DATE DUE

OCT 8 1991 5541	
MAR 17 1992	
BRODART, INC.	Cat. No. 23-221

Regional Stratigraphy of North America

Regional Stratigraphy of North America

William J. Frazier
David R. Schwimmer

Columbus College
Columbus, Georgia

Plenum Press • New York and London

Library of Congress Cataloging in Publication Data

Frazier, William J., 1946–
  Regional stratigraphy of North America.

  Bibliography: p.
  Includes index.
  1. Geology, Stratigraphic. 2. Geology—North America. I. Schwimmer, David R. II.
Title.
QE651.F75 1987   557   87-7019
ISBN 0-306-42324-3

© 1987 Plenum Press, New York
A Division of Plenum Publishing Corporation
233 Spring Street, New York, N.Y. 10013

Printed in the United States of America

To Sandra and Gabriele

Preface

An early reviewer of this book stated that he had difficulty assessing its marketability because it "falls between the cracks" of geological literature. We have designed this book to meet a need of modern geology: namely, a single source providing both detailed and synoptic stratigraphy of the various regions of North America, through geological time. Shortly after beginning work on such a book, we realized why it had not yet been written: it required six years of effort, assimilation of an incredible amount of information, and two years' additional work to cut the volume down to publishable size. Further, by the time the final chapter was written, the first few were already out of date.

Nevertheless, the book lies in front of you. It is intended to serve several purposes. As a textbook, it will serve the following courses:

- Regional stratigraphy
- Sedimentary tectonics
- Regional tectonics
- Advanced historical geology
- Survey-level paleontology

Obviously, not all portions of the book are relevant to all of the above courses. We assume the reader will retain this book after the particular course is done, and will use it as a reference book. Hopefully, others will obtain the book solely for reference purposes. We believe it will be especially useful for the working geologist or academic geologist seeking generalized and some moderately detailed information about a region or geological time interval which is unfamiliar. Discussions herein are detailed to the formational level on an exemplar basis: that is, not all formations for a time and region are discussed, but the characteristic strata are generally explained in a paragraph and equivalent units are noted. It is hoped that the reader can find the depositional history of virtually any North American unit of note in this text.

We omit discussions of basic principles because it is assumed the reader will have had courses in physical geology, historical geology, mineralogy, principles of stratigraphy, and at least a basic introduction to plate tectonics theory. A considerable amount of paleontology is included because many undergraduate curricula do not require paleontology and many discussions herein involve details of biostratigraphy. Then too, even a course in undergraduate invertebrate paleontology will not prepare the reader for the various elements of vertebrate history we have interwoven with the stratigraphy.

One may, of course, simply read the book for a thorough survey of North American stratigraphic history, exclusive of course or reference needs. Certain discussions will undoubtedly be heavy going, and for that reason we liberally provide overviews throughout. One may skip around the text and use the overviews to reorient or summarize detail one chooses not to read. A most valuable part of this book for many readers will be the citations. We opted for a compromise between the rigorous citation format of technical papers and the general lack of citations in textbooks. Rather than cite all arguments (as in journal articles), we typically cite both classical and very current references for major arguments, and assume the reader will check those sources and their citations for further in-depth reference. We have attempted to avoid citing obscure or archaic sources, unless there is an important point to be made by the nature of such reference (e.g., the original authorship of a major idea). Specialists in many areas will undoubtedly bemoan our omissions of crucial papers: we ask indulgence and consideration of how many we found.

An additional source of understandable culpability is in the scope of the book itself. Modern geology is inherently international, considering the importance of global tectonics. One may fairly ask how we dare create a sizable book dealing with the history of a single continent when all geologists realize that this slab of crust has been strongly influenced by the comings and goings of other continents, smaller terranes, and the expansions and contractions of seafloors. If we are myopic, it is for practical reasons: the book is already as large as possible. To provide a measure of the global perspective, we open most chapters with analyses of global paleogeography, showing especially which adjacent landmasses were affecting the margins of North America during the time in discussion. In addition, in subsequent discussions in text we analyze the effects of extracontinental influences on this continent's geological development.

William J. Frazier
David R. Schwimmer

Columbus, Georgia

Acknowledgments

During the long course of production of this book, a great many individuals have assisted in manuscript preparation, securing literature, providing illustrations, and performing technical reviews. In figure captions we acknowledge the sources of all illustrations, so at least one group will receive a portion of the credit due. Here, we wish to thank selected individuals. For manuscript typing: Karen Jackson, Daniella Horneck, Martha White, Bonnie Edwards, and Martha Kilgore. For reproduction of text figures: Jan Haney. For literature search and acquisition: Dr. Sharon Self and Fred Smith.

Many colleagues have shared ideas with us during the course of writing this manuscript, but we would especially like to thank Drs. Ronald S. Taylor, Thomas B. Hanley, Sydney W. Fox, William C. LeNoir, and William Birkhead for valuable critiques during the course of research. We also wish to acknowledge the material and financial support given by Columbus College during the long course of this effort.

Technical reviews were done by the following individuals, whose inclusion here does not necessarily indicate agreement with all material appearing in the book, but to whom we are grateful: Chapters 2 and 3, Dr. Kent C. Condie; Chapter 4, Dr. Juergen Reinhardt; Chapter 5, Dr. Peter W. Bretsky; Chapter 6, Dr. John M. Dennison; Chapter 7, Dr. Walter H. Wheeler; Chapter 8, Dr. Kenneth A. Aalto; Chapter 9, Drs. William J. Fritz and John Attig.

Contents

CHAPTER 1

Introduction

Stratigraphy lies at the heart of geology and we have set for ourselves the rather daunting task of describing, albeit in summary form, the regional stratigraphy of North America. Our purpose is to develop a stratigraphic framework against which to view the history of the continent and with which to test ideas about Earth mechanics. No such treatment can ever be complete, as ours is not, but we hope the attempt will provide the reader with at least a working summary of North American stratigraphy and geological history.

Through all the changes in geological understanding that have marked the history of geology, one of the primary goals of geological science has been the elucidation of the history of our planet. To that end, generations of geologists have measured, collected, analyzed, and theorized. It is our belief in writing this book that knowledge of the Earth can only come from careful analysis of the rocks themselves, that hypotheses and theories, though vital in the ordering of observations, must always be constrained by the data. One of the most important, and difficult, parts of dealing with stratigraphy is organization of available data in the most natural fashion. To this end, we employ in this book a system which blends the best of both time- and stratum-bounded units as stratigraphic subdivisions. The remainder of this chapter describes the format of all subsequent discussions.

The nine chapters (save this introductory one) are organized by two different types of geological units. Chapters 2 and 3 cover Precambrian time subdivided into the earlier (Archean) and later (Proterozoic) eons. Chapters 4 through 9 are organized in terms of cratonic sequences (Sloss, 1963), which are stratigraphic units of cratonic scale, bounded by major unconformities (see Fig. 1-1). Rationale for and definition of these cratonic sequences is presented in Chapter 4 (the first to employ the units). Here it will suffice to explain that the use of cratonic sequences is followed because the geological time units which subdivide so many

geology books are not suitable for our comprehensive stratigraphic analysis. The geological time units are almost entirely non-American in origin and they often divide similar strata artificially in North America. For example, it makes for poor organization to end a chapter on the Permian Period and begin a subsequent Triassic chapter stating "things have not changed much. . . ." By definition, cratonic sequences are North American units bounded by major depositional hiatuses which were created by significant events in the continent's history; in other words, they make very natural punctuation points in discussion.

The paleontological discussions, however, are not organized strictly around the cratonic sequences, because life history tends to more closely parallel the conventional geological time subdivisions. This is not coincidental but rather reflects the fact that geological time is generally delimited by natural changes in fossil assemblages in type regions in Europe. Even though North American fossil organisms did not necessarily parallel European taxa in evolution and extinction patterns, the relationships were generally close. We have attempted to follow the time bracketing a given cratonic sequence in discussion of the fossil organisms which lived during those times.

By title, this book treats the stratigraphy of North America region-by-region. In actuality, our approach is a twofold organization whereby most chapters subdivide the continent into cratonic versus marginal regions, within which we further subdivide discussions into a combination of geographic regions and physiographic provinces. Although this approach is not categorically pure, it does allow for practical-length collections of data to be presented without too much danger of having the broader pictures obscured.

Typically, chapters dealing with Phanerozoic events begin with an introduction featuring global paleogeography, and a few peripheral topics (often paleoclimatology). The

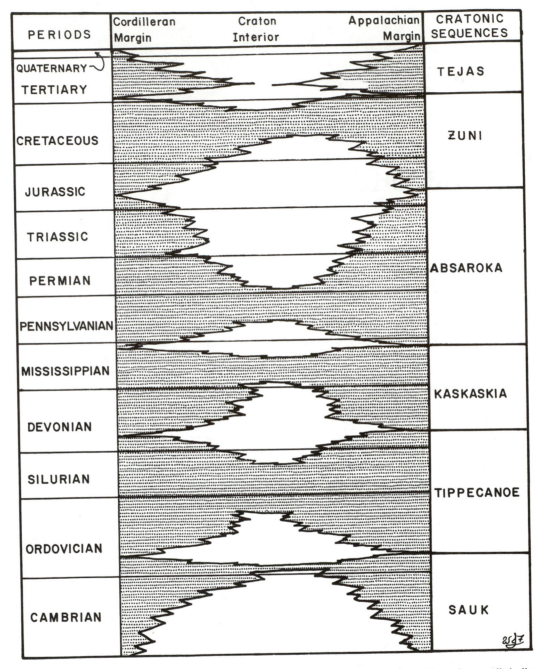

Fig. 1-1. Schematic diagram of cratonic sequences. Stippled areas represent sedimentary units; white areas indicate the absence of sedimentation. Geological time units on the vertical scale are presented in approximate proportion; however, no attempt is made to realistically proportion the horizontal relationships. (Modified and redrawn from Robertson and Marshall, 1975.)

second section is almost invariably a survey of the cratonic stratigraphy for the geological time interval dealt with by the chapter. The subsequent sections discuss events and rock units at the continental margins. The final sections of each chapter deal with the dominant life forms of the time during which the cratonic sequence was deposited.

The physiographic provinces of North America are typically defined and explained in a course in historical geology; they are the large regions of the continent which have a characteristic physical makeup and a unified geological development. Figure 1-2 presents a map which the reader may wish to use as a reference as we discuss the details of development of the various provinces through time. No single map can be sufficiently detailed to complement all of our coverages and therefore supplementary physiographic and/or geological maps will accompany parts

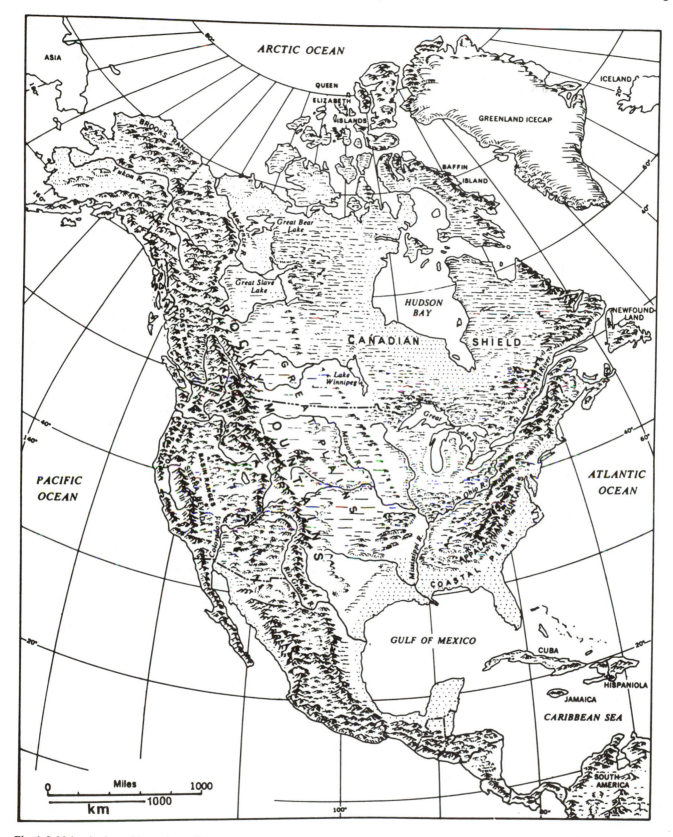

Fig. 1-2. Major physiographic provinces of North America. (From Kay and Colbert © 1965 by John Wiley & Sons, reprinted with permission of John Wiley & Sons.)

of the text. The reader should note that through geological time, the boundaries of various physiographic regions have changed; for example, the Rocky Mountains were part of the craton during much of the Phanerozoic Eon; the High Plains formed with and subsequent to the Rocky Mountains; portions of the Appalachian Mountains were marine and nearshore environments during much of the Phanerozoic, and even the "stable" Canadian Shield underwent a series of developmental stages with migrating ancient shorelines and rising and wasting mountain ranges during the Precambrian. The physiography presented in Fig. 1-2 is that of the modern continent and provides only glimpses of older provincial boundaries and geological characteristics.

Of necessity this book uses a variety of time, rock, and time-stratigraphic units in the descriptive contexts. We assume the reader is familiar with the use of all these descriptive modes [or, see Krumbein and Sloss (1963) for authoritative discussion]. Table I presents the time-stratigraphic units used in the text and which most efficiently delimit North American stratigraphic subdivisions. Although there are North American-derived series and stages for all geological systems of the Phanerozoic, we retain use of European terms for several systems. In addition, we use both European and North American nomenclature in some systems where it best explains or divides rock units. In general, we use North American stages and series in the Paleozoic Erathem, and European units in the Mesozoic. In discussions of the Upper Cretaceous, we use both (since a reasonably well-known western section is defined in North America). Our intent is the best communication possible, and we employ American terminology in preference to the international European-based terminology only where the former is firmly established and of continentwide usage. Table I is presented to smooth out any problems caused by jump-shifts in domestic to imported time-stratigraphic units.

Relating time and strata is the purpose of the time-stratigraphic units. However, mixed opinions exist about usages such as "Tuscaloosa time" (to describe the time during deposition of the Tuscaloosa Formation) or "Sauk time" (to describe the time during deposition of the Sauk cratonic sequence). We believe it is perfectly all right to describe the time of formation of a rock unit as "(unit name) time." Further, one may describe the time of deposition of a time-stratigraphic unit as "(time-stratigraphic unit name) time": e.g., "Conemaugh time." What is not correct is mixing references to stratigraphic positions and time units; e.g., the incorrect statement "this was formed during the Lower Cambrian." The time of deposition of the Lower Cambrian series is properly termed Early Cambrian time and there is no reason to eschew the proper terminology. Other seemingly aberrant "time" terminology may appear where a specific feature existed for an interval worth delimiting; e.g., we may refer to "Guadalupe reef time." The

intent is clearest communication and we hope to have achieved this.

A text such as this is characteristically a recitation of facts, hypotheses, and interpretations. Little overriding philosophy would seem to be necessary to such a purpose, and yet such a philosophy must be present to allow interpretations of past events from extant materials. We refer to the assumption of uniformitarianism, which is the principle that allows inferences to be drawn from ancient records by comparison with recent deposits and processes. Uniformitarianism is often stated as "the Present is the key to the Past." Given such an assumption, the analogy between, say, the sands of modern beaches and Ordovician sandstones, suggests the presence of Ordovician-age beaches.

Acceptance of uniformitarianism is widespread in the scientific world and few would question the general validity of the uniformity principle. However, in recent years, a modification of classical Huttonian uniformitarianism seems to be in order, following recognition of very different conditions at times during the Earth's past (especially the distant past). This modified assumption, derived from uniformitarianism (and certainly not a contradiction of it), is termed *actualism*, and is the precise guiding philosophy of this book. Actualism assumes constancy of the principles of science but does not require that physical and chemical conditions on Earth have always been the same.

Differences between actualism and uniformitarianism are subtle but important. As propounded by James Hutton in the mid-18th century, and as popularized by Charles Lyell in the famous early geology text *Principles of Geology* (1872), uniformitarianism is a one-to-one correspondence whereby ancient rocks and structures are interpreted as having formed in the same environments as are modern sediments, magmas, and so on. This assumption may be totally valid in a majority of cases (especially those dealing with younger events), but it obfuscates the point that assumptions of ancient-to-modern correspondences are themselves dependent on a more fundamental correspondence: that of the identity of natural processes between past and present. Thus, actualism implies that the same processes evidently operating on the present Earth have always been in operation. It does not automatically mean that present conditions are identical to those of the past, nor that unique events cannot occur.

As an example of where actualism may be quite distinct from uniformitarianism, consider that life on Earth might not be able to form today by means postulated to have been the original cause. The present atmosphere is strongly oxidizing whereas the postulated early reducing atmosphere probably favored formation of large organic molecules which are unstable in oxidizing conditions. Huttonian uniformitarianism would suggest a search for present cases of newly forming life whereas the actualistic approach is to

Table I. Phanerozoic (Cambrian to Holocene) Time-Stratigraphic Terminology Used in the Text[a,b]

	System	Series		Stage	
		European	(American)	European	American
CENOZOIC	Quaternary	Holocene			
		Pleistocene			Rancholabrean Irvingtonian
	Tertiary	Pliocene			Blancan Hemphillian
		Miocene			
		Oligocene			Jacksonian Claibornian Wilcoxian Midwayan
		Eocene			
		Paleocene		Danian	
MESOZOIC	Cretaceous	Upper	(Gulfian)	Maestrichtian Campanian Santonian Coniacian Turonian Cenomanian	Laramian Montanan Coloradoan Dakotan
		Lower	(Comanchean) (Coahuilan)	Albian Aptian Neocomian	
	Jurassic	Upper		Portlandian Kimmeridgian Oxfordian	
		Middle		Callovian Bathonian Bajocian Aalenian	
		Lower		Toarcian Pliensbachian Sinemurian Hettangian	
	Triassic	Upper		Rhaetian Norian Carnian	
		Middle		Ladinian Anisian	
		Lower		Scythian	
PALEOZOIC	Permian	Upper			Ochoan Guadalupian Leonardian
		Lower			Wolfcampian
	Pennsylvanian	Upper			Virgilian Missourian
		Middle			Desmoinesian Atokan
		Lower			Morrowan Springeran

(*continued*)

Table I. (*Continued*)

System	Series European	(American)	Stage European	Stage American
Mississippian	Upper			Chesterian / Meramecian / Osagian
	Lower			Kinderhookian
Devonian	Upper	(Chautauquan)	Famennian / Frasnian	Bradfordian / Cassadaga / Fingerlakesian / Taghanican
	Middle	(Erian)		Tioughniogan / Casenovian / Onesquethawan
	Lower	(Ulsterian)		Deerparkian / Helderbergian
Silurian	Upper	(Cayugan)		Keyseran / Tonolowayan / Salinan / Lockportian
	Middle	(Niagaran)		Cliftonian / Clintonian
	Lower			Alexandrian
Ordovician	Upper	(Cincinnatian)		Richmondian / Maysvillian / Edenian
	Middle	(Champlainian)		Trentonian / Blackriveran / Chazyan
	Lower	(Canadian)	Tremadocian	Whiterockian
Cambrian	Upper	(Croixan)		Trempealeauan / Franconian / Dresbachian
	Middle	(Albertan)		
	Lower	(Waucoban)		

(The system column is grouped under PALEOZOIC *rotated along the left margin.)*

[a]Data from Harland *et al.* (1982), Matthews (1984), Palmer (1983), and others.

[b]We have adopted North American series and stages in very wide application, but otherwise we have retained standard European terminology. The table is by no means exhaustive, and the absence of either European or American names for parts of the geological column does not mean there are no such terms; rather, it signifies they they are not used in this text. In parts of this composite geological column, several sets of nomenclature are given, reflecting common American usage (typically, these are both European and North American names).

assume the continuity of known chemical processes from the earliest times of living matter to the present.

As another example of the use of actualistic versus uniformitarian reasoning, consider the deposits of an ancient delta complex. Using uniformitarian reasoning, a geologist would try to draw a one-to-one comparison between the ancient sediment mass and a specific modern delta, e.g., the Nile delta or the Mississippi delta. Unfortunately, such an attempt is probably doomed at the outset because no two deltas are ever exactly alike. This is because the shape, sediment types, facies distributions, and historical development of a given delta are all functions of the chemical, physical, and biological processes which affect the delta. Since it is unlikely that all significant processes affecting two different deltas would be exactly the same, it is similarly unlikely that the two deltas would exactly resemble each other. Thus, the best one can hope for using uniformitarian thinking is an approximate correspondence between the recent and ancient. Actualism, on the other hand, presumes only that all deltas reflect the processes which operated to form them. Instead of trying to force a comparison between a modern and an ancient delta, actualistic reasoning seeks to compare the processes. Therefore, one may recognize as deltaic certain sedimentary deposits which do not exactly resemble any modern example. We believe that this is a more parsimonious assumption because it does not require exact analogs to demonstrate the general nature of a depositional environment.

A second philosophical approach that will be in evidence in this book is frequent application of "Occam's Razor." This is the philosophy that states, in essence, "given multiple choices of explanations, with equal or unknown probabilities, choose the simplest." Application of this principle will be seen in discussions of tectonics and in paleontology, especially in discussions of disputed topics such as mass extinctions and sudden appearances in the fossil record. Occam's Razor does not preclude the possibility that a complex explanation for a given problem is correct, but, rather, that given no preferred alternative, one chooses the simplest.

CHAPTER 2

The Archean

A. Introduction

The Archean Eon encompasses events from earliest Earth history to 2.5 Byr. ago. The rock record begins at approximately 3.8 Byr. ago and shows that the Archean Earth was strikingly different from that of the present day. For example, the atmosphere contained no free oxygen and the biosphere consisted solely of the simplest bacteria and blue-green algae. Continents existed during the Archean, but they probably did not consist, as do modern continents, of tectonically stable cratonic regions surrounded by tectonically active mobile belts; rather, Archean continents were *permobile* (i.e., they featured no stable areas at all but rather were everywhere susceptible to tectonism). Indeed, it is this tectonic permobility which distinguishes the Archean Eon as a separate interval of geological time, and it was the advent of stable cratons that marked the end of the Archean.

Detrital sedimentary rocks of Archean age do contain textures and structures which resemble those of younger deposits and which were probably affected by the same mechanical processes of transportation and deposition. On the other hand, some chemical sediments (e.g., chert-banded iron ores) are atypical of younger deposits and include varieties not encountered in strata of more recent vintage. Similarly, Archean igneous and metamorphic rocks contain some nonuniformitarian lithologies (i.e., rock types which apparently formed under physiochemical conditions not found today). Thus, it is clear that geological processes during the Archean were different from those today. Nevertheless, the present Earth evolved from the Archean Earth and today's crustal chemistry, tectonic patterns, and the geochemical balance between the atmosphere, hydrosphere, lithosphere, and biosphere all may be thought of as end products of processes traceable back to the Archean.

A.1. Distribution of Archean Rocks

A.1.a. Archean Rocks outside of North America

Rocks of Archean age occur on every continent but are usually confined to relatively small patches within Precambrian shields. A *Precambrian shield* is a large area of subaerially exposed Precambrian basement, often (but not always) forming the center of a craton. Typically, such areas are broadly convex upward, grossly resembling ancient military shields.

Some of the best-known Archean rocks are in southern Africa, such as the Swaziland Supergroup (Onverwacht, Fig Tree, and Moodies Groups) of South Africa's Barberton Mountain Land, and the Sebakwian, Bulawayan, and Shamvian Groups of Zimbabwe. Rocks of all these units have been subjected to extensive granitic intrusion, structural deformation, and low to moderate grades of metamorphism. Such deformed and slightly metamorphosed Archean strata comprise *greenstone belts*. Greenstone belts are commonly associated with areas of much-higher-grade metamorphic rocks, usually defined by extensive areas of granitic gneisses and amphibolites; these areas are termed *high-grade terranes*. In southern India, greenstone-belt rocks similar to those in southern Africa comprise the Dharwar System. In western Australia, Archean rocks comprise the Kalgoorlie System and in northwestern Scotland, rocks of the lower Scourie complex are considered to be of Archean age. Archean rocks occur in other regions as well, but the above-mentioned areas are among the best studied and are featured prominently in current tectonic models developed for the Archean.

A.1.b.   Archean Rocks of North America

North American Archean rocks occur in the Canadian Shield, the Wyoming Province, and the North Atlantic craton. In the following paragraphs, each of these areas will be briefly introduced.

i.   Canadian Shield.   The Canadian Shield occupies an area of approximately 4,828,000 km^2 in northcentral North America (Stockwell, 1976) and is subdivided into a number of geological provinces (see Fig. 2-1). Precambrian geological provinces are typically recognized on the basis of the timing of major deformations and regional metamorphism, as revealed by radiometric dating; boundaries between provinces are drawn where structural trends of one province are truncated by trends in another, either along major unconformities or orogenic fronts.

The Canadian Shield is bounded on all sides but one by younger geological provinces. To the south, it is overlapped by relatively undeformed, platform strata of the interior United States. Around its other sides are deformed continental-margin rocks: to the southeast is the Appalachian orogen; to the west is the Cordilleran orogen; and to the north is the Innuitian orogen. Only on the Labrador coast, facing the Greenland Shield, do Canadian Shield rocks occupy a continental margin.

Rocks of Archean age occur principally within Superior and Slave Provinces but are also found in areas of variable size within Bear and Churchill Provinces, where they exist in a structurally complex relation with Proterozoic rocks, possibly as the result of tectonic remobilization during the Proterozoic. A particularly large Archean exposure in western Churchill Province comprises the Kaminak craton of Hoffman *et al.* (1982). Additionally, Archean rocks comprise Eastern Nain Province of coastal Labrador. The other provinces of the Canadian Shield essentially represent mobile belts stabilized at various times during the Proterozoic and will be discussed in the next chapter.

ii.   Wyoming Province.   In Wyoming and adjacent states, Archean rocks occur in a series of Rocky Mountain uplifts (Fig. 2-2). Archean rocks of Wyoming Province, along with surrounding Proterozoic rocks, were deeply eroded and beveled off prior to the Cambrian Period and acted as continental basement, upon which Paleozoic and Mesozoic

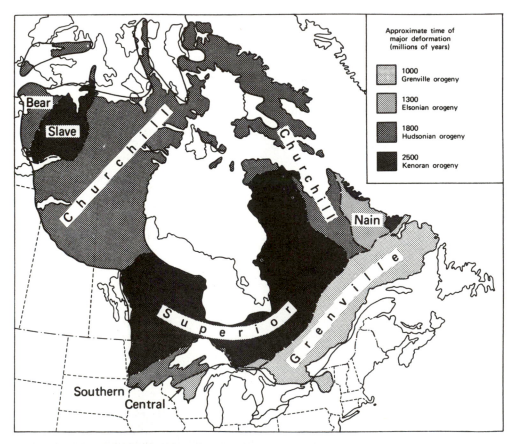

Fig. 2-1. The major geological provinces of the Canadian Shield and the ages of the last deformations that affected them. (From Stearn *et al.* © 1979 by John Wiley & Sons, reprinted with permission of John Wiley & Sons.)

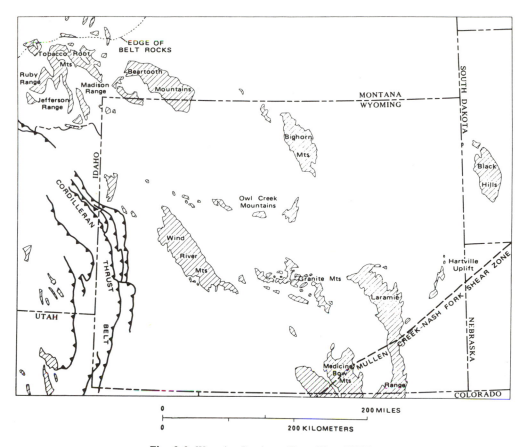

Fig. 2–2. Wyoming Province. (From King, 1976.)

strata were deposited. During the Late Cretaceous and Early Tertiary, however, the area was fragmented as a result of the Laramide orogeny into a series of crustal blocks which were jostled up and down, resulting in the exposure of Archean basement rocks at the base of the uplifted blocks (discussion of the Laramide orogeny is in Chapter 9).

iii. North Atlantic Province.

Along both eastern and western coasts of southern Greenland, exposed out from under the ice cap, is an extensive region of Archean rocks (see Fig. 2-3) containing some of the world's oldest rocks, such as the Amîtsoq Gneiss of the Godthaab region of southwestern Greenland (3760 ± 70 Myr.; Bridgwater *et al.*, 1974). Bordering this large Archean area on the north is an Early Proterozoic orogen termed the Nagssugtoqidian Mobile Belt. To the south of the Archean craton is the Ketilidian Mobile Belt, of Middle Proterozoic age.

These Archean and Proterozoic rocks of Greenland are very similar to rocks of eastern Labrador (technically, Eastern Nain Province of the Canadian Shield) and to rocks of the Scourian Complex of northwestern Scotland. These three areas were considered by Bridgwater *et al.* (1973) to have been, prior to Mesozoic rifting, parts of a single Arch-

ean province, which they termed the North Atlantic craton (see Fig. 2-3).

iv. North America as a Tectonic Collage.

Recently, a different breakdown of North America's Precambrian has been proposed. Hoffman *et al.* (1982) described North America's Precambrian crustal structure as a continentwide tectonic collage (see Fig. 2-4). The term *tectonic collage* was proposed by Helwig (1974) to describe a collection of diverse tectonic elements which originated independently of each other and which were subsequently accreted together by collisional tectonics. Helwig originally proposed the term to describe the character of an individual orogen, but Hoffman *et al.* (1982) broadened the term to characterize the entire Precambrian basement of the continent.

North America's basement collage consists of five distinct tectonic elements (shown in Fig. 2-4): (1) Archean cratonic blocks (the Superior, Kaminak, Slave, Wyoming, and North Atlantic cratons); (2) pre-1.9-Byr. Proterozoic orogens and reactivated Archean crust (the Committee, Rinkian, and Nagssugtoqidian fold belts); (3) Middle Proterozoic mobile belts (the Wopmay, Circum-Ungava, Rein-

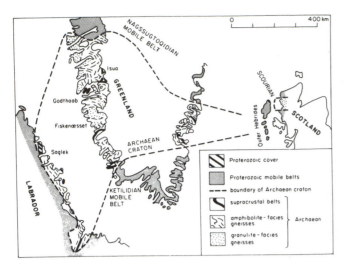

Fig. 2-3. General map of the North Atlantic craton. Note that the Greenland part of the craton is bounded to the south by the Ketilidian Mobile Belt and to the north by the Nagssugtoqidian Mobile Belt, both of which are of Proterozoic age. (From Windley © 1977 by John Wiley & Sons, reprinted with permission of John Wiley & Sons.)

deer, Ketilidian, and Penokean orogens); (4) intracratonic igneous and sedimentary rocks which formed between 1.6 and 1.3 Byr. ago; and (5) the Grenville orogen. A sixth Proterozoic element of the basement collage, not usually considered a separate geological province, is a major rift

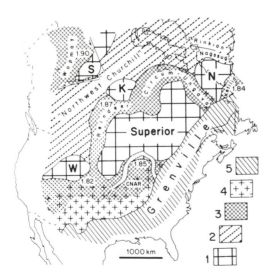

Fig. 2-4. Major Precambrian geological provinces of North America. Symbols are as follows: (1) Archean cratons (S, Slave; K, Kaminak; W, Wyoming; N, North Atlantic); (2) pre-1.9-Byr. Proterozoic orogens and reactivated Archean areas; (3) middle Proterozoic orogens; (4) 1.6- to 1.3-Byr. igneous intrusions and related volcanics; (5) Grenville orogen; CNAR, Central North American Rift. Greenland has been restored to predrift position by fitting adjacent continental slopes. (From Hoffman *et al.*, 1982, *D-NAG Spec. Publ. #1*, Fig. 2, p. 4.)

system extending to both the southwest and the southeast of the Great Lakes. Called the Central North American Rift (CNAR in Fig. 2-4), it formed during the later Middle Proterozoic in association with voluminous basaltic volcanism, possibly as the result of an incipient continental rifting.

A.2. Problems of Archean Geology

Archean rocks record an interval of geological time almost twice as long as the entire Phanerozoic Eon. Nonetheless, knowledge of the Archean is poor. This paucity of knowledge of the Earth's first 2 Byr. (including the time interval before the oldest rocks) is due to the following circumstances: (1) extant Archean rocks are limited in terms of both exposure and accessibility; (2) there are no index fossils for stratigraphic correlation; (3) Archean areas are structurally and petrographically complex; and (4) non-uniformitarian conditions may have existed during the Archean.

We have already noted that relatively few Archean areas are presently exposed. Although Archean rocks are known or inferred to be present beneath a cover of younger strata in other areas, knowledge of such rocks is obviously limited to information derived from remote sensing (e.g., geophysical data and drill cores). In addition, study of certain exposed Archean areas is made difficult by their remoteness and, in the case of Greenland and northern Canada, by the severity of their climates. Of course, some Archean areas (in particular, South Africa and Zimbabwe) have received attention out of proportion to their size because they contain important deposits of economic minerals, especially ores of Au, Ag, Cr, Ni, and Cu. Nevertheless, it is probably fair to say that limited outcrop extent is a major hindrance to the understanding of Archean events.

Well-preserved strata of Archean greenstone belts contain very simple fossils, notably algal stromatolites and putative bacteria (see discussion of Precambrian life in Chapter 3). But such fossils are rare and appear (at least in the light of the presently limited fund of data) to be essentially similar throughout the Archean. Thus, they are virtually useless at present for regional or interregional stratigraphic correlation, rendering impossible a presentation of detailed Archean geological history.

Of even greater significance is the fact that all known Archean rocks have been subjected to varying degrees of deformation and metamorphism. In many cases, primary textures and structures have been destroyed and original mineralogies have been altered. Then too, most Archean areas have been tectonized not once but several times, leading to the development of large-scale fold-interference patterns, and repeated episodes of metamorphism have completely altered the original character of the rock. Thus, interpretation of Archean rocks is predicated on palinspastic unraveling of deformed areas and determination of metamorphic *protoliths* (i.e., original, premetamorphic lithologies).

The permobile character of Archean tectonics has already been discussed but not its implication. Today, as throughout most of geological time since the Archean, crustal deformation is limited to narrow zones located where two rigid, lithospheric plates interact. The fact that Archean continents were subjected to intense deformation over their entire extent rather than in narrow mobile belts suggests that rigid lithospheric plates, at least as we know them today, did not exist during the Archean and that non-uniformitarian conditions were at work. Such a possibility has profound implications for interpretation of Archean geological processes. As Drury (1978) has pointed out, the only necessary constraint on Archean tectonic models (other than the empirical character of the Archean rocks and structures themselves) is that ''past events are the material basis for present processes.'' In other words, Archean geological processes *evolved into* Proterozoic and then Phanerozoic processes; therefore, any tectonic model developed for the Archean not only must be able to explain Archean processes but must also provide an evolutionary base from which to derive modern ones.

The purpose of this chapter is to introduce the reader to lithologies and structures typical of Archean regions and to present several different views on the nature of Archean tectonics. No attempt has been made to include all of the models that have been proposed, because the complexity of Archean tectonics has engendered quite a few; rather, we have chosen examples to illustrate the range of speculation. Further, because good Archean exposures are limited and because several important Archean tectonic models have been developed for areas of continents other than North America (especially the greenstone belts of southern Africa), this chapter deals with Archean areas from both within and outside of North America. (As such, this chapter represents a departure from the plan of the rest of the book, which focuses primarily on North American geology.) Readers who wish to pursue this subject are directed to Windley's (1984) text on continental evolution, McCall's (1977) collection of papers on Archean geology, Condie's (1981) symposium on greenstone belts, and Kröner's (1985) brief overview of Archean continental-crust evolution.

B.   Major Archean Lithologies

It is significant that Archean terranes are remarkably similar, at least in terms of their gross lithologies and structures, in almost all of their occurrences. This similarity is due partly to the fact that there are only two main lithological associations in Archean areas: greenstone belts and high-grade metamorphic terranes. Greenstone belts consist of thick and deeply infolded sequences of predominantly mafic volcanics and associated sediments which have been subjected to relatively low-grade metamorphism and to extensive intrusions of granite. High-grade terranes consist of a variety of granitic gneisses, amphibolites, and metasediments which have been affected by very-high-grade metamorphism, commonly up to the granulite grade, and by very strong deformation. Greenstone belts and high-grade terranes are commonly associated with each other but their structural relationships are obscure so that their relative ages are controversial. In this section, we will discuss these two associations and will offer some observations on the implications of certain lithologies to the character of the Archean environment.

B.1.   Greenstone Belts

Greenstone belts are large and complexly deformed terranes which contain the oldest well-preserved volcanic and sedimentary rocks on Earth. Their lithological associations differ from those of younger volcanic–sedimentary accumulations and their shape, structures, and relations to surrounding metamorphic terranes are all distinctive. They also contain the oldest fossils on Earth. Although there is much debate currently over their age relative to Archean high-grade terranes, there is no doubt that they contain clues to the earliest periods of Earth history and that the events which formed and deformed them were part and parcel of the formation of continents. They are of more than academic interest, however, because they also contain major deposits of Au, Ag, Cr, Ni, and other commercial and strategic metals as well as deposits of important nonmetals such as talc and asbestos.

Greenstone belts are typified by low to moderate grades of metamorphism, commonly ranging from the prenite–pumpellyite facies to the amphibolite facies (Condie, 1984). Several authors (e.g., Watson, 1978; Condie, 1981, 1984) have suggested that greenstone-belt metamorphism reflects alteration under relatively low-pressure conditions (perhaps at depths of as little as 20 km or less) in the presence of a geothermal gradient of 50–70° C/km. Such a gradient is comparable to those observed today in mid-oceanic ridge areas, but is considerably higher than modern continental gradients.

Glikson (1976) was the first to point out that most of the Earth's greenstone belts can be grouped into two categories based on age: one containing belts older than 3.0 Byr. and the other containing belts approximately 2.8 to 2.6 Byr. old. In addition to their ages, however, they also differ in their lithological associations. Condie (1984) characterized the older greenstone-belt group (> 3.0 Byr.) as: (1) containing larger proportions of komatiites than do greenstone belts of the younger group; (2) having few or no andesites; (3) having a greater proportion of chemical sediments, especially banded and carbonaceous cherts, silicified evaporites, and stromatolites, than younger belts; and (4) containing large amounts of pelitic deposits and, locally, quartzites, both of which are of only minor amount in the younger group. On the other hand, graywackes are rare or absent in the older group. Younger greenstone belts are characterized as having volcanics and sedimentary rocks of more highly variable composition than the older belts and as having few or no stromatolitic cherts and carbonates. Volcaniclastic sedimentary rocks are present in both groups, but volcaniclastics in the older group tend to be silicified. Greenstone belts of the younger group have massive sulfide deposits. Generally, the above-mentioned trends seem to indicate increasing abundance and influence of sialic crust throughout the Archean and increasing variation in water depths in which sediments accumulated (Windley, 1984).

The apparent age dependence of greenstone lithological associations in Archean rocks represents a contradiction of uniformitarian assumptions (i.e., that rock-types and associations are independent of age and may be found in rocks of all ages). Thus, Archean greenstone belts not only provide clues to an antique time in the Earth's past but also reveal the evolutionary nature of the Earth's history. We will return to this idea later in this chapter.

Although many lithological differences distinguish older from younger greenstone belts, it should be noted that both types are similar in their gross structures and their relations to Archean high-grade terranes. In the following discussion, we will describe greenstone belts in general, without distinguishing older from younger types.

B.1.a. Shape

Greenstone belts are characteristically synformal in cross section and vary in plan view from arcuate or cuspate, as in the Barberton Mountain Land of South Africa and the Rhodesian craton of Zimbabwe (see Fig. 2-5), to linear as in the Yilgarn craton of southwestern Australia or the Superior Province of Canada (see Fig. 2-6). They also vary in size from small areas such as the Barberton Mountain Land which is 40 km wide by 120 km long to the Abitibi greenstone belt of Canada's Superior Province, the largest continuous greenstone belt on Earth, which is 200 km wide by 800 km long. A possibly significant nonuniformitarian aspect of greenstone belts was noted by Engel and Kelm (1972) who observed that the length/width ratios of greenstone belts rarely exceed 5 : 1 while those of younger deformed belts (e.g., the Appalachian–Caledonian orogen or the Cordilleran orogens of North and South America) often exceed 100 : 1. Further, Windley (1984) has noted the tendency of

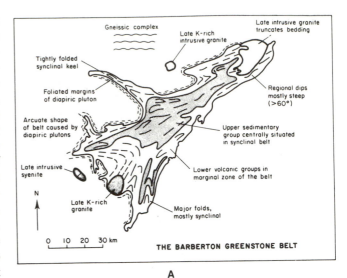

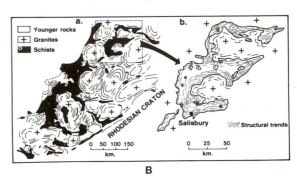

Fig. 2-5. Arcuate, cuspate, and irregular greenstone belts. (A) The Barberton greenstone belt in South Africa (from Windley © 1977 by John Wiley & Sons, reprinted with permission of John Wiley & Sons); (B) greenstone belts ("schists") of the Rhodesian craton (from Anhaeusser, 1975, reproduced, with permission, from the *Annual Review of Earth and Planetary Sciences,* Vol. 4, © 1975, Annual Reviews Inc.). Note that in both cases, the overall shapes of these greenstone belts are constrained by the diapiric "gregarious batholiths."

older greenstone belts (> 3000 Myr.) to be relatively smaller than the younger ones (< 3000 Myr.).

B.1.b. Structure

Greenstone belts are characterized by large dome-shaped granitic intrusions so numerous that they have been called "gregarious batholiths" (MacGregor, 1951). As may be seen in Fig. 2-5, the arcuate–cuspate shapes of many greenstone belts are due to these intrusions. The volcanic–sedimentary strata of greenstone belts have been folded into large, well-developed, variably plunging synclines. These synclines are commonly tight and isoclinal with steeply dipping limbs which may be overturned. In the Barberton Mountain Land, anticlinal folds are either absent,

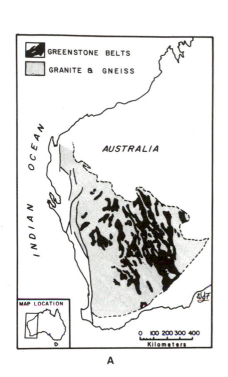

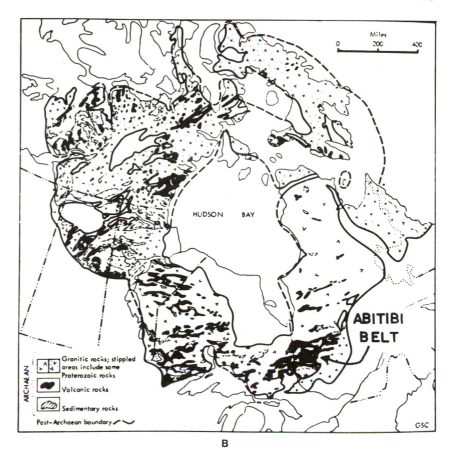

Fig. 2-6. Linear and subparallel greenstone belts. (A) Archean shields of western Australia showing striking parallelism of Yilgarn Shield greenstone belts (modified from Glikson, 1972, *Geological Society of America Bulletin,* Fig. 1, p. 3325); (B) Major Archean exposures of the Canadian Shield; note that greenstone belts in Superior Province show a subparallel pattern, somewhat less well developed than in the Yilgarn Shield. (From Goodwin, 1968, *Proceedings of the 23rd International Geological Congress,* Vol. 1, Fig. 2, p. 76.)

having been faulted off by high-angle faults (see Fig. 2-7), or poorly developed (Anhaeusser *et al.,* 1968). In the Abitibi greenstone belt, on the other hand, both synclines and anticlines, also isoclinal, are present (Fig. 2-8A) with granitic plutons occurring preferentially along anticlines and volcanic and sedimentary rocks (so-called *supracrustal rocks,* i.e., strata deposited on the crust rather than intruded into it) occurring along synclines (Goodwin and Ridler, 1970). Anhaeusser (1975) and Viljoen and Viljoen (1969), among others, interpreted the structural style of greenstone belts, especially those in southern Africa, to have been the result of simple downsagging of the crust without any significant compressive stress; they cite as evidence the relatively low grade of regional metamorphism, the near absence of penetrative structures (e.g., cleavage and schistocity), and the synclinal character of folds between high-angle faults. On the other hand, Ramsey (1963) and Wood (1966) argued that horizontal compression was the principal deforming stress in the Barberton Mountain Land.

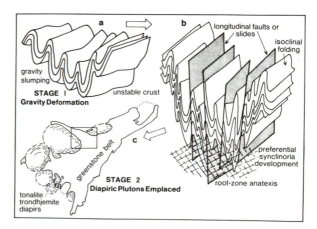

Fig. 2-7. Anhaeusser's description of folds in the Barberton greenstone belt. Note subvertically oriented faults and presumed root-zone anatexis; rectangle in *c* is the location of detailed structural analysis shown in Fig. 2-9. (From Anhaeusser, 1975, reproduced, with permission, from the *Annual Review of Earth and Planetary Sciences,* Vol. 4, © 1975, Annual Reviews Inc.)

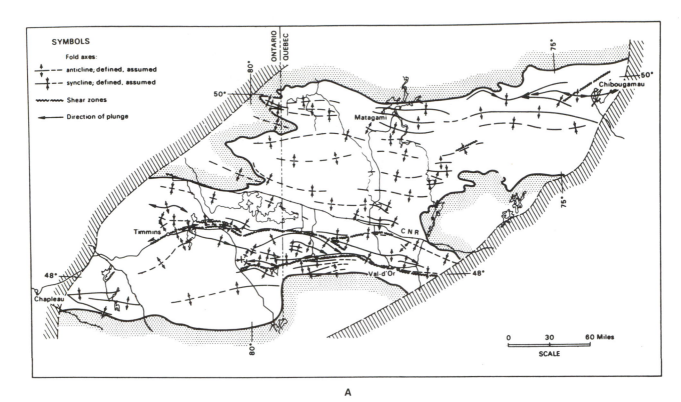

A

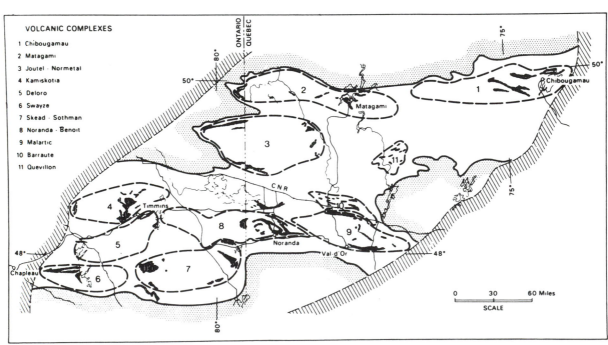

B

Fig. 2-8. Structural features of the Abitibi greenstone belt (location shown in Fig. 2-6B). (A) Major folds and shear zones; (B) major volcanic complexes, interpreted to represent originally circular stratovolcanoes which were deformed by horizontal compression. (From Goodwin and Ridler, 1970, reproduced with permission of the Minister of Supply and Services, Canada.)

Stowe (1974) and Coward *et al.* (1976) proposed a similar compressive tectonic style for the Rhodesian craton of Zimbabwe, and Wood (1973) estimated that approximately 75% of crustal shortening was experienced by some Rhodesian belts. Goodwin and Ridler (1970) described large, elliptical volcanic-complexes (see Fig. 2-8B) in the Abitibi greenstone belt which they interpreted as stratovolcanoes; presuming an originally circular shape, they used the ellipticity of the complexes to calculate that over 50% shortening had occurred, which they believed to have been due to horizontal compression. Of course, at least some of the structures of greenstone belts are due to stresses attendant on the diapiric rise of the "gregarious batholiths" (see Fig. 2-9) which was certainly a vertically directed event but, as can be seen from this brief discussion, the overall cause of Archean deformation is still highly controversial. (One may glimpse in this controversy the thread of the old vertical versus horizontal tectonics argument which goes all the way back to James Hutton's time.)

B.1.c.   Stratigraphy

Figure 2-10 is a composite stratigraphic column of the Swaziland Supergroup from the Barberton Mountain Land, which is an example of the older (> 3.0 Byr.) greenstone-belt group (Anhaeusser, 1973), and will serve as a general model for greenstone-belt stratigraphy. In the following discussion, we will divide Swaziland rocks into three lithological groupings: (1) a lower, ultramafic–mafic group (the Tjakastad subgroup of the Onverwacht Group); (2) a middle, mafic to felsic, calc-alkaline group (the Geluk subgroup of the Onverwacht); and (3) an upper, sedimentary group (the Fig Tree and Moodies Groups). While specific differences may exist between them, most older Archean greenstone belts are characterized by this same three-part sequence.

i.   Lower, Ultramafic–Mafic Group.

Due to the synclinal nature of greenstone belts, the lowest strata are found in outermost areas of the belt and have been strongly affected by granitic intrusions and metamorphic alteration. Indeed, in some cases these rocks are only present as inclusions within the bordering granites. These oldest strata are primarily interbedded ultramafic and mafic lava flows, often with well-developed pillow structures indicating extrusion underwater. Massive sills and layered igneous intrusions may also be present. Completing the picture are very small amounts of pyroclastic sediments (including felsic tuffs) and thin, discontinuous lenses of shaley and/or carbonaceous chert and banded iron formation (see below). However, sial-derived detrital sediments are generally absent. In the Barberton Mountain Land, these rocks attain a thickness of approximately 7 km.

A brief examination of the petrology of Onverwacht lavas is in order. Viljoen and Viljoen (1969) showed that Tjakastad subgroup rocks have a unique geochemistry and named them komatiites from the Komati River. Komatiites are recognized mainly by their high MgO content (> 9%, up to 32% in some), their high CaO/Al_2O_3 ratio (> 1.0), and their low K_2O and TiO_2 contents (both < 0.9%). Vil-

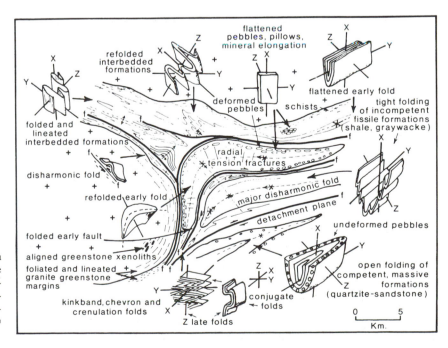

Fig. 2-9. Detailed structural analysis of small area (shown in Fig. 2-7) of the Barberton greenstone belt showing structures due to emplacement of marginal plutons. (From Anhaeusser, 1975, reproduced, with permission, from the *Annual Review of Earth and Planetary Sciences*, Vol. 4, © 1975, Annual Reviews Inc.)

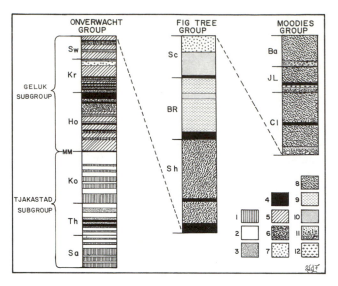

Fig. 2-10. Stratigraphy of the Swaziland Supergroup. Symbols for stratigraphic formations are as follows: Sa, Sandspruit Formation; Th,
Theespruit Formation; Ko, Komati Formation; MM, Middle Marker; Ho,
Hooggenoeg Formation; Kr, Kromberg Formation; Sw, Swartkoppie For-
mation; Sh, Sheba Formation; BR, Belvue Road Formation; Sc, Schoon-
gezicht Formation; Cl, Clutha Formation; JL, Joe's Luck Formation; Ba,
Baviaanskop Formation. Symbols for lithologies are as follows: 1, ultramafic lavas; 2, mafic lavas; 3, siliceous and aluminous felsic tuffs; 4, chert
with minor shale and limestone; 5, metatholeiites; 6, felsic lavas, tuffs,
agglomerates, and porphyries; 7, mafic pyroclastics, agglomerates, and
pillow breccias; 8, graywackes and shales; 9, shales; 10, tuffs; 11, conglomerate and quartzite; 12, amygdaloidal lava. (From Anhaeusser, 1978.)

joen and Viljoen divided komatiites into two main groups:
ultramafic or peridotitic komatiites and mafic or basaltic
komatiites. In addition to their geochemistry, komatiites are
characterized (at least in relatively unaltered samples) by
so-called spinifex texture, a texture defined by large skeletal
crystals of olivine or pyroxene. Spinifex texture (named
after a spiky Australian grass) apparently was caused by
very rapid cooling (quenching) of the magma, possibly by
submarine extrusion (a possibility supported by the common pillow-structures of these rocks). Komatiites are significant because of their "primitive" (i.e., relatively undifferentiated) character, indicating that they resulted from
high degrees of partial melting within the mantle. Green
(1972) estimated that mafic komatiites would have required
between 40 and 60% melting of the mantle and ultramafic
komatiites, up to 80% melting! Such high amounts of melting are unknown from modern environments but may be
explained by relatively shallow depths of melting within the
mantle and a higher geothermal gradient during the Archean
(Windley, 1984).

The layered igneous intrusions of the lower group are
also interesting. These intrusions are generally composed of
two members, the lower of which consists of alternating

layers of olivine and orthopyroxene (bronzite) cumulates
(i.e., early formed crystals which settled out of the magma
and accumulated on the floor of the magma chamber) and the
upper of which consists of serpentinized peridotites, pyrox-
enites, gabbros, norites, and anorthosites (Anhaeusser,
1978). The Tjakastad layered intrusions were crystallized
from a peridotitic magma rather than from a tholeiitic-
basaltic magma as were younger layered intrusions such as
the well-known Skaergaard complex. Thus, we see again
that Archean magmas were more "primitive," i.e., they
represented a higher degree of mantle melting, than younger
magmas. The implications of this will be discussed later in
this chapter.

**ii. Middle, Mafic to Felsic, Calc-alkaline
Group.** Overlying the ultramafic–mafic group is a sequence of successively more differentiated volcanic rocks
of the calc-alkaline group. In the Barberton Mountain Land,
the boundary between the two groups is drawn at a remarkably persistent, thin, chert horizon called the Middle Marker which separates the Tjakastad subgroup from the Geluk
subgroup (see Fig. 2-10). Geluk strata consist of a sequence
of cyclic volcanic units, each of which consists of a thick
zone of tholeiitic basalt overlain by a thinner zone of dacite
and rhyolite. Andesite is uncommon. The relative proportion of rhyolite in the units increases vertically as does the
average K_2O content. Pyroclastic debris is more common
than in the lower, ultramafic–mafic group. In Canadian
greenstone belts (which are of the late Archean type), a
similar pattern may be seen in the thick and extensive outpourings of tholeiite, called mafic platforms by Goodwin
(1968). These platforms are overlain by andesitic and
dacitic flows and pyroclastics and then by somewhat lesser
amounts of rhyolitic pyroclastics. These volcanic buildups
were termed felsic edifices by Goodwin and were probably
volcanic islands which underwent subsequent erosion and
detrital sedimentation. Other evidence for the island nature
of these felsic edifices was supplied by Henderson (1977)
who reported the presence of welded ash-flow tuffs within
the calc-alkaline group of Canada's Slave Province.

Sediments interlayered with the calc-alkaline flows
comprise only a relatively minor part of the section but their
character reveals much about environmental conditions during volcanic island formation. Lowe and Knauth (1977)
present an excellent discussion of Geluk subgroup sediments and much of the following is based on their observations. Most Geluk sediments are today composed of chert,
which earlier authors (e.g., Viljoen and Viljoen, 1969) interpreted as deposits of primary, colloidal silica gel. Lowe
and Knauth used careful field and microscope study to demonstrate that most of the chert is of secondary origin, due to
silicification of original volcaniclastic and carbonate sediments. Volcaniclastic sediment was composed of alkali

feldspar, volcanic quartz, pumice and volcanic-rock fragments, and glass shards. Detrital quartz is extremely rare,
making up less than 1% of the coarse, current-deposited
sediment. In addition, in the coarse conglomerates that
occur at the top of the Swartkoppie Formation (see Fig.
2-10), only chert pebbles are encountered; none of the clasts
indicate deep erosion of the sourceland. Some Geluk cherts
contain organic matter as either fine, intergranular material,
discrete particles, or thin, hairline laminae. Lowe and
Knauth speculated that some of this material may have originated as *in situ* organic mats or coatings. Some of the
oldest known putative microorganisms were recovered from
Swartkoppie cherts (Barghoorn and Schopf, 1966). Among
strata of the Bulawayan Group of the Rhodesian craton, also
a generally calc-alkaline volcanic unit, are limestones
which contain some of the oldest known stromatolites.

Sediments from the Hooggenoeg Formation (see Fig.
2-10) contain thin beds with erosional bases, graded bedding, small-scale crossbedding, and other features characteristic of turbidity-current deposits. They appear to have
been composed almost entirely of locally derived volcaniclastic sediment. Kromberg Formation sediments reveal
deposition in shallower-water environments. Primary carbonates (dolomitic in composition today) containing large-
and small-scale crossbedding were deposited as high-energy
sand bodies whose tops were occasionally exposed subaerially. Adjacent fine-grained volcaniclastic sediment records an environment in which low-energy conditions alternated with periods of low-velocity currents. Lowe and
Knauth referred to the depositional environment of these
sediments as ''problematical'' but, from their description,
they would seem to resemble deposits of modern tidally
dominated coastlines, such as tidal flats. The Swartkoppie
Formation is, according to Lowe and Knauth (p. 709), ''the
most lithologically and stratigraphically complex unit'' in
the Barberton Mountain Land. In southern areas, it is composed of silicified volcaniclastic and carbonate sediments
deposited in shallow water while, in northern areas, it is
composed of massive, structureless, black chert probably
deposited in deep water. In between, debris flow, turbidity
current, and low-energy (pelagic?) sediments alternate, suggesting deposition on a submarine fan at the foot of the
volcanic edifice. In his description of sediments in Slave
Province, Henderson (1977) described similar submarine
turbidite fans.

The mineralogy and texture of Geluk sediments suggest a local volcanic source—which is hardly surprising
considering the interlayered lava flows and pyroclastics.
The nearly complete absence of detrital quartz indicates that
sialic crust was not close enough to supply sediments to this
area. There is also a near absence of mudrocks within the
Geluk subgroup which suggests either that environmental
conditions were not right for chemical weathering and clay-

mineral formation or, perhaps more likely, that the volcanic
islands were not large enough to provide sizable amounts of
clay and that other sources, e.g., continents, were not significant contributors (at least in the Barberton area). Lowe
and Knauth's depositional/tectonic model for Geluk strata
is presented in Fig. 2-11.

iii. Upper, Sedimentary Group.
In the Bar-
berton Mountain Land, volcanics and sediments of the Onverwacht Group are overlain unconformably by two major
sedimentary units: the Fig Tree and Moodies Groups (see
Fig. 2-10). Taken together, their characteristics will serve
as a general model of Archean sediments.

The Fig Tree Group is over 2100 m thick and is composed primarily of rhythmically interbedded, thin layers of
graywacke and shale. Thin chert bands are a minor component and, near the top, felsic tuffs and agglomerates, lavas,
and thin beds of banded iron formation. Graywacke grain

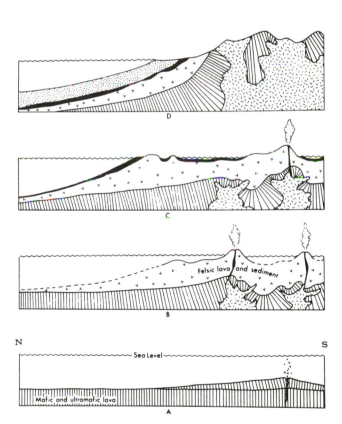

Fig. 2-11. Lowe and Knauth's model of sedimentological and tectonic
evolution of the Barberton Mountain Land during deposition of the Onverwacht and Fig Tree Groups. (A) Initial phase of submarine ultramafic–
mafic volcanism (Tjakastad Subgroup); (B) calc-alkaline volcanism during
deposition of the Hooggenoeg and Kromberg Formations; (C) deposition
of the Swartkoppie Formation (black); (D) deposition of Fig Tree Group
strata (stippled); note exposure in sourceland of granitic intrusions and
uplifted pre-Fig Tree lithologies. (From Lowe and Knauth, in *Journal of
Geology*, Vol. 85, p. 720, © 1977, University of Chicago Press.)

size increases gradually toward the top where occasional conglomerates are present. Sedimentary structures include graded bedding, flute casts, convolute bedding, ripple crossbedding, and load structures. Fig Tree detrital sediment is interpreted as turbidity-current deposits, possibly representing a submarine fan complex which grew horizontally and vertically with time, as suggested by the vertical increase in grain size. The sediment is texturally and mineralogically very immature, suggesting derivation on a high-relief landmass (Anhaeusser et al., 1968), rapid transportation, and burial in a rapidly subsiding basin. Graywacke composition implies a source area with a geologically varied terrane including sediments, volcanics, and granite (Lowe and Knauth, 1977). Condie et al. (1970) reported a vertical increase in the relative amount of granitic components which, they suggested, implied the unroofing of sialic crust, although it could also imply unroofing of a granite-cored orogen (see Fig. 2-11). The pyroclastics and lava at the top indicate that volcanic activity had continued from Onverwacht time. Similar turbidite deposits have been reported by Goodwin and Ridler (1970) from the Abitibi greenstone belt, where they are more than 2400 m thick, and by McGlynn and Henderson (1970) from Yellowknife Supergroup strata of Slave Province.

The Moodies Group is a thick (> 3100 m) mass of sandstone with shale or argillite beds, local banded iron formation, and a persistent basal conglomerate. Its sandstone compositions vary widely, from subgraywackes and subarkoses to quartz arenites. The conglomerates are polymictic, containing clasts of chert, granite, sandstone, shale, banded iron formation, various metamorphic lithologies, and even pebbles of conglomerate. Anhaeusser et al. (1968) recognized conglomerate lithologies derived from both Onverwacht and Fig Tree strata. Sedimentary structures include abundant crossbedding, especially trough crossbedding, ripple marks, and mud cracks. The Abitibi greenstone belt contains similar sediments, referred to by Goodwin and Ridler (1970) as the Timiskaming facies, characterized by coarse detrital sediments and by abrupt lateral facies changes and intraformational unconformities. Anhaeusser et al. referred to the Moodies as a molasse sequence and stated that it represented deposition in shallow marine and braided stream environments.

Chemically precipitated sediments, such as chert, carbonates, and banded iron formation, have been mentioned as occurring in some Archean strata but they generally comprise a relatively minor part of the section. Nevertheless, they are of interest because of the light that they shed on the geochemical nature of Archean surface environments. Chert is a common rock type in some Archean sediments, especially associated with iron minerals in banded iron formation. As mentioned earlier in the discussion of Geluk Group

sediments, some cherts are probably of secondary, i.e., replacement, origin. But many other cherts were probably primary, possibly deposited as amorphous silica on the seafloor. Siever (1977) argued that the early Precambrian ocean contained approximately 40–60 mg/liter of silica in solution, almost an order of magnitude greater than the silica concentration in modern seawater. The exact concentration was probably controlled inorganically by the balance between the rate of silica supply, from chemical weathering and from glassy volcanic materials, and the rate of silica removal, by precipitation of primary siliceous sediments and by formation of clay minerals. The greater amount of silica in solution in Archean seawater was probably due to the absence of biological removal, which is the principal control on silica concentration in the modern ocean. Siever pointed out that the above-mentioned silica-concentration level is somewhat below the solubility of amorphous silica (120 ppm) and he therefore hypothesized that hypersaline conditions would have been required for silica precipitation. This implies that much Archean chert was deposited in sedimentary basins which were at least partially restricted from the open ocean and where evaporative conditions pertained. The occurrence within the Gunflint Chert (Proterozoic) of halite pseudomorphs and of possible silica-replaced gypsum crystals in some Geluk Group cherts supports Siever's hypothesis. The commonly noted absence of detrital sediment in Archean cherts may be due to the presence of some barrier to sediment dispersal which separated the basin from a detrital sourceland or due to chert deposition around the margins of essentially base-leveled landmasses which supplied little coarse detritus.

Carbonate sediments are rare in Archean strata. Henderson (1977) described carbonate rocks of the Yellowknife Supergroup from Canada's Slave Province where they are typically located in basin-margin areas, transitional between felsic volcanic units and deep-water, turbidite deposits. Such a location suggests that these carbonates accumulated in shallow-water environments which fringed volcanic islands just as similar carbonates do today; of course, there would not have been any biohermal structures (i.e., organically constructed, wave-resistant structures, such as coral reefs) and carbonate textural elements would not have included bioclasts or fecal pellets. Dolostones are minor lithologies in the Geluk Group and in other Archean strata but they are not common. The reason for the paucity of Archean carbonates is controversial and relates to the chemical character of the Archean ocean. There are two proposed explanations: (1) carbonate sediments in general were not deposited because seawater chemistry was incorrect for inorganic carbonate precipitation, e.g., the pH was too low, the necessary conditions for concentration did not exist, and so on; or (2) carbonate sediments were deposited, accumulating to significant thicknesses, but were not, in general, preserved. Those who argue that carbonates were not deposited suggest that the partial pressure of CO_2 may have been higher in the Archean atmosphere than today, resulting in a somewhat lower seawater pH (Ericksson and Truswell, 1978). They point to the abundance of cherts in Archean sediments and suggest that high concentrations of dissolved silica are possible only at relatively low pHs and that, under those conditions, carbonates would probably have remained in solution. The opposition counters that silica concentrations of up to 120 ppm are theoretically possible in modern seawater (in the absence of biological or adsorptive removal mechanisms). Such theoretical concentration levels are higher than Siever's (1977) speculative Archean silica concentration of 40–60 ppm; thus, the greater amount of Archean chert does not necessarily imply any difference in the pH of seawater. Further, they present evidence that at least some Archean cherts were formed by silicification of preexisting carbonates. Such silicification may have been much more common than originally believed—Lowe (1980) described stromatolites in 3.4-Byr. cherts of western Australia and stated that they were silicified carbonate stromatolites. Since putative stromatolitic cherts are occasionally found in some Archean sequences, it is possible that considerable bodies of carbonate sediments did accumulate and that the present-day paucity of carbonates is artifactual.

Banded iron formation (BIF; also called taconite, itabirite, jaspilite, and so on) is a chemical sediment composed of layered iron minerals interbedded with chert or, occasionally, dolostone. James (1954) proposed a lower iron-content limit of 15% but most BIF deposits contain 20–35% iron. Four main mineralogical facies of BIF are recognized: *oxide* (hematite and magnetite); *carbonate* (siderite); *silicate* (primarily greenalite, a hydrated iron phyllosilicate with lesser amounts of minnesotaite, which is considered to be of diagenetic or low-grade metamorphic origin); and *sulfide* (pyrite).

BIF is virtually restricted to the Precambrian. Phanerozoic iron-rich sediments (commonly called ironstones) are also important sources of iron but their mineral associations differ from those of BIF in several important ways: (1) the iron oxide geothite is not found in BIF but is common in Phanerozoic ironstones; (2) chamosite, rather than greenalite, is the major iron silicate of ironstones; (3) glauconite, common in some ironstones, is unknown from BIF; and (4) marcasite also is not found in BIF but is present in some ironstones. These differences suggest that environmental conditions during the formation of BIF were different from those which attended the deposition of Phanerozoic ironstones but the nature of the differences is controversial. It may be noted, however, that both BIF and Phanerozoic ironstones tend to occur in the same two general associa-

tions: (1) the granular/oolitic facies, composed primarily of iron oxides and/or iron silicates which occur as crossbedded, rippled granular or oolitic deposits associated with mature quartz arenites, dolostones, shallow-water limestones, and other related sediments; this facies is sometimes said to be "miogeosynclinal," implying deposition on a relatively stable, shallow shelf; (2) the laminated facies, characterized by thinly bedded or laminated iron minerals, such as magnetite and iron silicates or pyrite interbedded with dark shales; this facies is considered to represent deeper-water environments and is sometimes called "eugeoclinal."

BIF is found in many Archean areas but usually in only minor amounts, much less abundant than in Proterozoic rocks. Archean BIF deposits, termed Algoma-type deposits, average 10–50 m in thickness with only a few exceeding 100 m; they range up to a maximum of 75 km in length along strike. By comparison, Proterozoic BIF, termed Superior-type deposits, may be well over 1000 m thick and more than 100 km long. Archean BIF is also more restricted in textural varieties than Proterozoic occurrences, being represented primarily by thinly bedded, finely crystalline varieties (i.e., the laminated facies). Mineralogically, Algoma-type deposits are composed primarily of oxides and carbonates although both sulfides and silicates are found in rare instances. Goodwin (1973) showed that the distribution of BIF mineral-facies in the Abitibi greenstone belt was directly related to bathymetry: the oxide facies was disposed around basin margins, carbonates were associated with deeper waters farther out into the basin; minor deposits of the sulfide facies represent the deep, basin-center area.

Veizer (1983) suggested that a principal cause of the differences between Algoma-type (Archean) and Superior-type (Proterozoic) BIF deposits was the difference in sedimentary–tectonic environment in which they accumulated. Veizer argued that, during the Archean, in the absence of stable cratonic blocks, continental shelves would have been narrower than during the Proterozoic. Algoma-type BIF deposits accumulated on these narrow shelves and the minor volume of such deposits, relative to Superior-type ores, as well as their limited mineral-facies occurrences are both due to this restriction of the shelf environment. With the onset of stable cratons and concomitant development of extensive continental shelves, the widespread Superior-type ores were possible.

B.2. A Digression: Weathering and Sedimentation during the Archean

At this point, we pause for a brief consideration of the general nature of sediment production, transportation, and deposition during the Archean. We begin with an overview of environmental conditions on the Archean Earth's surface.

B.2.a. Surface Temperature

Considering the abundant evidence for the presence of water on the Earth's surface (e.g., pillow structures in lavas, hydrodynamic sedimentary structures, and chemical sediments precipitated from aqueous solution), we may assume that average surface temperature of the Archean Earth was somewhere between 0 and 100°C. Furthermore, if the various putative Archean fossils are indeed of organic origin, then the average temperature may have been less than 30°C, i.e., similar to that of the present Earth. Today, the surface temperature is controlled by solar radiation with very little input from geothermal heat. While the Archean geothermal gradient was probably higher than today's, the major surface-temperature control was probably also the sun. However, most models of solar evolution suggest that the luminosity of the sun may have been slightly lower during the Archean, resulting in less total insolation (i.e., incoming solar radiation). Further, Siever (1977) postulated that the Earth's albedo (i.e., reflectivity) may have been somewhat higher due to increased amounts of atmospheric dust, caused by higher levels of volcanic activity, more frequent meteorite impacts and resultant ejecta dust, and the lack of terrestrial vegetation to hold down fine detritus from the erosive force of the wind. The implication of lower insolation and higher albedo is that the surface temperature may have been lower than today. On the other hand, if large landmasses were absent (as hypothesized in some Archean tectonic models), the greater surface area of the Earth covered by water would have increased the amount of heat absorbed by the Earth. In Siever's view, the average surface temperature of the Archean Earth was slightly lower than at present and large polar ice-caps were probably present.

Of course, the significance of insolation extends far beyond the average surface temperature. The difference between polar and equatorial temperatures (like voltage in an electrical circuit) provides the force to drive atmospheric and oceanic circulation. Differences in the actual magnitude of the polar/equatorial temperature-differential may, thus, have had important effects on the rate and even the configuration of surface circulations. Further, much absorbed insolation is in the form of latent heat (i.e., heat which causes phase changes rather than temperature increases). Latent heat is absorbed by evaporation, stored in atmospheric water vapor, and released by rainfall or snowfall. It is, thus, the ultimate source of energy for the erosion of land surfaces by running water and glaciers. For this reason, change in the total amount of insolation and absorption could have had effects on rates of erosion. Since landmass elevation and relief are at least partially a function of the ratio (rate of

tectonism + volcanism)/(rate of erosion), change in erosional rates could have affected the overall character of Archean topography. Again, we see that possible differences in insolation and absorption could have had significant effects on the rates of geomorphic processes.

B.2.b. Atmospheric Composition

A full discussion of the Precambrian atmosphere is beyond the scope of this book but a brief consideration of the probable composition of the Archean atmosphere is needed here. The reader who wishes to pursue this topic is directed to Walker et al. (1983). Although much controversy swirls around this subject, there is general agreement that the Archean atmosphere probably contained the following gases: H_2O (vapor), H_2, CO_2, SO_2, and minor amounts of CO and H_2S. There is much less agreement on the relative amounts (i.e., partial pressures) of these gases. For example, the Archean partial pressure of CO_2 (P_{CO_2}) has been the subject of considerable debate, with opinions ranging from very little atmospheric CO_2 to much more than today. Eriksson and Truswell (1978) argued that the paucity of Archean carbonate rocks was evidence for greater Archean P_{CO_2} since higher P_{CO_2} would have resulted in lower seawater pH. Siever (1977) opined that Archean P_{CO_2} was similar to that of today but the cooler surface temperature (see above) would have allowed more CO_2 to dissolve in seawater, again effecting a lowering of pH and less carbonate sedimentation. On the other hand, as discussed above, the paucity of Archean carbonates may be more artifactual than actual, blunting the thrust of both above arguments.

The reader will notice that O_2 is not included in the above list, an important omission when one considers the great significance of O_2 to contemporary weathering and sedimentation, not to mention biological processes. Of course, this point, too, is controversial and some authors, such as Dimroth and Kimberley (1976), have argued that abundant O_2 was available as far back as the oldest sedimentary rocks. They cite as evidence the presence of iron oxide as weathering rinds on andesite pebbles, as coatings on quartz pebbles, and as crusts on pillow basalts. But, as Blatt et al. (1980) pointed out, hematite is stable at very low P_{O_2} (as low as 10^{-68} bar). Thus, the presence of hematite as weathering rinds, etc. is insufficient evidence to claim an O_2-rich atmosphere. Further, the presence in alluvial (and therefore subaerial) sediments of rounded, detrital pyrite and uraninite grains has been cited by many authors as strong evidence for an O_2-deficient atmosphere. Of course, O_2 may have been generated, then as now, by photosynthesis so that shallow-marine environments with "clumps" of photosynthetic organisms may have had higher O_2 concentrations. Such a situation was discussed by Eriksson and Truswell (1978) as an explanation of the limited distribution of Archean BIF. But such "pockets" of higher O_2 would have released little O_2 to the atmosphere since most of it would have been used up in oxidation reactions involving reduced (ferrous) iron in solution. Further, any O_2 which did "leak out" to the atmosphere would have been taken up by oxidation of CH_4, CO, and other "reduced" gases. Thus, we may assume that P_{O_2} in the Archean atmosphere was very low.

B.2.c. Weathering Processes

We now have sufficient background to speculate on the nature of weathering processes during the Archean. Much of the following has been drawn from Siever's (1977) thoughtful review of early Precambrian weathering and sedimentation.

Chemical weathering of silicate minerals would probably have proceeded much as it does in modern environments. Feldspar alteration to clay is pH-controlled and thus depends today on P_{CO_2}. As we have seen, some geologists have speculated that CO_2 may have been present during the Archean in greater quantities than today, resulting in more-acidic meteoric waters and, consequently, an increased rate of feldspar weathering. Even if CO_2 were less abundant than today, other acid gases (e.g., SO_2, H_2S, HCl) would have been available. One interesting possible difference may have been the formation of an ammonium clay [e.g., $NH_4Al_3Si_3O_{10}(OH)_2$] by reaction with dissolved NH_4^+. The weathering of mafic minerals would probably also have been similar to present-day weathering because dissolution of mafic minerals does not depend on P_{O_2}. In modern weathering, Fe^{2+} released from the crystal lattice by hydrolysis is oxidized *in solution*. In an O_2-deficient atmosphere, Fe^{2+} would have been released in exactly the same manner but would not necessarily have undergone oxidation; thus, large amounts of Fe^{2+} could have been carried away in aqueous solution to the oceans.

If, however, the character of chemical weathering during the Archean was similar to modern processes, the *rates* of weathering probably were not. For one thing, if Siever is correct that the average surface temperature was lower, then chemical weathering, which is very temperature sensitive, would have been slower. Of greater significance would have been the lack of terrestrial vegetation, which acts today to retard runoff as well as to produce humic acids which accelerate chemical weathering. A rather exaggerated example of the effects of weathering under colder and apparently abiotic conditions can be seen in the Viking photographs from the Martian surface.

As will be seen later in this chapter, most students of Archean terranes believe that tectonic activity was greater then than now. If this were true, the dominance of physical weathering would have resulted in a rough and youthful topography characterized by jagged, bare rock exposures and deeply gullied, badlands topography. Abundant coarse sediment would probably have formed large alluvial fan and bajada systems at the bases of mountains, and streams would have been typically braided. Eolian transportation would probably have been extensive, even in moist climates, and eolian dunes and loess deposits were probably common. Briefly put, most terrains, even those in moist climates, would have looked very much like modern deserts.

B.2.d. Sedimentation

From the preceding discussion, one may conclude that texturally and compositionally very immature detritus, such as the polymictic conglomerates of the Moodies Group, was generated by Archean weathering and was transported to the sea via braided streams and glaciers. Since chemical weathering was probably of less significance than physical weathering, the gravel/sand and sand/mud ratios would have been somewhat higher than today because it is chemical weathering which tends to produce the finer grain sizes (Siever, 1977). Coarse arkoses, litharenites, and graywackes were the dominant early Archean sediment. By later Archean time, however, the existence of stable, sialic landmasses allowed the production of more quartzose sediments. Subarkoses, sublitharenites, quartz arenites, and quartz wackes have all been reported from later Archean deposits. In addition, volcaniclastic sediments were abundant, not only as detrital sediment but also, in the case of volcanic glass, as the raw material for chert.

Relatively few paleoenvironmental studies have been made of Archean sediments but three broad sedimentary facies may be recognized. (1) The turbidite/pelagic facies, usually characterized by rhythmically interbedded, graded volcaniclastic graywackes and shales with lesser amounts of dark chert, is referred to by several authors as the flysch facies and is considered to represent deposition in relatively deep-water environments as a result of turbidity currents and quiet-water settling of suspended loads. (2) The shelf/peritidal/terrestrial facies, a rather inclusive category, is characterized by coarse, crossbedded, heterogeneous sands, shales, and conglomerates deposited in a variety of shallow-marine to terrestrial environments. Siever (1977) suggested that most of today's major depositional environments, such as deltas, estuaries, barrier complexes, and shallow-marine sands, would also have been present during the Archean. This facies has been referred to as the molasse facies. (3) The starved-basin facies is characterized primarily by chemical sediments, especially chert and BIF, with very little associated detrital sediments. Sedimentation in these basins was controlled by the geochemistry of the Archean atmosphere and hydrosphere. The lack of detrital sediment probably points to the existence of some geographical barrier to sediment dispersal, such as a deep trench, or to the location of the basin adjacent to a low, peneplaned landmass which supplied little detritus.

This three-part division of major sedimentary facies into flysch, molasse, and nondetrital types is certainly not limited to the Archean! Indeed, it may be used with very little modification for much of the entire record of continental-margin sedimentation on Earth. In other words, while different, i.e., nonuniformitarian, sedimentological conditions may be recorded in some Archean sediments, they are differences of specific detail, controlled by the chemistry of Archean atmosphere and oceans, rather than of overall depositional framework, controlled by the physical results of tectonics.

B.3. High-Grade Terranes

The other major Archean lithological association is referred to as high-grade terranes or gneiss belts. As the term implies, these are areas of very-high-rank metamorphic rocks, predominantly ortho- and paragneisses, and are characterized by extremely complex structural deformation. Three main groups of metamorphic rocks are recognized in high-grade areas: (1) quartzofeldspathic gneisses, or "gray gneisses," composed to varying degrees of quartz, plagioclase feldspar (often with high Na/K ratios), potash feldspar, biotite and/or hornblende, and hypersthene; (2) metasupracrustal rocks, especially amphibolites, mica schists, quartzites, marbles, and metamorphosed BIF; and (3) metamorphosed layered igneous intrusions. Coward *et al.* (1969) and Condie (1984) observed that high-grade terranes typically contain a larger proportion of sediments than do greenstone belts, which are dominantly volcanic; especially common in high-grade terranes are pelitic sediments and quartzites. Metamorphic grade is typically of the granulite facies or, less commonly, the amphibolite facies. Windley (1984) pointed out that such high-temperature metamorphism, which is rare in younger metamorphic areas, was probably the result of steeper geothermal gradients within the Archean lithosphere. In some areas, retrograde metamorphism (i.e., mineralogical readjustment to pressure and temperature conditions lower than those which originally affected the rocks) occurred but the distinction between retrograde and prograde metamorphism in Archean rocks is often difficult to make.

High-grade terranes are among the most complexly deformed and structurally confusing areas on Earth. They have undergone multiple periods of strong deformation and are characterized by refolded isoclinal folds and large fold-interference patterns whose convoluted structures render meaningless the standard relative-age rules of strata in normal anticlines and synclines. In addition, all of the major lithologies occur in a seemingly "layer-cake" pattern due to tectonic intercalation, comparable to large-scale thrust sheets and nappes, and to concordant igneous intrusions "between the sheets." Bridgwater *et al.* (1978, p. 20) stressed that "virtually all contacts between rocks of different type . . . are either intrusive or tectonic" (see Fig. 2-12) so that, again, the conventional wisdoms of superposition, original lateral continuity, and so on are inapplicable. In the past, such structural and stratigraphic complexity allowed for very little detailed understanding of the geological history of high-grade terranes. Recently, however, careful studies in the North Atlantic craton (see Fig. 2-3) have disclosed "keys" to the structure so that a gross developmental history may be ascertained (e.g., McGregor, 1973; Bridgwater *et al.*, 1974).

In the following discussion, the main rock groups of the North Atlantic craton, especially in Greenland and Labrador, will be examined and will serve as a representative example of Archean high-grade regions. Figure 2-12 is a hypothetical vertical section which illustrates diagrammatically the usual relationships of the rock groups discussed.

B.3.a. Isua Supracrustals: The Oldest Rocks

The oldest known rocks of the North Atlantic craton are metamorphosed sediments and igneous rocks found as tectonic inclusions, or "rafts," contained within both the slightly younger Amîtsoq gneiss (3.75–3.6 Byr.) of West Greenland and the equivalent Uivak gneiss of Labrador (Bridgwater *et al.,* 1978). These inclusions vary consider-

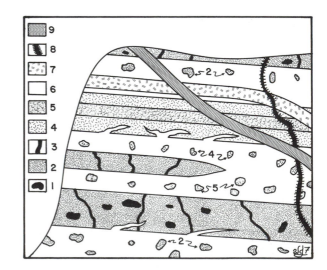

Fig. 2-12. Hypothetical vertical section through part of the Archean gneiss terrane of the North Atlantic craton. (1) Isua-type inclusions of amphibolites, metasediments; (2) quartzofeldspathic gneiss (Amîtsoq–Uivak gneisses); (3) Ameralik dikes; (4) Malene–Upernavik metasupracrustals; (5) layered anorthosite complexes; (6) younger quartzofeldspathic gneiss (Nûk); (7) Ilivertalik granite; (8) mafic to intermediate dikes; (9) Qôrqut granite. (From Bridgwater *et al.*, 1978.)

ably in size, the largest of which is called the Isua supracrustal belt of West Greenland and is approximately 40 km long and 2–3 km across. Most of the inclusions are composed of amphibolites believed to have originally been mafic and ultramafic volcanics (McGregor and Mason, 1977). Some mafic amphibolites are relatively fine-grained and have chemistries similar to those of modern oceanic basalts (although many authors have warned against taking geochemical comparisons too literally). Others appear to be compositionally and texturally layered (i.e., they were probably layered igneous intrusions) and have chemistries similar to those of komatiites (see section on greenstone belts). As the oldest volcanic rocks yet known, these amphibolites have been touted to be remnants of an originally simatic (i.e., oceanic) crust. Alternately, they may be volcanic extrusions onto an older (sialic?) crust that has not been preserved—at least in its original form. Debate over the nature of the primordial crust, and of the early Archean crust, is among the most hotly argued controversies of modern geology because the character of the crust is fundamental to any tectonic model. We will return to this topic in the next section.

Most metasediments in Isua-type inclusions are related to volcanic activity. For example, a major metaconglomerate within the Isua belt is composed of felsic volcanic clasts in a fine matrix of quartz, feldspar, and mica. The clasts are virtually unsorted and may represent a lahar, or volcanic mud-flow deposit, possibly derived from an adjacent massive felsic unit interpreted by Bridgwater *et al.* (1976) as a rhyolite flow or ignimbrite. Other, originally detrital sediments include quartz-pebble metaconglomerates, meta-siltstones, and biotite schists (probably metapelites). Rare but interesting are thin marbles from the Uivak gneiss; other marbles from high-grade Archean terranes were described by Windley (1984) and range in thickness from several meters up to several tens of meters. One particularly extravagant marble in southern India is over 250 m thick and 30 km long. Metamorphic BIF, composed of quartz (recrystallized chert) and magnetite, are rare components of some Isua strata, one occurrence being of sufficient tenor (i.e., grade) to mine. Finally, it is interesting to note that some biotite schists contain graphite; in some cases, the amount of graphite is considerable, as in the Graphite System of Malagasy (Besairie, 1967). The origin of these graphites is questionable as modern geochemical study of them has not been made; however, the possibility that they might be of organic origin makes them highly deserving of such study.

B.3.b. The Amîtsoq and Uivak Gneisses

Enclosing the Isua-type inclusions is quartzofeldspathic gneiss called the Amîtsoq in West Greenland and the Uivak in Labrador. Both are generally of granodioritic composition with occasional small bodies of more iron-rich gneiss. Complementary ages of 3.8–3.6 Byr. have been obtained using Rb/Sr, U/Pb, and Pb/Pb methods (e.g., Black *et al.*, 1971; Moorbath *et al.*, 1972). Typically, the gneisses are foliated and show compositional banding as well as augen structures in the more-iron-rich varieties.

The origin of these gray gneisses, and by extension the origin of gray gneiss in general, has been the source of much controversy. Their original lithology (protolith) has been variously interpreted as sedimentary (especially arkoses and graywackes), volcanic (of andesitic–dacitic composition), or intrusive granite. The argument for a sedimentary protolith is based primarily on the presence of Isua-type inclusions of demonstrable sedimentary origin; these were considered to represent patches of the original sedimentary mass which, for some reason, escaped full-scale gneissification (Smithson *et al.*, 1971). Proponents of a volcanic origin point to the general similarity (with a few exceptions) of the chemical compositions of the gneisses to those of greenstone-belt calc-alkaline volcanics. Of course, it is difficult to distinguish geochemically an extrusive from an intrusive lithology of the same original composition, especially after high-grade metamorphism and severe deformation. Windley (1984) elaborated on these discussions and concluded, based on geochemical and petrographic grounds, that the majority of Archean gray gneisses were originally plutonic granites, a conclusion shared by McGregor, Moorbath, Bridgwater, and others who have worked with the Amîtsoq/Uivak gneisses. The granodioritic gneisses mentioned above are, thus, believed to have been derived from originally calc-alkaline, syntectonic, plutonic rocks and the more iron-rich gneisses from tholeiitic intrusive rocks, possibly with some sialic contamination (Bridgwater *et al.*, 1978). The implication of this interpretation to the nature of the early Archean crust is fundamental. The oldest sialic masses yet discovered, i.e., the gray gneisses, were probably intrusive into simatic and volcaniclastic rocks, *rather than* underlying them as basement!

Amîtsoq gneisses are cut by thin mafic intrusions referred to as the Ameralik dikes; Uivak gneisses similarly contain the Saglek dikes. These dikes are not found in younger gray gneisses (see below) and their presence provided the "key" to unraveling the complex structural history of the Godthåb region of West Greenland (McGregor, 1973).

B.3.c. Malene and Upernavik Supracrustals

Up to 20% of the Archean gneiss complex of the North Atlantic craton is of metasupracrustal origin, only a small portion of which is accounted for by the very old Isua-type inclusions. The bulk of these rocks are younger than the Amîtsoq/Uivak gneisses but older than the extensive younger gneisses (e.g., the Nûk gneiss; see below). These supracrustal rocks are called the Malene supracrustals in West Greenland, where they are dominated by amphibolites derived from mafic igneous rocks, and the Upernavik supracrustals of Labrador, dominated by metasediments.

Amphibolites are generally massive except for those from several areas in West Greenland where pillow structures and pyroclastic, tuffaceous textures may be discerned. Associated with these rocks are detrital sediments containing coarse, poorly sorted rudites having clasts of plutonic, volcanic, and sedimentary rocks in a fine-grained matrix. These sediments are interpreted as possible slump breccias. Large bodies of ultramafic rocks with komatiitic chemistry and textures suggestive of spinifex texture (see section on greenstone belts) are associated with Upernavik rocks.

Metasediments include quartz-rich rocks which are interpreted as recrystallized chert and which contain garnet–quartz horizons, possibly representing minor BIF beds in the original succession. Metapelites such as biotite schists, some containing graphite, are abundant in Upernavik gneisses. Possibly-detrital quartzites occur as thin layers within some Scourian sequences in Scotland and may, in other Archean high-grade areas such as the Limpopo belt of southern Africa, reach thicknesses of 3 km. These quartzites are usually massive and structureless but, in a few places, may show traces of crossbedding.

B.3.d. Younger Quartzofeldspathic Gneisses

A major part (as much as 80%) of the gneisses of the North Atlantic craton is composed of gray gneisses which were metamorphosed and deformed *after* formation of the Malene/Upernavik supracrustals. This younger gneiss is called the Nûk gneiss in West Greenland and is similar in almost all respects to the earlier Amîtsoq/Uivak gneisses, except for the absence in the Nûk of the Ameralik dikes which serve as the key to separate the two gneiss complexes. The Nûk presents a new problem to petrologists: Was the Nûk formed by the metamorphism of new granites, intrusive after Malene time, *or* was it merely formed by the *reworking* (i.e., remetamorphism and diapirism, i.e., mobilization as an intrusive mass) of the Amîtsoq? It may be remembered that structural evidence indicates multiple periods of deformation so that reworking of old sialic material to make new sialic material is certainly possible. Moorbath (1977) has presented a summary of his elegant approach to this problem in an article in *Scientific American*. Moorbath's argument is based on the decay of radioisotopes contained within the Archean rocks and we digress briefly to follow his story.

As the reader may know, ^{87}Rb is radioactive and decays at a constant geometric rate to ^{87}Sr. ^{86}Sr, on the other hand, is not radiogenic. When an igneous rock crystallizes, the $^{87}Sr/^{86}Sr$ ratio of the magma at the time of crystalliza- tion is "fixed" as the initial $^{87}Sr/^{86}Sr$ ratio of the rock. As time passes, however, ^{87}Rb decays to ^{87}Sr while ^{86}Sr remains constant (we assume a closed system) and, consequently, the $^{87}Sr/^{86}Sr$ ratio increases while the $^{87}Rb/^{86}Sr$ ratio decreases (see Fig. 2-13A). Whole-rock $^{87}Rb/^{86}Sr$ dating recognizes the fact that all rocks are heterogeneous on a small scale and that different areas of a rock will have had slightly different initial amounts of ^{87}Rb. Since the ^{87}Rb decay rate is a function of the amount of ^{87}Rb present, a plot of $^{87}Sr/^{86}Sr$ versus $^{87}Rb/^{86}Sr$ for a group of randomly chosen samples from a given specimen, or outcrop, will define a straight line whose slope is directly proportional to the rock's age and whose ordinate-axis intercept is the initial $^{87}Sr/^{86}Sr$ of the rock (see Fig. 2-13B). Using such a method, Moorbath showed that the initial $^{87}Sr/^{86}Sr$ ratio of Amîtsoq gneiss was ~ 0.715. Now, if the Nûk gneiss (age ~ 3030 Myr.) were reworked Amîtsoq, its initial $^{87}Sr/^{86}Sr$ ratio would have to be the same as that of 3000-Myr.-old Amîtsoq gneiss, i.e., ~ 0.715. In point of

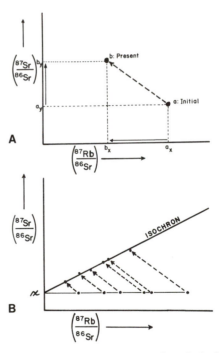

Fig. 2-13. The rubidium–strontium whole-rock method of radiometric dating. (A) Point "a" represents the initial ($^{87}SR/^{86}SR$ and $^{87}Rb/^{86}Sr$) ratios of a hypothetical rock sample; as ^{87}Rb decays to ^{87}Sr with time, the isotopic ratios change along the dashed line to point "b," which represents the present ratios. (B) The initial ($^{87}SR/^{86}SR$) ratios of randomly chosen samples from a given rock are all the same regardless of initial ^{87}Rb quantity; thus, a graph of initial conditions would result in a horizontal line through ($^{87}Sr/^{86}Sr$) = x (the initial Sr-isotopic ratio). Samples with more ^{87}Rb would decay more rapidly than samples with less ^{87}Rb so that a graph of the present isotopic ratios yields a straight line whose slope is proportional to the age of the rock. Note that the line, called the isochron, intercepts the ordinate axis at x, the initial value of ($^{87}Sr/^{86}Sr$).

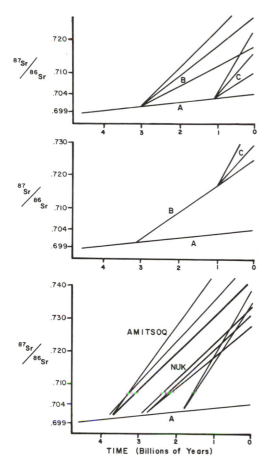

Fig. 2-14. Strontium-isotopic evidence on the origin of the Nûk gneiss of West Greenland. (A) Consider two hypothetical suites of granitic gneiss, B and C, with suite B being older than suite C; line A describes the change in ($^{87}Sr/^{86}Sr$) of the mantle with time and the converging lines labeled B and C represent the ($^{87}Sr/^{86}Sr$) changes of a group of related rocks with different original amounts of ^{87}Rb but all with the same initial ($^{87}Sr/^{86}Sr$), indicating that they formed at the same time. In graph A, suites B and C originated independently, with both derived from mantle materials; thus, their initial ($^{87}Sr/^{86}Sr$) ratios equal the strontium ratio of the mantle at the time of their formation. (B) If C forms from B, then C's initial ($^{87}Sr/^{86}Sr$) must be equal to B's strontium-isotopic ratio at the time of C's formation but *not* equal to that of the mantle at that time. (C) Graphs from Amîtsoq and Nûk gneisses are similar to graph (A) and the Nûk is therefore considered to have been derived directly from the mantle. (Redrawn from Moorbath, 1977.)

fact, however, its initial $^{87}Sr/^{86}Sr$ ratio is ~ 0.702, which is very close to the calculated $^{87}Sr/^{86}Sr$ ratio of the 3000-Myr.-old mantle (see Fig. 2-14). Thus, Nûk gneisses did *not* form from reworked Amîtsoq but rather from granites newly differentiated from mantle material. This is a profoundly important point because it provides strong evidence that sialic crustal material did not come into being all at once but rather was created by geochemical "milling" of the mantle over a period of time. Therefore, there must have

been less sial during the early Archean than in later geological periods.

The reader will not, by now, be surprised to learn that controversy exists over Moorbath's findings. Davies (1975) suggested that Moorbath's samples were biased in favor of rocks of well-preserved igneous affinities. Bridgwater and Collerson (1976) argued the more theoretical point that high-grade metamorphism may have caused considerable diffusion of both parent and daughter elements, thereby disallowing Moorbath's highly quantitative argument. Bridgwater *et al.* (1978) indicated that much of the younger gneisses of the Saglek Fjord area of Labrador are reworked Uivak but they agreed with Moorbath that Nûk gneisses of Greenland were probably formed from newly intruded calc-alkaline magmas. Of course, even if only a minority of the younger gneisses are of igneous origin, the point is still made: *sialic crust formed not all at once but over a long period of time.*

B.3.e. Late Granites

Completing the picture are a variety of relatively late granitic intrusions. For example, the Ilivertalik granite of Greenland (Rb/Sr whole-rock age $\sim 2760 \pm 30$ Myr.; Kalsbeek, reported in Bridgwater *et al.*, 1978) is a massive porphyritic granite emplaced under high-grade conditions *after* the main intercalation event but *before* the end of regional deformation. The youngest major Archean granitic intrusive rock is the Qôrqut granite (Rb/Sr whole-rock age $\sim 2520 \pm 90$ Myr.; Moorbath and Pankhurst, 1976). The Qôrqut is a massive K-rich granite emplaced after the end of granulite-grade metamorphism.

B.4. Relation of High-Grade Terranes to Greenstone Belts

It may be seen from the above discussion that Archean high-grade terranes generally represent a complex association of sialic rocks characterized by a long history of episodic strong deformation, granitic intrusions, volcanic activity, and sedimentation on a generally, but not exclusively, shallow-water shelf. Greenstone belts, on the other hand, represent a thick volcanic pile whose composition varies from ultramafic at the bottom to felsic at the top and which contains intimately interbedded sediments generally indicative of deep-water environments but showing a shallowing of environment through time. It remains to consider the relationship between these two major Archean lithological assemblages. Unfortunately, this is not easily done because deformation in Archean areas is such that present-day contacts between rocks of the two terranes (where they contact each other at all) are almost exclusively tectonic in nature, the relationship being obscured by massive younger igneous intrusions, large-scale faulting, or post-Archean reworking. This lamentable reality is the bane of Archean-tectonics research!

There are, of course, three possibilities: (1) high-grade terranes predate greenstone belts, implying that greenstone belts were originally ensialic; (2) greenstone belts are older, implying that the early crust was simatic; or (3) both terranes are essentially contemporaneous. We have shown that the oldest known supracrustal rocks (the Isua belt of West Greenland) are composed primarily of mafic and ultramafic rocks with sediments that did not have any appreciable sialic provenance. And we have shown that isotope geochemistry reveals that the amount of sialic crust has increased with time. On the other hand, some parts of greenstone belts lie unconformably on high-grade gneisses, as in areas of Superior Province (Ermanovics and Davison, 1976) and in Tanganyika (Haidutov, 1976). And considerable quantities of sial-derived sediments are present in greenstone belts which appear not to have any intrabelt sialic sources. Clearly, the evidence is contradictory.

The history of geological controversies teaches that in such situations, neither side of the debate is wholly right nor wholly wrong. In this case, the answer seems to lie in the middle: high-grade terranes and greenstone belts are contemporaneous. In advancing this idea, Sheckleton (1976) stated that there is structural and metamorphic continuity between the two types of terranes in Africa and India and that a gross stratigraphic correlation may be drawn between relatively young sediments of southern Africa greenstone belts and those of the high-grade Limpopo belt. As mentioned in the introduction of this chapter, Drury (1978) sagely advised that Archean conditions and processes must be considered the material basis for present conditions and processes. Thus, it seems reasonable that, even in the early Archean, there might have been areas of sialic crust, albeit smaller than today, *and* areas of simatic crust existing contemporaneously just as continental blocks and oceanic areas do today. A more specific analysis of their relations, however, depends fundamentally on the character of Archean tectonics and, naturally, opinions vary. We turn now to consider several of the most popular tectonic models.

C.   Hypotheses on Archean Tectonics

Speculation on Archean tectonic mechanisms is a stimulating exercise in interpretive geology but, as in any such exercise, the relative usefulness of a given tectonic model depends on the experience, skill, and bias of the modeler. Thus, it is not surprising that regions of inherent geological complexity such as Archean terranes would have engendered a variety of interpretations. Some tectonic models have been developed to explain the features of a particular

region and are therefore provincial in nature, with limited applicability to other regions. Other models are more broadly based, developed to account for general Archean tectonic evolution; but these models often gloss over specific regional details which do not fit the overall picture. Probably the most valid approach has been to incorporate local details into an interregional model. But here one encounters problems of bias: different geologists may view the same Archean terrane as having been the result of entirely different, even mutually exclusive, tectonic mechanisms. Thus, different schools of thought have developed whose proponents share the same philosophical orientation toward the interpretation of geological data.

In the following pages, we will discuss three types of tectonic models which have been chosen to illustrate the range of speculation: (1) "classical" models, developed by appeal to an almost strictly vertically oriented tectonic regime; (2) uniformitarian models, developed by analogy to modern plate-tectonics mechanisms and based on the strict-uniformitarian assumption that Archean tectonic mechanisms were nearly identical to today's; and (3) actualistic models, based on theoretical analyses of evolutionary conditions affecting tectonics, e.g., changes in the geothermal gradient, crustal composition, and lithosphere thickness. Naturally, all three types of models conclude at the same point, with the development, by the end of the Archean, of large, stable, sialic continental masses. But they certainly do not begin at the same point and debate over the nature of the primordial crust is a thread of contention which laces through all.

C.1.   "Classical" Models: The Downsagging Basin

Some of the first modern tectonic models for the Archean were developed by South African and Australian geologists (e.g., Viljoen and Viljoen, 1969; Anhaeusser, 1975) to account for the features of greenstone belts. These models begin with the presumption that greenstone belts preserve remnants of an originally simatic crust and that sialic material developed later in Archean history. They further presume that the weight of this simatic crust caused it to subside, or sag, into the mantle where it underwent partial melting, thereby generating large volumes of magma which rose diapirically into the crust near the surface. Glikson's (1972) discussion is a good example of such models.

Glikson argued that the oldest crust was simatic on the basis of the following evidence, some of which has been disputed by later authors: (1) ultramafic–mafic assemblages of greenstone belts are never observed to overlie granitic rocks unconformably (cf. Ermanovics and Davison, 1976); (2) Archean granites are always intrusive into greenstones; (3) gray gneisses of the Scourian complex in Scotland originated as mantle-derived volcanics rather than as reworked sial; and (4) Archean mafic rocks show no signs of geochemical contamination by sial, as might be expected if they had intruded a thick sialic crust. Thus, Glikson concluded, the earliest crust was a continuous, oceanic-type crust composed of ultramafic and mafic rocks such as the Tjakastad subgroup of the Onverwacht Group in the Barberton Mountain Land.

Figure 2-15 is Glikson's model of Archean tectonic evolution. The earliest events, termed the oceanic stage, were initiated by subsidence of linear zones of simatic crust to relatively great depths. The cause of this crustal subsidence, referred to as "megarippling" of the crust, was not discussed, but Anhaeusser (1975) stated that subsidence was caused by large-scale gravitational instability. He argued that the lower part of the crust would have been composed of eclogite (density = 3.5), which is the high-pressure-phase chemical equivalent of basalt; since peridotite of the mantle has a density of 3.3, the dense, eclogitic lower crust would have sagged down isostatically, giving rise to calc-alkaline magmas as the eclogite melted in the higher-temperature regime of the mantle. Also during the oceanic stage, chemical sediments (e.g., cherts, BIF) were deposited within subsiding areas of the crust; the near absence of detrital sediments was presumably due to the absence of emergent land areas to supply them.

Calc-alkaline magmas generated by partial melting of the sagging crust intruded the simatic crust as oval-shaped or elongated batholithic masses. This is termed the early plutonism stage. Resulting plutons are composed primarily of trondhjemite, a light-colored igneous rock consisting of sodic feldspar (oligoclase) and quartz with minor amounts of biotite. Green and Ringwood (1968) showed that trondhjemitic magmas could be produced by the partial melting of eclogite under wet conditions; thus, the presence of trondhjemites is easily explained by Glikson's model.

Areas of extensive trondhjemites rose isostatically while intervening areas of simatic crust subsided, leading to the volcanic–sedimentary stage. The result was a kind of topographic inversion where old subsiding areas (of the oceanic stage) became emerging sialic islands while simatic areas in between subsided isostatically, with consequent partial melting and calc-alkaline igneous activity. These newly subsided basins were to become the greenstone belts. In them were deposited calc-alkaline lavas and pyroclastics, chemical sediments, and, as time went by, detrital sediments derived from weathering of the sialic islands. Thus, Glikson's model explains the development of Geluk-type mafic-to-felsic volcanics over Tjakastad rocks and the subsequent deposition of deep-water Fig Tree flysch and shallow-water Moodies molasse.

The orogenic stage was a continuation of the previous stage whereby further subsidence of greenstone-belt basins

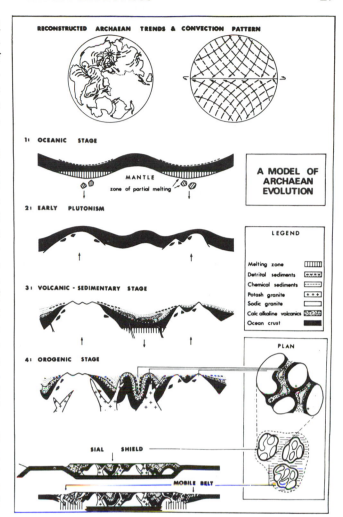

Fig. 2-15. Glikson's model of Archean tectonic evolution; see text for discussion. (From Glikson, 1972, *Geological Society of American Bulletin*, Fig. 12, p. 3338.)

led to the onset of vertical, isoclinal folding and associated faulting (see Fig. 2-7). Partial melting of the lower crust where it had subsided to great depths generated large volumes of potassic granitic magma, which intruded the already-deformed greenstone belts to form the "gregarious batholiths" mentioned earlier; stresses attendant on these diapiric intrusions further deformed greenstone-belt strata.

The result of all this sagging, intrusion, and deformation was the formation of small "granitic nuclei." These nuclei aggregated upon cooling to form sialic shields, a process on which Glikson did not elaborate.

There are some attractive aspects of this model, especially its explanation of greenstone-belt stratigraphy and structures. (Of course, it was developed specifically for greenstone-belt areas.) On the other hand, it has some serious weaknesses. It does not address, except in passing, the geology of high-grade terranes. Since some geochemical and petrological differences between the two types of Archean terranes are greater than may be explained by differences in metamorphic intensity alone, this is an unfortunate omission. Of greater significance, however, is the fact that there is, as we have shown, considerable evidence in high-grade terranes for horizontally directed tectonics; such horizontal tectonics during the Archean are implicitly denied by Glikson's model. Another problem is that many recent authors have argued convincingly that the Archean geothermal gradient was considerably higher than today's and, therefore, that eclogite would not have been stable in the crust due to the existence of much higher temperatures at relatively shallow crustal depths (Drury, 1978). Were eclogite not present in the lower crust, Glikson's model would be deprived of a fundamental element. A final objection lies in the fact that the downsagging-basin model does not provide the "material basis" for Phanerozoic plate tectonics, as Drury (1978) insisted that any Archean tectonic model must. In other words, Glikson's model does not show an evolutionary development culminating in horizontally moving plates of lithosphere.

C.2.   Uniformitarian Models

The set of theories referred to collectively as plate tectonics was developed initially to explain contemporary global seismicity and volcanism but its usefulness as an interpretive model for ancient tectonic events was realized rapidly and it has been used successfully to explain many aspects of Phanerozoic orogenesis. But arguments have developed over *when* modern plate tectonics mechanisms began. Advocates of the previously discussed "classical" models clearly believe that such mechanisms were not functional during the Archean but other geologists have seen in Archean terranes evidence for the workings of essentially modern lithosphere-plate tectonics. Thus, they make the strict-uniformitarian assumption that the very oldest extant rocks were formed in the same tectonic environments as are modern rocks. Moorbath (1975) said, ". . . the only really major nonuniformitarian geological event that we can discern with any certainty is the formation of the Earth itself." In this section, we will examine two different plate-tectonics models for Archean tectonics, one of which considers greenstone belts to have formed as rift zones between divergent sialic blocks and the other of which treats greenstone belts as marginal (i.e., back-arc) basins and high-grade terranes as associated magmatic arcs.

C.2.a.   Rift-Basin Model

Goodwin and Ridler (1970) described the Abitibi greenstone belt of Canada's Superior Province (see Fig. 2-8) and concluded that it may have formed as a rift basin within a sialic continental mass. This conclusion is clearly based on the assumption that sialic crust predated the greenstones. Henderson (1977) presented several lines of evidence from greenstone-belt strata (the Yellowknife Supergroup) of Slave Province to support this view: (1) quartz-rich detrital sediment is common in Yellowknife rocks and appears to have been derived from a granitic terrane; (2) judging from the great volume of these quartzose sediments, the size of the granitic source-terrane must have been very large; (3) unconformable relations are observed between underlying granite and overlying greenstone-belt strata; (4) basal conglomerates of some greenstone-belt rocks contain granite clasts; (5) granitic boulders, presumably derived from the underlying basement, are present in diatremes which intrude greenstone-belt volcanics; and (6) gravity measurements indicate that greenstone-belt strata are *not* underlain by extensive mafic rocks.

Goodwin and Ridler's rift-basin model is based on the spatial and chronological relationships of Abitibi rocks. They showed that major volcanic complexes (see Fig. 2-8B) lie close to the sialic forelands (i.e., high-grade terranes) which border the basin to the north and south. The location of younger volcanism shifted toward the center of the basin so that progressively younger volcanics occupy progressively more central positions. Furthermore, they also demonstrated that younger sediments occupy more central positions within the basin. The implication, of course, is that the earliest supracrustal rocks were formed on the margins of an initial rift; as the sialic blocks diverged, younger volcanics (associated with the "spreading center") and sediments were disposed in more central positions. The resulting relationship is very much like that shown by the relative ages of volcanic rocks in an oceanic ridge or a modern rifted basin such as the Red Sea. It is also interesting to note that Abitibi banded iron formation (see p. 21) indicates that the environment of the basin center, where the youngest rocks were deposited, was one of considerably deeper water than the environment under which older, marginal strata were deposited. This brings to mind the isostatic subsidence which accompanies contemporary continental rifting. One final line of evidence is that lower greenstone-belt volcanics are geochemically similar (but not identical) to modern oceanic tholeiitic basalts, suggesting that the Archean volcanics may also have formed at divergent boundaries.

One of the problems with this model is its failure to explain the origin of the sialic crust whose rifting began the sequence of events. Also, the model fails to account for areas, such as the Barberton Mountain Land, where evidence seems to suggest that sialic crust did *not* predate greenstone-belt formation. Both of these objections are probably unfair, however, because Goodwin and Ridler's ideas were not advanced as a general Archean tectonic model but rather are specifically applicable only to the

Abitibi area. A more valid objection lies in the fact that Superior Province greenstone belts (the Abitibi included) form a roughly linear array (see Fig. 2-6) separated by roughly linear high-grade terranes. If Goodwin and Ridler's model were strictly correct, the indicated explanation would be multiple riftings of long, thin slivers of sial. Such a situation is not observed in Phanerozoic tectonics. Finally, as Windley (1984) pointed out, Goodwin and Ridler's model is vague about the tectonic and structural controls on rifting, with no explanation of *why* rifting occurred where it did.

C.2.b. Marginal-Basin/Magmatic-Arc Model

The geochemistry of greenstone-belt volcanics has led various authors to propose analogies with recent volcanics. As mentioned above, the geochemistry of lower, ultra-mafic–mafic rocks has been compared with that of modern oceanic tholeiites, and stratigraphically higher volcanics have been related to modern magmatic-arc tholeiites because they are more differentiated. In both cases, however, the analogies were somewhat strained. Tarney *et al.* (1976) demonstrated a better geochemical similarity with the basalts of modern marginal basins, especially in terms of the relatively high amounts of K, Rb, Ba, Cr, and Ni. Based on this similarity, and on a striking modern analogue (see below), Tarney *et al.* proposed that greenstone belts are the remains of Archean marginal basins. Agreeing with this proposal, Windley (1984) pointed out that normal oceanic lithosphere is usually lost by subduction and is only preserved as highly deformed slivers of ophiolite within the suture zones of collisional orogens. Further, Windley argued, magmatic-arc volcanics are characteristically uplifted isostatically and eroded so that only their plutonic roots are preserved. Thus, of the various possible modern analogues, only marginal-basin lithologies stand a fair chance of being preserved in the geological record.

Another argument for the marginal-basin model is the existence of a striking Tertiary analog to Archean greenstone belts: the Rocas Verdes complex of southern Chile. The Rocas Verdes (which means "green rocks"!) complex is a synclinal basin whose dimensions, igneous rock associations, structural style, sedimentary rocks, greenschist metamorphism, and relationship to older sialic basement are all remarkably similar to Archean greenstone belts.

Figure 2-16 illustrates Tarney and colleagues' model of greenstone-belt development. Note that detrital sediments deposited within the basin were carried to it from sourcelands on either side. Such dual sediment sources have been observed in Archean greenstone belts by several authors (e.g., Goodwin and Ridler, 1970). Note also that this model accounts for the basin extension postulated by Goodwin and Ridler for the Abitibi belt.

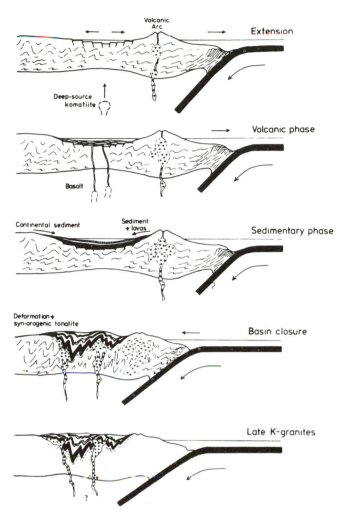

Fig. 2-16. Developmental model for Archean greenstone belts based on analogy with back-arc marginal basins such as the "Rocas Verdes" complex of southern Chile. (From Tarney *et al.* © 1976 by John Wiley & Sons, reprinted with permission of John Wiley & Sons.)

The counterpoint to the marginal-basin model for greenstone belts is the magmatic-arc model for high-grade terranes. Windley (1984) elaborated on this model, pointing out that Archean high-grade terranes find their closest analogs in the Mesozoic batholithic complexes of the North and South American Cordillera. These complexes were formed by intensive igneous intrusion and high-rank metamorphism at an active, convergent continental margin. He discussed the following lithological similarities: (1) the abundance of tonalites and granodiorites in both Archean and Mesozoic areas as well as their association with rarer potassic granitic intrusive rocks; (2) the presence, within Mesozoic complexes, of deformed and metamorphosed "rafts" of continental shelf and slope sediment, similar to Isua-type tectonic inclusions (see p. 24); and (3) the presence in both areas of layered igneous intrusions. Further, the common occurrence of major thrusting and tectonic intercalation in Archean high-grade terranes is mirrored by similar structural patterns in Mesozoic high-grade terranes. Based on these similarities, Windley argued that the Archean high-grade rocks were formed as the intrusive roots of magmatic arcs. The large horizontal component of tectonic stresses recorded in these rocks was due not only to the normal underthrusting characteristic of subduction zones but also to occasional arc/arc or arc/continent collisions. Isostatic uplift of the arc mass caused erosion of the upper stratigraphic levels of the arc complex, primarily involving volcanic and sedimentary supracrustals, so that only deep-seated, high-grade metamorphic and igneous rocks remained. This explains the observation by several authors (e.g., Goodwin, 1978) that high-grade terranes represent deeper levels of erosion than do greenstone belts.

Perhaps the most attractive aspect of the marginal-basin and magmatic-arc models is that they may be integrated well together into a single, coherent picture of Archean tectonics. Windley's integrated model is shown in Fig. 2-17. He speculated that magmatic arcs formed slightly *before* marginal basins. Somewhat later extension in back-arc areas initiated marginal-basin, i.e., greenstone-belt, development. Thus, the earliest marginal-basin volcanics may have formed on the thinned sialic crust of the adjacent magmatic arc (as in Canadian greenstone belts) or in a deeper, simatic environment (as in the southern Africa examples); the difference might have been due to different rates of

marginal-basin extension. Note that this avoids the question of primordial crustal type, implying that the answer depends on where one looks! Occasional arc/arc collisions and suturing resulted in the growth of sialic masses, so that by the end of the Archean, large, stable, continental masses had been formed. The episodic collisions of linear arcs and associated marginal basins may also explain the tendency of some greenstone belts and high-grade terranes to be arrayed linearly (as in Superior Province of Canada and the Yilgarn craton of Australia; see Fig. 2-6).

One last aspect of this model deserves mention. The distinction between the long, narrow orogens of Phanerozoic ages and the relatively short Archean orogens has been mentioned previously in this chapter and we have mentioned the permobile character of Archean tectonics. Both of these observations may be explained by the existence, throughout much of Archean time, of relatively small sialic masses which had not yet grown (by addition of new sial and by collision and suturing) into large, stable continents. Being small, these sialic "islands" could not have had long marginal orogens; as the size of the sialic masses increased, so did the length of the marginal orogens. Since the width of an orogen is constrained by the angle of subduction, and may be considered in a uniformitarian model to be relatively constant, the length/width ratio of orogens *must* increase with time as the sialic mass grows. Subduction-related events (e.g., intrusion, metamorphism) would have affected virtually the entire extent of such small, sialic

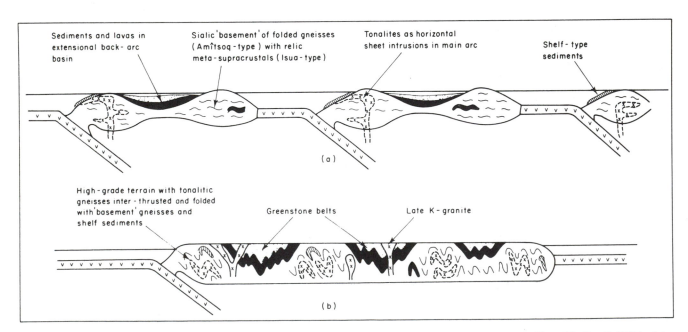

Fig. 2-17. Generalized plate-tectonics model of Archean tectonics incorporating the marginal-basin model of greenstone belts and the magmatic-arc model of high-grade terranes; arc/arc or arc/remnant-arc collisions result in formation of larger sial masses. (From Windley © 1977 by John Wiley & Sons, reprinted with permission of John Wiley & Sons.)

islands so that, after stable-continent formation, the appearance would be of a large terrane everywhere affected by tectonic activity.

Windley's integrated model explains many aspects of Archean tectonics and it deserves further, careful study. But the assumption upon which it is based, that of strict uniformitarianism, is questionable. It is interesting to note that Windley and others frequently mention the probability of a higher Archean geothermal gradient but they do not carry this idea to its logical conclusion. If the Archean geothermal gradient had been higher than today's, the lithosphere would probably have been thinner and the asthenosphere thicker (Drury, 1978); thus, the overall tectonic behavior of the Earth's plates may have been considerably different than now. This, of course, is not to say that it *was* different but rather that one cannot *assume* a uniformitarian position without careful analysis of the possible effects of different Archean conditions.

It is our orientation that planets, the Earth included, go through an evolutionary development [see, e.g., Kaula's (1975) article on the seven ages of a planet]. Thus, plate tectonics should be understood as an evolutionary process, i.e., changing slowly with time as the forces which drive it and the conditions which constrain it change. In the following section, we will consider one further model which attempts to develop an actualistic picture of Archean tectonics.

C.3. Development of an Actualistic Model

C.3.a. The Archean Geothermal Gradient

Since radionuclides decay at known, geometric rates and their present-day abundances can be estimated, it is fairly easy to calculate their abundances during the Archean. From such calculations, it can be shown that there were about eight to nine times more ^{40}K and about four times more ^{235}U and ^{238}U during the Archean and that these isotopes were the main sources of radiogenic heat at that time (McKenzie and Weiss, 1975; Tarling, 1978b). Naturally, the greater quantity of radionuclides resulted in greater amounts of internal heat; Lambert (1976) estimated that approximately three to four times more heat was produced during the Archean than today. It is instructive to consider the probable effects of this increased heat production on properties of the Archean asthenosphere, lithosphere, and crust.

i. Asthenosphere. The physics of convection is such that increasing the amount of heat at the base of a convecting cell results in an increase in the rate of convective overturn rather than in an increase in the thermal gradient within the cell (Tozer, 1972). This is because the

viscosity of the material is maintained approximately constant. Owing to this, the Archean asthenosphere was probably characterized by more rapid overturn than today but with a similar geothermal gradient (Tarling, 1978b). This is not to say that the Archean asthenosphere was not hotter than today's but rather that the gradient (i.e., the rate of temperature increase per increment of depth) was about the same.

A different thermal regime within the mantle raises the possibility that the geometry of convection may have been different during the Archean. Burke and Kidd (1978) suggested that the oceanic ridge system may have been longer at that time in order to dissipate the greater amount of internal heat. They also suggested that the rate of lithosphere spreading may have been greater as well. Of equal importance is the suggestion of McKenzie and Weiss (1975) that the actual nature of Archean convection may have differed from that of today. Within a layer such as the asthenosphere, whose thickness is small compared with its other dimensions, there are often two different but coexisting sets of convective motions: a large-scale convection affecting much of the layer and a small-scale convection superimposed on it (see Fig. 2-18). In the smaller-scale flow, heated material rises at a single point (a thermal jet, or plume) and spreads radially outward across the surface of the convection cell, where it loses heat, cools, and finally sinks to complete the cell. Since many small-scale cells would be present in the convecting layer, they would interfere with each other and acquire mutually convenient, polygonal boundaries. This dual form of convection is probably present in any thin, plastic layer to which heat is applied at its base, but the relative velocities of the two convection types depend on the character of the material and the actual amount of heat supplied. McKenzie and Weiss (1975) argued that under early Archean conditions, the small-scale flow would probably have been dominant. The implications of this for Archean tectonics are profound and will be considered later in this chapter.

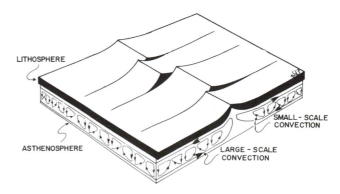

Fig. 2-18. Idealized sketch of large- and small-scale convection within the asthenosphere.

One final point: while convection would have dissipated heat within the asthenosphere so that large-scale melting did not occur, incipient melting (less than 1%) would probably have extended through a much larger part of the mantle (Tarling, 1978b). Thus, the Archean asthenosphere was probably thicker than today's.

ii. Lithosphere. While the geothermal gradient within the Archean asthenosphere may have been similar to today's, that of the lithosphere probably was not. Since the Archean asthenosphere was hotter than at present but the Earth's surface temperatures were similar, the near-surface gradient must have been steeper! The reason for this steeper gradient is that conduction, which is the principal heat-transfer mechanism of the rigid lithosphere, is much less effective than is convection. Based on the rare-earth geochemistry of komatiite flows in the Barberton Mountain Land, Green (1975) suggested an Archean near-surface gradient of 25°C/km. Determination of Archean near-surface gradients can also be made by examination of the mineralogy of metamorphic rocks; estimates based on this approach range from 60 to 15°C/km with several determinations around 25–35°C/km (Drury, 1978). Condie (1984, and references therein) states that metamorphic conditions in greenstone belts indicate a near-surface gradient of 50–70°C/km. Modern near-surface gradients in continental shield areas average about 12°C/km and in oceanic areas, about 16–20°C/km. Thus, there is both theoretical and observational evidence that Archean near-surface gradients were steeper than modern ones (see Fig. 2-19).

Perhaps the most significant effect of this steeper gradient on the nature of the Archean lithosphere is that any given temperature zone would have been closer to the surface than today. Since the lithosphere–asthenosphere boundary probably represents the solidus temperature of mantle rock (i.e., that temperature above which solid and liquid may coexist in equilibrium), the clear implication is that the Archean lithosphere must have been thinner. Of course, radiogenic heat production decreased at a geometric rate throughout the Archean so that lithosphere thickness probably increased with time as the actual amount of heat declined.

Estimates of Archean lithosphere thickness range considerably. Condie (1984) recognized three types of Archean lithosphere and speculated about their thickness. "High-grade" lithosphere, i.e., the lithosphere associated with high-grade terranes, was considered to have been the thickest, being around 75 km thick. Oceanic lithosphere, which is not represented by any extant lithological assemblage, was considered to have been about 15 to 25 km thick. And "greenstone" lithosphere, which Condie considered to have been situated between high-grade and oceanic lithosphere and to have been intermediate in thickness, was

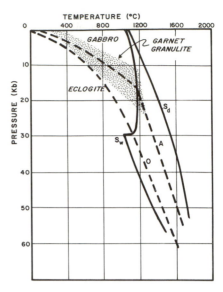

Fig. 2-19. Comparison of hypothetical Archean near-surface geothermal gradient (A) with near-surface geothermal gradient in modern oceanic regions (O). Solidus lines for mantle rock under anhydrous (S_d) and hydrous (S_w) conditions are shown for comparison as are stability fields of garnet granulite (patterned), gabbro, and eclogite. (Modified from Green, 1981.)

considered to have been about 50 km thick. The structural effects of a thinner lithosphere and a thicker asthenosphere will be considered later in this section.

iii. Crust. As we have seen, there has been considerable controversy over the nature of the earliest crust. This problem can also be examined in the light of high Archean heat production. In order for a sialic crust to exist, it must be buoyant enough to resist the downward pull associated with the sinking limbs of convective cells beneath it. If it is not sufficiently buoyant, it would be pulled down continually and stirred into the asthenosphere. Kaula (1975) stated that, in order for the crust to resist the downward pull of convection, the density difference between the crust and the asthenosphere must be comparable to the density differential driving the convection. There are two aspects to this problem: the nature of convection and the density difference between the crust and the mantle. We have already shown that Archean convection was probably more vigorous than today's; thus, the downward component of convective force would have been greater. By comparing the early Archean near-surface geothermal gradient with the pressure–temperature stability fields of common crustal rock-types, one may deduce which lithologies would have been present in the early Archean crust. Tarling (1978b) used this approach to show that, prior to 4.0 Byr. ago, asthenosphere convection would have been too vigorous for *any* differentiated crust to have been preserved. Tarling presented a hypothetical analysis of the geothermal gradient existing just before the beginning of the Archean (see Fig. 2-20) showing that the gradient at that time had not yet intercepted the stability fields of either gabbro, eclogite, or garnet granulite (i.e., rocks generally of basaltic composition). The implication is that no stable, differentiated crust was then feasible and that the lithosphere was, at that time, only rigidified asthenosphere, possibly with a thin but impermanent sialic frosting.

The beginning of the Archean, in Tarling's view, represents that point in the history of the geothermal gradient at which the gradient intercepted the stability field of garnet granulite (see Fig. 2-20B). This is a very interesting point. It suggests that rocks of much greater age than about 3.9 Byr. will probably not be found, because only by this time were rocks of sufficiently low density formed in significant quantities to be preserved from being remixed back into the mantle. Thus, the beginning of the Archean may not be just an arbitrary date representing the age of the oldest rocks yet to be found, but rather a real and significant event in Earth history: the beginning of the crust.

Of significance to our present discussion is the corollary of Tarling's arguments that the earliest Archean crust was composed of rocks freshly differentiated from the mantle such as ultramafics (like komatiites) and mafics (like basalt). Incidentally, the presence of komatiites is strong evidence for a higher near-surface geothermal gradient because they represent a high degree of mantle melting at a relatively shallow depth (Green, 1975). Since basaltic rocks were a major component of the early crust, their behavior is of some interest to Archean geology. Drury (1978) and Tarling (1978b) both noted that the Archean geothermal gradient was too high for eclogite to form (see Fig. 2-19). Instead, garnet granulite was the stable phase. This is a very important point because garnet granulite's density is approximately 3.0–3.2, which is *less* than that of the peridotitic mantle (3.3), whereas the density of eclogite (3.5) is *greater* than the mantle's! The significance of this will be seen later in this section.

C.3.b. The Primitive Stage

The preceding discussion of the Archean geothermal gradient provides the theoretical background upon which the following model of Archean tectonics is based. Since the character of tectonics is considered in this model to have changed during the Archean, we have divided the discussion into three parts: (1) the primitive stage, which describes the earliest Archean tectonic regime; (2) the permobile stage, which characterized most of Archean history; and (3) the Archean–Proterozoic boundary.

The beginning point of Drury's (1978) model is the presence, as discussed in previous paragraphs, of a thick,

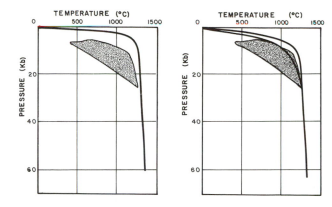

Fig. 2-20. Hypothetical near-surface geothermal gradients for pre-Archean and Archean times; the stability field of garnet granulite is patterned. (A) Pre-Archean gradient; only one line is shown because continents presumably did not then exist. Note that the gradient is nearly linear and almost vertical up to a depth of approximately 30 km where it bends and changes dramatically near the surface; the change in slope probably reflects the lithosphere–asthenosphere boundary. Note also that the gradient does not intercept the stability field of garnet granulite. (B) Archean gradient; the slight separation of the gradient into two lines above the granulite field reflects the presence of continents. (From Tarling, 1978b.)

convecting asthenosphere, a thin lithosphere, and an ultramafic–mafic crust initially containing little sialic material. Convection in the asthenosphere was dominated by the small-scale convection mentioned above so that heat and mantle-derived magmas rose at randomly oriented "hot spots." Three-armed rifts developed in the thin lithosphere over the thermal jet; spreading of the resulting three plates radially away from the three-armed rift led to the development of a ridge–ridge–ridge type of triple-plate boundary

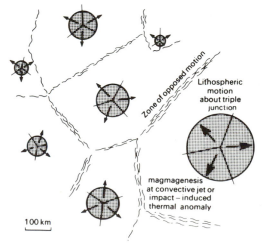

Fig. 2-21. Magmatism, lithospheric motion, and polygonal interference pattern for the primitive stage of Archean tectonics. (From Drury, 1978, reproduced with permission of Elsevier Publishing Company.)

centered at the hot spot. The plates moved across the surface until they met opposing plates associated with an adjacent convection cell (see Fig. 2-21). Since the convection cells were probably of approximately polygonal shape, the zones of opposing lithospheric motion were probably polygonally disposed about the hot-spot.

Exactly what happened at the zones of opposing motion would have depended on the thickness and physical properties (e.g., temperature, viscosity, ductility) of the lithosphere at that time. If the lithosphere were sufficiently weak, it may have buckled at these zones, resulting in the thickening of the lithosphere. On the other hand, the opposing plates may have been thrust one under the other. In Phanerozoic (and, presumably, Proterozoic) tectonics, when a simatic plate is pushed beneath another plate, a phase change occurs in the downgoing plate (as a result of increased pressure and temperature) in which basalt is changed to eclogite. Since eclogite's density is greater than that of the asthenosphere, the eclogitic part of the subducting plate pulls it steeply down into the mantle. But during the Archean, the near-surface geothermal gradient was too high for eclogite to have formed and garnet granulite was the stable lithology. Since granulite is of *lower* density than the asthenosphere, the underthrust plate would not have dipped steeply into the asthenosphere but rather would have been shoved beneath the overriding plate in the manner shown in Fig. 2-24. The reader should ponder this suggestion carefully for it is one of the major points of difference between this model and the more familiar model of Phanerozoic plate tectonics.

Calc-alkaline material could have been produced in one of several ways during this primitive stage (see Fig. 2-22). It may have formed as a result of partial melting of the simatic crust where its base subsided below the depth of the basalt solidus temperature. Such subsidence may have been due either to thickening of volcanic edifices at hot spots (Fig. 2-22a) or to lithosphere buckling at zones of opposing motion (Fig. 2-22b). Alternately, calc-alkaline material may have formed due to the partial melting of an underthrust plate (Fig. 2-22c). Regardless of the specifics, however, sialic material increased in quantity with time within the Archean crust.

C.3.c.   The Permobile Stage

Before long in the Archean, the small, three-armed spreading centers began to coalesce (see Fig. 2-23), possibly as a result of interference between adjacent convective jets to form a geothermal "saddle" between them or as a result of a slight increase in the thickness of the lithosphere as the geothermal gradient slowly declined. At any rate, linear zones of magmatic activity and lithosphere spreading developed and an Earth-surface geometry of divergent and

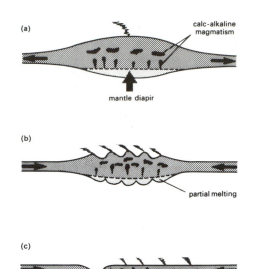

Fig. 2-22. Possible causes of secondary, calc-alkaline magmatism during the Archean; all three involve partial melting of ultramafic crust depressed below the basalt solidus (dashed line). (a) Depression of crust due to isostatic subsidence beneath a thick volcanic pile; (b) depression of crust due to buckling at zones of opposing lithospheric motion; (c) low-angle underthrusting at zones of opposing lithospheric motion. (From Drury, 1978, reproduced with permission of Elsevier Publishing Company.)

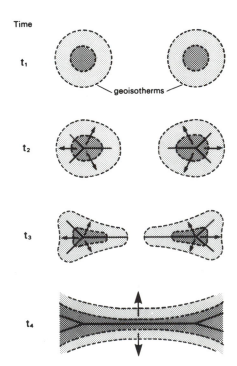

Fig. 2-23. Coalescence of radial-spreading centers into linear, spreading ridges. (From Drury, 1978, reproduced with permission of Elsevier Publishing Company.)

convergent plate boundaries ensued. Since small-scale convection still dominated asthenosphere motions, though its presence was only dimly perceived in lithosphere motions, plates were probably relatively small. Further, the higher Archean heat flow necessitated that spreading centers, where most heat is lost from the Earth's interior, be considerably longer (Windley, 1984). The low-angle underthrusting described above continued to characterize plate boundaries at subduction zones.

We have developed a picture of relatively small plates, some probably with a crust of sial and others of sima. Let us now consider what is thought to have happened when a plate with sialic and simatic crust was subjected to compression, possibly by a change in the nature of convection beneath it. We are primarily interested in the sialic block because it, unlike the simatic block, will ultimately be preserved in the geological record. It is at this point in the discussion that the significance of the much thinner nature of the Archean lithosphere is apparent, for the lithosphere probably responded to horizontal compression, at least at first, by buckling. The form taken by the lithosphere–asthenosphere boundary was probably a series of sharp cusps of low-viscosity asthenosphere material penetrating upward between broad downwarps, or lobes, of high-viscosity lithosphere (see Fig. 2-24A). This cusp–lobe pattern is suggested on the basis of experimental work done by Ramsey (1967) and is constrained by: (1) the magnitude of

the greatest compressive stress; (2) the relative thickness of the lithosphere and asthenosphere; and (3) the viscosity contrast between the lithosphere and the asthenosphere. The cusp–lobe pattern is another important difference from modern plate behavior and it has two significant consequences. First, lithospheric thinning over the cusps would have resulted in the isostatic subsidence of the surface in linear basins. These basins were to become the greenstone belts and this proposed origin, i.e., due to isostatic subsidence, fits the story argued by those geologists who stressed a vertically oriented tectonic regime. Paradoxically, the vertical tectonics of isostasy is here caused by horizontal compression! Note also that the presence of a series of cusps would result in the parallelism of linear greenstone belts observed previously in this chapter. It also provides sialic forelands on either side of the greenstone belts to supply detrital sediments bidirectionally into the sagging basins. Second, adiabatic uplift of asthenosphere material into the cusps would have resulted in considerable partial melting of the uplifted asthenosphere, thus supplying the voluminous ultramafic and mafic magmas which occur at the base of most greenstone-belt stratigraphic successions.

Lithosphere buckling would not, of course, have continued indefinitely and a new subduction zone would have developed, probably at the boundary between simatic and sialic crustal elements (see Fig. 2-24B). The low-angle subduction of the underthrust plate beneath the buckled sialic

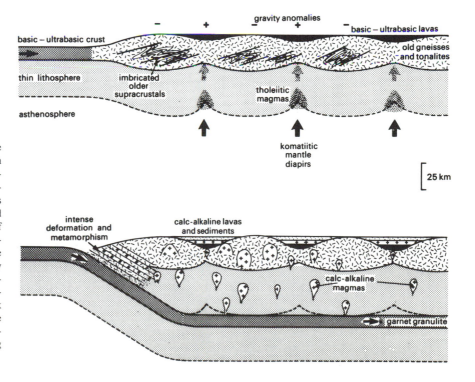

Fig. 2-24. Actualistic model of permobile Archean tectonics. (A) Compression between thin simatic and sialic lithosphere causes cusp-and-lobe configuration of the lithosphere–asthenosphere interface; thinning over sharp cusps and large-scale extrusion of komatiitic and tholeiitic magmas cause isostatic subsidence of the surface to form greenstone belts. (B) Shearing at sima–sial boundary leads to low-angle underthrusting of simatic crust, which partially melts beneath the sialic crust resulting in extensive calc-alkaline magmatism; the zone of tectonic intercalation, deformation, and high-rank metamorphism at the subduction zone forms the high-grade terranes. (From Drury, 1978, reproduced with permission of Elsevier Publishing Company.)

lithosphere is of importance here because magmas derived from its partial melting rose up to the surface over much of the area of the sialic crust, rather than only along its margin (as would have been the case with magmas associated with a steeply dipping subduction zone). Therefore, volcanic activity of the calc-alkaline type was spread over the whole area and calc-alkaline volcanics superseded ultramafic–mafic volcanics in greenstone-belt areas. Note that the buckling of the lithosphere and the intrusion of calc-alkaline magmas over large areas explain very nicely the permobile nature of Archean tectonics.

The actual convergent plate boundary, i.e., the area where the downgoing plate was underthrust beneath the sialic plate, was an area of intense activity involving major horizontal compression. Tectonic intercalation of lithologies would have been common as well as the intrusion of large masses of calc-alkaline magma. Very high grades of metamorphism would also have been common. Such convergent boundaries are thus considered to have resulted in the high-grade terranes.

C.3.d. The Archean–Proterozoic Boundary

The above discussion constitutes the main part of Drury's model. There remains only to describe the changes in the tectonic character of the lithosphere which ended the Archean and began the Proterozoic. Throughout the Archean, radiogenic heat production had decreased and the geothermal gradient had become more shallow. When the near-surface gradient dropped below 15°C/km, eclogite formed instead of granulite. This initiated the steeply dipping subduction that has characterized convergent plate-boundaries ever since. This was a geologically ''sudden'' event rather than an evolutionary one because only a small change in the gradient would have been necessary to cross the phase boundary from granulite to eclogite and thereby institute the dramatic shift in tectonic behavior. Without the low-angle underthrusting, plate margin deformation and intrusion rather than permobile deformation became the order of the day and the Proterozoic Eon had begun.

CHAPTER 3

The Proterozoic

A. Introduction

The beginning of the Proterozoic Eon is defined by the stabilization of large continental-crustal blocks and may have represented a significant change in the nature of terrestrial tectonic processes. The existence, from even the earliest Proterozoic, of platformal deposits, aulacogens, and sedimentary prisms formed on continental margins seems to imply geotectonic regimes identical to those of the Phanerozoic. There were, however, several differences, the most significant of which was the presence in Proterozoic rocks of (apparently) ensialic mobile belts involving large-scale remobilization and remetamorphism of Archean basement.

In our introduction to the continent's Precambrian geology in Chapter 2, we introduced the concept of North America as a continentwide tectonic collage. The obvious implication of the collage concept is that North America was assembled by the accretion of disparate tectonic elements during the Middle Proterozoic. Hoffman *et al.* (1982) implied that such assembly was the result of plate-tectonic collisional accretion in a manner similar to the assembly of Pangaea during the later Paleozoic. But other authors have argued that plate-tectonic mechanisms, at least as they are observed today, did not operate during the Proterozoic.

Discussion will begin with an examination of Proterozoic rocks of the Nagssugtoqidian and Ketilidian mobile belts associated with the North Atlantic craton. Following that, we will consider the history of the Circum-Superior belt (the string of early Proterozoic mobile belts which nearly surround the Superior craton; included in the Circum-Superior belt are rocks of Southern and Churchill Provinces as well as several areas within Grenville Province). We will then consider Bear Province of northwestern Canada and the Proterozoic strata which comprise the Wopmay orogen and the Athapuscow aulacogen and will conclude with an analysis of the Grenville orogen and of Precambrian exposures within the United States.

B. North Atlantic Craton

The North Atlantic Craton (Bridgwater *et al.*, 1973) was introduced in the previous chapter and is shown in Fig. 2-3. In Greenland, Archean rocks comprise the Pre-Ketilidian massif at the center of the craton and have undergone limited subsequent deformation. Proterozoic rocks of the North Atlantic craton in Greenland occur principally within the Nagssugtoqidian Mobile Belt to the north and the Ketilidian Mobile Belt to the south of the Archean craton. Along with the Laxfordian belt in Scotland and Eastern Nain Province of the Canadian Shield, these mobile belts form a roughly polygonal boundary around the relatively small Archean craton. Sutton (1978) pointed out that such a two-fold pattern of crustal structure (i.e., small blocks of little-deformed Archean blocks enveloped within a network of Proterozoic mobile belts) is generally characteristic of Precambrian areas.

The Pre-Ketilidian massif exhibits two distinctive types of structures formed during the Proterozoic. The first of these is vast swarms of basic dikes, intruded both prior to and during early to middle Proterozoic time. The Kangamiut dike swarm of western Greenland is one of the world's densest but is only 12.5 km across by 240 km long (Escher *et al.*, 1975). Other dike swarms are truly grand in scale: a swarm in northern Canada can be traced from Labrador to the Slave Province, a distance of over 2500 km; the Sudbury–Mackenzie swarm in southern Canada is 500 km across and at least 1500 km long (Windley, 1984). Proterozoic swarms of similar scale occur in other shield areas of the world; taken together, these dike swarms represent a remarkable tectonic development, totally distinct from either Archean or Phanerozoic tectonics. Escher *et al.* (1976) concluded that these dike swarms were intruded along conjugate shear fractures in intracontinental positions and Windley (1984) suggested that they represent early, abortive attempts to disrupt the young and not yet entirely rigid continents.

Second, Archean cratons are traversed by belts of intense shear deformation. These belts are typically steeply inclined and marked by conspicuous foliation and finer grain size than that of adjacent regions (Sutton, 1978). Shear belts are commonly aligned parallel to dike swarms, when the two features are juxtaposed. Metasomatic processes, such as amphibolitization, often accompanied shearing, possibly as the result of mantle-derived fluids connected with the generation of mafic magmas (Beach, 1976). From the above observations, it may be seen that virtually no area of the Precambrian continents was entirely free from some form of tectonic disturbance. Thus, whereas Proterozoic tectonic style was certainly different from the permobile style of the Archean, it also differed from that of the Phanerozoic in that intracontinental deformation and igneous activity were much more widespread than they have been subsequently.

B.1. Nagssugtoqidian Mobile Belt

The Nagssugtoqidian mobile belt (see Fig. 2-3) is a Lower Proterozoic orogenic belt composed today of remobilized Archean basement and highly contorted and metamorphosed Lower Proterozoic supracrustal rocks (Read and Watson, 1975). The Nagssugtoqidian belt was stabilized around 1800 Myr. ago, coinciding with the Hudsonian (Penokean) orogeny of the Canadian Shield. Much of the Nagssugtoqidian belt consists of reworked tonalitic and minor anorthositic gneisses, similar to rocks of the Archean Pre-Ketilidian massif. Within the main portion of the mobile belt, rocks have been raised to the granulite grade of regional metamorphism, which, along with the scarcity of late-orogenic granitic intrusions, implies a relatively deep-seated origin (Read and Watson, 1975). Supracrustal rocks of the mobile belt have also been metamorphosed and are seen today as mica schists, greenstones, and quartzites, some of which still retain primary sedimentary structures.

Near the southern margin of the mobile belt (i.e., the Nagssugtoqidian orogenic front) one may see the effects of tectonic regeneration of basement gneisses. Over a 5-km transitional zone, Archean-age gneisses undergo a retrogression from granulite grade in the Pre-Ketilidian massif to amphibolite grade in the mobile belt. Accompanying this retrogression is reorientation of gneissic foliation to conform with the generally east-northeast trend of the Nagssugtoqidian belt.

B.2. Ketilidian Mobile Belt

The Ketilidian mobile belt to the south of the North Atlantic craton (see Fig. 2-3) is similar to the Nagssugtoqidian belt but probably does not represent as deep a level of erosion. The degree of metamorphism ranges from granulite grade to virtually none, and late-stage granitic intrusions are common. North of the Ketilidian front, strata of the Ketilidian cover sequence are undisturbed and consist of detrital sediments interbedded with basic volcanics. Dolostones, cherts, and iron formation are minor components of the cover sequence. Passing southward across the Ketilidian front and into the mobile belt itself, the thickness of supracrustal rocks increases greatly. Similar thickness-trends typify many other Proterozoic mobile belts (e.g., the Huronian Supergroup of Southern Province of the Canadian Shield; see discussion later in this chapter). Sutton (1978) stated that such thickness trends may indicate continental-margin crustal thinning associated with extensional tectonics. In other words, Ketilidian rocks may represent deposition on an evolving passive continental margin. As with the Nagssugtoqidian belt, there is a transition in the Ketilidian mobile belt from undisturbed basement and cover sequence, north of the orogenic front, to strongly deformed and mobilized tectonites within the southern portion of the mobile belt. The Ketilidian area became stabilized around 1800–1500 Myr. ago, somewhat later than the Nagssugtoqidian belt to the north.

B.3. The Gardar Assemblage

In southern Greenland, a final area of Proterozoic rocks is found. Termed the Gardar Assemblage, these rocks are composed of an unmetamorphosed cover sequence called the Gardar Formation and associated alkaline intrusions and basic dike swarms. The Gardar Formation is composed of almost 3 km of sandstones, often red in color, as well as conglomerates, basaltic flows, and pyroclastics. The age of the Gardar Formation is bracketed between 1600 Myr. (date of stabilization of the Ketilidian Mobile Belt) and 1250–1000 Myr. (age of the igneous intrusions).

C. Circum-Superior Mobile Belt

Superior Province was stabilized during the Kenoran orogeny at the end of the Archean and, from the earliest Proterozoic time, acted as a stable, cratonic block. During the Early Proterozoic, supracrustal sequences developed in a series of elongate, geosynclinal troughs around the margin of the Superior craton. These geosynclinal assemblages were deformed and metamorphosed at the end of the Early Proterozoic in a series of orogenic events collectively called the Hudsonian orogeny. Today, these orogens comprise elements of Churchill, Southern, and Grenville Provinces and are separated from each other by the waters of Hudson Bay, by Phanerozoic cover sequences, and, in southeastern areas of Superior Province, by tectonic reworking as a result of the Grenville orogeny (see Fig. 3-1). That these areas are part of a single, continuous mobile belt is inferred from the presence of gravity and magnetic anomalies which, in almost all cases, connect elements of the mobile belt in those areas where they are obscured. Thomas and Gibb (1977) termed the series of deformed Lower Proterozoic rocks the Circum-Superior belt. Baragar and Scoates (1981) suggested that it represents a complete early Proterozoic plate margin surrounding the Superior plate. This interpretation is controversial, however, because application of plate-tectonics mechanisms to Proterozoic orogenesis is not agreed upon by all students of Precambrian geology. In the following sections, we will describe the various elements of the Circum-Superior belt and their tectonic development.

C.1.   Southern Province

We begin our discussion of Circum-Superior areas of the Canadian Shield with the Southern Province because it was among the first studied areas of the Shield. It contains major economic deposits, especially iron, and has remained an area of active geological investigation for well over 100 years. Within the Southern Province, rocks range in age from Archean to Upper Proterozoic with especially thick Lower Proterozoic sequences. Because work on Southern Province began prior to detailed study of most other Shield areas, it has provided a stratigraphic and tectonic framework to which the histories of other provinces can be referred.

Southern Province is divided into the Port Arthur Homocline northwest of Lake Superior, the Lake Superior Basin, the Penokean Fold Belt, which occurs in both the

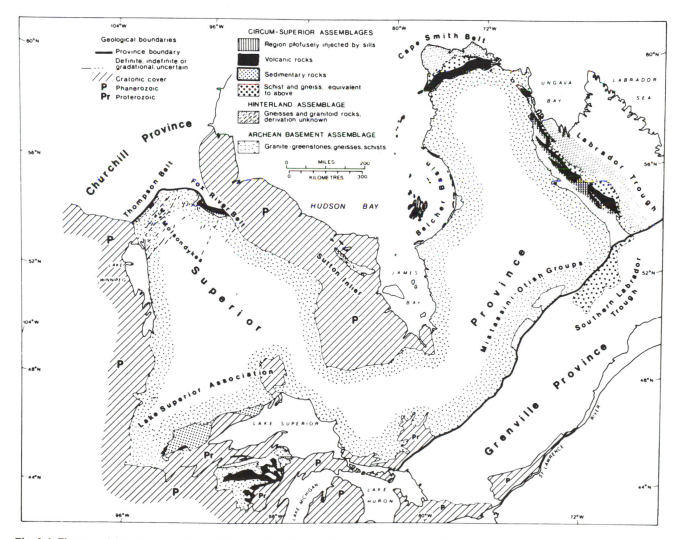

Fig. 3–1. Elements of the Circum-Superior mobile belt of the Canadian Shield. (From Baragar and Scoates, 1981, reproduced with permission of the authors and Elsevier Publishing Company.)

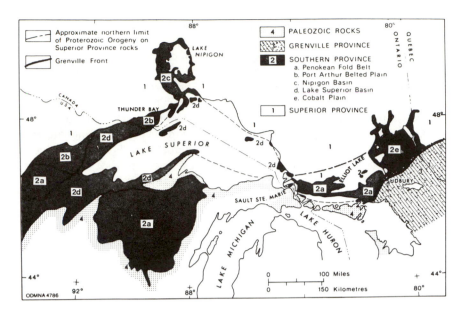

Fig. 3-2. Southern Province of the Canadian Shield and its tectonic subdivisions. (From Card *et al.*, 1972, reproduced with permission of the Geological Association of Canada.)

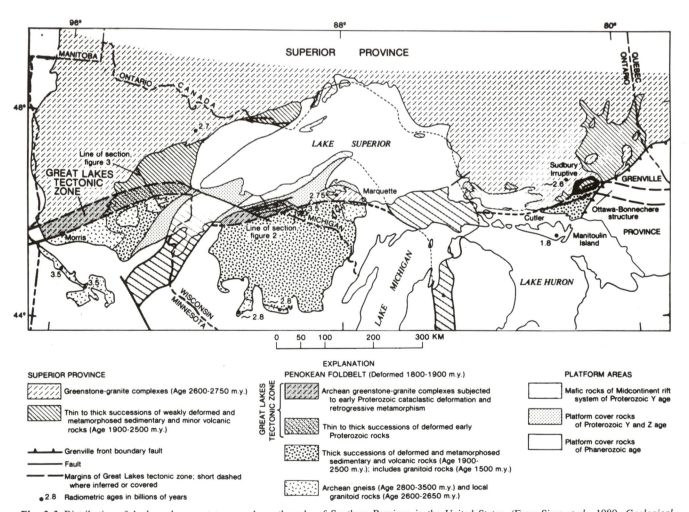

Fig. 3-3. Distribution of Archean basement terranes beneath rocks of Southern Province in the United States. (From Sims *et al.*, 1980, *Geological Society of America Bulletin*, Fig. 1, p. 691.)

United States and Canada, the Nipigon Basin, and the Cobalt Plain (see Fig. 3-2). The Archean basement of the Southern Province has been divided by Sims (1980) into two crustal elements (see Fig. 3-3). In Canada, northern Minnesota, northern Michigan, and adjacent parts of Wisconsin, the basement is an Archean greenstone-granite terrain similar to and continuous with rocks of the Superior Province just to the north. In areas farther south, the basement is dominantly an Archean high-grade terrane composed of gneiss and migmatite. The boundary zone between these two basement terranes is termed the Great Lakes tectonic zone (Sims *et al.*, 1980). Supracrustal strata of the Southern Province comprise a wedge-shaped, complexly deformed prism up to 11 km thick, thinning northward onto the southern flank of the Superior Province (Card *et al.*, 1972). The boundary between the Southern and Superior Provinces is an unconformity with Lower Proterozoic or younger rocks lying on deformed and metamorphosed Archean basement.

In northern parts of the Southern Province, strata are only gently deformed, but structural deformation increases in severity southward toward the Penokean Fold Belt, accompanied by increases in metamorphic rank and intensity of igneous activity. Actually, the exposed part of the Penokean orogen represents only the northern part of a linear orogenic belt of early Proterozoic age which runs from the area just north of Lake Huron in Canada southwestward into the Lake Superior district of the United States and thence beneath the cover of undisturbed Paleozoic rocks in the northern Midwest. Sims (1980) stated that the degree of Penokean deformation and metamorphism was at least partially controlled by the nature of the basement: where Lower Proterozoic strata overlie the greenstone-granite terrane, deformation is limited due to the relative stability of that crust; conversely, strata deposited on the gneissic–migmatitic basement of the south were much more strongly affected by Penokean events because that crust was strongly remobilized. Deformation and metamorphism also increase eastward toward Grenville Province, until, on the other side of the Grenville front, Lower Proterozoic rocks have been thoroughly mobilized and comprise areas of migmatite. These effects, however, were the result of late Middle Proterozoic Grenville orogenesis and will be discussed in a later section.

C.1.a. Lower Proterozoic Strata of Southern Province

Lower Proterozoic supracrustal rocks of the Southern Province comprise the Huronian Supergroup in the segment of the Penokean Fold Belt north of Lake Huron, the Animikie Group of the Port Arthur Homocline and adjacent Penokean areas of eastcentral Minnesota, the Mille Lacs

and Marquette Range Supergroups in the Penokean Belt south of Lake Superior, and the Whitewater Group in the area around Sudbury, Canada. Together, they form a southward-thickening wedge primarily composed of detrital sediments whose grain-size distributions and crossbed orientations indicate derivation from the north, i.e., from Superior Province. Windley (1984) pointed to the lack within Huronian and Animikie strata of sediment-distribution patterns typical of Cordilleran-type orogens (i.e., craton-derived detrital sediments followed by mobile-belt-derived sediments transported toward the craton), a lack of ophiolitic masses, and of syn- or post-orogenic deposits; on that basis, Windley argued that Southern Province represents deposition on a passive continental margin, similar to the modern continental shelf and slope of eastern North America. Such depositional settings do not seem to be represented in Archean deposits; thus, Lower Proterozoic rocks of the southern Canadian Shield record the onset of a new type of tectonic environment whose existence is further evidence for the presence of stabilized continental blocks during earliest Proterozoic time.

i. Huronian Supergroup. The Huronian Supergroup is among the best-known Proterozoic units of North America. The lowest and least extensive Huronian unit is the Elliot Lake Group whose strata occur in a nearly continuous belt from Elliot Lake north of Lake Huron to Sudbury, Ontario (see Fig. 3-4). Across this belt, Elliot Lake rocks show considerable lateral lithological variation. In westernmost parts of the belt, a thick sequence of immature, feldspathic quartzite (the Livingston Creek Formation) occurs at the base of the section and is overlain by more than 1000 m of flow basalt (the Thessalon Formation), which Robertson (1971) considered to be the result of fissure eruptions. In the Sudbury area near the eastern end of the Elliot Lake belt, a thick sequence of basaltic, intermediate, and rhyolitic lavas (Frood, Stobie, and Coppercliff Formations) comprises the lower part of the group and is overlain by up to 3000 m of graded graywacke, siltstone, and argillite (the McKim Formation). Throughout central parts of the belt, volcanic rocks are absent and the sequence is composed of coarse, immature arkoses and quartzites of the Matinenda Formation and overlying siltstones and argillites of the McKim. McGlynn (1976) stated that the Matinenda varied in thickness from only a few meters to almost 250 m; thicker sections occupy basement lows which correspond to valleys carved into the Archean bedrock. Matinenda and Livingston Creek sediments were interpreted by Card *et al.* (1972) to be of fluvial origin and paleocurrent data suggest a southward-flowing stream system (see Fig. 3-4). Deeper water conditions are inferred for the McKim, which Card *et al.* considered to represent turbidity-current deposits.

Fluvial deposits of the Matinenda and Livingston

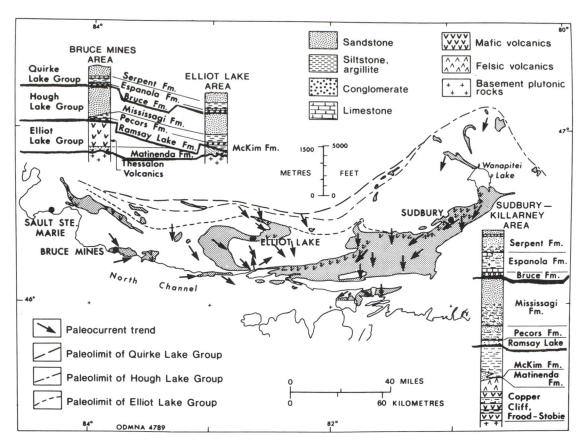

Fig. 3-4. Distribution, stratigraphy, and paleocurrents of the Elliot Lake, Hough Lake, and Quirke Lake Groups. (From Card *et al.*, 1972, reproduced with permission of the Geological Association of Canada.)

Creek Formations contain rounded detrital grains of pyrite and uraninite similar to those found in rocks of the correlative Witwatersrand System in southern Africa. In modern terrestrial environments, both pyrite and uraninite are rapidly and pervasively oxidized long before they have been transported far enough to have acquired any significant rounding; thus, such grains in the Huronian of Canada and the Witwatersrand of Africa are considered evidence of an anoxic atmosphere during the Early Proterozoic. Other evidence of reducing atmospheric conditions is the chemistry of paleosols beneath Huronian rocks (Robertson, 1963) and the absence of red color in undoubtedly terrestrial deposits (Roscoe 1973).

Volcanic rocks in the Huronian Supergroup are restricted to the Elliot Lake Group, but the great quantity of Elliot Lake lavas indicates a major extrusive episode. Frood, Stobie, and Coppercliff volcanics resemble the volcanic edifices of Archean greenstone belts (see p. 13) and, thus, may represent the last phase of the tectonic events which climaxed the Archean. Thessalon basalts, on the other hand, suggest the magmatic activity of divergent plate-margins, and their association with immature, feldspathic sandstones may indicate the development of a rifted continental margin during early Huronian time. The occurrence of McKim turbidites overlying older Elliot Lake strata suggests a deepening of the depositional environment, possibly due to subsidence associated with continued evolution of the rifted margin.

Younger Huronian strata comprise the Hough Lake Group (Ramsay Lake, Pecors, and Mississagi Formations), Quirke Lake Group (Bruce, Espanola, and Serpent Formations), and Cobalt Group (Gowganda, Lorraine, Gordon Lake, and Bar River Formations). Roscoe (1973) recognized within these strata three major depositional cycles, the first being recorded in Hough Lake rocks, the second in rocks of the Quirke Lake Group, and the third in Gowganda and Lorraine strata. In each case, the cycle begins with thick paraconglomerates (Ramsay Lake, Bruce, and Gowganda), followed by thinner deposits of argillites, siltstones, and fine quartzites (Pecors, Espanola, and lower Lorraine), and capped by coarse, crossbedded, arkosic and quartzose sandstones (Mississagi, Serpent, and upper Lorraine). Explanation of the cycles hinges on interpretation of the paraconglomerates.

Huronian paraconglomerates are best characterized as diamictites and are composed of polymictic gravels containing subangular to subrounded clasts of granite, amphibolite, siltstone, quartzite, and metavolcanics, supported in a graywacke matrix (McGlynn, 1976). Although generally massive, they are associated with varved argillites and quartzites containing dropstone boulders, especially in the Gowganda Formation; beneath the Gowganda in some areas are striated bedrock surfaces. All of these features suggest a glacial-till origin for Huronian paraconglomerates (e.g., Robertson, 1961; Young and Church, 1966).

Agreeing with the glacial hypothesis for Huronian paraconglomerates, Roscoe (1973) suggested that the cyclicity mentioned above was the result of glacial advance and retreat. Young (1973b) elaborated upon this, proposing that the three-part stratigraphy observed within each cycle was the result of crustal glacioisostasy. In Young's scheme, each cycle began with a major glacial advance causing crustal subsidence and deposition of tills and associated ice-front deposits (the paraconglomerates); during these periods, the shoreline was far to the south of present Huronian outcrops. Recession of the glaciers followed and, since the rate of isostatic rebound was probably lower than the rate of glacial retreat, rising sea level accompanying deglaciation caused flooding of the still-depressed crust, resulting in deposition of fine-grained, deep-marine sediments. Gradually, however, isostatic rebound did have an effect, causing shallowing and emergence of the crust and concomitant formation of shallow marine and fluvial sediments. Recurrence of glacial advance began the cycle anew. Young's hypothesis of back-and-forth migrating glaciomarine facies is illustrated in Fig. 3-5.

One further paleoenvironmental inference can be drawn from Huronian strata. Roscoe (1973) stated that glacially controlled depositional cycles ended with deposition of the Lorraine Formation; younger Huronian rocks record very different depositional settings. Associated with middle and upper Lorraine sandstones is an unusual suite of aluminous minerals, consisting of kaolinite, diaspore (a component of bauxite), andalusite, and kyanite. Kaolinite and diaspore are considered to have formed as the result of intense chemical weathering of feldspars. The metamorphic minerals andalusite and kyanite are probably not of detrital origin because they can be observed to have formed *in situ*

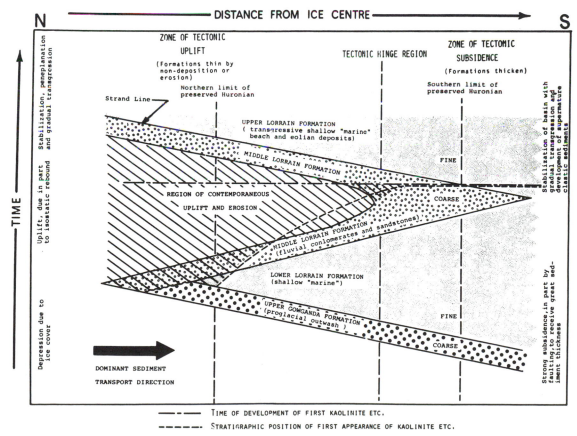

Fig. 3-5. Depositional model of upper Huronian strata in terms of migrating glaciomarine facies; whereas this model was developed specifically for Gowganda and Lorraine sediments, it is generally applicable for all three Huronian glacial-controlled cycles. (From Young, 1973b, reproduced with permission of the Geological Association of Canada.)

from kaolinite (Church, 1967). Young pointed out that ka- olinite and diaspore form today in tropical to subtropical climates and he reasoned that similar conditions were prob- ably responsible for Lorraine aluminous minerals. Thus, following Gowganda glaciation, the climate apparently un- derwent a distinct and rapid amelioration. Further, strati- graphic relations within the Lorraine suggest that climatic warming occurred contemporaneously throughout the area. From these observations, we may infer that Early Pro- terozoic North America had, by Lorraine time, drifted from a polar or subpolar position to more tropical latitudes.

ii. Animikie Group. The Animikie Group of the Port Arthur Homocline and adjacent parts of the Penokean Fold Belt is a thick sequence of detrital and chemical sedi- mentary rocks, especially notable for its iron formations, which supply the preponderance of iron mined from the Great Lakes region. Animikie strata occur in a discon- tinuous string of low ridges (locally termed ranges) which comprise the great iron-mining districts of southern Ontario and Minnesota (e.g., Gunflint, Mesabi, and Cuyuna ranges; see Fig. 3-6). For many years, Animikie rocks were consid- ered to be Huronian equivalents, but Van Schmus (1972) published radiometric data which showed that Animikie rocks, although Lower Proterozoic in age, were younger than Huronian rocks. Animikie formational terminology varies among the ranges (see Table I) but, in each case, exhibits the same three-part stratigraphy.

Lowest Animikie rocks (Kakabeka, Pokegama, Mah-

Table I. Stratigraphic Nomenclature and Inferred Correlation of Animikie Group Rocks in Minnesota and Adjacent Ontario[a]

Gunflint Range	Mesabi Range	Cuyuna Range	Eastcentral Minnesota
Rove Forma- tion	Virginia For- mation	Rabbit Lake Formation	Thompson Formation
Gunflint Iron Formation	Biwabik Iron Formation	Trommald Iron Formation	
Kakabeka Quartzite	Pokegama Quartzite	Mahnomen Formation	
		Unconformity	

[a]Modified from Morey (1973).

nomen Formations) are coarse, conglomeratic quartzites and associated detrital-sedimentary rocks which record gen- erally shallow-marine conditions (Morey, 1973). Thick- nesses range from less than 1.5 m in the Gunflint range to more that 600 m in the Cuyuna range. This southwestward- thickening trend is accompanied by a progressive decrease in grain size, leading Morey (1973) to infer that Kakabeka and Pokegama rocks represent basin-margin deposits, pos- sibly fluvial in origin, while Mahnomen strata represent a basin-interior position where shallow-marine sediments accumulated.

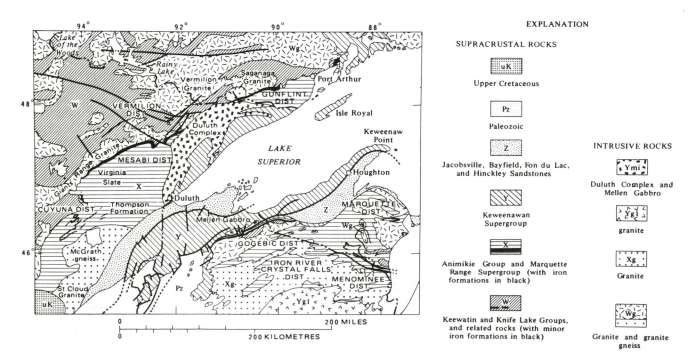

Fig. 3–6. Geological map of the western Lake Superior area. (From King, 1976.)

Middle Animikie units (Gunflint, Biwabik, and Trommald Formations) contain the major iron-bearing strata of the group. Thickness trends of these formations show less regularity than do lower units: Gunflint strata thicken northwestward from about 90 m in Minnesota to 120 m in Ontario; Biwabik rocks, on the other hand, are thickest (~ 250 m) in the middle of the Mesabi range and thin both east- and westward; Trommald strata vary from 14 m to around 50 m. The Gunflint consists of two nearly identical cyclic successions of algal chert, cherty carbonates, taconite (ore from banded iron formation), and pyritic, graphitic, tuffaceous shale (Card *et al.,* 1972). Barghoorn and Tyler (1965) reported the recovery of delicate filaments of blue-green algae from Gunflint cherts, among the oldest multicellular organic remains yet discovered (see discussion of Proterozoic life). The Gunflint is capped by a limestone–dolostone unit. Several authors (e.g., Goodwin, 1956) have argued that Gunflint chert–taconite cycles indicate fluctuating sea level; according to this interpretation, each cycle began with shallow-water conditions over most of the basin within which stromatolitic deposits were formed. Increasing water depth was apparently accompanied by explosive volcanic activity, resulting in the deposition of dark shales and tuffs.

Overlying chert–taconite strata of the Gunflint and correlative units are upper Animikie rocks of the Rove, Virginia, Rabbit Lake, and Thompson Formations. These rocks are generally characterized by a lower, argillaceous zone composed of dark mudstones and siltstones and an upper zone with thinly bedded, graded graywackes intercalated with carbonaceous argillites and clayey, micritic limestones with lesser amounts of quartzite and iron formation. Toward the south and southwest, these deposits thicken and show evidence of increasingly deeper-water deposition. For example, the southernmost of the units, the Thompson Formation, contains graphitic and pyritic black slates along with thin, very-fine-grained sandstones. Upper Animikie rocks were interpreted by Card *et al.* (1972) as having been derived from the granitic Archean craton in the north and carried southward by turbidity currents into deep waters of the basin center. Morey (1972) argued that deposition of these sediments commenced when the Animikie Basin was affected by rapid subsidence, perhaps associated with volcanism in the Michigan and Wisconsin area.

In northern areas of their exposure (e.g., in Gunflint, Vermilion, and Mesabi districts; Fig. 3-6), Animikie strata directly overlie Archean basement, whereas toward the south, into eastcentral Minnesota and adjacent parts of Wisconsin and Michigan, Animikie rocks overlie a sequence of highly deformed Lower Proterozoic (?) rocks. For example, in the Cuyuna district, Animikie strata lie with angular unconformity over a sequence of metamorphosed graywackes and volcanics termed the Mille Lacs Group (Morey, 1978). Although their age is not known with certainty, Mille Lacs

strata have been correlated with the Chocolay Group of the Marquette Range Supergroup in the Marquette and Menominee ranges, and with the Sunday Quartzite and Bad River Dolomite of the Gogebic range in Michigan (see Fig. 3-7).

iii. Marquette Range Supergroup. Composed of Chocolay, Menominee, Baraga, and Paint River Groups, the Marquette Range Supergroup comprises the Lower Proterozoic section of the Menominee and Marquette districts of Michigan and adjacent parts of Wisconsin (see Fig. 3-6). These strata are considered to be generally correlative with the Mille Lacs–Animikie sequence to the north and west (see Fig. 3-7). The Chocolay Group, lowest unit of the Marquette Range Supergroup, consists of two associations: (1) platformal strata characterized by supermature quartzites and carbonates, both containing evidence of shallow-water

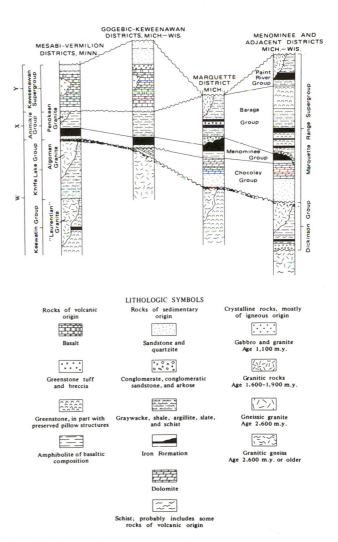

LITHOLOGIC SYMBOLS

Rocks of volcanic origin	Rocks of sedimentary origin	Crystalline rocks, mostly of igneous origin
Basalt	Sandstone and quartzite	Gabbro and granite Age 1,100 m.y.
Greenstone tuff and breccia	Conglomerate, conglomeratic sandstone, and arkose	Granitic rocks Age 1,600-1,900 m.y.
Greenstone, in part with preserved pillow structures	Graywacke, shale, argillite, slate, and schist	Gneissic granite Age 2,600 m.y.
Amphibolite of basaltic composition	Iron Formation	Granitic gneiss Age 2,600 m.y. or older
	Dolomite	
	Schist; probably includes some rocks of volcanic origin	

Fig. 3-7. Stratigraphic cross section of Lower Proterozoic rocks in the Lake Superior region. (From King, 1976.)

deposition, and (2) basinal strata composed primarily of dark slates (Larue, 1981). Basal conglomerates associated with Chocolay rocks have been interpreted by Larue and Sloss (1980) as alluvial fan deposits; stratigraphic relations and paleocurrent data suggest that these and related lower Chocolay sediments accumulated in elongate troughs. The fact that Chocolay basins were elongate rather than circular, and that these troughs were subparallel to younger (i.e., Penokean) structural trends led Larue and Sloss (1980) to argue that Chocolay basins were rift basins rather than cratonic basins (as had been suggested by Morey and Sims, 1976). They further argued that Chocolay rocks formed during early stages of the development of a rifted continental margin.

The Chocolay is overlain by rocks of the Menominee Group, characterized by metamorphosed quartz arenites, graywackes, mudstones, volcanics, and major iron formations. The Menominee Group is considered to be equivalent with strata of Biwabik, Gunflint, and Trommald Formations of the Animikie Group (King, 1976). Larue and Sloss (1980) stated that Menominee strata are more constant in their thickness trends than are Chocolay rocks and, therefore, represent essentially uniform deposition over the region rather than within discrete troughs. Thus, if Chocolay rocks are correctly interpreted as rift-valley deposits, then Menominee rocks probably represent deposition on the newly developed southern continental margin of the Superior craton following cessation of rift-valley-related subsidence. Early rifting is a feature seen in most of the other Circum-Superior mobile belts and is one of the lines of evidence that led Baragar and Scoates (1981) to postulate that the Circum-Superior belt represents a true plate margin. Exactly which crustal block or blocks may have rifted away from the Superior craton, how far away it went, and how such an event may be related to the Hudsonian (Penokean) orogeny (see below) are unknown.

Rocks of the Baraga Group are composed of great thicknesses of slates, argillites, graywackes, iron formation, and extensive volcanics. Baraga rocks are thought to be correlative with upper Animikie rocks and have been interpreted as turbidite deposits formed during major regional subsidence (Larue and Sloss, 1980).

From the preceding discussion, it can be seen that the lower three units of the Marquette Range Supergroup strongly resemble a rifted-margin prism, similar to younger examples, such as Atlantic Coastal Plain Province of eastern North America. The Paint River Group occurs only in the centers of the deeper downfolds and, consequently, its stratigraphic position is poorly known. The Paint River Group is composed of slates, graywackes, and more iron formation but was not considered by Larue and Sloss in their discussion of Marquette Range deposition.

The last major group of Lower Proterozoic rocks in Southern Province is the Whitewater Group, composed of Onaping, Onwatin, and Chelmsford Formations. Restricted to the Sudbury Basin, the Whitewater Group is considered to be an equivalent of the Animikie Group of the Port Arthur Homocline. The Onaping consists of nearly 1200 m of andesitic tuffs and breccias associated with minor flows; it has been interpreted in two different ways: (1) as a volcaniclastic deposit formed as the result of explosive volcanism (e.g., Stevenson, 1971) and (2) as a "fallback breccia" caused by a major meteorite impact (French, 1970). The Onwatin Formation consists of a thick sequence of carbonaceous and pyritic slates and siltstones with minor limestone, cherty carbonate, and graywacke. According to the meteorite-impact model of the Onaping, Onwatin strata should be interpreted as subaqueously reworked material from the Onaping ejecta blanket along with more "normal" marine deposits. Chelmsford strata are composed of up to 840 m of graded to massive graywackes and argillite with flute casts, convolute laminations, and other structures characteristic of proximal turbidites (Card *et al.*, 1972). Paleocurrent data suggest that Chelmsford deposits were derived from the north to northeast and were carried into the Sudbury basin area by turbidity currents.

C.1.b. The Hudsonian (Penokean) Orogeny

Lower Proterozoic rocks of Southern Province were deformed, intruded, and metamorphosed by a complex orogenic event named the Hudsonian orogeny in Canada and the Penokean orogeny in the United States. The Hudsonian (Penokean) orogeny is used as a convenient punctuation to separate Lower Proterozoic from Middle Proterozoic rocks. In the Canadian portion of Southern Province, Hudsonian deformation resulted primarily in the development of large but somewhat open folds trending nearly east–west. Toward the south, into the United States, the degree of deformation increases so much that, in the Cuyuna district (see Fig. 3-6), rocks of the Thompson Formation have been mistaken for the Archean Knife Lake Formation (King, 1976). In this same general region, Holst (1984) recognized at least one major, northward-directed nappe and suggested the presence of additional nappes, implying that major crustal mobilization must have coincided with the Hudsonian (Penokean) event, in at least southern areas of the orogen.

In both northern and southern parts of the deformed area, two phases of folding are recognized. The first fold-generation is characterized by isoclinal, recumbent folds with nearly horizontal axes. The second generation is more typified by open, upright to steeply inclined folds, also with subhorizontal axes. Associated with both generations of folding are axial-plane cleavages (Holst, 1984). In addition to folding, there are at least three generations of faults, the first of which probably developed during the main phase of folding; later episodes of faulting, however, occurred long afterward.

Igneous activity associated with the Hudsonian (Penokean) orogeny also occurred in two distinct phases, although they did not correspond with the two phases of folding. The first phase of igneous activity was characterized by mafic to felsic volcanism and is radiometrically dated at about 1859 Myr. ago (Van Schmus, 1980). Mafic volcanics and shallow intrusive rocks of this first phase in Canada comprise the Nipissing intrusions, individual plutons of which range in thickness from a few meters up to nearly 300 m (McGlynn, 1976b). Nipissing dikes are composed of tholeiitic gabbro with lesser amounts of pyroxenite and granophyre. In the Sudbury area, similar rocks comprise the Sudbury Gabbro. Intrusion of Nipissing and Sudbury igneous rocks occurred after the first phase of Hudsonian (Penokean) deformation but prior to major regional metamorphism (Card *et al.*, 1972).

The second igneous phase was characterized by tonalitic to granitic plutonism which occurred around 1840 Myr. to 1820 Myr. ago (Van Schmus, 1980). Major Hudsonian (Penokean) granitic bodies include the Cutler Granite, along the north shore of Lake Huron, which is a medium-crystalline, equigranular rock, and the Creighton Granite near Sudbury which is primarily a pink, porphyritic rock. In central Minnesota, rocks of the Animikie belt have been intruded by complex plutonic masses such as the Saint Cloud Granite (see Fig. 3-6). The McGrath Gneiss of the same area was considered also to be a Penokean intrusion, but recent work has indicated that it is probably an Algoman (i.e., late Archean) body with a Penokean overprint (King, 1976). South of Lake Superior, in northern Wisconsin, Precambrian exposures are limited because of the extensive blanket of Pleistocene glacial surficial deposits. Where exposed, however, the main body of basement rocks are relatively high-grade metamorphics, principally metasediments and metavolcanics, probably equivalent to the Marquette Range Supergroup in Michigan to the north (King, 1976). Intruding into these rocks are a number of granites, some of which may be unrelated to the Penokean Orogeny (e.g., the Wolf River batholith, which is considered to be an Elsonian intrusion).

An interesting and important igneous rock body of the Southern Province is the Sudbury Irruptive, an elliptical ring of intrusives 37 miles long by 17 miles wide which occurs within the Sudbury Basin. The body is composed of a lower zone of norite, an upper zone of granophyre, and a narrow transition zone (McGlynn, 1976b). The Sudbury Irruptive is of major economic significance because it is one of the largest accumulations of nickel ore in the world. The three-dimensional shape of the body is somewhat controversial; Thomson (1956) considered it to be a large ring-dike

but Wilson (1956) argued that it is a funnel-shaped mass similar in shape to a cone sheet and extends downward into essentially undifferentiated ultramafics. Both Thomson and Wilson considered the Sudbury Irruptive the result of "ordinary" igneous activity. Dietz (1964) suggested that the Irruptive may be a deformed astrobleme (meteorite impact scar) and that the magmatic activity associated with it may have been generated by the force of impact. In this view, the nickel ore of the Irruptive is derived from the material of the original meteorite and rocks of the Onaping and Onwatin Formations (of the Whitewater Group; see earlier discussion in this section) are ejecta materials. In their review of the structure of the Sudbury Basin, Brocoum and Dalziel (1974) concurred with Dietz's astrobleme hypothesis and stated that the basin-forming event was coeval with major folding in the region.

Metamorphism associated with Hudsonian (Penokean) deformation was mainly of lower greenschist grade in much of Canada. Metamorphic grade increases toward the south, however, coinciding with increase in structural deformation, and rocks of the Penokean fold belt in the United States attained metamorphic grades up to the amphibolite facies (King, 1976). The major metamorphic event of the Hudsonian (Penokean) orogeny occurred after the first phase of deformation and intrusion of Nipissing rocks, but prior to the second folding event.

It is important to understand that many geological events which occurred over a long period of time have been lumped together under the general heading of the Hudsonian (Penokean) orogeny. Some authors have attempted to recognize several different orogenic events, distinguishing between an earlier, Penokean event and a later, Hudsonian event. Others have pointed out that radiometric dates and field relationships indicate the southern margin of the Superior craton was tectonically active for an extended period of time. Such a long span of activity may be evidence that the area was an active plate margin during the Early Proterozoic (Cambray, 1978).

Several authors have proposed models for the Hudsonian (Penokean) orogeny. As with most tectonic models developed for Proterozoic events, they may be classified into two groups: intraplate models and plate-margin models. Agreement is general on several points: that Archean events had been dominated by compressional tectonics, that Early Proterozoic deposition occurred on a crustal area undergoing extensional tectonics (e.g., the Chocolay Group of the Marquette Range Supergroup), and that, from about 1900 Myr. to about 1850 Myr. ago, the area was affected by another episode of compressional deformation. Sims *et al.* (1980) argued that events associated with the Hudsonian (Penokean) orogeny were the result of alternating contraction and expansion of the gneissic basement block south of the Great Lakes tectonic zone, relative to the more stable, granite–greenstone basement block north of the tectonic zone (see Fig. 3-3). Alternatively, a tectonic model has been proposed by Cambray (1978) which explains Hudsonian (Penokean) events as the result of continental rifting followed later by a collisional event during which a north-facing magmatic arc was driven northward up onto the continental margin.

C.1.c. Middle Proterozoic Rocks of Southern Province

Following the Hudsonian (Penokean) orogeny, the Southern Province underwent long-term erosion as the orogenically active area gradually became stabilized. Later, during middle and later Middle Proterozoic time, another change in the tectonic regime occurred within the Province, leading to the development of a major continental rift within the interior of North America.

i. Sibley Formation and Its Correlatives. Oldest Middle Proterozoic rocks in the Canadian portion of the Southern Province compose the Sibley Formation, which occurs along the northern margin of the Lake Superior Basin and northward into the Nipigon Basin (Fig. 3-2). The Sibley is a red-bed sequence up to 150 m thick; it is composed of a basal conglomerate overlain by thin but laterally continuous units of sandstone, mudstone, stromatolitic chert, and limestone. It contains well-preserved sedimentary structures such as ripple marks, mud cracks, and intraformational conglomerates, all of which indicate a shallow-marine to supratidal depositional environment. Card *et al.* (1972) stated that the presence of corrensite and authigenic microcline in Sibley rocks was probably the result of clay diagenesis in a highly saline mudflat or restricted coastal basin, under hot, semiarid conditions. The Sibley accumulated within a grabenlike basin and may mark the beginning of the rift event which led to the formation of Keweenawan rocks (see below).

Southwest of the Lake Superior Basin, in southwestern Minnesota and adjacent South Dakota, is a broad, plateau-like terrace underlain by the Sioux Quartzite. Primarily a mature quartzite, the Sioux has a thickness of up to 900 m but thins southeastward into Wisconsin where it is exposed along the western flank of the Wisconsin Arch as the Barron Quartzite. The Sioux and Barron Quartzites have been warped into broad, open folds; to the south, in the subsurface of Iowa, the Sioux is interbedded with rhyolite (King, 1976). Like the Sibley Formation in Canada with which they are correlated, Sioux and Barron rocks seem to represent the initiation of a new phase of tectonic activity, a phase which led to the formation of the Keweenawan Supergroup.

ii. Keweenawan Supergroup. Much of Lake Superior and areas adjacent to the northwest, southwest, and south are underlain by rocks of the Lake Superior Basin (Fig. 3-2). Within this basin, little-deformed late Middle Proterozoic volcanics and red detrital sediments comprise the Keweenawan Supergroup. Keweenawan rocks are thickest (up to 15,000 m) within the axial portion of the Lake Superior trough but thin rapidly toward the basin margins; White (1966) suggested that Keweenawan rocks probably never extended much beyond the present-day northwestern and southeastern trough margins. The southwestern end of the Lake Superior Basin is overlapped by Paleozoic strata so that the original extent of Keweenawan rocks in that direction is obscured.

In Canada, Keweenawan rocks are represented by the Osler Formation. The Osler consists of two units: (1) a lower unit with up to 3000 m of vesicular and amygdaloidal basalt interbedded with thin layers of sandstones and mudstones, and (2) an upper unit composed almost entirely of tholeiitic, amygdaloidal basalt, andesitic and felsic flows along with a variety of shallow intrusive rocks (Card *et al.*, 1972). Osler rocks overlie the Sibley Formation disconformably but, because they are of nearly the same age, McGlynn (1976) placed both the Sibley and the Osler into the Keweenawan Supergroup. Radiometric dates for Osler lavas commonly fall in the range 1100 to 1000 Myr. (Card *et al.*, 1972).

In the United States, Keweenawan stratigraphy is more complicated but also can be generalized into two units: a lower, volcanic unit (called the North Shore Group northwest of Lake Superior in Minnesota and the Portage Lake Group south of the Lake in Michigan and Wisconsin) and an upper, detrital sedimentary unit (the Oronto Group). The lower, volcanic unit begins with a thin and discontinuous sandstone but is principally composed of vesicular and amygdaloidal basalts, andesitic flows, minor rhyolites and pyroclastics. Native copper occurs as amygdules in basalts and has been a major economic resource since it was first exploited by American Indians. Flow structures within basalts indicate that the main direction of flow was from axial regions outward toward trough margins. Thus, in their petrology and geometry, Keweenawan volcanics resemble similar ones which, today, form in continental rift-valley settings.

The Oronto Group is composed of the Copper Harbor, Nonesuch, and Freda Formations. The Copper Harbor Formation is a coarse conglomerate consisting of rounded volcanic clasts. Directly above it is the Nonesuch Shale, a dark, fissile mudstone containing organic compounds, microfossils, and hydrocarbons. And above the Nonesuch is the Freda Formation which is composed of a relatively thick sequence of red, arkosic sandstone with evidence of terrestrial deposition (e.g., alluvial fan and fluvial deposits). Paleocurrent data indicate that sediments were carried into the basin from both sides and petrographic evidence suggests derivation from basement rocks exposed along the flanks of the Keweenawan trough.

Intruding into the Keweenawan Supergroup are a variety of mafic plutons, the largest of which is called the Duluth Gabbro. The Duluth is a differentiated, layered lopolith, 240 km long and up to 1500 m thick at its center. It contains gabbroic, anorthositic, and granophyric phases. Like other mafic intrusions into the Keweenawan, rocks of the Duluth Gabbro are compositionally similar to those extrusive rocks which comprise the bulk of North Shore and Portage Lake Formations and probably represent ponded magmas, trapped during their upward migration to the surface.

Because they are dominated by voluminous, dense, mafic extrusive and intrusive rocks, Keweenawan rocks produce a strikingly large, positive Bouguer gravity anomaly which can easily be recognized on gravity maps of the central midcontinent. Termed the midcontinent gravity high, this anomaly extends in a discontinuous belt from Lake Superior all the way to central Kansas (Fig. 3-8).

Based on its discontinuous, offset geometry and its associated rock types, the midcontinent gravity high is considered to represent a major continental rift zone (Chase and Gilmer, 1973). In their geophysical analysis of its structure, Ocola and Meyer (1973) stated that the Keweenawan trough was probably most similar to the modern Red Sea rift and probably represents an incomplete continental-separation event in the center of Proterozoic North America. Because of this, it has been cited as unequivocal evidence that at least some form of lateral plate motion characterized the behavior of the Earth's lithosphere as far back as the late Middle Proterozoic (Chase and Gilmer, 1973).

C.1.d. Upper Proterozoic Rocks of Southern Province

The youngest Precambrian rocks of the Southern Province are of Late Proterozoic age and unconformably overlie Keweenawan rocks within the Lake Superior Basin. In Canada, only one unit, the Jacobsville Formation, is recognized. The Jacobsville is a flat-lying, nonmarine redbed composed of conglomerate, quartz sandstone, and shale. It is up to 200 m thick. In the United States, several units are

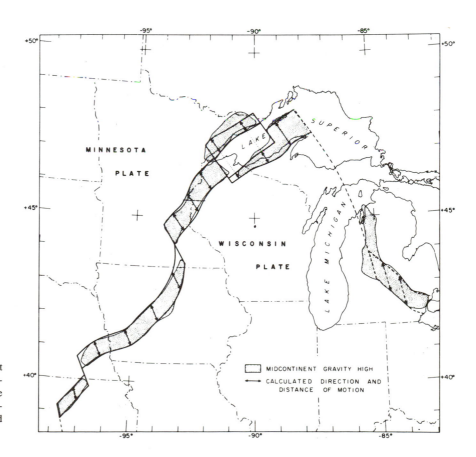

Fig. 3-8. Generalized map of the midcontinent gravity high, showing the inferred plate-tectonics geometry based on best-fit pole of relative motion. (From Chase and Gilmer, 1973, reproduced with permission of North-Holland Publishing Company, Amsterdam.)

recognized: the Jacobsville Formation (the same as the Canadian Jacobsville) in Michigan, the Bayfield in Wisconsin, and the Fon du Lac and Hinkley in Minnesota. These units are all similar in their lithologies and probably all represent terrestrial deposition. King (1976) said that paleocurrent determinations from these strata show that sediment-transport directions, like those from detrital Keweenawan sedimentary rocks, were toward the axis of the Lake Superior trough, suggesting that subsidence of the rift valley continued into Late Proterozoic time. Unlike Keweenawan rocks, however, Jacobsville and correlative sediments are more mature, being quartz arenites rather than arkoses and having less varied heavy-mineral suites. This higher degree of mineralogical maturity probably can be interpreted as the result of decreased levels of tectonic activity following Keweenawan rifting.

Also during the Late Proterozoic, the Southern Province was partially affected by the Grenville orogeny and Hudsonian (Penokean) rocks and structures underwent reworking, especially on the eastern side of the Grenville front (Fig. 3-2). We will discuss the Grenville orogeny later in this chapter.

C.2. Churchill Province, I: Circum-Ungava Mobile Belt

Churchill Province is a very large area in the central and northern part of the Canadian Shield composed primarily of gneisses and granitoid rocks of Archean and Lower Proterozoic age. In many areas, Archean rocks have been strongly reworked during Proterozoic tectonic events, but in other areas, they show only the effects of the Kenoran orogeny. Indeed, the Kaminak Craton (Chapter 2) was until recently considered a Churchill subprovince. In addition to these very-high-grade metamorphic rocks, the Churchill Province also contains linear to arcuate fold belts containing highly deformed Lower Proterozoic strata. Also part of the Churchill Province are several large, Middle Proterozoic sedimentary basins which have been only little deformed. Churchill Province, then, is a geological polyglot containing a number of different terranes; indeed, Davidson (1972) said that Churchill Province was the remainder of the Canadian Shield after the more easily distinguishable older and younger provinces had been delineated. In our discussion, we will consider Churchill Province in three sections: the first two will deal with those portions of the province which compose elements of the Circum-Superior belt (the Circum-Ungava belt, the Sutton Inlier, the Fox River belt, and the Thompson Nickel belt) and the third will treat the western part of the province.

That portion of Superior Province which extends from Hudson Bay to Labrador is termed the Ungava Peninsula and, during the Early Proterozoic, acted as a stable cratonic block. Surrounding the Ungava craton is an apparently continuous mobile belt which, during the early Early Proterozoic, was the site of active sedimentation. Called the Circum-Ungava continental-margin prism (or, geosyncline), this belt was deformed in the Hudsonian (Penokean) orogeny and today comprises a major segment of the Circum-Superior belt. The Circum-Ungava prism is composed of the following subdivisions, each of which will be discussed below: (1) the Labrador trough and its extensions into Grenville Province (i.e., the Southern Labrador trough and the Otish and Mistassini Basins); (2) the Cape Smith fold belt to the north of the Ungava craton; and (3) the Belcher fold belt along eastern Hudson Bay.

C.2.a. The Labrador Trough

Along the eastern side of the Ungava Peninsula is the Labrador trough (Fig. 3-1). Principally of Early Proterozoic age, the trough is bounded to the west by the Superior Province, on which strata of the Labrador trough rest with prominent nonconformity. To the east, the trough is bounded by the Labrador hinterland, whose high-grade metamorphics grade into Archean rocks of the Nain Province (a portion of the North Atlantic craton; see Chapter 2). Similar to those of other elements of the Circum-Ungava continental-margin prism, rocks of the Labrador trough near the Ungava craton are relatively unmetamorphosed but metamorphic grade increases away from the craton and rocks on the eastern side of the trough have been raised to the amphibolite grade; rocks of the hinterland are of granulite grade. Granitoid gneisses of the Archean basement complex beneath Labrador trough strata can be traced almost continuously around the northern end of the trough and can be seen (albeit highly reworked) in several localities to the east of the trough. These observations seem to indicate that the Labrador trough is an entirely ensialic orogen, a point stressed by those (e.g., Dimroth, 1972) who argue against plate-tectonics interpretations of these rocks.

Strata of the Labrador trough are termed the Kaniapiskau Supergroup and are divided into three units: a miogeoclinal sequence on the western side (the Knob Lake Group); a eugeoclinal sequence on the east (the Doublet Group); and intrusive igneous rocks, primarily of mafic composition (the Montagnais Group). Miogeoclinal strata are composed of orthoquartzites, dolostones, iron formation, and shales. Beneath the miogeosynclinal sequence are continental red beds composed of fluvial and alluvial detrital deposits which occur in normal-fault basins (Baragar and Scoates, 1981). Within the miogeoclinal sequence, Dimroth (1972, 1981) recognized evidence of three major cycles of sedimentation. The first cycle is present only in the southern part of the trough and is composed of basal quartzites and dolostones which grade upwards into flyschlike interbedded graywackes and shales associated in central parts of the trough with tholeiitic volcanics. The second cycle affected the whole trough, again beginning with quartzites and dolostones but also containing important deposits of iron formation (the Sokoman Formation); the second passes upwards into thick and extensive felsic pyroclastics. The third cycle is also seen throughout the trough and is grossly similar to the second except that it is capped by a very thick sequence of tholeiitic pillow-basalt. Two source areas have been recognized for detrital sediments of the trough: lower, quartzitic sediments of each cycle appear to have been derived from the Ungava craton whereas immature graywacke deposits of upper portions of each cycle were derived from eastern sources, perhaps from within the hinterland. During later stages of the trough's development, some sediments appear to have been derived from a rising sourceland within the trough (the geanticline of Dimroth, 1972). Reversal of sediment-transport direction during the evolution of a continental-margin sedimentary assemblage is a characteristic feature of Cordilleran-type mobile belts and has been cited by those authors who favor plate-tectonics interpretations of the Labrador trough (e.g., Windley, 1984).

The eugeocline is composed almost entirely of mafic flows, mafic pyroclastics, and shallow mafic and ultramafic intrusions. Dimroth (1972) estimated that as much as 80% of the eugeocline fill was composed of such rocks. Serpentinized peridotites are present in the form of large sills. There are also sills of gabbro and related phaneritic igneous rocks. Chemically, eugeocline volcanics of the Labrador trough are closely related to oceanic tholeiites (Baragar, 1970) and some authors have referred to these rocks as ophiolites. Davidson (1972), however, stated that such a classification is based only on their geochemistry and geometrical relationships with the rest of the basin fill, not on their structure or paragenesis; thus, Davidson recommended that the term *ophiolite* not be used for Labrador mafics and ultramafics. Although metasedimentary rocks comprise only a small volume of Labrador trough eugeoclinal rocks, two depositional cycles, somewhat similar to cycles described from miogeoclinal strata, have been recognized and correlations have been suggested with the upper two cycles of the miogeocline.

During the Hudsonian orogeny, rocks of the Labrador trough were deformed, metamorphosed, and thrust westward toward the Ungava craton (see Fig. 3-9). The principal structure in miogeosynclinal rocks of the western region are large, relatively high-angle, eastward-dipping thrust faults. Large-scale folds, overturned to the west, characterize central portions of the trough. Rocks of eastern areas are typified by intense, multiphase deformation, usually in the form of isoclinal folds. As many as five fold generations can be recognized in some areas (Donaldson, 1970). It is

significant that no Hudsonian intrusive igneous-rock masses have been recognized within the Labrador trough, although several large granitic bodies with Hudsonian radiometric ages have been identified in the Labrador hinterland.

Following the Hudsonian orogeny, the Labrador trough ceased to be tectonically active and there are only a few rock units younger than the Lower Proterozoic in the region. Probably the most important of these post-Hudsonian units is the Sims Formation which overlies folded strata of the Kaniapiskau Supergroup. The Sims is primarily a quartz arenite with local layers of pebbly vein-quartz and jasper conglomerates (Donaldson, 1976). In some areas, the Sims shows evidence of Grenville-age folding, which suggests a Middle Proterozoic age for deposition of Sims detrital sediments.

Toward the southeast, structural and lithological trends of the Labrador trough are truncated by the Grenville Front, across which the degree of both structural complexity and metamorphism increases dramatically. Nevertheless, near the Grenville Front, Labrador trough rocks can still be distinguished as a separated crustal entity; farther into the Grenville orogen, however, they lose their identity as their structure becomes increasingly obscured by Grenville overprinting and they blend in with migmatitic and intrusive rocks. Before losing its identity, the trend of the Labrador trough can be observed to swing southwestward and can be traced about 300 km (Baragar and Scoates, 1981). This extension into Grenville Province is called the Southern Labrador trough.

In the Superior Province, just west of and truncated by the Grenville Front, are several small basins containing Proterozoic rocks (see Fig. 3-1). Called the Otish and Mistassini Groups, strata of these basins are nonconformable on the Archean basement and are only gently folded. Chown and Caty (1973) described rocks of these basins and divided them into a lower sequence of arkosic sandstones and conglomerates derived from the Superior craton, and an upper sequence consisting of dolostones and iron formation. Although not precisely dated, Otish and Mistassini rocks are considered to be the platformal–miogeoclinal equivalents of Labrador trough rocks and so are included with them as part of the Circum-Ungava geosynclinal complex.

C.2.b. Cape Smith Fold Belt

Along the northern margin of the Ungava craton, Lower Proterozoic geosynclinal strata are found in the Cape Smith fold belt (Fig. 3-1). The Cape Smith belt is an obvious northward extension of the Labrador trough and its stratigraphy is grossly similar to Labrador rocks except that less of Cape Smith's miogeoclinal suite of rocks has been preserved; thus, most of the Cape Smith belt is dominated by mafic volcanics and shallow intrusions. Sialic basement

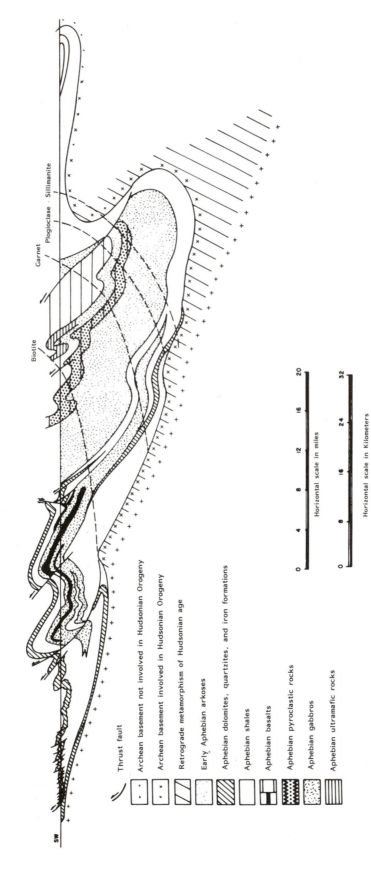

SW

Thrust fault

Archean basement not involved in Hudsonian Orogeny

Archean basement involved in Hudsonian Orogeny

Retrograde metamorphism of Hudsonian age

Early Aphebian arkoses

Aphebian dolomites, quartzites, and iron formations

Aphebian shales

Aphebian basalts

Aphebian pyroclastic rocks

Aphebian gabbros

Aphebian ultramafic rocks

Biotite

Garnet

Plagioclase - Sillimanite

Horizontal scale in miles

0   4   8   12   16   20

Horizontal scale in Kilometers

0   8   16   24   32

Fig. 3-9. Geological cross section across the Labrador trough. (From Dimroth, 1981, reproduced with permission of the author and Elsevier Publishing Company.)

beneath Cape Smith supracrustals can be traced continuously around the eastern end of the fold belt and for a considerable distance along the northern margin of the belt. For this reason, the Cape Smith belt, like the Labrador trough, is considered an ensialic trough. Also like the Labrador trough, the Cape Smith belt shows an increase in metamorphic grade outward from the Ungava craton, going from subgreenschist in the south to granulite in reworked Archean crust of the hinterland just north of the northern boundary of the Cape Smith belt proper.

Rocks of the Cape Smith belt have been divided into two sequences. The lower sequence is termed the Povungnituk Group and is composed of more than 300 m of mafic to intermediate flows, mafic and ultramafic sills, and minor sedimentary rocks. Povungnituk volcanics are primarily tholeiitic basalts, commonly pillowed and amygdaloidal, with lesser quantities of andesite. Lowest sedimentary deposits of the Cape Smith belt are thin, platformal sequences of quartz arenites, dolostones, shales, and minor iron formation; these are replaced upward in the succession by flyschlike graywacke and shale. As in Labrador trough eugeoclinal strata, two cycles of deposits are recognized but are not rigorously correlated.

The upper stratigraphic unit of the Cape Smith belt is the Chukotat Group and is composed of a thin basal assemblage of detrital sedimentary rocks followed by as much as 5000 m of pillowed basalts and andesites associated with gabbroic and peridotitic sills. Chukotat rocks were considered by Bergeron (1957) to be unconformable over Povungnituk strata, with the Povungnituk deformed and slightly metamorphosed prior to deposition of Chukotat rocks. Baragar and Scoates (1981), however, said that more recent work has failed to confirm the presence of an unconformity between the two groups and they treated the Povungnituk–Chukotat sequence as conformable.

C.2.c. Belcher Fold Belt

The third major element of the Circum-Ungava continental-margin prism is the Belcher fold belt which includes strata of the Belcher Islands of southeastern Hudson Bay and parts of the mainland just to the east of the Belcher Islands (see Fig. 3-1). Rocks of the Belcher area are thinner and contain much less volcanic material than rocks of either the Labrador or the Cape Smith area. In mainland parts of the belt, Belcher strata are divided into Pachi, Richmond Gulf, and Nastapoka Groups; in the Belcher Islands themselves, the sequence is thicker than to the east and is termed the Belcher Group.

Lowest strata of the Belcher fold belt are terrestrial arkoses associated with east–west-trending grabens. These are overlain by shallow-marine strata such as quartz arenites, dolostones, and iron formation. Paleocurrent data show that the principal source of detrital sediments in the

lower and middle parts of this succession was the Ungava craton to the east. In the Belcher Islands area, these craton-derived strata are succeeded by graywackes and shales which are, in turn, succeeded by red, terrigenous arkoses and conglomerates. The source area for these strata was in the west (Baragar and Scoates, 1981); thus, the same change of sediment-transport polarity seen in rocks of the Labrador trough and the Cape Smith belt also characterizes detrital deposits of the Belcher area. Also in the area of the Belcher Islands, sedimentary strata are associated with considerably more mafic, volcanic material, reflecting the fact that the Belcher Islands area was closer to the center of the mobile belt. Similarly, stratigraphic thicknesses of correlative strata increase from the mainland to the Belcher Islands.

C.3. Churchill Province, II: Western Circum-Superior Belt

The mobile-belt boundary of the Superior craton in the west is less obvious than similar boundaries in the east and north, because so much of the western region is obscured by Hudson Bay and by Paleozoic cover sequences. A patch of craton-margin rocks is exposed in the Sutton Inlier south of Hudson Bay; marginal rocks are also exposed in the Fox River and Thompson Nickel belts which form part of the boundary between Superior and western Churchill Provinces (see Fig. 3-1). Although these areas are separated by large covered intervals, their subsurface connection is indicated by major gravity anomalies which can be traced almost continuously from the Sutton Inlier to the Fox River area. In the same fashion, the Sutton area can be connected to the Belcher and Cape Smith areas by gravity anomaly patterns. In the following discussion, we will consider each of the above-mentioned areas.

C.3.a. Sutton Inlier

The Sutton Inlier is a relatively small region just south of Hudson Bay in which fold ridges of Lower Proterozoic and Archean rock poke up through a blanket of undisturbed Paleozoic sediments. Archean rocks are present only in southern parts of the Inlier and are composed of granodioritic gneisses and intrusive granites, similar to lithologies seen elsewhere in Superior Province. Proterozoic strata unconformably overlie Archean basement and consist of three units, the lowest of which is a shallow-marine carbonate. This is succeeded by a graywacke, shale, and iron formation unit. A third unit is formed by mafic sills which intrude the sedimentary strata. Comparison of the above description with those of Belcher, Cape Smith, and Labrador trough strata reveals a clear similarity among all areas mentioned and the correlation between them was first suggested by Bostock (1971).

C.3.b. Fox River Belt

Bordering the Superior craton for about 300 km in Manitoba is the Fox River belt, a deformed zone of Lower Proterozoic sedimentary rocks, mafic and ultramafic volcanics, and large, differentiated sills. This region is considered to be a western extension of geosynclinal rocks in the Sutton Inlier (Baragar and Scoates, 1981). Fox River sedimentary rocks consist of: a lower horizon containing laminated siltstones, argillites, and shales associated with sandstones, quartzites, dolostones, and minor iron formation; a middle horizon consisting of quartzose silt- and sandstone and feldspathic sandstone, all affected by contact metamorphism caused by the intrusion of the Fox River Sill, a huge, differentiated, stratiform sill; and an upper horizon containing argillites and shales. Volcanic strata associated with these sedimentary rocks are chiefly massive and pillowed basalts and komatiitic basalts. In some areas, they have been sheared and serpentinized; columnar jointing is observed in some flows as well as vesiculated and brecciated flow tops. Although Baragar and Scoates (1981) do not term these rocks ophiolites, that term would seem to apply.

C.3.c. Thompson Nickel Belt

The final element of the Circum-Superior belt is the Thompson nickel belt, which lies between Superior and Churchill Provinces in Manitoba. Supracrustal rocks of the Thompson belt are named the Ospwagan Group and consist of metasedimentary, metavolcanic, and ultramafic rocks. These rocks are intimately associated with highly reworked Archean basement gneisses and migmatites. Metasediments consist of laminated siltstones, shales, quartzites, dolostones, and iron formation along with minor chert and graywacke (Baragar and Scoates, 1981). Similar to volcanics of the Fox River belt, Thompson belt volcanics are basalts and komatiitic basalts. Ultramafics are common, as they are in the Fox River belt, and consist of serpentinized peridotite which occurs in pods and sill-like masses.

C.4. Tectonic Interpretation of the Circum-Superior Belt

The Circum-Superior belt presents a number of intriguing questions about the nature of tectonics during the Proterozoic. For example, if the connections inferred by Baragar and Scoates (1981) among all the above-mentioned areas are correct, then how did such an extensive, continent-scale, annular structure form? Certainly, continuity of the various orogenic belts suggests that the Circum-Superior zone was once a plate boundary; but the ensialic nature of most, if not all, of its orogens seems to imply that different tectonic styles may have been in operation. Of course, an ensialic setting for an orogenic zone does not necessarily rule out the operation of modern tectonic mechanisms in the development of that zone. Consider, for example, that continental collision can result in the development of a deformed belt of supracrustal deposits between two sialic blocks, one of which will extend beneath both the bulk of the deformed supracrustal belt as well as beneath the other sialic block (as a consequence of subduction) resulting in overthickened sialic crust. An example of this may be seen in the ensialic setting and associated crustal thickening of the modern Himalaya orogen, which is a suture zone between the colliding Indian and Eurasian plates. It is significant, therefore, that crustal overthickening has been reported to characterize the crust beneath several Circum-Superior areas, such as Southern Province (Sims *et al.*, 1980), the Labrador trough (Kearey, 1976), and the Cape Smith belt (Thomas and Gibb, 1977). For these reasons, the ensialic nature of Circum-Superior mobile belts is, by itself, insufficient evidence to rule out plate-tectonic mechanisms; other lines of evidence are needed to affirm or deny the operation of plate tectonics during the Proterozoic.

Additional evidence can be derived from the geometry and inferred history of individual mobile belts within the Circum-Superior belt. We have seen that all are characterized by depositional asymmetry, with miogeoclinal strata formed near the Superior craton and eugeoclinal strata farther away. Similarly, deformation and metamorphism are asymmetric in their effects across the belt, with greatest structural and metamorphic disturbance on the side of the belt away from the craton. Folds are overturned and verge toward the craton; reverse faulting was also craton-directed. Volcanic activity was also asymmetric. In addition to these geometrical observations, there are other suggestions of plate-type activity; for example, rocks very similar to modern ophiolites (i.e., oceanic crust and associated supracrustal pelagic sediments) are found in several of the belts. Also, at the base of Circum-Superior miogeoclinal sequences are normal-fault basins filled with coarse, red, locally derived detrital sediment deposited in terrestrial environments penecontemporaneously with faulting; in the Phanerozoic record, basins such as these are associated with the development of a rifted continental margin. All of the above-mentioned features are points of similarity between Phanerozoic mobile belts and those within the Circum-Superior belt.

Probably the strongest argument against plate-tectonic mechanisms in Proterozoic orogenesis has come from the study of paleomagnetism. Paleomagnetic pole positions, as determined from Proterozoic strata of different areas of the Canadian Shield, seem to indicate that little differential movement could have occurred between the various Shield areas (McGlynn and Irving, 1975). This is because paleopoles from the several provinces seem to be nearly identical (for the same interval of time) and, therefore, the provinces in question must have been part of a single, coherent piece of crust (Piper, 1976). But, such paleomagnetic data certainly do not rule out plate-tectonic mechanisms, since continental rifts may abort, even today. Even so, some authors have called into question certain of the paleopole determinations and some of the logic behind this interpretation. For example, Gibb and Walcott (1971) argued that the Superior craton was a separate sialic block from the Slave craton prior to the Hudsonian orogeny. They proposed that the northern portion of the Circum-Superior belt was a Proterozoic suture along which continental collision occurred between the Superior craton and the Slave craton, resulting in the Hudsonian orogeny. Cavanaugh and Seyfert (1977) agreed that differential movement had occurred between Superior and Slave but they chose to locate the suture between them in the interior of western Churchill Province rather than along the northern Circum-Superior belt.

Baragar and Scoates (1981) proposed the following tectonic model for the origin and history of the Circum-Superior belt (see Fig. 3-10). The scenario begins following the Kenoran orogeny at the end of the Archean, when Superior Province had been consolidated from greenstone belts and high-grade terranes. At that time it constituted a relatively thick, stable crustal block. Extension of the crust caused the development of tension fractures around the Superior block; these fractures continued to be active as crustal extension continued, resulting in the formation of continental rift basins into which terrestrial sediments were deposited in alluvial fans and fluvial plains. Further extension caused full continental separation, the opening of a marine trough along the locus of spreading, and the onset of mafic volcanism due to partial melting of the mantle. The trough opened along the northern margin of the Superior craton to approximately 300–500 km before crustal extension ceased. Further separation does not seem likely in view of paleopole positions from adjacent Proterozoic areas. Along the southern margin, however, extension apparently continued. Convergence followed cessation of extension and the northern chain of mobile belts was disrupted by compression. In the south, a subduction zone developed with oceanic lithosphere subducting northward beneath Southern Province, resulting in, among other things, the intrusion and extrusion of calc-alkaline rocks into Southern Province; no similar rocks are found along the northern belt of mobile belts.

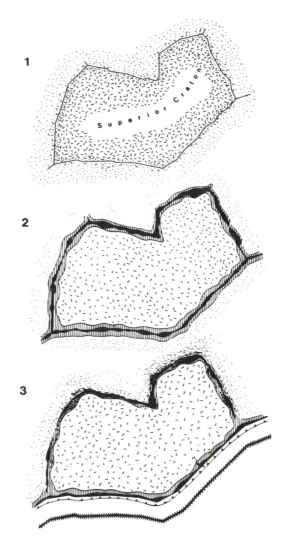

Fig. 3-10. Sequential model showing the development of the Circum-Superior belt. (1) Crustal extension causes development of fractures around the thickened crust of the Superior craton; (2) continued extension causes necking of the crust, opening of rift basins along the locus of spreading, and the advent of volcanism (black) within the rift trough; (3) spreading ceases along the northern segment of the belt while continuing along the southern segment; subsequent convergence causes closure of the northern troughs with concomitant deformation while a subduction zone develops on the southern margin of the Superior plate. (From Baragar and Scoates, 1981, reproduced with permission of the authors and Elsevier Publishing Company.)

D. Churchill Province, III: Western Churchill

The western part of Churchill Province is a vast, penaplaned wilderness veneered by glacial drift and obscured by thousands of lakes, swamps, and forests; consequently, details of its geology are poorly known. Even major structural features, such as fold belts, are indistinctly known or not known at all. Thus, Churchill Province has not been subdivided as carefully as have other provinces of the Shield.

Archean rocks are present in much of the western Churchill area, either as relatively undisturbed crustal rocks

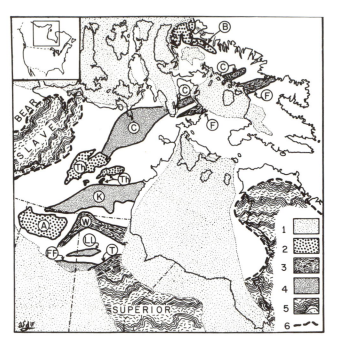

Fig. 3-11. Generalized map of western Churchill Province showing major geological regions. A, Athabasca Basin; B, Borden Basin; C, Committee fold belt; F, Foxe Basin; FF, Flin Flon region; K, Kaminak craton; LL, Lynn Lake region; T, Thompson nickel belt; Th, Thelon Basin; W, Wollaston Lake fold belt; 1, Phanerozoic cover strata; 2, relatively undeformed Proterozoic strata; 3, Proterozoic fold belts; 4, essentially unreworked Archean tectonic blocks included within western Churchill Province; 5, Archean rocks of Superior and Slave Provinces. (Modified from Davidson, 1972.)

formed during the Kenoran orogeny or as thoroughly remobilized crust, probably reworked as the result of the Hudsonian orogeny. In some areas, reworked Archean crust may resemble closely rocks formed during the Hudsonian event, making differentiation of them difficult. In other areas, lithologies and lithic associations of Churchill supracrustals are very similar to those of Archean greenstone belts so that mistakes in age assignments are easily made. In the ensuing discussion, we cover the following areas of western Churchill Province and their tectonic interpretation (see Fig. 3-11): Wollaston Lake, Foxe, and Committee fold belts, Kaminak craton, and Thelon, Borden, and Athabasca Basins.

D.1. Major Fold Belts

D.1.a. Wollaston Lake Fold Belt

The largest fold belt so far recognized within Churchill Province is the Wollaston Lake belt in the southcentral portion of the province, which extends nearly due westward from the southwest corner of Hudson Bay to Wollaston Lake on the eastern margin of the Athabasca Basin (Fig. 3-11). The

Wollaston Lake belt is typified by long, closely spaced, parallel isoclinal folds, often associated with cataclastic foliation (Davidson, 1972). Granitoid basement is exposed in the centers of some anticlines.

Along the northwest side of the Wollaston Lake fold belt, most of the metamorphosed supracrustal rocks are of miogeoclinal aspect (Money, 1968), but on the southeastern side of the fold belt, rocks of the lower Lower Proterozoic Sandfly Lake Group seem to be eugeoclinal. Davidson (1972) has stressed, however, that these rocks are not known well enough to be sure that they are correlative. Unconformably overlying the Sandfly Lake Group in western portions of the Wollaston Lake belt are quartzites, conglomerates, and mudrocks of the Meyers Lake Group, which is in turn overlain by arkoses, graywackes, and shales of the Daly Lake Group, both of which are probably of upper Lower Proterozoic age. At the eastern end of the Wollaston Lake fold belt, near Hudson Bay, a similar stratigraphy is observed: deformed strata, possibly correlative with Sandfly Lake rocks, are unconformably overlain by the Great Island Group, which may be correlative with Meyers Lake and Daly Lake strata.

Note that recorded within these strata are two major Lower Proterozoic stratigraphic sequences: an older sequence represented by the Sandfly Lake Group and its eastern correlatives, all of which have been intensely deformed and metamorphosed, and a younger sequence represented by Meyers Lake, Daly Lake, and Great Island rocks. If the upper sequence is actually Lower Proterozoic, then the orogeny during which Sandfly Lake rocks were deformed was a separate event from the Hudsonian orogeny in Southern Province, which occurred at the end of the Early Proterozoic. Whatever their actual tectonic history, Wollaston Lake rocks record a long and complicated geological history; Davidson (1972) postulated that these rocks may represent a major geosynclinal trough associated with the Kaminak craton just to the north.

D.1.b. Foxe and Committee Fold Belts

In northern Churchill Province, extending from Melville Peninsula across to central Baffin Island, is the Foxe fold belt (Heywood, 1967). The Foxe belt is composed of metasediments and remobilized Archean basement tightly folded around eastward- to northeastward-trending axes. The oldest sedimentary rocks along the northwestern side of the Foxe belt are widespread, mature quartz arenites and limestones of platformal or miogeoclinal aspect; these are overlain by immature graywackes, quartz sandstones, and mudstones, possibly representing a sediment source within a mobile belt to the southwest (Jackson and Taylor, 1972). On the southeastern side of the Foxe belt on Baffin Island, Jackson (1971) reported the presence of an ophiolite sequence consisting of metamorphosed mafic volcanics and ultramafic sills and dikes. From such little information, we may only suppose that the Foxe fold belt marks the site of another Churchill Province sedimentary basin, possibly a continental-margin prism, with a long and complex history.

To the northwest of the Foxe fold belt is the Committee fold belt, a partially reworked zone of greenstones and layered amphibolites associated with metasediments and iron formation (Davidson, 1972). Unlike most Archean greenstone belts, Committee belt supracrustals contain considerable amounts of quartzite. The age of these rocks is uncertain but may be Archean in part and Lower Proterozoic in part.

D.2.   Supracrustals of the Kaminak Craton

In the region of the Kaminak craton (Fig. 3-11), Lower Proterozoic strata form a cover sequence of only slightly deformed detrital sedimentary rocks overlying strongly deformed Archean basement. These strata have been divided into three units. The lowest of these units, termed the Montgomery Lake Group, is a coarse detrital sequence of local derivation and is found only in discontinuous basins (Davidson, 1972). Because of their petrology and geological setting, it is tempting to speculate that Montgomery Lake strata may represent fault basins formed during initiation of crustal extension and rifting following the Kenoran orogeny.

Above Montgomery Lake rocks are strata of the Hurwitz Group, subdivided into the Padei, Kinga, and Ameto Formations. These strata seem to represent a transition from a platformal tectonic environment to a geosynclinal environment. The Padei Formation is a coarse conglomerate and immature graywacke of variable thickness (Davidson, 1972), which grades upward into highly pure quartz arenites of the Kinga Formation. The Kinga is an extensive sheet of rippled and crossbedded quartz sand, spread out over more than 50,000 km^2. Overlying the Kinga across a minor disconformity are shales, siltstones, wackes, carbonates, and mafic flows of the Ameto Formation. Overlying the Hurwitz Group in the Kaminak area is the third major sedimentary sequence, an unnamed unit composed of arkose and conglomerate.

Strata similar to the Hurwitz Group in their petrology and stratigraphic position with respect to the basement are found in several areas between the Kaminak craton and the northern coast of mainland Canada. Whether these deposits actually represent outliers of an originally more-extensive Hurwitz cover sheet, or whether they are similar because they represent a similar tectonic–sedimentary environment has yet to be determined.

In his discussion of Kaminak Lower Proterozoic supracrustals, Bell (1970) interpreted Padei detrital sediments

as continental to shallow-marine deposits, representative of a stable, cratonic phase of sedimentation. Cratonization had occurred at the end of the Archean as a result of the Kenoran orogeny and was apparently followed by a period of local basin filling (represented by the Montgomery Lake Formation), perhaps the result of crustal extension and rifting after the Kenoran orogeny. The Kinga Formation, according to Bell, is the remains of a miogeoclinal sequence of marine deposits and the Ameto represents the beginning of orogenic activity in a mobile belt to the south or southeast, possibly a precursor to, or an early phase of, the Hudsonian orogeny.

D.3.   Thelon, Athabasca, and Borden Basins

In western Churchill Province, there are several large areas underlain by undisturbed or only gently disturbed Middle Proterozoic strata which overlie highly deformed Lower Proterozoic strata. They are predominantly terrestrial in their depositional character and are associated in some areas with igneous extrusions and shallow intrusions of mafic to intermediate composition. Originally, Middle Proterozoic strata may have covered much of the western part of the Shield [McGlynn (1976a) said that they may even have extended over into Slave Province] but subsequent erosion has removed much of this sedimentary veneer and Middle Proterozoic rocks are found today only in patches, scattered around the western part of the province. Most of these patches are relatively small but there are three areas within the Churchill Province where such rocks comprise large sedimentary basins: the Athabasca Basin of northern Saskatchewan, the Thelon Basin of the Northwest Territories, and the Borden Basin at the northwestern end of Baffin Island.

D.3.a.   The Thelon Basin

The Thelon Basin is located in the middle of western Churchill Province, just about halfway between Hudson Bay and Slave Province (Fig. 3-11). It is composed of strata of the Dubawnt Group, subdivided into six formations. The South Channel Formation is the lowest of the six and is composed of up to 1500 m of coarse conglomerate. It is overlain by the Kazan Formation, a terrestrial red bed approximately 2500 m thick. Both South Channel and Kazan strata were gently folded prior to the eruption of volcanics of the overlying Christopher Island Formation. These volcanics are composed of agglomeritic pyroclastics and flows of andesite, latite, and trachyte. The Pitz Formation is a second volcanic unit.

Unconformably overlying Christopher Island and Pitz volcanics are nearly flat-lying sands of the Thelon Formation, which comprise a blanket-shaped lithosome extending over much of the Thelon Basin. Thelon sands are composed of highly mature quartz arenites containing ripple marks, festoon crossbedding, and other primary sedimentary structures (McGlynn, 1976c). On the basis of its structures and stratification sequences, the Thelon is interpreted as a fluvial unit, formed on the stable, early Middle Proterozoic continent following cessation of the Hudsonian orogeny. The Thelon is overlain by stromatolitic dolostones of shallow-marine to supratidal origin and by thin basalts.

D.3.b.   The Athabasca Basin

The Athabasca Basin of northern Saskatchewan is composed almost entirely of the early Middle Proterozoic Athabasca Formation. The Athabasca is remarkably uniform in character over most of the basin, consisting of mature quartz arenites, quartz conglomerates, and rare, interbedded shale. It contains well-preserved tabular and trough crossbedding whose orientations indicate sedimentary transport toward the west and northwest. Like the Thelon, with which it is correlated, the Athabasca is considered to be the product of slow weathering within a tectonically stable source area, and deposited within a fluvial environment. Together with the Thelon, Athabasca sands are remnants of a blanket of terrestrial deposits which probably covered much of western Churchill Province.

The Athabasca is overlain by the Carswell Formation, a laminated dolostone with stromatolitic and brecciated zones. The Carswell is probably a shallow-marine deposit and may be correlated with similar marine dolostones overlying the Thelon Formation.

Two other units should be mentioned in this discussion although they are not parts of the Athabasca Basin. Just north of the basin, near Uranium City, are exposures of the Martin Formation, which consists of a minimum of 4000 m of coarse, immature sandstones and conglomerates associated with basaltic to andesitic flows and sills. The Martin is interpreted as a fluvial red-bed sequence deposited near the end of the Hudsonian orogeny within intermontane fault basins. Because of the similarity of their lithologies and geological settings, the Martin Formation and the South Channel Formation (of the Dubawnt Group in the Thelon Basin) may be correlative.

South of the Great Slave Lake are strata of the Nonacho Group, which are mainly red-colored, polymictic conglomerates, lithic and feldspathic sandstones, siltstones, and shales. Laminations and crossbedding are common in Nonacho sandstones; shales often reveal mud-cracked bedding surfaces, and adjacent sandstones contain mud chips in intraformational conglomerates. These rocks, too, probably represent fluvial deposition after the Hudsonian orogeny, but exact correlations with strata of the Thelon or Athabasca Basin are uncertain.

D.3.c.   The Borden Basin

The Borden Basin is smaller than the two above-mentioned Middle Proterozoic basins and is composed of a very different sequence of deposits. It appears to represent not a craton-interior setting, as do strata of the Thelon or Athabasca Basin, but rather a craton-margin mobile belt. At the base of the sequence, overlying Archean basement, is the Eqalulik Group, composed of a lower unit with about 300 m of massive and amygdaloidal basalt, quartzite, and tuff, and an upper unit consisting of greater than 1000 m of red arkose and quartzite. The Eqalulik is considered to be early Middle Proterozoic in age and is unconformably overlain by the upper Middle Proterozoic Uluksan Group, a conformable sequence over 6500 m thick containing detrital and chemical sedimentary deposits of a generally miogeoclinal aspect.

D.4.   The Hudsonian Orogeny in Western Churchill Province

If the preceding discussion seemed disjointed, it is because the western Churchill area is a geological jigsaw puzzle whose different areas have little or no real relation to each other. As pointed out earlier, the whole of Churchill Province is less a true structural/petrological province than it is a collection of tectonic leftovers after the other, obvious provinces had been defined. In the western Churchill region are large areas of nearly unreworked Archean rocks which seem to represent stable, microcontinental blocks prior to the end of the Early Proterozoic; there are also areas in which obvious Archean rocks have been thoroughly remobilized and remetamorphosed. Then too, there are extensive, but geometrically unrelated, Lower Proterozoic fold belts which appear to mark the sites of major sedimentary basins; although clearly ensialic in their modern geological settings, these Lower Proterozoic fold belts contain stratigraphic sequences which closely resemble those of Phanerozoic sedimentary masses whose origins as continental-margin assemblages are evident. Davidson (1972) claimed that it is debatable whether Churchill Province had been a distinct entity prior to the end of the Early Proterozoic, when the whole region, as well as adjacent parts of Superior, Southern, Bear, and Slave Provinces, were affected by the Hudsonian orogeny.

Diverse radiometric ages from rocks ascribed to the Hudsonian orogeny, as well as structural and stratigraphic relations among the tectono-stratigraphic units mentioned in the preceding paragraphs all indicate that the Hudsonian orogeny itself seems to have been a collection of events rather than a single tectonic entity. Some of these events were penecontemporaneous but in different areas, some were sequential in the same area, and some bear no apparent relationship to the others at all. One may infer from the above that Churchill Province is best thought of as a large tectonic collage composed of possibly unrelated Archean microcontinental blocks and Lower Proterozoic rifted-margin prisms, magmatic arcs, and cratonic basins. These disparate tectonic elements were gradually accreted together in a sequence of unrelated orogenic events and welded onto the growing margin of North America, the whole process being called the Hudsonian orogeny.

Cavanaugh and Seyfert (1977) suggested that areas in the northwestern part of Churchill Province and the Slave craton formed a single plate during the Early Proterozoic; they considered the rest of the Province to have been associated with the Superior craton, forming a second plate. Collision of these two plates resulted in the merging of two major portions of the Canadian Shield. Cavanaugh and Seyfert located the Superior–Slave suture between the Kaminak craton, which they claimed on the basis of paleomagnetic data to have been part of the Superior plate, and the Thelon Basin, considered to have been part of the Slave plate. In northern Canada, they located the suture between the Foxe fold belt to the southeast and the Committee fold belt to the northwest.

A somewhat different model for the Hudsonian orogeny in western Churchill was presented by Gibb (1983), who suggested that Hudsonian effects were due to a complex collision along the northern Circum-Superior line between the Superior and Churchill plates. In Gibb's model, the original shape of the Superior block is reflected by the present shape of the Circum-Superior belt. Thus, the northern area of the Superior continent featured two promontories, represented by the Thompson and Ungava salients, and a deep reentrant between them, in the area of southeastern Hudson Bay. Gibb proposed that these promontories acted as rigid protrusions which were forced into and "indented" the Churchill plate as it converged with and overthrust the Superior block. The resulting stress fields set up in the Churchill plate caused the development and propagation of thrusts, wrench faults, and belts of shear deformation, in the same manner as that suggested by Tapponier and Molnar (1976) for Asian continental fracturing in response to the Himalayan collision.

These models seem to answer some questions about western Churchill geology but other questions remain. For example, in the paper mentioned in the last paragraph, Gibb (1983) acknowledged that the western Churchill area may not have been a single continental plate but rather may have existed as independent microcontinents and island arcs. If this were true, it would seem to contradict his model, since the collision-caused stress fields he postulated would only have had the orientations he illustrated if they were imposed on a single, coherent plate. Of course, assembly of a single Churchill plate from smaller microplates could have oc-

curred prior to collision with the Superior block. We suspect, however, that geological relations and tectonic models in the Churchill area will become increasingly complicated as more knowledge is gained about the area and as the collage character of the region becomes more apparent.

E. Bear Province

The Bear Province is composed principally of Lower Proterozoic geosynclinal strata and associated metamorphic and igneous rocks; it also contains lesser amounts of Middle and Upper Proterozoic platform deposits. It is located between the Slave craton to the east, on whose Archean basement nearly undisturbed Bear strata rest nonconformably, and the MacKenzie Lowlands to the west, south, and north, whose Phanerozoic strata blanket Bear rocks (Fig. 3-12).

Bear Province can be divided into three tectonic elements (Fig. 3-12): (1) the Wopmay orogen; (2) Athapuscow and Bathurst aulacogens; and (3) the post-Lower Proterozoic Amundsen Basin. The Wopmay orogen is further broken down into three parts: (a) the Coronation margin, composed of the Epworth fold belt (or, Epworth Basin) along the eastern margin of the orogen and the Hepburn metamorphic–plutonic belt near the center; (b) the Great Bear magmatic belt (or, Great Bear batholith) on the western side of the orogen; and (c) the Hottah terrane just to the southwest of the Great Bear belt. In the following section, each of these areas and their tectonic implications will be discussed.

E.1. The Wopmay Orogen

E.1.a. The Coronation Margin

i. Epworth Fold Belt. Unmetamorphosed Lower Proterozoic sedimentary rocks are found on the eastern side of the Wopmay orogen, nonconformably overlying Archean basement. These rocks are termed the Coronation Supergroup and are considered to represent the origin, development, and destruction of a rifted continental margin, called the Coronation margin. The Coronation Supergroup is divided into three units (Fig. 3-13): (1) the Akaitcho Group, found only in isolated fault basins in western parts of the Coronation margin; (2) the Epworth Group, which is comprised of up to 3 km of miogeoclinal strata; and (3) the Recluse Group, a sequence of dark shales and graywacke flysch (in the western part of the Epworth Basin) and more quartz-rich molasse (in the easternmost portion of the basin).

Occurring only along the western boundary of the Hepburn belt, strata of the Akaitcho Group are composed dominantly of coarse detrital material grading upward into finer

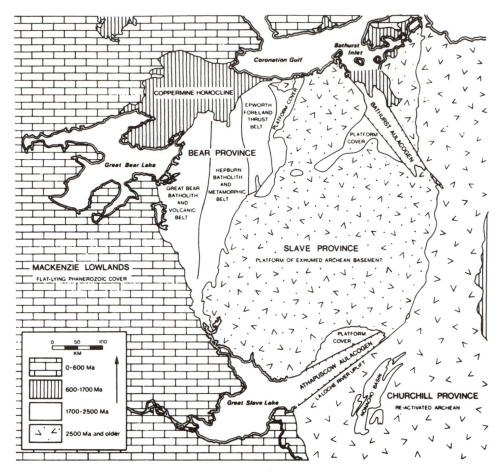

Fig. 3-12. Generalized map of Bear and Slave Provinces showing the Wopmay orogen, Athapuscow and Bathurst aulacogens, and Amundsen Basin (located within the Coppermine homocline); not shown are the Brock and Minto inliers on Melville Island to the north. (From Hoffman *et al.*, 1974, reproduced with permission of the Society of Economic Paleontologists and Mineralogists.)

deposits showing progressively greater marine influence. Lower Akaitcho strata (Zephyr and Drill Formations; Fig. 3-13) are very coarse, reddish conglomerates, lithic sandstones, arkoses, and mudstones intercalated with flows of continental tholeiites (Hoffman and Bowring, 1984); these lower Akaitcho strata are interpreted as alluvial-fan and fluvial deposits and probably represent the sedimentary fill of rift-valley basins. Above these continental red beds are strata of the Vaillant Formation, composed of thick layers of basaltic to rhyolitic lava-flows and tuffs which Hoffman and Bowring (1984) interpreted as coalescent shield volcanoes. Capping the sequence are dolomitic algal-reef complexes or restricted basinal muds.

Strata of the Akaitcho Group are interpreted to represent the earliest phases of continental rifting during which crustal extension led to development of block faulting and accumulation of terrestrial sediments in downfaulted areas. Continued extension produced further crustal subsidence as the thickness of the crust began to be significantly thinned;

this subsidence resulted in the incursion of the sea into the narrow oceanic gulf so that marine deposits are associated with lavas at the top of the Akaitcho.

Strata of the Epworth Group are composed of up to 3 km of primarily miogeoclinal sedimentary rocks. These rocks were thrust eastward over Slave craton basement during the orogenic destruction of the rifted margin. Figure 3-13 illustrates the inferred relationship of Epworth to Akaitcho strata; notice that Epworth rocks are inferred to overlie Akaitcho strata in western areas of the belt but, in eastern parts of the Epworth Basin, Epworth rocks rest directly on Archean basement. In both areas, however, Epworth strata are composed of shallow-water deposits. The lowest unit is the Odjick Formation, whose craton-derived, mature quartz arenites decrease in grain size westward, i.e., away from the Slave craton. Odjick strata represent the quartzite phase in the development of the Coronation margin (Hoffman, 1973). These detrital sediments pass upwards into massive and stromatolitic dolostones of the

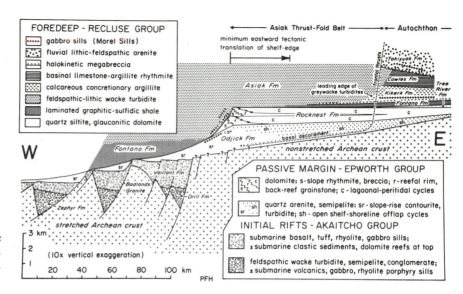

Fig. 3-13. Restored east–west stratigraphic cross section of the Coronation Supergroup. (From Hoffman and Bowring, 1984, *Geology*, Fig. 2, p. 70.)

Rocknest Formation, which represent the development of a shallow carbonate platform on the newly formed continental margin; this Hoffman (1973) called the dolomite phase. Around the edge of the platform are the remains of an algal reef complex which formed at the break in slope between the continental shelf and slope, and which acted as a rim around the platform. To the east, behind this algal rim, are deposits of a broad shelf-lagoon containing extensive stromatolites and storm deposits (Hoffman and Bowring, 1984).

To the west of the platform-edge algal rim are Epworth strata composed of siltstones, massive quartz beds, and graded graywackes; these are interpreted (respectively) as hemipelagic, contourite, and turbidite deposits formed on a continental slope and rise at the foot of the carbonate platform. Notice from Fig. 3-13 that deep-water Epworth deposits occur in the region where Akaitcho strata record major rift-valley development (and, therefore, major crustal separation); obviously, this area continued to subside rapidly, forming the lower slopes of the new continental margin.

The uppermost stratigraphic unit of the Epworth Basin is the Recluse Group, whose dark, deep-water deposits overlying shallow-water Epworth rocks have been interpreted as recording the development of a foredeep basin, i.e., a deep marine basin formed along the outer edge of an old miogeocline and deriving its sedimentary fill from a tectonic sourceland within a mobile belt (it may also be termed a deep foreland basin or a detrital wedge). Development of the foredeep basin was due to the foundering of the continental shelf, probably as the result of anisostatic subsidence related to incipient subduction of the continental margin beneath an approaching magmatic arc. Lowest units of the Recluse are black, pyritic shales (Fontano Formation)

which record the initial foundering of the shelf; these rocks represent the preflysch phase of the Coronation margin. Overlying preflysch strata of the Fontano are very immature feldspathic and lithic flysch of the Asiak Formation. These deposits are interpreted as turbidites and their grain-size trends and petrography indicate derivation from the west (i.e., from a tectonic source offshore of the continental margin). To the east, turbidite grain size decreases and the flysch facies passes into a facies dominated by fine, concretionary mudstones and argillaceous micrites of the Kikerk and Cowles Formations. These units are considered to have been starved-basin deposits along the eastern side of the foredeep basin and, therefore, to have been farthest from the area of detrital influx. Some calc-flysch within the Cowles Formation may have been derived from the carbonate platform. Capping the Recluse section in the east are coarse, lithic and feldspathic sandstones of the Takiyuak Formation which Hoffman and Bowring (1984) considered to represent a fluvial molasse facies formed penecontemporaneously with the onset of orogeny.

ii. Hepburn Metamorphic–Plutonic Belt. At the western edge of the Epworth fold belt is a region of highly deformed, metamorphosed, and intruded eugeoclinal rocks. At the bottom of the assemblage are feldspathic rocks related to the Akaitcho Group and overlying rocks are correlative with Fontano and Asiak Formations. Metamorphic grade increases westward across the Hepburn belt, from biotite grade in the east to sillimanite in the west. Intruding these rocks is granodiorite of the Hepburn batholith. The granodiorite has a distinct foliation paralleling that of gneisses within the country rock, indicating that intrusion was pre- or syn-tectonic.

E.1.b. Great Bear Magmatic Belt

The largest part of western Bear Province is comprised of intrusive igneous rocks and metasediments of the Great Bear magmatic belt (or, Great Bear batholith). The Great Bear belt is about 100 km wide and over 400 km long; its intrusive lithologies are dominated by nonfoliated granodiorites and adamellites associated with smaller masses of granite, tonalite, and diorite. These intrude a thick supracrustal section of subaerial basaltic to rhyolitic lava flows, welded tuffs, agglomerates, lahar breccias, and nonmarine sedimentary rocks (Hoffman, 1973). Thus, rocks of the Great Bear belt seem to represent a major volcanic arc; Hilderbrand (1982) presented rare-earth-element data on Great Bear plutons, showing that their geochemical enrichment trends and abundance patterns are more typical of continental magmatic arcs than of oceanic arcs. Since Great Bear plutons are not foliated, their emplacement must have been posttectonic.

Located within small, scattered basins at the top of the Great Bear batholith are Lower Proterozoic, red-bed deposits of the Echo Bay and Cameron Bay Groups. Echo Bay strata are divided into a lower unit composed of tuff, argillite, chert, arkose, and conglomerate with minor andesite flows, and an upper, volcanic unit of andesitic flows, breccias, tuffs, and agglomerates. The large amounts of tuffs, volcanic breccias, and other pyroclastics suggest a history of explosive volcanism. The Cameron Bay Group is made up of a conformable sequence of arkosic conglomerate, arkose, siltstone, and shale. Conglomerate clasts are mostly composed of volcanic rocks along with chert, argillite, and sandstone, suggesting derivation from Echo Bay strata. Taken together, Echo Bay and Cameron Bay strata are considered to represent the filling of late-orogenic basins cut into the top of the Great Bear batholith.

E.1.c. Hottah Terrane

The Hottah terrane is exposed only at the southeastern corner of Great Bear Lake (Fig. 3-13). It consists of a highly deformed suite of metavolcanic and metasedimentary rocks, intruded by plutonic masses of dioritic to granitic composition. Hoffman and Bowring (1984) stated that the Hottah has been deformed several times. The terrane is interpreted as a displaced microcontinental block, which was accreted to the Coronation margin as a result of collision.

E.1.d. Tectonic Interpretation

It is of great significance that Epworth and Recluse strata seem to preserve a record of the same kinds of tectonic events as do the deposits of much younger rifted continental margins. Indeed, Fraser *et al.* (1972) stated that the history of the Coronation sedimentary prism is very similar to that of the more familiar Appalachian continental-margin prism of Phanerozoic age. In both the Wopmay and Appalachian mobile belts, earliest deposits are continental, restricted to local fault basins, and associated with extensive outpourings of basaltic lavas. These rift-valley and proto-oceanic-gulf phase rocks are overlain by passive- (or, rifted-) continental-margin deposits which define the main part of the miogeoclinal assemblage in both areas. For the most part, detrital sediments of both the Appalachian and Coronation miogeoclines appear to have been derived from the craton and carried toward the mobile belt.

Following the passive-margin (or, miogeoclinal) phase, a "preflysch" phase ensued, marked by deposition of black, fissile shales. These are followed by development of an extensive detrital wedge composed of immature detrital sediments. These detrital wedge sediments were derived from within the mobile belt and were carried toward the craton; this change in sediment-transport polarity reflects either the development of an active plate margin by disruption of the old passive margin or the approach of a magmatic-arc terrane on a converging plate. Lower deposits of the detrital wedge comprise a flysch sequence, the detrital elements of which were derived from the west although some calc-flysch in eastern parts of the Epworth Basin were probably derived from the edge of the shallow carbonate platform and carried down into deep water by turbidity currents. Similarity of geological records suggests the possibility of similar formative mechanisms: in this case, continental rifting. Thus, strata of the Epworth Group lend strong support to the hypothesis that plate-tectonic mechanisms were in operation even during the earliest Early Proterozoic (Fraser *et al.*, 1972).

The strata of the Epworth Basin were deformed and intruded as the result of a series of tectonic events which began shortly after the development of the rifted continental margin. The first orogenic event, termed the Calderian orogeny by Hoffman and Bowring (1984), was characterized by the main phase of thrusting in the miogeoclinal assemblage, subsidence and filling of the foredeep basin, and intrusion of Hepburn peraluminous plutons. This event was considered by Hoffman and Bowring to have been the result of westward subduction of the leading edge of the Slave microcontinental plate beneath the advancing Hottah block. Probably, the earliest effect of this process was foundering of the continental shelf, ending Rocknest deposition and beginning the filling of the resulting foredeep.

The second tectonic event envisioned by Hoffman and Bowring was characterized by the development of northeast-trending, en echelon cross folds involving both the Slave basement and all Calderian structures. These structures follow the same trend as major fold and fault features in western Churchill Province to the east of Bear Province and were thought by Hoffman and Bowring to have formed as the result of collision between the Slave microplate and the Superior plate.

The third orogenic event spelled the demise of the Coronation area as an active plate margin. This event was characterized by development of large numbers of conjugate, transcurrent faults indicative of east–west crustal shortening. These faults crosscut both Calderian and younger fold structures and so must postdate them. This event was the terminal orogeny in the area and was held by Hoffman and Bowring to have been the result of a continental collision at the western, unexposed margin of the Hottah terrane.

From the above discussion, it is probably clear that plate-tectonic mechanisms can usefully be invoked to explain most of the structural and stratigraphic features seen in the Wopmay orogen. As we will show in the next section, the same is also true of Athapuscow and Bathurst areas of Bear Province. All of this is of great significance because it provides some of the best evidence that plate tectonics, in much the same manner as today, must have operated during the early Proterozoic. This is not to say that all other tectonic mechanisms should be ruled out for Proterozoic orogenesis everywhere. Rather, it suggests that the Proterozoic was not a strange eon when Earth processes were exotic or unknowable; just the opposite, observations from Bear Province (and others, of course) lead one to the conclusion that some supposedly modern tectonic mechanisms have been around a very long time indeed.

E.2. Athapuscow and Bathurst Aulacogens

Trending at high angles into the Slave Province are two long, narrow basins which taper slightly into the Slave craton (see Fig. 3-12). The basins are bordered on both sides by major normal faults along which Lower Proterozoic rocks have been downdropped into contact with Archean basement. Both basins have the same general form, history, and tectonic significance. The southern one, termed the Athapuscow aulacogen (also, the East Arm of Great Slave Basin), has been the most carefully studied of the two and will serve to exemplify both. Although not treated here, the Bathurst aulacogen is discussed by Tremblay (1968) and its relation to the Coronation and Athapuscow areas is considered in Fraser et al. (1972).

The Athapuscow aulacogen occupies the East Arm of Great Slave Lake and separates the Slave from the Superior Province. However, because its history is intimately linked with that of the main area of Bear Province, it is usually considered to be part of that province. The aulacogen is bordered on both sides by major normal faults, the southern of which is called the McDonald Fault and is revealed by a

distinctive, straight fault-line scarp over 400 km long. The northern border fault is covered by a cratonic cover sequence. These faults diverge gradually as they trend away from the center of the Slave block, and strata within the aulacogen thicken toward the Coronation area. Unlike rocks of the continental-margin prism, however, aulacogen rocks have not undergone any significant metamorphism. Figure 3-14 is an interpretative cross section of aulacogen units and their relation to strata of the adjacent Slave platform; it should be consulted during the following discussion.

Oldest Lower Proterozoic strata of the Athapuscow aulacogen comprise the Union Island Group which consists of a basal arkosic conglomerate containing boulders of Archean granite overlain by several hundred meters of dolostone. The sedimentary sequence is capped by thin beds of black shale. These strata are followed by thick layers of pillow basalt, volcanic breccias, and gabbroic sills (Hoffman, 1973). Union Island strata were tilted and truncated by erosion prior to the onset of aulacogen development.

The main rifting phase of the Athapuscow aulacogen is recorded by deposits of the Sosan Group, the lowest unit of which is a coarse, somewhat feldspathic quartzite up to 1600 m thick. This unit, called the Hornby Channel Formation, is considered to be equivalent to the Odjick Formation of the Epworth Group. Hornby Channel rocks coarsen toward the border faults and were probably deposited penecontemporaneously with faulting. Dolostones of the Duhamel Formation overlie Hornby Channel quartzites just as Rocknest dolostones overlie Odjick sandstones in the Coronation continental-margin prism. Duhamel dolostones are sandy and stromatolitic; they probably represent a continuation of the Rocknest carbonate platform of the Coronation margin into the Athapuscow area. Both Hornby Channel and Duhamel strata are restricted to the aulacogen and contain facies patterns which suggest that they were never much more extensive than they are today.

The upper unit of the Sosan Group, the Seton Formation, is composed of subaerial basaltic lava flows interbedded with mudstones and quartzites. The Seton differs from lower Sosan Group units not only by containing abundant volanic material, but also in that it extends from the aulacogen northward onto the Slave block, where it intertongues with and is replaced by fine, crossbedded quartz sands of the Kluziai Formation and overlying mudstones of the Akaitcho Formation (see Fig. 3-14). Together, these two units form the base of a cratonic cover sequence. In Hoffman's (1973) scheme, Seton strata are correlative with pre-flysch-phase rocks of the Coronation margin.

As discussed in the previous section, the Coronation margin foundered after its miogeoclinal stage resulting in the influx of immature, flysch-type sediments from a tectonic sourceland within the mobile belt. The Athapuscow aulacogen was also affected by foundering of the shelf but

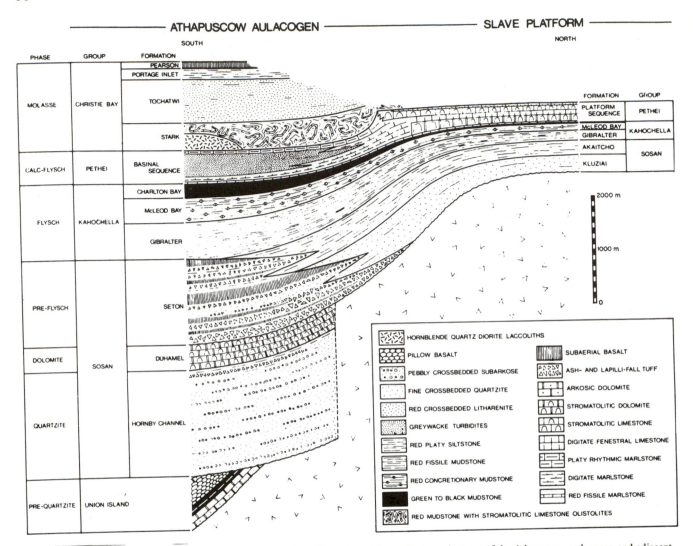

Fig. 3-14. Somewhat diagrammatic, restored north–south stratigraphic section of Lower Proterozoic strata of the Athapuscow aulacogen and adjacent Slave platform. (From Hoffman *et al.,* 1974, reproduced with permission of the Society of Economic Paleontologists and Mineralogists.)

it did not receive turbidites from the west as did the Coronation margin. Instead, flysch-phase deposits of the Athapuscow aulacogen are green and red mudstones, variously mud-cracked and concretionary, with calcareous beds and bentonite layers. These strata comprise the Kahochella Group, composed of Gibralter, McLeod, and Charlton Bay Formations. They are very similar to Kikerk and Cowles Formations of the Recluse Group with which they are correlated. Although Kahochella strata increase in thickness as they pass from the craton into the aulacogen, they undergo little change in sediment lithology, indicating that the aulacogen had ceased to be a separate, subsident feature and was behaving in a manner similar to that of the adjacent platform.

During the following interval of time, the aulacogen once again acquired a sediment facies different from that of

the platform. Pethei Group strata of the Slave cover sequence record a shallow carbonate platform with stromatolitic dolostones and limestones. Within the main part of the Athapuscow aulacogen, however, Pethei deposits are composed primarily of rhythmically bedded, calcareous mudstones, graywacke turbidites, and shaley limestones. At the edge of the Athapuscow Basin, Pethei carbonates record a platform-rim complex facing southward toward the deepwater environment of the basin center. Near the margin of the basin, at the foot of the carbonate platform, are debris-flow deposits with chaotic carbonate blocks.

The molasse phase in the history of the Athapuscow aulacogen is recorded in rocks of the Christie Bay Group. Sharply overlying Pethei basinal strata are extensive olistostromal masses containing angular blocks (up to 45 m thick and 1000 m long) of shallow-marine limestone and dolostone chaotically dispersed within a red mudstone matrix. These olistostromes comprise the Stark Formation and were probably caused by seismic instability of the platformal area, which accelerated as the Hottah terrane gradually closed on the Slave continental margin. Over the Stark are 850 m of red, laminated lithic sandstones, commonly cross-bedded and rippled, and mud-cracked siltstones and mudstones. Termed the Tochatwi Formation, these deposits represent a true molasse, deposited penecontemporaneously with the Calderian orogeny. They are overlain by red, evaporitic mudstones of the Portage Inlet Formation and columnar basaltic flows of the Pearson Formation.

The final chapter in the Athapuscow aulacogen's history is recorded in thick, massive conglomerates of the Etthen Group. Termed fanglomerates by Hoffman (1973), these rocks are up to 4000 m thick and contain rounded boulders up to 1 m in diameter and set in a calcite-cemented, lithic sandstone matrix. Sedimentation was fault-controlled and restricted to the aulacogen.

The tectonic significance of the Athapuscow aulacogen lies in the fact that aulacogens are generally considered to form during continental rifting as the third, or "failed" arm of a three-armed rift system. Initially, a three-spoked fracture pattern develops within continental crust uplifted above a hot spot. As the crust continues to be elevated, the fractures open into rift valleys accompanied by intrusion and extrusion of mafic magmas into axial regions of the rift. As rift valleys deepen, coarse, immature (often arkosic) detritus derived from weathering and erosion of the adjacent, uplifted crust is carried into the rifts and deposited in a variety of continental environments, especially alluvial fans and fluvial systems. With continued crustal extension, two of the three rifts continue to widen, resulting finally in the opening of an ocean basin. Following full continental separation, the third rift is shifted off the hot spot as the continental block in which it is located drifts away from the speading center. As a consequence of drift, the "failed" arm ceases to behave as a typical rift valley; instead, the area of the failed arm becomes the center of a cratonic basin as the dense, mafic core of the old rift pulls down on the surrounding sialic crust. Such a basin may have a long and varied history; because of its location striking into the cratonic interior, it is often the site of a major deltaic system. Later in its history, an aulacogen may undergo deformation associated with a continental-margin orogeny, but its setting within thick, sialic block (rather than on its margin) serves to protect it from the full effects of the orogeny. Aulacogens, therefore, are typically faulted and folded but not significantly metamorphosed.

Aulacogens develop very early in the history of a continental separation event. Therefore, they would seem to signify a style of tectonic response caused by the tensional stressing of rigid, lithospheric plates. For this reason, they are taken to indicate the operation of plate-tectonic mechanisms. And the Athapuscow aulacogen, being of Early Proterozoic age, indicates that such mechanisms have been at work since at least the beginning of the Proterozoic (Hoffman, 1973; Windley, 1984).

E.3. Middle and Upper Proterozoic Strata of the Amundsen Basin

Post-Lower Proterozoic sediments were deposited in the Amundsen Basin, remnants of which are now parts of the Coppermine homocline (see Fig. 3-12) and the Brock and Minto Inliers to the northwest of the main part of Bear Province. These sediments comprise three Middle Proterozoic sequences and one of Late Proterozoic age.

Lowest of these postorogenic deposits is the Hornby Bay Group, composed of up to 2000 m of sandstone, stromatolitic dolostone, shale, and minor conglomerate (Fraser et al., 1972). The Hornby Bay Group unconformably overlies deformed Lower Proterozoic basement and is itself unconformably overlain by the Dismal Lakes Group, which is also composed of sandstone, shale, and dolostone. Hornby Bay and Dismal Lakes strata are considered to be of fluvial and shallow marine origin and are very similar in both lithology and paleocurrent trends to rocks of the Thelon and Athabasca Formations of western Churchill Province. This similarity led Fraser et al. (1972) to suggest that all of these units are correlative and were originally parts of an extensive sheet of terrestrial sediments covering much of the western Canadian Shield.

Conformably overlying Dismal Lakes rocks are volcanics and associated sediments of the Coppermine River Group. Lowest Coppermine River strata are approximately 3000 m of flood basalts, individual flows of which range up to 70 m thick. Flows are usually massive but may be amygdaloidal at their tops. Upper Coppermine River rocks are red sandstones, siltstones, and shales interbedded with basaltic flows. These deposits cap the Middle Proterozoic section and were gently folded, faulted, and eroded prior to deposition of Upper Proterozoic strata.

While not part of the supracrustal sequence discussed above, the Muskox mafic–ultramafic intrusion is closely related to them. The intrusion is a highly differentiated, layered, igneous complex composed of dunite, peridotite, gabbros, and granophyres. Located along the boundary between the Epworth fold belt and the Hepburn metamorphic–plutonic belt, the Muskox intrusion is 118 km long and 13 km wide and has a funnel-shaped cross section. Fraser et al. (1972) interpreted it as having formed by the intrusion and fractional crystallization of a tholeiitic basalt magma; they further suggested that it is located within a major rift zone (much of which is beneath younger strata) and that its emplacement was related to the extrusion of Coppermine River flood basalts. In their model, Muskox magmas were intruded into the join between the Epworth and Hepburn areas and were trapped at the unconformity beneath strata of the Hornby Bay Group so that the fluid spread out into a funnel shape.

Upper Proterozoic rocks of Bear Province are sandstones, shales, argillites, stromatolitic dolostones and limestones, and some gypsum (Fraser *et al.*, 1972). These rocks comprise the Rae Group and rest unconformably on Coppermine River basalts.

F. Grenville Orogen

The youngest Proterozoic region of the Canadian Shield is the Grenville Province in eastern Canada. It extends almost 1500 km from the Atlantic coast of Labrador to Lake Huron and, at its widest point from Sudbury, Ontario, to the Adirondacks, is 675 km across (see Fig. 3-15). South of Canada, Phanerozoic strata obscure the basement so that the nature and extent of Grenville rocks are less certain; in addition, where exposed in the eastern United States, Grenville rocks have been strongly reworked by tectonic events associated with the Appalachian orogen so that their recognition is difficult. Nevertheless, rocks related to the Grenville orogen can be traced discontinuously all the way down eastern North America to the Llano Uplift and Van Horn Mountains in Texas.

The Grenville Province is characterized by extremely high grades of metamorphism in both basement terranes and supracrustal rocks of the Grenville Supergroup. Over much of the province, metamorphic grade is of the amphibolite rank and attains the granulite rank in some areas, such as the Adirondack Highlands. Strongly deformed Grenville gneisses are often associated with migmatites and evidence of metamorphic reworking and redeformation is common. Tight, asymmetric to overturned, northeast-trending, gently plunging folds, vergent to the northwest, constitute the typical structure in the province; within the interior of the orogen, they often take the form of mantled gneiss domes. Folds may be associated with large-scale faults, which are marked by distinct zones of cataclasis and mylonitization. Igneous rocks of Grenville Province comprise a large and varied group but the most common igneous lithologies are of the anorthosite–monzonite–norite suite, although many Grenville anorthosites may be reworked from Archean terranes, especially near Nain Province of Labrador.

Of its various aspects, the most significant to its tectonic interpretation is the apparently ensialic nature of the Grenville orogen. Much of the province is underlain by monotonous, homogeneous quartzofeldspathic gneisses which are interpreted as remobilized and remetamorphosed Archean crust. This interpretation is clearly indicated by the fact that structural and lithological trends associated with older, adjacent Shield provinces can be traced in the basement across the Grenville Tectonic Front and into Grenville Province. As a result, Grenville areas near older provinces often show evidence of multiple periods of deformation, indicated by the development of complex fold-interference patterns. Farther into the Grenville belt, the trends and lithologies of adjacent provinces gradually become less distinct as they are reoriented into concordance with Grenville trends and as remetamorphism renders them increasingly indistinguishable from Grenville rocks. Because reworking of older sialic crust is so apparent in Grenville Province, many geologists consider it to be a classic example of an ensialic orogen.

F.1. Subdivisions of the Grenville Orogen

In his discussion of Grenville Province, Wynne-Edwards (1972) recognized seven subdivisions, each of which is briefly described below and illustrated in Fig. 3-15. In addition, we include introductions to major Grenville areas of the United States.

F.1.a. Grenville Tectonic Front

The Grenville Front is the western boundary of Grenville Province, separating areas of metamorphosed Grenville rocks to the southeast from areas to the northwest which were essentially unmetamorphosed by the Grenville event. The Front is recognized as a wide belt of steeply dipping cataclastic and mylonitic rocks of granulite metamorphic grade and with distinct northeast–southwest-trending foliation. It is associated with a prominent negative Bouguer gravity anomaly. Seismic profiles indicate that the crust beneath the Front may be up to 8 km thicker than adjacent areas of the crust.

F.1.b. Grenville Foreland Belt

The Foreland Belt is a narrow zone immediately to the northwest of the Grenville Front in which largely unmetamorphosed Lower and Middle Proterozoic deposits rest on the crystalline basement of older provinces, e.g., the Mistassini and Otish Groups of Superior Province mentioned earlier in this chapter. The northwestern boundary of the Foreland Belt is the limit of Grenville-related deformation, i.e., folding, cleavage development, faulting.

The Foreland Belt is divided into four parts, corresponding to the four older Shield Provinces which adjoin the Grenville orogen in Canada. In the Southern Foreland zone, rocks of the Huronian Supergroup show effects of the Grenville orogeny and thicken slightly toward Grenville

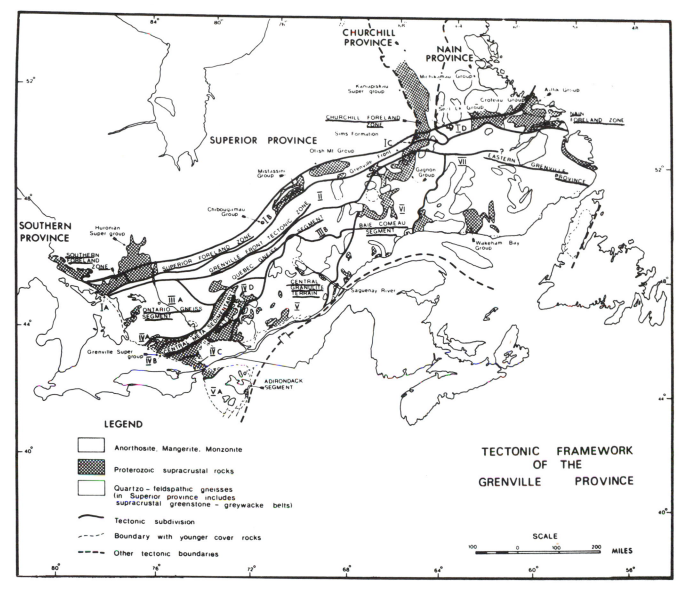

Fig. 3-15. Major subdivisions of the Grenville orogen in Canada. (From Wynne-Edwards, 1972, reproduced with permission of the Geological Association of Canada.)

Province. Huronian lithologies can be traced for some distance into the main part of the Grenville orogen. The Superior Foreland zone contains strata of the Mistassini, Otish, and Chibougamau Groups, all of which were undeformed by the Hudsonian orogeny but mildly affected by the Grenville event. As mentioned earlier, these units were probably deposited during pre-Hudsonian time, the result of sedimentation along the Circum-Superior belt; their lack of significant Hudsonian effects probably reflects the fact that they lay on stable sialic crust and, thus, were shielded from intense Hudsonian diastrophism. The Churchill Foreland zone contains rocks of the Kaniapiskau Supergroup (associated with the Lower Proterozoic Labrador trough) which were

deformed during both Hudsonian and Grenville orogenies. Post-Hudsonian, pre-Grenville stratigraphic units of the Churchill Foreland zone of the Grenville belt include the Sims Formation (mentioned earlier during discussion of the Labrador trough). The Nain Foreland zone contains several interesting units, particularly the Seal Lake Group which is interpreted by Wynne-Edwards (1972) as part of a late Middle Proterozoic, syntectonic detrital wedge (i.e., a molasse) associated with Grenville orogenic uplift.

F.1.c. Central Gneiss Belt

To the southeast of the Grenville Front, in the southern half of Grenville Province, is a belt of polydeformed rocks typified by quartzofeldspathic gneiss considered to be re-worked Archean crust. Some of these gneisses, however, may also be metamorphosed Grenville Supergroup strata. Typically, the gneisses have been raised to the upper amphibolite metamorphic grade. The strong northeast–southwest foliation of the Grenville Front becomes less distinct within the Central Gneiss Belt and obvious cataclastic zones become less apparent. The dominant structures in this region are well-defined, northeast-trending folds and gently dipping gneiss domes.

F.1.d.   Central Metasedimentary Belt

To the southeast of the Central Gneiss Belt is a region of metasedimentary and metavolcanic rocks termed the Grenville Supergroup. The thickest sequence of Grenville metasediments is in the Hastings basin of southern Ontario (indicated by IVB in Fig. 3-15). Metamorphic grade declines from amphibolite facies along the margins of the Metasedimentary Belt to greenschist in the center of the Belt.

F.1.e.   Central Granulite Belt

Northeast of the Central Metasedimentary Belt is a region of intensely metamorphosed gneisses and associated anorthosites. The boundary separating it from the Central Metasedimentary Belt is a zone of prominent cataclasis and mylonitization.

F.1.f.   Baie Comeau Segment

The Baie Comeau Segment adjoins the Central Granulite to the northeast but is of amphibolite metamorphic grade and is separated from it by a zone of cataclasis. Most of the Baie Comeau region is underlain by reworked basement gneisses, but up to 10% of the area is underlain by metamorphosed Grenville Supergroup rocks. In the northern region of the Baie Comeau Segment are polydeformed rocks of the Kaniapiskau Supergroup. As in above-mentioned divisions of Grenville Province, the Baie Comeau region contains numerous anorthositic massifs intruding crustal rocks; some of these anorthosites show evidence of metamorphic reworking while others appear to be fresh and retain their original, igneous textures.

F.1.g.   Eastern Grenville Province

This region is characterized by extensive terranes of quartzofeldspathic gneisses and anorthositic intrusions, all raised to the upper amphibolite grade of regional metamorphism. These rocks are probably reworked Archean

and/or Lower Proterozoic rocks of Nain Province just to the north. Middle Proterozoic supracrustals of the Eastern Grenville area are represented by the Wakeham Bay Group.

F.1.h.   Adirondack Mountains of Northern New York

The Adirondacks are surrounded and overlapped on all sides by Phanerozoic cratonic strata, except to the northwest where supracrustals have been stripped off and Precambrian basement is continuously exposed into Canada. The Adirondacks are divided into two regions, the northern of which is termed the Adirondack Lowlands and is characterized by amphibolite-grade metasedimentary rocks of the Grenville Supergroup. As such, it probably represents a southward extension of Wynne-Edwards's Central Metasedimentary Belt. The southern Adirondack segment, called the Adirondack Highlands, is underlain by granulite-grade gneisses and intrusive anorthosites and syenites. It is probably a southward continuation of the Central Granulite Belt. The boundary between the two Adirondack regions is called the Adirondack Highland Line and is recognized as a metamorphic boundary marked by the garnet–clinopyroxene isograd (Wynne-Edwards, 1976).

F.1.i.   Northern Appalachian Orogen

South of the Canadian border, exposures of Grenville rocks are obscured by younger supracrustals and by tectonic overprinting due to Appalachian orogenic events. Nevertheless, Grenville terranes probably constituted the basement upon which much Northern Appalachian strata were deposited. Grenville-age crust is found as basement in the Long Range of Newfoundland, the Green Mountains of Vermont, the Berkshires of Massachusetts, the Hudson Highlands of New York, and the Reading Prong of New Jersey (King, 1976). In most of these areas, Grenville rocks are paragneisses, marbles, and metaquartzites, similar to rocks of the Grenville Supergroup. Over much of the Northern Appalachians, Grenville rocks were raised to the amphibolite grade, except in the Reading Prong where granulite grades were obtained. In all areas, metamorphism related to Paleozoic tectonic events has caused retrogression of Grenville rocks and resetting of radiometric clocks; these effects have made recognition of Grenville terranes difficult. Consequently, other Grenville areas may yet be identified.

F.1.j.   Central and Southern Appalachians

Grenville lithologies of the Central and Southern Appalachians are exposed primarily in Blue Ridge and Piedmont Provinces but, as in the Northern Appalachians, probably form the basement beneath much of the Appalachian orogen, except possibly in the Piedmont, which may be underlain by exotic crustal blocks.

In western Appalachian regions, Grenville rocks comprise the core of the Blue Ridge. In Maryland and northern Virginia, Grenville basement is dominated by metamorphosed plutonic rocks such as hypersthene granodiorite and granite; one anorthosite body is recognized (King, 1976). This plutonic complex is continuous to the southwest into the North Carolina Blue Ridge, where it comprises the Cranberry, Beech, and Max Patch Gneisses. These units constitute the basement beneath Upper Proterozoic rocks of the Mount Rodgers Group and Ocoee Supergroup. To the northwest, however, Grenville paragneisses are exposed in the Stone Mountain thrusts, a series of low-angle reverse faults in southwestern Virginia and northeastern Tennessee. In this area, Grenville rocks constitute the Saddle Gneiss, a metasedimentary rock intruded by the Grenville-aged Striped Rock and Carsonville Granites and the Grayson Granodiorite. In the segment of the Blue Ridge in Georgia, Grenville rocks are represented by Whiteside Granite, the Corbin Gneiss, and the Salem Church Granite, all of which comprise basement beneath Ocoee strata.

During the early days of geological investigation in the Piedmont, most of the area (except for a few obviously posttectonic intrusions) was considered to be of Precambrian age. Increasingly, however, data have been acquired suggesting that much of the Piedmont is composed of Paleozoic rocks. Further, some areas clearly of Precambrian age are now thought to have been accreted to the continent during the Phanerozoic in the form of exotic, displaced terranes (see discussions of Appalachian history in later chapters). Consequently, in his catalog of Precambrian areas of the United States, King (1976) recognized only a few Piedmont rock units to be of Grenville age. These include the Baltimore Gneiss of Maryland and Pennsylvania, gneisses and schists of the Sauratown Mountain area of North Carolina's northwestern Piedmont, and the Woodland Gneiss and Jeff Davis Granite in the southwestern Piedmont of Georgia. Exactly how much more of the Piedmont is of Grenville age is currently being studied; there are some suggestions (especially as a result of COCORP seismic profiling; see Chapter 7) that much of the Piedmont's deep structure may be composed of Grenville crust, the surface of which acted as a décollement zone over which accreted terranes and sedimentary wedges were thrust during the Alleghenian orogeny.

F.1.k. Llano Uplift

In central Texas, Precambrian and lower Phanerozoic rocks are exposed in a basement uplift surrounded on all sides by much younger strata. Called the Llano uplift, this area also features rocks of Grenville age. The majority of Grenville rocks comprise the Valley Spring Gneiss and the Packsaddle Schist. The Valley Spring Gneiss is a felsic paragneiss, composed of pink, fine to medium crystalline, quartzofeldspathic gneiss. The Packsaddle is a more mafic assemblage, composed of schists of amphibole, graphite, and mica, marbles, quartzites, and calc-silicate rocks. Flawn and Muehlberger (1970) considered these two units to be of metasedimentary origin, originally comprising a sequence not less than 6500 m thick. These rocks were intruded by the Grenville-aged Big Branch Gneiss (a metamorphosed quartz diorite) and Red Mountain Gneiss (a metamorphosed granite), both of which are considered to be syntectonic. These rocks are similar to Grenville lithologies in the eastern United States and their radiometric ages agree well with Grenville ages (King, 1976). Although the local term "Llano orogeny" has been used to describe the tectonic event they record, Llano rocks probably represent an extension of Grenville orogenesis to the southwest across the Mississippi Embayment.

F.1.l. Van Horn Mountains

The last area considered by some geologists to be of Grenville age is in the Van Horn Mountains of the Trans-Pecos region of western Texas. In this desert region, exposures of Precambrian rocks comprise small, structurally complicated areas separated by wide areas in which younger strata obscure the basement. Consequently, the Precambrian geology is poorly understood. Principal Precambrian units of the Van Horn area include the following: the Carrizo Mountain Group, containing 13 metasedimentary units (primarily metamorphosed arkoses and quartzites with lesser schists, phyllites, and marbles) and three metaigneous units (metarhyolite, amphibolite, and granodiorite); the Allamoore Formation, composed of metamorphosed, commonly amygdaloidal, basaltic flows, pyroclastics, and limestones; the Hazel Formation, which is primarily a conglomerate (with clasts derived from the Allamoore) in its lower half and grading upward into fine, red, silty sandstone; and the Van Horn Sandstone, a massively bedded, coarse, arkosic sandstone with preserved crossbedding (Flawn and Muehlberger, 1970). King (1955) described the Van Horn Precambrian as representing a classic orogenic cycle, featuring initial, rapid accumulation of a thick section of supracrustal deposits, including both immature sediments and volcanics. There followed several phases of orogenic deformation and these events were followed by postorogenic block-faulting and tilting, then by peneplanation prior to deposition of Paleozoic strata. As in the Llano area, the inferred relation of the Van Horn area to the Grenville orogen in the east is based on the ages of metamorphism as revealed by radiometric dating and by a generally similar tectonic history. Because of its location near the southern end of the Rocky Mountains, however, the Van Horn area may be more related to the history of the Cordillera than to that of the Grenville belt.

F.2. Stratigraphy of Grenville Rocks in Canada

The bulk of Grenville sedimentary rocks are found in the Central Metasedimentary Belt and Eastern Grenville Province (see Fig. 3-15); additional strata related to the Grenville orogen are found in the Grenville Foreland Belt. In the following discussion, we will consider the strata of each of these areas; we will not further discuss Grenville-related strata of the United States because they are poorly understood at this time.

F.2.a. Grenville Supergroup

Metasedimentary rocks of the Central Metasedimentary Belt comprise the Grenville Supergroup, the thickest sequence of which occurs in the Hastings Basin. Correlatives are found in the Adirondack Lowlands, the Central Granulite Belt, and the Baie Comeau. The Grenville Supergroup is typified by coarsely crystalline, white marbles varying from calcitic to dolomitic over short distances (Wynne-Edwards, 1972). Grenville marbles are relatively pure, characteristically being composed of greater than 90% carbonate; impurities include graphite, diopside, phlogopite, and pyrite, among others. In addition, Grenville strata commonly contain white to gray quartzites and aluminous paragneisses. The total, original thickness of the Supergroup is uncertain but was probably somewhere around 10,000 to 15,000 m (Wynne-Edwards, 1972).

From bottom to top, the Grenville Supergroup is divided into the Hermon, Mayo, and Flinton Groups. Mayo rocks are dominated by marbles with minor quartzites and paragneiss, probably representative of a shallow-water sequence. The Hermon consists of metavolcanics, primarily pillow basalts and silicic pyroclastics. Hermon and Mayo strata are thought to be penecontemporaneous. The Flinton rests unconformably on the Hermon and consists of marbles, quartzites. thin slates, and conglomerates. Flinton rocks are considered to record deposition in continental or shallow-marine environments.

Wynne-Edwards (1972, 1976) has interpreted the Mayo and Hermon Groups as remnants of a miogeoclinal assemblage which began to form during the early Middle Proterozoic after the Hudsonian orogeny. The Flinton Group, he suggested, was formed as part of a detrital-wedge blanket during the late stages of the Grenville orogeny.

F.2.b. Wakeham Bay Group

Along the northern shore of the Gulf of St. Lawrence in Eastern Grenville Province are strata of the Wakeham Bay Group, which consists of around 7500 m of quartzite, often exhibiting ripple marks and crossbedding. Wakeham Bay rocks have only been raised to greenschist or lower amphibolite facies but are considered by Wynne-Edwards (1972) to be correlative with the Grenville Supergroup. Whereas this correlation is debatable, Wakeham Bay rocks, like those of the Grenville Supergroup, do seem to represent miogeoclinal deposition. In addition, they bear the same age relation to the basement as do Grenville rocks, and many of their intrusions are similar to Grenville Supergroup intrusions.

Assuming that Hermon, Mayo, and Wakeham Bay rocks do represent a miogeocline, it would be reasonable to ask if eugeoclinal strata have been found. Wynne-Edwards (1976) stated that none occur in the Grenville area as it is currently exposed; he did, however, speculate that such rocks may have been present to the southwest but have now been completely obscured by Appalachian deposition and metamorphic overprinting.

F.2.c. Grenville Foreland Belt Strata

One final assemblage of Grenville-related strata deserves mention. Within that portion of the Grenville Foreland associated with Nain Province are thick accumulations of quartzite, chert, slate, and volcanic rocks called the Seal Lake Group. Seal Lake rocks are about 6500 m thick and have been intruded by almost 5000 m of diabase sills. Seal Lake strata contain evidence of shallow-water to terrestrial deposition and Wynne-Edwards (1972) has suggested that the Seal Lake Group, like the Flinton Group of the Grenville Supergroup, represents the remnant of a once extensive detrital-wedge sequence formed penecontemporaneously with later stages of the Grenville orogeny.

F.3. Tectonics of the Grenville Orogeny

It is in the debate over tectonic mechanisms of the Grenville orogeny that the schism between groups of Precambrian geotheoreticians is most apparent. On one side are arrayed those who view Proterozoic orogenesis as having been substantially different from Phanerozoic orogenesis; these geologists consider the Grenville belt to typify the ensialic orogen and, therefore, to be the product of intracontinental, i.e., intraplate, diastrophism. Members of this camp we will call "actualists." The other camp is comprised of those who regard Proterozoic orogenesis as having proceeded from the same, i.e., plate-tectonic, mechanisms observed in operation today. These we will term "uniformitarianists."

Wynne-Edwards is, perhaps, the foremost advocate of the actualist camp and has argued strongly for the ensialic nature of the Grenville orogen. In 1976, Wynne-Edwards proposed a tectonic mechanism, termed the "millipede" model, for the Grenville event. It is based upon the hypothesis that ductile spreading of sialic crust, rather than plate-margin subduction, was the dominant process in Grenville orogenesis (see Fig. 3-16). In this scheme, continental crust is considered to have been hotter and, therefore, more ductile due to a higher level of heat flow. Because of this difference in the physical behavior of the crust, a different style of crustal response ensued upon being stressed by thermal upwelling in the mantle specifically, the warm crust above the thermal plume would have undergone ductile stretching, leading to a thinning of the crust and concomitant deposition of supracrustals. This would have been accompanied by partial crustal melting and by intrusion of mafic magmas from the mantle. The ductile crust was deformed as it was shifted off the mantle plume and compressed against cooler, more rigid crust which had not been affected to as great an extent by mantle heating. These ideas are appealing because they fit nicely into an evolutionary paradigm of terrestrial tectonics; unfortunately, their philosophical orientation is irrelevant to their acceptance.

In 1981, Baer proposed a different actualistic model for Grenville orogenesis. In contrast to Wynne-Edwards's ideas, Baer stated that there is no evidence for greatly increased crustal ductility in mid-Proterozoic times. Instead, Baer argued that Proterozoic tectonics, under the influence of a hotter mantle, involved thinner and more laterally extensive sialic crust but did not involve subduction. Were subduction not to operate, then the only movement allowed to the plates would be two-dimensional shifting about the spherical surface of the Earth so that the broader, thinner continental slabs would be "jammed" into one another. The result of such "plate jams" would have been lateral compression of the slabs and, according to Baer, concomitant up- and down-buckling within slab interiors, similar to the behavior of a throw rug when compressed laterally.

Uniformitarian, i.e., plate-tectonic, models of the Grenville orogeny primarily explain the event as the result of continental collision. In 1973, Dewey and Burke proposed that the Grenville orogeny was similar to events presently active beneath the Tibetan Plateau north of the Himalayas in Asia. Their model is presented in Fig. 3-17. Dewey and Burke suggested that the Grenville event resulted from the collision of a sialic block ("Grenvillia") against eastern North America. Partial underthrusting of the colliding block beneath the margin of North America resulted both in thickening of the crust and in high levels of both lithostatic and directed stress at the base of the crust. Under such extreme conditions, much of the lower crust was remobilized and underwent partial melting, resulting in the production of granitic melts and anorthositic residues. Granitic liquids rose through the crust to be emplaced at higher levels while anorthosites and related gabbros (and norites) remained in the lower crust. Long-term erosion following the orogeny stripped away much of the superstructure of the orogen revealing its roots, which would be characterized by reworked crust, minor remnants of greatly deformed and metamorphosed supracrustals, and anorthositic massifs, all of which typify the Grenville orogen.

In Dewey and Burke's model, the intense cataclastic zone marked by the Grenville Front is not a plate suture but rather a zone of movement more analogous to the deep crustal roots of a foreland thrust belt separating allochthonous crust of the orogen from essentially autochthonous crust of the foreland (see Fig. 3-17); note that this would explain the extension of basement structures into the Grenville orogen. The actual plate suture would be difficult to recognize because, at such a deep level of exposure, there would be little to identify such a demarcation other than a narrow belt of ultramylonites which itself might have been lost to tectonic reworking during later, Paleozoic events.

Dewey and Burke's model is attractive but faces several problems, the most serious being equivocal support (at best) from paleomagnetism. As mentioned earlier in this chapter, continental collision presumes that the colliding plates moved independently prior to collision so that their apparent polar-wander (APW) paths should differ for the interval before orogeny. Numerous workers have examined Canadian Shield rocks for their magnetic signatures and

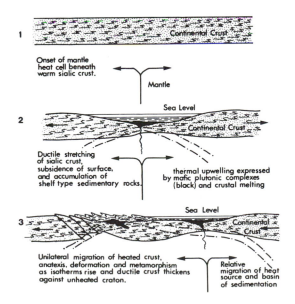

Fig. 3-16. Cartoon of Wynne-Edwards's model of Proterozoic ensialic orogenesis; see text for discussion. (From Wynne-Edwards, 1976, *American Journal of Science*, Fig. 5, p. 939.)

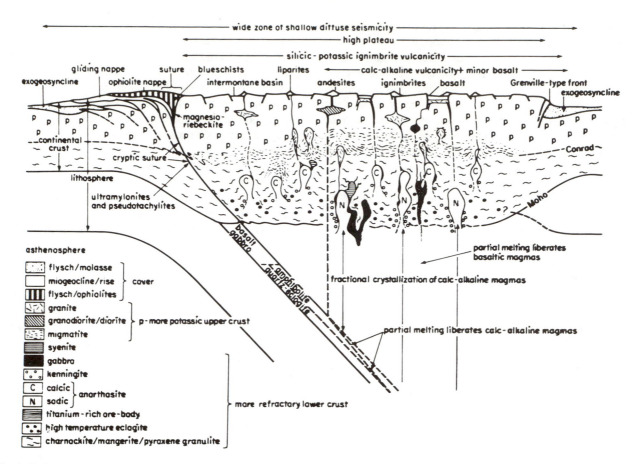

Fig. 3-17. Schematic illustration of a Grenville-type collisional orogen. (From Dewey and Burke, in *Journal of Geology*, Vol. 81, p. 689, © 1973 University of Chicago Press.)

have come to strikingly different conclusions. Some have stated that APW curves from "Grenvillia" in fact do differ from those of Superior Province and, thus, support the notion of a Grenville collision (e.g., Irvine and Findlay, 1972; Irving and McGlynn, 1976). Others (e.g., McElhinney and McWilliams, 1977; Piper, 1976) have argued that paleopoles from pre-Grenville terranes in both Grenville and adjacent provinces are the same, contradicting a Grenville collision. The heart of the problem lies in the fact that paleomagnetism is a delicate quantity and can be altered in any number of ways by subsequent geological events. Further, for very old rocks, radiometric dates are rarely better than plus/minus 50 Myr., which is not accurate enough to uniquely distinguish APW curves, especially if the plates in question were moving rapidly. [In 50 Myr., a plate moving at only 2 cm/yr. can travel 1000 km but Irving and McGlynn (1981) suggested an average mid-Proterozoic spreading rate of 5–6 cm/yr., occasionally as high as 10 cm/yr.] Even were dating more accurate, there is always the question of whether magnetism and isotopic differentiation were acquired at the same time. For these reasons, Burke *et*

al. (1976) argued that paleomagnetism is not yet able to aid significantly analysis of Precambrian tectonics. Irving and McGlynn (1981) agreed that paleomagnetic data from Precambrian areas can be problematic but stressed that they are not incompatible with a plate-tectonic interpretation.

But other problems seem to devil the Grenville collision model. As pointed out by Wynne-Edwards (1976) and Baer (1981), within the Grenville orogen, there are no apparent ensimatic terranes, no ophiolites, no blueschist mélanges, and no obvious plate sutures. On the other hand, at the present depth of erosion, blueschists, ophiolites, and other evidence of a suture could easily have been lost. Alternatively, such terranes could be hidden beneath younger cover sequences; indeed, Dewey and Burke's model predicts that the Grenville suture must be to the east of present limits of Grenville Province in Canada, covered by Appalachian supracrustals and obscured by younger tectonic events. In 1980, McLelland and Isachsen proposed that the Grenville suture may be represented by the Alabama–New York geophysical lineament of King and Zeitz (1978). They also hypothesized that the northeastern extension of the suture may lie beneath the Champlain Valley–Middlebury synclinorium or the Connecticut Valley.

Other, slightly different plate-collisional models have been suggested (e.g., McLelland and Isachsen, 1980; Seyfert, 1980). These models vary in the direction of subduction, the number of possible subduction zones, and the identity of the colliding block. They all, however, share the assumption that the postulates of plate tectonics (specifically, the Wilson cycle) are sufficient to explain Grenville orogenesis.

G. Proterozoic of the United States Craton

In most of the United States craton, Proterozoic rocks are known or inferred to form the basement beneath a veneer of Phanerozoic strata. Because of this cover, our knowledge of the Proterozoic basement comes mainly from a few deep drillholes and from geophysical studies. Principal exposures of Proterozoic rocks south of Southern Province of the Canadian Shield are in central and southern Michigan, the St. François Mountains of southeastern Arkansas, the Rocky Mountains, the southern part of the Great Basin, and the Transverse Ranges and Mojave Desert of southeastern California.

Recognition of Proterozoic provinces in the United States has been slow in development. Van Schmus and Bickford (1981) recognized four major Middle Proterozoic provinces (see Fig. 3-18): (1) the Penokean belt, characterized by dates ranging from 1900 to 1820 Myr.; (2) the Interior belt (here named; Van Schmus and Bickford chose not to name it) characterized by dates ranging from 1780 to 1690 Myr.; (3) the 1680–1610 Myr. Southwestern belt (here named; Van Schmus and Bickford termed it the Mazatzal belt); and (4) the 1480–1380 Myr. St. François belt (here named; Van Schmus and Bickford termed it the granite–rhyolite terrane; for reasons which will be discussed later in this chapter, this area might also be termed the Elsonian belt). The Keweenawan belt is also a major feature of North America's Proterozoic basement but was discussed earlier in connection with Southern Province and will not be discussed again. We will use Van Schmus and Bickford's (1981) province framework, modified slightly after Condie (1982) and Dutch (1983), to organize our discussion. Upper Middle and Upper Proterozoic rocks formed on the Cordilleran continental margin (the Belt and Purcell Supergroups, the Windermere Group, and related units) and on the Appalachian continental margin (Ocoee Supergroup, Mt. Rodgers Group, and related units) will be discussed later in this chapter.

Before beginning a discussion of Phanerozoic belts of the craton, two general trends should be mentioned. First, the Proterozoic belts of the U.S. craton lie to the south of the Superior and Wyoming cratons and extend northeasterly, resulting in a pronounced northeast–southwest "grain" to the basement. Second, the ages of the belts decrease progressively outward from the center of the continent; this relationship has been generally understood for many years and was the basis upon which 19th century American geologists such as James Dwight Dana proposed the concept of continental accretion.

G.1. Penokean Belt of the Northern Great Plains

In Section C of this chapter, we discussed the Penokean fold belt of Southern Province, showing that it can be explained as a passive-continental-margin prism formed by rifting along the southern margin of the Superior craton soon after the beginning of Early Proterozoic time and later deformed as the result of a plate collision (the Hudsonian, or Penokean, orogeny). Southwestward from the Wisconsin–Minnesota area, the Penokean fold belt is lost beneath Paleozoic cover. For many years, it was felt that the Archean Superior and Wyoming cratons were continuous in the subsurface and represented a single cratonic block; similarly, it was observed that the Penokean belt, if extended along strike, appears to coincide with folded, supracrustal, Proterozoic rocks in central Colorado to the south of the Wyoming craton (see Fig. 3-18). Consequently, Penokean and central Colorado Proterozoic rocks were considered to represent the same tectonic province.

Recently, however, Van Schmus and Bickford (1981) and Dutch (1983) have shown that a major post-Archean deformed belt runs northward in the subsurface through the western Dakotas and into Canada, separating the Superior and Wyoming blocks. This belt was considered by Van Schmus and Bickford to be an extension of Southern Province's Penokean belt and is, therefore, termed the Western Penokean belt. It represents the part of the Penokean belt which bends sharply northward in central South Dakota, around the southwestern margin of the Superior block. This suggests that the Superior and Wyoming areas were not originally a single tectonic entity and, therefore, that Penokean and central Colorado Proterozoic rocks are not necessarily correlative. In Dutch's view, the fold belt intervening between the Superior and Wyoming blocks is continuous in the subsurface with deformed Proterozoic rocks to the north in Canada and represents a continuation into the U.S. subsurface of western Churchill Province.

Rocks and structures associated with the Western Penokean belt are exposed in the Black Hills Uplift of western South Dakota. King (1976) stated that more than 12,000 m of Lower Proterozoic supracrustal strata in this area is steeply or isoclinally folded along north-south trending axes (i.e., paralleling the trend of the Western Penokean belt).

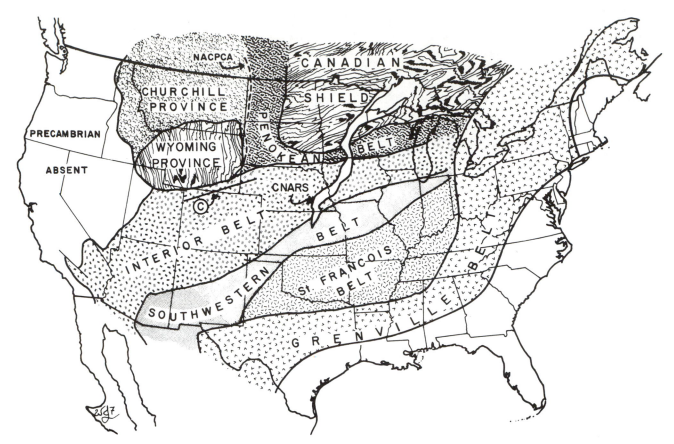

Fig. 3-18. Proterozoic provinces of the United States. (Based on data from Van Schmus and Bickford, 1981; Condie, 1982; Dutch, 1983; and Thomas *et al.*, 1984.)

The lower part of this sequence is composed of thick quartzite and conglomerate with associated iron formation, schist, and limestone. The upper part, exposed in the western part of the Black Hills, is composed of eugeoclinal rocks, originally deposited as graywackes, shales, cherts, and pillowed lava flows (King, 1976). This stratigraphic succession is suggestive of the subsidence of a rifted continental margin. If this interpretation is correct, Lower Proterozoic supracrustals in the Black Hills area may be genetically related to the Snowy Pass Supergroup in the Medicine Bow Mountains and the Red Creek Quartzite in the Uinta Mountains of Utah, both of which have been interpreted as rifted-margin deposits formed along the southern margin of the Wyoming craton during the Early Proterozoic (discussed in the next section). Later strong deformation and metamorphism may have been caused by the collision of the Wyoming and Superior blocks.

The ages of rocks within the Western Penokean fold belt range from 1900 to 1820 Myr. and, therefore, deformation and metamorphism were contemporaneous with similar (i.e., Hudsonian) events in both the eastern part of the Penokean belt and in Churchill Province. Assuming the strata of these three areas were once parts of a continuous belt, then they may represent the southwestern margin of

the Superior continent. Dutch (1983) proposed that the Western Penokean belt was deformed during the later Early Proterozoic by collision of the Wyoming continent against the Superior block. Camfield and Gough (1977) suggested that the suture between the accreted blocks may be represented by the North American Central Plains conductive anomaly (NACPCA in Fig. 3-18). This geophysical feature extends from the Cheyenne shear zone in the Laramie Range northeastward to the area just east of the Black Hills uplift. From the Black Hills, it bends northward becoming approximately coincident with the western borders of South and North Dakota and extending into Saskatchewan, Canada. The North American Central Plains conductive anomaly was considered by Dutch (1983) and Van Schmus and Bickford (1981) to represent a fundamental discontinuity in the continental crust.

Another result of the Superior–Wyoming collision, in addition to welding the two Archean blocks into a single crustal entity, was to cause metamorphism, intrusion, and deformation of supracrustal deposits and tectonic reworking of sialic crust along the margins of both colliding blocks. Naturally, the fold belt may also contain smaller, exotic terranes caught in the crunch between Superior and Wyoming continents. Western Penokean structures were shown by Dutch (1983) to extend southward almost to the South Dakota–Nebraska border where they appear to be truncated by younger, northeast–southwest–trending structures associated with the Interior belt. Therefore, the Wyoming–Superior collision must have occurred *before* the accretionary tectonic events along the southern margin of the Wyoming block discussed in the next section.

G.2.　Interior Belt

The zone of Proterozoic rocks and structures running diagonally across the United States, which we term the Interior belt, was once considered part of Southern Province of the Canadian Shield. As mentioned previously, however, Penokean structures appear to be truncated sharply along the boundary of the Interior belt. Further, although dates of rocks in both areas lie within the range of the "Hudsonian" orogeny as defined by some authors [e.g., 2000 to 1700 Myr., following Muehlberger (1980)], dates of Interior belt rocks are distinctly younger than those of western Penokean rocks. Most authors now choose to restrict the Penokean orogeny to the range 1900–1820 Myr., with peak activity at about 1840 Myr. (Van Schmus and Bickford, 1981); so defined, Penokean deformation is limited to the Southern Province Penokean belt and the western Penokean area. Ages of Interior belt rocks fall within the range 1780–1690 Myr. and so appear to represent an entirely different orogenic event.

Rocks of the Interior belt are seen in limited exposures in southern Wisconsin, the Rocky Mountains, and southern Basin and Range Province. We will discuss the Proterozoic rocks of each region separately.

G.2.a.　Wisconsin

Exposed Interior-belt rocks in Wisconsin comprise a granite–rhyolite terrane in the southern part of the state and a few scattered granitic plutons in the northern part. Southern Wisconsin granites are primarily of epizonal origin and are intrusive into genetically related rhyolitic flows and ignimbrites (Van Schmus and Bickford, 1981). Zircons from these rocks have been dated at 1760 Myr. and, thus, represent a probable post-Penokean event. Northern Wisconsin granites are more coarsely crystalline than those in the southern part of the state, probably reflecting a more mesozonal origin, but are geochemically similar and may represent the same magma series.

G.2.b.　Southern Margin of the Wyoming Craton

Along the southern margin of the Wyoming block, mildly disturbed Proterozoic strata lie atop Archean sialic basement. These strata constitute an allochthonous block

thrust northward over the margin of the Wyoming craton and are separated from more strongly deformed and metamorphosed correlative rocks of central Colorado by a belt of intense shear deformation, called the Cheyenne belt or the Mullen Creek–Nash Fork shear zone (see Fig. 3-18). In the Medicine Bow Mountains of Wyoming, rocks of the allochthon are termed the Snowy Pass Supergroup and in the Uinta Mountains in Utah, they are called the Red Creek Quartzite.

The Snowy Pass Supergroup consists of Deep Lake and Libbey Creek Groups. Deep Lake strata are composed of coarse, immature sandstones, shales, and conglomerates which have been interpreted as fluvial and glaciomarine deposits (Karlstrom et al., 1983). These are overlain by deposits of the lower Libbey Creek Group, whose detrital sediments record a marginal-marine setting typified by deltaic and related coastal environments. The upper Libbey Creek Group is dominated by stromatolitic, cherty carbonates and associated iron formation, dark shales, and greenstones; these rocks are interpreted as representing an open-marine environment. Because of their record of environmental change from terrestrial to coastal to open marine, and because of their marginal position with respect to the Wyoming cratonic block, Snowy Pass strata were considered by Karlstrom et al. (1983) to record the development of a passive-continental-margin prism, formed by early Early Proterozoic rifting along the southern margin of the Wyoming craton. In their model of Snowy Pass Supergroup deposition, Karlstrom et al. suggested the following developmental scenario: earliest episodes of rifting led to the development of a grabenlike trough into which were deposited freshly weathered alluvial and fluvial sediments. Correlative strata of the Sierra Madre range in southcentral Wyoming show greater marine influence than do Snowy Pass rocks in the Medicine Bow Mountains and, therefore, probably represent the open, marine end of the rift. The rift continued to widen as continental separation proceeded and, during later Deep Lake and early Libbey Creek times, was the site of a major glacier, which deposited coarse, polymictic, paraconglomeratic detritus. (It is possible that the glacial event recorded by these rocks is correlative with the major glaciation inferred for Southern province from strata of the Huronian Supergroup.) During middle Libbey Creek time, glacial influences waned and shallow-marine sedimentation began. By the end of Snowy Pass time, continental separation had been completed and open-marine deposition ensued.

Sears et al. (1982) described Proterozoic rocks of the Uinta Mountains west of Medicine Bow and Sierra Madre areas and suggested that the Red Creek Quartzite in that area is at least partially correlative with Snowy Pass strata. The Red Creek Quartzite consists of quartzose sandstones and conglomerates, pelitic rocks, and minor carbonates; although they have undergone metamorphism to varying degrees, some exposures of the Red Creek contain traces of primary structures such as cross and graded bedding. Sears *et al.* characterized these rocks as a miogeoclinal assemblage which probably began to form during the same rifting event which gave rise to the Snowy Pass Supergroup. Thus, Red Creek and Snowy Pass strata probably represent deposits of the same continental margin. In a similar fashion, supracrustal rocks of the Black Hills area (described in the previous section) may also have formed along the eastern margin of the Wyoming block.

G.2.c. Central and Southern Interior Belt

Within the Interior belt to the south of Wyoming Province, crustal rocks older than approximately 1800 Myr. are unknown, and geochemical studies of basement rocks from those areas suggest that they were derived from the mantle with no involvement of Archean rocks (Nelson and DePaolo, 1982). However, details of the geological history of this large area are conjectural at best; the following sketch is taken from the discussions of Hills and Houston (1979) as modified by Condie (1982).

Following the rifting event recorded in Snowy Pass strata, the next significant tectonic event to affect the southern margin of the Wyoming block occurred at approximately 1760 Myr. and featured the closing of the ocean basin and consequent collisional orogeny, which deformed Snowy Pass and Red Creek rocks. Since sialic rocks of an appropriate age are absent immediately south of the suture zone (the Cheyenne shear zone), the block which collided with the Wyoming craton is considered to have been a volcanic complex, perhaps represented by the Green Mountains Formation of southeastern Wyoming. COCORP (*co*nsortium for *co*ntinental seismic *r*eflection *p*rofiling) seismic profiling in the Medicine Bow and Laramie ranges of southeastern Wyoming shows that the Cheyenne shear zone is a fundamental crustal feature, defining the boundary between Archean basement to the north and Lower Proterozoic rocks to the south (Allmendinger *et al.*, 1982). The COCORP profile showed that the shear zone dips at approximately 55° to the southeast; over this shear surface, Snowy Pass and related rocks were thrust northwestward during the orogeny. Although this and related tectonic events in the Interior belt have been termed "Hudsonian" by some authors (e.g., Muehlberger, 1980), they were separated in time from the Hudsonian orogeny in its "type" area of the Great Lakes region by 50 Myr. Thus, the two events were probably unrelated.

Detrital sediments derived from the resulting mountains were deposited along the southern margin of the orogen and are represented by paragneisses of the Idaho Springs Formation in the Big Thompson Canyon area of the central Colorado Front Range. King (1976) proposed that

Idaho Springs gneisses originally were a thick eugeoclinal sequence of graywacke and shale. They are associated with interbedded lenses and layers of amphibolite (the Swandyke Gneiss) which may represent volcanism related to initiation of subduction along the new, Andean-type plate boundary.

Condie (1982) argued that a new cycle of continental-margin evolution, extending from 1760 to 1720 Myr., began as the volcanic arc formed at the end of the previous cycle began to migrate southward. In the resulting back-arc basin, rocks began to accumulate which represent a bimodal volcanic assemblage. The bimodal assemblage, represented by rocks such as the Dubois and Salida greenstone assemblages in Colorado, is characterized by mixed tholeiitic and felsic volcanic rocks, associated with volcaniclastic sediments; Condie considered the bimodal volcanic assemblage to have formed in shallow-marine to subaerial environments. In Condie's model, further arc migration led to the accumulation of similar bimodal-assemblage rocks in southern Colorado and northern New Mexico (e.g., the Gold Hill and Pecos greenstones in the Sangre de Cristo Mountains). Prograding over these bimodal-assemblage rocks were craton-derived sediments of Condie's quartzite–shale assemblage, characterized by crossbedded quartzites, local arkoses, shales, and conglomerates; these rocks were considered by Condie to have been deposited in nearshore marine, intertidal, and fluvial environments. Deposits of this type overlie Pecos greenstone in the Sangre de Cristo Mountains. Similar relations may also be represented by the mafic Brahma Schist and metasedimentary Vishnu Schist of the Grand Canyon area of northwestern Arizona.

The tectonic cycle represented by rocks of the Interior belt came to an end at approximately 1720 Myr. with closure of the back-arc basin and related Andean-type orogeny. Calc-alkaline volcanics from the Yavapai Supergroup in central Arizona were considered by Condie (1982) to represent volcanic-arc rocks, possibly associated with the terminal orogeny of the Interior belt. Igneous plutons, such as the Zoroaster Granite in the Grand Canyon, the Boulder Creek Granite of the Front Range in Colorado, and the Tenmile Granite of the Needle Mountains in southwestern Colorado, were probably all intruded in association with the orogeny.

Condie's (1982) scenario of the development of the Interior belt implies that the whole of the Interior belt formed as an accretionary terrane. As such, its history may have been considerably more complicated than his model suggests. Comparison with Phanerozoic accretionary terranes, such as the North American Cordillera, indicates that such broad areas have complex, multifaceted histories. The tectonic development of accretionary terranes often features numerous tectonic events such as the collisional accretion of displaced terranes and large-scale strike-slip faulting, rather than a simple, long-term, arc migration with a single tectonic climax at the end of the cycle.

G.3. Southwestern Belt

The Proterozoic area south of the Interior belt is called the Southwestern belt and extends from southeastern Arizona to the midcontinent region (see Fig. 3-18). Condie (1982) said that Southwestern belt rocks record the same type of tectonic cycle as do rocks of the Interior belt. Igneous and metamorphic activity associated with this cycle were widespread in both space and time, and radiometric ages from the Southwestern belt range from 1720 to 1650 Myr.

The entire southern half of the Lake Superior region was subjected to a thermal metamorphic event which reset most radiometric indicators at approximately 1630 Myr. Although no igneous bodies have yet been recognized in the southern Lake Superior region as having an age even close to 1630 Myr., Van Schmus and Bickford (1981) suggested that this metamorphic event was the foreland manifestation of the extensive Southwestern belt orogenic event to the south.

Igneous bodies which resulted from the Southwestern orogenic event occur in scattered areas throughout the midcontinent and the Southwest. For example, in Kansas and Missouri, several intrusive bodies have ages of 1625 ± 250 Myr. (Van Schmus and Bickford, 1981). In southern Arizona and central and southern New Mexico, metaigneous rocks of the bimodal volcanic assemblage (described earlier in this section) have ages ranging from 1610 to 1680 Myr. In the Ladron and Manzano Mountains of central New Mexico, strata of the quartzite–shale assemblage unconformably overlie rocks of the bimodal assemblage. In the Tonto Basin of Arizona, the quartzite–shale assemblage is represented by the Mazatzal Group, which conformably overlies bimodal volcanics of the Haigler and Alder Groups. And igneous intrusions associated with the Southwestern belt orogenic event occur within the Pinal Schist of southeastern Arizona. Granitic and granodioritic intrusions ranging in age from 1680 to 1650 Myr. occur in the Transverse Range of southern California and regional metamorphism affected rocks of the Mojáve Desert at about 1650 Myr.

The nature of the tectonic episode which affected the Southwestern belt is highly speculative. Silver (1978) favored an Andean-type orogeny associated with subduction beneath the continental margin along the southern Interior belt. In 1979, Condie and Budding suggested that Southwestern-belt tectonism was due to continental rifting, but later, Condie (1982) argued that Southwestern-belt rocks are similar to those of the Interior belt and represent the southward migration of a volcanic arc and concomitant growth of a back-arc basin, followed by Andean-type orogeny and resulting closure of the back-arc basin. Like the Interior belt, however, the Southwestern ''orogeny'' may not have been a single event, but rather a sequence of tectonic events, perhaps associated with the accretion of displaced terranes.

G.4. St. François Belt

In the southern midcontinent region, extending from the panhandle region of Texas to northwestern Ohio, is a basement region approximately 2000 km long and 500 km wide and composed almost entirely of rhyolitic volcanics and epizonal granitic intrusions. The principal exposure of these rocks is in the St. François Mountains on the crest of the Ozark uplift in southeastern Missouri. In the southwestern St. François Mountains, the dominant lithologies are stratified rhyolitic and related felsic flows associated with tuffs and volcanic breccia (King, 1976), while in northeastern areas, several varieties of granite occur as large, sill-like plutons intruding into the volcanics. The ages of these rocks range from 1480 to 1380 Myr. (Van Schmus and Bickford, 1981). Recently, Thomas et al. (1984) showed that the St. François belt may be subdivided into two parts: an older (1480–1440 Myr.) region stretching north- and eastward from southcentral Missouri and adjacent Arkansas and including the St. François Mountains themselves; and a younger (1400–1350 Myr.) region extending westward to eastern New Mexico.

The St. François belt is remarkable among Proterozoic (and Phanerozoic) regions for several reasons. First, over almost their entire extent, rocks of the St. François belt reveal a striking uniformity of petrographic and geochemical character (Thomas et al., 1984). Most of the volcanics are ash-flow tuffs with little or no penetrative deformation and with only minor, low-grade metamorphic effects (Van Schmus and Bickford, 1981). Second, because of their compositional uniformity, they constitute a remarkably large mass of silicic rock; Van Schmus and Bickford (1981) said that, if these rocks comprise only the upper 1 km of the crust in this region, they would still represent approximately 1 million km^3 of highly silicic and potassic rocks. Third, they are not associated with any appreciable mafic or intermediate igneous rocks. Such a large area containing but a single dominant igneous composition is unusual, to say the least.

In areas of the midcontinent and southern Rocky Mountains outside the main region of the St. François belt, a number of anorogenic granites (i.e., granites whose intrusion was not apparently related to an orogenic event) and related intrusions yield dates of 1380 to 1480 Myr. and, therefore, may be related to the St. François belt. Representatives of this suite of intrusives are exposed in the Rocky Mountain region, and include the St. Kevin Granite of Colorado (Van Schmus and Bickford, 1981), the San Isabel batholith in the Wet Mountains of Colorado (Thomas et al., 1984), The Eolus Granite of the Needle Mountain area

(King, 1976), and the Ruin and Oracle Granites of Arizona (King, 1976). There are also a few in northwestern Kansas. Whereas most granitic plutons within the St. François belt have petrographic features which indicate emplacement at shallow, epizonal depths, features of the anorogenic plutons suggest deeper level emplacement (Van Schmus and Bickford, 1981).

The fact that this large area is composed almost exclusively of volcanics and shallow intrusive rocks suggests that some other type of crust must underlie it. Thomas *et al.* (1984) suggested that 1700 Myr.-old crust may underlie the volcanics, but no evidence of this is yet available. Tectonic models for the St. François terrane are largely unable to account simply for all its characteristics. The lack of andesites, tonalites, and related igneous rocks would seem to mitigate against an origin associated with a convergent plate boundary. On the other hand, the absence of significant volumes of mafic volcanics gives little support to a rift-valley origin. Van Schmus and Bickford (1981) suggested that the continental crust of the southern and southwestern midcontinent area had been thickened by convergent plate tectonics during the Interior and/or Mazatazal orogenies. Sialic rocks at the base of the isostatically depressed continental crust would probably have been subjected to intense heating and would have partially melted, yielding highly silicic and potassic magmas. These magmas would have risen buoyantly to be emplaced at near-surface depths and extruded explosively to produce the flows and volcaniclastics seen, for example, in the St. François Mountains. If Van Schmus and Bickford's model is correct, it would provide an excellent example of ensialic tectonism, but at a plate-margin setting. At any rate, the St. François belt appears not only to be unique among Proterozoic terranes of North America but also to differ from modern plate-margin terranes.

To the south of the St. François belt are rocks and structures that formed during the Grenville orogeny, which was the last major tectonic cycle of the Proterozoic Eon. The Grenville orogeny of the eastern and southern areas of North America was discussed in Section F of this chapter and will not be considered further here.

H. Early History of the Cordilleran Continental Margin

Although in the preceding discussion we mentioned a number of areas which are part of the Cordilleran region of North America, their histories prior to approximately 1450 Myr. ago were related to tectonic events associated with an active, accretionary plate-margin along the southern side of the continent; there is no evidence of continental-margin sedimentation in the Cordilleran region while those events

were taking place. But, somewhere around 1450 Myr., tectonic events began which led to the development of the Cordilleran continental margin. The history of this earliest phase of the western margin is frustratingly fuzzy because many of the rocks that formed concomitantly with these events were lost to later erosion or hidden beneath younger strata and allochthonous structural blocks. On the other hand, most strata of the earliest Cordilleran margin are only slightly metamorphosed and, where exposed, contain much information about their depositional history.

The best preserved and exposed rocks of this age occur in northern Idaho, northwestern Montana, and adjacent Alberta and British Columbia. Related rocks are exposed in small, scattered areas of the United States, such as the Uinta Mountains of Utah, the Grand Canyon of the Colorado River, the Mazatzal Mountains of Arizona, and Death Valley, California. Far to the north, in the Mackenzie, Ogilvie, and adjacent mountains of the Yukon and Northwest Territories, there are much more extensive exposures of related rocks.

The tectonic environments represented by these strata are controversial; thus, the following discussion will deal first with the stratigraphy and sedimentology of the units, and will leave consideration of their tectonic interpretation to the end of the section.

H.1. Belt and Purcell Rocks and Their Correlatives

H.1.a. The Belt and Purcell Supergroups

The most extensively exposed Precambrian supracrustal sequence of the United States is the Belt Supergroup (King, 1976). Together with their extension into Canada, termed the Purcell Supergroup, they comprise an area of approximately 104,000 km² (see Fig. 3-19A). Belt–Purcell rocks are enormously thick, ranging from about 6100 m in thickness along its eastern margin to a probable maximum of 20,000 m in the center of the Belt Basin (King, 1976). Throughout much of their extent, Belt–Purcell rocks have been subjected only to lower grades of metamorphism and still retain their primary sedimentary structures. Indeed, their fresh appearance led early workers to conclude that they must be younger, perhaps even lower Paleozoic.

Belt–Purcell stratigraphy is complicated. Harrison (1972) recognized four fundamental units within the Belt Supergroup (see Fig. 3-19B): (1) an unnamed lower division composed of the Prichard Formation in central parts of the Belt Basin, which passes eastward into a coarse, conglomeratic facies represented by the LaHood Formation and overlying Chamberlain Shale and Newland Limestone (Atlyn Limestone of the Purcell Supergroup); (2) the Ravalli Group, composed of siltstones and argillites with minor

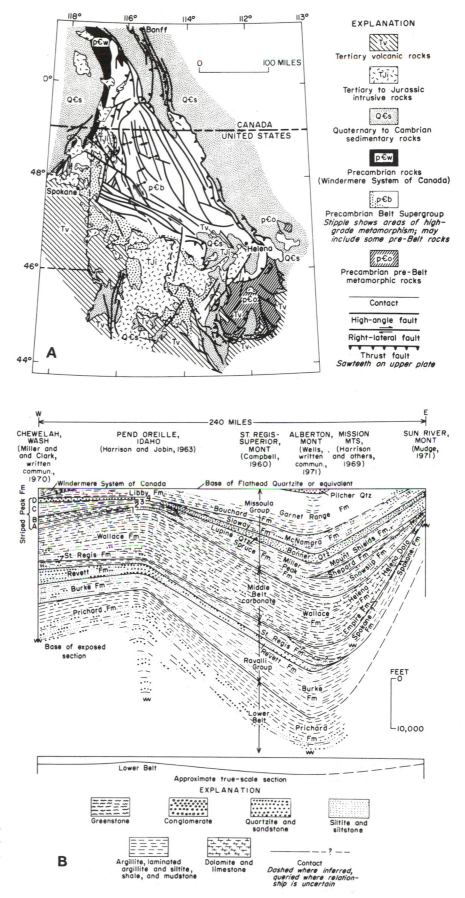

Fig. 3-19. The Belt and Purcell Basin. (A) Generalized geological map; (B) east–west stratigraphic cross section. (From Harrison, 1972, *Geological Society of America Bulletin*, Figs. 1 and 7, pp. 1216 and 1226.)

quartzites; (3) an unnamed middle group composed of carbonates and fine, calcareous detrital sediments and represented by the Wallace Formation in central basin areas, the Helena Limestone in the east, and the Siyeh Limestone in Canada; and (4) the Missoula Group, containing numerous stratigraphic units, especially the Purcell Lava at the base of the Group in Canada.

The Belt Supergroup is relatively monotonous in appearance, being composed mainly of siltstones and argillites with only occasional interbeds of medium-grained quartzite and dolostone. Throughout their entire section, Belt rocks contain evidence of shallow-water deposition, such as mud cracks, festoon crossbedding, salt-crystal impressions, and algal stromatolites. Red beds are common in eastern parts of the Belt Basin but decrease in abundance toward the west. Ross (1970) showed that Belt sediments accumulated in a complex, marginal-marine depositional system consisting of extensive mud flats with algal stromatolites and adjacent, very shallow marine waters. Depositional slopes were probably very low and wave energy was limited so that the shoreline was poorly defined and there were no distinctive littoral environments.

Thickness trends indicate that Belt sediments accumulated within a large trough striking northwestward from the area of southwestern Montana. Paleocurrents within the Belt trough ran generally parallel to trough axes (Harrison, 1972). Grain-size trends show that the principal source areas were in the east and southeast; it is significant that there is no indication of a western source. In addition to coarsening toward the east, most Belt units thin or pinch out in that direction, suggesting that their present limit is close to their original depositional limits. Ross (1970) suggested that the adjacent landmass was a broad, stable, flat-lying platform with very low relief and deeply weathered basement rocks.

Deposits of the Purcell are generally similar to those of the Belt but differ in several important ways. They, too, consist primarily of fine-grained, well-sorted detrital materials and they also contain evidence of an eastern, cratonic source (Gabrielse, 1972). In the Mackenzie Mountains, red siltstones and shales of the Purcell are associated with as much as 300 m of bedded gypsum. Purcell strata also contain minor but widely distributed andesite flows and gabbroic to dioritic dikes and sills, especially in the northern Rocky Mountains. In southern British Columbia and adjacent Montana, the Purcell Lava consists of 70–100 m of andesitic volcanics which date to approximately 1100 Myr. In the Purcell Mountains of southeastern British Columbia, deposits of the Aldridge Formation (equivalent to the middle and upper Prichard Formation) contain well-developed graded bedding, scour markings, and slump structures indicative of turbidite deposition and, therefore, suggestive of accumulation in a relatively deep-water environment. Over-

lying strata, however, have features considered to be shallow-water indicators, as do Belt strata to the south.

The time of deposition of the Belt and Purcell Supergroups ranged from 1450 Myr. to 850 Myr. (Harrison, 1972), an interval of time almost equal to the entire Phanerozoic. King (1976) said that even the great thickness of Belt–Purcell sediments was insufficient to account for the great spans of time embraced by their deposition, and he hypothesized the presence of an unknown number of diastems within the sequence.

H.1.b. Uinta Mountains Group and Big Cottonwood Formation

South of the Belt Basin, in Utah, correlatives of the Belt–Purcell sequence comprise the Uinta Mountains Group and the Big Cottonwood Formation. Strata of the Uinta Mountains Group are approximately 7600 m thick (Hansen, 1965) and form the core of the Uinta Mountains in northeastern Utah, where they lie nonconformably on the Lower Proterozoic Red Creek Quartzite (discussed earlier in this chapter). The Uinta Mountains Group is composed of a lower sequence containing as much as 1500 m of quartzite and arkose with interbedded thin shales and an upper sequence, termed the Red Pine Shale, which is primarily a dark shale and siltstone with minor quartzite (King, 1976). Thickness trends indicate that Uinta Mountain sediments accumulated in a narrow trough which extended almost due eastward toward the center of the craton. Sediments of the Uinta Mountains Group were derived from crystalline basement to the north and northeast of the Uinta Mountains basin trough and, within the trough, were transported westerly, paralleling the axis of the basin. Wallace and Crittenden (1969) stated that three major facies assemblages could be recognized in Uinta Mountain rocks: a continental facies with fluvial and alluvial deposits; a delta-plain facies; and a shallow-marine facies.

Due west of the Uinta Mountains, in the Wasatch Range just southwest of Salt Lake City, are exposures of the Big Cottonwood Formation. The Big Cottonwood is composed of 5000 m of quartzites and variegated shales which contain well-preserved ripple marks, crossbedding, and mud-clast (or, intraformational) conglomerates. These features are commonly thought to indicate shallow-water deposition. Stratigraphic correlations with other units are equivocal but the Big Cottonwood is considered to be correlative with both Belt rocks to the north and Uinta Mountain strata to the east.

H.1.c. Grand Canyon Supergroup

In the area of the Grand Canyon of the Colorado River, strata believed to correlate with the Belt–Purcell sequence comprise the Grand Canyon Supergroup, which is composed of up to 4600 m of primarily detrital-sedimentary rocks and is divided, from bottom to top, into the Unkar Group, the Nankoweap Formation, the Chuar Group, and the Sixtymile Formation (Elston and McKee, 1982).

The Unkar Group, lying on the eroded surface of the Vishnu Schist, is between 2000 and 3000 m thick and is divided into the basal Hotauta Conglomerate, Bass Limestone, Hakatai Shale, Shinumo Quartzite, Dox Sandstone, and, at the top, the Cardenas Lavas. Deposition of Unkar strata was essentially continuous except at a disconformity separating the Hakatai Shale from the Shinumo Quartzite (Elston and McKee, 1982). Throughout the Group are mafic sills and dikes, which were probably emplaced primarily during deposition of the Dox Sandstone. Cardenas Lavas have yielded radiometric dates of about 1100 Myr. (King, 1976), making them correlative with Purcell Lavas. Elston and McKee noted that Cardenas Lavas at the top of the Unkar sequence are interbedded at their base with the Dox Sandstone, indicating that no hiatus separates them.

The Nankoweap Formation, which King (1976) placed at the top of the Unkar Group, is a thin, red-bed unit. At its base is a discontinuous, ferruginous member which lies over erosionally truncated Cardenas Lavas (Elston and McKee, 1982). The ferruginous member is overlain by an upper member of red sandstone and siltstone of probably continental to marginal-marine origin.

The Chuar Group is dominated by dark, marine shales with local carbonate beds containing stromatolites. It is divided into the underlying Galeros Formation and the upper, Kwagunt Formation. Chuar strata contain small, circular, carbonaceous structures called *Chuaria*, once thought to represent primitive brachiopods but now interpreted as crushed algal spheres.

Chuar dark gray shales grade upwards into a dominantly red-brown unit called the Sixtymile Formation, which contains evidence of deposition concurrent with deformation. For example, it contains units of polymictic breccias intimately associated with normal faults (e.g., the Butte fault), whose activity appears to have been contemporaneous with sedimentation. Prior to deposition of the uppermost Sixtymile strata, folding and crenulation of its middle member occurred in response to activity associated with the Butte fault and the syndepositional Chuar syncline. In the center of the syncline, highest Sixtymile breccia deposits overlie folded strata of the middle member with abrupt local relief of 1.5 m (Elston and McKee, 1982). Tectonic effects associated with deposition of the Sixtymile Formation indicate that it was formed during the minor orogenic event called the Grand Canyon orogeny. This event was probably correlative with East Kootenay and Racklan orogenies of Canada (Elston and McKee, 1982).

Grand Canyon Supergroup strata seem to have accumulated in a deep, troughlike basin striking into the craton,

similar to Belt–Purcell and Uinta Mountain strata with which they correlate. This interpretation is controversial, however, because Unkar rocks are not as widely exposed as are rocks of the other two units mentioned and, consequently, their distribution is less well understood.

H.1.d. Crystal Springs and Beck Springs Formations (of Pahrump Group)

In the Death Valley area of southwestern California, Belt–Purcell correlatives make up the Crystal Springs Formation and the Beck Springs Dolomite, which, along with the overlying Kingston Peak Formation, comprise the Pahrump Group. The Crystal Springs Formation is 900–1200 m thick and contains quartzites and shales at both its bottom and top with a medial unit composed of limestones, dolostones, and chert (King, 1976). The Crystal Springs is extensively intruded by diabase. The Beck Springs Dolomite is composed of massive dolostone up to 300 m thick. It thins and pinches out both west- and southwestward. Crystal Springs and Beck Springs rocks define a relatively narrow belt extending approximately 180 km northwestward from the Kingston Range east of the Death Valley area to the Panamint Range; thus, they may also represent a late Proterozoic trough striking into the craton. Wright *et al.* (1974) proposed that this trough was an aulacogen, whose formation occurred prior to Crystal Springs deposition but whose main phase of activity was during Kingston Peak time. Stewart and Suczek (1977), however, said that evidence for a trough in the Death Valley area prior to Kingston Peak time is equivocal and argued that Crystal Springs and Beck Springs strata had been preserved only because they were downfaulted.

H.1.e. Other Belt–Purcell Correlatives

To the east and southeast of Grand Canyon and Death Valley areas, there are exposures of supracrustal rocks probably correlative with Belt-age strata. The most widespread of these is the Apache Group and overlying Troy Quartzite of central Arizona. The Apache Group is divided into three formations: a lower unit, termed the Pioneer Shale, which is primarily an argillaceous red-bed sequence; a middle unit called the Dripping Spring Quartzite; and the upper, Mescal Limestone. King (1976) stated that lithologies of these units are similar to the Unkar Group but they do not occur in the same stratigraphic order. For example, the Mescal Limestone of the Apache Group and the Bass of the Unkar Group both contain comparable stromatolitic horizons, but the Bass lies near the base of the Unkar sequence whereas the Mescal lies near the top of the Apache Group. In both units, however, mafic flows cap the sequences. The Troy Quartzite is a relatively clean quartz arenite, similar to, and considered by King (1976) to correlate with the Dox Sandstone of the Unkar Group. Compared to the previously mentioned Belt-age strata, rocks of the Apache and Troy are considerably thinner (380–490 m for the Apache Group and 360 m for the Troy; King, 1976). Further, Apache and Troy thickness trends do not suggest deposition in a troughlike basin but rather appear to be platformal strata. A similar, correlative quartz arenite called the Lanoria Quartzite occurs in northern New Mexico. Apache, Troy, and Lanoria strata may represent patches of an originally more extensive sand sheet in the Southwest. If these suggestions are correct, platform-type sedimentation involving mature sands occurred in the southwestern cratonic area contemporaneously with deposition in actively subsiding troughs farther west. Mafic sills in both the Apache and the Troy have been dated at 1150–1200 Myr. (King, 1976), i.e., only slightly older than Cardenas Lavas in the Grand Canyon area. Potassium–argon dating of these same mafic rocks reveals a date of about 800–900 Myr., which probably represents a later heating event (King, 1976).

H.2.   Racklan–East Kootenay–Grand Canyon Tectonic Event

At sometime around 800 Myr. ago (the event is poorly dated), rocks of the Belt–Purcell sequence and many of their correlatives were subjected to a relatively minor deformational and metamorphic event. This event has been termed the Racklan orogeny in the Mackenzie Mountains of the northern Cordillera, the East Kootenay orogeny in the Belt–Purcell Basin and vicinity, and the Grand Canyon orogeny in the Grand Canyon area. Elston and McKee (1982) advocated abandoning all of these local terms and calling the event the "Grand Canyon–Mackenzie Mountains disturbance." For reasons that we will discuss later in this section, this may not be a good idea.

Generally, the disturbance resulted in regional uplift and mild regional metamorphism (of greenschist to sillimanite grade; Burchfiel and Davis, 1975) accompanied in some areas by block faulting and in others by folding. Moderate amounts of igneous activity occurred in some areas. In the Belt–Purcell area, the disturbance resulted mainly in minor folding and regional uplift, recorded in the unconformity which separates Belt–Purcell strata from overlying Windermere rocks (Gabrielse, 1972). In the northern Cordillera of Canada, however, the event was more intense and was accompanied by tight folding about north-northeast-trending axes in the Selwyn Mountains and large-scale block faulting in the Mackenzie Mountains to the east. In the Grand Canyon area, the main effects were uplift and block faulting, with as much as 3.2 km of structural offset (Elston and McKee, 1982). Although only a few igneous

rocks are obviously associated with the disturbance, resetting of the potassium–argon clock mentioned earlier in connection with mafic intrusions in the Apache Group and Troy Formation may be related to this event. We will postpone tectonic analysis of the Grand Canyon–Mackenzie Mountains disturbance until after the following discussion of the Windermere Group and its correlatives.

H.3.   Windermere Rocks and Their Correlatives

H.3.a.   The Windermere Group

The Windermere Group is a westward-thickening mass of principally detrital sediments up to 6100 m thick. It is typically developed in the Purcell and Selkirk Mountains of southeastern British Columbia and overlies Belt–Purcell rocks across a slight unconformity. Unlike strata of the Belt and Purcell Supergroups, Windermere strata are exposed almost continuously along the length of the Cordillera in Canada (Gabrielse, 1972). Correlatives of the Windermere are also widespread in the United States segment of the Cordillera but are not continuously exposed; they were, however, considered by Stewart (1972) to have been continuous along the western craton margin during their formation (see Fig. 3-20). In Canada, strata of the Windermere comprise a tremendous volume of impure detrital sediments, the lower part of which consists dominantly of phyllite, slate, siltstone, and sandstone (Gabrielse, 1972). The most distinctive rocks are gritty, feldspathic sandstone with opalescent, bluish quartz. The finer-grained lithologies dominate in the lower half of the unit and quartzites dominate in the upper half; limestones and dolostones occur in some areas. Gabrielse (1972) said that the Windermere Group of Canada represents a continental-terrace wedge, whose depositional environments were primarily marine and whose main source area was the crystalline shield area to the east. Stewart (1972) said virtually the same thing about Windermere correlatives of the United States.

Locally, at the base of the Windermere is an extremely coarse, poorly sorted diamictite containing clasts of many different lithologies, including quartz, quartzite, slate, feldspar, and granite. These rocks comprise the Toby Conglomerate in Canada and the equivalent Shedroof Conglomerate in the United States, and are widespread at the base of correlative sections from the Mackenzie Mountains of the northern Canada Cordillera to southern California. Numerous authors (e.g., Aalto, 1971; Ojakangas and Matsch, 1980) have argued for a glacial origin for these late Proterozoic diamictites, based not only on their textural properties but also on the occurrence of striated, grooved, and polished bedrock surfaces beneath them in many areas and the presence of dropstones and striated clasts. Litholog-

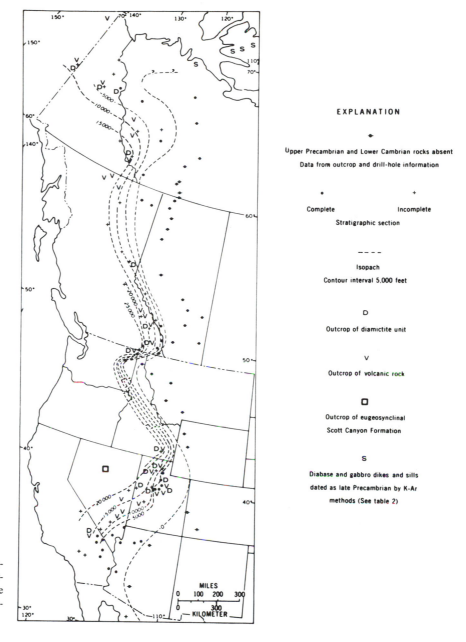

EXPLANATION

✦

Upper Precambrian and Lower Cambrian rocks absent

Data from outcrop and drill-hole information

•   +

Complete   Incomplete

Stratigraphic section

– – – –

Isopach

Contour interval 5,000 feet

D

Outcrop of diamictite unit

V

Outcrop of volcanic rock

▯

Outcrop of eugeosynclinal

Scott Canyon Formation

s

Diabase and gabbro dikes and sills

dated as late Precambrian by K-Ar

methods (See table 2)

Fig. 3-20. Generalized isopachous map of Upper Proterozoic (Windermere Group and correlatives) and Lower Cambrian strata in the Cordillera. (From Stewart, 1972, *Geological Society of America Bulletin*, Fig. 2, p. 1348.)

ically and genetically similar deposits are widespread not only in the Cordillera of North America but throughout upper Proterozoic sequences in many areas of the world.

Overlying the Toby and Shedroof Conglomerates are extrusive rocks of the Irene (Canada) and Leola (United States) Volcanics. These rocks are altered tholeiitic basalts, variously amygdaloidal and pillowed, and are associated with minor flow breccia and agglomerate (Stewart, 1972). In Washington, diamictites and volcanics are combined into the Huckleberry Formation. Overlying these volcanic rocks are fine-grained detrital sediments of the Horsethief Creek (Canada) and Monk (United States) Formations and then the Hamill (Canada) and Gypsy (United States) Quartzites, which comprise the main detrital mass of the Windermere Group. Speaking of both the Windermere and its correlatives to the south, Stewart (1972) said that unconformities occur at various horizons within these upper Proterozoic strata and within overlying Cambrian strata, but none can be traced for any great distance and the Lower Cambrian Series cannot be separated systematically from the upper Proterozoic rocks because they are lithologically similar and fossils are sparse.

H.3.b. Windermere Correlatives in Southeastern Idaho and Utah

In southeastern Idaho, there is a sequence of rocks termed the Pocatello Formation which is generally correlative with the Windermere Group. Near the base of the Pocatello is the Scott Mountain Member, composed of coarse diamictites similar to those of the Toby and Shedroof Conglomerates. Interbedded with Scott Mountain diamictite are altered basaltic flows and tuffs of the Bannock Volcanic Member, which are reminiscent of the Irene and Leola Volcanics intercalated with Toby and Shedroof diamictites (King, 1976). Pocatello rocks above the Scott Mountain Member comprise a sequence of shale, siltstone, and thick quartzites similar to Horsethief Creek and Gypsy strata. Near the top of the Pocatello is a thin limestone unit.

Windermere correlatives also outcrop in central and western Utah. For example, in the Big Cottonwood Canyon area of the Wasatch Range just south of Salt Lake City (about 300 km south of Pocatello), Windermere equivalents comprise the Mineral Fork Tillite and the Mutual Formation. The Mineral Fork lies on the eroded, striated, and polished surface of the Big Cottonwood Formation and is composed of up to 1000 m of coarse graywacke and diamictite. Ojakangas and Matsch (1980) divided the Mineral Fork into a lower member composed of thick, massive diamictites with minor lenses of shale–siltstone, sandstone, and conglomerate, and an upper member with more shale and siltstone with prominent laminations, graded bedding, and dropstone pebbles. The lower member Ojakangas and Matsch interpreted as glacial tills and outwash-plain deposits; the upper member they interpreted as glaciomarine strata. The Mutual Formation consists of up to 360 m of red and purple quartzites and red to green shales. It lies on the Mineral Fork across an erosional surface and is overlain, also unconformably, by the Lower Cambrian Tintic Quartzite (see discussion in Chapter 4).

Sixty or so kilometers to the west, in the Sheeprock Mountains of westcentral Utah, is a thick sequence of rocks also correlative with the Windermere. These rocks comprise the Sheeprock and Brigham Groups (Christie-Blick, 1982). The Sheeprock Group consists of three formations; from bottom to top, these are: the Otts Canyon Formation, made up of over 2000 m of lutites, diamictite, and quartzite, intruded near the top by diabase sills; the Dutch Peak Formation, composed of as much as 1750 m of diamictite, graywacke, sandstone, quartzite, siltstone, and shale; and the Kelley Canyon Formation, consisting of shale and siltstone. Dutch Peak diamictites contain clasts of various plutonic, volcanic, metamorphic, and sedimentary lithologies, some up to 3 m in size. The Brigham Group consists of four units, the uppermost of which is of Early Cambrian age (the Prospect Mountain Quartzite, a Tintic correlative). At the base of the Brigham Group is the Caddy Canyon Quartzite, which Christie-Blick (1982) described as consisting of at least 2000 m of quartzite with subordinate interbeds of siltstone, shale, and conglomerate, some of which contain large, angular, intraformational clasts. The Caddy Canyon is overlain by up to 145 m of slate with subordinate sandstone and quartzite called the Inkom Formation. Between the Inkom and the Prospect Mountain Quartzite is 500 m of red and purple feldspathic quartzite and conglomerate of the Mutual Formation, whose features are similar to those of the Mutual in the Big Cottonwood Canyon area. Taken together, strata of the Sheeprock and Brigham Groups attain a maximum thickness of greater than 7200 m (Christie-Blick, 1982); when compared to the correlative Mineral Fork and Mutual Formations of the Wasatch Range, they show the westward-thickening trend discussed previously for the Windermere Group.

Even farther west, along the Utah–Nevada border in the Pilot, Snake, and Schell Creek Ranges, strata comparable to the sequence just mentioned comprise the McCoy Creek Group. Rocks of the McCoy Creek Group include several persistent quartzite units separated by units of argillite, siltstone, and subordinate marble (King, 1976). McCoy Creek strata are greater than 2700 m thick but were considered by Christie-Blick (1982) to correlate for the most part with the Caddy Canyon Quartzite, again showing that these Windermere-correlative strata thicken progressively westward.

H.3.c. Kingston Peak Formation (Upper Pahrump Group)

The last important Windermere correlative in the United States is the Kingston Peak Formation, upper unit of the Pahrump Group in the Death Valley area. Kingston Peak strata are composed almost entirely of coarse diamictites similar to those described earlier. Kingston Peak diamictites range from 0 to over 1800 m across a few kilometers. This extraordinarily sudden thickness was considered by Stewart and Suczek (1977) to indicate that Kingston Peak strata were deposited in a fault-bounded basin. Near its margins, the formation contains gigantic blocks up to 300 m long, which have been interpreted to be slide blocks, perhaps caused by slumping along basin-margin fault scarps.

H.3.d. Rapitan Group

Far to the north, in the Mackenzie Mountains of Canada's Northwest Territories, uppermost Proterozoic strata equivalent to the Windermere comprise the Rapitan Group (Gabrielse, 1972). Rapitan strata lie with marked angular unconformity on Purcell rocks and have an aggregate thickness of 1800 m. The Rapitan is divided into three forma-tions, the lowest of which consists of diamictite, siltstone, shale, sandstone, volcanic ash and tuff, and banded iron formation. The middle formation is primarily composed of coarse diamictite and the upper unit is made up of shale, siltstone, and arkosic sandstone. The banded iron formation was said by Gabrielse (1972) to represent chemical pre-cipitation in a sediment-starved basin. Rapitan diamictites, like others in the Windermere and its correlatives, are con-sidered to be of glacial origin.

H.4. Late Precambrian Tectonics and Development of the Cordilleran Margin

Strata of the Belt and Purcell Supergroups, Winder-mere Group, and their correlatives are the oldest rocks that obviously formed in association with a continental margin in the Cordilleran area, but tectonic models of Cordilleran-margin development differ significantly. Gabrielse (1972) and Wheeler and Gabrielse (1972) suggested that the Cor-dilleran margin was formed by continental rifting around 1450 Myr. ago and that Belt–Purcell deposition occurred along the newly formed margin. The westward-thickening mass of Purcell strata in the Canadian Cordillera has many of the characteristics of a rifted-margin prism and is clearly of the appropriate thickness. The presence of thick, bedded gypsum deposits in the Purcell rocks of the Mackenzie Mountains area also supports a rift model of basin develop-ment. Further, as pointed out by Windley (1984), the pres-ence within Belt–Purcell strata of phosphorite deposits indi-cates deposition on a continental margin; this is because such deposits are characteristically developed in areas of ocean-margin upwelling. Strata of the Belt Supergroup in the United States formed within a rapidly subsiding trough which apparently opened northwestward toward the sea; several authors (e.g., Harrison et al., 1974) termed the Belt Basin a marine reentrant into the United States. Thus, the southern part of the Belt–Purcell Basin may be interpreted as an aulacogen formed within a failed arm associated with the Cordilleran rifting event. Strata of the Uinta Mountains Group and, possibly, the Grand Canyon Supergroup and the Crystal Springs–Beck Springs Formations may also be in-terpreted as aulacogens. Sears and Price (1978) extended this idea by suggesting that the continental fragment that was rifted away from the Cordilleran margin at about 1450 Myr. is represented by the Siberian platform, a large, dis-crete, older Precambrian craton surrounded by younger, continental-margin sedimentary prisms.

The Racklan and East Kootenay orogenies, in Wheeler and Gabrielse's (1972) model, are considered to have re-sulted from initiation of subduction along the Belt–Purcell margin. Subduction was short-lived, however, and the con-tinental margin soon reverted to its predisturbance condi-

tion. Following the disturbance, Windermere strata were deposited on the passive margin. The coarser and more immature nature of Windermere sediments would have been due to greater relief in the source area as the result of orog-enic uplift.

Stewart (1972, 1976) and Stewart and Suczek (1977) disagreed with the interpretation that Belt and Purcell strata represent a rifted-margin prism. They argued that evidence of an ocean associated with Belt–Purcell rocks is equivocal and suggested, instead, that these rocks formed in fault-bounded, epicratonic troughs similar to the Keweenawan trough of the cratonic interior (discussed in Section C of this chapter). If, however, rocks of the Belt, Uinta Mountains, and other basins formed as aulacogens, they would not necessarily be associated with an ocean in their immediate vicinity; rather, they would open westward toward the ocean adjoining the main continental-margin prism, which in the United States would be obscured by younger rocks and structures. Additionally, these authors' comments refer specifically to rocks in the United States; their objections do not seem to apply to western Canada, where Purcell rocks are more obviously miogeoclinal and, thus, probably asso-ciated with a continental margin during their deposition. Finally, the great thicknesses of Belt and equivalent strata in the United States would seem to militate against their depo-sition within simple epicratonic basins, because such basins typically do not subside so much so rapidly. Their closest corollaries in North America are the Ancestral Rocky Mountain basin-system of late Paleozoic age and the South-ern Oklahoma aulacogen (both of which are discussed in Chapter 7), and Triassic and Jurassic basins of the Atlantic Coastal Province (Chapter 8). All of these deep basins were probably formed in relation to extensional plate-margin tec-tonics. If the Belt, etc. basins are *not* related to continental-margin extension and rifting, then some other tectonic mechanism adequate to explain their thicknesses must be found.

Although opposed to the idea that Belt and Purcell strata represent rifted-margin deposits, Stewart (1972, 1976) and Stewart and Suczek (1977) did consider the Cor-dilleran margin to have developed by continental rifting. The difference between their interpretation and that of Gabrielse (1972) and Wheeler and Gabrielse (1972) is that they considered the rifting event to have occurred around 850 Myr. ago and to be recorded by rocks of the Winder-mere Group and correlatives. In their model, structural ef-fects associated with rifting define the Grand Canyon–Mackenzie Mountains disturbance (used in the inclusive sense proposed by Elston and McKee, 1982). Deformation due to the disturbance was relatively minor because the event was essentially extensional, rather than compres-sional, in origin. Early in the history of Windermere rifting, a major (continental?) glaciation event occurred, resulting in deposition of coarse diamictites along most of the newly formed margin. Evolution of the Cordilleran rifted-margin continued into the Early Cambrian.

Clearly, the two tectonic models mentioned above conflict. One should notice that the models were developed to explain stratigraphic and structural relations in two different parts of the Cordillera and were then extended to the other parts. Dickinson (1977) attempted to reconcile the two models by suggesting that both are correct in the areas for which they had been developed but neither can be extended much beyond the International border. In his model, Dickinson (1977) proposed that two different rifting events were involved in the evolution of the entire Cordilleran margin (see Fig. 3-21). The first of these rifting events occurred at about 1450 Myr. ago along the Canadian segment of the Cordilleran belt and led to continental-margin deposition of Purcell and northernmost Belt strata. In this model, the main part of the Belt Basin formed as an aulacogen at the southern end of the Purcell continental margin. The Racklan and East Kootenay parts of the Grand Canyon–Mackenzie Mountains disturbance may have been caused by a plate collisional event. Dickinson's proposal would explain why deformation associated with the Racklan event was so much more intense than that in the Belt Basin or farther south; presumably, tectonic effects in the Northern Cordillera would have been due to a full-scale collision, whereas the Belt Basin would have experienced less dramatic effects because aulacogens do not usually take the full brunt of collision. Following the Racklan–East Kootenay collision, continental-margin deposition resumed along the Canadian margin. At about 850 Myr. ago, the second rifting

event began to affect the U.S. Cordilleran area in much the same manner suggested by Stewart (1972). Structural effects in the Grand Canyon area would have been related to extensional tectonics associated with this rifting event. If the scenario is correct, it is probably unwise to link by a single name (i.e., Grand Canyon–Mackenzie Mountains disturbance) the tectonic events of the Racklan and East Kootenay events with those of the Grand Canyon area, because they did not develop in the same way.

Dickinson's (1977) fusion of Gabrielse's and Stewart's ideas seems to answer most of the objections raised to the original models. One difficulty remains: Dickinson, like Stewart, considered the Uinta Mountains Group, Grand Canyon Supergroup, and Crystal Springs–Beck Springs strata to have been deposited in epicratonic basins. We showed earlier in this section that Crystal Springs and Beck Springs rocks may not represent a significant depositional trough but, rather, may have been preserved because they were downfaulted during Kingston Peak time. Since the Kingston Peak Formation is a Windermere equivalent, its origin as an aulacogen is reasonable in terms of Dickinson's model. But the other two basinal areas mentioned still require explanation for their substantial and rapid subsidence.

I. Early History of the Appalachian Continental Margin

The long history of the Appalachian continental margin began at approximately 850 Myr. ago with a continental-separation event which led to the opening of the Iapetus Ocean. Knowledge of geological history in the eastern part of the continent during the interval between this continental-separation event and the Grenville orogeny (1000 Myr. ago; see discussion in Section F) is almost nonexistent; this is because there are (apparently) no extant rocks of that interval in the East. But with initiation of rifting in the Appalachian region began the deposition of sediments, first in rift basins and then along the new continental margin.

One of the more stratigraphically important questions presented by strata of the early Appalachian margin is where in the thick stack of strata to place the boundary between the Cambrian and the Precambrian. This is not a question unique to the Appalachian section; it enlivens discussions of most correlative stratigraphic sections around the world, with debate on the point being principally between those who favor biostratigraphy versus those who favor lithostratigraphy. We will elaborate upon this subject in the next chapter (see discussion in Section C of Chapter 4); suffice it to say here that the Cambrian–Precambrian boundary is usually placed somewhere within the Chilhowee Group in the Southern Appalachians. In the present

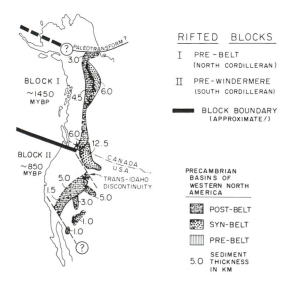

Fig. 3-21. Upper Precambrian deposits of the Cordilleran area. (Modified from Dickinson, 1977, reproduced with permission of the Society of Economic Paleontologists and Mineralogists.)

discussion, therefore, we will dwell primarily on pre-Chilhowee rocks.

Post-Grenville, Precambrian, supracrustal rocks are locally very thick but are present only in eastern portions of the Appalachian orogen (see Fig. 3-22). In western Appalachian regions, rocks of Cambrian or younger systems lie directly on granitic basement. Often, even where upper Precambrian rocks are present, they are obscured beneath younger strata or allochthonous structural blocks, or have been so thoroughly reworked in younger tectonic events as to be virtually unrecognizable. For these reasons, most information on the latest Precambrian in the East comes from a limited number of exposures in the Blue Ridge, western Piedmont, and western New England Highlands areas of the Appalachians. Upper Precambrian rocks are also present in portions of the Appalachian orogen to the east of these areas, but are probably parts of displaced terranes which formed elsewhere and were accreted to the Appalachian continental margin much later in its history. For this reason, we will defer discussion of them to Chapters 5 and 6.

I.1. Upper Precambrian Rocks of the Central and Southern Appalachians

The largest area of exposed upper Precambrian rocks in the East is in the Blue Ridge and western Piedmont of the Central and Southern Appalachians. Because upper Precambrian rocks of these areas comprise three different, but generally correlative, stratigraphic successions, we will divide the following discussion into three parts: (1) the Blue Ridge and adjoining Piedmont areas of northern and central Virginia; (2) the northwestern North Carolina–southwestern Virginia area; and (3) the Blue Ridge and Great Smoky Mountains of western North Carolina and eastern Tennessee.

I.1.a. Northern and Central Virginia

The major supracrustal stratigraphic unit of the northwestern flank of the Blue Ridge in northern and central Virginia is the Catoctin Greenstone, composed dominantly of altered basaltic flows up to 1500 m thick (King, 1976). Reed and Morgan (1971) said that Catoctin basalts are primarily tholeiitic; they contain well-developed columnar jointing in some areas and many are amygdaloidal. Associated with the flows are variable amounts of detrital sediment; in addition, rhyolitic volcanics and welded tuffs are locally present, e.g., at South Mountain in Pennsylvania (Rankin, 1975). Reed and Morgan (1971) argued that, based on geochemistry, Catoctin lavas were probably subaerial, a point supported by their amygdaloidal structure.

Locally at the base of the Catoctin Greenstone is a thin,

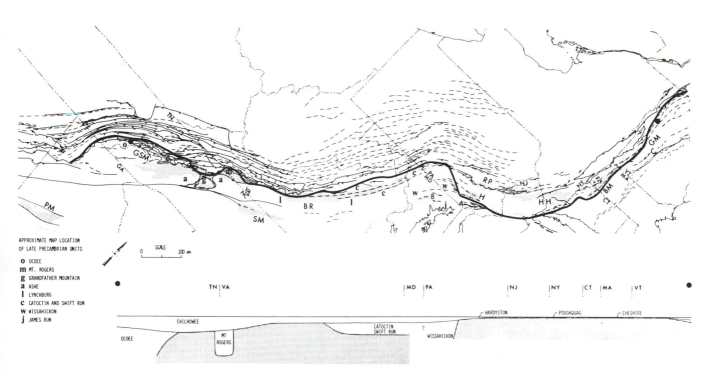

Fig. 3-22. Distribution of Upper Precambrian and Lower Cambrian strata in the Appalachian area. Heavy line on the map shows cratonward limit of Upper Precambrian rocks; shaded areas of the map represent exposures of Grenville-age basement. (From Thomas, 1977, *American Journal of Science,* Fig. 2, p. 1235.)

conglomeratic unit called the Swift Run Formation, whose thickness ranges from 0 to 30 m. The Swift Run contains fine- to coarse-grained conglomerate, arkose, quartzite, and tuffaceous slate or phyllite (King, 1970). The erosional surface on which the Swift Run and overlying Catoctin lie has up to 300 m of local relief.

Overlying the Catoctin Greenstone is the Chilhowee Group which, in the central Virginia area, is composed, from bottom to top, of the Loudoun, Weverton, and Harpers Formation and Antietam Quartzite. The lower part of this sequence is unfossiliferous except for the trace fossil *Skolithos;* within the Antietam Quartzite, however, is found the Lower Cambrian trilobite, *Olenellus.* Stratigraphic relations in the Blue Ridge area of central Virginia are shown in Fig. 3-23.

On the southeastern flank of the Blue Ridge, correlative rocks comprise a much thicker sequence composed of the Lynchburg Formation and the Evington Group. The Lynchburg Formation consists of a strongly metamorphosed sequence of layered detrital sedimentary rocks which originally were arkosic sandstones and siltstones with interbeds of arkosic conglomerate (King, 1970). In lower parts of the Lynchburg Formation, there are lenticular layers of mafic rocks. At the base of the Lynchburg is the Rockfish Conglomerate Member, a bouldery arkosic orthoconglomerate up to 30 m thick, which King (1970) considered to have formed in large alluvial fans. In upper parts of the Lynchburg Formation, there are amphibolites which represent the extension of Catoctin volcanics into the Lynchburg area. At the top of Catoctin correlatives in the Lynchburg are meta-

sediments of the Evington Group. These rocks are considered to be equivalent to the Chilhowee Group farther to the west, albeit finer-grained and of deeper-water origin.

I.1.b.   Southwestern Virginia and Northwestern North Carolina

Catoctin correlatives in the area of the North Carolina–Virginia border make up the Mount Rogers Formation. Like the Catoctin, the Mount Rogers is principally a volcanic unit but the nature of the volcanics is more varied and it contains more interbedded detrital sediments (Rankin, 1970). The formation can be divided into three parts. The lower part consists of intercalated sedimentary, basaltic, and rhyolitic rocks; the middle unit is composed entirely of thick layers of rhyolite; and the upper part is primarily sedimentary rock with minor basalt and rhyolite. The basalts of the lower member of the Mount Rogers Formation are similar to Catoctin basalts. Rhyolites are commonly associated with welded tuffs. Conglomerates within the lower Mount Rogers contain boulders of plutonic basement rocks (e.g., the Grenvillian Cranberry Gneiss); higher within the sequence, however, conglomerates contain only well-rounded quartz pebbles (King, 1970).

Within the upper member of the Mount Rogers Formation is a most interesting assemblage of sedimentary rocks, characterized by fine, laminated mudstone and diamictite. The laminated mudstones have a distinctly varved appearance, with fine to very fine arkosic wacke alternating in almost paper-thin layers with fine mud. The mudstones also

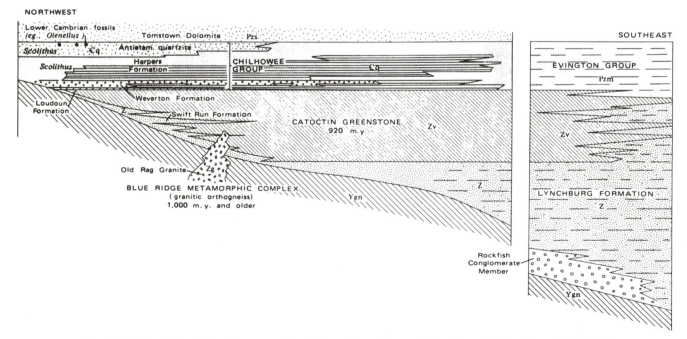

Fig. 3-23. Interpretative stratigraphic cross section of Upper Precambrian rocks in northern Virginia. (From King, 1976.)

contain abundant dropstone pebbles and the diamictites are very coarse and poorly sorted. Rankin (1970, 1975) considered these deposits to be tillites and other glacially related sediments. Since strata of the Mount Rogers (and Catoctin, etc.) units grossly correlate with Windermere Group strata of the Cordilleran continental margin (discussed in the previous section), it is interesting to note that both contain rocks interpreted as glacial in origin. Overlying the Mount Rogers Formation is the Chilhowee Group.

To the east and southeast of Mount Rogers strata, correlative rocks comprise the Ashe Formation and the Grandfather Mountain Formation. The Ashe is composed of high-grade metamorphics such as layered mica gneisses, amphibolites, and schists; it also contains granule and pebble metaconglomerates. Rankin (1975) called the Ashe a metagraywacke sequence and correlated it with the Lynchburg Formation of central Virginia.

The Grandfather Mountain Formation is similar to the Ashe and Lynchburg. It consists of metamorphosed arkose, siltstone, shale, and volcanic rocks and has a thickness possibly greater than 9000 m. King (1970) said that sedimentary components of the Grandfather Mountain Formation are similar to those of the Snowbird Group (of the Ocoee Supergroup) except that Snowbird strata contain no volcanic rocks; thus, the Grandfather Mountain sequence may constitute a transition between the sedimentary Ocoee rocks and the sedimentary–volcanic rocks in the east.

I.1.c. Western North Carolina and Eastern Tennessee

In the western North Carolina–eastern Tennessee area, the Blue Ridge broadens out into a large and rugged region called the Great Smoky Mountains, almost all of which is underlain by upper Precambrian rocks of the Ocoee Supergroup. The Ocoee is among the most impressive stratigraphic units of the Appalachians, being greater than 12,000 m thick and composed throughout its entire section of detrital-sedimentary rocks. The Ocoee is divided into three units, the relations among which are poorly understood because of the great structural complexity of the region. In western and northern parts of the Great Smokies, the Ocoee is composed of the lower, Snowbird Group (with five formations) and the upper, Walden Creek Group (with four formations). The Snowbird consists entirely of highly immature metasediments derived from an igneous–metamorphic terrane, probably to the east and southeast (Hadley, 1970). It is highly feldspathic and contains lenses of quartz-pebble conglomerate.

The overlying Walden Creek Group is lithologically very heterogeneous; it contains dark gray siltstones and argillite, feldspathic and quartzose sandstones, polymictic conglomerates, and even minor carbonates. King (1976)

stated that the Walden Creek Group formed on an unstable shelf and had a northwestern source.

The third Ocoee unit is the Great Smoky Group, which only occurs in eastern parts of the Great Smokies. Much of this unit is composed of medium- to coarse-grained, graded turbidites (King, 1976). The Great Smoky Group is considered to represent deeper-water deposition, possibly (but not definitely) correlative with Walden Creek shelf deposition. At the top of Ocoee strata are more mature, thinner strata of the Chilhowee Group. In eastern Tennessee, the Chilhowee is divided into the Cochran Formation, Nichols Shale, Nebo Quartzite, Murray Shale, Hesse Quartzite, and Helenmode Formation (see further discussion in the next chapter). Correlative strata in Georgia comprise the Weisner Formation.

Strata of the Ocoee are probably correlative with much more highly metamorphosed rocks on the eastern side of the Murphy Synclinorium, in the center of which is the Murphy Marble. The age of this marble unit is problematic but has been correlated with the Shady Dolomite (of Cambrian age) in the Ridge and Valley; therefore, some of the succession beneath the Murphy Marble may be of Cambrian age, too.

I.2. Upper Precambrian Rocks of the Northern Appalachians

Upper Proterozoic strata of the Northern and Maritime Appalachians are more poorly preserved and less exposed than in the south but reveal essentially the same history as their southern counterparts. For example, in the Newfoundland foreland, greater than 9000 m of coarse, detrital sediments composes the Fleur de Lys Supergroup (Williams and Stevens, 1974). Associated with these detrital strata are thick, tholeiitic volcanics of the Lighthouse Cove Formation and associated mafic feeder-dikes cutting across Grenville basement. At the base of Lighthouse Cove volcanics is a thick quartzite–conglomerate unit termed the Bateau Formation; the relation of the Bateau to the Lighthouse Cove resembles that of the Swift Run Formation to the Catoctin Greenstone in Virginia. A lack of pillow structures in Newfoundland mafic flows suggests that most of them are terrestrial, particularly in western exposures; some, however, like the Maiden Point Formation of the allochthonous terrane in Newfoundland, do contain pillowed flows and so were probably extruded under water. Williams and Stevens (1974) interpreted Lighthouse Cove volcanics as plateau basalts formed in an environment of continental rifting.

In Quebec, between Quebec City and Stratford, mafic flows of the Tibbit Hill Formation directly overlie Grenville basement with no intervening basal conglomerate. The Tibbit Hill is overlain by graywackes, slates, and dolostone of the Oak Hill succession (St. Julien and Hubert, 1975). Toward the east, these strata pass into a finer flysch succession represented by the Rosaire Group. Williams and Stevens

(1974) also considered the overlying, arenitic Caldwell Group to be part of the Upper Proterozoic sequence. Similar strata are also found in the Mendon Group and Pinnacle Formation of Vermont and the Nassau Formation of the New York allochthon.

In southeastern Pennsylvania and Maryland, an important series of upper Precambrian rocks is termed the Glenarm "Series," which lies over the Grenville-age Baltimore Gneiss. At its base, the Glenarm contains thin deposits of metamorphosed quartz sandstone, shale, and carbonate termed the Setters Quartzite and Cockeysville Marble. The bulk of the Glenarm is made up of the Wissahickon Formation, which is a flysch succession more than 6000 m thick and composed of metamorphosed graywacke, shales, and submarine slumps (Higgins, 1972). The depositional character of the Wissahickon and its great thickness suggest that it formed in relatively deep water, perhaps on the continental slope or continental rise. The age of the Wissahickon has been a matter of debate; it was once thought to be correlative with the Lynchburg Formation to the south, but Higgins (1972) showed that its deposition spanned Chilhowee time. Consequently, the Wissahickon is considered to be the off-shelf correlative of the Chilhowee Group, whose more mature, quartzose sediments formed on the continental shelf. The relation of the Wissahickon to the underlying Setters and Cockeysville is questionable, being either in normal, depositional sequence or in thrust contact.

I.3. Late Precambrian Rifting and Early History of the Appalachian Margin

Sometime around 850 Myr. ago, a continental-separation event began the history of the Appalachian continental margin. The identity of the continental block that had adjoined the prerift Appalachian is unknown. Rankin (1975), Thomas (1977), and Williams and Stevens (1974) all presented tectonic models of the origin of the Appalachian margin.

The earliest history of the event featured the development of a rift valley along the region that was to become the continental margin. The exact history and geometry of this rift are poorly known but more recent continental-separation events have typically begun in association with hot spots beneath the continental lithosphere. The first response of the crust is to dome broadly upwards above the hot spot with concomitant mafic volcanism. As the dome grows, a fracture pattern develops in the crust with rifts radiating away from the center of the dome in a three-armed pattern. As the rifts grow longer, they intersect rifts associated with other areas to produce a complex, interconnected rift-valley system similar to the East Africa rift system. At the same time, erosion of the domed area typically strips away much, if not all, of any previous supracrustal strata and cuts into the crust itself. Sediments derived from this erosion are carried into the rift-valley system, resulting in deposition of coarse, immature, arkosic sediments. Deposits of the Snowbird, Swift Run, Rockfish, and Bateau Formations probably formed in this way. Associated with these sediments were outpourings of basalts and lesser quantities of rhyolites. More recent riftings, directly associated with hot spots, feature the extrusion of much greater volumes of lava; thus, it can be suggested that Mount Rogers, Catoctin, and Lighthouse Cove volcanic units may have formed in association with such hot spots. Isostatic subsidence during this earliest phase was very rapid, as attested to by the great thicknesses of Ocoee and Fleur de Lys strata; sediments deposited at that time were buried before being significantly reworked and so are texturally and compositionally very immature. At least some of the diamictites mentioned in earlier discussions may have formed in this way, associated with alluvial fan systems along the margins of the rift valley.

The lack of volcanic components in the Ocoee Basin suggests that that area was not at the locus of main rifting. Thomas (1977) suggested that Ocoee rocks accumulated in a basin formed between the continental margin and a crustal block that was partially rifted away from the continent and then was stranded, Madagascar-like, by a shift in the locus of spreading (see Fig. 3-24). In this model, the stranded block is represented by the eastern Blue Ridge area (labeled G in Fig. 3-24), which is underlain by Grenville-age granitic basement. Note in Figs. 3-24 and 4-2 that other blocks, e.g., the Pine Mountain area of Georgia (labeled P), Sauratown Mountain block in northwest-central North Carolina (labeled S), and the Baltimore Gneiss (labeled B), were probably also left stranded, possibly by the same spreading shift. While Ocoee deposits accumulated in their protected basin, deposits of the Ashe and Lynchburg Formations were forming at the active continental margin.

Although most Appalachian geologists agree with the general model presented in this discussion, there is healthy disagreement over details. Especially since the publication of COCORP seismic reflection profiles across the Southern Appalachian orogen, questions have arisen about this model. This is because the profiles suggest that continental (i.e., Grenville-age) crust extends considerably farther to the east beneath central parts of the orogen than previously thought. Thus, Blue Ridge and related areas may not be near the edge of the Appalachian continental margin. If this is true, their formation could not have been related to continental-margin events. On the other hand, all parts of the Blue Ridge, Inner Piedmont, and related interior parts of the orogen appear to be allochthonous and to have been shoved far westward during the Alleghenian orogeny near the end of the Paleozoic. Consequently, rocks of those areas may have formed in continental areas after all.

Toward the end of Ocoee time, the newly opened

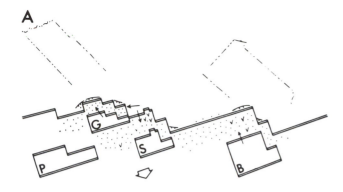

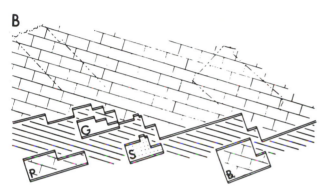

Fig. 3-24. Thomas's tectonic model of Late Precambrian rifting along the Appalachian continental margin; see text for discussion. (From Thomas, 1977, *American Journal of Science*, Fig. 11, p. 1267.)

fossils are the remains of single-celled organisms, and many early Precambrian fossils are of doubtful biotic origin (see Schopf, 1975). Yet, some of the earliest forms, such as stromatolites, are so similar to structures produced today by organisms that their biotic affinities are widely accepted. Most Precambrian fossils dating before 1.0 Byr. fall into the following categories:

1. Spheres and bacilliform structures (possible bacteria, blue-green algae, bacterial or fungal spores, fungi)
2. Filaments (possible algae)
3. Stromatolites (algal/bacterial structures)
4. Clusters of spheres (colonial bacteria or algae)
5. Spheres undergoing division (bacteria, algae, fungi, or other single-celled eukaryotes)
6. Irregularly shaped single-celled structures
7. Fossil carbon compounds

Multicellular *body fossils* (i.e., remains of the actual creature versus preserved traces termed *ichnofossils*) are known only in Ediacarian and younger rocks less than 800 Myr. old. These show creatures tremendously advanced over the earlier Precambrian organisms and they will be discussed in context with the earliest Phanerozoic sediments in Chapter 4.

Since the oldest known fossils approach the age of the oldest known unmetamorphosed sediments, one may not expect the age of the oldest life to be pushed back significantly earlier than the 3.5 Byr. range determined at present. The following discussion is arranged chronologically and is summarized in Table II.

Iapetus Basin was probably similar to the Red Sea and represented a proto-oceanic gulf. More mature deposits on the Appalachian margin from this time show the effects of slower deposition and greater amounts of reworking as the rate of margin subsidence slowed. By the onset of shallow-marine, Chilhowee deposition, a fully developed continental shelf had formed as well as a continental slope and rise, on which sediments of the Wissahickon were deposited. Subsequent evolution of the Appalachian margin occurred during the Cambrian and will be detailed in the next chapter.

J. The Fossil Record of Early Life

J.1. Introduction

The study of Precambrian paleontology has advanced enormously in the past three decades and the known fossil record has been pushed back more than 3.0 Byr. before the Cambrian Period (once regarded as the beginning of the fossil record). However, all but the youngest Precambrian

J.2. Archean Fossils: 3.5–2.5 Byr.

A majority of objects reported from the earliest part of the Archean are of dubious biogenic origin—an understandable phenomenon if one considers the extreme antiquity of the interval, and the nature of extremely simple, microscopic life forms. Nevertheless, and somewhat anomalously, the very oldest reported "fossils" are widely accepted as true organic remains and suggest that life had reached moderate complexity by 3.5 Byr.

The Warrawoona Group in Western Australia, in the vicinity of the so-called "North Pole" dome, has yielded several types of microfossils from cherts dated to 3.5 Byr. Although many are of dubious biogenicity [see Schopf and Walter (1983) for discussion of these and many other Archean "dubiofossils" and "nonfossils"], certain structures reported by Awramik *et al.* (1983) are widely accepted microfossils. These consist of radiating filaments showing cross-links (as if dividing between cells) and resemble certain modern cyanobacteria (blue-green algae). Schopf and Walter (1983) accept these as the only true biogenic structures predating 2.8 Byr. Other Warrawoona structures may

Table II. Chronology of Precambrian Fossil Reports[a]

Age (in Myr.)	Rock unit and locality	Lithology	Description of reported fossils
3556 (±32)	Warrawoona Series, North Pole, Western Australia	Carbonaceous stromatolitic chert	Filaments with cross-walls, microspheres
3540 (±30)	Onverwacht sediments, Swaziland System, South Africa	Carbonaceous cherts	Microspheres, filaments, hydrocarbons
3300 (+200)	Fig Tree cherts, Swaziland System, South Africa	Carbonaceous cherts	Microspheres, bacilliform structures, filaments, hydrocarbons
2768 (±24)	Tumbiana Formation, Fortescue Group, Western Australia	Cherts in stromatolitic carbonates	Filaments
2700 (+300, −35)	Bulawayo limestones, Rhodesia	Stromatolitic carbonates	Stromatolites
2700 (±?)	Soudan Iron Formation, Minnesota, U. Michigan	Siliceous ironstones	Microspheres, filaments, hydrocarbons
2600 (+200, −100)	Steeprock Group, Ontario	Stromatolitic carbonates	Stromatolites
2500 (±?)	Hamersley Group, Western Australia	Ironstones	Microspheres and subhedral structures
2300 (+100, −80)	Ventersdorp Group, South Africa	(From Schopf, 1975)	Stromatolites
2275 (+25, −55)	Wolkberg Group, Transvaal	(From Schopf, 1975)	Stromatolites
2250 (+50, −30)	Olifants River Group, Transvaal	(From Schopf, 1975)	Stromatolites
2100 (+120, −150)	Pretoria Group, Transvaal	(From Schopf, 1975)	Stromatolites
2000 (+185, −150)	Wyloo Group, Western Australia	(From Schopf, 1975)	Stromatolites
2000 (+200, −100)	Gunflint Iron Formation, Ontario	Cherts, some carbonates	Microspheres, filaments, bacilliform structures, irregular microforms, tubular and septate branching filaments, stromatolites, hydrocarbons; apparent blue-green algae, bacteria, other prokaryotes
1800 (+400, −50)	Belcher Group, Hudson Bay, Canada	Cherts in stromatolitic carbonates	Filaments, microspheres, tubular branching forms, large double-walled spheroids; apparent filamentous blue-green algae, bacteria, other prokaryotes
1700 (+300)	Frere Formation, Western Australia	Banded iron formation	Filaments, microspheres, irregular structures; apparent blue-green algae, bacteria, other prokaryotes
1650 (+100, −100)	Paradise Creek Formation, Queensland, Australia	Cherts in stromatolitic carbonates	Filaments; probably mat-forming blue-green algae
1600 (+100, −90)	Amelia Dolomite, Northern Territory, Australia	Cherts in stromatolitic carbonates	Filaments, tetrads of microspheres; apparent blue-green mat-forming algae and possible mitosing eukaryotic algae
1500 (+300, −400)	Bungle Bungle Dolomite, Western Australia	Cherts in stromatolitic carbonates	Filaments, microspheres with bound organelles(?); possible blue-green algae, possible eukaryotes
1300 (+100, −100)	Beck Springs Dolomite, Southern California	Cherts and carbonates in stromatolites and oncolites	Blue-green filamentous and coccoidal algae, large spheres, mitosing eukaryotes, Chlorophyta (?), Chrysophyta (?), other eukaryotes (?)
1300 (+100, −100)	Newland Limestone, Belt Supergroup, Montana	Shales	Filamentous blue-green algae, large microspheres; possible eukaryotic algae (Chlorophyta or Chrysophyta)
1000 (+340, −260)	Skillogalee Dolomite, South Australia	Cherts in algal laminae	Filamentous blue-green algae, branching algae (possible eukaryotic forms) with cross-walls
950 (+470, −110)	Myrtle Springs Formation, South Australia	Cherts in carbonates	Filamentous blue-green algae, large single-celled eukaryotes
900 (+470, −110)	Bitter Springs Formation, Central Australia	Cherts in algal laminae, carbonates	Filamentous and tubular blue-green algae, bacteria, eukaryotic algae, fungi (?), undetermined eukaryotes
900 (+440, −160)	Auburn Dolomite, South Australia	Cherts in carbonates	Filamentous blue-green algae, large single-celled eukaryotes
800 (+50(?), −150)	Worldwide		First appearance of metazoans

[a] Data from Schopf (1975), Schopf and Walter (1983), and others.

be organic fossils but their morphologies are not as compellingly lifelike.

The next oldest set of Archean microfossils is of nearly the same antiquity, but the dating is imprecise of the Onverwacht Group, in the Swaziland System in Transvaal, South Africa. The sediments lie thousands of meters below the Fig Tree Series, itself dated between 3.1 and 3.5 Byr. The putative microfossils come from carbonaceous black cherts and are themselves carbonaceous cell-like spheroids and filaments (Engel et al., 1968; Brooks and Muir, 1971; Muir and Grant, 1976). The sizes range from 1 μm to 55 μm in length and diameter for the longest filaments and the largest spheroids; the filaments and spheres may be blue-green algae and/or other forms of bacteria.

A variety of early stromatolites are reported in the literature, including those in the Warrawoona Group and other units approaching the 3.0 to 3.5 Byr. range of antiquity. Archean stromatolites are discussed in detail by Walter (1983), who expressed uncertainty about their biogenic nature. Modern stromatolites are known to be produced by combinations of bacteria, blue-green algae, and other algae in combination with sediment; hence, they are definitely biogenic structures. Nevertheless, Schopf and Walter (1983) consider most early stromatolites to be abiogenic and the product of exclusively sedimentary–crystalline origin. Walter (1983) listed 11 Archean stromatolite occurrences (see Table II) of which the Bulawayo Group in Zimbabwe is probably the best known. These stromatolites have a minimum age of 2.64 Byr., based on a crosscutting pegmatite dike which has been accurately dated, and if these stromatolites are biogenic, they offer evidence that photosynthetic life was present by that date.

The additional forms given the distinction of being true Archean microfossils, or at least "dubiofossils," by Schopf and Walter (1983), occur in the Fortescue and Hamersley Groups in Western Australia, and pyrites in the Witwatersrand complex showing petrographic textures resembling alga-formed boghead coal structures. The Fortescue Group fossils seem the most plausibly biogenic of these latter fossils, comprising filamentous structures in cherts from stromatolitic carbonates in the Tumbiana Formation, dated to approximately 2.77 Byr. The Hamersley Group fossils are like the Onverwacht spheroids but date to 2.5 Byr. and are considered questionable by Schopf and Walter; similarly, they classify the Witwatersrand textural evidence as dubious, and the dating is quite imprecise.

Fox and Dose (1977) reported the presence of microstructures and significant amounts of the amino acids characteristic of proteins (in quantities of 14 μg of amino acid per g of sediment) in sediments of the Witwatersrand System of South Africa. Fox and Dose (1977) also described hydrocarbons in the Soudan Iron Formation in upper Michigan and Minnesota (dated approximately 2.5 Byr.); of greatest significance is the presence of C_{21} isoprenoids

(branched hydrocarbons) which may have come from lycopene, a 40-carbon molecule that is the precursor of vitamin A. In addition, a peculiar branched C_{18} hydrocarbon found in the filamentous blue-green alga Nostoc was also found in the Soudan rocks (Calvin, 1969). In contrast to the molecular fossils, the microfossils in the Soudan rocks are not particularly impressive.

J.3. Fossils of the Early and Middle Proterozoic: 2.5–1.4 Byr.

The fossil record suggests that the beginning of the Proterozoic Era did not bring on a rapid change in the overall rate of organic evolution. It seems that only prokaryotes were present in the early Proterozoic, and the most notable biogenic structures found in the interval from 2.5 to approximately 2.0 Byr. are stromatolites (see Table II).

At approximately 2.0 Byr., an apparent change in the complexity of living forms is evident from fossils found in fossiliferous cherts near the base of the Gunflint Iron Formation, which outcrops along the Lake Superior shoreline in Ontario. Gunflint chert fossils may be entirely prokaryotic but they show a diversity of morphologies which include advances far beyond the rods and spheres of earlier life.

Classical study of the Gunflint biota was done by Barghoorn and Tyler (1965), who described taxa including tubular-branching, septate (chambered), colonial, stellate (star-shaped), and hydralike structures in addition to more prosaic rods and spheres. In addition, the Gunflint biota includes very-well-preserved stromatolites and strings of blue-green algae. Barghoorn (1971) also argued that the $^{12}C/^{13}C$ ratios in Gunflint organic matter show that much of the carbon was fixed by photosynthesis; in addition, the compounds pristane and phytane, both of which may be geochemical alterations of chlorophyll, are present in argillaceous rocks from the Gunflint Formation.

Two Gunflint organisms have been compared to obscure living counterparts; Barghoorn's "hydroid like" Kakabekia has been shown to be very similar to an organism found in soils from an old castle in Wales (Siegel, 1977). What is even more remarkable is that the living species of Kakabekia has also been found recently in Arctic polar localities and on the flanks of Alaskan glaciers. It may therefore be a cold-loving or cryophilic organism. To date, the taxonomic position of the modern Kakabekia is uncertain; it does not seem to resemble any other known living organism. The organism Eoastrion, a filamentous radial-structured form from the Gunflint cherts, has been suggested as a relative of a modern bacterium that has a metabolism based on the oxidation of iron and manganese (Barghoorn, 1971). This is an extremely interesting claim since 2.0 Byr. also marks the beginning of the widespread occurrence of red beds (colored by oxidized iron) including the Gunflint Formation itself.

At about the same time as the Gunflint cherts were forming, banded iron formations were being deposited in a number of areas. Moorhouse and Beales (1962) described fossils in the Animikie rocks of the Canadian banded iron formation, and they have suggested that the fossils were sponge spicules. This contention, if correct, would make these the oldest animal (in fact, the oldest eukaryotic) fossils known. Rutten (1971), however, rather strongly disputes the correctness of the identification of these putative animal remains.

Tappan (1976) suggested that two spherical forms from the Gunflint, *Huroniospora* (Fig. 3-25) and *Eosphaera,* because of their apparent reproductive modes and morphologies as evident in the fossils, are members of the red algae (or Rhodophyta). If this were correct, then these would also be the oldest known eukaryotic cells (except as noted about disputed forms in the banded iron formation of Canada).

A banded iron formation in Western Australia, the Frere Formation, has yielded microfossils which are somewhere between 1.7 and 2.0 Byr. old. These fossils occur in stromatolitic silicic iron deposits and resemble members of the Gunflint biota including representatives of *Kakabekia, Huroniospora, Eoastrion,* and a number of filamentous algal forms (Walter *et al.,* 1976). The striking similarity of their fossils despite the wide geographic distance between the Gunflint and Frere strata, suggests that Gunflint-type organisms may have been the dominant life forms of their times.

Following Gunflint times (2.0–1.8 Byr.) to approximately 1.4 Byr., the known fossil record has been confined largely to communities of microfossils in stromatolites from three continents (see Table II). Some of these are poorly known or described and all appear to be composed of pro-

karyotic forms as judged by their apparent simplicity. Many of the organisms in the rocks listed are filamentous and probably are blue-green algae very similar to modern mat-forming types. Fossils in the Belcher Group generally resemble those in Gunflint collections, including colonial bacteria, tubular algal forms, and a set of large (20 μm) double-walled spheroids with uncertain affinities (Hofmann and Jackson, 1969).

Eukaryotes are organisms which have membrane-bound nuclei, mitochondria, and other discrete organelles, and which undergo reproduction via mitosis. The age of the first eukaryotic cells is of considerable interest to Precambrian paleontologists; therefore, in these older Proterozoic fossils, many are tempted to seek out likely candidates for the distinction of "oldest eukaryote." The double-walled spheroids from the Belcher Group, tetrahedral groups of four small cells from the Amelia Dolomite (thought to be produced by mitotic division), and specimens from the Bungle Bungle Group which are well preserved and appear to have membrane-bound organelles, among others, have all been cited as possible early eukaryotes. It will be shown in the following discussion that the first widely recognized eukaryotic cells have been identified based on almost incontrovertible evidence.

J.4.   The Late Precambrian Fossils: 1.4–0.6 Byr.

The oldest late Precambrian fauna, which has received detailed description by Licari (1978), comes from the Beck Springs Dolomite of eastern California. It has been dated to 1.2–1.4 Byr. and lies approximately 1800 m stratigraphically below the Stirling Quartzite (which straddles the Precambrian/Cambrian boundary; see Chapter 4). The Beck Springs contains well-preserved stromatolites which have yielded very good fossil filamentous blue-green algae

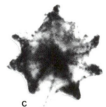

Fig. 3-25. Fossils from the Gunflint Cherts. (A) *Eoastrion,* about 15 μm wide; (B) *Huroniospora,* diameter about 20 μm; (C) *Kakabekia,* maximum dimension about 15 μm. (All from Barghoorn and Tyler, 1965, reprinted from *Science,* © the American Association for the Advancement of Science.)

5cm ~20μm

Fig. 3-26. Fossils and structures from the Beck Springs Dolomite. (Left) Polished section of an isolated stromatolite showing distinct laminations. (Right) Photomicrograph of filamentous algae, *Beckspringia,* from Beck Springs thin sections. (From Licari, 1978.)

(of the genus *Beckspringia*) from the laminae (see Fig. 3-26).

But chief interest in the Beck Springs fossils concerns the presence of several types of eukaryotic cells, belonging to two groups of plants: the Chrysophyta (yellow algae and diatoms) and the Chlorophyta (green algae). As evidence of the eukaryotic identity of the Beck Springs specimens, one can cite the following observations: at least some of the specimens of green algae have been observed undergoing a type of reproduction similar to that known in some groups of living green algae, and a majority of the eukaryotic Beck Springs forms show the presence of intracellular organelles (Licari found such in 134 out of 165 cells of green algae examined). In addition, a very large single-celled form is found in the Beck Springs (ranging in size to 62 μm); this organism characteristically contains intracellular organelles and the large size is far above that of most living prokaryotes (Schopf and Oehler, 1976). The Beck Springs flora is approximately coeval with the Belt Supergroup in Montana which contains shale-facies microbiota from the Newland Limestone, described by Horodyski and Bloeser (1978). These microfossils are rather similar to elements of the Beck Springs flora and include filamentous *Nostoc*-type blue-green algae and some large cells which may be eukaryotic green algae. It is interesting to note that the Belt fossils described are not from stromatolitic facies but resemble those from stromatolitic laminae (there are, however, many stromatolites known from the Belt Supergroup). One could conclude that the Beck Springs-type organisms were the dominant forms of their time. Fox and Dose (1977) describe Belt fossils (which they date to 1.1 Byr.) that re-

semble encysted protozoans and foraminiferans; however, it is very likely that the large blue-green algae described by Horodyski and Bloeser (1978) are the same forms described as protozoans by Fox and Dose.

It should be apparent that by the range of dates being addressed in this subsection, the Earth's atmosphere must have been oxygenated. It is impossible to determine the amount of oxygen present in the atmosphere at, say, 1.0 Byr. but the abundance and diversity of algal fossils which were undoubtedly photosynthetic compel one to believe that oxygen was rapidly accumulating in the atmosphere of the late Proterozoic.

The Skillogalee Dolomite of South Australia has been imprecisely dated (740–1340 Myr.; Schopf, 1975) but contains algae similar to some forms found in the Beck Springs Dolomite. The evidence of their eukaryotic nature consists of well-defined cross-walls in some of the filamentous forms. These cross-walls show that the structures are not the sheaths of blue-greens but rather were the cell walls of siphonaceous green or golden-green algae (Schopf and Oehler, 1976).

The youngest major pre-metazoan fossil find has been dated to span the interval between 790 and 1370 Myr., although a consensus of estimates for age suggests 900–1000 Myr. It comes from the Bitter Springs cherts in the Amadeus Basin of central Australia (Schopf, 1968; Barghoorn, 1971). The fossils represent five major types of organisms: filamentous and tubular blue-green algae of the *Nostoc* type, bacteria, green algae, two species of fungi (their first appearance in the fossil record), and two problematic forms. The most striking individuals of the flora are specimens of green algae preserved in the act of mitotic division. In these specimens, both daughter cells with their respective nuclei are visible. These nuclei undergoing mitosis provide indisputable proof that the organisms are eukaryotic. Another characteristic eukaryotic morphology found in the Bitter Springs algae is that of tetrahedral colonies of cells which also provide evidence of being products of mitotic division.

Two less-well-known fossiliferous deposits from Australia with approximately similar ages (740–1340 Myr.; probably close to 900 Myr.) are the Auburn Dolomite and Myrtle Springs Formation (Schopf, 1975). The fossiliferous lithologies in both are cherts from laminated carbonates. The fossils are similar to Bitter Springs specimens, with sheathlike filaments and large single cells (probably green algae) predominating. Within 100 Myr. after the probable average date for the Bitter Springs, Myrtle Springs, and Auburn fossils, the first metazoan fossils (also the first animals) appear in many areas of the world in Ediacarian rocks (see Chapter 4), and the apparent level of complexity for life on Earth took a broad jump during this interval.

CHAPTER 4

The Sauk Sequence: Ediacarian–Lower Ordovician

A. Conditions at the Beginning of the Phanerozoic

A.1. Overview

This chapter initiates the discussion of Phanerozoic history and, simultaneously, sets the format for the remainder of the book. The chapter divisions will be based on the cratonic sequences of L. L. Sloss, which reflect natural stratigraphic breaks in response to waxing and waning of the epeiric seas which have characterized much of North America's Paleozoic and Mesozoic history.

The Sauk Sequence was deposited during what were, for the Phanerozoic, somewhat unusual conditions; i.e., there was no significant tectonic activity taking place on either the eastern or western margin of the continent. In contrast, one or more continental margins were usually tectonically active throughout the remainder of geological history. The Sauk Sequence began in the latest Precambrian as the continental margins were gradually submerged and the sea transgressed onto both eastern and western platforms.

Basal Sauk units almost everywhere are mature sandstones, probably derived from reworkings of the Precambrian rocks of the Canadian Shield and the cratonic platform. Because of the long-term gradual submergence of the craton, basal Sauk sands range in age from late Precambrian to Late Cambrian.

By Middle Cambrian time, most areas of the craton had received some Sauk marine sediments, except for a sizable midcontinent region extending from Hudson Bay down through the upper Mississippi River Valley and reaching as far south as Kansas. By the Late Cambrian, marine units were being deposited in the upper Mississippi Valley

(notably Wisconsin and Minnesota), while elsewhere on the continent a characteristic three-part lithofacies pattern was firmly established. In the inner regions of the craton, detrital sands were deposited from sources in the remaining exposed Canadian Shield and from reworked older detrital sediments. Seaward (i.e., toward the continental margins) from these detrital lithofacies, a ring of carbonate shoals developed on shallow platforms under the epicontinental sea. Still farther out on the margins, a third zone of largely mixed carbonate/detrital sediments formed in deeper areas of the platform and on the continental slope. Sources here were largely from the craton (by sedimentary bypassing) but also include *in situ* carbonate formation and debris from slumps off the seaward edges of the carbonate shoals. In addition, extracratonic detrital sources may have been present in both eastern and western seas, possibly from magmatic arcs and/or microcontinents.

On the eastern margin, a rift had developed between North America and Ancestral Europe during the later Proterozoic (recall that they had sutured in earlier Proterozoic time; see Chapter 3). Eastern Cambrian sequences, now preserved largely in the Appalachian region, record development of a rifted margin prism.

In the latest Cambrian and Early Ordovician, most sediments deposited across the flooded craton were carbonates, commonly dolomitic. It seems that the supply of detritus from Precambrian uplands simply ran out or was covered by carbonates. Cambro-Ordovician units on the deeper platforms and continental shelves in the west were still mixed detrital sediments and carbonates. In the late Early Ordovician, the Sauk sea retreated. Most Sauk regressive sequences were stripped away by subsequent erosion but some regressive facies remain and they tend to be sandy or dolomitic, depending on the depth of erosion (i.e., whether

99

basal sands and Precambrian crystalline rocks were exposed as detrital sources).

During Sauk time, the first metazoans appeared (in Precambrian strata in Newfoundland, and in several latest Precambrian units on other continents); and at the base of the Cambrian, by definition, the first skeletonized organisms appear. By far the most important early organisms were trilobites, but inarticulate brachiopods, archaeocyathids, gastropods, several types of echinoderms, and a few rare additional forms also appear in the Lower Cambrian. Trilobites apparently occupied a variety of ecological niches, ranging from planktonic–pelagic to deepwater benthic infaunal. Archaeocyathids are the oldest known colonial reef-building animals, but they only survived until the early Middle Cambrian. Most carbonate buildups in Cambrian time were algal. One particular Middle Cambrian fossil fauna, that of the Burgess Shale, preserves a uniquely good record of soft-bodied organisms. By the Late Cambrian, nearly every major animal group was present except for coral, bryozoans, ammonoids, and most bivalves. At the Cambro-Ordovician boundary, trilobites lost their position of dominance and several major taxa became extinct; in their places, others appeared. Where fossils are absent, it is nearly impossible to separate uppermost Cambrian from Ordovician rocks.

A.2. Global Paleogeography

While this text is primarily oriented toward North American geology, we will digress briefly at the beginning of each Phanerozoic chapter to speculate on global paleogeography and continental reconstructions for each period of geological time. In particular, the global position of North America and environmental factors affecting it will be described.

Data used to approximate ancient continental positions are taken from apparent-polar-wander paleomagnetism studies, and from the distribution of climate-controlled lithologies, such as reef-related sediments, evaporites, paleosols, and tillites. Both types of data are "soft" (i.e., based on information drawn from rocks that may have been altered chemically or physically subsequent to their formation, using analytical techniques susceptible to considerable lack of precision), and therefore are subject to misinterpretation. Of greater significance is the fact that both sets of data can provide only information on paleolatitudes but not on paleolongitudes. Thus, global reconstructions presented here are hypothetical at best and may be expected to change as more and better data are acquired and increasingly more sophisticated analyses and interpretations are available.

Figure 4-1 is a paleogeographic map of the Earth during Late Cambrian time. North America was located at the Equator and was rotated about 90° so that the equator lay across the continent running from the present Arctic margin to the Gulf of Mexico region. The continent extended to approximately 30°N and 30°S latitude so that equatorial-to-subtropical climatic conditions prevailed over the whole continent. The eastern, or Appalachian continental margin (throughout the text, all compass directions are based on the present positions of the continents, unless otherwise stated) was in the Southern Hemisphere, in about the same global position as the northern and western margins of present Australia. The Appalachian margin of North America developed by continental rifting which sundered the Baltica block (Ancestral Europe) during the late Precambrian and created the Iapetus Ocean. The margin's general shape was probably due to the position of rift zones and major transform offsets (see Fig 4-2). Baltica's apparent movement relative to North America would seem to support the hypothesis that it rifted from North America during the late Precambrian. Gondwanaland also occupied an essentially equatorial position during the Late Cambrian, but its much greater size was such that it extended from around 50°N to 60°S latitude. North and south equatorial currents probably flowed westward as today so that surface currents flowing northwesterly (according to the Cambrian compass) along the Appalachian margin probably continued westerly, striking the Gondwanaland coast and being deflected southward.

A.3. Paleoclimatology

If the preceding discussion of global paleogeography is approximately correct, then we can make a few assumptions about North American paleoclimates during deposition of the Sauk Sequence. The climate must have been tropical to subtropical due to the equatorial position of the continent. Further, climate zones ran north–south (in terms of our present poles) and, therefore, areas such as the present Arctic North and the American Southwest showed the same tropical climate; the Appalachian area probably had a subtropical climate.

Tropical to subtropical latitudes would imply little seasonal variations in temperature, since the farther a point on Earth is away from the Equator, the more extreme is the climatic effect of the Earth's inclination. Therefore, an equatorial North America would be essentially seasonless.

Hamblin (1961) reported southerly to southwesterly paleocurrents in Upper Cambrian sediments from the upper Mississippi Valley, which data suggest that a southerly/southwesterly wind belt prevailed across the area. Examination of Fig. 4-1 shows that the upper Mississippi Valley lay in the zone of the Southern Trade Winds during Late Cambrian time, and that the prevailing wind direction by today's orientation should indeed have been south/southwesterly.

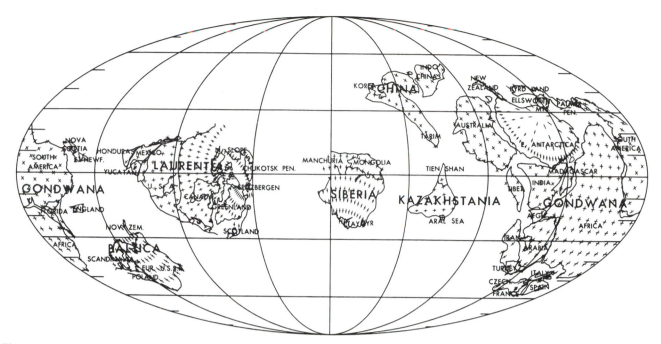

Fig. 4–1. Reconstruction of Late Cambrian paleogeography. Note that Laurentia (Ancestral North America and appendages now part of Europe, Asia, and Spitsbergen) is rotated approximately 90° clockwise with respect to the present. The existence of a Cambrian ''Kazakhstania'' is question- able and China may have been part of Gondwana, i.e., all landmasses shown east of Siberia may have been parts of Gondwana. (From Scotese *et al.*, 1979.)

B. Epeiric Seas and Cratonic Sequences

B.1. The Concept of Epeirogeny

We stated in Chapter 1 that the principle of actualism would be a guiding philosophy of this book (and, indeed, it is such for most geological work); the reader will recall that actualism is the interpretation of past events based on analogy with present processes. However, the following discussion will deal with past events which commonly have no direct modern analogues.

Epeiric seas were a dominant geological feature of the continent's history prior to the Holocene and they were especially widespread and long-lived during the Paleozoic Era. Characteristically, these seas were very shallow (< 100 m) and covered vast continental land areas through much of the Paleozoic. Transgressions and regressions of epeiric seas were caused by the interplay of two factors: eustatic sea-level changes, and *epeirogenic* (i.e., iso-static/tectonic) movements of continental crust. It is generally difficult to sort out the various contributors to movements of a past epeiric sea except where nearly global transgression occurred (in which case eustatic sea-level rise seems the likely cause). Because the present continents are all relatively high-standing, there are no true epeiric seas, and therefore no first-hand data on the behavior of such large shallow continental seas. Small areas are scattered

around the world, where some of the processes which took place in the great epeiric seas of the Paleozoic and Mesozoic Eras may be observed today on a smaller scale. In Shark Bay, Western Australia, the Bahama Banks, the Yellow Sea, and the Persian Gulf, one can study biological and sedimentological processes under conditions which proba-bly resemble those of the much larger epeiric seas of the past.

Conditions in epeiric seas were significantly different from those in true oceans and seas (i.e., marine areas under-lain by simatic crust and with water depths greater than several hundred meters). For example, relatively few areas on Earth at the present time are accumulating significant amounts of carbonate sediment, except in tropical, near-shore-shelf regions without nearby sources of detrital sedi-ments. This contrasts sharply with conditions inferred for epeiric seas of the past, where thousands to tens of thou-sands of meters of carbonate sediments accumulated during the lower Paleozoic across large parts of North America. Similarly, the marine fossil record shows that vast areas of the epeiric seas were inhabited by benthic organisms, whose relatives today live only in the nearshore, continental shelf environment.

At some time or another during Phanerozoic time, every part of the conterminous United States (i.e., the ''lower 48 states'') was flooded by an epeiric sea. Many areas have been underwater more than above it and, there-

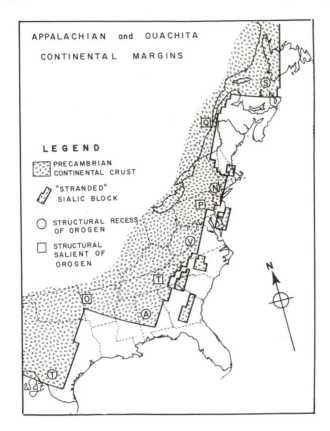

Fig. 4-2. Relation of salients and recesses of the Appalachian–Ouachita orogen to early Paleozoic configuration of the eastern margin of North America, as developed by late Precambrian rifting; see text for discussion. (From Thomas, 1977, *American Journal of Science,* Vol. 277.) Salients: O, Ouachita; P, Pennsylvania; Q, Quebec; T, Tennessee. Recesses: A, Alabama; N, New York; S, St. Lawrence; T, Texas; V, Virginia.

fore, it would appear that we are living with a somewhat unusual condition: a high, dry North America. Since epeiric seas dominated the history of North America for much of the Phanerozoic, our study of North American history from this point on will be dominated by examination of epeiric-sea sedimentation and benthic marine organisms.

B.1.a. Tectonic Interpretation of Epeirogeny

Tectonic activity in mobile belts (i.e., plate boundaries) has been recognized for many years; however, the "stable craton" has often been considered to be beyond such motions. Nonetheless, there is little debate over the fact that epeirogenic activity has been a repeating phenomenon at least through the Phanerozoic, and that epeirogenic movements involved sizable land areas moving several hundred meters vertically. The obvious question is, what is the cause of these epeirogenic movements?

An innovative approach to the problem of cratonic tectonics was proposed by Sloss and Speed (1974) who characterized the history of a typical craton by three types of

behavior: *emergence,* in which cratons are elevated above sea level; *submergence,* in which cratons are submerged below sea level; and *oscillatory,* in which cratons oscillate rapidly up and down near sea level. Further, Sloss and Speed demonstrated that in the submergent mode and during some phases of the oscillatory mode, there is a general increase in crustal instability and tectonic activity at cratonic margins. North America is currently in an oscillatory episode (Fig. 4-3) but for much of the Paleozoic, North America was submergent with intervening shorter emergent episodes.

Sloss and Speed proposed that the cause of this cratonic behavior is the presence of a molten asthenosphere beneath the continents. They assume that continents act as insulation to keep the Earth's heat from flowing easily away into space, as it does in oceanic areas. Thus, heat would build up slowly beneath continental areas, causing the melting of rocks deep below the surface and the slow accumulation of magma. Under "usual" conditions, the continent would ride low on the surface, resulting in submergent mode behavior; but buildup of magma and subsequent inflation of the asthenosphere beneath a continent, buoys up the continent, resulting in emergence. The slow buildup of asthenospheric magma ceases when its fluid pressure becomes great enough to cause the magma to be squeezed out from under the continent, causing the continent to subside (and also possibly increasing the rate of seafloor spreading by supplying increased quantities of magma to spreading centers). The oscillatory mode probably represents times of episodic inflation and deflation of the asthenosphere.

B.1.b. Dynamics and Characteristics of Epeiric Seas

One can infer some of the dynamics and characteristics of epeiric seas from their sedimentary and fossil records,

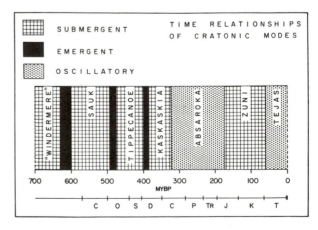

Fig. 4-3. Modes of cratonic tectonics related to geological time and cratonic sequences. (From Sloss and Speed, 1974.)

and one can reasonably guess at other characteristics by analogy with modern shallow-water shelf areas such as the Bahama Banks.

Since epeiric seas are usually described as being shallow, we will first consider their depth. In general, one finds epeiric-sea fossil collections populated with species whose closest modern equivalents (especially coral and bivalves) live in less than 100 m of water. One cannot consider these fossils as incontestable evidence of ancient epeiric-sea conditions, because there are no species known whose stratigraphic ranges extend from even the late Paleozoic to the Holocene. There are, however, long-ranging genera and families of organisms which appear to have similar Paleozoic and modern forms; it is their existence which allows the assumption that the shallow-water habitat of the modern forms resembles the habitat of their older cousins. Additional depth indications are fossil algae, which are photosynthetic and must live within the *photic zone* of the sea (i.e., the levels to which sunlight penetrates). This depth presently ranges down to approximately 100 m, and likely has not changed appreciably during the Phanerozoic. Since most fossil algae are found in what were probably the shallowest zones of epeiric seas, they do not really help determine the maximum depth of such seas.

The best information about the depth of epeiric seas comes from the sedimentary record. The patterns of carbonate deposition that one sees in many Paleozoic and Mesozoic formations are similar to those observed in recent carbonate-forming environments, such as the Bahama Banks, at depths generally less than 30 m. The presence of evaporites in older sediments has also been considered evidence for shallow water; however, some sedimentologists would argue that there are serious errors involved in modeling ancient evaporite basinal conditions from modern analogues (see Chapter 5). Overall, the consensus of opinions would argue for depths of near 50 m as average for epeiric seas, although deep (> 1000 m) intracratonic basins are known to have been present at times, resulting from normal faulting. Depths we are considering "average" are those of the broad areas of submerged sialic crust. Carbonate and evaporite deposits typically form at the shallow end of the range, whereas detrital sedimentation could occur to any depth.

Several conclusions can be reached about the dynamics of waves and currents in such shallow seas. It is not typical for large waves to propagate in shallow water, since waves begin to meet resistance by friction with the bottom when their wave length equals twice the depth. For example, in water of 50-m depth, the largest wave which could be propagated would have a wave length of 100 m. Examination of a broad shallow marine area, such as the northern Gulf of Mexico, will show that shallow-water environments usually contain small waves.

Large-scale movements of water in epeiric seas were probably due mainly to surface currents, driven by prevailing winds. In addition, winds probably caused episodic fluctuations of sea level due to *wind tides* (in which water is driven against a shoreline by strong winds and "piles up" there, causing water level to rise) or due to *storm surges* (which are slight elevations of the sea's surface beneath the low-pressure atmospheric center of a large tropical storm).

The temperatures of epeiric seas can also be inferred from the precipitation of carbonate sediments. In the modern ocean, carbonates generally form in subtropical to tropical waters at temperatures of 15 to 30°C; therefore, this range would be a reasonable estimate for the temperatures of past epeiric seas characterized by extensive carbonate deposition. In addition to inference based on their presence and quantity, several isotopic paleotemperature-measuring techniques based on carbonates have been discovered. These techniques include: the comparison of $^{18}O/^{16}O$ and $^{13}C/^{12}C$ isotopic ratios in carbonates (Lowenstam, 1961), and calculation of calcite/aragonite ratios in shells (Dodd, 1964). Probably the best paleotemperature indicator for ancient epeiric seas is the determination of paleogeographic position for such seas, as was discussed in the section on paleoclimatology.

Finally, to conclude the list of epeiric sea conditions, one must consider salinity. Average ocean water in the modern ocean ranges from 35 to 36 parts per thousand (35–36°/oo) dissolved material. Since evaporation tends to exceed precipitation in the modern ocean, it is presumed that epeiric seas with their large surface areas tended to be hypersaline except at the mouths of rivers (Friedman and Sanders, 1978). The extensive salt deposits mined today in the Silurian Salina Basin, the Devonian Williston Basin, and many other extensive evaporite beds worldwide, indicate that hypersaline conditions in epeiric seas were commonplace.

B.2. Cratonic Sequences

We introduced the concept of cratonic sequences previously so that the reader would understand both the chapter title and the reason for the use of the somewhat strange term "Sauk Sequence." We digress briefly to examine fully the units we will be using to divide up the remainder of the book.

Sloss (1963) proposed a new informal unit of stratigraphy, the sequence, which may be defined as a large-scale rock unit, consisting of genetically associated formations bounded by cratonwide unconformities. The sequence is an inherently natural unit because it is related to the processes of epeirogeny, eustasy, deposition, and erosion. All of the sedimentary rocks which form as a result of a single flooding of the craton by an epeiric sea constitute a single sequence. Sloss, in an earlier paper (Sloss *et al.,* 1949), observed four such sequences in the Paleozoic. In his 1963 paper, he recognized two more sequences which span the

Mesozoic and Tertiary intervals. His list of sequences is as follows, listed in proper order (i.e., youngest on top):

Tejas
Zuni
Absaroka
Kaskaskia
Tippecanoe
Sauk

The sequence names generally are derived from parts or features of the continent which contain characteristic exposures of each sequence's strata. Thus, for example, the Sauk Sequence is named for Sauk County, Wisconsin, the Absaroka for the Absaroka Mountains in Montana and Wyoming, Tejas is an old name for Texas, and so forth. Sloss (1984) presents a retrospective on the development of the sequence concept.

Figure 1-1 illustrates the approximate duration and extent of flooding that these sequences delimit. A point of strict terminology should be made here; sequences are time-free units in the sense that they are not bounded in theory or in practice by time planes. It is fairly easy to explain this principle if we examine the history of the basal facies of the Sauk Sequence. Marine sedimentation was under way in the eastern continental margin by the latest Precambrian; the Ocoee and Chilhowee Groups, among others, are preserved as part of the basal Sauk Sequence in the Appalachian Province. On the western margin, similar latest Proterozoic detrital sequences mark the basal transgressive facies of the Sauk sea. In Montana, farther inboard on the Cambrian craton, the lowest Sauk sediments comprise the Flathead Sandstone (Middle Cambrian), which itself becomes younger as one follows the outcrop cratonward into Wyoming. And, in the upper Mississippi Valley, high on the craton, the basal Sauk detrital sequences are of Late Cambrian age, such as the Dresbach Formation in Wisconsin. All of these sedimentary units are genetically related, comprising the basal Sauk detrital phase; however, they range in age diachronously from late Proterozoic to Late Cambrian. One should note that although the local or regional boundaries of a given cratonic sequence are variable, the total range of age for a given cratonic sequence can be readily defined, and these are the time spans as shown in Fig. 1-1.

Wheeler (1963) modified Sloss's sequence nomenclature because he recognized at least two more sizable Paleozoic regressions, one each within the Kaskaskia and Tippecanoe sequences. Wheeler's sequences are shown in Fig. 4-4. Nevertheless, we will use Sloss's terminology for our discussion because it is well known, and it has publication precedence. Wheeler's terminology does reflect two significant Paleozoic marine regressions; however, the mid-Tippecanoe regression, at least, appears to be of significantly smaller magnitude than the post-Sauk, post-Tip-

pecanoe, and post-Kaskaskia regressions. We believe that we can discuss the additional regressions Wheeler notes within the text and can thus avoid adding extra divisions to Sloss's list of sequences.

We shall designate herein Sloss's concept of the "sequence" by the term "cratonic sequence," simply because the word "sequence" is too useful as a general description for sedimentary relationships to confine it only to these informal units. For example, we will speak of "regressive sequences" and "lithological sequences" to mean successions of rocks. When we wish to designate one of Sloss's units, we will term it a "cratonic sequence" or use its proper name, such as the Sauk Sequence.

Although the nomenclature for cratonic sequences seems to be well established, the same cannot be said for the bounding unconformities. For example, the post-Sauk hiatus has been variously called the "Post-Knox unconformity," the "Pre-Bighorn erosion surface" (both from Lochman-Balk, 1971), and the "Owl Creek discontinuity" in the literature. Names have been given to the discontinuities, between most (not all) cratonic sequences, but not all such names are in common use. We will simply refer to all such erosion surfaces and intervals by reference to the preceding cratonic sequence name: thus, we will refer to the post-Sauk discontinuity, post-Tippecanoe discontinuity, and so on.

A final point about cratonic sequences: the Zuni and Tejas Sequences seem to be bounded by less widespread and pronounced regressions than are the Paleozoic cratonic sequences. In part, this may be due to the craton tectonics of Sloss and Speed (1974) discussed earlier; an examination of Fig. 4-3 will show that the Sauk, Tippecanoe, and Kaskaskia Sequences all may have resulted from the North American craton undergoing alternating submergent and emergent episodes of tectonic activity. The regular submergent–emergent–submergent pattern was broken by the irregular end-Kaskaskia regression, which initiated the Absaroka Sequence. Similarly, the end of the Zuni Sequence initiated the Tejas transgression which was ragged and which initiated another cratonic sequence which Sloss and Speed term "oscillatory."

C. A Digression, the Ediacarian: Do You Believe Rocks or Fossils?

The interval of Earth history from approximately 650 to 570 Myr. incorporates several interesting stratigraphic and paleontological problems. North America features substantial stratigraphic units deposited during this interval, located at the continental margins where the first Sauk seawaters transgressed onto the craton. In general, the late Precambrian basal Sauk units are unfossiliferous detrital

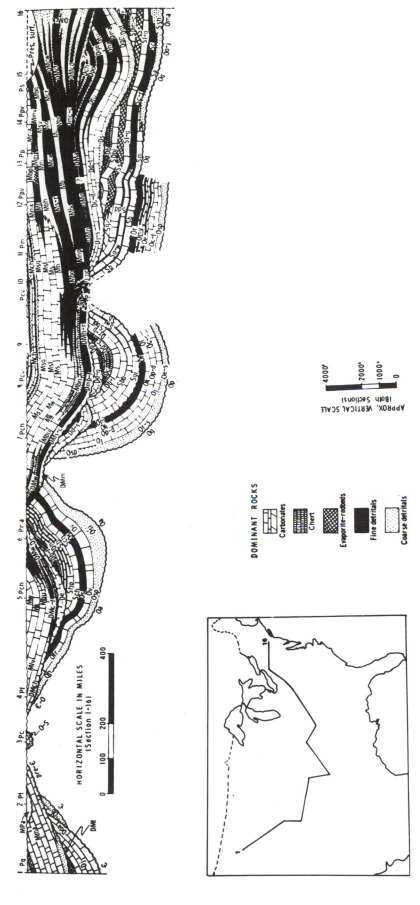

Fig. 4-4. Generalized cross sections of early and middle Paleozoic strata, showing cratonwide unconformities delimiting cratonic sequences. Inset shows the approximate line of section. (From Wheeler, 1963.)

formations which often grade upward conformably into fossil-bearing Cambrian strata. The basic problem is the placement of the Precambrian/Cambrian boundary. A further problem is understanding the reasons why organisms should leave abundant fossils only in upper portions of what appear to be uniform lithologies. Whereas the upper boundaries of these latest Precambrian units are commonly conformable, the lower contacts (with older Proterozoic or Archean rocks) are typically unconformable; thus, it underscores the distinctiveness of latest Precambrian conditions and has suggested to many authors that latest Precambrian time is more akin to the Paleozoic than to the Proterozoic.

A few latest Precambrian strata have provided workable fossils and these have helped clarify relationships among rocks. In North America, only latest Precambrian strata in Newfoundland contain good fossils, but contemporary strata from South Australia, England, Baltica, and South Africa have yielded excellent remains of soft-bodied organisms which are clearly several grades in organization higher than algae and bacteria. These animals are the oldest metazoan-grade forms (see Section C.3.a) and give a distinctively "advanced" sense to the times in which they lived. The presence of such organisms has led some authors to firmly place latest Precambrian time into the Paleozoic (e.g., Dott and Batten, 1981; Stearn et al., 1979). The validity of such placement will be discussed below. Typical fossils of this late Precambrian assemblage were originally described from the Pound Quartzite in the Ediacara Hills, South Australia, and the assemblage is thus termed the *Ediacaran* (see Glaessner, 1961).

C.1. Nomenclature for Latest Precambrian Time

Various geological time terms have been applied to the rocks and fossils in question. "Eocambrian" (meaning "dawn of the Cambrian") is among the first and clearest of such, but its use has often been disfavored because the original units so named turned out to incorporate strata in Scandinavia ranging from middle Proterozoic through Cambrian in age (see Broegger, 1900, in Salop, 1977). A similar fate befell the term "Vendian," as applied in the USSR for latest Precambrian-age rocks; "Infracambrian" was used in Africa but rarely applied elsewhere. Finally, in deference to the Australian fossils mentioned, many authors have used "Ediacarian" as a time unit for the latest Precambrian; see Cloud and Glaessner (1982) for formal designation of the stratigraphic usage of the term. We will use the term "Ediacaran" to designate the characteristic early metazoan fauna, and the term "Ediacarian" as the latest Precambrian time unit.

Additional complexity follows placement of the Ediacarian time unit within the hierarchy of the geological time scale; some authors consider it a latest Precambrian interval whereas others consider it part of the Phanerozoic Eon. Here, we beg the controversy and separate Ediacarian time from either eon, recognizing that the Ediacarian represents a truly transitional interval in life history. The fact that this discussion is incorporated with the Sauk Sequence underscores the close relationship between Ediacarian events and those of subsequent Sauk time.

C.2. Occurrence and Nature of North American Ediacarian Strata

Most Ediacarian rocks in North America represent continental-margin deposits laid down on stable platforms in the initial transgression of the Sauk sea. Such deposits are dominated by coarse detritus from the craton and eroding Proterozoic orogens, but also may include volumes of finer detrital sediments and platform carbonates. Three major regions contain the bulk of such platform Ediacarian deposits: the Great Basin, Alaska and the Rocky Mountains, and the central and southcentral Appalachian Mountains. Contemporary units are also present in Newfoundland, but these appear to represent deposits on an unstable margin; they contain some volcanics and considerable amounts of fine-grained materials, and the units are very thick. One must remember too that sizable parts of Newfoundland were not part of North America during Sauk time.

C.2.a. Southwestern (Great Basin) Ediacarian Strata

If a "typical" North American Ediacarian section were designated, it would likely be in the California–Nevada border region. Several apparently conformable Ediacarian/Cambrian sequences are known there, exposed in various ranges of the Basin and Range Province. The section with best paleontological control on the boundary interval is in the White–Inyo Mountains of eastern California, where the Reed Dolomite straddles the boundary. Other regional boundary units are the Stirling Quartzite and Wood Canyon Formation in the Spring Mountains–Death Valley area, the Prospect Mountain Quartzite farther north and east in Nevada into Utah, and the Mutual Quartzite and Uinta Group, both in Utah. Generally, these units are poorly fossiliferous and placement of the Precambrian–Cambrian boundary is tenuous.

C.2.b. Northern Rocky Mountain Ediacarian Strata

Ediacarian units exposed in the Rocky Mountains tend to be thick, widely exposed relative to the intermittent exposures in the Great Basin, and composed of westward-thickening wedges of sandstones and some argillites. The lithologies grade upward without discernible lithological change into fossiliferous Lower Cambrian units (North, 1971). Because of this gradational contact and because the Precambrian strata are generally unfossiliferous, their age assignment is very tenuous.

The probable Ediacarian outcrops along the northern and central Canadian Rockies, extending from the Columbia River south to the international boundary, are upper units of the Windermere Supergroup. These rocks are largely quartz arenites and platform carbonates. Discussion of the Windermere and its correlatives in Idaho and Utah is found in Chapter 3.

C.2.c.   Central and Southern Appalachians

A band of Ediacarian strata extends sinuously down the Appalachian Mountains in the Blue Ridge Province, from central Pennsylvania to northern Alabama (Rodgers, 1956). Outcrops are discontinuous and not all units include conformable Ediacarian/Cambrian sequences. The strata are largely assigned to a variety of formations in the Chilhowee Group, which has its type area in eastern Tennessee. As with the western Ediacarian, assignment of the era boundary in conformable Chilhowee sequences is problematic.

Chilhowee sequences contain both coarse and fine detrital sediments, and locally may include carbonates. In the type region, the formational sequence is as follows: Cochran (basal conglomerate and sandstone)/Nichols (shale)/Nebo (sandstone)/Murray (shale)/Hesse (sandstone)/Helenmode (calcareous with mixed detrital sediments) Formations (Whisonant, 1974). Formation names are quite variable along the Appalachians and the reader who wishes a synthesis of the Chilhowee makeup is referred to Palmer (1971b). Overall, Chilhowee deposition reflects transgression and sedimentation in a subsiding trough bounding the eastern continental margin. Rates of subsidence were variable: Whisonant (1974) stated that rates were relatively rapid in Cochran–Nichols time and that they slowed subsequently.

Assignment of ages to the various Chilhowee units is often difficult; *Skolithos* tubes are fairly widespread in upper units and Palmer (1971b) stated that the oldest fossil in the eastern United States is the paleocopid ostracode taxon *Indiana*, found in the Murray Shale. Mount *et al.* (1983) assigned this occurrence to the Tommotian assemblage (see Section C.3). Palmer (1971b), while not citing *Indiana* as a Tommotian fossil, tentatively placed the Precambrian/Cambrian boundary within the Murray Shale. Other Chilhowee regions probably feature only Cambrian-age formations; e.g., in northern Georgia and Alabama only the

Cambrian Weisner Quartzite, which contains lower Cambrian archaeocyathids, represents the group.

Ediacarian fossils (see next subsection) have been reported from metamorphosed sediments in the Carolina Slate Belt in North Carolina by Gibson *et al.* (1984). These fossils are assigned to the cosmopolitan genus *Pteridinium*, which is a form of pennatulid octocoral (like *Charnia* and *Rangea* from Old World Ediacarian faunas). Gibson *et al.* also revised an earlier report by St. Jean (1973) of a purported Cambrian trilobite from the Slate Belt; they showed that St. Jean's "*Paradoxides*" was in fact another occurrence of the sea pen *Pteridinium*. Cambrian fossils are, however, also known from the Slate Belt in South Carolina, and these occurrences underscore the evidence for the Slate Belt as an exotic terrane (see Section D.6.a).

C.2.d.   Newfoundland and Other Areas

Detrital sediments of late Precambrian age are abundant in Newfoundland, largely assigned to the Hodgewater Group, except in the Avalon Peninsula (easternmost Newfoundland), where probable equivalent units are the Musgravetown Group. The Hodgewater typically consists of over 4000 m of fine-grained detrital sediments, suggesting an orogenic facies rather than the more typical Ediacarian passive margin facies. The Ediacarian strata in Newfoundland also contain excellent fossils of medusoid cnidarians, best known from the Cape Kovy Formation of the Conception Group.

Salop (1977) stated that the earliest Cambrian shelled fossils (not including trilobites) appear in the upper Hodgewater Group, thereby placing the era boundary down in the Hodgewater. Crimes and Anderson (1985) reported sequences of trace fossils from southeastern Newfoundland, spanning latest Precambrian through Early Cambrian ages. Based on these data, they suggested the boundary lies within the Chapel Island Formation, which is a local unit that appears to be slightly younger than the Hodgewater Group (although this is uncertain).

An additional Ediacarian unit was described by Salop in Ellesmere Island, in the Canadian Arctic Archipelago, comprising the Ellesmere Group. These are detrital sediments, of mixed origins, which lie below biostratigraphically dated Cambrian rocks, and include a few *Skolithos* tubes.

C.3.   Ediacarian Life and the Significance of Metazoans

All fossils mentioned in Chapters 2 and 3 derive from single-celled plants and putative single-celled animals. Ediacarian time marks the first appearance of *metazoans*.

C.3.a.   The Metazoan Body Plan

Metazoans feature diversified cells and generally have their cells organized into *tissues* (i.e., groups of cells which perform a particular function) and *organs* (specialized grouping of tissues). Metazoan-grade organisms are distinguished from colonial single-celled organisms, since the latter are composed of more-or-less organized arrangements of large numbers of the same type of cells. The cells in metazoans are arranged in two (in the simplest forms) or three embryonic layers which evolve into all the body structures. These layers are termed *ectoderm, mesoderm,* and *endoderm.* The simplest metazoans, e.g., sponges and jellyfish, lack organized mesodermal tissue, whereas in higher forms the mesoderm contributes the bulk of the cells that form internal organs.

The most primitive metazoans are the *Diploblastica,* a group including several phyla which develop from two embryonic layers and which may have an undifferentiated third layer (termed a *mesoglea*). Among these (see the Appendix for a taxonomy of the major fossil groups) are Cnidaria (with mesoglea) and Porifera (without mesoglea). Higher metazoans belong largely to two groups which contain three embryonic layers but which differ in having either: (1) a cavity between the endoderm and mesoderm (the *Pseudocoelomata*), (2) a cavity within the mesoderm (the *Coelomata*), or (3) no internal cavity (*Acoelomata*). Virtually all important fossil taxa share the coelomate condition, and they may be divided further into *Protostomia,* including Arthropoda, Annelida, Mollusca, the lophophorate phyla, and others; and *Deuterostomia,* which includes Chaetognatha, Echinodermata, and Chordata. Figure 4-5

illustrates the developmental differences between the two subdivisions of Coelomata.

C.3.b.   The Stratigraphic Record of First Metazoans

As we have stated, the best of early metazoan records comes from the Pound Quartzite (Fig. 4-6) in the Australian Ediacara Hills (see Glaessner, 1961, 1971). Other summaries (see Cloud, 1976; Stanley, 1976; Fedonkin, 1981) have recognized 19 species of Ediacaran cnidarians, including both medusoid and pennatulate (i.e., like a sea pen, a featherlike octocoral) forms, five annelid worm species, a possible primitive arthropod, and two novel organisms, one of which, *Tribrachidium,* appears to be an echinoderm with a threefold symmetry rather than the conventional fivefold symmetry of all post-Cambrian echinoderm groups. Ford (1979) and Cloud and Glaessner (1982) summarized information on the appearance and affinities of Ediacaran metazoans and other Ediacarian fossils.

The type Ediacaran fauna is not subject to radiometric dating because the fossiliferous unit is not adjoined by igneous bodies. However, its Precambrian age is evident from its stratigraphic position. The fossils lie approximately 50 m below the top of the Pound Quartzite; pre-trilobite Lower Cambrian fossils are found approximately 175 m above the base of the next unit, the Ajax Limestone, and trilobites appear farther upward (Cowie, 1967). Fortunately, elements of the Ediacaran assemblage occur in several other sites and some have been dated radiometrically.

In Charnian beds near Leicester, England, pennatulate Cnidaria of the genus *Charnia* occur (Fig. 4-7); these rocks have been dated, based on coeval igneous rocks, at over 680 Myr. (Evans *et al.*, 1968, in Cloud, 1976). Cloud listed other dated occurrences of Ediacaran organisms: in northern Siberia, dated to over 680 Myr. (in common with the Leicester fossils); and in South-West Africa, where the age is bracketed between 700 and 550 Myr. (see also Jenkins, 1985).

McMenamin (1982) and Stanley (1976) noted additional occurrences of nondated Ediacaran fossils from China and northwestern Canada. An assemblage of Ediacaran fossils in Newfoundland has not been dated, but Cloud (1976) cited a probable age of 620 Myr., since the assemblage contains both medusoid and pennatulate Cnidaria and tends to resemble in part both the Australian and English assemblages.

As previously noted, Gibson *et al.* (1984) described the first Ediacarian-age body fossils from the United States. These are from the Carolina slate belt Albemarle Group (see also Chapter 3) and consist of two specimens assigned to the pennatulid cnidarian genus *Pteridinium.* These occurrences

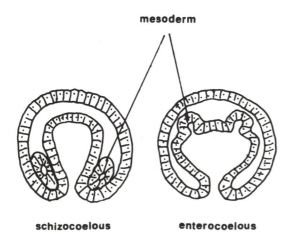

mesoderm

schizocoelous   **enterocoelous**

Fig. 4-5. Developmental differences between two coelomate blastulas. (Left) Schizocoelous blastula, characteristic of annelids, arthropods, mollusks, and lophophorate phyla (brachiopods, bryozoa, phoronids). (Right) Enterocoelous blastula, characteristic of echinoderms, chaetognaths, and chordates. (Modified and redrawn fom Hickman *et al.*, 1984.)

Fig. 4-6. Representative metazoan fossils from the Pound Quartzite, Ediacara Hills, South Australia. (A) *Spriggina*, a segmented vermiform animal, possibly an annelid; (B) *Dickinsonia*, another segmented worm; (C) *Mansonites*, a medusoid cnidarian. (Photographs courtesy of Dr. Mary J. Wade, Queensland Museum.)

Fig. 4-7. The cosmopolitan pennatulid cnidarian *Charnia*, from Ediacarian strata in Charnwood Forest, Leicester, England. (Photograph courtesy of Trevor D. Ford.)

are not subject to radiometric dating, but the fossils appear to be typical Ediacaran forms.

In addition to the cosmopolitan occurrences of soft-body fossils, many Ediacarian trace fossils have been reported. These consist largely of the burrows of wormlike organisms. Cloud *et al.* (1976) reported a particularly well-dated occurrence of J- and Y-shaped burrows in tuffs from the Carolina Slate Belt in the vicinity of Durham, North Carolina. These trace fossils date to 620 ± 20 Myr. Other non-Ediacaran contemporary and older Precambrian metazoan reports are generally inconclusive. Bassler (1941) reported an imprint assigned to the "jellyfish" genus *Brooksella*, from the Nankoweap Group in the Grand Canyon; other "jellyfish" have been reported from older Grand Canyon rocks (the Unkar Group). These reports probably describe sedimentary structures (Cloud, 1968), rather than organic remains. A variety of vermiform burrows have been reported from the Belt Series, but these too are doubtful organic remains. A particularly novel Precambrian fossil was reported from the Tindir Group in eastern Alaska by Allison (1975). The alleged fossil was interpreted to be a flatworm (phylum Platyhelminthes) less than a millimeter in

length. If Allison (1975) is correct, this occurrence of ancient Platyhelminthes is noteworthy, since the phylum is acoelomate and very near the roots of the metazoan phylogenetic tree. The specimen was assigned an age of over 850 Myr., making it the oldest North American metazoan. Cloud *et al.* (1976) disputed both its age and taxonomic assignment, although they were not certain the organism represented was not another metazoan (a hexactinellid sponge).

C.3.c. The Tommotian Skeletal Fossils

Historically, the base of the Cambrian section was designated by the first appearance of skeletal fossils, which happened to be trilobites. In fairly recent years, however, sizable faunas of pretrilobite skeletonized organisms have been described from many regions. These new reports make the traditional Precambrian/Cambrian boundary designation virtually meaningless and have engendered a host of further studies. One approach to the problem seems in current favor; i.e., the designation of a "Tommotian" Stage at the very bottom of the Cambrian System, below trilobite zones, but above the Ediacarian. This stage theoretically excludes faunas composed exclusively of soft-bodied metazoans; however, this separation is not as simple as might be hoped, because there are few absolute dates taken on the boundary sequences.

Among the important early North American Tommotian finds is that of Taylor (1966) who reported a mollusklike conical fossil named *Wyattia,* from the Reed Dolomite in the southwestern Great Basin, many hundreds of meters below the lowest trilobite beds. More recently, Mount *et al.* (1983) described additional shelled fossils in the overlying Deep Springs Formation, which still lies below the oldest regional trilobite beds. In the Mackenzie Mountains, northwestern Canada, Conway Morris and Fritz (1980) reported skeletonized microfossils, which include at least one protoconodont (these are enigmatic fossils, arguably ancestral to conodonts, which also are enigmatic but definitely left by metazoans with phosphatic hard parts). McMenamin *et al.* (1983) described a Tommotian assemblage from Sonora, Mexico, composed largely of worm-tubes and conical, enigmatic shells, lying some 900 m below the lowest trilobite beds. Palmer (1971b) noted the occurrence of the primitive ostracode *Indiana* in the lower Murray Shale in Tennessee, and Mount *et al.* (1983) consider this an Appalachian Tommotian fossil.

Outside North America, Tommotian fossils are widely distributed. A very important assemblage of Siberian Tommotian fossils was described by Zhuravleva (1970) and Matthews and Missarzhevsky (1975), and includes archaeocyathids, hyolithids (mollusks), and worms. In addition, there are approximately a half-dozen additional Tommotian faunas described from Mongolia, China, Australia, England, Scandinavia, and Baltica (see Stanley, 1976, and Mount *et al.,* 1983).

C.3.d. Ediacarian/Tommotian Paleoecology and Hypotheses for the Advent of Skeletons

McMenamin (1982) summarized the chronology of Ediacarian/Cambrian events as follows:

- Earliest trilobites—550 Myr.
- Base of Tommotian (Cambrian)—600 to 570 Myr.
- "Vendian, Ediacarian, or soft bodied fauna"—650 to 600 Myr.
- Upper Precambrian tillites—1000 to 650 Myr.

This sequence of events is fairly constant in all areas containing good Ediacarian/Cambrian stratal sequences. The sequence and its features bring up two major questions: Why did metazoans appear apparently suddenly, after nearly 2.9 Byr. of exclusively single-celled life on Earth? Why, similarly, did skeletonized organisms appear seemingly all at once?

Several explanations offer possible insights into the life and times of Ediacarian organisms. Stanley (1973) suggested that "cropping" (i.e., grazing by herbivores or predation by carnivores) is necessary for an ecosystem to maximize its diversity. Therefore, the appearance of an efficient herbivore at, say, 800 Myr. would open new ecological niches that might favor evolution of a novel form—a metazoan. Continuation of such a process might produce an explosive radiation of Metazoa and, finally, a similar explosive radiation of skeletonized forms.

A completely different model was championed by Cloud (1976), based on the presumed importance of atmospheric oxygen pressures in the metabolism of early metazoans. Cloud stated that the first metazoans may have arisen when oxygen partial pressures reached 6.2% of present levels, based on studies of tolerances of modern polychaete worms and jellyfish. Cloud further believed the thin, flabby morphology of early metazoans was a response to the necessity of diffusing meager oxygen supplies through tissues, and that appearance of skeletons also resulted from increasing oxygen partial pressures.

Many other explanations have been offered. A common idea for appearance of skeletons is that the evolution of an aggressive predator forced coevolution of protective skeletons. This is largely untestable; however, absence of such an organism in the fossil record militates against its acceptance. Most attractive to us is the parsimonious idea that routine evolution, and time, yielded the first metazoans. We note that the majority of Ediacaran forms are assignable to a single phylum (Cnidaria) and therefore may be products of a single diversification. Others, such as *Dickinsonia* and *Spriggina* (Fig. 4-6), seem to be annelid worms and may be descendants of the vermiform creatures which left some of the reported older Precambrian burrows; in other words, these annelids were fortuitously preserved members of a relatively long-standing lineage.

The current knowledge of Ediacaran fossils suggests that most oceanic habitats are represented, but planktonic–pelagic forms predominate. Pennatulate Cnidaria probably were attached benthos, and the presumed annelids were probably infaunal benthos. The modes of life of completely mysterious creatures such as *Parvancorina* and *Tribrachidium* are unknown. Presence of burrows and trails in Ediacarian rocks proves there was a benthic infauna, and tends to argue against Precambrian scenarios that feature low oxygen levels, since levels in bottom sediment would be even lower than in the water column.

Finally, we consider the mounting number of pretrilobite shelled fossils as clear evidence that there is no mystery to the appearance of skeletons. As Tommotian fossils show, the advent was actually gradual with several phyla taking part; Cambrian trilobites, although showy fossils, were merely additional skeletonized forms.

D. Phanerozoic Sauk Sedimentation

D.1. Overview of Cambrian Sedimentation

North America was quite different in overall form during Sauk time than during subsequent geological times. The main difference is the absence of known collisional tectonics along any margin of the continent, with the resulting lack of mountains and volcanoes. Thus, the continent received quite uniform bands of sediments across much of the margin and continental interior, at times forming virtual rings of similar sediments. This pattern persisted until interrupted by tectonic events during the Late Ordovician, after the Sauk sea had withdrawn.

The transgression of the Sauk sea continued without interruption from the Ediacarian. Progressive onlap onto the passive margin continued into the Late Cambrian, with minor offlap intervals, yielding sandy basal detrital sediments which young cratonward, largely overlain by finer detrital sediments and carbonates (see Fig. 4-8).

By the Late Cambrian, carbonate platforms or shoals virtually encircled the craton and separated inner and outer zones of largely detrital sedimentation. These carbonate shoals were generally broad and reached widths of 400 km. Palmer (1960, 1971a,b) designated the three parts of the classical Cambrian lithofacies as the "inner detrital belt," the "carbonate belt," and the "outer detrital belt."

The tectonic setting of the development of this three-part facies pattern (inner and outer detrital belts, and carbonate shoals) is generally clear. In the east, late Precambrian continental rifting led to the development of a rifted-margin prism. By the late Early Cambrian, the rapid-subsidence phase had ceased and the continental margin had entered the carbonate–shale phase of slower subsidence. Carbonate shoals developed on the continental shelf, beyond the reach of most cratonic detritus. These shoals ended near the continental slope, on which mostly fine sediments accumulated; these sediments, which constitute the outer detrital zones, were derived primarily from the craton by sedimentary bypassing. Sediments of the eastern inner detrital zone were dominantly shallow-water deposits whose source was clearly cratonic.

In the west, the situation was approximately the same except that the rifted origin of the western margin apparently occurred earlier than in the east and its history is more obscure. Nevertheless, a rifted-margin prism was developed there as well and the major facies patterns described in the east also developed in the west (Stewart and Suczek, 1977). In central and western Nevada, deep-water sediments (turbidites and pelagic deposits) accumulated. This region is interpreted as a deep marginal basin between the western continental margin and a magmatic arc complex farther to the west, where the ancestral Pacific plate was consumed by subduction beneath the ancestral North American plate.

D.2. The Craton during Sauk Deposition

The Early and Middle Cambrian craton was not a flat monotonous plain, but rather a land surface with mountains and valleys of Precambrian igneous and metamorphic rock. Among the major physiographic features was the Transcontinental Arch, a broad upraised feature 2250 km long by 960 to 1300 km wide running from Ontario to northern Mexico (Lochman-Balk, 1971). Crossing this uplift in a northwest-to-southeast trend was the Cambridge–Central Kansas Arch (Fig. 4-9). These features persisted through the Cambrian, and represent virtually the only part of Cambrian North America that did not receive marine sediments. This Transcontinental Arch system remained a strongly positive feature through the Early Ordovician.

Other cratonic features active in the Cambrian include basins in Michigan and Illinois, which may have been connected in the Cambrian and later separated by an arch (the Kankakee Arch); the Reelfoot Rift, a graben in the lower Mississippi River valley region, which may have been active since the late Precambrian (Ervin and McGinnis, 1975), and which may have been both the precursor of the Mesozoic–Cenozoic Mississippi Embayment and the reason for modern seismicity centering on New Madrid, Missouri; a large basin in western Canada and the northwestern United States, which has been called the Western Canada or Interlake Basin and which was the forerunner of the Williston Basin which features prominently in the later Paleozoic history (Chapters 6 and 7); and a number of positive features including the Findlay Arch extending down from the Canadian Shield and the Ozark and Montana Domes (Fig. 4-9).

To the west of the Transcontinental Arch lay a broad and roughly triangular coastal plain, narrowing southwestward into Arizona and bordered on the west by a deep marginal basin. East of the Transcontinental Arch was another broad coastal plain. Its geometry was complicated by continental rifting and the development of a rifted-margin prism in the Appalachian and Ouachita–Marathon regions at the shores of Iapetus, and by an aulacogen in the Arbuckle–Ouachita region whose activity was closely related to the continental rifting which resulted in the Iapetus Ocean.

D.3. Cambrian of the Eastern Margin

D.3.a. Lower Cambrian (Waucoban) Basal Detrital Sediments and Carbonates

Cambrian strata of the eastern continental margin outcrop almost exclusively along the Appalachian Mountains, as did the Ediacarian units described previously. Geographic subdivisions are made within the Appalachians, for purposes of grouping Cambrian exposures into relatable units,

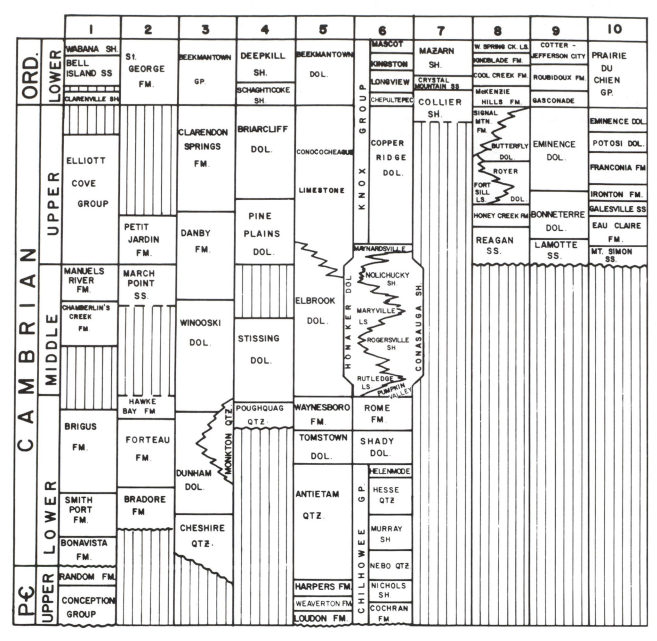

Fig. 4-8. Sauk stratigraphic columns. Location of columns as follows: 1, southeastern Newfoundland; 2, western Newfoundland; 3, western Vermont; 4, southeastern New York; 5, Shenandoah Valley, western Virginia; 6, eastern Tennessee; 7, Ouachitas; 8, Arbuckle and Wichita Mountains, Oklahoma; 9, Kansas; 10, northern Illinois; 11, Wisconsin and eastern Minnesota; 12, Death Valley; 13, central House Range, Utah; 14, eastern Tintic Mountains; 15, southcentral Montana; 16, Front Range, Colorado; 17, northern Park Ranges, Canadian Rockies; 18, Devon Island, northern Canada; 19, southeastern Franklin miogeosyncline; 20, East Greenland.

and these consist of the central and southern region (central Pennsylvania to Alabama), the northcentral region (western Massachusetts to eastern Pennsylvania), northern New England to Nova Scotia, and Newfoundland. The southern sections (see Fig. 4-10) are by far the most straightforward and provide the bulk of information on "representative" Cambrian history of the eastern margin. Northern Cambrian strata have commonly experienced considerably greater tectonic and other postdepositional deformation than southern units, and most formations are represented by disjunct slices of outcrop. In addition, there is an anomalously thin Lower Cambrian section in the northcentral region, which Palmer and Rozanov (1976) attributed to an intra-Lower Cambrian unconformity. A further, and quite interesting complication

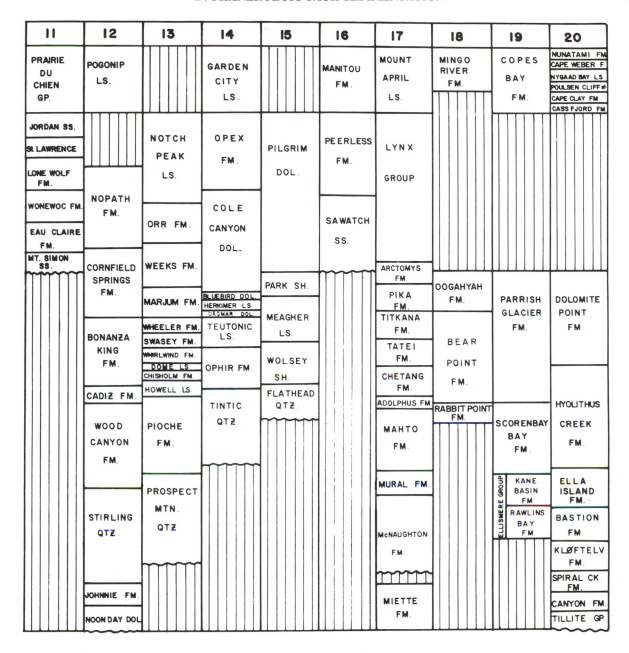

Fig. 4-8. (*Continued*)

is that Cambrian units east of the Appalachians in New England, Nova Scotia, and Newfoundland, show European affinities and were formed either within the Iapetus Ocean or on its eastern shore. The following discussion is in large part summarized from Palmer (1971b), Rodgers (1956), and North (1971).

Lower Cambrian strata in the Appalachians may be lumped into three lithosomes: a lower detrital unit, a middle carbonate unit, and an upper mixed fine-detrital and carbonate unit. However, some regions which experienced early

flooding by the Sauk sea developed carbonate environments in advance of other regions; thus, locally there may be carbonates far down in the section.

The lower detrital sequence consists of the various formations assigned to the Chilhowee Group in the central and southern region, and of a number of other units farther north, e.g., the Cheshire and Tyson Formations in Vermont, the Poughquag Formation in southeastern New York, the Rosaire Formation in Quebec, and the Bonavista Formation in southeastern Newfoundland. After late Chilhowee

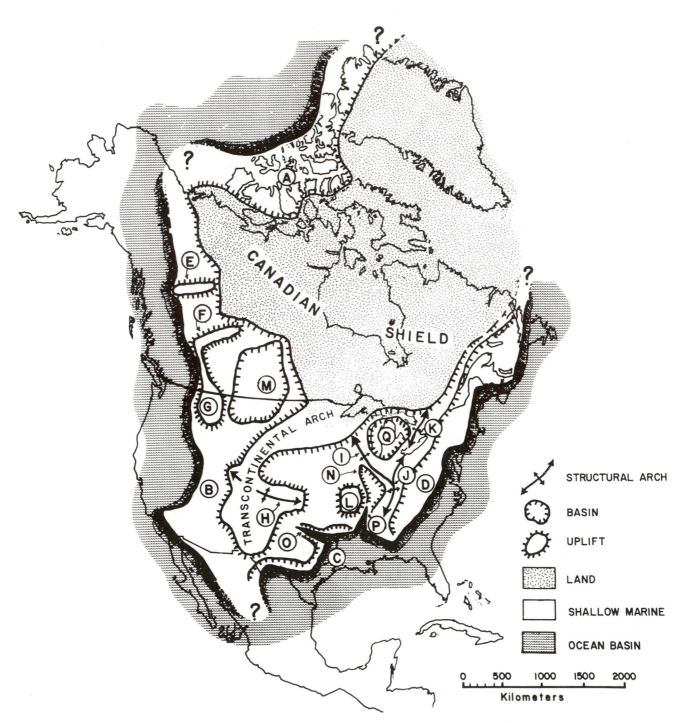

Fig. 4-9. Major physiographic and structural features of North America present during the early Paleozoic Era. Labeled features: A (in northern map), Innuitian Shelf; B, Cordilleran Shelf; C, Ouachita Trough; D, Appalachian Shelf; E, Tathlina Arch; F, Peace River Arch; G, Montana Dome; H, Cambridge–Central Kansas Arch; I, Kankakee Arch; J, Cincinnati Arch; K, Findlay Arch; L, Ozark Dome; M, Williston Basin; N, Illinois Basin; O, South Oklahoma Aulacogen; P, Reelfoot Rift (the Rome Rift may be a northeastern extension, not shown); Q, Michigan Basin.

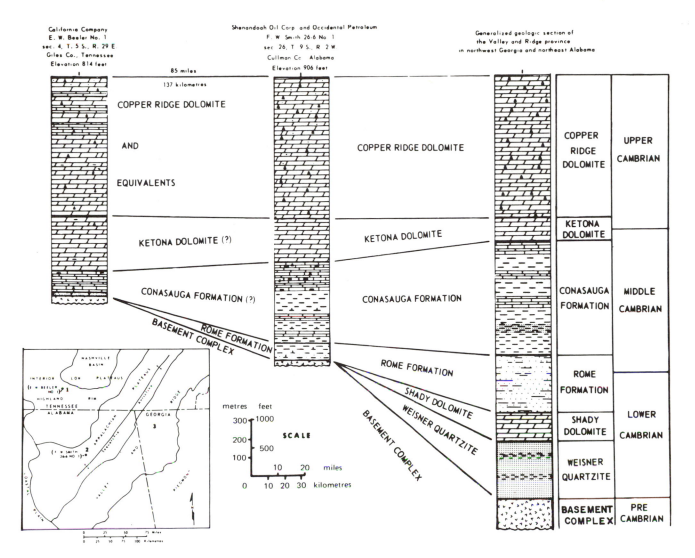

Fig. 4-10. Generalized Cambrian cross section in the southernmost Appalachians (southcentral Tennessee, northeastern Alabama, and northwestern Georgia). (Modified from Kidd and Neathery, 1976, *Geology,* Vol. 4(12), Fig. 1, p. 768.)

time, the foundering continental margin apparently stabilized and broad carbonate shoals developed in the regions previously receiving terrigenous sediments. In the southern and central Appalachians, these carbonates are represented by the Shady Dolomite (Virginia to Georgia) and the Tomstown Dolomite (Maryland to Pennsylvania).

A number of characteristic Lower Cambrian carbonates formed in the northern Appalachians, and, as mentioned in the opening discussion, locally even the basal Cambrian strata may be carbonates. Representative of the northern carbonates are the Dunham Dolomite in Vermont and the Forteau Formation in western Newfoundland. Detailed discussion of the units mentioned, and the many unmentioned, is beyond our scope. An analysis of the Shady Dolomite by Pfeil and Read (1980) may serve to charac-

terize the platform carbonates of both Lower Cambrian and the remaining Sauk sequence. The Shady Dolomite features shelf-edge algal reefs, which pass seaward (i.e., eastward) into a great variety of carbonate deposits, including considerable breccia derived from the algal reefs. At the seaward margin of the carbonate shoals, black, shaley limestones and other lithologies formed, and algal bioherms were absent.

An influx of terrigenous sand and (largely) mud occurred across nearly all the eastern platform, temporarily limiting carbonate deposition and producing widespread detrital and mixed carbonate/detrital strata. Typical are the Rome Formation from Tennessee to Alabama, the Waynesboro Formation from Virginia to parts of Pennsylvania, and the Monkton Quartzite in western Vermont. The representative Rome Formation consists of red and green shales with a few thin carbonate layers. The red shales show some mud cracks and suggest temporary regression with exposure of the platform.

A few eastern regions experienced dolomite deposition during the generally detrital phase of late Early Cambrian sedimentation. These include the uppermost part of the Dunham Dolomite in Vermont; locally, in central Pennsylvania the Ledger Dolomite; and in southeastern New York, the Stissing Dolomite.

Over most of the eastern margin, the detrital phase ended with renewed transgression of the Sauk sea and re-establishment of carbonate environments over an area similar to that of the Lower Cambrian. This event began Middle Cambrian time and initiated a metastable condition that prevailed through the remaining Cambrian stages.

D.3.b. Middle and Upper Cambrian (Albertan and Croixan) Lithofacies

The Sauk sea returned to roughly the same region it had transgressed during the carbonate phase of Early Cambrian time. Palmer (1971b) noted that the location of the main carbonate belt remained fairly stable through the Middle Cambrian and into the early Late Cambrian, although the carbonates did not extend as far eastward as they had previously.

The eastern carbonate shoals are composed generally of pure carbonates with abundant oolitic, stromatolitic, bioclastic, and other textures that suggest warm, shallow water conditions analogous to the conditions on the modern Bahama Banks. There were a variety of environmental conditions present, varying especially in intensity of wave or current action, water depth, and character of biotic communities. Typical of these Middle Cambrian carbonates are the Maryville and Rutledge Limestones in southern Virginia and southern Pennsylvania, the Pleasant Hill Limestone in central Pennsylvania, the Winooski Dolomite in western Vermont, and the March Point Formation in western Newfoundland.

The carbonate shoals of the Middle Cambrian were relatively narrow (~ 100 km wide) and even minor fluctuations of sea level brought areas out of the carbonate environments and into the reach of cratonic detritus [i.e., within the inner detrital zone of Palmer (1960)]. This pattern is distinctively recorded in the southernmost Appalachian region by the Conasauga Group. In easternmost Alabama and western Georgia, and a portion of Tennessee, the Conasauga consists predominantly of buff-colored shales with thin limestones. The detritus apparently is craton-derived and the trilobite faunas of Middle Cambrian age feature taxa characteristic of the cratonic assemblages observed in western inner detrital regions. In the subsurface in parts of Tennessee, and in more central areas of Alabama, the Middle Cambrian Conasauga consists predominantly of carbonates and includes the Rutledge and Maryville Formations as part of the carbonate belt sequence (see Fig. 4-11).

The third lithofacies which developed on the eastern margin and stabilized in the later Middle Cambrian, is termed the "outer detrital zone." Here, fine-grained (often black) detritus, and thin-bedded, argillaceous limestones, and other lithologies, were deposited. Reinhardt (1977) suggested that the presence of graded beds of carbonate detritus, deep-water conglomerates, and other evidence, shows that at least some Late Cambrian outer detrital units represent debris flows and turbidity currents on the continental slope and rise on the seaward side of the carbonate shoals. Relatively few such units are known to be of Middle Cambrian age, and all these occur in the central or northern Appalachians: they include uppermost units of the Kinzers Shale in Pennsylvania, the West Castleton Formation in the Taconic region of New York, and the lower Woods Corners Group in western Vermont.

The three-part facies pattern in the eastern margin became even more stable during the early Late Cambrian (Dresbachian), while the Sauk sea transgressed farther into the continent and began to leave significant cratonic strata. Most preserved Dresbachian strata in the Appalachians are carbonates: these include the upper Elbrook Formation in the central region, the Maynardville Limestone in southwestern Virginia and Tennessee, the Ketona Dolomite in parts of Alabama, the Pine Plains Dolomite in southeastern New York, the upper Danby Formation in western Vermont, the Petit Jardin Formation in western Newfoundland, and many others.

Palmer (1971b) noted that the sea apparently withdrew slightly during parts of Dresbachian time, as shown by several detrital units among the carbonates: these include the Nolichucky Shale overlying the Maryville and below the Maynardville in Tennessee, and detrital sediments in the Danby Formation in Vermont. Following this brief regression, the remainder of Cambrian time (Franconian and Trempealeauan) saw even greater cratonic transgression and widespread carbonate deposition in the Appalachians. Characteristic of the uppermost Cambrian carbonate sequences are the Copper Ridge Dolomite of the Knox Group in northeastern Alabama through Tennessee, the Conococheague Group in Virginia to southern Pennsylvania, the Gatesburg Dolomite in central Pennsylvania, the Briarcliff Dolomite in southeastern New York, the Clarendon Springs Formation in western Vermont, and the Saint George Formation in Newfoundland.

Few sediments of the inner detrital zone were deposited in the Appalachians during the Late Cambrian, since transgression of the Sauk sea shifted the inner detrital zone farther west on the craton. The Conasauga Formation in parts of northeastern Alabama and northwestern Georgia

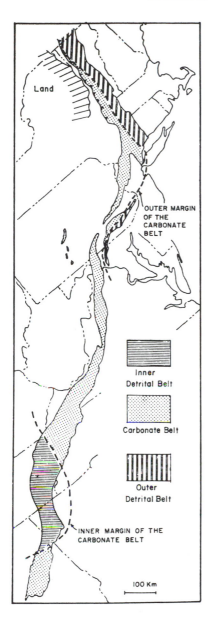

Fig. 4-11. Exposures of Middle and Late Cambrian strata on the eastern U.S. margin, showing distribution of lithofacies zones. (From Palmer © 1971 by John Wiley & Sons, reprinted with permission of John Wiley & Sons.)

contains mixed detrital and limestone beds of Late Cambrian age, which were probably deposited at the eastern edge of the inner detrital zone.

Outer detrital zone sediments are preserved largely in the northernmost regions. In northern Vermont, the Woods Corners Group and Gorge Formation were indicated by Palmer to be deposits on the seaward side of the carbonate bank. In the Taconic Mountains of New York, the allochthonous sequence of Cambrian outer detrital units (see Taconic orogeny discussion in Chapter 5) is capped by the Hatch Hill Formation, consisting of shale (metamorphosed to slate), carbonate conglomerates, and bedded carbonates.

In western Newfoundland, the Cow Head Breccia and Humber Arm Group are apparently outer detrital units. The Cow Head is renowned for its wide variety of carbonate clasts in a spectacular conglomerate texture. The carbonates range in age from Middle Cambrian to Middle Ordovician and seem to comprise debris from the outer edge of the carbonate shoals redeposited down the continental slope (Hubert *et al.,* 1977; James, 1981). The Humber Arm is a more conventional deep-water unit with graywackes, multicolored shales, and volcanigenic rocks (Karson and Dewey, 1978).

An interesting sequence of cryptic paleogeographic affinities is the Lévis Limestone Conglomerate, which occurs in Québec Province near Québec City. The Lévis conglomerates, like the Cow Head Breccia, contain fossiliferous clasts of Middle and Late Cambrian age; the deposit itself seem largely of Ordovician age (Rasetti, 1963). Carbonate blocks within the Lévis are generally fine-grained and white-colored (North, 1971), which distinguishes the material from any regional Cambrian units. Further, trilobites within the Lévis carbonate clasts are generally of the extracratonic assemblages (see discussion in E.7.) of Lochman-Balk and Wilson (1958), in common with the Cow Head fauna. Overall, the texture and nature of Lévis conglomerate clasts suggest deposition in a shallow basin environment rather than at the edge of the carbonate bank, but the trilobites are representative of the cosmopolitan (oceanic) forms typically part of the outer-detrital zones.

D.4. Cambrian of the Western Margin

Having described the eastern marginal facies first, we may present an abbreviated version of the contemporary western strata, since these followed a similar depositional history. As was observed in the discussion of the Ediacarian, many conformable Ediacarian/Lower Cambrian sequences are known in the west, almost entirely featuring thick, coarse detrital strata derived from the craton. Typical of such Lower Cambrian formations are the Prospect Mountain, Wood Canyon, and Stirling Formations in the Great Basin (Palmer, 1971a), the Gog Group in the main Rocky Mountain ranges of Alberta and British Columbia, the Cranbrook Formation in the western ranges of British Columbia, the Hamill Group in the Selkirk Mountains (north of Washington state in British Columbia), and a host of regional units farther north in Canada and Alaska (see McCrossan *et al.,* 1964, and North, 1971). These quartzites are generally unfossiliferous, but a few finer-grained Lower Cambrian detrital units are present, notably the Pioche Shale in Nevada, which include good fossil faunas dominated by olenellid trilobites (see Section E.2.b). The terrigenous detrital sequences typically thicken westward, reaching to 6000m in the United States (Stewart and Suczek, 1977).

As in the east, during the Middle Cambrian, the Sauk sea initiated a major transgression onto the craton and formed the three-part lithofacies pattern of Palmer (1960). The inner detrital zone migrated eastward onto the craton as the carbonate shoals formed, stabilized, and broadened on the western platform. Most western marginal rocks are either carbonates or outer-detrital mixed fine-grained detrital sediments and carbonates. Middle Cambrian inner-detrital rocks are found only in the Grand Canyon (Arizona and Colorado), Montana, northwestern Wyoming, and British Columbia. These are at the base of transgressive sequences and are all capped by carbonates representing the encroaching carbonate shoals. The progression, both geographically and temporally, from inner-detrital to carbonate environments is clearly shown in several western sequences.

In the eastern Grand Canyon of Arizona, the basal Lower Cambrian Tapeats (or Prospect Mountain) Quartzite is overlain by the Bright Angel Shale, also an inner-detrital unit, and was in turn overlain by the Muav Limestone as the carbonate shoals formed. A more extensive time-transgressive sequence is evident in Montana and adjacent Wyoming areas. Figure 4-12 shows that the Middle Cambrian basal sandstone, the Flathead, is overlain by an inner-detrital shale, called either the Gordon or the Wolsey, and subsequently overtopped by thick carbonates in westernmost regions (e.g., northwestern Montana), thinner carbonates at the peripheries of the carbonate shoals (e.g., in central Montana), or by shales in northwestern Wyoming. A brief regression is marked, near the top of many western sequences that were near the inner margin of the carbonates, by shale tongues at the top of the Middle Cambrian sequence; in Montana these are the Switchback and Park Shales.

Middle and Upper Cambrian carbonate shoals are represented in deposits stretching from east-central Alaska down through southern Nevada, and on into Mexico. The greatest areal extent of these carbonates was developed in

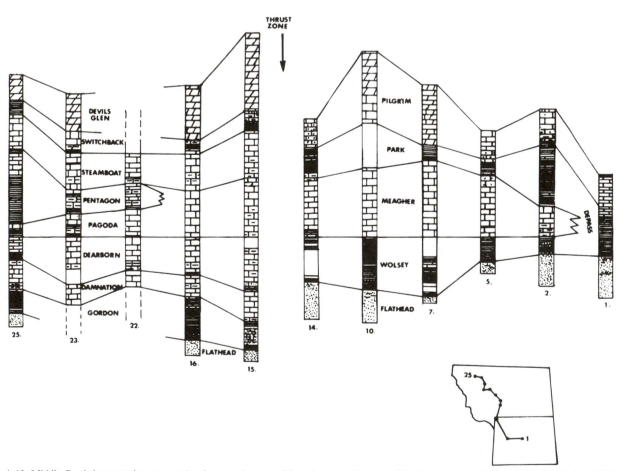

Fig. 4-12. Middle Cambrian strata in cross section from northwestern Wyoming to northwestern Montana (see inset for section line). The basal Flathead Sandstone is a diachronous unit, younging to the southeast, generally overlain by shales and carbonates. To the northwest, carbonates thicken appreciably as part of the carbonate bank lithofacies. (From Schwimmer, 1975.)

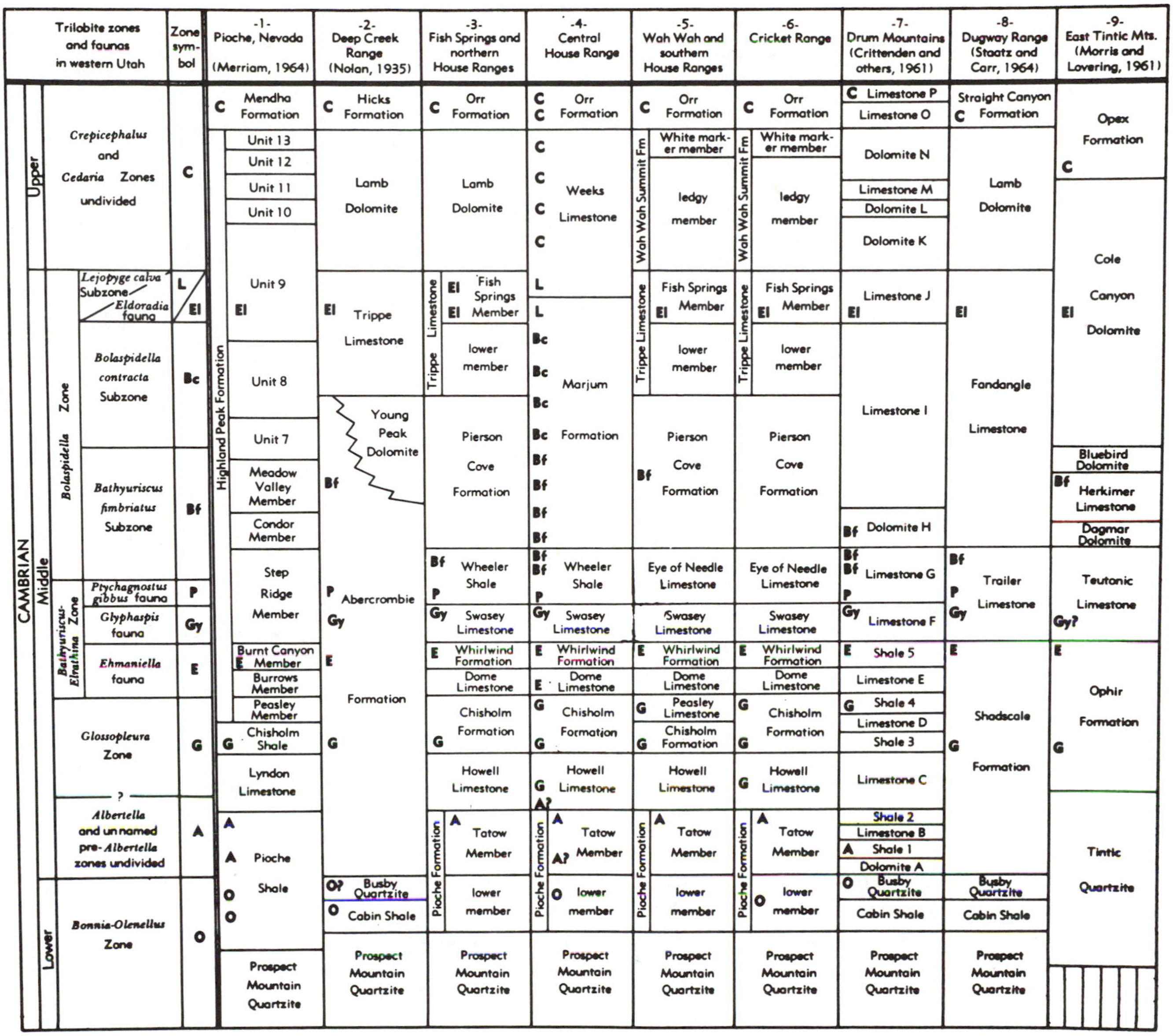

Trilobite zones and faunas in western Utah	Zone symbol	-1- Pioche, Nevada (Merriam, 1964)	-2- Deep Creek Range (Nolan, 1935)	-3- Fish Springs and northern House Ranges	-4- Central House Range	-5- Wah Wah and southern House Ranges	-6- Cricket Range	-7- Drum Mountains (Crittenden and others, 1961)	-8- Dugway Range (Staatz and Carr, 1964)	-9- East Tintic Mts. (Morris and Lovering, 1961)
(CAMBRIAN — Upper)	C	Mendha Formation	Hicks Formation	Orr Formation	Orr Formation	Orr Formation	Orr Formation	Limestone P / Limestone O	Straight Canyon Formation	Opex Formation
Crepicephalus and *Cedaria* Zones undivided	C	Unit 13 / Unit 12 / Unit 11 / Unit 10 *(Highland Peak Formation)*	Lamb Dolomite	Lamb Dolomite	Weeks Limestone	White marker member / ledgy member *(Wah Wah Summit Fm)*	White marker member / ledgy member *(Wah Wah Summit Fm)*	Dolomite N / Limestone M / Dolomite L / Dolomite K	Lamb Dolomite	Cole Canyon Dolomite
Lejopyge calva Subzone / *Eldoradia* fauna	L / El	Unit 9	Trippe Limestone	Fish Springs Member / lower member *(Trippe Limestone)*	Weeks Limestone	Fish Springs Member / lower member *(Trippe Limestone)*	Fish Springs Member / lower member *(Trippe Limestone)*	Limestone J	Fandangle Limestone	Cole Canyon Dolomite
Bolaspidella contracta Subzone	Bc	Unit 8	Young Peak Dolomite	Pierson Cove Formation	Marjum Formation	Pierson Cove Formation	Pierson Cove Formation	Limestone I	Fandangle Limestone	Bluebird Dolomite
Bathyuriscus fimbriatus Subzone	Bf	Unit 7 / Meadow Valley Member / Condor Member	Abercrombie Formation	Wheeler Shale	Wheeler Shale	Eye of Needle Limestone	Eye of Needle Limestone	Dolomite H / Limestone G	Trailer Limestone	Herkimer Limestone / Dagmar Dolomite
Ptychagnostus gibbus fauna	P	Step Ridge Member	Abercrombie Formation	Wheeler Shale	Wheeler Shale	Eye of Needle Limestone	Eye of Needle Limestone	Limestone G	Trailer Limestone	Teutonic Limestone
Glyphaspis fauna	Gy	Step Ridge Member	Abercrombie Formation	Swasey Limestone	Swasey Limestone	Swasey Limestone	Swasey Limestone	Limestone F	Trailer Limestone	Teutonic Limestone
Ehmaniella fauna	E	Burnt Canyon Member / Burrows Member / Peasley Member	Abercrombie Formation	Whirlwind Formation / Dome Limestone	Whirlwind Formation / Dome Limestone	Whirlwind Formation / Dome Limestone / Peasley Limestone	Whirlwind Formation / Dome Limestone	Shale 5 / Limestone E	Shadscale Formation	Ophir Formation
Glossopleura Zone	G	Chisholm Shale / Lyndon Limestone	Abercrombie Formation	Chisholm Formation / Howell Limestone	Chisholm Formation / Howell Limestone	Chisholm Formation / Howell Limestone	Chisholm Formation / Howell Limestone	Shale 4 / Limestone D / Shale 3 / Limestone C	Shadscale Formation	Ophir Formation
Albertella and unnamed pre-*Albertella* zones undivided	A	Pioche Shale	Busby Quartzite	Tatow Member / lower member *(Pioche Formation)*	Tatow Member / lower member *(Pioche Formation)*	Tatow Member / lower member *(Pioche Formation)*	Tatow Member / lower member *(Pioche Formation)*	Shale 2 / Limestone B / Shale 1 / Dolomite A	Busby Quartzite	Tintic Quartzite
Bonnia-Olenellus Zone	O	Pioche Shale	Busby Quartzite / Cabin Shale	lower member	lower member	lower member	lower member	Busby Quartzite / Cabin Shale	Busby Quartzite / Cabin Shale	Tintic Quartzite
		Prospect Mountain Quartzite	Prospect Mountain Quartzite	Prospect Mountain Quartzite	Prospect Mountain Quartzite	Prospect Mountain Quartzite	Prospect Mountain Quartzite	Prospect Mountain Quartzite	Prospect Mountain Quartzite	

Fig. 4-13. Middle Cambrian stratigraphy of the House, Wah Wah, and adjacent ranges in Utah. Boldface symbols designate known occurrences of fossil collections keyed to the trilobite zones shown on the left. (From Hintze and Robison, 1975, Fig. 3, p. 883.)

Late Cambrian (Croixan) time. There are a great many carbonate formations which comprise this belt of rocks (see Palmer, 1971a; North, 1971; Cook *et al.,* 1975; McCrossan *et al.,* 1964), with a wide variety of lithologies dominated by massive or thick-bedded limestones and dolostones, including many oolitic, oncolitic, bioclastic, and pelleted grainstones (Stewart and Suczek, 1977). Local ripples and desiccation cracks indicate periodic subaerial exposure of the shoals. These lithologies characterize the shoals for both Late and Middle Cambrian times; but in the Late Cambrian, the shoals migrated eastward, along with the increased transgression of the Sauk sea onto the craton, and the width of the carbonate facies belt increased concomitantly.

The outer-detrital belt in the west represents sedimentation on the continental shelf-edge and slope, as it does in the east, but strata are characterized by laminated shales and carbonates. In general, outer-detrital units are less abundant than are carbonate bank facies, but several examples may be noted. In the House Range, Utah, at least two Middle Cambrian strata have outer-detrital components and their associated trilobite faunas (which contain generally cosmopolitan taxa such as agnostids and eodiscids; see Section

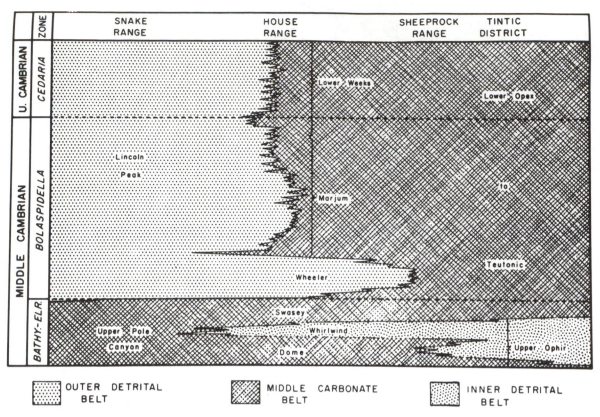

Fig. 4-14. Generalized cross section between eastern Nevada and central Utah, showing intergradation of upper middle and lower upper Cambrian lithofacies. Rock thicknesses are not to scale: datum horizon is at the biostratigraphic boundaries. Note eastern extension of outer-detrital lithologies (e.g., the Wheeler Shale) and western extension of inner-detrital lithologies (e.g., the Whirlwind Sandstone) into the House Range, Utah; compare with Fig. 4-13. (From Robison, 1964, Fig. 3, p. 999.)

E.2). Robison (1964) considered the Wheeler Shale and part of the Marjum Formation to reflect largely outer-detrital sedimentation (see Figs. 4-13 and 4-14). (The Wheeler Shale is uniquely fossiliferous and yields the abundant *Elrathia kingii* specimens included in most undergraduate fossil kits.)

Another typical but interesting outer-detrital zone section (see Fritz, 1971) is found in the vicinity of Kicking Horse Pass, at the British Columbia–Alberta border approximately 200 km north of the Montana–Canada border. There, the section has the Lower Cambrian Gog Formation (an inner-detrital basal sandstone), overlain by a mixed carbonate and outer-detrital shale Middle Cambrian sequence consisting of the Mt. Whyte (shale)/Cathedral (dolostone)/Stephen (shale)/Eldon (dolostone)/Pika (dolostone)/Arctomys (shale) Formations. Of great interest in this section are small lenses of dark shaley sediment within the Stephen Formation. These lenses probably represent small slumps of detritus off the outer shelf edge, and comprise the famous Burgess Shale (Fig. 4-15 and see Section E.6).

Three unusual western formations, the Scott Canyon, Paradise Valley, and Harmony Formations, all in north-central Nevada, show lithologies characteristic of deep

water and are interpreted to represent sedimentation in a deep marginal basin to the west of the continental margin. The Scott Canyon is composed of thick deposits of pelagic cherts, argillite, and greenstone (which originated, in part, as pillow lavas), with minor sandstone, quartzite, and limestones. The age of these rocks is ambiguous, but most fossils from the limestones indicate Early or Middle Cambrian age. The Scott Canyon is allochthonous, and apparently has been transported eastward 80 km or more (Stewart and Suczek, 1977). Fossils found in the cherts suggest that it may in part be admixed with deep-water sediments from intervals as late as the Devonian. The Upper Cambrian Paradise Valley chert in Nevada also seems to be a deep-water basinal sediment, but the origin of the overlying arkosic Harmony Formation is not as clear. Stewart and Poole (1974) referred to the facies characterized by both the Paradise Valley and the Harmony Formations as the "arkosic facies" and distinguished them from the "siliceous facies" comprising only the Scott Canyon Formation. The arkose in the Harmony may represent accumulations on the continental rise transported there by turbidity currents (Stewart and Suczek, 1977), originally derived from a topographically prominent feature in Idaho called the

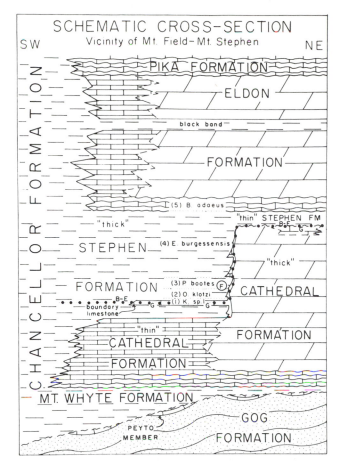

Fig. 4-15. Cambrian section in British Columbia, showing the transition from "thick" to "thin" Cathedral Formation strata. The Burgess Shale lens is at the letter "F" midway down the "thick" Cathedral section. (From Fritz, 1971.)

Lemhi Arch. This arch may have exposed Precambrian crystalline basement rocks, which may also be the source for a feldspar-rich quartzite unit in the St. Charles Limestone (Upper Cambrian) in northern Utah and southeastern Idaho.

D.5. Cambrian of the Craton

Central cratonic strata were deposited by the Sauk sea only from Middle Cambrian through Early Ordovician time, simply because transgression had not reached beyond the margins until then. Initial sedimentation on the craton was very much like basal sedimentation on the margins—coarse detrital materials, sometimes assigned the same formational names used in marginal areas. These basal quartz arenites are locally very well-sorted, rounded, and virtually pure; they are termed *supermature* and indicate multiple cycles of erosion and redeposition in high-energy intertidal zones. The Potsdam Formation in central New York is a good

example (also note similar units in the Middle Ordovician; see Chapter 5). Upper Cambrian basal sandstones are present as either surface or subsurface formations in a huge, roughly triangular area extending on the eastern craton from central New York to Oklahoma, and north to Minnesota. On the western craton, Upper Cambrian basal sandstones extend eastward from the Rocky Mountains as far as the eastern Dakotas and Kansas (see Fig. 4-16). Only small regions of the southwestern craton, Nebraska, and most of the Canadian Shield seem to have escaped some deposition during the Late Cambrian transgression.

Following the sandstone phase of transgression, carbonates were deposited across most of the transgressed cratonic regions, sometimes interbedded with fine detrital sediments. By the latest Cambrian, limestones and especially dolostones were forming across virtually the entire U.S. craton and much of the inner continental margins. The few regions which continued to receive sandy latest Cambrian sediments were probably at the receiving end of streams

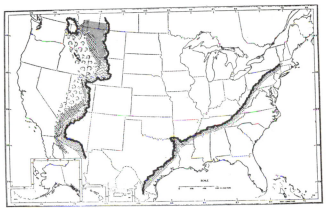

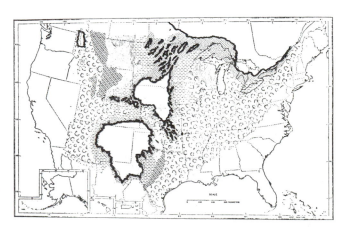

Fig. 4-16. Paleogeographic reconstructions of the U.S. craton. (Above) Middle Cambrian (*Bathyuriscus–Elrathina* time); (below) Late Cambrian (middle *Saukia* time). Areas shown unshaded were either above sea level (in the western interior) or under deep marine conditions (at the margins). (From Lochman-Balk © 1971 by John Wiley & Sons, reprinted with permission of John Wiley & Sons.)

washing detritus down from the exposed Canadian Shield; these areas include the northern and eastern Great Lakes region (northwestern New York, northern Michigan, northern Wisconsin, northern Illinois), and central North Dakota and nearby Wyoming.

The Sauk Sea transgressed the upper Mississippi River valley by earliest Late Cambrian time. The reference section for the Upper Cambrian is located in the region and includes the Dresbach, Franconia, and Trempealeau Formations (locally, Groups). In Wisconsin and Minnesota, the Dresbach is the lowermost Croixan unit and features three members (or formations): a basal Mt. Simon Sandstone, an argillaceous middle Eau Claire, and a largely sandstone upper Galesville. Locally the Eau Claire contains thin dolostones. The Franconia Formation overlies the Dresbach and is composed generally of sandstone with local dolostones. The Trempealeau Formation at the top of the Upper Cambrian section is largely sandstone and shale in the type region but it too may be locally dolomitic. Odom (1975) analyzed the sedimentology of Upper Cambrian arenites in the upper Mississippi valley.

Correlations of typical Croixan units outside the type region are generally uncomplicated, and trends toward increasing thickness and greater proportions of dolostone are evident away from the craton and toward the margins. During Late Cambrian time, several minor transgressions and regressions left a large number of well-defined strata across the craton. The reader is directed to Lochman-Balk (1971) for discussion of the many additional units and their interrelationships.

D.6. Peripheral Cambrian Strata

D.6.a. Eastern Cambrian Strata

As noted in the opening discussion, certain New England and Canadian outcrops of Cambrian rocks, east of the Appalachians, contain trilobites and other fossils of European affinity. Such rocks are known from southeastern Newfoundland, Cape Breton Island, southern New Brunswick (North, 1971), east central Maine, eastern Massachusetts (Palmer, 1971b), and Rhode Island (Skehan et al., 1978). Although it is generally accepted that these rocks formed in non-North American terranes, their precise origins are uncertain; current consensus suggests origins in oceanic magmatic arcs or microcontinents, appended to the eastern North American margin during closure of the Iapetus Ocean in the late Paleozoic. Williams and Hatcher (1982) and Skehan et al. (1978) include these exotic Cambrian terranes as part of Avalonia (see discussion of Appalachian tectonics in Chapter 5). Following this logic, when the eastern margin reopened during the Mesozoic with formation of the Atlantic Ocean, portions of the exotic terranes, including possible European crust, remained attached to North America.

North (1971) described the exotic eastern Cambrian units in Canada, mostly shales, slates, and thin limestones, many of which are fossiliferous and include trilobites of the Atlantic Province (i.e., of European affinities) such as *Callavia* (Lower Cambrian) and *Paradoxides* (Middle Cambrian). The Manuels Brook Formation in southeastern Newfoundland is a famous fossiliferous unit containing abundant, well-preserved *Paradoxides*. In the eastern United States, notable exotic Cambrian strata in eastern Massachusetts are assigned to the Lower Cambrian Weymouth and Middle Cambrian Braintree Formations (Palmer, 1971b). Both of the latter are marine shales which contain sparse Atlantic Province trilobites, especially *Callavia*. Skehan et al. (1978) described a Middle Cambrian *Paradoxides* trilobite assemblage, which appears to be a European fauna, from detrital strata near Newport, Rhode Island.

Cambrian fossils have been found in metamorphosed sediments of the Carolina Slate Belt in South Carolina; these occurrences confirm long-held suspicions that the Slate Belt represents exotic terranes appended to the continent during later Appalachian tectonic events (see also discussion of Ediacarian fossils in the Slate Belt, Section C.2.c). Cambrian trilobites were reported from southernmost South Carolina by Maher et al. (1981), and near Batesburg, South Carolina, by Secor et al. (1983). In both cases, the trilobites reported were taxa with non-American affinities, and may be placed in Middle Cambrian *Paradoxides* assemblages of the Atlantic faunal province. Secor et al. (1983) stated that the Carolina Slate Belt occurrences are sufficiently distinctive to suggest that they derive from an oceanic source separate from the more northern Avalonian terrane occurrences (described in the opening section of this discussion); however, Williams and Hatcher (1982) include the Carolina Slate Belt as a southern portion of the Avalon terrane. Further discussion of suspect terranes in the eastern margin will be given in Chapters 5 and 6.

D.6.b. Cambrian of American Arctic Regions

Cambrian strata are present across a geographically widespread area of the Western Hemisphere Arctic, extending from northeastern Greenland to Alaska; however, the actual amount of exposed Cambrian rock which has been identified and described is relatively limited compared with the lower latitudes. In addition to limited exposures due to ice cover, much of the American Arctic border consists of the Canadian Arctic Archipelago, which engenders difficulties in correlating strata across frozen ocean reaches. In Chapter 5 we present a synthesis of the early evolutionary history of the Arctic, with description of the physiographic regions and their tectonic development. Below is a brief discussion of representative Sauk units in the Arctic to show genetic relationships with the southern regions.

In general, the three-part lithofacies patterns described for the southern craton seem to have extended across the Arctic as well. It is possible that carbonate shoals literally encircled the entire continent by the late Middle Cambrian, and continued to exist through the remaining part of the period. In addition, after a (possibly local) erosional hiatus during the latest Cambrian, carbonate environments were again established during the late Early Ordovician and lasted until the overall post-Sauk erosion event, in concert with the southern areas. Although carbonates are hardly products of present Arctic environments, one must remember that the Cambro-Ordovician American Arctic region lay almost astride the equator (see Fig. 4-1).

Lower Cambrian rocks in the Arctic are composed of both detrital and carbonate sediments, but coarse detrital sediments seem to dominate the record. The latter include the Rensselaer Bay Formation and several possibly equivalent units in northwestern Greenland (Dawes and Peel, 1981), the Ellesmere Group in eastern Ellesmere Island, the Rabbit Point Formation in Devon Island, and the Gallery Formation in northwestern Baffin Island (the latter data from Cowie, 1971). Lower Cambrian carbonates commonly lie over the (presumably craton-derived) detritus. For example, in Inglefield Land, northwestern Greenland, Dawes and Peel (1981) report that Lower Cambrian sandstones are overlain by later Early Cambrian dolostones and limestones which grade upward into similar units with Middle Cambrian trilobites. Similar lithologies are present elsewhere in northern Greenland, and Dawes and Peel indicated that these can grade upward through the remaining Cambrian, and into the Ordovician, without significant depositional change. Although most Middle Cambrian units from the Arctic are dolostones, a few detrital sediments, such as the lithologically mixed Parrish Glacier Formation in Ellesmere Island, may have formed on the inside of the carbonate shoals.

Late Cambrian units are poorly known but the strata described suggest that dolomite deposition dominated the region. Dawes and Soper (1973) reported Upper Cambrian carbonates in Washington and Wolff Lands, northern Greenland, and Dawes and Peel (1981) discussed the carbonate sequence in the Bache Peninsula, northern Greenland, mentioned above, which ranges through the Cambrian and into the Lower Ordovician. Dawes and Peel also mentioned an unnamed Lower-to-Upper Cambrian carbonate sequence in western Peary Land, unconformably overlain by Ordovician carbonates, and a similar, > 1000-m sequence in southern Peary land.

In east-central Alaska, Palmer (1968) described limited Cambrian strata and trilobites ranging from Early through Late Cambrian age. Cambrian strata in Alaska are largely part of the Hiliard Limestone, which ranges through the period, but older detrital units include the Lower Cambrian Adams Argillite and Late Cambrian trilobites come from the Jones Ridge Limestone. Palmer (1968) observed that Alaskan Cambrian trilobites appear to have largely North American affinities, with minor portions of the assemblages favoring European or Asian faunas.

In summary, the Arctic Cambrian sedimentary history generally follows patterns exhibited by the rest of the continent, but the amount of information known is too small to be conclusive. Then too, evidence in northern Greenland suggests a Late Cambrian regression (Dawes and Peel, 1981) not characteristic of southern latitudes. It seems safe to assume carbonate shoals were extensive across the Arctic, and that sizable zones of detritus lay at least landward from these shoals. The presence of outer-detrital sedimentation is not evident but this may reflect an absence of information.

D.7. Lower Ordovician Sedimentation

The transition from Late Cambrian to Early Ordovician time passed without gross changes in the nature of cratonic and marginal sedimentation. The period boundary is, however, marked in North America (as it is in Europe where the boundaries were defined) by pronounced changes in fossil assemblages. Upper Cambrian faunas are dominated by a greater diversity of trilobites and phosphatic brachiopods than are Lower Ordovician faunas (see Chapter 5). In the Ordovician, a wide variety of new taxa, distributed among virtually all animal phyla, appeared along with new varieties of trilobites. In areas where fossils bounding the periods are absent, or where fossils are present but are indeterminable with respect to age (such as stromatolites, many ichnofossils, and some long-ranging species), strata from the interval are collectively called ''Cambro-Ordovician'' rocks. Such units are both common and generally dolomitic.

The carbonate shoals of the latest Cambrian persisted into the Early Ordovician (Canadian), and actually increased in areal extent through much of this time (Fig. 4-17). Ross (1977) showed the Ordovician paleoequator running approximately through the United States from the site of modern San Francisco to the northern tip of Idaho; and, thus, the western United States was in a favorable position for the development of very extensive carbonate shoals.

The Early Ordovician carbonates were very widespread around the continent, extending from Alaska to the Great Basin in the west, from the Arctic Islands to Greenland in the north, and from Greenland to Alabama and the Ouachita region in the east and southeast. They are also

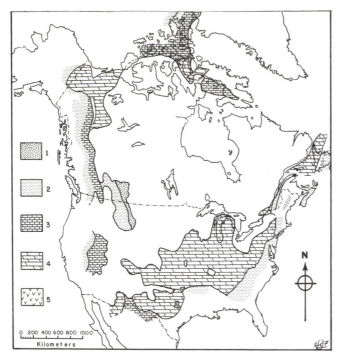

Fig. 4-17. Lower Ordovician lithofacies in North America. Map symbols are: 1, sandstone; 2, shale; 3, limestone; 4, dolostone; 5, volcanics.

present in both the surface and the subsurface over much of the midcontinent. Most of these carbonates are dolomitic, including the Cambro-Ordovician Knox Group in Ohio/Tennessee/Alabama/Georgia/Kentucky, the Lower Ordovician Prairie du Chien Dolomite in the Upper Mississippi Valley, the Franklin Mountain Dolomite in the Northern Interior Plains and the Yukon regions of Canada, the Beekmantown Group of New York, the Manitou Formation in Colorado, the Ellenberger and El Paso Formations in western Texas, and the Goodwin and Fillmore Formations in the Basin and Range region of the west. Some biohermal structures (see Chapter 5 for a discussion of "reef" structures) have been noted in western Lower Ordovician carbonates, in the Fillmore and Garden City Formations (Ross, 1977); these carbonate buildups were typically small and may have been formed by algae and sponges.

Carbonate deposition in the Early Ordovician occurred over most of the areas which were in the inner detrital zone of the Late Cambrian, but deeper water regions of the outer detrital zone continued to receive carbonate slump deposits, and siliceous and volcanic detritus. A siliceous/volcanic unit of the deep, marginal-basin environment on the western margin is the Valmy Formation, whose deposition may span the entire Ordovician. The Valmy Formation consists of literally thousands of meters of chert, shale, greenstone, and quartzite and is widespread through much of north-central Nevada. Other outer-detrital deposits of the Lower

Ordovician include a portion of the Cow Head Breccia of western Newfoundland (Section 3.b), the Collier/Crystal Mountain/Mazarn/Blakely Formations of the Ouachita region, the Bliss Sandstone of western Texas, and the sandy Descon and Deadwood Formation of (respectively) southeastern Alaska and the Williston Basin.

In the latest Cambrian and in the Ordovician and later times, deeper-water sediments, such as those named above, bear abundant graptolites and are thus termed the *graptolitic facies*. Such units are the result of open-ocean, deep-water sedimentation in which planktonic organisms yield virtually the only preserved fossils. They contrast with carbonate facies of the epeiric seas, which contained a high proportion of carbonate-shelled animals, such as articulate brachiopods, mollusks, and in the Middle Ordovician, bryozoans and coral. The carbonates have been collectively referred to as the *shelly facies* and represent generally shallow benthic environments. In the latest Cambrian and Early Ordovician, the graptolitic facies characterized deep continental-marginal environments, while the shelly facies dominated most submerged parts of the craton and continental shelves.

In post-Canadian and pre-Champlainian time (latest Early Ordovician to just before the Middle Ordovician), the Sauk sea withdrew from the craton. There is a widespread unconformity marking the interval in all but marginal areas of the continent, and in many sections in the west there is evidence of renewed detrital sedimentation from erosion of exposed basement rocks or Cambrian sandstones. In the Great Basin, withdrawal is indicated by deposition of the Eureka Quartzite in southeastern and central Nevada, the Kinnikinic Quartzite in Idaho, and detrital members of the Swan Peak Formation in northern Utah. In many parts of the midcontinent and the eastern craton, carbonates of the Knox, Prairie du Chien, Arbuckle, and Beekmantown units among others are overlain unconformably by Middle or Upper Ordovician sands or carbonates. The lack of regressive Sauk facies in these sections suggests either that post-Sauk erosion stripped such sediment away or, more likely, that there was no source of detritus exposed for detrital sedimentation to occur before the next sea transgressed onto the craton.

E. Earliest Animals with Skeletons: First Recorded Taxonomic Radiation

E.1. On the Appearance of Skeletonized Animals in the Lowermost Cambrian

It was shown in Section C that lower Sauk sediments can often be divided into Cambrian and Precambrian components only by the presence of skeletal fossils in the former. This phenomenon is nearly global in scope and led to the assumption of an "explosive" evolutionary event among skeletonized organisms at the Precambrian/Cambrian boundary.

The first Cambrian faunas to appear in the boundary sections are somewhat variable among continents with basal Cambrian sequences, but generally they are characterized by the following skeletal animal assemblages:

- Obolellid brachiopods
- Archaeocyathids
- Mollusk-like univalves
- Non-trilobite arthropods
- Olenelline or redlichine trilobites

The remainder of this chapter will be a brief characterization of the important Cambrian fossil groups. Actually, nearly all animal phyla including Chordata were extant by the latest Cambrian, but most were minor accessories in preserved Cambrian faunas. Each major group of organisms will be introduced in the context of its initial occurrence of stratigraphic importance in the fossil record.

E.2. Trilobites

The Cambrian sea was apparently full of trilobites and they dominate overwhelmingly the Cambrian stratigraphic record in North America. Trilobites are an extinct class of marine arthropods (see the Appendix for a synoptic taxonomy of fossil groups), which appeared in Early Cambrian time and became extinct in the Permian. They are characterized by having bodies flattened dorsoventrally (i.e., top-to-bottom), divided into three regions in cross section (two side portions, or *pleurae*, and an axis) and into a distinct tail (*pygidium*), body (*thorax*), and head (*cephalon*).

Their morphology and the strata in which they occur show that most trilobites were benthic, but some may have been nektonic or planktonic (see Cisne, 1974, and Robison, 1972). They were probably not predators since they lacked real mouthparts. Along with the stratigraphic appearance of trilobites is the appearance of multilimbed crawling (*Cruziana*) and resting (*Rusophycus*) tracks in Cambrian sediments. The nature of some *Cruziana* suggests that many benthic trilobites were grazers or detritivores, somewhat in the manner of horseshoe crabs, living and feeding just at the sediment–water interface.

E.2.a. Morphology

Because trilobites are the most widely used Cambrian index fossils, detailed studies of their morphology are numerous and a large number of morphological characteristics have been defined. Figure 4-18 shows the most important characters used in systematic description of the major taxa. Features specific to or most characteristic of trilobites include the *facial sutures*, which are zones of weakness in the head shield along which the exoskeleton could break predictably during molting; the *glabella*, a nose-shaped hump on the head shield which probably covered some sort of stomach or gut system; and the *hypostome*, a ventral (bottom on a trilobite) flap of exoskeleton located under the glabellar region and which may have covered the mouth.

Early classifications of trilobites used facial sutures for discrimination of higher taxa. This has been largely superseded by taxonomies based on multiple characteristics [see Harrington (1959) in the *Treatise on Invertebrate Paleontology*] but there is no completely natural classification system available to date.

The trilobite exoskeleton was composed largely of calcite, in several layers, with a sizable organic matrix in and between the layers. External surfaces feature a variety of sculpting, anastomosing lines, canals, tubercles, and other ornamentations. Margins of exoskeletons were commonly spiny (in some types featuring spines longer than the body), but many were smooth. The border of the cephalon tucks under the anterior and lateral margins on the ventral side in a structure termed the *doublure*. A similar marginal doubling was commonly present under the pygidium. These marginal areas, together with the hypostomata and a few smaller plates of exoskeleton under the head, comprised the ventral skeleton. Presumably, most of the remaining underside of trilobites was composed of legs and soft tissues.

Muscle scars, other shell structures, and, especially, a number of specimens preserved with soft anatomy intact (see Cisne, 1974, and discussion of the Burgess Shale, Section E.6) yield good information about the appendages and anatomy of trilobites. Limbs under the thorax and pygidium were *biramous* (i.e., double-branched) with both "walking" and gill portions. Appendages under the cephalon include a pair of antennae, a first pair of single-branched legs, and several biramous appendages. The first pair of cephalic appendages had enlarged basal segments which may have served as jaws, by means of rolling, crushing motions.

Trilobites are the oldest creatures known to have had eyes. Most feature compound eyes with lenses closely packed, as in modern insects; such eyes are termed *holochroal*. A major post-Cambrian group, the order Phacopida, featured a novel eye structure formed by individual round lenses and their internal accessories, usually much larger than the individual prisms of the holochroal eyes. The latter eye type is termed *schizochroal* and may represent an early evolutionary attempt at stereo vision (Clarkson and Levi-Setti, 1975). The lenses of schizochroal eyes consist of individual calcite crystals with the c axis oriented such as to eliminate the double-image obtained from any other orientation of optically clear calcite.

Several   groups   of   trilobites   were   apparently

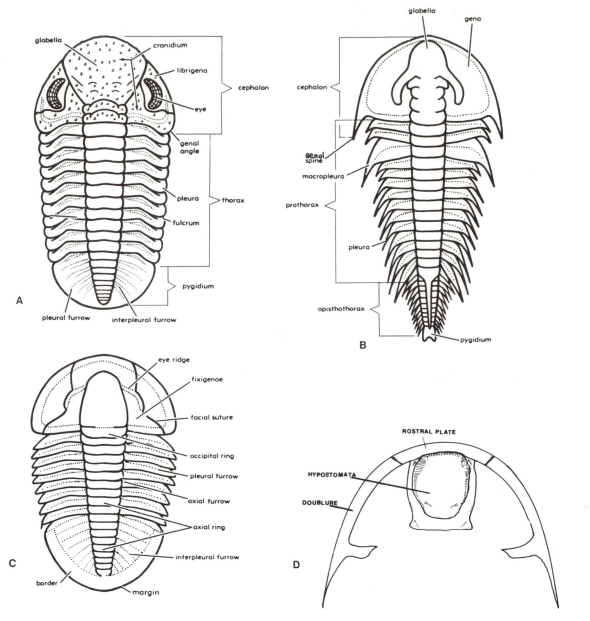

Fig. 4–18. Morphological features of trilobites. (A) An olenellid; (B) a phacopid; (C) a ptychopariid; (D) ventral features of a generalized trilobite cephalon. [A–C, from Moore (ed.), 1956, *Treatise on Invertebrate Pal-* *eontology,* courtesy of the Geological Society of America and the University of Kansas Press.]

planktonic/pelagic and show specializations for that niche; most were blind and small (e.g., the Agnostida), and some feature apparent flotation modifications such as frilled, wide borders (e.g., the Trinucleina) or very long spines (e.g., some corynexochids such as *Oryctocephalus*). At the other extreme, certain taxa were very large and heavy, and clearly could only crawl on the bottom (e.g., *Terataspis*) or, in the case of certain very smooth forms (e.g., the asaphinids such as *Isotelus*), could perhaps tunnel or "swim"

through bottom sediment. Most trilobites probably both crawled and swam, in a manner similar to modern horseshoe crabs.

Trilobites of several lineages were able to enroll in the manner of modern terrestrial isopods ("pill bugs"). Taxa with large tails when enrolled generally featured a tight fit between head and tail, with the thorax completely covered ventrally. Small-tailed (or *micropygous*) trilobites also were able to enroll, sometimes by means of a twofold pattern which tucked the tail under first, followed by bending of the thorax. Many post-Cambrian trilobites are commonly found enrolled (e.g., *Flexicalymene, Phacops*) whereas Cambrian taxa are rarely in that state.

The pattern of larval development is known for many trilobite taxa, and varies among groups (e.g., see Palmer, 1962). Typically, the newly hatched trilobite consisted of a round or oval single component, termed a *protaspid,* which increased in size at each molt and to which was added a tail as a furrow developed at the posterior part of the single shield. When the pygidium was free, the larva became a *meraspid,* and thoracic segments were subsequently added until the full adult number was reached. The last thoracic segment transformed the trilobite into a *holaspid,* which still needed several additional molts to complete adult shape.

E.2.b. Evolution and Taxonomy

The oldest trilobites, and presumably the most primitive and perhaps ancestral types, are assigned to the order Redlichiida (Harrington, 1959) and feature elongate, crescentic eyes, tiny pygidia, usually numerous segments (more than 15), spiny margins, and long, multisegmented glabellae. The two suborders of redlichiids vary in the nature of their cephalic sutures: the olenellines feature *ankylosed* (i.e., joined, inoperative) sutures, whereas the redlichines have *opisthoparian* sutures, which originate at the anterior corners of the head, and exit between the posterolateral corners (the *genal angles* in Fig. 4-18) and the *occipital ring*. Opisthoparian sutures are a common type, as will be noted below, and trilobites with this feature typically are found with the cephalon broken along the sutures. The central part of the cephalon, minus the lateral portions, is termed the *cranidium*. The lateral portions with the eye surfaces and the genal angles attached, are termed *librigenae* or "free cheeks."

Olenellines (Fig. 4-19) are common fossils at the bottom of the Cambrian section in North America, and are exclusively of Early Cambrian age, whereas redlichines are representative of the Lower and Middle Cambrian strata of Atlantic (European) Province. Middle Cambrian redlichines of the Atlantic Province are largely assigned to the superfamily Paradoxidacea, and *Paradoxides* species are largely the basis for the chronological and paleobiogeographical assignment of exotic Cambrian strata in southeastern Newfoundland, eastern New England, and the Carolina Slate Belt (see Section D.6).

Shortly after the olenellines appeared, several other trilobite groups joined Early Cambrian biotas. The most aberrant are the Agnostida, which differ from all other trilobites in being very small, having no more than three thoracic segments, and in having virtually identical pygidia and cephala. Of the two agnostid suborders, the Agnostina feature two thoracic segments and were always blind. The Eodiscina have three thoracic segments and in a single family (Pagetidae) have eyes. It is postulated (Robison, 1972) that many, if not all, agnostids were planktonic/pelagic. The phylogenetic relationships among most trilobite groups are obscure (Eldredge, 1977), but those between the agnostids and all other trilobites are especially unclear. Agnostines survived until well into the Late Ordovician, whereas eodiscines were extinct before the end of the Middle Cambrian.

Also in Lower Cambrian strata appear two additional trilobite orders: Corynexochida and Ptychopariida. Corynexochids are an exclusively Cambrian group which most closely resemble redlichines in having opisthoparian sutures and prominent glabellae; however, they also have a reduced number of thoracic segments (typically eight), and large, often spiny, pygidia. The group includes a wide variety of morphologies and may represent an artificial lumping of taxa. Ptychopariids include an even more diverse group of trilobites, typically featuring opisthoparian sutures and 12 to 14 thoracic segments. All post-Cambrian trilobites, save for surviving agnostines, probably descended from Cambrian ptychopariids. Primitive ptychopariids (Ptychopariina), of the Lower and Middle Cambrian, comprise a group of subtly varying morphologies which almost defy classification. Ptychopariines survived through the Ordovician but by the end of the Cambrian apparently had spun off virtually all post-Cambrian stem groups.

The post-Cambrian ptychopariid suborder Asaphina contains large, smooth, opisthoparian trilobites which may have lived infaunally. The suborder Trinucleina includes small, blind trilobites with tiny thoraxes and pygidia and large heads with a prominent marginal frill; these likely were planktonic. A suborder of slightly larger, sighted, probably benthic taxa is the Harpina, which also had cephala with marginal frills.

A major radiation of taxa occurred in the Early Ordovician and probably derived from ptychopariid ancestry. Most important among the post-Cambrian taxa is the order Phacopida, which dominated the Ordovician through Devonian trilobite biotas. They are distinguished by having a more highly derived form of facial suture, which cuts across the posterior cephalon ahead of the genal region, thereby forming a small free cheek. This *proparian* suture in time became vestigial and a radical molting pattern evolved, in which the entire head was shed in one unit. In addition, the schizochroal eye, mentioned previously, is a phacopid feature.

Another largely post-Cambrian order, the Proetida, evolved from the ptychopariids in the Late Cambrian and were relatively unchanged from the ancestors except with respect to the tails, which typically were large and multiseg-

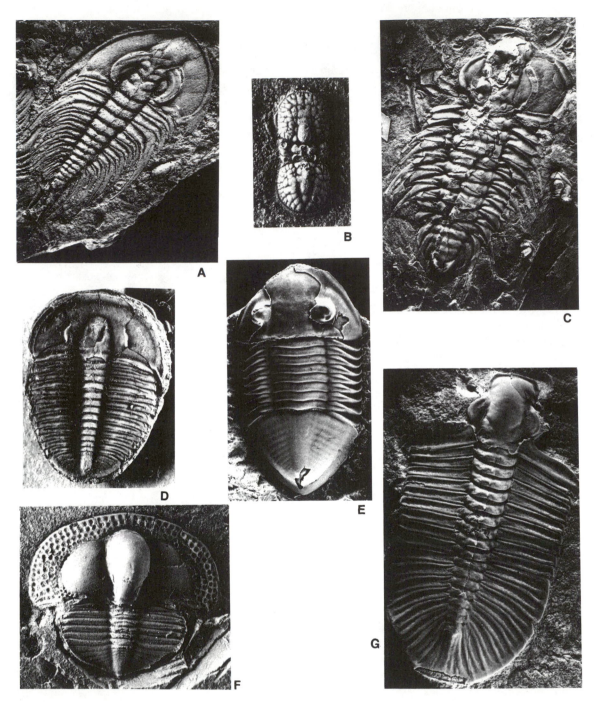

Fig. 4-19. Representative Cambrian and Ordovician trilobites. (A) *Olenellus* (Early Cambrian), Olenellida; (B) *Glyptagnostus* (Late Cambrian), Agnostida; (C) *Zacanthoides* (Middle Cambrian), Corynex-ochida; (D) *Elrathia* (Middle Cambrian), Ptychoparina, Ptychopariida; (E) *Isotelus* (Middle Ordovician), Asaphina, Ptychopariida; (F) *Cryptolithus* (Middle and Late Ordovician), Trinucleina, Ptychopariida; (G) *Orria* (Middle Cambrian), Corynexochida. (Specimens made available for photography by Dr. Fred Collier, U.S. National Museum of Natural History. DRS photographs.)

mented. Proetids include the last trilobites of the Paleozoic and are the only taxa found in post-Devonian rocks. Two additional, relatively minor orders, Lichida and Odontopleurida, also appeared in the Ordovician, disappeared in the Late Devonian, and probably descended from ptychopariid ancestors.

The end of the Cambrian represented a crisis in trilobite history, with extinction of the corynexochids as well as many lower taxa; and the end of Sauk time represents a second crisis. After Sauk time, trilobites never again constituted the dominant taxon in any preserved assemblage: one does not find mid-Paleozoic age "hash" composed of trilobite fragments, such as is common in many Sauk sediments.

E.3. Archaeocyatha

In several regions of the world, notably South Australia, Siberia, USSR, and China, archaeocyathids are dominant elements of the lower Cambrian faunas. They also have the distinction of being one of the very few (or, according to some taxonomies, the sole) extinct phyla.

In general, archaeocyathids were animals featuring erect, conical or tubular, calcitic skeletons, which typically were composed of two concentric cups. Between the cups were *septa* (i.e., dividing walls) and both the internal and external cups, as well as the septa, were usually perforated with numerous pores. The overall shapes varied considerably (see Fig. 4-20), and some colonial archaeocyathids formed bioherms; indeed, Rowland (1984) argued that archaeocyathids built true "framework" reefs (see Chapter 5 for discussion of the "reef" as a concept). Maples and Waters (1984) identified archaeocyathan patch reefs in the Shady Dolomite in Georgia.

Since they are extinct, and no soft-tissue casts are available to yield information such as is the case for trilobites, details of the soft anatomy are unknown. Their occurrence in carbonate-rich benthic sedimentary environments suggests a coral-like habit; however, their skeletal morphology clearly differentiates them from corals. They were almost certainly immotile except when shifted by currents. The porosity of the skeletons suggests that they were filter feeders and, in fact, the conical, open-topped shape of their skeletons may have evolved as a water-circulating mechanism.

The taxonomy of archaeocyathids is complex below the class level, and is based on details of the skeletal walls, the nature of the intervening space between cups (termed the *intervallum*), the nature of perforations in walls and septa, presence of horizontal (*tabulae*) and convex (*dissepiments*) elements in the skeleton, outgrowths on the skeleton, the nature of the basal structure or *holdfast*, the exter-

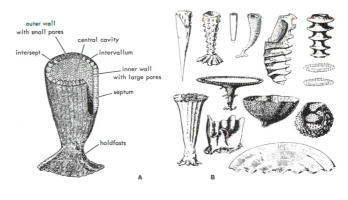

Fig. 4-20. Archaeocyatha. (A) Reconstruction of a regular archaeocyathid with general morphology labeled; (B) variations in external morphologies known among archaeocyatha; (C) reconstruction of an Early Cambrian benthic habitat with the inferred relationship between depth and growth form in archaeocyatha. (All from Hill, 1972, *Treatise on Invertebrate Paleontology,* courtesy of the Geological Society of America and the University of Kansas Press.)

nal shape, colonial or solitary nature, and other characteristics. The division into two classes, "regulars" and "irregulars," is based largely on whether during ontogeny the inner wall of the skeleton appears earlier (in the regulars) or later (irregulars) than the dissepiments. In addition, the regularity of pores in the walls and septa varies between the two classes as the names imply. Ordinal classification is based on the presence/absence of inner walls, septa, and radial tubules; and further taxonomic subdivision is based on the presence and types of tabulae (Hill, 1972). In general, the regulars are the more diverse class and include both solitary and colonial forms. Nearly all archaeocyathids are of Early Cambrian age, but a few taxa from both classes are found in early Middle Cambrian rocks, and a single, dubious genus is reported from the Upper Cambrian of Antarctica (Hill, 1972).

E.4. Inarticulate Brachiopods

As noted in previous discussions of the earliest Cambrian faunas, phosphatic inarticulate brachiopod shells are among the lowermost (although not absolutely lowest) Cambrian fossils found in many sections. The earliest brachiopods are largely from the order Obolellida, although several additional phosphatic inarticulate groups are characteristic of the Cambrian.

In general, brachiopods are bivalved marine animals which feature a distinctive feeding/respiratory mechanism called a *lophophore*. Two additional lophophorate phyla are known, Bryozoa and Phoronida (see the Appendix), of which the bryozoans comprise very important fossil taxa (see Chapter 6) and the phoronids do not. Brachiopods also feature a supporting/suspensory/or retractile structure termed a *pedicle*, and unique among bivalved organisms, one shell is modified to accommodate the lophophore structure whereas the other is modified to allow the pedicle to emerge. This yields the terminology of a pedicle valve and a brachial valve, because the lophophores are supported by skeletal structures termed the *brachia*.

Two subphyla, Articulata and Inarticulata, are defined by the shell mineralogy, nature of the shell articulation, musculature, and evolutionary grade. Inarticulates (see Fig. 4-21) are the more primitive and lack shell hingement mechanisms, generally have phosphatic shells, and have complex musculature which must both open and close the valves and hold them in relative position. Articulates, the more highly derived group, typically have calcareous shells, possess a hingement mechanism, and have musculature chiefly concerned only with opening and closing the valves. Although articulate brachiopods appeared in the Cambrian, their prominence in the fossil record begins in the Middle Ordovician, and we therefore will postpone extensive discussion until Chapter 5.

Four inarticulate brachiopod orders are commonly recognized. Most typical are the Lingulida, which feature small-to-large, *equilateral* (i.e., each valve symmetrical), *equivalved*, tongue-shaped, phosphatic shells, a long retrac-

tile pedicle, and, uniquely among brachiopods, tolerance for brackish water. Lingulids live infaunally, in slit-shaped burrows, and subsist by filtering seawater. The pedicle is used to rapidly pull the body into the burrow. The lingulid mode of life apparently has been very successful because members of the general group (i.e., the family) date from the Early Cambrian to the present.

Among the three remaining inarticulate orders, the Acrotretida are small, inequivalved brachiopods with phosphatic shells and reduced or absent pedicles. Fossil acrotretids are common in the Cambrian and tend to resemble small, rounded fish scales; however, in an uncompacted state, one valve is indeed flattish whereas the other is conical. Like lingulids, acrotretids are living fossils and survive to the present. Obolellida are an exclusively Cambrian order consisting of moderate-size (~ 0.5 cm) calcareous forms with subrounded valves featuring eccentric growth lines and a slightly flattened hinge line. The final order, Paterinida, range from the Cambrian through the Ordovician and show advances suggestive of their being ancestral to articulate brachiopods. Paterinids featured phosphatic, subrounded valves which generally resemble those of obolellids; however, they also feature distinctive musculature and shell specializations for the pedicle to emerge. In spite of their abundance in Cambrian strata, inarticulates are poor index fossils because of their morphological simplicity and long-ranging occurrences.

E.5. Cambrian Echinoderms

Echinodermata is an animal phylum which bears no known evolutionary relationship except with Chordata. The structure of echinoderms is based on a unique water-vascular system, and all but the primitive groups show characteristic pentaradial symmetry. Of the numerous fossil echi-

  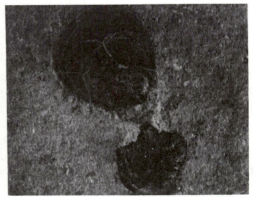

Fig. 4-21. Representative inarticulate brachiopods; generic identifications in quotation marks are uncertain. (A) "*Lingula*" (this specimen Middle Devonian), a typical lingulid; (B) *Obolus* (Late Cambrian), a lingulid with an atypical suboval outline; (C) "*Acrotreta*" (Late Cambrian), an acrotretid. (All are DRS photographs.)

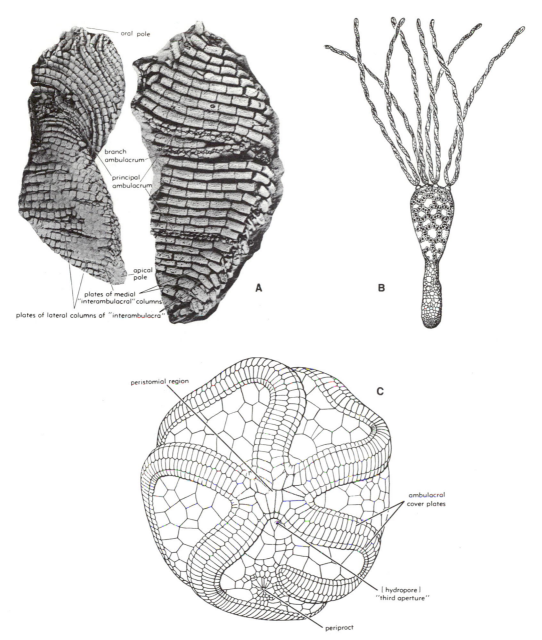

Fig. 4-22. Representative primitive echinoderms found in Cambrian strata (and younger strata except A). (A) *Helicoplacus*, a rare form, restricted to the Lower Cambrian of the western United States. (B) *Gogia*, an eocrinoid of primitive morphology, relatively common in western North American Cambrian rocks. Eocrinoids persisted until the Silurian and may be ancestral to crinoids (which, however, date back to the Early Ordovician).

(C) A representative edrioasteroid. This particular taxon, *Edrioaster*, is of Middle Ordovician age but other edrioasteroids range from Early Cambrian to Mississippian age. (All from the *Treatise on Invertebrate Paleontology*, courtesy of the Geological Society of America and the University of Kansas Press.)

noderms, three major groups appeared in the Cambrian (see Fig. 4-22) and all but one of those are extinct.

Edrioasteroids are classified (Regnell, 1966) as a class in the subphylum Echinozoa, which includes sea urchins and sea cucumbers. They featured saclike bodies, covered with small plates of calcite and surmounted with a starfish-shaped structure which was the external portion of the water-vascular system; this structure is termed an *ambulacrum* and is a common feature of many echinoderms. Edrioasteroids resemble a creature from the Ediacarian Pound Quartzite named *Tribrachidium* and may be its descendants; however, *Tribrachidium* featured a threefold symmetry in contrast to the pentaradial pattern of younger echinoderms. Edrioasteroids also share a few features with the stalked echinoderms (or pelmatozoans, see below); notable are details of the *hydropore,* a single opening to the water-vascular system, and in the makeup of plates. They are represented by a single Middle and Late Cambrian family, but new groups appeared in the Ordovician, and several survived to the Mississippian.

Several rare, primitive types of Pelmatozoa (stalked echinoderms) are known from the Cambrian; eocrinoids are the most common. These are primitive representatives of what subsequently was a major radiation in the Ordovician Period, distinguishable from true crinoids (see Chapter 5) by the presence of pores along sutures between plates in some taxa, and by several details of the body plates. A very exotic, extremely rare group of Early Cambrian echinoderms, called Helicoplacoidea, come from western North America (Durham and Caster, 1963). These football-shaped sacs of calcite plates are oriented in a spiral from what appears to be the oral end to what may be an anal end. Only a few specimens are known and their relationships to other echinoderms are unknown.

Although we will discuss chordates in virtually all subsequent chapters, it is pertinent here to note that echinoderms precede the earliest chordates only slightly, and very likely have some part in the origin of our phylum. A fossil fish report from the Upper Cambrian of Wyoming (Repetski, 1978) pushed the earliest vertebrate record back from the previously reported record in the Ordovician. It is possible that more extensive paleontology in the Cambrian may reveal older and better-preserved vertebrate ancestors and it may reveal transitional echinoderm/vertebrate ancestors.

E.6. Cambrian Miscellanea and the Burgess Shale Fauna

Discussions of "dominant" taxa for a given time tend to obscure the presence of many other, nondominant or unspectacular forms which may still have been significant parts of the biota. Thus, we have discussed for the Precambrian: jellyfish, other cnidarians, putative annelid worms, and miscellanea; and Cambrian trilobites, archaeocyathids, brachiopods, and echinoderms. The list below (data from Harland *et al.*, 1967) shows other major animal groups present by the latest Cambrian, all of which will be discussed in subsequent chapters.

- Foraminifera: allogromids
- Porifera: hyalosponges, demosponges
- Annelida: errant and sedentary polychaetes
- Onychophora

- Arthropoda: trilobitomorphs, merostomes (aglaspids), ostracodes
- Molluska: gastropods, monoplacophorans, nautiloid cephalopods, rostroconchs, and other rare bivalves
- Graptolites: dendroids
- Chordata and Hemichordata
- Priapulida

Conspicuously absent from the list are coral, bryozoa, most bivalves, diatoms, higher plants, and radiolaria: all of these appeared in the Ordovician. Several small gastropods are common elements of Early and Middle Cambrian rocks, including *Helcionella,* a simple cap-shelled form which may represent a distinct subclass, and *Hyolithes,* a small, conical shell of undetermined relationships. By the Late Cambrian, several more typical gastropod groups were common. Insoluble residues of Cambrian limestones may yield sponge spicules, suggesting that they were common parts of the biota. In this context, finally, we should remind the reader that algae, especially calcareous algae, were abundant in the Cambrian and were the major biohermal organisms as well as the dominant producers of carbonate sediment.

A large proportion of the exotic elements in the Cambrian biota are known from the Burgess Shale. In 1910, on a trip through Burgess Pass 5 km north of the town of Field, British Columbia, Charles D. Walcott discovered one of the greatest fossil finds in history. A slab of rock overturned by one of Walcott's horses contained beautifully detailed fossils of soft-bodied animals; Walcott discovered that this slab came from a 2-m-thick lens of black shale at the base of the Middle Cambrian Stephen Formation (Fig. 4-15).

The environment of deposition of the Burgess Shale may have been at the base of a very large, high-standing carbonate shoal or bank, perched above a deep-water basin (Conway Morris and Whittington, 1979). Although soft-bodied organisms are present in great abundance in the rock, their trails and burrows are notably absent, suggesting that they were rapidly buried (and, of course, killed) by one or more slumps of debris from the bank or a higher shelf. These slump sediments may have carried the invertebrate fauna into anoxic (and, therefore, poisonous) deep-water muds, thus preserving them. Rapid burial by such a mechanism would prohibit movement by the dying animals, hence the lack of trails or burrows; and, as the slump deposits compacted, the organisms were flattened, devolatilized, and preserved as perfect carbon-film residues.

The Burgess Shale shows us in detail the diversity of the Middle Cambrian fauna, and dramatizes how little of it is generally preserved. Among the fossils (Fig. 4-23) are several types of annelid and other worms, Onychophora

Fig. 4-23. Fossils of soft-bodied organisms from the Burgess Shale. (A) *Waptia*, a crustacean-like arthropod, showing external gills; (B) *Aysheaia*, an onychophore (a transitional phylum sharing characteristics of annelid worms and arthropods); (C) (center) *Canadia*, a polychaete annelid worm, (upper right) *Marrella*, an arthropod with undetermined affinities; (D) *Burgessia*, a form with undetermined affinities (possibly an arthropod), showing paired antennae and internal paired diverticula. (Specimens made available for photography by Dr. Fred Collier, U.S. National Museum of Natural History. DRS photographs)

(transitional forms between annelids and arthropods), jellyfish, sea urchins, pteropods, holothurians ("sea cucumbers"), trilobites and a number of trilobitelike forms called *trilobitomorphs,* and organisms which still cannot be accurately placed in higher taxa (Conway Morris and Whittington, 1979).

It will be noted in several subsequent discussions [e.g., concerning the Mazon Creek Pennsylvanian fauna (Chapter 7), the Jurassic Solenhofen Limestone and the Cretaceous Niobrara Chalk (Chapter 8), and the Eocene Green River Shales (Chapter 9)] how profoundly a perfectly preserved set of fossils can alter one's perception of past life. Where soft anatomy and soft-bodied organisms are preserved, one often receives a completely altered sense of the proportion of the ancient biota which is fossilized, and simultaneously an altered sense of dominance within past ecosystems.

E.7. Cambrian and Early Ordovician Paleobiogeography

It has been observed repeatedly in the literature that early Paleozoic trilobites are strongly patterned in their occurrences, and attempts to interpret such patterns have been made. Early work was done chiefly by Lochman-Balk and Wilson (1958) in North America, who recognized trilobite "realms" or biofacies which correspond to the Middle and Upper Cambrian detrital and carbonate belts described by Palmer (1960). Thus, the outer-detrital facies comprised an oceanic realm, with free access to the sea, featuring cosmopolitan-pelagic trilobites including numerous agnostids. In contrast, the inner-detrital facies comprised a cratonic realm, with *endemic* (i.e., geographically restricted) trilobite faunas lacking many agnostids and containing taxa quite different from those of the oceanic realm. The carbonate banks contain a third, mixed fauna which generally contains more oceanic than cratonic trilobites. Clearly, the oceanic faunas may be shared between continents, whereas the cratonic faunas are restricted to single continents; nevertheless, cratonic trilobites in North America may be wideranging. For example, nearly identical Middle Cambrian ptychopariid species may be recognized as far apart as the Conasauga Group in Georgia and the Meagher Limestone in Montana (Schwimmer, 1975, and personal observation). Even more curiously, those same species have no close kin in outer-detrital assemblages scant hundreds of kilometers away in the outer detrital zone.

There have been several attempts made to explain the distinctive separation of trilobite realms within the Cambrian ocean. Lochman-Balk and Wilson (1958) postulated tectonic influences (volcanism within the outer-detrital zones) as a major control over Middle and Upper Cambrian trilobite biofacies. Palmer (1969) suggested that topographic barriers, resulting from the elevation of carbonate buildups to nearly sea level, impeded migrations of benthic trilobites between the inner- and outer-detrital environments. Taylor (1977) proposed more purely environmental factors, such as water temperature and depth, to have been the cause of biofacies zonation.

Lower Cambrian trilobite biofacies patterns are considerably different from those of the Middle and Upper Cambrian. Before establishment of the zoned detrital and carbonate lithofacies, various regions of North America (and Ancestral Europe) had relatively restricted flooded areas, and the relative positions and relationships among such areas are not clearly demarcated in the stratigraphic record. In very general terms, an "olenellid" province was established in the Lower Cambrian of most of North America, and this was shared in part (albeit with generic differences) with northwestern Europe. Most of Asia, Australia, and Antarctica shared a different group of Lower Cambrian trilobites and comprise the "redlichiid" province, whereas southwestern Europe, northern Africa, and eastern USSR featured yet another characteristic Lower Cambrian assemblage and comprise the "*Holmia*" faunal province. Once again, the reader should note that these three faunal provinces are not internally uniform at the lower trilobite taxonomic levels and may be artificial groupings.

One may infer that northern Europe and eastern North America show related trilobite assemblages because the continents were barely separated by Iapetus during the Early Cambrian; thus, it is no surprise that nearly identical species of *Callavia* (Lower Cambrian) and *Paradoxides* (Middle Cambrian) may be found in the "Atlantic" facies in the Appalachian region (i.e., outer-detrital zone and suspect terranes) and in certain northern European strata. Definitive trilobite biozonation patterns will emerge only when reliable global reconstructions are available for the Cambrian, and when the environmental requirements of trilobites are thoroughly understood. As an example of working with the limited knowledge of the above elements of trilobite zonation, Jell (1974), in a controversial study, used quantitative techniques to delimit three global Middle Cambrian faunal realms; and from these, he attempted global reconstructions of Cambrian geography. One of his conclusions is that Cambrian Gondwanaland must have been crossed by a seaway which provided a migration route for trilobites.

Early Ordovician trilobite biogeography has also been examined. Whittington and Hughes (1972) observed four biogeographic provinces, with one including North America, Siberia, and northernmost Europe (exclusive of England and areas south). It was their conclusion that, as with Cambrian biofacies, both continental positions and ecological factors (especially temperature) controlled Ordovician trilobite distribution.

CHAPTER 5

The Tippecanoe Sequence: Middle Ordovician–Lower Devonian

A. Introduction

The second great Paleozoic epeiric sea, the Tippecanoe sea, transgressed on the craton beginning in the early Middle Ordovician and withdrew totally in the Early Devonian. During this interval a number of important events transpired, among the most significant being the Taconic Orogeny on the Appalachian continental margin. Beginning with Tippecanoe topics, and continuing virtually through the remaining text, we will split (somewhat artificially) discussion of cratonic events from discussion of marginal/tectonic events; this is a necessity due to the large variety of conditions which were present on the post-Cambrian continent.

A.1. Overview

A.1.a. Craton

Sedimentation in the Tippecanoe sea began in the Middle Ordovician and followed initially the pattern of the early Sauk sea. A widespread blanket of mature sand was deposited as the transgressing sea encroached on the older, eroded, cratonic surface; then, as the epeiric sea deepened and the shoreline stabilized, carbonate sedimentation occurred over most of the craton. Carbonate sediments comprise the basal Tippecanoe facies in regions of the craton isolated from sources of detritus. In the Late Ordovician (Trenton–Maysville time), the northeastern craton began to receive a heavy load of detritus from the emerging Taconic highlands, and by Richmond time (latest Ordovician) the influence of this detrital source had spread into what is now the Mississippi Valley.

Lower Silurian (Medinan) history is difficult to interpret because sediments from that time are lacking over much of the craton. A widespread regression may have occurred during this interval or Medinan deposits may have been stripped off in later times. The preserved rock record shows that the craton was undergoing some tectonic activity and beginning to form the basins of the present Great Lakes region (Michigan and New York) and the Williston (or Interlake) Basin located in Montana/Dakotas/southcentral Canada. These basins remained as prominent structural features through the Paleozoic Era. Most Middle Silurian (Clinton–Lockport) sediments on the craton are carbonates associated with coral/stromatoporoid/algal reefs (see Section D). The northeastern craton was still receiving detritus from the Taconic region, but the amount decreased throughout the Silurian.

The Upper Silurian (Salina) section of the Great Lakes region is characteristic and distinctive because it contains both evaporites and patch reefs; again, very little of this interval's sediments are preserved outside of cratonic basins due either to post-Salina erosion or nondeposition during Salina time. The notable development of saline conditions in Salina time suggests that seas were shallowing; therefore, nondeposition would seem to best account for the nature and distribution of sediments.

Cratonic sedimentation continued to be restricted to basins in the Great Lakes region and central New York, on into the Early Devonian, as the Tippecanoe sea withdrew completely to the craton margins. The sea left regressive sequences in some areas, especially in the New York region; and, finally, those basins emerged, marking the end of the cratonic sequence.

A.1.b. Continental Margins

Beginning in later Ordovician time, the continent experienced the first documentable marginal tectonic activity of the Phanerozoic Eon. Events in the *Appalachian* margin (i.e., the eastern continental border) are generally clearer than are those of the three additional "sides": the western *Cordilleran*), northern (*Innuitian*), and southern (*Ouachita*) margins. Each region will be introduced in this chapter save for the Ouachita margin which was first notably active in Kaskaskia time (Chapter 6).

The initial Tippecanoe transgression reestablished carbonate-platform conditions on the Appalachian continental margin but soon thereafter the margin was disrupted by the Taconic orogeny. Plate convergence initiated a subduction zone offshore which crumpled up and metamorphosed much of the rifted margin prism. Associated synchronous uplift of the old continental slope and rise occurred as the outer continental shelf foundered, forming a foreland basin into which turbidity-current sediments were carried while allochthonous blocks slid from the newly uplifted area to the east. During the Late Ordovician, further compression uplifted the whole continental margin, resulting in mountains out of which streamed a great quantity of sediments. These comprised the Taconic detrital wedge which grew across much of the eastern United States. In the south, orogenic activity also occurred as the result of plate convergence, resulting in the smaller Blount detrital wedge. Waning sediment supply in the Silurian caused a transgressive sequence as detrital-wedge facies migrated back eastward. By the end of the Silurian, detrital sedimentation related to the Taconic orogen was nearly finished and a major period of carbonate-sediment deposition occurred.

Early events (especially) in the northern (or Innuitian) continental margin are somewhat indeterminable because of the ice cover but, in general, the region contained, to the south, the stable craton's northernmost edge, which experienced sedimentation from Sauk and Tippecanoe seas not unlike that of the more southern regions; farther northward, the northern continental shelf which in Tippecanoe time contained a broad, carbonate platform with shelf-edge algal buildups and which dropped off into the deeper waters of a marginal basin; and, bounding this marginal basin, was a magmatic arc which shed flysch southward into the basin. In Late Silurian and Early Devonian time, probable collision of East Greenland with an as yet undeterminable landmass produced the Caledonian orogeny across the far northeast. This orogeny initiated a number of fold belts and uplifts which strongly affected subsequent sedimentation. A number of structural trends in the region run perpendicular to the Caledonian structures and probably were controlled by preexisting trends.

In Tippecanoe (and earlier) time, the western, or Cordilleran, continental margin apparently contained elements of two major geological provinces which parallel the coastline: in the eastern regions, a broad, relatively stable continental shelf bordered the craton while somewhere (indeterminably) to the west a belt of exotic terranes is indicated by a discontinuous series of lithologies. The continental shelf facies were relatively stable through deposition of the Tippecanoe Sequence, comprising widespread basal detrital units on the inner shelf in the Middle Ordovician, followed generally by extensive dolostone deposition with algal shelf-edge buildups through the Early Devonian. The outer shelf and continental slope experienced largely mixed-carbonate and fine-grained detrital sedimentation and there are some deeper-water (possibly continental rise) units which are siliceous and volcaniclastic. The far western terranes (referred to as the "insular belt" or "borderland terranes") are of an extremely complex nature with mélanges, ophiolites, and a wide variety of lithologies suggesting subduction and obduction events in some areas and the presence of magmatic arcs across most of the margin. It has been suggested that as many as six microplates existed off the western margin of the continent in the Lower Paleozoic, accounting for the variability of lithologies found and for the obvious tectonic activity indicated by the rock units.

We close the chapter with the introduction of many organisms which evolved in the Tippecanoe sea (or which left their first common fossils in its sediments). These can be categorized as: planktonic taxa, such as the graptolites; reef-building taxa, such as bryozoans, coral, stromatoporoids, and lithistid sponges; benthic taxa, occupying a variety of niches, such as the articulate brachiopods, pelmatozoans, post-Cambrian trilobites, gastropods, and bivalves; and the nektonic invertebrates such as the cephalopods. In Tippecanoe time too, the vertebrates and some of their (possibly) ancestral forms became parts of the freshwater or marine faunas; and the shelled microorganisms, such as foraminifera and ostracodes, became common. Finally, during the Ordovician and Silurian, plants and animals (respectively) made their first documentable appearances on the land, foreshadowing the great forests of the Carboniferous and the rise of insects and amphibians.

A.2. Global Paleogeography

Figure 5-1 is a hypothetical reconstruction of continental positions during the Middle Ordovician, i.e., at about the time of the initial Tippecanoe transgression. The Iapetus Ocean has been estimated as having a width of 1000 km, based on British paleomagnetic data (Briden *et al.*, 1984). North America was located squarely astride the Equator, reaching to 30°N and 30°S latitude. The climate experienced by much of the continent was, thus, probably tropical, as indicated by the ecologically diverse nature of fossil assemblages and the presence of bioherms in many areas of the craton and continental margin. Chemical weathering in such a climate would probably have been intense, as may be

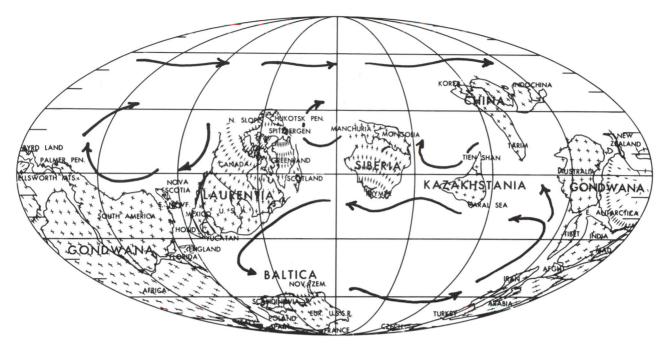

Fig. 5-1. Middle Ordovician paleogeography, showing hypothetical oceanic circulation patterns. (Modified from Scotese *et al.*, 1979.)

inferred from the great quantities of argillaceous sediments derived from detrital sourcelands of the Appalachian area. Lying at the Equator, the continent would have experienced prevailing, westward winds of the Northeast and Southeast Trades. That this may indeed have been the case is suggested by the distribution pattern of volcanic ash beds derived from the Taconic orogen of western North Carolina; these ash beds are found today in Middle and Upper Ordovician strata of Tennessee and Alabama, far to the west of the source volcanoes.

If the reconstruction given in Fig. 5-1 is reasonably correct, we may then postulate surface currents as shown. Speculations such as these on current patterns provide a rather stimulating exercise in applied oceanography but are of more than passing interest since surface currents control the distribution of planktonic organisms and of benthic organisms that have planktonic larval stages. They also control the climate of continental-margin areas by bringing relatively warmer or cooler waters in from other latitudes.

Figure 5-2 presents a reconstruction of the Middle Silurian Earth. The position of North America had changed relatively little since the Middle Ordovician and we will not elaborate further on environmental conditions. On the other hand, significant changes of other continental masses had occurred. Baltica had migrated to an equatorial position to the east of North America, while Siberia had moved relatively northwestward and stood off the northeast coast of North America. In the Southern Hemisphere, the South American–African coast of Gondwana had moved north-

ward, slowly diminishing the size of the Iapetus Ocean. A large part of Gondwana in the vicinity of the South Pole was covered by a continental glacier which left numerous tillite deposits in South America and Africa.

B. Tippecanoe Cratonic Sedimentation

B.1. The Post-Sauk Regression

The post-Sauk craton seems to have been an uneven surface even before erosion. Lochman-Balk (1971) noted that the cratonic surface had relatively high areas, including the Wisconsin Highlands, the Cincinnati and Findlay Arches, the western craton coastal plain, and uplifted southwestern and adjacent parts of Montana. In addition, the Transcontinental Arch, which was uplifted, perhaps initiating the regression, remained a strong positive feature on into Tippecanoe time and accounts for the absence of Ordovician exposures in most parts of the line of states from Minnesota to Arizona.

The topography of the cratonic surface after post-Sauk erosion was irregular, but of only modest relief. Fluvial erosion features are evident in many sections, showing a maximum relief of about 100 m to the bottom of channels, and many of the remaining Sauk carbonates show solution features.

Along the Transcontinental Arch, Hudson Bay, and parts of the Dakotas and adjacent Canada, all Sauk sedi-

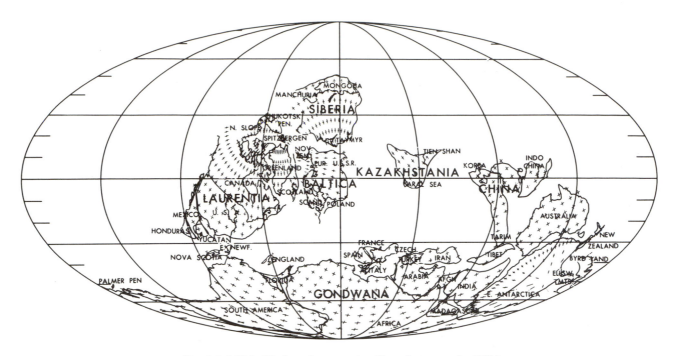

Fig. 5-2. Middle Silurian paleogeography. (From Scotese *et al.*, 1979.)

ments were eroded away exposing Precambrian basement rock or sediments. In other areas of the craton, where Sauk sediments were preserved, it is difficult to estimate the thicknesses of Sauk sediments lost to erosion and the precise time represented by the disconformity.

One characteristic of this erosional interval is the striking difference between the organisms in the strata above and below it, suggesting that the hiatus was significantly long, or that environments had changed significantly. The Tippecanoe sea was populated with bryozoans, coral, stromatoporoids, cephalopods, bivalves, a wide variety of articulate brachiopods, graptolites, and other forms which were either not present or were minor constituents of the fauna which inhabited the Sauk sea. The trilobites, so notable a feature of the Cambrian Period, became far less diverse in the Ordovician although they were locally present in considerable numbers. This fairly drastic change of faunas over the hiatus indicates that a considerable amount of evolution at all taxonomic levels had gone on at the margins of the continent, and that in fact the makeup of organic communities populating the shallow-water benthos had changed. Thus, one can conclude that the hiatus occupied a significant amount of time; or, alternatively, that the hiatus was brief but that the selection pressure due to confinement of benthic organisms to a shrinking environment (during regression) forced an explosive rate of evolution in the early Tippecanoe transgression.

B.2. Basal Tippecanoe Detrital Sedimentation

B.2.a. Blanket Sands: The St. Peter Sandstone

The Tippecanoe sea transgressed onto the broad, irregular, slowly submerging cratonic surface during Chazy time. As the shoreline migrated across very large areas of the craton, due to the low gradient of the submerging surface, the sea reworked sediments derived from exposed Cambrian sands and from newly eroded Precambrian basement rock. This detritus was redeposited in a thin, widespread sheet of sand which can be readily traced across much of the present upper Mississippi Valley, constituting the well-known St. Peter Sandstone and its correlatives (Fig. 5-3).

A study of the St. Peter (Dapples, 1955) showed that the prevailing transport of sands, as indicated by crossbedding, was to the southwest with the shoreline oriented northeast–southwest; thus, the St. Peter was interpreted as a product of longshore current deposition. A subsequent study by Dott and Roshardt (1972) suggests that at least locally, in Wisconsin, the St. Peter showed a less well-defined current direction. Interest in the St. Peter stems from several characteristics: it is a *blanket sand,* it is a supermature quartz arenite, and it is extensively mined for glassmaking.

Blanket sands are, as the term implies, widespread but thin. The St. Peter is no more than 130 m thick (and generally much thinner) but extends in an almost unbroken sheet from Colorado to Indiana (Krumbein and Sloss, 1963). In hand sample, it resembles granulated sugar, with each grain very well rounded, highly spherical, and slightly frosted. The sand is very well sorted, more than 99% pure quartz, and it features large- and small-scale trough cross-strata as well as tabular to concave-upward sets of cross-strata up to 15 m thick (Fraser, 1976). These large-scale cross-strata suggest submarine sand waves, and various authors have suggested that the St. Peter formed in one of several possible sublittoral depositional environments; it is possible that multiples of environments, all nearshore to sublittoral, may be represented by St. Peter deposition.

Friedman and Sanders (1978) note that in places, such as the Anadarko Basin of westcentral Oklahoma, the St. Peter sands in the subsurface are cemented by both quartz and calcite. The calcite was most likely supplied by leaching from overlying carbonate rocks.

The source of the detritus in St. Peter sands is not precisely known but we presume that much of it came from eroded Upper Cambrian sandstones in the Great Lakes region and southern Canada. These sands were probably brought down to the upper Mississippi Valley, along with detritus eroded from the granitic rocks of the Canadian Shield, as fluvial sediments in major river systems draining the central craton.

B.2.b. Other Basal Tippecanoe Detrital Sediments on the Craton

Outside of the present upper Mississippi Valley and areas to the west and south of the Transcontinental Arch, sands of approximately Chazy age seem to correlate with the St. Peter and probably represent transgressive facies of other shores of the Tippecanoe sea. These sediments include sands underlying (and perhaps part of) the Winnipeg Formation in the Williston Basin, the Harding Sandstone in Colorado and Wyoming, and sands occurring in at least one *diatreme* (a volcanic vent emplaced in sedimentary strata) in southern Wyoming (Chronic *et al.,* 1969). In the Oklahoma–midcontinent region (the Arbuckle Mountains), the formations of the Simpson Group overlay the post-Sauk hiatus; but it is difficult to interpret their lithostratigraphic correlations with the St. Peter because they contain varying lithologies (limestone, shale, and sands). They are apparently of Chazy–Black River age and thus were deposited roughly contemporaneously with the St. Peter, perhaps representing an offshore facies.

None of the sands listed above are as lithologically distinctive as the St. Peter, perhaps because they did not have the mature Upper Cambrian sands of the northeastern craton as part of their sediment supply; however, the Hard-

ing Sandstone was long famous as the formation containing the oldest vertebrate fossils, until the report of a Cambrian fish (Repetski, 1978) supplanted the distinction.

B.3. Chazyan Limestones: Basal Carbonates and an Introduction to Reefs

Our discussion of basal detrital sedimentation might lead to the conclusion that all first Tippecanoe sediments were sands; however, this is far from correct. Along the eastern cratonic margin, from New York to Alabama and as far west as Ohio, Kentucky, and western Tennessee, basal Tippecanoe strata are carbonates.

Apparently, the sources of quartz sand, which so strongly influenced the St. Peter and farther western depocenters, were isolated from the eastern margin in the early Middle Ordovician. Environments of deposition in the east were reestablished similar to those of latest Sauk time and, thus, carbonates formed just as they had in the Cambro-Ordovician. The first carbonates left by the Tippecanoe transgression contain evidence of deposition in supratidal to intertidal environments. These units include the lower (Head) member of the Day Point Formation (Chazy Group) in northeastern New York and vicinity, the Black River Group in central New York, the Row Park Limestone in Maryland, the Hell's Creek/Camp Nelson Formations in Ohio and Kentucky, the Lenoir Limestone in southeastern Tennessee, the Blackford/Holston Limestones in northwestern Tennessee, the New Market Limestone of Virginia, and the Stones River Formation of Alabama. Naturally, the precise character of each varies (as do all in collective groupings we use herein for purposes of managing huge amounts of data), but genetically they all record return of the sea to carbonate-favoring environments, relatively free of detritus.

As sea level continued to rise, intertidal environments were succeeded by shallow subtidal environments, and, since the local topography dictated the rate and area of changing environments, considerable heterogeneity follows in the succeeding strata. For example, in western Virginia, the New Market Limestone is overlain conformably by cherty, bioclastic strata of the Lincolnshire Limestone, whose brachiopod and bryozoan fossils suggest normal marine conditions. In northeastern Tennessee (Fig. 5-4), Ruppel and Walker (1984) describe the deposition of the Chickamauga Group (which includes the Blackford and Holston Formations at the base), in all its complexity, and they elaborate on the effects caused by local downwarping of the regional platform and by transport of terrigenous material into the basin as Middle Ordovician time progressed.

For the remainder of this discussion of the first Tippecanoe carbonates, we will focus on the reference section

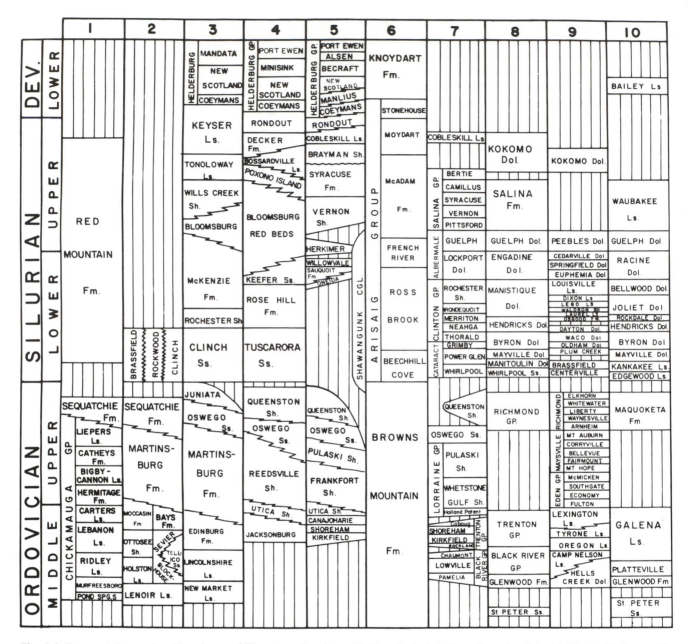

Fig. 5-3. Representative stratigraphic columns of Tippecanoe deposits; most columns are synthetic, having been assembled from multiple sources and from different localities within the area represented. Areas are as follows: 1, Georgia and Alabama; 2, Tennessee; 3, Virginia; 4, Pennsylvania; 5, eastcentral New York; 6, Nova Scotia; 7, western New York; 8, Michigan Basin; 9, Indiana, Kentucky, Ohio; 10, Illinois; 11, Kansas; 12, Arbuckle Mountains, southcentral Oklahoma; 13, Ouachita Mountains, Oklahoma and Arkansas; 14, central Colorado; 15, eastern Idaho; 16, northern Utah; 17, southeastern Nevada; 18, central Nevada; 19, northeastern Ellesmere Island (northern Canada); 20, northwestern Greenland.

in the Lake Champlain region of New York, Vermont, and Quebec. The type Chazy Group consists of the Day Point, Crown Point, and Valcour limestones, which overlie Cambro-Ordovician Beekmantown Limestones unconformably. Chazy limestones, and their correlative units to the south (Fig. 5-5), are commonly *bioclastic* (i.e., composed largely of fossil fragments); some Chazy limestones are also deposited in built-up mounds and may be termed *biohermal* (see discussion below). Although many Sauk carbonates feature a large proportion of trilobite fragments, these may be genetically distinguished from Chazy limestones in that the latter are composed of skeletal fragments from a diverse assortment of fossils (typically echinoderms, brachiopods, bryozoans, stromatoporoids, and algae).

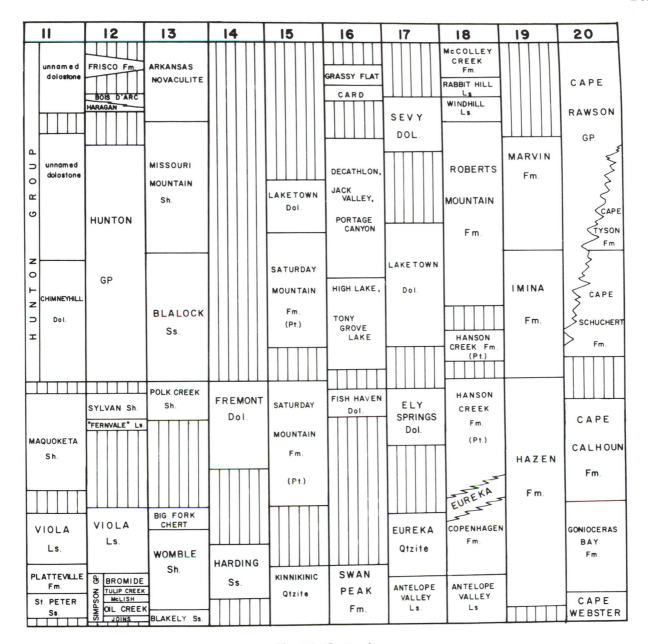

Fig. 5-3. (*Continued*)

B.3.a.   An Introduction to the "Reef"

Common for the first time in the Middle Ordovician, and reappearing time and again through the geological column, are structures which may be termed "multiorganism" reefs. They can be recognized typically by their apparent lack of stratification and by their rich collections of fossils.

The term "reef" originally derived from nautical usage and referred to any seabottom obstruction which could be hit by a ship. The term was lifted and applied to communities of organisms with a structural framework be-cause many such "reefs" were in fact hazards to navigation in tropical seas. In this book, we will adopt the following generalized but modern applications of the terms that describe the "reeflike" structures (Fig. 5-6) of ancient and modern seas:

- *Bioherm:* any mound-shaped structure built by skeleton-producing organisms
- *Reef:* a wave-resistant bioherm
  - *Patch reef:* an isolated, small reef
  - *Barrier reef:* a reef paralleling but separated from a shore

- *Fringing reef:* a reef paralleling and attached to a shore
- *Pinnacle reef:* a steep-sided, tapering patch reef
- *Atoll:* a ringlike bioherm surrounding a lagoon
- *Knoll:* a mound of carbonates below wave base
- *Biostrome:* a sheetlike accumulation of skeletal material

In addition, reefs allow the development of several characteristic sedimentary facies because of the different wave energies that the front, back, and sides of reefs experience.

In general, reefs develop around a framework produced by organisms (in this context called *framework builders*) that precipitate, bind, and/or retain carbonates as they grow upward toward the surface of the sea (Krumbein and Sloss, 1963). When the structure built up by such organisms reaches wave base, sizable amounts of material are eroded from the framework and become deposited on the reef flanks and at the back-reef region. Within the reef core, and the apron of debris on its flanks, there are innumerable voids present which become habitats for other organisms. These in turn contribute their skeletal materials and wastes to the total reef environment which is termed the *reef complex.* Thus, not only is the reef complex built up by the

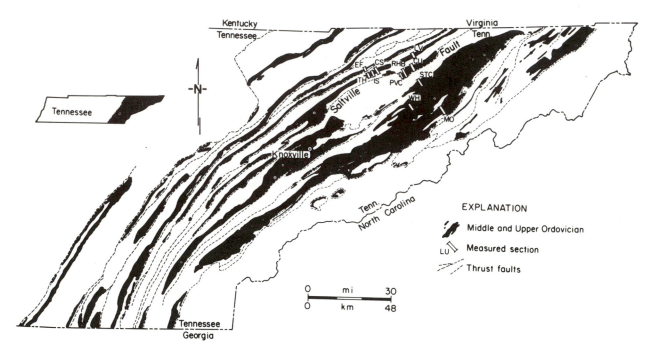

NORTH AMERICAN SERIES	NORTH AMERICAN STAGES	BRITISH SERIES	CONODONT ZONES Sweet & Bergstrom 1976	MIDDLE ORDOVICIAN OF E. TENNESSEE	
				Northwest of Saltville Fault	Southeast of Saltville Fault
CHAMPLAINIAN	TRENTONIAN	CARADOCIAN	Amorphognathus superbus	MARTINSBURG	
				?	BAYS
				MOCCASIN	
				?	
				WITTEN	?
	BLACKRIVERIAN		Amorphognathus tvaerensis	BOWEN	
				WARDELL	
				BENBOLT	SEVIER
			(P. gerdae)	ROCKDELL	
			(P. variabilis)	LINCOLNSHIRE	
	CHAZYAN	LLANDEILIAN	Pygodus anserinus	FIVE OAKS	BLOCK-HOUSE
				BLACKFORD	
		LLANVIRNIAN	Pygodus serra		
	WHITEROCKIAN	ARENIGIAN			

Fig. 5-4. Ordovician outcrops and correlations in eastern Tennessee. (Above) Outcrops of Middle and Upper Ordovician rocks along the Appalachian trend. (Below) General temporal relationships in the Ordovician rocks in the region. Note the differences between units on either side of the Saltville Fault (see map). (From Ruppel and Walker, 1984, *Geological Society of America Bulletin,* Vol. 95, p. 569, Figs. 1 and 2; note that most lettered symbols in the map refer to localities discussed in that source.)

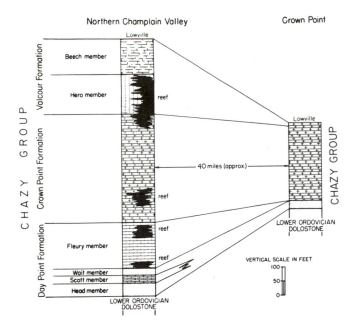

Fig. 5-5. Representative stratigraphic columns and correlations of the Chazy Group in the Lake Champlain region of New York and Vermont. Note positions of Chazy reefs. (From Shaw, 1969.)

interactions of multiple groups of organisms, but, in fact, the reef itself provides a number of different habitats because the various parts of the reef (i.e., the fore-reef, back-reef, reef-flanks, and reef-core; see Fig. 5-6) are subjected to differing wave energies and supply varying types of carbonate material to the local sedimentary environment. The reef actually changes its own original physical environment by modifying wave energies as it grows, and by forming a firm substrate for organisms to attach, thus effectively increasing the local area of benthic habitat.

The stromatolites discussed in Chapter 3 are reefs in the strict sense, but they were organically simple and rarely large individually; indeed, they lacked distinctive parts (i.e., reef-core, etc.) because of this simplicity. Similarly, the archaeocyathid reefs of the Early Cambrian (see Rowland, 1984) and the algal bioherms of the Middle Cambrian through Early Ordovician carbonate facies, were reefs technically but lacked many of the structural components and the complex organic interactions which characterized multiorganism reefs of Chazy and later times. The Chazy reefs were the first to be populated by recognizable *communities* of organisms, which may be defined as characteristic, interdependent groups of organisms which occupy a given environment during a given geological time. Typical marine benthic communities which form and occupy reefs include primary producers (i.e., single-celled photosynthetic plants at the base of food webs), consumers, predators, prey, structure-builders, structure-destroyers and occupiers, and peripheral species. Communities change with

changing environments (termed *community evolution*); in the broadest sense, one may say that all organisms live in communities with perhaps the exception of isolated colonies of autotrophic bacteria and algae.

Newell (1971) stated that modern (and, presumably, fossil) reefs are comparable to cities surrounded by more sparsely inhabited areas. The diversity of species (or *specific diversity*) in a modern reef probably exceeds by an order of magnitude that of the most nearly comparable marine community, such as the sublittoral marine environment.

Early Ordovician reefs are the earliest known communities with a sizable wave-resistant structure; such reefs are found in the Chazy Group of New York and Vermont and the Holston and Carters Formations in Tennessee. In addition, the Lower Ordovician El Paso Group in Texas contains structures made by a different set of organisms and lacking the structural integrity of the above, but which were nonetheless reefs. The Chazy, Carters, and El Paso structures are patch reefs, whereas the Holston structures may be the earliest known fringing or barrier reefs (Alberstadt *et al.*, 1974).

Externally, the Chazy reefs of New York are small but topographically prominent features, with heights from 2 to 8 m and diameters less than (and often much less than) 100 m. The earliest Chazy reefs, in the Day Point Formation, are smaller than those later occurring in the Crown Point and Valcour Limestones; and there seems to be a generally observable vertical increase in complexity within the reef communities.

Alberstadt *et al.* (1974) summarized the succession in Crown Point reefs as follows: development was initiated when stromatoporoids colonized a stable part of the seafloor. This was followed by additions of ramose (branching) bryozoans. The stromatoporoid–bryozoan assemblage was overlain by an assemblage of sponges, coral, bryozoans, and algae. This may have been the terminal stage or it may have been capped by stromatoporoids; or, as indicated by Finks and Toomey (1969), the reef surface was covered by a layer of sponges and stromatoporoids: see Fig. 5-6 and also Section F for discussion of the various reef-dwelling fossil taxa.

B.3.b. Skeletal Carbonates

As stated previously, bioclastic carbonates make up a large part of the volume of carbonates deposited in the Tippecanoe sea. The bioclasts in these sediments are abraded or whole, largely sand-sized particles derived from the skeletons of lime-secreting plants and animals which flourished in Chazy and later times. The list of Ordovician organisms includes, for the first time in geological history, crinoids, bryozoans, stromatoporoids, and coral; in addition, calcareous (articulate) brachiopods, and ever-present cor-

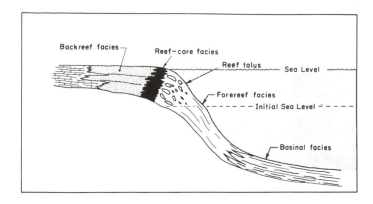

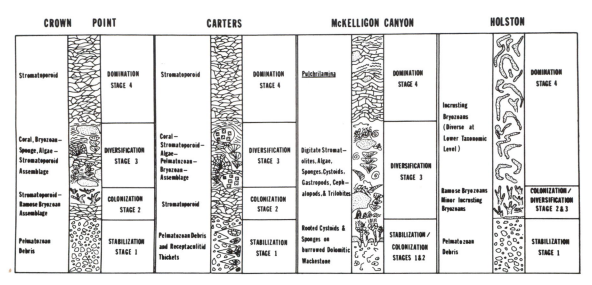

Fig. 5-6. (Above) Generalized reef morphology, showing the characteristic biolithofacies distribution. (Below) Organic succession in representative Ordovician patch reefs. (Above, from Friedman and Sanders © 1978 by John Wiley & Sons, reprinted with permission of John Wiley & Sons; below, from Alberstadt *et al.*, 1974.)

alline algae added considerable quantities of lime to Tippecanoe sediments.

Crinoids, and to a lesser extent, carpoids, crystoids, and blastoids yielded great quantities of bioclastic debris to Ordovician and later sediments. They had jointed stem-, body-, and arm-plates, all composed of thick calcite. These plates separated at the organism's death and thus each pelmatozoan became a small heap of bioclasts. Upon microscopic examination, it is easy to identify such echinoderm fragments because members of the phylum secrete each stem-, body-, or arm-plate as a single calcite crystal, regardless of its size or shape. Additional taxa which contributed bioclastic materials in sizable quantities to Tippecanoe carbonates include articulate brachiopods, bryozoans, stromatoporoids, and algae. The Late Ordovician rocks in New York, Ohio, and northern Kentucky, and the Silurian rocks in approximately the same areas (with the addition of several neighboring Appalachian regions) are among the most abundantly fossiliferous in North America (Fig. 5-7); indeed, this fact led to the term "shelly facies" (see Chapter 4) as a characterization of the cratonic-shelf sedimentary environment of the early Paleozoic.

B.4. Middle Tippecanoe Time: A Sea from Coast to Coast

In late-Middle (Trentonian) through Late Ordovician (Cincinnatian) time, the Tippecanoe sea probably covered most of the craton, and indeed may have covered virtually all of the continent except for the Canadian Shield and the Taconic Highlands (see Section C). The climate was warm, due both to the equatorial position of the continent and to the temperature-moderating influence of the seas themselves; therefore, the bulk of the sediments deposited were carbonates. However, the Taconic Highlands, which were being thrust up during this interval, initially spread a wide

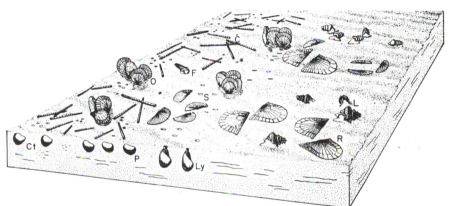

Fig. 5-7. Fossiliferous Late Ordovician rocks and paleoenvironments. (A) Cincinnatian shaley limestone from northern Kentucky (Richmond Formation) showing diversity of benthic taxa. (B) Reconstruction of Late Ordovician *Sowerbyella–Onniella* (both brachiopods) benthic community in Late Ordovician time. (Above, DRS photograph; below, from Bretsky, 1969, *Geological Society of America Bulletin*, Vol. 80(2), Fig. 4, p. 203.)

layer of mud and coarser detritus over the northeastern regions of the craton (New York/Pennsylvania) and later shed detritus which spread as far west as Illinois and as far south as Virginia and Tennessee.

The pattern of skeletal carbonate deposition described for the Chazyan continued on into Trentonian time, with an increasingly wide area undergoing carbonate deposition. This pattern is especially prominent in the central states, where the older detritus which produced the St. Peter sands apparently ran out and the basal sandstones gave way to carbonates.

The type Trenton section is at Trenton Falls, New York, where approximately 80 m of thin-bedded limestones is exposed in the gorge beneath the falls. In nearby areas of Pennsylvania, the Nealmont/Salona/Coburn Limestones are Trenton equivalents, as are many formations located across a broad area to the west of the Appalachian region. In the central states, the Trenton Stage is typified by the section in western Wisconsin where the St. Peter is overlain, in turn, by carbonates in the Platteville Formation and the Galena Dolomite. Elsewhere on the craton, Trenton carbonates include the uppermost Chickamauga Limestone of Georgia, Alabama, and parts of Tennessee, the Nashville Group in central Tennessee, the Lexington Limestone in Kentucky, and the Viola Limestone in the Arbuckle and Wichita Mountains of Texas and Oklahoma. There are also many other Trenton-age carbonates which clearly show that the prevailing marine environment in the middle Tippecanoe sea was warm and free from sizable quantities of detritus.

By the Late Ordovician (Cincinnatian), the Tippecanoe sea probably reached its greatest extent and may have spread over a wider area of the North American craton than did any previous or subsequent epeiric sea. During this time too, Ordovician benthic communities reached their peak in specific diversity and in the abundance of individual organisms.

Incredibly rich suites of Cincinnatian benthic marine fossils are found in several areas, especially Indiana, Ohio, and Kentucky. These rocks of the Eden, Maysville, and Richmond Formations comprise the type section for the Cincinnatian, and they are among the most fossiliferous sediments on Earth.

The Upper Ordovician carbonates on the central and southern portions of the craton are notably thin, with very limited areas of the craton containing thicknesses of carbonate rocks exceeding 300 m (Sloss *et al.*, 1966); the areas with greater thicknesses include southern Ohio and Kentucky. In general, Cincinnatian carbonate environments covered a broad area of the craton; but the rates of deposition were slow because (most likely) the craton was not submerging at rates sufficient to allow buildup of thick deposits. One may characterize the Cincinnatian carbonates as a veneer of sediment covering a wide area but with no great thickness.

Cincinnatian carbonates outside the type area include the Leipers/Arnheim/Fernvale Formations in Tennessee, the Fernvale in Oklahoma, the Maravillas ''Chert'' in western Texas, the Montoya Limestone in western Texas/New Mexico, the upper Fremont/Priest Canyon Formations in Colorado, the Bighorn Dolomite in Wyoming, the Stonewall/Stony Mountain Formations in the Williston Basin, and the Churchill River Formation in the Hudson Bay area.

B.5. The Taconic Influence: Upper Ordovician Detrital Sediments

Once again, not all the sedimentary facies for a given interval can be included in a single category. The Taconic Highlands, forming in present-day Massachusetts, New York, and Vermont, shed detritus westward in progressively increasing quantities through the Middle and Late Ordovician.

In Trenton through Maysville time, the Taconic influence on the craton is notable only in adjacent regions, including parts of New York, Pennsylvania, West Virginia, and Virginia. The Lorraine and Utica Shales in central New York and the Antes and Reedsville Shales in Pennsylvania are perhaps the most typical of these sediments; however, in many cross sections taken west-to-east in the above regions, the detrital influence can be clearly shown (Fig. 5-8). The Taconic-derived fluvial sediments of the northeastern craton grade eastward into the Martinsburg and Normanskill flysch facies in the Taconic region proper (see Section C).

By Cincinnati time, Taconic detritus was widespread across the eastern craton. Sections in western New York show the transition very well, where the Lorraine Shale

grades upward into the Oswego Sandstone and then to the Queenston Shale. The Queenston Shale (Fig. 5-8) is a red, silty lutite that forms the bed of Lake Ontario (due to its relative softness and scouring of Quaternary ice advances); the Queenston also underlies parts of southern Ontario. To the south, in Pennsylvania, West Virginia, Virginia, Maryland, and parts of Tennessee, Queenston-equivalent beds are named the Juniata Formation and tend to be composed of coarser clastics. All of these detrital sediments show clear evidence of their fluvial/deltaic environments of deposition (see Section C.2).

The Taconic detritus was apparently carried over very long distances by surface currents in the Tippecanoe sea; one finds Cincinnati-age detritus spread down through the present Mississippi valley and southwestward as far as Oklahoma. In the Mississippi Valley region, these Taconic-derived Upper Ordovician detrital sediments are found in the extensive Maquoketa Shale. In other areas of the craton, Taconic detritus fed the Upper Ordovician Polk Creek Shales in the northern Ouachita Mountains, the Sylvan Shale in the Wichita Mountains of Oklahoma, and the Sequatchie Formation in parts of Tennessee, Virginia, and the northwestern tip of Georgia. These detrital sediments were often admixed with carbonates similar to those discussed previously; overall, the ratio of detrital-to-carbonate sedi-

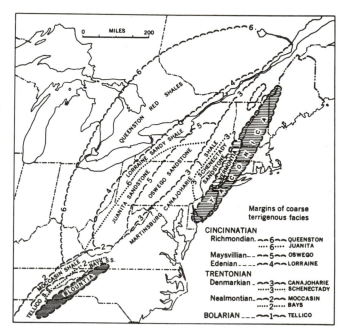

Fig. 5-8. Geographic extent of coarse terrigenous lithologies resulting from Taconic events. The map shows the approximate geographic positions of Taconica and Blountia (see Section C.3) positive areas, and the detrital wedges extending westward from these. (From Kay and Colbert © 1965 by John Wiley & Sons, reprinted with permission of John Wiley & Sons.)

ments deposited on the craton by the latest Ordovician may have been approximately 40 : 60.

A widespread series of *bentonites* are found in Champlainian rocks both on the craton adjacent to the Northern Appalachians, and in the Appalachians. Bentonites are rocks composed of altered volcanic ash in which most of the material is clay (chiefly montmorillonite). They have both interesting and occasionally troublesome properties, such as the ability to swell when wet (which causes problems in construction work) and the tendency to undergo plastic deformation. However, here we are interested in the information implicit in their mode of origin. In addition, volcanic ash beds provide time planes which can be traced across changing lithofacies. The difficulty of correlating Taconic flysch facies with "shelly" carbonate facies of the craton makes these bentonites especially useful in the Ordovician.

The precise source of the Champlainian bentonites on the eastern craton margin is unknown; however, the best guess would be from the Taconic Highland region (New York/Massachusetts/Vermont). Moore (1958) reported that lava flows in Pennsylvania have been associated with ash beds; in addition, volcanic rocks are associated with Middle Ordovician sediments in the far northern Appalachians (Canada and Newfoundland). Middle Ordovician bentonites are traceable as far south as Alabama and as far west as the present Mississippi valley.

B.6. Post-Cincinnati: The Sea Shallows

Events on the craton following the Cincinnatian Stage are rather difficult to interpret from the sedimentary record. The difficulties are threefold: first, the Silurian sediments have been stripped away from very large areas of the continent, leaving only patches of outcropping strata. Second, during the Silurian and subsequent periods of the Paleozoic Era, the craton was undergoing mild tectonic activity, the result of which was formation of clearly demarcated cratonic basins during the Silurian and Devonian Periods, and the formation of basin-and-dome structures (termed *yoked basins*) during the later Paleozoic. Silurian deposition in basins of the eastern craton, for example, need not parallel that of the Mississippi Valley region, nor were the post-Silurian effects of erosion the same between regions. The result is that correlation of Silurian strata generally involves basin-to-basin analysis without much help from intervening strata. Finally, the Appalachian region during the Silurian Period was in the final throes of the Taconic orogeny, and there are major differences in sedimentary style of proximal Appalachian environments versus sedimentation on the carbonate platforms of the craton farther to the west.

The Tippecanoe sea withdrew from most parts of the craton at the end of the Ordovician. This withdrawal, caused by emergence of the craton, did not affect all areas; thus, it was an incomplete emergent episode and does not constitute a cratonic sequence boundary. The nature of the contacts between Upper Ordovician and Silurian rocks also differs significantly among regions. For example, in Pennsylvania the contact between the Ordovician Martinsburg Formation (part of the Taconic retro-arc-basin flysch facies; see discussion in Section C) and the Silurian Medina Sandstone, is a very pronounced angular unconformity which denotes a significantly long hiatus (at least the entire Richmond Stage). In contrast, in the Niagara River Gorge in western New York, the lowermost Silurian rocks (the Whirlpool Sandstone) overlie Upper Ordovician Queenston shales with apparent conformity; the strata in the Niagara Gorge are perfectly parallel, with no real evidence of erosion present on the uppermost Queenston beds. One can conclude that there was in fact virtually no hiatus involved between deposition of the two units, or that the hiatus was extremely brief; this apparently conformable relationship between Upper Ordovician and Lower Silurian strata is not characteristic of the rest of the craton.

B.6.a. Lower Silurian Sedimentation

The Lower Silurian (Medinan) sediments of the eastern craton are mostly detrital ones, reflecting continued influence of the Taconic highlands and the uplift of Appalachian regions farther to the south. A number of formation names have been applied to the basal Silurian coarse detrital sediments of the eastern craton margin including the Shawangunk Conglomerate (New York to Pennsylvania), Tuscarora Sandstone (Pennsylvania to Virginia), and Clinch Sandstone (southwestern Virginia to Tennessee). In practice, these formations may represent different aspects of the same basic event: erosion of Appalachian uplifts deposited over the post-Ordovician erosion surface. The rocks may not be precisely time-equivalent across the entire region, nor do they overlie Ordovician rocks in a uniform manner. For example, the Tuscarora grades conformably down to the Juniata Formation in western Virginia, but farther to the east the Juniata had been stripped away and the Tuscarora lies unconformably over the Martinsburg, as described above. A paleocurrent analysis (Whisonant, 1977) demonstrated that the prevailing direction of sediment transport in the Tuscarora Sandstone was almost due northwest, i.e., from the Appalachians down to the alluvial plain below. The Tuscarora sands are mature, and may include eroded Cambrian and Ordovician sands from the uplifted areas among the sources of sediment.

Other basal Silurian detrital sediments on the eastern craton include the basal sands of the Red Mountain Group in Georgia and Alabama, the Grimsby Sandstone in central New York, and the Whirlpool Sandstone in the Niagara Gorge; and, farther to the west and away from the strong

Appalachian influence, the Centerville Shale under Ohio and eastern Kentucky.

Beyond the reach of Taconic detritus, the lowermost Silurian sediments, predictably, are carbonates. These carbonates are thinly and widely spread to virtually all corners of the craton; but they are conspicuously absent in much of the western cratonic interior and along a wide band of area adjacent to the Transcontinental Arch. Most Medina cratonic carbonate sediments occur in five general regions: Hudson Bay (the Severn River Limestone), the Williston Basin (the Interlake Dolomite), western Texas (the basal Fusselman Formation), Oklahoma and Arkansas (the Keel, Ideal Quarry, Cochrane, Chimneyhill, and Brassfield Limestones), and a large region extending from northern Michigan south to Tennessee, bordered on the east by the zone of Appalachian-derived detrital sediments, and extending west to Kansas (the Manitoulin, Brassfield, and Edgewood Limestones and equivalent strata). Small outcrops and subsurface occurrences on other areas of the craton suggest that Lower Silurian marine sediments were once widespread; diatremes in Colorado and Wyoming contain sediments with Lower Silurian fossils, just as they contained Ordovician rocks. Near Lake Timiskaming, on the Canadian Shield, an isolated Lower and Middle Silurian section containing carbonates shows that major parts of the Shield too may have submerged in Silurian time.

A noteworthy Medinan section is located on Anticosti Island, off the coast of Gaspé, Quebec; there, Lower Silurian strata correlate very well paleontologically with contemporary units in New York. The Lower Silurian Tippecanoe sea apparently occupied the Anticosti region at the very start of the Silurian, and transgressed down (southwesterly) across the Canadian Shield and into the New York and Great Lakes areas. Another arm of the sea, flooding the craton from the south around the southernmost reach of the Appalachian uplifts, transgressed northward and met the waters reaching down from the north by late Clinton time (Kay and Colbert, 1965). Evidence for the existence of two separate seaways is found in the similarity of fossil assemblages from southeastern New York and Anticosti Island in Late Clinton time; these contrast with the separate suites of fossils present in northern and southern areas from earlier Silurian times.

B.6.b. Middle Silurian: The Niagaran Section

During Middle Silurian (Clinton and Lockport) time, the distribution of lithofacies established on the craton in Medina time remained almost intact. Our discussion of Niagaran rocks will focus on the type section in the Niagara Gorge, below and downstream from Niagara Falls, and then we will briefly examine other parts of the craton. Figure 5-9 shows a generalized section along the Niagara Escarpment (an approximately east–west striking cuesta formed by resistant Niagara-age dolostones capping softer Niagaran units).

In the Niagara Gorge, the Middle Silurian contains sandstones at the bottom, which grade up to shale, limestone, limy shales, shaley limes, and ultimately into the massive Lockport carbonates. Eastward, to Rochester, New York, and beyond, the shaley layers thicken as the upper Clinton-age limestones pinch out (Grasso, 1973). The Rochester Shale thickens at Rochester, but the total section does not. To the west, at the escarpment in Ontario (see Fig. 5-9), the section thins significantly as almost all of the shaley and sandy rocks pinch out. In Ontario, the Niagaran section consists of little more than the Lockport and Reynales Formations, both carbonates; however, as the Niagaran section thins, the underlying Medina rocks (the Cabot Head and Manitoulin Formations) thicken. Farther west, to Manitoulin Island in northern Lake Huron, the section again changes as it thickens and becomes predominantly carbonate through both Lower and Middle Silurian. These lateral stratigraphic facies changes reflect fairly clear topographic and environmental differences between regions along the escarpment during the Silurian. Topographically, the Nia-

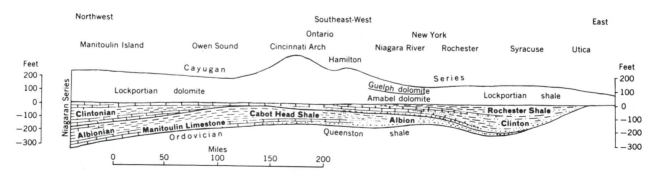

Fig. 5-9. Generalized Niagaran (Middle Silurian) cross section along the eastern Great Lakes region, between Manitoulin Island (Michigan) and Utica (New York). Many of the units shown are locally subdivided. (From Kay and Colbert © 1965 by John Wiley & Sons, reprinted with permission of John Wiley & Sons.)

gara Gorge and Rochester regions were on the edge of a subsiding retro-arc basin which deepened eastward toward the Taconic Highlands. The section under northern Michigan and Manitoulin Island was deposited on the edge of a cratonic basin in Michigan which was beginning to subside in the Middle Silurian and which underwent very dramatic subsidence in the Upper Silurian. Between these basins probably existed an up-arched region in Ontario, called the Algonquin Arch. The changing lithologies can be explained both by changes of sea level with the consequent migration of shorelines and zones of marine sedimentation, and by proximity to the Taconic detritus. The carbonate components increase westward, in general, and decrease eastward as the detrital components increase. The basinal sediments in New York terminate to the east of Herkimer, New York, at the overthrust zone of the Appalachians (see Section C); at the most easterly point, they are sandy, conglomeratic sediments considered to be molasse facies (Section C).

Of greater complexity than the lateral facies and thickness changes are the diachronous natures of some organism communities in the Silurian. One finds, for example, fossils in common between the carbonate rocks of Michigan and New York, even though these carbonates were deposited in Clinton and Lockport times, respectively. Similarly, the shaley Albion Group in Ontario contain organisms similar to the shaley (but younger) Clintonian fossils in New York. These seem to constitute classical examples of *facies fossils,* which are organisms whose habitats tended to limit their occurrences to specific lithologies. The graptolite/black shale association is a second classical example of facies fossils. In the case at hand, the temporal differences between Albion and Clinton times apparently were of less importance to the preserved faunas in terms of taxonomic differences, than were the shale-to-carbonate environmental differences.

The Lockport Dolomite forms the caprock at Niagara Falls and is the ridge-former along the cuesta. In the late Quaternary, the cuesta was scraped free of overburden and glacial material by the last ice advances and, consequently, the Lockport Escarpment now forms a topographic feature across western New York. The total thickness of the Lockport Group averages 50 m in the New York Basin. Eastward toward Syracuse, New York, the Lockport dolostones change lithology, become more calcitic, and comprise the Sconondoa Formation; and farther eastward in the area of Oneida, New York, the unit becomes shaley and is named the Illion Shale. Lockport carbonates are not highly fossiliferous because they are dolomitic: additionally, dolomitization of this massive unit in New York indicates that at least locally the Tippecanoe sea in Lockport time had restricted circulation with evaporitic conditions.

The Lockport Dolomites in the Niagara Gorge/Rochester area, although poorly fossiliferous, show the de-velopment of reefs which can be recognized as masses of unstratified carbonates, varying in size but often 10 m high and 30 m across. These reefs were probably formed in a manner similar to those we shall describe in detail in the Upper Silurian discussion. A final point of interest for the Lockport Dolomites in western New York is the presence of concentrations of metallic sulfides, especially those of iron, zinc, and lead. These metals are commonly present as crystalline vug-fillings and are often concentrated in voids in reefs. An explanation for these metal concentrations may be found in the concentration of metals observed in modern hypersaline reef environments due to thermohaline density stratification (Sonnenfeld *et al.,* 1977). These metallic sulfides provide additional evidence of the evaporitic conditions suggested for the Lockportian in New York (or perhaps in the interval shortly after the Lockport carbonates were deposited).

The discussion presented here has dealt with Niagaran rocks in New York, southern Ontario, and Michigan; we do not wish to imply that Niagaran sediments are found only in these areas. As our opening sentence of the subsection stated, Niagaran rocks are scattered around the craton in a pattern similar to the distribution of the underlying Medinan sediments. This pattern, once again, shows widespread carbonate sediments westward of the reach of Taconic detritus, with large areas between outcrops and subsurface occurrences. The best guess at the geographic extent of the Tippecanoe sea during Niagaran time would suggest that it covered virtually all the craton in the continental United States, and that it covered sizable parts of western Canada and the Canadian Shield. It may have been one of the widest-reaching epeiric seas, although it was certainly shallow everywhere. The total extent of Niagaran sedimentation will never be known because so much of the sedimentary record has been stripped away by erosion.

B.6.c. The Silurian Iron Ores

Rocks of Clinton age are economically important because they contain thin but persistent beds of iron ore, present in sections as widely spaced as Alabama and New York. The ores commonly occur as oolitic hematites, although hematite-cemented sands and oolitic chamosite (an iron-rich chlorite) are also widespread through the Clintonian section.

The source of all this iron is not precisely known; however, a very good guess would be from the same Taconic Highlands that contributed so much detrital material to the Tippecanoe facies in the east. Most of the iron is oxidized and it therefore indicates at least periodic subaerial exposure, perhaps during transport down from the highlands. The ores may contain marine fossils; Krumbein and Sloss (1963) state that the Clintonian ore formed by replacement of fossiliferous and/or oolitic limestones by hematite. Others (e.g., Hunter, 1970) suggest that the iron was transported in groundwater, with low Eh and Ph, which upon entering seawater with higher Eh and Ph initiated rapid precipitation of chamosite around sand or shell nuclei. This latter explanation would further suggest that the iron is a primary deposit rather than a replacement product, and that the hematite is an oxidation product of chamosite (see also Van Houten and Bhattacharyya, 1982).

The Clintonian iron ores have been mined on a large scale in regions near Birmingham, Alabama, where a sizable steel industry sprang up because of their presence. In Alabama, the iron is found in an 8-m-thick bed in the Red Mountain Formation, which contrasts with correlative iron beds in New York, such as the Furnaceville Hematite, which are very thin.

B.7.   Upper Silurian of the Michigan Basin

In Late Silurian (Cayuga) time, the Tippecanoe sea continued to shallow as the craton underwent localized downwarping to a greater degree than was evident at any earlier time. A lithofacies map of the continent for Middle Cayuga (Tonoloway) time would show sizable outcrops only on the western and northeastern craton margins, in Hudson Bay, in diatremes in Colorado and Wyoming (Chronic *et al.*, 1969) and in a large area centering on the Great Lakes and the Mississippi valley. By Latest Cayuga

(Keyser) time, almost all areas on the craton, except for the basins in Michigan and New York, had emerged from the sea. Once again, the inherent difficulty concerning interpretation of regressive sequences arises; were latest Silurian sediments deposited and stripped away by erosion, or were they never deposited? In most areas, it would seem more probable that they were not deposited; the evidence of evaporitic conditions even in the basins strongly supports extreme shallowing of the sea and the absence of strata over the emerging craton. However, the presence of Upper Silurian sediments in western diatremes, and putative Salinan (lowermost Cayugan) sediments in the Williston Basin, suggests that the late Tippecanoe sea left deposits in at least some areas of the craton for which we have no record. The reader may draw his or her own conclusions.

The record of Cayugan sediments in the Michigan basin deserves and will receive the major part of the subsequent Silurian discussion. We observed that the Michigan Basin was subsiding to some degree during the Niagaran interval; however, the center of this episode of subsidence was located in the present Upper Peninsula rather than in the center of Michigan. Niagaran sediments in Michigan are less than 250 m thick (Fig. 5-10). In the Late Silurian, by way of contrast, the Michigan Basin received over 1450 m of sediment, with basinal deposits laid down in a remarkably concentric pattern. The deposits are symmetrically thicker in the center of the basin and, as King (1977) points out, the configuration of Salinan sediments in Michigan and

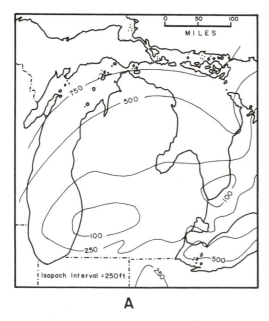

A

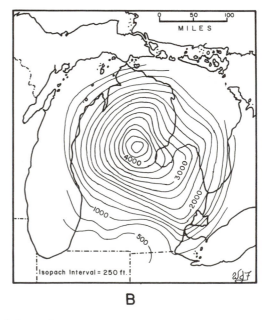

B

Fig. 5-10. Isopachs of Middle and Upper Silurian strata in the Michigan Basin. (A) Niagaran (Middle Silurian) isopachs; (B) Salina/Bass Island (Upper Silurian) isopachs. Observe the change in both degree of basin activity and location of the basin depocenter during this interval. There is

little post-Silurian structural deformation (not shown here), indicating that most basin activity took place during the Late Silurian. (Redrawn and modified from King, 1977.)

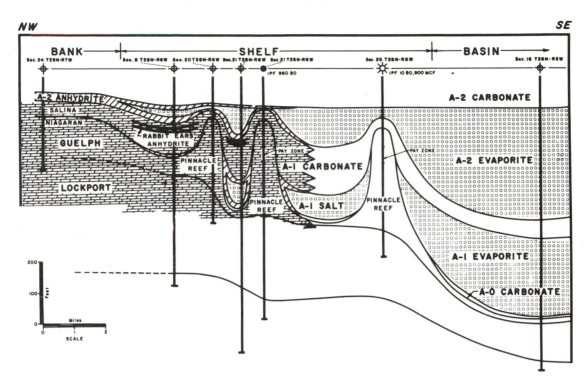

Fig. 5-11. Diagrammatic section of Silurian evaporite facies in the Michigan Basin, showing reefs and bank-to-basin transition. (From Huh *et al.*, 1977.)

the downwarping of the basin in Salinan time affect the outcrop pattern of deposits which pre- and post-date Salinan time. The extreme downwarping forced older sediments into the circular basinal configuration, and post-Salinan deposition too followed the pattern, presumably because the post-Salinan topography was basinal.

B.7.a. Evaporite Deposition in Salinan Time

What was being deposited and formed in the Michigan Basin during Cayuga time is as interesting as the cratonic tectonics which shaped the structure. The basin features a complex set of interbedded limestones, dolostones, and evaporites (Fig. 5-11); and, to add more spice to the mixture, both barrier and pinnacle reefs are present in and adjacent to the structure.

The sediments filling the basin in Salina time comprise the type section, and hence they are called the Salina Group. Halite, anhydrite, gypsum, and potash salts (sylvanite) have been mined in the region for many years and, therefore, the detailed stratigraphy of the Salinan sediments has been carefully studied. What has only recently been worked out is the origin of the evaporites in the Salinan section. Nurmi and Friedman (1977) present a brief historical sketch of the controversy surrounding the various evaporite models proposed by workers concerned with these deposits. The "evaporite problem" begins with simple calculations: It would require up to 1000 km (not meters!) of seawater to yield the total of 750 m of evaporites present in the Salina Group. Clearly, such a column of water never existed anywhere on Earth at any given time; and so the evaporites represent deposition from constantly replenished seawater. But replenishment implies flooding and the opposite of evaporation; and so the essential problem can be stated: "how does one model an evaporitic environment which must have plenty of water?" One can safely assume that the conditions were conducive to evaporation, i.e., hot and dry. Examination of the Silurian global geography (Fig. 5-2) can verify this assumption; Michigan was in the tropics in the Late Silurian.

Two models for evaporite accumulation can be applied to the Michigan Basin. First, the presence of considerable dolomite implies supratidal environments with splashover and seepage recharge (termed *sabkhas*). Sabkhas in the Persian Gulf today show thin crusts of salt due to this process, and it is conceivable that given time, such salt deposits could become quite thick and that other thick evaporites and dolomite could form. Sabkhas may explain some evaporite facies of the Salinan rocks but perhaps not all. Large circular areas of evaporites, such as are found in Michigan, may alternatively be explained by the model of a restricted basin with periodic recharge of seawater. This "barred-lagoon" model dates back to the last century (Bischof, 1875) but it may still be valid. Nurmi and Friedman (1977) discuss various evaporite facies in the Lower Salina Group (Fig. 5-11) and suggest a variety of conditions for the deposition of the different lithologies: laminated limestone and anhydrite may have formed in deeper hypersaline lagoonal conditions; nodular anhydrite probably formed in sabkhas.

B.7.b. The Problem of Silurian Reefs

Two types of reef structures are found in and around the Silurian Michigan Basin: numerous pinnacle reefs, located in a circular band around the center of the basin, and outside of this band a massive barrier reef, which also encircles the basin.

The first problem confronted in studying these reefs is one of synchrony (timing); when did these reefs live and die? They clearly began and developed in Niagara time, and they are similar to reefs found in Lockportian rocks in and around Michigan and in New York. The real question involves the second "problem": what is the relationship between these reefs and the Salinan evaporites? As Fisher (1977) states in the introduction to a symposium on reefs and evaporites, one group of geologists believes that all growth in Silurian reefs of the Michigan Basin ended in the Niagaran, whereas others claim that these reefs were still growing when initial carbonate and evaporite sedimentation was occurring in the basin during earliest Salina time. Still others claim that growth ceased during the first sequence of evaporite deposition and reinitiated with rising sea level and renewed carbonate-favoring conditions.

Reefs in Niagara/Medina time were not restricted to Michigan; both barrier and pinnacle types occur in Illinois, Indiana, Ohio, Wisconsin, Ontario, and New York.

The pinnacle reefs vary in height, becoming notably higher toward the center of the basin. Sears and Lucia (1979) report that heights range from 90 to 180 m, with an average width of 0.5 km. Their study mentions that over 500 such reefs have been discovered by petroleum industry drilling just in an area under 18 km wide and 160 km long; surely there are several thousand such reefs in the basin overall. The higher taxonomic makeup of these reefs is not qualitatively different from that of the Chazyan reefs. The specific composition of Niagaran/Medinan reefs does, however, vary significantly from that of reefs in the Chazyan section. Huh *et al.* (1977) described the evolution of Silurian pinnacle reefs and showed that it is not significantly different from the succession in Chazyan reefs described previously; however, according to their model, the influence of evaporitic conditions does seem to have stopped growth suddenly. The steep-sided "pinnacle" shape of these reefs apparently developed in an attempt by the reef community to maintain its position at wave base as the basin sank dramatically in earliest Salina or latest Niagara time. Fisher (1977) notes that if sea level lowered during the

carbonate sedimentation phase at the earliest Salinan, the taller reefs would stop growing (or be eroded down), while shorter ones would continue growing; and, therefore, differential apparent "ages" for Silurian pinnacle reefs result. What is fairly certain is that these reefs could not survive in hypersaline waters, and that contemporaneous deposition of salts and growth of reefs is highly unlikely. The sabkha model for evaporite deposition offers a possible answer to the problem of synchrony because evaporite deposition in sabkhas does not require poisonous levels of salinity for evaporites to form.

The barrier reef complex around Michigan, and similar reef facies in Indiana (the Fort Wayne bank) and Illinois, represent assemblages of organisms similar to those which make up the pinnacle reefs but which were living on the shelf edge adjacent to the area of rapid subsidence. Shaver (1974) claimed that almost 400 species are known from these reef complexes in Indiana, with diversity increasing through the Middle Silurian; the frame-building organisms remained approximately the same while the reef-dwelling organisms diversified intensely. Shaver also believes that the reefs are, in part, of Salina age.

Around the Michigan Basin the barrier reef complex shows differentiation of *biofacies* (organism-habitat groupings) which are analogous to those observed in present-day tropical reefs. Ingels (1963) presented a classic study of the Thornton Reef Complex in southern Chicago; and he reconstructed the final growth form of the reef crest (i.e., before growth ceased) and related the distribution of biofacies to wave and wind directions.

The barrier reef complex in Michigan is almost 18 m thick in places, and presents a steep inner face toward the center of the basin. As we have stated, the barrier complex virtually encircles the basin, and the presence of this barrier has frequently been cited as the restricting mechanism in the "barred-lagoon" model of evaporite deposition. Passes through the reef, and the probable presence of numerous openings in the reef subsurface, would allow periodic recharge of water in the basin; but, then again, how effective would such a permeable barrier be in restricting influx of seawater?

The foregoing discussion of Silurian reefs may have raised more questions than have been answered. The literature on the subject is voluminous, and a check into the reference list of most of the cited studies will lead the interested reader to a wide variety of other studies.

B.8. Final Tippecanoe Cratonic Sedimentation: Uppermost Cayugan and Lower Devonian

By the latest Silurian (uppermost Cayugan: Tonoloway and Keyser Stages) and through the earliest Devonian

(Helderberg Stage), the Tippecanoe sea had withdrawn to the eastern craton edge leaving sediments no farther west than Illinois, Tennessee, and Michigan. These last Tippecanoe sediments are predominantly carbonates. Since they extend far to the eastern edge of the craton in regions of New York, Pennsylvania, and Virginia, they indicate that the supply of detritus from the Taconic Highlands had finally diminished as the mountains had eroded down through the long depositional time of the cratonic sequence.

In the Illinois Basin, sediments of the latest Tippecanoe are found only in what was the deepest portion of the basin. These rocks comprise the uppermost Moccasin Springs (Silurian) and the conformably overlying (Devonian) Bailey Formations. A thin shaley unit which is unnamed is present at the very top of the Silurian section and may be part of the otherwise dominantly carbonate Moccasin Springs. A notable feature of Siluro-Devonian sedimentation in isolated parts of Illinois is the presence of Silurian reefs punching up through the overlying strata because of their topographic relief. In scattered localities, Devonian strata may be lower than or parallel to Silurian strata where reefs are present.

The Illinois Basin stratigraphic nomenclature (Moccasin Springs/Bailey Formations) is also used in southwestern Indiana, southeastern Missouri, and western Kentucky. In all these sections, where a Silurian–Devonian record is present, the boundary between periods is quite difficult to place; and in fact the base of the Bailey may be youngest Silurian age.

Sediments of the latest Tippecanoe sequence in the Appalachian Basin (the region westward and adjacent to the Appalachians) left a better record than is found in exposures farther to the west. In general, the Helderbergian and underlying rocks are dolomitic but many strata are composed of fossiliferous limestones. In Pennsylvania, West Virginia, and adjacent parts of Virginia and Maryland, the Upper Silurian rocks are named the Tonoloway and Keyser Formations (comprising the type sections for the stages) overlain in most areas by the Coeyman Limestone (locally called the "New Scotland" Limestone) or the Helderberg Limestone. The Devonian stratigraphic nomenclature of the region is complex and a number of other Helderbergian formations have been described in sections from Pennsylvania to Virginia. In Ohio and Michigan, the uppermost Silurian is named the Bass Island Limestone, and in Ohio the overlying Devonian is the Helderberg Limestone (which is absent in Michigan). In these sections too, the precise position of the Silurian/Devonian boundary is not placed with any certainty at the formation contacts.

The best-studied and best-exposed latest Tippecanoe strata are found in New York. These rocks outcrop over a widespread area and they are in places abundantly fossiliferous. In western New York, the post-Salina Silurian section is composed of the Akron and Bertie Formations, with the overlying Helderbergian absent (Buehler and Tesmer, 1963). Other authors (e.g., Ciurca, 1973) raise the Akron and Bertie to group status and describe a number of formations within the units. Of particular interest in these units are the thin "waterlimes," a term which defines impure limestones which in the past were burned to produce hydraulic cements (now replaced by the use of Portland cement). The environment of deposition for these waterlimes is not clear; however, the best interpretation seems to be brackish lagoonal with rather variable salinity. What is of unusual interest is that fossils of eurypterids, a group of arthropods most closely related to the modern horseshoe crabs, are found in the waterlimes. Eurypterids (see Section F.4) are nowhere common fossils, although they range stratigraphically from Ordovician to Permian; but in the Akron and Bertie Formations they are present in their greatest North American abundance. The conditions in which they lived, their prey (very likely primitive fish), and their somewhat odd and awesome appearance, with sizes to 180 cm, explain why they have aroused popular and professional interest for more than a century. Eurypterids may have been *euryhaline* organisms, i.e., forms which are tolerant of a wide range of salinities; this would explain their presence in the post-Salinan shrinking sea or lagoons adjacent to the sea. Salinities of both marine and lagoon environments in the Late Silurian were likely to have been highly variable, due both to the re-solution of bedded salts and to the shallowing of waters; and, thus, selection pressure may have favored animals tolerant of variable salinities over those adapted to ocean waters with a typically narrow salinity range (termed *stenohaline* organisms). Beds in the Akron and Bertie Formations, other than the waterlimes, tend to be dolomitic with the presence of casts of hopper salt crystals indicating at least sporadic continuation of evaporitic conditions into post-Salina time.

The Upper Silurian section in central/eastern New York and down the present Hudson River valley includes a change of nomenclature from that of western New York (the Buffalo/Niagara Falls area). Starting in the vicinity of Syracuse and Binghamton, the uppermost Silurian and the lowermost Devonian strata are assigned to the Rondout Limestone. The Rondout is overlain by the Helderberg Group (the type Helderbergian section) which is composed of poorly fossiliferous limestone and shale formations (the Manlius/Coeymans/Kalkberg/New Scotland) and which represents a classical time-transgressive marine sequence of nearshore environments [Ricard, 1962; Laporte, 1967; but also see Anderson *et al.* (1984) for an alternative model of Helderberg paleoenvironments]. These formations grade laterally and vertically into one another (Fig. 5-12), in a fashion which very well exemplifies Walther's law, and provided a good subject for Laporte's analysis of fossil

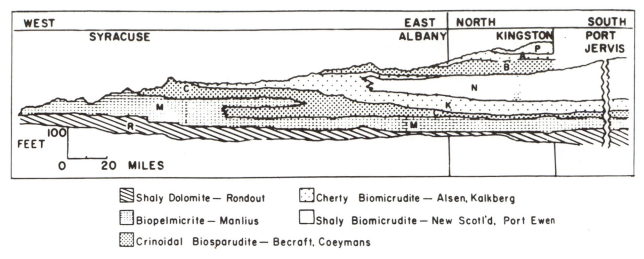

Fig. 5-12. Interpretation of lithofacies in the Helderberg Group, New York, exemplifying Walther's law. (From LaPorte, 1969, after Rickard, 1962.)

organism communities and sedimentary facies. Walker and Laporte (1970) showed that the communities present in various sedimentary facies of the Manlius Formation could be closely related to communities in the Middle Ordovician Black River Formation present in similar sediments; these comparisons apply at relationships above the generic level (i.e., organism communities with forms from the same families filling the same environmental niches, separated by perhaps 70 Myr.). Matthews (1984) noted that the relative thinness of the intertonguing Manlius and Coeymans Formations (under 50 m), and their environments of deposition interpreted from previously mentioned studies, suggests that a minor sea-level rise of less than 20 m would account for the deposition of the Helderberg sediments. This further suggests that that Helderberg Group in central and eastern New York represents a localized subsidence event in the final Tippecanoe Sequence as the rest of the craton emerged from the sea.

C. The Appalachian Continental Margin

The Appalachian orogen is one of the world's great mountain belts. Its stratigraphy and structure have served as a testing ground for most of the principal theories of tectonics. Although the Appalachians have long since ceased to be orogenically active, their history spanned the entire Paleozoic Era, and in the various strata and structures along their strike are recorded at least three major orogenic events and numerous minor ones, events which have left their marks printed over each other in complex and confusing patterns. The newest tectonic hypotheses (e.g., Williams and Hatcher, 1982) view much of the Appalachian orogen as a collage of "suspect terranes," i.e., scraps of crustal

litter accreted to the margin of North America by plate convergence. If this is so, then the Appalachians also contain the remnants of long-vanished ocean basins, microcontinents, and volcanic archipelagoes.

In the following pages, we present a brief introduction to the gross structure of the Appalachian orogen. After that, we examine more carefully the history of the Appalachian area during deposition of the Tippecanoe cratonic sequence, focusing especially on events of the Taconic orogeny. There are many sources of information available for further study of the Appalachians in general and the Taconic orogeny in particular. In the following discussion, we have leaned heavily on the following: Bobyarchick (1980), Bradley (1983), Cook *et al.* (1979), Hatcher and Zeitz (1980), Rankin (1975), Rodgers (1970, 1971, 1981, 1982), Rowley and Kidd (1981), and Williams and Hatcher (1982). Other views on the Appalachian orogen may be found in Dewey and Bird (1970), McKerrow and Ziegler (1971), and Robinson and Hall (1979).

C.1. Introduction to the Appalachian Orogen

The Appalachian orogen runs in a series of arcuate salients and recesses along the eastern margin of North America from Newfoundland to Alabama (Fig. 5-13), a distance of almost 3300 km. If continental reconstructions for later Paleozoic times are accurate, the Appalachian orogen was only one segment of a much larger mountain chain, which also included the Caledonian orogen of Greenland, England, and Scandinavia as well as the Ouachita–Marathon orogen of southern North America and the Mauritanides of northwestern Africa (Rodgers, 1982). The Appalachian orogen is divided into four segments: (1) the Maritime Appalachians, including Newfoundland, Nova

Scotia, New Brunswick, and the Gaspé Peninsula of Quebec Province; (2) the Northern Appalachians, including the New England states and adjacent parts of Quebec as well as New York state; (3) the Central Appalachians, running from Pennsylvania to Virginia; and (4) the Southern Appalachians, from Virginia to the Coastal Plain overlap in Alabama. Each of these segments is subdivided into provinces, containing essentially similar rocks and structures and having the same general history. In the following, we will discuss the province structure of the Northern and Southern Appalachians and will relate it to structural features in the Maritime and Central Appalachians.

C.1.a. The Northern Appalachians

In the following discussion, the major provinces of the Northern Appalachian orogen will be discussed. These provinces and their relation to adjacent geological regions are shown in Fig. 5-13.

In the New England area, the western boundary of the Appalachian orogen may be taken as the line of thrust faults separating the Appalachian foreland basin to the west from the Taconic allochthon and similar thrust nappes to the east (Rodgers, 1982). The foreland basin is composed of mildly deformed platformal strata of Paleozoic age. Lowest strata of the foreland basin consist of lower Paleozoic carbonates and associated detrital sedimentary rocks whose deposition recorded the formation and development of a passive continental margin on the eastern side of North America following late Precambrian rifting (Williams and Hatcher, 1982; see also discussions in Chapters 3 and 4). Overlying these lower Paleozoic strata are thick sequences of detrital deposits, derived from orogenic belts to the east and carried into the basin as part of great deltaic complexes during several intervals of the middle and later Paleozoic. The basement of the foreland basin is Grenville-age granitic crust (Rodgers, 1982).

The Taconic allochthon (Fig. 5-13) is a series of thrust slices composed of Ordovician graptolite-facies, deep-water shales, and graywackes with lesser carbonates (Rodgers, 1970). Emplacement of these thrust nappes occurred during middle to later phases of the Taconic orogeny. Similar, coeval thrust sheets include the Chaudiere nappes of Quebec, the Humber Arm Allochthon of Newfoundland, and the Hamburg klippe of Pennsylvania (Rodgers, 1971).

Genetically related to the Taconic allochthon and thrust over it at its eastern margin is a series of basement-cored nappes which comprise the Housatonic Highlands of western Connecticut, the Berkshire Hills of western Massachusetts, the Green Mountains of Vermont, and Sutton Mountain of Quebec. All of these areas comprise thrust slices of Grenville-age granitic crust and metamorphosed

supracrustal strata whose emplacement was also a Taconic event. Together, these areas comprise the Green Mountain–Sutton Mountain anticlinorium (Bradley, 1983).

To the east of the Green Mountain–Sutton Mountain structure is a post-Taconic basin composed of Silurian and Devonian metasediments and metavolcanics called the Connecticut Valley–Gaspé trough (Bradley, 1983). Strata of the Connecticut–Gaspé trough were deposited over eroded Taconic structures; thus, the trough may be considered a successor basin developed following the Taconic orogeny. Deposits of the Connecticut–Gaspé trough were deformed and metamorphosed during the Middle to Late Devonian Acadian orogeny.

Just east of the Connecticut–Gaspé trough, metamorphic grade rises considerably and rocks of the supracrustal sequence now consist of various phyllites, schists, and gneisses. Extending northward from west-central Connecticut is a complex, multiply deformed structure called the Bronson Hill anticlinorium (Fig. 5-13). The Bronson Hill structure extends northward into Maine where it is called the central Maine volcanic belt. The basement of the Bronson Hill structure consists of the highly metamorphosed roots of an uppermost Precambrian volcanic massif (Rodgers, 1970, 1982). Nonconformably overlying the Precambrian basement are Cambrian and Ordovician mafic and lesser felsic volcanics and volcaniclastics; associated with these volcanics are intermediate plutons. Thus, the Bronson Hill anticlinorium probably represents a volcanic arc massif whose formation may not have been related to the history of early Paleozoic North America and which was accreted to North America as part of the Taconic orogeny. Later, during the Silurian and Early Devonian, the same area was the site of renewed volcanic and plutonic activity, resulting in a feature termed the Piscataquis volcanic arc (Rankin, 1968). Piscataquis volcanics extend northward from the Bronson Hill structure into northcentral Maine, where they are associated with several genetically related plutonic bodies, such as the Katahdin batholith. Similar, coeval volcanics can be traced further northward to the Gaspé Peninsula (Rodgers, 1982). Piscataquis rocks were deformed by the Acadian orogeny.

Immediately to the east of the Piscataquis arc in Maine, New Brunswick, and Gaspé is a second successor basin formed after the Taconic orogeny. Called the Aroostook–Matapedia trough (Bradley, 1983), this feature is composed of metamorphosed sediments of Ordovician and Silurian age. The center of the Aroostook–Matapedia trough is marked by calcareous and noncalcareous turbidite deposits whereas strata of marginal areas appear to represent shallow-water environments which, during the middle Paleozoic, bordered the deep trough on either side. Bradley (1983) suggested that the Aroostook–Matapedia trough was formed during and as a consequence of the Taconic orogeny

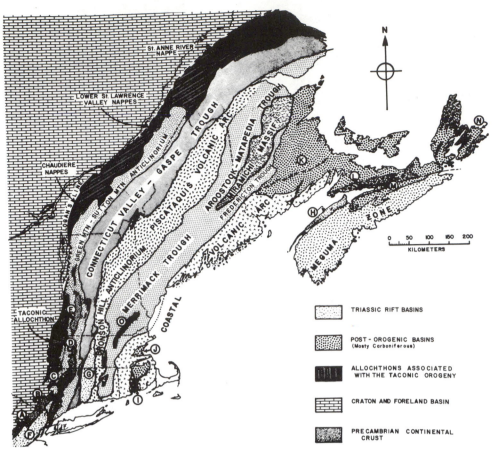

Fig. 5-13. (A) Generalized tectonic map of the Northern Appalachians. Symbols are as follows: A, New Jersey Highlands; B, Hudson Highlands; C, Housatonic Highlands; D, Berkshire Highlands; E, Green Mountains; F, Newark Triassic–Jurassic Basin; G, Connecticut Valley Triassic– Jurassic Basin; H, Fundy Triassic–Jurassic Basin; I, Narragansett Basin; J, Boston Basin; K, New Brunswick and Moncton Basins; L, Cumberland Basin; M, St. Marys Graben; N, Sydney Basin; O, Fitchburg pluton.

and persisted as an ensimatic embayment in the margin of North America until destroyed during the Acadian orogeny.

Bordering the Aroostook–Matapedia trough to the east is the Miramichi massif (Fig. 5-13), a complex igneous– metamorphic structure primarily of Ordovician age. It was interpreted by Rast and Stringer (1980) as an Ordovician volcanic arc whose accretion into the Appalachian tectonic collage accompanied the Taconic orogeny. The massif persisted as a land area during the Silurian and shed detrital sediments both westwards into the Aroostook–Matapedia trough and eastwards into the Merrimack–Fredericton trough, which was probably a major ocean basin at that time (see below). Silurian and Devonian volcanic and plutonic rocks of the Miramichi massif indicate that its history was similar to that of the Bronson Hill structure and its volcanic strata are similar to those of the Piscataquis arc, of which it is thought to have been a part (Bradley, 1983).

Extending all the way from the southern coast of Connecticut to the northern New Brunswick coast is the Merrimack–Fredericton trough. The southern (Merrimack) por-

tion of the trough is bordered on its western side by the Bronson Hill–Piscataquis structure; in New Brunswick, the northern (Fredericton) portion is bordered on the west by the Miramichi massif. The Merrimack–Fredericton trough is a major synformal structure composed of a thick sequence of polydeformed, metamorphosed turbidites and lesser black shales. These rocks are believed to represent deep-water deposition spanning at least the Silurian (Bradley, 1983; Rodgers, 1981) and to mark the site of a major ocean basin whose closure resulted in the Acadian orogeny of the Devonian Period.

Last on the menu of major geological provinces of the Northern Appalachians is the Coastal Volcanic Arc, a Silurian to Early Devonian belt of shallow-marine island-arc volcanics and related sediments. These rocks were deposited on a late Precambrian crustal block composed of polydeformed and metamorphosed arc-related volcanics and intrusives. The Precambrian deformation, metamorphism, and intrusion of similar rocks in Newfoundland has been termed the Avalonian orogeny, named for the Avalon Pen-

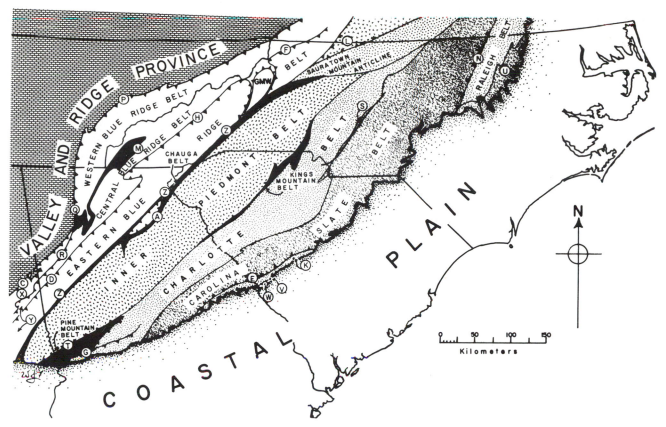

Fig. 5-13. (B) Generalized tectonic map of the Southern Appalachians. Symbols are as follows: A, Alto Allochthon; C, Talledega fault; D, Goodwater–Enitachopco fault; E, Modoc fault; F, Fries fault; G, Goat Rock fault; GMW, Grandfather Mountain Window; H, Haynesville fault; K, Kiokee Belt; L, Smith River Allochthon; M, Murphy Belt; N, Nutbush Creek fault; P, Great Smoky Mountain fault; Q, Cartersville fault; R, Hollins Line fault; S, Gold Hill fault; T, Towaliga fault; U, Eastern Slate Belt; V, Augusta fault; W, Belair Belt; X, Talledega Belt; Y, Ashland Belt (west of the Goodwater–Enitachopco fault) and Wedowee Belt (east of the Goodwater–Enitachopco fault); Z, Brevard Zone.

insula of Newfoundland (Lilly, 1966); subsequent authors have used this term to refer to the entire province, calling it the Avalon terrane (e.g., Williams and Hatcher, 1982) or Avalonia (Bradley, 1983). The Acadian orogeny may have been the result of the collisional accretion of the Avalon block against the margin of North America.

C.1.b. The Southern Appalachians

In the Southern Appalachians, too, the western boundary of the orogen may be drawn at the westernmost limit of structural deformation related to Appalachian orogenic events. Similar to the Northern Appalachians, rocks west of the structural front comprise a major foreland basin. Lowest strata of this basin are Cambrian and Ordovician shallow-shelf carbonates and related detrital sedimentary rocks, all of which reflect their origin along the passive continental margin which formed after late Precambrian rifting (Hatcher, 1978). Later detrital sediments were derived from the east and, like their counterparts in the north, were probably the result of rising tectonic sourcelands associated with an active continental margin which formed during the Ordovician.

Unlike the Northern Appalachians, however, the western limit of structural deformation is separated from metamorphosed rocks and basement-cored nappes by a wide belt of folded and faulted sedimentary strata, termed the Valley and Ridge Province. Strata of the Valley and Ridge range from Cambrian to lowest Permian and, therefore, record events spanning the entire Paleozoic Era (Colton, 1970). Valley and Ridge strata are typified by large, décollement-style thrust faults and associated rootless anticlines and synclines (see Chapter 7 for fuller discussion of Valley and Ridge structure). Most, but not all, of this deformation must have occurred at the end of the Paleozoic because even the youngest strata were involved; the structural event responsible for Valley and Ridge deformation is termed the Alleghenian orogeny.

Bounding the Valley and Ridge on the southeast is the Blue Ridge Province (Fig. 5-13), a belt of highly deformed and metamorphosed basement and supracrustal rocks thrust over Valley and Ridge rocks along a major fault system, represented by the Cartersville, Great Smoky, and Holston Mountain faults (see Fig. 5-13). The Blue Ridge may be divided into three zones (Hatcher, 1978): (1) a western belt composed of low-grade metasediments of latest Precambrian to Ordovician age and some Precambrian basement, compressed into a stack of imbricate thrust slices which dip generally eastwards; (2) a central core composed mostly of Grenville-age basement and high-grade metamorphic equivalents of supracrustal rocks in the western Blue Ridge belt; and (3) an eastern belt, separated from the central core by the Haynesville–Fries fault and composed of metasupracrustals, of which metavolcanics comprise a significant proportion. Although not strictly a part of the Blue Ridge, the Talladega and Ashland–Wedowee belts of east-central Alabama occupy approximately the same structural position with respect to Valley and Ridge strata as does the Blue Ridge. The Talladega slate belt is a thick sequence of low-rank metasediments whose gross stratigraphy resembles that of the adjacent Valley and Ridge, over which Talladega rocks have been thrust (Tull, 1978, 1982). At the top of the Talladega belt stratigraphic sequence is a metavolcanic unit called the Hillabee Greenstone of Lower to Middle Devonian age (Tull, 1978). Higher-grade (middle amphibolite) metamorphic rocks of the Ashland–Wedowee belt more nearly resemble Blue Ridge rocks and are thrust over Talladega rocks from the southeast (Fig. 5-13).

Forming the eastern boundary of the Blue Ridge is the Brevard Zone, a narrow belt of profound cataclasis showing the effects of at least three episodes of deformation (Hatcher, 1972, 1978). In addition to highly mylonitonized rocks, the Brevard Zone also contains exotic blocks of carbonate rocks which reveal the effects of only relatively low-grade metamorphism. The chemistry of these carbonates is similar to that of autochthonous, Cambrian–Ordovician platform carbonates of the Valley and Ridge, to which they have been related (Hatcher, 1971). Because these exotic blocks must have been caught within the cataclastic zone and dragged up from below, and because the Blue Ridge exhibits an overthrust relationship with respect to the Valley and Ridge, Hatcher (1971) suggested that the Brevard Zone is a major splay off a décollement which underlies the entire Blue Ridge (and, by implication, the Inner Piedmont as well). If this suggestion is correct, then both the Blue Ridge and Inner Piedmont are allochthonous and rest on top of relatively undeformed sedimentary strata which are continuous with those of the Valley and Ridge. (In Chapter 7, we will return to this subject and will discuss seismic reflection data which appear to confirm Hatcher's prediction.) Intimately associated with the Brevard Zone along its entire length is the Chauga belt, a narrow, multiply deformed synformal structure composed of low-rank metasediments (Hatcher, 1978).

To the northeast, into northcentral North Carolina, the Brevard Zone loses its identity. Just south of the Virginia border, the Brevard Zone appears to swing more eastwards around the south side of the Sauratown Mountain anticlinorium. The Sauratown Mountain structure is a highly deformed block of Grenville-age basement gneisses and associated metasupracrustals; its Grenville age sets it apart from surrounding rocks whose ages are primarily Ordovician but which range from latest Precambrian to Devonian (Rankin et al., 1973; Rankin, 1975). Because of these age relations, the Sauratown Mountain block may have been a separate crustal entity prior to collisional emplacement into the Appalachian tectonic collage. To the north, along the Blue Ridge of central Virginia, is another apparent displaced terrane represented by the Chopawamsic Formation, whose Lower Cambrian metavolcanics and related intrusive rocks were considered by Palvides (1981) to be Cambrian(?) island arc.

The Brevard Zone serves as a convenient boundary to separate the Blue Ridge physiographic province from the Piedmont physiographic province. The Piedmont province, however, is itself composed of a number of lithotectonic belts which must be discussed separately. Perhaps the largest of these, lying just to the east of the Brevard–Chauga structure, is the Inner Piedmont belt (Hatcher, 1978; Rodgers, 1982). The Inner Piedmont is a zone of very high rank (up to sillimanite grade) metamorphic rocks and associated plutonic intrusive rocks. It extends from the Coastal Plain overlap in Alabama to near the Virginia–North Carolina border, and used to be considered the most ancient, crustal core of the Appalachian orogen. More recent studies have shown that the Inner Piedmont is composed primarily of metamorphosed sedimentary rocks; radiometric dating of these rocks reveals a complex metamorphic history with a thermal peak during the later Devonian (i.e., probably associated with the Acadian orogeny; Butler, 1972).

Bordering the Inner Piedmont belt on the east is the Kings Mountain–Pine Mountain belt. The Kings Mountain belt of North and South Carolina is a major synformal structure composed of metasediments, especially quartzites and calcitic marbles, and metavolcanics (Williams and Hatcher, 1982). Its metamorphic rank, although high (up to amphibolite grade), is less than that of the Inner Piedmont belt to the west or the Charlotte belt to the east. The origin of the Kings Mountain belt has been highly controversial and suggestions as to its paleotectonic significance are numerous, including a collisional suture zone (Glover and Sinha, 1973) and a closed back-arc basin (Hatcher, 1978). Rocks of the Kings Mountain belt appear to become lost southwestward along regional strike but similar rocks outcrop in the Pine Mountain belt of southwestern Georgia and adjacent Alabama (Schamel et al., 1980). The Pine Mountain belt is separated from the Inner Piedmont by the Towaliga fault; it is separated from the Uchee belt to the southeast by the Goat

Rock fault (Fig. 5-13). Between these two faults, the Pine Mountain belt is composed of Grenville-age granitic basement and associated metasediments; quartzites and marbles of the Pine Mountain area are similar to rocks of the Kings Mountain belt with which they might be correlated.

The next major lithotectonic unit of the Southern Appalachians in the Charlotte belt. This province is composed of sillimanite-grade metasediments and metavolcanics which have been intruded by a large number of plutons. The compositions of these plutons vary greatly as do their ages, which range from Cambrian to Mississippian. Overstreet and Bell (1965) suggested that metasediments of the Charlotte belt are correlative with lower units of the Carolina slate belt just to the east. As can be seen from the map of the Appalachians (Fig. 5-13), the exact relationship between the Charlotte belt and the belts of southern Georgia and Alabama is problematical; Hatcher and Zeitz (1980) have proposed that the boundary between the Charlotte and Inner Piedmont belts, including the Kings Mountain belt, is a major continental suture which they termed the Central Piedmont suture. In this model, rocks of the Pine Mountain belt of Georgia and Alabama would not be correlative with those of the Kings Mountain area; indeed, because of its Grenville basement, the Pine Mountain block may have been a displaced crustal block similar to the Sauratown Mountain block in North Carolina (Thomas, 1977). Alternatively, the Pine Mountain belt may be a window through the allochthonous Piedmont nappe (Sears *et al.,* 1982).

The Carolina slate belt runs from southern Virginia to northeastern Georgia (Fig. 5-13) and consists of weakly metamorphosed volcanic and sedimentary rocks (Seiders and Wright, 1977; Black and Fullagar, 1976; Bearce *et al.,* 1982). In its northern part, Carolina slate belt volcanics are primarily calc-alkaline andesites and dacites with radiometric ages of about 650 to 550 Myr. More southerly portions of the slate belt are characterized by bimodal volcanic assemblages with lower rhyolites overlain by volcanic mudstones and lavas dating from about 570–520 Myr. Several authors (e.g., Butler and Ragland, 1979) have shown that the geochemistry of slate belt rocks is similar to that of modern volcanic island arcs. Dates of northern slate belt volcanics are similar to those of volcanics from the Avalon Peninsula of Newfoundland, leading Glover and Sinha (1973) to suggest that the tectonic event responsible for northern slate belt rocks (which they term the Virgilina orogeny) may have been related to the Avalonian orogeny of the Northern Appalachians. Williams and Hatcher (1982) proposed that the Carolina slate belt (as well as the Charlotte belt and several others discussed below) was part of a single, large Avalon block whose accretion to the margin of North America precipitated the Acadian orogeny during the Devonian.

The above-mentioned belts comprise the bulk of the Piedmont physiographic province of the Southeast. Several

other lesser lithotectonic belts should be mentioned, however. Adjoining the Carolina slate belt on the east are the Raleigh and Kiokee belts (Fig. 5-13), both of which are composed of plutonic igneous rocks and high-grade metamorphics (Bobyarchick, 1980). The Raleigh belt in North Carolina is separated from the slate belt by the Nutbush Creek fault; the Kiokee belt in South Carolina and Georgia is separated from the slate belt by the Modoc fault. Because the intervening area is obscured by a cover of Coastal Plain sediments, the exact relation between the Raleigh and Kiokee belts is unknown. Further, Snoke and Secor (1982) have argued that the Modoc fault is really a metamorphic front rather than a major fault boundary and that rocks of the Kiokee belt are merely high-grade equivalents of slate belt rocks. To the east of the Raleigh belt in North Carolina is the Eastern slate belt, whose rocks are similar to those of the Carolina slate belt; a similar relationship exists between the Kiokee belt and the Belair belt, only a small amount of which appears out from under the Coastal Plain cover. The Belair belt is separated from the Kiokee belt by the Augusta fault, a mylonite zone up to 200 m wide, upon which has been superimposed an unknown amount of brittle faulting. Although now in fault contact, the Kiokee and Belair belts have similar histories prior to late Paleozoic time and may have been part of the same tectonic element. South of the Goat Rock fault in southwestern Georgia is the Uchee belt (Schamel *et al.,* 1980). The Uchee belt is composed of relatively high-grade metamorphic rocks, principally gneisses and amphibolites. Although grossly similar to rocks of the Kiokee belt, the relation between them, if any, is unknown.

The faults mentioned in the preceding discussion, including the Nutbush Creek, Augusta, Modoc, Towaliga, and Goat Rock faults, have been grouped by Hatcher *et al.* (1977) into a single complex, which they termed the Eastern Piedmont fault system. Bobyarchick (1980) considered the Eastern Piedmont fault system to have been formed during the Alleghenian deformation, as the result of large-scale lateral displacement of crustal blocks due to continental collision. (The Alleghenian orogeny will be discussed in Chapter 7.)

C.1.c. The Appalachian Orogen as a Tectonic Collage

The subdivisions of the Appalachian orogen discussed in the preceding subsections are recognized primarily because of gross similarities of lithologies and structures within each one. Recently, however, new information and ideas have led to the view that the Appalachian orogen may best be considered as a tectonic collage into which lithotectonic entities, either singly or grouped together into larger terranes, were added, by collision or by transform faulting, throughout the Paleozoic Era. This view of the Appalachians was impelled by the somewhat earlier recognition of the significance of displaced, or "suspect," terranes in the history of the Cordillera; it has been strengthened by the results of seismic reflection profiling across the orogen which showed the presence of a deep, horizontal reflector horizon which has been interpreted as the master décollement over which Piedmont blocks were thrust during Appalachian orogenesis (see discussion in Chapter 7).

Williams and Hatcher (1982) applied the collage concept to the Appalachians, subdividing the orogen into lithotectonic assemblages which represent displaced terranes (Fig. 5-14). Notice that Williams and Hatcher recognize only a limited number of displaced terranes in the Appalachians. This is because several different lithotectonic belts may have originally been part of a single crustal entity whose accretion along the North American margin was a single event. For example, they lump the Bronson Hill–Piscataquis anticlinorium, the Aroostook–Matapedia trough, the Miramichi massif, and the Merrimack–Fredericton trough into a single feature, called the Gander terrane, which was accreted to North America as a result of the Taconic orogeny. Similarly, they consider the Charlotte, Carolina slate, Raleigh, Kiokee, Eastern slate, Belair, and Uchee belts all to be part of the Avalon terrane. Perhaps the biggest problem with Williams and Hatcher's model is the suggestion that most (but not all) of the terranes are continuous all along the orogen. This implication originates from a presumption that orogenic events must have been contemporaneous all along the continental margin. But detailed radiometric dating is beginning to show that such may not have been the case; rather, orogenic climaxes may have affected only small segments of the margin at any one time. They may have been diachronous along the margin, perhaps as the result of oblique convergence of colliding blocks, or may have had no counterpart at all in other areas of the margin. A glance at a map of the modern Pacific coast of North America, from the Aleutian volcanic arc of Alaska, to the continental arc of Washington and Oregon, to the faulted continental borderland of southern California and the Baja Peninsula should persuade the reader that a single continental margin may experience simultaneously a variety of essentially unrelated tectonic events. Indeed, even if events of different portions of the same continental margin were contemporaneous, there is still no guarantee that they were genetically related. Thus, there is no reason why events which affected the Southern Appalachians would have to parallel, either in type or time, those recorded in the north. We will return to this subject several times in succeeding discussions of the Appalachians.

C.2. The Taconic Orogeny

Having begun during the late Precambrian with a major episode of continental rifting, the passive margin of eastern North America was, by the Middle Ordovician, beginning to be affected by tectonic events which would result finally in its destruction. These tectonic events are termed the Taconic orogeny. In the following pages, we will discuss the Taconic orogeny in three sections: first, we will discuss the specific orogenic effects (i.e., deformation, intrusion, metamorphism) seen in the New England and Blue Ridge areas of the Appalachian orogen; second, we will discuss plate-tectonic speculations on the cause of the orogeny; and third, we will discuss the sedimentary record of the orogeny in the Appalachian foreland basin.

C.2.a. Nature of the Taconic Orogeny

It was believed at one time that the Taconic orogeny punctuated the end of the Ordovician and, thus, that its effects were of the same age throughout the Appalachians. Detailed analysis of unconformable stratigraphic relations and radiometric dating have since shown that Taconic events, though usually similar in kind and sequence, were

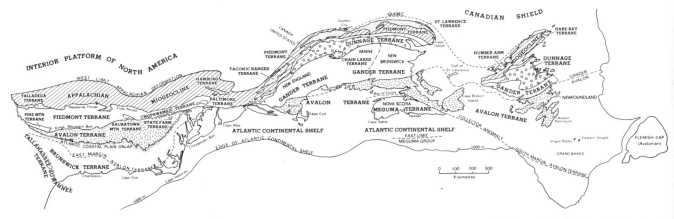

Fig. 5-14. Map of major suspect terranes of the Appalachian orogen; see text for discussion. (From Williams and Hatcher, 1982, *Geology*, Fig. 1, p. 530.)

not contemporaneous in all parts of the orogen (Rodgers, 1971). From such analyses, three or four major tectonic phases have been discerned which affected, to a greater or lesser extent, most of the eastern continental margin; these phases will now be discussed briefly. During the following discussion, the reader should keep in mind that the specific history of any given area of the Appalachian region varied considerably from that of other areas and that, even when two areas were affected by similar events, they were probably not affected simultaneously.

i. Tinmouth Phase. Several workers have recognized a period of moderate uplift of the shelf carbonate-platform that occurred at the end of the Early Ordovician. As we have seen, this was a time of lowered sea level over the whole continent, but additional tectonic uplift also seems to have affected the eastern edge of the continent, especially in the Northern Appalachian area where it exaggerated the extent of the post-Sauk disconformity. Williams and Stevens (1974) have stated that the Newfoundland carbonate platform was uplifted as much as 90 m at this time. Erosion did not apparently affect the deep-water deposits of the graptolite facies to the east of the Newfoundland shelf but Bird and Dewey (1970) suggested that red shales of the Blow-Me-Down Formation may have been derived from *terra rosso* soils of the carbonate platform; such soils today typify regoliths in karst areas. Erosion in the Quebec area stripped off all earlier strata, exposing Precambrian basement. Genetically related block faulting has also been recognized in these areas; in the New England area, for example, horst blocks were exposed and eroded while grabens underwent nearly continuous deposition (Williams and Stevens, 1974). In the Southern Appalachians, normal faulting has been postulated by Shanmugam and Walker (1978) as one explanation for the pronounced difference in subsidence rates of previously similar parts of the eastern Tennessee carbonate platform during the early Middle Ordovician.

ii. Penobscot (Oliverian) Phase. During the late Early Ordovician, a strong deformational event occurred which affected strata of the graptolite facies far to the east of the continental shelf. Local details of this event vary considerably from place to place but the general picture is of compressional tectonics with the development of major anti- and synclinoria, large westerly driven thrust faults, and strong metamorphism. In New England, for example, deformation within the Bronson Hill anticlinorium is related to this phase of the Taconic orogeny and occurred prior to deposition of the Middle Ordovician Shin Brook Formation and Ammonoosuc Volcanics (Bird and Dewey, 1970). In the northern Maine volcanic belt, basic volcanic rocks and related argillaceous strata of probable Cambrian age (such as the Grand Pitch Formation) were also strongly folded,

faulted, and cleaved before deposition of Middle Ordovician strata. Rodgers (1971) suggested that thrust-faulting and initial uplift of the Berkshire and Green Mountains anticlinoria of Massachusetts and Vermont may have occurred at about the same time; he also stated that coarse conglomerates in the lower Middle Ordovician Mictaw Formation of the southern Gaspé Peninsula (which contain metamorphic and granitic cobbles) may have been derived from island sourcelands which had been uplifted by that time. In Newfoundland, too, there is evidence of an early, strong tectonic event; there, basic volcanics of the Lush's Bright Group were completely thrust, folded, and metamorphosed before deposition of the Lower Ordovician Catcher's Pond Formation.

In the Southern Appalachians, an important tectonic event also occurred before the climactic phase of the Taconic orogeny. While the gross nature of this event differs significantly from that of the Penobscot phase in the north, some interesting comparisons can be drawn. In the Blue Ridge and western Piedmont of North Carolina, for example, a major episode of folding and westerly directed thrusting occurred during the Late Cambrian to Early Ordovician (Hatcher, 1978). At that time, mafic and ultramafic rocks with apparent mantle affinities were thrust into the deforming sediment pile that would become the eastern Blue Ridge. Several other major thrusts were formed at this time, such as the Greenbrier thrust which transects strata in the eastern Great Smoky Mountains. During the later stages of this event, granitic plutons were emplaced in the eastern Blue Ridge area and the western Inner Piedmont.

iii. Taconic (Vermontian) Phase. At some time following the events described above, the continental shelf of North America began to subside. That area had previously been a shallow carbonate platform on which "shelly" facies deposits accumulated but, by the middle Ordovician, deep-water, graptolite facies deposits were forming there. At the same time, farther out on the continental margin, thrust slices of deep-water muds began to be driven westwards, up and over rocks of the outermost continental shelf and slope. These thrust slices were to become Klippe of the Taconic allochthon, the Humber Arm allochthon, and related Ordovician thrust nappes. As the allochthonous slices were driven onto the continental margin, extremely chaotic masses of "blocks in shale," or *wildflysch,* were bulldozed in front of the slice. These chaotic masses were both pushed along in front of the allochthons and overridden by them; the mud matrix of the wildflysch contains fossils which allow close approximation of the time of emplacement of the nappes.

iv. Hudson Valley Phase. The final phase of the Taconic orogeny is the "classical" orogeny of early literature. It resulted in profound uplift of the outer continental shelf, producing a rapidly rising, mountainous landmass referred to as *Taconica*. Erosion related to this uplift caused the unconformity (seen by Mather and Rogers in the 1830s) at Becraft Mountain and at other localities in the Hudson Valley, northern New Jersey, and eastern Pennsylvania. Further, detritus derived from Taconica was carried westward to feed the deltaic complex of a great coastal plain which prograded into the foreland basin, ultimately spreading fine detrital sediment across much of the northeastern craton (the stratigraphy of this wedge-shaped mass of detrital sediments will be considered in a subsequent section).

Accompanying uplift was a period of folding which was most intense in New England but which affected strata as far west as Albany, New York. In the Green Mountains of Vermont, large isoclinal, recumbent folds with well-developed axial-plane cleavage were formed at this time. In eastern Pennsylvania, recumbent folds with wavelengths of tens of kilometers were developed. In addition, widespread regional metamorphism, up to sillimanite grade in southeastern New York, occurred at the same time.

All of the above-mentioned events of the "classical" Taconic orogeny in the Hudson Valley region and adjacent areas took place during the Late Ordovician but similar episodes of strong deformation and metamorphism occurred at other times in other regions. In some cases, these episodes were accompanied by emplacement of ophiolite sequences, such as the Thetford Mines complex of Quebec (St. Julien and Hubert, 1975) and the Bay of Islands complex of western Newfoundland (Williams and Stevens, 1974). In other cases, igneous activity was related to the deformation; examples include the Ammonoosuc Volcanics of Vermont and the Oliverian and Highlandcroft Plutonic Series of New England (Naylor, 1968). It is interesting to note that the four-phase subdivision of Taconic events is difficult to apply specifically in every area. In the case of the plutonic rocks mentioned above, for example, the timing of their emplacement is controversial, to say the least (some authors arguing for a Devonian age), so they may be related to the Penobscot phase instead of the Hudson Valley phase. On the other hand, since Oliverian plutons intrude Middle Ordovician Ammonoosuc Volcanics (which, in turn, overlie rocks deformed in the Penobscot phase), but were probably emplaced *before* the strongest metamorphic event (which was of Late Ordovician age in this area), the question of which phase is to be blamed seems rather moot.

In the Southern Appalachians, the climactic Taconic event was considerably older, probably beginning in the Middle Ordovician. Here, too, major uplift (and associated detrital sedimentation) was accompanied by strong deformation and metamorphism. Large-scale thrusting occurred along the Haynesville–Fries fault, which separates the eastern Blue Ridge from the core of the Blue Ridge (see Fig.

5-13). Major folding also occurred at this time, especially early isoclinal folding of Blue Ridge areas. Intense metamorphism was another effect of the Taconic orogeny in the Southeast. Butler (1972) stated that the peak of regional metamorphism in the Blue Ridge area occurred approximately 430 Myr. ago and is, therefore, related to the Taconic orogeny.

C.2.b. Plate-Tectonic Speculations

The Taconic orogen has engendered a number of tectonic models, including many plate-tectonic ones. Dewey and Bird's (1970) model was one of the first and still stands as a classic work of tectonic interpretation. In their model, Dewey and Bird suggested that the Taconic orogeny was the result of the initiation of westward-directed subduction beneath the margin of North America. The result was to cause a rising "thermal welt" along the continental slope and rise, i.e., directly above the newly developed subduction zone. This rising mass was the source of Taconic allochthons, which Dewey and Bird thought to be gravity-slide masses, slipping off the thermal welt. Subsidence of the continental shelf in this scheme was due to mass adjustments in the mantle in response to rising of the slope–rise area. By the Late Ordovician, the effects of thermal uplift were profound and affected the entire continental margin, producing the Taconic sourceland from which were derived the detrital sediments which flooded out into the foreland basin.

The scenario of Dewey and Bird (1970) has been modified considerably since its publication. A more recent view of Taconic tectonics may be seen in the discussion of Rowley and Kidd (1981). Probably the biggest single difference between the model of Dewey and Bird (1970) and Rowley and Kidd (1981) is the direction of subduction and the mechanics of Taconic-allochthon emplacement. Rowley and Kidd argue that the North American margin remained a passive margin within a larger plate, which was being subducted eastwards beneath a volcanic arc complex somewhere to the east (see Fig. 5-15). The volcanic arc, called the Ammonoosuc arc by Rowley and Kidd, is represented today by the Bronson Hill anticlinorium and its extension to the north, the central Maine volcanic belt. Probably also involved with the eastern arc complex was the Miramichi massif of New Brunswick; whether these arc complexes were all part of the same volcanic archipelago, or separate entities is not known. In their identification of suspect terranes of the Appalachian orogen, Williams and Hatcher (1982) referred to this offshore block as the Gander terrane (see Fig. 5-14).

Gradually, the outer margin of North America was brought into juxtaposition with the offshore arc (Fig. 5-15). The first apparent result was slight uplift of the outer conti-

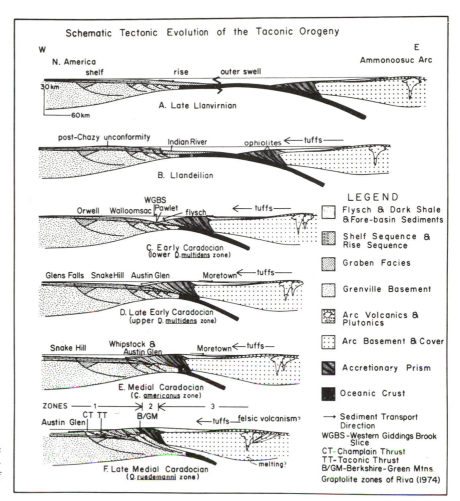

Fig. 5-15. Plate-tectonics model of the Taconic orogeny. (From Rowley and Kidd, 1981, *Journal of Geology*, Vol. 89, p. 212, © University of Chicago Press.)

nental margin (the Tinmouth phase), probably as the result of its passage through the outer-arc swell just to the west of the volcanic arc. As time passed, progressively more of the continental margin began to be pulled down into the subduction zone with several important effects. Subduction beneath the Ammonoosuc arc of increasingly thicker and more buoyant, sialic crust caused increasing stress within the arc; compressional folding and metamorphism of the Penobscot phase in the Bronson Hill anticlinorium and associated areas were probably caused in this way. Intrusions and related extrusions associated with the Penobscot phase within the colliding arc require no special explanation. The other major effect of subduction of the outer continental margin was the profound and rapid subsidence of the continental shelf (related to the Taconic phase); accompanying shelf subsidence was the thrusting of Taconic nappes onto the outer shelf. Both of these events may be thought of as part of the same process, i.e., the anisostatic drawing down of the shelf edge beneath the accretionary prism on the opposite plate. Thus, Taconic allochthons and related thrust nappes

are considered thick masses rucked up in front of and within the accretionary prism (Fig. 5-15).

As the accretionary prism acquired progressively more and thicker increments, it grew larger, finally forming an island mass. From this growing sourceland, detrital sediments were derived and carried into the peripheral foredeep basin, which had formed by anisostatic drawdown as well as by thrust-nappe loading of the outer shelf. At first, sediments were only fine, hemipelagic muds interspersed with rare turbidites; these deposits comprise the Normanskill Formation. Increase in the size of the sourceland resulted in increasingly greater sediment influx, reflected first in thick turbidity-current sedimentation of the Austin Glen Member of the Normanskill (Fig. 5-15).

As the subduction process continued, thicker continental crust began to be drawn into the subduction zone. The resulting stresses caused the granitic crust to be fractured into slivers (represented by the Green Mountain–Sutton Mountain anticlinorium) which were thrust up and backwards so that, today, they lie on top of and in thrust contact with the Taconic allochthon. Another result of the "hard-rock" thrusting of the Green Mountain–Sutton Mountain nappes was thickening of the crust at the suture zone so that it began to rise isostatically, producing the mountain belt which dominated the geology of the Late Ordovician Appalachians. From this mountain belt spread coarse detrital sediments into the foreland basin, forming the Taconic (or, Queenston) detrital wedge (whose stratigraphy will be discussed in the next section).

We now turn to the Southern Appalachians. Probably the greatest difference between Taconic events in the north versus the south was the presence in the south of a much larger crustal block offshore. Williams and Hatcher (1982) proposed that the Piedmont terrane, composed of the eastern Blue Ridge and the Inner Piedmont (see Fig. 5-14), was juxtaposed to the east of the margin of North America but riding on a converging plate. The North American continent occupied the other plate, which subducted eastwards beneath the Piedmont terrane. Beginning somewhat earlier in the south, perhaps as early as latest Late Cambrian or Early Ordovician, the margin of the continent began to interfere with the Piedmont block. As mentioned in the discussion of the Northern Taconic orogeny, the mechanical effects of that interference can explain most of the features considered to be of Taconic origin in the Southern Appalachians. The sourceland which resulted from collisional accretion of the Piedmont block supplied detritus to the Blount detrital wedge (see further discussion in the next section).

C.3. Stratigraphic Record of the Orogeny

The various structural effects of the Taconic orogeny described in the previous sections were limited to the edge of the continent but sediments produced by weathering and erosion of sourcelands uplifted by the orogeny were deposited over much of the eastern half of the continent. The resulting pile of detrital sediments is fan-shaped in plan view and wedge-shaped in cross section, being thickest near the sourceland and thinning westward toward the carbonate platform of the cratonic interior. Before we discuss the Taconic detrital wedge, we will digress briefly to discuss the nature of detrital wedges in general.

C.3.a. Nature of Detrital Wedges

When mountains are pushed up along a portion of a continental margin, they are immediately attacked by weathering and erosion, producing sediments which are carried away in stream systems to be deposited in subsiding areas on either side of the mountains. In the case of Cordilleran-type mountains (such as those of the Taconic orogen), the subsiding area on the ocean side would probably comprise a subduction prism and might be strongly deformed and metamorphosed in a subsequent orogeny. But the subsiding area on the continent side (i.e., a retroarc foreland basin, or *exogeosyncline*) would accumulate sediments that stood a good chance of being preserved in the stratigraphic record with only moderate folding or faulting.

Stream systems draining the mountains commonly flowed together to form one or two major deltaic systems so that sediments were introduced into the foreland basin at several distinct points. From these points of introduction, sediments prograded radially outward into the basin, which accounts for the commonly observed fan shape of detrital wedges. Occasionally found in the geological record are instances in which an extensive mountain system all along a continental margin supplied detritus to many fan systems which coalesced to form a great bajada-like coastal plain.

The individual detrital fan and the bajada-like coastal plain may be thought of as end members of a series; most real detrital wedges probably represent a situation somewhere between these two end members.

The internal stratigraphy of a detrital wedge is usually very complicated because of the multiplicity of sedimentary environments that were present as the great sediment pile grew into the foreland basin; nevertheless, several generalities can be made. Detrital wedges may be divided, for the sake of this discussion, into two types: those which grew into deep foreland basins and those which grew into shallow basins. In the first case, the lowest strata of the detrital wedge would be comprised of flysch sediments deposited by turbidity currents from the seismically active slopes of the foreland. The Normanskill Formation and its Austin Glen Member mentioned in the last section are examples. Flysch sedimentation would continue at a rate exceeding the rate of subsidence until the foreland basin literally filled up, at which time various shallow-water environments would grow over the now-shallow basin. Of course, the above discussion is very general, for deep and shallow environments would have been contemporaneous, with flysch accumulating in distal portions of the basin while shallow-water molasse was deposited in proximal areas. In the case of detrital wedges which prograde across shallow-water basins, such as an area dominated by a carbonate platform, molasse strata would grade outward into shallow epeiric-sea deposits with no lower flysch sequence.

As a detrital wedge prograded, nonmarine sediments would be deposited over shallow marine strata. Meckel (1970) proposed the following model of alluvial sedimentation in detrital wedges: during the early period of detrital-wedge growth, the rate of deposition would exceed the rate of basin-subsidence and all facies boundaries would migrate with time away from the sourceland, resulting in a regressive sequence. During this period, a traverse across a typical wedge would reveal coarse, pebbly sands, deposited in numerous fluvial channels, dominating the upper alluvial plain near the sourceland. Farther from the source, on the lower alluvial plain, sediments would be dominated by red muds deposited on large floodplains between stream channels. Even farther away from the sourceland, muddy red-bed strata would grade into shallow marine deposits. As long as deposition exceeded subsidence, red muds would be deposited rapidly with little or no reworking. At the maximum phase of regression, i.e., as the sedimentation rate began to decline, ending the outward-growth phase, a more-or-less static coastline would form along which muddy deposits would be reworked by the marine environment so that the nonred channel facies would grade directly into the nonred marine facies. Finally, when the sedimentation rate decreased to the point where it was exceeded by the subsidence rate, facies patterns would migrate back toward the source area, developing a transgressive sequence in which, once again, a red-mud facies would intervene between nonred marine strata and nonred stream-channel sands. In this case, however, the reason for red-mud deposition lies in the fact that, in a rising base-level regime, coarse sediments are trapped progressively farther upstream and only muds are supplied to the lower reaches of the alluvial plain. Meckel (1970) was able to show that this general model characterized the three major detrital wedges of the Appalachian Basin (Taconic, Acadian, and Alleghenian).

One complication to the general pattern described above should be mentioned. Occasionally, a large part of the detrital wedge would be "suddenly" overlapped by transgressive carbonate strata due to either a rapid rise of sea level or a momentary decrease in sediment supply. Such thin, nearly isochronous limestone units were referred to as *interwedge carbonates* by Thomas (1977). Probably the best example is the Tully Limestone of the Acadian detrital wedge which will be discussed in the next chapter.

With this general discussion of detrital wedges as background, we turn now to consider the two wedges associated with the Taconic orogeny.

C.3.b. The Blount Detrital Wedge

As discussed in the previous section, orogenic activity seen in the Southern Appalachians probably began in the Middle to Late Cambrian but at that time was primarily related to an eastward dipping subduction zone associated with the Sauratownia microcontinental block. Thus, it had no major effect on the main continental margin. By the earliest Middle Ordovician, however, the edge of the continental margin began to be affected by anisostatic subsidence as it was drawn down upon entering the subduction zone (see Fig. 5-14). Just after the Tippecanoe transgression had reestablished carbonate platform conditions, a sudden and profound subsidence dropped the outer shelf down into deep water at the geologically blistering rate of 1 m/1000 yr.

(Shanmugam and Walker, 1978). Areas of the outer shelf where intertidal carbonates of the Lenoir Limestone had been deposited shortly before were suddenly subject to the deposition of graptolitic shale and interbedded phosphatic and manganiferous limestones of the Whitesburg Formation, interpreted by Shanmugan and Walker as continental slope deposits. These are overlain by pelagic sediments of the Blockhouse Formation which appear to represent a starved-basin environment. Farther to the west, carbonate-platform deposition continued, apparently unaffected by events on the outer shelf. This dramatic foundering of the shelf is clearly similar to the "inversion of relief" in the Hudson Valley area described earlier but the large allochthonous slices seen in the north are not found here, possibly because the deep Southern Appalachian area was a peripheral basin and was therefore separated from the magmatic arc by the trench rather than being directly adjacent to the arc as was the retroarc basin of the Hudson Valley region.

Above the Blockhouse Formation, distal turbidites of the Sevier Shale appear in the section; farther to the southeast, proximal turbidites of the Tellico Formation occur at the same stratigraphic position, indicating that the sourceland lay offshore. Turbidite strata accumulated to a thickness of over 2000 m.

Toward the end of the Middle Ordovician, the subsidence rate had decreased and detrital sediments from the east were deposited in shallow-marine environments which prograded westward toward the carbonate platform which had not been, until then, significantly affected by detrital sedimentation in the east. The result was the Moccasin Formation of Alabama and Tennessee and the Bowen Formation of Virginia in which gray and reddish-gray, shallow-marine strata in the west intertongue with and grade into red, mud-cracked shales in the east. These deposits are believed to represent deposition on a subaerial mud flat at the front of the advancing detrital fan. To the east of these strata are red beds of the Bays Formation.

The whole mass of Blockhouse through Bays strata (about 2.5 km thick) comprises the Blount detrital wedge (named after Blount County, Tennessee). Strata of the Blount wedge grade radially outward into carbonates such as the Chickamauga Group of Alabama.

C.3.c. The Taconic Detrital Wedge

The Blount detrital wedge, while a significant feature of eastern Tennessee Ordovician strata, is of less overall importance in the development of Appalachian stratigraphy than the great detrital wedge which formed in the Northeast during the Middle and Late Ordovician and which lasted until the Middle Silurian. This great mass of sediment, over 4 km thick in eastern Pennsylvania near the center of the

Pennsylvania structural salient, has been referred to as the Queenston clastic (or detrital) wedge from the Queenston Formation of New York and northwestern Pennsylvania. However, we prefer the term *Taconic detrital wedge* to emphasize the relation of the stratigraphy to the causative orogenic event.

The history of the Taconic detrital wedge began with deposition in eastern New York of the Normanskill Shale whose Austin Glen Member represents turbidity-current deposits derived from the adjacent sourceland, Taconica. But these strata were deposited only on the outer shelf. By latest Trentonian time, however, black, graptolitic shales of the Canajoharie and then the Utica Shales had spread westward over almost all of New York and Pennsylvania, indicating that subsidence had affected this entire area and had initiated starved-basin conditions. A similar period of subsidence may be seen in northwestern Virginia where dark gray, graptolitic shales and deep-water carbonates of the Liberty Hall facies of the Edinburg Limestone overlie the shallow subtidal Lincolnshire Limestone. In eastern Pennsylvania, too, starved-basin conditions prevailed, represented by black, graptolitic shales, siliceous shales, and pelagic cherts of the lower Martinsburg Formation.

Turbidity-current deposits of the Martinsburg Formation began to be deposited on the eastern side of the deep foreland basin (or *foredeep*) adjacent to the eastern Taconic sourceland during the latest Middle Ordovician while black, graptolitic shales continued to accumulate in a starved-basin setting on the western side of the foredeep. The Martinsburg is one of the most widespread flysch units of the Appalachians, comprising a single, once-continuous sheet of turbidite strata from northeastern Pennsylvania to southeastern Tennessee (where it overlies the Bays Formation).

As more sediment was poured into the foreland basin, shallow marine environments developed in the east and turbidity-current deposition affected western parts of the foredeep. By the end of the early Late Ordovician, shallow marine strata of the upper Martinsburg were being deposited even in westernmost areas of the old foredeep; such shallow, upper Martinsburg strata graded west into the Reedsville Shale of western New York, Pennsylvania, West Virginia, and southwestern Virginia and south into shallow marine carbonates of the Inman and Leipers Formations of Tennessee and Alabama.

During latest Martinsburg time, nonmarine environments were growing ever westwards. The Oswego Sandstone of West Virginia and adjacent Virginia and correlative Bald Eagle Sandstone of Pennsylvania overlie shallow-water Martinsburg strata. The Oswego/Bald Eagle is a controversial unit because it has neither obvious marine features (especially, it lacks fossils) nor obvious alluvial features (especially, it lacks red coloration). The prevailing view of the moment (as espoused by Dennison and Wheel-

er, 1975) is that it is nonmarine and that the coloration is primarily a diagenetic phenomenon. At any rate, obviously alluvial strata of the Juniata Formation of the Southern Appalachians and the Queenston Formation of New York and Pennsylvania were deposited over much of the foreland basin during Late Ordovician time. The Queenston/Juniata is an extensive unit with well-developed fining-upward, fluvial sequences in some areas, thick floodplain deposits, and coastal tidal-flat deposits. The Queenston extends all the way across the Appalachian Basin, grading ultimately into cratonic strata of the upper Mississippi Valley, such as the Maquoketa Shale (see Section B.5). The Juniata grades south into the Sequatchie Formation whose red, gray, and greenish sediments represent a complex coastal-facies assemblage of mud flats, beach ridges, barrier bars, lagoons, and a muddy, marine shelf (Milici and Wedow, 1977). Taken altogether, these strata (i.e., Oswego/Bald Eagle, Juniata/Queenston, Sequatchie) comprise the regressive alluvial phase of Meckel's detrital wedge model.

The maximum regression phase of Meckel's model is represented in the Taconic detrital wedge by up to 120 m of light gray to white quartz arenitic sandstone named the Tuscarora Sandstone in western Virginia and areas to the north and the Clinch Sandstone in southwestern Virginia and areas to the south. The Massanutten Sandstone of northern Virginia's Shenandoah Valley and the Shawangunk Conglomerate of southeastern New York and adjacent Pennsylvania are equivalents of Tuscarora–Clinch strata in eastern outcrop belts. These units are all of very high textural and mineralogical maturity (composed of over 98% quartz with trace amounts of chert, zircon, and tourmaline) and are very tightly cemented. Because of these properties, they are the major ridge-forming sandstones of the Ridge and Valley. The Tuscarora–Clinch and its equivalents are unfossiliferous except for burrows such as *Arthrophycus*, which are generally considered to be of nonmarine origin. Based on this lack of fossils and on the presence of fining-upward sequences, Meckel (1970) interpreted the Tuscarora–Clinch Sandstone as being of fluvial origin. Paleocurrent data such as crossbed dip azimuths and regional decrease in grain size indicate that sediment dispersal was generally westerly to west-northwesterly. In western areas, Tuscarora–Clinch strata grade into fossiliferous marine strata such as the Whirlpool, Rumsey Ridge, and Cabot Head formations of western New York, the "Clinton Sands" of eastern Ohio, and the muddy and sandy Rockwood Formation of Tennessee. Even farther west, on the east flank of the Cincinnati Arch, equivalent strata are carbonates of the Brassfield Formation. Over much of its extent, the Tuscarora–Clinch lies conformably on Juniata–Queenston strata, indicating nearly continuous deposition. But on the extreme eastern side of the Appalachian Basin, the Tuscarora–Clinch–Massanutten–Shawangunk lies on eroded Middle Ordovician strata such as the Martinsburg Formation across an angular unconformity. Thus, uplift in the East had continued through latest Ordovician time and had begun to decrease during the Early Silurian. This probably accounts for why the Tuscarora represents the maximum regression phase.

By the beginning of the Niagaran Epoch, the rate of detrital influx from the Taconic sourceland had decreased significantly and facies boundaries retreated eastward as subsidence exceeded sedimentation. The resulting transgressive phase was marked by deposition of red, marginal marine strata of the Rose Hill Formation in the Central Appalachians. In the Southern Appalachians, however, which were much farther from the detrital source, subsidence greatly exceeded detrital influx and a turbidite basin developed whose deposits comprise the Red Mountain Formation. The event may be seen beautifully revealed in a series of outcrops in northwestern Georgia and southeastern Tennessee, where mud-flat deposits of the Sequatchie Formation are conformably overlain by lagoonal muds followed by proximal and then distal turbidites, all of the Red Mountain Formation. Further to the west, i.e., toward the craton, subsidence was considerably less and shallow-marine environments persisted; in this region formed fossiliferous iron ores such as those in the Birmingham, Alabama, Red Mountain strata (see Section B.6).

By later Niagaran time, two relatively small deltaic areas were active in the Central Appalachians: one was in western Virginia, where Massanutten Sandstone deposits continued to form; and the other was in southwestern Virginia, where the Keefer Sandstone developed. Before the close of the Niagaran, sea level again rose and flooded much of the Appalachian Basin, resulting in deposition of the McKenzie Formation in the eastern part of the basin and the Lockport Dolomite in the west (Dennison, 1970).

Another lowering of sea level (or slight rise of the sourceland) during early Cayugan time allowed the advance of a minor alluvial fan (the "last gasp" of the Taconic detrital wedge) represented by the Bloomsburg Formation, whose red shales and sands were deposited on broad floodplains and supratidal flats. The Bloomsburg fan was drowned in the middle Cayugan, and marine muds of the Wills Creek Shale accumulated over it. The Wills Creek and overlying carbonates of the Tonoloway Limestone of West Virginia and the Bossardville Limestone of Pennsylvania all contain evidence of occasional subaerial exposure, desiccation, and deposition of minor evaporites. For example, at Hively Gap, West Virginia, one of us (W.J.F.) has observed evaporite-dissolution breccias, well-developed mud cracks, and halite-crystal impressions in the Tonoloway. Similar features have been described at other Tonoloway localities (Ludlum, 1959). These evaporite-influenced units formed at the edge of the dwindling Taconic

coast, perhaps in sabhka-like settings, at the same time that major evaporite deposits of the Salina Group formed to the west on the craton (see Section B.6).

Detrital influx continued to wane and by the Early Devonian, carbonates of the Helderberg Group were deposited in Pennsylvania and New York (see Section B.8). Only very small amounts of mature quartz detritus came from the east, indicating that the old Taconic orogen had finally been worn down nearly to base level.

C.4. The Caledonian Orogeny in Greenland

The East Greenland fold belt consists of deformed Paleozoic strata which strike essentially parallel to the present-day coastline (see Fig. 5-16). The lack of extensive exposure due to the Greenland ice cap, the inaccessibility of much that is exposed, and the rigors of the climate all combine to limit knowledge of this large region. Thus, the following discussion is brief.

A rifted-margin prism was initiated during the Late Precambrian as a result of the continental separation that led to the formation of the Iapetus ocean; thus, the East Greenland rifted-margin prism developed at approximately the same time as the Appalachian prism far to the south. In eastcentral Greenland, the lower part of the rifted-margin prism is composed of the Eleonore Bay Group, containing up to 13 km of sediments ranging in composition from

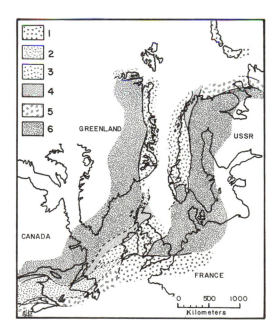

Fig. 5-16. Reconstruction of the Caledonian orogen prior to continental rifting. Shading patterns indicate the time of principal tectonic deformation: 1, Carboniferous; 2, Devonian; 3, Silurian; 4, Late Cambrian–Early Ordovician; 5, latest Proterozoic; 6, Grenville orogeny and earlier events. (From Roberts and Gale, 1978.)

coarse diamictites at the base, to immature graywackes and quartz sandstones in the middle, to interbedded shales and carbonates near the top. Associated with diamictites at the base of the Eleonore Bay Group is up to 1 km of greenstone, which Roberts and Gale (1978) have suggested may be penecontemporaneous with the Catoctin Greenstone of the Central Appalachians. The Eleonore Bay Group thus represents the early stages of rifted-margin-prism development, beginning with massive basaltic-lava flows and diamictites (alluvial fans?) of the rift-valley phase and the rapid accumulation of coarse detrital sediments which characterizes the basal clastic phase. Overlying Eleonore Bay strata is the Mrkebjerg Formation which contains tillites which may have formed during the hypothesized late Precambrian glaciation (see Chapter 3).

By the end of the Precambrian, the continental margin had ceased being affected by rapid subsidence and had passed into the carbonate–shale phase. The earliest Cambrian sediments of northeastern Greenland are transgressive quartz arenites with *Skolithos* tubes. In eastcentral Greenland, over 3 km of carbonates and shales was deposited from Early Cambrian to Middle Ordovician time. In both northeastern and eastcentral Greenland, a hiatus separating Middle Cambrian from Middle Ordovician strata probably represents the post-Sauk discontinuity.

During Late Silurian time, the East Greenland area was intensely deformed by the main phase of the Caledonian orogeny. The Caledonian orogeny was a major continental-collision event during which the western Scandinavian area of Europe was squeezed against East Greenland. Comparable orogenic activity occurred in England at the same time and was caused by the same continental collision (see Fig. 5-16). Thrust-sheets, some of which involved the crust as well as supracrustal strata, were driven in some cases over 100 km eastward in Greenland at this time. The collision began in Greenland but, as we will discuss in the next chapter, the North American and Baltican continental plates may have collided with a scissors-like motion so that the age of initial deformation decreases southward along the orogen, resulting finally in the Acadian orogeny (Late Devonian) of the Northern Appalachians.

D.   The Innuitian Continental Margin

The geological history of the northern, or Innuitian, continental margin has received relatively little attention thus far in this book because so little is known about this vast and inhospitable region. We begin our discussion of the northern continental margin in this chapter because Cambrian strata of the region are sparsely and poorly exposed (having been commonly subjected to varying degrees of metamorphism and deformation as well as being over-

lain, in much of northern Canada, by much younger strata) and therefore their relations are somewhat obscure.

D.1.   Major Geological Features

D.1.a.   The Canadian Shield

Intensely deformed and strongly metamorphosed rocks of Precambrian age are exposed on the southern part of the Innuitian continental margin. These rocks, whose structural development was discussed in Chapters 2 and 3, dip beneath Paleozoic and younger strata to the north just as they do to the south in southern Canada and the northern United States. Sialic continental crust is inferred to be present beneath younger strata as far north as the Innuitian fold belt (see below). There is no evidence of Precambrian sialic crust farther to the north.

D.1.b.   Arctic Platform

Overlying the Precambrian basement in the southern part of the Canadian Archipelago are essentially flat-lying Paleozoic strata which were probably deposited on a stable cratonic platform. This area, referred to by some authors as the Central Stable Region, is broken by a series of uplifts which expose Precambrian rocks, such as the Boothia, Wellington, Minto, and Coppermine Arches (see Fig. 5-17). These arches separate cratonic strata into three main intra-cratonic basins: the Foxe, Victoria Strait, and Wollaston Basins. The northern part of the Arctic Platform contains thicker cratonic strata of the Jones–Lancaster and Melville Basins which are separated by the Cornwallis fold belt, probably an extension of the Boothia Arch. These are interpreted as craton-margin basins, probably caused by slight subsidence of the stable craton accompanying the much greater subsidence of the continental margin farther to the north.

D.1.c.   The Innuitian Fold Belt

Folded and faulted Paleozoic strata occur in an arcuate belt from Melville Island to northern Greenland. This fold belt is divided into four parts. The *Parry Islands fold belt* includes folded Ordovician through Devonian strata on Melville and Bathurst Islands and is similar to the *Ellesmere–Greenland fold belt* which runs from northern Devon Island, along the southeastern part of Ellesmere Island, to the northern part of Greenland and which contains Cambrian through Devonian strata. On the northern coast of Ellesmere Island is a belt of deformed and metamorphosed strata of probable Late Precambrian to Devonian age referred to as the *North Ellesmere fold belt*, which is associated with igneous intrusions of several different types and

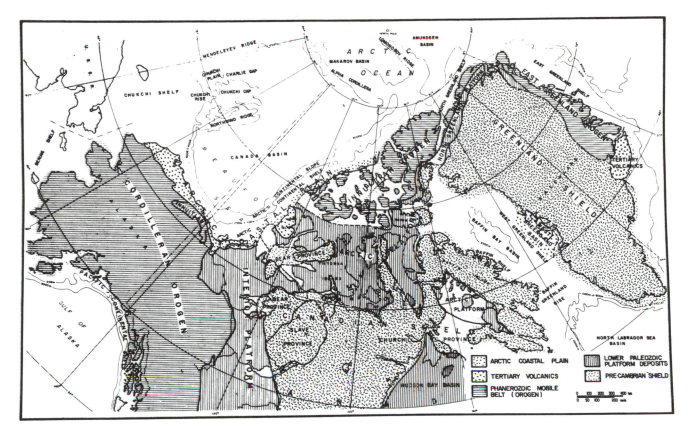

Fig. 5-17. Major geological provinces of the Innuitian margin and surrounding areas. (From Drummond, 1974, with permission of the author and Springer-Verlag.)

ages. Younger, less deformed strata of the Sverdrup basin (see below) overlie the area between the North Ellesmere fold belt and the Ellesmere–Greenland fold belt (as does a permanent ice cap) so that relations between the two belts are speculative. Taken together, the three fold belts described above constitute the main Innuitian fold belt and probably represent the Paleozoic continental margin.

Separating the Parry Islands fold belt from the Ellesmere–Greenland belt is the *Cornwallis fold belt* which is an extension of the Boothia Arch to the south. The Cornwallis belt is a curious structural feature in that the trends of its major structures run almost perpendicular to the Parry Islands and Ellesmere–Greenland trends (see Fig. 5-18). Some fold axes are bent through 90° from the Parry Islands belt to the Cornwallis belt; others are apparently truncated (McNair, 1961). As a relatively smaller fold belt which interrupts and trends perpendicular to a major fold belt, the Cornwallis belt is almost unique in the structural geology of North American orogens.

D.1.d. Brooks Range, Alaska

The Brooks Range is composed of several smaller ranges such as the DeLong, Baird, Endicott, and Romanzof Mountains of Alaska and the British Mountains of northwesternmost Yukon Province in Canada. The Brooks Range is extremely complex structurally because it bears the scars of multiple tectonic events which were associated with both Innuitian and Cordilleran continental margins and which were spread out over much of Phanerozoic time. Cambrian (?) through Devonian strata are represented as well as evidence of several different stages of igneous activity and metamorphism, probably related to events along the Innuitian continental margin.

D.1.e. Sverdrup Basin and Arctic Coastal Plain

The Sverdrup Basin in northern portions of the Canadian Archipelago contains a great thickness (over 1400 m!) of Upper Mississippian to Tertiary strata which were deformed during Tertiary time. These strata are of both marine and nonmarine origin; and it is interesting to note that the Tertiary fold structures and the trends of Paleozoic folds in underlying rocks are roughly parallel, suggesting that the Paleozoic structures had effects, long after their development, on Tertiary tectonics (Fortier, 1957).

The Arctic coastal plain contains undeformed strata of Jurassic, Cretaceous, and Tertiary ages. The coastal plain

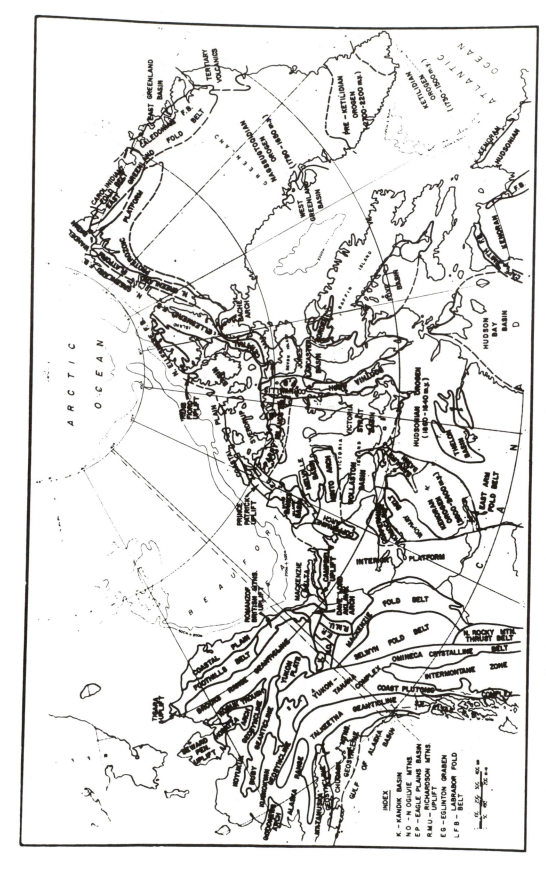

Fig. 5-18. Tectonic map of the Innuitian continental margin and surrounding areas. (From Drummond, 1974, with permission of the author and Springer-Verlag.)

runs from Ellef Ringnes Island along the extreme north of the Canadian Archipelago through Prince Patrick and Banker Islands to the mainland coast where it is associated with the Mackenzie River delta. It also extends northwestward along the northern coast of Alaska.

D.2. Early History of the Innuitian Continental Margin

The Phanerozoic history of the Innuitian area probably began with continental rifting in the Late Precambrian, during which time an as yet unknown continental mass was split off from North America. Dickinson (1977) pointed out that this rifting, dated at approximately 675 Myr. (from basaltic dikes, possibly formed during the prerift thermal arch or early rift-valley stage), occurred in time between the Cordilleran and Appalachian riftings and was, thus, part of the dismemberment of Pangaea I. Rifting led to the development of a Late Precambrian to Cambrian rifted-margin prism over 3000 m thick, composed of detrital and carbonate sediments (e.g., the Lower Cambrian Ellesmere Group) deposited in coastal and shallow-marine environments (Trettin, 1973).

At some time during the Early or early Middle Cambrian, the rifted-margin prism was disrupted by the initiation of plate subduction and subsequent formation of a magmatic arc to the north of the continental margin. In the North Ellesmere fold belt are gneisses, migmatites, and associated igneous rocks of the Cape Columbia complex which have been dated at 465 ± 19 Myr. (Wanless, 1969, in Trettin, 1973) and which represent the earliest direct evidence of magmatic-arc-related metamorphism in this area. Approximately correlative metamorphic events have been recognized in the Rens Fjord complex of northern Axel Heiberg Island and along the North Slope of Alaska. Drummond (1974) reports a maximum age of 547 ± 22 Myr. for some North Slope phyllites.

Indirect evidence of tectonic lands to the north is provided by the Grant Land Formation (Middle Cambrian) of the North Ellesmere fold belt. The Grant Land Formation contains more than 1100 m of slightly metamorphosed, immature, arkosic and quartzose arenites, conglomerates, and shales which were derived from gneissic terranes to the north and were deposited in deltas which prograded southward (Trettin, 1973). A similar tectonic setting has been inferred for lower parts of the Neruokpuk "Formation" (Cambrian through Devonian) in the Romanzof Mountains of Alaska.

That Grant Land Formation detrital sediments were not derived from the craton is further indicated by the presence of Middle Cambrian carbonates on eastern and central parts of the Arctic Platform. The Oogahgah and Bear Point Formations of Devon Island, Cape Wood Formation of the Bache Peninsula, and Turner Cliffs Formation of northwestern Baffin Island (Cook *et al.*, 1975) are parts of a once-continuous carbonate sheet which was deposited on a broad platform on the craton while Grant Land deltas offshore to the north grew southward from the magmatic arc.

We believe that this Middle Cambrian tectonic event was probably related to the onset of subduction during which an oceanic plate was thrust southward beneath the outer continental margin or adjacent oceanic crust (see Fig. 5-19). Thus, the Grant Land Formation may represent deposition in a back-arc tectonic setting, either in a "juvenile" marginal basin formed by back-arc spreading, or in a narrow, "trapped" marginal basin.

D.3. Tippecanoe History

By Ordovician time, a familiar paleogeographic picture had been established on the northern continental margin. The stable craton to the south passed northward into an unstable continental shelf; over both areas, a broad carbonate platform existed. The northern part of the carbonate platform dropped off precipitously into the deep waters of a marginal basin in which turbidites and pelagic sediments were deposited. This marginal basin is called the *Hazen Trough* (from the Hazen Formation, discussed below). Bounding the Hazen Trough on the north was a north-facing magmatic arc which was a positive topographic feature over a long period of time and supplied large volumes of sediment to the Hazen Trough. This positive area was named Pearya by Schuchert (1923) who considered it a "continental borderland"; later, it was referred to as the Pearya geanticline by Trettin (1973). We will use the term *Pearya magmatic arc* in our subsequent discussions.

D.3.a. Continental Shelf

Over the continental shelf and adjacent craton, a broad carbonate platform existed during the Early to early Middle Ordovician. This platform was similar to other upper Sauk carbonate facies of the shelf around the rest of North America. Examples of carbonate-platform deposits include limestones, dolostones, and evaporites of the Copes Bay, Baumann Fjord, Eleanor River, and Bay Fjord formations of southeastern Ellesmere Island.

Carbonate deposition also characterized Tippecanoe strata. Shelf-edge carbonate buildups, such as barrier reefs of the Cornwallis Formation and related bioclastic shoals, restricted circulation into the broad lagoonal area over much of the shelf, where fine, micritic carbonates were deposited and graded laterally into areas of evaporite formation (Drummond, 1974). Examples include the Read Bay and Allen Bay formations (Upper Ordovician) of Cornwallis and Ellesmere islands.

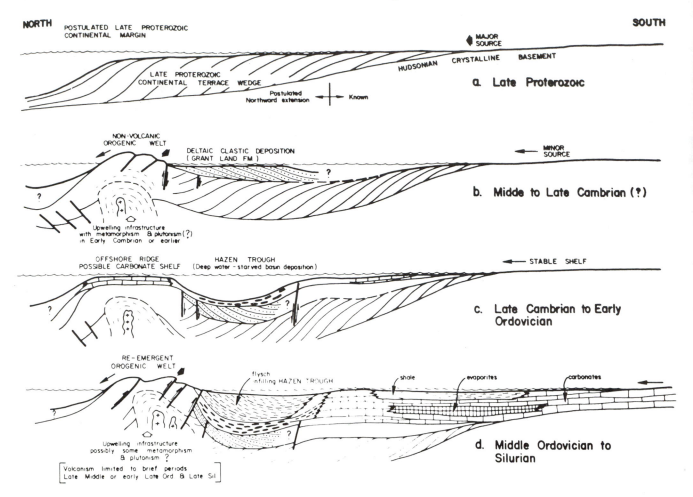

Fig. 5-19. Drummond's plate-tectonic model of the early Paleozoic history of the Innuitian margin. (From Drummond, 1974, with permission of the author and Springer-Verlag.)

During the Silurian, the outer shelf began to subside more rapidly, possibly related to effects of the Caledonian continental collision along the East Greenland continental margin (discussion follows). Some of the old bioherms grew upward, maintaining their relative bathymetric position on the subsiding shelf and resulting in great coral masses which stood up more than 200 m off the bottom. Normal carbonate deposition continued in southern areas throughout the rest of the Silurian.

D.3.b. Hazen Trough

The deep marginal basin to the north was bordered on the south by a continental slope characterized generally by the deposition of fine, graptolitic muds. Occasional slumps from the nearby shelf-edge carbonate platform created large debris flows which carried very poorly sorted carbonate sediments out onto the slope. These slumps were probably the cause of turbidity currents which carried carbonate mud

and silt farther out into the Hazen Trough. Graptolitic shales and debris-flow deposits of the continental slope comprise the Ordovician and Silurian Cape Phillips Formation.

The history of the main part of the Hazen Trough began during the Early Ordovician with deposition of the Hazen Formation which is characterized by radiolarian chert, graptolitic shale, carbonate turbidites, and rare debris-flow deposits. The fact that these deep-water sediments were deposited on top of deltaic sediments of the Grant Land Formation speaks eloquently of the rapid subsidence that accompanied the early development of the marginal basin. During early Hazen time, the trough was relatively narrow and the predominant type of sediments were carbonate turbidites with rare debris-flow deposits (Trettin, 1973). With time, the basin grew much wider and carbonate turbidites were restricted to the southern basin-margin, while cherts and pelagic shale dominated over most of the trough (see Fig. 5-20). This history (i.e., deltaic environment succeeded by a narrow trough which is succeeded by a wide basin) suggests to us that the Hazen Trough was formed by

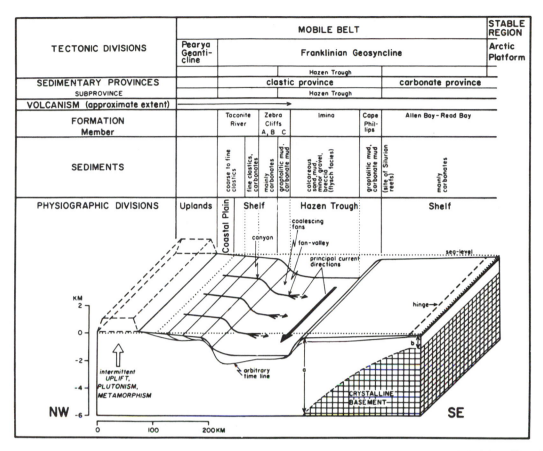

Fig. 5-20. Tectonic environments of sedimentary deposition along the Innuitian continental margin during the Late Ordovician. (From Trettin, 1973, reproduced with permission of the Association of Petroleum Geologists.)

back-arc spreading, such as has been inferred for modern marginal basins in the western Pacific.

By the Late Ordovician, the Hazen Trough was quite wide. The Pearya magmatic arc to the north was in an active phase and turbidity currents from there spread out into the marginal basin, resulting in extensive flysch deposits of the Imina Formation. Sedimentology and structures of the Imina suggest a complex of submarine fans growing out from the magmatic arc and coalescing to form a southwestward-prograding fan complex. This fan complex developed at the same time that deep, starved-basin cherts and graptolitic shales of the basal Ibbett Bay Formation were accumulating in the area of western Melville Island (Drummond, 1974). Thus, one may perceive a broad marginal basin which was being filled by turbidity-current sediments from the northeast while the southwestern part of the basin remained deep and relatively sediment starved.

D.3.c. Pearya Magmatic Arc

As we have noted, the Pearya magmatic arc had begun during the Cambrian and had supplied abundant detrital sediment to form the Grant Land deltas. By the Early Ordovician, however, uplift had slowed down so that continued subsidence in the marginal basin resulted in a "starved" basin (the Hazen Formation) rather than a flysch fan-complex as occurred later when arc activity picked up again (the Imina Formation). During the Early Ordovician lull, the Pearya arc was probably a narrow carbonate platform or an archipelago of low, carbonate islands.

By Middle Ordovician time, however, tectonic activity had increased again and tectonic lands again shed abundant sediments back into the Hazen Trough rather than into a relatively shallow, outer continent-shelf. The resulting sedimentary mass, which we refer to as a *back-arc detrital ramp*, is partly exposed in the North Ellesmere fold belt. This back-arc ramp is composed of three main parts: (1) alluvial and deltaic deposits, represented by the Taconite River Formation and detrital units of the Cape Discovery Formation, which constituted a narrow coastal plain; (2) shallow-marine carbonates, such as the Ayles and Marvin Formations and carbonate units of the Cape Discovery Formation, which may have formed during short-term pauses in local tectonic activity; and (3) a slope facies with debris-flow deposits and proximal turbidites, such as those of the Zebra Cliffs Formation, which grade southeastward into Imina flysch (Trettin, 1973).

D.3.d. Effects of the Caledonian Orogeny

During the later Silurian and Early Devonian, a major continental collision occurred along the East Greenland continental margin; this event, termed the Caledonian orogeny, had major structural and petrological effects throughout the northeastern part of North America [for further discussion, see the section (C) on the Appalachian continental margin]. While the main orogenic effects were experienced only in the area of the plate margin, collision-related stresses had effects far away from the orogen (cf. the effects of the Himalayan collision on intraplate tectonics of the Asian plate as discussed by Molnar and Tapponnier, 1977). During the Caledonian orogeny, the Boothia Arch and its northward extension, the Cornwallis fold belt, were positive features. The fold structures of the Cornwallis belt were produced at this time. The Rens Fjord Uplift of northern Axel Heiberg Island, whose trend approximately parallels that of the Boothia–Cornwallis Uplift, was also active during the Late Silurian and Early Devonian. While the trends of both the Rens Fjord and Boothia–Cornwallis uplifts were probably controlled by preexisting structures of underlying Precambrian rocks, the stresses to reactivate them were probably caused by the Caledonian orogeny.

E. Continental Margin and Magmatic Arc Assemblages of the Cordillera

E.1. Tippecanoe Cordillera: Overview and Summary of Regional Tectonics

The irregular emergence of the craton which terminated the Sauk sequence, and the more rapid, almost total submergence of the craton which initiated the Tippecanoe Sequence, did not radically alter the pattern of sedimentation on the western margin of the continent. After a period of late Early through Middle Ordovician (post-Whiterockian to Cincinnatian) detrital sedimentation, with sources probably from the craton, the western continental margin again contained a characteristic pattern of facies that formed on the continental shelf, slope, and rise in a probably unbroken belt from Alaska to Mexico. The facies developed along this margin are (from landward to seaward): shelf carbonates (and some quartz detritus); shale–limestone; chert units on the deeper shelf and upper slope; and shale–chert and siliceous and volcanic units on the slope and rise (modified from Stewart and Poole, 1974).

Additional rock units are found generally westward of the continental-margin sequences described above, scattered in "borderland terranes" (Churkin and Eberlein,

1977) which reflect very different tectonic environments from the continental-margin assemblages. They contain, among other lithologies, volcanic units, ophiolites, mélanges, and a variety of deep-water sediments and metamorphic rocks all of which suggest the presence of magmatic arcs at various times on the far western margin. The tectonic interpretation of these borderlands (also referred to as the "insular belt" by Monger et al., 1972) is complex and will be deferred until after we have described the marginal assemblages.

Within the Cordilleran region, and on the platform, were a number of structural trends which affected sedimentation both locally and over fairly large areas. Some of these include structures that may simply represent portions of the rifted margin; these structures in the past were considered to be a series of "troughs" within the "Cordilleran Geosyncline" (see, e.g., Gabrielse, 1966). Among such features which were "active" in the Late Proterozoic and lower Paleozoic are the "Selwyn Basin," the "Root Basin," the "Richardson Trough," and the "Porcupine Basin," all in the Northern Cordillera and which form part of an irregular but recognizable delineation of the continental margin (Fig. 5-20).

In the Great Basin, however, clearly located on the continent, are structures which seem to reflect intracratonic movements. Two intracratonic basins, the Ibex Basin in western Utah and eastern Nevada, and the Utah Basin in northwestern Utah, were separated by a strong, positive feature which affected sedimentation in the northeastern Great Basin, called the Tooele Arch (Fig. 5-21). These structures remained active until the Middle Devonian.

One final important tectonic aspect of the Paleozoic Cordillera is the presence of post-Tippecanoe crustal shortening which affects the interpretation of the sections discussed herein. We will document several such tectonic events in subsequent chapters but it should be noted here that reconstruction of Early Paleozoic Cordilleran sedimentation requires the use of palinspastic base maps which graphically undo *oroflexural* folding (crumpling of the surface without faulting) and which restore regions to prefaulting geometries. In the Great Basin, for example, the Roberts Mountains thrust may have involved a total of 145 km of eastward transport of pre-Upper Devonian strata (Stewart and Poole, 1974). In addition, several major right-lateral strike-slip fault zones, such as the Death Valley–Furnace Creek and Stewart Valley faults, may have resulted in comparable amounts of dislocation of strata in a general northwest direction.

E.2. The Continental Margin Assemblages

E.2.a. Middle and Upper Ordovician

Middle Ordovician sedimentation on the western shelf has been well studied in the Great Basin and to a lesser

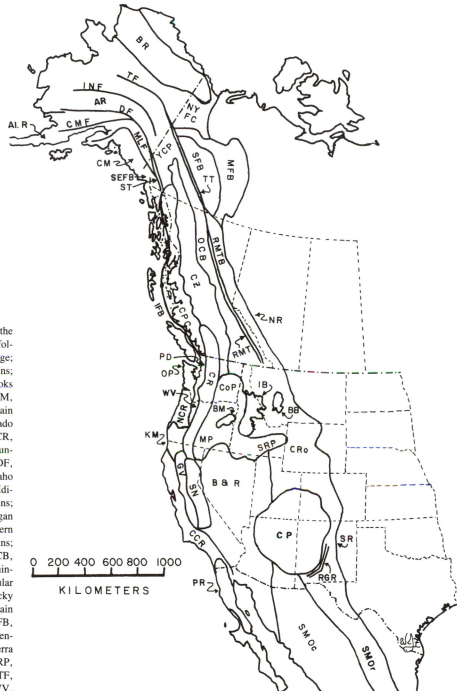

Fig. 5-21. Major geological features of the North American Cordillera. Symbols are as follows: Al.R, Aleutian Range; AR, Alaska Range; BB, Boulder Batholith; BM, Blue Mountains; B&R, Basin and Range Province; BR, Brooks Range; CCR, California Coast Ranges; CM, Chugash Mountains; CMF, Crystal Mountain Fault; CoP, Columbia Plateau; CP, Colorado Plateau; CPC, Coast Plutonic Complex; CR, Cascade Range; CRo, Central Rocky Mountains; CZ, Columbian Zwischengebirge; DF, Denali Fault; GV, Great Valley; IB, Idaho Batholith; IFB, Insular Fold Belt; INF, Ididarod–Nixon Fault; KM, Klamath Mountains; MFB, Mackenzie Fold Belt; MLF, Mt. Logan Fault; MP, Malheur Plateau; NCR, Northern Coast Ranges; NR, Northern Rocky Mountains; NYFC, Northern Yukon Fold Complex; OCB, Omineca Crystalline Belt; OP, Olympic Peninsula; PD, Puget Depression; PR, Peninsular Ranges; RGR, Rio Grande Rift; RMT, Rocky Mountain Trench; RMTB, Rocky Mountain Thrust Belt; SEFB, St. Elias Fold Belt; SFB, Selwin Fold Belt; SMOc, Sierra Madre Occidental; SMOr, Sierra Madre Oriental; SN, Sierra Nevada; SR, Southern Rocky Mountains; SRP, Snake River Plain; ST, Shakwak Trench; TF, Tintina Fault; TT, Tintina Trench; WV, Willamette Valley; YCP, Yukon Crystalline Platform.

degree in the Canadian Cordillera. In the Great Basin, detritus from the craton covered a broad area following post-Sauk emergence and early Tippecanoe submergence. The characteristic units formed by this event are the Eureka Quartzite in western Utah and most of Nevada, the quartzite member of the Swan Peak Formation in northern Utah, and the Kinnikinic Quartzite in Idaho. Equivalent strata in the northern Cordillera may include sandy units of the Sunblood Dolomite in the Northwest Territories.

During the Late Ordovician, these largely quartzose sediments gave way to widespread carbonates such as the Fish Haven Dolomite in Idaho and Utah, the Saturday Mountain Formation in parts of Idaho, the Hanson Creek Formation in parts of Nevada and Idaho, the Ely Springs

Dolomite in Nevada, the Mount Kindle Formation in the northern Yukon, the Whittaker Formation in the southern Yukon, and the Sandpile Formation in northern British Columbia. The Eureka and Swan Peak formations are absent over the Tooele Arch and there Upper Ordovician dolostones overlay directly Lower Ordovician units. Stewart and Poole (1974) state that these Upper Ordovician carbonates, like those of earlier times in the same region, were deposited in shallow, platform lagoons and contained massive bioherms built up largely by algae (such as *Girvanella*). The biohermal buildups tend to be located on the western fringe of the shallow carbonate platform, and, as was mentioned in the western Cambrian history, they may have restricted exchange of organisms with the inner-detrital zone. The dolomite in the Upper Ordovician may be dolomitized limestone, altered by magnesium-rich waters in the restricted lagoons (Stewart and Poole, 1974; Miller and Walch, 1977). The shelf carbonates contain shelly facies fossils and Ross (1977) noted that a complex combination of limestone and dolostone in central Nevada indicates a local deeper-water, well-circulated, more normal marine environment which supported a fauna rich in trilobites.

Westward, on the deeper shelf, beyond the algal buildups, much less dolostone and more limestone and shale were deposited in a zone Stewart and Poole describe as the "shale and limestone facies." These lithologies are preserved in limited areas of the Great Basin and in even more limited areas in other parts of the Cordillera. Stewart and Poole list these units in the Great Basin as comprising to the Aura, Perkins Canyon, Zanzibar, and Toquima formations, all in Nevada. Similar units in the northern Cordillera are largely absent, unnamed, or unrecognized.

The shale–limestone–chert and siliceous–volcanic facies in the Middle and Upper Ordovician are characteristically thick, widespread, and graptolitic. Minor limestone breccias, quartzites, and (notably) bedded barite are typical of these facies in the Great Basin. They are interpreted as representing deposition on the outermost continental shelf and upper continental slope. These rocks include the Middle Ordovician Vinini Formation in Nevada, shaley units of the Upper Ordovician Whittaker Formation in the Yukon, the Road River Formation in the northern Yukon and eastern Alaska, and the Valmy Formation in central Nevada. The Valmy comprises the siliceous–volcanic facies in the Great Basin and consists of several kilometers of cherts, quartzites, shale greenstones (pillow lavas), limestones, and barite. Stewart and Poole note that the Valmy contains much less shale and much more greenstone and quartzite than units of the "shale-and-chert" facies. They also suggest that the Valmy was deposited in waters with depths greater than 500 m, based on structural details of the pillow lavas.

The quantity of quartzite in the Valmy is odd for a deep-water lithology. It seems likely that the source for this quartz sand was extracratonic but it cannot be pinned down as yet. It is also interesting to note that the Valmy Formation was deposited continuously through the entire Ordovician, i.e., it was not affected by post-Sauk cratonic emergence.

E.2.b.   Silurian and Lower Devonian

In general, continental-margin facies of the upper Tippecanoe sequence of the Cordillera are similar to those of the Ordovician; however, the Silurian System is relatively thin in the Great Basin and limited aerially. Stewart and Poole (1974) recognize four facies on the shelf/slope/rise, similar to those of the Ordovician except as follows: inner-shelf quartzites are absent in the Silurian (but present in the Devonian); outer-shelf facies are characterized as laminated limestones in the Silurian and limestones and shales in the Devonian; and the deepest-water facies contained feldspathic sandstones in the Silurian and cherts in the Devonian. An analogous but simpler facies distribution for Silurian and Lower Devonian rocks (described for Copenhagen Canyon, Nevada, and extrapolated to the entire Great Basin) was proposed by Matti *et al.* (1974), consisting of an eastern dolomite suite deposited on the platform; a limestone-clastic suite deposited on the deep-subtidal to shelf-slope interface; and a slope and deeper volcaniclastic suite. They noted that the facies boundaries between the limestone-clastic and volcaniclastic suites are poorly understood because of regional thrusting during the Late Devonian–Mississippian Antler orogeny (Chapter 6).

Lower Silurian shelf carbonates include the upper parts of the Ely Springs, Hanson Creek, Saturday Mountain, and Fish Haven formations in the Great Basin (all of which are mostly of Ordovician age), and the middle units of the long-ranging Whittaker Formation in the Yukon. In the Great Basin there was an almost universal sedimentary hiatus during the late Llandoverian (Lower Silurian), which probably reflects cratonic uplift such as was observed across most of the northeast craton at the time. But in many sections, the rocks deposited subsequent to the hiatus are mapped as continuations of the underlying formations. These include the Ely Springs, Saturday Mountain, and Hanson Creek formations (Poole *et al.,* 1977).

Characteristic Middle and Upper Silurian shallow-shelf carbonates comprise the Laketown and Sevy dolomites in Nevada and Utah, the Roberts Mountain and Lone Mountain dolomites in Nevada, a number of unnamed formations in the Great Basin, and the upper Whittaker and Delorme dolomites in the Yukon. In addition, the Hidden Valley Dolomite, which has limited and rather poor exposure in southeastern California, has been dated tentatively by Miller (1978), using conodonts (see Section F.9), to range from the Early Silurian to at least the Early Devonian. The Sevy, Lone Mountain, and Roberts Mountain formations also persist into the Lower Devonian with no apparent change in lithology, but all shelf sedimentation apparently ceased briefly in the Great Basin in the mid-Lower Devonian. This hiatus marks the post-Tippecanoe emergence of the craton.

Stewart and Poole (1974) characterize the dolomitic facies in the Great Basin (particularly the widespread Laketown) as gray, thin- to thick-bedded dolostone with some dark-gray cherty units. Westward of the central Great Basin (onto the deeper shelf), these units change abruptly to dark, laminated limestones with abundant quartz and feldspar, graptolites, and carbonaceous material. Organisms characteristic of the "shelly" biofacies are found in the dolostones and are dominated by brachiopods, coral, and conodonts (Berry, 1977) whereas graptolitic biofacies are found in both deep-water lithologies and deeper-shelf carbonates.

As occurred during the Ordovician, Silurian and Early Devonian dolomite deposition probably occurred in broad, shallow lagoons fringed on the seaward side by biohermal mounds (however, the record of organic buildups is far poorer than that of the Ordovician). The dolomite may have been both primary and secondary in origin (Nichols and Silberling, 1977). The deeper-shelf laminated limestones deposited in Silurian time are reminiscent of the Middle and Upper Cambrian formations of the Great Basin deposited just seaward of the carbonate shoals; the same environment of deposition is probably represented, including carbonate debris carried by slumps and debris flows from the bioherms mentioned above.

The Silurian and Lower Devonian deep-basin lithologies (the chert–shale facies of Stewart and Poole) are represented by the Fourmile Canyon Formation and a number of unnamed units in the Great Basin, and by the Road River Formation in the Yukon. The Fourmile Canyon consists of chert, shale, argillite, siltstone, and minor sandstone; it may include some volcanics (however, these may be misidentified Valmy units). The "feldspathic sandstone" facies is represented in the Great Basin by the Elder Sandstone (mid-Lower to mid-Upper Silurian) in westcentral Nevada, and possibly by the upper Sandpile Formation (mostly Lower Silurian) in northwestern British Columbia.

The Elder Formation in a quartz sand unit containing a high percentage of potassium feldspar (15–25%) and muscovite (5%). Minor silt, shale, and chert are present as are traces of pumice and volcanic glass shards. This unit seems rather unusual, showing characteristics of both felsic plutonic and volcanic sources. In addition, it seems generally to be a deep-water deposit (based on the cherts) and, perhaps, locally a shallow-water deposit (evidenced by some possibly authigenic algal fragments). We will discuss en-

vironments of deposition and sources for the Elder Formation in the following subsection.

E.3. "Borderland Terranes" and "Insular Belt" Facies

In this final subsection dealing with the Tippecanoe sequence in the Cordillera, we encounter some very complex assemblages of rocks. In the previous discussion we examined the sedimentary assemblages of the continental shelf/slope/rise which formed generally in conventional patterns along the passive western continental margin. We also described a few unconventional formations (such as the Valmy and Elder) which were deposited close to the continental margin but which contain volcaniclastic materials and which seem to have extracratonic sources for their detrital sediments.

There are a number of other units of the Lower Paleozoic which were formed far west of the continental margin. These comprise the "insular belt" facies or "borderland terranes" introduced previously. They represent volcanic and plutonic igneous sources, associated either with a series of magmatic arcs separated from the continent by back-arc marginal basins or with parts of separate continental masses, or microplates, lying off the western shore of the American plate during the early Paleozoic. Included among these offshore facies are ophiolites and mélanges that represent the oceanic lithosphere and which are inferred to have formed in oceanic spreading centers and were emplaced in their present position by tectonic accretion in a subduction complex.

Dickinson (1977) and Churkin and Eberlein (1977) have proposed that several large crustal blocks were rifted from the various margins of North America during the Late Precambrian and early Paleozoic, and that some of these blocks were present off the western margin during Tippecanoe time. These blocks comprised the exotic borderland terranes as well as the sources of volcanics and western-derived detritus. Churkin and Eberlein propose that six or more separate lithospheric plates can be recognized based on the distinct basement assemblages which underlie some of these borderland terranes. Dickinson suggests that these "lost" blocks may have formed part of Eurasia (which always seems to be hard to track down in paleogeographic reconstructions) or part of Gondwanaland.

There seems to be little evidence for the presence of an insular volcanic arc environment in the Cordillera before the Silurian, except for volcanics and/or volcaniclastics in the Cambrian Harmony, Scott Canyon, and Valmy formations. The source for detrital sands in the Valmy, as well as the sands in the Silurian Elder Formation, may very likely have been a microcontinental mass; however, such a sialic igneous terrane, which yielded quartz and feldspar sands to the

Elder Formation, does not seem to be a likely source for the volcanics in these units. Monger *et al.* (1972) suggest that volcanic rocks in the eastern Cordillera (such as the Valmy) may "merely reflect local rifting (with associated volcanics) in an overall tensional environment."

In the Northern Cordillera, notably southeastern Alaska, there may have been an Ordovician/Silurian "orogeny," as deduced from local Ordovician granitic rocks and metamorphism, and from Silurian granitic rocks and granite-bearing Devonian conglomerates (Monger *et al.*, 1972). Gabrielse (1966) suggests that westward tilting of the margin accompanied this event and resulted in Upper Ordovician and Lower Silurian strata unconformably overlying progressively older strata eastward (due to beveling of the tilted surface). Monger *et al.* cite the rather startling conclusion that southeastern Alaska, with its somewhat aberrant Lower Paleozoic history, is a fragment of what is now California, emplaced in its present location by late Paleozoic or Mesozoic right-lateral strike-slip faulting! It is apparent, at least, that the Lower Paleozoic history of southeastern Alaska bears little resemblance to that of the continent.

During both Silurian and Early Devonian times, a number of borderland terranes developed somewhere west of the Cordilleran margin. Because of post-Tippecanoe faulting and overthrusting, it is difficult to restore palinspastically what may have been island arcs, or totally separate continental blocks, to their former positions. We will arrange a list of some of these distinctive, far-western units by their present locations.

From roughly south-to-north (Fig. 5-22), along the "insular belt," these units include: the Garlock Formation in southern California (Lower Devonian); the Elder and Fourmile Canyon formations (Silurian) and the Woodruff Shale and Slaven Chert in northcentral Nevada (in part Lower Devonian); the Phi Kappa Formation (Silurian) in central Idaho; the Shoo Fly Formation (Upper Ordovician through Lower Devonian?) in the Sierra Nevada; the Trinity Ophiolite (Ordovician), Moffett Creek Formation (Silurian), and Gazelle Formation (Upper Silurian through Upper Devonian?) in the Klamath Mountains in northern California; the Turtleback Complex (Ordovician) on San Juan Island; and a large group of volcanic/sedimentary/intrusive rock units along the Alexander Archipelago in southeastern Alaska, including the Ordovician/Silurian Descon Formation and the Silurian Bay of Pillers Formation.

In the eastern Klamath Mountains, the Trinity Ophiolite seems to represent a dismembered fragment of oceanic crust and upper mantle that outcrops in a large (75 by 50 km) area (Lindsley-Griffin, 1977). It has been interpreted as having been formed either by back-arc spreading within a marginal basin, or at a midocean ridge (Potter *et al.*, 1977; Irwin, 1977). It may have been uplifted during

the Middle or Late Ordovician and have formed the basement for a volcanic arc (Churkin, 1974). Associated with the Trinity is the Moffett Creek Formation, which is interpreted as a mélange composed primarily of sandstone, mudstone, and siltstone which are virtually unbedded and are highly sheared and broken into slabs in a shaley matrix. The Moffett Creek is probably a subduction complex and may have been part of a submarine fan system prior to its disruption (Potter *et al.*, 1977). The Gazelle Formation, which is associated with both of the above units, is a longranging collection of detrital units (units 1 and 2 are of Silurian through Middle Devonian ages). The middle unit contains volcaniclastic, massive sandstones; Potter *et al.* suggest that the Moffett Creek and Gazelle Formations derived from sources within or behind a volcanic arc and were deposited in an arc-trench environment. Again, the location of the eastern Klamath Mountains with respect to the North American continent in the Lower Paleozoic is unknown; but they certainly lay west of the continent and were probably separated from the continental margin by a back-arc basin or by an ocean basin. The Shoo Fly Formation is composed of several units, of which the middle and upper are part of the Tippecanoe sequence. The middle unit (roughly dated to Ordovician/Silurian) is a highly sheared shale and sandstone with slices of a variety of lithologies including carbonate clasts and volcanics. In the Taylorsville, California area, this unit is a mélange containing blocks of limestone and serpentine in a pelitic matrix (D'Allura *et al.*, 1977). The upper Shoo Fly (possibly Lower to Middle Devonian) is a light-gray chert, argillite, and shale with evidence of volcanic origins for at least some of the silica. The Shoo Fly probably represents deposition in a volcanic arc environment, possibly in a back-arc basin with detrital sources in the magmatic arc.

The Turtleback Complex contains a diverse set of basic and intermediate intrusive igneous rocks (gabbro, quartz diorite, and amphibolites) which may indicate an Ordovician collisional event for a crustal block off the western margin. Danner (1977) suggested that considerably more dating within the complex will have to be done before its origin and significance can be understood.

The Descon Formation in southeastern Alaska is a sandy, shaley unit that contains pillow structures, volcanic breccia, and tuffs in its upper beds. It overlays the metamorphic Wales Group (pre-Ordovician) which comprises the basement and may give evidence that the southeastern Alaskan orogenic event described previously had a long prehistory of volcanic activity. The Descon is over 7 km thick and is overthrust such that its upper contact is undeterminable (the thrusting probably dates to the Ordovician/Silurian boundary).

As mentioned, there are many other units that comprise the "borderland terranes" and "insular facies" along

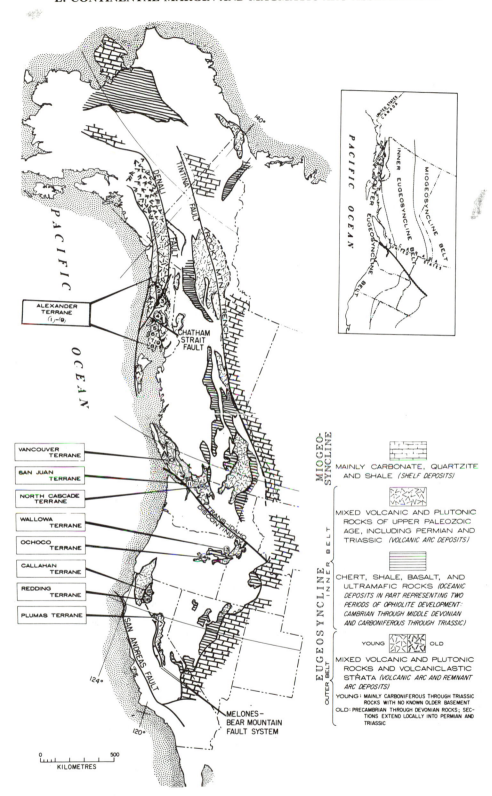

Fig. 5-22. Westernmost North American lithofacies of the "insular belt" (or "borderland terranes"). (From Churkin and Eberlein, 1977, Fig. 1.)

the far western margin of the continent. All give some type of evidence for either deep-water, volcanic, or tectonic origins as would be expected in a region which may have contained a series of magmatic arcs, back-arc and oceanic basins, and associated subduction zones (see Fig. 5-23).

An overriding question might be: "in which direction did subduction occur?" Churkin and Eberlein (1977) state that there is little agreement on the amount, direction, and type of movement that occurred in the emplacement of their borderland terranes. They further note that to evaluate the tectonics of the borderlands, it is first necessary to know whether a single lithosphere plate or a mosaic of microplates were involved (in the latter case, a series of relative plate motions and magmatic arc configurations would be possible and, in fact, likely). Churkin and Eberlein describe a number of possible tectonic histories for the borderland terranes, to which the interested reader is directed. In general, they suggest that, in the Paleozoic, multiple microcontinental plates moved toward and away from western North America, with a succession of marginal ocean basins opening and closing behind migrating arcs. They note that the faunas of these "borderlands" have more Asiatic affinities than do contemporary faunas of the continental margin assemblages.

Monger et al. (1972) state that the western terranes (their "insular belt") are probably not fragments of Asia but rather are a series of elements with diverse origins, notably island arcs above subduction zones with unknown polarity.

Regardless of the tectonics involved with these "borderlands," Stewart and Poole (1974) note that the long, relatively stable history of deposition for the continental margin assemblages suggests that volcanic arc activity may have been located sufficiently far off the coast such as not to significantly affect marginal sedimentation. They further suggest the possibility that a subduction zone and marginal sea did not develop before the Devonian and that the insular facies developed essentially completely remote from North America and were emplaced entirely after the Early Devonian.

F. Filling the Niches: Tippecanoe Life

F.1. Introduction

Throughout this chapter, several new groups of organisms were mentioned because of their initial occurrences in Tippecanoe strata, or because they characterize Tippecanoe fossil assemblages. Many of these organisms may have achieved importance because changing environments in Tippecanoe seas produced new and favorable habitats; a classic example would be the emergence of eurypterids as

the chief predators in the shrinking and occasionally hypersaline Late Silurian seas and estuaries. Another very important and characteristic assemblage of Tippecanoe fossils are early reef-building and reef-populating taxa: the stromatoporoids, bryozoans, tabulate coral, lithistid sponges, and others. As shown in Chapter 4, the Cambrian sea was inhabited largely by a limited number of skeletonized taxa, which radiated to fill a great number of niches. The Ordovician/Silurian sea hosted a renewed diversification of organisms, but this time, with a large variety of phyla occupying available habitats.

F.2. Above the Benthos

Planktonic and nektonic organisms yield virtually all the fossils found in deposits from deep ocean basins. There are abyssal benthic communities in modern oceans, but they are composed generally of species which would be preserved rarely as fossils [e.g., polychaete worms, holothurian and ophiuroid echinoderms, sea spiders (pycnogonids), and other soft-bodied or partly skeletonized animals]. Because the deep ocean bottoms are below the carbonate-compensation depth, organisms with carbonate skeletons cannot easily live there nor can carbonate parts from organisms in the water column above abyssal depths be preserved there. As we have mentioned in several earlier discussions, the deep marine facies off the continental margins received, during Tippecanoe time, so-called graptolite facies; therefore, we begin the discussion of fossil Tippecanoe plankton with graptolites.

F.2.a. Graptolites

Although taxonomic affinities of graptolites are unknown—the group is entirely extinct—most modern workers consider graptolites to be related to pterobranchs, which are modern marine colonial organisms placed in the phylum Hemichordata. This inferred association means that graptolites are primitive chordates and that they are really rather advanced organisms possessing at least rudiments of gills, a hollow dorsal nerve chord, notochord, and other features.

The gross appearance of the more typical graptolite is that of a tiny saber-saw blade. The "blade" may appear to have "teeth" on one or both sides, and several "blades" may be joined. The "teeth" are the cups (thecae) in which the individual animals (zooids) lived. The "blades" are colonies or stipes. Stipes typically are 1 to 15 cm in length but unusual specimens span over 1 m. The thecae that make up a graptolite colony are connected to an internal thread or stolon, which is a feature that is shared with the pterobranchs as is the outward shape and size. In the graptolites, however, the walls of the thecae are made up of two

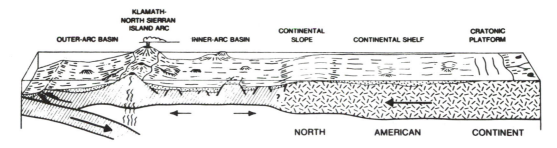

Fig. 5-23. Model of the Cordilleran continental margin during the Silurian and Devonian. Note that Klamath–North Sierran terrane is shown as a volcanic arc on the edge of the North American plate rather than on the other plate; also in this model, the deep-water environments between the arc and the continent are considered to be a back-arc basin rather than a true ocean basin. (From Poole *et al.*, 1977, reproduced with permission of the Pacific Section, Society of Economic Paleontologists and Mineralogists.)

layers whereas in the pterobranchs they are composed of a single layer (Bulman, 1970).

Most graptolites are preserved as carbonized impressions in carbonaceous shales. Some good-quality specimens have been prepared as acid leachates and it is these specimens that provide the known anatomical details. The taxonomy of the graptolites is based on variations in a number of morphological features present (or absent) in members of the various groups. Figure 5-24 illustrates several different types of graptolites; chief among the diagnostic features are: the *nema,* an extension of the stolon, which probably served as an attachment for the colony and which marks its base; the *sicula,* which is the first and presumably embryonic theca of the colony; the *dissepiments,* crossbars joining adjacent stipes in some types of graptolites; and the *rhabdosome,* which is another name for the entire graptolite colony. There are also three different types of thecae in the most primitive group of graptolites (the Dendroidea), which reflect their ontogenetic development. It has been suggested by Kozlowski (1949) that the male zooids secreted small cups (*bithecae*), the females made larger *autothecae,* and the third type of cups (*stolothecae*) gave rise to each successive group of cups.

Not all graptolites were planktonic. There are five generally recognized orders in the class of graptolites (pterobranchs comprise their own class of hemichordates), among which only the graptoloids were definitely planktonic. Three of the classes, Tuboidea, Camaroidea, and Stolonoidea, are not common fossils and are best known from the latest Cambrian (Tremadocian) of Poland. They were encrusting benthic organisms which may or may not have been ancestral to the dendroids and graptoloids.

The dendroids are the earliest group with a good fossil record. They appeared in the Late Cambrian and survived until the Mississippian but were overshadowed by the graptoloids during the latter group's heyday (Ordovician through Silurian). The most characteristic and widespread early dendroid species is *Dictyonema flabelliforme* (Fig. 5-24) which is found in the Tremadocian of Europe and North and South America. Dendroids may have been planktonic, with the colony suspended from floating objects by its nema, or they may have been epibenthic organisms which were suspended off the seabottom by their nema. Those forms that lacked nema (such as the bushy benthic form *Acanthograptus* from the Canadian Series) may have partly encrusted objects on the bottom and had part of their rhabdosome free.

In the younger group, Graptoloidea, there is an apparent trend toward simplification because these forms possess only one type of theca (and by implication, only one type of hermaphroditic zooid). They also lack preservable stolons. The graptoloids were very successful during Tippecanoe time and comprise the majority of graptolite fossils. Graptoloid morphology is highly variable but most types resemble variations of the saw-blade configuration, with between 1 and 100 stipes joined together in a great number of possible arrangements (Fig. 5-24). Some genera, such as the Ordovician *Diplograptus,* seem to have had a central float around which the stipes were suspended; and it is possible that many more graptoloid genera had floats which are not recognized in the fossil record. It is this morphological variability which makes graptoloids such valuable correlation tools.

F.2.b. Tentaculitids and a Few Other Miscellanea

Graptolites are not the only fossils from Tippecanoe strata which cannot directly be related to one of the modern phyla of animals. Among the apparent planktonic forms found in Tippecanoe rocks are the genera *Tentaculites, Orthotheca,* and several others. The fossils of these forms consist of small (0.5 to 4.0 cm) calcareou layered cones which show patterns of growth rings. The fossils are fre-

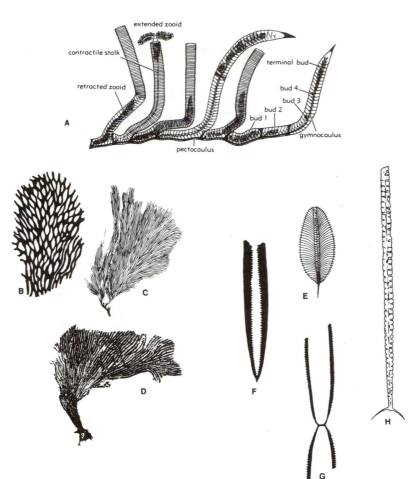

Fig. 5-24. Pterobranchs and graptolites. (A) Morphology of a living Rhabdopleurida (pterobranch) showing an extended zooid and details of the colonial cuticular skeleton. (B–D) Dendroid graptolites. (B) *Desmograptus* (L. Ordovician to Mississippian); (C) *Callograptus* (U. Cambrian to Mississippian); (D) *Dictyonema* (U. Cambrian to Mississippian). (E–H) Graptoloid graptolites. (E) *Phyllograptus* (L. Ordovician); (F) *Didymograptus* (Ordovician); (G) *Tetragraptus* (L. Ordovician); (H) *Climacograptus* (L. Ordovician to L. Silurian). (All from the *Treatise on Invertebrate Paleontology*, courtesy of the Geological Society of America and the University of Kansas Press.)

quently found in great numbers, although high concentrations of them are characteristic of the Middle and Late Devonian rather than Tippecanoe time; and they have been found in black shale facies, hence the implication that they were planktonic or nektonic. They seem to be most closely related to the pteropod mollusks, which are nektonic; however, some authors (e.g., Yochelson, 1963) suggest that the species assigned to *Tentaculites* may have been annelid worms, and Towe (1978) proposed that *Tentaculites* might be a specialized brachiopod taxon or a brachiopod structure (e.g., a spine), based on the microscopic structure of the calcite skeleton.

F.3. The Reef-Builders

F.3.a. Rugose and Tabulate Coral

The Ordovician and Silurian reefs that featured prominently in the discussion of Tippecanoe cratonic sedimentation, it must be reemphasized, were not coral reefs in the sense of modern reefs; but nonetheless the coral which first appeared in the Ordovician were important members of Tippecanoe reef communities and are essential framework builders of modern reefs.

Coral are a separate class (Anthozoa) of the phylum Cnidaria, which is one of the most primitive metazoan-grade groups of organisms and includes jellyfish, hydras, and coral. They may possess, among other features, specialized stinging cells (*nematocysts*), and during their life cycles various classes of Cnidaria may have a medusoid (jellyfishlike) and/or a polypoid (hydralike) morphology; Anthozoa are distinguished by not having a medusoid stage in their life cycle. The coral animal (or *polyp*) secretes a cup (*corallite*) in which it lives, and since many coral are colonial, the mass of cups of the individual polyps comprise the structure that is commonly called "coral." Within the corallites are structures which divide the animal's living space vertically and/or horizontally; these are termed, respectively, *septa* and *tabulae*. These hard parts are in turn secreted by *mesenteries* (partitions) within the soft tissue of the polyp. Some coral, especially members of the Rugosa, have diagonal or curved divisions in their corallites called *dissepiments*.

Not all living coral secrete compact aragonitic skeletons which would be preserved as fossils, and there are two living subclasses of coral (Octocorallia and Ceriantipatharia) which have slim fossil records, because they secrete skeletons of horny material or loose calcareous spicules. Octocoral are important in life history since many of the best Ediacaran fossils (Chapter 4) are "sea pens," a common term for gorgonacean octocorals. But in shelly facies faunas, such organisms are rarely preserved. The subsequent discussion will deal only with the two dominant Paleozoic fossil coral groups: Rugosa and Tabulata. A third Paleozoic order, Heterocorallia, is known from the Carboniferous of Europe and from rare specimens in the American Pacific region; but the group contains very few genera and serves no North American biostratigraphic function. Heterocoral apparently descended unsuccessfully from Rugosa. A fourth and final group of well-skeletonized coral, Scleractinia, appeared first in the Triassic Period and will be introduced in Chapter 7.

Unequivocally, both Tabulata and Rugosa appear first in the Ordovician, notably in the Chazyan reefs where the first tabulates appear. To date, the early evolution of coral has not been convincingly worked out and it is not possible to say which of these groups is primitive and which is ancestral. Rugose coral are commonly solitary (although many colonial genera existed) and they tended to form "horns" (Fig. 5-25) with a wide range of sizes (generally 1 to 18 cm). The external surface of the corallite tended to be rough (hence the name) and Wells (1963) showed that this roughness represents daily and annual growth increments. Rugosans built their corallites with various combinations of tabulae, septa, and dissepiments and these have provided the basis of their classification. They make excellent index fossils, although they are not always widespread specifically because they are sessile and benthic as adults (the larvae are free-moving and planktonic). In general, rugose coral comprised important elements of many Tippecanoe biofacies but they were commonly not reef-builders until the

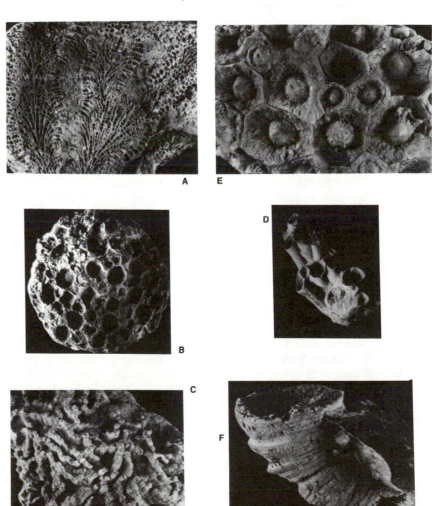

Fig. 5-25. Representative Paleozoic coral (A–D) Tabulata. (A) *Thamnopora* (Devonian), a colonial favositid featuring indeterminate-size, branching colonial form. (B) *Pleurodictyum* (U. Silurian to M. Devonian), a colonial tabulate with discoidal colonies and relatively determinate size range. (C) *Halysites* (M. Ordovician to U. Silurian), the colonial "chain coral" in surface view. (D) *Aulopora* (Ordovician to Permian), an encrusting colonial tabulate commonly found on brachiopods and other carbonate surfaces. (E, F) Rugosa. (E) *Lithostrotionella* (also *Lithostrotion* by some authors) (Mississippian), a colonial rugose coral featuring a prominent columella and hexagonal corallites. (F) *Zaphrentis* (Silurian to Mississippian but especially common in M. Devonian), a typical solitary rugosan. Accurate identification of most Paleozoic corals, especially rugosans, requires analysis of septa, tabulae, and/or dissepiment patterns as well as overall growth forms. (All are DRS photographs.)

later Devonian and Mississippian periods when colonial taxa such as *Hexagonaria* and *Lithostrotionella* became abundant.

Tabulate coral, in contrast, were almost always colonial and contributed significant structural elements to Ordovician and especially Silurian reefs. Tabulates, as the name implies, featured corallites characteristically divided only by tabulae (although some forms had dissepiments and some had vestiges of septa). This lack of septa may be interpreted either as a degenerative feature from the rugose morphology or as a primitive feature. Some tabulates had porous walls in the corallites which may have allowed contact between adjacent polyps. A hypothesis (Kazmierczak, 1984) suggests that at least certain tabulates (favositids) evolved directly from sponges and should in fact be classified as Sclerospongidae (discussed subsequently). This hypothesis is based on the microstructure of the corallite in favositids, and has not yet been thoroughly evaluated in the literature.

The tabulates which appeared in Chazy time either formed massive colonies or were members of a distinctive group of encrusting tabulate coral (the Auloporidae). Hill and Stumm (in Moore, 1956) note that the largest tabulates characterized the reef facies, and that the small and slender branching forms were more characteristic of colder- and deeper-water environments.

The Silurian tabulate fauna is dominated by genera of two groups: the favositids and halysitids which may be characterized by the respective genera *Favosites* and *Halysites* (Fig. 5-25). These taxa have been commonly called the "honeycomb" and "chain" coral for obvious reasons. Both genera appear in the Upper Ordovician and had worldwide distribution, but *Favosites* was abundant through the Middle Devonian while *Halysites* became extinct at the end of the Silurian. Many new tabulates appeared in subsequent Paleozoic periods but all died out by the Late Permian as did all Rugosa.

F.3.b. Bryozoans

Bryozoans are common organisms in modern oceans, but are far less known by most people than are coral, since modern bryozoans tend to form encrusting colonies or small attached colonies which may have no real skeletons. But during the Paleozoic Era, bryozoans often produced massive carbonate structures such as those comprising much of the Chazyan reefs.

Bryozoans superficially resemble colonial coral with very tiny openings for the individual animals; but there the resemblance stops because the groups are in no way related and, in fact, the skeletons are quite different on close examination. Bryozoans are far more complex structurally than cnidarians, and are related closely to the brachiopods. The simplest observational distinction between coral and

bryozoan skeletons is the lack of evident structures (such as septa or tabulae) within the cups that housed the organism (Fig. 5–26). In the terminology of bryozoan workers, these cups are called *zooecia* and the collection of such cups (although in bryozoans they are rather more like tubes), each separated from the other by an intervening wall, comprises the colony's skeleton and is termed the *zoarium*.

The animals (*zooids*) housed within the zooecia are somewhat irregularly shaped (Fig. 5-26) but they have one very distinguishing feature (which is shared with brachiopods): a feeding/respiratory organ termed a lophophore, composed of ciliated tentacies surrounding the mouth which, in bryozoans and brachiopods, are both extendable and retractable.

Colonies of bryozoans may grow by budding from the ancestral zooids. Bryozoans also feature sexual reproduction and in some groups there may be *polymorphic zooids*, i.e., specialized zooids which secrete tubes (*ovicells*) to house eggs and developing larvae. In addition, cheilostome bryozoans (see below) are strongly polymorphic and have additional specialized zooids in the colonies that perform protective functions; these include "bird's-head" individuals (*avicularia*) which remove sedimentary particles and encrusting organisms from the surface of the zoarium, and *vibracula* which serve similar functions by means of a whiplike structure.

The taxonomy of bryozoans is relatively simple at the highest level, with three classes recognized within the phylum Bryozoa of which one, the Phylactolaemata, is a freshwater group with few extant genera and an insignificant fossil record. Of the two remaining classes, the Stenolaemata incorporates four orders which form the bulk of the Paleozoic bryozoan record (and a fifth order of less importance), and the Gymnolaemata contains two orders including the dominant Mesozoic to present-day order (see discussion of cheilostomes in Chapter 7). The Appendix includes the current higher taxonomy and stratigraphic ranges of marine bryozoans (data from Boardman *et al.*, 1983).

The current taxonomy (Table I) of the stenolaemates is difficult to summarize because it does not represent a consensus of professional opinion, and because the divisions are based on detailed characteristics of the skeleton and inferred morphology of soft tissue (i.e., given that most taxa are extinct). In very general terms, trepostomes often formed massive colonies and are termed "stony bryozoans," whereas the cystoporates, cryptostomes, and fenestrates formed more delicate structures, frequently giving a lacy appearance to the skeleton. The Tubuliporata are very minor elements of Paleozoic fossil assemblages (although they alone of stenolaemates survived past the Paleozoic and became fairly important after the Triassic).

Evolution within the phylum Bryozoa is poorly understood, and the taxonomy is at least partly synthetic (termed

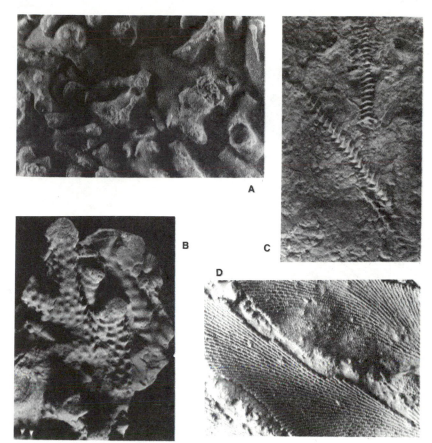

Fig. 5-26. Representative Paleozoic bryozoans. (A, B) Trepostomata. (A) *Dekayia* (= *Dekayella*) and other ramose trepostomes (this sample from U. Ordovician). (B) *Hallopora* (Ordovician–Devonian), another ramose trepostome. (C, D) Fenestrata. (C) *Archimedes* (Mississippian to L. Permian), a distinctively spiral fenestrate bryozoan. (D) *Fenestrellina* (Silurian to Permian), a generic name for a variety of fenestrate bryozoa, which may include the peripheral structures of *Archimedes* broken from the central spiral core. (All are DRS photographs.)

''polythetic'' in Boardman *et al.*, 1983). Since both classes of marine bryozoans have relatively minor orders which crossed the Paleozoic/Mesozoic boundary, it is not conclusive which group led to the modern, abundant Cheilostomata (although the ctenostomes would seem to be more closely related).

Table I. Current Classification of Marine Bryozoa

Class Stenolaemata	
Order Tubuliporata	(Ord. to present)
Order Trepostomata	(Ord. to Perm.)
Order Cryptostomata	(Ord. to Perm.)
Order Cystoporata	(Ord. to Perm.)
Order Fenestrata	(Ord. to Perm.)
Class Gymnolaemata	
Order Ctenostomata	(Ord. to present)
Order Cheilostomata	(Jur. to present)

F.3.c. Stromatoporoids

Among the important, extinct Paleozoic reef-building taxa are stromatoporoids (see Fig. 5-27), which were common, carbonate-secreting organisms capable of producing a variety of massive biohermal structures (see Kershaw, 1981); yet, their biological affinities are unclear. Hartman and Goreau (1970) suggested that stromatoporoids were related to an aberrant modern sponge group (Sclerospongiae), recently recognized in reef crevices near Jamaica.

The key morphological feature that suggests the interrelationship of sclerosponges and stromatoporoids is the presence in both groups of *astrorhizae* (see below) on their upper surfaces. An older, but still common view of the taxonomic affinity of stromatoporoids is that they are an extinct group of Cnidaria, based on features shared with the class Hydrozoa.

Stromatoporoid colonies take several general forms, but the most characteristic are encrusting or free-living laminar calcitic masses. Their growth seems to have been influenced by environmental conditions. In size they range from small nodules to sheets tens of centimeters thick by almost a meter in diameter (in some places, beds to 1 m thick are composed entirely of a single species of stromatoporoid and it is difficult to determine where one colony leaves off from the next). The main mass of the stromatoporoid colony consists of both tabular elements (*lamellae*) and vertical elements (*pillars*). On the surface of many stromatoporoids may be seen both prominent bumps (*mamelons*) and star-shaped systems of branching canals (*astrorhizae*). The sig-

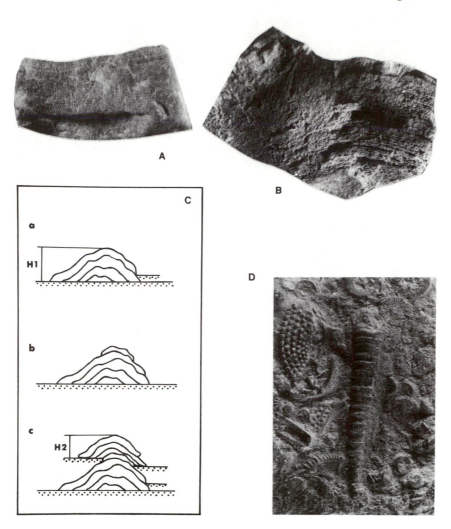

Fig. 5-27. Two anomalous Lower Paleozoic organisms. (A–C) Stromatoporoids. (A) *Stromatopora* (Silurian to Jurassic), a very common Paleozoic taxon; weathered specimen showing the tendency to split along laminae. (B) *Actinostroma* (Silurian to Jurassic), another common stromatoporoid featuring compact structure with prominent pillars, monticules, and astrorhizae (the latter evident only on the upper surface: these are lateral views). (C) Illustration of domal growth and its relationship with sedimentation rates, observed in the Middle Chazyan species *Pseudostylodictyon lamottense.* (a) First three growth layers reach down to sediment surface, fourth layer stops above surface; (b) sedimentation rate increases, growth stops except at uppermost-right; (c) as sedimentation rate decreases, new growth proceeds from the surviving region. (D) *Cornulites,* a Middle Ordovician to Mississippian fossil, especially abundant in Tippecanoe benthic carbonate assemblages: this photograph is considerably magnified (specimen about 6 mm long). *Cornulites* is reminiscent of tentaculitids, except the inferred habit of the former was benthic/encrusting whereas the latter were apparently planktonic and/or nektonic. (A, B, D, DRS photographs; C, from Kapp, 1974.)

nificance of both of the above features has not been determined but another feature, termed *latilaminae,* is a regularly spaced widening and thinning of the laminae in some species which seems to be a seasonal growth phenomenon. The actual living space of the organisms that made up the colony was probably fairly small but one cannot be sure whether they lived around or among the laminae and pillars.

Detailed taxonomic study of stromatoporoids requires thin-sectioning in both vertical and horizontal planes to observe the structure of laminae and pillars. There are five orders of stromatoporoids recognized (Stern, 1980), assuming that stromatoporoids are indeed in the class Sclerospongiae. The fossil sclerosponges (i.e., stromatoporoids) are of Cambrian through Cretaceous age but most commonly are found in Ordovician through Devonian limestones. Among the more common genera are *Stromatocerium* and *Cystostroma,* both principal Chazyan reef-builders. In overall environmental preferences, stromatoporoids seem to have lived in the same environments as Early Paleozoic coral and bryozoans and are most commonly found in fossiliferous limestones and early reefs.

F.3.d.   Calcareous Algae

Among the benthic algae are forms that become calcified, either wholly or in part, with the calcite (generally high-magnesian) or aragonite precipitated either in the cell walls, such as occurs in the red algae, or in and around the plant tissues such as occurs in the green algae (Bathurst, 1971). These red and green algae may collectively be called *calcareous algae;* a related term, *coralline algae,* is technically restricted to the family Corallinacea of the phylum Rhodophyta (red algae).

In Tippecanoe seas, several families of red and green algae were present and are presumed to have contributed a vast amount of carbonate to both the reef and nonreef benthic carbonate environments. The dominant coralline algae of Chazyan time was the genus *Solenopora,* a red alga of the family Solenoporacea; this family of rhodophytes was the only one known to have lived in the early Paleozoic. Thin sections of Ordovician carbonates show their characteristic large polygonal cells in samples cut parallel to the growing surface. Red algae are so named because of the reddish or purple color in their vegetative tissue (or *thallus*). They are still one of the major carbonate suppliers in the oceans.

Two families of green algae (or Chlorophyta) lived in the Tippecanoe seas: the Codiaceae and the Dasycladaceae. In codiaceans, the thallus is tubular and built up from interwoven branching tubes. The modern genus *Halimeda* is a typical codiacean which has cosmopolitan distribution in tropical seas. In ancient sediments, the carbonate debris from codiaceans can be recognized as carbonate needles which are the final decomposition product of their skeletons.

The second family of calcareous green algae which was present in Tippecanoe time are the dasycladaceans. In this group, the plant is basically a central stem with generally radiating slender branches. The carbonate secreted by these plants, which (as in the codiaceans) is aragonite, forms as a crust around the tissues of the plant (rather than within the cell walls as in the coralline algae) and thus the carbonate forms a mold of the alga.

Both chlorophyte families range from Cambrian to Recent in age, whereas the solenoporaceans began in the Cambrian but survived only until the Miocene. However, other red algae are abundant today as are members of several other important calcareous alga groups (such as the Coccolithaceae which appeared in the Jurassic and will be discussed in Chapter 8). Calcareous algae can serve as index fossils but it is often difficult to precisely classify ancient specimens for this kind of work; the dasyclads, however, do have reproductive bodies (*sporangia*) which are commonly preserved and help make them among the most useful algae for biostratigraphy.

F.3.e. Sponges and Spongelike Forms

The sponges (phylum Porifera), which are the most primitive animal group above the Protozoa, appeared in the Early Cambrian; some Late Proterozoic fossils too may be sponge *spicules* (skeletal elements) but these have not been proven to be such. Three common classes of sponges are generally recognized based on the composition of their spicular skeletons: the Demospongea have spicules made of *spongin* (which is a flexible organic material that gives bath sponges their characteristic ''squish'') and silica; the Calcarea have calcareous spicules; and the Hyalospongea (Hexactinellida, or ''glass sponges'') have siliceous spicules which are often interlocked in rather elegant basketwork arrangements. All three sponge classes range from Cambrian (or Precambrian for the hyalosponges) to Recent.

Sponges, as fossils, range in appearance from loose piles of spicules, or individual spicules, to discrete globular masses (see Fig. 5-28) and highly structured forms such as some glass sponges. The organism is basically an ''organized colony'' of specialized and generalized cells which secrete the spicular skeleton.

In Tippecanoe time, there were many living sponges but the forms most notable in Tippecanoe deposits are lithistid sponges, which were major contributors to Chazyan and other reef structures. Lithistid sponges were demosponges which formed strongly interlocked siliceous spicular skeletons.

Another enigmatic but important group of Paleozoic organisms have traditionally (but not currently) been assigned to Porifera. These are called receptaculitids (or ''sunflower coral'') but they are almost certainly not coral nor members of the same phylum as the sponges discussed above. The receptaculitids were calcareous-skeletoned, spicular organisms which are common in Ordovician through Devonian strata. They formed structures reaching 50 cm in diameter and can comprise considerable volumes of carbonate.

The simple identification of receptaculitids is made based on the unique and complex spiral arrangement of spicules (see Fig. 5-29) which rather resembles the arrangement of seeds in a sunflower head. Gould and Katz (1975) presented a geometrical analysis of this growth form and showed how the spirals can be generated using five variables in a computer simulation. Far more complex is the biological assignment of receptaculitids. Kesling and Graham (1962) attempted to prove they were dasycladacean algal structures; prior to that writing, they had been called virtually everything from pine cones to foraminifers. More

Fig. 5-28. Two common Paleozoic sponges. (A) *Astraeospongium* (Silurian to Devonian), a probable calcisponge (classification is uncertain) featuring hexagonal spicules which are barely distinguishable in portions of the photograph. (B) *Hindia* (Ordovician to Mississippian), a very common demosponge forming nearly spherical masses. B is magnified approximately three times greater than A. (Both are DRS photographs.)

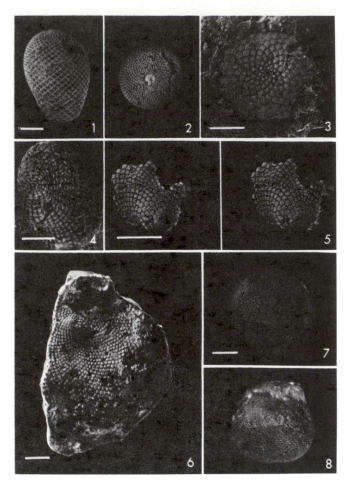

Fig. 5-29. Representative Receptaculitida. 1, 2, *Ischadites dixonensis* (Ordovician, Illinois); 3, *I. barrandei* (Silurian, Bohemia); 4, *I. koenigii* (Silurian, Great Britain and Illinois); 5, 6, *Fisherites occidentalis* (Ordovician, Québec); 7, 8, *Selenoides iowensis* (Ordovician, U. Mississippian). (From Fisher and Nitecki, 1982.)

recent studies, especially that of Fisher and Nitecki (1982), are typically noncommittal about their identity while tending to support the hypothesis of algal affinities (for example, Fisher and Nitecki refer to the body of receptaculitids as the ''thallus'').

F.4.　Euryhaline Organisms: Eurypterids

In the discussion of post-Salinan sedimentation, we mentioned the ecology of eurypterids; now we turn to their morphology and systematics. Eurypterids (Fig. 5-30) are a completely extinct group of the Chelicerata, a class of arthropods which also includes the fossil and living horseshoe crabs such as *Limulus*.

The body of eurypterids was distinctly segmented with a variable number of jointed legs originating from the *cephalothorax* (fused head and thorax). One of the anterior pair of legs was often specialized as pincers and the posteriormost pair of legs were generally flattened at the end into a

pair of paddles for swimming. In some species (such as *Mixopterus*), the front two pairs of legs had spines which may have served as weapons or grasping organs. It is presumed that most forms swam well using the paddles on their hind legs as oars; however, they may also have adopted the mode of swimming that young horseshoe crabs use, which consists of flopping over on their backs and propelling their bodies with the movable plates on the ventral side of the rearmost part of their body (the *mesosoma*). One eurypterid genus (*Stylonurus*) seems to have none of the swimming adaptations and may have marched about on the sea or lake bottom on its stiltlike legs.

They may have been fierce predators, if their appearance and armament are correctly interpreted, and it is possible that they preyed on or were protecting themselves from the newly evolving vertebrates in Tippecanoe lagoons. From the sediments in which they are found have also come the remains of vertebrates, nautiloid cephalopods, and a few other types of organisms; but rarely are eurypterids associated with coral, brachiopods, or trilobites and other constituents of the shelly facies. This has resulted in the conclusions previously presented that they inhabited an environment with widely fluctuating salinity.

Eurypterids are widely spread geographically but they are numerically rare generally. They range from the Ordovician to the Permian and are most common in Silurian and

Fig. 5-30. *Eurypterus* (Ordovician to Permian; this specimen *E. remipes*, U. Silurian), a relatively common eurypterid in the New York Salinan facies. (DRS photograph; specimen made available courtesy of Dr. Fred Collier, U.S. National Museum of National History.)

Devonian strata, probably because euryhaline conditions were most prevalent then. They may in part be rare because the exoskeleton was not as hard as that of trilobites, nor was it calcified. They may, in fact, have been common but were preserved only in selected sediments such as the Late Silurian waterlimes of New York. Their relative rarity makes them poor index fossils but they are certainly prizes for the collector, especially in the case of large specimens such as individuals of *Pterygotus* which range in size to over 2 m.

F.5. Success in the Tippecanoe Seas: Articulate Brachiopods

Although present in the Early Cambrian, brachiopods did not establish their dominance in the benthic marine environment until the Ordovician. They were the most common nonreef-building invertebrates of the times and may even have been the most common absolutely in terms of biomass. It is not yet determinable whether the articulates evolved from the inarticulates in the Late Proterozoic or Early Cambrian, or if both groups of brachiopods (and perhaps bryozoans) arose from a common ancestor.

Since the inarticulates were introduced with the Sauk fauna, the discussion here will be restricted to the Articulata. After Cambrian time, the inarticulate brachiopods apparently diminished in absolute population densities but persisted in a few conservative forms right up to the present.

There are seven major orders of articulate brachiopods as well as several minor groups which may deserve ordinal rank. Figure 5-31 shows typical external morphologies of the major groups and the inferred pathways by which they evolved. The orthids are the ancestral stock for the other articulates and gave rise to the pentamerids in the Middle Cambrian, and the strophomenids and spiriferids in the Ordovician. The subsequent evolution of articulates is less well defined but according to Rudwick (1970) the other three articulate orders evolved as follows: the pentamerids gave rise to the rhynchonellids in the earliest Middle Ordovician, and then to the atrypids late in the Middle Ordovician. In the Late Silurian, the atrypids gave rise to the terebratulids. Of the seven orders of articulate brachiopods, three survived to the present time (Rhynchonellida, Terebratulida, and Strophomenida); but this fact may be deceiving because among these orders there are many suborders and lower taxa which became extinct during or at the end of the Paleozoic (see Chapter 7). The Pentamerida survived only until the Late Devonian but they were especially abundant during the Silurian. The Atrypida survived until the Late Triassic (or possibly the Jurassic) and the Orthida became extinct at the Permo-Triassic boundary.

The shells of articulate brachiopods generally are not identical because one contains an opening for the pedicle (the *pedicle valve*) and one contains the brachial apparatus (the *brachial valve*). The line along which both shells hinge

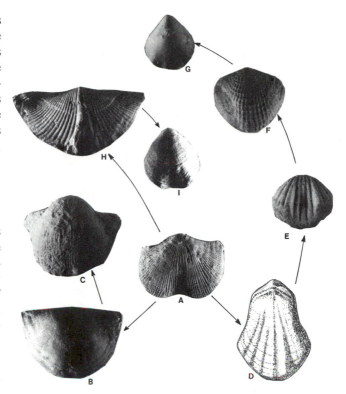

Fig. 5-31. Hypothetical evolutionary pathways among articulate brachiopod higher taxa. Various elements of this chart are very speculative, especially with regard to ancestry of rostrospiraceans, atrypaceans, and terebratulids. Brachiopod taxa represented are: (A) Orthida, (B) Strophomenida (Strophomenacea), (C) Strophomenida (Productacea), (D) Pentamerida, (E) Rhynchonellida, (F) Atrypida, (G) Terebratulida, (H) Spiriferida (Spiriferacea), (I) Spiriferida (Rostrospiracea). All figures show a view of the pedicle valve except D, which is viewed from the brachial valve. (Data in part from Rudwick, 1970; DRS photographs.)

is called, logically, the *hinge line* and the pointed part of the valves adjacent to the hinge line is the *beak*. Opposite the beak is an opening on one or both valves for the pedicle to emerge, termed the *pedicle foramen*. This opening may be round and located entirely in the pedicle valve, or commonly it is diamond-shaped and formed partly by both valves (but usually the major portion is in the pedicle valve). In the latter case, the *delthyrium* is the term applied to the pedicle opening in the pedicle valve, and the *notothyrium* to the smaller triangular opening in the brachial valve. The delthyrium may be partially closed by a pair of *deltidial plates*. The shell may have radial ridges that corrugate the entire thickness of the shell, usually extending from the beak to the opposite margin, called *plicae* or plication. If this sort of shell ornamentation does not extend to the inside of the shell, these radial lines are called *costae*. Many brachiopods also have a series of growth lines which parallel the margin and become smaller as they approach the hinge.

A few more diagnostic features can be found inside the valves and within the shell material itself. The various orders of brachiopods have a number of different types of brachia, ranging from simple loops or bracket-shaped devices to complex coils such as are featured by the spiriferids. (In some Middle Devonian *Mucrospirifer* which have been pyritized, notably from the Leicester Pyrite in western New York, these pyritized brachidia bear a remarkable resemblance to coiled springs.) Finally, the shells of brachiopods may be *punctate,* i.e., perforated with many small canals which pass from outside to the inside of the shell; *pseudopunctate* (which is a phenomenon that results from shells which have rods of calcite at right angles to the shell laminae in the outer layer only and which differentially weather in fossil forms to give the appearance of punctae); or *impunctate.*

Table II lists the diagnostic characteristics for each of the seven major orders of brachiopods using the above ter-

Table II. Diagnostic Features of Major Articulate Brachiopod Groups

Order	Overall morphology				Shell morphology			Miscellaneous
	Outline	Profile	Size	Interareas	Ornamentation	Structure	Brachia	
Orthida	Semicircular subround to elliptical, long hinge line	Unequal biconvex, relatively flattened	Small to medium	Both valves, delthyrium/notothyrium	Radially costate	Impunctate	Simple	Dalmanellacea are punctate
Rhynchonellida	Subtriangular to rounded, short hinge line, beaked	Unequal biconvex, globose, sulcus present	Small to medium	Small, on both valves	Strongly plicate	Impunctate	Absent (uncalcified)	Rhynchoporacea are punctate
Pentamerida	Oval to subround, short hinge line	Equal biconvex, globose	Medium to large	Small, on both valves	Smooth, some costate or plicate	Impunctate	Simple	Well-developed spondylia on pedicle valves
Strophomenida	Semicircular to subtriangular, very long hinge line	Unequal biconvex to concavoconvex, flattened to very flat	Small to large	Well developed, both valves	Smooth, costate, spiny (productids), annulate (e.g. *Leptaena*)	Pseudopunctate	Looped or spiral (usually unpreserved)	Productacea have one very globose valve, spines, weak or absent interareas Strophomenacea are nonspinose and flat
Spiriferida	Suboval to alate ("winged"), medium to long hinge line, beaked	Plano-convex to biconvex, globose, fold and sulcus	Small to large	On pedicle valve, can be large, triangular delthyrium with deltidia	Costate to plicate, some spinose, some concentric annulae	Impunctate	Spiral	Punctospiracea are punctate
Atrypida	Oval to subround, short hinge line	Unequal biconvex to concavoconvex, sulcus present	Small to medium	Small or absent, on pedicle valve	Smooth, costate, plicate, some concentric annulae	Impunctate	Spiral	Considered by some to be a taxon within Spiriferida
Terebratulida	Subcircular to oval, short hinge line	Equal to unequal biconvex, globose	Small to medium	Absent, round pedicle opening	Smooth to finely costate	Punctate	Looped	

minology. During the Middle Ordovician, brachiopod faunas were dominated by the orthids and strophomenids. In Upper Ordovician shelly facies, the rhynchonellids too are very abundant. In Silurian rocks, most of the orders except terebratulids, spiriferids, and atrypids are common, but pentamerids seem best to characterize Silurian benthic faunas in many areas. Brachiopods are, as mentioned, one of the most important groups of fossils used in Paleozoic biostratigraphy and they yielded dominance of the sessile epibenthic fauna to the bivalved mollusks only in the Mesozoic Era.

F.6. Early Paleozoic Mollusks

Among the earliest Cambrian shelled fossils are mollusklike taxa (e.g., *Hyolithes* and *Wyattia*; see Chapter 4), suggesting that the phylum is relatively old. Nevertheless, diverse and common molluskan fossils do not appear in the stratigraphic record until the Ordovician; and even then, the record suggests they were not close to their Mesozoic and Cenozoic abundances and diversification. One may speculate that brachiopods occupied the general niche now claimed by the Bivalvia (clams, oysters, pectens, and mussels), and that the modern role of gastropods (snails and sea slugs) may have been filled in early Paleozoic time by trilobites. Gould and Calloway (1980) suggested that the results of mass extinction rather than direct competition caused the apparent replacement of brachiopods with bivalves in younger strata.

Nevertheless, mollusks were present from the beginning of the Phanerozoic, and more varieties emerged during the Ordovician and Silurian. Although they never achieved any sort of predominant position in Tippecanoe faunas, they diversified during the interval.

The ancestral mollusks would be sluglike creatures with a head, a muscular foot, sensory appendages (such as the antennae of snails), and in most forms, the ability to secrete a calcareous shell by means of a layer of epidermal tissue called the *mantle*. Studies of a rare deep-sea mollusk, *Neopilina*, which is both a "living fossil" and a representative of the most primitive known group of mollusks (the Monoplacophora), confirmed the assumption that mollusks evolved from segmented organisms (possibly annelid worms) and that they have common ancestry with the arthropods.

One of the most primitive living group of mollusks is the class Amphineura. These are the familiar chitons of modern rocky intertidal zones, easily recognizable because most have overlapping segmented shells; but they are quite rare as fossils. Another more common, but stratigraphically unimportant, mollusk group is the Scaphopoda (or "tuskshells"), which appeared in the Ordovician. Scaphopod shells are simple tubes and do not provide the basis for easy specific identification.

There are three classes of mollusks which have left a major fossil record: the gastropods, bivalves, and cephalopods. In terms of the "basic" mollusk described, gastropods are a variant with the body twisted in a helical spiral and a spiral shell (although many gastropods are not shelled) and they are probably most like the ancestral mollusks. Bivalves are the most specialized; they have lost their heads, secrete a typically matching pair of shells, and have modified the walking-foot into a sort of "pushing"-foot (hence the former name for this group, Pelecypoda or "hatchet-foot"). The cephalopods are the nektonic variants on the mollusk theme, with a definite head surrounded by a ring of tentacles (hence the name, meaning "head-foot"), very large and capable eyes, and a unique and efficient jet propulsion system for rapid locomotion. A large number of extinct cephalopods (and a handful of living forms) secreted planispirally coiled shells which are very common fossils.

In Tippecanoe time, all three groups were present. The bivalves emerged in the Ordovician, the cephalopods in the Late Cambrian, and the gastropods (if *Wyattia* is a gastropod) in the earliest Cambrian.

The systematic description of both bivalves and gastropods is quite complex and requires an understanding of a great amount of detail about such features as hingement (with its tooth-and-socket structure) in bivalve shells, and the siphons and other soft anatomy in gastropods. In general, in Ordovician and Silurian strata, bivalves with affinities to modern pectens and mussels are present with no taxon more than locally abundant. A bivalve specialist can, of course, distinguish a lower Paleozoic bivalve community from a modern collection; but to the eye of the nonspecialist, the Paleozoic population might not appear strikingly distinctive except for a lack of oysters and siphonate clams.

The same holds true to a degree for the gastropods, except for the presence in Tippecanoe and later Paleozoic assemblages of a few characteristic extinct lineages. In modern oceans and freshwater, there are snails belonging to three major groups (which are of debatable taxonomic rank but probably should be subclasses) of which only one, the Prosobranchia, is commonly shelled and found in saltwater. Of the prosobranchs, only one order, the Archaeogastropoda, ranges back to the early Paleozoic. Archaeogastropods commonly resemble conventional snails with both high-spired and squat forms present in the group; one group of archaeogastropods, the macluritids, were distinctive, featuring fairly large shells (to over 20 cm in diameter) which were planispirally coiled and flattened on one side (Fig. 5-32). Another characteristic group of the early archaeogastropods were the euomphaliids which were also planispirally coiled but flattened on both sides.

An extinct "subclass" of gastropods were characteristic components of Tippecanoe and Devonian faunas.

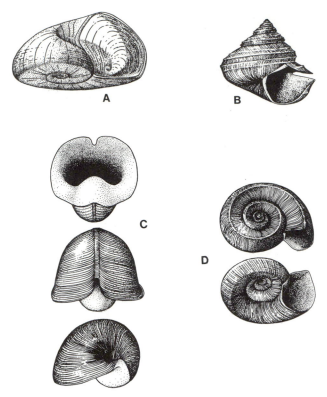

Fig. 5-32. Representative Paleozoic Archaeogastropoda: (A) *Maclurites* (Ordovician), a typical macluritid; (B) *Bembexia* (Ordovician to Pennsylvanian), a pleurotomarian; (C) *Bellerophon* (Silurian to L. Triassic), a typical bellerophontid shown in three views; and (D) *Euomphalus* (Silurian to M. Permian), a typical euomphaliid. (All from the *Treatise on Invertebrate Paleontology,* courtesy of the Geological Society of America and the University of Kansas Press.)

This group, the Amphigastropoda, included planispirally coiled snails which had a trumpet-shaped section on the outer coil of the shell that greatly enlarged the aperture in many species (and which yielded the superfamilial name for the largest group of amphigastropods, the Bellerophontacea, named for Bellerophon, the mythical trumpet-playing Greek hero who also rode Pegasus and slayed the Chimera). Bellerophontids range from the Ordovician through the Triassic.

The Cephalopoda are a diverse class of mollusks, represented in modern oceans by octopods and squids (including the giant squids of the genus *Architeuthis*), the externally shelled *Nautilus,* and the genus *Spirula,* which alone among living cephalopods has a coiled internal hard structure termed a *phragmocone.*

Starting in the Ordovician and continuing to the present time, cephalopods comprise important elements of marine invertebrate faunas. Various larger time units (e.g., the early Paleozoic, late Mesozoic) had characteristic cephalopod faunas: in the early through middle Paleozoic, "nautiloids" comprised most of the cephalopod popula-

tion. We shall return to other cephalopod groups as their times of prominence are discussed.

The informal term "nautiloid" is used to designate externally shelled cephalopods featuring tabular, simply curved, or slightly convolute *septa* between chambers of the shell; septa are the dividing walls between shell chambers, which appear toward the external part of the shell as lines termed *sutures.* Three subclasses, incorporating 11 orders, comprise the "nautiloids" but the most common, long-ranging, and characteristic among these is the formally designated subclass Nautiloidea. Among diagnostic features of "nautiloids" are the septal/sutural arrangement, the shape and position of the *siphuncle* (a calcareous tube internally connecting shell chambers), the presence and nature of internal calcareous deposits within the chambers (termed *cameral deposits*), the nature of the regions where the septa curve backward and join the siphuncle (termed the *septal neck*), the nature and presence of deposits within the siphuncle, and the nature of the overall shell (including size, shape, and the nature of the anterior shell opening). Shell shape engenders a detailed terminology itself: overall shape may be straight (*orthocone*), curved (*cyrtocone*), or coiled (*cochlear*); further, orthocones may be long and slender (*longicone*), or short and thick (*brevicone*) (see Fig. 5-33). Cochlear shells may have external coils covering earlier coils (termed *involute*) or they may have all earlier coils visible (*evolute*).

The "nautiloids" of the subclass Endoceratoidea range in age from Early Ordovician to Middle Silurian. They were typically medium-to-large, usually orthoconic forms with large siphuncles located in the ventral (outer) position, and possessing endosiphuncular deposits. There were two orders of endoceratoids and virtually all genera occur in the Ordovician (all taxonomy of "nautiloids" here from Teichert and Moore, 1964). Actinoceratoidea is a subclass with a single order, generally similar to the endoceratoids, but characteristically featuring cameral deposits and a number of specialized structures not found in other taxa. They were most common during the Middle and Late Ordovician but a number of genera may be found through the Pennsylvanian.

Nautiloids in the strict sense (the subclass Nautiloidea) include eight orders with a diversity of forms and features. Virtually the only features which they may not have are very large siphuncles and a few shapes to the septal necks. Six nautiloid orders are restricted to the Paleozoic. One survived into the Late Triassic, and a sole order among all "nautiloids" survived the range of geological time from the Paleozoic through the present. This latter order is the Nautilida, which also includes the majority of later Paleozoic taxa and virtually all cochlear "nautiloids."

Shelled cephalopods in general make excellent index fossils because they often are abundant, they preserve well, and most importantly, their mode of life was nektonic and

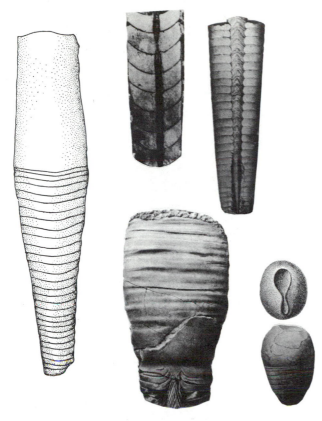

Blastozoa are the more primitive, and include the cystoids, a variety of primitive organisms variously called "eocrinoids" and "paracrinoids," and blastoids. The crinozoans are all placed in a single class, Crinoidea, which contains four orders.

Figure 5-34 shows typical morphologies of Paleozoic crinoids and blastoids. All have stalks composed of stacked, round (or rarely, star-shaped or oval) plates, each (as with all echinoderm skeletal parts) composed of a single calcite crystal. The stalked echinoderms feature a variable number of arms which extend above a body (*calyx*), of variable size and shape. In crinoids the arms may be long and multi-branched whereas in blastoids they are far less complex and small. Both subphyla feature strict pentaradial symmetry. Blastoids have a distinctive perforated plate system (*ambulacra*) covering a set of structures termed *hydrospires*. In addition, blastoids show five openings (*spiracles*) at the top of the calyx. These features reflect the water vascular system of the animal.

Crinoids (Fig. 5-35) lack obvious external expression of the water-vascular system; rather, the calyx is surmounted by arms which in most forms are at least as large as the calyx. Apparently, blastoids fed and respired by currents created in the hydrospires whereas crinoids used the arms as filtering organs, all with the intention of extracting nutrients and oxygen from seawater.

We have mentioned that pelmatozoan debris yielded

Fig. 5-33. Representative Paleozoic orthocone "nautiloids." Clockwise from left: *Actinoceras* (M. Ordovician to L. Silurian), a typical actinocerid; *Pleurothoceras* (U. Ordovician), an orthocerid shown in cut view to reveal simple curved septa and siphuncle; *Cameroceras* [M. Ordovician to (?) Silurian], an endocerid, shown in cut view to reveal tabular septa and details of more complex siphuncular structure; *Gomphoceras* (M. Silurian), an aberrant oncocerid, shown in ventral and aperture views; *Bathmoceras* [L. Ordovician to (?) U. Silurian], an ellesmerocerid. (All from the *Treatise on Invertebrate Paleontology*, courtesy of the Geological Society of America and the University of Kansas Press.)

their shells may float after death; thus, they may be widely dispersed. Although "nautiloids" are important index fossils of the Paleozoic, they are overshadowed by the Mesozoic ammonoids (see Chapters 6 and 8) which are considered by many biostratigraphers the overall best index megafossils of all geological times.

F.7. Diversification of Echinoderms

In Tippecanoe time, especially the Ordovician, echinoderms diversified and added at least eight classes to the edrioasteroids and other primitive forms mentioned in Chapter 4. Current classification [Sprinkle (1976) and the *Treatise*] recognizes five subphyla in Echinodermata. The focus in this discussion will be on two (formerly considered one: Pelmatozoa), termed the Blastozoa and Crinozoa. A third major subphylum, Echinozoa, will be examined in Chapter 8.

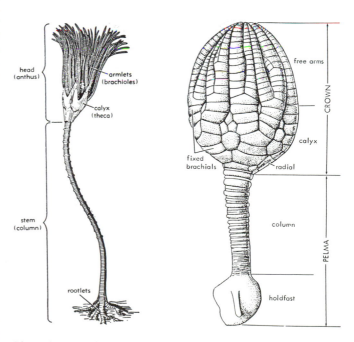

Fig. 5-34. Basic features of the dominant pelmatozoans. (Left) A Mississippian blastoid, reconstructed to show major body divisions and structures. (Right) A representative Silurian crinoid showing basic divisions and features. (Both from the *Treatise on Invertebrate Paleontology,* courtesy of the Geological Society of America and the University of Kansas Press.)

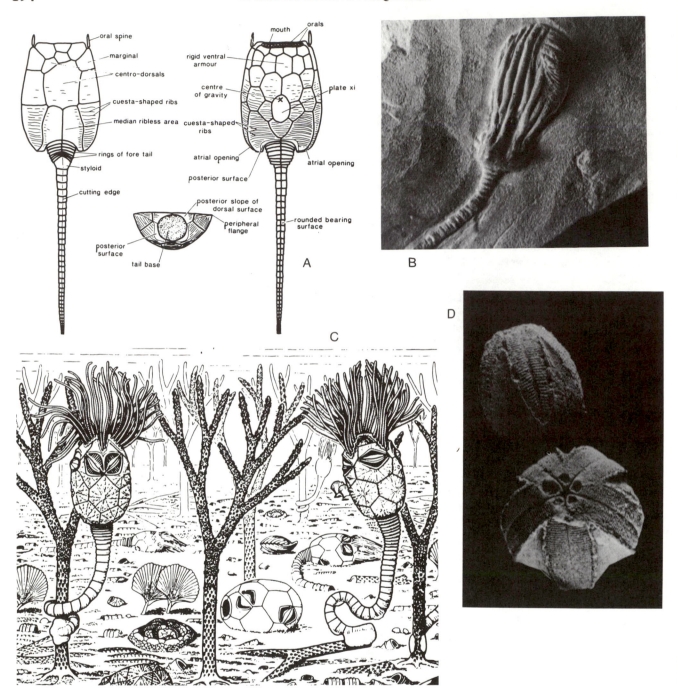

Fig. 5-35. Representative Paleozoic pelmatozoans. (A) *Ateleocystites* (Ordovician), a carpoid (= calcichordate) shown in dorsal (left), ventral (right), and posterior (bottom) views. (B) A Devonian camerate crinoid (genus undetermined), with partial stem and complete crown. (C) Reconstruction of a Late Ordovician seabottom in southern Indiana, featuring *Lepadocystis* cystoids in addition to other typical Cincinnatian fossils. (D) *Pentremites* (Mississippian to Pennsylvanian), a widespread, common blastoid genus shown here in anterior (oral) and side views. (A, from Jefferies, 1984; B and D, DRS photographs; C, from the *Treatise on Invertebrate Paleontology,* courtesy of the Geological Society of America and the University of Kansas Press.)

enormous volumes of carbonate clasts to many Tippecanoe facies. From their earliest appearance in the Ordovician, crinoids were very common members of Tippecanoe faunas, and in the later Paleozoic (especially the Mississippian), both groups dominate fossil assemblages. Most pelmatozoan limestones give evidence for clear-water conditions and it is believed that Paleozoic pelmatozoans were intolerant of turbid waters. Echinoderms were and are exclusively marine. In modern oceans, crinoids occupy a wide range of habitats, from shallow benthic to very deep ocean bottoms. Many modern crinoids lack stems (the stalked species are commonly called ''sea lilies'').

Blastoids survived in great numbers through the Mississippian, but began to decline in abundance soon after that time, disappearing at the Permo-Triassic boundary. Crinoids flourished through the Paleozoic and then underwent a rapid decline in the early Mesozoic; but they survived to the present. Two orders of crinoids (the Flexibilia and Camerata) became extinct in the Late Permian, and a third (the Inadunata) disappeared in the Triassic. All living crinoids belong to a fourth order (the Articulata) which appeared in the Triassic (Chapter 7).

F.8. The Shelled Microfossils

The modern practice of micropaleontology (which is essential to many economic and research applications of stratigraphy, paleontology, and oceanography) relies largely on the presence in fossil biotas of the remains of two groups of organisms: the Ostracoda (phylum Arthropoda) and the Foraminiferida (phylum Protozoa). In addition, the radiolarians, another group of protozoans, are less important stratigraphic fossils but are commonly found in ancient sediments. Both foraminifera and ostracodes are known with certainty to occur in Ordovician to Recent sediments. Foraminifera have also been reported in Cambrian sediments, but these reports are to date not widely accepted. Radiolarians have been more firmly reported from Precambrian cherts (they range in age to the present) and these reports are fairly convincing because radiolarians themselves are major contributors to modern siliceous deposits and very likely supplied silica for some Late Proterozoic cherts.

Foraminifera are probably the single most commonly and effectively used group of fossils in stratigraphic paleontology for all post-Mississippian geological periods. They are, in essence, amoebas with shells (*tests*), and occupy both benthic and planktonic environments. In some sediments they are almost as common as sand grains. Planktonic foraminifera serve as paleotemperature indicators in more recent pelagic deposits because many modern species, which range back in time, are zoned by temperature preferences and thus can be used to determine older

ocean temperatures in a given area by means of submarine cores (see Chapter 9).

The tests of foraminifera are composed of various substances, and this provides the basis of their taxonomy. According to Cushman (1950), there are four suborders of ''small'' Foraminiferida with the following distinctions (see Fig. 5-36): the foraminifera which make up tests by agglutinating detrital particles are the Textulariina; those which secrete porcelainlike calcareous tests are the Miliolina; those which secrete perforated calcareous tests are the Rotaliina; and those with membranous and pseudochitinous tests are the Allogromina (a primitive, early group).

Foraminifera, although present, were not highly diversified in Tippecanoe seas. During the Pennsylvanian Period a new suborder of ''large'' forams, termed fusulines, evolved, spread widely, and provide the earliest, widely used set of stratigraphic foram fossils. We will return to them in Chapter 7.

Ostracodes are not remotely related to forams except by their small size which classifies them as microfossils (a vague term which includes fossil forms that are small enough to be sampled statistically from drill cores). They are small crustaceans which secrete a tiny bivalvelike set of shells (Fig. 5-37). It is the presence of these paired shells that makes them infinitely more valuable as stratigraphic fossils than, say, crabs. The shells are calcareous and can be

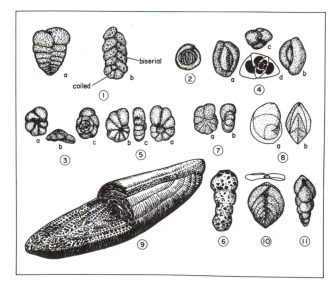

Fig. 5-36. Illustrations of representative foraminiferal morphologies. Representations 1, 10, and 11 are characteristic of textulariids, 2 and 4 are characteristic of miliolids, 3 and 5 are characteristic of rotaliids, 9 is a fusulinid, and 6 is characteristic of the allogromids. However, shapes illustrated in 1, 5, 7, 8, and 11 may be attributed to several groups. (From Tasch © 1980 by John Wiley & Sons, reprinted with permission of John Wiley & Sons.)

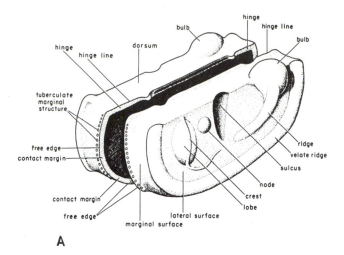

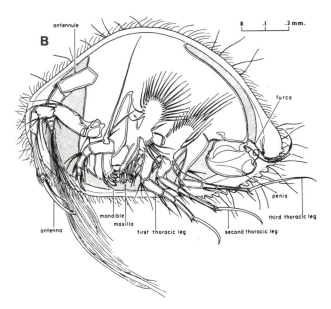

Fig. 5-37. Morphology of a representative Paleozoic ostracode carapace (A) and anatomy of the living ostracode *Bairdia* (B). (From the *Treatise on Invertebrate Paleontology,* courtesy of the Geological Society of America and the University of Kansas Press.)

ornamented and sculpted with a wide variety of lobes and depressions (termed *sulci*) all of which variation increases their biostratigraphic value. Ostracodes inhabit both fresh and saltwater. The taxonomy of ostracodes is too complex to present here (see Benson *et al.,* 1961); as a gross generality, Paleozoic/Mesozoic ostracodes tend to have deep lobes and sulci, in contrast to Cenozoic ostracodes which are smooth or which have a pitted, bubbled, or otherwise ornamented surface. Freshwater ostracodes tend to be smooth; but, again, there are very numerous exceptions to the above.

F.9. Early Vertebrates and Chordates

F.9.a. Earliest Chordates

The advanced animal phyla may be divided into three (or four) groupings, based on their larval development and inferred evolution. These are: the Metameria (annelids and arthropods), Molluscata (mollusks), Lophophorata (brachiopods, bryozoans, and phoronids or "feather worms"), and Deuterostomia (echinoderms, hemichordates, and chordates; classification in Valentine, 1977). Although Valentine separated the mollusks from Metameria, others consider them closely related because of the segmentation present in primitive mollusks such as the living monoplacophoran *Neopilina*.

Echinoderms and "chordates" (the latter is actually two phyla as described below) are differentiated from all other animals largely on the similarity of the larvae of enteropneusts ("acorn worms") and certain echinoderms. Certainly in gross appearance most echinoderms bear little resemblance to chordates: consider that the symmetry of chordates is usually bilateral (versus pentaradial in echinoderms) and echinoderms have external calcite skeletons versus internal phosphatic skeletons in chordates.

The phylum Chordata is subdivided currently into three subphyla: Urochordata, Cephalochordata, and Vertebrata. A fourth group, Hemichordata, is currently given phylum status and includes graptolites, pterobranchs, and enteropneusts. Enteropneusts have a fossil record dating back to the Middle Cambrian Burgess Shale (Chapter 4) but are generally absent from the fossil record.

Among the chordate subphyla, the Urochordata are interesting organisms, but not for our purposes since they are not known as fossils and because their relationships with higher chordates are undeterminable. Urochordates are commonly called "sea squirts" and "salps," comprising abundant elements of benthic and planktonic (respectively) modern marine faunas. All of the subsequent discussions on chordates in this book will focus on the two higher subphyla of chordates: the cephalochordates and the vertebrates.

F.9.b. Chordate Characteristics

Several morphological details discriminate chordates from other phyla. These include: a hollow nerve chord, usually in the dorsal (back) position; presence of a *notochord* (a rod of cartilage), usually ventral to the nerve chord; gill slits in the pharynx (throat); a generally high grade of tissue organization; and a tendency to have a body organized into segments. In vertebrates, the hollow nerve chord terminates in an expansion of the nervous tissue which constitutes the brain. Also in vertebrates, by definition, the notochord is encased in a series of bony or cartilaginous rings (vertebrae) and the notochord may be vestigial or absent.

F.9.c. Cephalochordates and Conodonts

Modern cephalochordates are a small group (two genera) of simple animals which superficially resemble sardines as they are ready for eating. They lack appreciable heads, have muscles visibly arranged in *myotomes* (segments in a chevron pattern), and lack separate fins, eyes, gill covers, and vertebrae. Typical members of the genus *Branchiostoma* (commonly termed "*Amphioxus*") are 4–5 cm long and live a semibenthic/seminektonic life. In cephalochordates, the notochord extends to the anterior tip of the animal, essentially where the head would be, and in current thinking this precludes these animals as vertebrate ancestors since the notochord extension precludes brain development. It is likely that modern cephalochordates and vertebrates have a distant common ancestor.

Cephalochordates as we know them have no major fossil record; however, a very important group of index fossils, conodonts, have often been attributed to chordate ancestry and may indeed be cephalochordate remains. Conodonts have also been attributed to a variety of other higher taxa, as will be discussed below. Superficially, conodonts have several characteristic morphologies including toothlike and bladelike forms. They are microscopic to submicroscopic in size, and typically composed of a calcium phosphate-complex material, very similar to the composition of vertebrate bone. They are fairly common in organic-rich shaley and sandy lithologies; indeed, they may be the only fossils found in many dark-colored sands. Conodont biostratigraphy is highly developed within their stratigraphic range of the Middle Ordovician to Triassic and conodont biozones are listed in most modern Paleozoic correlation charts (especially where subsurface work requires index microfossils). Youngquist (1967) presented an amusing narrative of the use of undetermined objects (specifically, conodonts) as index fossils, and used the analogy of a biostratigraphy of beer containers (glass quarts evolving to aluminum pop-tops) in human trash deposits.

It is clear that conodonts (Fig. 5-38) are neither jaws nor teeth of annelid worms; such fossils are well-known and called *scolecodonts,* with a known range of the Ordovician to Recent. Scolecodonts are usually composed of chitinous or siliceous material and are readily distinguished from conodonts. Conodonts, as mentioned above, are phosphatic and rarely (but can) show wear. The *Treatise* (Hass, 1962) states that what is known about the "conodont animal" is that they were soft-bodied (except for the conodonts *per se*), marine, pelagic, bilaterally symmetrical (because groups of conodonts are sometimes found in paired assemblages), and

small. Melton and Scott (1973) identified a "conodont animal" which appears to be a cephalochordate containing conodonts in the gut, presumably as a stomach support. Others have suggested conodonts to be vertebrate gill supports, dermal parts, mollusk parts, and even shark copulatory claspers. More recently, a case was made for their identity as chaetognath (a separate "worm" phylum) grasping spines; however, these were attributed to Cambrian "protoconodonts," which may not be true conodonts (Szaniawski, 1982). A strong, but cautious, case was made by Klapper and Bergstrom (1984) that some primitive Middle Ordovician conodonts are vertebrate teeth. These do show wear and internal structure analogous to true vertebrate teeth. A recent summary by Sweet (1985) included the following ideas: (1) protoconodonts probably are chaetognath grasping spines, but protoconodonts are probably not ancestral (nor closely related) to euconodonts (i.e., "true" conodonts); (2) euconodonts probably represent a distinct phylum, neither chaetognath nor chordate (although there is a likelihood that they are deuterostomes; see also Chapter 4 discussion of metazoan embryonic development); (3) conodonts are indeed food grasping and/or ingesting elements (i.e., they are what they appear to be).

F.9.d. Vertebrates of the Tippecanoe Sequence

The known early record of vertebrates extends back to the middle Late Cambrian, but Ordovician seas were probably the first to contain a sizable population and the Silurian saw the first abundant, diverse fish life. Until latest Silurian time, all vertebrates were jawless, fishlike creatures properly termed Agnatha ("without jaws"), a superclass in current taxonomy (after Moy-Thomas and Miles, 1971).

Classification of Agnatha follows a seemingly simple characteristic: presence of one or two nostrils. Taxa with double nostrils appear first in the record and, surprisingly, also gave rise to higher vertebrates (often it appears the earliest primitive forms also are the ancestral stock for advanced forms). Single-nostril agnathans appear in the Late Silurian, by which time agnathan life seems to have flourished and diversified quite extensively. Double-nostril agnathans, termed Pteraspidiomorphi, comprise two subclasses: Heterostraci (featuring solid dermal-armored heads and a projecting rostrum) and Thelodonti (lightly armored, somewhat flattened forms). Heterostracans are the oldest vertebrates known. Single-nostril agnathans comprise the class Cephalaspidiomorphi, featuring several subdivisions, one of which incorporates living lampreys and two extinct Paleozoic ostracoderm groups, and another subgroup which includes only the living hagfish. The Paleozoic cephalaspids include heavily armored, bottom-living forms called osteostracans, and lightly (or non-) armored forms called

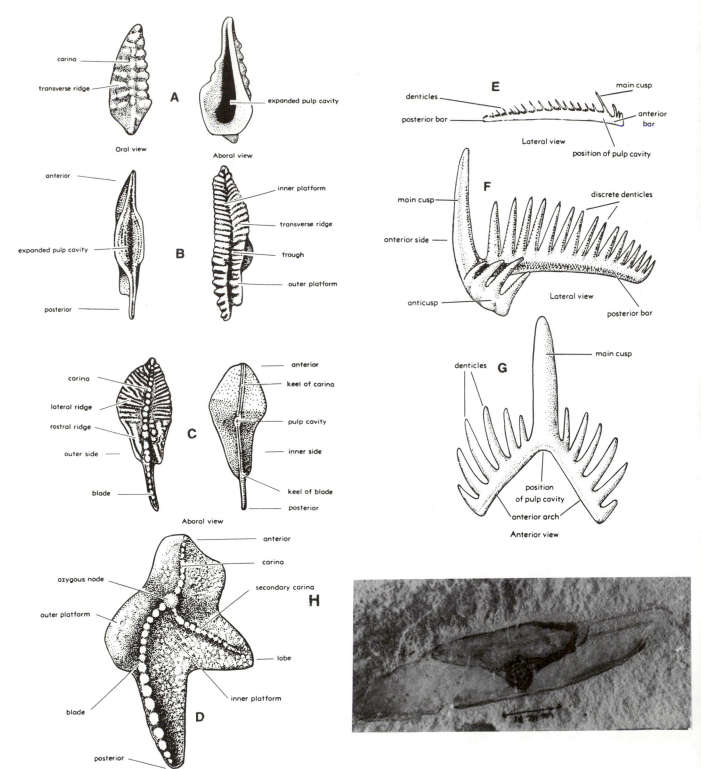

Fig. 5-38. Representative conodonts and a putative "conodont animal." (A, B) Platelike conodonts with expanded pulp cavities. (A) *Icriodus* (Devonian); (B) *Cavusgnathus* (U. Mississippian to L. Permian). (C, D) Platelike conodonts with lateral platforms. (C) *Siphonodella* (L. Mississippian); (D) *Palmatolepis* (U. Devonian). (E–G) Compound barlike conodonts. (E) *Hindeodella* (Silurian to Triassic); (F) *Ligonodina* (M. Ordovician to M. Triassic); (G) *Hibbardella* (M. Ordovician to M. Triassic). (H) A possible "conodont animal" from the Bear Gulch Limestone in Montana (U. Mississippian?). (A–G, from the *Treatise on Invertebrate Paleontology,* courtesy of the Geological Society of America and the University of Kansas Press; H, courtesy of William Melton, University of Montana.)

anaspids. Osteostracans typify the conventional image of an "ostracoderm" (Fig. 5-39).

Details of ostracoderm morphology are beyond our scope; a readable account is presented by Colbert (1980), and progressively more technical descriptions may be found in Romer (1966) and Moy-Thomas and Miles (1971).

Modern agnathans are parasitic and show probable degenerative features to accommodate that habit. For example, the mouth of typical modern Agnatha is reduced to a round sucker (hence the term "cyclostomes"). Paleozoic agnathans were far more fishlike in gross morphology and were almost certainly predatory or deposit-feeding rather than parasitic (besides, there were no other vertebrates to parasitize). Despite their bony armor, all internal skeletons of agnathans are cartilaginous.

The Upper Cambrian agnathan finds are from the Deadwood Formation in Wyoming (Repetski, 1978). This unit is of marine origin and adds fuel to arguments that the earliest Agnatha evolved in marine environments. Proponents of a freshwater origin for vertebrate life point out that agnathans are rare in normal benthic marine shelly facies. It is probable, based on the Cambrian specimens, that some significant evolutionary history preceded their known appearance, since even these oldest fossils are armored, and the oldest agnathans resemble closely forms from the Or-

dovician of northern Scandinavia (Spitsbergen) and a number of additional North American sites. The oldest agnathan specimens are assigned to the genus *Anatolepis*, which appears to be a heterostracan pteraspid; however, the fossils are only bits of dermal armor and biological judgments based on such evidence are clearly tenuous.

Ordovician agnathans are even more widely distributed, with the best known coming from the Harding Sandstone in Colorado. Many additional sites are known (see Darby, 1982), showing that agnathans were very abundant by the Middle Ordovician. The most representative Middle Ordovician agnathan is the heterostracan *Astraspis*, which reached sizes to 1.5 m. The next common appearance of agnathans is in the Middle Silurian of Europe, especially by *Jamoytius*, an elongate, lampreylike creature relatively lightly armored. By Late Silurian time, a diversity of agnathans were present, including the first cephalaspids. Many of these Upper Silurian specimens come from the eurypterid-bearing waterlimes and saline strata described in Section B, suggesting again that early vertebrates did not habitually occupy the normal marine environments. It has also been often proposed that there is a relationship between the armor of early fish and the presence of eurypterids (presumably as prey and predator, respectively). By Devonian time, agnathans were abundant globally, and simultaneously jawed fish appeared (in very latest Silurian time). Chapter 6 will continue discussion of the Devonian "age of fishes."

F.10. Ascent to Land

Thus far in this book, no mention has been made of land organisms. There are no definitive records of terrestrial plants or animals older than Silurian age; however, indirect evidence suggests that plants occupied the land as early as the Ordovician.

F.10.a. Vascular Plants

The algae which comprise most of the Precambrian and Cambrian fossil plant record are exclusively marine organisms, with a fairly simple makeup that has no need for root systems and tissues which can conduct fluids within the plant. Demands of life on the land are generally met by evolution of *vascular* (i.e., containing vessels) tissue to conduct fluids to the nonrooted portions of plants. Hence, "vascular plant" is a term virtually synonymous with "higher land plant." Nonvascular terrestrial plants include bryophytes and fungi; however, both of these groups reproduce by spores, unlike marine algae.

The oldest confirmed terrestrial plant life comes from the Middle Silurian of Wales, as described below, but re-

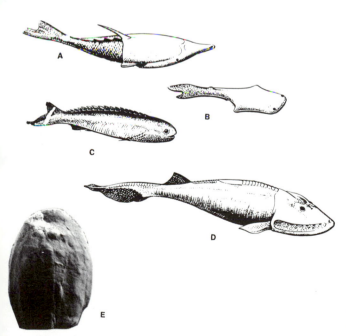

Fig. 5-39. Representative ostracoderms of the Silurian and Devonian. (A) *Pteraspis*, a pteraspid; (B) *Thelodus*, a coelolepid; (C) *Pterolepis*, an anaspid; (D) *Hemicyclaspis*, a cephalaspid; (E) the head shield of *Ariaspis*, a cephalaspid from the Upper Silurian of northern Canada. (A–D, from Colbert © 1980 by John Wiley & Sons, reprinted with permission of John Wiley & Sons; drawings by Lois M. Darling; E, DRS photograph.)

cent evidence of considerably older land plants has been reported from the Lower Silurian of the Southern Appalachians (Sequatchie Formation; Gray and Boucot, 1978), and from the Middle Ordovician of Libya (Gray *et al.*, 1982). In addition, Pratt *et al.* (1978) have reported the occurrence of terrestrial (or freshwater) plants from the lower Silurian of Virginia. These latter reports are based on indirect evidence rather than body fossils (i.e., plant material *per se*) and so are "unconfirmed." The indirect material consists largely of spores and tubes presumed to be part of the vascular systems.

However, the Middle Silurian specimens are the oldest to provide real insight into the earliest land plants. They have come from North America, Wales, and Czechoslovakia (Taylor, 1981) and are assigned to the genus *Cooksonia*, a member of the subdivision Psilopsida of the division Rhyniophyta (in botany, the "division" is approximately equivalent to the animal "phylum").

The psilophyte-type early plants were composed of not much more than a set of dichotomously branching, tapering stems each topped with a round spore-bearing fruiting body called a *sporangium*. The plants lacked true roots but were supported by structures shaped like the *rhizomes* (runners) of later plants.

In the Lower Devonian, a second group of vascular plants appear in strata widespread in Europe, Australia, and Spitsbergen. These plants had sporangia located on the sides of stems rather than at the tops. Like *Cooksonia*, representatives of these groups (such as *Zosterophyllum*) were leafless.

The most diverse set of early plant fossils come from the famous Rhynie Chert of the Lower Devonian of Scotland. Specimens of the characteristic genus *Rhynia* (Fig. 5-40) in these deposits have been very well preserved, perhaps in part because they may have been silicic as are some modern simple plants (such as the "scouring rushes"). By the Middle Devonian, the large number of vascular plant fossils from around the world shows that land plants were well established, with several groups present including the *Rhynia* (or *Cooksonia*) type, the *Zosterophyllum* type, early lycopods, horsetails, and several others. The latter groups mentioned and a wide variety of other Paleozoic plants will be discussed in Chapter 6, because they played a much more important sedimentological role in Mississippian and later times.

It is not clear which marine plant group gave rise to vascular plants but most current thinking assumes the Chlorophyta (green algae) are the ancestors, based on similarities of chlorophylls between the groups and the relative complexity (among algae) of the green algae. In addition, it is very uncertain whether the Ordovician housed the oldest vascular plants or if older fossils will appear.

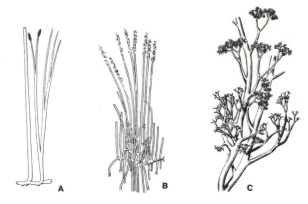

Fig. 5-40. Reconstructions of several Silurian–Devonian land plants. (A) *Rhynia*, a psilopsid genus featuring simple, dichotomous branching and terminal sporangia, with propagation from basal rhizomes. (B) *Zosterophyllum*, a more complex, dichotomously branching psilopsid, bearing sporangia at the sides of stems, and featuring side-extensions from the lower branches. (C) *Psilophyton*, an elaborate derivative of the *Rhynia*-type morphology, with persistent main-stems giving off dichotomously branching side-stems. Many excellent Devonian specimens of *Psilophyton* have been prepared by maceration techniques (see Stewart, 1983), showing a variety of complex morphologies derived from the simple psilophyte pattern. (A and B, from Banks, 1970; C, from Banks *et al.*, 1975.)

F.10.b. Land Animals

The earliest history of terrestrial animals is a subject of great controversy, mostly because the fossil record is very sparse, and because it is often difficult to prove a given organism did or did not dwell on land. For example, did a fossil found in shallow marine environments necessarily live in the water or could it have been washed into the sea? The same holds for nonmarine aquatic environments. Conversely, one may not be sure a fossil found in probable terrestrial sediments (e.g., red beds) was not an aquatic form which rarely ventured on land (like, say, a walking catfish).

For many years, it was claimed that scorpions were first on the land as far back as the Silurian; in more recent years, however, the consensus is that the early scorpions were aquatic, rather like eurypterids (see Shear *et al.*, 1984). Scant evidence of other Silurian terrestrial life exists, but what there is suggests that myriapods (millipedes) were probably the earliest land animals, followed by several additional arthropods by the latest Silurian (Rolfe, 1980). Two very dubious millipedelike forms occur in Silurian strata of England and Scotland, followed by somewhat more certain millipedes in the Old Red Sandstone in Scotland (where it is of Late Silurian age, rather than of Devonian age as in England). By the Early Devonian, one finds in the Rhynie Chert in Scotland several arthropod groups represented: collembolans (springtails; possibly ancestral to insects), mites, primitive (trigonotarbid) arachnids, and a possible primitive spider. Rolfe (1980) also discusses a Devonian arthropod fauna from Aiken an der Mösel, Germany, which includes many of the arthropods mentioned for the Rhynie Chert plus an arthropleurid myriapod (a gigantic group best known from the Pennsylvanian; see Chapter 7). The Aiken fauna also includes eurypterids and other arthropods which are aquatic; therefore, it represents a mixed terrestrial/aquatic fauna such as we mention above as a problem in paleoenvironmental interpretation. By Late Devonian time, insects probably appear (again, the fossils are controversial; see Chapter 6), and the first non-arthropod land animals are present (crossopterygian fish followed almost immediately in time by amphibians).

CHAPTER 6

The Kaskaskia Sequence: Middle Devonian–Upper Mississippian

A. Introduction

A.1. Overview

Post-Tippecanoe emergence and concomitant erosion erased much of the older stratigraphic record on the craton, especially in the area of the Transcontinental Arch. Thus, earliest returning Kaskaskian seas spread over a deeply eroded topography and resulted in the deposition of carbonate strata in basinal areas while uplifts were still emergent. By the end of the Onesquethaw, however, much of the eastern half of the craton was submerged; the western half remained a land area until finally being flooded by a major eustatic sea-level rise during the Taghanic Age. During the Late Devonian, the rate of detrital-sediment influx from adjacent orogens increased along the eastern, northern, and western margins of the craton. At the same time, much of the eastern half of the craton experienced the deposition of an extensive sheet of black, pyritic shale. On the western half of the craton, shallow-marine carbonates and mature quartz arenites continued to accumulate. Sea level fell at the end of the Devonian, resulting in an unconformity over much of the craton. Returning seas of the Mississippian Period left widespread carbonate deposits over much of the craton except along the northeastern craton, where detrital sediments from marginal orogens fed a large coastal-plain which repeatedly advanced into and retreated from the epeiric sea, producing repetitive strata. Cratonic stratigraphy of the Kaskaskian Sequence is complicated by a number of basins and uplifts which were active, but not in concert with each other, at the same time that true, eustatic sea-level changes were occurring (Fig. 6-1).

On the Appalachian continental margin, the oldest deposits of the Kaskaskian Sequence reveal that the Taconic orogen had been almost completely worn down and supplied only small amounts of mature quartzose sediments to the eastern margin of the Appalachian Basin, which was characterized in central and southern areas by deep-water environments. Detrital-sediment influx from the northeast increased during the Middle Devonian and increased even more during the Late Devonian, resulting in a major detrital wedge which prograded entirely across the northern and central Appalachian Basin and out onto the eastern craton. These sediments were derived from highlands along the northeastern margin of the United States uplifted by the Acadian orogeny. This orogeny was apparently caused by collision with Ancestral Europe and was thus a continuation and southwestward extension of the earlier Caledonian orogeny. The Acadian orogeny also caused high-grade regional metamorphism, large intrusions of granite and related igneous rocks, and intense deformation. The Acadian orogen provided detrital sediments in decreasing amounts to the Appalachian Basin throughout the Mississippian, thus supplying a large coastal-plain whose shoreline advanced or retreated according to complex interrelationships between sediment supply, basin subsidence, and eustatic fluctuations of sea level. A major rise of sea level occurred during the middle Mississippian which left carbonate deposits over virtually the entire area. Lowering of sea level and increase in detrital influx toward the close of the period caused the coastal plain to advance once more. In the Alabama area, the Middle Mississippian carbonate platform was invaded by deltaic molasse from the south, heralding the onset of tectonic activity on the southern continental margin.

Devonian strata of the Ouachita and Marathon regions reveal that the southern continental margin at this time was

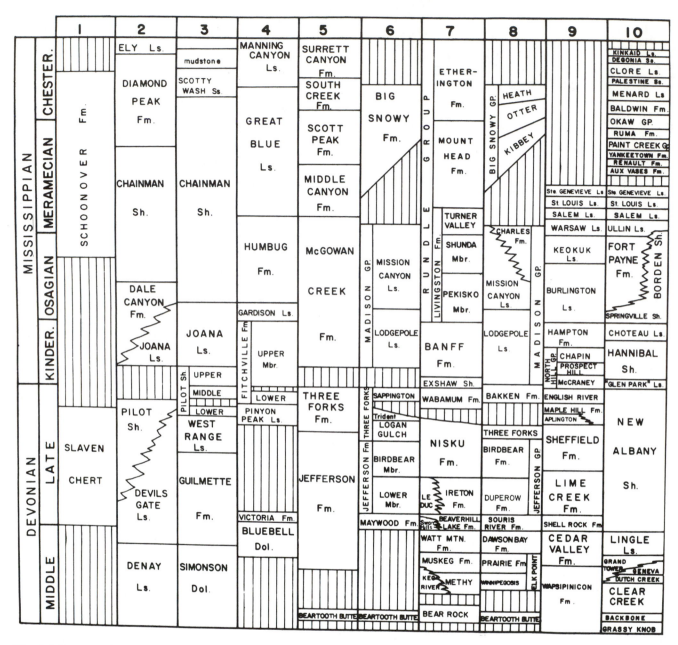

Fig. 6-1. Representative stratigraphic columns of Kaskaskian deposits; most columns are synthetic, having been assembled from multiple sources and from different localities within the area represented. Areas are as follows: 1, westcentral Nevada; 2, Diamond Mountains, Nevada; 3, Pahranagat Mountains, Nevada; 4, East Tintic Mountains, Utah; 5, central Idaho; 6, Montana; 7, Alberta, Canada; 8, Williston Basin; 9, Iowa; 10, Illinois Basin; 11, Michigan Basin; 12, eastern New York; 13, Pennsylvania; 14, Virginia; 15, Tennessee; 16, Georgia and Alabama; 17, Ouachita Mountains, Oklahoma and Arkansas; 18, Arbuckle Mountains, southcentral Oklahoma; 19, central Brooks Range, northern Alaska; 20, Bathurst Island, northern Canada.

passive and that a deep, starved, oceanic basin existed farther to the south. During Mississippian time, a north-facing magmatic arc advanced slowly from the south. This arc, which was probably oriented obliquely to North America's southern margin and which may have been associated with the South American coast of Gondwanaland, was relatively close to the southeastern Appalachian area by the end of the Mississippian, resulting in the molasse sediments of the Alabama area. Arc-derived sediments shed into the southern ocean-basin built a large submarine fan which grew westerly to northwesterly, resulting in the thick, shaley flysch of the Ouachita–Marathon region.

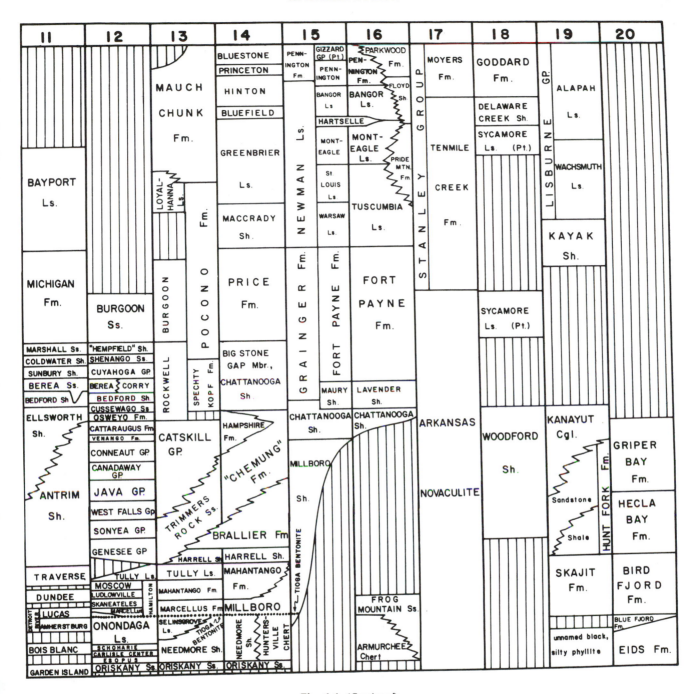

Fig. 6-1. *(Continued)*

On the Cordilleran continental margin, the lowest deposits of the Kaskaskia Sequence indicate that the continental shelf was the site of carbonate deposition on a broad, shallow platform which dropped off, on its western margin, into a deep marginal or oceanic basin. Carbonate turbidites, derived from the shelf, were deposited on the eastern side of this basin and graded into fine, pelagic deposits of the basin center. Farther west was a magmatic arc and related subduc-

tion complex which shed volcaniclastic sediment onto an archipelagic apron which grew eastward into the basin. During the latest Devonian, the eastern part of the basin was thrust up onto the continental margin. This event, termed the Antler orogeny, resulted in a tectonic landmass which was separated from the carbonate platform of the inner shelf by a deep, foreland basin on the outer shelf. The Antler tectonic land supplied coarse detrital sediments to the foreland basin throughout the Mississippian and the magmatic arc and intervening marginal basin continued to exist to the west. Carbonate deposition on the shelf persisted throughout the Mississippian.

The Innuitian continental margin in the Middle Devonian consisted of a carbonate platform which dropped off into a deep, marginal basin bordered to the north by a magmatic arc. Uplift in the arc resulted in a deep intra-arc basin and the deposition of thick flysch deposits in the marginal basin. During the Late Devonian, the Siberian plate collided with the magmatic arc, resulting in the first phase of the Ellesmere orogeny and the formation of a major detrital wedge which prograded onto the carbonate platform of the craton. The second orogenic phase occurred when the direction of subduction reversed following the arc–continent collision; the marginal basin was subducted northward beneath the deformed margin of the Siberian continental-block so that finally, in the Early Mississippian, the margin of North America collided with and partly underthrusted the Siberian plate. Gradual uplift following cessation of plate motion raised the entire northern part of the continent above sea level so that no sediments were deposited there throughout most of the Mississippian. In Alaska, however, Mississippian transgressive sequences reveal the flooding of the old, eroded orogen and the establishment of a carbonate platform over the whole area by Late Mississippian time.

In the history of life, too, Kaskaskian strata contain evidence of changes. The Devonian and Mississippian Periods saw the first great intertaxon extinctions and the succession of pelmatozoans to a place of prominence in the shallow marine realm. During this time, also, major evolutionary advances occurred in the vertebrates: primitive, bottom-dwelling ostracoderms and the earliest jawed fish evolved into many groups of cartilaginous and bony fish, which rapidly radiated to fill diverse ecological niches. By the end of the Devonian, some fish, such as the crossopterygians, had begun to venture onto the land and by the end of the Mississippian, amphibians had populated a major new niche. It was during the Devonian and Mississippian, too, that vascular plants took over the land, evolving rapidly from the simple Silurian forms to the arboreal lycopods, sphenopsids, and seed ferns of the Late Devonian and Mississippian.

A.2. Global Paleogeography

Figure 6-2 is a paleogeographic map of the Earth during the late Early Devonian (i.e., the time immediately after the first advance of the Kaskaskian sea). The main part of North America lay between 10°S and 30°N latitude and thus enjoyed a seasonless, equatorial climate, except in northernmost areas which were probably tropical to subtropical. Being directly at the equator, rainfall would have been heavy and chemical weathering intense in the northeastern region of the United States, which was the site of the Late Devonian Acadian Mountains; thus, abundant clays were available to a muddy coastal plain which advanced rapidly

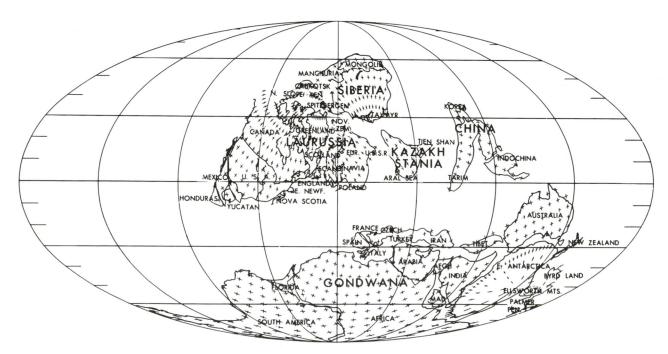

Fig. 6-2. A reconstruction of global paleogeography during the Early Devonian. (From Scotese *et al.*, 1979.)

into the epeiric sea. Atmospheric circulation would have been dominantly westward, a speculation supported by the distribution of volcanic ash beds, such as the Tioga Bentonite. Generalizations about wind patterns in northern North America are probably useless because the Caledonian orogen lay squarely across the belt of Northeast Trade Winds and would likely have greatly disrupted their regular pattern.

North America formed a single landmass, called Laurussia, in conjunction with the Ancestral Europe block with which it had collided during the Silurian. Gondwanaland was far to the south, separated from Laurussia, Kazakhstania, and Ancestral China by the Theic Ocean, which probably had a well-developed, westward-setting, surface current similar to the South Equatorial Current today. This current would have strongly affected the dispersal of organisms with planktonic juvenile or adult forms. Southern Africa and South America were probably covered by a large continental glacier at the South Pole. The North Polar region, on the other hand, was unconfined, open ocean.

Figure 6-3 is a paleogeographic map of the Earth during the Late Mississippian. The position of North America had changed very little since the Early Devonian and, thus, major environmental factors remained about the same. Perhaps the most important observation from the map is that Gondwanaland had moved considerably northward, almost completely closing the Theic Ocean. Other plate motions shown had little effect on North America.

B. Kaskaskian Events on the Craton

B.1. The Wallbridge Discontinuity

At the end of Helderbergian time, virtually the entire North American continent was emergent. Erosion stripped off earlier strata almost everywhere, resulting in the Wallbridge Discontinuity, which separates strata of the Tippecanoe sequence from those of the Kaskaskia sequence. Helderberg strata, having been deposited just prior to the emergence, were especially affected by subaerial erosion and are thus found only in basinal areas. On some uplifted areas, such as the Transcontinental Arch, the entire Silurian, Ordovician, and even Cambrian stratigraphic record was removed so that Kaskaskian units of these areas lie directly on Precambrian rocks. Even basinal areas were affected (though not as severely), with erosional unconformities underlying upper Lower Devonian strata in both the Michigan and Williston basins. Only in the Illinois Basin was deposition continuous across the Tippecanoe–Kaskaskia boundary. There, cherty limestone of the Bailey Formation (Helderberg) is conformably overlain by Deerpark-age Grassy Knob Chert (Collinson, 1967).

Over much of the craton, the main pre-Wallbridge lithologies were carbonates whose weathering yielded little coarse detrital sediment. Thus, returning Kaskaskian seas did not, in general, leave widespread basal, quartz-sand sheets as did the previous Sauk and Tippecanoe seas. Only

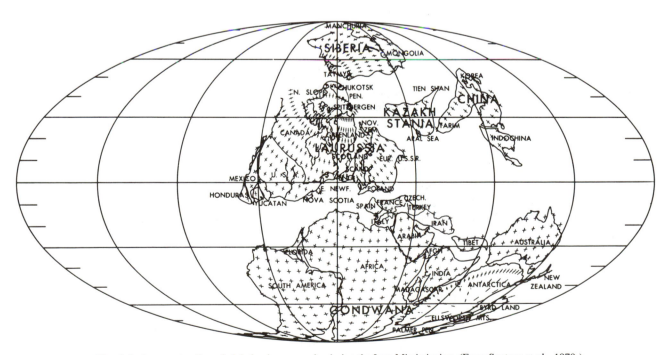

Fig. 6-3. A reconstruction of global paleogeography during the Late Mississippian. (From Scotese *et al.*, 1979.)

in the Appalachian Basin do important quartz-sand deposits (the Oriskany and Ridgeley Sandstones) lie at the base of the Kaskaskia Sequence.

B.2.   Structural Framework of Kaskaskian Cratonic Sedimentation

The distribution and facies patterns of Kaskaskian strata reveal that extensive epeirogenic activity affected virtually the entire craton throughout Devonian and Mississippian time. Of course, important cratonic basins and uplifts existed before deposition of the Kaskaskia Sequence, especially the Michigan, Illinois, Appalachian, and Williston (Interlake) basins and Findlay, Kankakee, Cincinnati, and Transcontinental arches. But, beginning in the Lower and Middle Devonian, the craton was subjected to increasingly greater warping, resulting in many new structural features which, along with the older ones, affected sedimentation in the Kaskaskian sea (Fig. 6-4). Some of the newer structures were only active during the Devonian, such as the Forest City Basin which barely lasted till the Mississippian. Others were significant features throughout the rest of the Paleozoic. Indeed, several of these features, e.g., the ancestral Uinta and Front Range uplifts, portended much greater activity during the Mesozoic Era.

B.3.   Early Devonian Events

The initial rise of the Kaskaskian sea began during Deerparkian time but strata of this age are very limited in their occurrence, being found primarily in basinal areas. For example, the Grassy Knob Chert of the Illinois Basin is overlain by the Backbone/Little Saline Formation, a cherty crinoidal limestone whose bioclasts were probably carried into the basin from shallow crinoid "meadows" which formed on the flanks of the Wabash Platform (a positive area formed by the intersection of the Kankakee, Findlay, and Cincinnati arches; Droste *et al.,* 1975). In the Arbuckle Mountains of Oklahoma, carbonate strata of the Frisco Formation are considered to be of Deerparkian age.

Rise of sea level continued into Onesquethawan time. In the Michigan Basin, the Bois Blanc Formation was deposited. Micritic limestone of the Bois Blanc is found in central and northern areas and cherty dolomicrite dominates around the eastern, southern, and western margins, where sediments were affected by various diagenetic processes associated with partial, intermittent exposure. Correlative strata in the Illinois Basin comprise the Clear Creek Chert. In northeastern Oklahoma, the Ozark region of Arkansas, and northern areas of the Ouachita Mountains, cherty, fossiliferous limestones and dolostones of Bois Blanc age have a variety of local names such as Sallisaw Formation, Pen-

ters, and Pinetop Cherts. The oldest unit of the Forest City Basin, the LaPorte City Chert is believed to be of early Onesquethawan age because of its stratigraphic position below the Wapsipinicon Formation and its cherty character.

The above discussion reveals that chert and cherty carbonates are a major component of lower Onesquethawan strata. This observation is strengthened by the occurrence of the Huntersville Chert of the same age in the western Appalachian Basin. We believe that this somewhat anomalous abundance of chert can be explained by reference to Knauth's (1979) model of silicification whereby subsurface mixing of fresh and marine waters provides the proper geochemical environment for the replacement of calcite by silica. Thus, lower Onesquethawan strata may contain abundant chert because rising seas had not yet drowned topographic highs, such as the Cincinnati Arch, which acted as catchment areas to supply freshwater to aquifer systems in which mixing and silicification occurred.

In the Illinois and Michigan basins, an unconformity separates Lower Devonian from Middle Devonian strata. This seems to have been caused by a brief retreat of the sea (Collinson, 1967). An unconformity occurs at the same stratigraphic position in western Canada, suggesting that the relative sea-level drop which resulted in these erosional surfaces might have been eustatic in nature. On the other hand, a correlative unconformity is *not* present in the Appalachian Basin (Dennison and Head, 1975). Such "disagreements" between the stratigraphic records of cratonic basins are not unusual; while the Kaskaskian was, as we have said, a time of common epeirogenic activity, basins (and uplifts) rarely acted in concert.

Throughout the entire Early Devonian, the western part of the craton continued to be emergent. Karst topography developed in regions underlain by carbonates while other areas were characterized by well-developed stream-drainage systems. Scattered, thin, terrestrial deposits of this age are termed the Beartooth Butte Formation in the United States and the Ashern in Canada.

Perhaps the thickest post-Wallbridge, Lower Devonian cratonic strata in North America occur in northwestern Canada in the Hay River and Elk Point basins (maximum thickness in the Elk Point Basin is approximately 375 m). To the west of this area, a large carbonate platform had developed during Helderbergian time (or even earlier) on the outer continental shelf. This platform restricted normal marine circulation to cratonic areas behind it, allowing the concentration of seawater into saline brines from which gypsum (or anhydrite) and halite were precipitated. These evaporitic deposits and associated dolostones and red beds comprise the lower Elk Point Group in the Elk Point Basin and the Bear Rock Formation in the Hay River Basin and on the Anderson–Great Slave shelf (see Fig. 6-5).

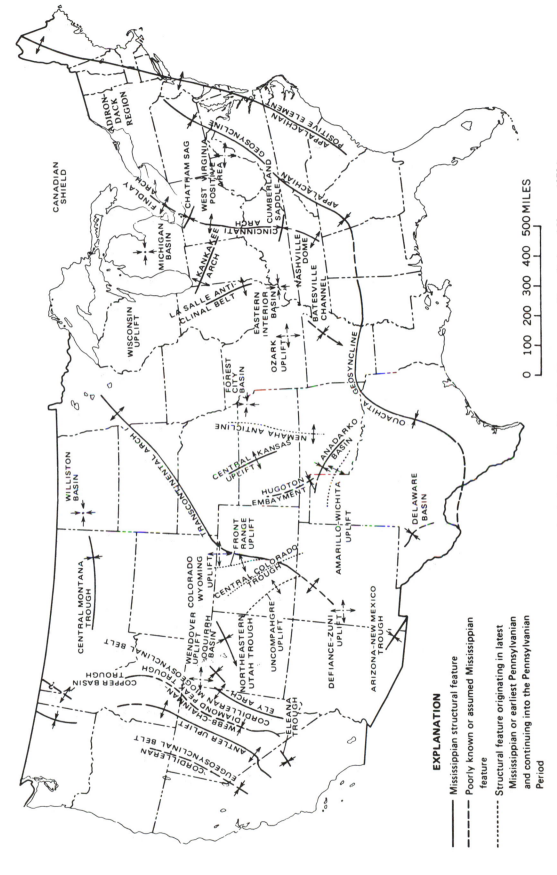

Fig. 6-4. Major structural features active during deposition of the Kaskaskian sequence. (From Craig and Varnes, 1979.)

EXPLANATION

—— Mississippian structural feature

— — — Poorly known or assumed Mississippian feature

········ Structural feature originating in latest Mississippian or earliest Pennsylvanian and continuing into the Pennsylvanian Period

0 100 200 300 400 500 MILES

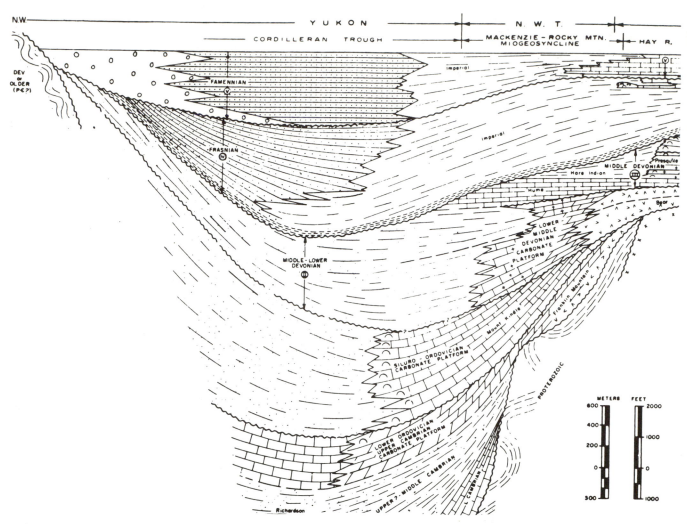

Fig. 6-5. Generalized stratigraphic cross section of Devonian strata in western Canada from Northern Yukon to Williston Basin. (From Bassett and Stout, 1967.)

B.4. Middle Devonian Events

By the beginning of the Middle Devonian (middle Onesquethawan), the sea had returned to the Illinois and Michigan basins (as well as to the western Canada basins; see below). The event was marked in the Illinois Basin by deposition of the quartz-arenitic Dutch Creek Sandstone, which is the lower member of both the Geneva Dolomite in the northern part of the basin and the Grand Tower Limestone in the southwest.

By late Onesquethawan time, continued marine transgression had flooded much of the eastern and central craton. Over this entire area, a continuous sheet of carbonate sediment was deposited. In the Illinois Basin, the Geneva Dolomite, Grand Tower, and Jeffersonville limestones were part of this sheet. The Jeffersonville Limestone is well-

known among students of North American historical geology because of its abundantly fossiliferous and well-exposed bioherms at the famous "Falls of the Ohio River" collecting locality. In the Michigan Basin, the upper Onesquethawan carbonate sheet is represented the Detroit River Group, consisting of Amherstburg and Lucas formations. Amherstburg facies patterns resemble those of the older Bois Blanc Formation except that large stromatoporoid bioherms (some up to 27 m thick and over 160 km² in area) developed along marginal areas of the basin associated with dolomitic strata. Continued growth of Amherstburg bioherm-complexes, especially in the southeastern part of the basin near the Chatham Sag (which functioned at that time as a connection between the Michigan and Appalachian basins), increasingly restricted free circulation with normally saline waters of the Appalachian Basin and thus re-

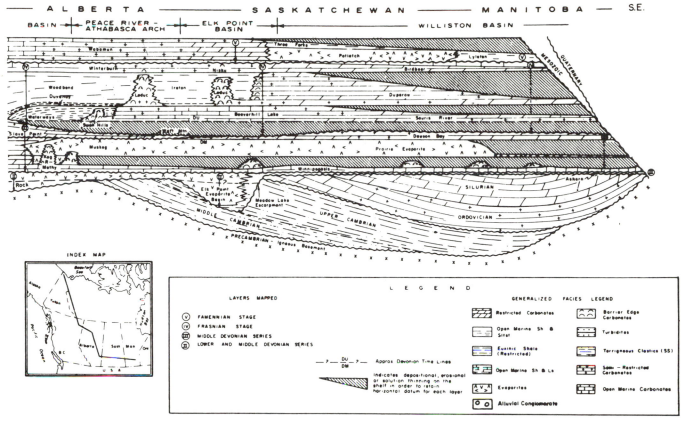

Fig. 6-5. (*Continued*)

sulted in deposition of evaporitic sediments of the Lucas Formation in the basin center. In the Forest City Basin, the Wapsipinicon Formation is of late Onesquethawan and Cazenovian age and is composed of a complex sequence of limestone, dolostone, chert, and commercially important gypsum deposits. Probably, the highly saline waters of the Michigan and Forest City basins were partially connected but were separated from the normally saline waters of the Illinois Basin by the Wabash platform and the Sangamon Arch (see Fig. 6-6).

Much of the western part of the craton continued to be emergent throughout most of the Middle Devonian (except for western Canada). This large land mass must have been relatively low because it supplied only minor amounts of mature quartz sediment to the continental margin.

The middle Onesquethawan transgression in the East was mirrored by a return of the seas in the West; as in the East, even more of the craton was flooded at this time than in the Early Devonian: the sea extended across the Meadow Lake escarpment and spread into the Williston Basin (see Fig. 6-7). It lapped far up onto the peninsula formed by the Peace River and Western Alberta uplifts whose erosion pro-

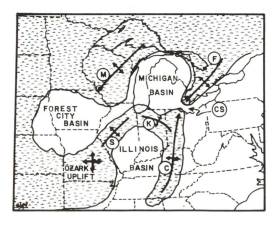

Fig. 6-6. Middle Devonian paleogeography of the eastern craton in the United States during deposition of the Lucas and Wapsipinicon Formations.

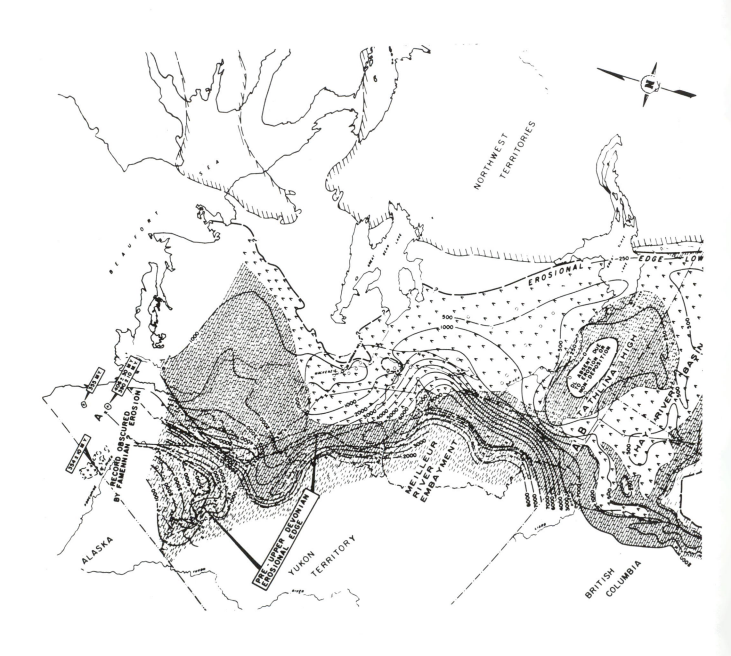

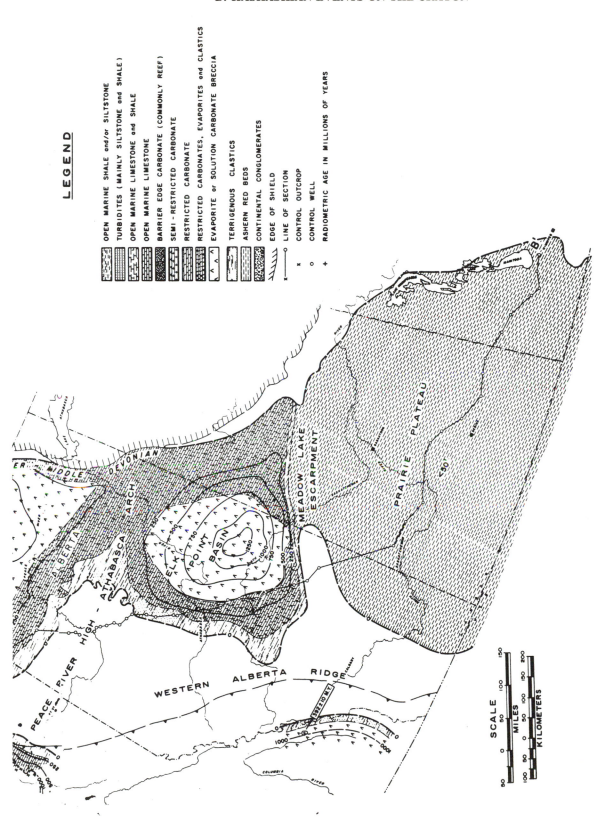

Fig. 6-7. Early Middle Devonian paleogeography of western Canada. (From Bassett and Stout, 1967.)

duced sediments which were transported both westward into the narrow sea on the continental shelf (see Section E on the Cordilleran continental margin) and eastward into the epeiric sea. These sediments built up fluvial and deltaic red-bed sequences named the Yahatinda Formation on the eastern part of the peninsula. Over much of western Canada, a broad shallow carbonate-platform was formed. North of the Peace River Uplift, platform strata are named the Hume Formation; south of there, they are called the Methy Formation (Alberta Basin) and Winnipegosis Formation (Williston Basin). Across this vast platform, circulation must have been good for there are relatively few evaporite deposits, even in the Williston Basin which was farthest away from the western ocean. Spectacular bioherms developed in several areas, especially in the Williston Basin, where they grew both along basin margins as well as in the basin center. Even larger bioherms grew in the area north of the Peace River Uplift, where they comprise the Keg River Formation.

The later Middle Devonian sea persisted, despite scattered epeirogenic activity, over most of the areas flooded in middle Onesquethawan time. In the Michigan Basin, the Dundee Formation unconformably overlies the Detroit River Group and consists of cherty limestones and dolostones whose facies patterns are similar to those of the older Bois Blanc Formation. In the Illinois Basin, deposition was continuous from Onesquethawan into Cazenovian time; limestone deposits (e.g., Lingle, North Vernon, Sellersburg formations) indicate nearly open marine conditions. In the Forest City Basin too, deposition was continuous and the Cedar Valley Formation (Tioughniogan and Taghanican age) overlies the Wapsipinicon conformably. The Cedar Valley wedges out against the Cambridge and Sioux arches and the Ozark Uplift but overlaps the Sangamon Arch, suggesting (at least by the end of the Middle Devonian) a partial connection between the Forest City and Illinois basins. The Cedar Valley is a complex unit with carbonates and evaporites divided into several members, all of which indicate deposition in a semi- to fully-restricted basin. Cedar Valley strata pass by facies change across the Wisconsin Arch into the Traverse Group of the western Michigan Basin, a sequence of argillaceous limestones which themselves grade eastwards into shales and mudstones of the Hamilton Group in the eastern part of the basin. These detrital sediments represent the massive introduction into the Michigan area of detritus derived by weathering of the Acadian Mountains far to the east and carried across the northern Appalachian Basin in the prograding Hamilton delta plain (see Section C on Kaskaskian events in the Appalachian Mobile Belt).

In western Canada, bioherm growth continued along the northwestern edge of the carbonate platform. The result was a shelf-edge barrier-reef complex (called the Presqu'ile barrier, Fig. 6-8) whose rapid growth began to close off inlets into the great epeiric sea behind it, restricting circulation and allowing the onset of evaporite deposition. The forward edge of the Presqu'ile barrier sloped steeply off into deep water in which argillaceous muds of the Hare Indian Formation were deposited. Behind (i.e., to the southeast of) the Presqu'ile barrier, progressively more restricted environments developed, ranging from nearly normal marine limestones immediately adjacent to the barrier in the Hay River Basin to limestones, dolostones, and evaporites in the Alberta and Williston basins. In the Williston Basin, there is a gradational contact between normal marine carbonates of the Winnipegosis Formation and thick sequences of anhydrite, halite, and potassium salts (such as sylvite and carnallite, whose presence indicates extreme aridity and evaporation) which comprise the Prairie Formation. These evaporites formed first in deep areas between pinnacle reefs of the Winnipegosis Formation (whose growth was probably terminated by the highly saline water) and then continued to be deposited, resulting finally in thicknesses of almost 200 m. Similar evaporite deposits of the Alberta Basin are named the Muskeg Formation. Following deposition of Prairie and Muskeg evaporites, the salinity of the epeiric sea decreased, possibly due to a minor rise of sea level; the result was a series of transgressive carbonate units, such as the Sulphur Point Formation near the Presqu'ile barrier, the rhythmically interbedded carbonates and evaporites of the Dawson Bay Formation in the Williston Basin. Sea level dropped once more near the end of the Middle Devonian, resulting in the erosional unconformity which occurs beneath Upper Devonian strata in most areas.

B.5. Late Devonian Events

The unconformity which separates Middle Devonian from Upper Devonian strata extends over much of the craton and therefore represents a major period of emergence. In many areas, erosion developed a pattern of stream channels which were flooded during rising sea-level at the beginning of Late Devonian time so that channel sands are overlain by estuarine deposits. In other areas, karst topography developed, such as in the Forest City Basin where a major cavern system, localized at the Wapsipinicon–Cedar Valley contact, formed during the pre-Upper Devonian emergence; these caverns were filled during the early Late Devonian by cave muds, which today comprise the Independence Shale.

B.5.a. The Eastern Craton

As will be discussed in the section on Appalachian geological events, the Late Devonian Epoch saw the progradation of a massive detrital-sediment complex (the Acadian detrital wedge) across most of the northern and central Appalachian Basin. Suspended muds from this sediment-dispersal system were carried far out into the epeiric sea so that Upper Devonian strata over most of the eastern half of the craton are composed largely of shales.

In the Michigan Basin, dark-gray to black, bituminous, pyritic shales of the Antrim Shale were deposited. In the Illinois Basin, similar deposits of black shale are named the New Albany Shale, which is continuous to the north with the Antrim and to the south and southeast with the similar Chattanooga Shale. Thus, these strata represent a thin sheet of dark mud which was deposited over most of the eastern craton. In the southern midcontinent area, too, most of the Upper Devonian section is composed of black shale, called the Chattanooga Shale in Oklahoma and Arkansas and the Woodford Shale in Texas. These units all have a restricted fauna composed primarily of conodonts and rare plant fragments; virtually no benthic organisms are present. In addition, these shales have a relatively high uranium content.

The restricted fauna of these shales, their dark, bituminous, pyritic nature, and the relatively high uranium content all indicate deposition in a strongly reducing (i.e., euxinic) chemical environment. Such an interpretation, however, presents a major problem for paleoenvironmental reconstructions because analogous chemical environments today are of only local extent and it is difficult to envision how an area as large as the eastern half of the craton could have sustained such oxygen-free conditions. One explanation of this dilemma is that the shales were deposited in very deep water but sediments from the western craton do not support this idea, and the requirement that the eastern craton be depressed deeply at the beginning of the Late Devonian and then released to pop back up at the end of the Devonian seems unrealistic. Another suggestion is that the Late Devonian epeiric sea was highly saline and therefore had a well-developed density stratification which did not allow overturn of the water column by wind-generated waves and currents and consequent oxygenation of the bottom. It has also been suggested that the Late Devonian sea was choked with planktonic marine vegetation which used up most of the oxygen dissolved in seawater so that anaerobic conditions existed even in a relatively well-mixed sea. This suggestion is intriguing but it does not explain why only Upper Devonian strata were thus affected. Further study of this fascinating problem is needed.

B.5.b. The Southwestern Craton

During the latest Middle Devonian and early Late Devonian, sea level rose and the sea flooded the craton from the west. This sea-level rise affected much of the western craton and has been correlated with the transgression which resulted in deposition of the Tully Limestone in the Appalachian Basin. This transgressive sequence, termed the Taghanic onlap by Johnson (1970), can be recognized in many parts of the craton and therefore probably records a major eustatic event. Transgressive near-shore and channel-fill sediments of the Maywood Formation in Montana and the Beckers Butte Sandstone of Arizona probably represent initial deposits of the Taghanic onlap. In the west, facies patterns were shifted eastward over 200 km so that the old, emergent western craton of the Lower and Middle Devonian was almost completely drowned. Initial transgressive deposits of Senecan age consist of channel sands with abundant fish remains overlain by estuarine deposits. With rising sea level, the sea soon flooded into large embayments in central Utah and in northern Mexico, resulting in carbonate sediments such as the Jerome Formation and the Temple Butte Limestone. In northwestern Arizona, sands and muds of the Chaffee Formation are interbedded in a cyclic repetition of beds, probably recording periodic, minor fluctuations of the sea. Bioherms were widespread in the warm, shallow sea and abundant marine fossils, especially coral and brachiopods, may be found in Upper Devonian strata of the western craton.

By early Chautauquan time, sea level began to drop and regressive sequences were deposited (e.g., the Ouray Formation of the Four Corners area of Colorado, Utah, Arizona, and New Mexico). In northern Utah, the Stansburg anticline was active and stood as a promontory or peninsula on the shore, jutting out into the western sea. Erosion of this peninsula produced coarse, immature sediments of the Stansburg Formation. Local epeirogenic activity increased during the late Late Devonian, possibly related to onset of the Antler orogeny on the western continental margin. Many small basins and uplifts were active and the resulting strata contain many minor, noncorrelative disconformities.

B.5.c. The Northwestern Craton

i. Senecan Events. The Taghanic onlap is also seen in deposits of the northwestern craton, where the returning sea spread even farther southward than it did during the Middle Devonian, depositing marine strata over almost all of western Canada and the northern Rocky Mountain region of the United States (see Fig. 6-9). In Montana, transgressing seas from the Williston Basin joined with a sea simultaneously transgressing eastward from the western continental margin. In the Hay River Basin, deep marine shales and carbonates of the Waterways Formation were deposited, grading northward into black, siliceous shales of

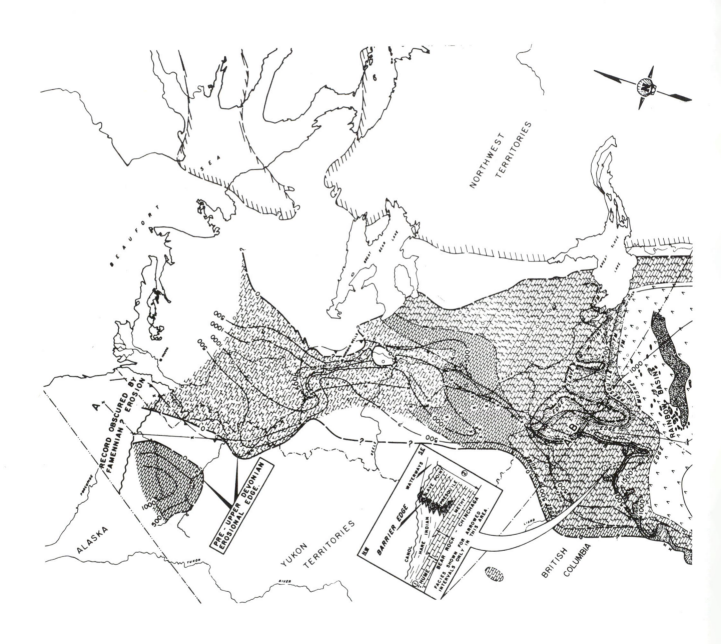

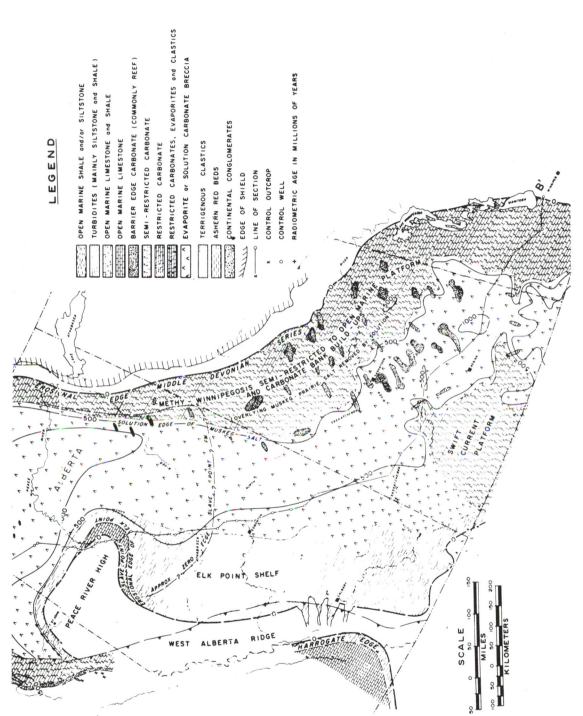

Fig. 6–8. Later Middle Devonian paleogeography of western Canada; insert shows inferred cross section of Presqu'ile barrier-reef stratigraphy. (From Bassett and Stout, 1967.)

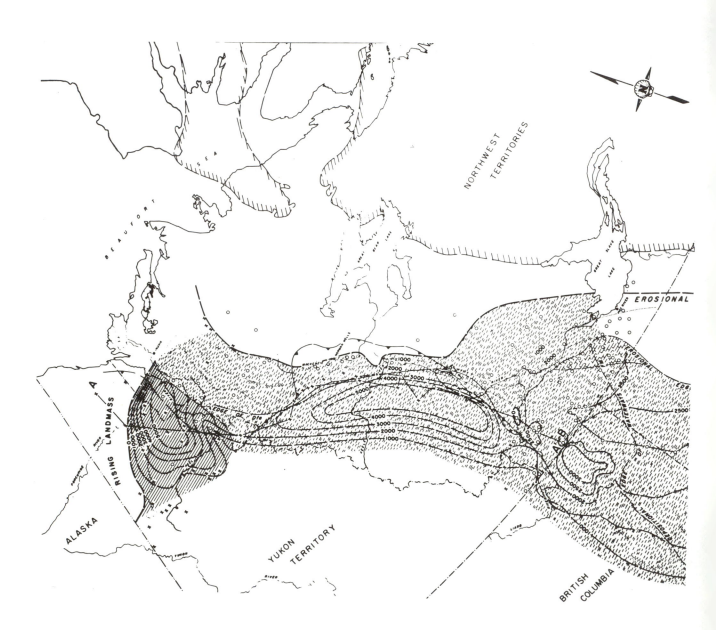

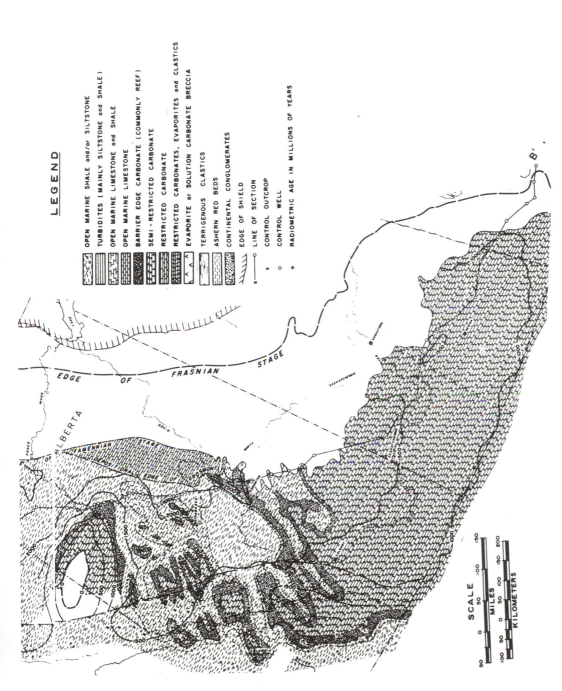

Fig. 6-9. Senecan paleogeography of western Canada. (From Bassett and Stout, 1967.)

the Canol Formation. Waterways shale is bounded to the south by Swan Hills Limestone in the subsurface of central Alberta, associated with the Peace River Uplift, and the Flume Formation which outcrops in the Front Ranges of Alberta's Rockies. The Swan Hills and Flume constitute a major bioherm complex which formed at the margin of a shallow carbonate platform extending far to the south, into the Williston Basin and adjacent Montana. Deposits of this platform are named the Beaverhill Lake Group in the Alberta Basin and the Souris River Formation in the Williston Basin and Montana. The Beaverhill Lake consists of essentially normal platform carbonates while the Souris River is composed of cyclic deposits of shale, limestone, dolostone, and anhydrite.

Following continued sea-level rise and drowning of the Swan Hills reef and Beaverhill Lake platform, a new, more extensive reef system was developed in central Alberta and around the Peace River Uplift. This large reef system, called the Le Duc reef, was initiated as relatively small bioherms associated with platform carbonates of the Cooking Lake Limestone. As sea level rose, bioherms grew upward rapidly, resulting in spectacular structures rising up through over 100 m of water. The lack of associated oolitic sediment probably indicates relatively quiet water with little tidal influence, a condition typical of epeiric seas. Where these bioherms crop out in the Canadian Rockies, they may be seen to change, within only a few hundred meters, from an obvious bioherm facies to a fore-reef facies containing micritic and bioclastic beds with coarse reef-talus debris (Wilson, 1975). These reef-talus beds extend several kilometers out into the basin, grading into dark siltstones and shales which themselves grade northward into distal turbidites of the Imperial Formation. Behind the Le Duc reefs, another broad and partially restricted carbonate platform was established whose deposits comprise the Duperow Formation in the Williston Basin and Montana. Like the Souris River Formation, the Duperow consists of cyclically interbedded carbonates and evaporites.

Near the end of the Senecan Age, regression resulted in the Birdbear Formation in the Williston basin and Montana, the Nisku Formation in the Alberta Basin, and the Winterburn Group in the Peace River uplift region. Continued lowering of sea level resulted in an unconformity which is present over most of the northwest craton area.

ii. Chautauquan Events. At the beginning of the Chautauquan Age, sea level rose once more, resulting in a transgressive sequence over much of western Canada and the northern U.S. Rockies. Detrital sands eroded from the Canadian Shield were carried westward and deposited at the shore of the transgressing sea, resulting in the Sassenach and Trout River formations in the Hay River and Alberta

basins. As during earlier transgressions, a shallow carbonate platform was developed in the Alberta Basin but large platform-edge or central-platform bioherms were absent. Rather, shallow platform carbonates in the western Alberta Basin (Wabamum and Palliser Groups) pass gradationally westward into deep-water shales of the Imperial and Besa River formations of the western continental margin and adjacent marginal basin. To the east, Wabamum and Palliser carbonates grade into restricted carbonate sequences and evaporite deposits of the Potlach and Settler formations in eastern parts of the Alberta Basin. These units are correlated with the Logan Gulch Member of the Three Forks Formation in Montana which contains approximately 150 m of evaporitic strata. Evaporites within the Logan Gulch are best developed in western Montana and grade eastward into silty dolostone and dolomitic siltstone and sandstone. This tendency for detrital influence to increase to the east is also seen in Canada where Potlach/Settler strata pass eastward into sands and shales of the Qu'Appele and Lyleton formations of the Williston Basin.

As in the Southwest, the end of the Devonian was marked by epeirogenic fragmentation of the area into numerous, small, restricted basins, especially in the northern U.S. Rocky Mountain region. While specific details of their stratigraphy vary from basin to basin, they all contain dark gray to black shales called the Exshaw Formation (in northwestern Montana, Idaho, and adjacent Alberta), the Bakken Formation (in the Williston Basin), the Sappington Member of the Three Forks Formation, and so on. These shales probably represent euxinic conditions in deep basins restricted from open-marine influence by lowering of sea level. A major disconformity separates Devonian from Mississippian strata.

B.6. Kinderhookian Cratonic Strata

Sea level had dropped at the end of the Devonian, resulting in an erosional unconformity over much of the craton, except in basinal areas such as the Michigan and Illinois basins and foreland basins associated with the Acadian and Antler orogenies. Almost immediately following this brief emergence, sea level rose again causing the deposition of transgressive sequences over most of North America. Figure 6-10 is a paleogeographic map of the United States during the Kinderhookian.

In the East, detrital sediments continued to be deposited on the northeastern craton and in northern portions of the Appalachian Basin. Much of this sediment was derived from the Acadian highlands to the east. But the Acadian detrital wedge was only one of three major sediment-dispersal systems related to continental-margin orogenies. Converging detrital wedges from the Caledonian orogen of East

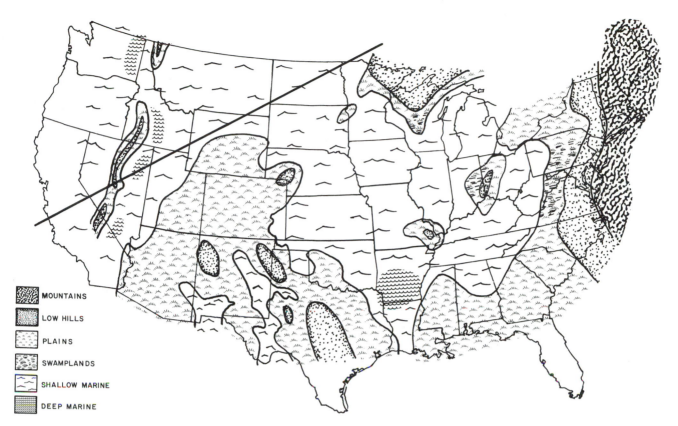

Fig. 6-10. Kinderhookian paleogeography of the United States. (Modified from Craig and Conners, 1979.) Straight, heavy line represents Mississippian equator.

Greenland and the Ellesmere orogen of the Innuitian continental margin as well as from the Acadian mountains had, by early Kinderhookian time, developed a great coastal plain on the entire northeastern part of North America. These three orogenic belts formed a roughly semicircular mountain-barrier along the northern and eastern margins of the continent and provided immature detritus which was carried in major river systems (such as the Ontario river) to growing deltas which prograded into the northern Appalachian Basin (the Red Bedford delta) and the eastern Michigan Basin (the Thumb delta; see Fig. 6-10). Sands of both deltas are named the Berea Sandstone, associated muds comprise the Bedford Shale. In the Michigan Basin, these deltaic sediments grade westward into contemporary, normal-marine deposits of the Ellsworth Shale and black muds of the Antrim Shale in the basin center. Falling sea level later in Kinderhookian time further restricted the Michigan Basin, and black, euxinic, Antrim-like muds of the Sunbury Shale were deposited over most of the basin.

Sediments deposited in the eastern epeiric sea were also derived from the Wisconsin uplift (the northeastern extension of the Transcontinental Arch which was a positive area throughout the Mississippian), the Ozark Uplift, and, to a lesser extent, the Cincinnati Arch. Wisconsin-area detritus was shed eastwards into the Michigan Basin, resulting in the Ellsworth Shale, and southwards into both the Illinois Basin and the Forest City Basin, resulting in the Hannibal Shale of western and central Illinois and Indiana and the English River Sandstone and siltstones of the North Hill Group in the Nebraska–Iowa region.

As seen above, most of the eastern Kinderhookian sea was dominated by detrital sedimentation. Nevertheless, several areas of carbonate deposition did exist at that time, especially in the southern Indiana–Illinois region which was shielded from eastern detrital-influx by the low-lying Cincinnati Arch. In this area, bioclastic and oolitic sediments of the Rockford and Chouteau Limestones were deposited, recording a shallow, current-washed platform.

Whereas detrital sedimentation characterized the eastern epeiric sea, being dominated by Acadian, Caledonian, and Ellesmere sediment-dispersal systems, carbonate deposition dominated the midcontinent region. A shallow, carbonate platform existed all the way from northwestern Illinois, where the McCraney and Starrs Cave limestones were deposited, to Oklahoma, where equivalent strata of the St. Joe Limestone Member of the Boone Formation are found. Environmental conditions naturally varied over such a large area but limestone deposits of this platform seem to indicate generally normal-marine conditions. Local detrital strata deposited on this platform were derived from the Ozark Uplift and Texas Arch, which shed sediments into the actively subsiding Anadarko Basin (see Fig. 6-10) and from the southern portion of the Transcontinental Arch.

As discussed above, the Wisconsin segment of the Transcontinental Arch was emergent during Kinderhookian time. Similarly, the southwestern segment of the arch was a positive feature during that time. Additionally, the Zuni-Defiance and Texas uplifts were also emergent, so that much of the Southwest constituted a low-relief landmass during the Kinderhookian Epoch.

Whereas northern and southern portions of the arch were emergent throughout the Kinderhookian, the middle segment was apparently not active at that time and allowed a connection between eastern and western epeiric seas. At the beginning of the Kinderhookian, virtually all of the Transcontinental Arch was emergent and erosion produced mature, quartzose detritus which was carried westward by streams. Transgressing seas covered the middle segment of the arch during the early Kinderhookian and reworked this detritus into a thin blanket which overstepped the karst surface developed on Upper Devonian carbonates during pre-Kinderhookian emergency. As the sea encroached further on the arch, carbonate deposition resumed on the west until, by latest Kinderhookian time, a broad carbonate platform existed over much of the area of Montana and Wyoming. On this platform were deposited shallow marine limestones and intertidal to supratidal dolostones of the Madison Limestone (and the Lodgepole Limestone of the Madison Group farther west). Relatively deeper-water conditions existed in the Central Montana trough which connected the Williston Basin to the western sea, allowing relatively free circulation.

The Williston Basin was a negative feature throughout the Kinderhookian in which were deposited carbonate and fine detrital sediments named the Lodgepole Limestone in the United States and the Souris Valley Formation in Saskatchewan. In the center of the basin, only very fine, organic-rich muds were deposited. Carbonate-platform strata surround these deep, starved-basin deposits.

In western Canada, too, Kinderhookian transgression brought a resumption of carbonate sedimentation to the western craton. The above-mentioned Souris Valley Formation grades northwestward into shallow-water oolitic and bioclastic limestones of the Banff Formation of western Alberta. Farther northwest, Banff strata become increasingly muddy; correlative strata in British Columbia are mostly calcareous shales.

B.7. The Osagian Stage

Sea level dropped slightly at the end of the Kinderhookian Stage and disconformities separate Kinderhookian from Osagian strata in southwestern Iowa, Kansas, western Illinois, and other stable cratonic areas. Deposition was continuous in most basinal areas, which were more active during the early Osagian than they had been during the Kinderhookian, but facies patterns indicate a shallowing of the sea. Rejuvenation of cratonic source areas such as the Wisconsin and Ozark uplifts and the Transcontinental Arch supplied fine detritus to the old, midcontinent carbonate platform.

As another result of lowered sea level, the great eastern deltas, whose sediment sources were in the old Acadian highlands, began to prograde rapidly westward across the Appalachian Basin. At the same time, Canadian Shield-derived sediments were deposited in deltaic systems which prograded across the northeastern part of the craton. In the Michigan Basin, for example, the Coldwater Shale conformably overlies the Sunbury Shale and indicates shallow marine and marginal marine environments whereas overlying strata of the Marshall Sandstone are composed of a complex lithological assemblage of deltaic and shallow marine deposits. Thus, a complex deltaic coast bordered the eastern Osagian sea on the north and east (see Fig. 6-11). Sediments of the Borden Formation (or Group) represent the progradation of deltaic sediments into the Osagian sea and are found not only in the Appalachian Basin but also in Indiana and Illinois. They contain the typical deltaic vertical-succession of bottomset, prodelta muds overlain by foreset, delta-slope deposits overlain by complex topset beds representing shallow marine environments, mud flats, barrier-island and lagoon systems, and marshes as well as the lobate deltas themselves.

The sea into which the Borden deltas prograded may have been as deep as 150 m (Craig and Varnes, 1979). In the deepest part of this sea, a sediment-starved basin existed into which intruded only an occasional turbidity current, caused probably by slumping on the eastern deltaic front. This deep basin may have been connected to the oceanic area to the south of the continent through a deep trough known as the Batesville channel (Glick, 1979). Swann (1964) suggested that the Illinois Basin during the Mississippian may have functioned as a "bypass" through which sediments derived from the Acadian highlands and far to the north in Canada were funneled into the southern ocean.

On the western side of the Illinois Basin, the old carbonate platform continued to thrive in the warm, shallow, normal-marine waters of the midcontinent Osagian sea. Conditions were ideal for life and the sea teemed with organisms; especially common were the pelmatozoan and fenes-

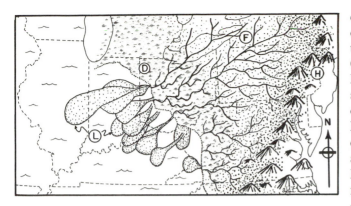

Fig. 6-11. Paleogeography of eastern interior region during Borden time; note inferred drainage from the Acadian orogen. (Modified from Kepferle, 1977.) Symbols are as follows: D, deltaic plain; F, fluvial plain; L, deltaic lobes; H, tectonic highlands.

trate-bryozoan ''gardens'' which supplied abundant bioclastic debris for large, current-washed shoals and banks, probably similar to sand bodies of the Bahama platform today (see, e.g., Ball, 1967). The Burlington and Keokuk limestones overlie the Fern Glen Formation and are composed of bioclastic limestones which record the shifting environments of the midcontinent carbonate platform. Small amounts of fine detrital sediment were introduced to this platform from the Wisconsin and Ozark uplifts but in most areas, the carbonates are remarkably pure. The eastern margin of the carbonate platform dropped off precipitously into the depths of the Illinois Basin sea.

Siliceous carbonates of the Ft. Payne Formation were deposited over a large area of the east central craton in the deep sea in front of the Borden delta-plain. The Ft. Payne thickens southward to where it comprises virtually the entire Osagian section in western Kentucky and Tennessee. Ft. Payne strata in the eastern interior are composed of dolo-siltite deposited with high initial dips in deeper water on the outer delta-slopes of the Borden complex (see Fig. 6-12). Ft. Payne sedimentation thus further reduced the size of the deep areas within the Illinois Basin. The last of these deep areas was filled in during late Osagian time by the Ullin Limestone so that by the end of the Osagian, the epeiric-sea floor was a relatively low-relief surface, implying that epeirogenic subsidence had decreased (Pryor and Sable, 1974).

During late Osagian time too, sea level rose rapidly and eustatically, causing one of the more spectacular floodings to be experienced by the North American continent (see Fig. 6-13). Detrital sediments supplying the growing Borden deltas were trapped in estuaries as the sea invaded the land and a broad carbonate platform was established over the old Borden delta-plain with the deposition of transgressive sequences of the lower Warsaw Limestone in Kentucky.

The southwestern part of the craton continued to be a low-relief land area during the early Osagian except for several areas of carbonate deposition. Near the end of the Kinderhookian, epeirogenic subsidence of the western craton margin initiated a marine transgression which resulted in deposition of the Whitmore Wash Member of the Redwall Limestone in Arizona. This transgression continued into the Osagian but shortly thereafter regressive sequences of the Thunder Springs Member recorded falling

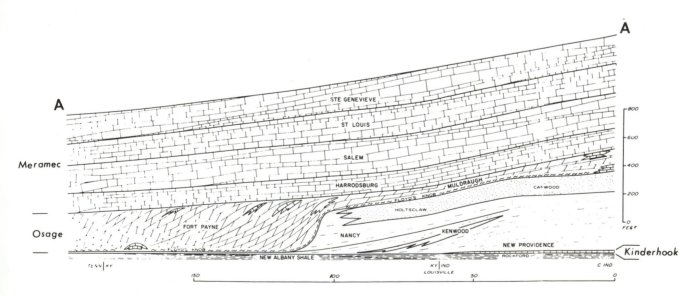

Fig. 6-12. Cross section of Mississippian strata from Indiana to Tennessee; note the clinothemic nature of the Fort Payne Formation. (From Pryor and Sable, 1974, Geological Society of America Special Publication No. 148, Fig. 6-A, p. 290.)

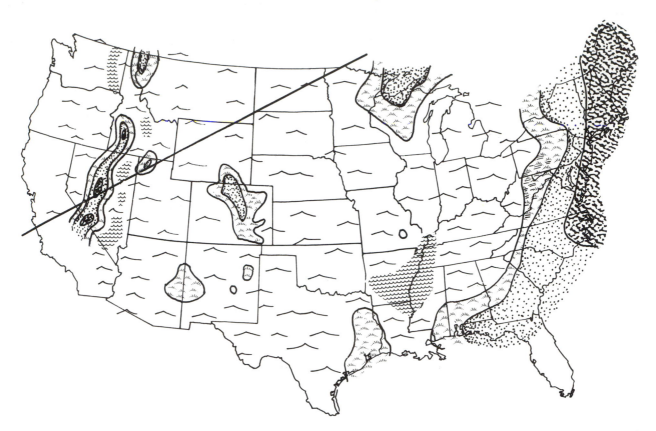

Fig. 6-13. Osagian paleogeography (during maximum transgression) of the United States. (Modified from Craig and Conners, 1979.) Symbols as in Fig. 6-10.

sea level (possibly related to the general Osagian regression). In westcentral New Mexico, an arm of the sea transgressed the craton from the south, resulting in the deposition of the Caloso Member of the Kelley Limestone.

During later Osagian time, virtually the entire area, except for the Zuni-Defiance and Pedernal uplifts, was flooded during the same eustatic rise of sea level recorded in correlative strata of the east. Major carbonate transgressive sequences were deposited over much of the southwestern craton. In southwestern Colorado, the Leadville Limestone represents shallow marine conditions; correlative strata in westcentral New Mexico comprise the Ladron Member of the Kelley Limestone. The Espiritu Santo Formation of New Mexico and the Mooney Falls Member of the Redwall Limestone in Arizona represent the same marine transgression.

The northern Rocky Mountain region also experienced a drop of sea level and concomitant development of regressive sequences during the early Osagian but the effects were not as sudden as they were in the East. Sea level apparently fluctuated rapidly (due to epeirogenic movements?) at that time, recording not a single regression but rather an overall regressive tendency superimposed on cyclic carbonate strata, such as the Woodhurst Limestone

Member of the Lodgepole Limestone of the Madison Group in southwestern Montana. Overlying strata of the lower Mission Canyon Limestone (also of the Madison Group) record generally regressive sequences in which dolostone dominates over interbedded limestone and which contains occasional evaporite deposits. Correlative strata in northwestern Montana comprise the Castle Reef Dolomite; in northcentral Utah and adjacent Idaho, the Brazer Dolomite and associated Deseret Limestone also formed at this time and contain a phosphatic member which was considered by Roberts (1979) to be the product of slow deposition in an area of the outer margin of the craton carbonate platform where cold waters upwelled from the deep Antler trough.

During the late Osagian, sea-level rise freshened the mildly hypersaline, western epeiric-sea and caused the deposition of transgressive sequences. Carbonate sediments of this time reveal the presence of a broad, shallow, current-agitated platform with oolitic and bioclastic shoals interspersed with somewhat deeper, protected environments in which carbonate muds were deposited.

Sedimentation in the Williston Basin was continuous from Kinderhookian to Osagian time and Osagian strata are almost ten times thicker than Kinderhookian strata, indicating an increase in depositional rate. Most Osagian strata of the Williston Basin in the United States are considered to represent Lodgepole and Mission Canyon limestones mentioned above but a widespread bed of anhydrite over 20 m thick is correlated with the Charles Formation of the Madison Group, which overlies the Mission Canyon. Similar strata are present in the Canadian part of the Williston Basin.

B.8.   Meramecian Events

The late Osagian high stand of sea level persisted into the early Meramecian and shallow marine sediments, primarily carbonates, were deposited over much of the North American craton. Actually, there were two separate transgressive events during the Meramecian and an intervening period of regression in which sea level was low, evaporites were deposited in shallow, restricted seas, and erosion removed earlier Meramecian (and older) strata from emergent areas. Figure 6-14 is a paleogeographic map of the United States during the Meramecian.

In the northeast, epeirogenic uplift along the Cincinnati, Findlay, and eastern part of the Kankakee arches resulted in a low landmass which separated the Appalachian Basin from the Michigan Basin. This, combined with con-

tinued emergence of the Wisconsin area, restricted the Michigan Basin from the main part of the epeiric sea and evaporite deposition there was frequent during the Meramecian. The resulting Michigan Formation is a complex association of stratigraphic units in which fine detrital sediments in the eastern and southern parts of the basin grade north- and westward into restricted carbonates and evaporites. Sea-level rise during later Meramecian time (the second transgressive event) freshened the Michigan sea and resulted in deposition of normal marine, cherty carbonates of the Bayport Limestone.

In the Illinois Basin and surrounding areas of the eastern interior, carbonate deposition predominated. Pryor and Sable (1974) have divided Meramecian strata of the East into three sequences, each of which reflects the different environmental condition which prevailed at that time. The lowest sequence consists of bioclastic and pelletal carbonates of the Harrodsburg (Illinois and Indiana) and Warsaw (Kentucky and Tennessee) limestones and overlying Salem Limestone (which is continuous over the whole area). These strata record a carbonate platform with environmental conditions ranging from quiet to strongly agitated and from clear to turbid water (near sources of detrital sediment such as the Cincinnati Arch and Ozark Uplift).

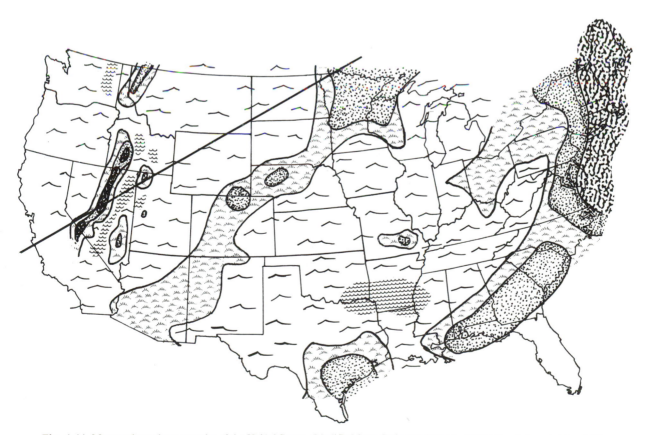

Fig. 6-14. Meramecian paleogeography of the United States. (Modified from Craig and Conners, 1979.) Symbols as in Fig. 6-10.

The second of Pryor and Sable's sequences is composed of fine, often carbonaceous, dolomitic carbonates and evaporites of the lower St. Louis Limestone. These deposits record the middle Meramecian regression and indicate deposition in shallow, restricted seas and in coastal sabkhas along the margins of emergent areas such as the Cincinnati Arch. (It is interesting to note that evaporites are present at this same stratigraphic position in both the Appalachian Basin and western cratonic strata, indicating that sea level was lowered over almost all of North America and that arid climatic conditions probably prevailed.)

The third, and last, Meramecian sequence in the East consists of oolitic and quartz-sandy carbonates of the upper St. Louis and Ste. Genevieve limestones and quartz arenites of the Spar Mountain Member of the Ste. Genevieve and the Aux Vases Sandstone in Illinois. These strata record a return to open marine conditions in the East following the second Meramecian transgression. Over almost the entire eastern half of the North American craton was a broad, shallow carbonate platform, generally characterized by large oolite shoals. Quartzose detrital deposits, such as compose the Aux Vases Sandstone, were probably derived from the north and were transported to the sea by the Michigan river system.

In the midcontinent region, too, carbonate deposition dominated, except for minor detrital materials deposited locally during the mid-Meramecian regression. In the northern part of the area (i.e., Nebraska, Iowa, and Kansas), the same stratigraphic units as in the eastern cratonic area are present and record generally the same geological history. Open marine carbonates of the Warsaw and Salem limestones indicate a widespread shallow sea; lowering of sea level during early St. Louis time resulted in emergence of northern and central Iowa (on the southern flank of the Wisconsin Uplift) and consequent erosional removal of Warsaw and Salem strata. Rising sea level during later St. Louis time reestablished carbonate deposition over most of the area.

In the Arkansas–Oklahoma region, carbonates of the Boone and Moorefield Formations are correlated with the Warsaw and Salem carbonates; these strata were partially eroded during the mid-Meramecian regression. Oolite shoals are recorded by the Short Creek Oolite Member of the Boone Formation in central Oklahoma; these shoals may mark the edge of the Meramecian carbonate platform, where it dropped off into the depths of the southern ocean where Stanley Group turbidites were accumulating.

Upper Mississippian strata of the western craton are very limited in their distribution due to numerous subsequent periods of erosion. In the Southwest, lower Meramecian strata, if they were ever present, were eroded away during the middle Meramecian. A significant exception is the Mooney Falls Member of the Redwall Limestone in Arizona, which represents the maximum eastward transgression onto the Transcontinental Arch of the western epeiric sea during Warsaw and Salem time.

In the northwestern cratonic area of the United States, too, the late Osagian sea persisted into the early Meramecian and carbonate deposition continued. As elsewhere, sea level dropped near the end of the early Meramecian, resulting in cyclically bedded, micritic limestones, dolostones, and evaporites of the upper Mission Canyon Limestone and the Charles Formation in Montana and Wyoming. Sea level continued to drop (or epeirogenic uplift continued to raise the area up) so that the sea drained completely off this part of the craton and karst topographic development was widespread. During later Meramecian time, probably penecontemporaneously with the upper St. Louis/Manuelitas transgression, sea level began to rise and returning seas flooded the outer continental shelf platform. The northern U.S. Rocky Mountain region, however, must have been epeirogenically uplifted at the same time because considerably less of the area was affected by rising seas than seems to have been the case in other areas of the craton. Major stream systems were established which drained highlands associated with the Transcontinental Arch in the Dakotas and eastern Wyoming. Deposits of these fluvial systems comprise the Amsden Formation of western and central Wyoming and the Big Snowy Formation of Montana.

B.9. The Chesterian Age: Final Retreat of the Kaskaskian Sea

The Chesterian is the uppermost stage of the Mississippian System and records the terminal withdrawal of the Kaskaskian sea from the North American continent; thus, regressive sequences are characteristic of Chesterian strata in most parts of the continent. Epeirogenic activity increased during the Chesterian; positive areas were uplifted and basins subsided more than during previous Mississippian ages.

In the eastern interior region, Chesterian history was dominated by a major deltaic complex which prograded southward into the broad, generally shallow sea which occupied southern cratonic areas (see Fig. 6-15). This delta system was supplied with sediments from the Canadian Shield (and, possibly, the old Acadian highlands) by the Michigan river system. The Michigan river probably arose far to the north, possibly on the southern flanks of the old Ellesmere Mountains north of the shield, and flowed southerly to southwesterly across Michigan to the area of active delta formation in Illinois and Indiana. The Michigan Basin was a land area, probably due both to increased rate of detrital influx and to a lower rate of subsidence.

Mississippian strata of the Michigan river delta provide one of the most interesting sedimentological studies in North American historical geology. Swann (1963, 1964) discussed the history of this area in detail and much of the

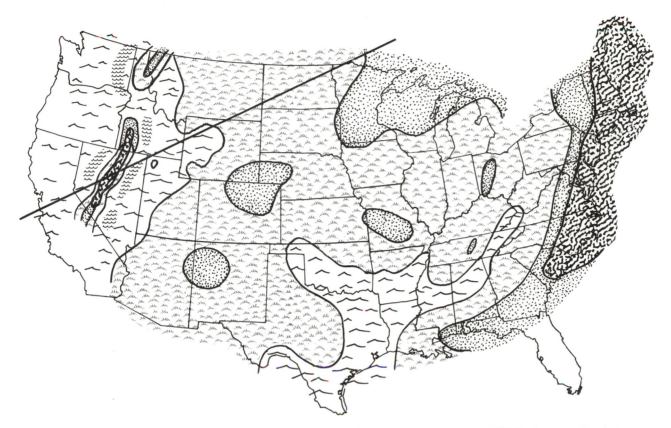

Fig. 6-15. Late Chesterian paleogeography of the United States. (Modified from Craig and Conners, 1979.) Symbols as in Fig. 6-10.

following is based on his ideas. Sandstones of the eastern interior region are thickest along a line from westcentral Indiana to southern Illinois and thin symmetrically to either side of this line, grading into mudstones and limestones. This line probably delineates the main locus of deltaic deposition (Fig. 6-16). But deltaic sands, both deeply channeled distributary-mouth bar deposits (i.e., "bar-finger" sands) and more tabular delta-front sands, as well as other, related topset deposits such as coaly marsh deposits and lagoonal muds, are distributed widely over most of the northern part of the Illinois Basin, clearly suggesting that the actual locus of deltaic deposition shifted back and forth over more than 300 km of the shoreline. In addition, sea level apparently fluctuated at the same time so that the shoreline moved back and forth over 1000 km across the low craton interior. The result is a complex sequence of transgressive and regressive strata in which deeply entrenched deltaic sands and associated deposits alternate with clean, shallow-water limestone (see Fig. 6-17). Over 20 different formations have been recognized in this complicated stratigraphic assemblage but most seem to be related to the same deltaic model discussed above.

Detrital sediments are also dominant in the southern midcontinent region of Arkansas and Oklahoma. Sediments were supplied to this area from the Illinois Basin through the Batesville channel as well as from the Ozark Uplift and

the Transcontinental Arch. In northwestern Arkansas and adjacent Oklahoma, shallow-water sediments of the Fayetteville Shale record deposition on a stable shelf south of the emergent Ozark Uplift. During the same time, detrital sediments associated with the Illinois Basin delta were car-

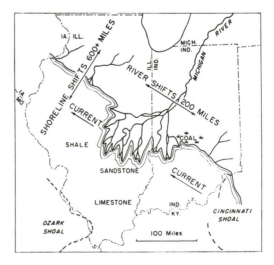

Fig. 6-16. Map of Michigan river and delta system during deposition of the Chesterian Stage. (From Swann, 1964, reproduced with permission of the American Association of Petroleum Geologists.)

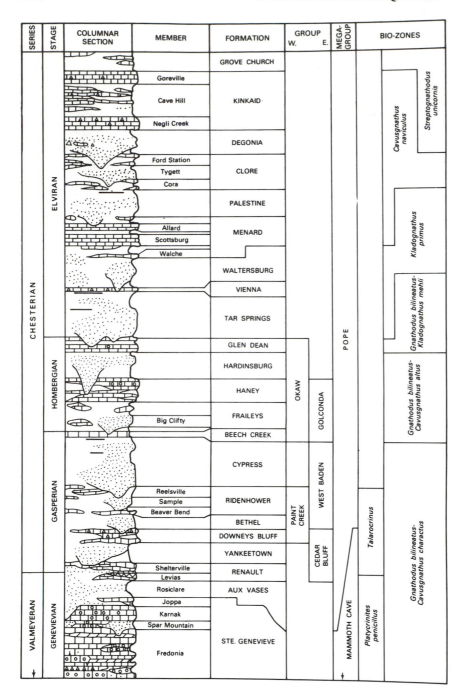

Fig. 6-17. Chesterian stratigraphy of the northeastern midcontinent region. (From Swann, 1964, reproduced with permission of the American Association of Petroleum Geologists.)

ried southward in the Batesville channel, resulting in the Batesville Sandstone. Near the end of Chesterian time, as the sea regressed nearly off the craton altogether, the Michigan River delta itself prograded all the way along the Batesville channel to disgorge its detrital load directly onto the southern continental margin. During middle and late Chesterian time, oolitic and bioclastic sediments of the Hindsville and Pitkin limestones accumulated on the southern part of the stable shelf area; scattered reef masses are present near the southern extent of the Pitkin carbonate platform. Occasional submarine debris flows, perhaps associated with slumping off the edge of the Pitkin platform, carried very coarse limestone blocks and shallow-water sediment into the deep Ouachita trough to be interbedded with Stanley Group turbidites.

Very few Chesterian strata are found in the Southwest even though they were probably widely distributed prior to post-Kaskaskian emergence and erosion. The presence of

Chester-age limestone in a thin erosional remnant in the Grand Canyon area suggests that detrital influx was limited and, thus, that the south end of the Transcontinental Arch was not actively uplifted at this time.

Central and northern parts of the Transcontinental Arch, on the other hand, were uplifted considerably during the Chesterian and detrital sediments derived from there dominated sedimentation patterns over the entire west-central and northwestern parts of the craton in the United States. The reader will remember that the northern U.S. Rocky Mountain region had been, during latest Meramecian time, generally emergent with well-developed drainage systems originating in the Transcontinental Arch region and on the Canadian Shield. This pattern persisted into the early Chesterian but soon thereafter, sea level began to rise, flooding the western craton margin. Along with sea level rise was renewed subsidence of the Central Montana trough which, with the Williston Basin, is referred to as the Big Snowy Basin. Rising seas flooded the Big Snowy Basin, leaving a transgressive detrital sequence over the entire area. This sequence comprises the lower part of the Big Snowy Group. During middle Chesterian time, the Big Snowy sea deepened and sands around the margins of the basin graded into dark gray and green muds in the basin center. Alternation of normal marine sediments, evaporites, and brackish-water deposits indicate the same kind of sea-level fluctuations as recorded in correlative strata of the East, suggesting a eustatic cause for the fluctuations. The Big Snowy Basin was separated from the Wyoming Basin by the Southern Montana Arch. During early Chesterian time, seas deposited transgressive sequences composed of fluvial, estuarine, and shallow-marine sands of the Darwin Sandstone Member of the Amsden Formation in the Wyoming area. The Darwin Sandstone varies considerably in thickness, probably as a result of the irregular, karst surface (developed during Meramecian emergence) over which it was deposited by the transgressing sea. In northwestern Utah, too, transgressive sequences were deposited whose sediments were derived from the eroding Transcontinental Arch.

Maximum transgression during later Chesterian time coincided with uplift throughout the Big Snowy Basin area, forming the Big Snowy Uplift. Recently deposited sediments were eroded and carried southward to be deposited in the Wyoming Basin. Detrital sediments were even carried far out onto the carbonate platform on the continental shelf, resulting in the Manning Canyon Shale. A general regression marked the end of the Chesterian Age.

A similar sequence of events may be inferred from extant Chesterian strata of western Canada. Transgressive marine detrital sediments, similar to the lower Big Snowy Group, are present in the Etherington and Stoddard Formations of Alberta and British Columbia. During later

Chesterian time, coarse detrital materials of the Tunnel Mountain and Taylor Flat formations were deposited in regressive seas.

C. Kaskaskian of the Appalachian Mobile Belt

C.1. The Wallbridge Discontinuity in the East

As discussed in Chapter 5, Helderbergian strata (Lower Devonian) record a broad, tectonically stable continental margin with a low carbonate platform in much of the Appalachian Basin and quartz-arenitic sands near the eastern margin. During early Deerparkian time, the seas withdrew from the entire craton, producing the Wallbridge Discontinuity (discussed in the preceding section on Kaskaskian cratonic events). In the East, erosion associated with emergence stripped away much earlier strata so that in many areas, especially on the margins of the basin, Lower Devonian, Silurian, and sometimes even Upper Ordovician strata were removed entirely. For example, in Tennessee, Upper Devonian Chattanooga Shale rests on Middle Ordovician Bays Formation. Such extreme stratigraphic truncation is not the norm, however, for the extent of erosion decreases toward the interior of the basin which was deep enough to be subaerially exposed by falling sea level only for a short period of time. Indeed, there is evidence that the deepest part of the basin was never completely drained and that sedimentation continued across the Tippecanoe/Kaskaskia boundary, unaffected by Wallbridge emergence (Dennison and Head, 1975).

Weathering and erosion during Wallbridge emergence produced quartzose detritus which was deposited as a thin, transgressive, blanket sand (similar in some aspects to the St. Peter Sandstone discussed in Chapter 5). This is the Oriskany (Ridgeley) Sandstone, a highly mature, fossiliferous, quartz arenite which forms an easily recognizable marker in much of the Central Appalachians and which is an important glass-sand deposit and petroleum reservoir. The Oriskany oversteps Cayugan strata in the northwestern part of the basin and Albion strata in the south (Dennison and Head, 1975) and, thus, may be considered the earliest Kaskaskian unit in the Appalachians.

C.2. The Onesquethaw Stage: Return of the Sea

During early Onesquethawan time, sea level began once more to rise. At first, only the center of the basin, which had never been drained, received sediments resulting in the Esopus Shale which occurs only in eastern New York, Pennsylvania, and adjacent New Jersey (Oliver *et al.*, 1967). Later, however, with further rise of sea level, sediments became even more widely distributed than in Oriskany time. In New York, the Carlisle Center and Schoharie Formations were deposited, showing evidence of a newly uplifted detrital source to the east. A narrow, discontinuous belt of sandstone (correlative with the Schoharie) extends all along the eastern margin of the Appalachian Basin and grades westward, in the basin center, into dark-gray muds of the Needmore Shale, which probably represent a deep prodelta environment associated with influx of Schoharie detrital sediments from the north (Dennison, 1971). Farther west, the Needmore grades into the Huntersville Chert which probably formed by the nearly complete silicification of a limy bioclastic sediment deposited in shallow water on the western margin of the Appalachian Basin.

Continued sea-level rise resulted in the deposition of the Onondaga Limestone in New York during the late Onesquethawan. Distribution of Onondaga-age strata throughout the East suggests that the sea covered nearly the entire area at this time, possibly even covering the Cincinnati Arch (Oliver *et al.*, 1967). Deposition of the Needmore continued in the central part of the basin, prograding over the Huntersville Chert in eastern West Virginia. The Frog Mountain Sandstone of Alabama and Georgia was formed at this time. The Onondaga Limestone is correlative with the Columbus Limestone of Ohio, the Jeffersonville Limestone of Kentucky, and the Detroit River Group in the Michigan Basin. These units probably represent parts of a once-continuous sheet of carbonates which extended across virtually the entire central and northeastern parts of North America during late Onesquethawan time. Clearly, the sea had returned and the broad carbonate platform of Helderbergian time had been reestablished after the interruption of the Wallbridge emergence.

The end of the Onesquethawan Stage is marked throughout the Appalachian Basin by the Tioga Bentonite, a volcanic-ash bed which is distributed from northerly exposures in New York to Tennessee in the south and to Ohio in the west. It has also been reported from the Illinois, Wisconsin, and possibly even the Williston basins; thus, the Tioga furnishes a superb isochronous surface throughout the eastern United States and allows for excellent stratigraphic correlations and paleogeographic reconstructions. Based on grain-size distributions and isopach data, Dennison and Textoris (1966) proposed that the source of the Tioga was somewhere near Monterey, Virginia, although more recent work suggests that the area around Fredericksburg is a more likely source area.

Soon after the beginning of the Middle Devonian, sedimentation patterns in the Appalachian foreland basin began to reflect greatly increased rates of detrital-sediment influx from the Northern Appalachian area. At first, these sediments affected only the area directly adjacent to the North-

ern Appalachians; by early Late Devonian time, however, they had formed a great deltaic complex which prograded far out onto the craton. This detrital-wedge complex was the result of the Acadian orogeny, to which we turn now.

C.3.   The Acadian Orogeny

The Acadian orogeny has been termed "the most important diastrophic event in the Northern Appalachians" (Zen, 1968). The Acadian orogeny is usually considered to have been a Middle Devonian event but its onset was probably diachronous, beginning during New Scotland time in Nova Scotia and reaching the latitude of New York only by Onondaga time (Boucot, 1962). It resulted in major structural deformation, widespread igneous activity, high-rank metamorphism, and a major mountain belt. From these mountains poured large volumes of detrital sediments which were swept westward to build the Acadian (Catskill) detrital wedge, comparable to but considerably more voluminous than the Taconic detrital wedge discussed in the previous chapter. The Acadian orogeny primarily affected the Northern and Maritime Appalachians but igneous activity, metamorphism, and at least minor deformation also occurred in the Southern and Central Appalachians at about the same time. In the following discussion, we will first describe the major effects of the Acadian orogeny and then discuss plate-tectonics models for the event.

C.3.a.   Effects of the Acadian Orogeny

Three different, direct effects of the Acadian orogeny may be distinguished: structural deformation, igneous activity, and metamorphism, each of which will now be discussed. The detrital-wedge complex mentioned above may be considered an indirect effect of the orogeny and will be considered after our analysis of the tectonics of the Acadian orogeny.

i.   Structural Deformation.   Along the western margins of the Appalachian orogen in Northern and Maritime areas, the effects of the Acadian orogeny may be seen in large-scale fold and thrust-fault patterns which overprint earlier, Taconic patterns as well as affecting post-Taconic deposits. For example, in the areas of the Hudson River valley, Lake Champlain, and the St. Lawrence seaway, large-scale, west-verging, asymmetrical and overturned folds and large-scale thrusts of Acadian age are observed. Closer to the center of the orogen, fold patterns reflect more intense and deep-seated stress; in the New Hampshire area, for example, the character of deformation suggests a plastic, almost fluid behavior (Thompson and Norton, 1968). In terms of deformational style, early, i.e., premetamorphic, folding was primarily of the recumbent to strongly overturned type whereas later, synmetamorphic deformation consisted of upright folding. The early phase of folding can be seen in large, recumbent, eastward-verging fold nappes along the western side of the Merrimack trough in Connecticut where it adjoins the Bronson Hill anticlinorium. Later, upright folding is observed in central Maine, where Acadian folds are tight, upright isoclines (Rodgers, 1981, 1982). Upright folding also reactivated and accentuated ancestral, i.e., Taconic, anticlinoria and synclinoria such as the Oliverian gneiss domes of New Hampshire (Bradley, 1983).

Of great importance in the tectonic interpretation of the Acadian orogeny are features considered to represent accretionary-prism complexes. Accretionary complexes form at zones of plate convergence and are composed of material scraped off the downgoing plate and welded onto the overthrust plate. It is important to note that an accretionary prism formed above a *westward-dipping* subduction zone should feature *east-verging* structures. With this in mind, consider that, within the Merrimack trough of northeastern Connecticut and adjacent Massachusetts, a wedge of metamorphosed turbidites is cut by a westward-dipping, eastward-verging stack of thrust faults which were considered by Rodgers (1981) to represent an accretionary complex developed above a westward-dipping subduction zone. Conversely, west-verging, recumbent nappes dominate Acadian structures of the Merrimack trough in New Hampshire and Maine. Thus, both east- and west-verging accretionary complexes occur within the Merrimack trough of New England. Bradley (1983) stated that the west-verging complex dominated the early, i.e., premetamorphic, structure in most of the Merrimack trough and structurally overlie the east-verging package. This is a significant point and we will return to it later.

ii. Igneous Activity. Igneous rocks related to the Acadian orogeny are present in several areas and reflect multiple phases of magmatic activity. Volcanic activity immediately preceded the orogeny in two distinct areas, the Piscataquis volcanic arc and the Coastal volcanic arc. Plutonic activity occurred in areas associated with both volcanic arcs and along the Merrimack–Fredericton trough in the center of the orogen in New Hampshire and Maine.

The Piscataquis arc (discussed in the introduction to the Appalachian orogen in the preceding chapter) had begun during the Early Silurian, probably as the result of the initiation of westward-directed subduction of oceanic lithosphere beneath the Bronson Hill arc massif (whose accretion to the North American margin during the Ordovician resulted in the Taconic orogeny). Earliest volcanics associated with the Piscataquis arc were of limited volume and sporadic occurrence; a marked increase in magmatism came during the Early Devonian with eruptions of thick volcanic piles all

along the arc from New England to Gaspé (Bradley, 1983). Volcanic rocks associated with the Piscataquis arc are composed mainly of andesites, basalts, dacites (in volcanic members of the Erving Formation in Massachusetts, the Littleton Formation in New Hampshire, the Piscataquis and Dockendorf units of Maine, and the Mount Alexandre volcanics of Gaspé), and rhyolites (Traveler and Big Spencer formations in Maine; Rankin, 1968). The presence, within these extrusive rocks, of welded tuffs, volcanic breccias, and interbedded, nonmarine, detrital deposits (e.g., Oriskany-age Tarratine and Matagamon formations; Boucot, 1968) all indicate that the Piscataquis arc was massive enough to form an archipelago of volcanic islands. Bradley (1983) said that the geochemistry of Piscataquis volcanics supports the idea that a pre-Acadian subduction zone dipped westwards beneath the North American mainland.

Between the Piscataquis arc and the North American mainland was a back-arc basin termed the Connecticut–Gaspé trough. The history of this basin began penecontemporaneously with that of the Piscataquis arc, probably as the result of back-arc spreading (Rodgers, 1981). Similar to igneous activity within the Piscataquis arc, subsidence within the Connecticut–Gaspé trough was localized and sporadic during the Silurian but increased markedly during the Early Devonian. Into this back-arc trough, turbidity currents carried detrital sediments from both the mainland to the west and the volcanic arc to the east. The resulting, monotonous mass of graywacke turbidites and black shales comprises the Gile Mountain Formation of Vermont, the Littleton of New Hampshire, the Seboomook of Maine, the Frontenac of Quebec, and the Temiscouata of Gaspé (Bradley, 1983).

The Coastal volcanic arc runs from eastern Massachusetts to southern New Brunswick; in fact, however, it is part of the Avalon terrane which extends at least from Newfoundland to Connecticut and which possibly can be extended to the Charlotte, Carolina slate, and related lithotectonic belts of the Southern Appalachians (see Figs. 5-13 and 5-14). Volcanic activity within the Coastal (Avalon) volcanic arc occurred at several different times, the oldest of which was during the late Precambrian, at which time occurred extensive subaerial outpourings of basalts and rhyolites with associated pyroclastics and detrital sediments (the Harbour Main Group of Newfoundland; Hughes, 1970). These rocks were intruded in Newfoundland by the Holyrood Plutonic Group, composed primarily of quartz monzonites and granites (these intrusions and related deformation and metamorphism are all related to the tectonic event termed the Avalonian orogeny; Lilly, 1966). Little is known of the origin of the Avalon block or its paleogeographic position during the late Precambrian but, by the Late Cambrian, it had apparently ceased to be volcanically active and shallow-water deposits accumulated with faunas quite unlike those of continental North America. Comparable rocks occur in Great Britain and Wales; Rodgers (1982) suggested that rocks of the Avalon block and the British Isles may all be related to the late Precambrian Cadomian orogenic belt in Normandy and Brittany. At any rate, volcanic activity resumed during the Silurian and continued into the Devonian, resulting in some areas in volcanic piles up to 8 km thick (Bradley, 1983). Volcanism of this episode defines the Coastal volcanic arc. In addition to its volcanics, the Coastal volcanic arc also contains fossiliferous Silurian and Lower Devonian sedimentary rocks deposited in deep-marine to nonmarine environments. Most workers have agreed to a volcanic island-arc origin for these extrusives (Bradley, 1983). Volcanic activity within the Coastal (Avalon) volcanic arc ceased during the Middle Devonian, at about the same time that it ceased in the Piscataquis arc.

Plutonic activity was associated with the volcanic arcs mentioned above, but in both cases it occurred earlier than extrusive activity. The Greenville plutonic belt is coextensive with the Piscataquis arc and is composed of gabbroic to granitic plutons (Hon et al., 1981). Volcanics of the Coastal (Avalon) volcanic arc were intruded by the Bays of Maine igneous complex, which is composed of foliated, ultramafic to granophyric plutons (Bradley, 1983).

The New Hampshire Plutonic Series, which principally intrudes rocks of the Merrimack trough, is among the most important igneous complexes of the Northern Appalachians and is composed almost entirely of granite and related felsic rocks. Older plutons of the New Hampshire Series occur as elongate bodies parallel to Acadian folds and have primary foliation; thus, they are considered to be syntectonic (Page, 1968). Younger plutons were injected parallel to regional foliation but are themselves unfoliated; they are considered to be posttectonic. Thompson and Norton (1968) argued that New Hampshire Series plutons were the result of anatectic melting at the base of the thickened continental crust within the center of the Acadian orogen. As evidence of this interpretation, Thompson and Norton cited the near absence of mafic rocks in the New Hamshire Series and the fact that the geochemistry of New Hampshire plutons can be fully accounted for by the partial melting of "eugeosynclinal" sediments (i.e., shales and graywackes).

iii. Metamorphism. Extremely high grades of metamorphism accompanied the Acadian orogeny. Highest-rank rocks are of sillimanite grade and occur in New Hampshire and adjacent Massachusetts, associated with anatectic bodies of the New Hampshire Plutonic Series (Thompson and Norton, 1968). Metamorphic rank decreases to the east, toward the Coastal (Avalon) arc, and westwards toward the Piscataquis arc. Another indication of the extreme degree of metamorphism is the plastic, almost fluidlike, character of folding associated with sillimanite-grade rocks.

C.3.b. Plate-Tectonics Model of the Acadian Orogeny

Among the earliest attempts to apply plate-tectonic concepts to the Acadian orogeny was the discussion of Dewey and Bird (1970). The authors suggested that the principal cause of the Acadian orogeny was the collision of the northeastern margin of North America with Ancestral Europe. However, with the recognition of numerous displaced ("suspect") terranes within the Appalachian orogen (see Fig. 5-14) has come the realization that the tectonic story is not that simple. The following discussion is drawn principally from the tectonic syntheses of Rodgers (1982) and Bradley (1983).

A brief review is necessary. The Taconic orogeny of the Middle and Late Ordovician was probably the result of the collision of a volcanic island-arc (represented by the Bronson Hill anticlinorium in New England and the Miramichi massif in New Brunswick) and the passive continental margin of North America. Following that collision, westward-directed subduction probably began along the eastern side of the collided block resulting in the onset of Piscataquis-arc volcanism. Opening of the Aroostook–Matapedia trough behind the Miramichi massif may have been the result of back-arc spreading. Associated with the beginning of subduction would have been the initiation of the east-facing accretionary prism which was to be preserved as the southern Merrimack trough.

The patterns developed after the Taconic orogeny persisted throughout the Silurian Period (see Fig. 6-18A). The floor of the Iapetos ocean basin subducted quietly beneath the eastern margin of North America, with no apparent tectonic effect other than Piscataquis-arc volcanism and incremental growth of the east-facing Merrimack–Fredericton accretionary prism. On the opposite side of the Iapetos ocean was the Avalonia microcontinental block. It, too, overrode Iapetos oceanic lithosphere with resulting arc volcanism manifested in the Coastal volcanic arc. At the leading edge of the Avalonian plate was a west-facing accretionary prism.

Caught between encroaching sialic blocks, the Iapetos ocean basin gradually shrank until, by the Early Devonian, the leading edges of the North American and Avalonian accretionary prisms were brought into contact with each other. The Avalonian accretionary prism is believed to have overridden the North American one, resulting in the complex, compound accretionary mass with both east- and west-vergent structures that was to become the Merrimack–

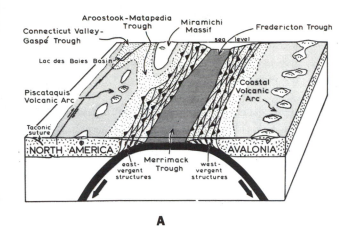

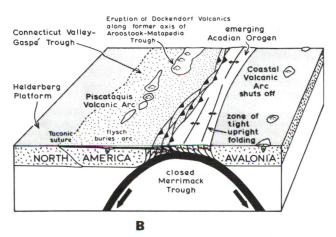

Fig. 6-18. Bradley's model of the Acadian orogeny. (A) Paleotectonic relationships during the Late Silurian; note that the Iapetus ocean basin (represented by the Merrimack–Fredericton trough) is flanked on both sides by oppositely verging subduction zones and associated accretionary prisms; fine stippling represents shallow-marine deposition. (B) Initiation of collision during the Early Devonian; note that the accretionary prism associated with the Avalon block has overthrust the North American prism. (From Bradley, 1983, *Journal of Geology*, Vol. 91, Figs. 5 and 6, p. 394, © University of Chicago Press.)

Fredericton trough (see Fig. 6-18B). As Avalonia overrode the leading edge of North America, it loaded the continental margin, causing it to subside. Into this newly subsident, peripheral foreland basin, turbidity currents carried detritus from the emerging Acadian orogen just to the east. These turbidite sediments first began to cover the Piscataquis arc and then to be swept into the Connecticut–Gaspé trough behind the arc; the resulting deposits were to become the Gile Mountain, Seboomook, and Littleton formations of New England and the Frontenac and Fortin formations of Canada. Sedimentary burial of the Piscataquis arc, however, did not extinguish volcanism there so that Littleton turbidites are interbedded with pillowed lavas above the axis of the old arc.

These events occurred during the Early Devonian, at the same time that Helderbergian carbonates were accumulating in the New York area with no appreciable detrital input from the east. By the earliest Middle Devonian, however, continued convergence of the North American and Avalonian plates had begun the growth of a modest orogenic welt and detrital sediments derived from this area were swept onto the old Helderberg platform, resulting in the Schoharie Formation of New York. But the rate of detrital influx was still relatively minor during Schoharie time and rising sea level during later Onesquethawan time produced a transgression during which the Onondaga Limestone was deposited (the Onondaga was part of a more extensive carbonate sheet deposited over much of eastern North America during Onesquethawan time; see discussion earlier in this section). Following the Onesquethawan transgression, detrital-sediment influx increased greatly in magnitude, reflecting the progressive growth of the Acadian orogen. By the Late Devonian, plate convergence had led to full-scale plate collision; partial subduction of the North American plate beneath the Avalonian block led to thickening of the sialic crust at the collision zone and the concomitant growth of a major mountain chain. From this orogen issued great river systems which transported detrital sediments to the epeiric sea; the resulting Acadian detrital wedge will be discussed in the next section.

All of the discussion thus far has focused on events in the Northern Appalachians and, indeed, the lack of a Devonian detrital-wedge complex in the southern part of the Appalachian foreland basin was long considered evidence that the Acadian orogeny did not affect southern areas of the orogen. Within Piedmont Province of the Southern Appalachians, however, there are numerous areas in which significant episodes of metamorphism occurred at approximately the same time as the Acadian orogeny in the Northeast. For example, Thomas *et al.* (1980) reported that the major dynamothermal metamorphic event to affect the Talledega belt of Alabama occurred penecontemporaneously with the Acadian orogeny. Seiders and Wright (1977) and Briggs *et al.* (1978) have stated that regional greenschist metamorphism within the Charlotte and Carolina slate belts was of Acadian age, although others (e.g., Black and Fullagar, 1976) have argued for a Taconic age. Thus, there is reason to suspect that some kind of tectonic event affected the Southern Appalachians at the same time as the Acadian orogeny in the North.

The recognition that lithotectonic provinces of the Piedmont and Blue Ridge in the Southern Appalachians may represent displaced terranes has led to new models which help explain some of the problems of Middle and Late Devonian tectonics. Williams and Hatcher (1982) proposed that all of the lithotectonic belts east of the Kings

Mountain belt (i.e., the Charlotte, Carolina slate, Raleigh, Kiokee, Eastern slate, Belair, and Uchee belts; see Figs. 5-13 and 5-14) were parts of the same Avalonian block whose collision with the Northern Appalachian continental margin caused the Acadian orogeny. In this scheme, the southeastern part of the Avalonian block was collisionally accreted to the eastern margin of the Inner Piedmont block at approximately the same time that it was being accreted to the outboard margin of the Bronson Hill–Miramichi block in the North. The Kings Mountain belt, thus, would represent the suture zone between the two tectonic elements.

This model has several virtues; for example, it explains the distribution of most observed Acadian metamorphic ages within the Piedmont. An exception is the Talledega belt, which lies on the western side of the Inner Piedmont. As Rodgers (1982) pointed out, this model also explains why detrital sediments of Upper Devonian age do not appear in the Southern Valley and Ridge. Prior to the Alleghenian orogeny of the late Paleozoic, during which major décollement-style thrusting shoved the Piedmont westwards, the accreted tectonic elements of the Piedmont were as much as 200 km farther to the east of the foreland basin than they are now. Thus, any sediments derived from the southern Acadian orogen were not carried into the Appalachian foreland basin but rather were trapped in basins associated with the various Piedmont blocks or swept either eastwards into the Rheic ocean basin behind the Avalon block or southwards toward ocean at the southern margin of the continent.

The suggestion that the single, Avalonian block was responsible for Devonian tectonics in the South has recently been questioned by Higgins *et al.* (1984). These authors questioned whether the terms Taconic, Acadian, and Alleghenian have any real meaning in the Southern Appalachians. They argued that these terms were defined primarily from the Northern and Central Appalachians and that there is no reason that orogenic events must have affected the entire continental margin. Instead, they suggest, metamorphism and deformation in the Southern Appalachians was a continuous, accretionary process beginning during the Ordovician and lasting until the terminal continental collision of the late Paleozoic that ended the history of the Appalachian continental margin. Even if orogenesis were penecontemporaneous in both northern and southern segments of the orogen, that is still not evidence that the colliding tectonic elements of the two areas were the identical block. In the scheme of Higgins *et al.* (1984), Piedmont Province should really be thought of as a tectonic collage composed of a large number of possibly unrelated thrust packages accreted against the southeastern continental margin throughout the middle and later Paleozoic. This idea is to be contrasted with that of Williams and Hatcher (1982),

as shown in Fig. 5-14, that only three or four displaced terranes comprise the Piedmont tectonic collage.

At any rate, the Appalachian orogenic area, both in the North and in the South, continued to be unstable for the rest of Kaskaskian time, as indicated by the continuous, albeit variable, influx of detrital sediments to northern and eastern parts of the Appalachian foreland basin throughout the Mississippian Period.

C.4. The Erian Stage: Initiation of the Acadian Detrital Wedge

As discussed above, the first stratigraphic evidence for the Acadian orogeny may be seen in the minor introduction of detrital sediments which resulted in the Schoharie Formation. This, however, proved to be only a harbinger of the massive influx of immature sediments that occurred during the Middle and Late Devonian. Sediments poured out of the rising Acadian highland and were deposited in the northern part of the Appalachian foreland basin. They formed a prograding complex of subtidal, intertidal, and terrestrial environments which has been referred to in general as the ''Catskill Delta Clastic Wedge'' (see Fig. 6-19). This is a slight misnomer, however, because the transition zone (between land and sea) was composed not of a single delta nor even a series of deltas but rather of a complex of mud flats, marshes, delta lobes, bays, barrier islands, and so on. We prefer the term ''Acadian detrital wedge'' which does not carry any sedimentological prejudices and which stresses the relation of the sediments to the orogenic event that supplied them.

C.4.a. Facies Patterns in the Foreland Basin

Before consideration of the Erian Stage in the Appalachian Basin, it will be useful to describe the overall framework of sediment deposition during both Middle and Late Devonian time. As detrital sediments poured into the foreland basin from the east, they built a westward-prograding wedge of sediments whose maximum thickness (more than 3.5 km; Thomas, 1977) occurs in eastern Pennsylvania. This sediment mass has been divided up into four idealized lithofacies which typically occur in regressive sequences but which are interrupted by thin transgressive sequences associated with interwedge carbonates (see discussion of detrital wedges in Chapter 5). The four major lithofacies of the Acadian detrital wedge are as follows.

1. *Prodelta lithofacies.* On the leading edge of the detrital wedge, turbidity current sediments were deposited. Composed of fine to very fine, slightly immature quartz wackes, rocks of this lithofacies are characterized by rhythmically interbedded, thin layers of sandstone and siltstone

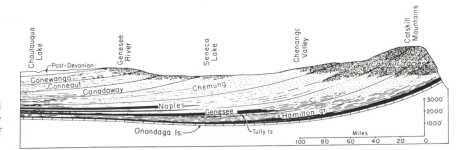

Fig. 6-19. Lithofacies of the Acadian detrital wedge. (From Dunbar and Rodgers © 1957 by John Wiley & Sons, reprinted with permission of John Wiley & Sons.)

or shale. Sole marks, e.g., flute casts, are occasionally found (McIver, 1970).

2. *Subtidal delta-plain lithofacies.* Overlying and interfingering with prodelta rocks are noncalcareous, shallow marine strata consisting of submature to immature graywacke associated with siltstone and gray shale. Many delta-plain lithologies are fossiliferous, some even being characterized as coquinites. Sands often contain low-angle, planar crossbedding, ball-and-pillow structures, megaripples, and horizontal laminations (Johnson, 1976; McIver, 1970). These sediments are considered to represent deposition on a marine delta platform or subtidal shelf.

3. *Peritidal lithofacies.* Eastern outcrops of strata previously considered to be of the delta plain type differ somewhat from typical delta-plain rocks in the west. They contain interbedded medium sandstone, massive siltstone, and mudstone with wavy, flaser, and lenticular bedding, mud clasts, wave and current ripples, bioturbation, plant debris, evidence of channeling, and local disconformities. Lenticular sand-bodies occur at the top of coarsening-upward sequences and associated gray-green shales contain a brackish-water fauna. This eastern aspect is termed the peritidal lithofacies and is considered to represent an intertidal shelf environment with tidal flats, barrier islands, lagoons, deltaic distributaries, and interdistributary bays.

4. *Coastal plain lithofacies.* In easternmost exposures of the Acadian detrital wedge, coarse-grained red beds with plant fossils are found. Pebble conglomerates, coarse sub-graywackes, and layers of siltstone, mudstone, and shale occur in fining-upward sequences associated with planar and tabular crossbedding, plant-rooted zones, mud cracks, and zones of calcareous nodules (Pedersen *et al.,* 1976). Multistory sand bodies and massive clay lenses, or "clay plugs," are also found. These red-bed deposits probably represent fluvial sedimentation on a low coastal plain with large meandering streams and broad, muddy floodplains.

While not strictly a part of the prograding clastic-wedge system, one other common lithofacies should be discussed at this time. In portions of the basin far away from the sources of detrital influx, sedimentation was very slow

and thus was unable to keep pace with subsidence. In such areas, a deep, starved basin developed where only very fine-grained, laminated, clayey, hemipelagic muds were deposited. These muds were dark gray to black and lacked a benthic fauna because of the anaerobic, reducing nature of the deep, stagnant basin. These dark gray shales are referred to generically as the "Black Shale lithofacies."

C.4.b. The Hamilton Group

With the above lithofacies terminology in mind, we turn to consider the initial development of the Acadian detrital wedge as recorded in the Hamilton Group of New York and Pennsylvania.

Directly after deposition of the Tioga Bentonite, there was a marked increase in detrital influx from the east accompanied by deepening of the basin center where black shales of the Marcellus Formation began to accumulate. In the northeastern part of the basin (around Kingston, New York), Marcellus equivalents record the development of typical Acadian detrital wedge stratigraphy (see Fig. 6-20): the Stony Hollow and lower parts of the Mount Marion formations contain prodelta and lower delta-slope sediments; the upper part of the Mount Marion has shallow marine lithologies of delta plain type. The Ashokan Formation contains peritidal strata and the Plattekill Formation represents fluvial, coastal-plain environments (Pedersen *et al.,* 1976). Thus, even in earliest Hamilton time, all of the major elements of the prograding Acadian coastal plain were present.

The same basic stratigraphic pattern, but with different rock-stratigraphic names, typifies virtually the entire Hamilton Group in New York, whose remaining formations (above the Marcellus) are Skaneateles, Ludlowville, and Moscow. Taken together, they record the progradation of the Acadian coastal plain westward across much of New York and Pennsylvania. However, superimposed on this regional regressive pattern are minor transgressive sequences associated with interwedge carbonates which represent fluctuations of shoreline. Such fluctuations may have

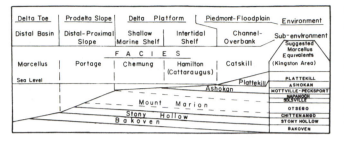

Fig. 6-20. Hamilton Group lithofacies in the Kingston, New York, area. (From Pedersen *et al.*, 1976, reproduced with permission of the New York State Geological Association.)

resulted from changes in the rate of sediment supply or locus of sediment discharge, changes in rate of basin subsidence, or eustatic sea-level changes. Probably the best example of an interwedge carbonate is the Tully Limestone of Taghanic age at the top of the Hamilton Group. The Tully is the most widespread limestone marker in the Acadian detrital wedge; it can be traced as far east in New York as the Susquehanna River and, on the western side of the basin, as far south as Bath County, Virginia, where it occurs as a zone of calcareous concretions within the Millboro Shale (Hasson and Dennison, 1974). Eastward of the Tully in New York are a series of detrital units (nicely described by Johnson and Friedman, 1969) which reveal the familiar delta-plain, peritidal, coastal-plain pattern of the detrital wedge. Thus, the Tully transgression did not destroy these environments but rather pushed them eastward, causing an abrupt facies shift of over 150 km. During sea-level rise, river mouths were drowned, stopping or drastically slowing the supply of detrital sediments to the shelf and thus allowing limy sediments to accumulate. Further, river-mouth drowning resulted in the preservation of estuarine sediments, which are rare in "normal" regressive sequences.

The Tully formed as a result of a major eustatic rise of sea level (Dennison and Head, 1975). This eustatic event is recorded in Taghanican-age strata from many parts of North America and has been termed the Taghanic on-lap by Johnson (1970; see further discussion in sections on cratonic and Cordilleran events in this chapter).

C.4.c. Erian Strata in the Southeast

Erian stratigraphic relations in the southern part of the Appalachian Basin are considerably less complicated than those in New York. This is because the area was far from the site of sediment influx and was, therefore, relatively deep. Its depth assured that it would not be affected by minor sea-level fluctuations.

South of the main body of the detrital wedge, prodelta deposits of the Mahantango Formation accumulated in

southern Pennsylvania, Maryland, and northern portions of the Virginias. The Mahantango prograded over Marcellus black shales, which had been deposited over virtually the entire basin excepting eastern New York and Pennsylvania. In southern Virginia, strata identical to and correlative with the Marcellus comprise the lower part of the Millboro Shale. Unlike the Marcellus, however, black shales of the Millboro continued to be deposited throughout Erian time, unaffected by the sedimentological excitement in the north. The southern part of the foreland basin was probably bordered on the east by a marginal land area, very close to sea level and of low relief.

C.5. Upper Devonian Stratigraphy and Paleogeography in the East

The tectonic and sedimentological patterns established in the East during the Middle Devonian persisted into the Late Devonian but continuation of orogenic activity along the continental margin had, by then, created an even larger orogenic belt from which detritus in ever-increasing quantities streamed westward onto the Acadian coastal plain. Evidence for such greatly increased sediment influx may be seen in the following observations: (1) Upper Devonian strata comprise more than two-thirds of the *entire* Devonian System in the Appalachian Basin; (2) during this time, the location of transitional, delta-plain environments shifted all the way to western New York, Pennsylvania, and West Virginia, extending as far south as Roanoke, Virginia; and (3) fluvial, red-bed lithologies dominate in most areas of the northern and central Appalachian Basin.

The two major fluvial units of the Upper Devonian are the Catskill Formation of New York and Pennsylvania and its southern equivalent, the Hampshire Formation of Maryland, Virginia, and West Virginia. Both units are texturally and compositionally immature, the predominant lithologies being muddy siltstone and fine, feldspathic graywacke. Heavy mineral suites reported from the Catskill Formation (Leeper, 1963) indicate an igneous-metamorphic source, suggesting the exposure of deep-seated rocks in the source area.

The transitional zone of the Acadian coastal plain during the Late Devonian was complicated by numerous different types of depositional environments but was, on the whole, muddy instead of sandy. Walker (1971) speculated that rapid progradation of the shore created a broad, low, alluvial coastal plain whose slow, low-gradient streams were competent to transport only fine sediments. Also, intense chemical weathering associated with equatorial regions provided abundant clays. Thus, sands carried down to the alluvial plain were trapped there and only mud was supplied to the shore. During much of the Late Devonian, the configuration of the coastline revealed the presence of four major delta lobes and three intervening bays as shown in Fig. 6-21.

The coastal plain has been divided by Woodrow *et al.* (1973) into lowland and upland parts. Red muds and fine sands comprise the lower coastal plain which was characterized by large, meandering streams and broad, muddy floodplains. Major stream divides on the lower plain were possibly very long-lived features and gross drainage patterns may have persisted for very long periods of time (Woodrow *et al.*, 1973). Such permanence of drainage may have allowed for the development of the four major deltaic lobes described above. The upland portion of the coastal plain was characterized by braided streams and coarser non-red sediments. Drainage basins were mostly small and stream divides evanescent. This upper surface may have formed from the coalescence of a series of alluvial fans at the break in slope at the foot of the Acadian Highlands where streams were no longer able to carry the coarser sediment as they began to wind slowly across the lower alluvial plain.

Sedimentary deposits of Late Devonian age are extremely uncommon in New England because this area was undergoing uplift and erosion, caught in the throes of the Acadian orogeny. Nevertheless, two small basins in northeastern Maine and adjacent New Brunswick (the Saint Andrews and Blacks Harbour basins) contain sediments with plant fossils identified as Late Devonian in age (Schluger, 1973). These sediments constitute the Perry Formation and are composed of very immature, polymictic, boulder and pebble conglomerates, arkoses, graywackes, and others, associated with basaltic lava flows. The sediments rest nonconformably on granite and represent deposition in alluvial fans and fluvial environments associated with high, intermontane fault basins.

In Greenland too, sediments of Devonian age were deposited in intermontane basins. These basins, however, were located in the old Caledonian mountains. Igneous activity continued in the Devonian with extrusion of both mafic and felsic volcanics and intrusions such as the Kapp Franklin Granite. Fluvial and lacustrine strata contain important vertebrate faunas including the first known group of tetrapods, the Ichthyostegalia, from the Upper Devonian Mount Celsius Series (Butler, 1961).

In southern and western parts of the Appalachian Basin, black muds continued to accumulate throughout the Late Devonian, resulting in the Chattanooga Shale, a fine-grained, black, carbonaceous shale with well-developed fissility and continuous, undisturbed laminations suggestive of hemipelagic deposition. The Chattanooga has a very limited fauna, consisting mainly of pelagic forms such as conodonts, which indicate an Upper Devonian to Lower Mississippian age. The Chattanooga is correlative with black shales of the eastern craton (e.g., Antrim, New Albany, and Woodford shales) whose stratigraphy and origin were discussed earlier in this chapter.

C.6. A New Period Begins: Kinderhookian Events in the East

Just before the end of the Late Devonian, sea level rose once more, flooding the great eastern deltaic complex and pushing the shoreline over 160 km eastward. The result, in addition to the extension of the southern black shale lithofacies into western parts of the basin (see above), was the deposition of the Oswayo Formation, a marine unit that overlies fluvial lithologies in western and central Pennsylvania, and the lowermost parts of the Pocono Formation in Maryland and West Virginia. These units span the Devonian–Mississippian boundary. In eastern Pennsylvania, correlative fluvial sediments of the Huntley Mountain Formation were deposited on a narrow coastal plain bounded on the east by coarser, conglomeratic sediments of the Spechty Kopf Formation. As sea level began to drop during mid-Kaskaskian time, all of the streams draining the Acadian Mountains were strongly rejuvenated, producing large quantities of sediment and causing the great deltaic complexes to prograde once again westward. From Pennsylvania to Virginia, large rivers built the Cussewago, Gay–Fink, Cabin Creek, and Carolina–Virginia deltas (see Fig. 6-22). Sediment influx was even greater than during the Late Devonian so that coarse, sandy, nonred sediment, which had been confined to the upper reaches of the coastal

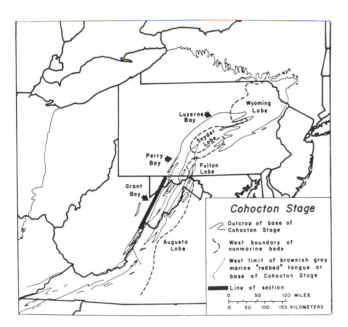

Fig. 6-21. Upper Devonian (Cohocton Stage) eastern shoreline of the Appalachian basin sea. (From Dennison and Head, 1975, *American Journal of Science*, Vol. 275, Fig. 12, p. 1112.)

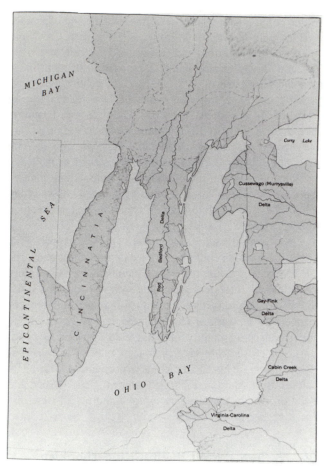

Fig. 6-22. Kinderhookian paleogeography of the northern Appalachian basin. (From Pepper *et al.*, 1954.)

ment of entrenched meanders which then filled with massive, channel sand. This period of rejuvenation was penecontemporaneous with the one postulated for the coastal plain on the eastern part of the Appalachian Basin. The Red Bedford delta grew rapidly southward but was almost completely flooded during late Kinderhookian time. Dark muds of the Sunbury Shale were deposited over most of the area.

Black shale deposition continued in southern parts of the Appalachian Basin throughout the Kinderhookian, resulting in the Big Stone Gap Member of the Chattanooga Shale in Virginia and eastern Tennessee, the Maury Shale in Georgia, Alabama, and the rest of Tennessee, and a thin stratum of clay shale with phosphate nodules in Kentucky (part of which may be correlative with the Sunbury Shale mentioned above). All of these units record a "starved" basin environment as did the Chattanooga Shale below them. Obviously, the new geological period brought little or no physical change to the southeast.

C.7. Osagian Events

As the Osagian Epoch began, the old Acadian coastal plain was once more prograding rapidly westward as a result of falling sea level. Rivers carried immature sediments out of the mountains of the East and deposited them in the four great delta systems of the Central Appalachians (Cussewago, Gay–Fink, Cabin Creek, and Carolina–Virginia deltas). These sediment dispersal systems dominated the geological history of the northern Appalachian Basin throughout most of the Osagian (see Fig. 6-13).

In Pennsylvania, Maryland, northern Virginia, and West Virginia, the Pocono Formation (Pocono Group in Pennsylvania and Maryland) is a coarse, immature sandstone whose composition ranges from arkosic to lithic arenite (Dennison and Wheeler, 1975). It is interbedded with thin layers of conglomerate, sandy shale, and coal. Plant fossils are common but some beds contain marine invertebrate fossils; thus, the Pocono records primarily a deltaic coastal plain with some associated marine strata.

In southwestern Virginia, the Price Formation contains marine strata at the base, overlain by conglomeratic sandstone (the Cloyd Conglomerate Member) and nearly 275 m of crossbedded sandstones, siltstones, and coal (Walker, 1964). Grain size decreases toward the west as does the number and thickness of coal seams. Mineralogical evidence points to a gradual change in provenance from sedimentary near the base to low-rank metamorphic in the middle to plutonic at the top (Dennison and Wheeler, 1975). Such a change in provenance is best explained by the "unroofing" of a rising orogenic source, in which the lowest strata in the depositional area record erosion of the highest rocks in the mountains; as erosion cuts deeper, successively younger strata contain the debris of progressively lower

plain (sometimes called the "Pocono" lithofacies of the Acadian detrital wedge), was carried all the way across the coastal plain to be interbedded with nonred marine sediments, with no red fluvial lithofacies intervening (Meckel, 1970). The result was the Pocono, Price, and Grainger formations of the Central Appalachians.

During earliest Kinderhookian time, a large deltaic complex was developed in Ohio whose source was to the north or northeast; this was the well-known Red Bedford delta described in the classic monograph of Pepper *et al.* (1954). They postulated an extremely lobate delta, fed by the "Ontario river," which prograded southward approximately parallel to the Cincinnati Arch, which separated the "Ohio bay" from the main part of the epeiric sea (see Fig. 6-22). The Bedford Shale, which records the delta's progradation, is a regressive sequence in which bluish-gray shale is overlain by reddish siltstone and sandstone. The Bedford is overlain by the Berea Sandstone which, in some areas, is strikingly channeled into the Bedford, possibly due to rejuvenation of the Ontario river and consequent develop-

levels within the orogen. Farther south, in eastern Tennessee, correlative strata are named the Grainger Formation but they are completely of marine origin (Hasson, 1971). There is no evidence of any significant eastern detrital source farther to the south in the Valley and Ridge.

As the eastern delta systems built westward, they began to affect western parts of the Appalachian Basin. Early in Osagian time in southwestern Ohio and eastcentral Kentucky, prodelta turbidity current sediments began to be deposited over Kinderhookian dark muds. Continued sediment influx developed delta slope and delta front environments, the resulting sedimentary complex comprising the Borden Formation, which is correlative with and grades eastward into the Pocono and Price formations. The high rate of sediment supply and the relatively low-energy conditions of the epeiric sea allowed the development of highly lobate deltas so that specific correlations within the Borden are difficult (Lewis and Taylor, 1979). Nevertheless, the whole unit demonstrates the development of a deltaic coast in central Kentucky by the late Osagian. The Borden Delta Front (see Fig. 6-11) is a line which separates Borden detrital sediments from autochthonous sediments of the Fort Payne Formation in western Kentucky and Tennessee.

Just as lower sea level had affected northern parts of the Appalachian Basin, so too were southern parts affected. Throughout northwestern Georgia, northern Alabama, western and central Tennessee, and western Kentucky, Osagian strata are cherty bioclastic limestones, dolostones, and replacement cherts of the Fort Payne Formation. Although the character of the Fort Payne varies, it represents, in general, the development of a broad, shallow carbonate platform across the south. Chowns and Elkins (1974) proposed that geodes, common in the Fort Payne, formed as pseudomorphs after anhydrite nodules; indeed, many of them still contain relict anhydrite. These evaporite minerals and the nature of dolomitization have led some geologists (e.g., Thomas and Cramer, 1979) to postulate sabkha-type environments for parts of the Fort Payne, possibly associated with low, ephemeral islands. Local, subaerial exposure might also explain the abundant chert: subsurface mixing of saltwater with freshwater lenses associated with islands would produce geochemical conditions in which silicification could take place. Only in the southern part of the Valley and Ridge of Georgia are Osagian sediments found which do not indicate deposition on a shallow carbonate platform. There, dark gray, calcareous shales and argillaceous line mudstones of the Lavender Shale Member of the Fort Payne Formation were deposited in a deep basin adjacent to the platform on the northwest.

Near the end of the Osagian, an apparently eustatic rise of sea level began which ultimately affected the entire Appalachian Basin. In the south, the shallow waters of the Fort Payne platform deepened; carbonate deposition continued but the amount of dolostone and chert steadily decreased as sea level rose and fewer islands were available to provide the chemical environments necessary for dolomitization and silicification. In eastern parts of the basin, Fort Payne lithologies on-lapped Grainger and Price strata as the delta plains were flooded. In Kentucky, too, marine transgression caused the deposition of carbonate sediments on the old Borden delta platform, probably because detrital sediments were trapped in estuaries on the coastal plains to the east. Sea level continued to rise as the Osagian Epoch ended and the Meramecian began.

C.8. The Meramecian Stage in the East

The eustatic rise of sea level which had begun in the Osagian resulted in the last major episode of carbonate deposition throughout the East. But important tectonic events also occurred during the Meramecian. Thus, this epoch is of significance to the overall geological history of the East.

In the southern part of the Appalachian Basin, rising sea level at the end of the Osagian ended deposition of the Fort Payne Formation and initiated deposition of the Tuscumbia Limestone of northern Georgia and Alabama and the Warsaw and St. Louis Limestones of Tennessee, all of which are primarily of lower and middle Meramecian age. They are characterized by bioclastic and micritic limestones with scattered chert nodules and occasional beds of dolostone (Thomas and Cramer, 1979; Milici et al., 1979), indicating that the platform was not yet completely submerged. Sea-level rise continued, however, and by the late Meramecian the whole area was covered with shallow water resulting in a broad, Bahamian-type carbonate platform (Ferm et al., 1972). The Monteagle Limestone of Tennessee, Georgia, and Alabama records this environment and contains evidence of well-developed oolite shoals, tidal bars, and spillover lobes of coarse bioclastic debris. Even here, however, there is evidence of occasional, ephemeral islands, in the form of brecciated caliche paleosoils and laminated, subaerial crusts such as reported from the Florida Keys by Multer and Hoffmeister (1968).

In western and central Kentucky, Meramecian strata are divided into Warsaw, Salem, St. Louis, and Ste. Genevieve limestones (Rice et al., 1979). The lower units record a single carbonate transgressive–regressive cycle, ending with tidal-flat deposits, supratidal dolostone, and scattered evaporites in the lower St. Louis Limestone. Transgression was renewed during late Meramecian time (contemporaneous with Monteagle deposition in Tennessee) with subtidal deposits of the upper St. Louis and Ste. Genevieve. The separation of Meramecian strata into two cycles is also observed in midcontinent strata and has been discussed earlier.

In Ohio, too, carbonate deposition took place throughout the Meramecian, resulting in the Maxville Limestone which spans both Meramecian and Chesterian Epochs (Collins, 1979). The Maxville is thin and discontinuous due to several periods of erosion during its history, one of which corresponds to the break between the two Meramecian carbonate depositional cycles. Final dissection occurred at the end of the Mississippian.

In eastern Tennessee, southwestern Virginia, and southern West Virginia, a transgressive detrital-sediment sequence named the Maccrady Formation was deposited during the early and middle Meramecian. Red-bed strata of the Maccrady are primarily marine except in a small area in Virginia (Montgomery and Giles Counties) where deltaic and fluvial strata are found (Dennison and Wheeler, 1975). Commercially important deposits of gypsum and halite occur in the Maccrady around Saltville, Virginia. Dennison and Wheeler (1975) noted that evaporite deposits of approximately the same age are widespread about the Appalachian Basin (and much of the craton as well) and probably indicate a period of regional aridity.

Overlying the Maccrady is the upper Meramecian Newman Limestone of Tennessee and Kentucky and the Greenbrier Limestone of Virginia, West Virginia, and Maryland. Both of these units are complicated by interbedded detrital sediments derived from the east and southeast but generally represent an eastward extension of the Monteagle/St. Louis–Ste. Genevieve/Maxville limestone sheet of western parts of the basin.

In Pennsylvania, too, carbonate deposition took place, resulting in the Loyalhanna Limestone. However, at the same time, epeirogenic uplift across the northeastern and northwestern parts of the state caused the erosion of Osagian deposits there (the Burgoon Formation of the Pocono Group) which produced sands that were carried southward into the northern Loyalhanna sea. Furthermore, immature detrital sediments continued to pour in from the great eastern mountains just as they had done virtually since the Erian Epoch of the Devonian. The result was the Mauch Chunk Formation, a detrital red-bed unit deposited in deltaic, mudflat, and fluvial environments similar to those of the coastal-plain lithofacies (see the section on the Erian Series). Thus, the Loyalhanna Limestone was restricted to a narrow basin between the uplifted land to the north and the Mauch Chunk delta plain to the southeast.

The last important aspect of Meramecian time in the East is the introduction into Alabama and Georgia of detrital sediments with a southerly to southwesterly source (Thomas, 1979). During the earliest Meramecian, the Fort Payne platform in the Black Warrior Basin of southwestern Alabama began to founder, subsiding more rapidly than its counterpart to the northeast, referred to as the East Warrior Platform. Coinciding with this subsidence was influx of sandy muds from the south and southwest, resulting in the

Floyd Shale, a dark-colored, clay shale with local calcareous and sandy beds and a brachiopod fauna. In the southwestern part of the Black Warrior Basin, the Floyd is about 200 m thick but it thins rapidly to the northeast as it approaches the East Warrior Platform. In northern Alabama, the Floyd is represented by a thin shale tongue called the Pride Mountain Formation which pinches out northward into Monteagle Limestone (see Fig. 6-23). The Floyd is also found in the Appalachian fold belt of eastern Alabama and western Georgia. Its thickness patterns there suggest that deposition may have been penecontemporaneous with early folding. Floyd sediments record distal prodelta, delta front, and marine bay environments which prograded northeastward across the southern part of the great Meramecian carbonate platform (Thomas, 1974).

An explanation of the Floyd Shale's southerly source was proposed by Thomas (1974) who pointed out that Meramecian strata are similar to those of the Ouachita Mountains in Arkansas and Oklahoma (see discussion of the Ouachitas later in this chapter). Thomas posited that both the Mississippian detrital sediments of the southern Alabama Valley and Ridge and those of the Ouachitas were derived from a southern source area. Sediments of Alabama, in Thomas's model, would represent a proximal, shallow-marine facies whereas Ouachita detrital sediments would represent distal, deep-water environments.

Graham *et al.* (1975) suggested that a north-facing magmatic arc existed to the south of the southern margin of North America during the Mississippian and gradually encroached upon the continent. Recent petrographic studies by Mack *et al.* (1983) showed that both Mississippian and Pennsylvanian sediments of the northeastward-prograding detrital wedge complex in Alabama contain epiclastic components indicative of derivation from a sedimentary and low-grade metamorphic fold-thrust belt, an arc complex, and probably a subduction prism. Thus, Mack *et al.* (1983) concluded that these deposits were probably derived from an orogenic uplift produced by arc–continent collision as the result of southward subduction of the North American continent. Whether this arc was an Andean-type arc on the leading edge of an approaching continent or microcontinent or, rather, a volcanic island-arc associated with oceanic lithosphere is unclear, although the abundance of quartz suggests the former possibility.

Also unclear is the relationship between this southern arc and the volcanic arc represented by the Hillabee Greenstone of the Talledega slate belt (Tull, 1982). Hillabee volcanics formed along the southeasternmost segment of the Appalachian continental margin during Early and Middle Devonian time, probably as the result of continent-directed subduction (Tull, 1982). Immediately following Hillabee volcanic activity, the area was subjected to a major phase of dynamothermal metamorphism, which Tull (1982) related

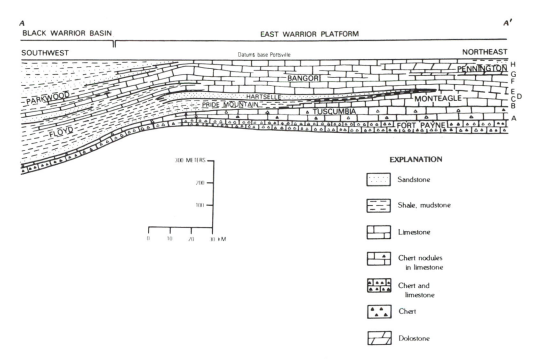

Fig. 6-23. Stratigraphic cross section of Mississippian rocks in Alabama. (From Thomas, 1979.)

to the Acadian orogeny. If Hillabee volcanism was caused by westward subduction of oceanic crust and the southern arc proposed by Graham *et al.* (1975) and Mack *et al.* (1983) indeed faced northward, then some kind of major transform fault must be postulated to connect them (now conveniently buried beneath the Coastal Plain). Alternatively, the southern arc may have faced southward above a north-dipping subduction zone, although this interpretation would not seem to fit the petrographic data of Mack *et al.* (1983). At any rate, by Meramecian time, the southern arc had been drawn close enough to the southern margin of North America for arc-derived sediments to be carried into the Black Warrior Basin, resulting in the Floyd Shale. (We will return to the question of the southern arc and its significance to North American geology in the next section.)

C.9. Chesterian Events: End of the Kaskaskian in the East

Conditions at the beginning of the Chesterian Age were very much like those at the end of the Meramecian. In the East and South, detrital sediments continued to be introduced, causing the development of deltaic systems which prograded across the broad carbonate platform. As time passed, the amount of sediments and their rates of supply increased on all fronts so that by the end of Chesterian time, detrital sedimentation dominated over almost the entire Appalachian Basin (see Fig. 6-15).

In Pennsylvania, Maryland, Virginia, and West Vir-

ginia, deposition of Mauch Chunk red detrital sediments continued on the eastern coastal plain. The thickest Mauch Chunk (greater than 1050 m) is in southwestern Virginia and adjacent West Virginia and was apparently deposited by the Virginia–Carolina delta (Dennison and Wheeler, 1975). The primarily fluvial and deltaic Mauch Chunk overlies Greenbrier and Newman limestones in western parts of its occurrence where it also contains marine beds which alternate with nonmarine beds in a cyclical manner (Dennison and Wheeler, 1975). The presence of coal in these strata suggests cyclothemic deposition, so prevalent in Pennsylvanian deposits of the Eastern Interior.

In the South, too, detrital influx increased during the early Chesterian. In Alabama, Floyd prodelta deposits are overlain by sandstones and shales of the Parkwood Formation which extends northward onto the southern carbonate platform (Thomas, 1974). The Parkwood is more than 700 m thick in the Black Warrior Basin but thins rapidly as it passes onto the East Warrior Platform, disappearing altogether farther northwest where it pinches out into the Bangor Limestone. Thomas (1974) interpreted the Parkwood as a deltaic complex with associated barrier islands, beaches, lagoons, and tidal flats.

In the northwestern part of the Appalachian Basin, sediments of the Chesterian Michigan river sediment-dispersal system were introduced onto the carbonate platform.

From the above discussion, it is clear that the broad Meramecian carbonate platform in the East was being progressively restricted to the Tennessee and Kentucky area.

There, carbonate deposition continued until virtually the end of the period. At the beginning of the Chesterian, carbonate deposition in Kentucky was interrupted by a period of widespread exposure, possibly caused by renewed activity on the Waverly Arch (Rice *et al.*, 1979). At the same time in eastcentral Tennessee, Georgia, and Alabama, a sandstone unit, referred to generally as the Hartselle Sandstone, was deposited. In Alabama, the Hartselle is thickest on the southwest part of the East Warrior Platform and thins gradually eastward and northward, pinching out between the Monteagle and Bangor limestones (see Fig. 6-23). The Alabama Hartselle was deposited as part of a barrier-island complex formed near the southern edge of the East Warrior Platform (Thomas, 1979).

Overlying the Hartselle is the Bangor Limestone, an impure and slightly cherty, bioclastic and oolitic limestone with some micrite and thin, interbedded argillaceous shale. It represents in most areas a shallow subtidal environment as did the Monteagle before it. In Kentucky, too, the Bangor is present but indicates deposition in intertidal and supratidal environments (Rice *et al.*, 1979).

Near the end of the Chesterian, detrital influx accelerated even more; sediments poured into the Appalachian Basin from the north, east, and south. Mauch Chunk deposition was extended even farther west and, in Tennessee and Georgia, the Pennington Formation was deposited. The Pennington is a complex unit with coarse deltaic sands, shales, conglomerates, and coal in its type area (Pennington Gap, southwestern Virginia) and dolostone, limestone, and shale on the west side of Tennessee's Cumberland Plateau. Basal dolostones represent supratidal deposition and contain evidence of Sabkha-type environments (Frazier, 1975). Shales and sandstones were deposited in barrier bars, lagoons, and deltaic environments which migrated westward near the end of Pennington time.

The increasing supply of sediments throughout the Chesterian, the initiation of sediment influx in Tennessee and Georgia from the southeast, and continuing syndepositional deformation all lead to the conclusion that tectonic activity was increasing on the southeastern and southern continental margin. The narrow remnant-ocean between eastern North America and Gondwanaland was almost completely closed and the southern arc was close enough for deltaic sediments, derived from the arc, to be deposited in the Black Warrior Basin and East Warrior Platform. By the Early Pennsylvanian, full-scale continental collision was taking place along the southeastern continental margin and coarse sediments were pouring westward into the southern Appalachian foreland basin.

The Mississippian–Pennsylvanian boundary is hard to place in the eastern part of the Appalachian Basin because deposition was continuous across the boundary. Several years ago, the boundary was drawn at the contact between Mauch Chunk/Pennington/Parkwood strata and the overlying Pottsville Group. Most recent workers (e.g., Edmunds *et al.*, 1979; Milici *et al.*, 1979; Thomas, 1979) have placed the boundary somewhere within the lower part of the Pottsville.

C.10. Mississippian Events in the Northern and Maritime Appalachians

At the northern end of the Appalachian orogen, sediments continued to be deposited in small, localized fault basins associated with the eroding Acadian mountains (similar to the Perry Formation of the Upper Devonian discussed earlier). But, during the latest Devonian, a large triangular rift-valley developed in northern Nova Scotia and Newfoundland. More than 3000 m of detrital sediments and extrusive igneous rocks was deposited within this rift valley, called the Fundy Basin, while only about 300 m of sediment accumulated on the more stable platform areas outside the rift (Belt, 1968). Rifting within an area of general compressive stress (associated with continental suturing) was explained by Belt (1968) as being caused by tension between northeast–southwest-oriented, en echelon wrench faults which were produced by east–west-directed compression. As rifting proceeded, a basin-and-range-type topography developed within the rift; extremely coarse, alluvial-fan sediment, associated with rising blocks, graded out into finer fluvial and lacustrine sediments in subsiding areas. Penecontemporaneous with sedimentation was the extrusion of over 600 m of rhyolitic, dacitic, and andesitic lavas. The resulting sequence of sediments and volcanics is named the Horton Group which was, itself, later deformed by northwest–southeast-directed compression. During Chesterian time, the whole area was flooded by rising seas and carbonate sediments were deposited on adjacent shelf areas as well as within the Fundy Basin. A few of the higher fault-block mountains within the rift remained as islands, which continued to shed coarse, fanglomerate sediments into the sea. Local evaporite deposits were left when the sea withdrew before the end of the Mississippian. The sedimentary complex of limestones, sandstones, conglomerates, and evaporites deposited as a result of this Chesterian sea is called the Windsor Group.

D. Ouachita Mountains, Marathon Uplift, and Related Areas

D.1. Introduction

The southern continental margin of Paleozoic North America has received little attention in previous chapters of this book because we believe that understanding of this large and complex area would best be served by a single, coherent description of its development and early history. After we have briefly discussed that early history, we will turn to the important changes in sedimentation and tectonism that occurred in this area during the Mississippian Period.

Today, Paleozoic rocks of the ancient southern continental margin are, for the most part, buried beneath Cretaceous and younger strata of the Gulf Coastal Plain. They are exposed from under this sedimentary blanket only in the Ouachita Mountains of southwestern Arkansas and southeastern Oklahoma and in the Marathon and Solitario uplifts of western Texas (see Figs. 6-24 and 7-40). The Ouachita Mountains are a region of folded and faulted strata whose main structural trend is east–west and whose structures resemble those of the Southern Appalachians. The Marathon region also contains deformed Paleozoic strata but their main structural trend is northeast–southwest. Major sedimentary and tectonic events in both areas were similar in their style and in the timing of their inception. In addition, deformed Paleozoic sediments of very different character are found in the Criner Hills, Arbuckle, and Wichita mountains of southwestern Oklahoma and adjacent Texas. While this area contains sediments and structures of very different types than those in the Ouachita and Marathon regions, its history is intimately linked with that of the rest of the southern continental margin.

Although these areas are separated today by thick and extensive deposits of younger sediments, they are continuous in the subsurface. Drilling in the Llano uplift region of central Texas has revealed Paleozoic rocks similar to those in the Marathon area. Further, drilling in the Fort Worth area has revealed the presence of Paleozoic strata whose petroleum-rich rocks also resemble those of the Ouachita and Marathon regions. These observations suggest that a single paleogeographic setting characterized all these areas. In addition, Thomas (1973) has shown that Appalachian structures and strata can be traced along strike beneath the Coastal Plains of Alabama and Mississippi and has demonstrated some continuation of major structural trends from the Ouachitas to the subsurface Appalachians. Finally, as we will see, several authors have commented on the similarity of geological events in the Ouachitas and the Black Warrior Basin of Alabama.

D.2. Earliest History

D.2.a. Cambrian Rifting

The Paleozoic history of the southern continental margin began in the Cambrian with a major continental rifting event. The rifting probably began with the development of a large prerift thermal dome centered in southern Oklahoma and adjacent Texas where the Ouachita–Marathon belt changes direction from northeast–southwest to east–west. During thermal doming, erosion stripped away a great deal

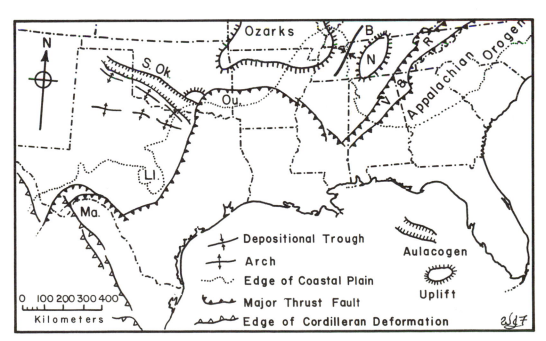

Fig. 6-24. Map of major structural features of the Paleozoic southern continental margin (based principally on King, 1977). Symbols are as follows: B, Batesville Channel; Ll., Llano Uplift; Ma., Marathon area; N, Nashville Dome; Ou., Ouachita area; S. Ok., Southern Oklahoma Aulacogen; V & R, Valley and Ridge Province of the Appalachian orogen.

of older rocks, unearthing the Tishomingo Granite, a large plutonic body of Elsonian age which forms the basement beneath Paleozoic strata in this area. As doming proceeded, a three-armed rift developed, two arms of which continued to open, forming the Paleozoic ocean to the south of North America. The Ouachita–Marathon belt represents sediments deposited in this southern ocean and along the new continental margin. Deposits from earlier stages of rifting along the continental margin (rift-valley, proto-oceanic gulf, and young-ocean-basin stages) are not today exposed but may be buried deep beneath large, imbricated thrust-sheets which were driven up onto the continental margin from the south and east during the Late Paleozoic orogeny that affected the entire southern area (the Ouachita orogeny; see Chapter 7).

D.2.b. Southern Oklahoma Aulacogen

The third arm of the rifted thermal-dome failed to continue opening and thus remained as a deep rift valley extending northwestward far into the craton. Into this deep rift, called the Southern Oklahoma aulacogen, were carried coarse, immature detrital sediments (e.g., the Tillman Graywacke) derived from uplifted blocks along the sides of the rift. These sediments were interbedded with thick basaltic and rhyolitic flows (and associated sills and dikes) to form a stratigraphic sequence over 5000 m thick.

By the latest Cambrian, the rate of subsidence associated with the rift-valley stage had decreased as the aulacogen passed into a stage of moderate regional subsidence, possibly isostatic in nature due to the weight of the great thickness of volcanic rocks along the rift's axis. During this time, the Reagan Sandstone was deposited. The Reagan, which is part of the extensive Upper Cambrian transgressive deposits of the midcontinent area, is thickest in the area of the old rift but extends outward, thinning over areas not part of the rift. The area of regional subsidence surrounding the old aulacogen is called the *Anadarko Basin*. By the end of the Cambrian, a carbonate shelf had developed over the whole southern continental margin (a part of the Cambro-Ordovician carbonate platform which existed over much of North America; see Chapter 4). In the Anadarko Basin, carbonates of this platform are called the Arbuckle Limestone.

Pre-Tippecanoe emergence resulted in the post-Sauk discontinuity (see Chapter 5) which, in the Anadarko Basin, underlies the Simpson Group. The Simpson contains a relatively thick sequence of detrital and carbonate sediments, the lower half of which (i.e., Joins, Oil Creek, McLish, and Tulip Creek formations) is correlative with the much thinner St. Peter Sandstone of the central interior; the upper part (the Bromide Formation) is correlative with the Platteville Formation (see Fig. 6-25). The relative thickness of the

Simpson indicates that the Anadarko Basin continued subsiding during the Middle Ordovician. Upper Ordovician carbonate sediments of the Corbin Ranch, Viola, and Fernvale formations and muds of the Sylvan Shale all thicken in the Anadarko area but generally record shallow-marine environments whose history mirrors that of adjacent cratonic areas.

During the Silurian and Devonian, the Anadarko Basin continued to subside slowly with limy sediments, such as the oolitic Keel Formation and the bioclastic Cochrane and Clarita formations, accumulating on a shallow carbonate platform. These strata contain very little detrital sediment (less than 5% insoluble residue; Amsden *et al.*, 1967) but the Upper Silurian Henryhouse Formation contains up to 20% insoluble detrital residue. The aerial distribution of this detrital material indicates that it was derived from a source somewhere to the south or southeast. Helderbergian strata are composed of dirty micritic limestones (the Haragan Formation) and well-washed bioclastic limestones (the Bois d'Arc Formation). These are overlain, across the Wallbridge discontinuity, by cherty bioclastic limestones of the Middle Devonian Frisco Formation and the Upper Devonian Woodford Shale (a Chattanooga Shale equivalent).

It will be seen from the foregoing discussion that the Southern Oklahoma aulacogen, which began as a rapidly subsiding rift valley and evolved into a normal cratonic basin, followed the typical developmental history of aulacogens.

D.2.c. Early History of the Ouachita–Marathon Continental Margin

Lower Cambrian deposits are not exposed in the Ouachita–Marathon belt and, thus, little is known of the early opening of the southern ocean. It is clear, however, that by the Late Cambrian, a relatively stable continental shelf was present in central Oklahoma and Arkansas on which accumulated carbonate platform sediments similar to Cambro-Ordovician deposits of the Anadarko Basin and the southern craton. This shelf area is referred to as the Arkoma Basin (see Fig. 6-24) and it continued to receive shallow marine sediments, primarily carbonates, throughout Early and Middle Paleozoic time. These generally shallow sediments of the Arkoma and Anadarko basins are referred to as the Arbuckle facies (Briggs, 1974).

The oldest strata of the Ouachitas are of Lower Ordovician age and record deposition in a deep, starved basin. Collier and Mazarn Shales are composed of dark, graptolitic clay-shales and distal carbonate turbidites, probably derived from the carbonate platform to the north. The Crystal Mountain Sandstone, which occurs between the above two shale units, consists of thick-bedded quartz sandstones interpreted by Morris (1974b) as proximal turbidites and de-

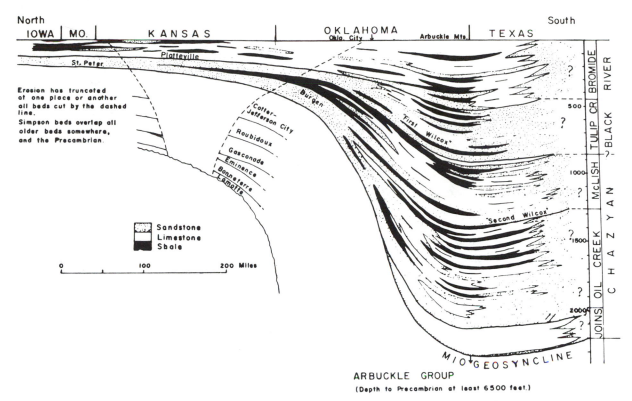

Fig. 6-25. Cross section of the Simpson Group and its relation to the St. Peter Sandstone, generalized from sections in the Arbuckle Mountains and areas to the north. (From Ireland. 1965.)

bris-flow deposits. The Blakely Sandstone of Middle Ordovician age conformably overlies the Lower Ordovician Mazarn Shale with no evidence of the Owl Creek discontinuity; the Blakely contains fine-grained quartz arenites, silts, and shales with less common diamictite beds, some of which contain blocks of granite and metamorphosed arkose up to 15 m in diameter. Overlying Ordovician strata (Womble Shale, Bigfork Chert, and Polk Creek Shale) are composed primarily of distal turbidites, graptolitic shales, and pelagic cherts, again indicating deposition in a deep, starved basin. Generally, sediments of this deep basin are referred to as the *Ouachita facies*.

From the above discussion, one may infer the presence of a deep, oceanic basin in the south and a shallow, continental shelf in the north. The nature of the continental-slope transition between these two environments is problematical because it has been obscured by thrust-faulting during the Ouachita orogeny. Nevertheless, some speculations can be made from observations of the character and petrology of diamictite beds within the Ouachita facies. Large boulders were derived by slumping on the upper continental slope and in submarine canyons and were transported into the deep basin by debris flows. The slope was probably relatively steep, judging from the great size of some of the blocks, and the granitic and metaarkosic composition of the blocks suggests derivation from basement exposures, possibly from steep cliffs along the sides of submarine canyons.

The Silurian Blalock Sandstone of the Ouachita facies recorded an important change in the sedimentation and tectonism of the deep basin. The Blalock is a shaley flysch deposit in which greenish-gray shale is interbedded with thinly laminated, poorly sorted, feldspathic graywackes. These immature, feldspathic sands were interpreted by Morris (1974b) as having been derived from a magmatic arc somewhere to the south of the area. The reader will recall that the Silurian Henryhouse Formation of the Anadarko Basin also contains evidence of detrital influx from the south or southeast, suggesting that adjacent continental areas also began to receive sediments from a new, southern source. If these sediments do, indeed, represent a magmatic arc to the south, one may surmise that opening of the southern ocean, which had begun during the Late Precambrian, had ceased by the Silurian and that convergence had begun.

Overlying the Blalock Sandstone is the Missouri Mountain Shale which is composed primarily of shale with thin, quartz-rich turbidites near the top. These turbidites were probably derived from the craton (Morris, 1974b). Thus, we see in Silurian strata the inception of multiple sediment-sources for the deep oceanic basin. Arc-derived sediments were texturally and mineralogically immature and came from the south while craton-derived sediments were quartz-rich and came from the north. This pattern will be repeated in Mississippian and Pennsylvanian deposits.

D.2.d. Devonian Events: The Arkansas/Caballos Novaculite

Devonian strata of the Ouachita–Marathon belt are composed of novaculite interbedded with lesser amounts of dark-gray, fissile shale, occasional red shale, thin sandstones, and rare zones of manganese nodules and manganiferous chert. Novaculite is a generally light-colored siliceous rock similar to chert except that it is composed of very finely crystalline quartz instead of chalcedony; it is usually well-bedded and massive. (Novaculite is also known as "Arkansas stone" and is prized for its utility as a sharpening stone.) In the Ouachitas, these strata are named the Arkansas Novaculite and in the Marathon region they are called the Caballos Novaculite. Fossils are rare in both units but conodonts, radiolarians, sponge spicules, and spores have been found (Sellars, 1967). Correlations based on conodonts suggest that novaculite deposition virtually spanned the Devonian Period and persisted into the Early Mississippian. In the Ouachitas, novaculite strata rest conformably on the underlying Missouri Mountain Shale (see above) and are overlain by the Stanley Group (Mississippian) across a gradational contact.

The origin of the Arkansas/Caballos Novaculite has been a source of controversy since the 19th century. Most workers have considered the novaculite to be a deep-water deposit but some have suggested a shallow-water origin. [The interested reader is directed to an excellent discussion of this controversy by Folk and McBride (1976, 1977).] In his defense of a deep-water origin for the Caballos, McBride (in McBride and Folk, 1977) presents the following observations: (1) Caballos sediments are very similar to modern oceanic siliceous oozes; (2) they are associated with red shales which he believes to represent deposits similar to modern deep-ocean red clays; (3) turbidite strata and manganese nodules are present in the Caballos, as in modern oceanic sediments; and (4) there is a complete absence of shallow-water fauna or obvious littoral or neritic sedimentary facies. In addition, the paleogeographic setting of the Ouachita–Marathon belt (discussed above) and the character of under- and overlying units also indicate a deep-water setting.

The origin of silica for these novaculites is also controversial. Some geologists (e.g., Folk and McBride) have suggested a biogenic origin for the silica, citing as evidence the presence of radiolarians and spicules. Lowe (1975) postulated that oceanic upwelling may have supplied silica-rich bottom water to near-surface zones, encouraging high productivity among silica-secreting planktonic organisms. Oth-

ers (e.g., Sellars, 1967) have argued that a biogenic source would not be sufficient to account for all the silica in the novaculite and have suggested a volcanic source, possibly as fine siliceous ash. It is interesting to consider the paleogeographic location of the Ouachita–Marathon region during the Devonian as shown in Fig. 6-2. Note that this area was in the Southern Hemisphere, just south of the equator and, therefore, squarely within the track of the Southeast Trade Winds, which would have been ideally situated to carry volcanic ash to the Ouachita–Marathon area from a southern or southeastern magmatic arc.

D.3. The Stanley Group: Development of a Mississippian Submarine Fan

Mississippian strata of the Stanley Group in the Ouachita area are over 3100 m thick and are divided into the Tenmile Creek, Moyers, and Chickasaw Creek formations. These strata consist generally of shaley flysch composed of dark-gray shale interbedded with fine-grained quartz and feldspathic wackes whose accessory grains include chert, quartzite, and schist with rare shale pebbles. Sands of these formations are weakly graded and commonly contain small-scale cross-lamination near the tops of sand beds. In the central Ouachitas of Oklahoma, sandy strata are thin and comprise only about 15% of the unit; they have been interpreted as distal turbidite deposits. In the southern Ouachitas, however, sand beds are thicker, make up more of the section, and are of slightly greater grain size. Sand also makes up more of the section in the eastern Ouachitas; in the region around Hot Springs, Arkansas, for example, the basal 150 m comprises a sandy flysch sequence with individual sandstone lenses up to 6 m thick. These deposits are interpreted as proximal turbidites. Associated with these sediments in the southern Ouachitas are volcanic ash beds composed of siliceous tuffs with coarse, euhedral plagioclase crystals, quartz grains, devitrified glass, and occasional fragments of altered basalt. In the northern part of the Ouachitas (known as the Frontal Ouachitas), the Stanley is shaley, contains thin beds of pelagic chert and much less tuff, and has massive diamictite deposits with exotic blocks of shallow-water limestone and quartz sandstone which were probably derived by slumping on the edge of the continental shelf to the north (e.g., the Pitkin Limestone).

Based on the above discussion, the Stanley is interpreted as a turbidite deposit whose proximal facies was in the east and south and whose distal facies occupied the central and northern part of the area. Paleocurrent directions of turbidity currents may often be inferred from the orientations of flute casts which were formed by scouring of the bottom by extreme turbulence in the head of the turbidity current as it rushed downslope and out onto the basin floor. Measurements of such features have been reported by sever-

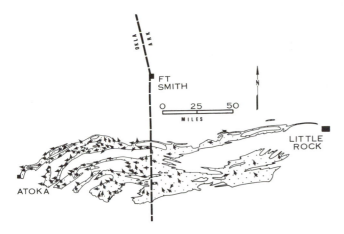

Fig. 6-26. Paleocurrent data from the Stanley Group. (From Morris, 1974b, reproduced with permission of the Society of Economic Paleontologists and Mineralogists.)

al authors (e.g., Johnson, 1968) and show that the predominant paleocurrent direction in much of the Ouachitas was toward the west except in the south where lower parts of the Stanley contain evidence of northerly to northwesterly directed currents (see Fig. 6-26).

Evidence presented above suggests that turbidity currents from somewhere to the southeast carried great quantities of muddy and fine-sandy sediment into the Ouachita–Marathon Basin, resulting in the Stanley Group in the Ouachitas and the Tesnus Formation in the Marathon area.

(We have not described the Tesnus because it is similar to the Stanley, except for being somewhat less sandy; it has also been interpreted as a turbidite deposit.) In an excellent analysis of Stanley paleoenvironments, Niem (1976) showed that basal parts of the sequence in Oklahoma represent deposition on an abyssal plain. Into this deep basin, submarine turbidite-fans prograded from the east and south (see Fig. 6-27). Proximal turbidite facies represent the upper part of the fan (characterized by well-defined distributary channels with levees along their sides) while the distal facies represents the outer part of the fan (characterized by a smooth surface which merged into the abyssal plain). Niem suggested that several different fans (or different lobes of the same fan) may have been active, accounting for the differences in paleocurrent direction discussed above.

A glance at the paleogeographic reconstruction of the southern continental margin for the Mississippian (Fig. 6-28) will convince the reader that sediments of the Stanley Group are closely related to those of the Black Warrior Basin in Alabama (see the section on Mississippian events in the Appalachians). There, thick sequences of shale and sandstone were deposited in delta complexes which prograded northward into the Black Warrior Basin and the East Warrior Platform beyond. Thus, sediments of both areas were derived from a tectonically active source area to the south of the continental margin. The presence in the Stanley Group of ash beds, euhedral plagioclase feldspar crystals, and altered grains of basalt implies that the source area was probably a magmatic arc. This arc was probably closer to

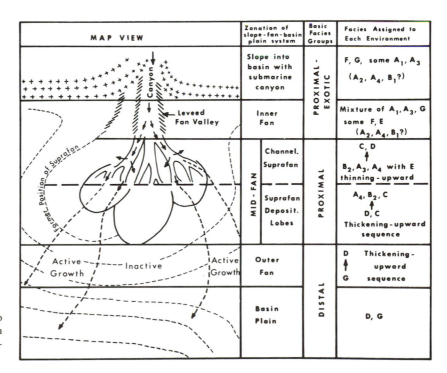

Fig. 6-27. Submarine-fan model of Stanley Group deposition. (From Niem, 1976, reproduced with permission of the Society of Economic Paleontologists and Mineralogists.)

Fig. 6-28. Paleogeographic reconstruction of the southern continental margin during Mississippian time. (From Chamberlain, 1978, reproduced with permission of the Society of Economic Paleontologists and Mineralogists.)

the Black Warrior Basin area and trended off to the southwest so that arc-derived sediments formed a great fan-shaped mass which sloped off into the deep Ouachita–Marathon Basin to the west. In this deep basin, flysch sediments were deposited while Floyd and Parkwood molasse sediments accumulated in Alabama. As time went on, the great sedimentary wedge grew westward with the molasse facies migrating over the flysch facies; here, we may see an explanation of the long-observed tendency for flysch sediments to be overlain by molasse.

At the beginning of the Mississippian Period, the arc was probably far enough away so that arc-derived sediments did not affect the Ouachita–Marathon area. However, plate convergence, which had probably begun in the Silurian, continued throughout the Mississippian with the effect that arc-derived turbidites were introduced in increasing quantity. The nature of this convergent boundary is highly speculative; global paleogeographic reconstructions for the Silurian and Devonian suggest that Gondwanaland was drifting northwestward toward North America and the hypothesized convergent boundary was probably related to ocean-closing associated with that northwesterly drift. Graham *et al.* (1975) and Wickham *et al.* (1976) have suggested that the convergent boundary consisted of a south-dipping subduction zone and a north-facing magmatic arc although other arc-configurations cannot be ruled out at present. Oceanic crust of the North American plate was subducted beneath the advancing Gondwanaland plate (assuming a north-facing arc as mentioned above), on whose leading edge was a magmatic-arc orogen. It is interesting in this regard to note that major orogenic event of Mississippian age is recorded in Venezuela and adjacent northwestern South America, where Pennsylvanian strata unconformably overlie deformed older strata (Walper and Rowett, 1972).

E. The Cordilleran Continental Margin

For almost 500 Myr., since late Precambrian rifting, a passive (i.e., intraplate) continental margin had existed on the western edge of the North American craton. Sediments of this margin were deposited in continental shelf, slope, and rise environments as well as within the deep, oceanic areas farther to the west; associated with the oceanic terrane today are volcanic-arc complexes which may be displaced terrances. This paleogeographic pattern persisted into Middle and Late Devonian time. Soon thereafter, however, a major change occurred which transformed the Cordilleran margin into an active continental margin and which established a tectonic land and adjacent flysch-basin along much of the western margin of North America. This tectonic event, termed the Antler orogeny, is frequently overlooked or quickly brushed over in historical geology textbooks because it apparently produced neither a major mountain belt nor an extensive detrital wedge. Nevertheless, the Antler orogeny marked an important change in the character of the western continental margin and, therefore, has great significance to the later history of the North American Cordillera. In this section, we will first describe the nature of the Cordilleran margin prior to the Antler orogeny and then discuss the nature, origin, and effects of this important tectonic event.

E.1. The Pre-Antler Continental Margin

E.1.a. Continental Shelf

Post-Tippecanoe emergence affected most of the western continental shelf, as it had much of the rest of the North American continent. (The reader will recall that the western part of the craton remained emergent until the major marine transgression of the Taghanic Age.) Quartz-arenitic sands, probably derived from weathering on the craton, are found at the base of Middle Devonian strata in most areas of the western continental shelf and constitute the oldest deposits of the Kaskaskian transgression in the Cordilleran region. Continued rise of sea level resulted in reestablishment, on the western continental shelf, of a carbonate-platform environment similar to the one that had existed there during the Silurian and Early Devonian. In central Nevada, for example, Middle Devonian limestones of the Nevada Formation (Group, in some areas) were deposited in a variety of subtidal marine environments; coeval deposits of the Simonson Dolomite in eastern Nevada and Utah are composed of sandy dolostones which represent shallow marine to supratidal environments (Poole *et al.*, 1977). A similar pattern may be seen in the Middle Devonian Jefferson Formation of Idaho, where limestones grade eastwards into sandy, dolomitic strata, again representing the change from an open-marine, shelf environment in the west to progressively more shallow-water conditions in the east. On the outer shelf and upper continental slope, thinly bedded limestones, mudstones, and lesser amounts of chert accumulated in a deeper-water environment to the west of the carbonate platform; one example of such outer-shelf deposits is the Denay Limestone of westcentral Nevada. The same general depositional pattern persisted into the early Late Devonian, with subtidal, marine carbonates of the Devil's Gate Limestone and Guilmette Formation of Nevada and western Utah passing eastwards into intertidal to supratidal carbonates of the Hyrum and Gilson dolomites of central Utah (Poole et al., 1977).

In western Canada, too, carbonate deposition dominated the shelf environment during the Middle and early Late Devonian (Bassett and Stout, 1967). Earliest carbonate strata (equivalent to the Elk Point Group of the western craton, discussed earlier in this chapter) are represented by limestones and dolostones of the Bear Rock Formation, which represent shallow, marine-shelf conditions. These deposits grade eastward into evaporitic strata, suggesting that much of the shelf interior was restricted from the western, open-marine shelf environment during the early Middle Devonian. Overlying subtidal limestones of the upper Middle Devonian Hume Formation reflect the advent of more open-marine conditions to the western Canada shelf. At the eastern edge of the carbonate shelf in southwestern Canada and adjacent Montana was a broad peninsula formed by the Peace River and Alberta uplifts (see discussion in the section of this chapter on the northwestern craton). Detrital sediments were shed westward from the Peace River–Alberta peninsula, resulting in an alluvial–deltaic red-bed complex which prograded out onto the western carbonate platform, where Hume-equivalent limestones of the Harrogate Formation were accumulating (Bassett and Stout, 1967). Continuing marine transgression toward the end of the Middle Devonian pushed the carbonate platform eastwards onto the craton margin of western Canada and allowed deposition of deeper-water, pelagic shales of the Hare Indian Formation on the outer shelf.

Bioherms were common on the shelf carbonate-platform, but the most spectacular and well-known examples occurred at the outer margin of the platform in western Canada. In that area, massive coral and stromatoporoid buildups formed the barrier between the broad, shallow platform of the shelf and western craton and the deeper, outer shelf and slope to the west (see Fig. 6-8). Some Late Devonian examples had greater than 100 m of relief and slope angles up to 10 degrees, e.g., the Miette Bank and the Ancient Wall complex (Wilson, 1975). (Further discussion of the western carbonate platform occurs in the section on Kaskaskian strata of the northwestern craton.)

E.1.b. Oceanic and Volcanic Terranes

To the west of the continental shelf and slope during the Middle and early Late Devonian was a deep, oceanic basin. Sediments of this basin consist mainly of radiolarian chert with lesser amounts of micritic carbonates, graptolitic shales, and thin, graded sandstones. An example of these deposits is the Slaven Chert of western Nevada (Poole et al., 1977). The present extent of the Devonian oceanic terrane is a matter of considerable discussion. There are a large number of simatic lithostratigraphic units within the Basin and Range physiographic province but many of them appear to be younger than Devonian or else were accreted to the continent later than the Devonian. Also, exactly which rock units one accepts as being related to the Devonian ocean-basin terrane depends on one's view of the Antler orogeny, as we will discuss later in this section.

Near the continental slope, deposits of the ocean-basin terrane contain extrabasinal limestones, characterized by graded bedding, flute casts, and other sole markings. These strata probably represent the deposits of turbidity currents which originated along the edge of the carbonate platform; thus, these deposits may represent a continental-rise prism formed at the foot of the continental slope. Examples of such strata include the Woodruff Formation of western Nevada and the Milligen Formation of western Idaho.

In areas to the west of the ocean-basin terrane, there are a variety of lithotectonic terranes containing volcanic rocks. In the Klamath Mountains of northern California, for example, Devonian lava flows and shallow intrusive rocks of the Copley Greenstone and Balaklala Rhyolite represent a magmatic-arc complex (Irwin, 1977; Poole et al., 1977). Overlying these rocks is the Kennett Formation which contains siliceous shale, rhyolitic tuff, and limestone, possibly deposited within an intra-arc basin. Correlative with these strata are Devonian flows and volcaniclastics of the Sylvester Group in British Columbia and unnamed volcanic graywackes, lava flows, and limestones in islands of southeastern Alaska. In northwestern Washington is a sequence of rocks called the Chilliwack Group which consists of Devonian carbonate buildups associated with a thick section of volcanic rocks (Danner, 1977). The overall picture that one gets from the above observations is of a sequence of volcanic islands and associated carbonate buildups lying somewhere to the west of the continental margin.

Associated with the volcanics and volcaniclastics of the Klamath Mountains is a series of metamorphosed and structurally complex rocks such as the Salmon Hornblende Schist and the Abrams Mica Schist (Irwin, 1977). These units are composed of a wide variety of lithologies, such as quartz-mica schist, micaceous marble, calc-schist, and amphibolite associated with large ultramafic complexes such as the Trinity ultramafic sheet. All of these lithologies have been chaotically intercalated, along with metamorphosed rocks of various ages from Ordovician to Devonian, by west-directed thrust faults into a massive mélange whose combined structural thickness is over 5 km. Poole *et al.* (1977) considered this mélange to represent a subduction complex associated with the Klamath volcanic arc.

Opinions differ on interpretation of the ocean-basin terrane. Poole *et al.* (1977) considered strata of this area to represent a back-arc, marginal basin which separated the continental margin from the volcanic-arc terrane farther to the west. This concept is based on the view that volcanic-arc and subduction-complex rocks to the west of the ocean-basin terrane have been part of North America's geological structure since their formation in the Ordovician or Silurian. Other authors (e.g., Dickinson, 1977; Speed and Sleep, 1980) have suggested that volcanic terranes west of the Devonian continental margin may be displaced terranes, i.e., lithotectonic terranes which formed in some other place and which were accreted to the continent later in their history. These two views of Devonian paleogeography are tied to contrasting interpretations of the tectonic mechanism of the Antler orogeny and cannot be further elaborated without discussion of the Antler's effects.

E.2. The Antler Orogeny: Destruction of the Cordilleran Passive Margin

Beginning in the latest Devonian and continuing through most of the Mississippian, an episode of tectonic activity, called the Antler orogeny, affected the Cordilleran continental margin. The Antler orogen is traced from west-central California, through central Nevada, to southcentral Idaho (Poole and Sandberg, 1977). The major effects of the Antler orogeny were regional metamorphism (of greenschist to amphibolite grade) and large-scale thrust-faulting, during which a thin mass composed of slivers of simatic, oceanic crustal rocks and associated deep-water sediments was thrust more than 100 km eastward onto the edge of the continental margin (Roberts *et al.*, 1958). The emplacement of such a mass of oceanic-facies rocks is termed obduction, and the Antler orogeny is best considered to have been an obduction event (Dickinson, 1977). It is significant that no magmatic activity occurred within the Antler orogen, either before, during, or after the orogeny. Indeed, volcanic-arc-derived detritus is apparently absent from the Antler allochthon (Dickinson, 1977). The key fault over which emplacement of the Antler allochthon took place is the Roberts Mountains thrust of northcentral Nevada and the overthrust sheet is termed the Roberts Mountains allochthon; within the Roberts Mountains allochthon, however, there are a large number of smaller thrusts, usually associated with intense deformation of the fine muds and cherts which comprise the bulk of the thrust mass.

The Roberts Mountains allochthon formed a tectonic land on the outer edge of the continent which supplied immature detrital sediments that were swept eastwards into an adjacent, deep foreland basin (a "foredeep" basin) developed on the outer edge of the continental shelf penecontemporaneously with the Antler orogeny. This basin, located between the carbonate platform of the inner shelf and the Antler orogen at the outer edge of the shelf, accumulated detrital flysch throughout the Mississippian Period. In western Canada, too, eastward-directed thrusting and large-scale detrital-sediment deposition indicate the onset of tectonic activity during the latest Devonian and Mississippian (Monger *et al.*, 1972). Thus, the Antler orogenic event apparently affected the whole western continental margin and heralded the beginning of tectonic activity which has persisted episodically right up to the present.

E.2.a. Plate-Tectonics Speculations

The nature of the Antler orogeny has been discussed by many authors (e.g., Roberts *et al.*, 1958; Burchfiel and Davis, 1972, 1975; Churkin, 1974; Poole, 1974; to mention only a few). Nevertheless, there is no universal agreement on a tectonic model for the event. The following discussion has been influenced by the ideas of Dickinson (1977), Poole and Sandberg (1977), and Speed and Sleep (1980).

There are two principal, competing tectonic models of the Antler orogeny. The first, championed by Poole and Sandberg (1977), suggests that obduction associated with the Antler orogeny was the result of back-arc spreading. These authors favor the view that the oceanic terrane west of the continental-margin carbonate platform was a marginal basin, formed behind a west-facing volcanic arc. With acceleration in the rate of seafloor spreading during the latest Late Devonian, the rate of subduction would have increased as well as the rate of back-arc spreading. The result, in this scheme, was to cause the crust of the oceanic (or, marginal) basin terrane to break along its eastern side, at the base of the continental crust, and to be shoved up and onto the margin of the continent. This process, termed back-arc thrusting, is illustrated in Fig. 6-29A. The reader should note that this model presumes that arc rocks west of the oceanic terrane were associated with the Cordilleran margin all through their history.

Probably the biggest problem with the back-arc thrusting model lies in the mechanics of emplacing such a thin, thrust slice, composed primarily of pelagic and other deep-water deposits, across more than 100 km of continental crust. As well as 100 km of lateral transport, the back-arc thrusting model requires that the allochthon be carried *up* the continental slope and out onto the edge of the shelf. Such tectonic transport would seem improbable without some kind of "grease zone" over which the allochthon

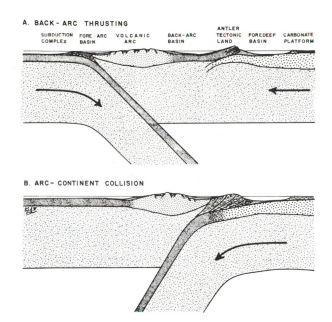

A. BACK-ARC THRUSTING

SUBDUCTION COMPLEX | FORE-ARC BASIN | VOLCANIC ARC | BACK-ARC BASIN | ANTLER TECTONIC LAND | FOREDEEP BASIN | CARBONATE PLATFORM

B. ARC-CONTINENT COLLISION

Fig. 6-29. Alternative plate-tectonic models for the Antler Orogeny; see text for discussion.

could glide with little friction. A second problem was pointed out by Dickinson (1977), who noted that the Roberts Mountains allochthon must have been emplaced at a rate of several centimeters per year, but such a rate is about an order of magnitude greater than strain rates associated with back-arc thrusting calculated from other areas.

The other major tectonic model for Antler obduction calls upon the collision of an east-facing magmatic arc with the western continental margin (Fig. 6-29B). In this model, the Roberts Mountains allochthon is considered to have been a subduction prism formed by the off-scraping of supracrustal deposits as oceanic crust of the North American plate subducted westward beneath the approaching volcanic arc (Speed and Sleep, 1980). The above-mentioned mechanical difficulties of emplacing the Roberts Mountains allochthon up and onto continental crust are avoided in this model because the western edge of the North American continent is considered to have *underthrust* the subduction prism rather than the prism overthrusting the continent. Further, Dickinson (1977) said that the rate of emplacement of the Roberts Mountains allochthon is comparable to those for ordinary subduction.

Dickinson (1977) pointed out that the arc-collision model seems to neglect evidence from the Klamath Mountains of California which indicates that eastward, rather than westward, subduction was associated with the western arc terrane. But, in the arc-collision model, the volcanic arc in question is considered to have been a displaced, or "exotic" terrane whose previous history was not necessarily related to that of the North American continent. Since the

Klamath arc is now also considered to be a displaced terrane, indications as to subduction polarity relating to the Klamaths probably have little significance to Antler tectonics. The actual identity of the arc massif responsible for the Antler orogeny is uncertain because, as we will discuss in the next chapter, post-Antler rifting apparently affected the Cordilleran continental margin, with the result that any arc-terrane which may have been there was shifted offshore and may now be hidden within the Cordilleran tectonic collage. An alternative possibility is that the arc ceased to be active following collision so that it cooled and subsided, forming the simatic basement upon which Pennsylvanian and younger strata were deposited in western Great Basin physiographic province. (We will return to this subject in the next chapter.)

E.2.b. The Cordilleran Continental Margin during the Antler Orogeny

i. Beginning of the Orogeny.
The time of onset of the Antler orogeny may be estimated from the first appearance of sediments derived from the orogen. In southern Nevada, for example, lower Senecan carbonate strata contain interbedded, mature quartz-arenites which were probably derived from the craton. Overlying these strata are deposits, also dated as Senecan, which contain immature sands derived from the west (Poole *et al.*, 1967). These sands would seem to announce the initiation of uplift along the continental margin. In eastern Nevada and western Utah, Chautauquan-age carbonate strata of the Devil's Gate Limestone are interbedded with and overlain by siltstones and mudstones of the Pilot Shale, whose turbidite and debris-flow sediments (termed "protoflysch" by Poole *et al.*, 1977) were derived from the embryonic Antler tectonic land and carried eastwards to be deposited on the outer continental shelf.

By the middle of the Mississippian Period, the Antler orogeny was fully in progress. It is instructive to consider a profile across the Cordilleran continental margin at this time in order to see the effects of the orogeny on the continent. We will begin our discussion of the Mississippian Cordillera at the westernmost part of the area and will work our way eastwards.

ii. Ocean-Basin and Volcanic Arc Terranes.
To the west of the Antler orogenic land were the ocean-basin and related magmatic-arc terranes, similar to those discussed earlier in this section. For example, in the northern Sierra Nevada, a subduction complex and magmatic arc are represented by strata of the Peale Formation. The Peale contains a lower member consisting of rhyolitic lava-flows and an upper member with chert, sandstone, shale, and volcaniclastics (Poole and Sandberg, 1977). In southeastern

Alaska, volcanic activity resulted in basaltic and andesitic flows of the Saganaw Bay Formation (Churkin, 1974). Again, however, it is important to remember that these rocks may represent events that occurred far from the margin of North America. Ocean-basin strata of Mississippian age west of the Antler land include the Bragdon Formation of northern California and the Coffee Creek Formation of Oregon. In addition, flysch sediments derived from the Antler land and carried westward into the ocean basin are found in the Schoonover Formation of western Nevada.

iii. Antler Tectonic Land.

The Antler tectonic land itself was a rapidly rising, structurally complex, mountainous ridge or string of islands, probably deeply dissected by erosion and heavily forested (based on the amount of plant detritus in Antler-derived sediments). Poole and Sandberg (1977) suggested that carbonate banks may have been present in some areas, perhaps as fringing or barrier reef complexes. By the late Mississippian, as the Antler orogeny began to wane, sediments overlapped up onto the Antler land, resulting in the deposition of subtidal to supratidal carbonates and coarse detritus.

iv. Foreland Basin.

On the outer continental shelf, just east of the Antler land, were three, deep, rapidly subsiding flysch basins: the Eleana trough in southcentral Nevada, the Webb–Chainman–Diamond Peak trough in northwestern Nevada, and the Cooper Basin trough in east-central Idaho. A submarine slope dropped steeply from the Antler land into these foredeep basins and was cut by submarine canyons, possibly associated with the mouths of major river systems on the tectonic land. From these canyons issued sediment gravity-flows that resulted in flysch sedimentation within the foreland basins. Poole (1974) provided an excellent description of Antler flysch-sedimentation. He showed that flysch sediments there may be divided into three groups based on their method of deposition: (1) *hemipelagic sediments,* which are fine muds and silts introduced into the basin as dilute suspensions (such as turbidity currents) but deposited by simple settling in quiet water. These deposits today contain disseminated pyrite and up to several percent organic matter, indicating a strongly reducing bottom-environment. (2) *Turbidite sediments,* which are characterized by dirty sands containing grains of quartz, chert, quartzite, and rare feldspar. These sands often contain graded bedding, flute casts, groove marks, and other turbidity-current-related structures as well as displaced shallow-marine fossils and even rare terrestrial plant remains. (3) *Debris-flow sediments,* which are composed of diamictites and conglomeratic sandstones whose clasts consist of chert, quartzite, rare carbonate pebbles, and intraformation ''rip-up'' clasts (some of which are up to 1 m in diameter). Poole considered these sediments to represent deposition in submarine fans

which grew out from the rising Antler land and which prograded into a quiet, euxinic, bathyal basin between the tectonic land and the shelf carbonate-platform (see Fig. 6-30). Flysch deposits of the Antler foreland basin comprise the Eleana Formation of southern Nevada, the Tonka and Diamond Peak formations of central Nevada, the Chainman Shale in eastern Nevada and western Utah, and the Scorpion Mountain Formation of central Idaho (Poole and Sandberg, 1977).

On the eastern side of the flysch basin, far removed from Antler-derived turbidity currents and cut off from craton-derived detrital sediments by the carbonate-platform-edge buildup, was a deep, sediment-starved basin. This basin was characterized mainly by phosphatic shales such as those in the upper McGowan Creek Formation in central Idaho, basal members of the Little Flat Formation in southeastern Idaho, Deseret Limestone and Woodman Formation in western Utah, and strata overlying the Joana Limestone in eastern Nevada (Poole and Sandberg, 1977). Into the eastern part of this starved basin were carried carbonate turbidites derived from the edge of the carbonate platform.

v. Continental Shelf.

Standing high above the starved eastern side of the Antler foredeep basin was the carbonate platform, which occupied the main part of the continental shelf. The carbonate platform had persisted on the shelf since its reestablishment during the Taghanican transgression, seemingly little affected by the orogenic excitement to the west. Near the end of the Devonian, however, the shelf and western cratonic margin were fragmented into a series of local uplifts and basins. This may have been caused in some way by Antler-related compression at the edge of the continental crust. In the Rocky Mountains region of the northwestern United States, the Central Montana uplift was reactivated, and thrusting occurred along the Cedar Creek anticline. Basinal areas between uplifts were sites of deposition of black, carbonaceous shales such as the

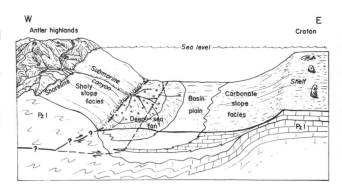

Fig. 6-30. Turbidite fans from Antler lands. (From Nilsen, 1977, reproduced with permission of the Pacific Section, Society of Economic Paleontologists and Mineralogists.)

Exshaw Shale, the Sappington Member of the Three Forks Formation, and the Leatham Formation, all three of western Montana and eastern Idaho. In northwestern Utah, the Stansbury anticline was uplifted, resulting in the Stansbury peninsula which shed detrital material westward onto the shelf.

At the close of the Devonian, the sea apparently withdrew almost completely from the cratonic margin and shelf, resulting in a disconformity over most of the area (except in the Antler flysch basin which was deep enough not to be affected by lowering of sea level). When the sea returned during Kinderhookian time, it left a transgressive sequence over most of the shelf. In Montana, for example, the Madison Group contains the record of this transgression (Smith, 1977). The lower Madison formation is the Lodgepole Limestone whose lowest member, the Cottonwood, records lagoonal and shallow-marine environments near the coast of the Madison sea. The overlying Paine Member contains dark, thin-bedded, argillaceous micrites interpreted as deep-water limestones. Thus, the Paine probably represents shelf-derived carbonate-turbidites deposited in the starved basin east of the Antler flysch basin (see above). Paine deposition ended with the progradation of shallow carbonate-shoals composed of crossbedded oolitic and bioclastic sediments which today compose the upper, Woodhurst Member of the Lodgepole Limestone. Farther to the east, correlative strata of the lower Madison Formation represent carbonates of the eastern shelf and craton margin. By Osagian time, a regressive carbonate platform was present over much of the shelf; this is represented by the Mission Canyon Limestone (the upper formation of the Madison Group) of Montana, the Brazer Dolomite of Utah, and part of the well-known Redwall Limestone of the Grand Canyon.

It is interesting to notice that these Osagian regressive carbonates were contemporaneous with regressive detrital-wedge strata far to the east, at the edge of the Acadian orogen (e.g., the Borden, Pocono, and Price formations to name only a few), and with the shallow-marine, cherty carbonates of the southern Appalachian Basin, such as the Fort Payne Formation. Thus, one can see that the early part of the Osagian Epoch was probably a time of eustatically lowered sea level.

At the end of the early Meramecian, epeirogenic uplift drained the sea from the carbonate platform, causing erosion and karst-development over much of the area. Returning seas probably reestablished the shelf carbonate-platform but much of the record from this time has been lost by erosion associated with post-Kaskaskian emergence. Only on the western margin of the platform are Meramecian and Chesterian carbonates preserved. They, like their predecessors, record shallow subtidal to intertidal environments. Examples include the Monroe Canyon Limestone of south-

eastern Idaho and the Ochre Mountain Limestone of Utah. Bissell and Barker (1977) suggested that the Great Blue Formation of northwest-central Utah was a deep-water limestone deposited on the downwarped edge of the shelf and, thus, probably represents continued carbonate-flysch deposition on the eastern side of the Antler flysch basin.

E.2.c. Final Stages of the Antler Orogeny

During the Late Mississippian, flysch sedimentation in the foredeep basin had reached its full development as very thick deposits of coarse chert and quartzite conglomerates accumulated in the channeled, upper-fan facies of the Diamond Peak Formation and sandy, shaley deposits formed in the middle- and outer-fan facies of the Chainman Shale. The sea-level fluctuations which affected the shelf carbonate-platform had little or no effect in the deeper water of the flysch basin.

But by later Chesterian time, the rate of sediment influx had apparently exceeded the rate of basin subsidence so that the basin began to fill up. Thus, upper Chesterian strata, such as the Scotty Wash Sandstone of southern Nevada, contain reworked flysch sediments deposited as crossbedded, lenticular strata associated with thin coal beds. Poole (1974) stated that the final, infilling stage of the flysch basin was characterized by shallow-marine sediments, generally deposited above wave base. These sediments contain a shallow-water fauna characterized by brachiopods, gastropods, pelmatozoans, and rare trilobites.

As the foreland basin filled up, reworked flysch sediments began to spill out of the basin and spread onto the western carbonate platform, resulting in detrital strata of the Indian Springs Formation in southern Nevada and the lower part of the Manning Canyon Shale in western Utah. Root structures (stigmaria) of *Lepidodendron* and similar lycopod plants at the base of the Indian Springs Formation probably indicate estuarine conditions (Poole, 1974).

Trough infilling continued across the Mississippian–Pennsylvanian systemic boundary, recording the waning phase of the Antler orogeny. We will return to these final, Antler-related events in the next chapter.

F. The Innuitian Continental Margin

To the Innuitian continental margin, too, the Devonian and Mississippian brought a major orogeny and concomitant detrital sedimentation. Indeed, it is interesting to observe that the Late Devonian was a time of major orogenic activity on Appalachian, Cordilleran, and Innuitian margins, all at the same time, a condition not matched at any other time in the Phanerozoic history of North America. In this section, we will describe the Innuitian continental margin prior to the Late Devonian and then discuss the effects and possible cause of the Ellesmere orogeny, which affected the entire northern margin. We will also discuss briefly the effects of this orogeny on the northern Cordilleran continental margin. The reader is reminded once again (as in Chapter 5) that knowledge of Innuitian stratigraphy is very scanty and that specific tectonic interpretations are thus tentative.

F.1. Middle Devonian Stratigraphy

The paleogeographic setting of the northern continental margin at the beginning of the Kaskaskian Sequence was very much like that revealed by strata of the upper Tippecanoe Sequence, described in Chapter 5. In the south was a stable craton and continental shelf on which existed a broad carbonate platform. North of this shelf, a well-developed continental slope facies led down into the deep-water environment of a marginal basin, the Hazen Trough. This basin was bordered to the north by the Pearya magmatic arc, which we have hypothesized to be a north-facing arc. We will now briefly examine the Middle Devonian stratigraphy of each of these regions.

F.1.a. Continental Shelf

The Tippecanoe sea apparently withdrew from the northern shelf, as it had from the rest of the continent, during the later Early Devonian, resulting in the post-Tippecanoe Wallbridge discontinuity which underlies Middle Devonian strata of the Innuitian shelf. During the earliest Middle Devonian, there was a widespread transgression which reestablished a broad carbonate platform on the shelf (see Fig. 6-31). The Blue Fjord Formation, which extends virtually the length of the Arctic Platform from Banks Island to Ellesmere Island, is a micritic limestone with fossiliferous beds containing well-preserved coral, stromatoporoids, brachiopods, nautiloids, pelecypods, and others (Miall, 1976), and which probably represents a broad, shelf lagoon behind shelf-edge organic buildups, such as those of the western Canada carbonate-platforms. On Banks Island, limestones of the Blue Fjord Formation pass downward gradationally into a micritic dolostone member. This dolostone is probably correlative with dolostones of the Disappointment Bay Formation, which underlies the Blue Fjord on Cornwallis Island. These dolostones probably represent coastal and supratidal conditions on land areas associated with the Coppermine and Cornwallis uplifts, which continued to be active from earlier times.

In northern Alaska, carbonates of the Skajit Limestone probably represent deposits of a similar carbonate platform. Further, Miall (1976) has pointed out that the Blue Fjord limestones were probably continuous with limestones of the

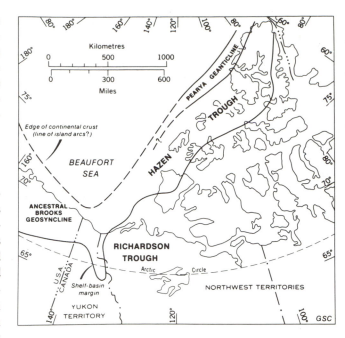

Fig. 6-31. Paleogeography of Innuitian continental margin during the Middle Devonian. (Modified from Miall, 1976, *Geological Society of America Bulletin,* Fig. 6, p. 1606.)

Gossage Formation of the northern Cordilleran continental margin. Indeed, one may easily imagine that Middle Devonian carbonate deposits of both the northern and western continental margins were deposited on the same carbonate platform which ran continuously from New Mexico up to the northern Yukon and on around to Ellesmere Island.

Lower Middle Devonian strata of the Arctic Platform on eastern Ellesmere Island are composed of evaporitic, quartzose red beds of the Vendom Fjord Formation which overlies the post-Tippecanoe disconformity. The source of Vendom Fjord detrital sediments was to the east, possibly the Precambrian Shield (Kerr, 1967).

F.1.b. Continental Slope and Marginal Basin (Hazen Trough)

To the north of the Blue Fjord Formation are contemporaneous, calcareous and silty shales, argillaceous limestones, and siltstones with interbedded, coarse bioclastic limestone. These deposits, which thin toward the north, comprise the Orksut Formation of the western Arctic Platform and the Eids Formation in the east (Miall, 1976). They are interpreted to represent a continental slope facies which formed a transition between the carbonates of the shelf and the deeper-water deposits of the marginal basin. The coarse bioclastic limestones had a steep initial-dip and were probably reef-talus deposits related to nearby shelf-edge bioherms.

At the western end of the Canadian Archipelago, the Orksut–Eids slope facies grades northward into the Nanuk Formation (Miall, 1973). The Nanuk consists of black siliceous shales and pelagic chert with minor, thin, carbonate turbidites. The Nanuk, like the similar Ibbett Bay Formation (Ordovician to Silurian) of the same area (see Chapter 5), is considered to have formed in a deep, starved, euxinic basin at the western end of the Hazen trough (see Fig. 6-31).

Toward the eastern part of the Hazen trough, on the other hand, Orksut–Eids slope sediments graded northward *not* into a deep, starved basin, but rather into a thick flysch sequence composed of calcareous graywackes and siltstones of the Cape Rawson Group (Kerr, 1967). Cape Rawson detrital sediments were derived from the north, in the Pearya magmatic arc, and poured southward into the marginal basin in a series of submarine fans which probably formed a fan complex that grew slowly toward the deep Nanuk Basin (see Fig. 6-31). Thus, the same pattern seen in the Ordovician and Silurian (i.e., Cape Phillips slope facies, Imina flysch facies, and Ibbett Bay starved basin facies) persisted into the Middle Devonian.

F.1.c. Pearya Magmatic Arc

The Pearya magmatic arc (see Fig. 6-31) had continued to be a positive feature since the Ordovician. During the Early Devonian, over 3750 m of coarse, red detrital sediment was deposited in the northern Axel Heiberg Island area. These deposits, named the Stallworthy Formation, contain interbedded conglomerate, crossbedded quartz arenites, and red siltstones; most of the pebbles from conglomerates are chert but include reworked volcanic materials and carbonate clasts, in part schistose, which Trettin (1967) inferred to have been derived from uplifted Rens Fjord complex (Cambrian?). These sediments were probably deposited in fluvial and deltaic environments and thus represent a back-arc detrital ramp, similar to the one discussed in Chapter 5. Clearly, the Stallworthy also represents considerable uplift of the arc terrane just to the north.

The Stallworthy is overlain conformably by approximately 3300 m of volcanics and immature sediments of the Svartevaeg Formation (Trettin, 1967). The Svartevaeg is divided informally into two members, the lower of which is composed of almost 1000 m of pillow lavas (mostly spilites), tuffs, volcaniclastic graywackes, and minor shale. The upper member contains over 2300 m of graded, volcaniclastic graywackes with lesser siltstones, diamictites, and rare lava-flows. The upper 60 m is composed of basaltic flows with fossiliferous-limestone fragments of Silurian age, some up to 4 m across. The Svartevaeg is interpreted as representing proximal turbidites and debris-flow deposits

associated with large-scale submarine volcanic activity; it probably indicates a deep-marine environment.

The tectonic implications of the Stallworthy and Svartevaeg are of considerable interest since they represent two extremely different depositional environments which are vertically gradational. This gradational contact obviously suggests rapid subsidence during earliest Svartevaeg time but the large amount of volcanic activity is not characteristic of a normal marginal basin. Indeed, the clear implication is that these sediments were deposited squarely within the magmatic arc. Yet, the great thickness of Svartevaeg deposits and their apparent deep-water environment would indicate an extreme rate of subsidence. For these reasons, we interpret the Svartevaeg Formation as having formed in an intra-arc basin. Intra-arc basins are usually related to tensile stress across the arc due not to extension but to uplift. It is thus possible that, paradoxically, the intense subsidence revealed by Svartevaeg strata may well be the first indication of the great uplifts which immediately followed in the Late Devonian. Middle Devonian uplift in the Pearya arc may also be inferred from the Cape Rawson flysch, which had a northerly source. In this regard, we might point out that in a large and complex magmatic arc, downdropped intra-arc basins are usually associated with uplifted blocks in other areas. The exact nature of the Devonian Pearya arc and its relations to strata on the southern margin of the Hazen trough are obscured by younger deposits of the Sverdrup Basin which overlie Devonian (and older) strata.

F.2. The Ellesmere Orogeny

As shown in the last section, uplift along the magmatic arc to the north of the Innuitian continental margin had accelerated during the Middle Devonian. By the Late Devonian, much of the northern part of the continent was affected by a major orogeny, termed the Ellesmere orogeny, which produced mountains out of which streamed massive quantities of sediments. These sediments accumulated in a large foreland basin on the northern part of the craton. Deposits of this detrital mass are found today all along the Innuitian fold belt and Arctic Platform; they are present in the Brooks Range of Alaska and even along the northern part of the Cordilleran margin. In this section, we will discuss briefly the stratigraphy of the Ellesmere detrital wedge and describe the nature of associated igneous activity, metamorphism, and structural deformation. We will conclude with a possible tectonic model for the Ellesmere orogeny.

F.2.a. Ellesmere Detrital Wedge

Uplift in the Pearya arc accelerated during later Middle Devonian time so that Cape Rawson flysch (see above) was supplied at a rate that exceeded the rate of Hazen trough subsidence; thus, Cape Rawson flysch literally filled the trough. An increase in the rate of later Middle Devonian sedimentation may also be seen in the western part of the Hazen trough, which had previously been a deep, starved basin (in which Nanuk Formation black shales and cherts were deposited; see above). By the later Middle Devonian, however, this region was receiving detrital turbidite sediments whose source was toward the northeast (Miall, 1976). These sediments comprise the Weatherall Formation, which is the lowest unit of the Melville Island Group. Miall considered Weatherall turbidites to be the distal fringe of a prograding delta complex, probably located to the northeast.

Deltaic and alluvial deposits of the Late Devonian Series are present throughout the Innuitian region. On the eastern Arctic Platform, approximately 3200 m of varicolored quartz arenites, sandy mudstones, shales, and thin coals of the Oska Bay Formation represents mainly alluvial and deltaic environments. These sediments were derived from the north and graded southward into dirty carbonates of the Bird Fjord Formation, which probably represent the shallow, distal fringe of the detrital wedge as it prograded over the old cratonic carbonate-platform. In western areas of the Arctic Platform, Weatherall deposition continued into Senecan time but was superseded during Chautauquan time by crossbedded quartz arenites of the Hecla Bay Formation, which is the middle unit of the Melville Island Group. These sediments probably represent delta-front sands which grew southward into the old Hazen trough. The Hecla Bay is overlain by argillaceous sands of the Griper Bay Formation (upper unit of the Melville Island Group) which probably formed in delta-plain and fluvial environments. Thus, considered together, sediments of the Oske Bay, Bird Fjord, and Melville Island units constitute remnants of a great molasse wedge deposited in the foreland basin behind the main Ellesmere orogen.

In Alaska, too, Late Devonian orogeny resulted in a detrital wedge which thickens and coarsens toward the north. In the Brooks Range, especially in the Romanzof and British mountains, very coarse gravels of the Kanayut Conglomerate grade laterally into sands and shales of the Hunt Fork Shale, which is present over much of the northern and western Brooks Range (see Fig. 6-32). These deposits are also deltaic and alluvial sediments of a molasse wedge.

Toward the southeast, conglomerates and sands of the Kanayut and Hunt Fork formations grade into deep-water deposits of the Imperial Shale, which comprises virtually the entire Upper Devonian section of the outer Cordilleran continental shelf in northwestern Canada (see Fig. 6-5). Northern sections of the Imperial Shale (in the Norman Wells regions of the Yukon Province, for example) consist of over 1600 m of graywacke with graded bedding, flute

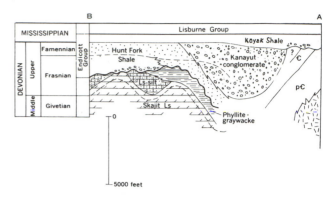

Fig. 6-32. Stratigraphic cross section of Hunt Fork and Kayanut formations. (From Brosgé and Dutro, 1973, reproduced with permission of the American Association of Petroleum Geologists.)

casts, groove marks, and other turbidite features. These sediments represent Innuitian-orogen-derived flysch deposited in the deep basin between the Innuitian detrital wedge, the Cordilleran outer magmatic arc of southeastern Alaska, and the western Canada craton. The nature of this basin is obscure because of extremely complex structural and stratigraphic relations in this area of intersecting orogenic trends. At any rate, the Imperial Shale thins and fines toward the southeast, gradually losing its turbidite character and passing into a deep, shelf facies where fine muds accumulated in front of the great carbonate platforms of the western Canada craton.

F.2.b.   Tectonic Effects of the Ellesmere Orogeny

Igneous activity and regional metamorphism associated with the Ellesmere orogeny affected much of the Innuitian area. In the northern Ellesmere and Axel Heiberg Islands, small plutons of granite and quartz diorite were emplaced during the Middle and Late Devonian. In the Romanzof and British Mountains area, larger granitic plutons, such as the Okpilak batholith, were emplaced at approximately the same time (Sable, 1977). Greenschist-grade regional metamorphism occurred in the northwestern part of Ellesmere Island, associated with some of the larger plutons. Regional metamorphism of higher grade affected the Romanzof Mountains region.

Structural deformation probably occurred in at least two phases. The first took place during the Late Devonian and affected pre-Chautauquan strata. It was this phase which was associated with the uplift of the Pearya arc region and resulting Ellesmere detrital wedge. Figure 6-32 illustrates the effects of this event in the Brooks Range area; notice that Senecan strata were folded prior to the deposition of the Hunt Fork and Kanayut formations. These

Chautauquan strata were deformed by the second folding event, of Mississippian age. It was this Mississippian event which had the greatest effect on continental margin strata of the Innuitian fold belt. During this event, supracrustal strata were displaced toward the south over a major décollement and were folded and faulted at the same time. This folding event was accompanied by general uplift of the entire Innuitian continental margin.

F.2.c. Plate-Tectonics Model of the Ellesmere Orogeny

The north-facing Pearya arc had been established during the early Paleozoic and subduction of oceanic crust beneath it had continued throughout the Ordovician, Silurian, and early Devonian (see Fig. 6-33A). Riding on this subducting plate was a continental mass, possibly the Siberian block (see Fig. 6-2). During the Middle Devonian, the outer parts of the Siberian passive continental margin began to enter the trench area and to interfere with plate subduction. Increase in crustal thickness beneath the Pearya are probably resulted in the first phase of uplift, associated with derivation of Cape Rawson flysch in the Hazen trough and

with the development of the Svartevaeg intra-arc basin (see Fi. 6-33B). By the Late Devonian, full-scale arc–continent collision was taking place, causing terminal uplift of the arc (which would have been thrust up onto the advancing Siberian continental margin) and emplacement of granitic plutons (see Fig. 6-33C). This collision was thus responsible for the Late Devonian phase of folding, granitic intrusion, and uplift which resulted in the Ellesmere detrital wedge. But plate convergence continued and a new subduction zone, with reversed polarity, was established beneath the Siberian collisional orogen. A second collision followed shortly thereafter, this time between the Siberian continental block and the Innuitian margin of North America (see Fig. 6-33D). This collision thrust supracrustal strata of the partially subducted North American continental margin back toward the craton, over the décollement mentioned above. Following the Ellesmere continental collision, plate convergence ceased and uplift occurred throughout the Innuitian area during the later Mississippian.

F.3. Mississippian Events

The general uplift of the northern margin caused the whole area to be eroded so that sediments derived from there were carried off in major drainage systems, ultimately being deposited in paralic environments of the craton south of the Canadian Shield. The result in the Innuitian area of Canada is that there are today no Mississippian sediments present. (The only exception is the existence of latest Mississippian strata at the bottom of the Sverdrup Basin but since the history of this feature rightly belongs with discussion of later geological periods, we will defer an examination of these deposits until the next chapter.)

In the Brooks Range area of Alaska, however, Mississippian strata are present and reveal the waning stages of the Ellesmere orogeny. In the eastern Brooks Range, Lower Mississippian shales of the Kayak Shale lie with striking angular unconformity on the Upper Devonian Hunt Fork and Kanayut formations (see Fig. 6-32). The Kayak Shale of the Romanzok Mountains consists of clean, well-sorted quartz arenites, shales (often with carbonaceous material), coal and conglomerate in its lower part, and interbedded shale and sandstone in its upper part (Sable, 1977). These sediments are interpreted as a transgressive sequence with upland, fluvial environments represented at the bottom, overlain successively by fluvial sediments of a lower coastal plain, paludal (i.e., swampy) deposits, and marginal marine sediments (e.g., lagoonal, estuarine, barrier island). Thus, deposits of the Kayak Shale reveal that basin subsidence in the south was no longer matched by equal or greater uplift of the orogen to the north. Kayak strata grade southward into and are overlain by Mississippian and Pennsylvanian carbonates of the Lisburne Group. The lower (i.e., Mis-

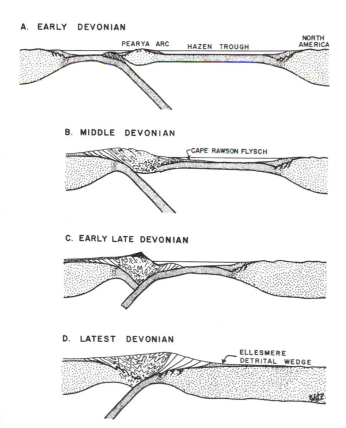

Fig. 6-33. Plate-tectonics model of the Ellesmere orogeny; see text for discussion.

sissippian) formation of the Lisburne Group is the Alapah Limestone, a partly silicified and dolomitized bioclastic and micritic limestone which becomes sandier toward the north. The Alapah probably represents deposition on a shallow carbonate platform with moderate current energy. Carbonate deposition began in southern parts of the Brooks Range area during the Early Mississippian but it did not begin in the main orogenic area until the Late Mississippian. Farther to the south, shallow subtidal strata of the Alapah pass into the "black cherty unit," an informal member composed of dark-gray to black mictitic limestone, chert, shale, and minor dolostone and phosphate. These sediments may have been deposited in deeper-water conditions to the south of the main Alapah platform.

Sedimentation was apparently continuous from the Mississippian to the Pennsylvanian on the Lisburne platform because no unconformity separates Alapah strata from Pennsylvanian Wahoo Limestone.

G. Revolutions among Invertebrates and the "Age of Fishes"

G.1. Invertebrate Revolutions

Painted with a broad brush, the fossil record seems to present patterns of events which characterize certain geological intervals. The current paleobiological literature pays great attention to these patterns, in a general attempt to understand causes of mass extinction and the reradiation of new forms. During the early and middle Paleozoic, several apparent "explosive radiations" are commonly recognized: specifically, in the Early Cambrian (trilobites and many primitive shelled animals; see Chapter 4), in the Early Ordovician (a wide variety of carbonate-shelled invertebrates; see Chapter 5), and in the Early Devonian (many new fishes, and a host of lower invertebrate taxa; see following subsection). Also, several mass extinctions apparently occurred during the early and mid-Paleozoic: in the Late Cambrian–Early Ordovician transition (loss of many trilobites and other stocks), in the Late Ordovician–Early Silurian transition (loss of many carbonate-shelled taxa), and during the Late Devonian (see next discussion). Data for these generalizations come from many sources but the reader is directed especially to Raup (1976), Signor (1978), Harland *et al.* (1967), Sepkoski (1979), and Boucot (1983), for a range of ideas on this controversial subject.

Theories abound for the nature and causes of sizable changes in past life. One overriding problem is in the nature of paleontology itself: dealing with imperfect (i.e., fossil) evidence engenders uncertainties about the quality of data; Sheehan (1977) and Raup (1977) particularly address this topic. Two critical uncertainties are (1) the possibility of loss of information in preservation (a concept analyzed in the science of *taphonomy*) and (2) the possibility of bias in the creation of new taxa by paleontologists. Since it is professionally advantageous to "discover" new taxa, there is a natural tendency to split taxa and erect new groups. Since analyses of past radiations and extinctions are largely based on numbers of taxa appearing and disappearing at various times, such studies incorporate both potential sources of paleontological error. Then too, the possibility of differential preservation of taxa (i.e., many forms are soft-bodied or otherwise rarely fossilized) may seriously skew the fossil record toward shelled forms.

Because these problems are so prominent, in our thinking, and because the literature is so rich in theories on extinction and diversification of past life, this book will largely shy away from discussion of these topics. We will, however, examine the mass extinctions of the Permo-Triassic and Cretaceous–Tertiary boundary intervals in some detail because of their evident magnitude and reality.

In Chapter 5, we introduced most invertebrate groups which have important Paleozoic fossil records. Table I presents a summary, using vernacular names, of the appearances and disappearances of invertebrates during Kaskaskian time. Some authors (e.g., House, 1967) consider the Late Devonian extinctions to have been very sizable, whereas many others are less impressed. The reader may survey Table I and make his or her own decision.

Reasons for such apparent (or real) changes of the Earth's biota of the past are commonplace in the literature, but many are difficult to discredit. With respect to the Late Devonian, it has been frequently suggested that the emergence of landmasses shrank marine habitats and led to intense competition among invertebrate groups. Other conjectures involve innovations in predators (e.g., among shell-cracking predators; see Signor and Brett, 1984), climate changes, and the new roles of vertebrates which were rapidly diversifying during the Devonian. In fact, any and all may be partial solutions to a multifaceted problem. An equally valid possibility is that the Devonian events reflect a natural reduction of overly diversified taxa, and that the protracted Late Devonian time of crisis may reflect a stabilizing effect of competition.

G.2. The Age of Fishes

In two prior chapters, the earliest vertebrates were shown to have appeared in the Late Cambrian and to have comprised a population of mostly jawless forms through the Ordovician and Silurian. Jawed fish appeared in the Silurian and reached their zenith in the Devonian. From Devonian times onward, *gnathostomes* (vertebrates with jaws) un-

Table I. Tabulation of Marine Invertebrate Changes from Middle/Late Devonian to Early Mississippian[a,b]

Devonian and older taxa lost	New Mississippian taxa
All receptaculitids	Four ammonoid superfamilies
Atrypids, pentamerids, four families of rhynchonellids, all but one family of orthid brachiopods	Thirteen families of rugose coral
	Four orders of blastoids
All primitive pelmatozoans (homalozoans, cystoids, others)	Numerous lower taxa (genera) of crinoids
	Numerous lower taxa of fenestrate bryozoans
Four ostracode families	Numerous lower taxa of spiriferid brachiopods
All phacopids, all but one ptychoparioid trilobite suborder	
Four orders of nautiloid cephalopods	
All graptoloids, all but one family of dendroid graptolites	
Fifteen families of rugose coral	
Seven ammonoid (goniatite) superfamilies	

Summary of the changes in "dominant" taxa in the Kaskaskia Sea

Silurian through Lower Devonian	Middle through Upper Devonian	Lower Mississippian
Pentamerid brachiopods	Phacopinid trilobites	Crinoids and blastoids
Trepostome bryozoans	Spiriferid brachiopods	Fenestrate bryozoans
Calymenid trilobites	Cryptostome and fenestrate bryozoans	Spiriferids
Tabulate and rugose coral	Atrypid brachiopods	Goniatites
Stromatoporoids	Crinoids and blastoids	
Crinoids	Tabulate and rugose coral	
Nautiloids	Goniatite cephalopods	
	Dendroids	

[a]Taxa are listed at various hierarchies used in text discussions.
[b]Data from Harland *et al.* (1967), House (1967), and other sources.

equivocally dominated the seas and ultimately the land, while Agnatha declined such that by the Mississippian, all jawless forms were reduced to parasitic lampreys and hagfish. In fairness, it should be noted that jawless ostracoderms were quite abundant during the Devonian, and may have even exceeded their earlier diversity; nevertheless, by the end of the period they were gone. The following discussion will be organized around current taxonomy of the Paleozoic gnathostome fish as presented by Moy-Thomas and Miles (1971).

G.2.a. Superclass Teleostomi (Bony Fish: Acanthodii and Osteichthyes)

There is an anomaly observed right at the outset of this discussion: although cartilaginous tissue is usually a precursor of bony tissue in ontogenetic and evolutionary lineages, here we find the oldest jawed fish to be bony, whereas more "primitive-looking" cartilaginous fish are later appearances. Acanthodians are the oldest jawed vertebrates, appearing first in Niagaran strata. The overall morphology is not striking: they are small, fusiform, and feature paired fins, single gill covers, scales rather than armor, large eyes, often large spines stiffening the fins (hence a common name "spiny sharks"), and ossification of the vertebrae, braincase, and other skeletal elements. Although they are quite distinguishable from Osteichthyes (advanced bony fish, including all Paleozoic through modern types), they hardly resemble their presumed agnathan ancestors. The group was apparently common in the Early Devonian, and declined gradually through the era, becoming extinct in the Early Permian. The most common fossils of acanthodians are the fin spines and scales.

Osteichthyes appeared in the Early Devonian and are divided into three infraclasses: Actinopterygii (ray-finned fish), Crossopterygii (lobe-finned fish), and Dipnoi (lungfish). All infraclasses date back to the Devonian and survive today; however, only actinopterygians are common at present.

Actinopterygii appear in the lowermost Middle Devonian (Onesquethawan) as two families, and expanded rapidly to 13 new families in the Mississippian. Their morphological characteristics are those of modern bony fish: possession of a swim bladder, paired fins, single gill covers, fins with a large membranous portion stiffened with flexible fin-rays, and scales of flexible material. Traditionally, three subdivisions of actinopterygians are recognized: Chondrostei (mostly Paleozoic but with sturgeons, paddlefish, and *Polypterus* as survivors), Holostei (mostly Mesozoic but with bowfins and gars as survivors), and Teleostei (common Cretaceous and Cenozoic bony fish). The features which distinguish the superorders are mostly details of skeletal organization and structure of the scales. Many consider these divisions to be grades of development rather than discrete taxonomic units. All Paleozoic Actinopterygii are chondrosteans except the Late Permian genus *Acentrophorus,* a holostean (Moy-Thomas and Miles, 1971). *Cheirolepis* (Fig. 6-34) is a typical early bony fish and represents well the Chondrostei; one may observe that it superficially resembles modern bony fish except for possession of a heterocercal tail (i.e., with top lobe containing supporting vertebrae and musculature).

Dipnoi are the true lungfish, which are rare at present

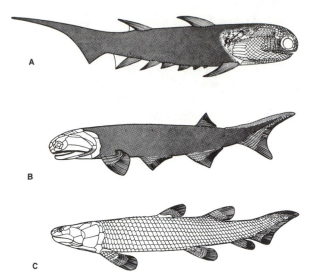

Fig. 6-34. Representative Paleozoic Teleostomi (bony fish). (A) *Climatius* (Upper Silurian to Lower Devonian), a typical, primitive acanthodian (not to scale; most specimens are approximately 10 cm long); (B) *Cheiroplepis* (Devonian), a typical paleoniscoid actinopterygian (ray-finned fish), approximately 30 cm long; (C) *Osteolepis* (Devonian), a typical crossopterygian (lobe-finned fish), about 15 cm long. (All from Colbert © 1980 by John Wiley & Sons, reprinted with permission of John Wiley & Sons.)

but were somewhat more common soon after their appearance in the Early Devonian. The modern Australian lungfish *Neoceratodus* shows typical features of the group: the swim bladder functions as a lung, the fins are modified as crawling limbs, the tails are symmetrical above and below the vertebrae (i.e., of the diphycercal form). Presumably, the appearance of Dipnoi shows adaptation to the emergent nature of many Devonian land areas, with shrinking lakes, nearshore swamps, and other water sources. Even at their earliest appearance, Dipnoi were so morphologically different from the earliest *tetrapods* (i.e., four-legged vertebrates) that they are definitely off the main stem of evolution to amphibians, notwithstanding their abilities to survive on land.

The remaining osteichthyan group, the Crossopterygii, contains the probable ancestors of land vertebrates. Crossopterygii appear in Lower Devonian strata of Germany, Spitsbergen, and Russia and were thought to have become extinct in the Late Cretaceous; however, a single genus, *Latimeria*, of the order Coelacanthidae, was discovered to live in deep waters of the Indian Ocean and thus it has become probably the best-known "living fossil." Crossopterygii are characterized by possession of "lobe fins" with well-developed fleshy lobes containing relatively strong bones arranged in a branching pattern. They also have two dorsal fins (in common with Dipnoi) and well-ossified skulls; the arrangement of skull bones is the tax-

onomic criterion used to distinguish the two main crossopterygian groups: rhipidistians and actinistians (or coelacanths). The rhipidistians are the more generalized forms, with the internal nostrils that enabled them to breathe without opening their mouths (and, thus, they could pump air into their air bladders). This feature suggests that they provided the gene pool from which amphibians evolved. Among the best-known taxa of Paleozoic rhipidistians is the genus *Eusthenopteron,* a top predator in Upper Devonian freshwater environments of Europe and North America (rhipidistians possessed large, sharp teeth and reached lengths to 130 cm). Because rhipidistian crossopterygians were undoubtedly ancestral to the earliest amphibians, discussion here will return to that group after examining the cartilaginous fish which also diversified in the Devonian.

G.2.b.   Superclass Elasmobranchiomorphi (Cartilaginous Fish: Placodermi and Chondrichthyes)

Present taxonomy places two very different types of fish in a single higher taxon because they both feature cartilaginous skeletons. In all other respects, placoderms and sharks are as different as one could conceive. It is very likely that there is no direct evolutionary relationship between the two groups.

The Placodermi were probably the most unusual group of Paleozoic vertebrates and are simultaneously the most distinctive elements of Devonian assemblages. They survived as a class only until the latest Devonian, except for two orders which persisted until the earliest Kinderhookian; but, during their span of existence, they probably comprised the dominant vertebrate taxa. The group was extremely diverse and one description cannot possibly cope with all known forms: indeed, the class itself is probably artificial. The most characteristic placoderms are in the order Arthrodira featuring armored head and trunk shields connected by a unique ball-and-socket joint (Fig. 6-35), paired pectoral and pelvic fins, heterocercal tails, unarmored posterior parts with scales present in some taxa, some ossification of the vertebrae, and very simple jaws consisting of not much more than projecting, shearing dental plates derived from dermal bones. Overall, the heavy armored head and body and the head/thorax joint are the distinctive arthrodire features.

Some arthrodires reached very large size, notably *Dunkelosteus* (''*Dinichthyes*''), a Middle Devonian genus, which reached 10 m and more in length. Their size is made even more impressive by the massiveness of their heads and the impressive size and efficient cutting of their teeth; very likely, arthrodires such as *Dunkelosteus* were the top predators of the Devonian.

Other placoderms varied quite strikingly from the ar-

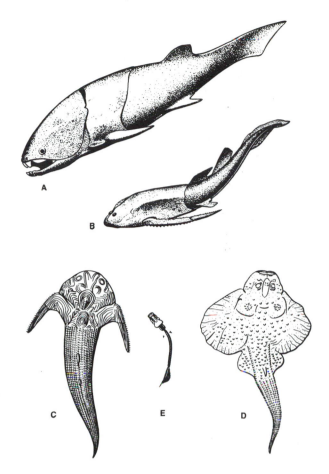

Fig. 6-35. Representative placoderms (all of Devonian age). (A) *Coccosteus,* an arthrodire, about 25 cm long; (B) *Bothriolepis,* an antiarch, about 20 cm long; (C) *Lunaspis,* a petalichthyid, about 12 cm long; (D) *Gemuendina,* a stegoselachian (rhenanid), about 12 cm long; (E) *Paleospondylus,* a unique taxon with uncertain affinities, about 2 cm long. (All from Colbert © 1980 by John Wiley & Sons, reprinted with permission of John Wiley & Sons. Drawings A and B by Lois M. Darling.)

throdire morphology. One suborder, the Antiarcha, were bottom-crawling small forms (Fig. 6-35) which featured very heavy head and trunk armor, reduced jaw structures, and the development of large pectoral spines which probably were the locomotor organs for crawling. The antiarchs apparently reversed the common evolutionary trend of spines evolving into fins and this specialization probably adapted them to the deposit-feeding benthic niche. An additional placoderm group evolved morphology almost perfectly parallel to skates and rays (which are Chondrichthyes). These are called Rhenanidae.

Chondrichthyes are familiar to most students of nature and need little introduction here. There are two infraclasses, Elasmobranchii (sharks and rays) and Holocephali (chimaeras). Both groups appeared in the Devonian (Early and Late, respectively, for the infraclasses).

Elasmobranchs were abundant by the Middle Devonian and have changed relatively little since then. Changes through time have largely centered on improved jaw suspensions, variations in teeth (notably with specializations for crushing in certain lineages of the Mesozoic), and development of the elongate rostrum characteristic of most modern sharks. Chondrichthyes, in general, can be distinguished from other fish by the absence of lungs or air bladders (indicating exclusively marine origins), entirely cartilaginous skeletons (with hard teeth), internal fertilization in (at least) modern and a few fossil taxa, and in most forms, the development of dermal denticles ("teeth") in the skin as counterparts of fish scales.

The elasmobranchs are by far the better-known chondrichthyans in both fossil and recent faunas. The Upper Devonian Cleveland Shale has yielded numerous fossils of *Cladoselache,* a greater than 1-m-long shark that featured very large pectoral fins, with broad stout bases, a strongly heterocercal tail (in common with most sharks), and a stout spine projecting upward behind the head (Fig. 6-36). Sharks, including *Cladoselache,* maintained the primitive gnathostome gill slits (five plus the spiracle marking the hyomandibular apparatus). The teeth of sharks are common fossils found because each shark produced replacement teeth throughout its life and has several rows of teeth present at any one time.

The cladoselachian sharks dominated the Paleozoic chondrichthyan fauna and then were replaced in the early Mesozoic by the hybodontoids. An evolutionary side branch of elasmobranches evolved in the Late Devonian, called xenacanths. These were freshwater sharks with long thin bodies, elongate dorsal fins, and two-cusped teeth (Fig. 6-36), with diphycercal tails producing a very un-sharklike appearance. They are poorly known and were far from the main stem of shark evolution, becoming extinct in the Triassic.

Rays and skates (classified as the superorder Batoidea by Compagno, 1973) appeared in the Jurassic and flourished from Cretaceous to modern times. The remaining chondrichthyan group, Holocephali, are rarely observed in modern oceans because they largely inhabit deep waters. They are rather bizarre in appearance: the skin lacks dermal denticles, the tail is an elongate structure with the fins often constricted to appear jointed, and there is a hooklike head structure in males used to secure to the seabottom during mating; the heads of all feature a very long projection (rostrum) which presents a "rat-nosed" appearance. Two families of Holocephali appeared during the Late Devonian, five more were added in the Mississippian, but only a single family survives at present. Chimaeras are relatively rare fossils except in the upper Mesozoic, where cephalic hooks and jaws (largely cartilage but partly ossified) are common marine fossils.

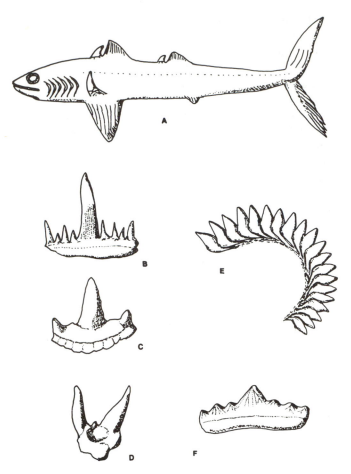

Fig. 6-36. Paleozoic chondrichthyes. (A) Reconstruction of *Clado-selache,* a Late Devonian shark, about 60 cm long; the spines shown extending from the pectoral and dorsal fin regions are disputed by some specialists. (B, C) Cladodont shark teeth, common Devono-Mississippian fossils, consisting of a large central cusp and two or more accessory cusps. The roots of cladodont teeth are irregular generally and easily distinguished from those of Mesozoic and later sharks. (D) Tooth of *Xenacanthus,* an unusual but common freshwater shark of the Mississippian through Permian Periods. (E) The tooth whorl of *Helicoprion,* of Permian age, presenting something of a mystery with regards to the position and function of the whorls. (F) Tooth of *Protacrodus,* an Upper Devonian shark which may be an early hybodontid. (Modified and redrawn from Moy-Thomas and Miles, 1971.)

G.3. The First Land Vertebrates

The transition from crossopterygian fish to the earliest amphibians occurred almost certainly during the middle Late Devonian, because rhipidistians were at their zenith during that time and the oldest amphibians are of latest Devonian age. Three taxa of early amphibians are known from the classical site in East Greenland, all assigned to the Ichthyostegalia, of which *Ichthyostega* (Fig. 6-37) is by far the best known. Additional early ichthyostegids are known from Australia, and Thomson (1980) states that these are

earlier occurrences than the Greenland material (but much poorer-quality specimens).

There are no series of transitional forms showing half-modified structures, such as would be expected in the classical gradual model of evolution; rather, the rhipidistian-to-ichthyostegid evolution probably occurred in geologically brief time, more in accord with the "punctuated equilibrium" model of Eldredge and Gould (1972). Nevertheless, the evolutionary relationship between the advanced fish of the rhipidistian family Osteolepidae, and the Ichthyostegalia, is unquestionable based on similarities listed in Table II. In general, although *Ichthyostega* and relatives are fully developed tetrapods, they possess vestiges of tail fins, skull bones directly carried over from the fish, weak limbs, and rhipidistian-like vertebrae and teeth. It is also no coincidence that the earliest amphibians bear the most fishlike characters.

Anatomically, one may show direct correspondence between the bones in the fins of certain crossopterygii such as the osteolepid *Sterropterygion,* and the limbs of tetrapods (see Rackoff, 1980). A more difficult transition to demonstrate is the modification of the hyomandibular bone from the fish jaw suspension into the stapes of the amphibian (and higher vertebrate) middle ear (see Carroll, 1980). The spiracle, a relict from the first gill arch of jawless fish, became the eustachian tube and ear canal (Romer, 1966). Development of the limb girdles (pelvic and pectoral) may

Table II. Similar and Distinct Features of Advanced Rhipidistians and Ichthyostegids

	Rhipidistian	Ichthyostegid
Similarities		
Teeth	Labyrinthine enamel	Labyrinthine enamel
Vertebrae	Wedge-shaped inter-centrum, reduced pleurocentrum	Wedge-shaped inter-centrum, reduced pleurocentrum
Jaw	Nine bones	Nine bones
Limbs	All bones of tetrapod limb present, plus several additional fin elements	Full tetrapod limb bones
Differences		
Occiput	4-part	Single surface with one occipital condyle
Braincase	2-part	Single
Gill operculum	Present	Replaced by otic notch
Tear ducts	Absent	Present
Limb girdles	Weak	Fully load-bearing
Hyomandibular	Only serves in jaw suspension	Modified to stapes
Spiracle	Present	Modified to ear canal and eustachian tube

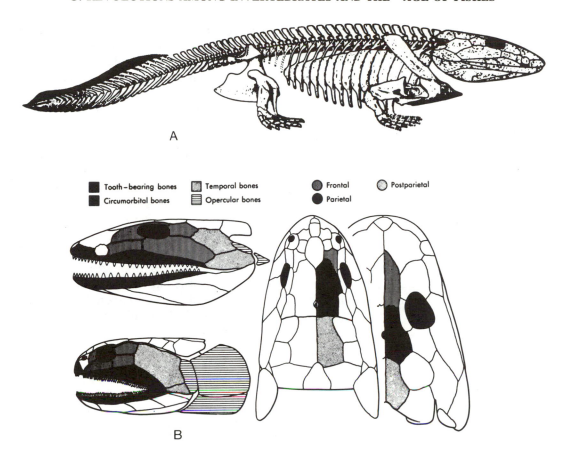

Fig. 6-37. *Ichthyostega,* and homology of advanced crossopterygian and ichthyostegan skull bones. (A) Reconstruction of *Ichthyostega* (Upper Devonian), about 90 cm long; (B) comparison of the skull bones of *Ichthyostega* and *Osteolepis* (illustrated also in Fig. 6-34). Note the homologous bones in ventral and dorsal aspects of both primitive amphibian and rhipidistian crossopterygian fish, although the proportions of various bones are modified from fish to amphibian. (A, from Romer, A. S., *Vertebrate Paleontology,* 3rd ed., © 1966, University of Chicago Press; B, from Colbert © 1980 by John Wiley & Sons, reprinted with permission of John Wiley & Sons.)

be reasonably described from the before-and-after fossils (e.g., Rackoff, 1980), but intermediate steps are unknown and ichthyostegids appear with well-developed limb girdles already established.

The Ichthyostegalia are considered a primitive superorder in the amphibian subclass Labyrinthodontia (the name derives from the complex infolded enamel in their teeth—a direct inheritance from the rhipidistians). By Mississippian time, amphibians had diversified and dominated the terrestrial environment throughout the Mississippian-to-Early Permian interval. We will discuss the dominant amphibian taxa in Chapter 7.

G.4. The Literal Rise of Terrestrial Plants

G.4.a. Devonian Flora

As we showed in the previous chapter, the early record of terrestrial plants is very poor, but it is clear that by earliest Devonian time (Helderbergian), at least two groups of Psilopsida were present, generally referred to as *Rhyni*- and *Zosterophyllum*-type psilophytes. An additional set of primitive plants are known to have developed no later than the Helderbergian: *Psilophyton* (also referred to as the genus *Sawdonia;* Taylor, 1981) and its relatives; and the lycopods. *Psilophyton* (Fig. 6-38) is the oldest plant to feature alternate branches off the main stem, which contrasts with the dichotomously branched psilopsids. The psilophytons were short-lived (to Onesquethaw time) but could be the ancestors of many or most higher plants.

Lycopods also date to the Early Devonian, and Taylor (1981) suggests *Zosterophyllum* as their ancestor. They are characterized by having dichotomous branches (thus are not likely descendants of *Psilophyton*), which bear small, true leaves, and by having sporangia borne in the axil or on a specialized leaf. Lycopods are still extant and are commonly called "club mosses." Of more interest at hand,

Paleozoic lycopods became arborescent (i.e., treelike) and in fact comprised many of the largest forest trees of the Late Devonian through Permian interval. Arborescent lycopods are commonly termed lepidodendrids, after the common genus *Lepidodendron*.

Also in latest Early Devonian time, a second arborescent group emerged, this too represented in the modern flora by small herbaceous forms. These are termed Sphenopsida, known at present from the ''horsetail'' *Equisetum*. Sphenopsid characteristics include straight, jointed stems (becoming trunks in arborescent species), with joints at regular intervals and whorls of branches bearing leaves at the joints. Modern sphenopsids include stems bearing a terminal cone with its own hanging whorl of sporangia, and sterile shoots bearing only leaves: presumably, Paleozoic sphenopsids developed similarly.

During the Late Devonian, several important events in terrestrial plant history occurred. Both *Rhynia*- and *Zosterophyllum*-type psilopsids became extinct, as did the psilophytons, but their descendant taxa flourished. One of these descendant groups is the gymnosperms, which appeared in primitive forms in the Late Devonian (many paleobotanists refer to these as progymnosperms because of their extreme primitive nature relative to later gymnosperms). Ferns made their appearance in either latest Middle Devonian (Taghanican) or earliest Late Devonian (Fingerlakesian) time. And, to complete this summary of mainstream Paleozoic plant evolution, the early gymnosperms differentiated into seed ferns (pteridiosperms) in the Early Mississippian and also provided the main stock for evolution of advanced gymnosperms (such as conifers) in the later periods (Fig. 6-38).

Ferns, or pterophytes, are a diverse group that has been divided loosely into ''preferns'' and filicales or ''true ferns.'' The general morphology of ferns is sufficiently familiar not to require definition here; however, the most important morphological feature that discriminates ferns from the primitive plants thus far described is their relatively large leaves borne on fronds. True ferns, and probably most preferns, bear spores in sporangia located variously on the pinnules (leaf stems), but always on the undersurfaces. The ferns probably exhibit more morphological variation than any other group of vascular plants, living or dead.

The late Paleozoic has often been thought of as the ''age of ferns'' because a great bulk of the preserved material from Mississippian coal swamps resembles fern imprints. It is true that ferns were very common and volumetrically important in the early land floras; but the fossils may be misleading both because other common Paleozoic plants generally resemble ferns (such as some lycopods) and, more significantly, because seed ferns were probably

as common as true ferns. Pteridiosperms have foliage and general structures which resemble those of true ferns but they also produced seeds on the main stems and pinnules. The distinction between spores and seeds is this: spores are generally single cells which can give rise to a new plant either in the *homosporous* condition (in which each sporangium produced gametophytes of both sexes) or in the *heterosporous* condition (featuring separate male and female sporangia). Spores can remain dormant for long time periods. *Rhynia, Psilophyton,* and some other, similar plants of the Siluro-Devonian were among the few homosporous forms, and thus the homosporous state is clearly primitive. Seeds, in contrast, are more complex structures that contain the embryonic plant within a protective covering and which also contain stored nutrients for the embryo. The evolution of seeds from spores occurred during the Devonian, but it is interesting to note that although seed-bearing is an advance over spore-bearing, seed ferns became extinct in the Mesozoic whereas true ferns survive to the present.

The final major plant group to be introduced in this chapter are the early gymnosperms, comprising the first representatives of a very diverse plant group which includes the living conifers (such as pine, cypress, sequoia, and many others), the ginkgoes, and many extinct taxa. Actually, following some taxonomic schemes, seed ferns could be considered primitive gymnosperms since the taxonomic description of gymnosperms incorporates a large group of plants having the seeds borne on open scales, usually in cones, which also lack true vessels in the woody tissue. The remaining group of vascular plants, the angiosperms, appeared late in the Mesozoic and will be introduced in Chapter 8. They are distinguished from the gymnosperms by having seeds borne in ovaries, having true flowers, and by the presence of tracheae (or vessels) in their xylem.

Paleobotany suffers a great deal from the nature of decomposition of dead plants (especially trees). When large plants die, the leaves and bark usually fall away from the stems and trunk, the trunk may topple from the root tops, seeds obviously drop or are otherwise separated from the plant, and in general the single plant almost inevitably becomes a set of separate components. If a large number of plants are living in close association (a common phenomenon), it becomes virtually impossible to reconstruct a complete individual of many extinct types of plants. Consequently, ''form-genera'' are a necessity in plant (and some animal) taxonomy, whereby foliage, roots, stems, trunks, or other parts are classified without formal knowledge of the interassociations between parts of plants. Some examples from the Paleozoic are fernlike foliage form-genera such as *Neuropteris* and *Pecopteris* (which probably belonged to both seed ferns and true ferns); lycopod stumps and roots

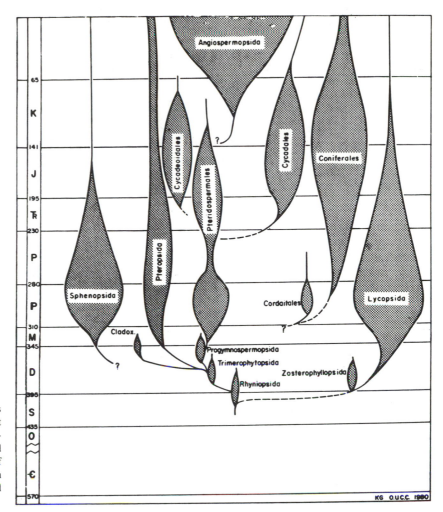

Fig. 6-38. Generalized phylogenetic relationships among major vascular plant groups. (Note that Ginkgoales are omitted; their ancestry is undetermined but as gymnosperms they would be placed within the cycad–conifer–pteridiosperm region of this diagram. Earliest ginkgoes date to the Permian and a sole genus survives at present.) (From Knoll and Rothwell, 1981.)

termed *Stigmaria;* rings of foliage from sphenopsids termed *Annularia;* and a large number of seed and spore form-genera which cannot be related to other plant parts.

In summary, by the end of the Devonian, most plant higher taxa, except for coniferous gymnosperms and angiosperms, were well established. It is unclear how extensive or tall Late Devonian forests were but several moderate-sized arborescent forms are known including lycopods such as *Cyclostigma* and *Protolepidodendron,* and progymnosperms such as *Callixylon* and *Aneurophyton.* A variety of herbaceous lycopods, ferns, seed ferns, and unassignable plants are also known, suggesting generally lush vegetation.

G.4.b. Mississippian Flora

The Mississippian Period (the Lower Carboniferous Period of Europe) initiated the interval of probably the most successful plant diversification in all geological history.

The lushest plant life in North America and Europe developed in the Pennsylvanian (Upper Carboniferous) but the onset of terrestrial plant radiations clearly occurred by Kinderhook time.

The Bradford/Kinderhook transition among plants was gradual but with renewed marine transgression on the craton in the Early Mississippian, large forests apparently developed along the shores and embayments of the cratonic sea. These forests are now represented in coal seams, notably in the Appalachians, northern Alaska, the Sverdup basin in the Innuitian continental margin, and western Europe. Gradual shift in dominance of plant groups occurred with the evolution of arborescent seed ferns, large lepidodendrid lycopods, and the first and most successful arborescent sphenopsids.

The spenopsids underwent the most dramatic evolution with the appearance of the genus *Calamites,* a 15-m-tall variety of horsetail, and related forms. *Calamites* is commonly found preserved in two forms: the trunk is often preserved with the external surface missing, such that it presents a vertically striated and jointed appearance resulting from traces of the primary rays of the wood (*Arthropitys* is a form-genus describing such wood fossils); the second common *Calamites* fossil form comprises leaf-whorls generally called *Annularia*.

The lepidodendrids were probably the most characteristic Lower Carboniferous large trees, including the genera *Lepidodendron, Cyclostigma,* and *Lepidodendropsis.* They evolved a mode of growth in which the leaves left behind rhomboidal leaf-scars, after falling from both stems and the trunk, resulting in the characteristic ''lizard-skin'' pattern, on their bark (Fig. 6-39). Large casts of the pseudo-root structures of lepidodendrids have been found in the number of places and such fossils are generally placed in the form-genus *Stigmaria.*

Among Lower Mississippian fossils are a great number of fernlike impressions placed in various form-genera. It is clear that arborescent seed-fern taxa existed at the time as evidenced by large seed-bearing foliage fossils; but these are difficult to associate with wood fossils and are often placed in one of the fern form-genera such as *Neuropteris.* In general, both true ferns and seed ferns of the Lower Mississippian were very abundant, comprising both herbaceous and shrublike forms.

Cordaites (Fig. 6-40) and its relatives were important and very large trees of the Late Mississippian and Pennsylvanian and they are probable descendants of earlier gymnosperms such as *Callixylon.* The most frequent fossils found are its long, slender leaves with lengths to 1 m, but associated flowering units, cones, pollen, wood, stems, and roots have been described and have been placed in form-genera. *Cordaites* had both pollen- and seed-bearing cones and its wood and stems have characteristic arrangements of cells within the pith area which make identification relatively easy.

True coniferous gymnosperms appeared during the Early Mississippian, along with the Cordaitales (which are gymnosperms but which are separated from the conifers). Foliage fossils bearing small spiral leaves and associated cones are known and suggest morphologies not unlike modern yew or arborvitae (white cedar).

By the Late Mississippian, the terrestrial forests seem to have matured and stabilized with a number of new genera but few new types of arborescent plants. *Sigillaria,* a large lycopod, joined *Lepidodendron* in the coal swamps, as did new large sphenopsids. One major change occurred during the Late Mississippian among ferns: primitive true ferns of the Coenopteridopsida were replaced by advanced groups, notably the Marattiales in the Late Mississippian and modern-type ferns (Filicales) in the Pennsylvanian. Among the Marattiales were large arborescent genera such as *Psaronius,* which reached to 10 m in height. Filicales probably were smaller trees during the Paleozoic but they undoubtedly formed much of the understories of forests. The genus *Botryopteris* is a common Pennsylvanian filicale which reached tree sizes.

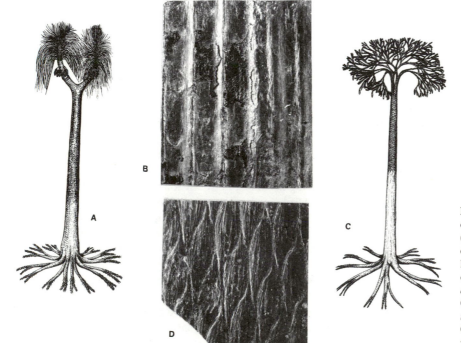

Fig. 6-39. Representative arborescent lycopods of the Mississippian and Pennsylvanian Periods. (A) Reconstruction of *Sigillaria;* (B) photograph of *Sigillaria* bark, in coal from the Pottsville Formation (Pennsylvanian); (C) reconstruction of *Lepidodendron;* (D) photograph of *Lepidodendron* bark, also from Pottsville coal. (A and C, from Stewart, W. N., 1983, *Paleobotany and the Evolution of Plants,* with permission of Cambridge University Press; B and D, DRS photographs.)

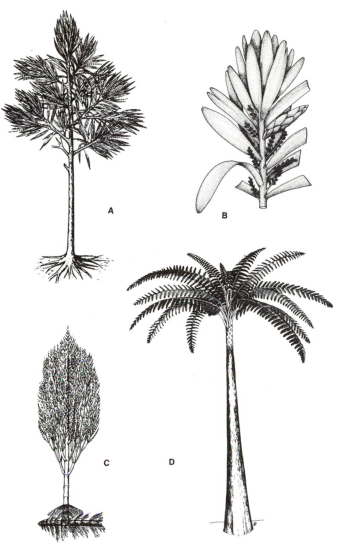

Fig. 6-40. (A) Reconstruction of a cordaitale tree; (B) detail of the leaves and inflorescences of a cordaitale; (C) *Calamites,* a typical sphenopsid of the Mississippian and Pennsylvanian (note the illustration shows a partial specimen with a portion of the lower trunk across the bottom); (D) *Psaronius,* a Pennsylvanian marattiaceous tree-fern. (A–C, from Stewart, W. N., 1983, *Paleobotany and the Evolution of Plants,* with permission of Cambridge University Press; D, from Andrews © 1961 by John Wiley & Sons, reprinted with permission of John Wiley & Sons.)

G.4.c. Pennsylvanian and Permian Floras

Terrestrial floras did not change substantially in the Pennsylvanian, except for even greater numbers of lower taxa among all major groups, and evidence of especially high densities of plant life in the cyclothem-producing environments which characterize the Pennsylvanian System. The "coal age" was dominated by large lycopods, large sphenopsids, seed and true ferns (arborescent and herbaceous), and early gymnosperms of the Cordaitales and

primitive Coniferales. There was little evident change in the Permian flora, until the end of the period, at which time drastic events occurred among plants (just as with marine invertebrates).

At the era boundary, all arborescent sphenopsids, all coenopterid ferns, most lepidodendrids, and most cordaitales disappeared. During the Late Pennsylvanian and Permian conifers, ginkgoes, and cycads had begun diversification and these came to dominate the upper stories of Mesozoic forests, while filicales occupied the lower stories.

G.5. Early Insects and Related Late Paleozoic Arthropods

In Chapter 5, we discussed early land arthropods, notably the Siluro-Devonian forms in the Rhynie Chert fauna and other assemblages dating very close to that time. These fossils co-occur with aquatic arthropods, suggesting life very near the water's edge or a mixed preservation of terrestrial and aquatic arthropods. Among the Rhynie Chert arthropods are collembolids (or "springtails") which are simple, tiny creatures, in many respects suggestive of the ancestors of insects. Collembolids are common in the modern world (Fig. 6-41) living in soil, and they may have survived as living fossils from Devonian time.

The Mississippian record of terrestrial arthropods is quite poor, but this is undoubtedly preservational because older (i.e., Devonian) taxa are evident and younger taxa are very well known: clearly, arthropods did not disappear for a geological period. In the Pennsylvanian, we encounter several good terrestrial arthropod assemblages from North America and Europe, which suggest that a considerable amount of evolution occurred among arthropods during the Mississippian, despite the absence of fossils (see Carpenter and Burnham, 1985). One very important advent was the evolution of flight and true insects.

Among the living and fossil winged insects are two broad groups (which are not taxonomic ranks) referred to as Paleoptera and Neoptera. Paleopteran insects have wings which are always extended and generally are of equal size; the dragonflies represent advanced paleopterans. Neopterans comprise most other insects with retractable wings (including wingless forms which have lost their wings through evolutionary modification). Insect wings are formed by flattened, double layers of the body walls with nerves and circulatory system components within. The evolution of insect wings is inferred to have followed from selective advantage gained by wingless forms that had lobate extensions of the prothorax which may have facilitated gliding (as a precursor of true flight); this is, however, conjectural. In addition, it is not totally established that the collembolids are insects and therefore it is possible that they are not ancestral to paleopterans; thus, it is possible that

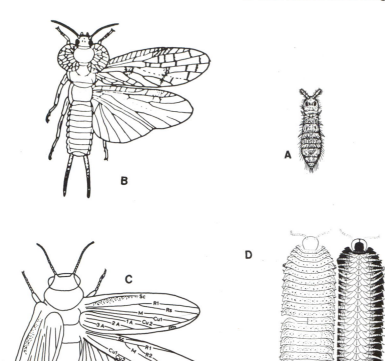

Fig. 6-41. Primitive insects and *Arthropleura*. (A) A modern collembolid, *Achorutes*, which resembles creatures from the Rhynie and other Siluro-Devonian terrestrial faunas; (B) *Lemmatophora*, a paleopteran from the Permian Wellington Shale in Kansas; (C) *Protelytron*, a neopteran insect, also from the Wellington Shale; (D) reconstruction of the gigantic Pennsylvanian aquatic arthropod *Arthropleura*. (A–C, from Tasch © 1980 by John Wiley & Sons, reprinted with permission of John Wiley & Sons; D, from the *Treatise on Invertebrate Paleontology,* courtesy of the Geological Society of America and the University of Kansas Press.)

flight in insects evolved primarily without leaving fossil evidence. Collembolids are classified in the class Hexapoda (all insects) because they are six-legged; but the hexapod condition could have evolved independently among several groups.

Insects today comprise almost two-thirds of described animal species (there are approximately three-quarters of a million insect taxa) and, logically, their taxonomy is complex. Within the scope of this book, we will describe only essential parts of their early record. In the Pennsylvanian and Permian, the paleopterans comprise the common fossil forms. These included giant (and other) dragonflies, of the order Odonata, with a notable genus (*Meganeura*) of the Late Pennsylvanian having wingspans to 65 cm. Paleopteran insect wings are generally more common fossils than are body parts of complete specimens. Upper Paleozoic deposits contain many wing form-genera classified by characteristic patterns of venation and overall shape. Among the nondragonfly paleopterans are the Palaeodictyoptera, a group of (probable) heavy-bodied insects, totally extinct, which are known from many wings and a few body-fossils; and the Megascoptera, which were light-bodied, probably predaceous paleopterans not very dissimilar from modern damselflies.

Also in Pennsylvanian and Permian strata are found

fossils of several neuropteran orders which appeared and diversified in the coal forests. These include: the blattoids (cockroaches), which were an abundant, primitive, and (unfortunately) highly successful group; the orthopteroids (grasshoppers, locusts, and crickets), also common in the upper Paleozoic and representing a basic modification of the primitive insect plan comprising adaptation of the hind pair of legs for springing; the Coleoptera (beetles) which appeared in the Permian and were not nearly as common in the pre-Tertiary as they are today (comprising more than half of the known insect taxa); the Ephemeroptera (mayflies), which are known from nymphs found in Permian rocks; and primitive members of the mecopteroids (a large group that includes flies, butterflies, and moths). The known Permian mecopteroids are four-winged flies which seem to be ancestral to the later two-winged taxa in the order Diptera. No true moths or butterflies are known from strata older than the early Tertiary nor do two-winged flies predate the Triassic.

Along with the insects, both arachnids and myriapods are known from the upper Paleozoic. A fossil millipede of the genus *Xylobus* has been found comprising a cast in a Pennsylvanian *Sigillaria* stump, suggesting that then, as now, they were vegetarians. Fossil centipedes, which are today carnivorous, are also known from the Pennsylvanian, especially the Mazon Creek fauna; however, nothing can be deduced about the feeding habits of the earliest forms. Other groups of myriapods are known from the Late Paleozoic including the extinct Arthropleurida, a group of 2-m-long giant myriapods (millipedes) that seem to share some limb structures with trilobites, leading to previous speculation of a direct relationship (which has been shown to be unfounded). They became extinct in the Late Pennsylvanian and left no descendants; their fossils are associated with coal deposits and this suggests that they were adapted to the coal swamps of the Pennsylvanian and could not adapt to the changing climate of the latest Pennsylvanian, with its decreasing flora.

CHAPTER 7

The Absaroka Sequence: Lower Pennsylvanian–Lower Jurassic

A. Introduction

A.1. Overview of Absarokan Events

The Absaroka Sequence was deposited through the longest of Phanerozoic geological intervals to be considered in this book. It took more than three geological periods (Pennsylvanian through Early Jurassic time) for the complete sequence to be laid down and, during these roughly 145 Myr., some fundamental changes occurred on Earth and on the continent. Among these changes were (1) assembly of Pangaea II, completed by the Permian Period, and the initial breakup of the global landmass during Triassic–Jurassic time; (2) the first sustained episode of nonvolcanic tectonic activity within the western craton; (3) the most prolonged interval of subaerial exposure of the craton in Phanerozoic time; (4) uniquely symmetrical and pervasive deposition of coal-bearing cyclical deposits during the Pennsylvanian and Permian Periods (also occurring on other continents); (5) appearance by latest Absarokan time of all known phyla of animals including all classes of vertebrates; (6) along with the above, appearance of a uniquely diverse arborescent flora and the first diverse terrestrial fauna; and (7) the most pervasive mass extinction event in the history of the marine biota, from which, it is claimed by some authors, less than 5% of all species of marine animals may have escaped!

Absaroka time is part of two eras; it incorporates all of the late Paleozoic and approximately one-third of the Mesozoic. One deals with a few complications concerning time and time–stratigraphic nomenclature in the Absaroka owing to European/American differences; immediately at the beginning of the sequence is the Pennsylvanian Period, an American unit which is, of course, the Upper Carboniferous of Europe. In North America, several time–stratigraphic systems have evolved for different cratonic basins and more than one are in wide use. The Appalachian Basin stratigraphic nomenclature has been commonly used because of the economic importance of Appalachian coal; however, this section does not work nearly so well as does the mid-continent section for use in dividing up Pennsylvanian strata (and time) across the central and western craton. For this book we will adopt the latter. For the Permian System we will use the widely recognized terminology based on the west Texas standard section.

The Triassic System presents a real dilemma for American terminology because there is no widely recognized American standard section. A similar situation also exists for the remaining Mesozoic eras and we will of necessity use European-based stages in our presentation.

A.2. Paleogeography

North America during the late Paleozoic occupied an equatorial position on the globe (Fig. 7-1). Several lines of evidence have been presented to support this hypothesis, including paleomagnetic data (on the direction toward the magnetic poles and on the approximate magnetic latitude) and paleoclimatic data. Of the paleoclimatic data, perhaps the most convincing come from study of the fossil plants which had flourished during the Pennsylvanian in the lush coastal marshes and lowland forests of North America and Europe. In these environments accumulated great thicknesses of organic muck and litter destined to become coal.

Also of interest in this regard is the occurrence of large evaporite deposits in the Sverdrup Basin of the Innuitian area (Otto Fiord and Mount Bayley Formations), which bespeak an environment of considerable aridity. According

271

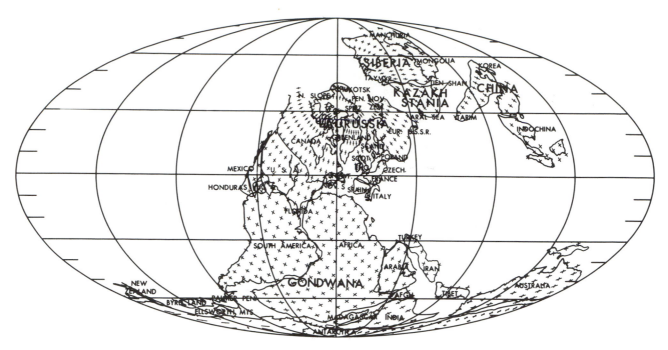

Fig. 7-1. Paleogeographic reconstruction of the Middle Pennsylvanian Earth. Notice that the Gondwana continents are shown to have collided with North America, resulting in a single landmass from the South Pole to approximately 40°N latitude. (From Scotese *et al.*, 1979.)

to the reconstruction in Fig. 7-1, the Sverdrup Basin area lay at approximately 30°N latitude. It was, therefore, in about the same paleogeographic position as is the Persian Gulf today, a region well known for its recent evaporites and carbonates.

Away from the coal swamps of the east and the evaporitic basin of the Innuitian continental margin, much of western North America during the Pennsylvanian experienced deposition of extensive platform carbonates. On the western continental margin (the "miogeosyncline"), a carbonate platform existed from Mexico all the way to northwestern Canada and the Brooks Range area of northern Alaska.

The final assembly of Pangaea II occurred during the late Paleozoic. Figures 7-1 and 7-2A present the relative positions of the other major continents during the Pennsylvanian and Permian. Note that the Gondwanaland continents are shown to have collided with the Appalachian–Ouachita margin of North America to produce a single landmass from the South Pole to at least 40°N latitude (during the Pennsylvanian) and, with the Permian collision of eastern Europe and Kazakhstania/Siberia, possibly even to the North Pole. If this reconstruction is reasonably accurate, we might imagine the effects that such a landmass would have had on oceanic surface currents (see Fig. 7-2B). The major drift currents would probably have established in the unbroken expanse of Panthalassa a nearly ideal pattern of east–west flows associated with global wind currents. This may have helped to occasion the restriction of global climates to lati-

tude-defined zones. Along the margins of Pangaea, the east–west currents would have been deflected into a series of north–south boundary currents.

Abundant evidence exists for the presence, during the late Paleozoic, of a major continental glacier on the Gondwanaland continents around the South Pole. In South America, southern Africa, India, Australia, and Antarctica, there are well-documented glacial deposits, such as the Dwyka Tillite of southern Africa, and glacially striated bedrock surfaces. This southern ice cap is of more than passing interest because it probably acted as a fundamental control of sea level and therefore profoundly affected sedimentation everywhere on Earth. The Pennsylvanian was a time of marked sea-level fluctuations as revealed by cyclical sedimentary sequences of several different types (e.g., coal-bearing cyclothems, carbonate–evaporite cycles). These fluctuations were frequent and rapid. There are up to 60 cyclothem sequences in the North American midcontinent region; and, since the Pennsylvanian lasted approximately 30 Myr., one may infer an average of 500,000 yr. for each episode of sea-level rise and fall. Further, the sea-level fluctuations were clearly eustatic: cyclic strata of Pennsylvanian age have been reported from almost every continent. There are several mechanisms which can account for eustatic sea-level changes, but the only one that can easily explain the remarkable periodicity of Pennsylvanian sea-level fluctuations is the regular advance and retreat of continental glaciers.

On the perimeter of the southern continental glacier

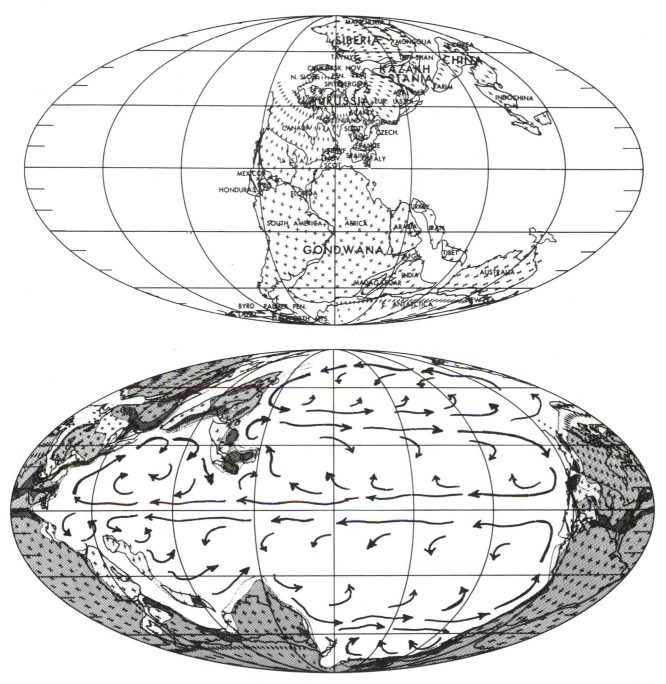

Fig. 7-2. Reconstruction of the Permian Earth. (A) Pangaea during the Late Permian; notice that the landmass extends literally from North to South Poles. (B) Paleogeography of the Panthallassa Ocean, showing hypothetical surface circulation. (Both, exclusive of current arrows in B, from Scotese *et al.,* 1979.)

was another major floral province, greatly different from the lycopsid flora of North America. This southern plant-fossil assemblage is termed the *Glossopteris* flora after its main constituent, a large tree-fern. The *Glossopteris* flora has relatively low specific diversity but a high density (i.e., there were few different species but a great many individuals of the same species); such an ecological situation is typical of northern-temperate to subpolar environments. Prominent growth rings in *Glossopteris*-flora plants indicate a seasonal climate, also typical of middle to high latitudes. Like the lycopsid flora, the *Glossopteris* flora formed large coal swamps but these probably represent a greatly different climatic situation.

Global paleogeography during the Permian was

grossly similar to that of the Pennsylvanian but several aspects of North America's paleoclimate were quite different. For one thing, the Allegheny Mountains formed a barrier to the Trade Winds and therefore blocked moist air from much of the eastern half of the continent. One may infer that southeastern slopes of the Allegheny Mountains supported a lush rain forest watered by moist winds rising up the southeastern side of the mountains and dropping torrential rains in the cooler air. In the rainshadow of the mountains, much of the eastern half of the continent was hot and dry. Red beds predominated in continental and some shallow marine environments and evaporite deposition characterized several major epicontinental basins.

Throughout the Permian, sea level slowly dropped, and by the end of the period most of the Earth's continents were high-standing and dry. There are few or, perhaps, no areas on Earth where conformable sequences of marine sediments were deposited across the Permian–Triassic boundary. Conversely, continental deposits, especially red beds, are very common in both Permian and Triassic strata. Additionally, the retreat of the sea to the very edges of continents must have played an important part in the mass organic extinctions which mark the end of the Paleozoic.

The cause of this major eustatic drawdown of sea level is not known, but a change in the shape of the ocean basins as a result of the assembly of Pangaea is a plausible cause. It is believed that a rearrangement of global tectonic features (e.g., midocean ridges, trenches) and plate motions followed the assembly of Pangaea. Ultimately, the new seafloor spreading regime split apart the supercontinent and directed the movements of the resulting plates; the pattern of global tectonics seen today was initiated soon after the beginning of the Mesozoic. Since the elevation and lateral extent of midocean ridges are controlled by the rate of seafloor spreading, one may suggest that a deepening of the ocean basins could have occurred during the time between Pangaea assembly and the advent of the new plate-tectonics regime, when older ridges were sinking following their abandonment and newer ones were not yet fully active. An alternate possibility is that assembly of Pangaea resulted in the compression of old continental-margin prisms (and other sedimentary masses) into high-standing mountain belts. The effect would be for continental crust to occupy relatively less area. Since the volume of continental crust would not have been significantly different, the result would have been an increase in the volume of the ocean basins and, therefore, a lowering of sea level. Other possibilities, such as glacial control, may also be invoked. At present, there is a lack of sufficient data to evaluate these possibilities; we also may not have realized the right combination of factors to explain late Paleozoic sea-level changes.

The Triassic was the most arid geological period of all Phanerozoic time (Habicht, 1979). It was a time of extensive red-bed deposition over much of North America, Europe, South America, Africa—indeed, almost the entire Earth! Global paleogeography during the Triassic was similar to that during the Permian and many of the climatological ideas discussed earlier apply equally to the Triassic. One significant difference is that Pangaea began to break up toward the end of the Triassic, with North America the first wanderer of the new tectonic regime. The earliest signs of continental rifting were rift valleys, block-fault mountain ranges, and extensive mafic volcanism along the southern and eastern margins of North America. Some rifts became major evaporite basins, especially on the southern margin where Triassic evaporites and red beds occur at the bottom of the thick Gulf of Mexico coastal-plain prism. At the end of Triassic time, separation of North America from Pangaea was not quite complete. By the time that the Jurassic Period was well along, however, full-scale separation had occurred and rifted-margins were evolving on both eastern and southern continental margins.

B.   Absarokan Cratonic Sedimentation

B.1.   Overview

Strata of the Absaroka Sequence record a change from predominantly epicontinental marine sedimentation to mixed marine, paralic, and (especially in the later Permian and Triassic) predominantly terrestrial sedimentation (Fig. 7-3). In many respects, this change in pattern at the Kaskaskia/Absaroka boundary may mark a more significant environmental event in North America than does the Paleozoic–Mesozoic Era boundary, which also falls within Absaroka time and which marks a profound worldwide change in marine faunas.

Absaroka strata are the youngest exposed in the eastern half of the craton (with the exception of continental outliers of the Late Jurassic in Michigan, and of course Quaternary glacial deposits in the North). Pennsylvanian and Lower Permian strata in the eastern and central regions are predominantly cyclical deposits which characteristically contain coal, especially in the Appalachian Basin where paralic and terrestrial sedimentation prevailed. Pennsylvanian cyclothems on the eastern craton contain great quantities of detrital sands, derived primarily from the Appalachian Mountains. In the later Pennsylvanian and Permian, the Ouachita Mountains also supplied detritus to southeastern cratonic sections.

The Absaroka Sequence featured the first Phanerozoic episode of oscillatory cratonic–tectonic behavior. In addition to the cyclical up-and-downwarping of the central and eastern craton, in Pennsylvanian and Permian times the southeastern and central western craton underwent especially strong epeirogenic activity which produced sizable uplifts and adjacent basins (termed *yoked basins* or *zeugogeosynclines*). These basins characteristically received very thick detrital sediments derived from the adjacent uplifts (hence the "yoked" term), but they also formed evaporites during times of restricted marine flow. While these basins were forming thick marine sequences, the surrounding stable cratonic platforms received at best thin veneers of detrital or carbonate sediments. Many of these yoked basins yield important quantities of hydrocarbons (e.g., the Paradox Basin in the "Four Corners" area). The uplifts in the southwest too have importance in oil prospecting; for example, the Nemaha Ridge in eastern Kansas and Oklahoma was active during the Early Pennsylvanian and resulted in structural traps being formed due to its activity and subsequent truncation.

Lower and Middle Pennsylvanian strata on the western craton also show cyclicity but the cyclothems are dominated by marine units, often containing evaporites. Pennsylvanian sedimentation (Morrow–Desmoines age) in the east was confined largely to four basinal areas: the Appalachian, Illinois, Michigan, and the very large, homoclinal, Midcontinent Basin. By Desmoines time, the transgression of the sea from the western craton may have overtopped the Transcontinental Arch and covered the entire midcontinent. It is probable that some Pennsylvanian strata of the east were nearly or totally continuous between cratonic basins and with marine carbonates of the west, and have been eroded off intervening uplifted areas. West of a line from Iowa to Oklahoma, Pennsylvanian units are present almost everywhere in the subsurface. King and Beikman (1976) note that eastern and western subdivisions of the Pennsylvanian System apparently correlate well across barren regions, suggesting the earlier presence of continuous strata.

In late Pennsylvanian and Early Permian (Missouri–Virgil) times, the eastern craton uplifted and all sedimentation ceased except in the Dunkard Basin within the north-central Appalachian Basin, where, for a brief time in the earliest Permian, terrestrial deposits were laid down conformably over the youngest Pennsylvanian.

On the western craton, however, the Late Pennsylvanian through Permian were times of extensive marine transgression, although the seas were frequently restricted and left large masses of evaporites. The Midcontinent Basin, which in the Permian had its depocenter in Kansas, received generally cyclical carbonate-shale formations until Leonard time, and then received admixed marine muds, carbonates, alluvial plain deposits, and in the Early Triassic, continental red beds, as the sea regressed from the region. In west Texas and adjacent New Mexico, sharply differentiated basin and platform environments developed in the later

Pennsylvanian and persisted through the latest Permian. The basins were very deep and "starved," featuring thick deposits of bituminous sands and muds while the platforms received veneers of carbonates. Periodic restriction of seawater flow on the platforms produced evaporites while at the basin margins, sizable reefs developed, composed of algae, sponges, and bryozoans and populated by a diverse set of shelly invertebrates including many bizarre brachiopods. The Permian Basin of west Texas became a broad platform by the late Ochoan and, by Permo-Triassic time, was above sea level.

The youngest Absarokan strata were deposited on the far western craton. In the extreme northwest, a deep basin parallel to the Cordilleran margin contained mostly siliceous, shaley, and carbonate sediments of probable deep-water origin; but it also formed phosphorite and phosphatic detrital units due to probable upwelling from the basin depths onto a marginal platform. Eastward through Montana, Wyoming, and the Dakotas, marine shales, limestones, and sandstones were deposited through the Permian, followed by evaporites and red beds in the Late Permian and Early Triassic. Triassic strata across the northern craton comprise a complexly intertonguing, diachronous sequence grading from marine units in the basin in western Montana and eastern Idaho, through shelf sediments eastward and southward, to continental red beds at the far eastern region in Wyoming and northern Colorado. By the Middle Triassic, the sea had shallowed over the region generally and a sequence of red beds was deposited throughout the remainder of the period. These red beds are punctuated periodically by thin, wide-ranging limestone tongues which probably correlate with limestones formed in the deeper waters at the Cordilleran margin. By latest Triassic time, the sea had withdrawn, and fluvial/eolian sands generally cover the last marine units. Some of these sands may have been deposited as late as the Early Jurassic.

The southwestern craton has the very last Absarokan sediments. In the region now called the "Colorado Plateau," which includes the Grand Canyon, Lower Permian limestones were deposited in broad seaways while at the margins of those seaways, continental red beds formed with sources of detritus coming from the eroding uplifts which dotted the region. Extensive sand deposition followed in the early Middle Permian (Leonard time), followed by renewed transgression and the widespread deposition of carbonates. Regression across the region in the later Permian eroded many sections, but in the Early Triassic the sea again flooded much of Arizona, New Mexico, Colorado, and Utah. The sediments deposited grade from marine carbonates in the west to marginal-marine and continental deposits in the east. The Absaroka sea again withdrew, this time permanently, and Middle Triassic erosion was followed by massive deposition of continental deposits.

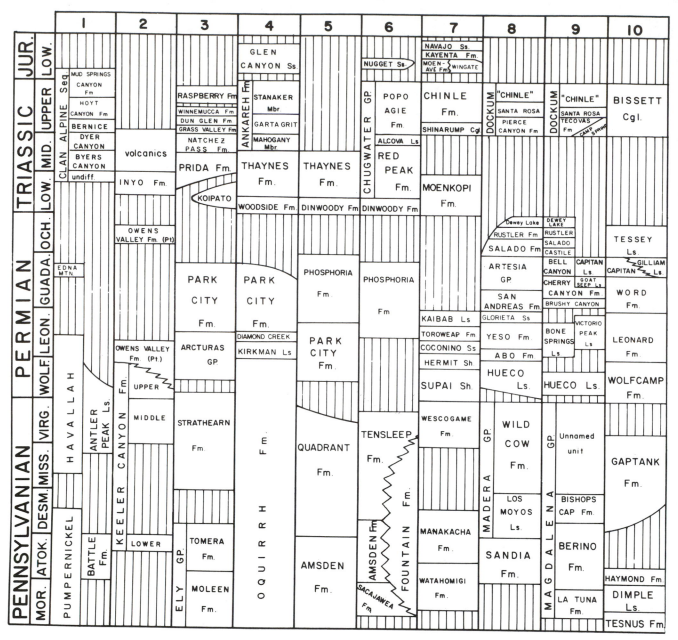

Fig. 7-3. Representative stratigraphic columns of Absarokan deposits; most columns are synthetic, having been assembled from multiple sources and from different localities within the area represented. Areas are as follows: 1, central Nevada; 2, southeastern California; 3, northeastern Nevada; 4, Utah; 5, southwestern Montana; 6, Wyoming; 7, northern Arizona; 8, New Mexico; 9, Permian Basin, northwestern Texas; 10, Marathon area, southwestern Texas; 11, central Texas; 12, Arbuckle Mountains, southcentral Oklahoma; 13, Nebraska; 14, Black Hills, South Dakota; 15, Illinois Basin; 16, West Virginia; 17, Eastern Triassic–Jurassic Basins; 18, Georgia and Alabama; 19, eastern Canada; 20, northern Alaska; 21, Sverdrup Basin, northern Canada.

Upper Triassic strata of the western craton feature brightly colored mixed detrital lithologies including widespread basal conglomerates, considerable volcanic material, petrified wood, and important uranium ore (carnotite) deposits. In the latest Triassic, eolian sands were deposited, followed briefly by deposition of sandy, shaley fluvial sequences, and then through the end of the Triassic, and well into the Early Jurassic, massive eolian sands formed, marking the end of Absarokan cratonic sedimentation.

B.1.a. The Post-Kaskaskia Erosion Surface

Lowermost Absarokan strata overlie Chesterian and older units unconformably across most of the craton; how-

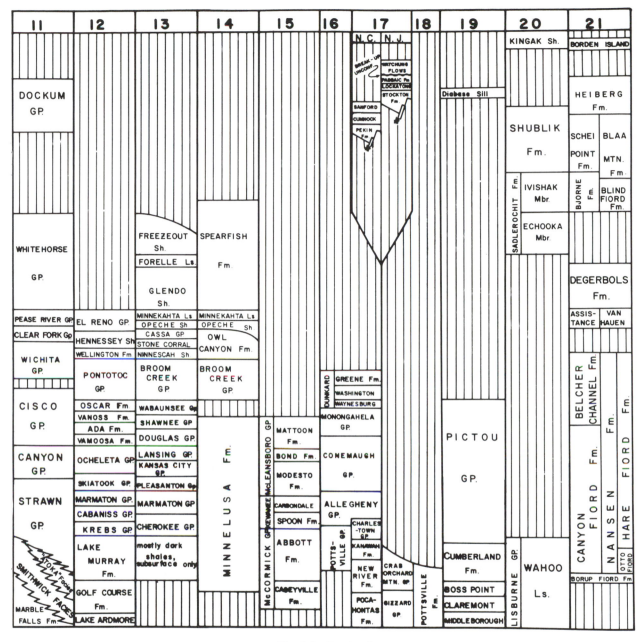

Fig. 7-3. (*Continued*)

ever, within areas occupied by deep or rapidly subsiding foreland basins at the Appalachian, Ouachita, and Cordilleran margins, Chesterian/Morrowan conformable sequences do occur over relatively large areas (e.g., Pottsville Group strata in the Appalachian area from Alabama to Pennsylvania). Generally, and as might be expected, the magnitude of the Mississippian/Pennsylvanian unconformity increases westward and northward in the Appalachian Basin, away from the Acadian uplifts and foreland basins, and it is more pronounced and involves greater angular unconformity in regions adjacent to uplifts and arches in the midcontinent and western craton.

The retreat of the Chester sea left most of the craton emergent and undergoing extensive, locally spectacular erosion. In some places the post-Kaskaskia erosion surface is extremely rugged. In the Illinois Basin, for example, the southward-retreating Chester sea fostered the strong development of the southwestward-flowing drainages that eroded deep valleys reported to be up to 145 m deep and over 30 m wide (Bristol and Howard, 1974). These valleys experienced considerable mass wasting and have been shown by subsurface work to contain sizable slump blocks along the buried valley walls, due in part to the failure of several swelling-clay units in the Mississippian section.

In the southwestern United States, as well as other regions across the continent, many Mississippian limestone units developed rugged karst surfaces during the hiatus. In central Colorado, the Lower Pennsylvanian Molas Formation occurs principally as cave-infilling in the Mississippian Leadville Limestone/Dolomite (Tweto and Lovering, 1977). Elsewhere around the craton, many basal Absarokan strata fill and/or cover Kaskaskian karst as would be expected subsequent to an interval of subaerial exposure of extensive carbonate strata.

B.1.b.   Index Fossils in the Absaroka Sequence

In the following brief discussion, we will break slightly with our format and cover some aspects of Absarokan fossils early in the discussion rather than at the end of the chapter. Absarokan strata differ from those of most earlier sequences in at least one respect: units on either side of several time boundaries are conformable and contain virtually the same lithologies. In addition, because of the cyclical nature of Pennsylvanian and some Lower Permian strata, members of cycles may more closely resemble lithologically similar members of younger or older cyclothems than they do different lithologies within the same cycle. For these two and other reasons, the use of biostratigraphic index fossils is absolutely essential to upper Paleozoic and Triassic stratigraphy and, therefore, we wish early in the discussion to acquaint the reader with the fossils involved.

In Pennsylvanian and Permian strata, the foremost set of index fossils are fusulinid foraminifera (see Chapter 5 for a general description of foraminifera). Fusulinids (more correctly, fusulinaceans) are a distinctive superfamily of calcareous foraminifera which reached large sizes (in gigantic species, up to 6 cm but generally 2–6 mm in length and 1–3 mm in width). They also featured distinct shell structure, composed of microgranular calcite, sometimes packed with detrital particles, and they had unique shell shapes, being generally fusiform (American football-shaped) overall, coiled in sagittal (cross) section, and layered almost like an onion in axial (long) section (Fig. 7-4).

Fusulinids existed from the Late Mississippian into the Late (but not latest) Permian, and thus were around for what will be shown to be the part of Absaroka time that included major marine sedimentation on the craton. They were also common; in fact, so common that they may exceed in volume all other Late Paleozoic marine fossils (Thompson, 1964). Within the general fusulinid morphology described, there was much variation among taxa, especially in details of the septa, *axial fillings* (deposits of dense calcite), and the *spirotheca* (spiral shell walls, above the chambers). Identification of fusulinids is generally done by making sagittal and axial sections of specimens and comparing

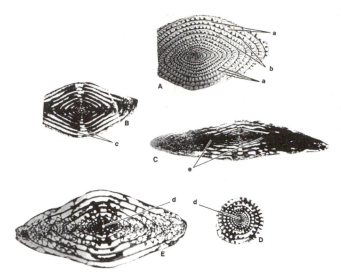

Fig. 7-4. General morphology of fusulinids. (A) *Neoschwagerina* (U. Permian), in oblique axial section; (B) *Yangchienia* (U. Permian), in axial section; (C, D) *Wedekindellina* (M. to U. Pennsylvanian), respectively in axial and sagittal sections; (E) *Triticites* (U. Pennsylvanian to L. Permian), in axial section. Labeled features are: a, transverse septa; b, parachomata; c, chomata; d, proloculus; e, axial fillings. (From the *Treatise on Invertebrate Paleontology,* courtesy of the Geological Society of America and the University of Kansas Press.)

structures with the long list of described taxa (i.e., in the *Treatise,* Thompson, 1964). Most Pennsylvanian and Permian stratigraphic boundaries, including the series and even the systemic boundary, are based on fusulinid zones; for example, the genus *Pseudoschwagerina* is the index fossil for the lowermost Permian beds in North America, Europe, Asia, and Arctic regions, whereas species of *Triticites* characterize the Upper Pennsylvanian. In many conformable sections, the Penno-Permian boundary is marked by the presence or absence of these genera, as are other intervals of the late Paleozoic marked by other fusulinid taxa.

In Pennsylvanian stratigraphy, many terrestrial deposits have been put into time perspective by use of the abundant plant fossils. Plants similarly are present in Permian continental deposits but not nearly to the extent in the Pennsylvanian. The boundary between Pennsylvanian and Permian in some continental sequences (e.g., in the Appalachian region) has been defined by index plant species, notably the conifer *Callipteris conferta* (Wanless and Bell, 1975). In the Permian, with generally sparser floras in the arid environments implied by deposits in the southwestern craton, correlation and delimiting of continental strata are difficult where plants may be absent. Some work has been done with amphibian and reptile fossils, but these too can be rare; in fact, more so than plants. Since, as will be shown, some Permian units represent environments ranging from fluvial to marine, stratigraphers have been able to correlate some unfossiliferous nonmarine facies by assuming they were deposited synchronously with their correlatable marine facies. This practice leaves room for possible error where such marine/continental units were deposited diachronously.

The Triassic was similarly a time of extensive continental as well as marine deposition. Since fusulinids became extinct at the end of the Permian, as did a vast host of other shelly marine invertebrates, they cannot be used for Triassic index fossils. Fortunately, in the Triassic, and to a much greater extent in the later Mesozoic, ammonoid cephalopods underwent dramatic radiation. These creatures were not only accommodating enough to diversify at a time when the seas were seemingly depauperate of their shelled biota, but they also were nektonic and cosmopolitan so they spread around the globe as quickly as they evolved. They thrived in the Mesozoic seas (while the nautiloids declined to a single family by the late Triassic), and diversified, taking on a great (and sometimes bizarre) variety of shapes which makes their value as index fossils even higher. Among the variations of Mesozoic cephalopods were the complex suture patterns evolved in most taxa. These alone can be used for taxonomy and serve to make even a fragment of some ammonoids a diagnostic tool. We will discuss the group at greater length in Chapter 8.

Nonmarine Triassic strata can be correlated by plant, reptile, amphibian, fish, and freshwater mollusk fossils, where such are found. In general, the fossil record of reptiles in the Triassic is notably better than that in the Permo-Pennsylvanian, but not nearly as good as the later Mesozoic. Dinosaurs were living in Late Triassic time but had not yet reached near their zenith of success. Most Triassic reptile fossils used for correlation are not dinosaurs; one group of aquatic reptiles, phytosaurs, is fairly common in lake deposits found especially in western cratonic sections. We discuss Mesozoic reptiles in detail in both this and the next chapter. Overall, Triassic correlations in nonmarine environments are difficult and may have been inconclusive (especially with respect to the location of the systemic Triassic–Jurassic boundary; see Pipiringos and O'Sullivan, 1978).

B.2. Early Pennsylvanian Time on the Eastern Craton

B.2.a. Terrestrial and Paralic Facies in the Appalachian Basin

The geological history of the Appalachian Basin during the late Paleozoic is deceptively simple. To the east and southeast of the basin was a rising highland, the embryonic Allegheny Mountains, whose erosion supplied sediments almost continuously to the basin. Less significant sources of sediment lay to the north of the basin and to the south. Large quantities of detrital sediment were carried down from the highlands and out onto an alluvial plain which sloped off toward the sea in the west and northwest. Large delta systems and related barrier-island complexes comprised the coastal environments, in which a variety of sediments accumulated. Farther west, in the quiet waters of the epeiric sea, marine muds and limestones were deposited. In essence, the above sketch may be applied, with only a few modifications, to almost all Pennsylvanian and Permian sediments of the Appalachian Basin.

Of course, as is so often the case, the simplicity of the first glance becomes the complexity of careful examination. The Pennsylvanian and Permian Periods were times of rapid, cyclic migration of facies back and forth as shorelines oscillated in response to sea-level fluctuations, alternations in rates of sediment supply, or some other factor. Superimposed on these cyclic fluctuations were longer-term changes in the rate and locus of basin subsidence and sourceland uplift and the effects of penecontemporaneously growing structures. In this section, we will discuss deposition of the Pottsville Series, the Appalachian Basin chronostratigraphic equivalent of Morrowan and Atokan strata of the midcontinent. Further discussion of Appalachian Basin history will appear later in the elaboration of Absarokan cratonic history.

i. Mississippian–Pennsylvanian Boundary Problem. As discussed in the previous section, a definite erosional unconformity separates Kaskaskian from Absarokan strata throughout most of the midcontinent area. In western parts of the Appalachian Basin, too, an unconformity separates Mississippian from Pennsylvanian rocks. But on the eastern side of the basin, stratigraphic sequences appear to be conformable and recognition of the systemic boundary is difficult at best. In Pennsylvania for example, coarse, alluvial-plain detrital sediments of the lower Pottsville Formation (Pennsylvanian) were accumulating in the eastern part of the state at the same time that upper Mauch Chunk deltaic red beds (Mississippian) formed in central and western parts of the state (Edmunds et al., 1979).

We may conclude then that a significant unconformity at the Mississippian–Pennsylvanian boundary is probably present in most of the Appalachian Basin except along its extreme eastern and southeastern edge, where subsidence greatly exceeded that of more western areas of the basin and deposition was essentially continuous across the boundary.

ii. Stratigraphy of the Pottsville Series. We now turn to an overview of Pottsville (i.e., Morrowan–Atokan) strata in the Appalachian Basin. The Pottsville Formation is named from exposures in the Southern Anthracite field of eastern Pennsylvania, which is also the type area for the Pottsville Series. There, it is divided into three members: the Tumbling Run and Schuylkill, both of which are of Morrow age, and the Sharp Mountain, which is now considered to be of Desmoines age (Wanless, 1975). Atokan strata are either absent or comprise the uppermost Schuylkill and lowermost Sharp Mountain. All of these units are composed of coarse sands and gravels with little shale or coal. Grain size and bed thickness both increase toward the east and southeast. In western Pennsylvania and adjacent areas of southwestern New York, Ohio, and northern West Virginia, an area referred to here as the western Pennsylvanian bituminous field, equivalent sediments comprise the Pottsville Group, of which the Olean (New York and Pennsylvania) and Sharon (Ohio and West Virginia) Conglomerates are of early Morrow age. Both units coarsen to the north and were probably derived from exposures of older rocks in southeastern Canada. The Pottsville section above Sharon–Olean Conglomerates contains an almost overwhelming complexity of named stratigraphic units, each unit being one lithological stratum within the cyclically repeating sequence of sandstones, shales, coals, and limestones. This complexity of terminology will be encountered in all Pennsylvanian strata of the Appalachian Basin (and elsewhere) and it will be our policy to neglect this detailed stratigraphy except in the case of important or well-known units.

In southern West Virginia, southwestern Virginia, and eastern Kentucky, Morrowan strata comprise the Pocahontas and New River formations and Atokan strata comprise the Kanawah Formation. These units, too, are characterized by cyclically repeating sequences and contain coals, a few of which are of commercial importance. Pocahontas coals (e.g., Pocahontas No. 3 coal) are among the principal sources of low-volatile coals used to manufacture metallurgical coke. The Pocahontas and New River formations also occur to the east of this region, looking very much as they do in southern West Virginia. But farther to the west, the Pocahontas is absent and the Lee Formation rests unconformably on Mississippian strata. The stratigraphic position of the unconformity rises to the east so that Lee pebbly sands overlie Pocahontas sands and shales in the central region and grade into New River sediments in eastern parts (Englund, 1974). Above New River strata in Virginia are sediments of the Atokan-age Norton, Gladeville, and Wise formations. In eastern Kentucky, the Lee Formation comprises the Morrow Series and Atokan strata are the Breathitt Formation.

In Tennessee, lowest Pennsylvanian strata comprise the Gizzard, Crab Orchard Mountains, and Crooked Fork groups. Wanless (1975) considered all three to be of Atoka age, with Morrowan strata absent, but, as discussed above, others (e.g., Milici *et al.*, 1979) have demonstrated continuous sedimentation from Mississippian into Early Pennsylvanian (Morrow) time.

In Alabama, the Pottsville section thickens considerably toward the south, reaching a thickness of over 3000 m at the southern terminus of exposed Appalachian Basin strata where they are overlapped by Mesozoic sediments of the Gulf Coastal Plain. Well data in the subsurface of Mississippi, beneath the Coastal Plain, show over 3600 m of Pennsylvanian strata. Based on plant fossils, some of the upper Parkwood Formation (mainly Mississippian; see Chapter 6) may be Pennsylvanian, but the main Pennsylvanian section is referred to as the Pottsville Formation, divided into upper and lower parts. The above-mentioned southward thickening and the disruption of Pottsville strata by Alleghenian deformation naturally divides the Alabama Pennsylvanian into four areas of coal-mining activity: the Warrior field in the Black Warrior Basin and the Plateau field to the north on the Cumberland Plateau, and the Cahaba and Coosa fields within the Ridge and Valley province. The Cahaba field contains the thickest Pottsville section. Pennsylvanian deposits of Alabama are significant for a number of reasons, not the least of which is the great quantity of coal they contain. Of more academic interest is the fact that the regional Pottsville paleoslope was westward toward the Ouachita foredeep basin. Further, there is evidence that sediment in the Warrior and Cahaba fields came from the south. The great thicknesses of Alabama (and Mississippi) Lower Pennsylvanian strata indicate the extreme rate of subsidence that affected the southern continental margin at that time.

iii. Tectonic Controls on Appalachian Basin Sedimentation.

The general tectonic framework of Pennsylvanian sedimentation in the Appalachian Basin was established during Pottsville time. Thus, it will be instructive to consider the basin tectonics which affected sedimentation.

The main source of sediments in the Appalachian Basin was a rising orogenic highland to the east and southeast of the basin. This is borne out by the relative southeastward increase in detrital-sediment grain-size and thickness of beds, paleocurrent data, and facies patterns. The cause of uplift was the Alleghenian orogeny, in its early stages (see Section C). Sediments from the Appalachian Highlands were probably supplied all along the eastern side of the basin from Philadelphia to Georgia (Wanless, 1975). But the area of maximum subsidence, and, consequently, maximum sedimentation, was in southwestern Virginia and southern West Virginia.

The Olean and Sharon Conglomerates in northern parts of the basin apparently derived from the north. The Olean contains pebbles similar in lithology and fossils to Devonian strata farther to the north in New York.

In Alabama, as mentioned above, sedimentation from a southern, Ouachita source is inferred. The volcanic and low-grade metamorphic character of the provenance, as inferred from sandstone mineralogy, allows the hypothesis that these sediments were derived from a magmatic arc juxtaposed to the south due to closure of the southern ocean, caught between North America and the approaching Gondwanaland.

Subsidence in the Appalachian Basin was greater on its eastern side (i.e., near the sourceland) than on its western side, but the increase in rate of subsidence was not uniform from the west to the east. Rather, there was a *hinge line* (i.e., flexure zone) between a relatively stable platform in the west (Ohio, northwestern West Virginia, Kentucky, and Tennessee) and a rapidly subsiding trough in the east (see Fig. 7-5). This hinge line has been interpreted as one or more *growth faults* (i.e., faults which are active penecontemporaneously with sedimentation) by Ferm and Cavaroc (1968), or as a monoclinal fold by Dennison and Wheeler (1975). We consider the subsiding trough to be a retroarc foreland basin associated with the Allegheny orogen to the east.

iv. Pennsylvanian Sedimentology in the Appalachian Basin.

The sedimentology of Pottsville strata (and Pennsylvanian strata generally) is complicated because almost every type of coastal and near-coastal depositional environment is represented somewhere in the Appalachian Basin. Barrier islands, tidal inlets, tidal deltas, tidal mudflats, rivers and river deltas, bays, marshes, lakes, and so on were all present and their deposits grade into one another laterally, according to the changing paleogeography of the time, and vertically, according to cyclic oscillation of sea

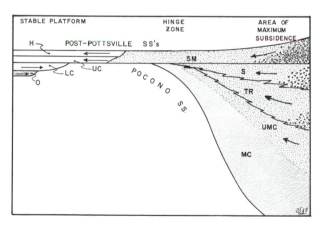

Fig. 7-5. Generalized stratigraphic cross section of Pottsville strata in Pennsylvania. Note the thickening and coarsening of units toward the Anthracite field of eastern Pennsylvania, and the southerly direction of sediment transport in the Olean Conglomerate and the lower Connoquenessing Sandstone of southwestern New York. Symbols are as follows: H, Homewood SS.; UC, upper Connoquenessing SS.; LC, lower Connoquenessing SS.; O, Olean Congl.; SM, Sharp Mountain Fm.; S, Schuylkill Fm.; TR, Tumbling Run Fm.; UMC, upper member, Mauch Chunk Fm.; MC, main body of Mauch Chunk Fm. (Modified and redrawn from Meckel, 1967.)

level. Below, and briefly, we will summarize a few general features of Appalachian deltaic sedimentation of the Pennsylvanian [the interested reader is directed to Ferm (1974) and Ferm *et al.* (1972) for greater detail]. Our subsequent discussions on the origin of coal and the nature of "typical" midwestern cyclothems will amplify topics introduced here.

Alluvial plains bordering the Appalachian Highlands were made up of large stream complexes separated by broad floodplains with lakes and swamps, in which accumulated great amounts of organic detritus destined to become coal. Streams debouched into the sea, building out large deltaic complexes with a variety of environments: marshes, lakes, distributary channels, levees, distributary-mouth bars, and interdistributary bays. Frequently, but not always, the delta grew into a lagoon bounded on the seaward side by a barrier-island complex, also with a variety of environments: tidal inlets, tidal marshes, tidal deltas, washover fans, foreshore. Shoreline oscillation pushed all of these back and forth over one another, developing the complex stratigraphic patterns of the Pennsylvanian System (Fig. 7-6).

Marine deposits are better developed in western parts of the Appalachian Basin than in eastern areas. These deposits contain fossiliferous limestones and shales and record a variety of offshore environments, ranging from shallow, agitated platforms to quiet, low-energy environments (possibly in deeper water). During regressive phases, these marine deposits were buried by the sediments of advancing coastal environments. Coal beds represent marshy areas around the margin of the lagoon. All of these environments owe their existence to the presence of a barrier island. The barrier itself, composed of mature quartz arenite, owed its existence to paleogeography, since it could form only where coastal areas fronted on a sea both wide enough and deep enough for waves to rework immature detritus brought down to the shore by rivers.

In southwestern Virginia and southern West Virginia, major delta complexes migrated northwestward over marine deposits at the same time that the barriers mentioned above were active in southern Tennessee (Fig. 7-7). Englund (1974) showed that the Pocahontas Formation represented several rapidly prograding delta lobes and that, during periods of transgression, Pocahontas delta-front sands were reworked into barrier islands. This was apparently a common situation in Pottsville time and long stretches of the coastline must have been characterized by delta complexes which grew out into lagoons bordered seaward by barriers; in some cases, the barriers may even have been attached to the delta and fed directly by delta distributaries.

B.2.b. Deltaic/Marine Sequences of the Illinois and Michigan Basins

Outside of the Appalachian Basin, Pennsylvanian units are classified by time–stratigraphic nomenclature based on

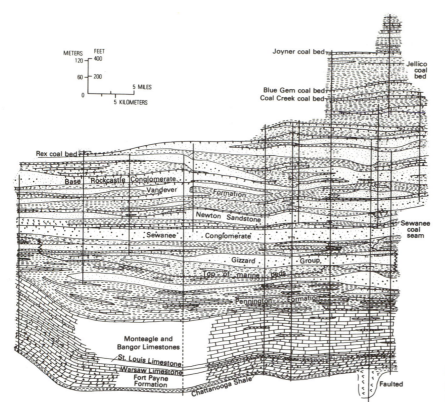

METERS   FEET
120 — 400

60 — 200   5 MILES

0   5 KILOMETERS

Joyner coal bed

Jellico coal bed

Blue Gem coal bed
Coal Creek coal bed

Rex coal bed

Base Rockcastle Conglomerate

Vandever   Formation

Newton Sandstone

Sewanee coal seam

Sewanee   Conglomerate

Gizzard   Group

Top of marine beds

Pennington   Formation

Monteagle and
Bangor Limestones

St. Louis Limestone
Warsaw Limestone
Fort Payne
Formation

Chattanooga Shale

Faulted

Fig. 7-6. Stratigraphic cross section of Mississippian and Pennsylvanian strata in eastern Tennessee. (From Milici *et al.*, 1979.)

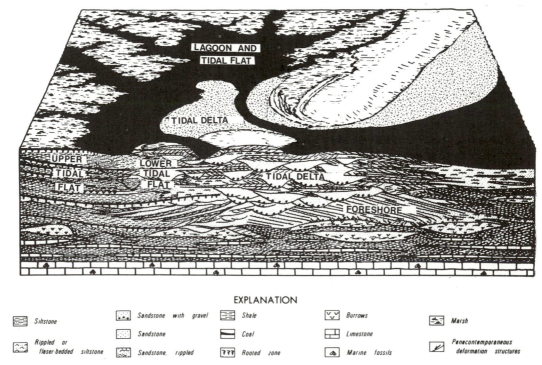

LAGOON AND TIDAL FLAT

TIDAL DELTA

UPPER TIDAL FLAT

LOWER TIDAL FLAT

TIDAL DELTA

FORESHORE

EXPLANATION

Siltstone	Sandstone with gravel	Shale

Rippled or flaser-bedded siltstone   Sandstone   Coal   Burrows   Marsh

Sandstone, rippled   Rooted zone   Limestone

Marine fossils   Penecontemporaneous deformation structures

Fig. 7-7. The barrier-island model of Upper Mississippian and Lower Pennsylvanian strata in southern Tennessee, showing typical sedimentary structures and stratigraphic relations. (From Ferm *et al.*, 1972.)

type sections in Arkansas (Morrow, Atoka), Iowa (Des-moines), Missouri (with a series of the same name), and Oklahoma (Virgil). Fortunately, units in the Appalachian, central and western cratonic basins can be readily correlated. Pennsylvanian units west of the Appalachian Basin contain proportionately more marine and less continental sediment, with marine sediments predominant in the Mid-continent Basin and western platform strata. We will first examine Lower Pennsylvanian (Morrowan and Atokan) units of the basins east of the midcontinent line, and then shift the focus to the western craton in a subsequent section.

The Illinois Basin (previously discussed in Chapter 6) occupies much of the state of Illinois, a sizable part of southwestern Indiana, the "western coal region" of Kentucky, and small areas of adjoining states. It was subsident through most of the Mississippian and the earlier two-thirds of the Pennsylvanian and during those times was bounded on all sides by positive cratonic elements.

Despite being surrounded by positive elements, it is almost certain that the Illinois Basin was connected with the Appalachian and Midcontinent basins at least periodically during Pennsylvanian time, and that it had further connection with the southern ocean through the Ouachita region. Basal sands in channel fills and marine limestones both show continuity between basins; and outliers of Pennsylvania rocks between basins indicate that strata were once more widespread and perhaps were continuous. Indeed, it will be shown subsequently that sands deposited in flysch sequences in the Ouachita margin may have been transported all the way down from the Canadian Shield and the old Acadian Highlands through the Illinois Basin.

At the close of the Mississippian, the region encompassed by the Illinois Basin was a southwest-dipping coastal plain with sizable, linear stream valleys. Initial Pennsylvanian sedimentation consisted primarily of alluvial infilling of the stream channels, as the Absaroka sea transgressed northward, followed by burial of the unconformable surface under a series of prograding deltas which intermittently retreated and reestablished, thus initiating typical Pennsylvanian cyclicity. Within the total Pennsylvanian section of Illinois, there are more than 50 successive delta sequences preserved. Marine units are somewhat more common in Illinois than in the Appalachian Basin but the Morrowan through Desmoinesian section characteristically represents delta plains, distributaries, salt marshes, and shallow marine platforms (Pryor and Sable, 1974).

As stated, cyclical sedimentation began in Chester time and continued with the first Pennsylvanian sediments above the channel-fills, but the characteristic, complete ten-part "typical Illinois cyclothem" did not develop until Atoka and Desmoines times (see Section B.3). Early units tend to be sandy with rare and thin coals and very restricted or absent marine units in cycles.

Morrowan units in the Illinois Basin comprise the Caseyville Formation in Illinois and western Kentucky, and the Mansfield Formation in Indiana. Sandstone, shale, silt shale, and siltstones (in descending order) are the dominant Caseyville lithologies; there are several thin coals present but only one, the Gentry Coal Member, is important enough to have been named (Atherton and Palmer, 1979). Sandstone units within the Caseyville are notable for containing abundant, well-rounded quartz pebbles. The Mansfield Formation in Indiana is similar to the Caseyville; however, in the lowermost Mansfield, near French Lick, Indiana, is a bed of clay-bonded siltstones with exceptionally smooth surfaces. These have been mined for many years as sharpening stones (the Hindostan Whetstone beds; Gray, 1979). The environment represented by these whetstone beds may have been a varved lake or lagoon. Minor marine transgressions reached the basin during late Morrow and early Atoka times, leaving the Fulda and Ferdinand Limestone members of the Mansfield Formation and unnamed equivalent limes in western Kentucky.

Atokan strata in Illinois are the Abbott Formation, whereas in Indiana equivalent units are the Brazil Formation, and in western Kentucky comprise the lower part of the Tradewater Formation. The Abbott and Caseyville units are quite similar. Coals are more numerous, although not nearly as abundant as in younger sections; there are at least six named coals in the Abbott in Illinois. The Brazil Formation in Indiana and the Tradewater in western Kentucky are apparent continuations of the Morrowan environments into Atoka times. Coals in the Brazil are more prominent than in the older beds and some have been mined. The most persistent and thick coals in Kentucky are in the upper Tradewater (Rice et al., 1979); yet these units still do not show the full complement of associated deposits developed in later cyclothems nor are the coals and marine deposits as volumetrically important in the Atokan as they are in Desmoinesian and later units. In Kentucky, a prominent limestone member of the Tradewater, the Curlew, indicates that marine conditions prevailed temporarily and were centered there in Atoka time; in both Illinois and Indiana, marine units are virtually absent in cyclothems from the late Morrowan until the Desmoinesian.

The Michigan Basin was active during the Pennsylvanian, though to a far lesser degree than during the Silurian (Chapter 5). Bounded westward by the Wisconsin Dome, northward by the Canadian Shield, and southward and southeastward by the Kankakee and Findley arches, the Michigan Basin received largely continental but also some marine deposits in the Early Pennsylvanian. Since the basin was active through most of the Paleozoic, it is logical that these youngest units, of Pennsylvanian age, are located in the basin center surrounded by Mississippian and older bedrock.

All Pennsylvanian strata in Michigan are placed in the Saginaw Formation and the overlying Grand River Formation. The Saginaw is a cyclical unit with considerable vertical variability. At the base is the Parma Sandstone Member, present locally, which may represent channel-fill on the rugged post-Chesterian unconformity. Lower units of the Saginaw above the base consist of sands, shales, minor coals, and limestones. A notable limestone unit near the top of the Saginaw Formation, termed the Verne Limestone Member, may correlate with the lowermost Desmoinesian Curlew and Seville Limestone members of the Spoon Formation in Illinois. It has been suggested that these limestones may be parts of a once-continuous deposit left by an extensive marine transgression (Ells, 1979). The Saginaw ranges in age diachronously from Middle–Late Morrowan to Desmoinesian and the Grand River seems to be Desmoinesian.

B.2.c. Marine Facies of the Midcontinent Basin

Westward from the Illinois Basin, and separated from it by the Mississippi River Arch and the Ozark Dome, was a very large basin that received relatively thin sediments through much of Pennsylvanian and Permian times. This basin, like the Illinois, has been referred to by several names but we will adopt the most descriptive: the Midcontinent Basin. (Chief among the alternative names are the "Forest City" and "Western Interior" basins.) At the maximum extent of the Absaroka sea, in Desmoines–Missouri times, this basin extended from Iowa to Oklahoma including much of Kansas, Nebraska, northcentral Texas, and Missouri; and it had an east–west-trending "tail," extending through Arkansas and Oklahoma, which impinged on the Arkoma Basin and the Ouachita margin.

Structurally, the basin was not a symmetrical, subsiding trough but, rather, a very gently westward-dipping platform with local depressions. Strata deposited during the Pennsylvanian dip westward at low angles and thin to a feather edge eastward. Units within the region show cyclicity, in common with Pennsylvanian strata described elsewhere, but marine rocks dominate the cyclothems. Much of the Midcontinent Basin did not receive Morrowan–Atokan sequences because the Transcontinental Arch, which both bounded the basin westward and was itself within the post-Atokan basin, remained above sea level until the Desmoinesian. Most conformable or reasonably complete Morrowan–Atokan sequences are in Arkansas or Oklahoma; there the continental shelf dropped off into the deep Ouachita trough and Lower Pennsylvanian units deposited are transitional between sands, shales, and limestones on the platform and deep-water units, largely flysch, in the southern ocean. The hinge line between Ouachita and cratonic environments shifted north during the later Pennsylvanian (Atokan–Desmoinesian), as the Arkoma Basin

underwent tectonic activity with uplifting. Strata at this hinge line thin drastically northward from the Ouachita trough to the platform edge.

Morrowan sediments at the Ouachita-platform hinge (Fig. 7-8) were deposited when the Arkoma Basin was still the shallow-water northern platform of the Ouachita mobile belt (Briggs (ed.) 1974). The Chester–Morrow unconformity is present at the hinge and to the north; but it is undeterminable or absent southward into the deep marine basin. The Morrowan units on the platform are exposed in the Ozark region (Arkansas) and comprise the Hale Sandstone and the Bloyd Shale, which are penecontemporaneous with the flysch facies of the Springer and Wapanulka formations in the deeper Arkoma Basin, and the upper Johns Valley Formation in the Ouachitas. The Hale and Bloyd represent relatively normal marine transgressive units deposited on the platform edge, and together comprise the Morrow Group, which is also the type series for the Lower Pennsylvanian in northwestern Arkansas.

West of the type region, in northeast Oklahoma, Morrowan strata have been called the Sausbee and McCully

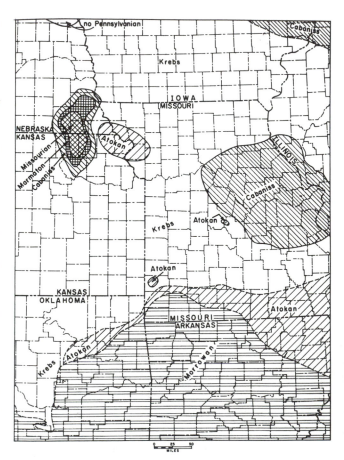

Fig. 7-8. Lower Pennsylvanian lithofacies distribution on the southcentral craton, near the Ouachita hinge line. (From Branson, 1962a.)

formations. These units are similar to the Hale and Bloyd except that the quartz arenites are supplanted by carbonate muds, calc-arenites, and detrital shales (Manger and Saunders, 1980).

Above the areally restricted Morrowan strata in the southeastern Midcontinent Basin is the widespread Atoka Formation, the type section for the Atokan Series. At the type locality in Oklahoma, the Atoka Formation consists of dark shales with numerous sandstone tongues and lenses (Branson, 1962). During Atoka time, the sea clearly had transgressed over a very broad area of the southern midcontinent; but detrital influences still predominated over much of the southeastern platform. The Atoka covers a wide area, including large regions in Oklahoma and Arkansas, and smaller areas in Missouri. Thicknesses increase drastically into the Arkoma and Ouachita basins; a thickness of over 3000 m is reached in the subsurface at the Ouachita hinge line. Briggs (1974) cites evidence that the Atoka near the Ouachita front in Arkansas attains a thickness of over 6400 m, but with irregular, steplike increments southward, possibly due to a series of parallel normal faults. These are interpreted as growth faults as the Ouachita Basin underwent increasingly rapid subsidence during deposition of the Middle part of the Atoka Formation.

In the western Arkoma Basin, the Atoka contains several sandstone tongues which are oil-bearing, each capped with a shale. Northward, the Atoka thins and contains more calcareous rock and a few limestones.

It is interesting to consider the probable source of all the sand in Morrowan and Atokan strata in the southeastern sections described above; and, surprisingly, the detritus seems to have come all the way down from the Canadian Shield and the Northern Appalachians (the eroded Acadian lands), through the Michigan, Illinois, and parts of the Appalachian basins. In Chester through Desmoines times, a major drainageway called the "Michigan River" (Chapter 6) flowed southward through the Illinois Basin, carrying sands derived from the Shield and old Acadian lands, depositing them in a series of southward-prograding deltas. These deltas comprised a large part of the deltaic sands in Illinois Basin cyclothems of the times, but as the deltas extended to the extreme south of the Illinois Basin, Graham *et al.* (1975) suggested, an orogenic land lying beyond the exposed end of the southern Appalachian belt blocked movement of sediment southward and shunted sands westward into the Ouachita–Arkoma region.

Lower Pennsylvanian units are also present in other parts of Oklahoma and in Texas and eastern New Mexico. These sections fit neither into the context of "craton" nor "continental margin" because areas within the region intermittently functioned as platforms or oceanic troughs during much of the Pennsylvanian. The best physiographic term for the region is the "Texas Foreland System" and it in-

cludes the Ardmore Basin, Criner Hills, Arbuckle Anticline, and, in general, the Wichita orogen and the adjacent Anadarko Basin. These features will be examined subsequently with events on the continental margin.

B.3. Cyclic Sedimentation and the Formation of Coal

As we have already shown, cyclic sedimentary sequences in which nonmarine and marine deposits alternate vertically are common in Pennsylvanian strata and they will be encountered throughout most of our discussion of subsequent Absarokan history. Such sequences comprise the bulk of Absarokan strata on the eastern craton and are of great economic importance because they contain enormous reserves of coal.

Sedimentary sequences containing vertically repeating strata are relatively common in the geological record but they are not all considered to be cyclic because some lack regularity of repetition. We define a *cyclothem* as a sequence of strata which records the variation of depositional environments due to a single seaward and landward oscillation of a shoreline. As originally used (Weller, in Wanless and Weller, 1932), the term cyclothem referred specifically to deposits of Pennsylvanian and Permian ages but it is a very useful term and has been generalized (e.g., Duff and Walton, 1962) to include similar deposits of any age. Each distinguishable element of the succession (e.g., the coal bed, the nonmarine shale) is called a *phase*.

B.3.a. The "Ideal" Cyclothem

In 1930, Weller elucidated the concept of cyclic Pennsylvanian sedimentation and described the "ideal" cyclothem. This generalized sequence, which is nowhere completely exposed in one outcrop, is composed of ten phases and is illustrated in Fig. 7-9. The sequence begins at the erosional surface (a local disconformity) beneath massive to thin-bedded sandstones which sometimes contain plant fossils and clasts derived from the underlying mudstone; the sandstone, which is considered to be phase 1, may be deeply channeled into the underlying sequence or it may rest on the lower sequence across an even, sharp surface. The second phase is sandy shale, also with sporadic plant fossils and sometimes with fossils of terrestrial invertebrates. Phase 3, which occurs mainly as nodules, lenses, and discontinuous layers, is a freshwater algal limestone and is generally unfossiliferous. Phase 4 is underclay, or seat earth, a nonfissile claystone or siltstone which underlies coal, phase 5. The sixth phase is gray shale with terrestrial plant fossils, often beautifully preserved, pyritic nodules, and ironstone concretions at the base. Phase 7, which is often absent, is argillaceous, micritic, fossiliferous

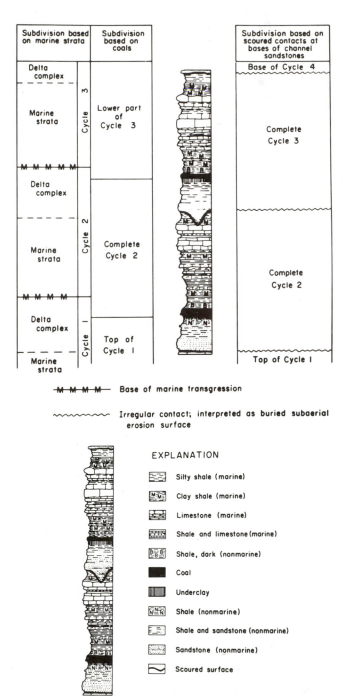

M—M—M—M—   Base of marine transgression

〜〜〜〜   Irregular contact; interpreted as buried subaerial erosion surface

EXPLANATION

- Silty shale (marine)
- Clay shale (marine)
- Limestone (marine)
- Shale and limestone (marine)
- Shale, dark (nonmarine)
- Coal
- Underclay
- Shale (nonmarine)
- Shale and sandstone (nonmarine)
- Sandstone (nonmarine)
- Scoured surface

Fig. 7-9. Generalized stratigraphic sequence of an "ideal" western-Illinois-type cyclothem, showing three ways of subdividing the repeating sequence. The right-hand subdivision, based on erosional surfaces, is the traditional method; subdivisions based on deltaic and marine stages are newer and probably in better accord with modern sedimentology. (Modified from Friedman and Sanders, 1978.)

marine limestone. Phase 8 is dark-gray to black marine shale. This shale is commonly laminated, very organic-rich, and often radioactive; it has a very limited fauna, almost exclusively of planktonic or nektonic habit. Some black shales contain limy beds with a more normal marine faunal assemblage, featuring coral, bryozoans, crinoids, brachiopods, fusulines, and others. Phase 9 is clean, fossiliferous marine limestone. The final phase consists of bluish-gray shale which becomes increasingly more sandy toward the top. This shale phase contains a marine fauna near its base but scattered plant debris and very few marine fossils in its uppermost layers.

The breakdown of cyclothem phases described above is by far the most common way to subdivide the repetitive sequence and it appears in most textbooks. But it is certainly not the only way to subdivide the sequence. A second way is shown on the left side of Fig. 7-9. This breakdown considers the bluish-gray shale above the fossiliferous limestone (i.e., phase 10 above) to be the beginning of the cycle and each phase is numbered accordingly. The reasons for this different subdivision are related to the environmental interpretation of the cyclothem and will be discussed below.

It should be noted that the stratigraphic sequence discussed above is a generalized one, an ideal model. Real cyclothems vary somewhat, but never grossly, from the ideal. For example, strata are sometimes missing due to erosion or nondeposition; strata may be repeated out of sequence such as the occurrence of two coal beds separated by a gray shale within a single cyclothem. Nevertheless, the correlation of individual phases within a given cyclothem over extremely large areas is, perhaps, the most striking aspect of cyclothems (other than their remarkable cyclicity). Moore (1950) stated that some phases could be traced over 640 km along strike and over 480 km downdip into basins. Furthermore, it is also possible to correlate individual cyclothems all the way from western Kansas to the Appalachian Basin. Thus, Pennsylvanian cyclic sedimentation affected a truly enormous area of the continent!

B.3.b.   Walther's Law and the History of a Cyclothem

Walther's law states that in a vertical stratigraphic sequence unaffected by unconformities, strata vertically adjacent to each other also formed in laterally adjacent depositional environments. This principle is particularly pertinent to the interpretation of cyclothems, for they clearly record the migrations of a set of interrelated sedimentary environments as shorelines oscillated back and forth. Walther's law provides a constraint on the types of depositional environments one may postulate for any specific phase; for example, black shale may form in deep-water, basinal conditions but such an environment cannot reasonably be inferred for the black shales of cyclothems because they lie between shales with nonmarine fossils and shallow-water limestones. Therefore, the depositional model that one chooses must explain not only the characteristics of any one cyclothem phase but also the occurrence of the phase within the vertical sequence.

The choice of the erosional surface as the base of the sequence seems a logical way to separate the cyclic strata into repeating units, especially when one realizes that this choice was made at a time in the history of stratigraphy when unconformities were used to define most major chronostratigraphic units. But such a choice was unfortunate because it forced geologists to make depositional sense out of a rather unnatural pattern. The erosional surface is now recognized as a result of channeling of deltaic distributaries into the silts and muds of the prodelta. Therefore, choosing the erosional surface as the base of the sequence split up the deltaic record into two parts and made its recognition difficult. This is the reason why cyclothems are now subdivided such that shallow-marine limestones mark the end of one cyclothem and the advent of prodelta muds marks the beginning of the next.

The depositional history of a cyclothem begins at a time of maximum marine transgression. As sea level began to drop, or the rate of sediment influx to increase, deltas began to prograde westward and southward into the epeiric sea (see Fig. 7-10). The first sediments of the new regressive regime were very fine detrital muds carried far out into the sea and deposited, like a blanket, over the carbonate platform. As the deltas prograded, the area of the old carbonate platform was invaded, first by lower prodelta-slope muds and then by upper prodelta-slope silts and sands; the result was the coarsening-upward sequence typical of delta progradation. These prodelta deposits comprise the bluish-gray shales at the base of the deltaic stage in Fig. 7-9. Continuation of the regression allowed distributaries of the subaerial delta to cut across the prodelta deposits, producing a channeled erosional surface. Sands of the distributary channels and distributary mouth bars were sometimes reworked by the weak waves and currents of the sea into a sand sheet overlying the erosional surface; more often, distributary sands were left unreworked and remain as bar-finger sands, characteristic of "bird's-foot" deltas such as the Mississippi Delta. Deltaic sedimentation constructed a platform at sea level and on this platform were deposited a variety of sediments such as the fining-upward sequences of stream point-bars and the muds of overbank deposits which comprise the sandy shales above distributary sands. Nodular algal limestones formed in freshwater lakes on the delta plain. Underclays probably represent soils developed on the subaerial delta surface. As the area subsided to or below the water table, swamps ensued. Coal swamps were vast and within their stagnant waters accumulated thick masses of

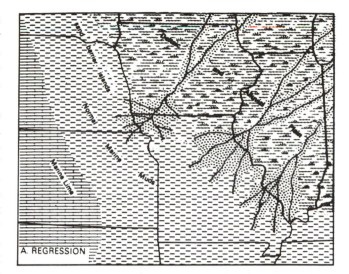

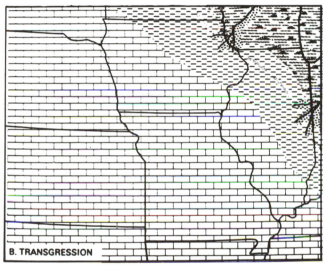

Fig. 7-10. Midcontinental paleogeography, reconstructed during a regressive phase (A) and a transgressive phase (B) of cyclothemic deposition. (From Dott and Batten, 1981, with permission of McGraw-Hill Book Company.)

peat which were to become coal. The coal swamps received little, if any, detrital sediments and as the organic detritus underwent compaction, the level of the swamp subsided further so that lakes were established in which were deposited the gray shales which have preserved so well the flora and fauna of the times. These shales probably represent the maximum extent of regression.

Rising sea level, or decrease in the rate of sediment supply, allowed the sea to transgress the delta platform. Since the surface of the delta platform was nearly flat and at or very near sea level, only a slight rise of the sea would have been required to flood the whole area and establish an extensive, shallow body of water. In most places, the first sediments were black, organic-rich muds. The presence within the resulting black shales of only planktonic and nektonic fossils suggests that surface conditions were conducive to life while bottom conditions were stagnant and euxinic. As water depth increased and the shoreline retreated before the advancing sea, detrital sediments were trapped on the coastal plains by alluviation. Soon only limy sediments accumulated in the epeiric sea, now fully open to the effects of waves and currents and well-oxygenated all the way to the bottom. Benthic faunas flourished as they had before delta growth had begun the new cycle. The carbonate platform was reestablished.

B.3.c.   Regional Variations and Paleogeography

The generalized cyclothem discussed above is based on sequences in western Illinois and therefore is not representative of cyclic Pennsylvanian strata elsewhere on the craton. Figure 7-11 is a comparison of three general cyclothems: from western Illinois (the "ideal" cyclothem), from the southern Appalachians, and from Kansas. Not only is the typical sequence in the Appalachian area (the Piedmont cyclothem of Wanless and Shepard, 1936) considerably thicker overall than the western Illinois sequence, but individual phases are also thicker. The average grain size of Appalachian-sequence sediments is coarser, with conglomerates at the base of the distributary-channel phase. The sequence is dominated by deltaic and terrestrial sediments and marine phases are subdued or absent altogether.

The Kansas-type (or neritic) cyclothem, on the other hand, is thinner than the western Illinois sequence and is dominantly marine. Deltaic strata comprise less of the Kansas sequence and subaerially deposited strata, such as underclays and coal, may be absent. Limestones comprise a much greater portion of the Kansas cyclothem and, in many cases, record far-offshore environments.

The paleogeographic picture that may be inferred from the nature of cyclic Pennsylvanian strata is shown in Fig. 7-10. The continental interior was low, nearly at sea level, and relatively flat. Thus, very slight changes in relative sea level or in rate of sediment supply would have had dramatic effects on the position of shoreline, which is believed to have oscillated over a region 800–1000 km wide. Sediments supplied to the advancing deltas were derived from the rising Appalachian Highlands in the east and from the Canadian Shield in the north. Potter and Siever (1956) showed that eastern cyclothems contain sands derived from metamorphic terranes of the Appalachians whereas western cyclothems were supplied with multicycled sands derived from earlier Paleozoic strata in Canada.

B.3.d.   Causes of Shoreline Oscillation

In the discussion so far, we have been deliberately vague as to the causes of cyclothems, other than to say that the stratigraphic record shows an oscillation of shorelines. We now turn to this subject of causes. The reasons behind cyclic shoreline-oscillation have been the subject of much debate. Two reasonable basic hypotheses are: (1) epeirogenic up- and down-warping of the craton and (2) eustatic sea-level fluctuations.

Rapid, cyclic up- and down-warping of the craton is called upon by Sloss and Speed (1974) to explain cyclothems. Their model supposes that quivering of the craton surface and associated large-scale block-faulting of cratonic areas (the oscillatory mode of craton tectonics) was a transitional stage following the extreme submergence of Kaskaskian time.

The second hypothesis, that of eustatic sea-level fluctuations, is currently the most widely accepted one. In addition to the appealing simplicity of the idea, there are two other reasons for its wide acceptance. First, there is some evidence from the strata of other continents that depositional cyclicity was global in nature. And, second, a cause of sea-level fluctuations is readily at hand: during Late Mississippian through Permian times, a continental glacier covered large parts of Gondwanaland, which, at that time, occupied a South Polar position; advance and retreat of the ice sheet, responding to the same dynamics as the Pleistocene ice sheets, would have caused cyclical variations in the amount of water in the ocean basins and, therefore, cyclical fluctuations of sea level.

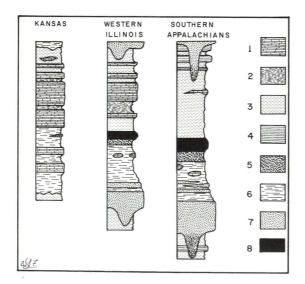

Fig. 7-11. Comparison of cyclothemic deposits in three regions. Legend of map symbols: 1, marine limestone; 2, carbonaceous shale; 3, marine shale; 4, marine sandstone; 5, underclay; 6, nonmarine shale; 7, nonmarine sandstone; 8, coal. (Redrafted using data from Wanless and Shepard, 1936.)

B.4. Early Pennsylvanian of the Western Craton: Shallow Seas, Uplifts, and Yoked Basins

B.4.a. Morrowan–Atokan Transgression on the Cordilleran Platform

On the western cratonic platform, the Absaroka sea initially transgressed eastward out from the old Antler foreland basin, to which it had retreated in late Chester time (Chapter 6), and spread widely across the platform in Montana, the western Dakotas, and Wyoming. Subsequently, it spread southward down to the "Four Corners" area (junction of New Mexico, Arizona, Colorado, Utah) but the progress was quite irregular as shown by variable hiatuses in regional strata.

West of the Cordilleran foreland basin lay the Antler tectonic lands, traceable in a north–south belt from southcentral Idaho to southeastern California, and still farther west were a diverse set of exotic terranes including island arcs and back-arc basins. It is not known whether waters in the foredeep basin communicated with those of the marginal basins across the Antler lands.

The initial transgression on the northwestern U.S. platform left the Amsden Formation in Montana and eastcentral Idaho. This unit spans the Chesterian–Morrowan for its time of deposition and contains thin sandstones and mudstones, probably with a western (Antler) provenance (Sando et al., 1975). Farther south, the platform received less detritus and included a number of low emergent regions. Carbonate sedimentation prevailed. Morrowan–Atokan carbonates in Nevada and Utah are largely assigned to the Ely Limestone but in southeasternmost Nevada and northwest Arizona, extensive oolite shoals developed on the shallow platform and formed the Callville Limestone (which ranges in age through the Virgilian). An additional unit in southcentral Nevada is the Bird Spring Group, deposited in a small basin located approximately 200 km south of Ely. The Bird Spring formed through Pennsylvanian time and on into the Leonardian (Middle Permian), and contains mostly carbonates but also significant shales and sandstones, the latter suggesting a nearby source.

In southwestern New Mexico and southern Arizona, Pennsylvanian strata overlie Precambrian to Chesterian rocks with disconformable to angular-unconformable relationships. The Pennsylvanian units there range from probable later Morrow to Virgil ages with thicknesses reaching approximately 3500 m (but generally less). These units in southeastern Arizona and adjacent New Mexico are generally assigned to the Horquilla and the lower Earp formations, and in southcentral Arizona they are the Naco and the lower Supai formations (Kottlowski, 1962). The dominant lithology is limestone, deposited on the passive shelf, with considerable detritus interbedded containing sand to clay-size particles. The detrital materials came mostly from the Zuni landmass but some may have derived from the Florida and Pedernales uplifts. In most localities, the basal beds are of early Atoka age but some exposures yield Morrowan fusulinids. Younger (Desmoinesian) strata in the region are more cherty than are the Morrowan and Atokan units, and in Missouri time, units became more shaley and some big reefs developed. In northcentral Arizona, the Supai red beds developed as early as Missouri time, representing delta deposits prograding south-southeastward across Arizona. The source for these deltaic detrital sediments may have been the Kaibab Arch on the northcentral Arizona–southcentral Utah border.

Eastward from the above, basal Absarokan strata vary tremendously from place to place across the western craton because of the development of basins and uplifts through the Pennsylvanian, and continuing into the Permian. Most such activity, notably of the yoked basin format, occurred in New Mexico, Arizona, Colorado, and Utah; but in Texas, northern Mexico, and Oklahoma, a series of uplifts and associated yoked basins were present, comprising the Wichita and Texas Foreland systems. All of the above features will be the topic of the next subsection.

By Desmoines time, the Absarokan transgression from the Cordilleran foredeep basin had breached the midcontinent, and the sea from the west met the sea in the central basins. However, not all central and western cratonic areas were flooded. Portions of northernmost Montana, Wyoming, the Dakotas, Minnesota, Wisconsin, and in general the upper Mississippi valley do not contain Pennsylvanian units. These regions were uplands of the central platform and Canadian Shield and remained emergent during Absarokan and virtually all later times.

B.4.b. Uplifts and Their Adjacent Basins: The Enigma of Cratonic Tectonics

The up- and down-warping of the southcentral and western craton in Absarokan time has no apparent parallel in the North American geological record. Certainly basins were active in other times (recall the Michigan Basin in the Silurian, the Williston Basin in the Devonian, and so on), but the profound activity of basins, coupled with apparent sharp uplifts of nearby arches which occurred in Penno-Permian time, may have been unique.

The most pronounced development of uplifts and yoked basins centered around the Four Corners area (Fig. 7-12) with development of the uplifts which together comprise the "Ancestral Rockies." Lying largely in Colorado were the Front Range and Uncompahgre uplifts (both active by the latest Mississippian), between which developed an irregularly deep, narrow basin called the Colorado Trough [we are adapting here the terminologies of Eardley (1951)

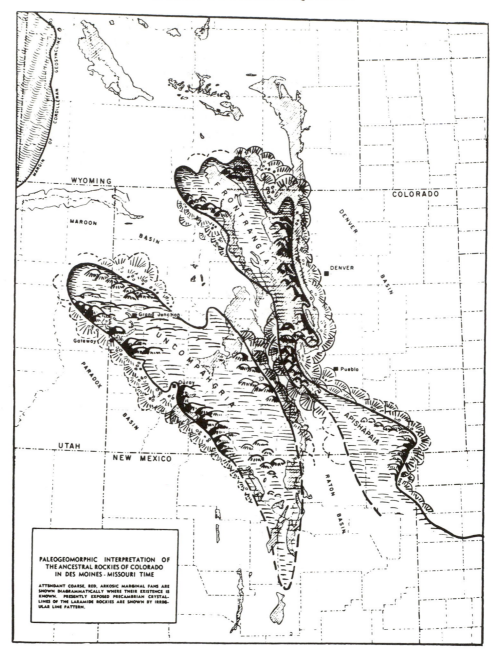

Fig. 7-12. Paleogeomorphology of the Ancestral Rockies of Colorado, during Desmoines–Missouri time. Arkosic alluvial fan deposits are diagrammatically shown peripheral to uplifts. (From Mallory, 1958.)

and King (1977), each in part]. These Ancestral Rockies uplifts may have been topographically very high structures and were almost certainly formed by some sort of block-faulting. Their presence on the craton indicates significant tectonic activity but of a nature imprecisely known.

Southwest of the Uncompahgre Range, and apparently dropping precipitously from the normal-faulted boundary, was the Paradox Basin lying equally in Colorado and Utah

(Fig. 7-12). On the border between New Mexico and Arizona, still farther southwest, was the Zuni Uplift. The Zuni may have been very low-lying, and it may have been occasionally awash in the shallow sea (Kottlowski, 1962), but its areal extent was large and it seems a product of simple uplift rather than fault-block activity. In nearby northcentral Utah, the Oquirrh Basin developed in Morrow time and received over 4500 m of Pennsylvanian sediment (with an additional 3000 m in Permian time) over a relatively small area; it was probably a very sharply downdropped graben on the platform. In central, southern Idaho, a much larger basin developed at the foothills of the Antler lands, called the Wood River Basin, and it remained active through the Pennsylvanian.

In the southernmost, western corner of the craton, the Pedernal Uplift developed in central New Mexico and it may have been a southern extension of the Uncompahgre Range; however, its precise nature has been difficult to define. A long finger of land called the Sierra Grande–Las Animas Uplift may have connected the Pedernal and Ancestral Rockies. Also in New Mexico, the Florida Uplift and Diablo platform emerged (and lay partly in northern Mexico) and east of these lay the Texas Foreland System. The Texas basins and uplifts feature prominently in our discussion of Permian events and will be discussed in great detail there; they lay north of the Marathon orogenic belt and included a complex of arches and three major basins: the Delaware, Midland, and Marfa basins, which generally accumulated less than 1 km of sediment in the Pennsylvanian (most in Morrowan–Atokan times) but which were notably active in the Permian.

North of the Texas Foreland System, in northern Texas and Oklahoma, was the Wichita System of uplifts and basins (Fig. 7-13, also see Fig. 7-40) including a long, narrow series of uplifts, which together comprise the Wichita Range (or Anadarko Uplift), and to the north of this, the Anadarko Basin. The Anadarko was active in the latest Mississippian and remained active through the Pennsylvanian and into the Permian. To complete the regional overview of basins and uplifts, north of the Wichita System, on the central stable platform, were a number of small or topographically minor basins and arches active during the Pennsylvanian; but most important were two uplifts: the Nemaha Ridge in Kansas and Oklahoma, and the Central Kansas Arch (part of the Transcontinental Arch System). The Nemaha Ridge was a very narrow, very sharp upland which was active only briefly during the early Pennsylvanian. The Central Kansas Arch, by comparison, was a low, broad uplift which may have been a reactivation of the Devonian Ellis Arch.

B.4.c. Analyses of Selected Upper Paleozoic Cratonic Basins

i. The Oquirrh Basin.   The basin was centered in the location of the present-day Wasatch Mountains, east of Provo, Utah (Morris *et al.,* 1977); however, its eastern and western limits are obscured and, additionally, it is likely that post-Paleozoic tectonics have transported Oquirrh basinal facies eastward as much as 160 km, thus truly obscuring the original location. [The interested reader is directed to Bissell (1974), which presents a summary history of attempts to locate the original basin's position.] Units within the Oquirrh, ranging in age from the later Mississippian through the Permian, show that the greatest activity was in the Permian; however, throughout the latest Morrow

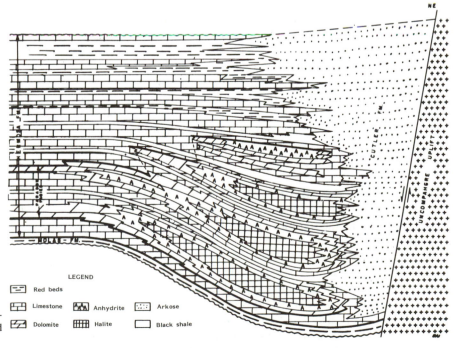

Fig. 7-13. Halite-bearing cycles in the Pennsylvanian of the Paradox Basin. (From Herman and Barkell, 1957.)

LEGEND

- Red beds
- Limestone
- Dolomite
- Anhydrite
- Halite
- Arkose
- Black shale

through Wolfcamp times, the basin underwent profound subsidence. Bissell (1974) considers the Oquirrh a classical example of *hypersubsidence* and states that the basin was the site of the deepest down-warp of crust within the Cordilleran ''miogeosynclinal belt'' (i.e., on the western platform-edge).

The lowermost Pennsylvanian units within the basin overlie the Chester-to-Morrow-age Manning Canyon Formation unconformably. Since the Manning Canyon does not thicken appreciably in the basin, it is assumed that activity began in the latest Morrowan, with the lower units of the Oquirrh Group spanning Atoka through Wolfcamp times. Approximately 5000 m of the group was deposited in the Pennsylvanian. The Oquirrh Group is composed primarily of clean, fine-grained quartz arenites, interbedded with limestones.

During the Permian, the depocenter shifted westward and the term ''Oquirrh–Sublette'' Basin denotes this changed configuration (Bissell, 1974). In southcentral Idaho, the ''Sublette'' portion of the basin received more than 6750 m of Permian detritus while the extension of the original basin in Utah received approximately 3000 m of additional sediment.

The units deposited in the Permian (the upper Oquirrh Group) were described by Bissell as ''one of the finest suites of orthoquartzites and calcarenaceous orthoquartzites which we have on record'' (1974, p. 91). At about the same time as the upper units of the Oquirrh Group were deposited (Wolfcampian), in the considerably less spectacular, adjoining Arcturus Basin in eastcentral Nevada, approximately 3000 m of a similar but more complex mix of quartz arenite, limestone, dolostone, shale, and evaporites accumulated in the Arcturus Group. The Arcturus Basin may have been under detrital influence from the western Antler erosion remnants or from the Sonoma orogen.

Activity in the Oquirrh Basin ceased in Wolfcamp time with deposition of the relatively thin Kirkman Limestone (suggesting that the basin had filled to relatively shallow, carbonate-shelf depth) whereas activity in the Arcturus Basin continued well into Leonard time with deposition of the Park City Formation, preserved in the Eureka district in Nevada and in northern Utah.

ii. The Paradox Basin. The Paradox Basin is located adjacent to the southwestern side of the Uncompahgre Uplift, within the southwest corner of Colorado and the southeast corner of Utah (see Fig. 7-12). It had a relatively short span of activity, Atokan through Virgilian, and it appears to have featured restricted circulation with the adjacent shelf areas subsequent to its initial activity. Within the Paradox region, the basal Pennsylvanian units (mostly Atokan in age but in many sections showing diachronous boundaries through the lower Pennsylvanian) comprise the relatively thin Molas Formation. This detritus-and-carbon-

ate mud unit may have incorporated sediments eroded from exposed units in Chester time and probably represents coastal deposition on the shallow platform. In later Atoka time, a shallow basin developed with restricted circulation to the western sea (remember that the Uncompahgre Uplift was active since Chester time, blocking eastward marine flow). This basin initially received carbonates and some red, gray, and greenish muds, but soon received dark, organic-rich gray muds, suggesting eutrophy and a closed basin (Shawe, 1976). These lithologies comprise the lower limestone member of the Hermosa Formation and the initial activity of the Paradox Basin (Fig. 7-14).

The basin was fully active by the Middle Pennsylvanian (Desmoinesian–Missourian) and continued to be restricted from adjacent seas. The lower unit of the Paradox Member (Hermosa Formation) deposited at this time contains a mix of organic shales, carbonates, and other sediments in the first few hundred feet, overlain by over 1000 m of cyclically stratified, predominantly evaporitic strata, chiefly halite, indicating extensive evaporation. The basin was probably a graben, bounded by normal faults at the foot of the Uncompahgre Uplift; the basin deepens toward the land with no evidence of shore facies.

Upper beds of the Hermosa Formation contain progressively fewer evaporites and are capped by almost 600 m of marine limestones interbedded with mixed-color shales and minor sandstone. In the latest Pennsylvanian, the thin Rico Formation, which locally (in the San Juan Mountains region) intertongues with the uppermost Hermosa Formation, contains some evaporites and some shales; Shawe (1976) suggests that by this time the basin had reestablished connections with the open sea. By the earliest Permian, the basin was inactive and Uncompahgre detritus flooded the area, causing localized shallowing and even emergence of areas within the region and the development of deltas.

iii. The Central Colorado Basin. This long, irregularly deep trough located between the Uncompahgre and Front Range uplifts, was active through the late Paleozoic and received predominantly coarse, feldspathic detritus during its period of activity. The basin was irregularly subsident and we will describe units from the northernmost depocenter, in Eagle County, Colorado, which has been called the ''Eagle Basin.''

Early Pennsylvanian units in the basin comprise a thin bed of the Molas Formation, which here (as it was locally in the Paradox Basin) was largely an infilling of the pre-Absarokan karst surface. The Molas is locally bright yellow and red regolithic silt and clay with abundant chert (Tweto and Lovering, 1977). In cave fillings, the Molas has been bleached by hydrothermal activity. Above the Molas is the relatively thin Belden Formation (0–61 m) which is fossiliferous and composed generally of dark-gray to black shale, thin-bedded sandstones, and black limestones. It is

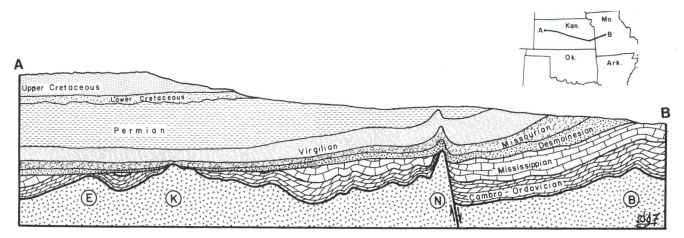

Fig. 7-14. Structural/stratigraphic cross section across the Nemaha Granite Ridge and vicinity. Map symbols: E, Ellis Arch; K, Kansas Arch; N, Nemaha Ridge; B, Bourbon Arch.

the oldest of the central Colorado Basin clastic assemblages and it was clearly deposited during a period of very moderate activity; the presence of limestones and the absence of red, coarse units show that the basin was not topographically deep during Belden time.

Above the Belden, however, approximately 600 to 2000 m of arkosic conglomerate, coarse sand, thin-bedded carbonates, micaceous shales, and mixed detrital sediments forms the Minturn Formation, which denotes both activity in the basin and topographic prominence in the uplifts. In northwestern parts of the Eagle Basin and other areas of the Colorado trough, the Minturn Formation contains as much as 500 m of evaporites. The evaporite basin was at the western edge of the Front Range highland, geographically in the present-day Gore Range, and represents localized but profound restriction in the basin, which (as stated previously) was highly variable in bottom topography and activity.

The remaining basinal unit is the Maroon Formation, consisting of up to 1280 m of red sandstones, conglomerates, and mixed detrital sediments. The Maroon represents predominantly arid, subaerial erosion from the partly eroded uplifts and correlates in part with the Weber Sandstone in other Cordilleran cratonic sections. The Maroon is poorly fossiliferous but probably was deposited from Middle Pennsylvanian through Early Permian times. After Maroon time, the basin was inactive and Middle Permian through Middle Triassic strata are absent.

iv. Synoptic View of Other Cratonic Basins.
The stratigraphic sections in the basins discussed thus far are important, tectonically significant examples of Absarokan Cordilleran events, but they are geographically small regions. It would be impractical to describe sections

from each basin in the west at length but here we will broadly cover several additional areas.

Among the largest down-warped cratonic areas was the Wood River Basin, located in southcentral Idaho and occupying the former site of part of the eroded Antler lands. It was active from Desmoinesian through Virgilian times and received several thousand meters of largely quartzitic and calcareous sandstone. The basal Wood River Formation is over 160 m of massive, brown conglomerate, overlain by dark limestones and then thick, monotonous sequences of calcareous sands and sandy limes (Williams, 1962). Rich (1977) suggests that the source for the detritus in the Wood River Formation was from an emergent terrane in western Montana derived from uplift of the Copper Basin (a small basin active in Morrow and Atoka times). Stratigraphic relations in the Wood River area are difficult to reconstruct due to post-Paleozoic faulting associated with the Rocky Mountain orogeny.

On the flanks of the Colorado uplifts (east of the Front Range Uplift and southeast of the Uncompahgre Uplift) are thick detrital units clearly derived from the uplifts. The Fountain Formation contains up to 1200 m of pink, red, and orange crossbedded sands, conglomerates, and mudstones deposited probably in Desmoines through Wolfcamp times. The redness in Fountain, Sangre de Cristo (below), and related units comes from the presence of both orthoclase and ferric oxides, indicating oxidizing conditions and rapid deposition. These units also contain mud cracks and channel-fill features which, together with the redness, point to terrestrial sources (Van Horn, 1976; Mallory, 1958). Mudstones within the Fountain represent floodplain or lake deposits from intervals of relatively slow erosion of the highlands. In the Golden Quadrangle, Colorado, and elsewhere, the Fountain is overlain by the Permian Lyons Sandstone, which in part interfingers with the Fountain lithology; but the Lyons contains less red material and has been considered a supratidal deposit on the strand of the transgressing Permian sea (Van Horn, 1976).

The Sangre de Cristo Formation, which outcrops in a mountain range of the same name, was deposited east of the southeastern end of the Uncompahgre Uplift. Like the Fountain Formation, it consists of a thick wedge of coarse, red, bouldery sediments, derived from the adjacent uplifted mass (King, 1977). The Sangre de Cristo Basin occupies the southern end of the Central Colorado Trough and received terrestrial sediments from the Late Pennsylvanian through the Early Permian.

To end the basin tour, we will briefly examine the Anadarko and Ardmore Basins associated with the Wichita uplifts in the Texas panhandle and Oklahoma; these features will be reexamined subsequently because their position close to the Ouachita and Marathon margins involved tectonics from those sources. The Anadarko Basin was a large, hypersubsident feature, active through the Pennsylvanian and Permian. It lay across the north flank of the Wichita Uplifts (and see Fig. 7-40) and received over 6 km of sediment during its time of activity. Particularly thick are the Morrow and Atoka formations which together comprise almost 2 km of the total section in the basin. Most units in the basinal facies are shaley and sandy, derived largely from the uplifts. The Anadarko Basin deepened precipitously south to the uplifts, and northward, into northern Oklahoma and Kansas, the basin units thin to a feather edge. The Morrow and Atoka units in the Anadarko Basin pinch out completely northward, where Desmoinesian and younger carbonates overlie Mississippian rocks. Within the basin, Desmoinesian units overlie Morrowan–Atokan units disconformably, suggesting some sort of orogenic activity in late Atokan time. The Ardmore Basin is located east of the Anadarko Basin and the Wichita Uplifts; and it connected with the Anadarko and may have been closely related tectonically. It was a much smaller basin, occupying a narrow space between the Arbuckle and Criner uplifts (and see Fig. 7-40), but it received a thick set of Pennsylvanian sediments. As in the Anadarko, Ardmore units are shaley and sandy, and over 4000 m thick. The Ardmore units are largely confined to the Pennsylvanian, including the Springer, Dornick Hills, Deese, and Hoxbar groups, but in the basin center is a relatively thin conglomeratic sequence of the Pontotoc Group which records postorogenic coarse detritus from the emergent Ouachita tectonic lands.

B.4.d. Analyses of Pennsylvanian Cratonic Uplifts: The "Ancestral Rockies," Nemaha Ridge, and Others

Unlike the previous discussion on cratonic basins, we have no sections to describe from the positive features on the Absarokan craton, because uplifts do not receive sedi-

ments but rather yield them. The stratigraphic record of, say, the Ancestral Rockies Uplifts lies in the coarse detritus of the Paradox, Frontier, Central Colorado, and Sangre de Cristo basins, among others. Yet, some vestiges of ancient cratonic uplifts may remain on-site in the form of the igneous cores of structures and in the topographic reversal produced by the tectonic uplifting of thick sedimentary sections.

One Pennsylvanian uplift has left a definite structural record: the Nemaha Granite Ridge in Kansas (which, strictly speaking, is a midcontinent feature but we will discuss it here in the context of uplifts). This was an apparently sharp, long, straight positive feature, active in the Early Pennsylvanian, which was uplifted sufficiently to expose Precambrian crystalline basement rocks at its core after erosion (hence the term "granite ridge"). The eastern flank was probably a normal fault scarp and sediments deposited on that side include considerable detritus from the ridge itself; Eardley (1951) reported that there is over 1200 m of relief within the ridge (Fig. 7-15). The ridge was totally buried beneath Late Pennsylvanian debris. None of the Precambrian core of the ridge is exposed at present, and in fact the entire ridge is subsurface; yet, its presence was detected early in the history of midwestern petroleum drilling. As the Nemaha Ridge continued to uplift during pre-Missouri time, sediments deposited on the western flank formed a strongly westward-dipping monocline (the ridge may not have been faulted on the western flank). Then, as the ridge eroded through Missouri time, it received relatively impermeable shaley units which overstepped the older Pennsylvanian rocks and formed excellent stratigraphic petroleum traps.

The neighboring Central Kansas Arch, as well as the Zuni Uplift, Pedernal Uplift, and a number of others featured in Fig. 7-40 probably were not bounded by faults and almost certainly had far less relief than did the Nemaha Ridge and the Ancestral Rockies. They are not easy to reconstruct topographically because their relatively lower elevations and gentle slopes, in contrast to the faulted margins of the above uplifts, did not yield characteristic coarse red sediments in restricted areas at their flanks but rather shed detritus across broad areas.

B.4.e. Causes of Yoked Basins and Other Intraplate Features

It is not clear whether the appearance of all the described western cratonic features during the late Paleozoic shows an interrelationship of their causes, nor is it clear how they relate to marginal tectonics of the times. Several arguments may be reasonably given for the western events: (1) response to Ouachita–Marathon marginal tectonics; (2) response to Appalachian marginal tectonics; (3) effects of sediment loading from the Antler and Sonoma orogenies in

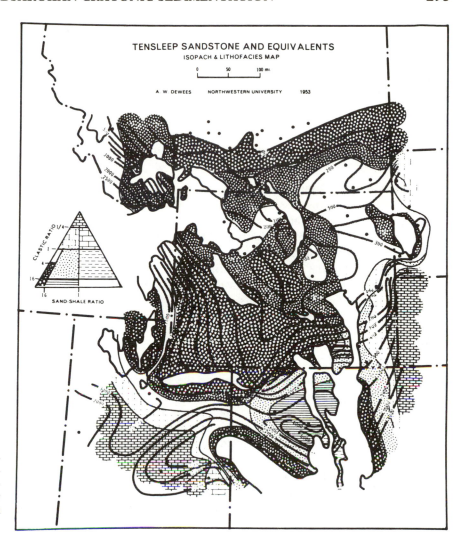

Fig. 7-15. Lithofacies–isopach map of the Tensleep/Quadrant/Weber sandstones on the northwestern craton, showing sand–shale and carbonate–clastic ratios. (From Sloss *et al.* © 1966 by John Wiley & Sons, reprinted with permission of John Wiley & Sons.)

the Cordillera; and (4) a generalized response to marginal tectonics, caused by deflation of the asthenosphere.

Explanation No. 1 is at least superficially appealing because of the impressive set of marginal events occurring in the western continent. Kluth and Coney (1981) proposed a model specifically for the Ancestral Rockies, due to collision and suturing of North America and the South American/African landmass. However, we are cautious about accepting this model because the distance between the margin and cratonic uplifts and basins is so great, and because intervening regions commonly show little evidence of tectonic disturbance. For this same reason (and more so), arguments based on relationship with Appalachian tectonics (Explanation No. 2) seem tenuous, given the extreme distance of, say, the Pedernales Uplift and the Appalachian margin.

Argument No. 3 is reasonable but virtually impossible to confirm or deny because the sedimentary mass involved has long since been reworked, transported, and/or involved in younger orogenic events. Argument No. 4 is essentially

that of Sloss and Speed (1974) whose model for cratonic tectonics has been previously cited. According to their model, an "oscillatory condition of the craton" (e.g., as in cyclothemic behavior) results from erratic but severe recovery from an episode of severe deflation of the asthenosphere, such as characterized the later Mississippian. The consequence is extreme retention of asthenospheric melt and net uplift of the craton to high elevations (such as we find in much of the late Pennsylvanian through Triassic and modern times), but with periodic extraction of asthenosphere from under the craton as an adjustment to reach equilibrium. Because of the high uplift during such oscillatory events, and assuming that the continental crust and underlying simatic crust retain their continuity, the sialic crust at the craton margin undergoes normal, high-angle faulting as an inelastic response to the tension of rapid and erratic uplift from the drastically deflated state. Thus, block-faulted uplifts and adjacent dropped basins develop.

Sloss and Speed (1974) also noted that the Absaroka Sequence contains no notable volcanic or shallow plutonic units on the craton (in contrast with the Tejas Sequence; see Chapter 9) and, therefore, no melt was expelled except at the margins. Since major tectonic events were occurring during this time, it seems logical that the absence of shallow cratonic magmas would correspond with topographic disturbances due to deep motions of the asthenosphere.

B.5.   Middle Pennsylvanian across the Craton (Desmoinesian–Missourian)

Desmoines and Missouri times saw the most widespread late Paleozoic transgression on the craton. It is probable that seas encroached on the western margin from the northeast in Montana and Wyoming, from the southwest in Arizona, and from the southcentral west in Texas (around the uplifts), reaching the Midcontinent Basin some time during the Desmoinesian. Since this was a west-to-east transgression, we will describe events accordingly.

B.5.a.   The Western Craton

Desmoinesian lithofacies east of the Antler belt show that three marine environments prevailed on the western craton: a basinal region in eastern Nevada through western Utah and extending up to central Idaho (encompassing the Oquirrh, Wood River, Ely, and other basins previously described) which collectively has been called the Hogan Basin (Rich, 1977); a broad carbonate shelf from northern Mexico to western Montana (encompassing deposition of the Callville, and equivalent carbonate units); and northeastward in Wyoming, Colorado, and central and eastern Montana, a broad shelf with local detrital sources supplying abundant sand.

The carbonate shelf sediments are largely assigned to the upper Ely Limestone and the middle parts of the Callville Limestone and Bird Spring Group (all described previously). Their lithologies are relatively uniform through the Lower and Middle Pennsylvanian but, in many sections in eastern Nevada and western Utah block-fault ranges, the upper Ely is overlain by the basinal, detrital Hogan Formation. Farther east and north, Rich (1977) described the Oquirrh Formation as a transitional unit between the Hogan Basin and the detrital shelf (termed the Weber–Quadrant Shelf).

B.5.b.   The Northwestern Platform

Sandy, detrital, and carbonate units on the northwestern platform in the west are largely assigned to the widespread Weber Sandstone (northeast Utah, southeast Idaho, and northern Colorado), the Tensleep Sandstone (southeasternmost Montana and Wyoming), and the Quadrant Sandstone (northwest Wyoming and Montana) (Fig.

7-16). These units have diachronous boundaries but formed approximately in Desmoines through Virgil times. The characteristic lithologies are clean, crossbedded, coarse-grained, cliff-forming quartz arenites and calc-arenites, featuring minor dolostones and evaporites. Associated with these sandy units are fringing carbonates which are transitional to the carbonate shelf to the southwest. In addition, there are gradational lithological changes within the sandy units: for example, the Weber at its type locality, Weber Canyon, Utah, is of marine origin, probably a barrier bar or submarine dune unit, whereas eastward, toward the Uinta Mountains and Dinosaur Monument, it appears to be an eolian dune sand.

The sea on the Tensleep/Quadrant/Weber shelf was certainly shallow, with a bottom and shoreface nearly featureless, to have allowed such widespread deposits of relatively uniform lithology to form. In addition to detrital sources from the Front Range Uplift and the Canadian Shield northeast of Montana, sediments may have come from the Transcontinental Arch east of the shelf and from small uplifts on the shelf.

B.5.c.   The Williston Basin Region

Eastward and slightly northward of the sandy shelf is the Williston Basin region, which in Desmoines and Missouri times received largely dolomitic sediments and some

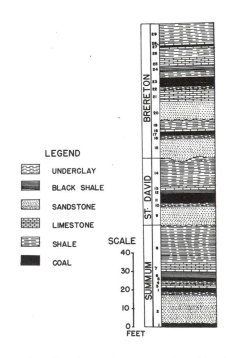

LEGEND

- UNDERCLAY
- BLACK SHALE
- SANDSTONE
- LIMESTONE
- SHALE
- COAL

SCALE

40
30
20
10
0
FEET

Fig. 7-16. Stratigraphic column of a representative midcontinental Desmoinesian section. (From Wanless *et al.*, 1963.)

detrital sands. The basin was not notably active in the Pennsylvanian and only a small, down-warped area at the junction of Wyoming, South Dakota, and Nebraska (termed the Lusk Embayment) received over 300 m of sediment (Bates, 1955). Units within this region are assigned generally to the Minnelusa Formation, but in the region between the Laramie Mountains in Wyoming and the Black Hills of South Dakota, penecontemporaneous strata are assigned to the Casper and Hartville formations as well as to the Minnelusa. Exposures of the Minnelusa are described only from the Black Hills; elsewhere it is subsurface. The Minnelusa section in the Black Hills contains cyclical sands, shales, and carbonates (largely dolostone) in the following order: (bottom) carbonate, shale, sand, shale, carbonate, . . . (top). Some Minnelusa cycles also include evaporites and, even more distinctively, thick breccias. The breccias and evaporites may be related: Bates (1955) suggests that brecciation may have occurred when significant quantities of anhydrite were altered to gypsum by hydration. The resulting volume change caused fracture and heaving of adjacent beds, thus allowing the transgressing sea to redeposit the disturbed material as breccia.

Toward the later Pennsylvanian (Missourian–Virgilian), the region of Minnelusa deposition apparently began spasmodic uplift. The proportion of sand in the upper Minnelusa increased and, because of the proximity of the region to the Shield, it seems likely that much of the detritus came from the north with the Minnelusa platform becoming part of the continental uplands by Wolfcamp time. Some of the detritus in the Minnelusa may have also come from the Weber/Quadrant/Tensleep shelf.

B.5.d. The Midcontinent Basin

Continuing southward from the Dakotas is the beginning of the homoclinal Midcontinent Basin which by Desmoines time connected with the western craton and received predominantly marine cyclical units. In Iowa, Kansas, Nebraska, Missouri, and northern Oklahoma, Desmoinesian cyclothems are divided into two groups: the Cherokee and the overlying Marmaton. Within these groups are very numerous units including a few thin but persistent coals.

On the Oklahoma platform (parts of Oklahoma adjacent to the Arkoma Basin and northeastern Arkansas), Desmoinesian–Missourian units are assigned to the Krebs, Cabaniss, Skiatook, and Ocheleta groups (in ascending order; Branson, 1962). Despite the name changes between regions, Oklahoma platform deposits in the Desmoinesian–Missourian were also markedly cyclical, with repeated sequences of sandstone, coal, limestone, and shale, in common with the more northern midcontinent groups. (See also related strata in the Ouachita mobile belt.)

Desmoinesian sections in southeastern Kansas and northeastern Oklahoma (where most western Midcontinent Basin units outcrop) are predominantly shaley and sandy, with continental units notably thin. Wanless et al. (1963) note that marine detrital sediments dominated there while sections in easternmost Kansas and Missouri contain more marine limestones within their cyclothems. Cherokee units in Kansas are noteworthy for being characteristically very thin and shaley. The most persistent elements are coals and black shales which are quite fissile. Many of these shales in Kansas form caps to structural and stratigraphic petroleum traps over the Nemaha Ridge.

Coals, underclays, and black shales, which are essential characteristics of Illinois and Appalachian Basin Desmoinesian cyclothems, are only present sporadically in midcontinent cycles. Where terrestrial deposits are present, they tend to be laterally persistent and help in correlations with eastern basinal sequences.

In addition to the generally "shaley-eastern, limy-western" pattern of midcontinent Desmoinesian cyclothems, are units of generally red, deltaic and floodplain muds which are more common in later Desmoinesian (Brereton Formation) cyclothems in northern Missouri and Iowa. Wanless et al. (1963) present a detailed series of paleoenvironmental maps for the Desmoinesian of the Illinois–Indiana and Midcontinent basins to which the interested reader is referred. Figure 7-17b presents a representative column of Desmoinesian units in the Midcontinent Basin, and Fig. 7-17a is a lithological cross section of the Summum, St. David, and Brereton cyclothems.

The widespread black shales present in midcontinent (and Illinois) cyclothems are interesting because they imply the presence of widespread, eutrophic conditions. In addition to high percentages of carbon, such strata are generally somewhat radioactive (making them easy to trace in the subsurface by gamma-ray radioactivity well logs; Moore, 1950). As discussed in Chapter 6 ("the Chattanooga Shale problem"), it is difficult to understand how such areally extensive, anoxic conditions could have prevailed.

Missourian strata of the Midcontinent Basin are approximately the same as those of Desmoines age, except that the rhythms of cycles became more regular and limestones are thicker and more widespread. Unconformably overlying the Desmoinesian Marmaton Group is the Pleasanton Group, with a sandy base and shaley upper unit at the type region. Above the Pleasanton is the Kansas City Group, composed predominantly of marine limestones and shales. At the top of the typical Missourian section is the Lansing Group, predominantly of limestone. Locally, an additional limestone unit, the thin Peedee Group, overlies the Lansing. The craton was apparently warping upward to the east and downward in the midcontinent during this time and examination of younger units will show that progressively more marine sediments followed.

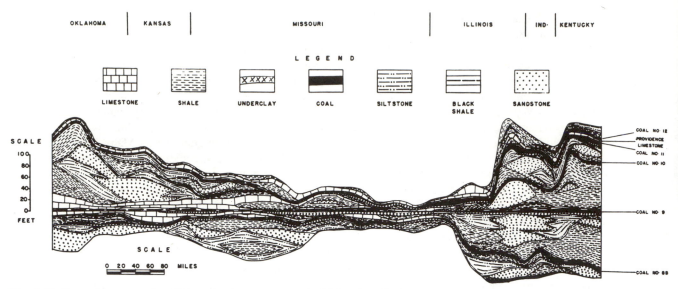

Fig. 7-17. Composite cross section of Desmoinesian strata across the Midcontinent Basin and into adjacent Illinois and Appalachian basins. (From Wanless *et al.*, 1963.)

B.5.e.   The Illinois and Michigan Basins

Strata of the Illinois Basin probably were continuous with those laid down in the midcontinent during Desmoines and Missouri times. When the Absaroka sea rose (or the craton fell) to levels sufficient to breach the Transcontinental Arch, it also overtopped the Mississippi River Arch which had formerly separated the basins. In Illinois, Desmoinesian strata are assigned to the Spoon and Carbondale formations of the Kewanee Group and the lower Modesto Formation of the McLeansboro Group. Overlying Missourian strata are assigned to the upper beds of the Modesto Formation, the Bond Formation, and the lower half of the Mattoon Formation (Fig. 7-18).

These Desmoinesian and Missourian units in Illinois correlate with the section in Indiana, on the eastern part of the basin, in which strata are assigned to formations of the uppermost Raccoon Creek Group, the Carbondale Group, and the McLeansboro Group (Gray, 1979).

The "classical" Pennsylvanian cyclothem was defined in Illinois and the cyclothems of the Desmoinesian in Illinois are probably the most regular. Desmoinesian strata in Illinois are markedly more marine than those of earlier intervals, denoting regular transgressions and regressions. The Kewanee Group, and notably the Carbondale Formation, contains over 99% of the mappable coals in Illinois, with more than 20 traceable in the Desmoinesian section of southwestern and southeastern Illinois (Atherton and Palmer, 1979). The ten-part classical cyclothem of the Illinois Basin Desmoinesian section (Fig. 7-9) contains more marine strata than correlative units in the Appalachian Basin, and thinner coals.

In Indiana too, four of the five most productive coal beds are in the Desmoinesian Carbondale Group. The Pennsylvanian area of outcrop of Indiana is relatively limited compared to Illinois, and less than one-fifth of the state is underlain by Pennsylvanian strata. In contrast, equivalent units outcrop or lie near the surface under two-thirds of Illinois.

Once again, recognizing the difficulty of simply describing paleogeography during deposition of cyclical units, we may generalize that the Desmoinesian in the Illinois Basin represented a change from the irregularly fluvial–deltaic environments of the Morrowan–Atokan to more marine dominated conditions with regular transgressions and regressions. However, periodic emergence still (and notably) produced black shales and coals.

In early Missouri time, emergence of the eastern craton had begun and the Illinois Basin underwent extensive subaerial erosion between cycles, with numerous deep channels developed on exposed surfaces. Sequences of successive deltas again formed, consisting of muds and sands derived from the north or northeast. However, periodic marine transgressions continued through the Missourian, maintaining an apparent rhythm of strong emergences and submergences; and the Missourian sequences contain the thickest Pennsylvanian limestones within the basin, notably to 15 m thick in the Bond Formation.

In the Michigan Basin, the Desmoinesian represents the last gasp of Paleozoic activity with only the lower Desmoinesian represented. Desmoinesian strata are assigned to upper beds of the Saginaw Formation, except where locally derived, uppermost Paleozoic sands are termed the Grand River Formation, of questionable collective age and gene-

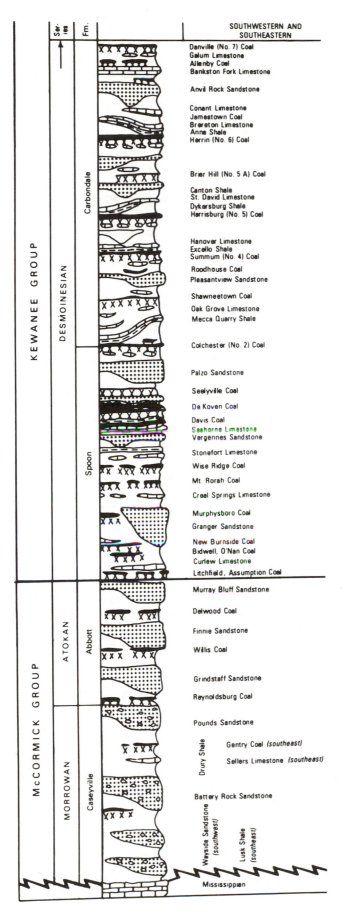

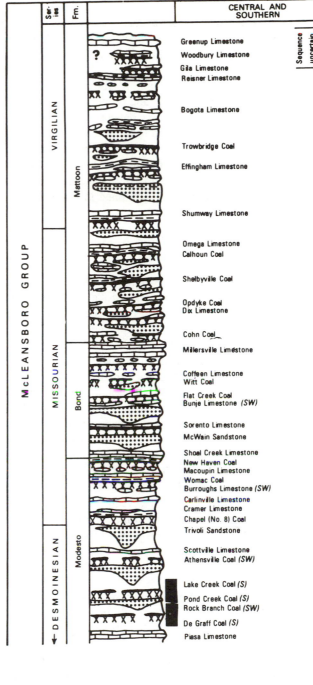

Fig. 7-18. Representative column of Upper Pennsylvanian strata in Illinois, illustrating the McCormick, Kewanee, and McLeansboro groups. (From Atherton and Palmer, 1979.)

sis. As with lower Saginaw strata, upper beds are generally fluvial sands formed by southwest- or west-draining streams flowing across a broad alluvial plain, located substantially landward from the sea in Illinois and the west (Shideler, 1969). Pennsylvanian sands in Michigan were derived consistently from the northeast and probably represent multicycle erosion of older Paleozoic detrital arenaceous units. The total Pennsylvanian section in the Michigan Basin is 230 m thick; over the Pennsylvanian, locally, are terrestrial or fluvial sediment considered to be of Jurassic age. These latter represent the youngest pre-Quaternary units on the eastern craton and will be examined in Chapter 8.

B.5.f. Desmoinesian and Missourian Strata of the Appalachian Basin

In the Appalachian Basin, geological conditions that had begun during Morrow and Atoka times persisted into Desmoines and Missouri times. In general, a major detrital source, the tectonically rising Appalachian Highlands, supplied abundant sediment to the eastern margin of the basin. Three major depositional environments characterized the Appalachian Basin area: an eastern alluvial plain adjacent to the highlands, a deltaic coastline associated in some areas with barrier islands, and a shallow marine environment on the western side of the basin. Cyclic shoreline oscillations and variations in the locus of river discharge caused these to fluctuate back and forth resulting in piedmont-type cyclothems which typify most Desmoinesian and Missourian units of the Appalachian Basin. Desmoinesian and Missourian strata are not preserved south of Kentucky and southwestern Virginia so the following discussion pertains only to central and northern parts of the basin.

i. Desmoinesian Events.

Desmoinesian strata in the eastern Pennsylvania anthracite fields are assigned to the Sharp Mountain Member of the Pottsville Formation and to the overlying Llewellyn Formation, from which most of eastern Pennsylvania's coals are mined. In the western Pennsylvania bituminous field, Desmoinesian deposits comprise the Allegheny Group composed (in ascending order) of the Clarion, Kittanning, and Freeport formations. Arkle (1969) separates the relatively fine detrital sediments of the Allegheny Group in northern West Virginia from coarser sediments in southern West Virginia which he calls the Charleston Group. Dennison and Wheeler (1975) interpreted the Charleston Group as the upstream alluvial facies of Allegheny deltaic deposits to the northwest. Equivalent strata of eastern Kentucky are placed in the upper part of the Breathitt Formation. And in southwestern Virginia, they are termed the Harlan Sandstone.

Figure 7-19A is a paleogeographic reconstruction of the Appalachian Basin during Desmoines time. Marine deposition characterized western and northern areas of the basin while deltaic and alluvial sedimentation dominated in the southeast. Sediment to feed the deltas was derived from the Appalachian Highland, especially in the Piedmont and Blue Ridge regions of the Carolinas and Virginia, and was carried northwestward down the regional paleoslope by major river systems such as the Virginia–Carolina River which supplied the great deltas of West Virginia. The alluvial plain stretched from southern West Virginia to central and eastern Pennsylvania and on this swampy expanse accumulated thick deposits of peat and other organic sediments, resulting in important deposits of coal (e.g., the Mammouth Coal of the eastern Pennsylvania anthracite fields, which is up to 24 m thick). For this reason, Alleghenian strata were referred to in the 1800s as the "Lower Productive Measures." Associated with alluvial deposits of the Harlan Sandstone in southwestern Virginia is the only Pennsylvanian bentonite yet recognized in the Appalachian Basin.

The southeastern coastline extended (on the average) from eastern Kentucky to western Pennsylvania but the actual locus of deltaic sedimentation shifted along the coast in response to changes in the course of the river. Such "hopping" of the delta mass back and forth along the coast resulted in a greater number of cyclothems than in the Illinois Basin just to the west. In eastern Kentucky, a barrier-island complex persisted down-coast from the great West Virginia deltas.

Fine, carbonate sediments were deposited occasionally during high stands of sea level. Three particularly widespread limestones within Allegheny strata are (in ascending order) the Putnam Hill, Vanport, and Columbiana limestones.

ii. Missourian Events.

Missourian strata in most of the central and northern Appalachian Basin comprise the Conemaugh Group composed of the Glenshaw and Casselman formations. The upper part of the Llewellyn Formation in the eastern Pennsylvania anthracite fields is probably also of Missouri age (Wanless, 1975). Conemaugh strata reveal greater marine influence than other Pennsylvanian strata of the Appalachian Basin; further, they contain more red-bed strata and less minable coal (hence their older classification as the "Lower Barren Measures"). With these exceptions, however, Conemaugh strata still reveal the same overall depositional framework of alluvial and deltaic sedimentation in the southeast and marine deposition in the west. Red-bed strata are primarily shales and are limited in their extent to southeastern parts of the basin. They grade laterally into freshwater limestones and are commonly marked with mud cracks, raindrop imprints, and other indications of occasional subaerial exposure. Dennison and Wheeler (1975) interpreted the red beds as representing very shallow coastal environments or broad, lower delta plains.

During Conemaugh time, the history of Appalachian Basin sedimentation became more complicated because of increasing rates of detrital-sediment supply from the east. Wanless (1975) contended that delta progradation in West Virginia periodically isolated the northeastern arm of the Appalachian Sea from the main epeiric sea to the west (see Fig. 7-19). As a result, the Appalachian Sea periodically became brackish. This accounts for the observation that some Conemaugh strata contain features characteristic of marine deposition but lack a normal marine fauna. Ferm (1974) proposed that the red coloration of some lower delta-plain shales was another result of restriction of the Appalachian Sea. He pointed out that offshore barrier complexes, which were common on earlier Pennsylvanian coastlines, protected back-barrier environments from the effects of waves and strong currents, thereby preventing overturn of the water mass and retarding consequent oxidation of organic matter.

B.6. Late Pennsylvanian and Conformable Penno-Permian Sequences (Virgilian–Wolfcampian)

B.6.a. Terminal Sedimentation in the Appalachian Basin

By Virgil time, uplift associated with the Allegheny Orogeny had become intense and great quantities of detrital sediment poured westward. Subsidence of the Appalachian Basin was not rapid enough to handle the ever-increasing load of detritus and the basin began to fill up. Missourian strata of the Conemaugh Group reveal significant marine influence but strata of the overlying Monongahela Group contain no marine deposits at all, its sediments being primarily of marsh, lake, delta, or stream origin. And the Dunkard Group of probable Permian (Wolfcamp) age is dominantly alluvial. Thus, one may see in these units the final retreat of the sea from the Appalachian Basin, never to return.

i. Final Stages of Deltaic Sedimentation: Monongahela Group.
Strata of Virgil age in the Appalachian Basin include the upper part of the Conemaugh Group, the Monongahela Group (composed of Pittsburg and Uniontown Formations), and (probably) the Waynesburg Formation of the Dunkard Group. Each of these units reveals the now-familiar pattern of coarse alluvial sediments grading into finer deltaic sediments but the regional paleoslope was almost due north by Virgil time. Arkle (1959) designated three lithofacies, the ''red,'' ''transitional,'' and ''gray'' for the alluvial-to-deltaic transitions.

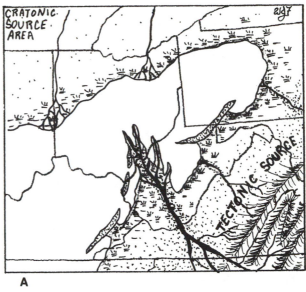

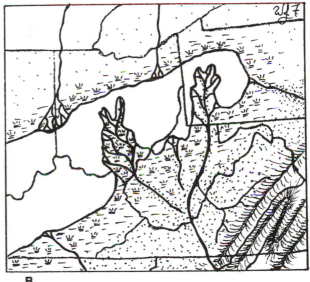

Fig. 7-19. Middle Pennsylvanian paleogeography of the central and northern Appalachian Basin: (A) Desmoinesian and (B) Missourian. (Redrawn from Donaldson, 1974.)

At the end of Conemaugh time (i.e., early Virgilian), the sea withdrew from the basin, never to return. Probably this was due to the combination of decreased rate of basin subsidence and increased rate of sediment supply. The sea's retreat left a vast, low plain on which developed large swamps and lakes (Wanless, 1975). Freshwater limestones deposited in the lakes are very well developed (e.g., the Benwood and Fishpot limestones), some as much as 25 m thick. They are among the most extensive freshwater limestones in the United States. Of great importance are the coals which formed from peat deposits of the swamps; today, they constitute some of the most important coal deposits of the Appalachian Basin (e.g., the Pittsburg Coal, the most extensively mined coal in the Appalachian field). Because of these abundant coals, Monongahela strata used to be known as the "Upper Productive Measures."

Deltas invaded the lakes and pushed distributaries over the swampy areas, only to be abandoned when a shift in river discharge developed a new delta lobe somewhere else. Donaldson (1974) was able to delimit formation and degradation of 12 such Monongahela deltas. With time, sediments of the upper alluvial plain (the "red facies") grew northward; the swamp plain decreased in size and its lakes became smaller. There was no break in sedimentation as deposition of the Dunkard Group began.

ii. Alluvial Deposits of the Dunkard Group.

The Dunkard Group, composed (in ascending order) of the Waynesburg, Washington, and Greene formations, is the youngest stratigraphic unit of the Appalachian Basin, but its actual age is controversial. Originally called the "Upper Barren Measures" due to their paucity of coal, Dunkard strata were considered by Fontaine and White (1880) to be of Permian age. Romer (1952) argued for an Early Permian age based on vertebrate assemblages which contain, for example, *Eryops*, a common labyrinthodont of Permian red beds in Texas. But Beerbower (1963) defended a Late Pennsylvanian to Early Permian age for the Dunkard because of the presence of the amphibian *Diplocerapsis*. The majority opinion today, stated by Dennison and Wheeler (1975), is that the Dunkard spans the systemic boundary, with the Waynesburg probably being Upper Pennsylvanian while Washington and Greene strata are probably Lower Permian.

While the Dunkard's age may be controversial, its regional stratigraphy and depositional environments are not. The Dunkard is similar to underlying Monongahela strata in its complete lack of marine deposits; but the "red," alluvial facies and the "transitional" facies are shifted farther to the north, indicating the greater significance of alluvial sedimentation in Dunkard time (see Fig. 7-20). Dunkard sediments coarsen and their sandstone/shale ratio increases toward the south; freshwater limestones and coal are more common in the north, associated with the gray facies, but even there they are thin and poorly developed. These data, as well as paleocurrent data shown in Fig. 7-20, indicate a southern source and generally northward sediment transport.

The Dunkard represents a broad alluvial plain in the south merging northward into a swampy plain, probably very near sea level and dotted with lakes. The mudstones and argillaceous limestones of the swampy plain contain a vertebrate fauna consisting primarily of fish and aquatic amphibians; possibly, there was too little dry land in the

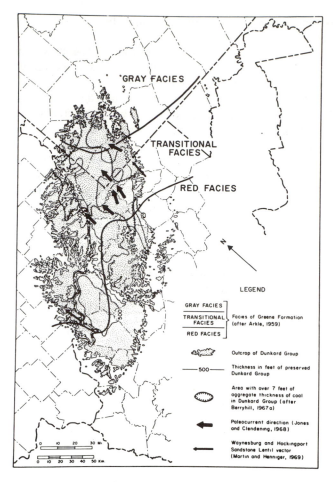

Fig. 7-20. Outcrop and isopach map of the Dunkard Group, showing distribution of the red, gray, and transitional facies of the Greene Formation. (From Dennison and Wheeler, 1975.)

area for reptiles or terrestrial amphibians to flourish (Romer, 1952). Invertebrate faunas contain principally freshwater forms such as ostracodes and branchiopods. *Lingula* fossils are found in one dark shale horizon in West Virginia, probably representing a brief incusion of brackish water into the peat swamp (Cross and Schemel, 1956). Younger Dunkard strata contain progressively more conglomeratic and sandy alluvial deposits and fewer lake and swamp deposits.

The top of the Dunkard is an erosional surface. We can only assume that general regional uplift associated with the main phase of the Allegheny Orogeny caused the cessation of deposition in the Appalachian Basin.

B.6.b. Latest Paleozoic in the Illinois Basin

Incomplete sections of Virgilian rocks are present in Illinois, comprising the upper Mattoon Formation, in Indiana, where youngest McLeansboro Group beds may be of

Virgil age (however, most Indiana sections do not contain post-Missourian units), and in western Kentucky where Virgilian strata are assigned to the Sturgis Formation.

These Illinois Basin Virgilian strata are cyclical and include the whole complement of lithologies characteristic of the older cyclothems. Apparently, depositional environments in the region persisted largely unchanged from Missouri to Virgil times, featuring strong fluvial and deltaic channeling on exposed terrestrial surfaces and numerous marine transgressions and regressions which emplaced marine limestones on delta and prodelta deposits. Atherton and Palmer (1979) note that at least seven marine transgressions in the basin occurred during Virgil time.

Uppermost Virgilian and all Permian strata are absent in the Illinois Basin, except as noted below. The southern part of the basin was strongly uplifted sometime after deposition of the latest Pennsylvanian units.

In western Kentucky, Rice *et al.* (1979) report that fusulinids of Early Permian age have been recovered from drill cores in a single locality. The units sampled are from an unnamed limestone zone at the top of the Sturgis Formation and are conformable with the remaining Virgilian Sturgis units. These limestones may represent the youngest Paleozoic strata in the Illinois Basin and suggest that a marine transgression was the last preserved depositional event to occur.

B.6.c. Pennsylvanian–Permian Transition in the Midcontinent

As the Appalachian and Illinois basins emerged in the latest Pennsylvanian and earliest Permian, the midcontinent and western craton remained generally submergent and locally experienced considerable down-warping. Thus, the emphasis of post-Missourian cratonic stratigraphy must be on sections from the midcontinent westward. Penno-Permian time was an interval of virtually continuous sedimentation from Nebraska to Oklahoma and westward on the craton; the systemic boundary is often difficult to determine in many western regions because conformable sequences are present.

Virgilian strata in Iowa, Nebraska, Missouri, Kansas, northwest Arkansas, and northern Oklahoma are assigned generally to the Douglas, Shawnee, and Wabaunsee groups (which have type sections in Kansas). On the Oklahoma platform, contemporary strata are the Vamoosa and Vanoss formations.

The Douglas Group is a generally shallow-water, cyclical sequence of sandy shales and sandstones, with some limestone tongues and a few coals of local extent (Branson, 1962). These lithologies are fairly constant across the midcontinent but southward, toward the Oklahoma platform, they grade into the Vamoosa Formation which characteristically contains red beds with a basal conglomerate, sandstones, claystones, and cherts.

The overlying Shawnee Group is a sequence of four limestone formations with intervening shale units. It contains thin, local coals and generally reflects repeated marine onlap and offlap. Again, southward to the Oklahoma platform, the group grades into red beds of the upper part of the Vamoosa and the lower Vanoss formations.

Final Pennsylvanian units in the midcontinent are Wabaunsee cyclothems, containing sequences of shale, limestone, and thin coals (Branson, 1962). The limestones are apparently thin and disappear toward Oklahoma, where they "shale out." On the Oklahoma platform, Wabaunsee units, like those below, grade into red beds, assigned to the upper Vanoss Formation. Uppermost Wabaunsee strata are limestones and are overlain by a sandstone, the Indian Cave Member of the Admire Group, which represents the probable lowermost Permian unit. The Penno-Permian contact across most of the midcontinent is conformable and Mudge (1967) notes that the uppermost limestone below the lower Admire sand contact is arbitrarily used as the systemic cutoff. Admire strata above the sand are generally red and gray muds and so can be readily distinguished from the Pennsylvanian limestones.

Permian units in the midcontinent outcrop in eastern and central Kansas and southeastern Nebraska. Kansas has a large number of oil wells drilled to tap Permian oil and therefore subsurface strata have been described with considerable control. Wolfcampian units in Kansas and Nebraska are assigned to the Admire, Council Grove, and Chase groups; in northcentral Oklahoma, the same names are used for equivalent strata except that the Admire is not elevated to group status.

In Wolfcamp time, the structure of the Midcontinent Basin changed somewhat from that of the Pennsylvanian. Parts of the midcontinent, notably Missouri, Iowa, and Arkansas, do not have determinable Permian sections, whereas Kansas apparently down-warped into a shallow cratonic basin receiving largely carbonate strata. The Kansas Permian basin shallowed westward across western Kansas, western Nebraska, and eastern Colorado, all of which areas shared similar styles of sedimentation. Farther west in Colorado, platform sediments grade into red, arkosic coarse detritus which poured out of the Front Range uplands and were deposited as the Fountain Formation.

Wolfcampian strata are very similar to those of the underlying Virgilian units in the midcontinent but are strikingly different from the overlying red beds (Branson, 1962). Within the Admire–Council Grove–Chase sequence, Mudge (1967) lists 24 almost perfectly alternating shale and limestone formations, among which are over 39 differentiated, alternating shale and limestone members; the rhythm of shale–limestone–shale . . . is fantastically precise; and, clearly, cyclicity in sedimentation characterizes the Lower Permian of the midcontinent. Many of the limestones are cherty and Mudge notes that the distribution of cherts roughly parallels the configuration of the Kansas Permian basin; thus, he concludes that the chert was formed penecontemporaneously with the limestones. On the western edge of Kansas, the chert is absent on the platform.

In general, Wolfcampian units represent intertidal and high-subtidal marine environments in central and eastern Kansas, and deltaic/alluvial environments on the western platform. The Kansas Permian basin subsided periodically, producing the cyclicity evident in the alternating shales and carbonates, with carbonates forming in deeper waters of the basin. Some of the limestones are biostromal and it may be these buildups (largely algal) that are the source or reservoir rocks for the important midcontinent oil fields. The shelly, invertebrate fossils within some Wolfcampian limestone units are very abundant, reflecting diverse faunas. Imbrie (1955) did a classical biofacies analysis of the Florena Shale Member of the Beattie Limestone in the Council Grove Group. Some fossil plants have been recovered from Kansas Wolfcampian strata, indicating that marginal marine swamps or coastal forests existed periodically.

B.6.d.　Penno-Permian of the Northwestern Craton

Conditions on the sandy, northwestern shelf remained relatively constant from Desmoines through Wolfcamp times but a major regional regression occurred at the Permian boundary. Most formations deposited in the Desmoinesian–Missourian either continued to form through the Virgilian–Wolfcampian, with a Penno-Permian hiatus present, or are overlain unconformably by Permian strata, with the Virgilian section absent.

In the western Dakotas and eastern Montana, the Minnelusa Formation was deposited through the Penno-Permian boundary. In central and southern Montana and across Wyoming, the Virgilian section is characterized by the Tensleep and Quadrant formations, and a number of equivalent units (e.g., the Hartville and Casper formations). There is an unconformity generally present within these formations separating the beds of the different periods. Maughan (1967) states that the base of the Permian System is believed to be unconformable across the region except in southeastern Wyoming (Tensleep Formation). In addition, Maughan notes that a prominent red mudstone, called the Red Marker Bed, overlies the boundary unconformity across southeastern Wyoming and southwestern South Dakota, distinctly marking the systemic break.

Maughan (1967) also notes that regional Permian units differ from those of the Pennsylvanian in being more red-

colored, containing more evaporitic strata, lacking radioactive black shales and limestones (although dolostones are present in the Permian), and a few additional minor characters. These changes point toward more terrestrial-dominated sedimentation and, especially, toward hotter and more arid environmental conditions such as we will see mirrored elsewhere on the continent.

The Pennsylvanian/Permian unconformity in these northwestern cratonic sections seems to represent local uplift of the shelf, with greatest activity centered in south-central Wyoming. This uplift may be related tectonically to late stages of the Front Range activity. Missourian and Virgilian strata are absent in central Wyoming and it seems likely that they were not deposited due to this uplift; apparently, uplift began there in latest Demoines time but did not become regionally significant until Virgil or earliest Wolfcamp times.

Farther south, the Pennsylvanian–Permian boundary is also disconformable but units above and below suggest that uniform conditions prevailed throughout the interval. In northeast Utah and northwest Colorado, the Weber Sandstone was probably deposited through Virgil time and in at least some Utah localities, upper Weber beds contain Wolfcampian fusulinids (Bissell and Childs, 1958, in Sheldon et al., 1967). In the latest Pennsylvanian, detrital sands were deposited across a very wide area of southeastern Idaho (the Wells Formation), southwestern Wyoming, northeast Utah, and northwest Colorado. Sheldon et al. (1967) suggest that these were of shallow-water shelf origin but the presence of a Penno-Permian unconformity would suggest that, locally, some sands were deposited in supratidal dunes.

A somewhat confusing arrangement of stratigraphic units occurs in the Wolfcampian and Middle Permian sections of southwestern Montana and southeastern Idaho. Sheldon et al. (1967) assign Lower Permian strata in southeastern Montana (overlying upper beds of the Quadrant Sandstone at the top of the Pennsylvanian section) to the Grandeur Member of the Park City Formation. The Grandeur contains a lower dolostone unit with interbedded sands and upper beds of tan sandstone and red and tan mudstone. In southeastern Idaho, however, the "Grandeur Tongue" of the Park City Formation is placed by Sheldon et al. (1967) and Stevens (1977) in the upper Leonardian section. The strata are poorly fossiliferous and the assignment of beds in Montana to Wolfcamp time is based on a few diagnostic fusulinids. To compound the problem, elsewhere (e.g., Nevada) the Park City "Group" contains most of the Upper Permian section including the important Phosphoria and Kaibab formations, among others.

Overall, during Virgilian–Wolfcampian times, sandy conditions predominated across Wyoming and most of Montana, and deeper-water shelf conditions were present in southwestern Montana/southeastern Idaho. Examination of contemporary units farther to the south and west (Section B.8) will show that the sea was present almost continuously in portions of Colorado, Nevada, and western Idaho during this time. Wyoming and adjacent areas represented the highest reaches of transgressing waters, and, simultaneously, were the first to shallow out during regression on the northwestern craton.

B.6.e. The Central–Western and Southwestern Craton

The region from central Colorado westward to Nevada and southward to Arizona probably contains the most complex set of Penno-Permian strata known, as this area includes most of the uplifts and basins which were featured in our earlier Pennsylvanian discussion. In previous discussions, we traced the depositional history of several Pennsyl-vanian–Permian basins in the region and, therefore, we will tax neither ourselves nor the reader to review those. In general, carbonates were deposited in deeper waters on the platform in eastern Nevada, western Utah, and southeastern California, comprising the upper beds of the Callville, Ely, Bird Springs, and Arcturus formations. In northern Colorado, most nonbasinal strata are either Weber Sandstone beds (toward the northeast), or equivalents of midcontinent cyclical units (toward the east-southeast). At the eastern margin of the Front Range uplift, the red arkosic Fountain Formation (Fig. 7-21) grades laterally into Weber or midcontinent units (Admire/Council Grove/Chase groups). Collectively, they show that environmental conditions were stable across the central-west from roughly Desmoines through Wolfcamp times, with the Late Pennsylvanian withdrawal and return of the sea as the most notable event.

Farther south of the major area of uplifts and basins,

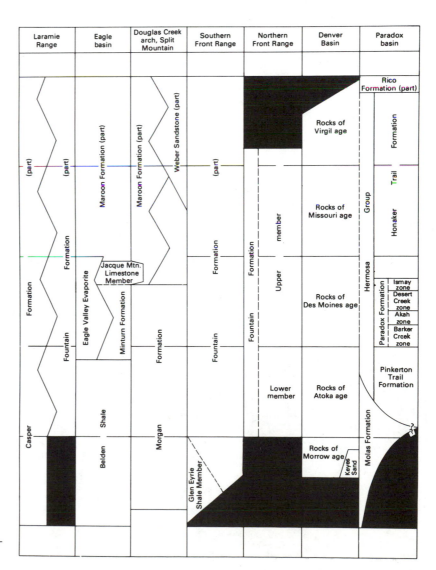

Fig. 7-21. Pennsylvanian correlation chart for Colorado. (From Chronic, 1979.)

the systemic-boundary section includes two widespread, red detrital units. In southwestern Colorado and northcentral New Mexico (including parts of the Paradox Basin which ceased activity in the Permian) is the Lower and Middle Permian Cutler Formation, and in northeastern Arizona/northwestern New Mexico and adjacent southern Utah is the long ranging Pennsylvanian through Middle Permian Supai Formation. The Cutler is a highly variable formation, characteristically composed of red conglomerate, sandstone, and mudstone but also containing members with gypsiferous beds (the Cedar Mesa Member) and some probable marine carbonate and mud beds at its western extent. The Supai is also red, sandy, and muddy. Some sands within both units show crossbedding, clearly indicative of eolian deposition, and, together, they suggest the presence of a coastal area with well-developed dune systems. The red conglomerates and sands in these units are probably of terrestrial origin and Wood and Northrop (1946, in Hallgarth, 1967) believed portions of the Cutler to be fluvial. Clearly, the Absaroka sea had shallowed in regions of the southwestern craton by the Early Permian and, by the latest Wolfcampian, the red lithologies in Cutler and Supai strata predominated over the probable dunal lithologies, suggesting even greater emergence of the region. Within the Cutler is a limestone (the Elephant Canyon Member), in eastcentral Utah (Hallgarth, 1967), showing that a carbonate marine environment still prevailed in Wolfcamp time north of the area of red-bed deposition. It is likely that Elephant Canyon limestones were deposited on a platform roughly similar to the Callville and Ely depocenters to the west, and similar to the Hueco and Magdalena basins to the extreme south (see below).

In southern Arizona, southwestern New Mexico, and northcentral Mexico, a number of structural features were present on the craton in Penno-Permian times. Some have been discussed previously but many are too localized to have been included at that level of discussion. Figure 7-22 shows an interpretation of structures on the southwestern craton in the early Wolfcampian; as discussed previously, interpretation of geological events in a region such as this, featuring block-fault mountain exposures and thick Tertiary and Quaternary alluvium in basin-fills (see Chapter 9), is very difficult, and definitive stratigraphy is almost impossible. Many regional Pennsylvanian units are undescribed and undifferentiated. We will discuss below only a few better-known sequences in two basins.

In the "Orogrande Basin" (Fig. 7-22), Pennsylvanian units are generally called either the Madera Formation (northern basin) or the Magdalena Group (southwestern basin). Many additional unnamed or local strata exist in the basin too. Using the most broadly applicable units, the Magdalena Group contains a lower limestone, the Sandia, which may be as old as the Devonian, and an upper limestone, the Madera, which was probably deposited in the Desmoinesian through Virgilian. Above the Madera Limestone in many sections are red beds generally called the Bursum and Abo formations, of probable Wolfcamp age. The Bursum lithology includes limestones interbedded with red detrital sediments and suggests intermittent terrestrial and marine conditions within the basin. The Penno-Permian boundary may be at the Madera–Bursum contact but the sequence appears to be conformable.

In southeastern Arizona (the "Pedregosa Basin"), another Penno-Permian series of strata includes the Horquilla

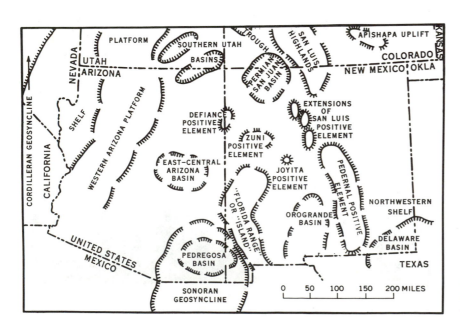

Fig. 7-22. Structural features of Arizona, New Mexico, and surrounding regions during Wolfcamp time. (From McKee, 1967.)

and Earp formations, which range down to the Lower Pennsylvanian. These units extend locally up to the Wolf-campian and are predominantly limestones with some de-trital deposits. In some localities too, the Earp extends through the Wolfcampian and contains some interbedded pink detritus with the limestones (Kottlowski, 1962). In other places, much of the Wolfcampian section is assigned to the Hueco Limestone, which either underlies or inter-tongues with the Abo in Arizona and western New Mexico (McKee, 1967); however, fusulinid data show that most of the Abo is of late Wolfcamp age and is more typically underlain by strata assignable to the Bursum Formation.

To help explain what are confusing stratigraphic rela-tionships, it is assumed that a series of small basins, large uplifts, and intervening carbonate platforms, sandy shelves, and uplands existed in the extreme southwest craton during Penno-Permian times. The precise locations and determina-tions of these features will likely never be known because strata are exposed intermittently in fault-bounded mountains and cannot be traced laterally. Most regional units are lime-stones but many red beds too were formed in Wolfcamp time. Sedimentation continued apparently uninterrupted from the Pennsylvanian through the Early Permian as shown by conformable basin sequences but, in the later Permian, continental red beds encroached broadly across the region as the sea withdrew totally.

B.7. West Texas and Eastern New Mexico: The "Permian Basin"

Of all regions in North America containing Permian strata, probably none is more important economically than the west Texas area: the so-called Permian Basin. Major oil and gas fields have been discovered in several parts of this region and it is one of the most intensely drilled areas on Earth.

B.7.a. Pennsylvanian History

The study of Pennsylvanian strata in west Texas and eastern New Mexico is largely a subsurface problem since less than 5% of the exposed bedrock is of Pennsylvanian age (Adams, 1962). In addition, few subsurface units can be traced province-wide and named formations are limited to shelf areas. Most correlations within the subsurface Pennsylvanian are done biostratigraphically, using index fusulinid species because these microfossils are abundant in cores from the generally limy units.

Prior to the Pennsylvanian, a broad arch system appar-ently existed in central Texas, and effectively separated a large lowland interior region in the west from the Oklahoma–east Texas embayment on the east-northeast. The latter area includes much of the region discussed pre-

viously with our history of the Oklahoma platform and events in the Ouachita Foreland System. The following dis-cussion will focus on the western basin.

During the Early Pennsylvanian, a number of arches bordered the large west Texas area on the north, northeast, and west (Fig. 7-23). None of the positive features seems to have been sizable enough to comprise a significant detrital source, and most Early Pennsylvanian units within the re-gion are carbonates.

The Mississippian/Pennsylvanian boundary is confor-mable locally, within basins, but over much of the interior lowland it is absent. Morrowan strata, where present, are largely undifferentiated and unnamed; a notable exception is in the Llano Uplift area of central Texas where the Mor-rowan Marble Falls Limestone outcrops.

The Marble Falls is a dark gray and black, fos-siliferous, siliceous unit, containing some black shale and grading eastward into a shaley facies (Plummer, 1950, in Crosby and Mapel, 1975). Elsewhere in the Texas interior, unnamed subsurface units of probable (but undifferentiated) Morrow and Atoka age are largely limestone and fine-grained detritus with some interbedded black shales. In the Early Pennsylvanian, the sea apparently transgressed slowly in from the Oklahoma and Arkansas shelf into the Texas interior. The interior region may have been generally emergent except locally where down-warps created small basins.

In later Morrow time, more intense tectonic activity occurred, including block-faulting. Some yoked basins formed and were filled in quickly by coarse detritus. By the early Atokan, extensive marine flooding reworked older detritus leaving shaley deposits in the Texas interior, es-pecially in basins, and formed broad carbonate platforms.

In late Atoka time, the major west Texas basins formed. These basins, which will feature prominently in the Permian discussion, are the Delaware, Midland, and Marfa basins (see Fig. 7-23). A broad platform between the basins, the Central Basin Platform, will also be a prime player in the Permian history. While most of the interior resumed limestone deposition in the later Atokan, the Dela-ware and Midland basins were too deep for lime-secreting, benthic invertebrates, and were too remote to receive de-tritus from a positive source; therefore, they primarily re-ceived siliceous shales (i.e., they were "starved").

On the broad carbonate platforms, bioherms formed in many places and at least one, the very large Horseshoe Atoll (Fig. 7-23) in the northern Midland Basin, reached thick-nesses over 170 m. This reef remained as a positive feature through later times.

From middle Desmoines time onward, the Midland and Delaware basins, as well as others, deepened while uplifts (Diablo, Central Basin platforms, and Bend Arch) remained weakly positive or were slightly submerged

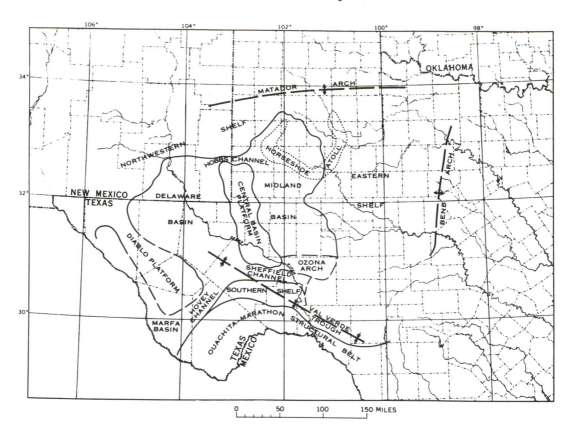

Fig. 7-23. Late Paleozoic tectonic features in the Permian Basin and vicinity. (From Oriel *et al.*, 1967.)

(Crosby and Mapel, 1975). Most detritus in the region was derived from the Ouachita mobile belt while interior basins were generally starved.

Carbonate platform and starved basin environments continued regionally through the latest Pennsylvanian. General lack of exposure makes detailed description of Pennsylvanian units difficult; however, in the Llano area, the Big Saline Formation, a cherty gray limestone, overlies the Marble Falls. Elsewhere, most Atokan units are assigned to the Atoka Formation but in the far northern part of the region, almost in Oklahoma, equivalent rocks are assigned to the Dornick Hills Formation. Two lithological units of relatively wide distribution are present in west Texas sections and characterized later Pennsylvanian deposition. The Strawn Limestone dates paleontologically to the Desmoinesian and contains several hundred meters of dark-gray mudstones and limestones. It is a major part of the later Pennsylvanian sections in the Midland and Delaware basins and the Val Verde Trough. In the Midland Basin, above the Strawn is a dark-gray, unnamed mudstone of Missourian–Virgilian age; in other less-extensively drilled areas the section is unexplored. Elsewhere, some uppermost Pennsylvanian units have been assigned to the Cisco and Canyon

Groups and these too are "starved basin" deposits of thin, dark limestones and siliceous, dark mudstones.

B.7.b. Wolfcampian

By Wolfcampian time, the regional tectonic setting was approximately that presented in Fig. 7-23. It should be remembered that west Texas was not yet isolated from marine waters. A broad seaway existed in New Mexico, connecting the Permian Basin with the ocean to the southwest, while northward the Permian Basin was connected with the epeiric sea in Oklahoma, Kansas, and Nebraska. It will be seen, however, that in post-Leonard time the Permian Basin did become restricted from normal marine circulation.

The Pennsylvanian–Permian systemic boundary is indistinct lithologically (but determinable by fusulinid biostratigraphy) in most west Texas cratonic basins, but recognizable in many structurally positive areas because of a minor regression which left a basal detrital layer adjacent to uplands. An angular unconformity at, or just above the systemic boundary is present in some sections, including those in the southern part of the Pedernales Uplift and in the Glass Mountains. It seems likely that localized tectonism

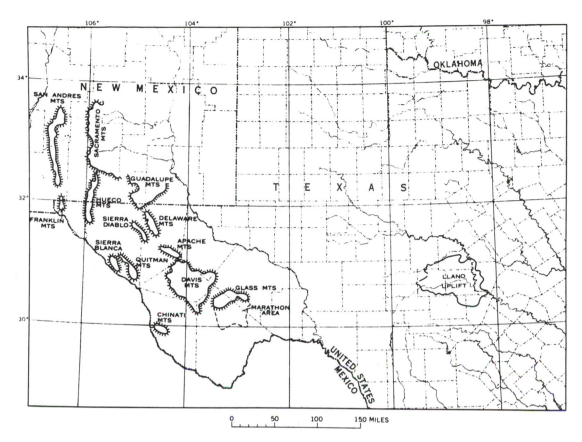

Fig. 7-24. Modern physiographic features of the Permian Basin and vicinity. (From Oriel *et al.*, 1967.)

affecting positive areas occurred at the period break but this activity did not notably affect shelves or basins established in the Pennsylvanian.

In Wolfcamp time, the contrasts between basins and platforms increased, largely due to enhanced subsidence of basins. The Ouachita–Marathon mobile belt to the south culminated its orogenic activity at about this time, resulting in northward thrusting of newly formed tectonic lands toward the Permian Basin. Rapid erosion of the mountains shed considerable detritus into the rapidly subsiding Val Verde Trough (Oriel *et al.*, 1967).

The Wolfcampian section in west Texas, in common with both younger and older units, reveals a wide range of lithologies in shelf/margin and basin facies. Most regional Wolfcampian units are assigned to the Wolfcamp Formation, the type unit for the series, but the Wolfcamp is extremely variable. It was named for exposures in the present-day Glass Mountains, in the area occupied in Permian time by part of the Val Verde Trough, and there it consists of laterally variable, thin, gray limestones and dark-gray, sandy and muddy detrital sediments. In the basin areas, Wolfcampian strata include the Third Bone Spring Sand and the Dean Sandstone (the former largely in the Delaware Basin and the latter in the Midland Basin) which are mixed, fine- and coarse-grained detrital units with argillaceous, cherty limestones. These basinal detrital sediments are characteristically dark-gray; this color contrasts with that of Wolfcampian strata deposited in shelf areas which tend to be either lighter-colored limestones or red detrital sediments. Shelf units include the Hueco Limestone in westernmost areas, locally the Wolfcamp Limestone, and on the eastern shelves, the Pueblo/Moran/Putnam formations.

Overall, one could summarize by saying that in the Wolfcampian, the shelves of west Texas received mostly shallow-water, fossiliferous, and bioclastic limestones as the dominant sediment while, simultaneously, the major basins received dark-gray to black organic-rich limestones and shales, reflecting relatively strong subsidence. As the relief between shelves and basins increased with time in the Permian, the zonation of organisms became more evident as the basins sank to inhospitable depths and soon the basin margins were to become the sites of massive reef limestone formation while the basins became virtually devoid of life but received the debris of the fringing reefs.

B.7.c. Leonardian

Whereas Wolfcampian tectonics within the Permian Basin enhanced the topographic differences between basins and platforms, Leonardian tectonic events, largely subsidence of the entire region, affected basins and shelves uniformly. Oriel *et al.* (1967) state that Leonardian deposits are of relatively uniform thickness (despite marked facies changes) across basins and shelves. This is in contrast to Wolfcampian units which drastically thicken in basins, as proper sedimentary sequences generally do. However, the relief between basins and shelves remained great and their conclusion is that basins were ''semistarved, or at least not well nourished'' (p. 45), while shelves featured rapid deposition of carbonates.

The Marfa Basin, which we have largely ignored because its behavior and extent in the Permian are relatively poorly known, was a negative element through Leonard time. Sediments from the region, assumed to be basinal in the Leonardian, are largely detritus with probable sources from the Diabolo Uplift to the northeast. Oriel *et al.* (1967) note that the Marfa Basin may have extended southwest into Mexico as shown by Leonardian deposits in the Placer de Guadalupe area.

Classical patterns of Permian sedimentation in west Texas (so-confusing to early workers in the region) really began in the Leonardian but climaxed in subsequent Guadalupian time. The riddle posed by the strata was eloquently phrased by King (1977), in a description of his early years in west Texas:

> The [drillers'] logs were sufficiently baffling, as one would report thousands of meters of limestone, and another only a few kilometers away indicated great thicknesses of salt and gypsum or a great thickness of sandstone. . . .

In fact, this and related riddles had two equally important answers. First, was the presence of strongly differentiated basins and uplifts in the region, especially the Midland and Delaware basins separated by the Central Basin Platform, which was not known to early workers (who believed the area to contain a single basin). Second, was the new element in our discussion, the presence of spectacular limestone buildups and reef structures on the margins of the Delaware and Midland basins.

Leonardian strata in the Permian Basin are outstandingly varied among environments of deposition (basin/ margin/shelf) and, even, among basins. Some of these lateral variations are marked by separate formation names but many striking facies changes occur within single units, notably the type unit for the series, the Leonard Formation. In the western Glass Mountains, the Leonard is composed of siliceous shales, sandstone, and thin- to thick-bedded limestones. To the east-northeast, however, these units inter-

tongue with biohermal limestones which, in turn, change eastward into thin-bedded, lagoonal limestones.

In the Delaware Basin, Leonardian units are assigned to the Bone Spring Limestone, a predominantly black, bituminous, argillaceous, sparsely fossiliferous limestone with some sandstone and chert beds (King, 1948). In the Midland Basin, typical units include the Sprayberry Sandstone, composed of very fine sands and silts with interbedded dark limestone and mudstones, and the overlying Clear Fork Group, composed largely of limestones and dolostones. A large number of other units have been described from the Delaware and Midland basins, again representing drastic facies changes over relatively short distances.

The basin-margin environments of the Texas Permian featured massive limestone buildups. At the margins of the Delaware Basin, the Bone Spring Limestone grades into the Victorio Peak Limestone, which is a light-gray, thick-bedded fossiliferous deposit with minor chert and sandstone beds. Oriel *et al.* (1967) believe the Victorio Peak to be a carbonate bank deposit rather than an organic reef and they cite the lack of hermatypic organisms in the assemblage as evidence of this. They suggest that small reefs may have been present around basin margins in the Delaware and Midland basins as well as limestone buildups, all of which may have virtually encircled the basin. Basinal sediments of the Midland Basin are not as well exposed but, in general, they grade into shelf limestones and dolostones of the Yeso Formation without the clear presence of intervening massive carbonates.

B.7.d. Guadalupian

In Guadalupe time, activity in the Permian Basin centered around the Delaware Basin and its margins. The Delaware Basin continued to subside and remained semistarved, receiving deep-water deposits under euxinic conditions. Around the basin margins, spectacular organic reefs grew and contributed reef talus to the basin edges. The Midland Basin, however, ceased its subsidence and gradually filled through Guadalupe time, becoming a shelf such as the areas around it (e.g., the Central Basin Platform).

The growth of reefs around the Delaware Basin began to restrict marine circulation in the early Guadalupian and considerable deposition of evaporites occurred in shelf areas, the Midland Basin, and platforms. The euxinic Delaware Basin received lithologically mixed, dark sediments of the Delaware Mountain Group (see Fig. 7-25). King (1948) noted that structures such as crossbedding in the coarser sandstones, oscillation ripples, and oriented fusulinids in the Brushy Canyon Formation suggest agitated, circulating water. The two overlying units (Cherry Canyon and Bell Canyon formations) seem to have been deposited in quiet water conditions.

The lithofacies change that confounded early workers in the region occurs between the upper two formations in the Delaware Mountain Group and the reef limestones on the basin margin. The Cherry Canyon grades laterally and abruptly into the Goat Seep Limestone, and the Bell Canyon grades into the Capitan Limestone, spectacularly exposed in El Capitan, a peak at the eastern front of the Guadalupe Mountains (and the highest point in Texas) (see Fig. 7-26). Reef sediments of both units are thick, massive-bedded, highly fossiliferous limestones with a diverse fauna dominated by sponges, algae, and bryozoans (the framework builders) and containing a host of brachiopods, fusulinids, echinoderms, and rugose coral.

The reef limestones dip into the basin at a steep angle, and what is perhaps most intriguing about this section is that the dip of these limestones approximates the paleoslope from the reef front to the original basin. Thus, erosion has approximated the original topography and a climb of 600 m down from El Capitan today follows approximately the same 25° slope that a submarine traveler down from the reef to the basin depths would have experienced in Capitan time.

During the later Guadalupian, reef growth was more extensive laterally than vertically; thus, subsidence did not keep pace with sedimentation on the basin margins. However, the reefs were more than high enough to restrict circulation to the adjacent shelves once a continuous sill had developed (beginning in Goat Seep time) around the Delaware Basin. On the shelves, Guadalupian units include the San Andres Limestone and the overlying Artesia Group (formerly the Carlsbad Limestone in King, 1948). The San Andres is mostly dolostone with some limestone and chert beds. Westward, away from the Delaware Basin in the Guadalupe Mountains, the San Andres grades into evaporites. It also thins eastward across the Midland Basin and grades into increasingly greater amounts of evaporitic strata. The Artesia Group is a mixed limestone and evaporite unit which also contains anhydrite deposited near the end of Guadalupe time. In general, evaporite deposition was occurring on the distal portions of shelves in lower Guadalupe time and encroached toward the reefs by the late Guadalupian. At the northern and eastern extremities of the Permian Basin, far from the Delaware Basin, arkoses and red beds were developing by late Guadalupe time and these, along with the ever-increasing evaporites, denote the impending end of the sea in the region.

B.7.e.   Ochoan and Post-Permian

Latest Permian in west Texas was a time of evaporite and red bed deposition as the marginal reefs drastically restricted circulation while subsidence continued, maintaining the evaporitic pan. Ochoan evaporites in Texas are among the thickest in North America and are assigned to

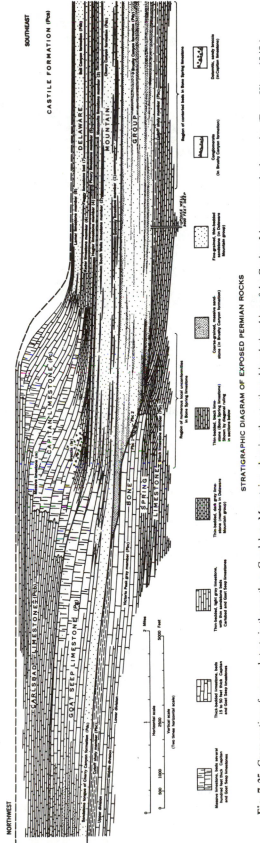

Fig. 7-25. Cross section of exposed strata in the southern Guadalupe Mountains, showing the stratigraphic relationships of the Capitan Limestone and other units. (From King, 1948.)

Fig. 7-26. El Capitan, west Texas, formed by the resistant Capitan Limestone reef structure. The modern erosional slope closely reflects the submarine paleoslope of Late Permian time, when the Capitan reef stood at the shelf edge above the deep basin in which the Delaware Mountain Group (foreground) was deposited. (DRS photograph.)

three formations: the Castile, Salado, and Rustler. These are overlain by red beds of the Dewey Lake Formation. An additional marine unit, the Tessey Formation, is present locally in the Glass Mountains and there it is a massive, poorly bedded dolostone which may grade northward into the Salado and Rustler formations in the Delaware Basin.

The Dewey Lake red beds are exposed in the former Delaware Basin and marginal areas (by Dewey Lake time, the area was filled in). The unit is a uniform, orange-red siltstone with abundant feldspar, some sandstone beds, and abundant gypsum. The feldspars suggest uplift of a nearby granitic or lacustrine deposit on an arid, hot, windswept plain. The age of the Dewey Lake in uncertain, since it is unfossiliferous, and it may be Triassic. However, it overlies the Rustler conformably and is itself overlain disconformably by the Upper Triassic Dockum Group. Dockum strata are extensive continental red beds that probably represent part of a laterally intertonguing sequence including the upper Chugwater Group in Wyoming and the Chinle Formation in the southwestern craton (both discussed subsequently). The disconformity between Dewey Lake and Upper Triassic strata is considered evidence of tectonic stability at the Permo-Triassic boundary and it marks the hot, dry end to the history of regional cratonic activity.

B.8. The Midcontinent Permian Sea

Permian strata are present (including the surface and subsurface) from central Texas to Nebraska and west to the Front Range in Colorado, in a virtually unbroken and laterally traceable sequence. In prior discussions, we examined lowermost Permian units in the region (largely three Wolf-

campian groups: the Admire, Council Grove, and Chase groups) and observed that the units were cyclical and seemed to relate genetically to the underlying Pennsylvanian strata. Permian strata are generally quite different.

A broad, shallow epicontinental sea occupied the region through most of later Permian time, although its chemistry and environments of deposition varied widely through the interval. During the Wolfcampian, cyclical deposits formed as shallow subtidal conditions to the east intertongued with and overlapped deltaic and alluvial environments to the west, with sediments in eastern Colorado coming from the Front Range Uplift (Mudge, 1967). In Leonard time, more marine conditions prevailed and, although considerable vertical and lateral facies changes are preserved, the units deposited do not show regular cyclicity as do those of Penno-Permian time. The sea in the midcontinent was often restricted through the post-Wolfcampian and many units are at least in part evaporitic.

Leonardian units vary slightly around the midcontinent, reflecting multiple detrital sources from the Front Range in Colorado, the postorogenic uplands and Anadarko Basin in southern Oklahoma and northern Texas (Arbuckle and Wichita mountains), and the inland basin located in Kansas. But, in general, similar styles of sedimentation prevailed regionally through the Permian. The following discussion is extracted largely from Mudge (1967).

The lowermost regional Leonardian unit is the Wellington Formation of Oklahoma, eastern Nebraska, and Kansas, which in Kansas and eastern Nebraska is overlain by the Ninnescah Shale. Locally, especially in western Nebraska, the Ninnescah occupies the entire lower Leonardian section. Both units are predominantly red mudstones containing large amounts of anhydrite. The Wellington in Kansas contains the greatest amount of evaporite sediments in the midcontinent, with a thick halite unit in the middle.

Toward eastern Colorado, the Wellington/Ninnescah (and most lower Leonardian formations) grade laterally into the Lyons Sandstone which represents probable beach deposits in the lower portion and dune deposits in the upper. The Lyons was clearly derived from weathering of the Front Range Uplift at the shores of the inland Permian sea; it is characteristically composed of about 300 m of well-sorted, subangular quartzose detritus, and it shows crossbedding, ripple marks, and some raindrop impressions. It is not notably arkosic except in lenses to the south, suggesting that the Front Range had been worn down considerably by Leonard time.

In Oklahoma, the Wellington is overlain by the Hennessey Shale, consisting of 100–200 m of red mudstone with small quantities of evaporites (MacLachlan, 1967). In the Oklahoma section, the Wellington thickens greatly toward the Anadarko Basin, reaching over 350 m. In the Oklahoma and Texas panhandles (northernmost Texas, westernmost Oklahoma), lower Leonardian units are the Wichita Group, which contains sediments similar to the Wellington/Ninnescah/Hennessey formations but which also contains considerable dolostone and green mudstone along with the anhydrite. Locally in northern Texas, the Wichita Group is dominantly sandstone, siltstone, and mudstone.

More widespread dolostone and anhydrite deposition followed in eastern Colorado, Kansas, and southern Nebraska as shown by the Stone Corral Formation in those areas, overlying the Ninnescah or Wellington Formation. The Stone Corral is a laterally persistent but thin (2–35 m) unit which outcrops in Kansas as a dolostone and dolomitic mudstone, but which is almost entirely composed of anhydrite downdip to the west in Colorado. Stone Corral deposition took place in an extensive, but restricted shallow sea.

Mudge notes that, following Stone Corral time, much of the region was a broad alluvial plain with brackish conditions. Between the Stone Corral and the uppermost Leonardian units are a series of predominantly red mudstones, sand and sandy mudstones, assigned to the Harper/Salt Plain/Cedar Hills/Flowerpot formations in Kansas and eastern Nebraska. In western Nebraska, the Cassa Group is probably equivalent to the Harper–Cedar Hills section and the Opeche Shale correlates with the Flowerpot. Locally, in western and southwestern Kansas, up to 100 m of halite and anhydrite is present within the upper Leonardian section.

In Colorado, the Lyons Sandstone intertongues with Harper–Flowerpot lithologies, and in Oklahoma, equivalent units comprise the lower part of the El Reno Group (Duncan Sandstone and Flowerpot Shale). El Reno lithologies include mixed evaporites, mudstones, and dolostones with deltaic sands and muds derived from probable eastern (Ouachita or southern Appalachian) sources (MacLaughlan, 1967).

The end of Leonard time was marked by widespread restricted-marine conditions, and gypsum and anhydrite were deposited in Kansas and eastern Nebraska. The Blaine Formation is a well-known evaporite unit in the subsurface of Kansas, Nebraska, and eastern Colorado, which is largely composed of evaporites, and it is overlain by the Dog Creek Shale which is a mudstone and dolostone unit marking the latest Leonardian in the Kansas Basin. In outcrops in Kansas, the Blaine also includes lithologies similar to the Dog Creek, and the two units are not easily separable. In northeastern Colorado and western Nebraska, an equivalent unit is the Minnekahta Limestone.

In Oklahoma, gypsiferous Blaine and Dog Creek strata are described in the upper Leonardian section, suggesting that conditions there were more similar to those in the rest of the restricted midcontinent sea at that time; but in the Texas panhandle, uppermost Leonardian strata are assigned to the Clear Fork Group and the overlying Glorietta Sandstone. These latter show strong detrital influences with sediments derived from several probable sources: the Wichita Uplifts, the Ancestral Rockies, and local arches such as the Sierra Grande.

Guadalupian and Ochoan midcontinental strata are generally lumped together, because neither series is readily correlated with units in the type areas in the Permian Basin. Conditions in the midcontinent sea presumably continued relatively unchanged from the Leonardian, with intermittent restriction of the Kansas Basin, and with deposition of considerable amounts of fine-grained reddish detrital sediments. Upper Permian units in Kansas and eastern Nebraska are dominated by red mudstones, producing the characteristic "Permian red-bed" look. It is likely that the shallow sea periodically gave way to ephemeral lakes and to very low-lying alluvial plains throughout the interval, so that admixed marine, freshwater, and intermittent brackish conditions were present.

In eastern Colorado, the Upper Permian section is assigned to the Lykins Formation, which contains several members but which has about the same overall mix of lithologies as the Kansas–Nebraska Upper Permian; however, Mudge notes that the Lykins may have been deposited rapidly and under more uniform and normal marine conditions. It contains at least two widespread limestone members, the Falcon and Glennon, and three shale members, the Harriman, Begern, and Strain, all lithologies absent elsewhere in the midcontinent.

In Oklahoma, the Whitehorse is elevated to group status and includes two dominantly sandstone formations. The Cloud Chief Formation, which is a mudstone with dolomite and gypsum lenses above the Whitehorse, correlates with the Day Creek in Nebraska/Kansas/Colorado. The youngest units in Oklahoma deposited during Absaroka time are members of the Quartermaster Formation which is a predominantly red mud unit deposited under conditions similar to those of the Whitehorse: i.e., a mud flat bordering on the sea (MacLaughlan, 1967). In the Texas panhandle, the Upper Permian is assigned by Dixon (1967) to undifferentiated Whitehorse and Quartermaster formations, with an indeterminate age for the uppermost units.

Across most of the midcontinent, Guadalupian or Ochoan strata are the youngest Absarokan deposits present, and these top beds are generally red-colored mudstones. Units overlying the Permian vary widely in age across the midcontinent but only in southeastern Colorado and bits of western Nebraska do they include deposits laid down during later Absaroka time. There, three Triassic units are present: the upper beds of the Lykins Formation, which are conformable on Permian Lykins at the foot of the Front Range in central Colorado; in westernmost Nebraska, rocks of the Lower Triassic Spearfish Formation apparently overlie the Freezeout Shale conformably; and in southeastern Colorado, Upper Triassic red beds of the Dockum Group overlie the Taloga Formation unconformably. These latest Absarokan beds are red, fluvial-channel, ephemeral-lake, and other continental deposits which indicate by the presence of conformable contacts that no major tectonic events occurred during the post-Permian interval of Absaroka time in the region. Further, their dominantly terrestrial character indicates that the midcontinent Permian sea was almost totally gone by the beginning of the Triassic.

B.9. Permo-Triassic of the Northwestern Craton

Across the northwestern craton, there were two principal regions of upper Absarokan deposition: one incorporating eastern Montana, Wyoming, and the western Dakotas, comprising an area which had much in common environmentally with the midcontinent to the southeast; and the other including the present eastcentral Rocky Mountains and eastern Great Basin states, containing southwestern Montana, western Wyoming, northern Utah, and northeastern Nevada. Both regions received marine deposits from the Absaroka sea in the Permian but, by the Triassic, the region was under admixed marine and continental environments.

B.9.a. Permian of the Eastern–Northern Craton

After Wolfcamp time, the sea withdrew from eastern Wyoming and the western Dakotas due to epeirogenic uplift centered in Wyoming. The sea returned in Leonard time but it did not flood quite as extensive an area as that previously transgressed. The returning marine waters resulted in deposition of the Owl Canyon Formation over the Minnelusa Formation in South Dakota, and over the Casper/Tensleep/Hartville and correlative sandy shelf units in Wyoming and Montana. Owl Canyon strata are absent locally, especially in parts of southern Montana, northern Wyoming, and much of the Dakotas (except the Black Hills); these regions were probably still emergent in early Leonard time. The Owl Canyon contains sandstones and siltstones, which thicken to the north, and evaporites which increase to the south. The unit represents a variety of paralic environments formed at the margin of the transgressing Permian sea, including dune, delta, littoral, and subtidal environments. The detritus was largely derived from reworked older units (e.g., Tensleep, Casper, Minnelusa; Maughan, 1967). Some sands in the extreme northwestern and southwestern reach of the Owl Canyon have cross-stratification suggestive of eolian conditions and these grade laterally into deltaic and lagoonal muds.

Later Leonard time in the area is marked by deposition of the Opeche Shale and overlying Minnekahta Limestone. The Opeche correlates with the Flowerpot Shale, and the Minnekahta with the Blaine Formation of the midcontinent. The similarity of units between the northern craton and the midcontinent in uppermost Leonardian time indicates that the inland Permian sea occupied a very wide area. The Opeche represents more restricted marine conditions than does the Minnekahta; therefore, it is believed that the sea shallowed somewhat or regressed partly from the northern craton during Opeche time, leaving behind red beds and evaporites including thick halites. Minnekahta time probably saw renewed transgression; however, the Minnekahta is poorly fossiliferous and includes local evaporites and widespread dolostones, indicating that the sea was somewhat hypersaline (Maughan, 1967).

In the Late Permian (Guadalupian and Ochoan), the northern craton again received strata which correlate with and in most cases are given the same names as midcontinent units. Across most of the area, the Glendo Shale (previously described in western Nebraska) is at the bottom of the Upper Permian section. The Glendo in the north is a thin (20 m) mudstone and siltstone with local evaporites, and grades southward into limestones and dolostones. Maughan (1967) describes the Glendo as having been deposited in a shallow sea during warm, arid conditions. Above the Glendo is the Forelle Limestone, a thin algal limestone which correlates with Day Creek strata in the midcontinent. Forelle deposition marked a brief return to almost normal marine conditions but to the south, in the midcontinent, and to the east on the northern shelf, it grades into evaporites, showing that restricted local basins with evaporite pans still existed.

The uppermost Permian section regionally includes

three intertonguing lithologies: in the Dakotas and eastern Montana is the Pine Salt Formation, a red mudstone and halite facies; elsewhere, the section contains the Freezeout Shale and the overlying Ervay Formations. Freezeout strata are red mudstones whereas Ervay strata are beds of gypsum and anhydrite. In the Black Hills, the entire Guadalupian and Ochoan section is assigned to the Spearfish Formation, and the Glendo, Forelle, Pine Salt, Freezeout, and Ervay are assigned member status. In central Wyoming, the uppermost Permian strata are carbonates, assigned to the ''Ervay Carbonate/Rock Member'' (of the Spearfish Formation).

B.9.b. Permian of the Northcentral Rockies Region

In a previous discussion, we traced the Penno-Permian transition in western Wyoming, Montana, and eastern Idaho; regional Wolfcampian strata were shown to include middle sandy units of the Wells Formation in western and central Montana.

During Leonard time, a relatively deep marine basin formed in Idaho and western Montana, which adjoined shelf areas eastward in eastern Montana and Wyoming, northward in northern Montana, and southward toward Colorado (Fig. 7-27). Coarse detritus and carbonates were deposited primarily on the shelves while in the basin, organic-rich, fine detritus, minor carbonates, and, notably, silica and phosphorites were deposited (Sheldon *et al.*, 1967). These basinal deposits form the widespread Phosphoria Formation whereas shelf deposits are included in the Shedhorn Sandstone, Park City, and Goose Egg formations.

The Phosphoria is composed characteristically of chert, phosphorite, phosphatic shale, dark to black mudstone, and sandstone. It is generally 100 m thick in Montana but it thickens fivefold westward toward the Great Basin. Phosphoria deposition spanned Leonard through Guadalupe times but most of it is of Guadalupe age. Phosphorite is a relatively rare lithology which is seen forming today only in areas of upwelling from deep ocean waters. Presumed presence of a deep basin off the edge of the Montana/Idaho continental shelf probably explains the deposition of persistent, thin beds of phosphorite but the actual source of the phosphorus in the Phosphoria is unknown.

Sheldon *et al.* (1967) state that the whole Leonardian section regionally (and by extension, the whole Upper Permian section) forms one big transgressive–regressive cycle; therefore, all regional units are interrelated and grade into one another according to Walther's law. While the deep marine dark mudstone, chert, and phosphorite facies are assigned to the Phosphoria, shallow-water limestone units are assigned generally to the Park City Formation, coarse shelf-derived detrital sediments are assigned to the

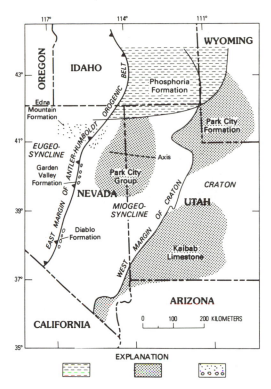

Fig. 7-27. Tectonic and stratigraphic relationships of Permian Park City and equivalent units in the central-western craton. (From Wardlaw *et al.*, 1979.)

Shedhorn Sandstone, and nonmarine or shallow-marine red mudstones and anhydrites are assigned to the Goose Egg Formation.

The same Phosphoria/Park City/Shedhorn/Goose Egg nomenclature representing the basin-to-shelf-to-shoreline transition persists through the Upper Permian section, with similar intertonguing relationships. Sheldon *et al.* state that a north–south high developed in Guadalupe time off the eastern edge of the Phosphoria Basin, and a shallow submarine ridge developed on the seafloor, separating the open ocean far to the west from the basin. These structural changes in the Phosphoria Basin may have enhanced phosphorite precipitation because Guadalupe-age Phosphoria beds contain greater proportions of phosphate rock.

Tongues of Phosphoria lithology are perhaps the most extensive and recognizable marker beds in the Upper Permian of the northwestern craton, ranging across the entire region during intervals of the transgressive–regressive cycle. Extensions of Phosphoria tongues onto the shelves imply periodic overturning of deep basin waters to spill out onto adjacent shallow environments.

King and Beikman (1976) note that many marine fossils in the Phosphoria are similar or identical to those found in the Guadalupian section of the Texas basins. Wardlaw *et al.* (1979) used brachiopod biostratigraphy to show that the

Retort Phosphatic Shale Member locally at the top of the Phosphoria, was deposited initially in two depocenters: one in southwestern Montana and the other in adjacent southern Idaho. This suggests that the basin was irregular and that it may have been segmented in an as yet undeterminable fashion.

The regional Permo-Triassic contact is generally at the top of the Phosphoria (Retort Shale and/Tosi Chert Members)/Park City ("Ervay Carbonate" or Franson Members)/Goose Egg/Shedhorn formations, with the apparent truncation of beds suggesting a regressive offlap sequence. The lateral gradation of beds around the region makes both description and determination of the nature of this boundary sequence rather difficult. Permian strata are generally overlain by Triassic strata with slight disconformity; but, locally, up to 650 m of Permian units was beveled by erosion. In the Montana–Idaho Basin, some Triassic units may overlie Phosphoria strata conformably but biostratigraphic evidence is generally absent since the boundary units are poorly fossiliferous; certainly, a major interval of erosion did not take place since the contact separates uppermost Guadalupian or Ochoan from Lower Triassic rocks and since the contact does not show deep fluvial channeling.

B.9.c.   Triassic of the Northern Craton

Lower Triassic strata across the northern craton comprise an intertonguing set of thick basinal sequences to the west, which thin across the basin edge in eastern Idaho–western Montana and grade eastward into shaley red beds across Montana and Wyoming and southeastward into Utah and Colorado (Fig. 7-28). These Triassic beds show the continued presence of the deep basin which influenced Permian Phosphoria deposition, located eastward of the Cordilleran tectonic axis in Idaho and Nevada and off the edge of the main continental shelf in Utah, Wyoming, and Montana. Much of the following discussion is extracted from Kummel (1954).

The most widespread lowermost Triassic unit is the Dinwoody Formation, representing basinal and deep-water shelf environments, with the type area in the Wind River Mountains, Wyoming. The Dinwoody in Wyoming is generally less than 30 m thick and consists of olive-to-tan siltstones and shales. Westward into southern Idaho, the Dinwoody contains calcareous siltstones and gray limestones and it thickens to between 250 and 800 m. Its upper and lower boundaries are diachronous, including biostratigraphically younger and older beds in the thicker sections. Dinwoody strata thin toward the edge of the shelf in Wyoming and northern Utah, but they do not pinch out until they reach extreme eastern Wyoming and northcentral Utah.

Above the Dinwoody in most of the northern craton, but at the base of the Triassic section in Utah, is the Woodside Formation, with type area in northeastern Utah; there it consists of over 300 m of maroon and red shaley siltstones. The Woodside and Dinwoody have approximately intertonguing thicknesses; thus, where thick Woodside is present, thin Dinwoody exists. But, to further complicate matters, a third and even a fourth basal Triassic unit is present regionally to intertongue with the Dinwoody/Woodside lithologies. These units are the Chugwater Formation in southcentral Wyoming and the Moenkopi Formation in northeastern Utah, extending across a wide area of the southwest. "Chugwater" is used to designate all red sandstones and siltstones which occur above the Dinwoody in Wyoming and which overlie Permian units in parts of Wyoming. The Chugwater consists of three members, the Red Peak, Alcova Limestone, and Popo Agie, of which only the Red Peak is definitely Lower Triassic. In Wyoming, the Red Peak section ranges up to 300 m thick.

Above the Woodside, especially toward the western shelf, occurs a widespread, thick limestone tongue named the Thaynes Limestone, with type section in Utah. Typically, the formation is over 350 m thick and composed of limestone, calcareous sandstone, green-gray shale, and a middle red shale. It contains an abundant ammonoid fauna and, as Kummel notes, is the most fossiliferous marine Triassic unit in the region.

The complex of interrelated marine units described above represents the largest wedge of sediments in the Triassic section of North America. Limestones such as the Thaynes and those in the Dinwoody, along with the light-colored siltstones, were deposited in relatively deep basin and shelf waters whereas the predominantly red Chugwater and Woodside lithologies are shallow-shelf deposits. The complex interfingering of lithologies shows that changes of seawater depth occurred over broad regions but that the major depocenters for each lithology remained relatively constant. The dominant red color of the eastern shelf detritus must represent nonmarine oxidation in continental environments east of Wyoming and Utah.

A major regression occurred after Triassic time, and subsequent units are progressively less marine-dominated and overwhelmingly red. In Idaho, a fairly complete Upper Triassic section is present where Lower or Middle Triassic beds are overlain by Upper Triassic red detrital sediments that are assigned generally to the Ankareh Formation (or Group). In Montana, on the other hand, units above the Thaynes/Woodside and below the post-Absarokan Jurassic are absent; a Middle Triassic depositional hiatus may be represented across much of the northern region, except for areas that have Middle Triassic Thaynes units (which are difficult to determine because of a lack of index fossils). Thus, it is apparent that the regression of the Absaroka sea in the Middle Triassic was widespread and long.

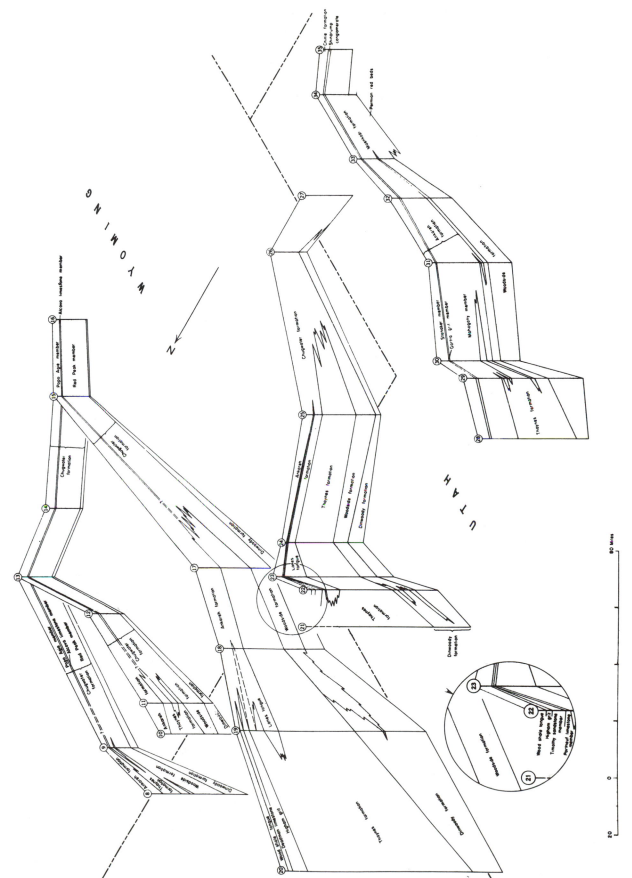

Fig. 7-28. Fence diagram of Triassic strata in the northwestern United States. The inset details relationships in the northeastern Utah corner. (From Kummel, 1954.)

Eastward into central Wyoming, the Ankareh grades into the upper beds of the Chugwater Formation, comprising the Alcova Limestone and Popo Agie members. The Alcova is a very persistent, thin (5 m) gray-to-pink thin-bedded micrite, which is notable for having yielded fossils of a nothosaur (Kummel, 1954), which is rare in North America, and an early representative of the long-necked marine reptiles which flourished during the Mesozoic. The Alcova is widely present as a resistant cap to Triassic sections in Wyoming and it is a classical marker bed in the middle of an otherwise-red sequence. Following Alcova time in Wyoming, Popo Agie red beds were deposited with very mixed components, including oolitic claystones, conglomerates, shales, and silty sandstones. Chugwater lithologies probably grade southward into Wyoming but stratigraphic sections are absent between areas where Chugwater is described and where Moenkopi–Chinle units make up the Triassic section (discussed in the next subsection).

Environments of deposition as indicated by the Ankareh/upper Chugwater lithologies show increased continental influences but within persistent marine carbonate environments. The sea in the Late Triassic was certainly less widespread and less deep than in the Early Triassic (or Permian). By the latest Triassic, complete regression of the Absaroka sea began on the northwestern craton and the widespread Nugget Sandstone was deposited over older Triassic strata. Nugget sands are generally considered to be of Early Jurassic age but in parts of central Wyoming they may date to the latest Triassic. Since the unit is unfossiliferous and the Nugget overlies the Ankareh/ Chugwater units with apparent conformity, Nugget time has generally been classified as Triassic(?), Jurassic(?). The formation consists of a lower Bells Springs Member, of red and gray, ripple-marked sandstone and red, green, and purplish-red siltstones, and an upper part of yellow-white to pink crossbedded sandstone. The Bells Springs is partly fluvial while the upper part is eolian (Pipiringos and O'Sullivan, 1978); clearly, the sea was withdrawing in Nugget time. This unit is but one of several to be described subsequently (Kayenta and Navajo sandstones) which comprise a widespread series of continental/eolian deposits formed as the very long stand of the Absaroka sea in the west finally ended and winds swept across the exposed landscape, eroding older Absarokan marine detritus.

B.10. Permian through Lower Jurassic of the Southwestern Craton: Final Retreat of the Absaroka Sea

We now follow final Absarokan cratonic history on the southwestern craton, where the sea made its final stand and left a sizable stratigraphic record of withdrawal. The study area here includes Arizona, New Mexico (west of the "Permian Basin"), Utah, and western Colorado. It will be recalled from previous discussions that this region encompassed a complex of late Paleozoic basins and uplifts, several of which were active during the interval to be examined. The following account does not consider lithofacies or events within the basinal areas since they have been described and since their histories are commonly different from the adjoining shelves. Positive areas will be included because they contributed detritus to shelves as well as yoked basins.

B.10.a. Permian History

By the latest Wolfcampian, limy marine conditions persisted only in western Utah, while most of the southwest was under mixed nearshore, intertidal, eolian, and continental influences, with large volumes of arkosic, red sediments being supplied by the Ancestral Rockies and other uplifts. The Leonardian brought renewed marine transgression across much of Arizona, New Mexico, and small parts of western Colorado and Utah. The basins in western Colorado (Central Colorado Basin) and western Utah (the foredeep basin west of the Cordilleran continental margin) were probably active throughout the Early and Middle Permian; the region around the Four Corners was probably above sea level (McKee, 1967). With renewed flooding, a wide variety of depositional environments left a diverse set of lithofacies (Fig. 7-29).

In western Utah during the early Leonardian, marine-to-supratidal sandstones were deposited as the upper Supai Formation (also called the Queantoweap Sandstone). In extreme southwestern Utah, related sands are called the Coconino Sandstone. The Coconino is a lighter-colored deposit, ranging to almost pure white, and is divided into the White Rim and Cedar Mesa members (which have locally been assigned formation status). The Queantoweap and Coconino sands represent supratidal, eolian, subtidal bar and intertidal deposits. Regionally, vast quantities of sand make up the lower Leonardian section (e.g., the Queantoweap is generally 250 m thick) and this sandstone lithosome extends widely with lateral gradations in color, grain size, and composition. Into northwestern Colorado, basal Leonard-age sands comprise the Schoolhouse Sandstone, which shows large-scale, high-angle cross-stratification, and these sands may correlate with Weber Sandstone beds even farther north.

Southward into northeastern Arizona and northwestern New Mexico, sands of the De Chelly Sandstone Member of the Cutler Formation correlate with the Schoolhouse/ Coconino/Queantoweap formations. Farther south, in Arizona and New Mexico, several intertonguing subtidal to deep-water marine units were deposited, including the Fort Apache Limestone Member of the Supai Formation, and the

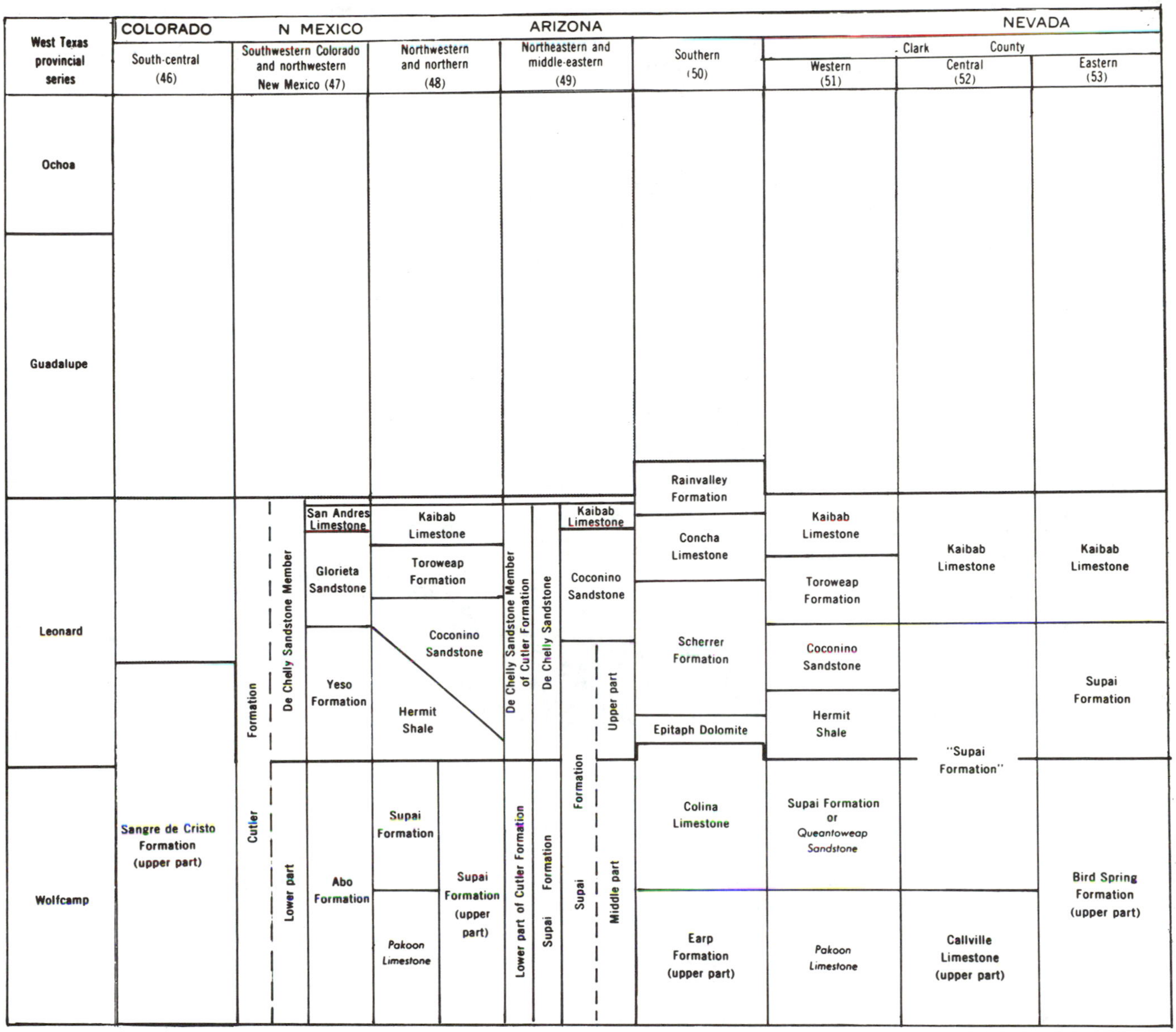

West Texas provincial series	COLORADO	N MEXICO	ARIZONA			NEVADA		
	South-central (46)	Southwestern Colorado and northwestern New Mexico (47)	Northwestern and northern (48)	Northeastern and middle-eastern (49)	Southern (50)	Western (51)	Clark County — Central (52)	Eastern (53)
Ochoa								
Guadalupe								
Leonard	Sangre de Cristo Formation (upper part)	Cutler Formation — De Chelly Sandstone Member; San Andres Limestone; Glorieta Sandstone; Yeso Formation; Hermit Shale	Kaibab Limestone; Toroweap Formation; Coconino Sandstone; Hermit Shale	De Chelly Sandstone Member of Cutler Formation / De Chelly Sandstone; Kaibab Limestone; Coconino Sandstone; Lower part of Cutler Formation; Supai Formation (Upper part / Middle part)	Rainvalley Formation; Concha Limestone; Scherrer Formation; Epitaph Dolomite	Kaibab Limestone; Toroweap Formation; Coconino Sandstone; Hermit Shale	Kaibab Limestone; "Supai Formation"	Kaibab Limestone; Supai Formation
Wolfcamp		Cutler Formation (Lower part); Abo Formation; Supai Formation; Pakoon Limestone	Supai Formation; Supai Formation (upper part); Pakoon Limestone		Colina Limestone; Earp Formation (upper part)	Supai Formation or Queantoweap Sandstone; Pakoon Limestone	Callville Limestone (upper part)	Bird Spring Formation (upper part)

Fig. 7-29. Representative correlations of the uppermost Permian units on the southwestern craton. (From McKee *et al.*, 1967.)

Hermit Shale of the Grand Canyon section in northwestern Arizona (Fig. 7-30).

Since the basal Leonardian sands were deposited by transgressing seas, the times of deposition for units above the Coconino/Queantoweap and equivalents vary. In western Nevada and northwestern Arizona (Grand Canyon), the Hermit Shale is overlain by Coconino beds and in turn overlain by the Toroweap and Kaibab formations. The latter two are marine deposits, with the Toroweap containing mixed lithologies of sandstone, carbonates, and evaporites, and the Kaibab comprising a marine carbonate sequence with a shelly invertebrate fauna. The sea in Toroweap time was restricted and, too, despite the name "Kaibab Limestone," the Kaibab sea deposited dolomitic limestone, suggesting some continued restricted circulation. The Kaibab is a widespread unit which occupies the top of the Leonardian section across northern Arizona and western Utah. In central and southern Utah, Kaibab-equivalent limestones may be of lower Guadalupe age and overlie Coconino sands that were deposited through the Leonardian. In areas toward central Arizona, the upper Leonardian beds are De Chelly sands of the Supai Formation and the Kaibab is absent; it most likely was eroded off in post-Leonard time.

Guadalupe and Ochoa times probably featured relative

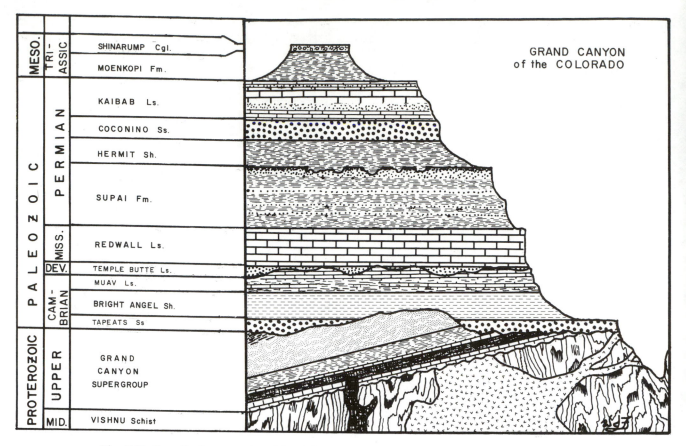

Fig. 7-30. Generalized cross section of Grand Canyon strata. (Redrawn and modified from Eardley, 1951.)

stability on the southwestern craton. In central and southeastern Utah, the Kaibab limestones described above extend into the post-Leonardian, and they are overlain by marine limestones assigned to the Plympton and Gerster formations. All together, the post-Leonardian Permian section there exceeds 700 m and may have been deposited as late as the Ochoan.

The Maroon Formation (Section B.4), in the Central Colorado Trough, was deposited generally through the Guadalupian; but in the Eagle Basin portion of the trough, Maroon strata are overlain by Guadalupian units termed the State Bridge Formation. These are thin mudstones with a thin fossiliferous limestone upper unit that correlates biostratigraphically with the Phosphoria Formation farther to the north and west. Where uppermost beds of the Maroon Formation are of Guadalupe age, they may exceed 1500-m thickness; State Bridge rocks were deposited in less active basin sections and are approximately 50 m thick.

In general, the presence of relatively normal-marine mudstone, limestone, and basinal detrital strata in northwest Colorado shows that the area was relatively stable through at least Guadalupe time and that the sea there probably connected with the waters in Utah where Kaibab limestones were forming. Equivalent Guadalupian sections may have been deposited in Arizona and New Mexico.

B.10.b.   Triassic History

Triassic rocks overlie Permian rocks unconformably across the entire region, but with variable lengths of time represented by the erosional surface. The sea withdrew almost totally from the region in the late Guadalupian or Ochoan, and its return is marked generally by a change from carbonate and clean sandstone deposition, characteristic of the later Permian, to the generally red mud deposition of the Early Triassic (Hallgarth, 1967). The most widespread regional basal Triassic unit is the Lower and Middle Triassic Moenkopi Formation; but in northeastern Arizona and northwestern New Mexico, the Moenkopi is absent and the Upper Triassic Chinle Formation overlies Leonardian strata. Locally in the southwest, and generally in southern Arizona, Triassic beds are totally absent and Permian strata are overlain by Cretaceous sediments.

The Moenkopi Formation is a variable unit, containing

characteristically red- and brown-colored beds of mudstones, siltstones, some sandstones, thin-bedded limestones, calcareous mudstones, and some thin, pure gypsum
beds. The unit is gypsiferous in many sections and some
halite crystal impressions have been found. Moenkopi strata
extend across the entire Colorado Plateau region and westward into southeastern Nevada and extreme southeastern
California; the overall structure of the unit is a northwest-
ward-thickening wedge of sediments, which thins to a
feather edge in eastern Arizona, Colorado, and western
New Mexico. At its thickest, along a line drawn north–
south through eastern Nevada at the Cordilleran margin, the
unit is approximately 1000 m thick. Beds in the west are
dominantly marine and contain the bulk of the carbonates
and evaporites, while the eastern beds are entirely continental; marine-to-continental lithologies intertongue in northcentral Arizona and southcentral Utah. The eastern margin
of the Moenkopi Basin formed two embayments, north and
south, separated by a projecting ridge in Arizona. The
northern embayment extends up into Colorado, while the
southern extended into New Mexico (Fig. 7-31). Moenkopi
units in the southern embayment show coarser and less
weathered detritus and probably were closer to sources
(McKee, 1954).

The Moenkopi correlates with Lower Triassic formations to the north including the Dinwoody, Thaynes, and
Woodside formations, and the Chugwater Group (see Fig.
7-28). The age of the Moenkopi has been a problem because
of the complex intergradation and diachrony of Lower Triassic units in the western craton, and because of the appar-

ently long hiatus between Kaibab and Moenkopi times. Upper Moenkopi beds are almost everywhere of continental
origin and bear only vertebrate fossils. These are generally
less precise biostratigraphic tools than are marine invertebrates but upper beds seem to be of latest Early Triassic or
early Middle Triassic ages. The environments of Moenkopi
deposition represented include shallow sea, deltaic, lake,
playa, floodplain, and tidal flat (Reif and Slatt, 1979).
Transgression into the region was accompanied by extensive fluvial deposition on the eastern part of the Moenkopi
Basin with detritus derived from the Uncompahgre, Front
Range, and Defiance uplifts, among others. Along the eastern margin of the depositional basin, fluvial, deltaic, and
mud-flat deposits mix with continental eolian sands; the red
mud flats comprise perhaps the most areally extensive beds.
Structures in many beds, including mud cracks, raindrop
impressions, vertebrate tracks, and salt casts, show that
many shaley siltstones and sands were deposited subaerially. The climate in Moenkopi time was probably arid
and tropical. Vertebrate fossils, although not abundant, are
generally those of large, probably, sluggish warmth-loving
amphibians and reptiles.

Regression of the Moenkopi sea was gradual, marked
by minor transgressions and regressions, and involved repeated basin filling and renewed subsidence. Eastern sands
and finer detrital deposits were borne by rivers over the
western basin as regression progressed and as deltas prograded across much of the area. Upper Moenkopi beds of
the Upper Red Member are largely continental across the
entire region, and mark the beginning of another major

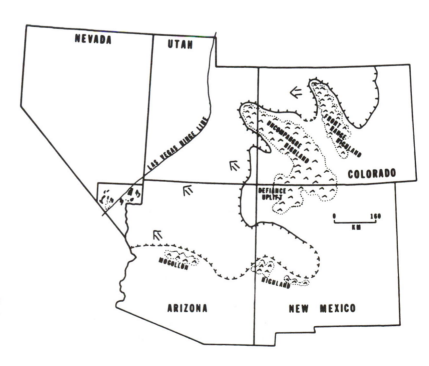

Fig. 7-31. East-southeastern source areas of the
Moenkopi Formation, and the paths of sediment disper-
sal; the hachured line shows the limits of deposition.
(From Reif and Slatt, 1979.)

depositional hiatus in the Absaroka Sequence and the last stand of the Absaroka sea on the craton. As a final note, the Moenkopi is a poorly resistant unit in general and characteristically forms the steplike slopes of mesas and buttes; together with its bright colors, the erodibility of Moenkopi strata makes it stand out strikingly in the arid landscape of the southwest.

After a hiatus which spanned most of the Middle Triassic, during which time extensive erosion produced what is one of the most widespread and conspicuous unconformities in the Triassic and Jurassic sections of the west (Pipiringos and O'Sullivan, 1978), renewed deposition of continental deposits began in the Colorado Plateau region. This depositional sequence was probably initiated by renewed uplift of the Front Range and other eastern sources. Basal deposits of the detrital wedge consist of approximately 35 m of coarse conglomerates which characteristically comprise gravel filling of channels in eroded Moenkopi beds. The roughly 350 m of continental deposits above the conglomerates, comprising the bulk of the Upper Triassic section on the southwestern craton, is the famous Chinle Formation. The conglomerate at the base is the Shinarump Member and it is uniquely widespread as well as remarkably thick for this type of lithology.

The Chinle is a brightly colored, shaley siltstone, dominantly red but also purple, maroon, and gray, with a number of conglomerate beds above the Shinarump. Among the exotic lithologies in the Chinle are extensive and economically important beds of carnotite (the principal uranium mineral), beds of volcanic ash which are present both as bentonites and as material interbedded with sandstones (Shawe, 1976), and large quantities of petrified coniferous wood. The Chinle Formation itself makes up the attraction in the famous Painted Desert, where the bright colors of the continental deposits are displayed spectacularly. In most completely exposed sections of the Upper Triassic in the region, the Shinarump is a prominant ledge-forming unit while the overlying Chinle beds form slopes.

The lime cement and limestones present in the Chinle are apparently from lakes which developed on the floodplains and lowlands during Chinle deposition. Much of the unit probably represents deposits on pediments and playas, which may account for the concentration of gypsum and calcite cement as caliche. Volcanic ash in sandstones, bentonites, and zones of tuff indicate that volcanism, probably from island arcs to the west in the Cordilleran mobile belt, was increasing during Chinle time relative to older Triassic and Penno-Permian times.

The Chinle, like the Moenkopi, is a widespread and interesting unit. Detailed stratigraphy and environmental analysis are beyond the scope of this book but the interested reader is directed to Stewart *et al.* (1972) and Poole and Stewart (1964) for monographic treatment of the formation.

The fossils in the Chinle deserve attention before we move up in the section. In the Petrified Forest, tree trunks up to 3 m in diameter by 40 m in length are preserved by beautiful replacement of the wood with agate and jasper. These fossils in the National Park are protected by law and can be observed laying about on the surface in the manner that streams emplaced them in Chinle time. Most, if not all, of the tree trunks were stream-transported and it is not unreasonable to assume they grew predominantly in uplands to the east rather than by the lakes, channel banks, and bajada–playa environments of Chinle deposition. The assemblage is dominated by the conifer *Araucarioxylon,* but several other arborescent groups are represented including cycads and ferns (Andrews, 1961), common in Mesozoic floras. Other plant fossils have been found in the Chinle, indicating that brush and scrub growth was present beneath the large conifers. Fossils of early dinosaurs, notably light, agile bipedal forms such as *Coelophysis,* were found in terrestrial Chinle beds in New Mexico and, in Chinle lake sediments, the remains of phytosaurs (crocodilelike thecodont reptiles; see Fig. 7-77) are found, along with other reptiles, amphibians, and some freshwater invertebrates.

B.10.c. Jurassic History

The uppermost beds of the Chinle contain some probable eolian sands, indicating the beginning of the end. The upper surface of the Chinle is an unconformity marking a relatively short hiatus before massive eolian sands swept across the Colorado Plateau. On the outcrop, this upper Chinle contact appears conformable, but at map scale, it is not. The position of the Triassic–Jurassic stratigraphic boundary is not certain; it lies above the Chinle but it may be within or above the overlying Glen Canyon Group (Wingate Sandstone Formation), or it may be between the Chinle and the Wingate (Pipiringos and O'Sullivan, 1978).

The Glen Canyon Group of continental, largely eolian deposits overlies the Chinle (Fig. 7-32) and consists regionally of four formations in ascending order: the Wingate/ Moenave/Kayenta/Navajo formations. These are further divided into members, representing fluctuating dune/fluvial/lake/playa environments across the southwest craton. The Wingate consists of an average 100–200 m of pale-reddish-brown sand in two members: the lower Rock Point, a slope-forming unit which may closely resemble Chinle beds and which generally lacks cross-stratification; and the upper Lukachukai Member, of similar-colored, cliff-forming, massive, crossbedded, fine, quartzose sand of probable eolian origin.

The Moenave is less than 100 m thick and may be equivalent to parts of the Wingate and Kayenta formations. It too contains two units: the Dinosaur Canyon and Spring-

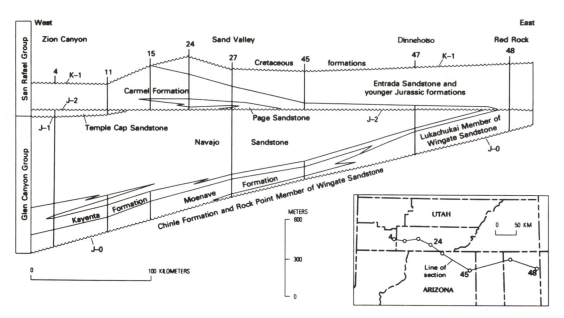

Fig. 7-32. Schematic cross section of the uppermost Absarokan and lowermost Zuni strata, along a section line near the Utah–Arizona border (see inset map). (From Peterson and Pipiringos, 1979.)

dale Members. The Dinosaur Canyon is confined to Navajo Co., Arizona, and is composed of mixed fluvial and eolian sands. The Springdale is ripple-marked and contains shale and mud lenses in a pale-red sandstone. It is crossbedded and ledge-forming, and represents lake and fluvial environments. Some fish fossils have been found in the Springdale. Over most of the region, the Springdale comprises the entire Moenave section.

The Kayenta is a thin (30 m), fluvial sediment that was deposited generally on an eroded Wingate surface or conformably on the Moenave, depending on local, lateral facies and thickness variations across the Colorado Plateau. Typically, it is a pale-red, crossbedded, calcite-cemented sandstone that thickens westward into Arizona–Utah and thins to a feather edge in eastern Arizona–Colorado. Where present, it is a ledge-former and contains fossils of freshwater mollusks, dinosaurs, and mammal-like reptiles.

At the end of Kayenta time, conditions reverted to those present during the eolian phase of Wingate deposition. The final unit of the Absaroka Sequence on the craton is the massive, cliff-forming, spectacularly large-scale crossbedded Navajo Sandstone, which is exposed to the fullest effect in Zion National Park, Glen Canyon, and on the Navajo Reservation in Utah and Arizona, where it comprises over 200 m of cliff-forming sands. The Navajo is a light-buff to tan-colored unit, composed of frosted, rounded, feldspathic sandstones and minor freshwater mudstones and limestones, which are easily recognized in outcrop be-

cause the unit shows virtually no horizontal bedding but exposes huge, high-angle, tabular-planar, wedge-planar, and trough-shaped crossbeds from 5 to 15 m thick (although one set measured to 34 m; Peterson and Pipiringos, 1979). The Navajo thins eastward and is beveled off near the New Mexico and Utah–Colorado state lines by the post-Absaroka unconformity. It thickens to the west in Arizona, southeastern Nevada, and western Utah, and may be truncated by the Cordilleran orogenic front. It overlies conformably and intergrades with the top of the Kayenta.

The paleoenvironment of the Navajo Sandstone is the subject of a long-term controversy. Classically, the unit was considered a continental eolian formation, with detrital sources from various areas around the western craton (see Shawe, 1976). Alternatively, the relatively recent observation that large-scale crossbedded sands form in aqueous environments has suggested to several workers (Stanley et al., 1971; Doe and Dott, 1980) that the Navajo, among other sandstones, may incorporate coastal dune and nearshore aqueous sands, in addition to purely terrestrial sands. Further, assuming subaqueous conditions, the orientation of crossbeds in the Navajo shows a predominantly southward direction of transportation, suggesting the sea lay close to the west of the craton in the Early Jurassic. Regardless of the model of deposition, after Navajo time, sedimentation ceased briefly on the southwestern craton (as it had ceased earlier elsewhere on the craton) and was reinitiated during the later Jurassic with the next cratonic sequence.

C. The Allegheny Orogeny: Tectonic Culmination in the East

C.1. Evidence for the Allegheny Orogeny

Late Paleozoic deformation is ubiquitous in strata of the Appalachian Valley and Ridge province and is characterized by large fold and thrust-fault structures traceable for hundreds of kilometers along strike (see Fig. 7-33). As a generality, folds dominate in the Central Appalachians (the "folded Appalachians") whereas thrusts dominate in the Southern (or "faulted") Appalachians. These structures, which affect even the youngest strata in the Appalachian Basin (the Permian Dunkard Group), were recognized very early in the history of Appalachian studies, but their relation to overall Appalachian history was not understood. Early geologists at first believed (incorrectly) that all deformation seen in the Appalachians occurred at the same time, as the result of a single, major tectonic event, which they termed the *Appalachian Revolution*. Later, after the deformations associated with the Taconic and Acadian orogenies were understood, the "Appalachian Revolution" was demoted to

Fig. 7-33. Map of major Appalachian structures showing distribution of miogeoclinal (M) and eugeoclinal (E) sequences. Bv, Brevard fault zone; W, window; CP, Coastal Plain; C, continental shelf; CS, continental slope; PM, Pine Mountain fault; WLF, western limit of thrusting. (From Harris and Bayer, 1979, *Geology*, Fig. 1, p. 569.)

the "Appalachian orogeny," but it was still considered to have been the major tectonic event in Appalachian history. Of course, since there had been other, previous tectonic events, one may not necessarily equate all Appalachian structures with the Appalachian orogeny. This semantic problem was resolved by Woodward (1957) who proposed substitution of the term *Allegheny orogeny*. With more detailed field studies and mapping in the crystalline Appalachians (i.e., the New England highlands, the Blue Ridge, and the Piedmont) and with accumulation of many radiometric age determinations, doubts began to appear that the Allegheny orogeny had actually been the greatest orogenic event of Appalachian history. Not only did major periods of igneous intrusion, metamorphism, and large-scale shear folding precede the Allegheny orogeny but also relatively few rocks were unquestionably formed or metamorphosed by it. By the early 1970s, some geologists were suggesting that the Acadian orogeny (see Chapter 6) represented the continental collision between North America and Africa, which ended the plate-tectonic history of the Appalachian orogen; the Allegheny orogeny was considered a modest affair, possibly the result of gravity tectonics.

In the past several years, stimulating new evidence has been developed that appears to reaffirm the view that the Allegheny orogeny may have been a major tectonic event.

C.1.a.   Debate on the Cause of Deformation in the Valley and Ridge

The great fold and fault systems of the Valley and Ridge have generated some of the most famous and theoretically significant debates in the history of structural geology. Two questions have commanded the attention of Appalachian geologists: (1) whether or not the great thrust faults of the Valley and Ridge are rooted in the basement (i.e., continue down into crystalline rocks below the sediments) and, more recently, (2) whether the main cause of deformation was horizontal compression or gravity sliding.

i.   Thick–Skinned versus Thin–Skinned Tectonics.   Before one can comment intelligently on causes of thrusting, one must understand the relation of the faults to the strata they affect. Of course, the character of faults at the surface is readily ascertained but their behavior at depth has been controversial. One school of thought, championed by Byron Cooper (e.g., 1968), held that major Appalachian thrusts extend down into the crystalline basement and, consequently, that large slices of the basement are thrust up into the sequence of supracrustal rocks. This has been called *"thick-skinned"* tectonics. The other side of the debate, first proposed by Rich (1934) and championed more recently by John Rodgers (e.g., 1970), held that most of the

major thrusts are *lystric* in nature (i.e., that they are concave-upward surfaces which are inclined steeply at the surface but which approach the horizontal at depth). In this *"thin-skinned" tectonics,* lystric faults are all rooted along a common, major décollement (or detachment surface).

In 1964, Gwinn published a summary of oil-company data, obtained from drilling and seismic profiling, which confirmed the thin-skinned model for the Central Appalachians. In 1976, Harris presented similar data from eastern Tennessee which also revealed the thin-skinned style of deformation. Since then, most analyses of Valley and Ridge structures presume lystric thrusts rooted in a master décollement.

There are two important corollaries to this thin-skinned model. The first involves the location of the master décollement. Within the strata of the Appalachian Basin are layers with varying degrees of shear strength: carbonate rocks and well-cemented sandstones have relatively high shear strengths and behave rigidly under stress (they are said to be *competent*); mudrocks, on the other hand, have very little shear strength and behave plastically when subjected to stress (they are *incompetent*). Thus, stress directed parallel to bedding does not affect every layer in the same way; greatest deformation occurs along the zones of lowest strength. In eastern parts of the Southern Appalachian Valley and Ridge, the master décollement lies primarily within the Rome Formation of Lower Cambrian age (the Lower Level Décollement of Fig. 7-34). The fault moves westward up a *tectonic ramp,* or higher-angle surface, to the level of the Chattanooga Shale (the Upper Level Décollement of Fig. 7-34). Tectonic ramps occur where the fault is deflected upward through competent rock units, possibly due to increased frictional resistance along the original horizon. The geometry of thin-skinned thrusting is thus described as a series of décollements in incompetent layers and tectonic ramps in competent ones.

The second corollary of thin-skinned tectonics involves the interrelationship between folding and thrusting. It has long been known that most Appalachian folds are asymmetrical with their axial planes dipping toward the southeast (they are said to *verge* toward the northwest). Based on an analysis of the nature of thin-skinned faulting, Rich (1934) proposed that anticlines (such as the Powell Valley and Pine Mountain anticlines in Fig. 7-35) were located over major tectonic ramps whereas synclines (e.g., the Middlesboro syncline, Fig. 7-35) occur over décollement surfaces. Harris and Milici (1977) noted that eastern anticlines are larger, i.e., of greater amplitude, than those farther to the west because the wedge shape of the sediment mass necessitated that a greater thickness of rock be duplicated by faulting in the east. Structures such as the Powell Valley anticline are said to be surficial or *rootless* because they do not extend down beneath the décollement surface.

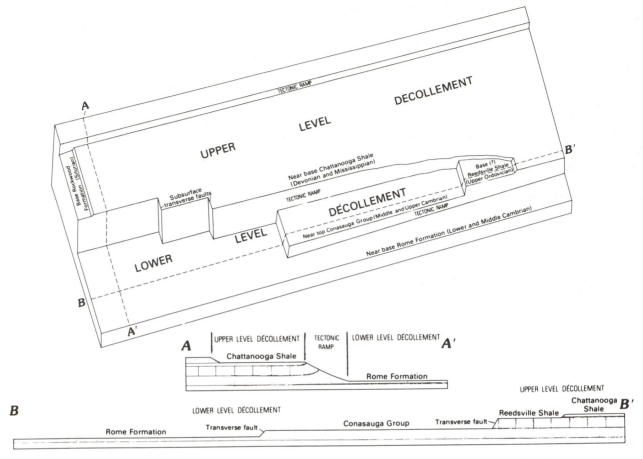

Fig. 7-34. Distribution of component parts of the Pine Mountain thrust of northeastern Tennessee showing the relation of décollements to incompetent horizons and of tectonic ramps to competent horizons. (From Harris and Milici, 1977.)

ii.   Horizontal Compression versus Gravity Sliding.   Once the character of Valley and Ridge thrusts and rootless folds was understood, debate over the cause of deformation could be joined. The traditional view was that all of these structures were produced by horizontal compression in which Appalachian strata were squeezed by a northwest-directed force. The other, and somewhat more recent, view is that uplift of the southeastern part of the orogen produced a northwestward slope; when the slope angle had increased to the degree that the component of gravitational force parallel to bedding exceeded the shear strength of the least-competent strata, thin sheets of rock broke free along décollement horizons and slid down the slope (see Fig. 7-36). Milici (1975) used a geometrical analysis to demonstrate that the relative ages of east Tennessee thrusts decrease from west to east, which he argued was due to their emplacement by gravity sliding combined with horizontal compression. The uplift to set the thrust slices in motion, as well as the associated horizontal compression, was thought by Milici to have been generated by rising and spreading of the thermal core (i.e., the deep-seated, hot, plastically de-

forming center) of the Appalachian orogen, represented today by the Piedmont. Milici considered the décollement to root *beneath* the Blue Ridge, implying that it, too, had moved westward.

These arguments left some geologists unconvinced. For example, Hatcher (1978) contended that the thrusts, which are clearly thin-skinned in the Valley and Ridge, ultimately propagated out of a deeply rooted fault zone in the Piedmont (see Fig. 7-37). This is an intriguing idea because it suggests, as shown in the illustration, not only that the thrusting was compressional in nature but also that both the Blue Ridge and the Piedmont may be allochthonous! We will now evaluate this suggestion.

C.1.b.   Are the Piedmont and Blue Ridge Allochthonous? Evidence from Seismic Profiling

Several authors (e.g., Root, 1970) have discussed the possibility that the entire Blue Ridge province is allochthonous. Evidence for this view may be seen in the

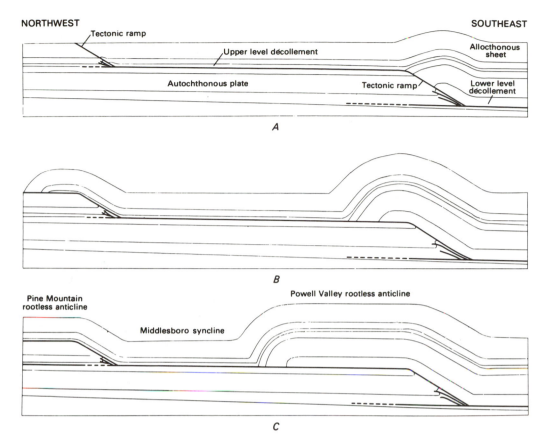

NORTHWEST

Tectonic ramp

Upper level décollement

Autochthonous plate

Tectonic ramp

SOUTHEAST

Allocthonous sheet

Lower level décollement

A

B

Pine Mountain rootless anticline

Powell Valley rootless anticline

Middlesboro syncline

C

Fig. 7-35. Sequential development of the Powell Valley Anticline of northeastern Tennessee. (A) Subsurface splay faulting initiates development of the anticline as the allochthonous sheet is pushed up the southeastern tectonic ramp; (B) northeastern displacement of the allochthonous sheet pushes the folded strata up the tectonic ramp and across the upper décollement, resulting in duplication of the stratigraphic section; (C) further northeastward displacement enlarges the Powell Valley Anticline while decreasing the width of the Middlesboro Syncline; the result is a broad, rootless anticline–syncline pair. (From Harris and Milici, 1977.)

presence of several large windows through the high-grade metamorphics of the Blue Ridge which expose sedimentary and low-grade metamorphic rocks. Rodgers (1970, p. 173) stated, "If the master fault [at the foot of the Blue Ridge in east Tennessee where it is thrust over younger strata] . . . finds its real roots at the back of the Grandfather Mountain window, then the entire Blue Ridge province here, basement and all, is floating over some unknown footwall." Hatcher (1978) noted the existence of virtually unmetamorphosed slices of carbonate rocks (which he correlated with the Shady Dolomite; see Chapter 4) within the Brevard zone of western North Carolina. He argued that these slices had been brought up from Paleozoic strata which hypothetically underlie allochthonous crystalline rocks of the Blue Ridge at least as far to the east as the Brevard zone and possibly even farther, i.e., beneath the Piedmont.

From these few references, one may see that several geologists were led by surface-geological relations to propose that (at least) parts of the Blue Ridge and Piedmont are allochthonous. But this historical sketch is brief because few geologists accepted the idea that such a large mass of crystalline rocks could have been thrust any appreciable distance. Cloos (1972), for example, believed that the Blue Ridge anticlinorium in northern Virginia and Maryland was essentially autochthonous, produced by large-scale shear folding. Nevertheless, thoughts on this subject have recently begun to change with the compilation of new data from seismic-reflection studies of the deep structures of the crystalline Appalachians.

In 1979, Cooke *et al.* reported the results of a seismic-reflection study conducted by the Consortium for Continental Reflection Profiling (COCORP) along a transect from east Tennessee (west of the Great Smoky fault, which separates the Valley and Ridge from the Blue Ridge) to the Coastal Plain overlap of the Piedmont just north of Augusta, Georgia. A summary sketch of the resulting seismic-reflection profile is shown in Fig. 7-38. Cooke *et al.* drew several important interpretations from these data. First, they postulated that the horizontal reflector-horizon deep below the

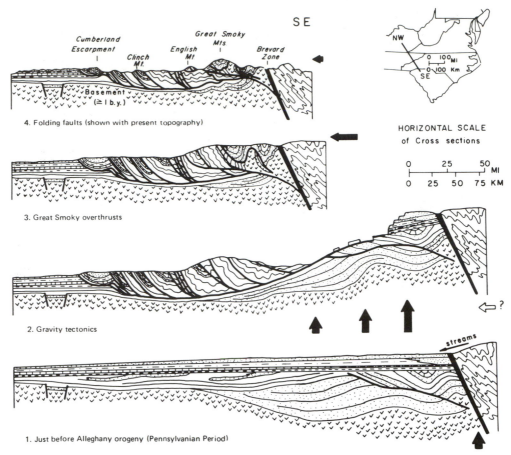

Fig. 7-36. Gravity tectonics as an explanation of Alleghenian deformation in the Valley and Ridge. (From Dennison, 1976, *Geological Society of America Bulletin,* Fig. 2, p. 1472.)

surface represents undeformed sedimentary rocks, an extension of Valley and Ridge strata beneath rocks of the crystalline Appalachians, just as Hatcher and the others had predicted. Cooke *et al.* calculated that the allochthonous crystalline rocks were thrust westward a *minimum* of 260

km! From these interpretations, they inferred that the edge of Paleozoic North America was not just east of present-day easternmost exposures of continental shelf sediments, as has been previously presumed, but *at least* 260 km farther to the east. In this regard, it is interesting to note that deep

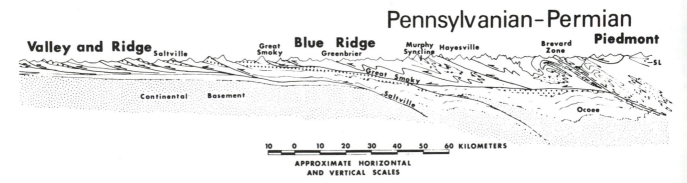

Fig. 7-37. Model of Alleghenian deformation stressing compressional, rather than gravity, tectonics. Notice that thin-skinned faults in the Valley and Ridge become thick-skinned (i.e., they root in the basement) below the eastern Blue Ridge. (From Hatcher, 1978, *American Journal of Science,* Fig. 5, p. 293.)

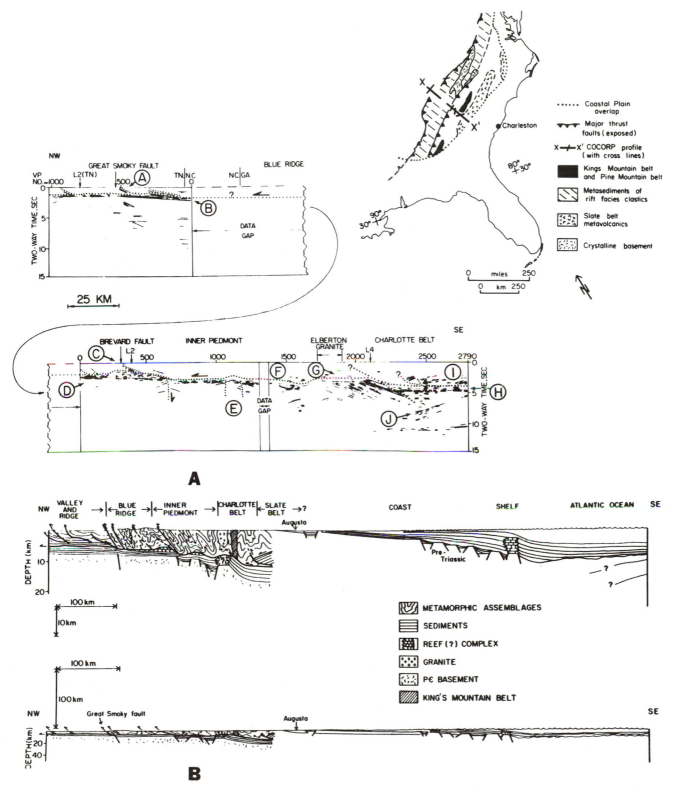

Fig. 7-38. (A) Sketch-tracing of the COCORP seismic profile of the Southern Appalachians; the section is drawn with a vertical exaggeration of 5×. (B) Geological interpretation of the profile and its relation to the geological structure of the continental margin of North America; this cross section is not vertically exaggerated. (From Cooke *et al.*, 1979, *Geology*, Figs. 3 and 6, pp. 564 and 566.)

beneath the Charlotte belt (at point G in Fig. 7-38A) are eastward-dipping reflectors which are considerably thicker than the horizontal reflectors to the west. These were interpreted as either the Paleozoic continental slope-and-rise sedimentary prism or a thrust-imbricated mass of sediments. If they are, indeed, a continental slope-and-rise prism, then the locus of the slope-break between the western horizontal reflectors and the eastward-dipping reflectors, near the location of the Kings Mountain belt, may be the edge of the Paleozoic North American continent and the eastward dip would represent the original slope of the continental margin. This interpretation was apparently preferred by Cooke *et al.* and is illustrated on their geological cross section (Fig. 7-38B).

The significance of these suggestions is profound and should not be obscured by the forest of detail: the claim is that the Blue Ridge, Inner Piedmont, Kings Mountain belt, Charlotte belt, and Carolina slate belt (see Fig. 5-13) *all are allochthonous,* pushed westward over a deep master-décollement! If this hypothesis is correct, it will force the rethinking of much Appalachian historical geology and, indeed, may greatly affect our view of the nature of orogens in general.

C.2. Plate Tectonics and the Allegheny Orogeny

Figure 7-39 is a plate-tectonics model of the Allegheny orogeny modified from Cooke *et al.* (1979). Gondwanaland is shown as a passive rider on the plate being consumed beneath eastern North America. The existence of a west-dipping subduction zone to the east of the presently exposed Piedmont is suggested by the continuation of folding and metamorphism, as well as the intrusion of numerous plutons, in the Carolina and Eastern slate belts and the associated Raleigh and Kiokee belts (see Fig. 5-13). The late Paleozoic plutons of the eastern Piedmont are of interest for two reasons: first, their presence is evidence that igneous processes (presumably related to subduction) were continuous from Early Mississippian time (after the Acadian orogeny) into Permian time. Second, their location near the Coastal Plain overlap hints that more Alleghenian intrusives may lie hidden beneath the cover of younger sediments to the east and, therefore, that the North America–Gondwanaland suture may also be buried beneath the Coastal Plain (as suggested by Hatcher, 1978). Several good reviews of the Alleghenian orogeny in the Piedmont have recently been published (e.g. Secor *et al.,* 1986; Dallmeyer *et al.,* 1986; and Secor, Smoke, and Dallmeyer, 1986) to which the reader is directed for further information.

At some time, the advancing edge of Gondwanaland collided with North America, resulting in the Allegheny orogeny. The exact timing of the collision is difficult to fix.

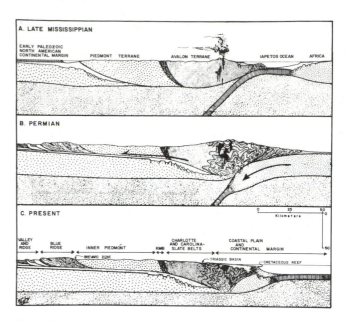

Fig. 7-39. Plate-tectonics model of the Allegheny orogeny. (A) Tectonic elements of the eastern North American continental crust after the Acadian but prior to the Allegheny orogeny; (B) collision of North America and Gondwanaland develops an intensely deformed and metamorphosed orogen and shoves a crystalline thrust-sheet backwards over the early and middle Paleozoic foreland basin, which is concomitantly affected by thin-skinned deformation; (C) configuration of tectonic elements of modern eastern North America. (Modified considerably from Cooke *et al.,* 1979.)

Evidence suggests that the plates began to interact as early as the Early Pennsylvanian because it was then that major detrital influx was renewed in the Appalachian foreland basin, signaling strong uplift to the east and south. The rate of sediment supply from uplifted Alleghenian terranes increased throughout the Pennsylvanian and Early Permian, after which time the stratigraphic record ends. After deposition of the Dunkard Group, the area was deformed by the faulting and folding discussed earlier in this section. Folding, faulting, and retrograde metamorphism of Alleghenian age also occurred (Kish *et al.,* 1978). We can put an upper age limit on Alleghenian compressive tectonics by noting that large-scale rifting occurred all along the Appalachian orogen during the Late Triassic (see Section E on eastern Triassic basins) indicating that tension rather than compression characterized the Appalachian region at that time.

Perhaps the most profound effect of the Allegheny continental collision was the thrusting of the entire crystalline Appalachians, which amounts to a considerable portion of the upper continental crust, back more than 260 km over relatively unmetamorphosed strata. A mechanism for emplacement of large, crystalline thrust sheets was proposed by Armstrong and Dick (1974); they suggested that, in a region of steep geothermal gradients such as a rear-arc area where heat-flow rates are high, compressive stress could displace the cooler, rigid upper level of the crust, to depths of 5–10 km, over the thermally softened base. At first, displacement would be by plastic deformation but with continued compressive stress, movement would become localized along a single fault zone, the master décollement, which would parallel the isotherms (i.e., equal-temperature contours) within the crust. The relevance of this model to the Allegheny orogeny can be appreciated by looking again at Fig. 7-39. The Piedmont area prior to continental collision was located above a westward-dipping subduction zone and, thus, was in an area of high heat flow. When the collision occurred, the tremendous compressive stress that resulted dislocated the rigid, upper crustal level from the more plastic base and shoved it backwards away from the line of collision. Armstrong and Dick pointed out that fluids released from the overridden rocks could have acted to partially reduce friction on the thrust plane after the crystalline sheet had been pushed up and away from the plastic base. They also suggested that these fluids would have resulted in retrograde metamorphism in the overthrust sheet, a well-known feature of the Allegheny orogeny (Tull, 1978). As thrusting proceeded, fault displacement occurred *within* the allochthonous sheet, primarily along old zones of weakness such as the Brevard zone.

D. Events on the Ouachita–Marathon Continental Margin

Just as in the Appalachian area, events during the Pennsylvanian Period in the Ouachita area approached a tectonic climax. But the tectonic events which affected the Southern continental margin were considerably different from those in the east as well as (apparently) somewhat less complicated. Reasons for these differences probably lie in the previous history of the two areas, going back to the continental rifting which initiated the Paleozoic histories of both areas. Unlike the Appalachian continental margin, rifting along the Southern margin apparently did not leave offshore microcontinental blocks which could later be shoved against the continent. Thus, the Ouachita–Marathon orogen is relatively narrow whereas the Appalachian orogen with its various belts and provinces spans a width of more than 400 km (even more if its extent beneath the Atlantic Coastal Plain is considered). Furthermore, subduction was never initiated beneath the Southern continental margin as it had been on the Appalachian margin; therefore, the Ouachita–Marathon continental margin remained passive for almost its entire history and its sedimentary record is, consequently, more straightforward than the Appalachian stratigraphic record.

In this section, we will describe early Pennsylvanian sediments of the Southern continental margin and will show

that conditions were little changed following the period boundary. We will then show that a major tectonic event, the Ouachita orogeny, occurred shortly thereafter and that it caused not only much of the structural deformation seen in the Ouachita–Marathon region (and Anadarko Basin region) but also resulted in a large detrital wedge which spread westward from the orogen across the foreland basins of Texas and Oklahoma. Figure 7-40 is a map showing the regional tectonic setting of Absarokan sedimentation in the Ouachita–Marathon and Southern Oklahoma aulacogen areas as well as related cratonic basins and uplifts.

D.1. The Southern Continental Margin before the Ouachita Orogeny

D.1.a. Continental Shelves: The Arbuckle Facies

Lower Pennsylvanian (Morrowan) strata of the continental shelves bordering the Southern ocean reveal deposition under tectonically stable conditions. In the Arkoma Basin area, for example, widespread, open-marine, carbonate deposition prevailed in the seas which returned following post-Kaskaskia emergence, transgressing an erosional surface with at least 24 m of relief cut into the Upper Mississippian Pitkin Limestone. These open-marine deposits are termed the Hale Formation in Arkansas and the Sausbee Formation in Oklahoma. The lower parts of these units consist primarily of interbedded oolitic–bioclastic grainstones and quartz arenites whose crossbedding and grain-size parameters indicate deposition on a turbulent, open-marine shelf fully exposed to wave and current activity (Sutherland and Henry, 1977). Later in Sausbee time, a large, complex carbonate-mudbank developed over much of the southwestern shelf (represented now by the Brewer Bend Limestone Member of the Sausbee) while detrital sediments accumulated in deltaic and shallow marine environments to the northeast, on the flanks of the Ozark Uplift from which the detritus was derived.

In central Texas, a broad carbonate platform, called the Caddo shelf, developed during early Morrowan time on the Concho platform and continued to exist until late Atokan time (see Fig. 7-41). Limestone strata of the Caddo shelf are primarily located today in the subsurface except along the north flank of the Llano uplift where the Marble Falls Limestone is exposed. Marble Falls carbonates were deposited on the eastern side of the Caddo shelf and are characterized by a complex facies-assemblage of cherty, shallow-water limestone types (mudstones, wackestones, packstones, grainstones) and coral–algal bioherms (Kier *et al.*, 1979). Marble Falls carbonates extend in the subsurface to the western side of the Fort Worth Basin (which is entirely a subsurface feature) but are not present in the eastern Fort Worth Basin. There, Morrowan strata are primarily

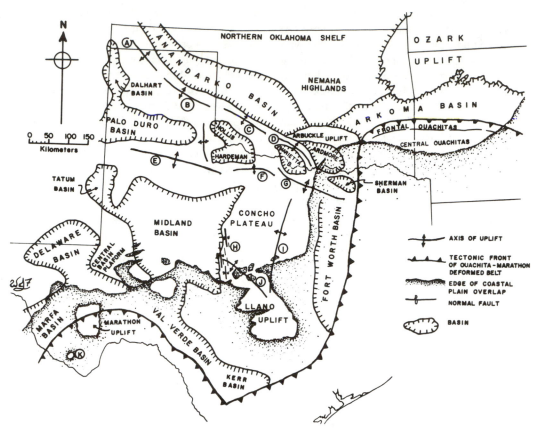

Fig. 7-40. Generalized tectonic map of the Ouachita–Marathon continental margin and related areas. Symbols are as follows: A, Cimmaron Uplift; B, Amarillo Uplift; C, Wichita Uplift; D, Criner Uplift; E, Matador Uplift; F, Red River Arch; G, Muenster Arch; H, Fort Chadbourne fault system; I, Bend Arch; J, Concho Arch; K, Solitario region.

thin-bedded, fissile, spiculiferous, black shale called the Smithwick Formation. These sedimentary rocks record relatively deep, starved basin on the outer continental shelf. This pattern of shallow shelf carbonates deposited in the west penecontemporaneously with deeper, outer-shelf muds in the east persisted into Atokan time.

In the Southern Oklahoma aulacogen, the Pennsylvanian stratigraphic record reveals a complex facies-assemblage of detrital and carbonate sediments whose deposition was related to the uplift of large fault-blocks along the southern margin of the aulacogen. For example, in the panhandle region of Texas, long and relatively thin blocks were uplifted along high-angle reverse faults resulting in the Amarillo and Wichita uplifts. Sediments derived from these uplifts were carried northward into the subsiding Anadarko Basin where they graded into clayey mudstones being deposited in the relatively deep, quiet waters of the basin center. Lesser amounts of detrital sediment were supplied to the Anadarko Basin from the north, probably derived from the low, emergent Nemaha highlands. The resulting strata in the basin, sandy on the sides and clayey in the center, is referred to as the Springer Formation. In the southern part

of the Ardmore Basin, adjacent to the Criner Uplift (which is an extension of the Wichita Uplift), a conglomerate up to 9 m thick and containing cobbles of Mississippian and Devonian lithologies is overlain by various carbonate lithologies and northward-thinning wedges of similar, coarse conglomerate. The whole unit is termed the Jolliff Limestone.

The overall picture, then, is one of rising positive areas, the Amarillo, Wichita, and Criner uplifts, which supplied abundant detrital sediments to the southern side of the Anadarko–Ardmore Basin; these sediments formed miniature detrital wedges, with conglomerates fringing the uplifts and grading basinward into sands and then to muds in the basin center. Uplift in some areas may have been extreme: some of the detrital sands are arkosic, suggesting that granitic basement had been exposed to erosion; Tomlinson and McBee (1959) indicated that nearly 1.6 km of uplift occurred on the Criner uplift. Fluctuations in the intensity of uplift and/or fluctuations of sea level caused variations in the rate of detrital-sediment influx with the result that sand and shales alternate with carbonates, especially in southeastern parts of the aulacogen where it adjoins the Arkoma and Fort Worth continental shelves. Morrowan uplifts in the

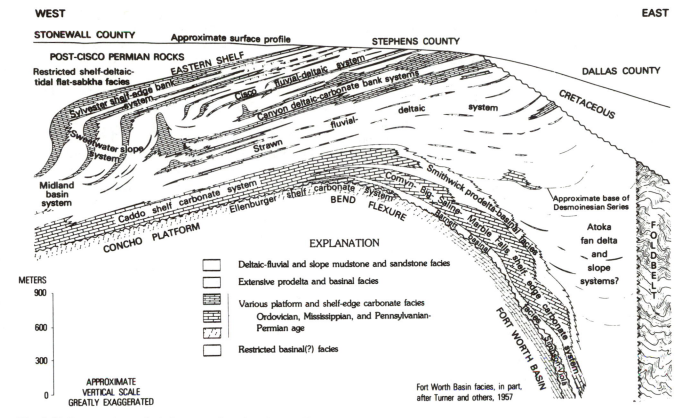

Fig. 7-41. Interpretative geological cross section of northcentral Texas showing the relation of strata in the Fort Worth Basin to those of the Concho Platform. (From Kier *et al.*, 1979.)

Southern Oklahoma aulacogen have been termed the *Wichita orogeny* (Fay *et al.*, 1979) but they are similar in kind to uplifts which periodically affected the aulacogen throughout the Pennsylvanian.

D.1.b.   Outer Shelf and Slope: The Frontal Ouachitas

Only in the Frontal Ouachitas, just south of the Choc-taw fault, do strata outcrop which appear to represent the outer continental shelf and upper continental slope. These strata lie geographically between shallow-water carbonates of the Arkoma Basin to the north and deep-water flysch of the central Ouachitas to the south. Further, their sedimentological properties are transitional between those of shelf and deep-water sediments and some of their lithologies and biostratigraphic zones can be traced either northward to the shelf or southward to the abyss, demonstrating their contemporaneity.

The lower Morrowan unit in the Frontal Ouachitas is the Springer Formation (which is correlative with the Springer of the Anadarko–Ardmore area but probably not indicative of the same environment). The Springer consists

primarily of dark-gray silty shale. In northern areas of the Frontal Ouachitas, the Springer is overlain by the Wapanucka Limestone, a sequence of interbedded carbonate mudstones, spiculiferous shales, and rare oolitic grainstone. The Wapanucka seems to have formed in a shallower-water environment than did the Springer (Morris, 1974a) and so may represent the outward growth of the carbonate platform during later Morrowan time.

Farther to the south, the Springer is overlain by the Chickachoc Chert, predominantly a dark-gray shale with layers of dark-gray *spiculite*. The Chickachoc, which is up to 215 m thick, is considered to be a deeper-water facies of the Wapanucka and probably represents, like the lower Smithwick Formation of the eastern Fort Worth Basin with which it is correlated, the outer shelf–upper slope environment.

D.1.c.   Deep-Water Environments: The Ouachita Facies

The final area of the Southern continental margin to be examined here is the continental rise and adjacent abyssal plain. Strata formed in these deep-water environments outcrop only in the Ouachita Mountains of Oklahoma and Arkansas and in the Marathon Uplift of southwestern Texas. Morrowan units of these areas are the Jackfork Group and Johns Valley Shale (Ouachitas) and the Tesnus Formation and Dimple Limestone (Marathon area). We will now consider each of these units.

i. Jackfork Group. The Jackfork Group is a sandy flysch composed of up to 1980 m of monotonous, interbedded, graded, quartzose sandstones and shales. It lies conformably on the Mississippian Stanley Group and for a long time was also considered to be of Mississippian age because of fossils. But Gordon and Stone (1973) demonstrated that these Mississippian fossils were reworked from older units on the adjacent continental shelf and subsequently transported by turbidity currents and redeposited within the Jackfork. Paleocurrent indicators, such as flute casts, reveal an almost-uniformly westward current regime during Jackfork time. Thus, the turbidity currents which transported and deposited Jackfork flysch traveled westerly, parallel to the axis of the Ouachita trough.

ii. Johns Valley Shale. Most of the strata of the Johns Valley Shale are composed of rather unexciting gray shale and thin, turbidite sand layers. The Johns Valley is geographically restricted, outcropping only in western Arkansas and southeastern Oklahoma. Nevertheless, it is one of the best-known stratigraphic units of the Ouachita Mountains. Its fame derives from the fact that it contains a thick zone of spectacular, chaotic megabreccia in which a great variety of *exotic* (i.e., extraformational) blocks occur in an extremely contorted shaley matrix. The exotic blocks vary lithologically from fossiliferous limestones and sandstones to massive chert and black shale; they range in size from granules and pebbles up to enormous blocks many tens of meters long! Shideler (1970) used fossils within the blocks to demonstrate an age range from the Cambrian to early Morrowan. The exotic blocks are believed to have been derived by large-scale slumping on the edge of the continental shelf, perhaps set off by violent earthquakes, and carried down the continental slope in giant debris flows. Such large masses of debris-flow-emplaced, chaotic masses containing exotic blocks are called *olistostromes*.

iii. Tesnus Formation. The Tesnus Formation of the Marathon Uplift conformably overlies the Caballos Novaculite (see Chapter 6) and is of Late Mississippian and Early Morrowan age. It consists of interbedded dark shales and fine sandstones and ranges in thickness from 100 m in western exposures to almost 2000 m in the southeast. At the base of the Tesnus is a relatively thin sheet of black, fissile shale. This black shale is of nearly uniform thickness over the whole area and comprises most of the Tesnus Formation

in the west. In the east, however, almost 1600 m of flysch reveals westward-directed currents and the sandstone/shale ratio increases toward the east (Kier *et al.*, 1979). These data all suggest the presence of a rising detrital source area to the east from which large debris flows and turbidity currents originated, moving down a westward-directed paleoslope and out into a deep, starved basin.

iv. Dimple Limestone. The Dimple Limestone marks a rather striking interruption of detrital flysch sedimentation in the Marathon area, occurring as it does between the Tesnus and the overlying Haymond Formation, which is a detrital flysch unit similar to the Tesnus. The Dimple overlies the Tesnus conformably and its strata have been divided by Thomson and Thomasson (1969) into shelf, slope, and basin facies. The shelf facies is composed of crossbedded, cherty oolitic and bioclastic grainstones and limestone-pebble conglomerates. Fossils of the shelf facies include most of the common, benthic, invertebrate assemblage of later Paleozoic seas, including crinoids, bryozoans, echinoids, and brachiopods. The slope facies is marked by chert and limestone-pebble conglomerates occurring in scoured depressions associated with graded calcarenites, probably proximal turbidites, and with coarse, ungraded deposits, some with "floating" (i.e., matrix-supported) pebbles, which may represent debris-flow deposits. Slump structures and contorted bedding are common in slope carbonates of the Dimple and a 10-m-thick olistostrome is present in eastern parts of the slope facies (Kier *et al.*, 1979). The basin facies is composed of rhythmically interbedded calc-arenites and carbonate mudstones which show graded bedding, sole marks, and other features characteristic of turbidity-current deposits. These basinal limestones are considered to represent distal carbonate turbidites derived from shelf areas to the east, as indicated by directional studies of flute casts and paleocurrent indicators.

The problem of the Dimple's origin is twofold: (1) what caused the temporary cessation of detrital flysch sedimentation and (2) what caused the development of a shallow-water bank where, only shortly before, there had been a deep, starved basin? Thomson and Thomasson (1969) suggested that thrusting in the Ouachita–Marathon belt to the east penecontemporaneous with sedimentation uplifted several long, ridgelike areas of the seafloor which then acted as sediment dams to block eastern-derived detrital sediment from the Marathon region. The surfaces of these upthrust ridges would probably have been essentially free of detrital sediment and, therefore, would have been reasonable sites for the development of carbonate banks.

v. Sediment Sources and Paleogeography. In the preceding paragraphs, we have shown that Ouachita–Marathon basinal facies consisted of a variety of flysch and debris-flow deposits. The source of these sediments is a very important question because of the constraints the answer would place on the paleogeography of the Southern continental margin and on acceptable models of tectonic development. Unfortunately, it is not an easily answered question. Three different areas have been suggested as the source of the detrital flysch: (1) the North American craton, especially the Illinois Basin which may have been connected to the Southern margin; (2) the Appalachian orogen, especially where it intersects the Southern margin in Alabama; and (3) a tectonically active, offshore source to the south and east, possibly a magmatic arc area. We now consider each of these three possibilities.

The connection between the Illinois Basin and the Ouachita area has been suggested by several geologists (e.g., Swann, 1964) because of their geographic proximity. They are thought to have been connected by a shallow seaway (or "sluiceway," Potter and Pryor, 1961) between the Ozark uplift and the Cincinnati Arch which we have termed the Batesville channel following Glick's (1979) usage. Cratonic-basin sands, already reworked several times, were eroded during post-Kaskaskia emergence and were carried southward to be dumped off the Southern margin through submarine canyons somewhere east of Little Rock, Arkansas, where the Gulf Coastal Plain overlaps the Ouachitas. Sea-level control of this process was stressed by Glick (1975) who indicated that Jackfork sands may have been introduced into the Ouachita trough primarily during times of falling sea level. Of course, the Pennsylvanian was a time of greatly increased detrital influx to the craton from the Appalachian area and a considerable quantity of Appalachian-derived sediment may have been detoured into the Ouachita trough via the Illinois Basin "bypass."

An Appalachian source for Ouachita flysch has been postulated by Graham *et al.* (1975) based on analogy with the present-day Himalaya Mountains–Bengal submarine fan system in the northeastern Indian Ocean. They suggested that Appalachian detritus was carried southward into the area of the Black Warrior Basin and from there into the Ouachita trough, diverted westward by the magmatic arc south of the Alabama continental margin (see Fig. 7-42). This is a very attractive model, illustrating well the use of actualistic reasoning to solve geological questions. Unfortunately, it is probably wrong. Paleocurrent indicators in the Black Warrior Basin show a generally northerly to northwesterly sediment-dispersal, rather than southerly to southwesterly as indicated by Graham *et al.* Furthermore, facies patterns in the Black Warrior Basin suggest that deltas entered the central Alabama region from the south and east (Hobday, 1974; Thomas, 1979).

The third possible source area is an offshore, tectonically active terrane to the south and east. The idea of a mountainous sourceland to the south of the Ouachita and

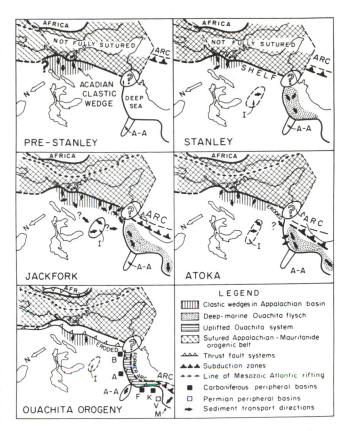

Fig. 7-42. Interpretative sketches of southeastern North America showing inferred Carboniferous evolution of Ouachita system. Features include: A–A, Anadarko–Ardmore Basin; A, Arkoma Basin; B, Black Warrior Basin; F, Fort Worth Basin; I, Illinois Basin; K, Kerr Basin; M, Marathon region; V, Val Verde Basin. (From Graham *et al.*, 1975, *Geological Society of America Bulletin*, Fig. 5, p. 280.)

Marathon regions is not new. In older literature (e.g., see Dumple, 1920), this source was called *Llanoria* and was considered to have been a geanticline, or continental borderland south of the Ouachita geosyncline. Today, the "Llanorian" source is believed to have been a magmatic-arc terrane, possibly associated with the northern margin of Gondwanaland. The quartzose character of the sediment may have been due to erosion of older sedimentary rocks uplifted within the arc orogen. Such a model would explain the small but ubiquitous sand-fraction consisting of feldspar, volcanic detritus, and metamorphic rock fragments. Another argument for a southern source is the fact that Ouachita turbidites decrease rapidly in grain size from proximal areas in Arkansas to distal areas in Oklahoma. And yet, flysch in the Marathon region is as coarse or, in specific instances, even coarser than Oklahoma flysch. If the majority of the sand came from either the Batesville channel or the Black Warrior Basin, then Marathon deposits should be much finer than Oklahoma deposits. One could rebut this argument by proposing that the flysch submarine-fan grew westward with time (as it certainly did) so that coarse Marathon sediments would be younger than similar sediments of the Ouachitas. However, as we have shown, Tesnus flysch and Jackfork flysch (both of very similar aspect) accumulated penecontemporaneously.

In the final analysis, the question of flysch provenance will probably not be answered by a single source. Indeed, a cursory look at modern marginal-ocean basins (e.g., the Gulf of Mexico) shows that many sediment sources of very different character interact to yield a single, complex deposit. The same must have been true of the Ouachita–Marathon marginal ocean. For example, Dennison and Wheeler (1975) proposed that both Appalachian and Ouachita (''Llanorian'') sources supplied sediments to the Black Warrior Basin. By extension of this idea, we suspect that cratonic sources, Appalachian sources, introduced through both the Illinois Basin and the Black Warrior Basin, and ''Llanorian'' sources may all have been significant.

D.1.d. Change in Sedimentary Dynamics during Atoka Time

The conditions described above continued for a brief time into the Atokan Epoch but, shortly thereafter, there occurred a major change in the tectonic and sedimentological character of the entire Southern continental margin. This change can best be seen in the Atoka Formation of Arkansas and Oklahoma.

The Atoka Formation (type unit of the Atokan Epoch) is a large and complex unit. In the Arkoma Basin, the Atoka is composed of a very thick sequence of deltaic and shallow marine detrital sediments along with commercially important coal beds. Paleocurrent data and evidence from facies patterns show that Atokan deltas came from the Ozark region and grew southward all the way across the shelf to the edge of the Ouachita trough. Arkoma Basin subsidence was the result of large-scale normal faulting on a series of down-to-the-south growth faults.

In the deep Ouachita trough, the Atoka is almost 6100 m thick and is composed of two distinctly different lithological assemblages. The lower part consists of monotonously interbedded, thin layers of sandstone and shale, a typical flysch similar in many ways to the Jackfork group. Paleocurrent studies of lower Atoka turbidites in the Ouachita trough reveal a westward-oriented current system, also similar to previous deposits. But the upper part consists of cyclic deposits of thick, channel-form sandstones, shallow-marine shales, and thin coal beds. These deposits are of deltaic origin and, with correlative Atoka strata of the Arkoma Basin, originally formed an extensive sheet of deltaic sediments from the Ozark Uplift to southernmost exposures of Ouachita structures. Paleocurrent studies of these sediments in the Ouachita trough demonstrate that the deltas

grew southward, with sediments derived from the Ozark area and other, nearby cratonic uplifts such as the Nemaha highlands.

From the above, we can reconstruct the Atokan history of the Arkoma Basin and Ouachita trough. During early Atokan time, southward-prograding deltas and associated marine environments dominated the continental shelf while a large submarine-fan complex grew westward within the Ouachita trough. By the middle Atokan, however, flysch sediment had virtually filled the Ouachita trough, implying that the rate of sediment influx outstripped the rate of subsidence. Into the now-shallow trough prograded large deltas from the shelf carrying great quantities of sediment. Sedimentation in the Arkoma area accompanied and largely kept pace with major normal faulting of the continental shelf. It is significant that the main sources of detrital sediments to feed these deltas lay to the north, east, and northwest but *not* to the south. Deltaic sedimentation continued across the epoch boundary in the Ouachita and Arkoma areas where deposits of the Desmoinesian Krebs Group record a southward-growing delta system.

In the Fort Worth Basin, an eastward-thickening wedge of coarse detrital sediment referred to as the ''Atoka facies'' (Kier *et al.*, 1979) first appears in eastern portions of the basin at this time. These sediments consist of sandstones, mudstone, and coal beds; they appear to represent deltas which grew westward across the Fort Worth Basin. Detrital grain size increases eastward, as does the relative amount of sand, indicating that sediments came from a generally eastern source, *not* a cratonic source as we described for the Oklahoma Atoka. ''Atoka facies'' sediments grade westward into black prodelta muds of the Smithwick Formation (see above) which, in turn, grades westward into the Marble Falls Limestone. As the Atoka Epoch progressed, these facies migrated westward (see Fig. 7-41). In the northern Fort Worth Basin and in northern areas of the Concho platform, Atokan strata record a system of southwestward-prograding deltas which carried sediments from the rising Red River and Muenster arches.

In the Marathon region, Dimple deposition continued into the early Atokan but was soon superseded by deposition of the Haymond Formation which represents a renewal of eastern-derived detrital flysch sedimentation. The Haymond, like the Tesnus of early Morrowan age, consists of interbedded thin layers of sandstone and shale. In eastern exposures, a 330-m-thick olistostrome occurs in the upper Haymond and contains exotic blocks up to 40 m long.

From the above discussion, it can be seen that important changes had occurred during the Atoka Epoch: (1) the Ouachita trough ceased to be a deep flysch-basin with an eastern source and had become a shallow marine basin dominated by deltas growing from the north; (2) deltaic deposits first appeared in the Fort Worth Basin, recording a westward-prograding shore associated with a source terrane within the Ouachita–Marathon belt; and (3) detrital flysch sedimentation returned to the Marathon area and continued into early Desmoinesian time.

In addition to these sedimentological events, several important structural events occurred, including major activity on the large growth faults of the Arkoma Basin and the development of a series of horsts and grabens (called the Fort Chadbourne system) near the western edge of the Concho platform (see Fig. 7-40). Strong uplift continued in the Southern Oklahoma aulacogen with major activity along the Criner, Wichita, Muenster, and Red River uplifts. Thus, while the main part of the Ouachita orogeny occurred during middle Desmoinesian time, tectonic events leading up to it, beginning as far back as the early Morrowan, were occurring throughout the Atokan.

D.2. The Ouachita Orogeny

The Ouachita orogeny was the major tectonic event to affect the southern continental margin during the Paleozoic and, indeed, the *only* compressive tectonic event there since the Precambrian. One might expect, then, that knowledge of this event would be good. Unhappily, this is not the case. As discussed previously, sediments deposited on the Paleozoic Southern margin and deformed in the Ouachita orogeny may be observed today only in a few, relatively small areas where they protrude from under the covering strata of the Gulf Coastal Plain. What is known suggests that the Ouachita orogen is much narrower than the Appalachian orogen, principally because there are no accreted tectonic blocks such as the displaced terranes which comprise the Southeast's Piedmont province.

There is an apparent lack of significant igneous activity associated with the Ouachita–Marathon belt. Neither surface nor subsurface Ouachita-related plutons have been reported. Metamorphism, too, is limited, being mainly of the penetrative type (e.g., slaty cleavage) in exposed areas of strong deformation.

The observation that igneous and metamorphic activity was of only minor importance may lead the reader to consider that the Ouachita orogeny itself was only a minor event. This, however, would be incorrect. In exposed stretches of the Ouachita–Marathon belt, structural deformation is intense. For example, in Arkansas and Oklahoma, the Frontal Ouachitas are characterized by large, subhorizontal, lystric thrust faults which dip southerly (see Fig. 7-43). The Choctaw fault, which separates the Ouachitas from the Arkoma Basin, is the farthest northward exposed thrust but others are present in the subsurface farther north. In the Central and Southern Ouachitas, the thrusts themselves have been deformed, in some areas so strongly that they are completely overturned and dip northerly associated with inverted stratigraphic sequences. As implied in Fig. 7-43, these thrusts and fold structures are considered to have been the result of thin-skinned tectonics, the entire sedimentary mass supposedly having been shoved northward over a master décollement.

The timing of the Ouachita orogeny is difficult to specify but must be some time after deposition of the lower Desmoinesian Krebs Group because it, too, was affected by the faulting. On the other hand, uppermost Krebs detrital deposits (Thurman and Stuart formations) were apparently derived from a southern source (rather than from a northern source as lower Krebs sediments had been); thus, they may imply the initiation of uplift in the Ouachita area. Further, as we will discuss later in this section, the rate of detrital influx in foreland basins adjacent to the Ouachita–Marathon belt increased dramatically during the middle Desmoinesian. Thus, we infer that uplift associated with the Ouachita orogeny and probably the main stage of deformation (at least in the Ouachita area) began during the middle Desmoinesian. Several authors (e.g., Glick, 1975) have pointed to the presence of middle and upper Desmoinesian strata, apparently deposited in a marine environment and overlying deformed strata of the Ouachita facies, in deep wells in southern Arkansas and northern Louisiana, arguing that these strata preclude a Desmoinesian age for Ouachita uplift. We will return to this argument later.

In the Southern Oklahoma aulacogen, tectonic events, primarily uplift of fault blocks along large, high-angle reverse faults, continued throughout the Pennsylvanian. As we have discussed, the Wichita orogeny of Morrowan–Atokan age resulted in as much as 1.6 km of uplift in the

Fig. 7-43. Structure cross section of Ouachita orogen in Oklahoma. (From Wickham *et al.*, 1976, *Geology*, Fig. 1, p. 173.)

Criner area and lesser, but still significant, uplift of the other positive tectonic elements of the region. Uplift of most of these areas continued, somewhat fitfully, through Desmoinesian and Missourian time reaching a peak in Virgilian time with the *Arbuckle orogeny*. During the Arbuckle event, strong compression affected the Ardmore, Marietta, and Anadarko basins and spectacular uplift occurred in the Arbuckle Mountains area.

One of the earliest attempts to explain the Ouachita orogeny by appeal to plate tectonics was proposed by Keller and Cebull (1973); Fig. 7-44A presents their model of Ouachita history. They suggested that the Southern rifted-margin prism had been transformed into a convergent boundary during the Late Ordovician, coinciding with subduction-induced collapse of the Appalachian rifted-margin prism (i.e., the Taconic orogeny). Keller and Cebull considered that the southern tectonic land (i.e., the "orogenic welt") rose actively from Early Mississippian through Permian time, supplying sediments to fill the Ouachita trough and, later, as the continental shelf foundered, the Arkoma Basin. Continued growth and horizontal expansion of the orogen's mobile core caused thrusting and folding of the Ouachita area during later Pennsylvanian time. There are several serious problems with Keller and Cebull's scenario. First, the great thicknesses of sediment which filled the Ouachita trough and Arkoma Basin were derived primarily from the east and north as indicated by paleocurrent studies, rather than from the south. Eastern-derived flysch of the Stanley and Jackfork Groups may well have been derived from an arc terrane (although the high percentage of quartz would suggest an Andean-type arc rather than a western-Pacific-type as shown in the model) but the large volumes of Atoka and Krebs sediments are clearly of cratonic derivation and, thus, do not fit Keller and Cebull's picture. Second, no Paleozoic igneous activity or high-temperature metamorphism is known from the Ouachita region, although both should be present if the area had overlain a subducting plate. Third, the deep wells of southern Arkansas and northern Louisiana mentioned previously, which should have encountered igneous and metamorphic rocks of Keller and Cebull's mobile core, encountered instead only unmetamorphosed Pennsylvanian sediments overlying deformed strata of the Ouachita facies.

Several collisional models have been proposed for the Ouachita orogeny. Morris (1974b) proposed that a north-facing volcanic island-arc collided with the Southern continental margin but, as above, the highly quartzose character of Pennsylvanian detrital sediments in the Ouachita trough seems to argue against the involvement of a simple island arc. A better model, discussed by Graham *et al.* (1975) and Wickham *et al.* (1976), supposes a collision involving a continental landmass, possibly South America. In this model (see Fig. 7-44B), the southern margin of North America is considered to have been passive from its inception until the Pennsylvanian collision, during which a north-facing, Andean-type magmatic arc was thrust against southern North America, driving thrust sheets northward toward the interior of the craton. Anisostatic subsidence of the leading edge of southern North America, as it was pulled down into the trench, is considered in this model to have been the cause of continental-shelf foundering along the normal growth-faults mentioned earlier. The same mechanism may have been responsible for faulting and earthquakes on the outer continental shelf which resulted in the large olistostromes of the Frontal Ouachitas. This model explains the lack of major igneous and metamorphic activity and it agrees well with the conclusions about flysch provenance discussed earlier. It does not, however, adequately explain the presence of marine, Pennsylvanian sediments overlying deformed Ouachita-facies strata in southern Arkansas and northern Louisiana. Further, this model presumes complete closure of the Gulf of Mexico but, as we will show in the next paragraph, some geologists now argue that Paleozoic oceanic crust underlies the northern Gulf coastal plain.

Comparison of Ouachita and Appalachian orogens suggests the presence of oceanic crust beneath Mesozoic–Cenozoic strata of the Gulf Coastal Plain. This interpretation is based on the behavior of seismic waves in the lower crust and upper mantle and has been championed by a team of geologists at Texas Tech University (e.g., Shurbet, 1964; Keller and Shurbet, 1975). They have argued convincingly that the Southern Ocean Basin was never fully closed at the end of the Paleozoic and that the oceanic crust beneath Coastal Plain strata is of late Paleozoic age, an unsubducted remnant of a formerly much-larger ocean-basin floor. Should this interpretation prove correct, the clear implication would be that continental collision such as occurred along the Appalachian margin probably did not affect the Ouachita–Marathon margin.

We have now argued ourselves rather neatly into a box. We have seemingly dismissed all three plate-tectonics mechanisms of orogeny: the Cordilleran model, the arc–continent collision model, and the continent–continent collision model. One possible way out of this box was suggested by Walper (1980) who proposed a model of incomplete continental collision (see Fig. 7-44C). In this model, as in the other collisional models discussed above, the southern continental margin is considered to have been passive until the Pennsylvanian. As the leading edge of the southern margin began to enter the trench at the boundary with the approaching Gondwanaland, anisostatic subsidence caused normal faulting of the continental shelf, affecting at first only the shelf edge and causing the slumping that produced the olistostromes such as in the Johns Valley Shale. As more of the continental margin was affected,

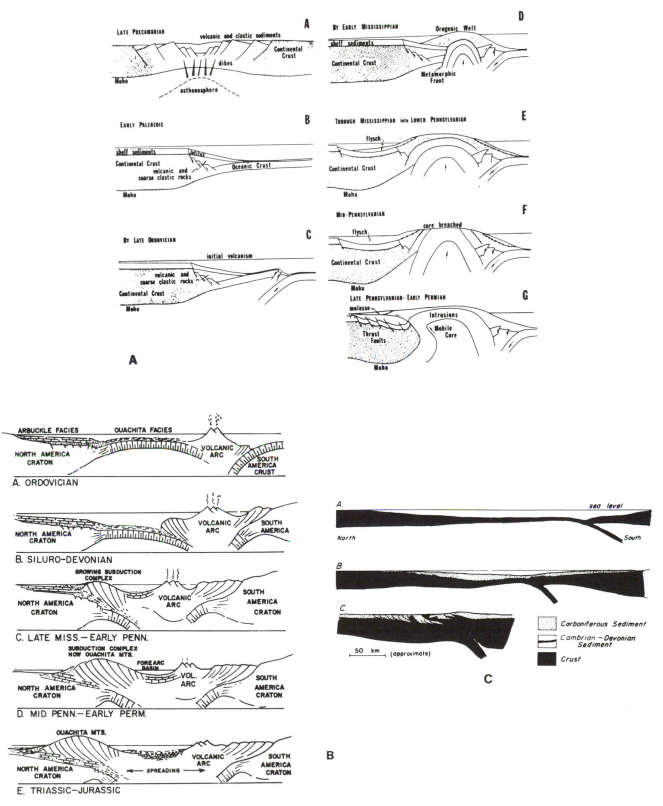

Fig. 7-44. Tectonic models of the Ouachita orogeny: (A) Keller and Cebull (1973, *Geological Society of America Bulletin*, Fig. 3, p. 1662); (B) Wickham *et al.* (1976, *Geology*, Fig. 3, p. 175); (C) Walper (1980). See text for discussion.

normal faulting affected progressively more of the shelf, resulting in rapid subsidence of the Arkoma Basin during Atokan time. In addition to these effects, as the thin edge of the continental margin actually began to be subducted beneath the Gondwanaland plate, first continental-rise and then continental-slope deposits were underthrust beneath the subduction prism. But, being thicker than abyssal plain deposits and relatively rigid, they were sheared across the bottom and shoved backwards rather than being intercalated into the subduction mélange. At length, the growing thrust mass, a sort of subduction prism, was pushed backwards over the continental shelf. At some point during this history, the subduction mass grew large enough to interrupt the dispersal of arc-derived sediments across the plate boundary and then large enough to act itself as a tectonic sourceland. Behind the subduction prism, however, the fore-arc basin persisted and shallow-marine sediments were deposited on top of the deformed thrust slices of Ouachita-facies sediment (in the same way that fore-arc basin sediments today overlie upturned and internally deformed subduction prism strata).

To this point in the story, there is little difference between Walper's model and the other collision models discussed above; the difference lies in what happens, or rather what does not happen, next. The reader will remember that South America was not an independent crustal element but, rather, was part of Gondwanaland. During Pennsylvanian time, the African margin of Gondwanaland was beginning to collide with eastern North America. By latest Pennsylvanian and Early Permian time, the collision had resulted in the Allegheny orogeny, and the Gondwanaland plate had nearly ground to a halt, a sialic, irresistible force opposed by a sialic, immovable object. Of course, the interaction of these two rigid continental plates in one area must have affected their interactions in other areas. Therefore, Walper suggested, the Gondwanaland plate was stopped *before* continental collision completely closed the Southern Ocean. For this reason, oceanic crust south of the Ouachita area was left unsubducted and fore-arc basin sediments were largely unaffected by tectonism. These fore-arc deposits remain today as the Pennsylvanian strata penetrated by the wells in southern Arkansas and northern Louisiana mentioned several times earlier in this discussion. The geometries of plate margins involved in the near collision are difficult to reconstruct accurately but the somewhat-earlier advent of arc-derived sediments in the Fort Worth Basin (the Atoka facies discussed earlier) and the more intense deformation of subsurface Ouachita-facies rocks in east Texas may reflect the close approach of crustal promontories while the later and less intense tectonic effects seen in the Ouachitas and in the Marathon region may have resulted from the presence of recesses on the continental margin (Thomas, 1977). This is an intriguing model.

A final aspect of southern continental margin tectonics requires attention. The vertically directed movements of the Southern Oklahoma aulacogen do not seem to fit easily into the picture developed above. Hoffman *et al.* (1974) used the Southern Oklahoma aulacogen as one of their type examples of typical aulacogens and stated that the compressive stage of aulacogen development (e.g., the Wichita and Arbuckle orogenies) was caused by major compressive events on the continental margin such as would accompany continental collision. This, however, does not seem to work for the Southern Oklahoma area. Vertical movements there began during latest Chesterian time and reached a peak in the Morrowan–Atokan (i.e., the Wichita orogeny), which was *before* any strong tectonic activity in the Ouchita–Marathon belt except for the thrusting hypothesized to explain cessation of the detrital influx in the Marathon area during Dimple time. It is possible that small-scale vertical adjustments of the crust occurred as the continent approached the trench, even when it was still some distance away; such adjustments would have been particularly strongly felt in the Southern Oklahoma aulacogen because aulacogens constitute fundamental fractures through the crust and are therefore free to move at the slightest disturbance. This seems inadequate, however, to explain the 1.6 km of movement described for the Criner area. Another problem is the fact that the time of greatest continental margin activity (i.e., the middle Desmoinesian) does not seem to have been a time of equally climactic activity in the aulacogen. The point that we are making is that the tectonics of the aulacogen seem, at the moment, to be somewhat difficult to reconcile with that of the continental margin. Until a single, composite model for the whole area is developed, uncertainty is probably the best stance to adopt toward tectonic models for either area.

D.3. Molasse Sedimentation in the Ouachita Foreland System

D.3.a. The Pattern Begins: Desmoinesian Strata

Regardless of problems with plate-tectonics interpretations of the Ouachita orogeny, there is no question that it was fully in progress by middle Desmoinesian time because great volumes of molasse sediment were deposited at that time in foreland basins all along the Ouachita–Marathon belt. We have grouped these foreland basins, including the western Arkoma, Fort Worth, Kerr, Val Verde, and Marathon basins (see Fig. 7-40), under the heading *Ouachita foreland system;* the final part of this section on the Ouachita–Marathon region will deal with its stratigraphy.

In the western Arkoma Basin and eastern parts of the Northern Oklahoma shelf, strata of the lower Desmoinesian Krebs Group, as mentioned previously, record a series of delta-dominated environments fed detrital sediments from the Ozark and Nemaha areas farther north. The *sedimentary strike* (i.e., the direction of elongation of sedimentary facies) of Krebs strata is approximately east–west, reflecting their northern deviation. Overlying the Krebs Group are sandstones, shales, coals, and limestones interbedded cyclically and revealing a generally north–south sedimentary strike (Frezon and Dixon, 1975). These strata comprise the Cabaniss Group, whose detrital deposits were derived from the east and south and, because they overlie folded strata of the Krebs whereas they are themselves not folded, testify to the onset of the Ouachita orogeny.

In the Southern Oklahoma aulacogen, the Amarillo–Wichita Uplift continued to rise, shedding coarse detrital materials both northward into the Anadarko Basin, where they are termed the Cherokee Group, and southward into the Hardeman and Palo Duro basins, where they are correlated with the Strawn Group of the Concho platform (see below). The Criner Uplift was apparently not active during this time and may have been submerged (Frezon and Dixon, 1975). The Arbuckle Uplift, on the other hand, was active and supplied coarse, conglomeratic sediments (termed the Deese Formation) to the Ardmore Basin. Coarse conglomerates at the base of the Strawn Group bordering the

Matador and Red River arches indicate that these uplifts, too, were active, at least during the early Desmoinesian.

In central Texas, Desmoinesian strata provide an excellent record of Ouachita molasse sedimentation. During early Desmoinesian time, or even latest Atokan, the rate of detrital influx from the east increased dramatically and deltaic–fluvial sediments prograded rapidly westward. After first filling the Fort Worth Basin, detrital sediments spread out onto the eastern Concho platform, overwhelming carbonate depositional systems and establishing a complex of elongate deltas, interdelta bays, and coal swamps (see Fig. 7-45). Here and there on the platform, carbonate deposition persisted, especially in western areas on the margin of the Midland Basin, which began to subside during the Desmoinesian. This great mass of detrital and minor carbonate strata is termed the Strawn Group (Kier *et al.*, 1979). Most Strawn detrital sediments were derived from the Ouachita orogen to the east except in northern and northeastern areas of the Concho platform where some detrital sediments were also derived from the Arbuckle and Wichita areas. Thus, Desmoinesian Strawn strata herald the initiation in central Texas of molasse sedimentation in a foreland-basin-type tectonic setting.

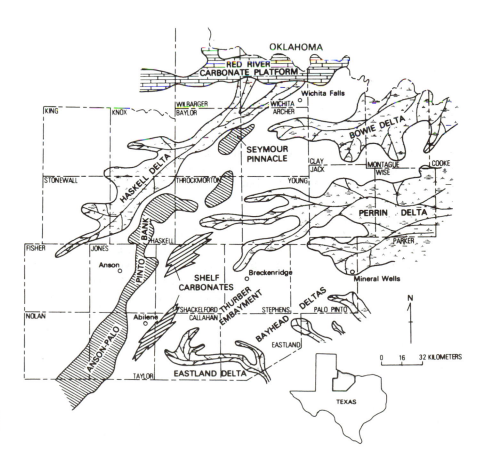

Fig. 7-45. Paleogeographic reconstruction of Strawn Group strata in northcentral Texas. (From Kier *et al.*, 1979.)

In the Marathon region during the early Desmoinesian, deposition of Haymond flysch continued without interruption. Relatively early in the Desmoinesian, a thick mass of shallow-water sediments composed of gravel, sand, mud, and some limy, bioclastic debris was deposited on an uplifted shelf area in the southeast. At the same time, deep-water conditions persisted in the northwest and are recorded in turbidites and pelagic shales. These deposits are called the Gaptank Formation. Whereas Gaptank detrital deposits represent the advent of molasse sediments in the Marathon area, a major difference of geological· circumstance exists between them and molasse deposits of the Fort Worth Basin: the Marathon region was actively undergoing deformation penecontemporaneously with sedimentation. Thus, the Gaptank contains clasts of Tesnus, Dimple, and Haymond lithologies derived from the eastern uplifted areas and, thus also, clasts of Gaptank lithology are found in sediments of overlying Wolfcampian strata, indicated that uplift continued following Gaptank time.

A quick review of the preceding paragraphs will convince the reader that the Desmoinesian was, indeed, a time of major activity on the Southern continental margin. Uplift, folding, and thrusting occurred in the Ouachita–Marathon belt and sediments derived there were spread northward, northwestward, and westward resulting in a complex deltaic–paralic molasse referred to in the various areas affected as the Cabaniss, Deese, Strawn, and Gaptank. The pattern of molasse sedimentation persisted through the rest of the Pennsylvanian and into the Permian. Strata formed during these later epochs will be described below.

D.3.b. The Pattern Persists: Missourian and Virgilian Strata

During the Missourian Epoch, a decrease in the rate of detrital-sediment influx affected much of the area. In the central Oklahoma region, Missourian strata are composed primarily of limestone and fine detrital materials except for areas adjacent to uplifts, e.g., the Amarillo, Wichita, and Arbuckle areas, from which sands and minor conglomerates were derived. The Ozark uplift also continued to be positive and to feed deltaic systems in eastern Oklahoma. In the Ardmore, Muenster, and Marietta basins, fine detritus accumulated under relatively deep-water conditions except for areas adjacent to uplifts. Strata of these regions are termed the Hoxbar Formation.

In central and northern Texas, the decrease in detrital influx is marked by the Canyon Group, a complex of deltaic deposits, shallow marine mudstones, and carbonates. Following Strawn time, carbonate deposition returned to much of the Concho platform, especially to the shallow areas left as delta lobes were abandoned. Sea-level fluctuations re-sulted in cyclic repetitions of deltaic and carbonate strata; the paleogeography of these cyclothems is shown in Fig. 7-46. Notice in this illustration that detritus was supplied not only from the principal, Ouachita source but also from the uplifts to the north, referred to as the Oklahoma Mountains. Detrital influence on the western Concho platform was relatively minor and a large carbonate shelf-edge bank developed there which persisted throughout Canyon time (see Fig. 7-46). The Midland Basin, just to the west, had been actively subsiding since Strawn time and contained, by the Missourian, a sea as deep as 300 m (Crosby and Mapel, 1975). The western edge of the Canyon carbonate bank was a reef complex which dropped off precipitously into the deep waters of the basin.

Tectonic activity increased again during Virgilian time both in the Ouachita area and in the Southern Oklahoma aulacogen (where intense folding, faulting, and uplift during the Virgilian is referred to as the Arbuckle orogeny). Renewed uplift in both areas resulted in another pulse of detrital sedimentation. In Oklahoma, cyclic strata of the Vamoosa, Lecompton, Ada, and Vanoss formations record oscillating, deltaic shorelines whose detrital sediments were derived from the Arbuckle and Wichita uplifts to the south and from the Ozark uplift to the east. In the Arbuckle area itself, coarse boulder conglomerates in the Vanoss reveal rapid uplift and deposition in alluvial fans and on fluvial plains which sloped northward toward the central Oklahoma sea.

In central Texas during the Virgilian, rejuvenation of the source area led to a resumption of detrital-dominated molasse sedimentation over much of the Concho platform. These sediments, referred to as the Cisco Group, are the last fluvial–deltaic deposits of the Concho platform. Cisco strata are cyclic in nature and record the familiar back-and-forth oscillation of deltaic and fluvial environments over shallow marine environments. Facies patterns tend generally to shift westward with time (neglecting the small-scale cyclic fluctuations) so that deltaic deposits in the lower Cisco pass upward into fluvial deposits near the top. While the fluvial–deltaic environments of the eastern Concho platform oscillated back and forth, a large carbonate bank prevailed on the western part of the platform at the edge of the Midland Basin. Coarse, bioclastic carbonates and reef-facies debris on the margin of the basin graded downslope into sand and silts of the slope facies and then into thin-bedded black shales and dark siliceous limestones of the basin floor. Outward growth of the entire carbonate-bank and slope facies during latest Pennsylvanian and Early Permian time created the Sweetwater slope system and the Sylvester shelf-edge carbonate bank system. The Pennsylvanian–Permian systemic boundary occurs within the Cisco Group and upper Cisco strata are of Wolfcampian age.

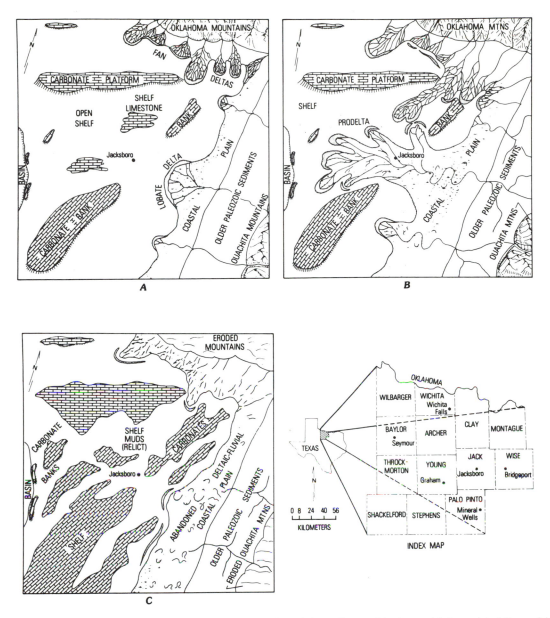

Fig. 7-46. Paleogeographic evolution of Canyon strata as recorded in a single cyclothem. Contrast this picture with the model of Strawn deltas shown in Fig. 7-45. (From Kier *et al.*, 1979.)

D.3.c. End of the Pattern: Permian Strata

It is often as difficult to specify when an orogeny ceased as when it began. In the Ouachita–Marathon area, for example, tectonic activity had apparently ceased in northern stretches of the orogen by latest Pennsylvanian time while deformation continued in the Marathon area into the Wolfcampian. In the following paragraphs, we will consider post-Cisco deposits of the Concho platform as an ex-

ample of tectonic end-game and then describe the final stages of orogeny in southwest Texas.

During post-Cisco Wolfcampian time, the rate of detrital influx began to decline on the Concho platform (which is referred to in Permian literature as the *Eastern shelf*), probably indicating that tectonic sourcelands were being eroded down. Fine detrital materials continued to be introduced from the east and fluvial–deltaic environments persisted but shallow marine environments became increasingly more extensive with time. Wolfcampian deposits of the Eastern shelf include Pueblo, Moran, and lower Putnam formations of the Wichita Group and are dominated by marine mudstones and limestones; thin, deltaic sandstones occur in eastern areas associated with minor coal beds. In northern parts of the shelf, coarser, arkosic sand and cherty gravel were deposited in subaerial and paralic environments fringing the uplifts of the Southern Oklahoma aulacogen (Oriel *et al.*, 1967).

In lower Leonardian strata of the Wichita Group are found cherty limestones, dolostones, and gray mudstones, thus indicating detrital-sediment supply had continued to wane and that subaerial environments occupied progressively less area as the Leonardian Epoch passed. During later Leonardian time, strata of the Clear Fork Group were deposited. They include a variety of limestones, dolostones, evaporites, and red beds. In their sedimentological characteristics and facies distribution, one can recognize very little tectonic-sourceland effect at all. From that time on, the Eastern shelf behaved sedimentologically as did the other shelf areas of the craton surrounding the west Texas Permian basins.

Lower Permian strata of the Marathon region comprise the Wolfcamp Group (type unit of the Wolfcampian Series) which is divided into the Neal Ranch and overlying Lenox Hills formations. Thrusting and folding continued penecontemporaneously with sedimentation. Neal Ranch strata overlie Gaptank strata disconformably but the Lenox Hills overlies both Neal Ranch and Gaptank strata with striking angular unconformity. Wolfcampian uplift of the Ouachita–Marathon belt in the southwest not only shed detritus into the Marathon area but also into the Val Verde Basin where a thick sequence of molasse deposits accumulated.

By the end of Wolfcampian time, uplift had apparently ceased. The eastern Marathon region became, in Leonardian time, a carbonate platform on which strata of the Skinner Ranch Limestone were deposited on top of deformed Wolfcampian and older rocks (Ross, 1978). To the west, Skinner Ranch shelf carbonates grade into deeper-water shales of the Cathedral Mountain Formation. As in foreland areas to the north, no further compressional tectonic effects or detrital-wedge-type sedimentation is noted in the Marathon region after Leonardian time. Further history of the region may be found in the discussion of the west Texas Permian basin.

E. Eastern Triassic–Jurassic Basins and the Gulf of Mexico: Onset of Continental Rifting

As a result of the continental collision discussed in the previous two sections, eastern and southern North America

were, at the beginning of the Triassic, the sites of a major Himalayan-type orogenic belt which stretched in great salients and recesses from the Marathon region of Texas to the Ouachitas of Oklahoma and Arkansas, to the Hercynides of Europe. Continental reconstructions for this time vary somewhat but there is general agreement that this mountain system was one of the most magnificent in history; remnants of it are distributed today on four different continents! By later Triassic time, however, erosion had brought low the great mountains and Pangaea had begun to rift asunder, splitting more or less at the "weld" along which the Paleozoic continents had been sutured. North America's separation began during the Triassic, relatively early in the history of continental disassembly, but was not fully complete until the Cretaceous. Most of the strata that record this continental separation are present today as the Atlantic and Gulf Coastal Plains but the earliest sediments, formed during the first phase of rifting, occur in a series of long, narrow fault-basins exposed along eastern North America and buried deeply beneath younger Coastal Plain strata. In this section, we will examine the sedimentology and structures of these fault basins, called *Eastern Triassic–Jurassic basins* (or, Newark fault basins), and will consider their relevance to continental separation. We will also discuss Late Triassic and Early Jurassic deposits of the Gulf Coastal Plain.

E.1. Eastern Triassic–Jurassic Basins

Figure 7-47 is a map showing the locations of most of the known Eastern Triassic–Jurassic basins, except for some of the smaller ones. Commonly referred to in older literature simply as "Triassic basins," they occur principally in the crystalline provinces of the Appalachian orogen and tend to be elongated parallel to the structural "grain" of the surrounding rocks. Locally, these basins may cut diagonally across regional structures such as along the western side of the Gettysburg Basin where the fault bordering the basin truncates the Blue Ridge anticlinorium, separating it from the Reading Prong, which represents a continuation of Blue Ridge structures to the north. (It is interesting to notice that the Susquehanna River flows eastward out of central Pennsylvania through just this area, very likely because the hard-rock barrier of the Blue Ridge had been cut out by Triassic faulting.) Triassic–Jurassic basins buried beneath the Coastal Plain are also recognized and some very large basins containing Triassic and Jurassic strata lie beneath the sediments of the Atlantic and Gulf continental shelves. Thus, we may say with some justification that most of the eastern continent was affected by basin-formation during the Triassic. A final point can be made about the distribution of Triassic–Jurassic basins; not only are they present throughout the eastern United States

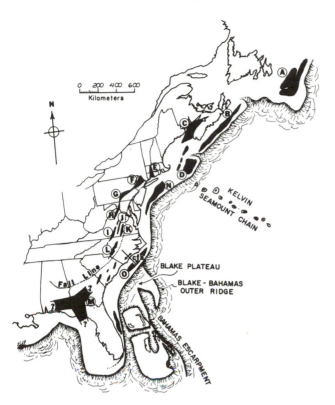

Fig. 7-47. Location map of major Eastern Triassic-Jurassic basins and related features. Symbols are as follows: A, Grand Banks Basin; B, Scotia Shelf Basin; C, Fundy Basin; D, Georges Bank Basin; E, Connecticut Basin; F, Newark Basin; G, Gettysburg Basin; H, Culpeper Basin; I, Danville Basin; J, Farmville Basin; K, Richmond Basin; L, Deep River Basin; M, South Georgia Basin; N, Baltimore Canyon Trough; O, Carolina Trough; cfa, Cape Fear Arch.

and Canada, but similar fault basins are also found in northwest Africa (Manspeizer *et al.*, 1978), which was adjacent to the present eastern margin of North America during the Triassic. Clearly, the forces which led to fault-basin development affected a very large area.

E.1.a. Basin Structure

The structure of Triassic–Jurassic basins is relatively simple but, nevertheless, has generated long-standing controversies. Triassic strata lie nonconformably on crystalline rocks last deformed during the Alleghenian orogeny and, as a general rule, Triassic basins are bordered on one side or the other by a normal fault toward which bedding dips monoclinally. In some cases, both sides of the basin are faulted but, in such cases, one of the border faults is of considerably greater vertical displacement than the other one. The strata are commonly intruded by diabasic dikes and sills and, in a few areas, have been mildly folded. This much is not disputed. But the gross cross-sectional shape and the original interrelationships of the basins have been

quite controversial and we will briefly consider each of these problems (Fig. 7-48).

There are two views of Triassic–Jurassic basin shape and structure. The first view, which goes back to Davis (1886) and has been argued more recently by Stose and Stose (1944) and Faill (1973), holds that the cross-sectional shape of the basin is typically synformal with one side faulted off. The implication is that the basin formed first by crustal down-warping and sedimentation; later, following cessation of subsidence and sedimentation, one side of the basin was faulted up and removed by erosion. The border faults, in this scheme, are not considered to be of particularly great displacement; indeed, Faill stated that, of the northwestern margin of the Gettysburg Basin (the margin usually thought to have a large-displacement fault), only 35% is actually faulted—the rest, he argued, is a simple overlap contact.

The other view of Triassic–Jurassic basin structure, as put forth by Barrell (1915), Longwell (1933), and Klein

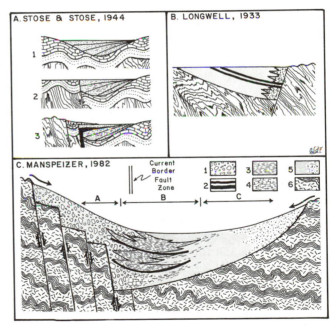

Fig. 7-48. Comparison of structural models of Eastern Triassic–Jurassic basins. (A) The model of Stose and Stose (1944) showing their interpretation of the basins as having formed by crustal down-warping unrelated to faulting; faulting and magmatism are considered in this model to have occurred later and fanglomerates related to the border fault are considered to be of only minor importance. (B) Longwell's (1933) model showing his interpretation that the basins had formed by normal faulting and that sedimentation was penecontemporaneous with faulting. (C) Manspeizer's (1981) model showing basins as having been originally more extensive, as suggested by Stose and Stose, but formed by faulting contemporaneous with deposition, as indicated by Longwell; symbols are as follows: A, marginal border-fault facies; B, central basin facies; C, marginal piedmont facies; 1, subaerial fan deposits; 2, lava flows; 3, lacustrine deposits; 4, playa deposits; 5, fluvial–deltaic deposits; 6, crystalline basement.

(1969) among others, is that the cross-sectional shape is generally triangular with the border fault comprising one side of the triangle and that the developmental history was a continuous, one-step process of faulting and penecontemporaneous sedimentation. In other words, the basins are thought to have formed as rift-valleys, i.e., grabens or half-grabens. Advocates of this view point to the presence of impressively large alluvial-fan deposits (*fanglomerates*) located directly adjacent to the border faults. Some of these fanglomerates contain angular to poorly rounded boulders up to several meters in size and are considered to have formed analogously to alluvial-fan deposits along the margins of modern block-fault basins in the Basin and Range Province of the southwestern United States (e.g., Death Valley, California). The fanglomerates, thus, are taken to be evidence of active fault-scarps bordering the basins during the Triassic and Jurassic. If this view is correct, then strata within the basin must dip continuously down toward the border fault at the same angle seen at the surface; consequently, displacement on the faults must be considerable, on the order of 5000 to 10,000 m!

The rift-basin model is today the most widely accepted view. As we will show later in this discussion, sedimentological evidence from the Triassic basins indicates depositional environments which resemble those of the great rifts of east Africa and the Basin and Range. Studies in the Wadesboro Basin (a sub-basin of the Deep River Basin) have clearly shown that faulting and fanglomerate-deposition were penecontemporaneous (Randazzo *et al.,* 1970); similar synchronous faulting and sedimentation have been argued for other Triassic basins (e.g., Cloos and Pettijohn, 1973; Thayer, 1970). Geophysical studies have revealed that displacements on some border faults are great, agreeing with the prediction of the rift-basin model, and have also indicated the presence of graben-form basins beneath the Coastal Plain. A final reason for wide acceptance of the rift-basin model is that it fits well into the emerging picture of global plate-tectonics history. Interpreted as rifts, Triassic basins imply the presence of a tensile-stress regime during the Late Triassic, which is the time, according to theory, when Pangaea began to rift apart.

A second long-standing controversy involves the original interrelationships among the several basins. Again, there are two views: (1) *the "broad terrane" model,* which holds that the basins were originally much larger, perhaps part of a single, continuous rift valley like the east Africa rift valley today; and (2) *the "local basin" model,* stating that the basins formed as separate, distinct rifts isolated from each other. The broad-terrane model (Russell, 1878; Sanders, 1963) is based on the observation that there is a gross symmetry in the distribution of the basins, their structures, and their facies. Strata in the western "string" of basins (e.g., Newark, Gettysburg, Culpeper, Danville) dip westerly toward their border faults and associated fanglomerates, whereas strata of the eastern string (e.g., Connecticut, Deep River) dip toward eastern border faults and fanglomerates. The interpretation is that the present two strings of basins represent the two sides of an originally much larger rift-valley whose middle has been warped up and removed by erosion. But this view is too simplistic; a careful examination shows that several basins do not fit the picture: the Richmond Basin, for example, is in the eastern string but has a western border fault; the Farmville Basin, with a western border fault, lies midway between the two strings. Further, the broad terrane model disregards the presence of basins buried beneath the Coastal Plain.

The local-basin model has recently been championed by Klein (1969) who used paleocurrent and provenance data to argue convincingly that each basin was isolated from the others during deposition. He showed that sediments were supplied from local source-areas virtually surrounding each basin. The implication of this interpretation is that widespread block-faulting, possibly due to tensile stress related to crustal extension as the plates began to pull apart, created a terrane of rotating fault-blocks, subsiding basins, and rising mountain ranges in the area of eastern North America and western Africa.

E.1.b. Basin Geology

i. Stratigraphy. Strata of the Triassic–Jurassic basins are referred to as the Newark Supergroup, from type sections in the Newark Basin. As a general rule, the sedimentary fill of the basins may be divided into three stratigraphic units; in the Newark Basin these are the Stockton, Lockatong, and Brunswick formations. Both the Stockton and the Brunswick are composed of red-bed strata characterized by immature sandstones, conglomerates, shales, and fanglomerates near the border faults. By contrast, intervening Lockatong strata are predominantly gray to black shales, mudstones, and siltstones. Coarse Stockton and Brunswick sediments were deposited along the margins of the basin, i.e., relatively near source areas, whereas finer Lockatong sediments were deposited near the basin center, far away from detrital sources. The three units are shown on most stratigraphic columns as overlying each other; this is not, however, because they represent layer-cake strata but rather because the monoclinal dip of the strata makes them appear to be so. In fact, they are facies of each other (see Fig. 7-49). Because the geometries of most Triassic–Jurassic basins are similar, the same three-part stratigraphic subdivision of coarse, immature red beds above and below a dark, finer-grained unit may be established for other basins (e.g., New Haven, Meriden, Portland in the Connecticut Basin; Pine Hall, Cow Branch, Stonesville in the Danville Basin; Pekin, Cumnock, Sandford in the Deep River

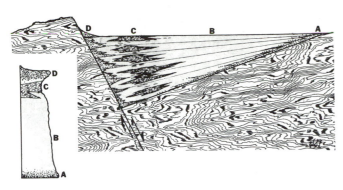

Fig. 7-49. Very generalized geological cross section of a typical Eastern Triassic–Jurassic basin. Notice that a stratigraphic column derived only from surface information yields a three-part "layer-cake" interpretation with lower and upper sandstone units and a middle mudstone unit. An analysis employing basin geometry as well as surface information, however, shows the various deposits to have formed as facies of each other.

Basin). It should be understood, however, that the similarity of stratigraphic descriptions for the various basins does *not* necessarily imply chronostratigraphic equivalence of the units nor synchroneity of climactic or tectonic factors; such similarity is merely a function of the geometry of basin strata as they are exposed today.

The age of Newark strata was considered until recently to be entirely Late Triassic. Since basin sediments are entirely of terrestrial facies, they contain no marine fossils and, therefore, cannot be related to standard biostratigraphic reference-sections. Further, the faunas that are present (see below for further discussion) are primarily composed of long-ranging taxa which provide little help in making precise age assignments. In 1975, however, Cornet and Traverse demonstrated the presence of Early Jurassic palynomorphs in upper portions of some northern-basin sections. Further, Olson (1975) showed that vertebrate assemblages of the Connecticut Basin are of Norian to Hettangian age. But Olson also showed that faunas of the southern basins (i.e., Deep River, Danville, and Richmond) are older (of Carnian age). The idea that southern basins are older than northern basins has now gained support from a variety of studies. Radiometric dating of volcanic flows from the Newark and Connecticut basins shows that they were extruded during the Early Jurassic (Hettangian and Sinemurian time) but volcanics in the Culpeper Basin are considerably older (DeBoer and Snider, 1979). This gradual northward decrease in age reflects progressive northward opening of the rifts, a process referred to by DeBoer and Snider (1979) as the "zipper effect." It appears, however, that the Fundy Basin contains older deposits than do basins to the south. Thus, it may not belong to the group of Triassic–Jurassic basins shown in Fig. 7-47; perhaps it belongs to a group of basins now located primarily beneath the Coastal Plain overlap, part of a different "zipper."

ii. Sedimentology. Composed primarily of immature detrital sediments, Triassic–Jurassic basin strata provide excellent material for studies of provenance, sediment dispersal, and terrestrial depositional-environments. In the following paragraphs, we will first consider the general character of the sediment and then discuss its paleoenvironmental and paleoclimatic interpretation.

Triassic–Jurassic basin rudites are divided into two groups: fanglomerates and ordinary conglomerates. The distinction between them is genetic, fanglomerates theoretically having been deposited as alluvial fans whereas ordinary conglomerates are primarily of fluvial origin. On outcrop, however, the terms are often used descriptively, with fanglomerate referring to extremely poorly sorted, angular debris and conglomerate referring to better sorted, well-rounded rudaceous sediments. Fanglomerates are typically found adjacent to border faults whereas ordinary conglomerates are usually found farther out into the basin, commonly comprising the lower parts of fluvial, fining-upward sequences. Both groups of rudites contain lithologies recognizable in nearby areas of the sourceland. In the Wadesboro subbasin, for example, argillite pebbles and cobbles were obviously derived from the adjacent Carolina Slate Belt (Randazzo *et al.*, 1970). Along the western side of the Gettysburg Basin, the side with the main border fault, thick fanglomerate deposits contain clasts derived from strata of the Appalachian Valley and Ridge Province, with representative lithologies of Precambrian to Devonian age. An example is the well-known "Potomac marble" in Maryland, which is not a marble at all but rather a Triassic–Jurassic carbonate–lithoclast conglomerate. Whereas some of the western fanglomerates are quite spectacular, the main source of sediment in the Gettysburg Basin was probably from the south and southeast (Glaeser, 1966). In other words, the majority of sediments came from the dip-slope of the rotated fault-block whereas only small amounts of sediment were derived along the fault scarp (see Fig. 7-50).

Triassic–Jurassic basin sandstones are typically immature arkoses, lithic arkoses, and litharenites. Feldspars in the sand are dominantly Na-plagioclase, reflecting their Piedmont derivation as does the presence of metamorphic rock-fragment grains such as phyllite and schist. Sandstones are often crossbedded and occur in fining-upward sequences characteristic of fluvial deposition.

Fine-grained sedimentary rocks predominate in the middles of the basins in formations such as the Lockatong or the Cow Branch. The most abundant lithology is massive, gray mudstone, often with thin laminae of carbonaceous material and, in some instances, small amounts of the zeolite mineral analcime. Other lithologies include dolomitic mudstone, black, fissile shale, and thin-bedded siltstone with ripples, small-scale crossbedding, and minor graded bedding. These mudstones and shales may be fos-

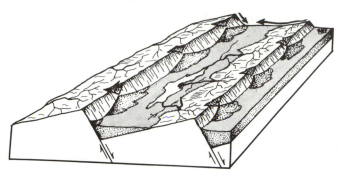

Fig. 7-50. Simplified model of sedimentation within an Eastern Triassic–Jurassic basin. Notice that the main sediment-transport system (large arrow) brings sediment into the basin from the dip-slope of the fault block, while only minor amounts of sediment, primarily deposited as fanglomerates, come from erosion of the fault scarp.

siliferous. Olson *et al.* (1978) described a well-preserved, diverse, mixed vertebrate and invertebrate fauna from the Cow Branch Formation of the Danville Basin to contain: eosuchian reptiles, phytosaurs, five genera of fish, and well-preserved insects (such as water bugs, March flies, true dipteran flies, crane flies, and beetles). Olson *et al.* concluded that the insect faunas reflect two major components: a lacustrine population and a forest population.

Rudites, sandstones, and shales are by far the most abundant lithologies in Triassic–Jurassic basins but several minor lithologies are important in the interpretation of depositional environments. Limestones have been reported from several of the basins and range in character from algal-laminated micrite (in the Connecticut Basin; Sanders, 1968) to oolitic limestone (in the Culpeper Basin; Carozzi, 1964). Caliche deposits have also been reported and appear to exist in almost all the basins. Commonly associated with these carbonates is chert, either as a replacement after limestone or as a primary, inorganic precipitate formed in playalike lakes (Wheeler and Textoris, 1978). Analcime is present in some gray shales and there are rare salt-crystal impressions on bedding planes associated with mud-cracked surfaces. In the southern basins, small amounts of coal have been found and, in a few cases, mined (e.g., the Cumnock Coal of the Deep River Basin; Reinemund, 1955). Indeed, the first coal deposits discovered in North America were near Midlothian, Virginia, within the Richmond Basin.

In terms of their paleoenvironmental interpretation, Triassic–Jurassic basin sediments can be divided into three groups: (1) alluvial fan deposits; (2) fluvial deposits; and (3) lake and swamp deposits. Deposits of alluvial fans, as we have discussed earlier, are principally composed of fanglomerates. The extreme angularity of the clasts, the low degree of sorting, the highly polymictic character of the sediments, and the large size of some of the boulders all

indicate that these deposits accumulated very near their source. The fanglomerates typically lack distinctive bedding except for crude size-grading and intraformational lenses of sand or mud. These features suggest deposition from mudflows, which are the principal agent of transportation on alluvial fans. In several instances, mapping of individual fanglomerates reveals a fan-shaped plan-view; in cross section, they wedge out away from the border faults, grading basinward into either stream or lake sediments.

Fluvial sediments typically contain conglomerates, sandstones, and thin shales. Both the average grain-size and the relative proportion of coarser sediment decrease toward the center of the basin. Commonly colored reddish-brown to brick red, the sediments are arranged vertically in a series of fining-upward sequences, with each sequence channeled into the one below it. Such sequences are considered to represent deposition on migrating point bars of meandering streams. Braided streams may be represented by very conglomeratic deposits near fanglomerates.

Typically, lake and swamp deposits occur in the fine-grained middle unit of Triassic–Jurassic basin strata. Sediments of these environments are the most heterogeneous within the basin, including both detrital sediments (e.g., mudstone, siltstone) and chemical sediments (e.g., limestone, caliche, chert, dolomite, analcime). The occurrence of these sediments in vertically repeating, cyclic strata has been noted by many authors, especially Van Houten (e.g., 1964) who recognized in the Newark and Gettysburg basins the presence of two different types of cycles. The repeating unit of the *detrital cycle* is composed of a coarsening-upward sequence with fissile, pyritic, black shale at the base which grades upward into massive, bioturbated mudstone, often with mud cracks. The *chemical cycle* is composed of dark, laminated shale at the base, commonly containing dolomite and calcite lenses, grading upward into massive dolomite- or analcime-rich mudstone, also often with mud cracks. Similar cycles have been recognized in the Connecticut Basin (Hubert *et al.*, 1976), the Danville Basin (Thayer, 1970; Olson *et al.*, 1978), and the Deep River Basin (Wheeler and Textoris, 1978). Both types of cycles are interpreted to represent fluctuations of the size of valley-floor lakes. During periods when lakes were extensive, laminated muds were deposited, their dark color and pyritic nature suggestive of reducing conditions along the bottom of a thermally stratified lake and their rhythmic laminae representative of seasonal varves (Van Houten, 1964). As the size of the lake decreased, extensive muddy flats became exposed, resulting in the upper, mud-cracked deposits. The difference between detrital and chemical cycles, Van Houten suggested, was geomorphological: during periods when through-flowing stream-drainage predominated, lake water remained essentially fresh and sediments of detrital cycles were deposited; but, during other periods of time, internal drainage prevailed within the basins and lakes were playas, with high ionic concentrations and consequent evaporite deposits, resulting in the chemical cycles. The presence of apparently authigenic analcime (of either primary or diagenetic origin) bespeaks very high Na concentrations in lake water, clearly reflecting the weathering of Na-rich source such as the metamorphic rocks of the Piedmont. The cyclic nature of both detrital and chemical sediments was explained by Van Houten as being due to short-term climatic fluctuations related to the 21,000-yr. precession cycle (i.e., the periodic variation of obliquity of the Earth's rotational axis to the plane of the ecliptic), an explanation supported by varve-counts of the laminated strata of each sequence.

All of this brings us to the most interesting and hotly debated aspect of Triassic–Jurassic basin studies: their paleoclimatology. We have noted the presence of sediments deposited in alluvial fans, braided streams, and playa lakes. We have also shown that other basin deposits record meandering streams, freshwater lakes with diverse organic communities, and forests. Dolomite, analcime, primary silica, and salt-crystal impressions are found; coal seams in minable quantity are also found. As the reader can see, the climate represented by Triassic–Jurassic basin strata is, to say the least, enigmatic. Barrell (1908) and Cloos and Pettijohn (1973) have argued for an arid climate. Krynine (1950) and Klein (1962) have opined in favor of a humid climate. Since there is evidence for both views, one is forced to seek a compromise (although, at first glance, a compromise between "wet" and "dry" yields only "damp," which satisfies none of the evidence). One escape from the dilemma lies in the concept, mentioned above, of climatic cycles, possibly controlled by the 21,000-yr. precession cycle. Since there is some consensus that such climatic fluctuations did occur, it seems a small step to suggest that those evidences of a humid climate formed during one part of the cycle while arid-climate indicators formed during the other. This point of view has been advanced by Wheeler and Textoris (1978). This suggestion, however, does not easily explain the climatic differences (if they were climatic) that led to variations between external and internal drainage as recorded by the detrital versus chemical cycles. Perhaps longer-term climatic cycles might be involved, such as Milankovitch proposed for the Pleistocene, or variations in solar activity, or, more mundanely, temporal variations in the rate of block-faulting. The question, at any rate, remains finally unanswered.

iii. Igneous Rocks. In addition to sedimentary rocks, Triassic–Jurassic basins also contain igneous rocks, mainly diabasic flows and sills, which are confined to the basins, and dikes, which are widely distributed throughout the Piedmont and Northern Appalachians. The diabases range considerably in age, but most, especially the dikes, are lower Jurassic. Again, however, one may observe a regional trend in ages: the oldest dikes are found in the Carolinas and dike ages decrease progressively to the north. The sizes of the dikes and sills are variable, with thicknesses ranging from 1 or 2 m up to approximately 550 m (for a sill in the Gettysburg Basin), and lengths up to 100 km (for the Conshohocton dike in eastern Pennsylvania). Some well-known American landmarks, such as Seminary Ridge and Big and Little Round Tops of the Gettysburg battlefield and the Palisades along the Hudson River, are formed of Triassic diabasic rocks. (The Palisades sill is equally famous among geologists for its well-developed columnar jointing and its classic exemplification of magmatic differentiation by fractional crystallization and crystal settling.) Dikes in the Southeast trend generally northwestward, those in Virginia, Maryland, and Pennsylvania trend approximately northward, and those in the Northern Appalachians trend roughly northeastward. Clearly, the dike trends crosscut the "grain" of regional structure instead of following it roughly as do the border faults, whose displacements led to Triassic–Jurassic basin formation. Therefore, the stresses which fractured the crust in association with Triassic magmatism in the East do not seem to have been disposed spatially in the same manner as the basin-forming stresses. It is interesting in this regard to notice that similar dikes of Triassic–Jurassic age are present on those other continents which, prior to Mesozoic continental drift, were adjacent to North America's east coast; when placed back into their predrift position, these dikes exhibit a gross but distinct radial pattern centered roughly on the Blake Plateau of North America's continental margin (May, 1971). We will discuss the significance of this pattern later in this section.

In terms of their petrology, the diabases are relatively simple, being composed primarily of labradorite and augite with accessory magnetite–ilmenite and apatite (Weigand and Ragland, 1970). Either olivine or quartz is also a major component but the two never occur together in the same rock. The geographic distribution of quartz-normative versus olivine-normative diabase reveals an interesting pattern: quartz-normative dikes occur throughout Piedmont and Northern Appalachian areas *except* in North and South Carolina, where olivine-normative dikes predominate (almost exclusively). Carolina diabases contain the highest total Fe and the lowest TiO_2 contents of all eastern Triassic–Jurassic diabases and are, thus, more primitive (i.e., less differentiated). They also differ from other Triassic diabases of the East in their magnetic character: they exhibit the highest-amplitude magnetic anomalies and have the strongest remanent magnetism, probably because of greater amounts of magnetite and, thus, more Fe in the rock. De-Boer and Snider (1979) interpreted these observations to mean that the Carolinas area was at or near a mantle hot-spot during the Late Triassic. Were this true, it would explain the more-primitive chemistry of Carolina diabases as being due to deeper partial-melting within the mantle because of higher heat-flux.

E.1.c. The Opening of the North Atlantic Ocean

The preceding discussion allows speculation on the tectonic events which accompanied continental disassembly during the Late Triassic and Early Jurassic. Perhaps the first question to be addressed is the cause of the lithospheric spreading which broke North America away from Pangaea. This question must be viewed from its geologic-historical perspective: throughout much of the Paleozoic, the eastern margin of North America had been a convergent plate boundary and had, at the end of the Paleozoic, experienced a major continental-collisional orogeny. Yet, only a (geologically) short time later, the same area was experiencing the effects of tensile stress (e.g., block-faulting) and the development of a divergent plate boundary. What caused this fundamental change in the tectonic regime of eastern North America? We do not know. Actually, this question is only part of a much larger problem, for collisional orogenies are commonly followed by rifting. There are many speculations, of which we will discuss three, but there is no general agreement. It may be argued, for example, that collision itself ended convergent tectonics along the East because sialic crust does not subduct very well, being relatively buoyant; when driven down into a subduction zone beneath an opposing sialic block, continental crust acts like a wrench caught in the gears—it stops the motion. But, since plate tectonics apparently plays a major role in dissipation of heat from the Earth's interior, cessation of sea-floor spreading in one configuration must cause onset of spreading in another configuration, a global reshuffling of the Earth's tectonic cards. In this scheme, Pangaea was disassembled by the new configuration of spreading, breaking apart along old zones of weakness, e.g., collisional "welds." Another, but not necessarily contradictory, view suggests that the thickened continental crust of the Appalachian orogen was a more efficient thermal insulator than "normal," i.e., thinner, continental crust and, therefore, held in more of the Earth's heat, causing the mantle and adjacent crust beneath the orogen to heat up. This process is considered to be responsible both for "softening" the crust (i.e., making it behave more plastically) and for the development of "hot spots" which could initiate the process of rifting. Yet another idea (also not necessarily contradictory) supposes that isostatic rebound of the over-thickened orogen, following cessation of convergence and accompanied by erosion and tension-faulting, caused the

progressive thinning of the crust and the development of the fault basins. Clearly, no agreement as yet exists on the underlying reason why rifting so often follows continental collision.

Agreement is general, however, that a series of normal-fault basins began to open along the Appalachian orogen during the Late Triassic and that a rifted continental-margin existed on the eastern side of North America by the Late Jurassic. The exact sequence of rifting is still being studied but, as we have shown, southern Triassic–Jurassic basins are probably older than northern basins and a hot spot may have existed around the area of the Carolinas during the Late Triassic. DeBoer and Snider (1979) suggested that this hot spot may have led to the development of a ridge–ridge–ridge-type triple juncture in the area of southern Georgia where a very large and deep Triassic–Jurassic basin exists beneath the Coastal Plain (Chowns and Williams, 1983). The three arms of this triple juncture are today represented by subsurface Triassic–Jurassic basins in southern Alabama and Mississippi (the western arm), by Triassic rifts in the Bahamas (the southern arm), and by the line of Eastern Triassic–Jurassic basins up the Appalachian orogen (the northeastern arm). DeBoer and Snider likened their proposed Carolina hot spot and Georgia triple juncture to the Galapagos area of the eastern Pacific where the Galapagos hot spot lies east of the Pacific–Cocos–Nazca triple juncture.

The decreasing age of the basins toward the north probably indicates that the crustal tear, begun in the South, propagated up the orogen. Far to the north, however, in the area formed by the Canadian Maritime Provinces, Morocco, and southwestern Europe, rifting began at about the same time as it had in the Southern Appalachians. According to Manspeizer et al. (1978), extensive prerift doming took place in this northern area and consequent erosion stripped away not only the pre-Late Triassic stratigraphic record but also a large part of the continental crust! Manspeizer et al. argued that gaps in continental reconstruction in the area of Canada's Maritime Provinces and Morocco are not an error of continental fit but rather a *real* gap, produced by intense uplift of the crust and concomitant erosion. They suggested that the cause of uplift may have been isostatic rebound of the orogenically thickened crust but the association of diabasic intrusions similar to those of the Southern Appalachians suggests the possibility of hot-spot-related uplift, perhaps similar to the prerift thermal doming observed in east Africa. At any rate, doming and erosion were followed by rapid subsidence of large rift basins in which were deposited both red-bed strata (e.g., the Eurydice Formation of the Grand Banks), similar to red beds of the other Triassic basins, and very thick sequences of salt and associated evaporites. Jansa and Wade (1975) proposed that the Argo salt of the Grand Banks and Scotia shelf, as well as the thick evaporites of the Essaouira Basin in Morocco, the Lusitanian and Algave basins in Portugal, and the Aquitaine Basin in southern France were all formed at this time in a large and complex rift basin (see Fig. 7-51). This basinal area may have been supplied with saltwater from the Tethys sea to the east; Manspeizer *et al.* cited palynological evidence for a major Tethyan marine transgression into the evaporite basin during the Late Triassic and Early Jurassic.

We are led by the foregoing discussion to speculate that (at least) two hot spots developed along the Appalachian–Hercynian orogen, roughly at about the same time, one in the area of the Carolinas and one in the eastern Canada–Morocco area. Rifts and transform faults associated with these hot spots began the process of continental disassembly. Unfortunately, the history of continental rupture is extremely complex and much information has been obscured by the blanket of younger sediments. Thus, we are left only with a few tantalizing hints of the separation. For example, the three-armed rift which developed in south Georgia did not cause continental separation to occur and so the rift failed to open farther; instead, for reasons that we do not understand, spreading "jumped" out to a position just

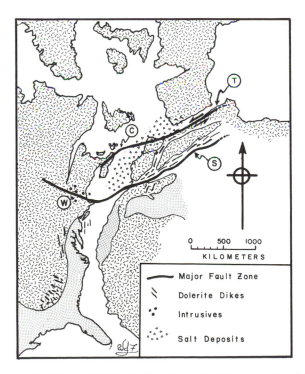

Fig. 7-51. Relation of Cobequid–Chedabucto Fault Zones and the White Mountains magma series to major fault zones in northwestern Africa; note the position of Late Triassic salt basin. Symbols are as follows: C, Cobequid–Chedabucto Fault Zone; T, Tethys Fracture Zone; W, White Mountains magma series; S, South Atlas Fault Zone. (Modified from Manspeizer *et al.*, 1978, *Geological Society of America Bulletin*, Fig. 7, p. 914.)

east of the Blake submarine plateau where it finally did result in continental separation.

E.2. Early Rifting on the Southern Continental Margin

Along the southern continental margin, too, rift basins formed during the Late Triassic. Strata of these basins are entirely in the subsurface, buried in several cases to depths of 3–4 km. Thus, they are known only from geophysical, especially seismic, studies and from a few deep wells. The fault basins begin at the large, buried Triassic–Jurassic basin beneath the south Georgia Coastal Plain and extend almost due westward from southern Alabama all the way to eastern Texas, closely paralleling the trend of the Ouachita fold belt in the subsurface. Strata of these basins are referred to as the Eagle Mills Formation and are composed primarily of detrital red beds of probable fluvial origin (Scott *et al.*, 1961; Rainwater, 1967). They are associated with diabasic flows and shallow intrusions.

From their sedimentological description, their association with diabasic igneous rocks, and their occurrence in normal fault basins parallel to regional, structural grain, Eagle Mills strata can easily be related to Newark Supergroup strata of the Eastern Triassic–Jurassic basins. Further, similar red detrital sediments and associated mafic igneous rocks of Late Triassic age exist in northeastern Mexico and are called the La Boca Formation. La Boca strata also occupy fault basins. Thus, it may be seen that tensional faulting extended across nearly the entire northern part of what is now the Gulf of Mexico.

F.   The Absaroka Sequence in the Cordillera

F.1. Between Orogenies: Paleogeography during the Pennsylvanian and Permian

The paleogeography of the Cordilleran region at the beginning of the Pennsylvanian Period was very much like that of the preceding Chesterian Epoch (see Fig. 7-52). The western cratonic platform was largely occupied by a shallow epeiric sea in which mainly carbonate sediments accumulated. Eastern parts of the continental shelf (the "miogeocline") apparently lay under somewhat deeper water but were also dominated by carbonate sediments. The western shelf continued to act as a foreland basin in front of the low but still emergent Antler orogenic belt. West of the Antler sourceland was a deep-water basin floored by oceanic crust; this deep basin has been interpreted as either a back-arc basin, similar to the Sea of Japan, or as a true ocean basin. Somewhere to the west (the location with respect to North America is debatable) was at least one vol-

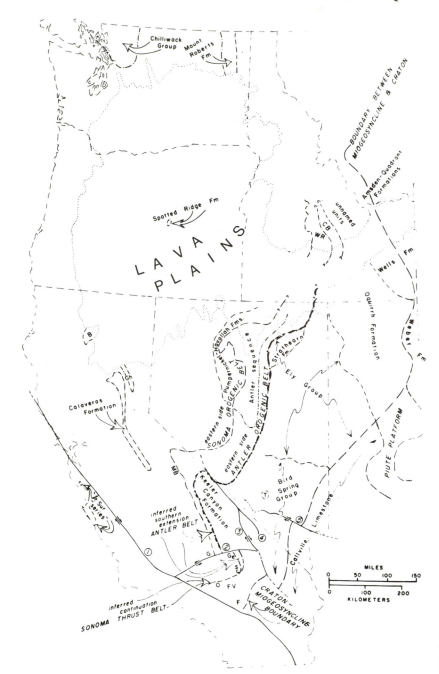

Fig. 7-52. Major structural and stratigraphic features of Pennsylvanian age in the Cordilleran region of the United States. (From Rich, 1977, reproduced by permission of the Pacific Section, Society of Economic Paleontologists and Mineralogists.)

canic-arc terrane. This general paleogeographic pattern persisted, with several important modifications, until altered at the end of the Permian by the Sonoma orogeny. In the following section we discuss in more detail the stratigraphy and paleogeography of each of these regions for the interval from the beginning of the Pennsylvanian to the end of the Permian. We will then consider the effects and tectonic models of the Sonoma orogeny.

F.1.a. Antler Orogenic Belt and Foreland Basin

The tectonic sourceland on the western edge of the continental shelf was still an effective barrier between the foreland-basin sea and the western ocean and, during Early Pennsylvanian time, it still provided detrital sediments to the foreland basin. In Nevada, conglomerates in the uppermost Diamond Peak (Tonka) Formation (central part of the state) and in the Moleen Formation (northern part of the state) were probably derived from the Antler-land and deposited primarily on the western margin of the foreland basin. Over most of the basin, referred to as the Ely Basin, moderately deep-water conditions prevailed in which dark micritic carbonates were deposited. In eastern Nevada these carbonates are called the Ely Limestone; in southern Nevada they are the lower Bird Springs Group; and in southeastern California they comprise the lower Keeler Canyon Formation (Ketner, 1977). On the eastern side of the Ely Basin, correlative sediments of the Oquirrh Formation record extreme rates and amounts of subsidence (the Oquirrh Basin). Farther eastward, shallow-marine and supratidal environments of the craton are represented by the Callville Limestone (southeastern California to southern Utah), Weber Formation (northern Utah and Colorado), and Amsden Formation (Montana and Idaho). (Further discussion of these units may be found in the section on Absarokan stratigraphy of the craton.)

During the Atoka Epoch, a mild pulse of uplift affected central and northern parts of the Antler orogen. In north-central Nevada, coarse conglomerates from Antler-lands comprise the upper Moleen and overlying lower Tomera Formation on the west side of the Ely Basin. Exposed in outcrops to the northwest are over 220 m of coarse, conglomeratic detrital materials of the Battle Formation, which were deposited on a narrow strand on the west side of the Antler sourceland (Rich, 1977). Battle conglomerates grade westward into shallow-water deposits of the Highway Limestone which probably formed on a narrow carbonate platform at the westernmost edge of the continental shelf (see Fig. 7-53). Note that Battle sediments were transported westward toward the western-ocean basin while Tomera sediments were carried eastward into the Ely Basin. However, Atokan uplifts were of only local importance in most areas because, over most of the Ely Basin, carbonate deposition continued as before with little apparent increase in detrital influx.

During the Desmoinesian Epoch, an even stronger pulse of uplift affected the area of the Antler orogen and areas just west of it. One result was the deposition of large volumes of detrital sediment along the western side of the foreland basin. Among the most interesting units of this detrital mass is the Wood River Formation (Desmoinesian to Wolfcampian) in central Idaho. Wood River sediments accumulated on the eroded surface of the old Antler orogen, indicating that the formerly positive area was, during the Desmoinesian, the site of active subsidence. Skipp *et al.* (1979) argued that basal gravels of the Wood River Formation were derived from uplifted Copper Basin strata on the *eastern* side of the Wood River basin but most sand in the Wood River was probably *not* derived from the Copper

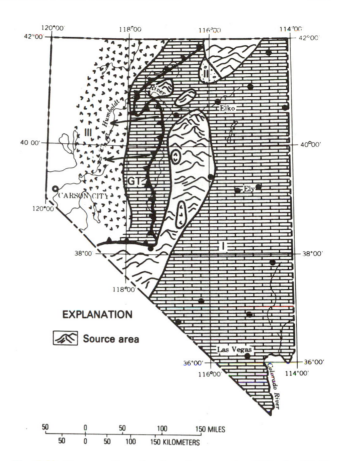

Fig. 7-53. Morrowan–Desmoinesian paleogeography of Nevada. (Modified from Larson and Langenheim, 1979.)

Basin Uplift. Paleocurrent data reported by Thomasson (1959) suggest derivation from a western source area (such as the Humboldt Highlands; see below) but the strongly quartzose mineralogy would seem to suggest a cratonic source (Hall *et al.*, 1974).

In northeastern and central Nevada, coarse detrital sediments of the upper Tomera Formation also reflect renewed uplift in the west. An unconformity at the top of the Tomera may imply that even the western part of the foreland basin was affected by uplift by the latest Desmoinesian. In eastern Nevada, the Ely Limestone is succeeded conformably by the Desmoinesian Hogan Formation, whose sediments are considerably more argillaceous than Ely sediments. Thus, conglomeratic and arenaceous sediments of the western part of the basin (e.g., the Tomera) passed eastwards into muddy and sandy strata of the basin middle (e.g., the Hogan).

Ketner (1977) termed this sharp pulse of uplift the *Humboldt orogeny* for the Humboldt River in northern Nevada; he called the uplifted western sourceland the Humboldt highlands. The tectonic nature of the Humboldt event is difficult to specify because so much of the data have been lost. There are two tantalizingly different tectonic models. Skipp *et al.* (1979) postulated that the Humboldt orogeny was a continuation of back-arc thrusting and resultant obduction of back-arc-basin sediments (see Fig. 7-54). The reader will recall that this mechanism was advanced by Burchfiel and Davis (1972, 1975) for the Antler orogeny. The Humboldt highland itself is considered in this scheme to have been an obducted sediment mass similar to the earlier Antler orogen. An alternate, and vastly different, model was suggested by Dickinson (1977) who proposed that rifting during the Pennsylvanian broke up the western shelf into a series of crustal blocks (see Fig. 7-55). In this scheme, the Humboldt Highlands are regarded as a crustal horst block which subsided less than the adjacent graben blocks. This would explain the foundering of the old Antler orogen in Idaho to form the Wood River Basin. The Copper Basin Uplift is considered by the Dickinson model to be another horst block. One attractive aspect of this model is its ability to explain the penecontemporaneous basin-and-uplift activity of the western craton; basins, such as the Oquirrh Basin, are considered rift-related grabens whose

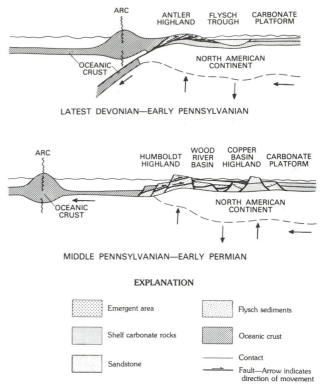

Fig. 7-55. Rifting model for the Humboldt orogeny. (From Skipp *et al.*, 1979.)

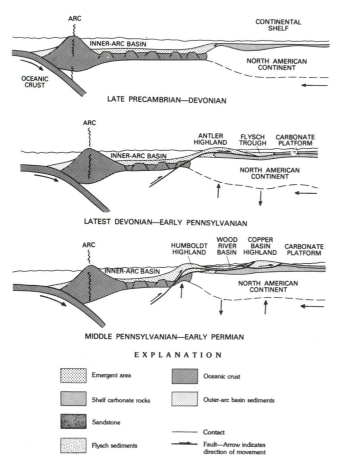

Fig. 7-54. Back-arc thrusting tectonic model of the Humboldt orogeny. (From Skipp *et al.*, 1979.)

total subsidence is comparable to that of aulacogens. Another corollary of the rift model is its explanation of the termination of the Antler orogeny. In most cases, a foredeep basin in which flysch is deposited evolves over time into a molasse basin with complex fluvial, deltaic, and shallow marine facies (e.g., the Taconic orogeny of the Appalachians, where Martinsburg flysch is succeeded by Juniata–Queenston molasse). In the case of the Antler belt, however, the molasse phase never quite got started before the orogen itself began to be overlapped by detrital sediments, e.g., the Wood River Formation. Dickinson considered this interruption of the ''normal'' orogenic scenario to have been due to rifting, which cut down the Antler orogen ''in its prime.'' It might be noted, however, that our knowledge of orogenesis is probably not broad enough at this stage to allow the development of rigorous propositions on the basis of negative evidence. Furthermore, it is unclear what, if anything, was rifted away from the western continental margin, although Skipp and colleagues' illustration (our Fig. 7-55) implies that it was the magmatic arc which was sundered from the continent, perhaps by an episode of back-arc spreading. If this was the case, the whole middle and late Paleozoic history of the Cordillera could be described as a sort of ''accordion-tectonics'' in which the western arc moves back and forth with respect to the continental margin. Such behavior would be bizarre but, as we noted several times, stranger things have happened. Speed and Sleep (1980) have suggested that the arc, which they believed collided with North America to produce the Antler orogeny, instead of rifting away, may have cooled and subsided following the collision, becoming part of the western-ocean-basin floor (see below). Perhaps Humboldt rifting was related to that subsidence, a kind of crustal relaxation following cessation of compressive strain. None of these ideas are entirely satisfactory and much more study is needed before the final story of the Humboldt orogeny is written.

Regardless of its origin, the Humboldt Highlands apparently persisted as a geographic entity throughout the rest of the Pennsylvanian Period and much of the Permian. Late Pennsylvanian deposits reveal the same general patterns of sedimentation as do those of the Desmoinesian. In Idaho, Wood River detritus continued to be deposited. In northeastern Nevada, the Missourian and Virgilian Strathern Formation is composed of chert-pebble conglomerates, silty limestone, and sandstones which rest disconformably on the Tomera Formation. In southeastern California, upper Pennsylvanian strata in the Inyo Mountains comprise the upper Keeler Canyon Formation. These deposits consist primarily of deep-water calcareous turbidites; they contain paleocurrent indicators which record a general flow direction of west-southwest, suggesting that the turbidity currents which deposited them came from a dominantly carbonate area to the east and north. This hypothesis is borne out by the gradation of the turbidite facies eastwards into fossiliferous shallow-water carbonate facies in the Panamint Mountains. The uppermost part of the Keeler Canyon is of Wolfcampian age (see below).

In the area of the Antler orogen in northcentral Nevada, Upper Pennsylvanian and Lower Permian deposits consist of limestone/chert pebble conglomerates and pebbly limestones (Wildcat Peak Formation and Antler Peak Limestone). Rich (1977) suggested that these deposits formed in a variety of detrital and carbonate environments associated with a low archipelago of islands, the last remnants of the Antler-land. The series of stratigraphic units which formed over the subsided Antler orogen is informally known as the *Antler overlap sequence* (Stevens, 1977).

During the Permian, the Antler orogen continued to be overlapped. Upper Antler Peak sediments in northcentral Nevada consist of both carbonates and detrital materials and were deposited in straits and embayments in the Antler archipelago during the Wolfcampian Epoch. A major unconformity separates these sediments from Guadelupian limestones of the Edna Mountain Formation, which comprise the top of the overlap sequence in northcentral Nevada. In southwestern Nevada, the orogen is covered by approximately 100 m of shallow marine sediments of the Diablo Formation. Speed *et al.* (1977) showed that the Diablo actually consists of two very different parts: an autochthonous part consisting of the shallow-water sediments and an allochthonous part consisting of deep-water strata. Allochthonous, deep-water Diablo sediments were probably deposited on oceanic crust at the base of the western slope of the Antler archipelago. Autochthonous, shallow-water sediments are fossiliferous, locally crossbedded calcarenites, limy quartz-chert arenites, and chert-pebble conglomerates. These deposits lie with strong angular conformity on the cherty, pelagic, Ordovician strata which comprised the Antler orogen. Thus, autochthonous Diablo sediment strata represent the Antler overlap sequence in the southwestern Nevada, southeastern California area. The Diablo is overlain conformably by the lower Triassic Candelaria Formation.

The old Ely/Hogan Basin had been nearly filled with sediments by the beginning of the Permian and, over most of its extent, shallow-water detrital and carbonate facies developed. The resulting units (the Lower Permian Arcturus Group and the Upper Permian Park City Group) are described in more detail in the discussion of cratonic stratigraphy. Only on the extreme western part of the basin, adjacent to the Antler archipelago, did a deep trough persist. Called the Dry Mountain trough (Stevens, 1977), this area continued to receive detrital sediments from the Antler archipelago; in eastcentral Nevada, these sediments comprise the Buckskin Mountain Formation. The Dry Mountain trough deepened and widened to the southwest. In southeastern California, carbonate turbidites of the Keeler Canyon Formation continued to accumulate in the deep waters between the carbonate shelf on the east and shoal areas associated with the old Antler orogen on the west. Keeler Canyon sediments (as mentioned above for Pennsylvanian strata) contain both shallow, carbonate-platform facies in the Panamint Mountains and a deep, turbidite-basin facies to the west in the Inyo Mountains. Farther westward, the upper Keeler Canyon grades into the Owens Valley Formation which consists of distal carbonate turbidites. By the end of the Guadelupian, the stratigraphic record of the Antler/Humboldt orogen and its foreland basin as separate tectonic entities disappears and, from Triassic time on, the area behaved as a single, stable crustal area.

F.1.b. Ocean Basin Terrane (West of the Golconda Thrust)

To the west of the Antler orogen (Fig. 7-52) during the Late Paleozoic was a sedimentary basin of probable oceanic depth in which a variety of deep-water sediments accumulated associated with basaltic pillow lavas and minor volcanic ash. Speed (1977c) referred to these rocks as the *ocean basin terrane* and we will adopt this usage. We use this term, however, in a descriptive rather than a genetic sense and the reader should keep in mind that these rocks have been interpreted alternately as deposits of a true ocean basin or as deposits of a back-arc, marginal basin. During the latest Permian to Early Triassic Sonoma orogeny, deposits of the ocean basin terrane were thrust eastward over the western margin of the Antler orogen in a manner very similar to the thrusting which accompanied the Antler orogeny. The main thrust surface over which the Sonoma allochthon was displaced is called the Golconda thrust (see Fig. 7-56) which today separates the overthrust ocean-basin-terrane rocks from autochthonous rocks of the Antler belt.

One of the better known units of the ocean basin terrane is the Havallah sequence in central Nevada. These strata were originally described in 1951 by Ferguson *et al.*, who recognized a lower, primarily greenstone unit with associated bedded chert and dark argillite, called the Pumpernickel Formation, and an upper unit composed of bedded chert, limestone, quartz arenite, agrillite, and local conglomerates, called the Havallah Formation. Stewart *et al.* (1977), however, observed that Pumpernickel-like and Havallah-like lithologies occur at several different horizons within the sequence and suggested that the two are probably not discrete stratigraphic units. Indeed, Silberling (1975) claimed that the sequence was not a depositional unit at all but rather a tectonic complex of thrust sheets intercalated together prior to or during Sonoma thrusting. These rocks are thus referred to informally as the Havallah sequence. In the Sonoma Mountains of Nevada, Havallah strata are composed of: (1) bedded, pelagic chert often containing radi-

olarian fossils; (2) detrital sediments such as conglomerate, graded quartz sandstone, and argillite; and (3) basaltic volcanic rocks (oceanic tholeiites, to be exact) with well-developed pillow structures. Detrital sediments are primarily of continental origin (Speed, 1977c) and appear to have been deposited from turbidity currents and other debris flows. The most likely source for such sediment was the Antler orogen along the western margin of the continental shelf which fed detritus both eastward (as discussed in the preceding section) and westward into the deep basin. Volcanogenic debris is rare and usually found in association with basalts, suggesting local derivation. Trace fossils, primarily of the *Nerites* assemblage [Seilacher's (1953) deep-water fossil assemblage], are infrequently found on Havallah bedding planes. Based on these lines of evidence, Havallah strata are generally considered to have been deposited in a deep-water environment, possibly of oceanic depth. Tholeiitic pillow lavas may represent scraps of oceanic crust caught up in the Sonoma thrusting and transplanted with the Sonoma allochthon. The Havallah is not notably fossiliferous but has yielded conodonts and fusulinids whose ages range from lower Wolfcampian to upper Leonardian; the Havallah is usually mapped as Pennsylvanian–Permian (Stewart *et al.*, 1977).

Several authors (e.g., Speed, 1971; Schweichert, 1976) have suggested that the Calaveras Formation of the western Sierra Nevada foothills in eastern California might originally have been part of the ocean basin terrane which was displaced westward during the continental fragmentation of the latest Permian. Thus, the Calaveras and the Havallah may be consanguineous. However, Speed changed his mind in 1977 and argued against the identity of

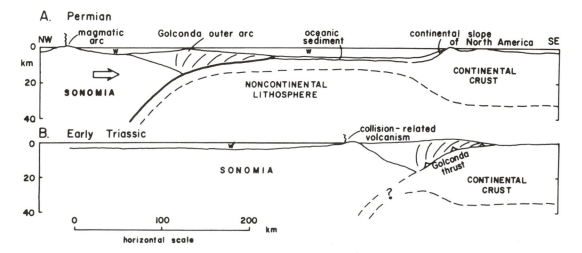

Fig. 7-56. The arc–continent collision model of the Sonoma orogeny. (From Speed, 1979, *Journal of Geology*, Vol. 87, p. 285.)

the two units because of uncertainties about the age of the Calaveras and lithological differences between the two units.

In northwestern Washington and adjacent parts of British Columbia are Pennsylvanian and Lower Permian (?) strata classified as the Mount Roberts Formation (Little, 1960). Whereas the lower part of Mount Roberts strata contains andesitic volcanics and shallow-water limestones (and, therefore, may be part of the volcanic arc terrane discussed in the next section), the upper part contains siliceous argillite, chert, graywacke-turbidites, and minor dark, micritic limestone. These upper Mount Roberts rocks, thus, bear a marked similarity to Havallah deposits and led Rich (1977) to suggest that the conditions reflected in Havallah sediments might have extended in a curvilinear belt, west of the Antler orogen, into northeast Washington (see Fig. 7-52).

In the case of all units mentioned above, definitive biostratigraphic data are rare and age assignments are vague, usually indicating only Penno-Permian age. The assumed relationship of these units is based on their common structural relation to the Golconda thrust and their lithological similarity. They all appear to represent deep-water environments associated with oceanic crust and were all thrust up onto the edge of the continent at the end of the Paleozoic. Other western Cordilleran units of Pennsylvanian and Permian age may also represent the ocean basin terrane but, because they also contain evidence of island-arc volcanism and/or deposition in a fore-arc or intra-arc basin, they are considered in the next section.

F.1.c. Volcanic Arc Terrane

Penno-Permian rocks to the west of the Golconda allochthon are exposed only in scattered and diverse localities including the Klamath Mountains of northern California and the Sierra Nevada Mountains of eastern California. Despite the distances which separate them, however, they share roughly similar lithologies, typified by andesitic volcanics, pyroclastics, volcaniclastic detrital deposits, and minor shallow-water limestones. These rocks all formed within the tectonic and sedimentary complex of an active volcanic-island arc. We will follow Speed (1977c) in referring to these rocks as the *volcanic arc terrane.*

Before beginning the analyses of individual units, the reader should understand the great complexity of the volcanic arc terrane and the problems that it presents to paleogeographic reconstruction. Rocks of this terrane are so widely separated in their occurrences that accurate correlations are very difficult. Similarly, there are a multiplicity of stratigraphic names which, like the trees that obscure the forest, considerably complicate a relatively simple pattern.

Of greater significance is the fact that three separate faunal provinces, defined on the basis of fusulinids are recognized: (1) a North American fauna, distinguished by generalized schwagerinids such as *Schwagerina* and *Parafusulina;* (2) an eastern Klamath fauna, characterized by specialized schwagerinid types like *Cuniculinella* and *Eoparafusulina;* and (3) a Tethyan fauna (i.e., similar to faunas commonly found in areas, such as Asia and southern Europe, which bordered the Tethys Ocean) defined by verbeekinid fusulinids such as *Yabeina.* The implication to be drawn from these three distinct faunas is that they probably represent three geographically isolated areas which have subsequently been telescoped together to form a structurally complex, composite arc terrane. Such a terrane is termed a *tectonic collage* (Helwig, 1974) and the individual tectonic elements of the collage are called *displaced terranes* (also called exotic or suspect terranes). These displaced terranes played a significant role in the history of the western Cordillera throughout later Paleozoic and later periods. In the following discussion, note that one implication of collage tectonics is that events recorded in the rocks of displaced terranes may not necessarily have occurred in the vicinity of North America, but instead *may* have occurred literally on the other side of the Earth.

i. Western Great Basin. In the western part of Nevada's Great Basin province are several major areas in which arc terrane rocks are exposed. In the region around Mina in southwestern Nevada, arc terrane rocks comprise (among other units) the Black Dyke and Mina Formations. The Black Dyke is composed of andesitic lavas and agglomerates, minor dacite, mafic intrusions, and a variety of volcaniclastic sediments including conglomerates, sandstones, and turbidites. Clearly, the Black Dyke represents a volcanic island arc. The Mina Formation is composed primarily of volcaniclastic turbidites, chert, argillaceous rocks, and fine quartz sandstone. These sediments are considered to represent a submarine fan complex immediately adjacent to the Black Dyke arc, whose igneous lithologies are mirrored almost exactly by Mina volcaniclastics. It is interesting to notice that the quartz sands of the Mina were probably not derived from the arc because such a mature sediment would not ordinarily be supplied by a tectonically active, andesitic terrane; Speed considered that the quartz was derived from a far-distant (cratonic?) source and carried to the Mina area by deep bottom-currents. The age of the Mina and Black Dyke formations is Middle to Late Permian based on radiometric dates obtained from hornblendes and confirmed by fossils.

In the northwestern corner of Nevada, outcrops in the Black Rock Desert expose rocks of the Happy Creek volcanic series. These volcanic rocks are virtually unlayered andesitic lavas and breccias with minor basalt, dacite, and sparse volcaniclastics. As in the Mina area, both shallow-water strata (limestones, conglomerates) and deeper-water deposits (turbidites) are found, indicating a tectonically active environment. An overlying, unnamed sequence of limestones, volcaniclastics, and lesser pillow lavas suggests a decline in volcanic activity and consequent erosion of the volcanic edifice.

Volcanic rocks of the Koipato Group occur in the Humboldt Range of northwestern Nevada. The Koipato is generally considered to be of Triassic age but Speed (1977c) had suggested that its lower unit, the Limerick Greenstone, is actually part of the late Paleozoic arc terrane. Limerick rocks are porphyritic and vesicular lavas of basaltic to andesitic composition and associated sediments are entirely volcanogenic. Younger (i.e., Triassic) Koipato rocks consist of thick rhyolite flows and ash-flow tuffs, indicating a major change in the tectonic character of the region.

ii. Sierra Nevada Mountains.

In the northern Sierra Nevada, volcanic arc terrane rocks comprise a variety of stratigraphic units. The Peale Formation consists of a lower, volcanic member of probable Mississippian age and an upper, Mississippian to Permian member with abundant chert, slate, and volcaniclastic sediments. Peale rocks are overlain by the Permo-Triassic volcanic sequence which consists of Goodhue, Reeve, and Robinson formations. The Goodhue is principally composed of olivine-basalt to andesitic basalt-breccias, pyroclastics, and pillow lavas. The overlying Reeve is mainly pyroclastics and marine limestones. The Reeve is, in turn, overlain by the Robinson Formation which is composed of volcaniclastic conglomerates, breccias, sandstones, slate, and minor limestone. These units are probably of Permian age but they are overlain by the Lower Triassic Hosselkus Limestone which is a very fossiliferous, shallow-marine limestone. Thus, the whole sequence records the volcanic origin and erosional destruction of an arc terrane and the development of a carbonate bank on the relatively stable remnants.

South of the Melones fault (see Fig. 7-59) are rocks of the Calaveras Formation consisting of argillites, cherts, greenstones, and metamorphosed limestones. The Calaveras was mentioned previously because Schweichert (1976) claimed that it was an equivalent of Havallah ocean-basin terrane rocks but later was widely separated by large-scale rifting. Others, however (e.g., Speed, 1977c; Davis *et al.*, 1978) have voiced opposition to this model and Speed (1978) has suggested that the Calaveras may have been part of a subduction complex containing rocks of different ages and lithologies. One's interpretation of the Calaveras Formation is, thus, predicated on the tectonic and paleogeographic model that one accepts for the southwestern Cordillera.

iii. Klamath Mountains.

In the Klamath Mountains of northern California and southern Oregon is an extremely complex terrane of tectonically intercalated rocks with ages ranging from Ordovician to Jurassic. Rocks considered to be part of the late Paleozoic volcanic-arc terrane are found primarily in the Hayfork terrane (Irwin, 1972) which largely contains andesite with associated volcaniclastic conglomerates, sandstones, chert, mafic intrusions, and pods of limestone with Permian fossils of the Klamath fusulinid fauna. Permian rocks are also found in the other Klamath terranes of the Western Paleozoic and Triassic belt (e.g., Rattlesnake Creek and North Fork) but are not so obviously of volcanic-arc origin and may, as in the case of the Rattlesnake Creek Terrane, represent sediments and basement rocks of oceanic crust juxtaposed with Hayfork terrane rocks much later, possibly during the Late Jurassic or Early Cretaceous.

In the area around Shasta Lake in northern California, to the southeast of the Klamath Mountains, are several more arc-terrane units. The Baird Formation is a Pennsylvanian unit consisting of tuffaceous sandstone, mafic lava flows, and pyroclastics. It is overlain conformably by the Permian McCloud Limestone, an extremely fossiliferous unit which probably represents a carbonate bank that developed on a quiescent volcano. Unconformably overlying the McCloud is the Nosoni Formation containing up to 1850 m of tuffs, volcanic breccias, andesite, basalt, volcaniclastic sandstone, and lesser chert. The Nosoni clearly represents a renewal of volcanic activity following the lull during which McCloud carbonates formed. The Late Permian Dekkas Formation caps the sequence and consists of andesitic tuffs, tuffaceous sedimentary rocks, basalt, and black shale. Conformably overlying Dekkas andestic rocks are Early Triassic flows of the Bully Hill Rhyolite.

iv. Other Areas.

The areas and units mentioned above are probably the best-known arc-terrane rocks but several other units should be mentioned to complete the picture of late Paleozoic volcanic-island arc terranes. In eastcentral Oregon, volcaniclastic detrital sediments comprise the Spotted Ridge Formation. The Spotted Ridge represents a variety of sedimentary environments associated with an island arc including both terrestrial and marine facies; it contains Morrowan-age plant fossils deposited in terrestrial or estuarine environments (Merriam and Read, 1943). In western Idaho are Middle Permian rocks consisting of a variety of volcanic flows, pyroclastics, and volcaniclastics. These rocks are termed the Seven Devils Group and are considered to be fragments of a volcanic arc (Stevens, 1977). In northwest Washington, some rocks of the Chilliwack Group may be volcanic-arc remnants of late Paleozoic age. The Chilliwack, however, is a unit which contains rocks of many ages and types; further, they contain Tethyan faunas. In northeast Washington and adjacent areas of British Columbia are volcanic and volcaniclastic rocks of the lower Mount Roberts Formation which probably also represent a volcanic archipelago.

Late Paleozoic volcanic-arc terranes are also found in the Cordillera of Canada and in southeastern Alaska. Monger *et al.* (1972) described such rocks within the Omineca Crystalline belt of the Columbian orogen and within the Coast Plutonic belt and Insular belt of the Pacific orogen. One such unit is the Thompson assemblage (Davis *et al.*, 1978) in British Columbia. The Thompson assemblage is an informal grouping of rock units such as the Mount Roberts Formation, Arachist Group, Mission Argillite, and Palmer Mountain Greenstone. The presence of andesitic flows and breccias, pyroclastic and volcaniclastic sediments indicates that Thompson rocks represent a volcanic arc and well-preserved fossils date the arc to the Early to Middle Permian. To the west of Thompson arc rocks are cherts, argillites, basalts, and ultramafics of the upper Paleozoic Old Tom, Independence, and Shoemaker formations which probably represent late Paleozoic oceanic crust. Probable oceanic crust and associated sediments are also represented by the Cache Creek Formation in southern British Columbia. Cache Creek strata range in age from Mississippian to Triassic and contain verbeekinid fusulinids of the Tethyan fauna. Similar arc-related rocks are found in southeastern Alaska. The Saginaw Bay Formation contains a lower volcanic member overlain by black shale, volcaniclastics, and, at the top, a shallow-water limestone. Correlative carbonate complexes of Pennsylvanian and Permian age include the Peratrovich Formation, Klawah Formation, and Ladrones Limestone. These units are all considered to represent a displaced arc terrane called the *Alexander terrane* by Churkin and Eberlein (1977).

v. Synthesis. As we stated earlier, the complexity of formation-name stratigraphy obscures the similarities among these various volcanic and volcanogenic units. The real complication lies not in nomenclature but in the identity of the arcs where several are represented. We begin by assuming that rocks with a Tethyan fauna were formed far away from North America and that their accretion to North America was probably a Mesozoic event. Klamath faunas were probably deposited on an arc somewhat closer to North America because there is some evidence of North American influence but the Klamath arc was definitely not directly adjacent to the continent. Even closer to the continent was the arc represented by rocks of the Black Dyke and Mina formations and the Happy Creek volcanics in western Nevada. As we will discuss in the next section, it was probably rocks of this arc-terrane which interacted with

North America to produce the Sonoma orogeny at the end of the Permian.

F.2. The Sonoma Orogeny

At the end of the Permian, the paleogeography of the western Cordillera as discussed in the previous section was fundamentally altered by a series of tectonic events whose close timing suggests a causal relationship, although the exact nature of that relationship is highly speculative. The first of these tectonic events was the thrusting of ocean-basin terrane rocks, such as the Havallah Sequence, over the western margin of the old Antler orogen. This thrusting and other, closely related events are termed the Sonoma orogeny. In this section, we will briefly describe some effects of the Sonoma orogeny and then discuss three tectonic models which have been offered to account for Sonoma events.

F.2.a. Timing and Effects of the Sonoma Orogeny

As originally defined (Silberling and Roberts, 1962), the Sonoma orogeny affected only the area of the Golconda allochthon, but more recent investigations (e.g., Davis *et al.*, 1978; Speed, 1979) have expanded the use of the term to include correlative tectonic events in the Klamath Mountains, where intensive volcanic activity occurred (as seen in the Dekkas Formation and Bully Hill Rhyolite), and even in British Columbia, where volcanic-arc and ocean-basin rocks were deformed, subjected to low-grade regional metamorphism, and uplifted concurrently with thrusting in Nevada.

The timing of the Sonoma orogeny is hard to pin down exactly but seems almost to punctuate the end of the Paleozoic Era. The thrusting was probably completed before extrusion of Early Triassic rhyolites and ash-flow tuffs of the Koipato Group which lie unconformably on deformed Havallah rocks. Volcanics of the Klamath area, which Davis *et al.* (1978) related to the Sonoma orogeny, range from Late Permian to Early Triassic in age. And in the British Columbia area, deformed Permian rocks, such as the Mount Roberts Formation, are overlain with angular unconformity by Middle and Upper Triassic strata.

The thrusting of Havallah Sequence rocks, which defines the "classic" Sonoma orogeny, was not the result simply of movement of a rigid sheet over a single fault surface. Instead, allochthonous Havallah rocks are complexly deformed and contain a number of internal thrust surfaces, themselves deformed, which repeat the stratigraphic sequence many times. This is the reason that Pumpernickel-like greenstones are found throughout the

Havallah sequence and not just at the bottom. A significant observation in this regard was made by Stewart *et al.* (1977); they noted that the youngest fossils yet found in the Havallah sequence (upper Leonardian) come from Pumpernickel-like rocks *near the base* of the allochthon.

Although the Golconda allochthon was clearly uplifted and eroded prior to Koipato time, the surface area and relief of the resultant landmass must not have been great because little or no detrital sediment was deposited in the foreland basin to the east. Collinson and Hasenmueller (1978) noted that few Early Triassic sediments in eastern Nevada or western Utah were derived from the west other than some chert-pebble conglomerates, up to 30 m thick, at the base of the Triassic section where it overlaps Permian rocks in central and southern Nevada. Then too, in other areas of the western continental margin, Late Permian to Early Triassic detrital sediments of western provenance are rare. It is possible that some turbidites in the upper Candelaria Formation overlying volcanic-arc terrane rocks in the Mina area of Nevada may have been derived from the Golconda area but this is debatable since much upper Candelaria sediment is volcaniclastic rather than epiclastic (Speed, 1977c). Clearly, then, uplift related to the Sonoma orogeny was moderate at best and few detrital sediments were derived.

F.2.b. Tectonic Models

The Sonoma orogeny, viewed from its effects, was hardly a major orogenic event. It did not involve large-scale metamorphism (except for the low-grade metamorphism in British Columbia); it did not cause deformation over large regions nor intrusions of large granitic plutons as did other Cordilleran orogenic events. Indeed, it did not even result in mountains. Nonetheless, the Sonoma orogeny has occasioned considerable controversy over its causes. This is primarily because it was the first of those events which mark the change in regional tectonic setting from the Paleozoic pattern of stable continental margins to the Mesozoic pattern of unstable, convergent-plate continental margins. Thus, understanding the Sonoma orogeny is very important and much study has been devoted to it. We will now describe the three major tectonic models so far proposed.

i. Closure of a Back-Arc Basin. Burchfiel and Davis (1972, 1975) presented the case for a back-arc-basin closure model for the Sonoma orogeny. Their model presumes that Havallah rocks represent the sediments of a back-arc basin which were obducted due to basin closure at the end of the Permian. In discussing this model, Dickinson (1977) pointed out that the Havallah back-arc basin may have been an unobducted remnant of the Devonian back-arc basin whose (partial?) closure was postulated by Burchfiel and Davis (1972) to have caused the Antler orogeny. Alter-

nately, the Havallah Basin may have been formed anew by back-arc spreading and westward arc-migration following the Antler orogeny (the "accordion tectonics" mentioned earlier).

Burchfiel and Davis (1975) suggested that basin closure was due to a type of "flake tectonics" in which the crust of the back-arc basin was subducted eastward beneath the continent while the sediment fill of the basin, instead of subducting too, was thrust back over the continent like veneer sliced off a rotating log. This mechanism is required because there is no oceanic crust represented in the Golconda allochthon. Some have argued, however, that such a thin "flake" of marine sediments would not have had enough strength to be thrust in this manner from 30 to 60 km up and over the edge of the continent. Eastward subduction and partial melting of the back-arc crust do however, provide an origin for Koipato magmas which were extruded following obduction of Havallah rocks; the relatively small volume of Koipato volcanics is considered by Burchfiel and Davis to be evidence that only a small amount of subduction occurred, such as would be expected from subduction of the crust beneath a small, back-arc basin.

The main volcanic-island arc, whose approach to the continent closed the basin, is believed to have been the east Klamath arc where Nosoni, Dekkas, and Bully Hill volcanism took place almost continuously from the Middle Permian to the Early Triassic. This continuity of volcanism is cited by Davis *et al.* (1978) as strong evidence that eastward-directed subduction beneath the arc did not cease after the Sonoma orogeny and, therefore, as evidence that the Sonoma orogeny was not the result of the collision of the continent with an east-facing arc as suggested by Speed (1977c; see below). The result of basin closure, in addition to Sonoma effects, would have been to weld the still-active volcanic arc to the western margin of the continent, thereby transforming the continental margin from passive to active and explaining the difference between Paleozoic and Mesozoic tectonic styles.

A serious problem with this model is the lack of abundant craton-derived sediments in Havallah rocks. If they formed in a back-arc basin, they should have an appreciable amount of continentally derived sediment, at least on the eastern side of the basin. Of course, a geographic barrier at the edge of the continent, the Humboldt Highlands for instance, might have cut off the transport of quartzose detritus to the west. Nevertheless, one would expect more evidence of so major a sediment source as the North American continent.

A second problem is the existence of Late Paleozoic volcanic-arc rocks, such as the Black Dyke–Mina and Happy Creek volcanics of western Nevada. Davis *et al.* (1978) mention these rocks briefly as representing an accreted Paleozoic arc-terrane but they do not discuss how such rocks relate to the Golconda allochthon or to east-dipping subduction in the east Klamath area.

ii.   Volcanic-Arc/Continent Collision.

The second model of Sonoma tectonics was presented by Speed (1977c) who postulated that Havallah sediments were ocean-basin sediments tectonically intercalated into a subduction prism and thrust over the edge of the Antler belt as it was partially subducted westward beneath an advancing volcanic arc (see Fig. 7-56). This volcanic arc is not represented by the east Klamath area but rather by volcanic rocks of western Nevada. This model does not have to rely on complex "flake tectonics" to explain the absence of oceanic crust in the Havallah because it supposes that the Havallah is essentially a mass of sediment scraped off a subducting plate prior to arc/continent collision.

Note that this idea is supported by the observation that the lowest Havallah rocks contain the youngest fossils; this is in accord with the geometry of modern subduction prisms. The absence of mélange-type structure and blueschist metamorphism would seem to constitute a problem for this model; Speed suggested that the depth of exposure may yet be insufficient to reveal such features. On the other hand, this model does offer a reasonable explanation for the general lack of deformation in the autochthonous terrane beneath the Golconda allochthon. It also provides a simple mechanism for the thrusting of the thin sheet of seafloor sediments over the continental margin: in essence, the theory is that the anisostatically depressed continental margin underthrust the Golconda mass! After arc/continent collision ceased in the Early Triassic, cooling of the arc and back-arc areas caused subsidence of the arc terrane. The subsidence resulted in the formation of a successor basin in which Middle Triassic and younger sediments were deposited. Also following the arc collision, plate convergence began somewhere to the west of the arc terrane. Davis *et al.* (1978), as we have mentioned, argued that the continuity of volcanism in the east Klamath area was evidence against Speed's model. But if the Klamath arc were not involved in the collision at all but were still somewhere off to the west of the continent, this argument would have less weight. A more serious problem is presented by Koipato volcanics whose origin is not adequately explained by Speed's model.

iii.   Collision of the Sonomia Plate.

In 1979, Speed presented a somewhat altered and expanded version of his arc-collision model. In his new version, Speed considered the arc-terrane rocks of western Nevada, the eastern Sierra Nevada, the eastern Klamaths, and southern Oregon to represent a single laterally continuous plate which he termed *Sonomia* (see Fig. 7-57). Sonomia is considered to be composed of sequential Paleozoic arc-terranes, the youngest being the active arc and associated Havallah subduc-

tion-complex which collided with North America to produce the Sonoma orogeny as discussed above. The significance of this new version is the addition of the third dimension, shown in Fig. 7-58. Speed considered that the Cordilleran region during the late Paleozoic was composed of three lithospheric plates: the North American plate, Sonomia, and Plate X, an oceanic plate of unknown extent. Notice in the figure that the North American continent is shown to extend farther to the southwest during the Permian than it does now. The Sonomia plate, moving southward, overrode the North American plate as well as Plate X. The subduction complex between Sonomia and North America is represented by the Havallah Sequence whereas the subduction complex between Sonomia and Plate X is represented by the Calaveras Formation. This idea rather neatly explains the general similarities of these two formations while also explaining their specific differences. Note that this model also explains the eastward (or actually northeastward) subduction inferred in the Klamath area as being the result of the subduction and partial melting of Plate X beneath Sonomia. With the collision of Sonomia and North America, movement of Sonomia ceased but Plate X continued to underride Sonomia so that Klamath-area volcanism continued. Plate X also continued to subduct beneath the hypothesized southwestern extension of North America, leading to the tectonic events of the Early Triassic. Indeed, one of the most attractive aspects of Speed's new model is that it integrates the tectonics of the Sonoma orogeny and Triassic events into a single, coherent plate model.

F.3.   Early Mesozoic Tectonics of the Western Cordillera

The Sonoma orogeny heralded the onset of tectonic pandemonium all along the North American Cordillera. Following closely after Sonoma deformation came a series of accreted terranes, transform faults, and volcanic arcs. The extant evidence from these events allows many, contradictory interpretations and in the following discussion we will be able to present only a few of the tectonic models which have been suggested for the post-Sonoma Cordillera. It will be clear, however, even from a cursory overview that extremely complicated plate interactions were probably involved and that the resulting styles of deformation differed greatly from Paleozoic styles. One major factor in this change may be the fact that it was during the Triassic that Pangaea began to break up. As a result, Mesozoic Cordilleran tectonics may be grossly related to the movement of the North American plate away from Pangaea. And, in so moving, the Cordilleran margin "swept up" the various displaced terranes which littered the Panthalassa Ocean floor. In this section, we have divided the Cordilleran

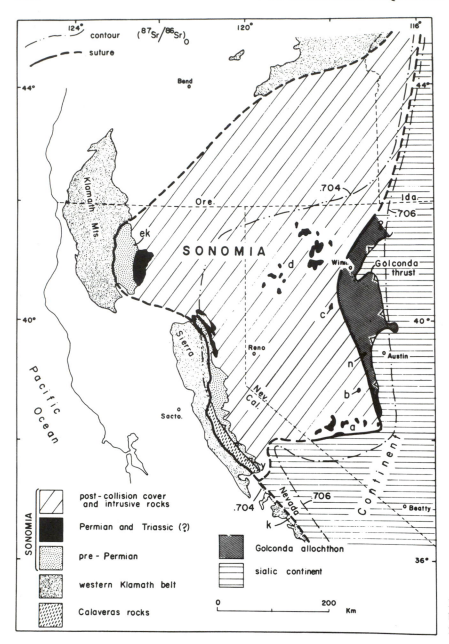

Fig. 7-57. The Sonoma microplate as envisioned by Speed (1979, *Journal of Geology*, Vol. 87, p. 280).

orogen into two segments both for ease of discussion and because the two segments had very different histories. The Southern Cordillera we define to be the area south of the Klamath Mountains, including the Sierra Nevada Mountains, the Great Basin, and regions to the south. The Northern Cordillera is defined here to include the Klamaths and areas to the north.

F.3.a. The Southern Cordillera

i. Sedimentation in the Post-Sonoma Great Basin. Following the emplacement of the Golconda allochthon in earliest Triassic time, the arc terrane of the western Great Basin subsided to become a large marine basin. Speed (1977c) called this new marine area a successor basin and suggested that subsidence was the result of cooling of arc and back-arc regions after the Sonoma collision and consequent cessation of westward subduction beneath the Sonomia plate. Speed (1978) recognized two major sedimentary provinces within the Triassic successor basin: an eastern, shallow shelf adjacent to the Golconda allochthon and a western, deep-water environment. The eastern shelf was occupied by a carbonate platform during the Middle Triassic time; these carbonates are now represented by the Star Peak Group in the northern area of the shelf and by the Luning Formation in the southern area In

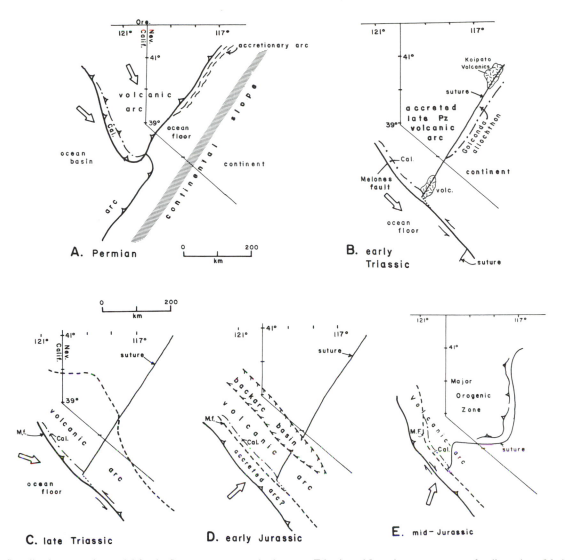

Fig. 7-58. Speed's plate-tectonics model for the Sonoma orogeny and subsequent Triassic and Jurassic events; see text for discussion. (Modified from Speed, 1978, reproduced by permission of the Pacific Section, Society of Economic Paleontologists and Mineralogists.)

the western basinal area were deposited fine, dark muds, distal turbidites, and pelagic cherts represented by the Clan Alpine sequence in central Nevada and by the Byers Canyon, Dyer Canyon, and Bernice formations in the Black Rock Desert area of northwestern Nevada. As more sediment was deposited in the basin, the carbonate platform prograded over basinal muds during the later Middle Triassic, capping the Clan Alpine sequence with shallow-water carbonates. The western part of the basinal area, however, continued to be a deep-water environment throughout the rest of the Triassic.

The difference between shelf and basin areas was probably due to different amounts of subsidence in the two areas. The shelf province lay along the western side of the uplifted Havallah Sequence and therefore was very close to the edge of the Paleozoic continental crust. Thus underpin-

ned by sial, the eastern shelf was not allowed to subside to any appreciable extent. On the other hand, the western area, being underlain only by mafic to intermediate arc-volcanic rocks of the Sonomia plate, subsided more rapidly.

ii. Early Triassic Truncation of Antler and Sonoma Trends. Consideration of Fig. 7-59 will reveal that the trends of major Paleozoic elements of the Southern Cordillera (e.g., the edge of the Paleozoic shelf, the axes of Upper Devonian and Mississippian foreland basins, and the strikes of the Roberts Mountains and Golconda allochthons) are all roughly parallel to each other. But major Mesozoic structures (e.g., the Foothills suture and Triassic–Jurassic fold belts in eastern California) and igneous bodies (e.g., the Sierra Nevada batholith) trend at high angles *across* the trends of Paleozoic structures. Burchfiel and Davis (1972)

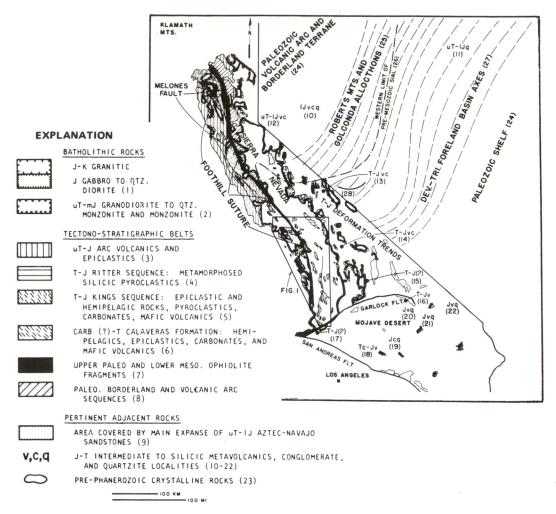

Fig. 7-59. Generalized geological map of Sierra Nevada region and its relation to Paleozoic structural and tectonic trends. The Foothills Suture (which represents the Melones Fault in the northern Sierra Nevada and the $^{87}Sr/^{86}Sr = 0.706$ contour line in the south) is the inferred line of conti- nental truncation and subsequent suture between Paleozoic rocks and ac- creted rocks of the Kings–Kaweah Belt. See text for discussion. (From Saleeby *et al.*, 1978, reproduced by permission of the Pacific Section, Society of Economic Paleontologists and Mineralogists.)

proposed that this relationship was caused by some kind of continental truncation event in the Early Triassic which re- sulted in the breaking off of an indeterminately large piece of North America. The break cut across the Paleozoic struc- tures so that subsequent tectonic activity took place along a beveled continental margin.

The idea of a continental truncation event is now well established but debate continues over the specific cause of the truncation and over the identity of the displaced (mis- placed!) continental fragment. There are two possibilities: (1) the truncation was caused by continental rifting along a newly formed divergent plate-boundary; or (2) the trunca- tion was caused by transform faulting. We will discuss each of these views briefly.

Schweichert (1976) proposed the rifting hypothesis and subsequently (1978) elaborated his views into a general model of Triassic and Jurassic plate motions in the Cor- dilleran region. He postulated that the truncated continental fragment is represented by rocks of the Alexander displaced terrane in Alaska and, possibly, of Wrangellia. The general scenario of the rifting model is shown in Fig. 7-60. Follow- ing rifting and concomitant opening of a new oceanic basin (the "Triassic Ocean floor"), a change of plate motion is believed to have initiated *dextral* (i.e., right-lateral), trans- form motion between the Tethyan Ocean Floor and the Triassic Ocean Floor. In this manner, the Alexander and Wrangellia terranes were moved northward. At the same time (Late Triassic–Early Jurassic), a subduction zone de- veloped along the truncated North American margin. This subduction led to volcanic and plutonic activity which char- acterized the Mesozoic Southern Cordillera and also al- lowed the Triassic Ocean Floor to plunge conveniently off

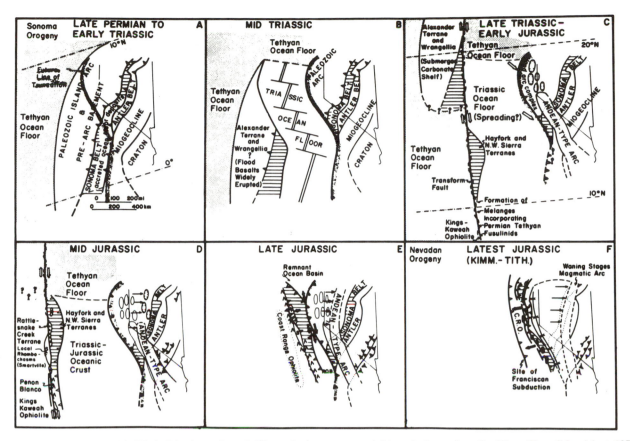

Fig. 7-60. Schweichert's model of Early Triassic continental rifting and subsequent tectonic history in the southern Cordillera. (From Schweichert, 1978, reproduced by permission of the Pacific Section, Society of Economic Paleontologists and Mineralogists.)

into obscurity. Schweichert's ideas have found few supporters. Davis *et al.* (1978) referred to the rifting model as ''imaginative, but unsupportable'' because they disagreed with Schweichert's correlation of some units in the Klamath area with other units of the Antler orogen. Similarly, as we have seen, Speed (1977c) disagreed with Schweichert's correlation of Havallah with Calaveras rocks.

On the other hand, several authors have concurred with Schweichert's identification of the Alexander terrane as the displaced fragment of California. Saleeby *et al.* (1978) made the same observation but they proposed that the truncation and displacement was caused by a major transform fault. If this is correct, the offset of the fault would have to be dextral in order to displace a piece of California up to Alaska. And it may be stated that dextral slip has been an important component of Cordilleran plate interactions at least since the Jurassic (see below); indeed, the San Andreas fault today is a dextral transform fault. But that certainly does not require the hypothetical Early Triassic transform to be dextral. In fact, majority opinion at the moment is that such transform faulting was *sinistral* (i.e., left-lateral). Davis *et al.* (1978) suggested sinistral faulting to account

for the truncation event and pointed to apparently offset Paleozoic detrital sequences in the El Paso Mountains of southern California which they correlated with similar detrital sequences of the shelf area to the northwest. Van der Voo *et al.* (1977) argued that the displaced Antler belt is now part of mainland Mexico and suggested that the transform faulting was genetically related to the opening of the Gulf of Mexico rather than to specific plate interactions in the Cordillera *per se*. Speed (1978, 1979) also argued for sinistral transform truncation. Speed's view may be seen in Fig. 7-58; note that following the Sonoma orogeny, Plate X is shown to have continued its southeasterly movement. As Speed put it (1978, p. 269), Plate X ''found means of detaching a fragment of the North American continent while maintaining essentially constant relative velocity.''

iii. Development of Andean Volcanic Arc. During the latest Middle Triassic, shortly after the truncation event, a subduction zone developed along the newly established continental margin. The result was the development of a volcanic arc across the southwestern part of the continent. Rocks of this arc are found in southwestern

Nevada and adjacent southeastern California where they comprise a number of unnamed local units of andesite, volcanic conglomerate, tuff, and volcaniclastic sediment. For example, the Gold Range Formation of Speed (1977b) consists of ash-flow tuff and fluvially transported volcaniclastics, thus indicating a subaerial volcanic arc in the southeastern area of the Triassic continental margin (i.e., the part of the margin underlain by sialic crust). In areas to the northwest, however, rhyolitic flows and breccias are associated with marine sediments, such as fossiliferous carbonates, suggesting that a submarine segment of the arc formed along the part of the margin underlain not by sial but by Sonomia mafic to intermediate rocks.

Detrital sediments derived from uplifted arc terranes were carried by streams to the old eastern carbonate shelf where they were deposited as the Late Triassic Grass Valley Formation. Toward the end of the Triassic, activity of the arc decreased considerably and the position of the arc migrated southwestward. The former arc terrane became a back-arc basin in which Lower Jurassic carbonate sediments were deposited over older, Upper Triassic volcanics (see Fig. 7-58). Speed related the continental truncation event, the development of the arc, and the southwestward migration of the arc to a steady, continuous clockwise rotation of North America with respect to the motion of Plate X (or, equivalently, to a steady counterclockwise rotation of Plate X with respect to North America). Further, Speed assumed that this rotation continued into Jurassic time during which it led to a complete shift in the sense of plate motion along the continental margin, leading to dextral slip where previously there had been sinistral slip. Thus, Speed related most of the complexities of Triassic Southern Cordillera tectonics to a relatively simple plate-tectonics scheme. Before awarding accolades, however, it should be noted that this model does not account for observations made on rocks of the Kings–Kaweah belt along the western side of the Sierra Nevada batholith.

iv. The Kings–Kaweah Belt.

Along the western side of the Sierra Nevada batholith and within country-rock roof-pendants hanging down into the batholith are strongly deformed and intensely metamorphosed rocks of late Paleozoic and early Mesozoic age which represent the prebatholith crust of the region (see Fig. 7-59). Correlation from pendant to pendant and from one side of the batholith to the other is difficult at best. Nevertheless, Saleeby *et al.* (1978) have recognized a variety of lithological units in these rocks and have presented an excellent analysis of their implications to Southern Cordillera tectonics.

Saleeby *et al.* described the *Kings–Kaweah* ophiolite belt as a "tectonic megabreccia" (i.e., a large-scale mélange) consisting of mappable slabs of ophiolite, termed the *Kings River ophiolite,* which occur within a serpentinite matrix. In the area south of Kings River, a similar unit is composed of ophiolite blocks up to 3 km in length within a serpentinite matrix; this unit is termed the *Kaweah Serpentinite mélange.* Metasediments and metavolcanics are also recognized and comprise the *Kings sequence,* composed of Upper Triassic and Lower Jurassic quartzites, subarkosic flysch, carbonates, metavolcanics, and olistostromal masses.

Saleeby *et al.* believed that the Kings–Kaweah ophiolite belt represents late Paleozoic oceanic crust which was accreted to the North American continent during the Late Triassic or Early Jurassic. They observed that the structure and petrology of the ophiolite are considerably more complex than "normal" oceanic-crust ophiolite and postulated that it had undergone strong deformation and hydrothermal metamorphism near its point of origin, resulting in much serpentinite alteration of the mafic rock *before* deposition of thick pelagic sediments upon it. From these observations, Saleeby *et al.* concluded that the Kings–Kaweah belt represents a major fracture zone complex.

The motion of the Kings–Kaweah fracture zone brought it progressively closer to the continent. This is documented by the sequence of sediments deposited on it. At first, only purely siliceous cherts were deposited when the area was still far removed from the continent. Overlying these are argillaceous cherts which probably represent the influx of hemipelagic sediments when the fracture zone area was in closer proximity to the continent. At the top of the sequence is flysch derived from basaltic–andesitic rocks of the Late Triassic marginal volcanic arc.

Perhaps the most interesting aspect of the Saleeby *et al.* model is its description of the accretion of the ophiolite mass to the continent. They argued that the essentially strike-slip, transform motion which had resulted in continental truncation had evolved into oblique subduction due to changes in relative plate motions, such as the rotation mentioned earlier. The resulting plate boundary, referred to as a *transpressive boundary,* is characterized by both arc volcanism and transform faulting. Oblique convergence continued and the resulting volcanic arc (the same one discussed earlier) began to be active along the southwestern continental margin. At some point in time, probably toward the end of the Late Triassic, the Kings–Kaweah fracture zone was brought into proximity with the trench. But fracture-zone rocks were composed to a significant extent of serpentinite, which is less dense than normal sial, and therefore resisted subduction. Thus, it was accreted to the edge of the continent instead of being lost down the trench. The effect was to cause the locus of subduction to shift to the other side of the Kings–Kaweah mass, explaining the apparent lull in volcanic activity in the southern Great Basin area and the subsequent southwestward shift of the arc during the Early Jurassic.

The newly accreted Kings–Kaweah ophiolite belt then functioned as basement for fore-arc sediments (see Fig. 7-61). The Kings Sequence contains a variety of sediment types and depositional environments associated with the complex ophiolitic basement. Figure 7-61 illustrates well the degree of sedimentological complexity expected in such a tectonic situation.

A final observation on this model relates to the presence of extensive quartz sands associated with arc-derived volcaniclastics. Such a large amount of quartz is anomalous within a volcanic arc environment. Saleeby et al. suggested that the entire Kings–Kaweah belt was farther south during the Early Jurassic than it is today, thus placing it near the southern end of the craton and, thus, adjacent to the area where quartz sands from the cratonic interior were being

carried off the continent by longshore and offshore currents. Jurassic sandstones of the western interior, such as the Navajo and Aztec sandstones, were parts of this southward to southwestward dispersal system. Being near the end of this sand dispersal system, the Kings–Kaweah belt received much submarine quartz detritus which was mixed with arc-derived sediments to comprise the complex Kings Sequence.

F.3.b. The Northern Cordillera

As we have seen, the Southern Cordilleran margin of the continent was, during Late Triassic to Early Jurassic time, a transpressive plate boundary. Farther north along

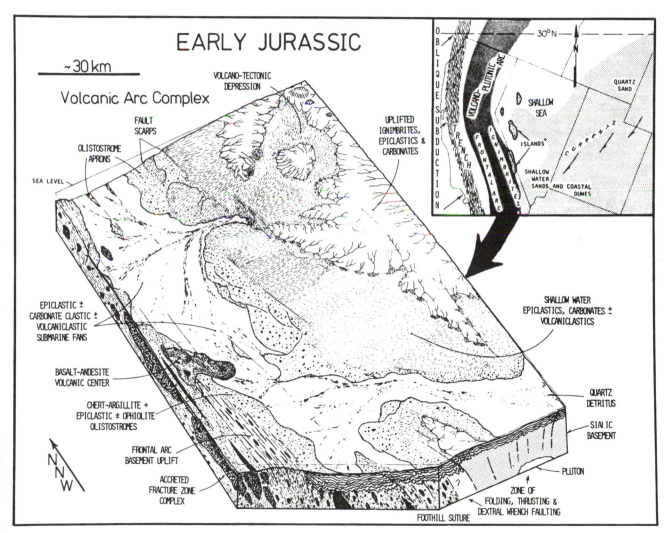

Fig. 7-61. Tectonic and paleogeographic model of Kings–Kaweah Belt rocks in the Sierra Nevada region, showing inferred depositional environments of Kings and Ritter sequences and their relation to the accreted Kings–Kaweah ophiolites; insert shows hypothetical paleogeographic interpretation of Kings–Kaweah area to cratonic sedimentary facies. (From Saleeby et al., 1978, reproduced by permission of the Pacific Section, Society of Economic Paleontologists and Mineralogists.)

the Cordillera, however, the geometry of plate motions was such that more nearly normal subduction occurred. Accreted terranes typically bear thrust-faulted relations to adjacent units rather than the essentially transcurrent-faulted relations of the transpressive boundary. The Northern part of the Cordillera is therefore considered to be an aggregate of accretionary terranes. It is to these accretionary terranes that we now direct attention.

i. Klamath Mountains. In the Klamath area there are several ophiolitic terranes underthrust beneath each other in a structural pattern of striking complexity. These accreted slices of ophiolite record an almost continuous history of plate convergence. Figure 7-62 illustrates the major terranes, their structural relationships, and their relative ages. Notice that the Central Metamorphic belt is underthrust beneath the Trinity ophiolite (early Paleozoic in age) and that the Hayfork–North Fork terrane similarly is underthrust beneath the Central Metamorphic belt. The Hayfork, as we have seen earlier, probably represents a Permian volcanic arc and the North Fork ophiolite may be a scrap of the oceanic crust which subducted beneath the Hayfork arc: In the western area, the Hayfork–North Fork sheet is underthrust by the younger Rattlesnake ophiolite terrane which may also represent a sliver of oceanic crust. Even this brief description shows that the Klamath area experienced a series of thrust intercalations during Later Triassic and Early Jurassic time and is therefore considered to have been part of the accretionary terrane of the Northern Cordillera rather than part of the Southern Cordilleran transpressive terrane.

ii. Central Oregon and Western Idaho. Inliers of Mesozoic arc-related rocks are exposed out from under the cover of Tertiary volcanic rocks in central Oregon and western Idaho. In the southeastern part of the area, Upper Triassic volcanic flows and volcanic breccias, with associated volcaniclastic sedimentary rocks, comprise the *Huntington arc terrane* (Dickinson, 1979). In the central part of the area is a strongly deformed assemblage of rocks composed of Permian ophiolites referred to as the *central mélange terrane*. And in the northern part of the area is a pile of volcanics and volcaniclastics called the *Seven Devils terrane* which is now considered (e.g., Jones *et al.*, 1977) to have been part of Wrangellia and, therefore, a displaced terrane whose accretion occurred later; we will not consider the Seven Devils terrane further at this time. Between the Huntington arc terrane and the central mélange is a sequence of detrital strata which comprise the *Mesozoic clastic terrane;* Dickinson (1979) proposed that these strata were fore-arc basin deposits and that they had accumulated in the area between the central mélange terrane and the Huntington arc.

iii. British Columbia. In British Columbia, Upper Triassic and Lower Jurassic strata are exposed in three areas: the Front Ranges and Foothills of the Southern Canadian Rockies, where a shallow-marine assemblage of cratonic detritus and carbonates probably represents a continental shelf area; along the southern end of the Omineca crystalline belt, where argillites, siltstones, volcaniclastics, and volcanic flows are interpreted as deposits of a back-arc basin; and the Okanagan area of southern British Columbia where volcanic-arc and putative oceanic-crustal rocks are exposed. In the following discussion (taken largely from Davis *et al.*, 1978), we will consider the rocks of the Okanagan region in more detail.

The orogenic event during Late Permian to Early Tri-

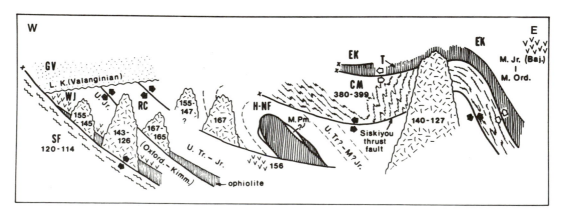

Fig. 7-62. Diagrammatic representation of stratigraphic and radiometric age control on thrust faulting in the Klamath Mountains, California and Oregon. Length of section is approximately 200 km. SF, South Fork Mountain Schist; GV, Great Valley Sequence; WJ, western Jurassic subprovince; RC, Rattlesnake Creek terrane; H-NF, Hayfork–North Fork terrane; CM, central metamorphic subprovince; T, Trinity Ophiolite; EK, eastern Klamath subprovince. Thrust faults with black relative-motion arrows are post-Paleozoic in age. (From Davis *et al.*, 1978, reproduced by permission of the Pacific Section, Society of Economic Paleontologists and Mineralogists.)

assic time (penecontemporaneous with the Sonoma orogeny) resulted in the destruction of the Thompson volcanic arc. During the Late Triassic, a new volcanic arc developed on the deformed remains of the Thompson arc. Deposits of that new arc in the Okanagan area are represented by the Nicola Group, which consists of almost 5000 m of alkaline–basaltic and andesitic flows, tuffs, volcanic debris-flow deposits, local conglomerates, sandstones, and biohermal carbonates. These extrusions and related rocks are intruded by comagmatic diorites, monzonites, syenites, and granodiorites. The Nicola arc, thus, was the site of diverse igneous activity. Farther north along the Cordillera, correlative volcanics of the Takla Group are very similar to Nicola rocks.

Takla–Nicola arc rocks are associated with ophiolitic units referred to as the Cache Creek and Bridge River Groups. These units probably represent the oceanic crust whose subduction resulted in Takla–Nicola arc volcanism. In the Okanagan area, the two units comprise a single ophiolitic body but, farther to the north, there are two separate ophiolite belts. The eastern belt, associated with Takla volcanics, is represented by the Cache Creek Group whereas the western belt is represented by Bridge River rocks.

In between the two ophiolite belts is a major displaced terrane referred to as the *Stikine block* or *Stikinia* (Monger, 1977). Composed of upper Paleozoic calc-alkaline volcanics and various carbonates, the Stikine block probably acted as a distinct microplate like Sonomia. The Stikine block, as it moved toward its rendezvous with the Cordillera, was bordered on the east by a volcanic arc; this arc, represented by Stuhini arc rocks, was probably caused by westward subduction of the Cache Creek oceanic plate, trapped between Stikinia and the Takla arc.

The Stikine block was emplaced into the Cordilleran collage during the latest Triassic, terminating activity along the Takla and Stuhini arcs. The deformation and intrusions associated with the collisional accretion of Stikinia have been termed the *Inklinian orogeny* (Douglas *et al.*, 1976). Activity of the Nicola arc, which was south of the accreted plate, continued into Middle Jurassic time.

iv. Alaska. Earlier in this section we mentioned that late Paleozoic rocks in southern and southeastern Alaska consist dominantly of volcanic-arc-related rocks. These volcanic-arc rocks are considered by Churkin and Eberlein (1977) to represent one or more displaced terranes, such as the Alexander terrane or Wrangellia. Churkin and his coworkers at the USGS have continued to analyze Alaskan tectonics; they concluded (1979, 1980) that much of central and southern Alaska is composed of a surprisingly large number of displaced volcanic-arc terranes, all complexly accordioned together. Figure 7-63 is a "pull-apart" map (i.e., a kind of palinspastic map in which displaced terranes

have been separated from each other for comparison) of eastcentral Alaska. Notice that a wide variety of microplates are represented: continental blocks such as the Yukon–Tanana terrane and oceanic blocks such as Wrangellia. Central and southern Alaska is a virtual antique store of crustal memorabilia.

During our earlier discussion of Triassic events in the Sierra Nevada, we presented evidence for northeastward movement (i.e., toward the northern Cordillera) of the western oceanic plate. Disposed at the end of this oceanic conveyor-belt, Alaska and western Canada were the targets of a diaspora of arc slivers and crustal slabs. Obviously, all these accreted crustal-fragments imply that somewhere there is an original continental margin which functioned as a "backstop" against which the displaced terranes were squeezed and molded by episodic plate collisions. Churkin *et al.* (1979) proposed that the Paleozoic continental margin is today approximately coincident with the southernmost extent of Lisburne Group carbonates in the southern Brooks Range. The Lisburne Group represents a shallow-water carbonate platform formed on the edge of the continent. It is believed to have graded southward down the continental slope into dark, argillaceous muds, chert, and carbonate turbidites of the Kagvik structural sequence (i.e., a sequence of structurally juxtaposed lithologies which once represented a coherent stratigraphic sequence). Kagvik rocks range in age from Carboniferous to Triassic and are considered to be deep-water deposits, possibly even an ocean-floor deposit. Thus, the Lisburne Group represents the approximate location of the continental margin against which the displaced terranes piled up.

The concept of the tectonic antique store with scraps and slivers of ancient crustal fragments from faraway places is an immensely intriguing romance and can be derived logically from the central ideas of plate tectonics. It seems to be able easily to explain problems previously enigmatic and is supported by paleomagnetic data from arc volcanics and by fossil assemblages in associated sedimentary rocks. We will return to this topic in the next chapter.

G. Upper Paleozoic and Lower Mesozoic Strata of the Innuitian Continental Margin

We turn finally to the strata and structures of the Innuitian continental margin. The geological history of this region during the late Paleozoic and early Mesozoic seems much less complicated than the histories of the other continental margins discussed in this chapter; there were apparently no major orogenic events and no accretion of displaced terranes such as those which clutter the history of the

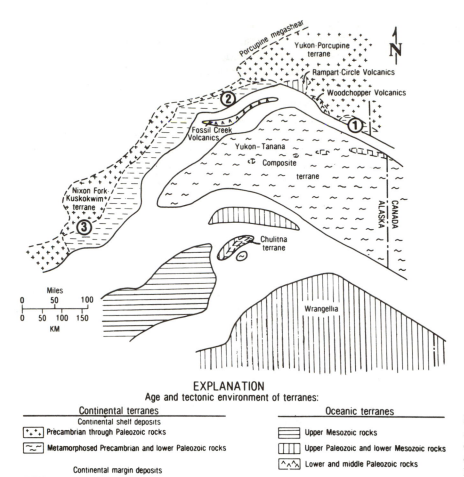

EXPLANATION
Age and tectonic environment of terranes:

Continental terranes
———————————
Continental shelf deposits
[+.+.] Precambrian through Paleozoic rocks

[~~] Metamorphosed Precambrian and lower Paleozoic rocks

Continental margin deposits
[=-=] Lower and middle Paleozoic rocks

Oceanic terranes
———————————
[≡] Upper Mesozoic rocks

[III] Upper Paleozoic and lower Mesozoic rocks

[^^^] Lower and middle Paleozoic rocks

Fig. 7-63. "Pull-apart" map of east central Alaska showing the inferred microplates (or, displaced terranes) which comprise the tectonic collage in the Alaskan Cordillera. (From Churkin *et al.*, 1980, *Geological Society of America Bulletin*, Fig. 6, p. 653.)

Cordilleran margin. (We remind the reader, however, that knowledge of this desolate and difficult area is poor. It is often the case that geological interpretations made on the basis of sparse data are simple but become more complex as the quantity of data is increased.) Nevertheless, Innuitian strata of these ages have yielded several fascinating sedimentological studies of shallow- to deep-water facies relations and of the formation of deep-water evaporites and gravity-flow deposits. These strata also contain, in Alaska, one of the major hydrocarbon reservoirs of North America.

G.1. Pennsylvanian Strata in Northern Alaska

At the end of the Mississippian, shelf carbonates of the Alapah Limestone, lower unit of the Lisburne Group, were being deposited over much of northern Alaska. Alapah deposition had been time-transgressive, beginning in the southern Brooks Range area during the Early Mississippian but not commencing in the north until the Late Mississippian. Lower Alapah beds are principally composed of arenaceous limestone whose quartz-sand fraction was de-

rived from a detrital source to the north (Sable, 1977). This sourceland is referred to in the literature of the North Slope as the Beaufort Uplift (Morgridge and Smith, 1972). The upper Alapah is primarily a dark, micritic, partly dolomitized limestone suggestive of deposition under relatively restricted marine conditions.

Overlying light-gray carbonates of the Lower Pennsylvanian Wahoo Limestone, upper unit of the Lisburne Group, contrast markedly with dark Alapah strata although the contact between them is apparently conformable. In the Romanzof Mountains of northeastern Alaska, the Wahoo consists of two members. The lower unit is composed of light-gray, micritic limestone with minor lenses of crinoidal biosparite and, near the base, several beds of bryozoan and brachiopod coquina. The upper member is massive crinoidal limestone. Thus, Wahoo strata appear to indicate a more-open marine environment than do Alapah strata.

Lisburne carbonates apparently represent progressively deeper-water conditions toward the south and are lost altogether just south of the Brooks Range. Churkin *et al.* (1980) have interpreted the southern end of Lisburne strata to represent the edge of the Paleozoic continental margin. Farther to the south, platform sediments are believed to have graded down the continental slope into deep-water deposits such as the black shales and radiolarian cherts of the Kagvik structural sequence. Churkin *et al.* (1979) considered the Kagvik sequence to represent deep, ocean-basin deposits correlative with Lisburne and younger strata of the Brooks Range area. Exact paleogeographic relationships of the two areas are problematical but the presence of carbonate turbidites within the Kagvik was cited as evidence of a nearby carbonate source, probably the Lisburne platform on the continental margin.

Following Lisburne deposition, which ended during or after Atokan time, the entire Brooks Range area was subjected to erosion. It apparently remained a low land area until the Early Permian.

G.2. Pennsylvanian and Lower Permian Strata of the Sverdrup Basin

The Late Devonian–Early Mississippian Ellesmere orogeny resulted in a major mountain system all across the northern part of Canada. Erosion of the Ellesmere orogen supplied great volumes of detrital sediments to depositional areas far to the south and southwest (e.g., the Michigan, Illinois, and Appalachian basins). Since the Innuitian region during Early Mississippian time was an actively eroding mountain belt, it is not surprising that Lower Mississippian through lower Upper Mississippian strata are entirely absent. But during the later Late Mississippian, a new phase in the geological history of the Innuitian area began with the onset of subsidence within the heart of the Ellesmere orogen. The resulting successor basin is named the *Sverdrup Basin* (see Fig. 7-64) and contains a virtually conformable sequence of strata ranging in age from Late Mississippian to Early Tertiary! Throughout the long history of the Sverdrup Basin, its axis coincided roughly with the axis of the mid-Paleozoic Hazen trough, thus suggesting that both subsident structures were related to some feature of the crustal basement. The Sverdrup Basin ceased to be an active tectonic element following the Early Tertiary Eurekan orogeny, during which Sverdrup Basin strata were strongly deformed.

G.2.a. Mississippian Beginnings

The oldest strata of the Sverdrup Basin are nonmarine detrital sediments of the Emma Fiord Formation, which occur only in scattered areas along the northeastern margin of the basin in northern Axel Heiberg and Ellesmere Islands (see Fig. 7-64). Emma Fiord strata lie with striking unconformity on upturned and beveled rocks of the Ellesmere orogen and consist mainly of carbonaceous siltstone and fine sandstone with lesser shale and coal. Plant fossils associated with the coal seams were shown by Bamber and Copeland (1976) to be of middle to late Chesterian age.

Overlying the Emma Fiord is the much more extensive Borup Fiord Formation, another predominantly nonmarine unit consisting of coarse, detrital red beds. The Borup Fiord is considered to be of latest Mississippian (Namurian) age (Trettin, 1972). As shown in Fig. 7-65, the Borup Fiord underlies most of the Sverdrup Basin except in the southeast where similar but much thicker sedimentary rocks are named the Canyon Fiord Formation. Clearly, nonmarine conditions prevailed over most of the Sverdrup Basin area just prior to the onset of intense subsidence during the Pennsylvanian.

G.2.b. Pennsylvanian and Early Permian Facies Patterns

The rate of basin subsidence apparently increased drastically at the beginning of the Pennsylvanian Period, outstripping the rate of sedimentation so that the Sverdrup Basin evolved rapidly into a marine basin whose axial region was the site of relatively deep-water conditions. The result was the development of a series of shallow-shelf facies around the margins of the basin and a deep-water facies in the center.

The main source of detrital sediments was the North American craton to the south; therefore, terrestrial and marginal-marine sedimentary facies composed dominantly of red, quartzose detritus occur along the southern margin of the basin. These deposits, referred to by Trettin (1972) as the detrital belt, are represented by the Canyon Fiord Formation. Canyon Fiord rocks grade northward into a belt of mixed detrital and carbonate rocks represented by the Belcher Channel Formation. Still farther north, similar mixed detrital and carbonate strata comprise the Antoinette and Tanquary formations which are separated from each other by evaporites of the Mount Bayley Formation. The juxtaposition of Mount Bayley rocks with shallow-water carbonates suggests that its evaporites may have formed in sabkha-like coastal embayments on the shelf.

Beyond the influence of detrital-sediment influx, clean, limy sediments of the Nansen Formation were deposited on a shallow, current-agitated, carbonate platform. The Nansen is composed of various carbonate lithologies including bioclastic micrites, calc-arenites, and oolitic limestones. Its carbonate petrology is similar to that of the Bahama platform today and Nansen strata probably represent the same type of environment. Cyclic carbonate sedimentation (e.g., alternating sequences of micritic and sparry carbonates) was recognized by Davies (1977b); of course, evidence of cyclicity in Pennsylvanian strata is hardly a novelty. In the upper part of the section, Nansen rocks are

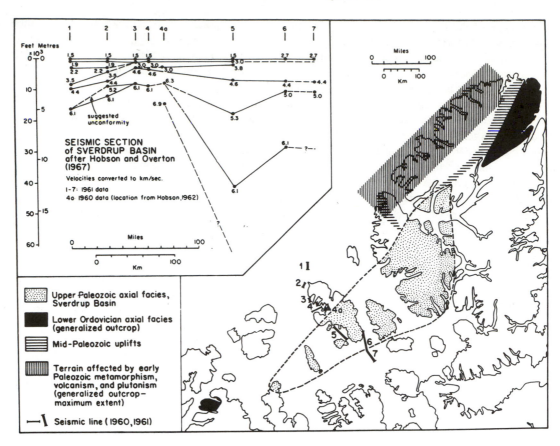

Fig. 7-64. The Sverdrup Basin. Fine stippled pattern indicates location of axial facies of the basin; dot-and-dash pattern outlines the approximate extent of shallow-water facies; coarse stippling indicates the location of covering strata of the Arctic Coastal Plain. (Modified from Trettin, 1972, reproduced by permission of the Geological Association of Canada from GAC Special Paper No. 11, Variations in Tectonic Styles in Canada.)

dominated by phylloid–algal reef masses. The Nansen ranges in thickness from 1200 to 2200 m, being thickest where it abruptly grades into deep-water deposits of the basin center. The Nansen platform partly surrounded the basin-center trough and Nansen carbonates are present today on both northern and southern margins of the basin. Nansen deposition ranged from Early Pennsylvanian to Early Permian time.

In extreme northeastern portions of the Sverdrup Basin, Nansen carbonates and underlying Borup Fiord red beds are interbedded with approximately 560 m of basaltic

flows and pyroclastic debris. These volcanics are termed the Audhild Formation (Trettin, 1972). Small amounts of detrital sediment are found in Nansen carbonates in northern and northwestern parts of the basin. These sediments were probably derived from basement uplifts just to the northwest of the basin.

Deep-water deposits of the basin center comprise two formations: a lower unit, the Otto Fiord Formation (Lower and Middle Pennsylvanian), composed of interbedded limestones, evaporites, and shales; and an upper unit, the Hare Fiord Formation (Upper Pennsylvanian to Lower Permian),

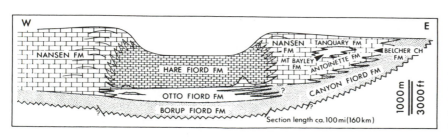

Fig. 7-65. Generalized east–west stratigraphic cross section of Upper Mississippian to Lower Permian units of the Sverdrup Basin in northwestern Ellesmere Island. (From Davies, 1977b, reproduced by permission of the Society of Economic Paleontologists and Mineralogists.)

composed of carbonate turbidites, debris-flow deposits, argillaceous rocks, and chert. These units are exposed in spectacular outcrops on the walls of Otto and Hare fiords and van Hauen pass on eastern Ellesmere Island. In these outcrops one can directly observe the vertical and lateral relations of Nansen, Otto Fiord, and Hare Fiord strata. Such natural stratigraphic cross sections invite detailed sedimentological analysis; in the following paragraphs we discuss several such studies.

i. Otto Fiord Evaporites. The Otto Fiord Formation consists of cyclically interbedded layers of anhydrite (occurring in both bedded deposits and massive, mosaic deposits formed by the coalescing of anhydrite modules) and fine, micritic limestones. In areas near the southern pinch-out of the Otto Fiord, detrital shales and crossbedded sandstones may also be present within the cyclic strata. Exposures along the walls of Otto Fiord reveal three major groupings of strata: (1) shallow-water, platform deposits of the Nansen Formation which define an *undathem* assemblage (i.e., strata deposited on a generally horizontal and shallow surface); (2) Nansen sediments deposited on the slope of the basin, sometimes with slope angles up to 40°, which represent a *clinothem* assemblage (i.e., strata deposited on a sloping surface); and (3) deep-water deposits of the Otto Fiord, which constitute a *fondothem* assemblage (i.e., strata deposited on a deep, relatively horizontal surface). Individual beds of the undathem may be traced visually in outcrop into beds of the clinothem and down to correlative beds of the fondothem. In this way, one can determine that the floor of the Otto Fiord trough was almost 300 m deeper than the edge of the Nansen carbonate platform. Clearly, anhydrite of the Otto Fiord Formation did not form in sabkha-like coastal embayments as did evaporites of the Mount Bayley Formation.

Davies and Nassichuk (1975) proposed that Otto Fiord anhydrites are "deep-water evaporites," i.e., formed by the precipitation and settling-out of evaporite minerals from a very hypersaline water mass. They argued that the cyclic character of the deposits was due to alternation of basin-water salinity possibly caused by sea-level fluctuations. This scenario presupposes that the Sverdrup Basin sea during Otto Fiord time was only connected to the open ocean via shallow channels through which free interchange was possible only during times of relatively high sea level. During times of lower sea level, the basin would have been partially or completely cut off from the open ocean; evaporation rapidly led to the very high salinities necessary for evaporite mineral precipitation. If the basin were only partially cut off, a constant influx of seawater would allow water level in the basin to remain essentially constant. On the other hand, complete basin isolation would lead to "evaporative drawdown" whereby the basin's water level

would have dropped. Davies (1977a) opted for the latter model to explain Sverdrup Basin evaporites and argued that algal mounds associated with carbonate beds of the Otto Fiord Formation were periodically affected by subaerial erosion, indicating that evaporative drawdown had lowered water level sufficiently to expose them. During these periods of lower sea level, craton-derived detritus was transported across the southern shelf and accumulated in prograding deltas which grew into the central-basin trough. Of course, such remarkable amounts of seawater evaporation require not only restriction or isolation of the basin but also a properly hot and arid climate. Paleogeographic reconstructions show that the Sverdrup Basin was located during the Pennsylvanian and Permian at approximately 30°N latitude. This places it in almost exactly the same global position and climatic zone as the Persian Gulf and northern Sahara Desert today.

ii. Gravity-Flow Deposits in the Hare Fiord Formation. Deposition of Otto Fiord evaporites ended during the late Desmoinesian, probably as the result of the opening of a deep access-way to the ocean. The Nansen carbonate platform continued to develop around the margins of the Sverdrup Basin while deep-water conditions persisted in the center, now dominated by pelagic and turbidite deposition. Davies (1977b) estimated that by the Early Permian, water depths in the central trough may have been as great as 1200 m. Tongue of the Ocean, the deep trough in the Bahama platform just east of Andros Island, is approximately 1280 m deep at its rear end where it is surrounded on three sides by shallow platform areas with current-swept shoals of oolitic and bioclastic sand. Therefore, Tongue of the Ocean and the adjacent Bahamas platform provide an excellent modern analogue for the Late Pennsylvanian to Early Permian Sverdrup Basin.

The presence of such great bathymetric relief and of steep slope-angles (up to 40°) provided the conditions necessary for the generation of a variety of subaqueous gravity flows ranging from bouldery debris flows to low-density turbidity flows. Deposits of these flows comprise the bulk of strata of the Hare Fiord Formation.

G.2.c. Observations on Basin Tectonics

Enough information has now been presented to allow some speculations about plate-tectonics mechanisms. Davies (1977b) pointed out that the association of basaltic volcanics and red beds at the base of the sequence, overlain by marine evaporites and then by shallow to deep "normal" marine strata, suggests that the Sverdrup Basin originated as a rift basin. Since continental rifting is theorized often to occur along the position of a former orogenic suture-zone, development of the Sverdrup Basin over the

Ellesmere orogen also fits the rift hypothesis. And paleo-geographic evidence suggests that the Siberian block (whose mid-Paleozoic collision with North America caused the Ellesmere orogeny) was in fact rifted away from North America during the late Paleozoic. These data seem to make a good case for the rift model. But the Sverdrup Basin did not develop beyond the proto-oceanic gulf phase of continental separation. In other words, rifting did not continue to evolve a new continental margin and the Sverdrup Basin remained a marginal rift basin throughout the Mesozoic Era. The reasons for this are unknown. Perhaps rifting "jumped" from the Sverdrup Basin area to some new rift zone farther to the north which did result in continental separation.

At any rate, the rapid-subsidence phase of the Sverdrup Basin ceased sometime during the Lower Permian. Later Permian and younger strata reflect lower rates of subsidence and more subdued bathymetric relief. The cessation of rapid subsidence may have coincided with the shift of rifting activity. A minor episode of folding in northwestern Melville Island and faulting on southwestern Ellesmere Island also occurred during the Early Permian. This event, called the Melville disturbance, may have been penecontemporaneous with the extrusion of basaltic lavas on northwestern Ellesmere Island, resulting in the Essayoo Formation (Trettin, 1972). Plauchut (1973) shows the Essayoo to have formed during the hiatus recorded over most of the Sverdrup Basin by the post-Nansen unconformity. Perhaps both Melville folding and Essayoo volcanic activity were related to the shift of active spreading to the north.

G.3.   Permo–Triassic Strata of Northern Alaska

A major erosional unconformity separates the Lisburne Group from overlying strata. Middle and Upper Pennsylvanian strata are missing entirely and Lower Permian rocks are only present locally. In southwestern areas of the Brooks Range, the Wahoo Limestone is overlain by siltstones, cherts, and shales of the Siksikpuk Formation (Lower Permian; see Fig. 7-66). In the Romanzof Mountains and on the North Slope, Lower Permian strata are absent and Wahoo carbonates are overlain by Upper Permian and Lower Triassic detrital deposits of the Sadlerochit Formation (see Fig. 7-66).

The Sadlerochit is present over much of the Brooks Range area and the North Slope and is one of the more important units of northern Alaska stratigraphy. In the Brooks Range, the Sadlerochit is divided into two members: the lower, Echooka Member (Upper Permian) composed of quartz sandstone, dark, siliceous siltstone, and shale, and the upper, Ivishak Member (Lower Triassic) composed of a lower, dark shale and an upper, clean, quartz sandstone (Sable, 1977). The Echooka probably represents an initial marine transgression followed by deposition of shelf muds below average wave base (Brosgé and Dutro, 1973). Overlying coarser sediments of the Ivishak were probably deposited as a prograding deltaic system (Detterman, 1973). Detrital sediments of both members were probably derived from the Beaufort Uplift. On the North Slope, the Sadlerochit is not formally divided but does contain two rather different lithofacies which probably are correlative with Echooka and

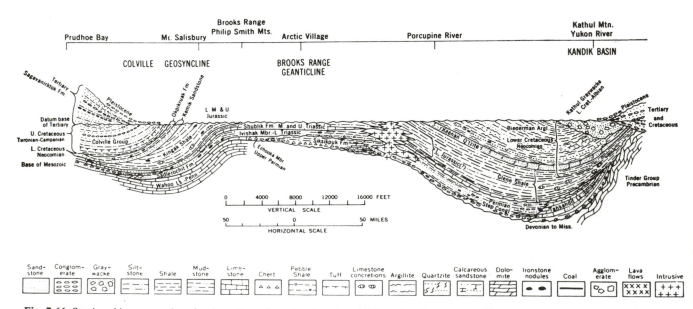

Fig. 7-66. Stratigraphic cross section of major sedimentary units of northern Alaska. (From Detterman, 1973, reproduced by permission of the American Association of Petroleum Geologists.)

Ivishak rocks of the Brooks Range. The lower part of the North Slope Sadlerochit contains sooty-black shales and siltstones which Morgridge and Smith (1972) interpreted to be prodelta deposits. Middle parts of the formation are deltaic sands, probably deposited in distributary-mouth bars. And upper Sadlerochit deposits represent a braided-stream environment. These environments correlate well with Brooks Range exposures of the Sadlerochit; for example, upper, alluvial strata in the Prudhoe Bay area correlate with Ivishak deltaic deposits in the Romanzof Mountains (Sable, 1977).

The Sadlerochit contains the principal petroleum reservoir strata of the famed Prudhoe Bay oil field, among the ten largest oil fields on Earth. In particular, braided-stream deposits of the upper Sadlerochit have porosities of 20–24% and permeabilities up to several darcies. These Sadlerochit braided-stream sands contain most of the recoverable hydrocarbons of the Prudhoe Bay field (Morgridge and Smith, 1972).

Overlying the Sadlerochit is the Upper Triassic Shublik Formation (see Fig. 7-66) which, in the Romanzof Mountains, is dominantly a dark-gray to black shale with phosphatic material and fossils. In the Prudhoe Bay area, the Shublik is much sandier, as befits its closer proximity to the Beaufort uplift. The Shublik appears to represent a transgressive sequence formed as rising sea level drowned the Sadlerochit delta plain. The Shublik is overlain by the Jurassic Kingak Shale whose deposition spanned most of Jurassic time; we therefore will postpone discussion of Kingak strata until the next chapter.

G.4. Permo-Triassic Strata of the Sverdrup Basin

The rapid-subsidence phase of the Sverdrup Basin had ceased by Upper Permian time and the central trough had gradually filled. Thus, later Permian and Triassic strata within the basin record generally much shallower environments than do earlier strata. Nonetheless, deposits are mainly marine and contain evidence of at least moderate water depths in the basin's axial region.

Upper Permian strata of the Sverdrup Basin are composed of three unconformity-bounded sequences. The lowest of these, found only in the southern half of the basin, comprise the Sabine Bay Formation. Sabine Bay deposits are primarily nonmarine and resemble Canyon Fiord Formation rocks. Unconformably overlying beveled Belcher Channel and Tanquary strata, the Sabine Bay probably represents the progradation of southern terrestrial facies northward across the shelf as a result of falling sea level.

Unconformably overlying the Sabine Bay are deposits of the sandy Assistance Formation in the south and correlative detrital muds of the van Hauen Formation in the central and northern parts of the basin. The Assistance is a glauconitic, highly fossiliferous, marine sandstone, probably deposited in relatively shallow water. van Hauen sedimentary rocks are primarily black siltstones and shales with a depauperate fauna, implying euxinic conditions and, possibly, moderately deep water. Assistance–van Hauen strata are unconformably overlain by the Trold Formation, a glauconitic detrital sand restricted to the south of the basin and the correlative Degerbols Formation whose cherty, fossiliferous limestones in the basin center imply more oxygenated environmental conditions than inferred from the underlying van Hauen.

For the most part, Triassic strata of the Sverdrup Basin record the same type of geological setting as do Late Permian deposits. The Early Triassic Bjorne Formation disconformably overlies Permian strata and consists of crossbedded, quartzose sandstones and conglomerates along the extreme southern part of the basin (see Fig. 7-67). The Bjorne Formation is mainly a nonmarine unit. Correlative with Bjorne strata are rocks of the Blind Fiord Formation in the basin center. The Blind Fiord consists of fossiliferous green and gray shales, siltstones, and sandstones which are interpreted as marine deposits (Thorsteinsson and Tozer, 1976). Overlying these deposits are silty and limy siltstones and sandstones of the Schei Point Formation. These sediments are inferred to grade basinward into shales of the Blaa Mountain Formation. Both Schei Point and Blaa Mountain are considered to be marine deposits.

During later Triassic time, a large volume of detrital sediment poured into the Sverdrup Basin from the south and literally filled the basin so that nonmarine conditions prevailed in most places. These nonmarine deposits are termed the Heiberg Formation (see Fig. 7-67). The source of Heiberg detritus, and of other Mesozoic units of the Sverdrup Basin as well, is believed to have been located some distance away from the basin, such as the western slopes of the Appalachian Mountains far to the south and east (Thorsteinsson and Tozer, 1976).

H. Late Paleozoic and Triassic Marine Life: A Study in Diversification, Extinction, and Repopulation

H.1. The Late Paleozoic Invertebrate Biota

The following discussion will deal only with taxa which underwent distinctive radiation or adaptation during the later Paleozoic.

H.1.a. Brachiopods

During the Pennsylvanian and Permian, brachiopods produced a host of new forms, especially among the stro-

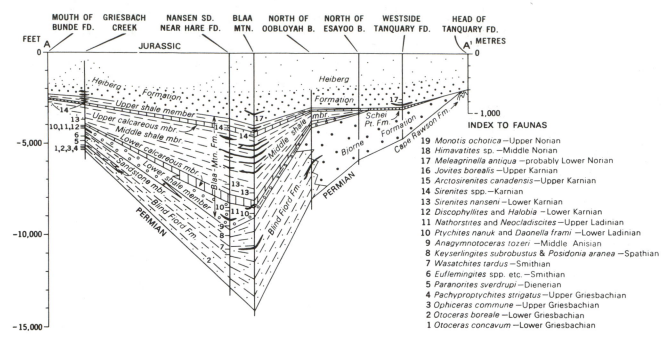

Fig. 7-67. Stratigraphic cross section of Triassic units of the Sverdrup Basin. (From Thorsteinsson and Tozer, 1976, reproduced by permission of the Minister of Supply and Services, Canada.)

phomenids of the suborder Productacea. Productids characteristically have very globose (rounded, high) pedicle valves, concave brachial valves, and well-developed spines on the posterior margin or over much of the shell. Like most strophomenids, they had long hinge lines, weak interareas, and pseudopunctate shells. They also developed extended anterior shell margins, producing "*trail*," a winglike portion of the shell that does not contribute to the living space even though both valves are incorporated. Although the group appeared in the Devonian, it did not achieve statistical importance in the shelly biotas until the Mississippian.

The productids underwent extreme diversification in late Paleozoic time, especially in the Permian, and some of the strange forms that emerged have been cited variously as examples of "racial senescence" (a once-popular concept that long-lived groups of organisms lose genetic vigor shortly before extinction and undergo bizarre evolution), convergent evolution, and adaptations to unique environmental conditions. The spines of productids may have evolved as a response to the need to anchor and stabilize the brachiopods within very-soft-bottom substrates (Rudwick, 1970).

A bizarre group of Permian productids (Fig. 7-68) are volumetrically important components of the Texas Permian platform carbonates. These are characterized by the genera *Scacchinella, Hercosia,* and *Prorichthofenia,* among others, and have horn-shaped pedicle valves oriented upright, in many species held up by spines. The brachial valves tend to be caplike and have spines directed inward. Some taxa

appear almost amorphous, and led to the suggestions of racial senescence; overall, these brachiopods more closely resemble solitary rugose coral than brachiopods, although they were not colonial (since they reproduced sexually rather than by budding) but rather lived commonly in tightly packed communities. Their coral-like appearance and community habit have been considered evidence for convergence toward the colonial structure-building habit of Paleozoic tabulates, which are nonexistent in the American Upper Permian.

Among the productids are also the largest known brachiopods, species of *Gigantoproductus* from the Mississippian, with individuals measuring 30 cm across the hinge line. The productids (and, in fact, strophomenids as a group; see below) seem to have had great abilities to evolve wide variations in shape and size. The absence of pedicles in all adult strophomenids and the spiny morphology of productids undoubtedly preadapted many taxa to the upright, corallike habit (Williams and Hurst, 1977); yet, despite their apparent genetic flexibility, productids disappeared at the end of the Permian.

Other articulate brachiopod taxa were common in the late Paleozoic including terebratulids, orthids (Dalmanellacea), spiriferids, rhynchonellids, and strophomenaceans. In addition, inarticulates were present during the late Paleozoic, as during all Phanerozoic time, and include the long-ranging and common genera *Lingula* and *Crania.*

A very specialized Permian strophomenacean genus,

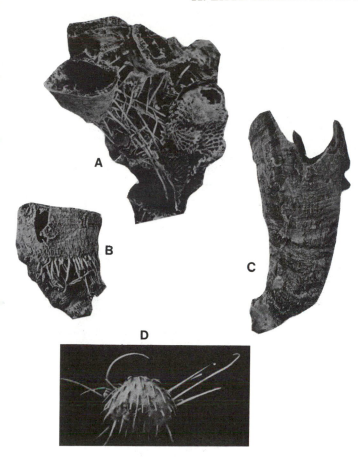

Fig. 7-68. Permian productid brachiopods from west Texas, showing bizarre, reef-forming morphologies and elongate spines. (A) Cluster of *Hercosia uddeni,* with interlocking spines and corallike morphology; (B, C) *Scacchinella titan,* small and large individuals, showing only the pedicle valve with corallike morphology; (D) *Echinauris parva,* a small productid with typical outline and elongate spines. (By permission of Smithsonian Institution Press, from Cooper and Grant, 1975, plate 281, figs. 8, 10, plate 286, fig. 2, plate 332, fig. 20.)

Leptodus, evolved an oysterlike mode of life. *Leptodus* species featured pedicle valves cemented to the bottom substrata or their fellows, and they built low bioherms much like those of the modern Atlantic oyster *Crassostrea virginica. Leptodus* also featured distinctive ridges and slits on respective brachial and pedicle valves, which may have had a relationship to the shape of the lophophore (Moore *et al.,* 1952). The genus seems to have adapted well to the oysterlike niche in west Texas but it is not found in midcontinent deposits, in contrast with the coral-like forms which are distributed widely over the western craton. Neither group of attached productids were reef structure-builders but they added considerable carbonate detritus to Permian reefs. *Leptodus,* like the productids, did not survive the Permian. Monographic treatment of Permian brachiopods from west Texas has been accomplished (Cooper and Grant, 1972 ff.) and the interested reader is directed there.

H.1.b. Late Paleozoic Radiation of Bivalves

Bivalves first diversified during the Ordovician and are elements of most subsequent Paleozoic shelly fossil assemblages but they never reached the population size and taxonomic diversity of brachiopods before the Mesozoic. In latest Paleozoic faunas (Pennsylvanian–Permian), signs of things to come do, however, appear in the form of diversification of the bivalves in the seas occupying the continental interior.

A very characteristic Penno-Permian group are the myalinids, mussels with very asymmetric valves (Fig. 7-69); *Myalina* is placed in the order Mytiloida, to which the modern edible mussels belong. Mussels are an ancient (Ordovician) bivalve lineage, from which both pectens and oysters may have evolved (Cox, 1960, in Tasch, 1980); they are characterized as having generally thin, prismatic, and *nacreous* shells (nacre is "mother-of-pearl," composed of thick aragonite with thin interstitial organic layers), an outer organic covering (*periostracum*), and a series of threadlike

Fig. 7-69. Two characteristic Penno-Permian North American fossils. (Above) *Girtyocoelia,* a common calcisponge which resembles "pop beads." (Below) *Myalina,* a distinctive mussel whose evolutionary history has been studied in great detail, with evolutionary stages traced from the typical mussel's subovate morphology, to the "winged" shape illustrated here. (Both are DRS photographs.)

attachments to their substrate (*byssus*). Whether myalinids had all the modern mussel attributes is uncertain, but they are common Pennsylvanian and Permian fossils in marine sections.

Not all bivalves are or were marine. In continental and estuarine strata of the Pennsylvanian are found series of bivalves, including the common mussel *Naiadites;* Moore *et al.* (1952) note that this genus is found in freshwater shales above coal, associated with the clams *Anthracomya* and *Carbonicola.*

Also common in Mississippian through Permian times were nearly modern-looking pectinids. These are bivalves with generally symmetrical, equal valves, strongly ribbed, winged (i.e., bearing projections at the beak end), and which can move rapidly above the sea bottom by contraction and relaxation of their adductor muscle, producing a ''flying'' motion. The wings aid in this means of locomotion and their presence in fossil taxa suggests similar life habits. A variety of pectinids were present in the later Paleozoic including two common forms from the western interior region, *Acanthopecten* and *Aviculopecten* which resemble closely (to the nonspecialist) the modern ''sea scallop'' *Aquipecten.*

Marine and freshwater clams were common but not generally distinctive faunal elements of the Pennsylvanian and Permian. Many derived from much older lineages virtually unchanged (e.g., *Nucula, Nuculana, Schizodus*). At least one notable clam taxon, *Aviculopinna,* appeared in the late Paleozoic and may represent the forerunner of the razor clams.

Another ancestral bivalve stock appeared in the Permian and foreshadowed the oncoming Mesozoic diversification of the class. These were oysterlike taxa characterized by the genera *Pseudomonotis* and *Prospondylus.* Most bivalves, especially clams, prefer soft substrates in which they can burrow. However, some, specifically the mussels, pectens, and oysters, live epifaunally and are commonly attached to hard substrates; the oysters also take the final step and attach to each other. During the Permian, a group of oysterlike bivalves apparently evolved from inequal-valved pectinids, like *Pseudomonotis,* and developed the ability to cement one valve to their substrate, like later true oysters; the pectens from which they emerged may have attached in similar fashion by their byssus (Newell and Boyd, 1970). These Permian forms were not true oysters because they possessed byssuses (absent in oysters) but, otherwise, they were very much like oysters. The genus *Prospondylus* includes large (10 cm) forms that almost certainly had evolved the habit of cementing a bottom valve to their substrate. These are common fossils in west Texas Permian strata on the platform regions and, curiously, are commonly found encrusted with productid brachiopods like *Prorichthofenia.* True oysters are generally considered to

have appeared in the Middle Triassic; however, Newell and Boyd (1970) cite examples of Late Permian, apparently true oysters from Japan.

H.1.c.   Sponges

In our discussion of the Permian of west Texas, we noted that the massive basin-margin reefs were built primarily by sponges, algae, and bryozoans. This trio of taxa (with the addition of stromatoporoids) also largely built the Silurian reefs of Michigan and the Devonian reefs of the Williston Basin, among others.

The Penno-Permian genus *Girtyocoelia* is a favorite of distributors of beginning-geology fossil kits because it is abundant in marine deposits from Texas and the western interior (and cheap to obtain); it is a favorite of students because it is easy to recognize, resembling a series of ''pop beads'' (Fig. 7-69). This genus is a calcisponge classed in the order Sycones, designating that it is thin-walled (as opposed to the order Pharetrones, which is thick-walled). *Girtyocoelia* was not, however, a reef-building taxon. A somewhat similar, and almost identically named genus, *Girtycoelia* is found only in the Upper Pennsylvanian of west Texas. *Girtycoelia* is also a thin-walled calcisponge, and features round, small individual chambers but these are fused rather than separated as in *Girtyocoelia.*

A number of other calcisponges are common Upper Paleozoic taxa including *Talpaspongia,* a Lower Permian index fossil from Texas, but probably more important as reef-builders were lithistid sponges (a group of demosponges, described previously in our discussion of the Silurian). Lithistids feature spicules that interlock and therefore allow the formation of rigid structures. Characteristic Guadalupian lithistid sponges include *Actinocoelia,* which ranges from the southern Guadalupe Mountains in west Texas to Wyoming, and *Guadalupia,* present in great numbers and in numerous species in the Capitan Limestone in west Texas (King, 1948).

H.1.d.   The Mazon Creek Biota

In Chapter 4 we discussed the Burgess Shale and showed that it was one of the few units which preserved soft-bodied organisms. Fortunately, another such unit is in the Absaroka Sequence, comprising the Francis Creek Shale Member of the Carbondale Formation (Desmoinesian) in northeastern Illinois, exposed along Mazon Creek.

Mazon Creek fossils occur in spheroidal and ellipsoidal ironstone (siderite) concretions, typically 1 to 30 cm long (Nitecki, 1979). Burial of the organisms occurred both rapidly and under anaerobic conditions; Nitecki notes that both phosphatic and organic remains were unchanged by preservation but that calcareous shells were largely dissolved. The fossils have been known and collected for over 100 years and coal mining operations along the creek have provided much material for study (Fig. 7-70).

The biota is generally divided into the marine Essex and nonmarine Braidwood components. Because of the wide attention the biota has received (more than 200 scientific papers), the described fauna and flora are enormous: in the nonmarine component are found centipedes, millipedes, spiders, scorpions, insects, fish, shrimp, amphibians, and of course a wide variety of mollusks and plants. The preservation is often excellent and in some insect specimens wing venation and setae (hairs) are visible. Many terrestrial organisms have first occurrences in the Mazon Creek biota at the generic or even much higher taxonomic levels. The marine component of the biota is also large and diverse, including (in addition to the common Pennsylvanian assemblage of mollusks, brachiopods, echinoderms, and other invertebrates): the only known fossil lamprey, hydra, and priapulid; the first occurrences of marine isopods, chaetognaths, and nemertine worms; holothurians; annelid worms, medusoid coelenterates, sharks (including the cartilaginous skeletons), marine shrimps, horseshoe crabs, and many others. On some fossils, original color patterns are still visible (although the coloration itself is altered by the siderite).

Fig. 7-70. A decapod crustacean, *Arthrapalaemon gracilis,* from the Francis Creek Shale at Mazon Creek, Illinois. (DRS photograph, specimen made available courtesy of Dr. Fred Collier, U.S. Museum of Natural History.)

The Francis Creek was almost certainly a delta-plain deposit, with the site of the nonmarine to marine transition located within the area of outcrop, generally parallel to Mazon Creek. Parts of the biota include mixed nonmarine and marine fossils and these may represent transport by storm surges or periodic distributary flooding of marine environments (which consequently both poisoned marine organisms with an influx of freshwater and simultaneously transported freshwater organisms to the marine environment; Baird, 1979).

H.2. The Permo-Triassic Boundary and Mass Extinction

One of the most enduringly controversial topics of paleobiology is the cause, nature, and reality of mass extinction as a natural process. The Permo-Triassic mass extinction of invertebrates is among the three most widely discussed paleobiological events [the others being the Cretaceous–Tertiary boundary extinctions (Chapter 8) and the Pleistocene large-mammal extinctions in North America (Chapter 9)].

Schindewolf (1950, 1953, 1954, in Newell, 1956) did classic studies on the Permo-Triassic and emphasized the catastrophic appearance of the faunal change at the era boundary. He analyzed global Permo-Triassic faunas and focused on regions that showed no apparent stratigraphic break between periods (notably in the Salt Range, West Pakistan); his studies showed that 22 categories of marine and continental taxa above the familial level disappeared at the close of the Permian, while 11 new higher taxa appeared in the Early Triassic. However, the terrestrial animals generally were less affected by the apparent mass extinction than were the marine taxa.

Subsequent studies tend to support the magnitude of the change described by Schindewolf. Raup (1979) analyzed the size of the "Permo-Triassic bottleneck" at the lower taxonomic levels and concluded that extinction of perhaps 88 to 96% of all marine vertebrate and invertebrate species may have occurred [based on a statistical technique called "rarefaction" which is designed to compare species richness in collections having differing numbers of individuals; see also Tipper (1979) for a critique of rarefaction as used in paleoecology].

Yet, others believe the Permo-Triassic extinctions to have been not unusually severe and suggest that the event has been overdramatized. Rhodes (1967) noted that sponges, gastropods, bivalves, and fish, among other marine stocks, remained largely unaffected. He also pointed out that the Permian was not a time of generally dwindling marine stocks but rather one of diversification among much of the marine biota. Rhodes suggested the nature of the Permo-Triassic event was one of slow Triassic replacement of routinely lost taxa, rather than unusual mass mortality.

H.2.a.   The Nature of the Faunal Changes at the Permo-Triassic Boundary

Before delving further into this problem, let us tabulate some of the taxonomic changes at the era boundary. The following list (data from Harland *et al.,* 1967) is not exhaustive but should provide some perspective on the change of marine faunas so that the reader can join in the extinction game; note that the groups enumerated represent various taxonomic hierarchies:

Taxa extinct at the end of the Permian	New taxa in the Early or Middle Triassic
Trilobites	
Eurypterids	
Tabulate and rugose coral	Scleractinian coral
Cryptostome, fenestrate, trepostome, and cystoporate bryozoans	
Fusulinids	Five families of foraminifera
Orthid and productid brachiopods	
Inadunate, flexible, and camerate crinoids	Articulate crinoids
Blastoids	
One suborder of echinoids	Five suborders of echinoids
Ten superfamilies of ammonoids	Ten superfamilies of ammonoids
Eight families of ostracodes	
Five families of elasmobranchs (sharks)	
Eight families of bony fish (Chondrostei)	Fifteen families of bony fish (nine Chondrostei, six Holostei)

It must be emphasized that this list is potentially deceiving in several ways: first, some taxa, especially fish, represent questionable occurrences of poorly preserved and possibly misinterpreted fossils; second, many groups not included had sizable taxonomic changes at hierarchies below those included in Harland *et al.;* third, and most important, the changes did not all occur in the latest Ochoan but rather were chosen for the entire Guadalupian–Ochoan interval; fourth, taxa that did not change appreciably are not listed.

H.2.b.   Evaluation of Conformable Boundary Sequences

Part of the thrust of arguments in favor of a significant boundary event relies on the few stratigraphic sequences on Earth that appear to be conformable through the Permo-Triassic. Kummel and Teichert (1970) narrowed down the sections with putative continuous marine sedimentation from the Late Permian into the Early Triassic to five areas: southern China; Kashmir, India; northern West Pakistan (Salt and Trans-Indus Ranges); a region on the border of Iran and Armenia, USSR; and Northeast Greenland. Rhodes also included the Phosphoria/Woodside formations in eastern Idaho and parts of Indonesia and Malaysia to the possible conformable sequences. In addition, a region in Timor, Netherlands East Indies, has a unique latest Permian fauna which bears on the question as discussed below.

In fact, none of the sections above may be proven conformable. Schindewolf examined the Salt Range sections and based much of his "catastrophic extinction" conclusion on the assumption that the sections he examined were conformable; he observed drastic turnover of the marine biota within apparently gradational sequences. However, a symposium on the problem of the Permo-Triassic in the Salt Range (Kummel and Teichert, 1970) yielded slightly different conclusions. In the lead-off paper, Kummel and Teichert showed that the Permo-Triassic boundary in sections they examined included a sharp lithological change and was not conformable but rather paraconformable. Previous studies had placed the boundary either above or below their chosen horizon; in addition, Kummel and Teichert show that locally several Permian lineages of brachiopods, nautiloids, and other taxa persisted briefly into Triassic time.

The Late Permian fauna from Timor adds another insight which will be considered in our analysis. Blastoids disappeared from virtually all marine environments globally in Leonard (or equivalent) time, except for remarkably diverse assemblages in Timor (in the Basleo Formation of approximate Guadalupian age) and scattered occurrences of two families in Australia and the USSR (Harland *et al.,* 1967). The Timor fauna contains not only 32 species among 14 genera of Permian blastoids, but it also contains 239 species among 75 genera of crinoids, of which 105 species were new and restricted to the area. To put this in perspective: all known post-Leonardian blastoids (and all but two occurrences of post-Lower Pennsylvanian blastoids) and more than half the post-Leonardian crinoids are from one formation in Timor.

H.2.c.   Causes of the Permo-Triassic Extinctions

Schindewolf suggested that cosmic events, such as heavy influx of cosmic rays or solar radiation, were a possible explanation for the catastrophe he considered to have occurred. Newell (1956), among others, handily refuted this idea by pointing out that terrestrial animals were far less affected by Permo-Triassic extinctions than were marine animals. Since seawater would shield the marine biota effectively, were cosmic agents involved, the reverse of the observed events would have transpired.

A more general thesis for mass extinction by Bretsky and Lorenz (1970) proposed that extinctions follow intervals of environmental stability in which taxa lose genetic variability. The consequence of this would seem to be less adaptability to changed environments in subsequent periods of stress. In contradiction of this thesis, first, it is not clear that the Permian was a period of low environmental stress in the marine benthos, and second, Valentine *et al.* (1973) effectively showed that in at least one modern studied organism, the "killer clam" *Tridacna*, living in a very stable tropical environment and able to control much of its microenvironment because it contains zooxanthellae in its tissues, genetic variability is high.

Other theses for the event have been proposed, including both drastic reduction and "blooms" of pelagic phytoplankton species, crowding of the marine benthos, climate changes, racial senescence, and accelerated competition [see Raup and Stanley (1978) for discussion and bibliography]. All these (except racial senescence) may have some validity but they are probably impossible to prove, especially since no deep ocean sediments dating to the Permo-Triassic are known. The thesis we most strongly support has several elements that are borne out by the data presented.

Three lines of evidence seem most compelling to us:

1. Permo-Triassic events largely affected the marine biota.
2. The apparent refuge in Timor suggests effects were not uniform globally or the fossil record yields a biased view.
3. Absence of good (or any) marine boundary sequences and the very sparse Lower Triassic marine record suggest an additional component of missing data.

These lines of thought point most clearly to global reduction of the benthic environments, probably due to eustatic sea-level changes; and to some terrestrial upheavals, probably due to continental assembly in the Late Permian. As we have shown in previous discussion, North American Permian and Triassic strata are dominated by terrestrial sequences. Those areas which do not have Permian and Triassic strata were almost certainly exposed land; clearly, North America was higher and drier during the Permo-Triassic than at any other time in the Paleozoic. And the general absence of Permo-Triassic boundary sequences globally speaks well for the assumption that lands were generally emergent. Emergent landmasses would reduce the availability of subtidal benthic habitat, especially, because broad epeiric seas were nearly absent; reduced habitat would impose severe stresses on most elements of the benthic fauna and would cause extinction of the less adaptable groups (especially those highly specialized such as the productids, requiring suitable bottom substrates, and pelmatozoans, requiring low turbidity).

Valentine and Moores (1972), in a broad discussion of the effects of plate tectonics on ancient communities, noted several additional conditions that would contribute to the lowering of organic diversity. They show that in times of fluctuating environments (as would occur along with regressions and accelerated orogenesis), fewer *trophic levels* could be supported in ecosystems, i.e., fewer links would be present in food webs. Therefore, global continental assembly would drastically reduce or eliminate barriers to dispersal of organisms, and as a result, would decrease the potential for allopatric speciation.

H.3. Triassic Marine Faunas

H.3.a. General Makeup of the Biota

In our discussion of Permo-Triassic extinctions, we necessarily mentioned elements of the change in faunas between Permian and Triassic times and listed some of the higher taxa which emerged in the Triassic. In this section we will discuss the general nature of the Triassic marine biota; however, certain groups, notably the ammonoids, marine reptiles, free-swimming articulate crinoids, and oysters are better exemplified by the later Mesozoic fossil record and will be detailed in Chapter 8.

Aside from questionably dated outcrops in Transcaucasis, USSR, Lower Triassic marine units worldwide contain no foraminifera, no bryozoans, no crinoids, and, most notably, no coral. Since some of these groups are represented in younger strata, including the cyclostome and ctenostome bryozoans, terebratulid brachiopods, and many families of foraminifera, we must presume that nonpreservation has produced this seeming depauperization of the biota rather than actual absence of these taxa.

What are abundant in Lower Triassic marine rocks are gastropods, bivalves, ammonoids, nautiloids, and crustaceans; in this we see the beginning of a "modern" appearance to the invertebrate biota, dominated by mollusks and arthropods. Bivalves completed the process begun in the Permian and replaced the brachiopods as the dominant shelled benthic filter- and deposit-feeders while gastropods expanded their role as the dominant benthic grazers and invertebrate carnivores.

During the Middle Triassic, the Scleractinia (coral) appeared as three families, and by the Late Jurassic had expanded to 19 families. At about the same time, numerous families of bony fish emerged, and although most chondros-

tean families were short-lived and did not survive the Triassic, the more advanced Holostei which were part of the
radiation thrived and evolved the Mesozoic populations we
will discuss in the next chapter. Also in the Middle Triassic,
from a single superfamily, Otocerataceae, which survived
the Permo-Triassic extinction, nine superfamilies of ammonoids emerged in the Middle Triassic while a second
surviving Paleozoic superfamily, Medlicottaceae, became
extinct. Nautiloids survived the extinction as four superfamilies, and a new one, the Nautilaceae, appeared in the
Middle Triassic. By late Middle Triassic time, all but the
newly evolved nautilids became extinct and this group comprises the sole surviving nautiloids of the remainder of geological time.

Brachiopod populations were decimated in the Permo-
Triassic event but at least one rhynchonellid family survived
and, in the Middle Triassic, four additional families appeared, possibly from the one. Spiriferids were affected by
the Permo-Triassic extinction but at least four families survived only to become totally extinct in the Middle Jurassic.
In the Middle Triassic too, three terebratulid families appeared after virtually all known Paleozoic groups became
extinct in the Permo-Triassic, as mentioned above.

H.3.b.   Scleractinia

For roughly the past 215 Myr., all coral have been
scleractinians. It is precisely because of attributes of the
scleractinians that one thinks of coral as being the primary
reef-builders and why we had to emphasize that Paleozoic
tabulates and rugosans were not reef-structure-builders.

The scleractinian polyp (Fig. 7-71) is generally cylin-
drical with a flattened base, and features a ring of tentacles
on the upper (termed *oral*) end. The base or *aboral* end is
attached by a disk to the substrate in solitary forms, and
attached to either the substrate or other corallites in colonial
taxa. The mesenteries and, consequently, the septa in the
corallite are in patterns of sixfold radial symmetry; hence
the general name for scleractinians, *hexacorals* (versus
tetracorals for rugosans). Of paramount importance in
scleractinian morphology is the so-called *edge zone,* consisting of part of the polyp lying outside the walls of the
cup; in colonial forms, it lies between adjacent polyps. It is
this edge zone that secretes aragonite which tremendously
strengthens the connections between corallites in colonial
forms and enables them to produce strong, wave-resistant
structures. Chamberlain (1978) described mechanical properties of coral skeletons and the adaptive significance of
such properties. Scleractinians invaded the high-energy marine environments which were probably deadly to tabulates
and rugosans and, as such, they not only expanded their
niche and habitable area, but they also created new marine
habitats by erecting reefs. Modern coral reefs are built gen-

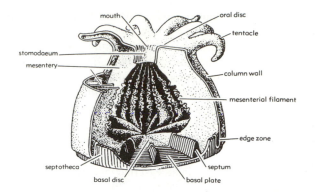

Fig. 7-71. (Above) Morphology of a generalized scleractinian coral
polyp, showing both soft anatomy and the makeup of the corallium. (Below) Representative modern hermatypic scleractinians, showing the variety of colony patterns: 1a,b, *Caulastrea;* 2, *Favia;* 3, *Goniastrea;* 4,
Leptoria; 5, *Hydnophora;* 6, *Pavona.* (All from the *Treatise on Invertebrate Paleontology,* courtesy of the Geological Society of America and the
University of Kansas Press.)

erally in the active wave zone of the nearshore shelf and,
consequently, by erecting a wave-resistant structure, coral
create a number of different environments. At the front of
the reef is the high-energy zone, commonly populated by
branching, staghorn coral (*Acropora*) and other organisms
that seek agitated water. On the surface of the reef are coral
and organisms that also seek agitated waters but with some-

what less energy; behind the reef is a quiet water zone and behind that, typically, is a lagoonal area or embayment which would have been wave-swept except for the presence of the reef. Back-reef areas are typically crawling with invertebrates as are many parts of the reef complex, including numerous taxa which live nowhere else.

As mentioned previously, reef-building (or hermatypic) scleractinians contain symbiotic algae (*zooxanthellae*) in their tissues. Part of their impetus to build reefs up as wave erosion and reef-boring organisms tend to tear them down is the need to maintain position in the photic zone. Scleractinians can be roughly divided into two groups, hermatypic and ahermatypic, depending on the respective presence or absence of zooxanthellae in their tissues. Ahermatypic scleractinians inhabit generally cold-water and deep-water environments, whereas hermatypic ones are typically tropical to subtropical and shallow. Stanley (1981) proposed that Scleractinia acquired zooxanthellae in the Late Triassic, which may account for the 25-Myr. hiatus between the first appearance of scleractinians and their prominence as reef-builders. Once they did become established, scleractinian coral essentially completed the "modernization" of marine benthic invertebrate faunas and represent probably the most notable change from Paleozoic marine life to that of the Mesozoic.

I. Early Amphibians and Reptiles

I.1. Dominant Amphibia of the Late Paleozoic

The first land animals, primitive amphibians, were introduced in the previous chapter. For roughly 55 Myr. after their appearance, amphibians dominated the terrestrial environment and shared it only with arthropods and plants. Further, for the next 50 Myr. (until well into the Early Permian), even after the appearance of reptiles, amphibians still maintained apparent control of the land (McFarland *et al.*, 1979). Late Paleozoic and Triassic amphibians were a large and diverse group, including forms which filled far more ecological niches than do modern amphibians. Indeed, modern amphibians are considerably different from the majority of amphibians which have appeared over the span of geological time; as a single example, many Paleozoic taxa reached lengths over 2 m. In the following discussion, we make no attempt to exhaustively describe the dominant amphibians; a general survey of Romer (1966) will fill in a bit more background. We will highlight only the better-known taxa, and the reader is reminded that discussion in Chapter 6 introduced the basic taxonomy of Paleozoic amphibians. Most discussion will focus on the Labyrinthodontia, which includes two important orders: temnospondyls and anthracosaurs.

I.1.a. Temnospondyli

The Temnospondyli include two suborders: Rhachitomi and Stereospondyli. The rhachitomes include the best-known and most common Permian amphibians such as *Eryops* of the Texas Permian red beds (Fig. 7-72) and a related form, *Cacops*. These rhachitomes featured heavy bodies with large, and in the case of *Cacops*, proportionately enormous heads. Both were armored and *Cacops* featured thick plating, suggesting a defensive purpose (probably from pelycosaurian reptiles, commonly found in the same strata). *Eryops* was probably a traditional amphibian, living partly in and out of water in the manner of modern crocodiles. *Cacops*, on the other hand, appears to have been an exclusively terrestrial form (excluding the larval stages which, as in all amphibians, must have been aquatic); the high, compact body, rugged legs, and short tail of *Cacops* all point toward adaptation to terrestrialism. Some

Fig. 7-72. Two characteristic Permian amphibians from Texas. (Above) *Seymouria*, an advanced anthracosaur, showing many characteristics suggesting this form is close to the transition to reptiles; this specimen approximately 80 cm long. (Below) *Eryops*, a common rhachitomous temnospondyl; this specimen approximately 1.2 m long. (Both are DRS photographs.)

rhachitomes were far better adapted to aquatic habitats, notably taxa such as the Permian *Archegosaurus* and the Triassic *Trematosaurus*. These were long-snouted forms, undoubtedly fish-eaters, and may have been exclusively aquatic. *Trematosaurus* is one of the few amphibians known from marine beds and Romer (1966) notes that they may have followed the Triassic migration of ray-finned fish from continental to marine habitats.

Advanced temnospondyls comprise the suborder Stereospondyli, which differ from *Eryops*-type rhachitomes in details of the skull and, especially, the vertebrae. Stereospondyls include the most common Triassic amphibian genus, *Capitosaurus,* and the largest labyrinthodont, *Mastodonsaurus,* with a skull length of over 125 cm. Stereospondyls tend to have proportionately smaller and flatter skulls than do rhachitomes and they feature double *occipital condyles* (articulating surfaces between the vertebral column and the skull); this double condyle is also a characteristic of mammals (but, curiously, not of reptiles).

I.1.b. Anthracosauria

The second order of labyrinthodonts of importance in the Permian is the Anthracosauria, including four suborders of which we will examine only one: the seymouriamorphs. Romer notes that the anthracosaurs have a known history quite different from that of the temnospondyls in that they were never diverse and became extinct early (in the Permian). Distinction of anthracosaurs from temnospondyls is based largely on details of the skull and vertebrae but in addition, anthracosaurs have five digits on front limbs whereas temnospondyls had four. *Seymouria,* the typical seymouriamorph, is found in Lower Permian red beds of Texas. It was a small creature that closely approximates the link between amphibians and reptiles. In many respects (e.g., single occipital condyle, structure of the shoulder girdle, five-toed hands, and others), it appears to have been a reptile, but it also possesses unquestionable anthracosaurian characteristics. Thus, it may be one of the best examples of an interclass transitional taxon in the vertebrate record. Another seymouriamorph, *Diadectes,* has characteristics also shared by both reptiles and amphibians but in addition it was a "first" of another kind: it had specialized teeth and palate suggesting that it was the first terrestrial herbivore. If it was an amphibian, it is the only known herbivorous amphibian; if a reptile, it was the first herbivorous reptile. The actual "proof" of the affinities of animals like *Seymouria* and *Diadectes* is their manner of reproduction, whether via unprotected eggs laid in water or via the self-contained amniote egg. Since very few fossil eggs of any sort are found, evidence of this nature will likely never emerge.

I.1.c. Lepospondyli

So far we have described labyrinthodont amphibians; there was another Paleozoic subclass, the Lepospondyli, which left a sizable Absarokan fossil record from Late Mississippian through Late Permian times. Lepospondyli have distinctive vertebral structure and tended to be specialized animals. One order, Aistopoda, includes mostly limbless, snakelike taxa such as *Ophiderpeton,* generally less than 1 m in length. The second order, Nectridea, contains a more varied assemblage of amphibians characteristically bearing a very wide, flat, "horned" skull which protruded at the posterolateral corners. *Diplocaulus,* a highly specialized Permian nectridean, was the last of its group and one of the largest, approximately 60 cm long.

A final group of Paleozoic amphibians, the Microsauria, are taxonomic orphans, classified as lepospondyls by some (e.g., Romer, 1966; Carroll and Gaskill, 1978) and as a distinct subclass by others (e.g., Panchen, 1977). In overall form (Fig. 7-73) they were reptilelike, but with relatively long bodies and weak limbs, not unlike salamanders. They possessed a distinctive skull/vertebral articulation and three toes on the forelimbs which, along with skull details, differentiates them from other amphibians and reptiles. McFarland *et al.* (1979) note that one greatly elongate form, *Lysorophus,* had more than 70 trunk vertebrae and very small limbs, thus apparently converging on the snakelike morphology of the aistopods. Microsaurs are relatively common in Late Pennsylvanian terrestrial deposits, surviving only until the Early Permian and, when found, can confuse the nonspecialist because of their resemblance to early reptiles.

All amphibians surviving into the Triassic were labyrinthodonts but during Late Triassic time, they too became extinct. It seems reasonable to suppose that the decline of amphibians in the latest Permian resulted from causes related to the Permo-Triassic mass extinction among the invertebrates, especially the lowering of migration barriers among Pangaean continents. In addition, the radiation and apparent competitive advantages of reptiles in the arid environments

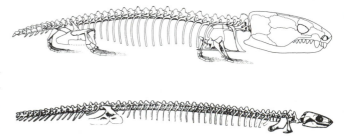

Fig. 7-73. Microsaurs. (Above) *Pantylus* and (below) *Micraroter,* from the Lower Permian of Texas. (Both from Carroll and Gaskill, 1978.)

of the later Permian probably contributed greatly to amphibian decline; certainly the reduction of swampy conditions in North America, which was a characteristic part of Pennsylvanian and Early Permian environments, would select against aquatic amphibians and those which spent most of their time in freshwater. The Triassic reradiation of labyrinthodonts is less easy to explain but it may represent replacement of the depauperate terrestrial fauna following extinction of a great many amphibian and reptile taxa in the Permo-Triassic. The final decline of labyrinthodonts may be the consequence of their competitive disadvantages in the face of the Triassic radiation of reptiles.

I.1.d. Lissamphibia

Lissamphibians are familiar today, comprising very numerous, widespread, relatively small animals which are most abundant in tropical to temperate climates. The most abundant are in the order Anuria, commonly called frogs and toads. They are distinguished by their reduced trunk vertebrae, elongate femurs and ilia, reduced ribs, and absent tails (in adults). The fossil record of frogs and toads shows a slow Jurassic diversification followed by accelerated radiation in the early Tertiary, continuing to the present. A single Triassic fossil found in Madagascar, assigned to the genus *Triadobatrachus,* appears to be a primitive frog and suggests that the group was present long before it left a good fossil record. The other major modern lissamphibian group is the order Urodela, or salamanders. These have a fossil record extending back to the Late Jurassic, but they are far more common in Early Tertiary strata.

I.2. Origin, Morphology, and Early Radiations of Reptiles

I.2.a. Differences between Reptiles and Amphibians

Since we introduced the probable ancestors of reptiles in the preceding discussion, we will plunge directly into reptile morphology and the characteristics that distinguish them from amphibians. First and foremost is the amniotic egg, which provided a means whereby reptiles (and, later, birds and primitive mammals) could reproduce without dependence on bodies of water.

Shown in Fig. 7-74 is a schematic drawing of the development of a typical vertebrate larva in an amniotic egg. The significant parts are three membranes, the *amnion, chorion,* and *allantois,* the first two of which develop from outgrowths of the embryo itself. Amniotic eggs also feature hard or leathery shells as a physical barrier to damage. The combination of these internal structures and the shell allows amniotic eggs to be deposited terrestrially and allows for large-size eggs. The waste products of embryogeny are left behind in the amniotic egg in contrast to the "reprocessing" required for wastes of anamniotes. Reproduction via amniote eggs also engenders internal fertilization and the potential for larvae to be totally independent of water bodies.

Table I lists additional major differences between the reptile- and amphibian-grade of development. Many of these point toward evolutionary advances allowing reptiles to exploit insects as food; small size, fast-snapping jaws (pterygoideus muscle development), upright limb posture, and freedom from water, all select for insectivorous ability (R. Carroll, in McFarland *et al.,* 1979).

I.2.b. Reptile Systematics and Temporal Fenestrae

Classification of the higher reptile taxa (subclasses) is based on the presence and position of skull openings termed temporal fenestrae. This classification may not reflect real evolutionary relationships in all cases. The lower portion of Fig. 7-74 shows schematic illustrations of the fenestrae characteristic of the various subclasses. The Diapsida are currently subdivided into two subclasses, both of which have the same type of temporal fenestrae but which are overall quite different in many respects. Lepidosauria are the diapsids with mobile skull bones (e.g., the expandable skulls of snakes) and the group includes snakes, lizards, and a few extinct taxa; and Archosauria are crocodilians, flying reptiles, and dinosaurs.

The function of temporal fenestrae is at least twofold: the first and probably most basic function is to provide room for the contraction of large temporalis jaw muscles, without necessitating a protruding bulge at the side of the skull. The existence of two versions of single openings (euryapsid and synapsid) suggests that this function can be accomplished in more than one way. A second function is lightening of the skull with minimum loss of strength. One may observe this in the extreme in certain carnivorous dinosaurs, where the diapsid skull condition allows enormous skulls to be constructed from thin stringers of bone (see Chapter 8).

I.2.c. Anapsida and the First Reptiles

The earliest reptiles feature skulls lacking temporal fenestrae, and therefore are placed in the Anapsida (Fig. 7-75). However, the interrelationship of advanced amphibians and early reptiles is so close that there is professional discord on the placement of various subgroups between classes. Heaton (1980) discusses this topic in detail and points out that of three suborders originally assigned to the "reptile" order Cotylosauria (once called "stem reptiles"

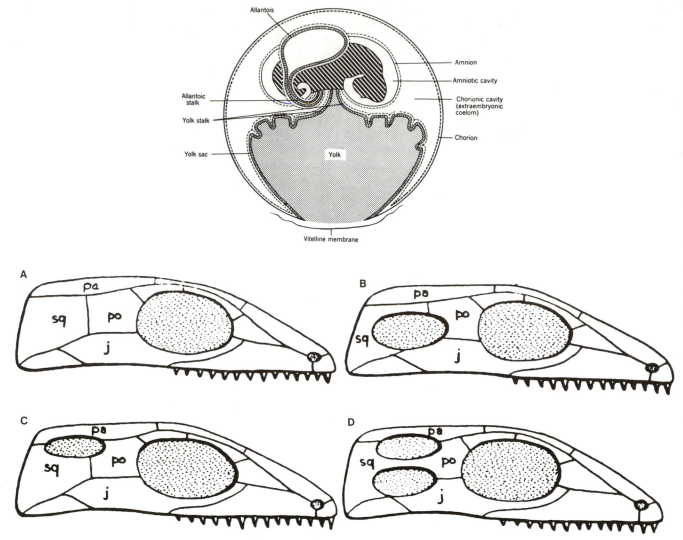

Fig. 7-74. (Above) The amniotic egg, showing the membranes which are formed by outgrowths of the embryo. (Below) Temporal fenestrate conditions characteristic of the four major subdivisions of Reptilia: (A) anapsid, (B) synapsid, (C) euryapsid, (D) diapsid (present in both Archosauria and Lepidosauria). The skull bones labeled in the lower portion of the figure are: j, jugal; po, postorbital; pa, parietal; sq, squamosal. (Above, from Torrey © 1967 by John Wiley & Sons, reprinted with permission of John Wiley & Sons; below, from Romer, A. S., *Vertebrate Paleontology*, 3rd ed., © 1966, University of Chicago Press.)

due to a misunderstanding of the derivation of the name), one suborder is currently assigned in total to the amphibians (Seymouriamorpha), the second (Diadectomorpha) contains some amphibians and some reptiles, and the third suborder (Captorhinomorpha) contains only primitive reptiles and may be a wastebasket classification into which all truly primitive anapsids are placed.

An additional taxonomic problem arose due to the original namesake of the Cotylosauria, which Heaton (1980) notes is an amphibian: that is, the "type" cotylosaur reptile is an amphibian!

Nevertheless, primitive reptiles, which we will term captorhinomorphs, date back to the Early Pennsylvanian and are represented by the oldest, *Hylonomus*, whose remains are best known from fossils found in hollow tree stumps (lycopods) from Nova Scotia. Primitive captorhinomorphs clearly show amphibian ancestry but they additionally developed higher and narrower skulls, single occipital condyles, and a number of diagnostic reptilian skeletal features. Among the captorhinomorphs, too, were very lizardlike forms, referred to collectively as romeriids (from the genus *Romeria*), which probably were early predators on terrestrial invertebrates. Another anapsid order, the Mesosauria, have an unclear relationship with the stem rep-

Table I. Discriminating Characters, Amphibians versus Reptiles

	Amphibian	Reptilian
Occipital condyles	Single or double	Single
Jaw adduction	Temporalis muscle	Temporalis + pterygoideus
Vertebrae	Variable: usually intercentrum + pleurocentrum	Pleurocentrum with reduced or absent intercentrum
Skull morphology	Flattened	Vaulted
Postparietal bones in the skull	All present	Reduced
Limb posture	Sprawling	More upright
Sacral vertebrae	Single	Two or more
Eggs[a]	Water-lain	Amniote
Larval development[a]	Aquatic	Terrestrial
Skin[a]	Smooth	Scaly

[a]Characters not determinable from fossils.

tiles but possess evident adaptations to aquatic life, such as long, narrow jaws filled with sharp, forward-projecting teeth that probably functioned to catch fish. Typical mesosaurs are 1 m long and widely distributed in Upper Pennsylvanian–Lower Permian marine strata from the Southern Hemisphere; they appear to have been an early, specialized offshoot from the stem reptiles, and they left no descendants.

A diverse and probably loosely interrelated group of stem reptiles are placed in the "order Cotylosauria," including some highly specialized forms that were quite common in the Triassic. One of these groups was the pareiasaurs, which includes large (2.5 m long), heavy-bodied forms with lumpy, grotesque skulls and the general appearance of modern rhinoceroses. Other "cotylosaurs," such as the procolophonians, were small and relatively unspecialized; these reptiles may be near the ancestry of both Diapsida and turtles (the only extant anapsid group).

Some "cotylosaurs" showed initial evolution of temporal fenestrae; for example, the genus *Millerosaurus*, from the Upper Permian of South Africa, shows a well-developed synapsid-type fenestra. This taxon was too young to have been ancestral to other reptile subclasses, but it points to the probable format of evolution by which in natural experiments, early reptiles independently evolved the fenestrated skulls of more advanced taxa.

Direct links between the anapsids and other subclasses are unknown but it seems that the romeriid captorhinomorphs are closer in overall morphology to the idealized ancestor than are any other known groups. In the Late Pennsylvanian, a great radiation of reptiles began with representatives of all five subclasses probably in existence before the Permian.

I.2.d. Synapsida

The synapsids feature the temporal fenestrae at the lower position on the side of the skull, and include two orders: Pelycosauria and Therapsida. Although the distinction between the suborders is partly abiological (pelycosaurs are largely Permian-age North American forms, whereas therapsids are Triassic African forms), the differences between extreme members are very great: indeed, the most advanced therapsids are virtually indistinguishable from (or may be!) mammals.

The simplest pelycosaurs are barely different from the romeriid captorhinomorphs except for the temporal fenestrae and a few other skeletal features. By the Late Pennsylvanian, primitive pelycosaurs were sparsely represented in terrestrial, coal-bearing strata and by the earliest Permian, they were common in the terrestrial fauna of Texas. Most studies have recognized three groups of pelycosaurs, at subordinal rank, of which the primitive, simple forms comprise the Ophiacodontidae. This group includes carnivorous, unspecialized pelycosaurs which may be the root lineage for other synapsids. Two taxa will serve to represent ophiacodonts; the Lower Permian genus *Varanosaurus*, a relatively small (under 2 m long), primitive form, and the contemporary *Ophiacodon*, a 4-m-long, relatively fierce genus. Among the primitive features of *Varanosaurus* are numerous, almost undifferentiated teeth (which may, however, have been a specialization for fish-eating) and apparently clumsy short limbs. *Ophiacodon*, also a probable fish-eater, had fewer, larger teeth, with greater size differentiation, longer and probably more efficient legs, and more massive and mechanically efficient jaws.

Probably from the ophiacodonts a second pelycosaurian lineage evolved: the Edaphosauria. These appear in the uppermost Pennsylvanian and are common in Lower Permian beds of Texas; they are the oldest-known herbivorous reptiles. The most representative genus, *Edaphosaurus* is large (4 m), and small-headed, with relatively small teeth, suggesting that it ate soft vegetation. *Edaphosaurus* also featured very large neural spines on its vertebrae that extended far above the body and were covered by vascularized tissue. These projections also had crossbars, giving it, as Romer phrased it, "somewhat the effect of a full-rigged ship." More advanced edaphosaurids, such as the Early Permian genera *Casea* and *Cotylorhynchus*, lacked the "sails." Another distinctive feature of *Edaphosaurus* was the chewing apparatus within the normal rows of teeth. This was apparently an early evolutionary experiment toward molarization of chewing; but, instead of adapting existing teeth toward flattened morphology, the group evolved plates of bone inside the mouth, studded with small teeth, as its way of processing vegetation.

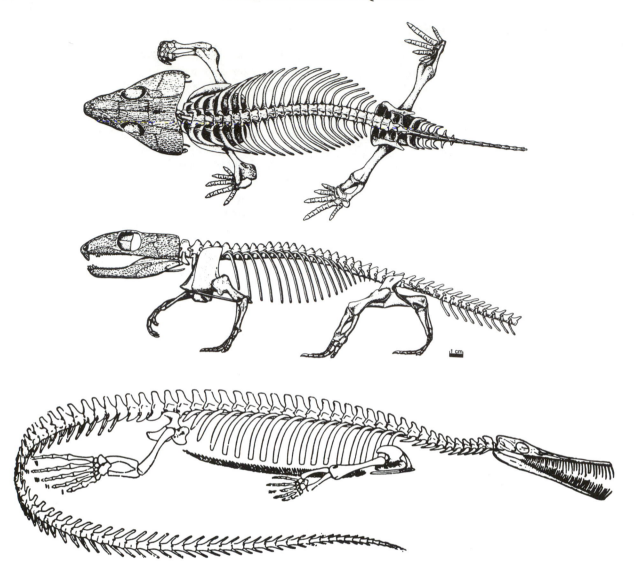

Fig. 7-75. Anapsida. (Above) Restoration of the Early Permian captorhinid anapsid *Eocaptorhinus* (length about 35 cm), showing typical morphology of the so-called "stem" reptiles: specimen shown in dorsal and left-side views. (Below) *Mesosaurus,* an aquatic anapsid (length about 40 cm) found in Upper Pennsylvanian–lower Permian deposits on most Gondwanaland continents. (Above, from Heaton and Reisz, 1980; below, from Romer, A. S., *Vertebrate Paleontology,* 3rd ed., © 1966, University of Chicago Press.)

The third group of pelycosaurs, the Sphenacodontia, were the most spectacular. These were large, predatory reptiles that are also the most common tetrapods of the Lower Permian of the western craton. The genus *Dimetrodon,* found in the Wichita and Clear Fork groups of north central Texas, is probably the best-recognized Permian reptile, featuring a distinctive sail formed from extensions of the neural arches of the vertebrae (as in *Edaphosaurus*), and also featuring a massive skull with curved (almost smiling) jaws and carnivorous dentition. Mature *Dimetrodon* reached to 4 cm in length (approximately the size of *Edaphosaurus*) and

Edaphosaurus may have been the main prey of *Dimetrodon* species.

There are some curious aspects to the paleoecology of these pelycosaurs: for example, why did both *Dimetrodon* and *Edaphosaurus* evolve sails while contemporary (but geographically separated) pelycosaurs, *Sphenacodon* and *Casea,* lived apparently quite well without "sails?" The distribution of *Dimetrodon* and *Sphenacodon,* which are morphologically almost identical except for the sails, is noteworthy: as stated, *Dimetrodon* is found primarily in central Texas, in deltaic strata which formed in marginal marine environments, probably not unlike conditions today in the Mississippi delta. *Sphenacodon,* without "sails," is found widely in the Abo red beds of New Mexico, and it lived in more arid, upland environments. The similarities between the genera indicate common ancestry; yet, the populations must have been geographically separated before the Early Permian. Pelycosaurs declined rapidly after the Early Permian and disappeared in the Middle Permian.

Dimetrodon, or a related form, was the likely ancestor of the Therapsida, which appear chiefly in Middle Permian through Lower Jurassic strata of the Gondwanaland continents (especially South Africa in the Karoo Series) but are known from scattered occurrences in North America. The physiological differences between pelycosaurs and therapsids are gradational, so that primitive therapsids are clearly recognizable as both reptiles and synapsids; whereas advanced therapsids, especially the suborder Theriodontia, are extremely mammalianlike and include animals which appear to be at the razor's edge of transition between reptiles and mammals. Figure 7-76 shows an advanced theriodont; note that the temporal fenestra of the synapsid reptile has opened into the mammalian jugal bar (cheekbone). Then too, the teeth are differentiated, a secondary palate is present, the carriage is inferred from the skeleton to be erect, and the jaw is composed almost exclusively of a single dentary bone. In the most advanced therapsid taxa, the sole reptilian features remaining in the skeleton are the absent bones of the middle ear (which are still present as jawbone vestiges). It is notable that the earliest mammals, of Late Triassic age, overlap the theriodonts both in time and in ear–jawhinge morphology (see discussion in Chapter 8).

I.2.e. Euryapsida and Triassic Turtles

Euryapsida is probably a *polyphyletic* (i.e., from many sources) assemblage and largely contains marine reptiles. Most euryapsids feature a temporal fenestra in a high position on the skull, and nearly all show specializations for life in water.

The sauropterygians comprised a euryapsid group of large, long-necked reptiles that began in the Triassic with the relatively unmodified "nothosaurs" (Fig. 7-77), found in marine deposits in Europe, Asia, and Africa. Sauropterygian evolution culminated in the Jurassic–Cretaceous with the plesiosaurs, enormous reptiles (lengths to 18 m) with necks that contained up to 76 vertebrae. In most sauropterygians, the evolutionary trends were toward long necks, small heads, and large, paddle-shaped limbs with many joints in the manus (a condition referred to as *hyperphalangy*); some sauropterygians, however, went the short-necked route and evolved enormously long heads. Sauropterygians are also examined in Chapter 8.

Turtles are anapsids whose ancestry among the reptiles is uncertain since the oldest known turtle still featured considerable specialization of form. *Proganochelys,* an Upper Triassic turtle from Europe, the earliest fossil turtle, already contained all of the shell structures of modern turtles (in addition to some that have since been lost). It had a horny beak, like modern turtles, but it also had some teeth inside the palate. In addition, like some turtles of later Mesozoic time (but unlike modern turtles), the head could not be withdrawn into the shell. Although the ancestry of Triassic turtles is unknown, their striking specializations suggest

Fig. 7-76. Mammallike reptiles (Therapsida). (Left) *Diademodon* (L. Triassic, South Africa), complete skeleton, about 1 m long, showing extremely mammalianlike morphology. (Right) *Cynognathus* (L. to M. Triassic, South Africa), skull length about 50 cm, a typical, relatively primitive therapsid. (Both are DRS photographs.)

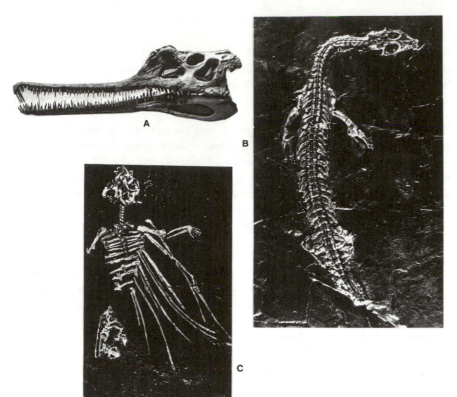

Fig. 7-77. A potpourri of Triassic reptiles. (A) *Rutiodon,* a large phytosaur (Archosauria) commonly found in nonmarine aquatic strata on the western craton; this skull approximately 1 m long. (B) *Pachypleurosaurus,* a nothosaur from the Middle Triassic of Europe; overall length approximately 20 cm. (C) *Icarosaurus,* an early lizard which also was adapted to gliding flight; overall length approximately 10 cm. (All are DRS photographs.)

that they evolved early from anapsid ancestors and that, like marine invertebrates, their Late Permian and Early Triassic record may have been lost due to the scarcity of marine strata from those times.

A final euryapsid lineage to mention in passing are the ichthyosaurs, the most highly specialized marine reptiles of all time. These are very dissimilar from other Euryapsida and probably arose independently, again, from unknown ancestors. Since the apex of icthyosaur evolution was reached in the Jurassic, with species as big as medium-size whales, we will discuss them in Chapter 8.

I.2.f. Diapsida

In this final section on reptiles we introduce two groups (considered subclasses), the Lepidosauria and the Archosauria, which have in common skulls with two temporal fenestrae. Both groups are more common and diverse in post-Triassic strata and will be dealt with in detail in the next chapter. Lepidosauria include ancestral diapsids, squamata (lizards and snakes), and rhynchocephalians, whereas the Archosauria include crocodilians, flying reptiles, two dinosaur orders, and an important Triassic group, the thecodonts.

The origins of the Diapsida are unknown but the ear-

liest found, lepidosaurians of the order Eosuchia, resemble romeriid anapsids and may point the direction of ancestry by this resemblance. The oldest eosuchians are from the Upper Pennsylvanian of Kansas, the genus *Petrolacosaurus* (Colbert, 1980), a small, slender, lizardlike form. A slightly younger eosuchian, *Youngina,* is a more common genus from the Upper Permian and Lower Triassic of South Africa and shows many morphological similarities with the cotylosaurs.

It is not known whether eosuchians are ancestral to Archosauria and younger Lepidosauria, or whether the archosaurs had independent origins. By Triassic time, advanced eosuchians with aquatic specializations, such as *Thallatosaurus* from North America, were present as well as very specialized gliding lizards, such as *Icarosaurus* from the Upper Triassic of New Jersey.

The Archosauria or ''ruling reptiles'' will dominate our discussion of Chapter 8. They appear as rare individuals in the latest Permian, and may be recognized by two features: a tendency toward bipedal walking (evident in elongate hind legs and reduced forelimbs) and a tendency toward enlargement of the skull with proportionately large temporal fenestrae. In Lower Triassic strata of South America, one finds fairly common fossils of *Euparkeria,* of the archosaurian order Thecodontia. *Euparkeria* was under 1 m

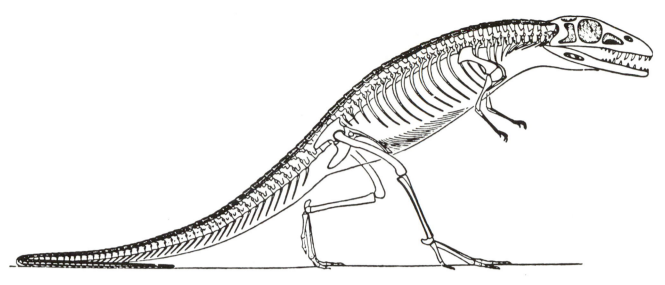

Fig. 7-78. *Saltopsuchus,* a Late Triassic thecodont from Europe which shows bipedalism and suggests the ancestry of dinosaurs. (From Romer, A. S., *Vertebrate Paleontology,* 3rd ed., © 1966, University of Chicago Press.)

in length; but slightly younger thecodonts, such as *Saltopsuchus* (Fig. 7-78) reached 1.5 m and suggest a trend toward dinosaur morphologies (see Chapter 8).

Common in North American strata from the western craton are fossils of aquatic thecodonts called phytosaurs. These were abundant in the Late Triassic and probably were top predators in lakes and streams. Phytosaurs resemble (but are not ancestral to) crocodiles and reached over 5 m in length, with extremely long, slender jaws. They can be readily distinguished from crocodiles by the position of their nostrils near the eyes (rather than at the tip of the snout). True crocodiles appeared in the latest Triassic, developing a secondary palate as an aid to breathing while submerged and hidden in water.

Latest Triassic time seems to represent a period of mass extinction of both labyrinthodont amphibian and many reptilian groups. However, one should be skeptical because the vagaries of fossilization and erosion of strata can turn a fossil record into an apparent hiatus. For example, Olson and Galton (1977) suggest that the Triassic–Jurassic tetrapod extinction is artificial. They base this on tetrapods from studies in Newark basins and further conclude that the apparent "event" is more due to a poor Lower Jurassic record than to a real crisis in reptile affairs.

CHAPTER 8

The Zuni Sequence: Middle Jurassic–Upper Cretaceous

A. Introduction

A.1. Overview

Craton. The Zuni cratonic sequence contains strata ranging in age from Early Jurassic (Sinemurian) through latest Cretaceous (Maestrichtian). The principal areas of Zuni sedimentation were: in the western craton and Cordillera, reaching from Alaska to Mexico; on the eastern continental margin, where the Atlantic Coastal Plain first developed during this time; and on the Gulf Coastal Plain, along with the birth of the Gulf of Mexico.

At the beginning of Zuni time, the Cordilleran orogen, as we know it today, did not exist; there was no Atlantic Ocean, no Gulf of Mexico, no Arctic Ocean. The Appalachian orogen stretched from the Oklahoma–Arkansas region (or, possibly, even farther to the south) in a series of arcuate sweeps to New England and thence eastward to Europe, which was adjacent to northeastern North America. Rifting within the Appalachian mountain system had begun during the Triassic, resulting in the eastern Triassic basins and, during the Early Jurassic, true continental separation began. But the Appalachians were still a major orogenic area, supplying detrital sediments which were carried by great river systems westward across the craton to the Pacific Ocean. From the Early Jurassic to the end of the Cretaceous occurred a sequence of events which resulted in a major Andean-type orogenic belt in the west and broad, open oceans to the east, south, and north of the continent.

Cratonic sedimentation commenced with Early Jurassic marine transgression in the west, followed by several regressions and transgressions through the remainder of the period, and culminating in the latest Jurassic with major marine regression and deposition of vast alluvial and other continental sediments on the western craton. In the Early Cretaceous, further limited transgressions and regressions ensued on the western craton, followed in the Late Cretaceous by the most extensive epeiric sea of post-Paleozoic history. This Late Cretaceous interior sea left an enormous sedimentary record. In the latest Cretaceous, the sea withdrew irregularly and, in the end, totally from the continental interior.

Atlantic and Gulf Coastal Province. Although rift-valley formation and the earliest phases of continental breakup had occurred during latest Absarokan time, it was during deposition of the Zuni Sequence that full-scale continental separation finally pulled North America away from Pangaea II, opening both the Atlantic Ocean and the Gulf of Mexico. On both newly-formed continental margins, rifted-margin prisms developed; western parts of this sedimentary prism were dominated by detrital sediments, primarily derived from the continent, whereas eastern areas were formed of carbonate sediments associated with a shelf-edge carbonate platform. At the base of the prism are evaporites, especially halite, formed during the proto-oceanic gulf phase of rifted margin development; these salt deposits are particularly thick in the Gulf of Mexico region, where they are responsible for the spectacular development of salt diapirs. During the Early Cretaceous, carbonate platforms were extensively developed all along the outer edge of the Atlantic continental margin, over the entire Florida–Bahamas area, and almost completely rimming the Gulf of Mexico.

The Late Cretaceous sea-level rise flooded all of these continental-marginal areas, pushing the position of shore far inland to the area of the modern Fall Line. This transgression is recorded in detrital sediments of the Salisbury Embayment, Cape Fear Arch, and eastern and western Gulf

Provinces. The same sea-level rise restricted carbonate-platform development to peninsular Florida and the Bahama Banks; concomitantly, the deep channels which cut through those shallow-water areas today (e.g., Florida Straits, Tongue of the Ocean) were formed by current scour and/or differential reef-development. Following the initial transgression, smaller-scale sea-level fluctuations and local tectonic and/or sedimentological effects resulted in back-and-forth migration of coastal and shelf environments, leaving a complex stratigraphy of interbedded sands and muds. In the Gulf of Mexico area, another major lithological constituent is chalk, which forms impressive, white outcrops on both eastern and western Gulf coastal plains.

North American Cordillera. The tectonic history of the Cordilleran continental margin during Zuni time is one of almost continuous activity. In the southern Cordilleran region, an Andean-type magmatic arc formed during the Middle or Late Jurassic; the result was an episode of deformation, magmatism, and uplift known as the Nevadan orogeny. Penecontemporaneously, in the northern Cordillera, a series of successor basins developed following the Triassic collisional accretion of the Stikine block. The succeeding magmatic arc, termed the Hazelton–Nicola Arc, supplied coarse detrital and volcaniclastic sediments both westward to the forearc basin (e.g., the Tyanghton–Methow Trough) and eastward to the Fernie–Laberge Basin on the edge of the continental margin.

During the Cretaceous, a classical magmatic-arc setting developed in the California area. The main igneous arc-massif today comprises the Sierra Nevada Batholith and related areas; to the west are forearc-basin deposits (the Great Valley Sequence) and subduction-complex deposits (the Franciscan Sequence). Behind the Sierra Nevada Arc was a broad foreland basin (the Upper Cretaceous detrital wedge of the western craton, mentioned above) and an intervening foreland thrust belt, termed the Sevier orogen. Large-scale thrusting occurred in this area, moving with time from west to east as the Cretaceous progressed, and comprised the Sevier orogeny. In the Northern Cordillera, too, arc-related thrusting occurred in a foreland thrust belt, termed the Columbian orogen, which was continuous with the Sevier belt to the south. Deformation associated with the Columbian orogeny, however, was partially interrupted during the Middle Cretaceous by an apparent cessation of subduction activity. This is believed to have been caused by the collisional accretion of Wrangellia, a major displaced terrane, which acted somewhat like a monkey wrench in the cogs of subduction. Development of a new subduction zone to the west of the Wrangellia block during Late Cretaceous time led to a new magmatic arc (the Coast Plutonic Complex of Canada) and a new forearc-basin complex (e.g., the Georgia Basin of Vancouver).

Innuitian Continental Margin. Zuni events of the far north are very poorly known except in northern Alaska (where extensive exploration for petroleum has elucidated regional stratigraphic patterns). In northern Canada, Zuni history was apparently dominated by continued subsidence of the Sverdrup Basin in which were deposited a variety of marine and nonmarine detrital sediments. The rate of Sverdrup Basin subsidence during the Jurassic was less than that of the Cretaceous. In addition, extensive fissure eruptions of basaltic lavas occurred in the northern part of the basin during the Cretaceous. These events are interpreted as having been related to the opening of the Amerasian Basin, i.e., the western half of the Arctic Ocean.

In northern Alaska, the history is more complex. Much of the Brooks Range and North Slope areas were, during the Jurassic, basinal areas in which thick marine shale sequences were deposited. Like the Triassic detrital sediments beneath them, Lower Jurassic portions of this sequence were apparently derived from the north. But at some time during the Jurassic (the timing is uncertain), a detrital sourceland to the south began to furnish sediment, at first only as fine muds. By the Cretaceous, however, massive quantities of coarse, immature detritus were being deposited in the northern Alaska area. The source of these sediments was an orogenic belt developed by the Brooks Range orogeny and the basin in which they were deposited is termed the Colville Foreland Basin.

The cause and evolutionary history of the Arctic Ocean are also poorly known and, as yet, quite speculative. Most models of the Arctic Ocean require an episode of seafloor spreading but the details vary considerably. Especially controversial are the nature and origin of the Alpha–Mendeleyev Ridge which has been considered alternatively as an ancient spreading ridge or an ancient volcanic arc complex. Nonetheless, by the end of the Cretaceous, the Amerasian Basin was a broad ocean basin.

Zuni Life. Also, during Zuni time, fantastic reptiles inhabited both land and sea. These organisms are the hallmark of the Mesozoic Era in popular fancy but, to an even greater degree, certain marine invertebrates characterize the Mesozoic. These latter include ammonite cephalopods and several types of bivalves. Terrestrial floras of Zuni time were equally distinctive and include the earliest angiosperms. Because of the large amount of Mesozoic strata preserved globally (and especially in North America), there are enormous numbers of Mesozoic fossils.

A.2. Zuni Paleogeography

As mentioned above, the Zuni cratonic sequence was deposited during the disassembly of Pangaea ll. Figures 8-1 and 8-2 show global paleogeographic reconstructions for the Triassic, Jurassic, and Cretaceous, illustrating the progress of that disassembly. Notice especially that North

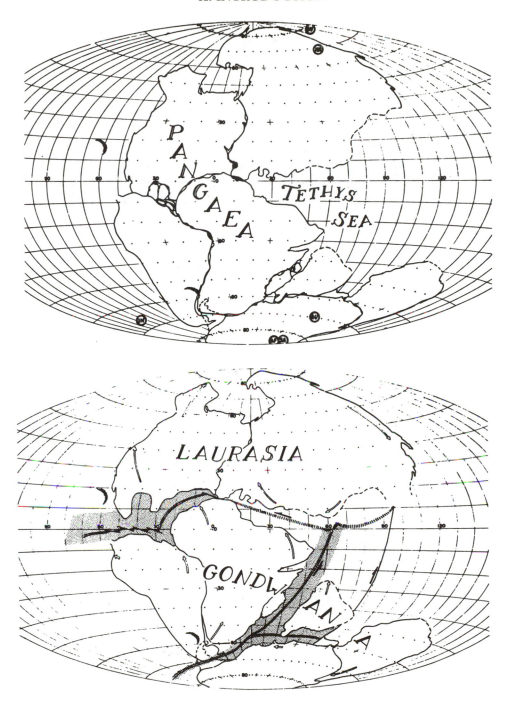

Fig. 8-1. Global paleogeography during the breakup of Pangaea II. (Above) Pangaea during the Late Permian, as a point of reference; (below) Late Triassic reconstruction, showing initial rifting at the eastern and southern North American continental margins, as well as in the southern hemisphere. (From Dietz and Holden, 1970.)

America was the first to split away from the assemblage during the Jurassic. South America was sundered off Pangaea II during the Cretaceous but final breakup of the great late Paleozoic supercontinent did not come until early in the Cenozoic. Figures 8-1 and 8-2 are interpretations of these events. During the Middle Jurassic, the principal connec- tion from the young North Atlantic to the rest of the world ocean was through the Tethys Sea between Europe and Africa. Although not shown in the figures, a second con- nection, from the Gulf of Mexico across southern Mexico, may also have been present. The current pattern developed at this time was essentially a single, clockwise gyre in the

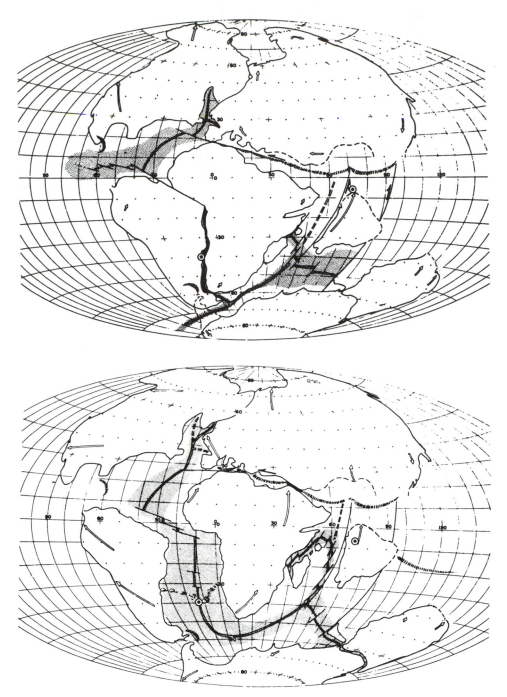

Fig. 8-2. Later Mesozoic paleogeographic reconstructions. (Above) Late Jurassic; (below) Late Cretaceous. (From Dietz and Holden, 1970.)

middle of the Atlantic Ocean. Note that no connection from the North Atlantic is shown to have existed between Greenland and Scandinavia. By the end of the Jurassic, a much more elaborate current system must have developed, at least partially because the North Atlantic Ocean was larger and apparently had deep-water access to the Pacific Ocean be-

tween Central and South America and to the northern ocean through a deep channel east of Greenland. The same pattern persisted in the North Atlantic during the Early Cretaceous, but continental drift of South America away from Africa had opened the South Atlantic by this time and a deep, relatively free connection existed between the North and

South Atlantic. Significantly, a surface current moved into the Gulf of Mexico from the south and continued along, either crossing southern Mexico to the Pacific or being shunted northward into the epeiric seaway of the western North American craton.

During the Late Cretaceous, a well-developed pattern of current gyres existed in both the North and South Atlantic. Closing the connection of the Gulf of Mexico, either westward to the Pacific or northwest to the epeiric sea, would have caused water to pile up in the Gulf, forcing the initiation of the Florida Current flowing around the southern tip of Florida and then northeastward to join the Gulf Stream. The Florida Current had profound effects on the subsequent development of the southeastern continental margin.

The Gulf Stream carried warm, tropical water far to the north in the Atlantic Ocean and, consequently, subtropical climates extended all along North America's eastern margin. Chalks, indicative of warm-water conditions, were deposited in the Labrador Sea, along coastal Greenland, and off Denmark. Glaser (1968) described the plant spores and pollen of Cretaceous strata from the Salisbury Embayment in Maryland; he concluded that the flora was characteristic of a warm, temperate to subtropical rain forest, similar to some New Zealand forests today. Other paleoclimate studies have been made of Cretaceous strata or fossils from the Atlantic and Gulf Provinces; these studies agree that conditions ranged from temperate to subtropical to even tropical.

By the Cretaceous, the North American continent had moved far enough north to have encountered the belt of prevailing westerly winds. One effect was probably the onset of seasonality (which does not strongly affect areas near the equator). A second effect of the position of the Cretaceous continent with respect to the westerly winds was the supply of abundant moist air to the Andean-type mountains of the western Cordillera. No doubt extensive rainfall greatly accelerated the rate of chemical weathering and the production of clay minerals. Most of these trends continued into the Cenozoic and we will return to them in the next chapter.

The history of sea-level changes through the Mesozoic and Cenozoic Eras has been studied by Vail and his colleagues. Figure 8-3 is a detailed graph showing fluctuations of sea level during that time. Notice that a long-term rise of sea level is shown from Sinemurian time to Maestrichtian time. Second-order sea-level oscillations complicate the trend (and third-order ones, too, although they are not shown in the graph) but the overall effect was the drowning of successively more and higher areas until, by the Late Cretaceous, a large part of North America (and other continents) was inundated. This long-term sea-level rise defines the Zuni cratonic sequence.

The cause of the Zuni transgression is debatable but the coincidence of sea-level rise with Pangaea II disassembly is obvious and has been noted by many authors (e.g., Valentine and Moores, 1972). Thus, the reduction of ocean-basin volume as a result of continental disassembly and rifted-margin prism development is possibly a major contributing factor in global eustasy. Changes in the rate of seafloor spreading and corresponding changes in the volume of spreading ridges, as suggested by Hays and Pitman (1973), may also be a factor (although the relationship between these two factors is as yet unknown). At any rate, by the Late Cretaceous, epeiric and broad pericontinental seas covered large areas of formerly dry land. A slight regression at the end of the Cretaceous marked the close of the Mesozoic.

B. Zuni Cratonic Sedimentation

B.1. The Post–Absaroka Unconformity

The unconformity between uppermost Absarokan and lowermost Zuni strata can be traced across virtually the entire western craton, where it lies generally above the Navajo, Nuggett, Popo Agie, or Spearfish formations. However, locally and regionally, beveling and truncation of exposed surfaces during the long time of Absarokan regression has resulted in Zuni strata overlying Moenave, Wingate, Chinle, or older Absarokan (and pre-Absarokan) units.

The depositional hiatus marked by the post-Absarokan unconformity varies greatly among regions. In "typical" sequences, Early Jurassic regressive sandstones such as the Navajo or Nuggett are overlain by Middle Jurassic marine sandy strata such as the Gypsum Springs Member of the Twin Creek Limestone (Fig. 8-4). But earliest Zuni strata are often younger than the Gypsum Springs and these sequences broaden the time span represented by the hiatus. During Cretaceous time, the most extensive of post-Paleozoic seas flooded the western one-half of the continent and in this great sea sediments were deposited in regions which had not received marine units since pre-Mississippian times. Consequently, at the easternmost reaches of the Late Cretaceous transgressions, one may find Upper Cretaceous strata overlying Pennsylvanian and older Paleozoic strata (especially in much of the eastern plains of Canada and in Eastern Kansas, Nebraska, and Iowa); in Minnesota, Wisconsin, and Manitoba, one may even find Upper Cretaceous strata overlying Precambrian rocks (Cobban and Reeside, 1952a; Pipiringos and O'Sullivan, 1978).

B.2. Jurassic Marine Sequences in the Western Interior

B.2.a. First Zuni Transgression

During the earliest Middle Jurassic (Bajocian time), an unusually shallow sea flooded the western craton across

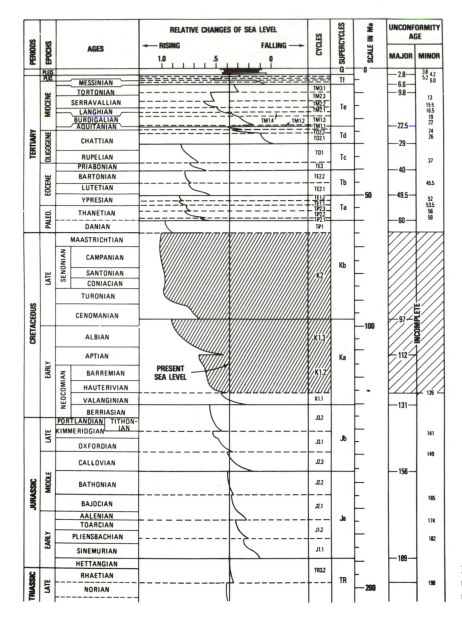

Fig. 8-3. Eustatic sea-level curves for Jurassic through Pleistocene time. (From Vail and Mitchum, 1979.)

Idaho, Wyoming, and northern Utah. It spread up to the Williston Basin areas of the Dakotas, Manitoba, and Alberta, and down to southwestern Utah and eastern Nevada (Fig. 8-5). The earliest Middle Jurassic sea was the first of three transgressions to occur during the Jurassic on the western craton; each subsequent transgression covered a wider area at its maximum. However, all Jurassic transgressions were bounded to the west (and their deposits end abruptly) along the tectonically active, rising Cordilleran front, which was especially active by the Late Jurassic. Interior seas were also bounded on the other three sides by nonsubmerged land; and it is almost certain western cratonic seas did not communicate at any time during the Jurassic with waters occupying the Mexican Basin.

Detritus coming to the Jurassic interior seas was derived from several sources; Hallam (1975) suggested that southeastern and southwestern detrital sources alternated with western (Cordilleran) sources through the Jurassic, with the high-standing Cordillera predominating later in the Jurassic. Greatest quantities of sediment formed in a major north–south trough at the foot of the Cordillera (Fig. 8-5) and most units thin eastward and northward, suggesting the western–southern source areas.

Deposits thicken in the Williston Basin, showing renewed activity of the ancient feature. In addition, several positive elements on the shelf were present during the Jurassic including the large Belt Island in central Montana (tectonically related to the younger Sweetgrass Arch), and the reactivated Uncompahgre Uplift in Colorado. These may not have been major detrital sources compared with the Cordilleran Highlands, but they did affect sedimentation by providing shorelines and submerged highs, yielding sedimentary wedges and pinch-outs (Hallam, 1975).

The first Zuni sea was commonly hypersaline and deposited gypsum, some halite, red and green silts and muds, and a mix of other lithologies. Much of the following description of Jurassic sedimentary lithofacies is from Imlay (1980) and Hallam (1975).

In the Williston Basin region, initial transgressive Zuni units comprise the Nesson Formation, now almost entirely in the subsurface. The oldest Nesson member, the Poe Evaporite, consists of 36 m of halite and gypsum with some red shale and dolostone. The Poe is overlain by the thin, dark-red, shaly Picard Shale Member, also containing some gypsum, which is, in turn, overlain by the variably thick (24–58 m) Kline Member containing a thick, locally oolitic limestone and an upper dolomitic unit.

The same early Bajocian sea deposited the Gypsum Springs Formation in western and central Wyoming and the Temple Cap Sandstone in southwestern Utah (Fig. 8-6). The Gypsum Springs ranges from roughly 15 to 65 m in thickness, and typically contains a massive gypsum bed at the base with overlying red siltstones and claystones, interbedded with lighter-colored limestones, dolostones, and fine detrital deposits. The Temple Cap Sandstone was formerly considered a member of the Navajo Formation but is now placed as a distinct unit between the Navajo and the overlying Carmel Formation in southwestern Utah (Peterson and Pipiringos, 1979). In Zion Canyon (the type area), the Temple Cap contains two members: the Sinawava Member, a slope-forming sandstone, silty sandstone, and mudstone with some gypsiferous beds; and the White Throne Member, a cliff-forming, fine-grained, cross-stratified sandstone.

A virtually pervasive unconformity separates the Nesson, Gypsum Springs, and Temple Cap formations from overlying Jurassic strata in the western cratonic interior, indicating that the shallow sea had withdrawn completely by middle Bajocian time (Pipiringos and O'Sullivan, 1978). Thus, in regions where the first Bajocian sea transgressed, a distinct, thin, evaporite–carbonate–silt sequence was deposited; elsewhere in the western craton, first Zuni sediments were deposited with the second or subsequent Jurassic transgressions.

B.2.b. Second Jurassic Transgression

In the middle through late Bajocian, the sea returned to roughly the same western cratonic areas but spread more widely than it had in the first transgression. Imlay (1980) noted that the rapid retreat and return of the sea over such a large region indicates that the western interior was at a very low elevation and relief. It was, therefore, easily influenced by minor up- or down-warping (or eustatic sea-level change). The second sea lasted longer than did the first and did not retreat completely until the late middle Callovian.

The second sea (which has been called the "Sundance Sea") had more normal marine waters. Basal deposits include the Sliderock Member of the Twin Creek Limestone in western Wyoming and adjacent northeastern Utah, the Piper Formation in central–western Montana, and the Page Sandstone in central–eastern Utah. These are generally normal-marine (or in the case of the Page, marginal-marine) lithologies: The Sliderock is a thin limestone with basal sandstones; the Piper is a variably thick, red and green shale with sandstones, gypsum beds, and oolitic and pebbly limestones; and the Page is a massive, cliff-forming crossbedded sandstone which resembles the Navajo and which may reflect the same marginal or nonmarine eolian environments of deposition. The various limestone units in the Sliderock and Piper formations are commonly fossiliferous, containing ammonites and the oyster *Gryphaea*.

The sea apparently deepened and stabilized during the latest Bajocian since nearly uniform lithologies, dating to that time, are regionally widespread. Imlay (1980, p. 111) stated that: "oolitic to dense highly fossiliferous gray lime mud was deposited over nearly the entire seaway from Idaho to South Dakota and from Saskatchewan to southwestern Utah." Units within this sequence include: the Sawtooth Formation in western Montana, middle units of the Piper Formation in central and eastern Montana, the lower member of the Shaunavon Formation in Saskatchewan, the Rich Member of the Twin Creek Limestone in easternmost Idaho, westernmost Wyoming, and northeastern Utah, lowermost members of the Sundance Formation near Jackson Hole, Wyoming, and lowermost members of the Carmel Formation across most of Utah except the extreme northeast. As Hallam (1975, p. 124) stated pointedly: "One reason why the great lateral persistence of many stratigraphic units has been obscured in the literature is that they have been given different formation names both within and between states, even though the lithology does not necessarily change significantly."

The sea shallowed subsequently and in the early Bathonian some evaporites and red beds were deposited, especially in the southern end of the trough fronting the Cordillera (i.e., in Utah) and in northeastern Montana. Hypersaline deposits include the gypsiferous, red-gray-green sandy and silty Banded Member of the Carmel Formation in southern Utah and, in northeastern Montana, the red, shaley Bowes Member (uppermost beds) of the Piper Formation. Normal marine units deposited around the western interior during the early Bathonian include: the upper, siltstone and limestone beds of the Sawtooth Formation in western Montana; sandy and limy uppermost Piper Formation units in

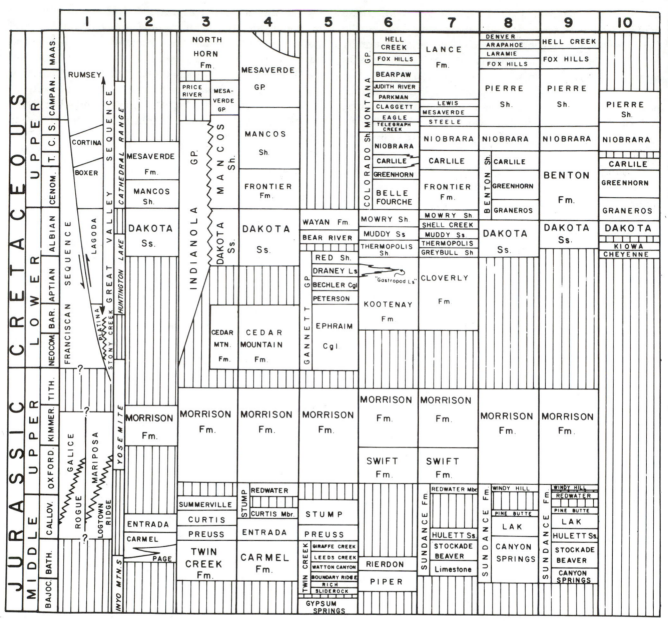

Fig. 8-4. Representative stratigraphic columns of Zuni deposits; most columns are synthetic, having been assembled from multiple sources and from different localities within the area represented. Areas are as follows: 1, California; *, igneous intrusive episodes in the Sierra Nevada Mountains of eastern California and western Nevada; 2, northern Arizona; 3, central Utah; 4, northeastern Utah; 5, southeastern Idaho; 6, Montana; 7, central Montana; the uppermost shale and limestone member of the Shaunavon Formation in the Williston Basin in Canada; the red and green siltstone, gray limestone, and sandy limestone Boundary Ridge Member of the Twin Creek Limestone in northeastern Utah/eastern Idaho/western Wyoming; lower-middle gray to red shaley and limy members of the Sundance Formation in central Wyoming; Wyoming; 8, Colorado; 9, South Dakota; 10, Kansas; 11, Alberta; 12, Cook Inlet, Alaska; 13, northern Alaska; 14, Sverdrup Basin, northern Canada; 15, Scotian Basin; 16, Salisbury Embayment, Maryland; 17, Cape Fear Arch, North Carolina; 18, Georgia and Alabama; **, Cretaceous series recognized in western Gulf Coastal Province; 19, eastern Texas; 20, northeastern Mexico.

and gray to red sandstone and siltstone units in the Carmel Formation in central Utah.

This same sea persisted through the Bathonian and withdrew in the late middle Callovian. During this interval, many specific shallowings and deepenings, restrictions, and freshenings occurred. We will not rigorously follow further details of sedimentation in this sea (the interested reader is

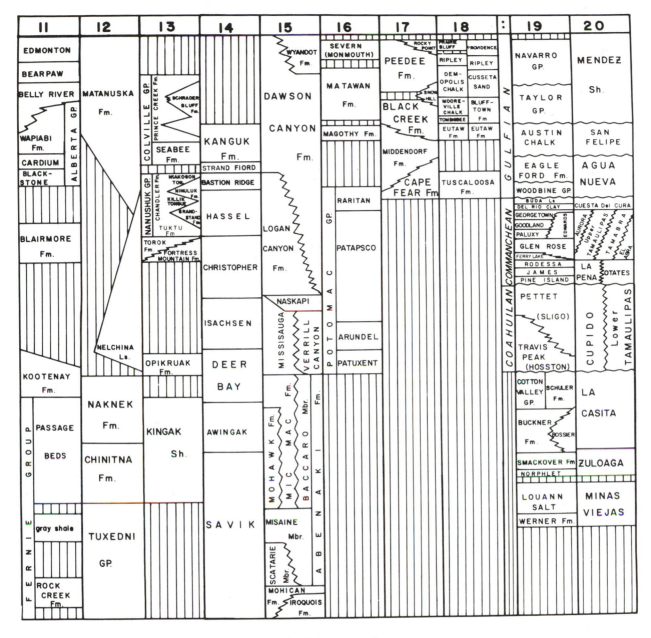

Fig. 8-4. (*Continued*)

directed to Imlay, 1980) but will highlight notable events and strata. Late Bathonian saw widespread deposition of lime mud in the widest transgression of this second Zuni sea. During that time, however, in southwestern Colorado and southeastern Utah, tidal flat deposits formed on the margins of the sea comprising the massive, white-to-orange, crossbedded Entrada Sandstone, which paraconformably overlies the Navajo Sandstone and represents similar but more proximal-marine environments.

In much of central Montana and in southern Alberta (Weir, 1949), the Upper Bathonian–Callovian interval is represented by the Rierdon Formation, containing generally 10 to 100 m of gray limestone, gray shale, silty shales, and sandy shales. However, in sections in between, on the Sweetgrass Arch, middle Bathonian through middle Oxfordian (Upper Jurassic) strata are conspicuously absent due to emergence of the Belt Island during that time. Across a broad area covering most of Wyoming and northern Colorado, the entire second Zuni transgression is represented by the Sundance Formation (from which the sea received its name). The Sundance contains seven members and is typically 60 to 125 m thick; locally, members pinch, swell, and

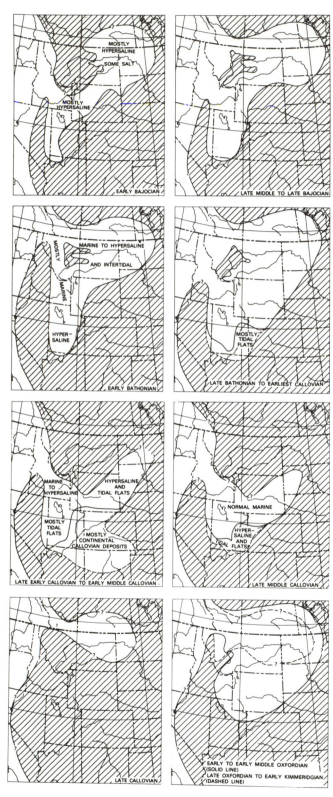

Fig. 8-5. Various stands of the Jurassic western interior seaway. (From Imlay, 1980.)

even disappear. The Sundance is mostly a normal-marine deposit, although highly variable, and includes a range of lithologies from glauconitic limestones in both basal and uppermost members, to red siltstones and shales in the middle Lak Member. Both upper and lower boundaries of the Sundance are diachronous and uppermost beds in many sections were deposited by the final Jurassic interior sea, with an unconformity separating members within the formation. The Twin Creek Limestone is a correlative formation in the region to the east of Sundance deposition, also containing seven members (including the Gypsum Springs deposited by the first Zuni transgression). The Twin Creek contains more limestone than does the Sundance but the two formations are genetically related and might be given a single name. Twin Creek strata are generally considered to have been deposited through the early Callovian and are overlain by the Preuss Sandstone, a relatively thick (to 400 m) pale-red, fine-grained ripple-marked sandstone which contains marine fossils and which probably represents shallow marine environments (thus, it may represent subaqueous environments adjacent to the Entrada shoreline).

B.2.c.   Final Jurassic Transgression

The sea withdrew almost completely from the western interior in the late Middle Callovian. The unconformity formed by the late Middle Callovian regression characteristically separates the Redwater Shale and Pine Butte members of the Sundance Formation, and splits the Curtis Formation in the Uinta Mountains of Utah into two parts.

The returning Late Jurassic sea in the western interior was present from late Callovian through Oxfordian and it entered the interior through an Arctic seaway, transgressing a trough parallel to the Cordilleran front and widening in the Williston Basin region and the western United States (Brenner and Davis, 1974). Its late Callovian extent was rather limited, encompassing only the trans-Canadian seaway and the Williston region; but in Oxford time, the sea expanded south across Wyoming, west across Idaho, and east into the Dakotas. During its maximum extent in middle Oxford time, it covered more than 1 million km^2, apparently reaching northern New Mexico and Arizona (Brenner and Davies, 1974; Hallam, 1975). [Imlay (1980), however, states that the Oxfordian sea did not extend south of northern Colorado and northern Utah.]

Two widespread sequences were deposited by this returning epicontinental sea: in northern Montana and the southern Alberta plains, the Swift Formation, and in Saskatchewan the Masefield Shale, overlie the Rierdon Formation. The Swift contains generally 10 to 60 m of glauconitic sandstone with a basal conglomerate, commonly containing siltstone and limestone beds. The Masefield is more shaly and thicker, but it becomes sandier to the south and proba-

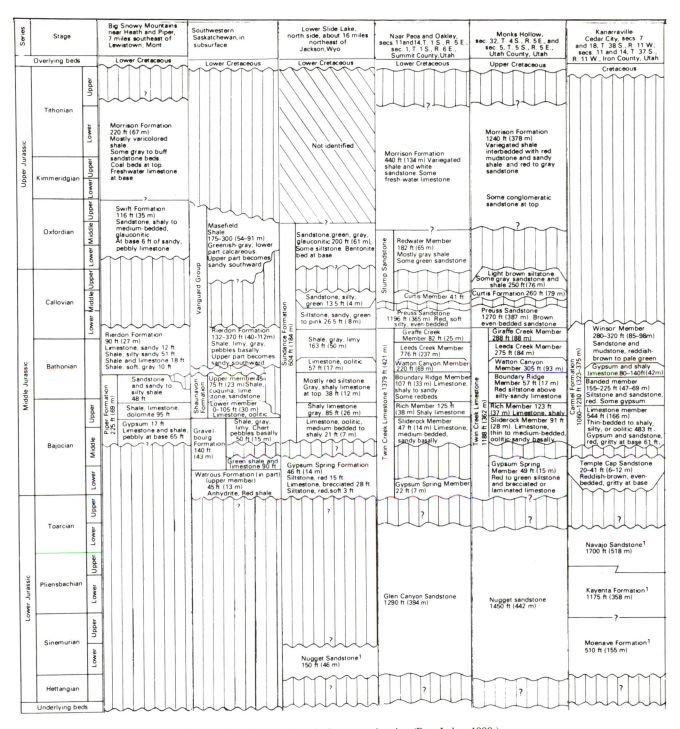

Fig. 8-6. Jurassic correlations in the western interior. (From Imlay, 1980.)

bly intertongues with Swift glauconitic sands. Swift units represent fairly normal transgressive basal conglomerates overlain by shallow marine sands and silts; in many places the Swift contains ripple marks. The uppermost beds of the Swift Formation grade conformably into the continental Morrison Formation, with the transition occurring approximately (but indeterminately) in earliest Kimmeridgian time.

The late Jurassic marine sedimentary unit includes the Stump Sandstone in southeastern Idaho, western Wyoming and northeastern Utah, and the Redwater Shale Member of the Sundance Formation in northwestern Colorado and central Wyoming. Typically, the Curtis Formation in central Utah contains green shale, glauconitic sandstone, and, locally, a basal conglomerate. The overlying Summerville Formation contains darker, generally maroon, brown, or reddish, even-bedded shales with thin-bedded sandstones and local gypsum deposits. The Stump Formation is similar to the Curtis but typically contains more glauconitic sand (Imlay, 1980; Pipiringos and Imlay, 1979).

The above strata demonstrate considerable homogeneity across the broad region. Marine sediments deposited from latest Callovian through middle Oxfordian times contain mostly green and gray silt and clay to the east and glauconitic, calcareous sand, sandy silt, and coquinoid mud to the west. The basal contact to the west is marked by a conglomerate of chert or quartzite pebbles, highly glauconitic sands, and abundant belemnites (see discussion in Section F.2); to the east the contact is marked by similar common belemnites, fewer coarse detrital sediments, and much less glauconitic sand (Imlay, 1980). A few localized gypsiferous beds are the only notable exceptions to what would appear to be a normal, mostly shallow-marine sedimentary suite across the entire region.

The last units deposited by the third Jurassic western interior sea vary lithologically around the region, just as does the probable age of the lower beds of the continental Morrison Formation which overlies these last Jurassic marine units. As mentioned above, Morrison strata overlie the Swift Formation in Montana with the transition occurring at approximately the Oxfordian–Kimmeridgian boundary. In central Utah, there appears to be a sizable unconformity between upper Oxford-age Morrison and underlying Summerville strata. In northeastern Utah, the Morrison overlies the Redwater Member of the Stump Formation with no sign of unconformity, but the age of the boundary is apparently middle Oxfordian. In most of Wyoming, parts of South Dakota, northcentral Colorado, and adjacent areas of other states, the uppermost member of the Sundance Formation, the Windy Hill Sandstone Member, is overlain conformably by the Morrison with the boundary dating from late Oxford to earliest Kimmeridge times.

B.3. Jurassic Continental Deposits

B.3.a. The Morrison Formation

After final retreat of the third Zuni sea by latest Oxford time, the entire western craton probably lay very near sea level. This huge region, just emerged from the sea, was a surface of very low relief. Onto this broad, featureless plain descended quantities of alluvium from the west, southwest,

and northeast. These sediments built up an enormous alluvial wedge represented by the Upper Jurassic Morrison Formation.

As described in the preceding sections, the Morrison generally overlies marine Jurassic strata conformably; such conformable marine-to-continental transitions could only have occurred without determinable breaks if the marine waters and land surfaces were at nearly the same elevations and if nearshore marine and marginal nonmarine environments intergraded. Hallam (1975) noted that although many Jurassic units in the west are given names which vary among regions for no real reason, the name "Morrison" is used everywhere (although its subdivisions do not persist laterally for any great distance). The reason is that the Morrison represents a mosaic of environments superimposed on a uniquely large system of alluvial plains. Indeed, the Morrison is the most extensive nonmarine unit in North America.

One might envision the overall Morrison environment to resemble an area of coalescing alluvial plains such as may be found today in the Hwang-Ho and Yangtze-Kiang Rivers in China (Mook, 1916, in Dodson *et al.*, 1980). On these alluvial plains, lying at almost unmeasurable low angles, numerous lakes developed and meandering streams flowed sluggishly to the sea which had withdrawn to the west. Morrison streams had wide floodplains through which they meandered. Oxbow lakes, cutoff meanders, and the backstream areas became sites of coal swamps.

The Morrison Formation covers approximately 1 million km^2 on the western interior and ranges to over 300 m in thickness. Detritus came mostly from the west, in the tectonically active Cordilleran orogen, but sources to the southwest in Arizona and to the northeast in the midcontinent (probably a reactivation of the Transcontinental Arch system) also contributed great quantities of material. The formation is divided locally into members with the most commonly recognized subdivisions comprising the basal Salt Wash Sandstone and the Brushy Basin Shale Members in southwestern Colorado.

Dodson *et al.* (1980) used a system of lithofacies to discriminate among the mosaic of Morrison units as part of their paleoenvironmental study (see Fig. 8-7). Among the laterally variable lithologies they describe, pastel-colored mudstone and siltstone are the most widespread and best typify the formation.

In central–western Montana, the Morrison contains commercially exploited coal in the upper 50 m. Far overshadowing the coal in economic importance, across a large portion of the outcrop, uranium ore (carnotite) is mined from the variegated mudstones, comprising the largest American uranium resource. An additional exotic lithology, bentonite, is common to the Morrison (as it is in many

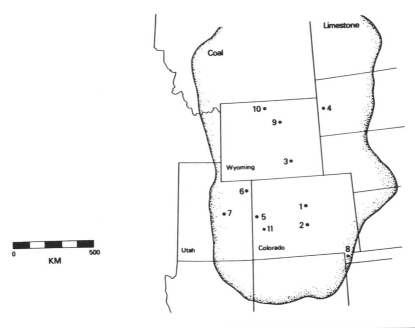

Lithofacies	Lithology & Texture	Color	Structures	Unit Dimensions & Shape
A. Coarse-Clastic and Sand-Lens	Sand, fine to coarse with mud and carbonate clasts common. Sand units usually well-sorted and clean; coarse clastic units poorly sorted.	White, brown-gray, yellow-gray, rarely pink-brown to red.	Cross-stratification; ripple laminae; irregular partings.	Lenticular, shoestring. Thickness: 0.5–5.0 m, usually 1–2 m. Lateral extent: <1 km, usually 50–100 m.
B. Variegated Mudstone	Dominantly clay, with variable % silt; some thin sand units, carbonate nodules and glaebules dispersed or in distinct zones, occasionally associated with angular mudclasts. Sorting: variable, good in clay units.	Red, red-brown, purple, light to dark green, dark brown.	Lacks well-defined macro-bedding; mottled, burrows and/or root traces common, ped structures (clay-coated granules), some fine laminae.	Tabular; thickness highly variable: 0.5–6.0 m. Lateral extent: usually >100 m, <1 km.
C. Drab Mudstone	Clay usually dominant, variable low % silt, CaCO₃ finely disseminated or concentrated in zones of nodules with angular mudclasts. Sorting variable.	Gray-green, yellow-gray, gray-brown, dark brown.	Lacks distinct macro-bedding, fine laminae locally common, burrows and/or root structures common, mottling, occasional mudcracks.	Tabular; thickness: 2.0–8.0 m, typically 5 m. Lateral extent: usually >1 km.
D. Limestone-Marl	CaCO₃ micrite, marl, silt and clay, occasional fine sand, bioclasts, mudclast lenses. Sorting generally good.	Light gray, gray, gray-brown, gray-green.	Thin interbedded limestone and clay, irregular but well-defined bedding. Limestone in discrete units, occasionally nodular. Mudcracks, abundant fine laminae, footprints, possible algal structures.	Tabular; thickness: 0.2–0.8 m. Lateral extent: 10 to 150 m, usually >50 m.

Fig. 8-7. The Morrison Formation distribution and lithologies. (Above) Distribution of the Morrison Formation, with noteworthy fossil localities labeled: (Colorado) 1, Morrison; 2, Canon City, Cope and Marsh quarries at Garden Park; 5, Grand Junction, Riggs quarry; 11, Potter Creek, Uncompahgre Plateau; (South Dakota) 4, northeast Black Hills, piedmont *Barosaurus* quarry; (Wyoming) 3, Como region; 9, Red Fork of Powder River, southern Bighorn Basin; 10, central Bighorn Basin, Howe quarry; (Utah) 6, Dinosaur National Monument; 7, Cleveland–Lloyd quarry; 8, Kenton region, Stovall site. (Below) Sedimentology of four major Morrison lithofacies. (From Dodson *et al.*, 1980.)

Cretaceous strata from the west) and its presence is evidence of increasing tectonic activity in the Cordillera, with accompanying volcanism.

Fossils in the Morrison include freshwater ostracodes, bivalves, gastropods, arthropods, and dinosaurs. The age of the Morrison was determined by correlation of its vertebrate fauna with that of African regions containing both marine and continental strata spanning Morrison time; certainly the most celebrated aspect of the Morrison is its rich suite of latest Jurassic dinosaurs.

Morrison dinosaurs are well-represented in museums, following the intense collecting in the late 19th century done for the Yale Peabody Museum by O. C. Marsh, and for the Philadelphia Academy of Natural Sciences by E. D.

Cope. An intense rivalry developed during their years of collecting (see Lanham, 1973). An apocryphal story claims that Marsh invented the term *coprolite* (i.e., fossil excrement) to "honor" Professor Cope. The Morrison dinosaur fauna has been made vivid in museum tableaux such as those painted by Charles Knight, showing huge "brontosaurs" pursued by grinning, bloodthirsty carnivorous theropods (see discussion of dinosaurs in Section G.1).

Although leaving discussion of Morrison reptiles for subsequent sections, we would like to elaborate here on the presumption held by many that the presence of large dinosaur faunas in the Morrison "proves" that the climate was humid in Morrison time. The largest sauropods in the Morrison fauna have commonly been presumed to have lived a semiaquatic life, using the buoyancy of water to overcome their enormous weight. Because the bones of such giant herbivorous reptiles are relatively common in some Morrison outcrops (especially at Dinosaur National Monument, northeastern Utah, the type area at Morrison, Colorado, the Cope and Marsh quarries at Garden Park, Colo., and other areas in central Colorado, Wyoming, and northeastern Utah; see Fig. 8-7), it has been presumed that water was abundant. Morrison plant life, while poorly represented as fossils, also seems to indicate moist conditions because cycads and giant rushlike plants were present. Additionally, the presence of coals and the obvious need for 30-plus-metric-ton herbivorous reptiles to feed themselves have pointed to lush vegetation, and, presumably, humid conditions. However, Dodson *et al.* (1980) cite evidence to suggest that Morrison environments were generally dry and periodically very arid. In their view, a strongly seasonal, periodically dry climate forced reptiles to wander across the western craton, resulting in wide distribution of species and also in occasional mass mortality (yielding the rich bone beds). In addition, they do not believe that the sauropods were semiaquatic (in the manner of the modern hippopotamus) but rather that they lived like modern elephants in dry uplands and sought ponds periodically for drinking water and cooling.

In southern Canadian parts of the Williston Basin, Morrison-age strata contain marine sands and carbonates which are assigned generally to the Fernie Group. These sediments represent isolated, regressive facies left by the last remnants of the Jurassic Zuni sea which retreated northward and westward out of the conterminous United States. Morrison deposition ended regionally in the latest Tithonian time but locally uppermost Morrison beds may have been deposited during the earliest Cretaceous. The first post-Morrison units regionally are nonmarine sands which overlie the Morrison unconformably. Pipiringos and O'Sullivan (1978) state that the hiatus represented was brief (3 Myr.), but other authors believe the time of erosion was relatively long, ranging from 10 to 40 Myr. (see Dodson *et al.,* 1980).

B.3.b. Jurassic Deposits of the Continental Interior

Outside of the western interior, only one cratonic region, the Michigan Basin, contains definite Jurassic-age strata. Continental deposits there are present in the basin center, ranging from 30 to 130 m in thickness and consisting of mixed red clay, shale, sandstone, and gypsum (Imlay, 1980). These sediments are generally referred to as the "Upper Red Beds," lying apparently isolated from related Jurassic strata; Imlay noted that they were deposited penecontemporaneously with the Morrison Formation (based on evidence from plant spores). It is paradoxical that a region stable for so long (post-Pennsylvanian to Late Jurassic) should undergo alluviation simultaneous with the long-active western interior and with similar styles of sedimentation.

Elsewhere on the midcontinent, continental strata overlying Pennsylvanian strata are reported from north-central Iowa (the Fort Dodge Gypsum) and in the subsurface in southern Canada and adjacent northern Minnesota (King and Beikman, 1975). These deposits have not been dated paleontologically but they overlie Pennsylvanian beds with the same relationship as do Jurassic units in Michigan. Conceivably, these and the Michigan continental units may be remnants of an alluvial deposit that extended across the entire northern United States–southern Canada border as far east as Michigan and with sources westward in the Transcontinental Arch (in the eastern Dakotas or Nebraska).

B.4. Early Cretaceous in the Western Interior

B.4.a. Introduction to the Cretaceous System

After Morrison time, i.e., at the Jurassic–Cretaceous time boundary, the nature of cratonic sedimentation changed rather dramatically from the patterns present since the Mississippian. In the Early Cretaceous, hesitant but eventually very sizable marine transgression took place between the Cordillera and the midcontinent; and during the Late Cretaceous, a Paleozoic-style epeiric sea flooded so much of the continent that a broad seaway was formed extending from the Arctic Ocean to the Gulf of Mexico (Fig. 8-8), and including the newly rifted Coastal Plain margins (see Section C). North America was divided into two landmasses by this most extensive sea, with very complex consequences for geological environments in the various parts.

Marine deposits of the Cretaceous seas are noteworthy for the sizable chalk deposits laid down, mostly during the late part of the period. The term "Cretaceous" derives from the Latin *creta,* meaning chalk, and the original use of the term was based on the famous chalks in southeastern England (the "white cliffs of Dover"). In America, Upper

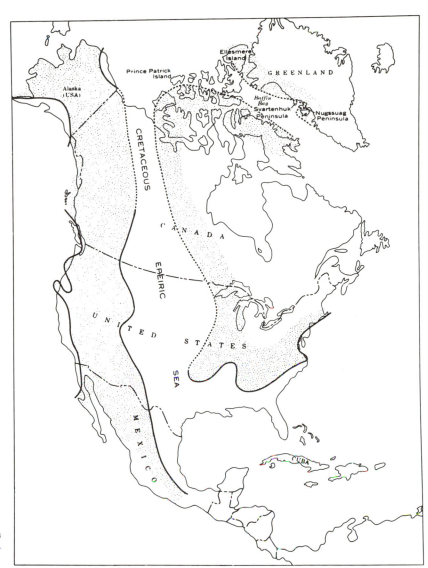

Fig. 8-8. Approximate distribution of land and sea during a maximum transgressive interval (late Campanian). (From Gill and Cobban, 1966.)

Cretaceous chalks of wide recognition include the Niobrara (Kansas and vicinity), Austin (Texas), and Selma (Alabama and Mississippi) formations. These chalks are largely the products of foraminifera, algal debris, and, most especially, the minute calcareous plates of the chrysophyte algae called coccolithophorids (see discussion in Section F.4).

We will continue in this discussion in the use of European stage nomenclature (as we have with Triassic and Jurassic discussions, in contrast with the Paleozoic). This practice is necessary because North American units are defined only regionally and are virtually never uniform in their boundaries. For example, in the Gulf Coast, three series are recognized: the Coahuila, Comanche, and Gulf. Three are also recognized in the western interior: the Gannett (or sometimes Dakota), Colorado, and Montana. However, the Gannett was deposited well into Comanche time, the Gulf

Series formed through most of Colorado and Montana times, and none of these subdivisions except the Colorado Series (which began at the beginning of the Late Cretaceous) correspond with European series, which are simply the Lower and Upper Cretaceous. To accommodate these diversities, we will refer to European series and stages for time references, while using regional terminology in appropriate discussions for organization of concepts. It should be noted, however, that although the nomenclature varies, interregional correlations of Cretaceous strata are quite good because the index fossils available for the Cretaceous are abundant and precisely dated.

Beginning here with discussion of the Cretaceous history of the western craton, we will integrate events in the Cordilleran Mountains closely with cratonic events, because from Cretaceous time onward the Cordilleran Mountains became the dominant (and often exclusive) sources of detritus for the western half of the continent. Williams and Stelck (1975) stated that by the late Neocomian the Cordilleran Mountains became continuous from Alaska to Mexico, bordering the North American plate and, except for occasional minor breachings, they prevented Pacific waters from entering the continental interior. Preserved Cretaceous strata in the interior thicken and coarsen westward to the margin of the Cordilleran Mountains, at which they are truncated by the beginning of the Rocky Mountain thrust zones. In far western sections, Cretaceous rocks reach over 10 km in thickness while eastward, across the hinge line that marks the edge of the Cretaceous shelf, the same composite Cretaceous section thins by an order of magnitude and eventually tapers to a zero edge in the midcontinent east of Minnesota and Iowa.

B.4.b. Lower Cretaceous Nonmarine Strata of the Central–Western Interior

No strata are known which can be dated right at the Jurassic–Cretaceous boundary interval, but several coarse detrital sequences from earliest Cretaceous time are known. These are interpreted generally as fluvial deposits shed from streams draining the rapidly rising Cordilleran Mountains, laid down on the very low, flat, post-Morrison surface. Several of the units to be discussed feature marine strata in their upper beds, reflecting incursion of marine waters into the whole length of the western interior by the late Albian.

Starting in Idaho, northeastern Utah, and far westernmost Wyoming, one may observe a set of coarse detrital rocks, dating from Neocomian through Aptian, and comprising a sequence averaging 1 to 1.5 km thick. This set of continental units is best represented by the Gannett Group (the reference sequence for the western interior Lower Cretaceous) described in Idaho. At the base of the Gannett is the very thick Ephraim Conglomerate, containing conglomerate, red-and-purple mudstones, and coarse-grained sandstone, clearly derived from the west. Above the Ephraim are the Aptian-age Peterson Limestone, Belcher Conglomerate, Draney Limestone, and an unnamed uppermost red shale. The Gannet Group is entirely of continental origin, and the Draney and Peterson Limestones were lake deposits formed during pauses in Cordilleran tectonic activity (King and Beikman, 1976). The Belcher Conglomerate marks renewed mountain-building and the uppermost red shale likely represents fluvial deposition during a relative pause in orogenesis. Above the Gannett Group is the Bear River Formation, composed of mixed marine and (mostly) nonmarine sandstone with shale and siltstone. The Bear River is approximately middle Albian age, and its marine beds indicate at least a temporary marine influx; however, above the Bear River is the Wayan Formation, a

nonmarine unit. The Wayan contains a basal, lacustrine limestone which grades upward into reddish siltstones and sandstones. The total Wayan section is up to 1 km thick and it shows that marine conditions were short-lived regionally during the Early Cretaceous.

In other localities abutting the Rocky Mountain thrust zone, thick, Lower Cretaceous coarse detrital units are called variously the Beckwith Formation (in southwestern Wyoming), and the Kelvin Conglomerate (northeastern Utah; data from Stokes *et al.*, 1955). Both units are probable equivalents of the entire Gannett Group and the Beckwith may include parts of the overlying Bear River Formation.

Eastward in Montana, and farther from the Cordilleran mountains, basal Cretaceous strata overlie the Morrison Formation unconformably but with almost similar lithology. The Kootenay Formation, present in Montana and adjacent Canada, is probably Aptian age and contains typically a basal conglomerate or sandstone overlain by sequences of coal, shale, and sandstone (Suttner, 1969). The Kootenay contains relatively less carbonate than does the underlying Morrison, probably reflecting less lacustrine influence, but at the top is a widespread limestone containing freshwater gastropod fossils. In contrast with the Morrison, which had mixed Cordilleran and eastern detrital sources, the Kootenay detritus derives entirely from eastern Idaho and extreme northwestern Montana.

Leaving Montana, one begins to encounter sections of Lower Cretaceous strata designated ''Dakota,'' either of group or formational status. Unfortunately, this encumbers the ''Dakota problem'' (King and Beikman, 1976) which is, basically, that the term ''Dakota'' designates many things in the Cretaceous of the western interior. In the type region for the Dakota Formation, eastern Nebraska near Dakota City, the name designates two nonmarine to marginal-marine sequences of transgressive sandstones: the Terra Cotta Member, a resistant, cross-stratified sandstone and mottled claystone, and the overlying Janssen Member containing distinctive crossbedded sand lenses in a generally clayey matrix. The Jansen Member also contains marine units near the top in Kansas. More importantly, the type Dakota Sandstone was deposited in the latest Albian and early Cenomanian, i.e., it is a transitional Lower–Upper Cretaceous unit.

However, as geological nomenclature is occasionally wont to do, the resistant quartzites of the typical Dakota were used to characterize other sections and, since most such strata are not fossiliferous, the name ''Dakota'' was applied too widely. In South Dakota, Wyoming, Colorado, and parts of Kansas, ''Dakota'' strata are entirely of Early Cretaceous age while elsewhere, they are entirely early Late Cretaceous. Even worse, in areas of Colorado and Wyoming, the ''Dakota Group'' comprises the entire Lower

Cretaceous section, intervening between the Morrison strata below and the Mowry shales above.

"Dakota" sandstones commonly form the ledges and dip-slope beds in many parts of Colorado, Utah, and Wyoming; these strata are by their physical nature well-exposed and have been extensively studied. Franks (1975) analyzed and described the Dakota Formation in Kansas as the lenticular clays and sands deposited by southwestward-flowing streams in an alluvial plain–deltaic complex by the shores of the shrinking late Albian–earliest Campanian seaway. Deposition of these Dakota lithologies continued until renewed, regional, Late Cretaceous transgression effectively ended marginal marine sedimentation. Karl (1976) noted that as the Cretaceous (Cenomanian) sea transgressed eastward, Dakota-bearing streams became less competent due to base-level change, and deposited fine-grained, lens-shaped sandstones where they had deposited tabular crossbedded sandstones previously (thus, the differences between the lower, Terra Cotta and the upper, Janssen members) (Fig. 8-9). Karl and Franks each note that some uppermost Dakota strata in Kansas are marine (Fig. 8-10) and grade into overlying marine shales and limestones of the Upper Cretaceous section. Uppermost Dakota beds in Kansas have also been shown to contain marine and brackish-water mollusks (Hattin, 1967), whereas most lower Dakota beds are nonfossiliferous.

As an example of an exclusively Lower Cretaceous "Dakota" unit, the Cloverly Formation in central Wyoming has been called the "Dakota Sandstone" by many authors, being a resistant, hogback-forming thin sand. It is of Aptian age and is entirely nonmarine. In eastern Wyoming adjacent to the Black Hills, and in South Dakota, lowermost Cretaceous strata are the Inyan Kara Group, which includes a single Lower Cretaceous unit, the Lakota Formation. Lakota Rocks are probably older than the Cloverly, but they may intercalate in the subsurface. The Lakota is a nonmarine, yellow-red-brown sandstone with some multicolored siltstone and claystone (Sohn, 1979). Lakota facies probably were deposited much farther from the shore than were Cloverly "Dakota" units and represent more continental influences.

In the Colorado Plateau, the Lower Cretaceous section is represented by two nonmarine units overlying the Morrison disconformably: the Cedar Mountain and Burro Canyon formations. The Burro Canyon type area is in extreme southwestern Colorado whereas the Cedar Mountain section is in central, eastern Utah. The two units are genetically similar, containing a basal conglomerate overlain by a highly varied lithology of varicolored sandstone, mudstone, siltstone, and shale (Stokes, 1952).

Above the Burro Canyon and Cedar Mountain formations in the Colorado Plateau is a ledge-forming, coarse detrital unit which seems to fit well into the "Dakota" type

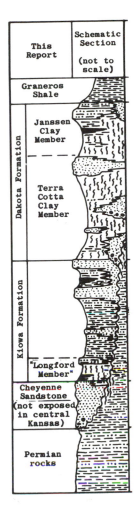

Fig. 8-9. General column of Lower Cretaceous rocks in Kansas. (From Franks, 1975, reprinted with permission of the Geological Association of Canada.)

lithology. It is light yellow to gray, of generally thin but variable thickness and in places coarse textured or cherty.

B.4.c. Lower Cretaceous Paralic and Marine Strata in the Southern-Western Interior

Early Cretaceous transgression on the southern craton was halting and irregular; thus, one finds extensive transitional continental-to-marine deposits across a wide area. We will trace such sequences from the extreme southwest up through the central-west although the seas did not follow quite such simple directions of transgression and regression; rather, transgression followed a generally northward course but regression commonly occurred eastward.

In southeastern Arizona, southwestern New Mexico, and northernmost Sonora and Chihuahua (Mexico), Lower Cretaceous rocks of the Bisbee Group represent thick (to 3000 m) sequences of piedmont-to-marine strata with most

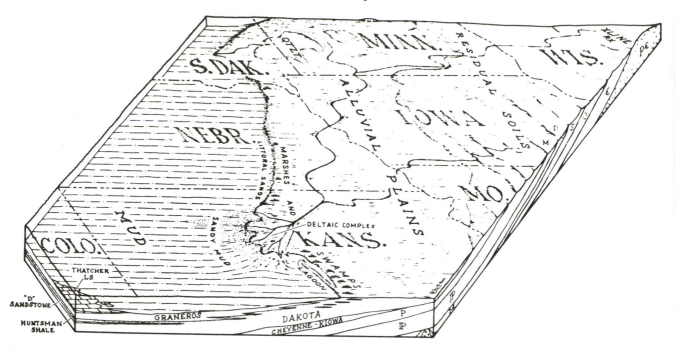

Fig. 8-10. Paleogeographic reconstruction of central Kansas and vicinity during late Dakota time. (From Hattin © 1967, University of Kansas Press.)

of the area receiving deltaic or marine sediments by the middle Albian. In southeastern exposures, Bisbee strata contain predominantly marine limestones, whereas in western areas of its deposition, the proportion of limestone decreases as the quantities of sand and arkose increase (Hayes, 1970b).

The basal Bisbee unit is the Glance Conglomerate, which represents fanglomerates from a positive area in southeastern Arizona, shed toward the sea in Mexico (see Fig. 8-11). Hayes stated that the seaway may have advanced from the Gulf to northern Chihuahua by the late Neocomian and by the Aptian, it may have advanced to Arizona. Most Bisbee strata date to Aptian–Albian with some Glance or equivalent sediments dating to the late Neocomian. Volcanics are present in Bisbee strata, probably traceable to sources in Arizona. Above the Glance is the Morita Formation, consisting of approximately 90 m of pink-gray to pale-red feldspathic sandstone, with siltstones and mudstones. These represent deposition on a delta plain, with some oyster beds in upper units showing intermittent flooding from the sea advancing from the southeast (Hayes, 1970b).

The maximum regional marine advance occurred during the early Albian, when brackish to marine strata were deposited over a broad area, comprising respectively the Apache Canyon and Mural formations. The Apache Canyon contains thin-bedded, platy, argillaceous limestones with a brackish-water fauna, whereas the Mural is a widespread, thick-bedded limestone with marine fossils including rudist bivalve bioherms.

The sea withdrew eastward in the late Albian–Santonian, leaving the Cintura Formation over the Mural and equivalents. The Cintura is interbedded sandstone, siltstone, and shale and correlates with arkoses to the west. It shows fluvial channeling and clearly represents regressive, alluvial deposits on the emerged continental surface. Above the late Albian–earliest Cenomanian Cintura is an erosional surface which regionally represents a hiatus to the early Campanian.

In western Kansas, eastern Colorado, Oklahoma, and parts of New Mexico, marginal-marine to marine units are present below "Dakota" sands of Cenomanian age. These Lower Cretaceous strata are assigned generally to the Cheyenne Sandstone, which in the type area (Kansas) is a light-colored, fine-to-coarse-grained, crossbedded sandstone which is generally thin (under 28 m) and contains shale beds. The overlying Kiowa is composed of dark-gray or olive-gray shale with brown sandstones and a variety of other lithologies [including calcite and siderite nodular beds, and coquinoid limestones (Scott, 1970; Franks, 1980)]. The Cheyenne and parts of the Kiowa formations represent terrestrial, coastal-plain environments whereas most Kiowa strata represent lagoonal, barrier bar, and open marine environments.

The Early Cretaceous seaway in the southern craton opened across the Oklahoma and Texas panhandles, almost certainly connecting with the Gulf Coast region and extending, in the early Albian, across Kansas. Franks (1975) showed that the basal Longford Member of the Kiowa in

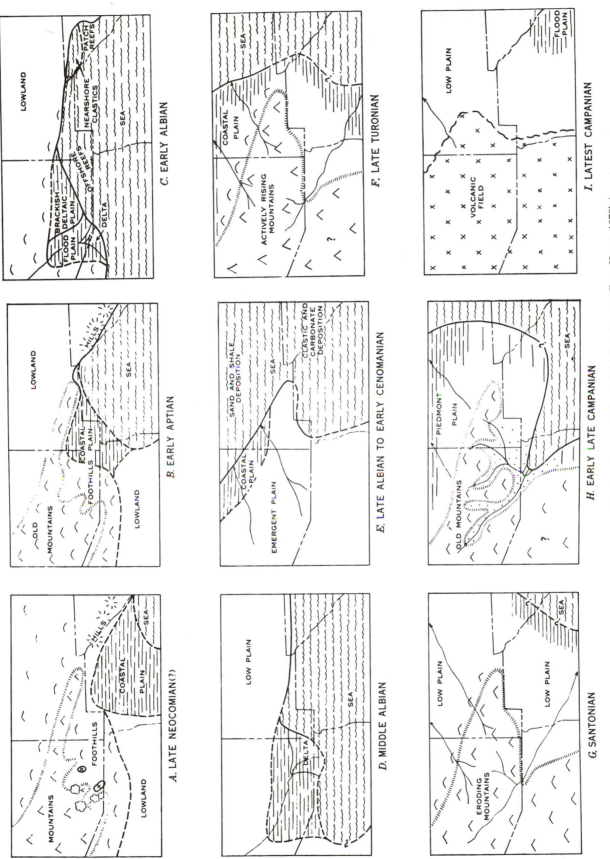

Fig. 8-11. Cretaceous paleogeography of southeastern Arizona and adjacent areas. (From Hayes, 1970b.)

Kansas represented fluvial and fluvial/estuarine realms while the overlying shale and topmost sandstone beds represented (respectively) open sea or open bay and tide-influenced barrier bar systems. Franks observed that a near-textbook model of marine transgression is preserved in the Kiowa.

Elsewhere in the southern cratonic interior, basal Cretaceous strata are the Purgatoire Formation in the Front Range in Colorado, the southern Colorado plains, Oklahoma, and northeastern New Mexico, and the Tucumcari Formation in parts of New Mexico. The Purgatoire contains basal sandstone and overlying gray shale members, and has been generally considered to be correlative with both the Cheyenne and the Kiowa. The Tucumcari Formation too is a probable genetic equivalent of the Purgatoire and Cheyenne/Kiowa formations, consisting of a basal conglomerate overlain by thin limestones and thin sandstones within a dark-gray shale. It represents open-sea environments and contains few or no fluvial and paralic deposits.

B.4.d.   Marginal Marine and Marine Units of Western Canada and Northernmost United States

Cretaceous strata of interior western Canada have received great attention because they contain the Western Hemisphere's largest known accumulation of hydrocarbons. Unfortunately, this oil is largely "heavy oil" or, more technically, bituminous sand, which is expensive to produce by current means. The total known Canadian resource comprises 1350 billion barrels (Mossop, 1980). One large deposit, the "Athabaska Tar Sands," has received popular attention; others include deposits in the Peace River, Wabasca, Cold Lake, Bonnyville, and Lloydminster areas (Fig. 8-12). All of these petroliferous units are included in correlatives of the Mannville/Blairmore Group, deposited during the Aptian and Albian.

The Mannville Group is defined in the northern Great Plains of western Saskatchewan, where it is almost exclusively in the subsurface. In central Alberta, too, Mannville strata are recognized, but westward to the Rocky Mountain foothills, correlative strata are called the Blairmore Group. In northeastern British Columbia, contemporary strata comprise the Fort St. Johns Group and far to the southeast in western Manitoba, contemporary units are the Swan River Group.

The lower Mannville and equivalent units are largely continental sandstones of fluvial origin, which are present across the broad region from western Manitoba to northeastern British Columbia, roughly paralleling the Cordilleran Mountains. In parts of this region, however, Lower Cretaceous marine shales were deposited within the sequence marking the initial southward transgression by the

Arctic Sea. In late Mannville time, the Arctic waters reached well into Alberta and sand deposition gave way to deposition of marine shales of the Clearwater Formation. The transgression was irregular and the intergradation of sandstone and marine shale units adds much to the complexity of the regional Cretaceous section.

Generally coarse, detrital units in the lower Mannville and correlatives include the McMurray Formation in central and northeastern Alberta, the Success and lower Centaur formations in the subsurface of Saskatchewan (Christopher, 1975), the Swan River Group in Manitoba, and the Gething Formation in many areas of Alberta. Of these, the McMurray in Alberta bears most of the bituminous deposits, but oil and gas resources are abundant across the regional sands. Thicknesses of these lower Mannville equivalents are generally less than 300 m; but since these represent stream-transported sands and muds distributed across a subaerially eroded landscape, thicknesses vary, especially in channeled areas. The lithofacies in British Columbia assigned to the Blairmore Group and the Gething Formation are generally more muddy than are Mannville strata. These are considered to be deltaic deposits which prograded into the southward-transgressing Arctic Sea; thus, sand deposition in the southern Canadian interior was synchronous with deltaic and marginal marine sedimentation in the northwest. Basement structural control is important regionally, with the Sweetgrass and Peace River arches especially controlling attitudes of basal sand sequences. For example, Stelck (1975) showed that the greater thicknesses of lower Blairmore strata coincide with the west Alberta Arch within the west Alberta Basin, whereas equivalent strata grade into marine units across the Peace River Arch.

In late Mannville time (early and middle Albian), marine waters transgressed far southeastward across western Canada and sand deposition generally gave way to deposition of marine shales such as the Clearwater Formation. The transgression was very irregular and the relationship between McMurray and Clearwater units and their equivalents is very diachronous. Vigrass (1968) showed that the southward-advancing sea was the probable source of early Mannville-age Canadian marine units but in late Mannville time, Arctic waters met Gulf waters approximately at or north of the United States–Canada border and from that time, until retreat of the sea near the Early–Late Cretaceous time boundary, sediments were derived largely from the southern sea.

Upper Mannville-equivalent strata consist generally of interbedded marine shales and subtidal-to-marginal marine shales and sandstones. Many of these strata, especially the sands, contain heavy oil accumulations. Upper Mannville equivalents include the Clearwater and Grand Rapids formations of Alberta, the Buckinghorse, Sikanni, and Sully (among others described locally) formations in the Fort St.

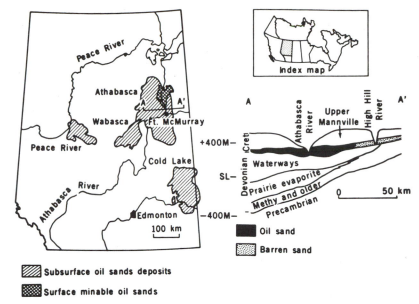

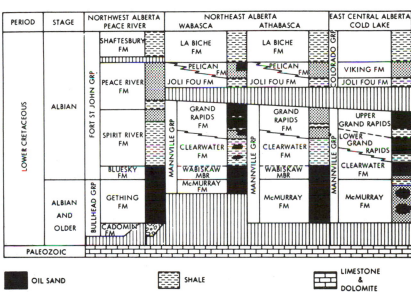

Fig. 8-12. Geology of the western Canadian oil sands. (Above) Map of Alberta, showing sites of major oil sand accumulation, and a schematic cross section of the Athabasca deposits. (Below) Correlation chart of the Mannville Group in Alberta, with oil-rich units indicated. (From Mossop, in *Science,* © 1980, American Association for the Advancement of Science.)

Johns Group of northeastern British Columbia, the Pense Formation in southwestern Saskatchewan, and upper units of the Swan River Group in eastern Saskatchewan and western Manitoba. The proportion of marine beds to nonmarine beds within these strata is very large and despite internal variations within the upper Mannville marine shale lithosome, one can say that relatively stable marine conditions ensued across western Canada through middle Albian time.

Before continuing on up the section, we should comment on the source for oil in Canadian Cretaceous deposits of the western interior, certainly a subject of no small im-

portance. Two hypotheses are in current favor: (1) oil migrated down from marine beds such as the Clearwater to be held in the porous, relatively pure-quartz-sand McMurray and equivalent units; or (2) oil migrated upward from Devonian and Mississippian rocks underlying the Cretaceous. There are mechanical problems with both hypotheses, largely involving the nature of reservoir structures and caps, but it is certain that the oil did not form in the fluvial sand bodies. Thus, it must have migrated into them from above or below.

Above upper Mannville strata are widespread, uppermost Lower Cretaceous marine units in Canada which correlate well with units in Montana and the Williston Basin. In large areas of Canada, this sequence contains either the Joli Fou and Viking formations (western Saskatchewan–eastern Alberta), the lower and middle parts of the Ashville Formation (eastern Saskatchewan and western Manitoba), the upper Sully Formation in northeastern British Columbia, or more localized units such as the Paddy Member of the Peace River Formation near the Peace River Arch in Alberta, or the Ft. Nelson Conglomerate. These and other strata comprise the basal part of a transgressive marine cycle which deposited the lower parts of the Colorado Group, one of the three widely recognized reference sequences of the western interior Cretaceous. The Joli Fou Formation is composed mostly of mudstones and siltstones. It is generally 50–55 m thick and includes bivalve shell-hash, and fish-bone and fish-scale beds; the basal 10 m of the unit is commonly a zone of reworked underlying units (Simpson, 1975). The marine, argillaceous Joli Fou lithology intertongues eastward on the shelf in Saskatchewan with the fine-grained, glauconitic Spinney Hill Sandstone, interpreted as having been deposited in an estuarine delta. Southeastward into Manitoba, shales of the Joli Fou pass into similar shales of the lower Ashville Formation.

The Viking, the middle sand member of the Ashville, the Bow Island Sand Member in Alberta, and other regional units represent a regressive phase of the upper Albian seaway in Canada. This sand forms a westward-thickening wedge and with it we can see the initiation of the consistent east–west intergradation of transgressive lime/shale and regressive sand/silt tongues that characterizes the Upper Cretaceous sections across the western interior.

Strata above the Viking and equivalents in Canada correlate with the Mowry Shale in the United States (discussed in the next subsection); but first, we will examine American equivalents of the Mannville–Viking sequences. In Montana, marine strata of the Blackleaf Formation overlie the nonmarine Kootenay Formation disconformably, demonstrating again that seas reached the northernmost craton in the United States by the middle Albian. The Blackleaf contains four members (Flood, Taft Hill, Vaughn, Bootlegger) which can be correlated with the Mannville-through-Viking/Bow Island strata in Canada and with other units in the northern United States. Uppermost Blackleaf strata are equivalent to the Mowry Formation as discussed subsequently. Cobban *et al.* (1976) noted that the Blackleaf section thickens southward, whereas the unconformably overlying Marias River Formation of the Upper Cretaceous thickens westward. One can interpret that the sea in Blackleaf time was still transgressing northward and that the international boundary region was that last to be inundated on the western craton.

In Wyoming, marine strata from middle and late Al-

bian comprise the Thermopolis Formation and the Mowry Shale. In the northern Black Hills and adjacent bits of Montana, South Dakota, and western Nebraska, Albian-age marine equivalents of the Thermopolis are the Fall River and overlying Skull Creek, Newcastle, and Mowry formations. The Fall River Sandstone is a fine-grained, flaggy, tan-weathering unit less than 30 m thick. It contains *Ophiomorpha* burrows (a trace fossil indicative of marine beach and bar environments), and generally resembles the Flood in Montana. The Skull Creek and most of the Thermopolis contain gray-black shale with rare marine fossils and some bentonite beds. Toward Montana, these units grade into the Taft Hill glauconitic sand and bentonitic shale lithology. The Newcastle is a sandstone unit with gray shales, glauconitic sands, and bentonites.

B.4.e. The Mowry Shale and Equivalent Latest Early Cretaceous Strata

Late-late Albian time saw the restriction of the western interior sea and the development of *endemic* (i.e., localized) marine faunas. The connection between the western interior sea and the Gulf of Mexico was broken briefly during the Early–Late Cretaceous boundary interval and, simultaneously, the connection between interior waters and Arctic waters may have become very narrow with relatively little exchange. Within the virtually isolated sea that resulted, large volumes of volcanic material and fine-grained continental detritus were deposited, producing the widespread Mowry and Aspen shales (Fig. 8-13) and their equivalents such as the Bootlegger Member of the Blackleaf Formation in Montana, and related Canadian units such as the Shaftesbury, "Colorado," Ashville, and Fort St. Johns.

A most characteristic element of Mowry and equivalent units is the abundance of fish scales and disarticulated fish parts within the rock. In many areas of the west, a so-called "fish-scale marker" is present in Mowry-equivalent units and these prominent layers of fish remains mark the latest Albian. Aside from the fish scales, bones, and teeth, fossils are generally rare in units deposited in Mowry time but faunas containing the ammonite genus *Neogastroplites,* the index taxon for the late-late Albian, are found in some Mowry strata. Reeside and Cobban (1960) showed that the Mowry may also contain freshwater invertebrates, wood, fern fossils, and, in general, a rare but diverse marine-to-freshwater biota. The Mowry thus represents a mixed set of environments of deposition.

The Mowry Shale at the type section in Wyoming is 60–100 m thick and consists of hard, siliceous, light-gray shale with numerous bentonites; locally, especially in Montana, the unit thickens to over 300 m. Some beds in the typical Mowry are very hard and are called porcelainites. The volcanic material which makes up much of the Mowry,

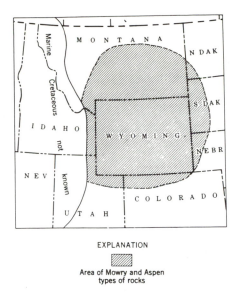

Fig. 8-13. Distribution of the Mowry Shale and related formations. (From Reeside and Cobban, 1960.)

with high silica and phosporous contents in the volcanic ash, apparently enhanced preservation of vertebrate materials such as the fish-scale beds.

The distinctive characteristics of the Mowry Shale have resulted in widespread use of the name; Mowry strata are defined in Wyoming, Montana, Colorado, northeastern Utah, and western Nebraska and the Dakotas (Fig. 8-14). In northwestern Montana, on the Sweetgrass Arch, part of the Colorado Shale is of Mowry age; there, the lower non-marine member is called the ''red-speck zone,'' containing prominent layers of heulandite crystals.

In western Wyoming, the Aspen Shale comprises the latest Albian section. The Aspen is slightly less silicic and sandy than the Mowry but in other respects it is the same; it would seem that the main distinction is semantic. In Canada, the volcanic ash-falls which yielded the Mowry were limited or absent. Simpson (1975) showed several unnamed shale beds in Alberta, within the Colorado Group, to comprise the *Neogastroplites* zone, and Reeside and Cobban (1960) indicated that in British Columbia, shale units of the Sikanni Formation of the Fort St. Johns Group were deposited in Mowry time.

While the Mowry was being formed in the United States, on the borders of the Mowry sea were streams depositing fluvial and alluvial sands referred to as ''Dakota'' units. Therefore, synchronous with Mowry deposition in the region shown in Fig. 8-14, zones of neritic-to-lagoonal sand and mud, and inland floodplain sands, ringed the Mowry sea and left ''Dakota'' units in the central Dakotas, Kansas, Nebraska, southern and central Colorado, and a narrow region of eastern Idaho and Utah. The latest Albian Dakota units form the ''type'' Dakota in Nebraska and

probably were deposited through the latest Albian into the Cenomanian as base level rose, ending fluvial deposition.

B.5. Late Cretaceous in the Western Interior

B.5.a. Overview of Upper Cretaceous Sequences and Sedimentation

The general pattern of Upper Cretaceous sedimentation in the western and west-central craton consists of a series of east–west intertonguing lithosomes which thicken to 6000 m at the front of the Cordilleran thrust zones, and which thin to roughly 600 m on the eastern shelf. The overwhelming bulk of this sediment is dark-colored marine shale, but there are two major limestones present: the Greenhorn and Niobrara, both of the Colorado Group. The only pervasive sand unit is the ''Dakota'' at the base of the section. Alone of the detrital units, the ''Dakota'' has a primarily eastern provenance. During latest Cretaceous (Maestrichtian) time, sands were shed from the Cordillera and spread widely, resulting in several regressive sandstone units; but these are by no means as widespread as the ''Dakota'' sands.

A number of coal-bearing, predominantly terrestrial units are present toward the far western craton. These coals, like Pennsylvanian coals, probably represent cyclical deposition resulting from transgressive–regressive shorelines

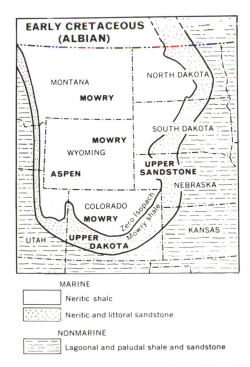

Fig. 8-14. Paleogeographic reconstruction of Upper Dakota–Mowry (Albian) time in the western interior. (From Kay and Colbert © 1965 by John Wiley & Sons, reprinted with permission of John Wiley & Sons.)

(Ryer, 1983); however, the precise site of coal formation in these Cretaceous cyclothems is controversial (see Fassett's comment, *Geology* **12**: 736). Regardless of their origin, the Cretaceous coal largely motivated early survey of the upper Missouri River valley by F. B. Meek and F. V. Hayden, resulting in designation of the Coloradoan and Montanan stratotypes.

As shown in Fig. 8-15, the Upper Cretaceous epeiric sea occupied a north–south elongated trough. During much of the Late Cretaceous, a broad coastal plain was present on both western and eastern edges of the sea, with transitional coast-to-shelf deposits formed symmetrically about the deeper regions. Such gradations of lithofacies characterize the stratigraphic record and, since numerous transgressions and regressions occurred, one finds elaborate lateral and vertical arrangements of lithologies, in accord with Walther's law.

The dominant sources of detritus were in the west, as mentioned, even though land was exposed to the east (but virtually without relief anywhere within reach of the western sea). During the maximum Cretaceous transgression, in the later Cenomanian through Turonian, the interior sea was some 1400 km across, and it is a sign of the intense tectonic activity of the time that even eastern strata across this vast seaway were fed from the west.

The Upper Cretaceous cratonic strata show four major transgressive units, deposited in east-to-west sequence as water from the prevailing eastern basin flooded the western coastal plain (see Fig. 8-15). Regression proceeded from west to east, with fluvial and deltaic deposits prograding over marine shales and chalk. The consequence of this cyclical east–west transgression and west–east regression is a very thick, well-preserved set of lithofacies.

Five dominant lithofacies associations may be recognized in the Upper Cretaceous section (Weimer, 1959): (1) gray to black marine shale, deposited in neritic environments (e.g., the Mancos, Pierre, Claggett, and Lewis shales); (2) limestone, chalk, or marlstone, deposited in deeper neritic environments (e.g., the Greenhorn and Niobrara formations); (3) tan or buff, massive or crossbedded sandstone, deposited in paralic, fluvial, and alluvial environments (e.g., the Dakota, Fox Hills, and Eagle formations); (4) dark-gray clay, tan lenticular sandstone, carbonaceous, coal, and lignite-bearing strata, freshwater

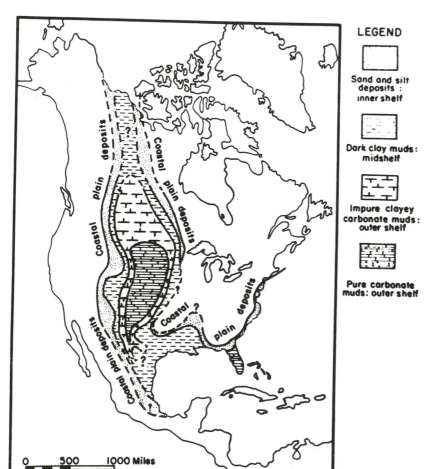

LEGEND

Sand and silt deposits : inner shelf

Dark clay muds : midshelf

Impure clayey carbonate muds : outer shelf

Pure carbonate muds: outer shelf

0   500   1000 Miles

Fig. 8-15. General distribution of sediments in the Rocky Mountain seaway during the Greenhorn transgression (early Turonian). (From Hancock, 1975, reprinted with permission of the Geological Association of Canada.)

marlstone and red beds, deposited in paludal, lagoonal, and lacustrine coastal plain environments (e.g., the Mesaverde and Laramie strata); and (5) coarse sand and bouldery conglomerate, deposited in Cordilleran piedmont environments (e.g., the Indianola and Price River strata). The reader should remember that the unusually broad and heterogeneous seaway meant that all those lithologies could be simultaneously present somewhere on the western craton, and that all may be represented in a single formation or pair of intertonguing formations.

B.5.b. Lower/Upper Boundary Sequences: "Dakota" *Redux*

The end of Early Cretaceous sedimentation is denoted by prominent fish-scale marker beds covering a wide area, even in regions far from the Mowry/Aspen depocenters. In many localities and regions, the later Albian and earliest Cenomanian were periods of marginal marine sedimentation which formed "Dakota"-type sand bodies; these early Late Cretaceous Dakota sands are in fact the typical Dakota strata, such as are found near Dakota City, Nebraska. These Dakota sequences derived from eastern cratonic sources and represent lower-fluvial, alluvial, and back-beach environments in transgressive sequences in front of Mowry and post-Mowry transgressions.

As the first group of the Upper Cretaceous sequence, Dakota units are present in Kansas and Nebraska unconformably overlying the Kiowa Shale, in northern and eastern Colorado unconformably overlying the Purgatoire Formation, in parts of Utah generally at the base of the Cretaceous section, in New Mexico unconformably overlying the Tucumcari or Purgatoire formations or located at the base of the section, and in Iowa and Minnesota at the base of the Cretaceous section (data from Cobban and Reeside, 1952a).

B.5.c. The Colorado Group

Where originally defined, the Colorado Group in Colorado Springs is entirely of Late Cretaceous age and consists of the Graneros (or Belle Fourche)/Greenhorn/Carlile/ and Niobrara formations, deposited from the early Cenomanian through the Santonian. However, in Montana, northwestern Wyoming, Saskatchewan, Alberta, and the Black Hills region, the Colorado Group has been defined to extend down the section to encompass Early Cretaceous units including the Fall River Formation in the United States and the Mannville (Blairmore) lower sands in Canada (Simpson, 1975). Further, the "Colorado Shale" is defined locally in Montana to parallel roughly the stratigraphic range of the Group as it is defined regionally. Other long-ranging units of approximate Colorado age are common in the western interior

and these too have diachronous boundaries; among such are the Frontier, Benton, Cody, and Mancos groups or formations, defined from Arizona to Montana.

i. The Typical Colorado Group. In post-Mowry (latest Albian) time, on the eastern shelf, readvancing seas initially induced "Dakota" Sandstone deposition; but in the former Mowry region (chiefly Wyoming and northern Colorado), widespread neritic environments became established and formed Belle Fourche (and Graneros) Shales. As the sea deepened, Greenhorn Limestone deposition followed (see Fig. 8-16).

Typically the Belle Fourche is a dark-colored, fissile shale, with ironstones and iron concretions, some hard siliceous beds, and numerous bentonites. Its thickness varies, reaching to over 200 m and thinning to less than 20 m. Some Belle Fourche strata (and Graneros units, which are identical) are entirely marine in origin, while other sequences grade from nonmarine near the bottom (Hattin, 1965) to marine on the top (with an ammonite and normal-marine bivalve fauna).

Greenhorn (and subsequent Carlile) deposition denotes a high-water mark with stable marine conditions over most of the northern, western American craton. The Greenhorn Limestone is roughly 50 to 125 m thick, but it becomes much thicker westward of the type area and consists of light-gray calcarenite, limestone, marl, shaley limestone, and dark noncalcareous shale in addition to numerous minor lithologies. The rocks are typically fossiliferous, containing ammonites, oysters, and *Inoceramus* (an important large Cretaceous bivalve; see F.3.b), which yield dates of early to middle Turonian for the unit.

The Carlile Shale represents regressive facies of the first Upper Cretaceous marine cycle. The formation is generally 200 m thick with several units: a basal, dark-gray shale; a middle, gray, sandy member; and an upper dark-gray shale member. Marine fossils are abundant in the Carlile, especially in the lower member, and the entire forma-

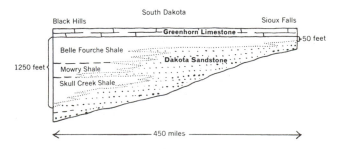

Fig. 8-16. Schematic cross section of the Great Plains Dakota-through-Greenhorn sequence, showing the intergradational nature of units up to the Greenhorn transgression. (From Kay and Colbert © 1965 by John Wiley & Sons, reprinted with permission of John Wiley & Sons.)

tion is of marine origin. Carlile lithology shows a return to predominantly neritic, shallow marine environments as the seas withdrew after the Greenhorn high-water mark. The formation was deposited during the middle through latest Turonian, with its upper boundary diachronous, as described below. Following Carlile time was a period of erosion or nondeposition, marked by a disconformity between the Carlile and Niobrara formations.

The Niobrara transgression that followed the Greenhorn–Carlile regression represents the last general transgression of the Cretaceous; above the Niobrara are very long-ranging and thick sequences of regressive facies, with minor transgressive units breaking up the pattern. The lower Niobrara boundary is diachronous but generally of the earliest Coniacian and the upper boundary is of the late middle Santonian. The Niobrara Chalk typically contains a lower, Fort Hays Limestone Member, and an upper, Smoky Hills Chalk Member. The Fort Hays is approximately 20 m thick, composed of thick-bedded, light-gray, olive, yellow, or white, chalky limestone with a low-diversity fauna of oysters and *Inoceramus* (Hattin, 1965). The Smoky Hills Member is much thicker (averaging 200 m), composed of olive-gray to very light-gray shaley chalk, with abundant spheroidal calcareous blebs. Hattin suggests that Fort Hays deposition in Kansas reflects agitated, well-circulated waters whereas Smoky Hills chalks are deeper, quieter-water deposits.

In Montana, near the Black Hills, the Niobrara contains numerous (perhaps over 100) bentonite beds (Cobban, 1951), and there the unit is typically thin (65 m). By way of contrast, in the Sand Wash Basin of northwest Colorado, Haskett (1959) described the units as over 400 m thick and composed of calcareous shales with four members. Among those members are some largely detrital beds and bentonites; the northwest Colorado Niobrara is also an oil-producing unit.

Life in the limy, enormously broad Niobrara Sea may have been unusual by many standards. Some of the most spectacular of megainvertebrate fossils are found in Smoky Hills strata, including gigantic specimens of *Inoceramus* with individual valves measuring approximately 1 m²! Hattin (1965) speculated that the poorly oxygenated waters of the deep Niobrara sea may have required development of massive respiratory tissue in such bivalves, with concomitant development of huge shells. There were also giant vertebrates in the Niobrara sea, including 18-m-long mosasaurs, 22-m-long plesiosaurs, and (somewhat) smaller ichthyosaurs. Niobrara chalks sometimes produce fossils which show preserved impressions of soft anatomy; among the soft-tissue features discovered in Niobrara chalks are the fishlike tails of ichthyosaurs, and the wing-membranes of pterosaurs (see Section G for discussion of Cretaceous vertebrates).

The upper contact of the Niobrara is unconformable. In Kansas, Colorado, Nebraska, and most of the Dakotas, there is an upper Santonian through middle Campanian hiatus, caused by regression of the Niobrara sea (Hattin, 1975). However, in areas to the west and north, where clastic sediments were supplied by uplifts in the Cordillera, sedimentation through the interval may have been continuous. Such continuous sections are represented in western and southern Colorado, parts of Utah and New Mexico (where the Mancos and Mesaverde sections are recognized), in Wyoming, where several units (Steele Shale and Cody Shale) span that interval, in Montana and the southern Canadian plains regions, where Cody and upper Colorado shales and Eagle sandstones were deposited in the interval, and in the Canadian Rockies, where Lea Park and Milk River formations span the interval in Saskatchewan and Alberta, and the Hanson/Chunge and Belly River formations span the time in British Columbia. By convention, the division between Coloradoan and Montanan reference sequences is drawn at the age of youngest Niobrara strata in Colorado.

ii. Canadian Coloradoan Strata. The Colorado Group is recognized in Saskatchewan and Alberta and contains units which can be correlated with American Coloradoan units. At the base of the sequence is the Fish Scale Marker unit (Fig. 8-17), which we noted previously to equate with the uppermost Mowry; it also represents early Belle Fourche time and is overlain by an erosional surface representing the remainder of Belle Fourche time. The Fish Scale Marker includes 3–35m of phosphatic, sandy, muddy, apparently storm-worked detritus. Fish scales and bones are, of course, abundant as are bioclasts and abraded fossils. Simpson (1975) suggested that the deposit represents storm-worked material from underlying units. Elsewhere in Canada, Belle Fourche correlatives either are termed "Upper Colorado Shales" or they comprise lower beds of the Favel Formation, containing 20 to 60 m of gray shale and bentonite. The Favel reaches eastward across the plains to Manitoba (Wickenden, 1953) and lies above the Fish Scale Marker, whereas westward, in Alberta and British Columbia, shales of Graneros–Belle Fourche age comprise the Alberta Shale Group, the Smoky Group, or simply are unnamed. In northeastern British Columbia, Cenomanian-age strata comprise the Dunvegan Formation (Stott, 1975), a massive conglomerate also containing sandstone, carbonaceous shale, and coal. Dunvegan strata vary laterally with the coarsest conglomerates located near Cordilleran tectonic lands, and sandy beds representing alluvial and deltaic environments. The formation decreases in thickness (from nearly 300 m) southeastward, grading from carbonaceous sandstone to siltstone and into marine shales with a shelly fauna.

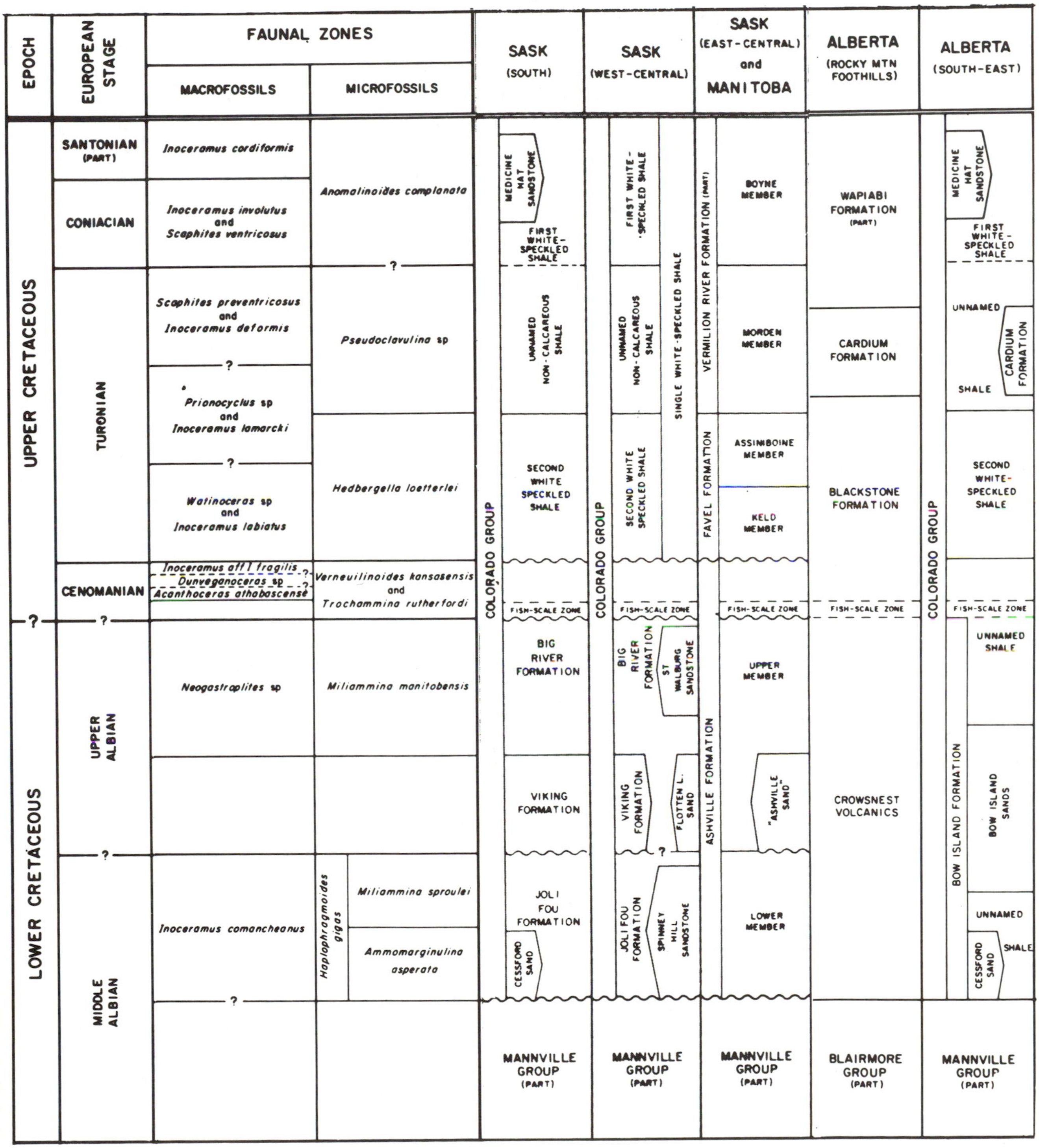

EPOCH	EUROPEAN STAGE	FAUNAL ZONES		SASK (SOUTH)	SASK (WEST-CENTRAL)	SASK (EAST-CENTRAL) and MANITOBA	ALBERTA (ROCKY MTN FOOTHILLS)	ALBERTA (SOUTH-EAST)
		MACROFOSSILS	MICROFOSSILS					
UPPER CRETACEOUS	SANTONIAN (PART)	Inoceramus cordiformis	Anomalinoides complanata	MEDICINE HAT SANDSTONE	FIRST WHITE-SPECKLED SHALE	BOYNE MEMBER (VERMILION RIVER FORMATION (PART))	WAPIABI FORMATION (PART)	MEDICINE HAT SANDSTONE
	CONIACIAN	Inoceramus involutus and Scaphites ventricosus		FIRST WHITE-SPECKLED SHALE				FIRST WHITE-SPECKLED SHALE
	TURONIAN	Scaphites preventricosus and Inoceramus deformis	Pseudoclavulina sp	UNNAMED NON-CALCAREOUS SHALE	UNNAMED NON-CALCAREOUS SHALE	MORDEN MEMBER	CARDIUM FORMATION	UNNAMED SHALE / CARDIUM FORMATION
		Prionocyclus sp and Inoceramus lamarcki		SECOND WHITE SPECKLED SHALE	SECOND WHITE SPECKLED SHALE (SINGLE WHITE-SPECKLED SHALE)	ASSINIBOINE MEMBER (FAVEL FORMATION)	BLACKSTONE FORMATION	SECOND WHITE-SPECKLED SHALE
		Watinoceras sp and Inoceramus labiatus	Hedbergella loetterlei			KELD MEMBER		
	CENOMANIAN	Inoceramus aff. I. fragilis / Dunveganoceras sp / Acanthoceras athabascense	Verneuilinoides kansasensis and Trochammina rutherfordi	COLORADO GROUP — FISH-SCALE ZONE	COLORADO GROUP — FISH-SCALE ZONE	FISH-SCALE ZONE	FISH-SCALE ZONE	COLORADO GROUP — FISH-SCALE ZONE
LOWER CRETACEOUS	UPPER ALBIAN	Neogastroplites sp	Miliammina manitobensis	BIG RIVER FORMATION	BIG RIVER FORMATION / ST WALBURG SANDSTONE	UPPER MEMBER / "ASHVILLE SAND" (ASHVILLE FORMATION)	CROWSNEST VOLCANICS	UNNAMED SHALE / BOW ISLAND SANDS (BOW ISLAND FORMATION)
				VIKING FORMATION	VIKING FORMATION / FLOTTEN L. SAND			
	MIDDLE ALBIAN	Inoceramus comancheanus (Haplophragmoides gigas)	Miliammina sproulei	JOLI FOU FORMATION	JOLI FOU FORMATION / SPINNEY HILL SANDSTONE	LOWER MEMBER		UNNAMED SHALE
			Ammomarginulina asperata	CESSFORD SAND				CESSFORD SAND
				MANNVILLE GROUP (PART)	MANNVILLE GROUP (PART)	MANNVILLE GROUP (PART)	BLAIRMORE GROUP (PART)	MANNVILLE GROUP (PART)

Fig. 8–17. Correlation chart for the Colorado Group in the Canadian Great Plains region, including the diagnostic mega- and microfossils. (Modified from Simpson, 1975, reproduced with permission of the Geological Association of Canada.)

Greenhorn-equivalent units in Canada are generally shaley, showing that shallower conditions prevailed. In eastern Saskatchewan and Manitoba, however, the Keld Member of the Favel Formation contains clean limestones and correlates with the Greenhorn (Williams and Stelck, 1975). Across much of Saskatchewan and Alberta, the so-called "Second White-Speckled Shale" contains limy equivalents of the Greenhorn, but there the limestone component is reduced to white calcareous zones within a predominantly shaley unit. In southern Saskatchewan, both the Second and First White-Speckled shales, with an unnamed noncalcareous shale in between, represent the entire upper Colorado section (the Second White-Speckled Shale represents the Greenhorn, the intervening shale is of Carlile age, and the First White-Speckled Shale correlates with the Niobrara; Simpson, 1975). Locally in Saskatchewan, the noncalcareous shale is absent and speckled shales make up the whole upper Coloradoan section. In the Foothills of British Columbia, the Kaskapu Formation of the Smoky Group features dark-gray shales which are contemporary with the Greenhorn Formation, but Kaskapu shales are generally noncalcareous and, in fact, apparently grade from marine-to-nonmarine through the section. The Vimy Member of the Kaskapu contains calcareous shales which correlate with the "white specks" to the south but above the Vimy are the increasingly continental Haven and Opabin members (Stott, 1975).

Carlile equivalents are almost entirely composed of dark-gray shales, except in British Columbia and parts of Alberta; there, the oil-bearing Cardium Formation contains sandstone, conglomerate, and shale beds, with coals and carbonaceous zones, reflecting lagoonal, alluvial, paludal, and marginal marine environments. The Cardium grades southeastward into dark-gray marine shales under the Great Plains in Saskatchewan and eastern Alberta. In far eastern Saskatchewan and Manitoba, the Morden Member of the Vermillion River Formation is of Carlile age, containing dark-gray, noncalcareous marine shales. Elsewhere, equivalent strata are the unnamed, noncalcareous unit between the white-speckled shales.

Above the Carlile–Niobrara interval, which we discussed previously, are the First White-Speckled Shales in Saskatchewan and Alberta, the Boyne Member of the Vermillion River Formation in Manitoba and eastern Saskatchewan, and the Medicine Hat Sandstone in southeast Alberta and southwestern Saskatchewan. The upper speckled unit, like the lower, consists largely of gray marine shale with white calcareous blebs in prominent zones. The Boyne Member is more calcareous than the speckled shales, composed generally of calcareous shale with bentonite and some noncalcareous gray shale beds (Wickenden, 1953). The relatively greater carbonate component clearly shows that the unit was deposited closer to the Niobrara sea depocenter, in

deeper water than other Canadian Niobrara equivalents. Simpson (1975) describes the Medicine Hat Sandstone as a silty mudstone and shelly sandstone unit with considerable fish-skeletal debris.

In British Columbia, Niobrara time is represented by the Muskiki, Bad Heart, and part of the Puskwaskau formations of the Smoky Group. British Columbian units of the Smoky Group show irregular but progressively greater deepening of the sea, starting with neritic dark-gray shales of the Muskiki, followed by largely nonmarine, sandy and shaley, coal-bearing strata of the Bad Heart, and followed by entirely marine dark-gray shales of the Puskwaskau Formation. The overlying Wapiti Formation of early Montana age is a nonmarine conglomerate, in common with the pattern described for American Niobrara–Pierre sequences.

iii. Coloradoan Sequences in Wyoming, Montana, and Parts of Utah.

There are many alternately named, local sequences spanning Colorado time described from sections in the northern United States. Among these are several with relatively wide recognition which will be covered in the following discussion. On the Sweetgrass Arch of northwest-central Montana, the Marias River Shale overlies the Lower Cretaceous Blackleaf Formation and can be readily correlated with typical Coloradoan strata. The Marias River contains four members: the Floweree, a variably thick, dark-gray shale and lighter-colored shaley siltstone (which correlates with the Belle Fourche); the Cone Member, a thin, calcareous shale and chalk marl (Greenhorn equivalent); the Ferdig Member, a dark-blue-gray, noncalcareous shale, 68 m thick (Carlile equivalent); and the Kevin Member, the most widespread and thickest (188 m) unit, composed generally of dark-gray marine shale but containing calcareous concretionary zones and bentonites (Niobrara equivalent; data from Cobban *et al.*, 1976). The Sweetgrass Arch section in general is quite similar to the type Coloradoan, except that it is thinner and Niobrara equivalents are much less calcareous, showing that the limestones of Colorado lens out through Montana even before reaching Canada.

In western Wyoming, the Frontier/Hilliard and Adaville formations represent the Coloradoan section; however, "Frontier" also has been used to delimit strata with very wide spans of ages outside the type region. Typically, the Frontier is of Cenomanian through Turonian age. The lower one-half contains nonmarine sandstones, siltstones, volcaniclastics, and bentonites. These strata were deposited west of the Belle Fourche sea in marginal environments and fluvial systems, and they also reflect a sizable, western detrital source. This lower unit is typically 300 m thick and is overlain by Greenhorn–Carlile-age marine sandstones, shales, and calcareous shale. The upper marine unit is approximately 300 m thick and includes some sizable sand bodies such as the Oyster Ridge Sandstone Member, which shows that the detrital source was still active during the Turonian. At the top of the Frontier is a massive sandstone unit with early Niobrara (Coniacian) marine fossils but in some areas this upper sand is nonmarine and bears mudstones and some coal (the Kemmerer Coal zone). The Hilliard Shale overlies Frontier strata, and Cobban and Reeside (1952b) stated that in the type Frontier region, the Hilliard contains almost 2000 m of marine shale and is, in turn, overlain by the nonmarine Adaville Formation; the Hilliard, Adaville, and the uppermost part of the Frontier (above the Kemmerer Coal) represent Niobrara time but with obviously very different lithology.

Sizable volcanic components in the typical Frontier, both as volcaniclastics and as bentonites, show that extrusive activity was ongoing from Mowry time. The bentonites comprise one of the economically important Frontier resources; the others are the coal, oil, and gas-bearing beds.

In nontypical areas of Wyoming and Utah, "Frontier" units extend far up the section to include beds of Santonian and Campanian ages. Locally, such extended "Frontier" strata are very thick, such as in the Wasatch Range in Utah, where Stokes *et al.* (1955) show the "Frontier Formation" to be 2700 m thick. In such sequences, virtually the entire Colorado Group and parts of the Montana Group are lumped together under the name "Frontier."

The name "Benton" is applied to the lower Colorado section over a wide area, notably the Benton Formation in Minnesota, North Dakota, Kansas, and Colorado, and the Benton Group in parts of New Mexico and many local areas on the Great Plains. "Benton" is approximately the equivalent of the "Colorado" Group below the Niobrara; thus, the typical "Benton" Group includes the Belle Fourche through Carlile sequence. As a historical point of interest, the term *bentonite* derives from the Benton Formation, since beds of altered volcanic ash are common in much of the typical Benton section.

An additional, widely applied, Colorado-equivalent name in the American section north of Kansas is the Cody Shale, commonly used in Montana. In the Bighorn Basin of Wyoming, the type Cody region, the unit is of Coniacian to middle Santonian age and correlates with the Niobrara and parts of over- and underlying units. But in some localities, especially in Montana, the Cody ranges all the way down the section to include the Belle Fourche as a member, and it extends well up into the Montanan section. Cody shales are typically like other Coloradoan units, including predominantly dark-gray marine shales and Niobrara-type carbonates.

iv. The Mancos Shale. In the southwestern craton, the Mancos Shale encompasses the thick marine shale section between Dakota sands and overlying Mesa-

verde nonmarine strata. The type section of the Mancos comes from Mancos, Colorado, where the formation correlates with the entire Colorado Group plus lower parts of the Montana Group. The Mancos is defined in sections from western Colorado, Utah, New Mexico, and Arizona and it differs from Coloradoan units farther north and east by containing a much greater proportion of marine strata. In general, lower Mancos lithology is similar to the Belle Fourche: poorly fossiliferous, dark-gray shale. Higher in the section, the gray-shale lithology persists but portions of Niobrara age are more calcareous and bear an *Inoceramus* fauna (Kent, 1968).

Monotonous as the sediment of the Mancos may be, the formation is distinctive in its strikingly intertonguing relationship with the Dakota Formation (see Fig. 8-18A) to the degree that Dakota and Mancos units actually repeat in single vertical sections. These reflect minor transgressions and regressions that destroyed and reestablished paralic and shallow marine environments at the Mancos shoreline. In addition, the Mancos is remarkable for rapid, lateral facies changes. Spieker (1949) noted that on the east side of the Wasatch plateau, the lower Mancos is all marine shale whereas 35 km westward across the plateau, contemporary strata are sandstones and conglomerates and 9 km farther west, the section is all rubbly conglomerate. The conglomerate comprises the nonmarine Indianola Group, containing generally very coarse sediments but with some sandstone, shale, and minor limestone, and reaching thicknesses up to 500 m. The Indianola represents alluvium from a nearby mountain region which was active in the Cenomanian through the Turonian.

The Mancos is a diachronous unit, with its time relationships largely dependent on the presence of the Mesaverde Shale as a westward-thickening wedge in the middle of the Mancos section. Where the Mesaverde is present in the section, its nonmarine sands and coal-bearing units divide the marine shales of the Mancos into a lower (Mancos) unit and an upper (Lewis Shale) unit. The ages of Mesaverde strata range from thick sections spanning the middle Turonian through upper Campanian, down through thin sections of only middle Campanian age. The overlying Lewis Shale is typically of the late Campanian but locally it can be much older. Where thick Mesaverde sections are present, Mancos deposition ended generally during the late Turonian as nonmarine detrital sources producing the Mesaverde apparently overwhelmed Mancos marine environments.

B.5.d. The Montana Group

The upper Upper Cretaceous reference sequence in the interior is somewhat complicated nomenclaturally because three Montana "reference" sequences are in widespread

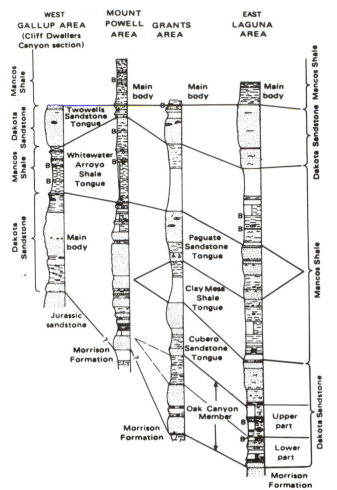

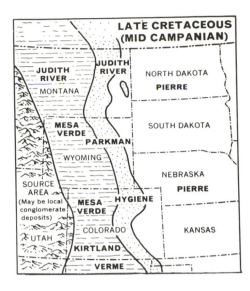

Fig. 8-18. (A) The complexly intertonguing relationships of the Dakota Sandstone and Mancos Shale, in westcentral New Mexico; note that the same formation names appear above and below the interfingering units. (B) Paleogeographic reconstruction of Mesaverde–Pierre time (middle Campanian) in the western interior; for legend to deposits, see Fig. 8-14. (A, modified from Cobban, 1977; B, from Kay and Colbert © 1965 by John Wiley & Sons, reprinted with permission of John Wiley & Sons.)

use. Fortunately, the rocks themselves are less complex and basically represent a continuation of the previous Coloradoan pattern of deposition. Lithosomal relations of Montanan strata consist of east–west-intertonguing detrital wedges with transgressive marine shales thinning westward and with regressive, prograding sandstones and volcaniclastic sediments thinning eastward toward the Pierre sea which occupied the eastern part of the western craton. Gill and Cobban (1966) stated that throughout much of Late Cretaceous time, the epicontinental (Pierre) sea was separated from the western Cordillera by a wide coastal plain and piedmont area traversed by streams and rivers which delivered only the finer fraction of their load to the sea— mainly fine sands, silts, and muds. Locally, in western Montana, the coastal plain was very narrow or absent, and marine waters at one time or another inundated hilly volcanic terrain. Montanan strata in essence represent a continuation of the slow regression from the Greenhorn/Niobrara high-water marks; but the regression is masked by intervening, rapid transgressions and renewed

regression. The Montana sequence over the entire western craton is devoid of carbonates, except near the base, where some persistent Niobrara environments show up, and in deepest water deposits of the Pierre Shale.

The Mesaverde nonmarine wedge in the southern craton thickens westward. Within the Mesaverde is a fluvial/ deltaic sandstone tongue, the Castlegate, which extends far to the east and shows that very rapid mountain-building occurred during latest Cretaceous time, associated with the Laramide orogeny. The Castlegate is but one of many detrital tongues shed from the Cordillera during Montana time. It appears that all detritus in the Montana sea came from the west and in addition to fluvial and deltaic sources, volcaniclastics make up a very sizable bulk of Montanan strata, especially in the vicinity of the Elkhorn Mountains, Montana.

i. Typical Montana Strata. In central Montana, there is no post-Niobrara unconformity; rather, Colorado Shale is overlain with apparent conformity by the Telegraph Creek Formation and the Eagle Sandstone. Together, these formations comprise a regressive sequence that was punctuated at least once by a minor, rapid transgression. The base of the Telegraph Creek is diachronous, younging eastward across western Montana and Wyoming. The unit is a thick, silty sandstone, representing shallow marine deposits of the Niobrara sea. Eastward toward the Missouri River in the Dakotas, Telegraph Creek silt beds grade into the lower beds of the Gammon Shale in the Cedar Creek Anticline in Montana, and, ultimately, into marly beds in the uppermost Niobrara. One can thus observe the transitional setting of Telegraph Creek strata as well as the penecontemporaneity of lowermost Montanan and uppermost Coloradoan strata in western and eastern sections.

The Eagle Sandstone contains two distinct units and it too is diachronous. In western Montana, the basal Virgelle Sandstone Member of the Eagle consists of white marine sandstones, representing beach and barrier-bar deposits which are younger eastward and grade into marine shales of the middle Gammon Shale in southeastern Montana (Fig. 8-19). The thickest part of the Eagle Formation is the unnamed upper member, containing dark, coal-bearing, shales and sands which represent fluvial and lagoonal environments. Gill and Cobban (1973) stated that a brief interval of transgression occurred during late Eagle time, marked by a widespread, black, chert-pebble horizon in Montana and southern Canada. The upper member of the Eagle Formation grades eastward into Virgelle-type white marginal and shallow marine sands and silts and, in southeastern Montana, into upper Gammon shales.

Telegraph Creek–Eagle regression lasted through the latest Santonian and early Campanian. The cause of regression seems to be related to volcanism and tectonism in far western Montana, centered around present-day Helena. Gill and Cobban noted that this activity lasted approximately 5.5 Myr., as deduced from ammonite biostratigraphy, and that during this time over 1000 m of Elkhorn Mountains Volcanics accumulated in western Montana.

Continued, explosive volcanism in the Elkhorn Mountains region may have resulted in subsidence of low-lying, near-coastal areas of western Montana. The Claggett Shale is a transgressive, westward-thinning tongue of marine silt and clay which overlies the Eagle Sandstone. At the base of the Claggett are thick, laterally persistent bentonites which correlate with middle units of the Elkhorn Mountains Volcanics; Gill and Cobban state that approximately 1250 km³ of bentonite is preserved in the Claggett and its equivalents and, very likely, much more was present originally. In the Claggett too are some porcelainite beds which, as described in the Mowry Shale, are siliceous, very hard, fine-grained sediments produced by volcaniclasts falling in shaley environments. Eastward toward the Dakotas, the Claggett grades into the Sharon Springs Member of the Pierre Shale

and is represented by organic-rich beds (Schultz et al., 1980). Westward, the Claggett lenses out into Judith River nonmarine volcaniclastics in the vicinity of the Dearborn River, northwestern Montana.

The Parkman Sandstone is a thin, white, marginal-marine sandstone unit that overlies the Claggett locally and represents initial stages of the Judith River regression. Although not recognized in the type area of the Montana Group, it is fairly widespread in Montana and is almost identical with the Virgelle Member of the Eagle Formation. The Parkman represents a return to beach and barrier-bar environments as the Claggett sea shallowed; and the formation intertongues diachronously on top and bottom with the Claggett and Judith River formations.

The Judith River regression began in the approximate middle Campanian but the formation is diachronous and older westward. Lithologically, the unit is variable, with basal, marginal-marine sands, overlain by brackish-water deposits, in turn overlain generally by friable, light-colored bentonitic sands, coals, well-indurated sandstones, bentonitic shales, and a great amount of volcanic debris. Much of the volcanic material present is coarse because typical Judith River deposition took place close to volcanic sources. The Judith River lithology resembles generally that of the Hell Creek Formation at the top of the Cretaceous section; because of this resemblance, it was formerly considered the uppermost Cretaceous unit (Waage, 1975). In outcrop, the Judith River weathers easily and forms extensive badlands. In Alberta and Saskatchewan, the Belly River Formation is identical to the Judith River and upper beds of the Belly River, the Oldman Member (a separate formation in some usage), are famous for having yielded one of the most diverse and sizable Cretaceous dinosaur collections in North America. Judith River beds also contain dinosaurs but the American strata have not been as fully explored.

The Judith River and Belly River formations are quite thick (averaging over 650 m) as are most regressive facies in the Upper Cretaceous. Schultz et al. (1980) noted that Judith River deposition involved unusually rapid regression from the Claggett high water and that this regression may have been in some part caused by the influx of volcanics on the western shore near the Elkhorn Mountains.

In extreme northwest Montana and adjacent parts of Alberta, Judith River equivalents thicken and comprise the Two Medicine Formation, which intergrades bottom and top (respectively) with the Eagle Sandstone and the St. Mary River Formation. The total Eagle/Two Medicine/St. Mary River section comprises a nearly complete, nonmarine Montana sequence. The Two Medicine is especially volcanic-rich, even in comparison to the Judith River.

The final Montanan transgression, the Bearpaw transgression, widened the interior sea to its maximum extent subsequent to Niobrara time. This transgression left a marine

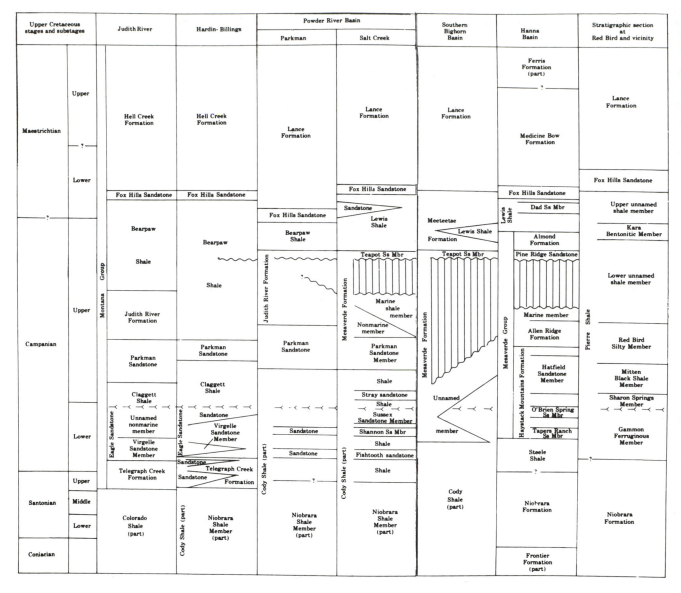

Fig. 8-19. Correlations in the Montana Group, Montana and Wyoming. (Modified from Gill and Cobban, 1973.)

record as far east as the Dakotas and over much of the craton, containing several hundred meters of predominantly dark-gray, fissile shales with abundant concretions. Upper units of the Bearpaw Shale represent an initial regressive sequence containing 50 to 100 m of silts and sands, which grade into the overlying Fox Hills Sandstone regressive sequence. Once again, volcanics are common in the Bearpaw, and Gill and Cobban (1973) suggested that their source was from the Elkhorn Mountains. Bearpaw transgression was apparently slow, and lasted roughly 3.5 Myr. beginning in the late Campanian and persisting through the latest Campanian.

Westward from the Judith River area, Bearpaw shales grade into shallow marine sandstones of the Lennep and Horsethief formations. These were deposited in environments like those of the Virgelle and Parkman, except that they represent westward-thinning marine tongues and they grade northwestward into the nonmarine Two Medicine Formation and upward into the nonmarine St. Mary River Formation.

Concretions in the Bearpaw have yielded invertebrate marine fossils, which are among the youngest marine fossils from the cratonic interior. Bearpaw beds have also yielded spectacular marine vertebrate fossils; William Melton of the

University of Montana Museum found a pair of sharks in the Bearpaw that died and somehow were preserved *in copula!*

While the sea advanced in Montana with Bearpaw deposition, it apparently retreated across eastern Wyoming, suggesting a relationship between crustal subsidence in the west and uplift in the east. Marine regression was well under way at the close of Bearpaw time (approximately earliest Maestrichtian) and initiated Fox Hills Sandstone deposition. The regression lasted some 2.5 Myr., dated by bentonites and ammonite zones, and was slow and regular early in Fox Hills time but rapid near the end. The Fox Hills Formation is a thin (100 m), nearshore and marginal-marine, light-colored, often massive sandstone unit which is very widespread and recognized far outside of the type Montanan area (including North Dakota, Wyoming, South Dakota, and Colorado). A lower, crossbedded, fossiliferous unit contains common bivalve and gastropod fossils, which Feldman and Palubniak (1975) reported showed evidence of strong storm tossing in sections from North Dakota. Above the shell beds are *Ophiomorpha*-burrowed sands which formed in intertidal zones, probably in barrier-bars. The uppermost part of the Fox Hills represents back-barrier and beach deposits and commonly contains ripple-marked sands.

During Fox Hills regression, the shoreline oscillated east and west to some degree but the shallowing moved generally eastward and reached the Dakotas by earliest middle Maestrichtian. There, the Pierre sea persisted so that, while Montana and sections in western Wyoming received continental deposits, marine shales were deposited over large eastern areas until the late Maestrichtian. In central and northwestern Montana, the Lennep and Horsethief sandstones are coextensive with the Fox Hills Sandstone and represent additional marginal-marine environments. Fox Hills regression produced a strandline that was very irregular, as Gill and Cobban illustrate (Fig. 8-20); a large delta system, called the Sheridan Delta, in southeastern Montana and the western Dakotas initially brought nonmarine sediments to an area of the Pierre sea.

ii. The Pierre Shale. As noted previously, there are three commonly recognized western interior Montanan reference sections; we have described the far western facies and now will examine the central and eastern facies.

The Pierre (pronounced "pier") sea lay east of the Montana–western Wyoming coastal plain and was bounded on the east by the cratonic shelf. Since virtually all sediments were derived from the Cordillera, far to the west,

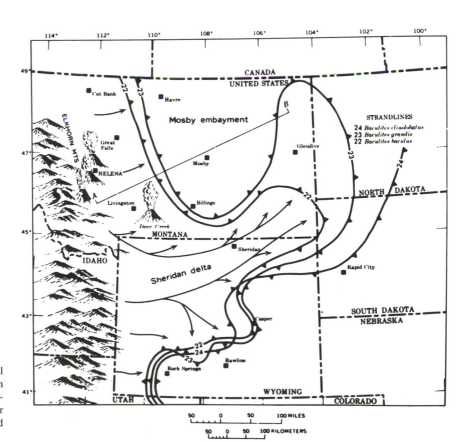

Fig. 8-20. Approximate strandlines during final phases of the Fox Hills regression. Numbers on the three strands correspond to fine-scale biostratigraphic time zones as indicated in the upper right (these are ammonite species). (From Gill and Cobban, 1973.)

detritus in the Pierre is exclusively fine-grained. Westward, Pierre lithology grades into the Montanan sequence described in the preceding subsection. Precisely how far west the interdigitation of lithologies extended for any given time depended on the transgressive and regressive events we described; thus, during maximum transgressions, such as in Claggett and Bearpaw times, Pierre-type sediments extended into western Montana (Fig. 8-21) while during maximum regressions, such as Eagle and Judith River times, western coarse detrital sediments and volcanics reached across Montana and Wyoming to the western shores of the Pierre sea.

The Pierre Shale was originally defined in the upper Missouri River valley, but Gill and Cobban (1966) sensibly proposed to use the very complete section at Red Bird, Niobrara County, eastern Wyoming, as a reference. There, the Pierre contains seven members: the Gammon Ferruginous/Sharon Springs/Mitten Black Shale/Red Bird Silty/unnamed lower/Kara Bentonite/unnamed upper members. Typically there is an unconformity at the base of the Pierre Shale, which represents Telegraph Creek time in central Montana, and latest Niobrara time in the Dakotas. The predominant Pierre lithology is gray marine shale, with limestone and siderite concretions, bentonites, and siltstones as major accessories (see Gill and Cobban, 1966). The total Red Bird Pierre Shale section is over 1000 m thick.

As in central Montana, the Pierre Shale in Wyoming grades upward into the Fox Hills Sandstone. Near Red Bird, the Fox Hills averages 135 to 165 m thick and consists of yellow and white, massive to thin-bedded sands which represent shallow marine environments concomitant with regional regression. The Fox Hills at Red Bird is one ammonite zone younger than at Lewistown in Montana and it represents the end of Montana time (the earliest Maestrich-

tian). Above the Fox Hills in Wyoming is the Lance Formation, a Laramian unit which will be discussed subsequently. Paleogeographic interpretation of the Pierre section in Wyoming is relatively clear: all units except the lower two members are marine and contain fossils largely of bivalves and ammonoids. Gill and Cobban (1966) inferred from paleontological and sedimentological indications that the sea in Wyoming was less than 65 m deep during Mitten Member through the end of Pierre times and somewhat deeper during Gammon and Sharon Springs Member times; but it was never very deep regionally. Gill and Cobban also suggested that the western strandline was never closer than 100 km west of Red Bird during Pierre time and generally it was twice that distance to the west.

The Pierre in western Kansas is thinner and contains four members: the Sharon Springs Shale/Weskan Shale/Lake Creek Shale/Salt Grass Shale Members. The total section measures approximately 430 m thick and the formation overall is largely dark-gray shale containing large quantities of limestone, ironstone concretions, and bentonites (Hattin, 1965). In Colorado, near Fort Collins, the so-called Hygiene Sandstone Member of the Pierre is a notable feature since the formation generally is devoid of coarse detritus. The unit lies approximately one-third up from the base of the Pierre and is also present in other Pierre sections in Colorado. Correlation is unclear but Hygiene sands probably represent an extension of the Mesaverde detrital wedge (discussed below); however, the sandstones in the Hygiene in the Fort Collins area are younger than typical Mesaverde strata (Scott and Cobban, 1959).

iii. The Mesaverde Section. As we have mentioned several times previously, the Mesaverde Group or Formation represents a major detrital wedge in the southwestern craton and comprises another major Montanan lithofacies. The Mesaverde is recognized in the type region of western and southern Colorado, in most of eastern Utah, western and southern Wyoming, and in the San Juan Basin of northern New Mexico. We have already noted the extremely diachronous nature of the Mesaverde and the fact that it commonly divides marine shales of the Mancos type into a lower unit called Mancos and an upper unit called the Lewis Shale. In general, the Mesaverde is a nonmarine, sandy sequence of approximate Judith River age, but because it is a variably thick regressive wedge, its age varies. Gill and Cobban (1966) showed, for example, that the Mesaverde Group in Durango, Colorado, is of Eagle–Claggett age (latest early Campanian), whereas in the Salt Creek Anticline in Wyoming the Mesaverde Formation is of early Bearpaw age (late Campanian).

At Durango, the type locality, the group contains two sandstone units, the Cliff House and Point Lookout, which represent back-beach and floodplain environments, with a

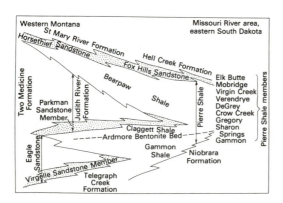

Fig. 8-21. Schematic cross section of the Pierre Shale and equivalents in the northern Great Plains. Stippled units are predominantly marine sandstone; otherwise, predominantly marine shale below the Horsethief/Fox Hills Sandstones, and nonmarine units above. (From Schultz *et al.*, 1980.)

nonmarine, coal-bearing sand and shale unit in between (the Menefee Formation). This type section represents more marine-influenced environments than do most Mesaverde sections described. In northeastern Utah, one finds a number of local, interfingering, nonmarine units making up the Mesaverde Group but with the Blackhawk, Castlegate, and Price River formations the most widespread. The Blackhawk is a coal-bearing floodplain and swamp deposit which grades southeastward into marine Mancos shales; the Blackhawk also contains some marine strata but in general it denotes the first Mesaverde regression in northern Utah (Hale and Van de Graaf, 1964). The Castlegate is a uniquely persistent sand unit in Montanan sections, deposited at the approximate beginning of Pierre time, and representing fluvial and deltaic incursions eastward into the Pierre Sea from a rapidly rising highland in the Cordillera (probably near the present Wasatch Range; Van de Graaf, 1972). Eastward, the Castlegate pinches out into the upper Mancos (however, it persists far into Colorado), and westward, the Castlegate pinches out into the Price River Formation after becoming coarser near the source area. The Price River Formation is a variable nonmarine unit, consisting of coarse, red conglomerate where it merges with Castlegate equivalents far to the west, and consisting of coarse sand and shale farther to the east in Price River Canyon. Price River environments were regressive floodplains and alluvial plains. Shortly after Price River time, renewed, rapid transgression deposited the Buck Tongue of the Mancos Shale over the Price River. Following deposition of the Buck Tongue, a series of minor regressions and transgressions in Utah deposited marine shales and nonmarine sands of the upper Price River Members.

Most Mesaverde sections are overlain by the transgressive Lewis Shale. Once again, we observe a very diachronous unit; for example, in Durango the Lewis is of early and middle Pierre age, whereas in Wyoming it is of latest Pierre to Fox Hills age (Gill and Cobban, 1966). The Lewis is essentially a Mancos-type, marine, dark shale. In Wyoming, it is approximately 350 m thick and contains an early Maestrichtian ammonite fauna (Roehler, 1979b).

In the Laramie Basin of central Wyoming, below the Mesaverde Group and above the Niobrara Formation, is the Steele Shale which is equivalent biostratigraphically to the Telegraph Creek–lower Claggett units in the reference Montanan area. In that same region, the Mesaverde Group consists of largely marine-to-brackish-water formations: the Haystacks Mountains/Rock River/Allen Ridge/Pine Ridge/Almond. These grade laterally eastward into the Pierre section (with the Lewis completing the upper Pierre; data from Gill et al., 1970).

Across most of the western craton, youngest marine regressive sandstones either lie above the upper Montanan marine units or comprise the upper Montana section. Fox Hills, Lennep, and Horsethief are the names generally ap-

plied to these units, where preserved, in the northern craton (with the term "Fox Hills" by far the more common one). Weimer and Land (1975) state that the Fox Hills in southwestern Wyoming, overlying neritic shales of the Lewis Formation, represents shallow-water deposits formed in barrier-bar and estuarine environments. The Fox Hills is generally a tan, fine- to medium-grained sandstone with minor associated siltstones and shales. As noted previously, the ages of Fox Hills strata and equivalents vary because the timing of the final Cretaceous regression varied around the western craton.

In the southwest, especially Durango (Colorado) and the San Juan basin (New Mexico), a distinctive regressive set of formations is defined, comprising the Picture Cliffs, Fruitland, and Kirtland formations. This sequence is of probable late Campanian through late Maestrichtian ages, but dating of nonmarine fossils in the upper units is incomplete. The Picture Cliffs is a marine sandstone which overlies the Lewis Shale and is penecontemporaneous with the Bearpaw Shale. The Kirtland and Fruitland are nonmarine fluvial and swamp deposits. The Fruitland contains coals within generally conglomeratic, sandy, and shaley lithologies whereas the overlying Kirtland does not. Both units have yielded important very-late Cretaceous vertebrate collections, including rare mammal fossils (L. S. Russell, 1975). It is uncertain whether these formations are Laramian equivalents or older but they are overlain by conglomerates called the Ojo Alamo Formation, which was originally considered of youngest Cretaceous age but is now considered to be of Tertiary age. After Kirtland time, but still in the latest Cretaceous, minor tectonic activity in the basin deposited the McDermott Member of the Animas Formation in small areas of the basin. Tilting of large parts of the basin in earliest Paleocene time yielded pebble- and boulder-sized detritus to newly forming rivers, producing the Ojo Alamo Formation.

The Upper Cretaceous strata of the San Juan Basin are famous for several attributes. The Picture Cliffs Sandstone is so named because it bears numerous *petroglyphs* (Indian rock carvings) on outcrop faces. The Kirtland and Fruitland strata are rich hydrocarbon and uranium sources, and the Fruitland yields commercial coal (Fassett and Hinds, 1971). In addition, a relatively rich dinosaur fauna containing some of the youngest taxa is present in the Kirtland, including the latest of sauropods, *Alamosaurus*.

iv. Montanan Volcanic Units and Southeastern Arizona Conglomerates and Volcanics. Before examining the Montanan sections in Canada, we should mention two exotic volcanic lithologies in the United States. The first is in southeastern Arizona, where the Lower Cretaceous Bisbee Group is overlain by approximately 300 m of nonmarine conglomerate, sandstone, and shale comprising the Fort Crittenden Formation. The Fort Crittenden is, in turn, overlain by volcanic rocks of the Salero Formation. The Salero is dated radiometrically to the late Campanian–Maestrichtian (Hayes, 1970a) whereas the Fort Crittenden is obviously older but contains nonmarine fossils which cannot be assigned to precise intervals. The history of the region is relatively simple: following deposition of the Bisbee Group, in marine, deltaic, and coastal plain environments, the region was an eroding land area until the Campanian during which deltas and alluvial plains developed, feeding detritus into the sea and producing the Fort Crittenden. During the latest Campanian, the region became a lava field, featuring andesitic and rhyolitic lavas comprising the Salero Formation.

The western Montana volcanics have been mentioned previously in passing in our discussion of the type Montanan section. In the Elkhorn Mountains, at or near the site of the volcanoes that strongly influenced typical Montanan units, two distinct sequences overlie the Colorado Formation: the Slim Sam Formation and the Elkhorn Mountains Volcanics. The Slim Sam is about 400 m thick and consists of a lower, quartz and chert sandstone with much tuff and some black shale, and an upper unit containing crystal-lithic tuff and some sedimentary tuff with minor siliceous mudstones (Klepper *et al.*, 1957). The Slim Sam lithology is present in a very restricted area and shows strong local volcanic influences. The overlying Elkhorn Mountains Volcanics consist of up to 3500 m of tuff, breccia, and lapilli, with some deposits apparently water-lain and most representing ash-falls and mudflows. A few lava flows are present and, clearly, these units were formed within a zone of major volcanic activity. The Slim Sam Formation contains an imperfectly dated plant fauna of approximate Judith River age, whereas the Elkhorn Mountains Volcanics to our knowledge have not been dated.

v. Montanan Strata in Canada. Fortunately, much of the Canadian upper Upper Cretaceous strata can be correlated directly with the Montana Group in the American reference section. In the western Saskatchewan and eastern Alberta plains, Campanian and Maestrichtian sequences show the same eastward-thinning nonmarine sands and silts and the westward-thinning marine silts and clays as were described for the type Montanan and Pierre sections; thus, in the Canadian plains, the Milk River and Belly River formations represent, respectively, the Eagle and Judith River detrital wedges borne by rivers from highlands in the west and pinching out eastward; Lea Park and Edmonton (or Bearpaw) formations represent the transgressive marine units, Claggett and Bearpaw, respectively (North and Caldwell, 1975; Shaw and Harding, 1949).

In the eastern plains of Saskatchewan and Manitoba, the Upper Cretaceous is far better exposed than in the western plains; therefore, more distinct local nomenclature was developed (North and Caldwell, 1975). The Telegraph Creek Formation of the American reference section is there represented by a depositional hiatus (Fig. 8-22), as it is in the western plains of Canada, and the Eagle/Claggett interval is represented by transgressive marine clays and silts of the Pembina Member of the Vermilion River Formation. (Recall that the Boyne Member of the Vermilion River is equivalent to the Niobrara.) Above the Pembina is the Riding Mountain Formation, which represents both Bearpaw and Judith River times, and which is largely a marine shale sequence with an ammonite and *Inoceramus* fauna, suggesting that the eastern plains were well-inundated by the Pierre sea during the late Campanian and early Maestrichtian. The youngest Montanan units are the thin, marine-to-continental sands of the Eastend or lower Boissevan formations, which correlate with Fox Hills and equivalent paralic sandstones; these are, in turn, overlain by nonmarine, coal-bearing, fluvial, swamp, and lake deposits of the Whitemud or upper Boissevan formations (data from North and Caldwell, 1975, and Wickenden, 1953).

In northeastern British Columbia, marine conditions were far less pervasive and the upper Upper Cretaceous sequence is not nearly as well developed as it is in the southern Canadian plains. The Puskwaskau Formation, of the marine Smoky Group, mentioned previously as the youngest Coloradoan unit, may be in part early Montanan in age (Stott, 1975); it is a gray, marine shale containing the index ammonite fossil *Scaphites vermiformis*. Above the Puskwaskau is the regressive Wapiti Formation, consisting of eastward-thinning wedges of largely continental sandstones, shales, and coal, reaching a thickness of over 1500 m near the foothills (Stott, 1975). The Wapiti contains dinosaur fossils, although they are rare, and is probably of latest Campanian to Maestrichtian age. The Wapiti is also the youngest Cretaceous unit in the northern foothills and is overlain by Tertiary strata with apparently conformable contact; the nature of stratigraphic contacts between Cretaceous and Tertiary strata will comprise part of our subsequent Laramie discussion.

B.5.e. The Laramie Formation and Equivalents

The youngest Cretaceous units of the western interior are nonmarine and tend to grade upward into similar strata with Paleocene-age fossils. In earlier days, this similarity in lithology and origin for Cretaceous and Tertiary units generated the "Laramie problem" because it was presumed that the change from Cretaceous to Tertiary would be marked by a change from marine to freshwater conditions. This misconception was compounded by so-called "layer-cake geology" which ignored facies changes, and by paleontological misunderstandings. The paleontological problem was, in essence, that dinosaur fossils, especially tyrannosaurs and *Triceratops,* in the Laramie and equivalent formations, said

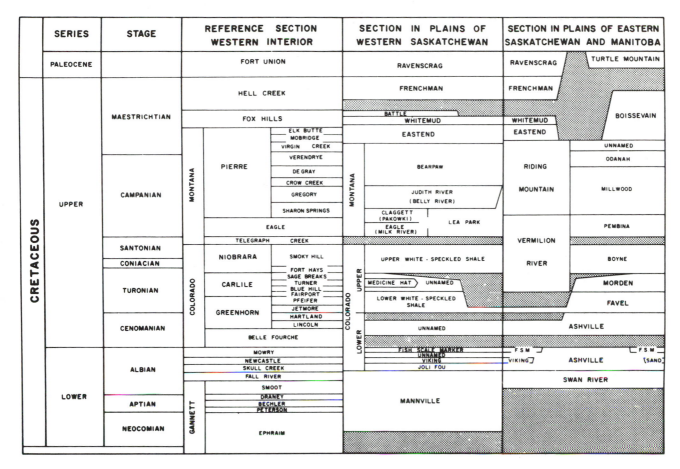

Fig. 8-22. Correlations of Cretaceous strata in Saskatchewan and adjacent regions. (From North and Caldwell, 1975, reproduced with permission of the Geological Association of Canada.)

"Cretaceous" while the plant fossils therein were like those in the Fort Union strata of Tertiary age (Waage, 1975; King and Beikman, 1976). Resolution of this problem rested on the following: recognition of the Paleocene Series with a transitional flora, acceptance of diachronous facies in the boundary sequence, realization that nonmarine Cretaceous strata were common, and a general understanding of the complex interfingering of Cretaceous strata in opposition to "layer-cake" models.

The uppermost Cretaceous reference section in current use is the Laramie/Arapahoe/lower Denver sequence in the northern foothills of the Front Range, Colorado. These overlie the Fox Hills Sandstone generally conformably and record the sequence of final retreat of the Zuni sea from the craton, and the development of delta and fluvial systems prograding into the retreating sea (Fig. 8-23). The Laramie Formation is generally 200 to 350 m thick and typically has two units: a lower member of tan and gray sandstone, siltstone, and claystone with generally thin coals (Weimer and Land, 1975), but with some reaching to almost 5 m in thickness (Van Horn, 1976); and an upper member composed primarily of varicolored claystone. The Laramie contains *Triceratops* fossils and is therefore of Maestrichtian age. The Arapahoe is an arkosic sandstone and claystone unit with thick basal conglomerates in several subunits; the conglomeratic beds rest with sharp scour surfaces on Laramie beds (Weimer and Land, 1975). Dinosaur fossils have also been recovered from the Arapahoe, showing that it, too, is of latest Cretaceous age. The Denver Formation consists of green-brown, clayey sandstones with a suite of amphibole and pyroxene clasts that suggest very youthful stages of erosion from uplifts. In addition, the Denver contains considerable volcanic material and three lava flows in the upper unit (Van Horn, 1976); the presence of volcanics is the criterion used to distinguish the Denver from the Arapahoe. The Denver is of Maestrichtian to Paleocene age, containing dinosaur material in lower beds and Paleocene plants in upper beds; thus, its deposition spanned the era boundary.

In Wyoming, the uppermost Cretaceous sequence contains an analogous but somewhat different set of strata assigned to the Lance Formation, which overlies the Fox Hills

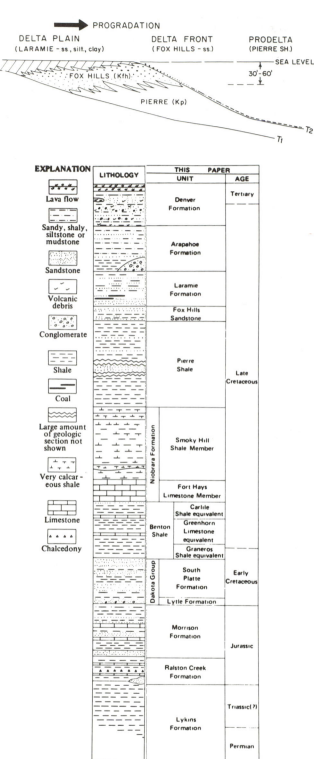

Fig. 8-23. (Above) Schematic diagram of the deltaic model of latest Cretaceous deposition in the Great Plains. (Below) Correlations of Permian through Tertiary formations in the Golden area, Colorado. Note the position of the Denver Formation straddling the Cretaceous–Tertiary boundary. [Diagram (above) from Weimer and Land, 1975, reproduced with permission of the Geological Association of Canada; chart (below) from Van Horn, 1976.]

and is, in turn, overlain by the Fort Union Formation, of Paleocene age. The Lance, therefore, was deposited through most of the Maestrichtian. The Lance is composed of gray siltstones and shales, with interbedded fine-grained sandstones and much coal. Some gray, silty dolomitic beds are also present and the formation ranges in thickness from 30 to over 800 m (Roehler, 1979b; Estes, 1964). Fossils from the Lance include a very diverse reptile assemblage, including numerous remains of dinosaurs, freshwater sharks, and bony fish (Estes, 1964); also included are important suites of Cretaceous mammals (L. S. Russell, 1975). Above the Lance are coal-bearing, economically important Fort Union strata, which are terrestrial deposits to be discussed in Chapter 9. An unconformity is present at the Lance–Fort Union contact with a 1–3° angular discordance (Roehler, 1979b). A paleoenvironmental analysis by Weimer and Land (1975) showed the Lance to contain all of the following nonmarine, nondeltaic environments: channel and overbank fluvial, marsh, lagoonal, and swamp.

Laramie- and Lance-age strata across most of the remaining American craton are assigned largely to either: the Hell Creek Formation of Montana, the Dakotas, and Nebraska; the lower North Horn Formation in Utah; or the McDermott and Animas formations in southern Colorado and the San Juan Basin, New Mexico. In general, these units are similar to the Lance and Laramie, consisting of nonmarine beds overlying marginal marine sequences. The Hell Creek is identical to the Lance, containing coal-bearing sandy and shaley sequences with abundant dinosaurs of the *Triceratops* assemblage. It is overlain generally by Fort Union Paleocene deposits but in Montana the Cretaceous–Tertiary boundary sequence is apparently gradational. The North Horn Formation in the Wasatch Plateau of Utah was deposited across the Late Cretaceous–Tertiary boundary and consists of red and variegated fluvial deposits shed from nearby Laramie uplifts. The unit is very thick (to 750 m) and contains mixed coarse conglomerates, sandstones, and lacustrine limestones in an eastward-thinning wedge (Hale and Van de Graaf, 1964). As with other era-spanning units, the North Horn contains dinosaurs of the *Triceratops* zone in lower beds, and Paleocene mammals in upper beds. The McDermott and Animas formations are conglomeratic, shaley, and sandy Lance equivalents that locally lie over the Kirtland Shale. They have complexly intertonguing relationships with the Kirtland and Fruitland and represent nonmarine, fluvial and alluvial environments. Above these units is the Ojo Alamo Formation.

To end our discussion of the Cretaceous cratonic regressive sequences (and simultaneously to end our discussion of Zuni and Mesozoic cratonic sedimentation), we will briefly mention Canadian Laramie equivalents. In southern Canada, these consist of the Frenchman Formation in the western plains of Saskatchewan, and the upper beds of the

Boissevan Formation toward the east in Manitoba. These units are entirely nonmarine and of fluvial and lacustrine origin. The upper Boissevan is a nonfossiliferous, thin, detrital sandstone (Wickenden, 1953), which is imprecisely dated but overlies the Riding Mountain Formation and therefore is presumed to be post-Montanan. The Frenchman is an interbedded sand and shale unit which uncomformably overlies two Fox Hills equivalents: the Eastend and Whitemud formations. The Frenchman is probably correlative with the Hell Creek in Montana.

Farther north in British Columbia, the Wapiti Formation is overlain apparently conformably by the Tertiary Paskapoo Formation and therefore spans the Maestrichtian. The Wapiti is a thick detrital unit derived from the Cordillera to the north and west. Farther north still, and in many regions of Canada, the uppermost Cretaceous section appears to have been eroded during Late and post-Cretaceous intervals of uplift and is absent.

C. Zuni Sequence of the Atlantic and Gulf Coastal Province

In the preceding chapter, we discussed the stratigraphy and structure of Triassic and Lower Jurassic deposits of the Eastern Triassic basins, and of correlative basin-deposits now buried beneath younger strata of North America's eastern and southeastern margins. We showed that those basins are most accurately described as grabens and/or half-grabens that formed in response to the initial phases of continental separation during which North America began to rift away from Pangaea. As rifting continued, the rift-valley phase of continental separation passed gradually into the proto-oceanic-gulf and then young-ocean-basin phases, during which a major sediment-mass, termed a rifted-margin prism, developed on the newly formed continental margin. The rifted-margin prism of eastern and southeastern North America is referred to as the Atlantic and Gulf Coastal Province; its stratigraphy and developmental history during Jurassic and Cretaceous time are the subject of the present discussion.

C.1. Introduction to the Atlantic and Gulf Coastal Province

The Atlantic and Gulf Coastal Province is the last major geological province to be introduced in this book. Since we have not previously discussed this area, a brief introduction and guided tour are in order (see Fig. 8-24).

The Atlantic and Gulf Coastal Province, although a single entity, is commonly subdivided in several different ways. Perhaps the simplest division is based on the position of sea level: the province is divided into subaerially exposed

and submarine parts. Obviously, the two parts meet at the present-day shore. The subaerially exposed part is loosely termed the *coastal plain,* but, as Murray (1961) observed, that term is a geomorphological expression relating to the surface character of the province and not to the sediment mass beneath it. By the same token, the terms *continental shelf, slope,* and *rise,* which are used to refer to the submarine portion of the Coastal Province, are geomorphological and, strictly speaking, should refer to bathymetric features of the continental margin rather than to the sedimentary mass upon which those features are impressed. It is our view that differentiation of the Coastal Province into subaerial and submarine parts is somewhat artificial since the boundary between the parts is dependent only on the position of sea level (which clearly fluctuates through time). Therefore, in the following discussion we will treat the Coastal Province as a single physiographic feature. Nonetheless, to the extent that different methods must be employed to study the two areas, and because the technical literature differentiates between them, correct terms are needed for these subdivisions. In this book, we will refer to the strata beneath the coastal plain as the *coastal-plain assemblage* and to the submarine part of the province as the *continental shelf-, slope-* and/or *rise-assemblage.*

A second way to subdivide the Atlantic and Gulf Coastal Province is geographically. The literature commonly distinguishes between the Atlantic Coastal Province and the Gulf Coastal Province. Again, it is our feeling that the Coastal Province is best thought of as a single entity but we also feel that such a subdivision is valuable to communication and we follow that usage here. Different writers have variously included Florida and/or the Bahamas with either the Atlantic or the Gulf Coastal Province; in order to avoid any confusion, we will treat Florida and the Bahamas as a separate subdivision of the Coastal Province. In addition, although it is certainly *not* part of North America's Coastal Province, we will also in this section discuss the Gulf of Mexico; this is because its history is intimately related to the history of North America's margins.

Deposits of the Atlantic Coastal Province occur in either thick, narrow accumulations, called basins or troughs, or in thinner masses termed platforms. Basins are thick, elongate sedimentary masses which generally parallel the edge of the continental shelf (Fig. 8-24). Major basins include: Labrador Shelf Basin, Northeast Newfoundland Shelf Basin, Grand Banks Basin, Scotian Basin, Georges Bank Basin, Baltimore Canyon trough, Carolina trough, and Blake Plateau Basin. The Southeast Georgia embayment, which underlies both the Georgia coastal plain and continental shelf, is a wedge-shaped prism of strata, triangular in plan view, which thickens continuously into the Blake Plateau Basin. Basins are usually bordered on the landward side by block-faulted continental crust and on the

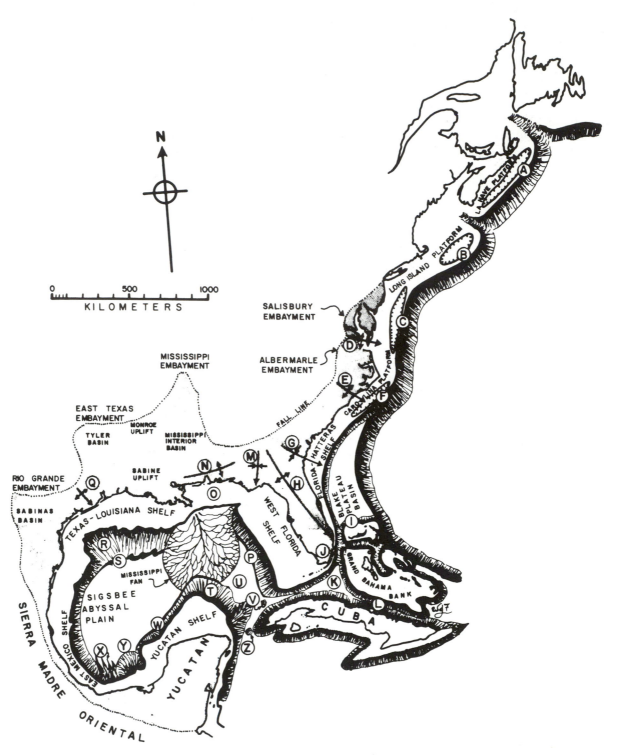

Fig. 8-24. Generalized map of the Atlantic and Gulf Coastal Province showing major structures of the province and features of the modern continental margin. A, Scotian Basin; B, Georges Bank Basin; C, Baltimore Canyon Trough; D, Monroe Arch; E, Cape Fear Arch; F, Carolina Trough; G, Southeast Georgia Embayment; H, Central Georgia uplift (in Georgia) and Peninsular Arch (in Florida); I, Little Bahama Bank; J, South Florida Basin; K, Florida Strait; L, Old Bahama Channel; M, Southwest Georgia Embayment; N, Wiggins Arch; O, Mississippi–Alabama Shelf; P, west Florida Escarpment; Q, San Marcos Arch; R, Texas–Louisiana Slope; S, Sigsbee Escarpment; T, Banco de Campeche Slope; U, Florida Abyssal Plain; V, Jordan Knoll; W, Campeche Escarpment; X, Mexican Ridges; Y, Golfo de Campeche Slope; Z, Pinar del Rio Knoll.

seaward side by the *East Coast magnetic anomaly* (a major positive magnetic anomaly which parallels the continental slope from Canada to the Blake Plateau). Longitudinally, the basins end at sharp offsets of continental crust which were interpreted by Kiltgord and Behrendt (1979) to represent fracture zones formed during continental rifting.

Platforms of the continental margin are much thinner sedimentary masses, usually having the shape of a very thin triangular prism, overlying continental crust which has not been significantly block-faulted. Platforms of the Atlantic Coastal Province include the La Have, Long Island, Carolina, and Florida platforms.

A final point about stratigraphic terminology is necessary. Since strata of the Coastal Province have not been grossly deformed, they tend to dip at relatively low angles away from the continent and toward the ocean. Similarly, isochronous beds and biostratigraphic markers also dip toward the continental margins. Although postdepositional changes (e.g., compaction, isostatic adjustments, epeirogeny) may have had local effects, one may consider the present dip-direction to agree grossly with the paleoslope direction. Naturally, the character of the rocks within a given time-parallel stratum will vary with position with respect to the paleoshore (i.e., it will show major facies changes perpendicular to strike). Consequently, it is convenient to distinguish different areas of Coastal Province stratigraphic units by specifying their location with respect to the dip of the unit. Thus, one may distinguish *up-dip* areas, i.e., areas high up on the dipping stratum, from *down-dip* areas, which are lower on the dipping stratum. As a general rule, up-dip areas were originally closer to the continental interior whereas down-dip areas were farther away, i.e., were more marine.

C.2. Jurassic Events

C.2.a. Atlantic Coastal Province

Murray (1961) virtually dismissed Jurassic stratigraphy of the Atlantic Coastal Province because no Jurassic rocks were then known definitely to exist in the East. Since that time, study of the Atlantic continental margin of North America (in the search for new petroleum resources) has led to the recognition of extensive Jurassic deposits in basins of the outer shelf and slope. Indeed, it is now believed that the bulk of basin-fill deposits of eastern continental-margin areas is of Jurassic age. In this section, we will consider these deposits and their sedimentary–tectonic significance.

At the base of the continental-margin prism occur Triassic–Lower Jurassic strata correlated with and lithologically similar to exposed strata of the Eastern Triassic–Jurassic basins (see Fig. 8-25). These strata tend to occur in normal-fault basins beneath platform areas and as essentially tabular

units beneath basins (e.g., in the Baltimore Canyon trough; see Fig. 8-24). In some COST (Continental Offshore Stratigraphic Test) wells on the Georges Bank, red beds, evaporites, and shallow marine carbonates of Triassic–Lower Jurassic age were encountered. On the Nova Scotia shelf, similar red-bed strata of the Mohican Formation are of Lower Jurassic age. Within correlative strata in the Baltimore Canyon Trough, designated Unit A by Schlee (1981), is an obvious angular unconformity called the *breakup unconformity*. Schlee considered that this unconformity separates rift-valley strata (tilted concomitantly with basin-subsidence) from on-lapping strata of Lower Jurassic age (which probably represent deposition penecontemporaneous with continental breakup). A somewhat similar circumstance is observed in the Southeast Georgia embayment where the breakup unconformity separates Triassic–Lower Jurassic strata in rift basins from Lower Jurassic lava flows. This unconformity can be traced westward (i.e., up-dip) to where Lower Jurassic volcanics directly overlie continental basement (Buffler *et al.*, 1980; see Fig. 8-25).

Continental-margin basins differ from smaller rift-basins beneath platform areas primarily in that they are larger and their activity continued longer. Schlee (1981) said that the Baltimore Canyon Trough originated during the Triassic as a much smaller basin located on the New Jersey shelf but gradually was enlarged as a result of postrifting subsidence. It is possible that a similar history may characterize other basins as well. The widths of continental-margin basins vary considerably; Klitgord and Behrendt (1979) said that basin width may be a function either of the rifting process itself or of later (i.e., postrifting) sediment-loading. They argued that if the rifting process were primarily responsible, then the basin-width pattern on one side of the Atlantic Ocean should be related to that on the other, i.e., the broadest basin on the African margin should match with the narrowest basin on the North American margin and vice versa. Figure 8-26 is a reconstruction of the North Atlantic at the time of the Blake Spur magnetic anomaly (approximately 175 Myr. ago). Note that the Blake Plateau Basin is opposite a segment of the African coast with no margin basin, that the large Senegal Basin is opposite the narrow Carolina trough, and that the Baltimore Canyon Trough is opposite another segment of the African coast with no margin basin. Notice also that the ends of the Blake Plateau, Senegal, and Baltimore Canyon basins match up remarkably well, supporting the hypothesis that basins end at transform offsets of continental crust formed during continental separation. Thus, the data seem to suggest that basin width was controlled (at least to a first approximation) by the rifting process.

As the Triassic passed into the Jurassic, continental separation gradually caused subsidence of the prospective continental-margin area until, at some time during the Early

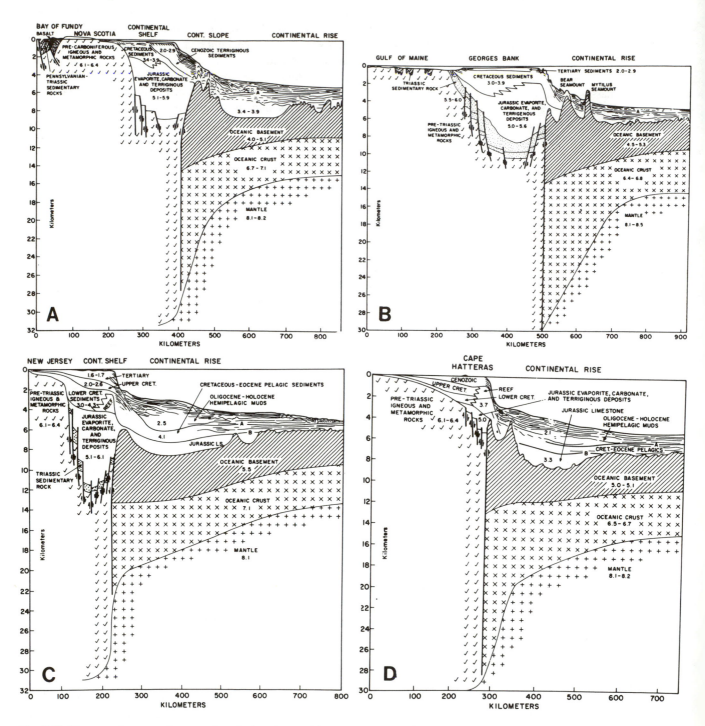

Fig. 8-25. Diagrammatic structural cross section of selected areas of the Atlantic continental margin. Sections are drawn perpendicular to the continental margin at the following areas: (A) Bay of Fundy; (B) Gulf of Maine; (C) Cape May, New Jersey; (D) Cape Hatteras, North Carolina; (E) central Florida. (From Sheridan, 1974, reproduced with permission of the author and Springer-Verlag.)

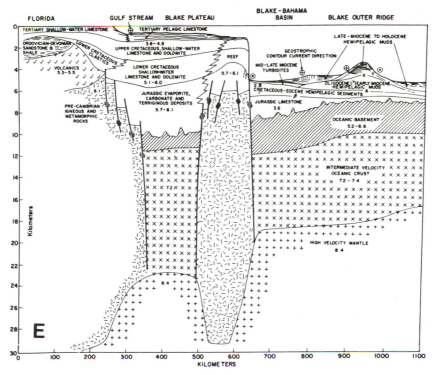

Fig. 8-25. (*Continued*)

Jurassic, oceanic crust began to form between the diverging continental blocks. This initial oceanic crust was unusually thick (Dickinson's "quasioceanic" crust) and is seen today as the basement high (marked by the East Coast magnetic anomaly) which bounds the Carolina Trough and other basins farther north on their seaward side. An alternate model for development of the outer basement high was proposed by Schuepbach and Vail (1980), who observed that outer basement highs are relatively common on rifted continental margins. They postulated that during the later part of the rift-valley phase of continental separation, a narrow, irregular uplift forms along the center of the broad graben. Erosion across the top of the uplifted block rapidly strips away its sedimentary veneer, the detritus being carried into the still-subsident parts of the graben to either side. Basaltic intrusions would be extensive within the uplifted block so that its average density would be intermediate between that of continental and oceanic crust. Schuepbach and Vail considered the medial uplift to be a thermal welt whose formation immediately preceded formation of true oceanic crust.

Subsequent plate-divergence causes a split running commonly, but not always, through the center of the medial uplift so that parts of the block are retained at the outer margins of both continents. At about 175 Myr. ago, a spreading-center jump took place, shifting the locus of divergence to the position of the Blake Spur anomaly.

Klitgord and Behrendt (1979) considered that the Blake Spur anomaly was formed continuously with the West Africa magnetic anomaly (Fig. 8-26). Possibly, the spreading jump may have been related to the greater separation-velocity along the southern (as opposed to the northern) portion of the break, as suggested by basin-width data. Sheridan et al. (1981) suggested that westward-dipping strata beneath the breakup unconformity of the outer Blake Plateau may represent a splinter of "African" continental crust stranded by the spreading-center jump. (Evidence for this interpretation may be seen in Fig. 8-25.) In any event, related to the Early Jurassic spreading jump was a change in the stress regime along the North American margin which produced a major episode of basaltic volcanism, resulting in the Lower Jurassic flows seen near the base of the Southeast Georgia embayment. Quite possibly, this is the same volcanic/intrusive episode recorded in the Eastern Triassic–Jurassic basins (p. 349) and sometimes referred to as the "Palisades orogeny." Small-scale folding of some Eastern Triassic–Jurassic basin strata may also have been caused by this alteration of regional stress regime.

As separation continued, the newly formed continental margin subsided rapidly. The margin basins, which had received primarily terrestrial sediment prior to breakup, now saw the influx of seawater. Within these still-shallow, narrow basins, circulation was restricted and evaporite sediments were deposited. As far as is known, evaporites

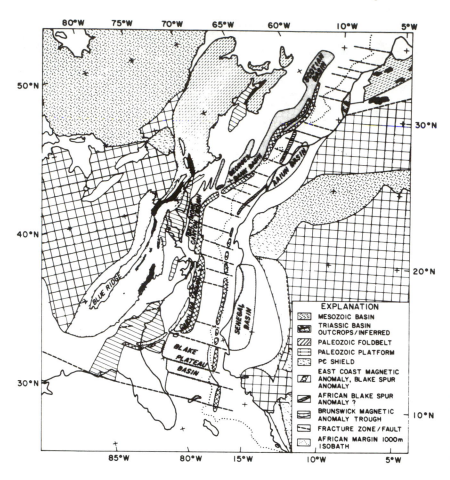

Fig. 8-26. Klitgord and Behrendt's illustration of the North Atlantic at the time of the Blake Spur magnetic anomaly (approximately 175 Myr. ago). Note that continental-margin basins on both sides of the Atlantic are clearly related to each other, both in terms of width and of location of the apparent transform faults which terminate them. See text for further discussion. (From Klitgord and Behrendt, 1979, reproduced with permission of the American Association of Petroleum Geologists.)

(mostly salt) are present near the bottoms of all the margin basins but the relative amount of salt varies from basin to basin. The Argo Salt of the Scotian and Grand Banks basins (Figs. 8-25 and 8-27) is extremely thick and is the source of diapirs which rise partially through superjacent strata; it is of Early Jurassic age (Pliensbachian; Imlay, 1980). Thick evaporites also apparently underlie the southern part of the Blake Plateau and the Carolina Trough; diapirs have been noted both beneath Exuma Sound in the Bahamas (Sheridan, 1974) and in the area between the Southeast Georgia embayment and the Cape Fear Arch (Grow *et al.*, 1983).

Differences in the quantity of salt from basin to basin may reflect differences in the degree of restriction among the individual basins. As discussed in the last chapter, salt deposits of the Grand Banks and Nova Scotia area are clearly related to similar deposits in basins of western Africa and Europe. Possibly, earliest continental separation in that area opened a narrow seaway between the Tethys and the rift valleys along the northern part of the break, so that they had restricted, shallow, highly saline seas while basins farther south (e.g., the Baltimore Canyon Trough) were still dominantly terrestrial. (The resulting circumstance may not have

been greatly different from the situation observed today between the Red Sea and the East Africa rift.) Similarly, as we will discuss in the next subsection, Gulf and Bahamian salt deposits may have been related to a marine incursion from the Pacific Ocean. Thus, basins at northern and southern ends of the continental break may have been opened to saltwater invasion first. By the time that seawater flooded basins of the middle margin, a proto-oceanic gulf existed all along the developing continental break, joining the Tethys Ocean tongue to the Pacific Ocean tongue and allowing relatively free circulation so that less evaporite material was deposited. If this suggestion is correct, oldest evaporites in middle-margin basins should be younger than oldest evaporites to either the north or the south.

As subsidence of the young continental margin continued, basins acquired progressively more open-marine conditions as a direct result of margin subsidence and eustatic sea-level rise. Thus, evaporite deposition ceased as basin waters freshened. On the landward side of the basins, continent-derived detritus was deposited in a variety of nonmarine and shallow marine environments (e.g., the Iroquois and Mic Mac formations of the Nova Scotia shelf; see Fig.

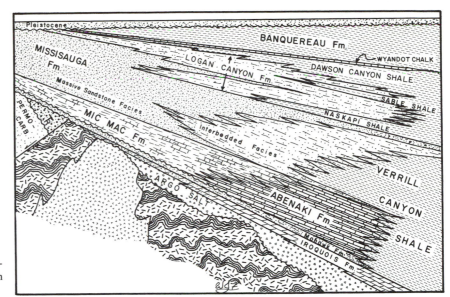

Fig. 8-27. Stratigraphic cross section (dip section) across the Scotian basin. (Modified from McIver, 1972.)

8-27) but, on the seaward side of the basin, carbonates began to accumulate (e.g., the Abenaki Formation; Fig. 8-27). The basement high formed by "quasioceanic" crust of the East Coast magnetic anomaly apparently formed an offshore bank between deeper waters of the margin basins to the west and the growing ocean basin to the east. Waters of the bank were probably warm, shallow, and clear (because they were far enough from the continent so that large quantities of detrital mud could not reach them). Consequently, the banks were ideal for the inception of carbonate deposition. By the Middle and Late Jurassic, a carbonate platform extended discontinuously all along the eastern margin of North America in much the same fashion as the Great Barrier Reef of present-day eastern Australia. In some areas, the carbonate buildup was represented by a narrow, Bahamas-like bank (the Abenaki Formation of the Scotian shelf is a good example because it is composed of a variety of carbonate lithologies including oolitic and related grainstones as well as various micritic lithologies), while in other areas, it was represented principally by a large reef complex.

The timing of carbonate-platform onset is somewhat arguable since the bottom of the carbonate complex has nowhere been penetrated by drilling. On the other hand, upper parts have been sampled, e.g., in COST B-3 well (off New Jersey) where rocks of Middle Jurassic (Callovian?) age consist of interbedded limestone, in places oolitic, and calcareous shale (Poag, 1980). In the Scotian Basin, limestones of the Abenaki Formation are of Middle Jurassic age. Thus, the Middle Jurassic is considered to be the time of platform development in most places. It should be noted, however, that specific timing probably varied from basin to basin.

During the Early Jurassic (Sinemurian) and persisting, with episodic fluctuations, throughout the rest of the Jurassic and into the Cretaceous, was a long-term eustatic rise of sea level. In addition to causing the rapid upbuilding of the shelf-edge carbonate mass, this sea-level rise conspired with post-rifting margin-subsidence to flood the new continental margin. Continent-derived sediments were deposited on a shelf whose depositional base-level was subsiding so quickly that little or no reworking probably took place. The result was that a very thick mass of sediment comprises much of the margin-basin fill. Grow *et al.* (1979) said that almost 70% of all subsidence beneath the New Jersey shelf took place during the Jurassic! This estimate fits nicely with Dickinson's theoretical analysis of rifted continental-margin development in which it was shown that nearly all of the isostatic subsidence of the margin due to crustal thinning and cooling of the lithosphere would be accomplished shortly after continental separation.

C.2.b. Gulf Coastal Province

Jurassic deposits of the Gulf Coastal Province are much better known than those of the Atlantic Coastal Province, not only because they are exposed in fold belts of eastern Mexico (e.g., the Sierra Madre Oriental) but also because, in the northern Gulf area, they have been extensively cored and have yielded very large quantities of petroleum. Thus, even though they do not outcrop in the United States, their sedimentary characteristics and facies patterns are relatively well known.

Rifting associated with Pangaea disassembly affected the Gulf region in a similar fashion as it had the Atlantic continental margin, beginning with the development, during the Triassic, of rift valleys whose strata (e.g., the Eagle Mills Formation of the U.S. Gulf Province) are similar to those of Eastern Triassic–Jurassic basins. Eagle Mills strata are truncated by a major erosional unconformity which may be equivalent to the breakup unconformity recognized above Triassic strata of the Atlantic continental margin. Jurassic strata above the post-Eagle Mills unconformity all share several general characteristics; for example, they are far more extensive than Eagle Mills strata, persisting with relatively little change from Florida to Texas and on into Mexico. They do, however, exhibit marked down-dip facies changes. In up-dip areas, Jurassic strata are predominantly coarse, red detrital-deposits, probably of terrestrial environments, with subordinate carbonates and evaporites. Farther down-dip, they are primarily evaporites or shallow marine carbonates. In extreme down-dip areas (at least, to the limits of drilling), Jurassic strata are typically composed of dark mudstones and pelagic limestones.

Directly overlying the post-Eagle Mills unconformity is the Werner Formation whose red, sandy, conglomeratic lower member rests on the Eagle Mills in parts of southern Arkansas and on Paleozoic rocks in other areas of southern Arkansas and northeastern Louisiana (Imlay, 1980). The upper member is composed mainly of anhydrite. Thus, the Werner probably represents the first introduction of seawater to the Gulf area as well as the onset of evaporitic conditions within the newly formed sea. The age of the Werner has been a matter of controversy because of its general lack of fossils. Imlay (1980) argued for a Callovian age based on inferred correlation with strata in Mexico which indicate that seawater first invaded the Gulf area during the Callovian, transgressing eastward across southern Mexico from the Pacific Ocean.

Overstepping the Werner throughout its extent is the Louann Salt, one of the most scientifically and economically important units of the Gulf Coastal Province. The Louann, whose maximum thickness has been estimated to be from 1500 to 3000 m, is the source of salt for the many diapiric structures which puncture the Gulf sedimentary prism. The age of the Louann has also been a matter of controversy, with estimates ranging all the way back to the Permian. Today, there seems to be general agreement that it is Middle Jurassic, probably Callovian to lower Oxfordian (Imlay, 1980). There are other Jurassic salt masses in the Gulf region, especially the salt associated with the Campeche knolls of the Golfo de Campeche slope and the Sigsbee knolls of the Sigsbee abyssal plain. In addition, a thick mass of salt and anhydrite, termed the Minas Viejas Formation, extends from the Coahuila Province of Mexico southeastward into the Tamaulipas Province. These deposits are all considered to be correlative with the Louann (as are salt strata beneath the southern Blake Plateau Basin). In addition, salt appears to be a component of the Challenger

seismic-stratigraphic unit which has been traced in seismic profiles to extend over much of the deep Gulf basin (refer to discussion in the next section).

Above the Louann is the Norphlet Formation (Oxfordian) which is a sequence of sandy, shaley red beds with local conglomerates and anhydrite. The Norphlet overlies the Louann disconformably in its up-dip areas and oversteps it to the north (Imlay, 1980). The Norphlet is interpreted to represent an arid-region coastal-sedimentary complex with fluvial, deltaic, sabkha, and eolian facies. In northeastern Mexico, Norphlet-equivalent rocks comprise the La Joya Formation (Murray, 1961).

Above the Norphlet is the Smackover Formation, whose stratigraphy is considerably better known because it is a major petroleum reservoir. In up-dip areas, e.g., southern Arkansas (the type area), the Smackover–Norphlet contact is disconformable (Moore and Druckman, 1981), but in down-dip areas such as Mobile County, Alabama, the contact is gradational (Mancini and Benson, 1980). By this point in our discussion, a general trend may be apparent: contacts between stratigraphic units in up-dip areas are typically unconformable, but become conformable in the down-dip direction. This corresponds with the well-known stratigraphic dictum that unconformities are common around basin margins but die out toward the basin center.

The Smackover is a thick sequence of carbonates (approximately 600 m in the type area) which can be traced from Florida to southern Texas. It is correlative with the Zuloaga Limestone of northern Mexico and is of upper Oxfordian age. Over its whole extent, the Smackover's lithological characteristics are essentially constant. The lower Smackover is a dense, laminated limestone, typically composed of pelletal wackestones and packstones, which probably represents a low-energy, shallow, subtidal environment formed by a marine transgression, which began during late Norphlet time (Mancini and Benson, 1980). In southern Arkansas, the lower member is similar, but contains more detrital clay and silt, suggesting that the area was closer to a low, terrigenous source. The Upper Smackover member consists of coarse, crossbedded carbonate grainstones, primarily oolitic, with lesser oolitic/oncolitic packstones and wackestones. Such lithologies have been described in southwestern Alabama, southern Arkansas, Texas, and elsewhere. They represent an extensive zone of wave- and current-washed calc-arenitic shoals which must have extended approximately parallel to the Jurassic coast all around the northern margin of the Gulf. The initial porosity of these sediments must have been relatively high but, in many areas, porosity and permeability have been enhanced by solution and dolomitization (Moore and Druckman, 1981). Thus, the upper member is a splendid reservoir-rock and is responsible for most Smackover oil production.

Mancini and Benson argued that a carbonate ramp model, rather than the carbonate platform model, best described the depositional setting of the Smackover. Major facies seem to be elongated parallel to the Jurassic coastline (with some variations due to subsurface structural features), rather than the more "patchy" distribution of facies on a broad platform behind a shelf-margin reef. Perhaps a carbonate platform did not develop because the Gulf waters were too hypersaline to allow the normal growth of hermatypic organisms so that shelf-margin reefs could not form (Nurmi, 1978). The overall depositional environment of the Smackover was considered by Mancini and Benson to have been similar to modern conditions along the southwestern coast of the Persian Gulf. Numerous shoreline fluctuations are recorded in the upper Smackover member but the general trend was regressive.

In its up-dip areas, the Smackover is conformably overlain by the Buckner Formation, which is of lower Kimmeridgian age and represents a continuation of the marine regression begun during later Smackover time. Toward the south, both the upper Smackover and the Buckner grade into dark, calcareous shales of the Bossier Formation (see below). The Buckner is lithologically complex; in its up-dip facies, it is primarily a detrital red bed, containing shales, siltstones, and sand, frequently with nodular anhydrite. Down-dip from the detrital facies, especially in southwestern Alabama and adjacent Mississippi, the Buckner is composed dominantly of anhydrite with interbedded dolostone; it may also contain salt. In northern Mexico, the Olvido Formation, which conformably overlies the Zuloaga Limestone, is lithologically identical with the Buckner (Imlay, 1980).

The uppermost unit of the Jurassic Gulf Province is the Cotton Valley Group, divided into the lower, Bossier Formation and the upper, Schuler Formation. As previously mentioned, the Bossier is a dark, calcareous, argillaceous shale with local limestone. It grades northward into both the upper Smackover and the Buckner formations; thus, it probably represents a relatively deeper-water (below wavebase) environment. The overlying Schuler Formation is a complex detrital unit with a number of different lithofacies. Its up-dip strata are coarse, red gravels and sands which grade down-dip to red sands and shales, all of which apparently represent the same kinds of terrestrial and transitional-marine environments encountered in previous units such as the Norphlet Formation. Still farther down-dip, Schuler strata grade into darker, calcareous muds. The relation between the Bossier and the Schuler is uncertain; originally, geologists in the area felt that the Schuler–Bossier contact must be unconformable because in up-dip areas where Schuler had overstepped the Bossier, Schuler strata locally lie on Smackover rocks with no intervening Buckner. It has been suggested, however, that the Schuler–Bossier contact

is, in fact, conformable *except* where the Buckner is missing. In those areas, the Buckner is assumed to have been uplifted across the top of rising salt diapirs and consequently eroded off. It is interesting to note that salt tectonics had already begun by Schuler time (even earlier in some places), indicating that salt mobilization does not necessarily require great thicknesses of overburden! In northern Mexico, similar strata occur. Bossier-like, dark, limy shales are called the La Caja Formation; they are relatively thin and lie basinward of correlative sandstones of the La Casita Formation. The La Casita is somewhat similar to the Schuler except that it is gray to black instead of red, contains coaly beds, and commonly is marked by a basal conglomerate where the Olvido (= Buckner) is absent. Imlay considered the Cotton Valley–La Caja/La Casita units to span the remainder of the Kimmeridgian Stage and all of the Tithonian.

C.2.c. Jurassic Stratigraphy and Tectonic History of the Gulf of Mexico

Figure 8-28 is a profile from the Sigsbee to the Campeche escarpments, showing the main seismic-stratigraphic units recognized throughout the deep Gulf basin and Table I presents their primary seismic characteristics and interpretations. Of these, only the Challenger unit is considered to contain Jurassic strata. Not shown on the profile is the Viejo unit, which is a weak, discontinuous reflector horizon which occurs in some areas between the basement and the Challenger unit. The Viejo apparently represents scattered areas of thin, detrital strata resting directly on oceanic crust (Watkins *et al.*, 1978). The Challenger unit is defined by a strong reflector horizon over a seismically transparent zone overlying either the Viejo or basement. The upper reflectors probably are Lower Cretaceous carbonate turbidites (equivalent to the carbonate banks which rimmed the Gulf at that time and which were sources of the turbidites) and the lower, transparent zone contains Jurassic salt (Worzel and Burk, 1979). Challenger salt is considered to be the source of the salt in diapirs of the Campeche and Sigsbee knolls. The Challenger is about 1 km thick throughout much of its extent, except in the Sigsbee–Campeche area where it may be more than 2 km thick (Watkins *et al.*, 1978). Assuming that approximately 40% of the Challenger unit is salt, Watkins *et al.* said that it must contain about 4×10^5 km^3 of salt (or approximately 20 times as much salt as is contained in Gulf waters today). Note, however, that this estimate is based on the assumption that the amount of salt in the Challenger unit is the same throughout the entire Gulf of Mexico—an assumption which is now highly questionable (see below).

The salt of the deep Gulf presents some important questions about the early history of the area. For example,

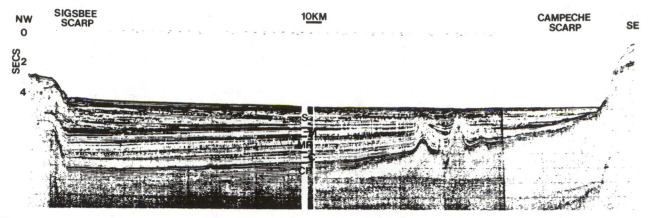

Fig. 8-28. Multichannel seismic profile from the Sigsbee escarpment to the Campeche escarpment. Note that the sediment mass and most units within the mass are wedge-shaped and thin toward the south; note also several salt diapirs (part of the Sigsbee knolls) north of the Campeche escarpment. (CH, Challenger seismic-stratigraphic unit; C, Campeche unit; MR, Mexican Ridges unit; M, Cinco de Mayo unit; S, Sigsbee unit.) (From Worzel and Burk, 1979, reproduced with permission of the American Association of Petroleum Geologists.)

Table I. Seismic-Stratigraphic Units of the Gulf of Mexico[a]

Unit	Seismic characteristics	Lithology	Age
6. Sigsbee	Closely spaced, evenly layered, strong, laterally persistent reflectors, unconformity at base	Silty clay with turbidite sands and clays	Pleistocene
5. Cinco de Mayo	Weak, laterally persistent reflectors	Silty clays with thin turbidites	Upper Miocene–Pleistocene
4. Mexican Ridges	Strong, closely spaced, discontinuous reflectors broken by numerous faults of small throw	Turbidites, more clayey at the bottom, sandier at the top	Lower Tertiary–Middle Miocene
3. Campeche	Largely transparent	Homogeneous, fine-grained sediments	Predominantly Cretaceous
2. Challenger	Strong upper reflector sequence, transparent in lower section	Fine turbidites at top and evaporitic sequence at bottom	Middle Jurassic to Lower Cretaceous
1. Viejo	Weak, discontinuous, pre-Challenger stratiform reflectors	Detrital rocks(?)	Pre-Middle Jurassic

[a]From Watkins *et al.* (1978) and Worzel and Burk (1979).

is the Challenger salt correlative with the Louann Salt and, if so, were they formed as a single, continuous layer or as separate deposits? Also, did the salt form in a shallow- or deep-water environment? The way that one answers these questions depends upon one's view of the tectonic history of the Gulf. We will examine several of the answers that have been proposed.

Humphris (1978) presented the model shown in Fig. 8-29. He demonstrated that the salt unit beneath the Sigsbee escarpment is considerably higher stratigraphically than the Challenger unit and he inferred that the salt unit of the shelf-slope area is continuous with the Louann of the inner-belt salt basins. Humphris observed that Challenger salt in the southern Gulf region was clearly the source of Sigsbee-knoll diapirs but he noted that no diapirs have ever been observed between the Sigsbee-knoll area and the Sigsbee escarpment. Since thickness of overburden in that area is even greater than in the area of the knolls, he argued that diapirs would have formed there were sufficient salt present. From this reasoning, Humphris deduced that salt deposits thick enough to form diapirs occur along the northern and southern margins of the Gulf but not in the middle. He then proposed the following historical sketch (Fig. 8-29). Louann and Challenger salt formed in a restricted, shallow-marine environment within a broad, subsiding basin divided approximately in half by a central (thermal?) uplift. The salt and related deposits would thus be considered the proto-oceanic gulf phase of continental rifting. (Note that this implies that thickest salt overlies transitional, not oceanic, crust.) Later Jurassic and Cretaceous seafloor spreading fully opened the Gulf and separated the northern (Louann) and southern (Challenger) salt masses with normal oceanic crust in between. Sediment loading of the Louann mass from the north during early Tertiary time caused southward-directed salt flowage during which the salt was thrust into

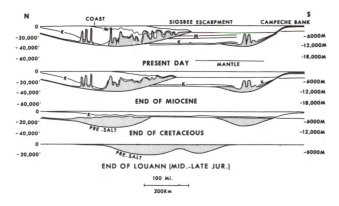

Fig. 8-29. Humphris's model of the tectonic evolution of the Jurassic salt deposits of the Gulf region; note the southward mobilization of the salt beneath the Cenozoic detrital-sediment mass. (From Humphris, 1978, reproduced with permission of the American Association of Petroleum Geologists.)

progressively higher stratigraphic levels, finally carrying the southern margin of the salt as much as 100 km farther south than its original location and several thousand meters higher stratigraphically than the salt-poor mid-Gulf Challenger unit at the base of the Sigsbee wedge (Fig. 8-29).

Buffler *et al.* (1980) presented seismic evidence that thick salt indeed occurs primarily over transitional crust except in the Texas–Louisiana areas where post-Jurassic salt flow has carried the edge of the salt southward beyond the limit of transitional crust (see Fig. 8-31). Their only

refinement of Humphris's ideas was the recognition that the central uplift of the proto-oceanic gulf (which divided the salt into two subequal basins) may have been the predecessor of the outer basement high observed at the edge of transitional crust north of the Campeche escarpment. They suggested that this central uplift formed in the manner described by Schuepbach and Vail (1980) and discussed earlier in this section.

We now turn to the tectonics of the opening of the Gulf of Mexico. The model described below has been assimilated from the discussions of Buffler *et al.* (1980) and Dickinson and Coney (1980); Fig. 8-30 illustrates the four major phases of Mesozoic Gulf history. As along North America's Atlantic margin, continental separation began with development of rift valleys in which were deposited red strata of the Eagle Mills Formation (Fig. 8-30A). With time, and with gradual thinning of the crust concomitant with separation, subsidence began to affect the entire area, leading to the deposition of lower Werner Formation detritus. If Imlay's (1980) assignment of a Callovian age for the Werner is correct, then regional subsidence probably began during the middle or late Middle Jurassic. At some stage in the subsidence, a breach was opened to the Pacific Ocean, allowing ingress of seawater from the west (at that time, the Atlantic Ocean was itself hardly more than a proto-oceanic gulf). The climate in the area of the resulting, broad, shallow sea was probably hot and arid; the Gulf area was, at that time, at about the same latitude as is the Persian Gulf today and had to its east (i.e., upwind with respect to

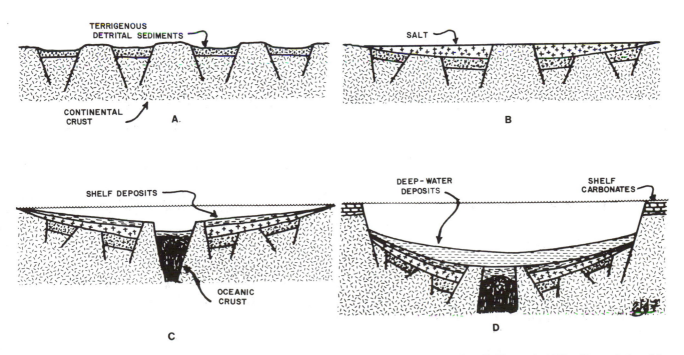

Fig. 8-30. Model of the tectonic development of the Gulf of Mexico; see text for discussion. (Redrawn from Buffler *et al.*, 1980, with permission of the Louisiana State University School of Geosciences.)

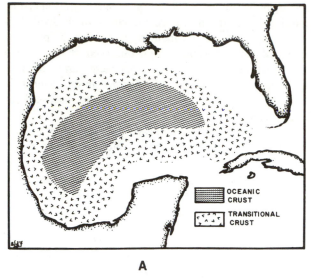

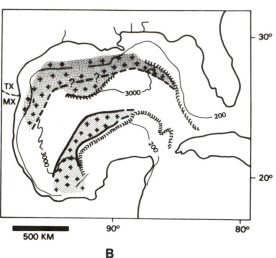

Fig. 8-31. (A) Inferred distribution of oceanic and transitional crust in the Gulf region; (B) generalized distribution of salt in the deep Gulf basin. (From Buffler *et al.*, 1980, reproduced with permission of the Louisiana State University School of Geosciences.)

the northeast Trade Winds) only the vast, dry expanse of Pangaean Africa. Thus, evaporitic conditions developed quickly and salinity in the shallow sea increased rapidly. This may be seen in the vertical progression from red beds in the lower Werner to upper Werner anhydrite and, finally, to Louann salt. Throughout the rest of the Callovian and into the Oxfordian, basin subsidence continued and connection with the Pacific remained restricted enough for salt precipitation to persist. Near the middle of the rapidly subsiding basin was the central uplift (Fig. 8-30B), which may have shed small amounts of detrital sediments to either side. By the late Jurassic, full continental separation had occurred and oceanic crust was forming as the continental blocks on

either side moved apart (Fig. 8-30C). Subsidence of the newly formed continental margin was rapid, as may be seen by the fact that strata of the coastal plain assemblage overstep one another throughout the Upper Jurassic and Lower Cretaceous Series. The probable cause of this subsidence was cooling of the crust as it moved away from the spreading center. Contemporaneously with continental separation in the Gulf, the seaway to the Pacific must also have been widening: salt deposition began to wane so that only thin salt strata were deposited on the new oceanic crust. By the middle Late Jurassic, evaporites deposited around the margins of the Gulf were principally anhydrite (reflecting a lower salinity) and were restricted to those coastal areas where barrier islands, uplifts (due to basement features or to salt tectonics), or other geographic barriers were present to restrict circulation. Deposition of carbonates on the continental shelves became common, resulting in widespread limestones of the Smackover and Zuloaga formations. Seafloor spreading continued in the Gulf region until some time during the Early Cretaceous, when it ceased. The margins of the Gulf were left by cessation of spreading more or less as they remain today, modified only by subsequent sedimentation and erosion. During this final tectonic stage, subsidence was initially rapid but slowed with time as the temperature of the oceanic crust decreased (Fig. 8-30D).

Figure 8-32 shows Dickinson and Coney's (1980) paleogeographic reconstructions of the Gulf area during the events discussed above. Notice that the orientation of the spreading ridge is shown to be parallel to the coasts of Texas and Yucatan and that the inferred transform offsets are parallel to transforms on the southern Atlantic margin of North America. (This is, of course, altogether reasonable since transform faults on the same plate are parallel to lines of spreading latitude, i.e., small circles drawn about the spreading axis, and must, therefore, parallel each other.) In this reconstruction, Yucatan is shown to have lodged against Texas and to have moved away in unison with the rest of Mexico, slipping past the "United States plate" along the Sonora–Mojave (also called the Silver–Anderson) megashear (marked "M" in Fig. 8-32). We must stress that the role played by Yucatan in this scenario is the most tentative and controversial aspect of the entire model. Other authors (e.g., Walper, 1980) have argued that Yucatan was a separate micro-continental block broken off the southern tip of North America and brought to its present position much later than the events described here.

The cessation of seafloor spreading in the Gulf is shown in the lower panel of Fig. 8-32 to have been caused by a "jump" of spreading to the Caribbean Sea. Presumably, this new episode of spreading broke South America away from Yucatan, opening a narrow seaway from the Pacific to the Atlantic. The cause of this spreading "jump" is highly speculative; it may be noted, however, that conti-

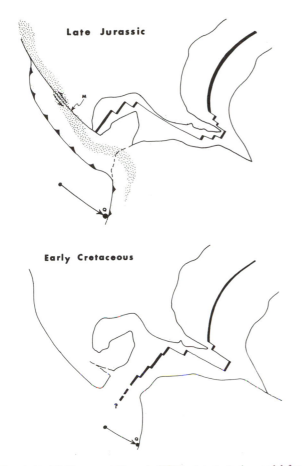

Fig. 8-32. Dickinson and Coney's (1980) plate-tectonics model for the opening of the Gulf of Mexico; see text for explanation. (Reproduced with permission of the Louisiana State University School of Geosciences.)

nental rifting broke South America away from Africa during the middle Cretaceous (a marine connection from the North to the South Atlantic had been formed by the early Turonian; Berggren and Hollister, 1974). Perhaps the spreading jump was related to events affecting the South American plate.

A final aspect of Dickinson and Coney's model involves the position of Cuba. According to the model shown in Fig. 8-32, Cuba at that time did not exist where and as we know it today. Rather, it is considered to have been part of an arc–trench complex somewhere between North and South America, perhaps a continuation of the Cordilleran magmatic arc system. Cuba was shown by Dickinson and Coney to have been brought to its present location by prograde arc-migration during the early Tertiary. (We will further discuss this subject in the next chapter.)

C.3. Cretaceous Events

We have seen that the Jurassic was a period of beginnings on the Atlantic and Gulf Coastal Province and that the Jurassic stratigraphic record is much fuller than previously expected. Nevertheless, it is the Cretaceous rocks of the Coastal Province that continue to attract the interest of geologists. Cretaceous rocks are definitely of commercial importance. In the Gulf area, Cretaceous strata contain important hydrocarbon reservoirs; on the Atlantic continental shelf, the hope of major oil discoveries has led to intense geophysical and drilling studies. But Cretaceous rocks also have the advantage of being exposed on the coastal plains and are, therefore, easily amenable to detailed stratigraphic study. Cretaceous strata outcrop in four major areas of the Atlantic and Gulf Coastal Province: (1) the Salisbury embayment and surrounding areas of the northern Atlantic Coastal Province; (2) the Cape Fear Arch; (3) the eastern Gulf Coastal Province from Georgia to Tennessee; and (4) the western Gulf Coastal Province from Arkansas to southern Texas and Mexico. During the Cretaceous, there were very few tectonic events to affect the Coastal Province; therefore, in this section we will focus attention on the stratigraphy of these four areas and on the subsurface stratigraphy of the continental margins and the Florida–Bahamas area.

C.3.a. Lower Cretaceous Strata—Atlantic Coastal Province

On the Atlantic Coastal Province, Lower Cretaceous rocks are known or inferred to be present beneath the continental shelf and slope from Newfoundland to Florida and the Bahamas. They are exposed, however, only in the Salisbury embayment where they comprise about 75% of the embayment's sedimentary fill. Along other portions of the coastal plain, Lower Cretaceous rocks have been overstepped by younger (primarily Upper Cretaceous) strata. As a general rule, Lower Cretaceous strata of the Atlantic Coastal Province are composed of detrital sedimentary rocks of nonmarine to shallow marine origin. On the eastern sides of most continental-margin basins are Lower Cretaceous reef complexes or narrow carbonate-platform strata. These rocks are continuous with underlying Jurassic rocks of similar character and probably represent shelf-edge carbonate buildups.

i. The Scotian Basin. Lower Cretaceous strata of the Scotian Basin are shown in Fig. 8-27 and are cited here to exemplify the stratigraphy of basins along the northern portion of the continental margin. Notice that carbonate rocks of the shelf-edge platform (represented by the Jurassic Abenaki Formation) are not observed in the Scotian Basin. Instead, Lower Cretaceous strata there are composed entirely of detrital sedimentary rocks which show a continuous lateral gradation from massive sandstone to interbedded sandstone, siltstone, and shale to dark, clay shale. The sands and interbedded sands and muds are termed the Missisauga Formation and the eastern clay shales comprise the Verrill Canyon and Shortland shales (McIver, 1972; Poag, 1978). (Note that Jurassic portions of the Verrill Canyon Shale occur eastward of the Abenaki platform and may, therefore, represent an upper continental slope facies.)

The obvious changes of grain size in Missisauga and Verrill Canyon detrital sediments from west to east probably reflect the development of a graded continental shelf. The concept of a graded shelf is useful in our discussions of Atlantic and Gulf Coastal Province strata and we digress briefly to consider it. On the gently and uniformly sloping surface of a continental shelf, the farther an area is from the shore, the deeper it will be and the finer will be the sediments that accumulate there. The reasoning runs as follows: sands will be held primarily near the shore because they are transported mainly as bed load. Shore sands may be shifted laterally parallel to the coast by longshore currents and may be shuffled on- and offshore by seasonal variations of wave characteristics. But sufficiently strong, offshore-flowing, bottom currents which might move large quantities of sand perpendicularly offshore do not exist. (This concept obviously neglects turbidity currents, subaqueous debris flows, contour currents, and so on, but as these do not generally affect shallow, shelf areas, the omission is probably not damaging to the overall argument.) Silts and clays, on the other hand, are carried principally as suspended load and, therefore, are not held to the shore but, rather, are moved by offshore-flowing, geostrophic (i.e., gravity-induced) surface currents. The size grading of shelf muds is a function of the settling velocities of the particles: silts settle more rapidly and, hence, are deposited relatively closer to land whereas clays settle more slowly and are carried farther out. Thus, sediments of a typical continental shelf (at least, of an Atlantic-type margin) should grade from nearshore sand to inner-shelf silts to outer-shelf clays. Such a sedimentological situation is termed a graded shelf. These ideas were stated as far back as 1932 (by F. P. Shepard) but, more recently, were dismissed when it became clear that the modern shelves are definitely *not* graded. The problem, of course, is that many of the sedimentary features of modern shelves were not shaped by marine processes but, rather, by nonmarine processes when the shelves were emergent during the Wisconsin glacial epoch. Thus, modern-shelf sediments are best considered to be relict. Curray's (1965) study of shelf sedimentology led him to state that a graded condition is now slowly returning to the shelves. Consequently, the graded-shelf concept seems to be on good theoretical footing and has been used by several authors (e.g., Swift *et al.*, 1969) to explain detrital-facies distributions of the Atlantic Coastal Province.

Returning to the Scotian Basin, its Early Cretaceous history may be described as a general marine regression

reflected in shoaling-upward, graded-shelf facies, interrupted by one or two minor transgressions. Missisauga facies prograded eastward throughout the early Early Cretaceous until an apparent rise of sea level (or decrease in the rate of detrital influx) caused a brief transgression. This transgression is seen in the westward shift of graded-shelf facies correlative with the Naskapi Shale. Following Naskapi time, regression resumed with deposition of the Logan Canyon Formation. The Lower–Upper Cretaceous boundary lies within the upper Logan Canyon; overlying units (Dawson Canyon Shale, Wyandot Chalk, and Banquereau Formation) record the great marine transgression during the Late Cretaceous.

ii. Central Atlantic Coastal Province. Lower Cretaceous rocks of the central Atlantic Coastal Province also record a general regression but a graded shelf did not develop because the carbonate bank on the eastern edge of the continental shelf continued to influence the sedimentary patterns of the shelf. With the exception of these shelf-edge carbonates, most of the rest of the Lower Cretaceous Series of the mid-Atlantic area is composed of detrital sediments. Lower Cretaceous strata are exposed in the Salisbury embayment of Delaware, Maryland, and Virginia; they are divided into Patuxent, Arundel (sometimes absent), and Patapsco formations, which (with the Upper Cretaceous Raritan Formation) comprise the Potomac Group. The Patuxent Formation is a nonmarine unit, probably deposited in a fluvial setting. It is characterized by quartz gravels, coarse to medium quartz sands, and lenticular clays, all arranged in fining-upward, point-bar sequences (Glaser, 1968a). The Patuxent also contains abundant trough crossbedding, local intraformational disconformities, and clay-pebble conglomerates, all of which are typical of fluvial deposits. Plant fossils, especially palynomorphs, indicate a late Neocomian (probably Barremian) age. Other than these plant remains, the Patuxent is unfossiliferous.

The Arundel Formation is a lignitic clay with siderite concretions (Glaser, 1968a). It contains palynomorphs similar to those of the Patuxent and so is probably of nearly the same age. In addition, the Arundel is known to contain fragmentary remains of eight species of dinosaurs, a crocodile, a turtle, a garfish, and assorted mollusks. From these fossils and from the presence of siderite concretions, the Arundel is inferred to be of primarily freshwater origin, possibly deposited in broad alluvial backswamps.

The Patapsco Formation, like the Patuxent, is composed of sandy to muddy sediments interbedded in fining-upward, disconformity-bounded sequences of probable fluvial origin. The Patapsco flora is extensive and well-preserved, containing numerous ferns, gymnosperms, and the first definite angiosperms of the Atlantic Coastal Province. Animal fossils, however, are sparse, consisting primarily of a few freshwater bivalves. Based on its flora, the Patapsco is considered to be of Albian age.

Although they are of nonmarine origin where exposed, Potomac Group strata grade down-dip into rocks of increasingly marine aspect. In coastal Maryland, for example, wells have penetrated fossiliferous marine strata of Lower Cretaceous age (Richards, 1967). Similar relations exist in Delaware.

Similarly, Lower Cretaceous strata in the Baltimore Canyon trough appear to be of mixed marine and nonmarine origin. Pre-Albian Cretaceous strata of the trough were included by Schlee (1981) in his seismic-stratigraphic Unit C; they contain interbedded sandstone, siltstone, and shale with local beds of limestone, lignite, and coal. Based on both seismic stratigraphy and well data, Schlee proposed the following historical scenario for the Baltimore Canyon trough: during the earliest Cretaceous, the western area was terrestrial. On the eastern edge of the shelf, a carbonate platform existed and in between the platform and the land were coastal and shallow marine environments. Through most of the Early Cretaceous, deltaic and related environments prograded eastwards in a major regression, correlative with the regression discussed earlier for the Nova Scotia area. By the end of the Albian, detrital sediments had been carried far enough eastward to bury the carbonate platform, terminating it as an active topographic entity and allowing the development of a normal graded shelf with outer shelf–upper slope deposits accumulating where the platform used to be.

The Early Cretaceous regression is also seen in strata of the Southeast Georgia embayment and northern Blake Plateau Basin. Graded-shelf facies, which developed relatively early in the northern Blake Plateau area because of the absence of an offshore carbonate buildup, prograded eastward. The top of the shelf was built up nearly to and maintained at sea level as shown by numerous disconformities in the Lower Cretaceous section (Dillon *et al.*, 1979). At the same time, terrestrial environments extended far out on the Florida–Hatteras shelf; Lower Cretaceous strata penetrated by a well on the Florida–Hatteras shelf offshore from the Georgia–Florida state line (Paull and Dillon, 1980) were unfossiliferous, sandy red beds with beds of limestone, calcareous shale, and coal (note the similarity to Lower Cretaceous strata of the Baltimore Canyon trough). Dillon *et al.* (1979) stated that, in the absence of an outer carbonate buildup, the position of the shelf edge was determined by the interplay of sedimentary accretion, erosion, and subsidence.

We have shown that during the Early Cretaceous a major marine regression characterized the Atlantic continental margin of North America from Newfoundland to the northern Blake Plateau. This regression was typified by eastward progradation of nonmarine, coastal, and marine-

shelf detrital facies. By the end of the Albian, this regression had stifled the shelf-edge carbonate buildups and a normal, graded shelf ensued. It may be seen, however, that Vail and Mitchum's (1979) eustatic sea-level curves (Fig. 8-3) indicate rising sea level during the Early Cretaceous; thus, a transgression, rather than a regression, might have been expected. The apparent contradiction may be resolved by considering the subsidence history of the Atlantic continental margin. During the Jurassic, subsidence of the margin had been rapid, probably due to crustal cooling following continental separation. Since the transgressive/regressive regime of an area is controlled by the balance between rates of sedimentation and subsidence, the very rapid initial subsidence of the Atlantic margin caused marine transgression. By the end of the Jurassic, however, the subsidence rate had declined considerably so that sedimentation outstripped subsidence and transgression gave way to regression. The unstated assumption, of course, is that sedimentation rate remained constant throughout this time. Such a situation would certainly be unlikely; sediment production from the old Appalachian orogen must have declined too. But all that is required is that the rate of sediment supply decline more slowly than the rate of subsidence. At that point where rate of sedimentation exceeded that of subsidence, facies would cease to migrate landward and begin to shift back out again.

C.3.b. Lower Cretaceous Strata—Florida and the Bahamas

During the Early Cretaceous, the history of Florida and the Bahamas began to digress from that of the Atlantic Coastal Province. Whereas the Atlantic continental margin continued to be dominated by detrital sedimentation, the Florida and Bahamas area was isolated from detrital sediments and experienced the development of a large carbonate platform. Relatively little is known about the early history of the Florida–Bahamas area. Lower Cretaceous units on the west side of the Peninsular Arch are named for correlative strata of the Gulf Province (which they generally resemble) whereas strata east of the arch are named for exposed units in Cuba.

At the beginning of the Early Cretaceous, the Peninsular Arch was a narrow finger of land extending south-southeastward from Georgia. Sandy and muddy red beds were deposited in predominantly nonmarine environments fringing the arch and grading outward in gray muds of probable marine origin (see Fig. 8-33, section D). The red beds are termed the Hosston Formation whereas gray shales are the Sligo Formation (both of these terms have been "imported" from Louisiana). The Sligo conformably succeeded the Hosston in a normal transgressive progression as rising sea level pushed both facies farther up onto the arch.

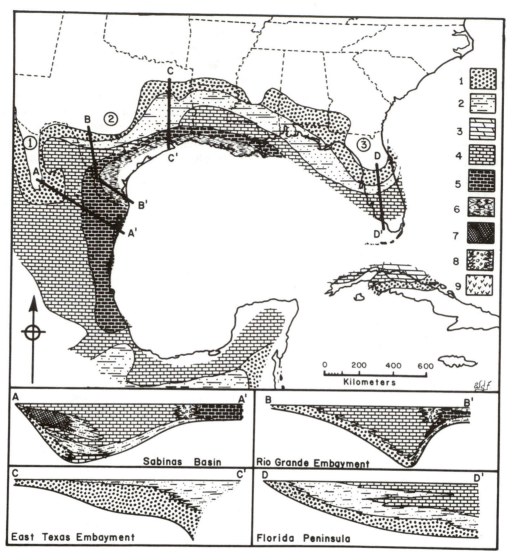

Fig. 8-33. Coahuilan lithofacies of the Gulf and southern Atlantic Coastal Provinces. Lithological symbols are as follows: 1, nonmarine detrital sediment; 2, shallow-marine detrital sediment; 3, dolostones; 4, shallow-marine limestones; 5, deep-marine limestones; 6, deep-basinal, detrital sediment, primarily muds; 7, evaporites; 8, biohermal structures; 9, volcanics. Numbers in circles are as follows: 1, Coahuilan Peninsula; 2, San Marcos Arch (associated with the basement structure called the Llano Uplift); 3, Peninsular (Ocala) Arch.

In southern Florida, far from detrital sources, platform carbonates developed and migrated up the arch as sea level rose. Termed the Pettet and Black Lake formations, these earliest carbonates interfinger with and overlie Sligo muds. They are continuous with limestones of the southern Blake Plateau Basin and, therefore, represent an already-large platform.

By Albian time, rising sea level had flooded all but the northernmost part of the Peninsular Arch, which supplied few detrital deposits to the carbonate areas around it. Over the entire Florida peninsula and shelves, carbonate sediments were deposited on a shallow platform which extended as far as the west Florida escarpment (see Fig. 8-34). The western edge of the platform was bordered by a large reef complex which may have been continuous both westward across the northern margin of the Gulf and southwestward to the platform-margin reefs of the Campeche escarpment (see Fig. 8-35). Antoine *et al.* (1974) thought that the west Florida reef was algal but Wilson (1975) considered it to be a rudist reef similar to contemporaneous reef complexes of West Texas and Mexico. The reef complex separated waters of the Gulf of Mexico from the shallow, lagoonal areas behind it and formed a barrier sufficiently restrictive to allow development of hypersaline conditions. Evaporites, such as the Ferry Lake Anhydrite (which is widely distributed in the northern Gulf region behind the reef), formed

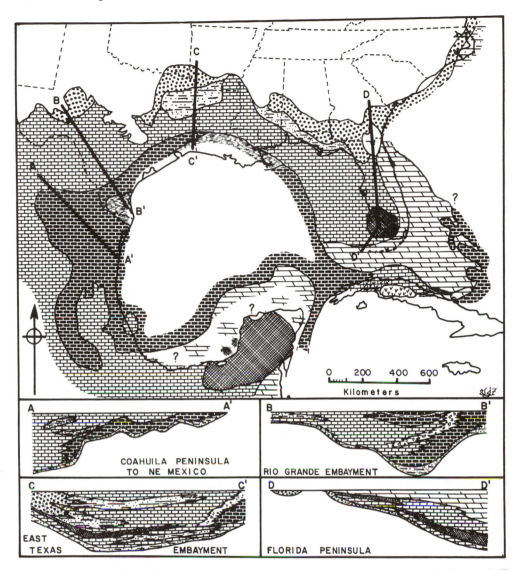

Fig. 8-34. Comanchean lithofacies of the Gulf and southern Atlantic Coastal Provinces. Lithological symbols are as in Fig. 8-33.

due to this restriction as did numerous local areas of dolomitization. Over most of the central and southern Florida area, however, normal shallow-water limestones were deposited (Fig. 8-34, section D). The character of these sediments suggests the kinds of depositional environments which exist today on the Great Bahama bank. The rate of subsidence increased toward the south but carbonate sedimentation kept pace, so that over 2400 m of shallow-water limestones comprises the Lower Cretaceous section of the South Florida Basin.

In the southern Blake Plateau area, the Jurassic reef complex of the Blake escarpment continued to grow during the Early Cretaceous and behind it developed a broad, shallow, carbonate platform which extended without interruption to the Florida platform. Shoals of carbonate sand shifted with the currents over deeper, carbonate-mud bottoms. The result was oolitic–skeletal calc-arenites and chalky micrites such as were sampled in a deep well on the northwestern corner of the Great Bahama bank. In addition, dolomitic rocks and evaporites are also present in Lower Cretaceous strata of the southern Blake Plateau Basin. This suggests that evaporitic conditions were locally present, perhaps in Sabkha-like environments associated with supratidal areas.

Figure 8-34 shows that by the end of the Early Cretaceous, a single, continuous carbonate platform extended from the west Florida escarpment to the Blake escarpment, a distance of over 11,000 km. The extension of shallow-water carbonate environments over such a large area is clearly a regressive trend and, therefore, probably reflects

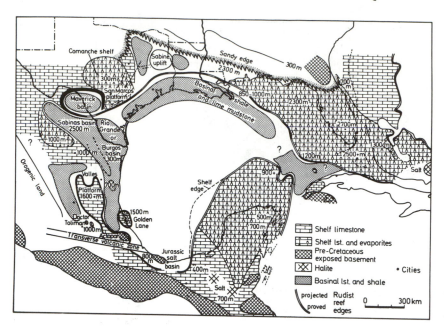

Fig. 8-35. Distribution of rudist-reef complex and associated carbonate platforms in the Gulf region during the Early Cretaceous. (From Wilson, 1975, reproduced with permission of the author and Springer-Verlag.)

the decreased rate of continental-margin subsidence mentioned earlier. The result was a vast, shallow marine province of warm, clear water, tropical climates, shifting shoals, and scattered islands and cays.

C.3.c. Lower Cretaceous Strata—Gulf Region

As in the Florida–Bahamas area, the other coastal regions surrounding the Gulf of Mexico were dominated during the Early Cretaceous by the development of extensive carbonate platforms (see Figs. 8-33 and 8-34). Only in northern areas of the U.S. Gulf coastal-plain assemblage did detrital facies persist. By the late Albian, a nearly continuous rim of carbonate platforms encircled the Gulf. Indeed, Worzel and Burk (1979) have suggested that even the deep-water corridor between the west Florida and Yucatan shelves may have been the site, during the Early Cretaceous, of a shallow-water reef complex. Since the rim of platforms surrounding the Gulf cut it off nearly completely from significant detrital influx, the deep Gulf was essentially a starved basin, receiving only pelagic sediment and occasional carbonate turbidites. These deep-water deposits comprise the upper Challenger and Campeche seismic-stratigraphic units (see Fig. 8-28).

The development of such large, platformal areas was probably due to the growth of barrier reefs along the margins of the continental shelves. These reefs, where exposed (in eastern Mexico) or sampled by drilling (in Texas and Louisiana), are composed of extinct, aberrant bivalve mollusks called rudists and lesser numbers of other invertebrates. The onset of reef growth during the earliest Cretaceous probably coincided with decrease of detrital influx

to the outer shelves because rudists (especially Early Cretaceous forms) were generally intolerant of turbid water. Behind the long, rudist reefs, a large variety of carbonate sediments accumulated in complicated facies-patterns which reflect the sensitivity of carbonate sediments to variations of environmental conditions. For this reason, stratigraphic relations of platform deposits are typically complex; we will try to avoid most of this complexity by stressing the historical development of the platforms rather than their formational stratigraphy.

As in other parts of this chapter, we follow European stage terminology. It turns out, however, that Cretaceous depositional sequences in the Gulf region do not coincide exactly with European sequences. For this reason, students of Gulf stratigraphy have devised a local reference section (based primarily on Texas strata). Since this Gulf-based terminology accurately reflects Gulf depositional sequences, we will use it as an organization tool to divide our discussion of Lower Cretaceous rocks into two parts: the *Coahuilan* (or lower Cretaceous) which corresponds to the Neocomian and lower Aptian stages and the *Comanchean* (or middle Cretaceous) which equals the upper Aptian, Albian, and lower Cenomanian.

i. Coahuilan Series. Rocks of the Coahuilan Series are present in the subsurface in most parts of the Gulf Coastal Province (Fig. 8-33) but are exposed only in extreme northwestern and western areas. Along the northern Gulf Coastal Province, lowest Cretaceous strata are detrital red beds, sandy at the bottom and gradationally more argillaceous toward the top. These strata are named the Hosston Formation and are recognized from west Florida (as mentioned earlier) around to the Rio Grande embayment of south Texas. Coarse, red Hosston detrital sediments apparently represent terrestrial depositional environments; they grade down-dip to dark argillaceous–calcareous muds with marine fossils. Conformably overlying the Hosston are marine deposits of the Sligo Formation. The Sligo is lithologically variable, being primarily a sandy mudstone with layers and lenses of sandstone and oolitic limestone in its type area (in Louisiana) but primarily a limestone farther west in Texas.

Consideration of Hosston–Sligo stratigraphic relations in their several areas reveals a general historical pattern: an initial Cretaceous rise of sea level, also recorded in Florida rocks, resulted in transgression of the Hosston coastal plain by Sligo shallow marine environments. In eastern regions, Sligo sediments were dominantly detrital in character because that area was relatively near a detrital source (the Appalachian orogen), while in the west, Sligo sediments were mainly carbonate-platform deposits because detrital sources were less important.

In northeastern Mexico, virtually the same pattern is seen (Fig. 8-33, section A–A'). Lowest strata are coarse, red, Hosston-like detritus of the San Marcos Formation. These grade Gulfward to dark, argillaceous and calcareous shales of the Taraises Formation and then to dark, basinal, pelagic limestones of the Tamaulipas Formation. Conformably overlying the San Marcos are Barril Viejos Formation muds; these, in turn, are overlain by shallow-water carbonates of the Cupido Formation. Both Barril Viejos and Cupido rocks are similar to the lithologically variable Sligo. The Cupido, especially, is similar to shallow-water Sligo carbonates not far to the north in Texas. Compare sections A—A' and B—B' and notice on both sections that a reef complex developed at the outer edge of the Sligo–Cupido platform and graded down to basinal Tamaulipas pelagics. During mid-Cupido time, evaporitic conditions developed at the back of the Cupido platform, leading to deposition of anhydritic deposits (La Virgen Formation). Again, the general pattern glimpsed in all of this is that of an initial transgression whereby marine sediments overrode terrestrial detrital sediments, followed by the development of an outer-margin reef and back-reef platform.

In eastcentral Mexico, Coahuilan strata are mainly basinal, pelagic carbonates of the Tamaulipas Formation.

ii. Comanchean Series.
Rocks of the Comanche Series are present in the subsurface in virtually all Gulf Coastal Province areas (see Fig. 8-34); they are exposed in large parts of central Texas and Mexico. Northernmost Comanchean strata in the subsurface from Georgia to Louisiana and at the surface in Arkansas, Oklahoma, and northeast Texas are composed of detrital sedimentary rocks, especially red beds, similar to the Hosston Formation (of

the Coahuilan Series). The mid-Aptian drop of sea level apparently exposed most up-dip areas to erosion so that lowest Comanchean (i.e., upper Aptian) strata unconformably overlie Coahuilan rocks. In northeast Texas, for example, detrital sediments of the Travis Peak Formation unconformably overlie Sligo carbonates. The source area of Travis Peak sediments continued episodically to supply detritus throughout the Early Cretaceous. The result was a stratigraphic pattern of detrital sand and mud tongues from the north interfingered with platform carbonates of the south; these tongues thin and become finger-grained in the down-dip direction. Perhaps the best example is the Paluxy Formation of northeast Texas and Louisiana (Fig. 8-34, section C–C'), whose varicolored sands and muds prograded far out onto the carbonate platform of the Comanche plateau during middle Albian time.

During the late Aptian, perhaps as a result of lowered sea level, a considerable volume of fine detrital muds was spread out over the shelves surrounding the Gulf of Mexico and interrupting, briefly, the deposition of carbonate sediments. (These detrital muds were penecontemporaneous with coarser sediments of the Travis Peak Formation.) In the northeast Gulf area, sediments deposited at this time are termed the Pine Island Formation, and consist of interbedded dark shales and micritic limestones. In the Rio Grande area of Texas, correlative black shale and dark micrite is termed the Pearsall Formation; in northeast Mexico, similar strata comprise the La Peña Formation, and in eastcentral Mexico they are called the Otates Formation. All of these units lie at the base of the Comanchean section and collectively represent a major interval of dark-mud deposition around the Gulf's margins.

By the beginning of the Albian, essentially clear-water conditions had returned to the shelves and carbonate platforms were reestablished surrounding the Gulf. Earliest carbonate deposits in the northeastern Gulf region are sandy, oolitic, coquinoid, and slightly anhydritic limestones of the James and Rodessa formations (Fig. 8-34, section C–C'). Overlying the Rodessa is the Ferry Lake Anhydrite, whose massive anhydrite, averaging 60–75 m in thickness, represents a significant restriction of circulation to the shelves, due possibly to a slight lowering of sea level or to extensive reef-growth on the platform margin. After Ferry Lake time, platform waters freshened and carbonate deposition returned, resulting in the Mooringsport Limestone, which is traced from Louisiana to western Florida.

In Texas, the entire carbonate section above the Pearsall is termed the Glen Rose Formation (Fig. 8-34, sections B–B' and C–C'). Together, the Travis Peak and Glen Rose formations comprise the Trinity Group. The Glen Rose is a highly variable platform-carbonate which contains, near its base, a massive anhydrite member correlative with the Ferry Lake Anhydrite. In down-dip areas, the Glen Rose contains relatively open-marine limestones, but in up-dip areas, it consists of intertidal and supratidal carbonates associated with detrital sands and shales. In some areas, the up-dip Glen Rose contains fossilized cycad stumps, abundant and well-preserved dinosaur tracks, and other indications of supratidal environments. Overlying the Glen Rose in northeastern Texas and Louisiana is the Paluxy Formation (mentioned earlier) which represents detrital influx from the north and west.

Overlying the Paluxy Formation on the Comanche plateau, and the Glen Rose on the Edwards plateau are strata of the Fredericksburg Group. Lower Fredericksburg formations, the marly Walnut Clay and chalky Comanche Peak Limestone, define a marine transgression after the lowered sea level of Paluxy (i.e., mid-Albian) time. The overlying Edwards Limestone is a thick and very widespread platform carbonate with abundant patchlike rudist mounds. The uppermost Fredericksburg unit, the Kiamichi, is a dark argillaceous shale similar to Pine Island–Pearsall–La Peña–Otates strata mentioned earlier. The Kiamichi is recognized over most of the northern Gulf region; thus, it may represent increased amounts of mud supplied to the shelves, possibly as a result of lowered sea level.

Strata of the Fredericksburg Group are overlain by the Washita Group which, in Texas, is divided into a complicated array of formations, most of which represent various carbonate or mixed carbonate–detrital facies. In general, Washita strata represent persistence of the Comanchean carbonate platform. Near the top of the Washita Group is the Georgetown Limestone, whose dark, pelagic micrites represent a major rise of sea level during latest Albian time. The shelf-edge reef complex and much of the platform behind it was submerged rapidly, terminating reef growth. Over Georgetown strata are dark, argillaceous shales of the Del Rio and Grayson formations (which are similar to Pearsall and Kiamichi shales, already mentioned). Overlying these are thin, fossiliferous and micritic carbonates of the Buda Limestone, which defines the top of the Washita Group and Comanchean Series.

The Comanchean carbonate platform extended without interruption into Mexico (see Fig. 8-35) where the term "Aurora Limestone" is applied to all the strata from Glen Rose equivalents to the top of Washita equivalents (Fig. 8-34, section A–A'). Near the base of the Aurora are anhydritic strata of the Acatita Formation which are correlative with the Ferry Lake Anhydrite. Carbonate platforms also existed in eastcentral Mexico during this time but we will defer discussion of them for the moment.

From the above discussion of Comanchean carbonate platforms of the northern and northwestern Gulf region, several generalities may be derived. During most of Comanchean time, detrital-sediment sources were probably low and supplied little sediment, except during times of depressed sea level. Eustatic sea-level changes are inferred to have been a major control on both detrital-influx rate and carbonate sedimentation because of the apparent synchrony of sedimentation events from Mexico to Florida. If this inference is correct, then the sea-level curve of Vail and Mitchum (Fig. 8-3) for the Cretaceous must represent the general trends of eustasy but not relatively minor fluctuations. Since detrital input was so limited, we may further infer that orogenic or epeirogenic uplifts around the Gulf were very limited.

At the outer edge of the Gulf Carbonate-platform complex was a remarkably long rudist reef. The reef extended, virtually continuously, from the west Florida escarpment to southern Louisiana, Texas, and northeastern Mexico and from the eastern Yucatan shelf westward and then southward to southeastern Mexico and thence northward to the Valles platform (see Fig. 8-35). Indeed, if Worzel and Burk (1979) are correct, the reef may have been continuous from Florida to Yucatan. The only exception would have been in the eastcentral Mexico area where a deep-water connection north of the Valles platform may have linked the Gulf to the Pacific. The only place where the reef is actually exposed is in the Rio Grande region south of the Edwards plateau. There, the reef is termed the Devils River Limestone and consists of a massive framework of biohermal rudists (notably caprinids; see Fig. 8-66) with lesser contributions from colonial coral, stromatoporoids, encrusting red algae, benthic foraminifera, and so on (Wilson, 1975). These organisms tended to construct large, wave-resistant mounds rather than continuous ramparts. Between the mounds were tidal inlets floored with coarse, highly crossbedded carbonate sands derived from breakup of the mounds and washed by strong tidal currents. The general structure of Lower Cretaceous rudist reefs and their relation to adjacent facies are shown in Fig. 8-36.

In the subsurface of Texas and Louisiana, the rudist reef is termed the Stuart City reef trend (Wilson, 1975). This (and, by extrapolation, the rest of the long reef as well) was remarkably narrow, only 5–10 km wide. Figure 8-36A shows the relationship of the Stuart City reef to the platformal carbonate facies. Perhaps the most interesting area of Comanchean rudist-reef development was in eastcentral Mexico. During the Neocomian, that entire area had been a deep-water environment in which only pelagic micrites (the Lower Tamaulipas Formation) were deposited. During the later Aptian (following deposition of the Otates Formation), however, rudist reefs began to grow, leading to the development of carbonate platforms which rose up from deep water just as the Bahama Banks do today. The major platforms are the Toliman, El Doctor, Actopan, Valles, and Faja de Oro (or, Golden Lane). The stratigraphy of eastcentral Mexican carbonate platforms is relatively simple. The reef itself and associated platform carbonates are termed the El Abra For-

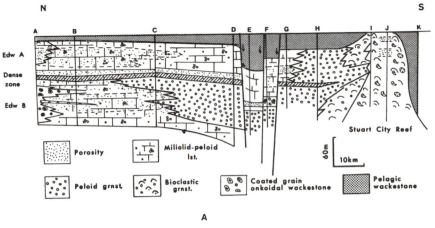

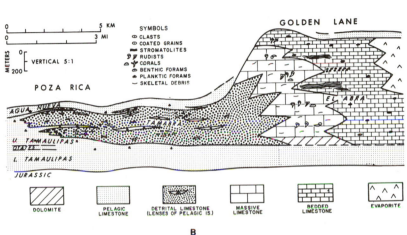

Fig. 8-36. Stratigraphic cross sections of Upper Cretaceous rudist-reef complexes and their relation to adjacent facies. (A) The Stuart City Reef in eastern Texas; note the complexity of carbonate-facies patterns on the back-reef platform; pelagic wackestones which blanket both reef and platform deposits represent the Georgetown Limestone of the Washita Group. (From Griffith *et al.*, 1969, reproduced with permission of the Society of Economic Paleontologists and Mineralogists.) (B) Simplified facies cross section of the Golden Lane platform; note that the Tamabra, which is interpreted as a deep-water reef-talus deposit, is shown to interfinger with basinal, interreef micrites of the Upper Tamaulipas Formation. (From Enos, 1977, reproduced with permission of the Society of Economic Paleontologists and Mineralogists.)

mation. Forereef deposits comprise the Tamabra Limestone and basinal deposits are the Upper Tamaulipas Formation (see Fig. 8-36B). The Tamabra represents reef talus which accumulated in deep water at the foot of the reef. Carrasco-V. (1977) suggested that as much as 1000 m of contemporaneous relief separated El Abra reefs from deep-water deposits adjacent to them.

C.3.d. Upper Cretaceous Strata—Atlantic Coastal Province

Throughout the time represented by the Zuni cratonic sequence, global sea level had been gradually rising, interrupted only briefly by short intervals of falling sea level (see Fig. 8-3). This long-term, eustatic trend culminated during the Late Cretaceous in a great marine transgression of North America. Never again would sea level be so high nor so much of the continent inundated: on the western craton, an epeiric sea stretched from the Arctic Ocean to the Gulf of Mexico, and on the Atlantic and Gulf Coastal Province, the shelf sea flooded at least as far inland as the present Fall Line and probably farther. Thus, Upper Cretaceous rocks

are widely distributed and widely exposed. In almost every case, the stratigraphic patterns of Upper Cretaceous strata of the Atlantic Province record a rapid, initial transgression followed by a stabilization of marine environments, subject to relatively small-scale, back-and-forth facies-migration in response to minor sea-level fluctuations.

i. Northern Atlantic Coastal Province. In the Scotian Basin (Fig. 8-27), the Late Cretaceous transgression is marked by the striking extension of outer-shelf muds, represented by the Dawson Canyon Shale, westward over coarser, inner-shelf strata of the Lower Cretaceous Logan Canyon Formation. The Wyandot Chalk above the Dawson Canyon Shale is a very-fine-grained carbonate composed of pelagic, biogenic debris, such as coccolithophores, and lesser amounts of hemipelagic clay. Uppermost Cretaceous strata in the Scotian Basin constitute the lowest part of the Banquereau Formation and are composed primarily of fine, clayey shales.

On Georges Bank, Upper Cretaceous rocks have been identified from dredge samples; drilling has penetrated over 400 m of Upper Cretaceous strata before bottoming in Campanian (?) rocks (Poag, 1978). Coniacian and Santonian parts of the section are composed primarily of dark to light olive-gray claystones, calcareous in some horizons, with scattered layers of detrital siltstone and sandstone. Campanian and Maestrichtian rocks are principally dark, laminated to massive siltstones with lesser sands and clays. Most of these strata were deposited under bathyal conditions as indicated by their microfauna.

ii. Baltimore Canyon Trough and Salisbury Embayment.

The same general pattern may be discerned along the central Atlantic Coastal Province. Lower Upper Cretaceous strata (Cenomanian to Coniacian) in the Baltimore Canyon trough are termed Unit D by Schlee (1981) and represent a graded, open marine shelf. In parts of the trough toward the land, shallow-water environments are recorded by fossiliferous, calcareous sandstone, siltstone, and shale with subordinate limestone. Farther from the land, the predominant sedimentary rock of Unit D is dark brown to gray calcareous mudstone indicative of moderate water depths.

These open-shelf strata grade westward to correlative nonmarine rocks in the Salisbury embayment where they are exposed in a narrow belt from New Jersey to Virginia (e.g., see Owens et al., 1970). The lowest of these Upper Cretaceous strata is the Raritan Formation, upper unit of the Potomac Group. The Raritan is a fluvial unit composed of interbedded sands and clays arranged in continuously gradational, fining-upward, point-bar sequences. Exposed Raritan strata contain few animal fossils but do have fossil plant material which indicates a Cenomanian age. Consequently, the Raritan is probably correlative with the Woodbine of Texas.

The Raritan is overlain disconformably by the Magothy Formation of approximately late Cenomanian and early Turonian age. The Magothy is a more complex unit than the Raritan, varying from coarse, fluvial gravels and sands in western areas to lagoonal and estuarine deposits in the east (Glaser, 1968). In the subsurface of southern Maryland, the Magothy contains strandline sands, whose high permeability makes them an excellent aquifer. Magothy strata have yielded a scanty vertebrate fauna including the fragmentary remains of crocodiles, turtles, and dinosaurs.

Rising sea level continued to push facies-patterns westward as the initial transgression continued into the Coniacian. Unit E (Coniacian through Maestrichtian) of the Baltimore Canyon trough reflects even deeper-water conditions on the continental margin while shallow-marine and peritidal environments are recorded by Salisbury embayment units. Schlee (1981) reported that Unit E strata contain calcareous, silty mudstone with thin beds of glauconitic sandstone and a faunal assemblage indicative of an outer shelf–upper slope environment.

In the Salisbury embayment, outcropping units correlative with Schlee's Unit E comprise the Matawan and Monmouth Groups (Formations in Maryland). The Matawan in Maryland is typically a dark green to black, micaceous, glauconitic, silty clay with little internal stratification (Glaser, 1968). Fossils include elements of the Exogyra ponderosa zone, indicating a Campanian age. Glaser said that the Matawan is the oldest wholly marine unit of the Maryland coastal-plain assemblage and represents the final stage of the initial Late Cretaceous transgression. In Delaware, the Matawan Group is composed of two units: the Merchantville Formation, whose general features resemble those of the Matawan in Maryland, and the overlying Wenowah Formation which is a reddish-brown, crossbedded sand typified by numerous Ophiomorpha burrows. Ophiomorpha typically occurs in well-sorted, highly crossbedded littoral sands. Weimar and Hoyt (1964) described nearly identical burrows produced today by the ghost shrimp Callianassa major, which inhabits a littoral to very shallow neritic environment. Thus, Wenowah sands overlying Merchantville glauconitic muds must represent a marine regression. In New Jersey, the Matawan contains several units (Woodbury Clay, Englishtown and Marshalltown formations) between the Merchantville and Wenowah but the same general regressive trend is noted. Several Upper Cretaceous units in New Jersey have yielded sparse but very important vertebrate faunas. The various Matawan units, for example, contain both herbivorous, "duckbilled" dinosaurs and carnosaurs (Horner, 1979).

The Monmouth Formation in Maryland is similar to the Matawan there except for being slightly coarser. It is a fine, micaceous, clayey sand with a glauconite content ranging from only a trace to over 50% of the whole sediment (Glaser, 1968). It contains a rich and well-preserved marine fauna with fishes, crabs, and over 140 mollusks, including Exogyra costata. In Delaware, the Monmouth Group is composed of the Mount Laurel–Navesink Formations and the Red Bank Sand. The Mount Laurel–Navesink is a sandy, clayey, glauconitic, fossiliferous deposit similar to the Monmouth of Maryland. The Red Bank is a reddish-brown to yellow, coarse sand. In New Jersey, the same units are observed but they are generally thicker than they are in Delaware. The New Jersey Monmouth section is capped by the Tinton Formation (Richards, 1967) which contains a very late Maestrichtian fauna. The Navesink in New Jersey has yielded a particularly fine assemblage of marine turtles, including some giants with shells nearly 3 m in length (Baird, 1978; Gaffney, 1975), mosasaurs (Baird and Case, 1966), and several types of dinosaurs (Horner, 1979).

The sedimentary pattern seen in all of the above stratigraphic details is one of an initial marine transgression from Raritan to early Matawan time, followed by the back-and-forth oscillations of coastal facies as recorded in strata of the upper Matawan and Monmouth Groups. The factor(s) controlling these facies oscillations is difficult to isolate but was probably not eustasy acting alone because the specific oscillations do not exactly match those recorded in the eastern Gulf Coastal Province.

iii. The Cape Fear Arch.

Upper Cretaceous strata are widely exposed in southeastern North Carolina and adjacent parts of South Carolina across the Cape Fear Arch. Because these strata are exposed over such a large area and dip at a relatively low angle, they offer a virtually unique opportunity to study in outcrop up-dip–down-dip facies relations of coastal-plain assemblage strata. One such study is that by Swift et al. (1969) and is recommended to the interested reader. The Upper Cretaceous of the Cape Fear Arch area is divided generally into four units: the Cape Fear, Middendorf, Black Creek, and Peedee formations. In the following paragraphs, we will discuss the main features of each.

Lowest of the four units, the Cape Fear Formation, which lacks macrofossils, unconformably underlies the other three, and was originally considered to be of Lower Cretaceous age (Heron and Wheeler, 1964). More recently, however, Christopher et al. (1979) showed that the Cape Fear Formation contains palynomorphs of Upper Cretaceous age. The sandy and muddy Cape Fear is probably of continental origin.

The Middendorf Formation lies unconformably on the Lower Cretaceous Cape Fear Formation and grades down-dip into the Black Creek Formation. Middendorf strata consist of quartz sandstone, some with a clay epimatrix (i.e., matrix due to intrastratal alteration of detrital feldspars), and interbedded mottled mudstone. These sands and muds are arranged in fluvial fining-upward sequences. The Middendorf Formation is Santonian in age and is, therefore, correlative with the Eutaw Formation of the eastern Gulf Coastal Province.

The Black Creek Formation is similar to, but not correlative with, the Magothy Formation in Maryland in that it is a complex unit with several different lithofacies. In rare instances, vertebrate remains are found in Black Creek strata, including fragmentary bones of dinosaurs, mosasaurs, turtles, and crocodiles as well as sharks' teeth. Thus, the Black Creek is similar to the Magothy in its general features and, like the Magothy, is considered to have estuarine, lagoonal, and littoral facies. The Black Creek contains Exogyra ponderosa and is, therefore, considered to be of Campanian age.

The Peedee Formation rests disconformably on the Black Creek. This disconformity, however, does not represent a significant time gap but is, rather, due to erosion at the shoreface as it was pushed backward by the advancing sea. Such an erosional surface is termed a *ravinement* (Swift, 1968) and should be understood to represent *no* subaerial exposure. The Peedee is typically a dark, very muddy sand whose primary sedimentary structures have been substantially destroyed by intense bioturbation. It is quite fossiliferous, containing numerous Exogyra and other oysters, belemnites, and rare marine-vertebrate remains. The Peedee is considered to be the deposit of an open-marine, graded shelf.

Near the approximate eastern limit of exposed Upper Cretaceous strata on the Cape Fear Arch, the uppermost Peedee contains a very different lithology from typical Peedee rocks described above. Called the Rocky Point Member, this unit is a fossiliferous limestone with abundant bivalve mollusks. At first glance, the Rocky Point resembles the Eocene Castle Hayne Formation which immediately overlies it and for many years was unrecognized as being of Cretaceous age. Its fauna, however, contains *Hardouinia,* an Upper Cretaceous echinoid, and *Exogyra costata.* The Rocky Point Member apparently represents a carbonate environment probably located far from the western, detrital source.

From west to east, the latter three Upper Cretaceous units of the Cape Fear Arch represent deposition in increasingly deeper-water environments. Taken together, they reveal the kinds of sedimentological features encountered along a traverse from an alluvial plain (Middendorf) to coastal marshes, estuarines, and lagoons near the shore (Black Creek) to a marine shelf (Peedee). Each sedimentary unit overlies the adjacent unit in a typical transgressive sequence exactly as predicted by Walther's law.

iv. Southeast Georgia Embayment.

In the Southeast Georgia embayment and northern Blake Plateau Basin, too, the Late Cretaceous was a time of marine transgression. Perhaps the most important effect of the transgression in this region was westward shift of the shelf edge from its previous position near the seaward side of the Blake escarpment to the present position of the Florida–Hatteras slope (Buffler et al., 1979b). This dramatic shift must have occurred during the Late Cretaceous (Cenomanian?) because Upper Cretaceous strata in most of the Blake Plateau Basin are deep-water deposits, including mostly chalks and calcareous shales (Poag, 1978). The Blake Plateau area has remained a deep-water environment, resembling a step between the deep ocean-basin and the Florida–Hatteras shelf, from the Late Cretaceous to the present, although the position of the shelf edge in the Southeast Georgia–Blake Plateau Basin region was controlled by the balance between deposition, erosion, and subsidence. The rapid Late Cretaceous sea-level rise perturbed that balance, causing the westward shift of the shelf edge and allowing the onset of deep-water conditions over the plateau. The actual location of the shelf edge was probably dictated by the newly formed Florida current which flowed northward in much the same fashion as does the Gulf Stream today (Buffler *et al.,* 1979b).

On the continental shelf itself, facies patterns were affected by two principal mechanisms. The first was small-scale fluctuation of sea level which caused back-and-forth migration of coastal and shelf lithologies as well as occasional emergence of much of the shelf itself, resulting in major unconformities. The initial Late Cretaceous transgression apparently lasted until some time during the Coniacian because Turonian and Coniacian strata are truncated by a regional unconformity above which are Coniacian–Santonian sediments (Paull and Dillon, 1980). Santonian strata in this region represent a progradational period during which the shelf edge grew eastward over deep-water deposits of the Blake Plateau. Another unconformity truncates the uppermost Santonian sequence, separating it from Campanian–Maestrichtian strata, which were deposited on a broad, graded shelf. Maestrichtian deposits of this shelf were sampled in a well drilled offshore from Charlestown, South Carolina, and are composed of soft, gray, silty clay similar to strata of the Peedee Formation on the Cape Fear Arch.

The second factor influencing facies patterns in the Southeast Georgia embayment was carbonate-platform deposition in the southern part of the area. The reader will recall that most shelf-edge carbonate areas of the Atlantic continental margin had been overwhelmed by the flood of detritus which prograded across the shelf during the Early Cretaceous regression. In the southern Southeast Georgia embayment, however, carbonate deposition continued, because it was far from rivers draining eastern slopes of the Appalachians. The Late Cretaceous transgression lowered stream gradients, thereby reducing detrital-sediment input to nearshore environments. Consequently, the carbonate platform increased in size, growing both seaward over deep-water environments and landward. According to Buffler *et al.* (1979b), virtually all Upper Cretaceous strata in the southern part of the embayment are carbonates. Paull and Dillon (1980) recognized a carbonate bank of Campanian and Maestrichtian age south of 30°N latitude. Conditions favorable for carbonate deposition were apparently widespread during the Maestrichtian, extending even to the inner shelf in some areas, as may be seen in the Rocky Point Member of the Peedee in North Carolina.

C.3.e. Upper Cretaceous Strata—Florida and the Bahamas

The Late Cretaceous history of the Florida–Bahamas area was also dominated by the eustatic sea-level rise discussed above. However, unlike the Atlantic Province where detrital sedimentation prevailed, the Florida–Bahamas region continued to be an area of carbonate deposition. Perhaps the most profound and far-reaching effect of the transgression was flooding of the outer margins of the carbonate platform at a rate which outstripped the rate of carbonate-sediment production. The result was initiation of deep-water conditions over areas which, during Albian time, had constituted a shallow carbonate platform. For example, the great barrier-reef complex of the Blake escarpment was terminated at this time by rapidly rising waters; southern portions of the Blake Plateau Basin began to receive only pelagic, calcareous sediment (Sheridan *et al.,* 1981). Concomitantly, the shelf edge shifted westward to a position just landward of its present location opposite the Florida Straits. The exact location of the shelf edge was probably controlled by the balance between rates of sea-level rise, carbonate-sediment production, erosion, and crustal subsidence. (This shelf-edge shift occurred in concert with the similar shift just to the north in the Southeast Georgia embayment, as discussed earlier.) At the same time, the western margin of the Florida platform was also flooded in a similar fashion. Mitchum (1978) stated that lowermost Upper Cretaceous strata from the West Florida slope are shallow-marine molluskan limestones, probably representative of the continuation of Early Cretaceous conditions. Abruptly overlying these, however, are unconsolidated, slightly argillaceous, foraminiferal–coccolith muds of apparently deep-water origin. By the end of the Cretaceous, the West Florida slope area was submerged beneath almost 900 m of seawater (based on paleobathymetric interpretation of foraminiferal assemblages; Mitchum, 1978).

Not only were marginal areas of the Florida–Bahamas region affected by the Late Cretaceous transgression but also the entire province. It was during this time that the entire Peninsular Arch was finally inundated and shallow-marine carbonate deposition extended over virtually the entire area of the modern Florida peninsula. Like earlier units of the Florida–Bahamas area, Cretaceous strata are entirely in the subsurface. Cenomanian through Santonian carbonates are separated from overlying Campanian–Maestrichtian strata by a major regional unconformity which apparently is correlative with the unconformity recognized by Paull and Dillon (1980) on the Florida–Hatteras shelf. For the most part, Upper Cretaceous carbonates of the Florida area are very pure, with little detrital contamination. The lack of detrital input to the carbonate platform during this time was not only due to the final drowning of the Peninsular Arch, which had furnished detritus locally during the Early Cretaceous, but also due to a prominent marine channel oriented NE–SW across northern Florida which separated the peninsular platform from detrital sources in Georgia. This channel is called the Suwanee Strait and is believed to have funneled strong currents from the Gulf to the Atlantic. Detrital sediments from the north which were introduced to the Suwanee Strait were probably shunted to either the Atlantic or Gulf basin, conceivably by turbidity currents or some other type of gravity-transport mechanism such as grain flow. It has been suggested (e.g., Mitchum, 1978) that the De Soto Canyon first began to form at this time, perhaps by the scouring effects of turbidity currents. The De Soto Canyon lies at the southwestern end of the Suwanee Strait and may have channeled turbid flows to the deep Gulf. The effectiveness of the Suwanee Strait as a barrier to southward detrital dispersal may be seen from the fact that almost pure carbonates were deposited along the southern side of the Strait.

Fortunately for tourists and carbonate petrologists, several large carbonate areas on the southern Blake Plateau were maintained at or near sea level, probably by rapid upbuilding of coral reefs, despite the Late Cretaceous transgression which submerged most of the rest of the plateau. The result was the beautiful and much-studied Bahama Banks. Controls on the exact positioning of the banks are controversial; some have argued for fault control whereas others have stressed organic control, i.e., a complex interaction of nutrient-bearing currents, reef progradation, and erosion.

The shallow carbonate platform of the Florida peninsula was separated from the Bahama Banks by the Florida Strait. This deep-water channel was due neither to faulting nor to offshore-bank development *per se* but, rather, to erosion as a result of the Florida current. The origin of the Florida current was discussed by Sheridan *et al.* (1981) who presented the following scenario. Prior to the drop of sea level which followed the Early Cretaceous, an arm of the North Equatorial current flowed from the Atlantic Ocean into the Gulf of Mexico and from there to the Pacific Ocean; the Gulf–Pacific connection may have been across south-central Mexico, possibly just north of the Valles platform (Fig. 8-35). With the Cenomanian drop of sea level, however, the Pacific connection was closed and waters in the Gulf backed up, creating the hydraulic head necessary to force an eastward-flowing current south of the Florida peninsula. This east-flowing current must have encountered the North Atlantic current gyre and was diverted northward to form part of the western Atlantic boundary current (now called the Gulf Stream). As it flowed northward over the newly flooded Blake Plateau, it eroded a channel between the Florida platform and the upbuilding Bahama Banks; indeed, the Florida current may have delineated the western boundary of bank development. By the same token, scour associated with the Florida current probably controlled the position of the entire Florida–Hatteras continental slope. It is interesting to speculate that the piling up of water in the Gulf hypothesized by Sheridan *et al.* to explain the Florida current may also have been responsible for producing the

current which swept through the Suwanee Strait mentioned above.

C.3.f. Upper Cretaceous Strata—Gulf Region

i. Eastern Gulf Region. The Gulf area, too, was affected by the Late Cretaceous sea-level rise and most detrital facies were shifted far landward of their Early Cretaceous positions. Thus, Upper Cretaceous strata typically comprise the basal outcropping units of the eastern Gulf coastal-plain assemblage. To a first approximation, the stratigraphic pattern revealed by coastal-plain-assemblage strata resembles that shown in the Salisbury embayment of the Atlantic Coastal Province. Specifically, in both areas a widespread fluvial unit is overlain unconformably by a sequence of peritidal units which records smaller-scale shoreline fluctuations. The sedimentological details differ between the two areas but such differences would be expected; not necessarily predictable, however, is the fact that timing of regional unconformities differs between the eastern Gulf and the Salisbury embayment. Thus, at least some component of smaller-scale shoreline fluctuations must be due to local factors (e.g., sediment-supply rate, basin subsidence) rather than to eustatic sea-level variations. In this section, we will discuss the stratigraphy and sedimentology of outcropping Upper Cretaceous units of the eastern Gulf region and show how a local history of shoreline fluctuations may be discerned from these data.

The lowest outcropping unit of the eastern Gulf coastal-plain assemblage is the Tuscaloosa Formation, which is exposed in a continuous belt from western Tennessee through northeastern Mississippi and central Alabama to western Georgia. In central Georgia, the Tuscaloosa (and the rest of the Cretaceous sequence) is overstepped by Cenozoic strata. Generally, the Tuscaloosa resembles the Raritan Formation of the Atlantic Coastal Province. Unlike the Raritan, however, which is correlative with the Woodbine of Texas, the late Cenomanian Tuscaloosa is correlative with the Texas Eagle Ford Group (Christopher, 1982). In its type area in Alabama, the Tuscaloosa is considered to be a group and is composed of the Coker and Gordo formations. In Mississippi and Tennessee, the Tuscaloosa is much thinner than in Alabama and is considered to be a formation but retains its generally terrestrial character. Its gravels are exclusively composed of chert and its sands also contain a considerable chert fraction; all of it derived from Devonian and Mississippian strata of surrounding areas (Marcher and Sterns, 1962). In the Chattahoochee Valley of eastern Alabama and western Georgia, the Tuscaloosa is also considered a formation and is characterized by fining-upward sequences. Its gravels, however, are composed entirely of quartz and quartzite pebbles, some of which were clearly derived from quartzite units of the

Piedmont just to the north (Frazier, 1982). In addition, Tuscaloosa sands in the Chattahoochee Valley region are arkosic, also reflecting a Piedmont provenance. From the above discussion, one may infer that the Tuscaloosa represents a fluvial plain between eroding upland areas to the north and the sea farther to the south. Variations of pebble and sand composition probably reflect a local origin for most Tuscaloosa sediments. Glauconite in the lower member of the Coker Formation and a somewhat more marine aspect for the Coker in general suggest that the present exposure belt does not exactly parallel the original sedimentary strike.

Unconformably overlying the Gordo Formation in westcentral Alabama is the McShan Formation, which is a glauconitic, rippled and crossbedded, very-fine to fine sand with thin beds of light-gray clay. McShan sediments closely resemble those of the lower member of the Coker Formation except for the McShan's lack of fossils.

Overlying the McShan in westcentral Alabama and the Tuscaloosa both to the west into Tennessee and east into Georgia is the Eutaw Formation. The Eutaw is the first unmistakably marine unit of the region and, in most areas, is composed of greenish-gray, crossbedded, glauconitic sandstones and associated dark, clayey mudstones. Fossils are common, especially bivalve mollusks such as *Ostrea cretacea* which in eastern Alabama occurs in large, reeflike masses. Numerous other fossils, such as *Gryphaea, Hardouinia, Exogyra,* sharks' teeth, and assorted fish vertebrae, indicate a strong marine influence on Eutaw depositional conditions. The Eutaw's fauna indicates a Santonian age (Reinhardt, 1980); thus, the unconformity separating the Eutaw from the upper Cenomanian Tuscaloosa Formation represents all of Turonian and Coniacian time. No doubt the duration of the exposure represented by the unconformity decreases toward central Alabama where the McShan intervenes but, even there, the post-Tuscaloosa unconformity is still the most significant hiatus in the Upper Cretaceous Series of the eastern Gulf region. In western Georgia, the Eutaw consists of a coarse to very-coarse, high-angle crossbedded sand facies in up-dip areas; this facies contains numerous evidences of deposition in a tidally dominated environment.

Abundant *Ophiomorpha* burrows in these sands indicate littoral to shallow sublittoral conditions. Therefore, these sands are interpreted as tidal-delta and associated back-barrier deposits formed during the early, transgressive phase of Eutaw time (Frazier and Taylor, 1980). During later Eutaw time, sea level apparently dropped and a regressive phase commenced, resulting in development of a new, offshore barrier system to the south of the Eutaw's present outcrop belt in eastern Alabama and western Georgia. Consequently, most upper Eutaw deposits in this area are of back-barrier, i.e., lagoonal, marshy, estuarine, origin

(Reinhardt, 1980). In the upper Eutaw of central Alabama are crossbedded sands of the Tombigbee Sand Member. These sands were interpreted by Frazier and Taylor (1980) as tidal-delta deposits; thus, they may represent the offshore Eutaw barrier mentioned above. This possibility is strengthened by the observation (made earlier) that the present-day outcrop belt of Cretaceous strata is probably not parallel to depositional strike so that central Alabama deposits represent more basinward conditions than do deposits of penecontemporaneous units to either the east or the northwest.

We have seen in earlier discussions that Santonian deposits of both the Atlantic and Florida–Bahamas provinces are truncated by a major regional unconformity. In the eastern Gulf, too, an unconformity occurs at the top of the Eutaw Formation; lowest Campanian strata represent a major marine transgression. Facies patterns during this transgression were shifted even farther landward than they had been during the early Eutaw transgression. The result was deposition of offshore-marine chalks in the Alabama–Mississippi area (represented today by the Mooreville and Demopolis Chalks of the Selma Group) and of marginal-marine detrital sediments in both Tennessee (the Coffee Sand and related units) and Georgia (the Blufftown and Cusseta formations).

The Mooreville Chalk consists of very calcareous, locally glauconitic marl and clayey chalk (Copeland, 1972). Mooreville faunas are somewhat limited, being composed of thin-shelled mollusks and sparse sharks' teeth along with the foraminifera, ostracodes, and other microorganisms which comprise the chalk. In the upper Mooreville is a thin, laterally persistent, fossiliferous limestone termed the Arcola Limestone Member. The Arcola contains abundant bivalves, especially *Exogyra ponderosa;* it characteristically is perforated by numerous borings and, therefore, may represent a carbonate hardground. Major collections of mosasaur fossils come from the Mooreville as well as from other Upper Cretaceous units in Alabama ranging from the Eutaw to the Ripley Formation (see Russell, 1967).

The Demopolis Chalk disconformably overlies the Mooreville and extends considerably farther to the northwest, reaching all the way into Tennessee. The Demopolis is a light-gray, fossiliferous chalk. Its lower part consists of thinly bedded, marly chalk, somewhat purer than the underlying Mooreville; its upper member is relatively pure chalk, up to 90% calcium carbonate in some outcrops (Copeland, 1972). In western Alabama, the pure chalk facies grades upward to a massive, abundantly fossiliferous, clayey marl called the Bluffport Marl Member.

In western Georgia and adjacent eastern Alabama, the detrital equivalent of the Mooreville Chalk is the Blufftown Formation. At first glance, the Blufftown resembles the Eutaw Formation which it overlies. The reason is that it is the product of deposition in the same suite of depositional environments. Much of the Blufftown represents a regressive sequence, but near the middle of the unit is a transgressive sequence, somewhat rare in barrier-island systems, in which back-barrier deposits are overlain by poorly preserved barrier-bar sands at the top of which is a poorly sorted, somewhat disrupted-looking sediment considered by Reinhardt (1980) to represent a ravinement caused by reworking of barrier deposits by the retreating shoreface. Above the ravinement unit are inner-shelf muds containing an abundant and diverse marine fauna and glauconite. Because the Eutaw and Blufftown apparently consist of similar lithofacies, they may be considered to be the deposits of the same suite of environments which migrated back and forth with fluctuating sea level. Reinhardt elaborated this idea and applied it to the entire Upper Cretaceous Series in the Chattahoochee Valley region (see Fig. 8-37).

The Blufftown is abruptly overlain by the Cusseta Sand, which is correlative with the Demopolis Chalk. The type Cusseta is composed of coarse, highly crossbedded, *Ophiomorpha*-burrowed sands and is, therefore, similar to barrier-related sands of the Eutaw and Blufftown formations. The transition from the Blufftown represents rapid shoaling of the environment due either to deltaic progradation or to a drop of sea level. Reinhardt (1980) favored the latter view and this hypothesis is given some credence by the Arcola Limestone hardground of central Alabama, which is approximately correlative with the Blufftown–Cusseta contact and which probably represents a period of much-decreased rate of carbonate-sediment production such as would occur during a brief lowering of sea level.

In western Tennessee, Mooreville-equivalent strata comprise the lower Coffee Sand (see Fig. 8-38) which is composed of fine to very-fine, crossbedded sand and brownish-gray, lignitic clay (E. E. Russell, 1975). Coffee Sand lithologies represent a suite of environments from fluvial and marginal marine in northern areas, to inner-shelf muds gradational into the calcareous deposits of the outer shelf to the south. A prominent tongue of Coffee Sand (the Tupelo Tongue; Fig. 8-38) represents the regression at the end of Mooreville–Blufftown time and separates the Mooreville from the Demopolis in Mississippi. Between the Demopolis and the Coffee Creek is the Sardis Formation

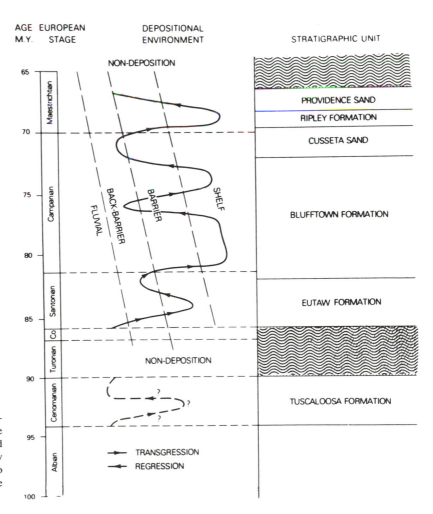

Fig. 8-37. Reinhardt's (1980) model of Upper Cretaceous transgressions and regressions for strata of the Chattahoochee Valley region. Notice that only a limited number of environments are hypothesized; these few environments migrated back and forth in response to sea-level fluctuations to produce the observed sequence of strata.

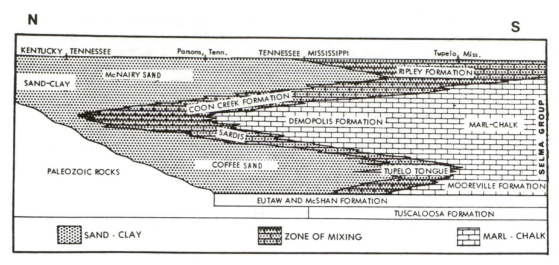

Fig. 8-38. Diagrammatic cross section showing the major lithostratigraphic relationships in the Selma equivalents of western Tennessee. (From E. E. Russell, 1975.)

which is a highly glauconitic sand, becoming progressively more argillaceous toward the top. The Sardis is a transgressive transitional unit between the peritidal deposits of the Coffee Creek and the outer-shelf chalks of the Demopolis. Overlying the Demopolis in this western area is the Coon Creek Formation, which is a regressive version of the Sardis Formation. The Coon Creek is abundantly fossiliferous and contains some of the best-known collecting localities for Cretaceous faunas of the entire eastern Gulf Coastal Province.

Overlying Cusseta–Demopolis–Coon Creek strata is the Ripley Formation. In Tennessee, Ripley-equivalent deposits are termed the McNairy Sand but in Mississippi the McNairy is considered a member of the Ripley. In western Georgia, the Ripley is a massive, bioturbated, micaceous, glauconitic, fine sand. It is very fossiliferous and locally contains indurated calcareous ledges or horizons of calcareous concretions. This description generally characterizes the Ripley in Alabama and Mississippi as well, except that Ripley strata in those areas are somewhat finer-grained and more marly. The Ripley contains both *Exogyra costata* and *E. cancellata* and is, thus, correlative with the Peedee Formation of North and South Carolina. The Ripley in Georgia was considered by Reinhardt (1980) to be transgressive (see Fig. 8-37) but is apparently regressive in western areas, especially in Tennessee and Mississippi where it is associated with McNairy detrital sediments (see Fig. 8-38). E. E. Russell (1975) recognized three lithological units in the McNairy: (1) a regressive, nearshore sand; (2) a middle wedge of coarser sands and lignitic clay believed to be fluvial, deltaic, and possibly estuarine in origin; and (3) an upper, transgressive sand which is only present in a

limited area. It is entirely possible that differences in the character of Ripley–McNairy deposits reflect variations in detrital-sediment supply rates.

Uppermost deposits of the Upper Cretaceous Series in central Alabama comprise the thin Prairie Bluff Chalk, a massive, compact, white chalk which contains varying amounts of quartz sand and abundant fossils (Copeland, 1972). To the west, the Prairie Bluff interfingers with highly fossiliferous, sandy muds of the Owl Creek Formation and, to the east, Prairie Bluff chalks grade into the Providence Sand of western Georgia. The Providence consists of two members, of which only the lower one (the Perote Member) is named. The Perote consists of laminated carbonaceous clay and silt. It is fossiliferous in some exposures and contains leached shells of *Exogyra costata, Turritella*, and others. The upper, unnamed member thickens eastward, reaching about 45 m in its type locality of Providence Canyons, Georgia. This member is mostly a high-angle crossbedded, quartz sand with abundant *Ophiomorpha* and clay clasts. In down-dip exposures, the upper Providence member is fossiliferous with bivalves and gastropods. At the top of the Upper Cretaceous Series throughout its outcrop belt in the eastern Gulf region, the Series is truncated by a major unconformity, the relief on which reaches up to 5 m in some eastern areas.

ii. Western Gulf Region. Upper Cretaceous strata in the western Gulf Coastal Province are exposed in a narrow, continuous belt from southwestern Arkansas to Mexico. The Upper Cretaceous section in Texas has been cited as a standard for all Gulf Province areas and defines the Gulfian Series, divided (from bottom to top) into Woodbine, Eagle Ford, Austin, Taylor, and Navarro Stages. As in our earlier description of Lower Cretaceous rocks in this area, we will use these Gulf Province-based stages to organize the following discussion.

The Woodbine Stage is defined by the Woodbine Group, which is principally a nonmarine to brackish-water unit deposited during the Cenomanian. The Woodbine is, therefore, correlative with the Raritan of the Atlantic Province and represents the early Late Cretaceous rise of sea level, during which coastal-plain stream systems deposited widespread alluvial sediments; it also represents, during later Woodbine time, the return of the sea to the East Texas plain. In its type locality, the Woodbine is divided into the Dexter Sand and Lewisville Formation. The Dexter is a typical fluvial unit, containing fining-upward sequences with coarse, pebbly sand at the base grading upward to massive or crossbedded sands with red and green mottled shale at the top. The Lewisville, which lies conformably on and oversteps the Dexter, consists primarily of laminated to massive, dark-gray, micaceous, carbonaceous siltstone and shale with thin lignite beds. Whereas the Dexter has no animal fossils, Lewisville strata contain a brackish to marine fauna with oysters, ammonites, and other mollusks. Thus, the Lewisville represents the initial flooding of the Woodbine alluvial plain by the sea. Toward the south, sandstone beds disappear so that nearly the whole section consists of dark, noncalcareous, clayey strata termed the Pepper Shale. Murray (1961) considered the Pepper to be a continuation of Lewisville strata, thus defining an overstep relation with the Dexter.

In southwestern Arkansas and adjacent Texas, Woodbine sedimentary rocks contain volcanic detritus including altered volcanic rock fragments and tuffaceous sands. Small amounts of volcanic debris have also been reported from Woodbine and Tuscaloosa deposits in the subsurface of Louisiana, Mississippi, and the Monroe uplift area of Arkansas. Genetically related to these volcanic materials are small plutonic masses in southwestern Arkansas. For example, southeast of Little Rock are two nepheline syenite masses, and in the Ouachita Mountains to the west are numerous, smaller bodies of similar lithology. In addition, near the coastal-plain-assemblage overlap in Arkansas are four diamond-bearing periodotite pipes. (These are the source of the only abundant diamonds, most of industrial grade, in the United States.) The origin of these somewhat unusual igneous plutons and their tectonic significance are poorly understood.

The Eagle Ford Group is more widely distributed than is the Woodbine Group beneath it and is of marine origin throughout its extent. The Eagle Ford is of upper Cenomanian and Turonian age and, therefore, is correlative with the Magothy Formation of the Salisbury embayment and the Tuscaloosa Formation of the eastern Gulf area. In the sub-

surface, Eagle Ford correlatives are widely recognized as dark calcareous shales and lesser chalks throughout southwestern Mississippi and most of Louisiana. In its Texas outcrop, the Eagle Ford is generally a dark, fossiliferous, marine shale; however, a variety of lithologies have been recognized and provide the basis for a somewhat complicated stratigraphy.

In the Dallas area, near the center of the East Texas embayment, the Eagle Ford Group is about 144 m thick and is divided into three formations, the lowest of which is the Tarrant Formation consisting of light-gray, calcareous sandstones and limy flagstones and siltstones. The Britton Formation is the middle Eagle Ford unit and consists primarily of dark-brown shale with many interbedded bentonite layers. At the top of the Eagle Ford in the Dallas area is the Acadia Park Formation which is a limy shale with subordinate siltstone. In the subsurface to the east of Dallas, these units can be traced into the Tyler Basin (which is the deepest part of the East Texas embayment). From the Tyler Basin, the Eagle Ford becomes sandier in the up-dip direction toward Arkansas and contains increasing amounts of volcanic detritus. On outcrop in Arkansas and southeastern Oklahoma, sandy Eagle Ford equivalents consist in large part of volcaniclastic sediments.

Southward from Dallas, the Eagle Ford thins considerably toward the San Marcos Arch (Fig. 8-24). Consequently, in the Waco area it is about 60 m thick. Furthermore, Dallas-area stratigraphic units are not recognized in Waco; instead, the Eagle Ford there is divided into the Lake Waco Formation and overlying South Bosque Formation. The Lake Waco is itself subdivided into a basal flagstone unit (the Bluebonnet Member), a dark, silty shale (the Cloice Member), and an upper, silty limestone with prominent bentonite beds (the Bouldin Member). The South Bosque is a dark, laminated-to-marine mudstone and shale with lesser limestone containing a variety of small ammonites, *Inoceramus*, and other mollusks. The Lake Waco and South Bosque are also recognized in the Austin area, near the center of the San Marcos Arch, where the total Eagle Ford thickness is only 12.5 m (Pessagno, 1969). In Austin, the Lake Waco is primarily a limy shale and the South Bosque is a concretionary, dark, argillaceous shale. Farther south, moving off the San Marcos Arch toward the Rio Grande embayment, the Eagle Ford increases in thickness and, again, changes its stratigraphic nomenclature. In the region of the Rio Grande River, the Eagle Ford consists of only one unit, the Boquillas Flagstone, which is a limy, gray flagstone and calcareous siltstone overlain by marls and chalks. The Boquillas unconformably overlies the Buda Limestone (of the Coahuilan Series) with no intervening Woodbine unit and is overlain conformably by the Austin Chalk. In Mexico, too, Eagle Ford equivalents overlie Coahuilan or older rocks. The major Eagle Ford equivalent in Mexico is the Agua Nueva Formation, which is composed of dark shale, shaley limestone, and flagstones similar to the Texas Eagle Ford.

From the above stratigraphic complexity, one may sift several general observations. First, the Eagle Ford represents a continuation of the Upper Cretaceous transgressive sequence and records the onset of open-marine shelf conditions over much of the western Gulf Province. Second, the San Marcos Arch continued to be a relatively positive feature, primarily influencing thickness trends. And, third, volcanic activity, principally centered in Arkansas, continued to supply volcaniclastic detritus and windblown ash to a large part of the northwestern Gulf.

The Austin Chalk is a widespread marine unit composed generally of white, fossiliferous chalk and lesser marl. The Austin Chalk is but part of the thick and extensive chalks of Upper Cretaceous age found in many parts of North America (and, indeed, the Earth). It is at least partially correlative with Selma Group chalks of the eastern Gulf Province and the Niobrara Chalk of the western interior. Petrographically, all of these chalks are relatively simple, being composed primarily of micritic carbonates made up of foraminifera, coccolithophores, and assorted other microscopic organisms; accompanying these pelagic components are varying (but usually small) amounts of insoluble materials, principally very fine clay. Thus, the Upper Cretaceous chalks represent wide and enormously fecund seas of clarity and warmth. The Austin Chalk is of Coniacian through lower Campanian age, but there has been disagreement on the age of the upper Austin. It rests with slight disconformity on the Eagle Ford in most localities. Although considered a formation, the Austin is as stratigraphically complicated as is the Eagle Ford Group, being comprised of numerous members (Pessagno, 1969). In this description, we will describe the Austin's major lithologies and avoid the biostratigraphic and lithostratigraphic complexities.

In its type area, the Austin is composed of the following members (from bottom to top): (1) Atco Chalk (thickly bedded, indurated, white chalk with marl interbeds); (2) Bruceville Chalk Marl (white to gray chalk and buff-colored marl); (3) Vinson Chalk (massive chalk); (4) Jonah Limestone (chalky, bioclastic limestone); (5) Dessau Chalk (gray chalk with interbedded marl); (6) Burditt Chalk Marl (buff to gray marl, similar to marls in the Dessau); and (7) Big House Chalk (indurated, phosphatic chalk). As may be seen in Fig. 8-39, marls increase in number and thickness northward along the outcrop belt, i.e., toward the center of the East Texas embayment, where apparent water-depth during deposition was greatest. In the Austin and San Antonio region, coarser, more bioclastic chalk predominates, indicating a shallower, higher-energy environment associated with the San Marcos Arch. Pessagno (1969) used the Aus-

tin's microfauna to argue that almost the entire Austin Chalk is of Coniacian and Santonian (i.e., Austinian) age. Others, however (i.e., Young, 1963), have suggested that Dessau and younger units are of Campanian age based on ammonite biostratigraphy. The unconformity shown in Fig. 8-39 within the Austin (separating Dessau rocks from underlying units) may represent the post-Santonian lowering of sea level which has been observed elsewhere in the Gulf Province.

Volcanic activity in the south Arkansas area continued during Austin time and began in central Texas. In the subsurface of Louisiana and Arkansas, the Tokio Formation (Coniacian) consists of tuffaceous sands and shales; the volcanic fraction of the Tokio increases and coarsens up-dip (Murray, 1961). In the Austin and San Antonio area, volcanic detritus and bentonite occur in the Austin at several different horizons; volcanic activity in the same area continued into Taylor time. Associated with these volcaniclastic sediments in central Texas is a group of small volcanic plugs and laccoliths composed of nepheline basalt and phonolite (King and Beikman, 1975) which intrude rocks up to Austin and Taylor age. Near Austin is the Pilot Knob intrusion which has been termed a fossil volcano. As stated earlier, the origin and significance of these igneous rocks and volcaniclastic sediments are uncertain.

In Mexico, the main Austin equivalent is the San Felipe Formation which consists of gray, green, or blue micritic limestone with interbedded clays and volcaniclastic sandstones. The San Felipe is almost 600 m thick in the Tamaulipas Province of northeastern Mexico but thins both westward and southward.

The principal Campanian unit of the western Gulf Province is the Taylor Marl, which is composed of highly fossiliferous marine marls, calcareous shales, and chalks. In its type locality, the Taylor is composed of lower and upper marl members separated by the Pecan Gap Chalk. Both the "Lower Taylor" and "Upper Taylor" marls are buff-weathering, fossiliferous, gray to dark-gray, calcareous mudstones with numerous, well-preserved *Exogyra ponderosa* fossils. The Pecan Gap Chalk Member is an abundantly fossiliferous, bluish-gray, somewhat bituminous, argillaceous chalk with a fauna containing *E. ponderosa, Inoceramus, Baculites,* and many other mollusks. To the west of San Antonio, Taylor marls are partly or wholly replaced by massive reef deposits of the Anacacho Limestone. Volcanic activity continued during Taylor time in the Austin–San Antonio area and volcaniclastics in the Taylor are associated with the nepheline basalt intrusion mentioned earlier for the Austin Chalk. All of the Taylor units described above may be traced down-dip into the subsurface; the Pecan Gap, especially, is widely recognized.

In southwestern Arkansas, the lowest outcropping Taylor equivalent is the Brownstown Marl whose dark-

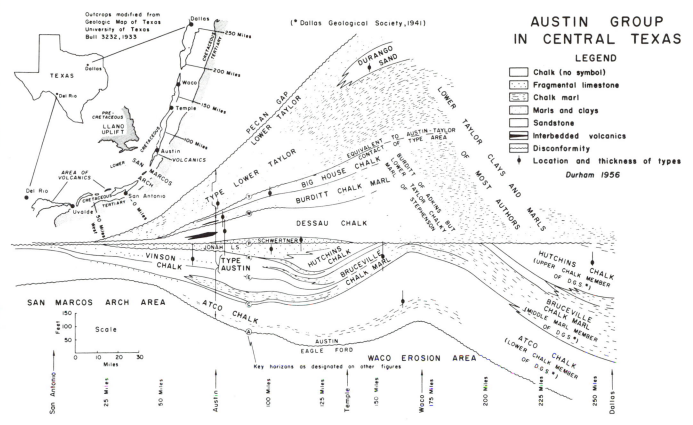

Fig. 8-39. Stratigraphic relations of Austin Chalk members along strike from Del Rio to Dallas; the unconformity through the middle of the Austin Chalk (shown as the datum) may represent the Santonian–Campanian boundary. (From Durham, 1957.)

gray, calcareous mudstones rest disconformably on the Tokio Formation. The Brownstown is richly fossiliferous with a similar fauna as that in the "Lower Taylor" Marl. Overlying the Brownstown disconformably is the Ozan Formation which is a micaceous, calcareous sand and sandy marl with considerable glauconite in its lower half. The Ozan is correlative with the Wolfe City–Buckrange sands in Texas. At the top of the Campanian Stage in southwestern Arkansas is the Annona Chalk, which is a massive, white chalk conformably resting on the Ozan. The Annona is correlative with the Pecan Gap Chalk and "Upper Taylor" Marl. In the subsurface of southeastern Arkansas, western Mississippi, and adjacent Louisiana, there is a considerable amount of water-lain volcanic detritus, indicating that volcanic activity continued there, too. Of particular interest in the subsurface is the presence of large reef complexes atop the Monroe uplift and Jackson (Miss.) dome. These reefs were initiated during Taylor time but persisted well into Navarro time. They are termed the Jackson Gas Rocks and Monroe Gas Rocks because they contain a large accumulation of natural gas.

In the Sabinas Basin of northern Mexico (Fig. 8-24), Taylor-correlative strata comprise the Difunta Formation (Murray, 1961). These strata consist of coal-bearing sandstones and shales which grade upward into detrital red beds and then to shallow-marine calc-arenites. The Difunta, thus, seems to define a generally transgressive sequence formed during rising sea level in the Campanian. Toward the southeast, in the Monterrey, Mexico area, the Difunta passes into the Mendez Shale which seems to represent a somewhat deeper, more open-marine environment.

The upper Gulfian stage is the Navarro, defined by the Navarro Group in Texas which is a complicated association of detrital and carbonate formations that vary widely in character (and terminology). In the Austin area, Navarro Group strata are relatively thin and are divided into only two formations: the lower, Corsicana Marl and a greenish, glauconitic mudstone termed the Kemp Clay. Both units are fossiliferous with a Maestrichtian fauna. In Waco, north of Austin, the Corsicana and Kemp are also present but several additional units are present beneath the Corsicana. The lower unit, termed the Neylandville Marl, is a greenish-gray to buff, fossiliferous, sandy, calcareous mudstone with scattered concretions. Overlying it (and beneath the Corsicana) is the Nacatosh Sand, which is a greenish fossiliferous sand with abundant glauconite; the Nacatosh is a very persistent unit, although somewhat variable in its character, and has been recognized in outcrop as far as southwestern Arkansas. In the subsurface, the Nacatosh is the most important oil- and gas-producing unit of Navarro age in the entire tristate (i.e., Texas, Arkansas, Louisiana) area (Murray, 1961).

Beneath the Nacatosh in Arkansas (and, therefore, equivalent to the Neylandville in Texas) is the Marlbrook Marl and the Saratoga Chalk (Pessagno, 1969). The Marlbrook conformably overlies the Annona Chalk and may be of Austin (i.e., Campanian) age. The unconformably overlying Saratoga is a hard, sandy, slightly glauconitic chalk with beds of soft, clayey sand. Above the Nacatosh is the Arkadelphia Marl which correlates with both the Corsicana and Kemp of Texas.

In the Sabinas Basin of the Rio Grande embayment are sedimentary rocks very different from the marly, chalky, marine-shelf strata of the Navarro Group in Texas. These are predominantly detrital and record a general marine regression in northeastern Mexico following the Campanian transgression (as seen in Difunta Formation strata discussed earlier). At the bottom of this Maestrichtian sequence is the Upson Clay, which is a dark, greenish-gray, glauconitic mudstone interpreted by Pessagno as having formed on a shallow, marine shelf. Conformably overlying the Upson is the San Miguel Formation, a shallow neritic to littoral sequence of crossbedded sand and calcareous mudstone containing an abundant molluskan fauna. The San Miguel is followed by the Olmos Coal, which is a rather typical nonmarine sequence with fluvial sands, marsh clays, and coal seams. At the top of the Maestrichtian sequence is the Escondido Formation which is a lagoonal to shallow-marine unit composed of crossbedded sandstones, some with reptile bones and clays containing large masses of *Ostrea cortex*, probably representing lagoonal oyster reefs or mounds. The greater amount of detrital sediment in Sabinas Basin strata may reflect the greater proximity of northeastern Mexico to the western Cordillera, which was an active Andean-type arc during the Late Cretaceous.

This discussion of Upper Cretaceous strata in the western Gulf Coastal Province should lead the reader to conclude that most sediments of the region were argillaceous and/or calcareous, marine-shelf deposits whose variations in lithic character and facies stratigraphy generally reflect small-scale sea-level fluctuations such as we described from eastern Gulf Province areas. In a way, the influx of coarse detritus to the Sabinas Basin during late Late Cretaceous time was a harbinger of the massive flood of detritus which would, shortly thereafter, permanently alter the sedimentology of the western Gulf region. Those detrital sediments, derived from areas affected by the Laramide orogeny, poured into the western and north Gulf Province soon after the

beginning of the Tertiary and will be discussed in the next chapter.

D. Zuni Sequence in the Cordillera

By the beginning of the Jurassic Period, an Andean-type magmatic arc had been established along nearly the entire Cordilleran continental margin following a variety of microplate collision–accretion events. Zuni-sequence rocks of the Southern Cordillera record the long-continued history of this magmatic arc throughout much of Jurassic and Cretaceous time. In western California, for example, sedimentary rocks of the Franciscan and Great Valley sequences and igneous rocks of the Sierra Nevada batholith preserve one of the best-known ancient convergent-margin suites on Earth. In the Northern Cordillera, several major displaced terranes, such as Wrangellia, were added to the continental margin. Indeed, as North America moved westward to northwestward following the breakup of Pangaea II, the leading margin "piled up" arc-crustal litter as if in front of a bulldozer blade. Some areas, e.g., southern Alaska, are practically "antique stores" of tectonic scraps (see Fig. 8-40). These accretion events and the deformational–sedimentational records they left comprise a major portion of Northern Cordilleran tectonic activity during Zuni time.

Before beginning our detailed look at Zuni Cordilleran tectonics, a brief cautionary note should be sounded. The late Mesozoic history of the North American Cordillera is replete with names of orogenies. A partial list includes: Inklinian, Alaska Range, Nassian, Nevadan, Sevier, Columbian, and Laramide. Each of these terms was coined to refer to a specific set of tectonic events which occurred during a given interval of time and which affected a specific area. In addition to being somewhat confusing, this plethora of names also creates a significant misconception about the nature of orogenesis. Namely, they *seem* to imply that orogenic events occur in neat spatial/temporal "packages" and, therefore, that the tectonic history of continental margins is episodic. But, along a typical convergent plate-boundary, subduction and related tectonic events (e.g., subduction-prism accretion, fore-arc basin subsidence, arc magmatism) are continuous, perhaps interrupted occasionally by collision with another arc or some other tectonic element. The point is that much of Cordilleran history (especially the late Paleozoic to Recent) is most accurately understood in terms of a continuum of tectonic activity, complicated, at times, by collisional events, shifts in the locus of subduction, and reorganization of relative plate-motions. In the following discussion, we will use the standard orogeny names because of their value as aids to communication but they should be understood as representing

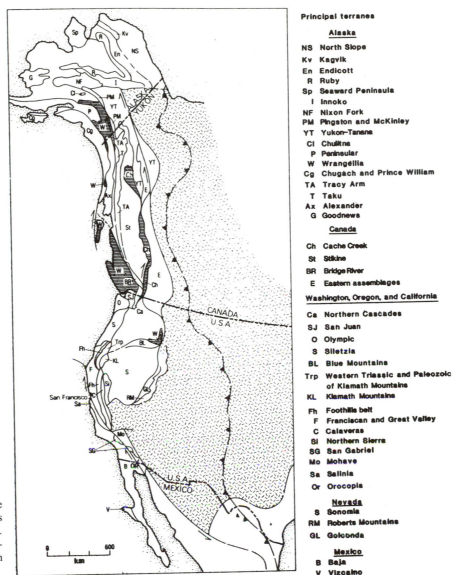

Principal terranes

Alaska

NS North Slope
Kv Kagvik
En Endicott
R   Ruby
Sp Seaward Peninsula
I   Innoko
NF Nixon Fork
PM Pingston and McKinley
YT Yukon-Tanana
Cl Chulitna
P   Peninsular
W   Wrangellia
Cg Chugach and Prince William
TA Tracy Arm
T   Taku
Ax Alexander
G   Goodnews

Canada

Ch Cache Creek
St Stikine
BR Bridge River
E   Eastern assemblages

Washington, Oregon, and California

Ca Northern Cascades
SJ San Juan
O   Olympic
S   Siletzia
BL Blue Mountains
Trp Western Triassic and Paleozoic
    of Klamath Mountains
KL Klamath Mountains
Fh Foothills belt
F   Franciscan and Great Valley
C   Calaveras
Si Northern Sierra
SG San Gabriel
Mo Mohave
Sa Salinia
Or Orocopia

Nevada

S   Sonomia
RM Roberts Mountains
GL Golconda

Mexico

B   Baja
V   Vizcaino

Fig. 8-40. The major displaced terranes of the North American Cordillera; barbed line indicates limit of Mesozoic–Cenozoic deformation. (From Ben-Avraham *et al.*, reprinted from *Science*, Vol. 213, p. 53, fig. 7, © 1981, American Association for the Advancement of Science.)

tectonic climaxes (perhaps related to times of collisional accretion of displaced, or exotic, terranes) rather than bursts of activity between long periods of inactivity. (Indeed, in some cases, an "orogeny" may not even reflect a tectonic climax but rather may be an artifact of intense study by generations of geologists in certain areas while other, less interesting or less hospitable areas remain poorly known.)

D.1. The Southern Cordillera: The Andean-Type Arc

D.1.a. The Jurassic Arc: Nevadan Orogeny

At the end of our discussion of Absaroka-sequence rocks in the Southern Cordillera, we described the accretion of the Kings–Kaweah ophiolite belt to the western margin of North America. One result of this event was a shift of subduction from an eastward position near the California–southern Nevada border to the western side of the Kings–Kaweah belt. During the Early and Middle Jurassic, a trench and related subduction complex probably existed in the approximate region of the Sierra Nevada foothills; magmatic activity associated with this subduction zone occurred along the eastern side of the Sierra Nevada region in what is today the Inyo Mountains (and vicinity). In that area, relatively small plutonic masses ranging in composition from intermediate to felsitic define the Inyo Mountains intrusive epoch (dated radiometrically at 180–160 Myr.; Evernden and Kistler, 1970).

The area behind the new arc was submerged beneath a shallow shelf-sea which was probably continuous with the western epeiric sea of the Jurassic discussed earlier in this chapter. Thus, most of the arc was separated from the North American mainland in the same way as the Sunda arc (Java and Sumatra) is today separated from the Southeast Asian mainland by the generally shallow South China Sea.

The Early and Middle Jurassic location of the subduction zone in the Sierran foothills was considerably eastward of its latest Jurassic and Early Cretaceous position. Thus, one must explain how the subduction zone jumped so far to the west. Schweichert and Cowan (1975) suggested that an east-facing arc collided with the west-facing Andean arc, resulting both in the westward jump of subduction (to the back side of the colliding arc) and in deformation and plutonism referred to as the Nevadan orogeny. Scraps of intra-oceanic-arc rocks juxtaposed against the Melones fault (part of the Foothills suture; see Fig. 7-59) were cited by Schweichert and Cowan as evidence of this collision. These rocks include pillow lavas, volcanic breccias, and other volcaniclastics of the Logtown Ridge Formation and slates, tuffs, and graywackes of the Mariposa Formation. Logtown Ridge rocks were considered by Schweichert (1978) to represent the colliding arc's volcanic base and Mariposa rocks to represent arc-related sediments, perhaps deposited in a back-arc basin. Apparently underlying these supracrustal rocks is an ophiolitic block, probably oceanic crust, called the Smartville Ophiolite. The whole sequence (Smartville, Logtown Ridge, Mariposa) is thought by Schweichert to be a displaced terrane accreted to the continental margin during the Late Jurassic.

All of this makes, as Davis *et al.* (1978) have said, a "conceptually pleasing" story but, as the reader might expect, there is doubt. Davis *et al.* believed that evidence of arc collision in the Lower Jurassic is weak and suggested instead that the Sierran arc was a single, broad feature (perhaps associated with a relatively low-angle subduction zone?). They said that large thrust faults associated with arc volcanics could be explained by intra-arc deformation; in other words, arc volcanics may not be exotic at all but, instead, may be thrust slivers of the original Foothills arc telescoped together by intra-arc compression. This view was elaborated by Saleeby *et al.* (1978) who said that the Nevadan orogeny may be considered the culmination of Jurassic intra-arc tectonics.

In the Klamath Mountains, too, the same dispute exists. Schweichert and Cowan (1975) and Irwin *et al.* (1978), among others, have argued the case for arc collision. Davis *et al.* (1978) have argued the opposite. The controversy in the Klamaths centers on interpretation of rocks in the Western Jurassic terrane, which form the upper thrust sheet over rocks of the Franciscan sequence just to the west. Of importance in this discussion are three units: the Rogue Forma-

tion, a sequence of intermediate to silicic metavolcanics which probably represents a volcanic arc; the Galice Formation, a metasedimentary unit composed mainly of phyllites, slates, and metagraywackes; and the Josephine ophiolite, primarily a pillowed spilite.

The Josephine–Rogue–Galice sequence is similar to the Smartville–Logtown Ridge–Mariposa sequence in the western Sierra Nevada and was considered by Schweichert (1978) to be correlative units of the same collided arc. The arc-collision school of thought presumes the Josephine to be a sliver of oceanic crust caught between the collided plates (similar to Cache Creek rocks in British Columbia) and the Rogue–Galice sequence to represent the collided, exotic arc. The Galice is considered to have been deposited within a back-arc basin. Davis *et al.*, on the other hand, argued persuasively that these units were part of a single, broad arc developed across a sutured plate boundary. The Josephine ophiolite, in this scheme, is considered to represent marginal back-arc basin crust and Rogue–Galice rocks to represent the continental margin arc. The whole mass is thought to have been telescoped together by intra-arc tectonics. This dispute has not yet been arbitrated.

D.1.b. The "Classic" Cretaceous Arc

Regardless of the mechanism of its inception, by latest Jurassic time a trench and related subduction complex had developed in the area occupied today by California's Coast Ranges (see Fig. 8-41). Activity along this subduction zone continued steadily throughout the Cretaceous and into the Early Tertiary. Strongly deformed rocks of the subduction complex are known as the Franciscan Sequence. Magmatic activity related to Cretaceous subduction is recorded by the great mass of plutonic igneous rocks collectively termed the Sierra Nevada batholith. In between the Sierran arc and the Franciscan trench was a broad fore-arc basin in which were deposited detrital sediments derived from the Sierras. These fore-arc basin deposits are termed the Great Valley Sequence. Detritus shed cratonward from the Sierran arc accumulated in the Cordilleran foreland basin whose complex strata (e.g., Indianola Group, Mancos Shale) were discussed earlier in this chapter. On the western side of the foreland basin is a major fold-thrust belt which was active penecontemporaneously with sedimentation. Deformation associated with this belt is termed the Sevier orogeny.

i. The Franciscan Complex. Much of California's Coast Ranges are composed of rocks generally referred to as the Franciscan Sequence (see Fig. 8-41). These rocks extend northward into Oregon, where they are known as the Dothan Formation, and southward into Baja California. The Franciscan Complex contains some of the most complexly deformed and interesting lithologies of North

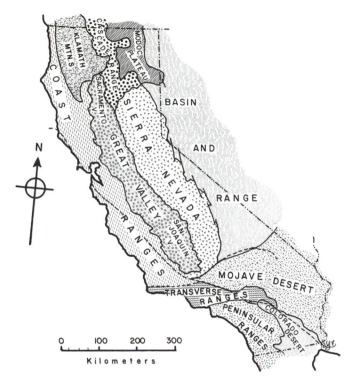

Fig. 8-41. Geological province map of California; dashed lines indicate major faults.

cases, broken formations are not really "blocks" but rather tectonic or depositional intercalations within mélanges (Aalto, 1981).

Franciscan rocks occur in discrete, sheetlike units separated from each other by major thrust faults (Cowan, 1974). Adjacent sheets may bear neither lithological similarity nor apparent stratigraphic relationships to each other; indeed, they often appear to represent greatly different depositional environments and histories of deformation and metamorphism. In his "anatomy" of the northern California Coast Range, Maxwell (1974) illustrated this complex structural pattern and showed that each sheet was a separate thrust sliver intercalated into the stack. Of significance is the fact that oldest rocks are generally disposed in *upper* parts of the stack whereas younger slivers are *lower*.

A final observation is needed before attempting a sedimentary–tectonic interpretation of Franciscan rocks: they have no visible basement. At every exposure, the lowermost contact appears to be faulted rather than depositional. It is generally presumed that oceanic crust exists somewhere beneath Franciscan rocks but it is nowhere exposed.

The sedimentary–tectonic model agreed upon by most workers (e.g., Dickinson, 1976) is that Franciscan rocks represent oceanic crust and various oceanic sediments (both pelagic and turbidites) which were intercalated together at a trench. Sediments and pieces of the crust were "scraped off" the eastward-subducting oceanic plate and underthrust beneath older, previously "scraped-off" sheets. The whole mass of deformed, "off-scraped" rock (the subduction complex) comprised the inner slope of the trench and is associated with trench-slope deposits. In some areas today, e.g., the Andaman Islands of the Indian Ocean, the subduction complex has grown so thick that its top is subaerially exposed as a chain of islands (called the outer arc) lying seaward of the volcanic arc. Turbidites may have been derived from the inner trench-slope itself (obviously an area of intense seismic activity). Large submarine slumps and slides would probably also have occurred periodically along the inner trench slope, resulting in olistostromal masses. Some exotic blocks of the tectonic mélange may have originated as blocks within an olistostrome. Other exotic blocks must have been tectonically emplaced into the mélange, perhaps by dynamic (rather than thermal) convection within the subduction complex itself as suggested by Cloos (1980). Much of the sand in Franciscan turbidites was probably derived from the Sierran arc to the east and funneled out onto the ocean floor through submarine canyons cut across the subduction complex.

ii. The Great Valley Sequence. Underlying much of California's Great Valley (whose northern part is called the Sacramento Valley and whose southern part is the San Joaquin Valley; see Fig. 8-41) is a thick mass (up to

America. Indeed, for a long time their exact nature was highly speculative and controversial. We will not discuss previous hypotheses but will consider only contemporary thinking on Franciscan rocks (a good discussion of the history of thought on Franciscan rocks may be found in Bailey *et al.*, 1964).

The Franciscan Sequence is a chaotic assemblage of graywacke, shale, siltstone, pillow basalt, and radiolarian chert in which are embedded exotic blocks of serpentinite, gabbro, and other lithologies commonly associated with ophiolites. Exotic blocks may be many tens of meters to up to several kilometers in size and are usually embedded in a pervasively sheared, fine-grained, argillaceous matrix. In most areas, the whole mass has undergone low-grade dynamic metamorphism. Commonly, exotic blocks have been subjected to even higher grades of metamorphism, also of the load type but containing minerals such as lawsonite, pumpellyite, and glaucophane; these minerals define the upper, blueschist facies of load metamorphism. Because of their higher grade of metamorphism, Cowan (1974) stated that these exotic blocks were probably metamorphosed *prior* to emplacement within the chaotic Franciscan mass. In some instances, exotic blocks up to several kilometers in size were hardly metamorphosed or deformed at all; such blocks may contain relatively simple stratigraphic sequences and are called "broken formations." In many

15,000 m) of sedimentary rock referred to as the Great
Valley Sequence. Great Valley strata range in age from
Upper Jurassic to Lower Tertiary. Upper Jurassic rocks
comprise the "Knoxville" Formation (in quotation marks
because it has lost its formal status as a result of recent
mapping which has shown the original definition of the
"Knoxville" to be unsound); Cretaceous rocks form Pas-
kenta, Horsetown (both Lower Cretaceous), and Chico (Upper Cretaceous) formations. These names are old now and
are not in general use because detailed studies have greatly
increased their terminological complexity.

   Most of the Great Valley Sequence is composed of
thin, graded graywacke and siltstone interleaved with dark,
fissile shale. Such rocks are well known to the reader by
now as flysch, probably of turbidity-current origin. In their
eastern exposures, lower Great Valley strata appear to lie
depositionally on ophiolitic oceanic crust. To the west,
however, they rest on upturned Coast Range ophiolite but
the nature of the contact is questionable. While much of the
evidence suggests a depositional contact, most authors
(e.g., Maxwell, 1974) have suggested that it is a faulted
contact, along which Great Valley rocks have been thrust
over the Franciscan Sequence. Great Valley strata overlie
coeval Franciscan rocks but (as we described earlier) Franciscan rocks decrease in age downward (due to subduction-
related underthrust-intercalation); further, Franciscan rocks
contain faunas of the same approximate age as those of the
Great Valley Sequence. Thus, Franciscan and Great Valley
rocks are probably penecontemporaneous even though
Great Valley strata are today structurally higher than Franciscan strata. Stratigraphic relations with Sierra Nevada
rocks in the east are much less complicated: Great Valley
strata rest depositionally on and progressively overstep
eroded Sierran rocks to the north and east. It was originally
felt that Great Valley rocks were everywhere younger than
Sierran rocks but agreement now is general (e.g., Ingersoll,
1978a) that the two units are penecontemporaneous, with
the eroding arc providing detritus for the slowly transgressing Great Valley Sequence.

   As mentioned, most Great Valley strata appear to be
turbidites with related debris-flow deposits. Ingersoll
(1978b) has presented an elegant interpretation of Great
Valley rocks as the remains of a submarine-fan complex
which grew out from the Sierran arc into the fore-arc basin.
During the Late Jurassic, the Franciscan subduction complex had not yet grown sufficiently large to act as a sediment barrier; thus, Sierra-derived, volcaniclastic-rich sediments were incorporated into Franciscan strata (see Fig.
8-42). Later, during the Cretaceous, almost all Sierra-derived sediments were trapped in the fore-arc basin. By the
Late Cretaceous, more arkosic, fluvial–deltaic sedimentary
wedges in the east prograded westward from the arc into the

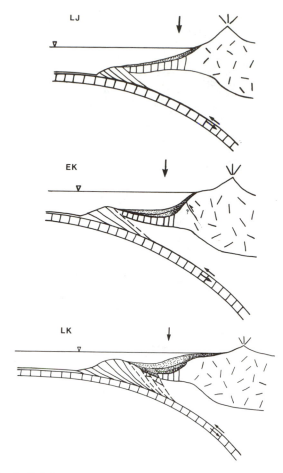

Fig. 8-42. Paleotectonic evolution of Franciscan, Great Valley, and Sierra
Nevada rocks during the Late Jurassic and Cretaceous. Note the progressive enlargement of the Great Valley fore-arc basin as the locus of
subduction shifted westward and arc magmatism shifted eastward. (From
Ingersoll, 1978a, reproduced with permission of the Pacific Section, Society of Economic Paleontologists and Mineralogists.)

fore-arc basin while deep-water sedimentation continued in
the western part of the basin. By the Early Tertiary, the
fore-arc basin was nearly filled so that shallow-water to
terrestrial sediments accumulated over much of the area.

iii.   The Sierra Nevada Batholith.   The Sierra
Nevada Mountains of eastern California (see Fig. 8-41) are
composed predominantly of intrusive igneous rocks ranging
in age from Jurassic to Late Cretaceous. This mass of igneous rocks is called the Sierra Nevada batholith but the term
is misleading because the "batholith" is actually a complex
of many smaller plutons with a wide variety of shapes and
sizes. Igneous rocks of virtually every composition from
gabbro to granite are present in the Sierra Nevada batholith
but the dominant lithologies are quartz monzonite, granodiorite,   and   quartz   diorite   (all   essentially   un-

metamorphosed). Plutonic rocks of Jurassic age are disposed both east and west of the high, central region of the Sierra (which is composed of Cretaceous plutonic rocks forming a continuous igneous mass 40 km wide and running the entire length of the range). Table II shows the timings of major plutonic events in the Sierra Nevada and related areas.

Jurassic plutons to the *east* of the Sierras were emplaced from 180 to 160 Myr. ago during the Inyo Mountains intrusive epoch, probably associated with subduction along the western Sierran foothills trench. Late Jurassic plutons to the *west* of the Sierra axis were emplaced between 148 and 132 Myr. ago as part of the Yosemite intrusive epoch. These plutonic bodies are relatively small, equidimensional bodies in their western occurrences but become larger toward the east, merging finally (as in the lower end of the Yosemite Valley) into large, continuous plutons such as the El Capitan Granite. For the most part, Yosemite-phase rocks are slightly less felsitic than Cretaceous rocks, being dominantly quartz diorites and granodiorites.

Cretaceous plutonic rocks of the Sierra Nevada were emplaced during two main intrusive epochs. The Huntington Lake phase (121–104 Myr; Evernden and Kistler, 1970) was responsible for emplacement of lesser masses of quartz monzonite and granodiorite. The main body of the Sierra Nevada batholith was emplaced during the Cathedral Range phase (90–79 Myr.; Evernden and Kistler, 1970). Rocks of the Cathedral Range phase are termed the Tuolomne Intrusive Series and are generally granitic to granodioritic (King and Beikman, 1975).

Plutonism of Yosemite, Huntington Lake, and Cathedral Range phases was probably due to magmatic activity

Table II. Comparison of Timings of Major Plutonic Events in the Southern Cordillera[a]

Age (Myr.)	Peninsular Range	Sierra Nevada	Klamath Mountains	Idaho batholith
75		75 ⌐ Cathedral Range 90 ⌐		70 ⌐
100	Baja 100 ⌐ 109 ⌐ ⌐115 S. Calif. 120 ⌐	104 Huntington Lake 121 ⌐		100 ⌐
125		127 132 ⌐ Yosemite 148 ⌐	140 ⌐ 145 ⌐ 155 ⌐	
150				
175		160 ⌐ Inyo Mountains 180 ⌐	165 ⌐ 167 ⌐	
200				

[a]Columns are arranged from most southerly (on the left) to most northerly (on the right). Plutonic events are formally named only in the Sierra Nevada region. See text for discussion.

associated with subduction at the Franciscan trench. Earliest magmatism (Yosemite) primarily affected the western Sierra Nevada area. During the Cretaceous history of the arc, the locus of plutonism gradually moved eastwards (see Fig. 8-43). This eastward growth may be seen in both Huntington Lake and Cathedral Range rocks, and parallels the general widening of the arc-trench gap (discussed later in this section).

Before leaving the subject of Cretaceous magmatic activity in the Southern Cordillera, mention should be made of two other major batholithic masses: the Peninsular Range batholith and the Idaho batholith. The Peninsular Range batholith, like the Sierra Nevada batholith, is composed of many plutons of composition ranging from gabbro to granite. Emplacement apparently occurred during a single epoch. Peninsular Range plutons intrude rocks of up to Albian age but are overlain nonconformably by rocks of Campanian age. Radiometric dates of Peninsular Range rocks from southern California (see Fig. 8-41) range from 120 to 109 Myr. whereas dates from Baja California are 115–100 Myr. (Armstrong and Suppe, 1973). Thus, they correlate well with the Huntington Lake phase of Sierra Nevada plutonism.

The Idaho batholith is a huge mass of granodiorite and quartz monzonite, similar to other batholiths discussed

above. Unlike the others, however, it is *not* composed of numerous smaller plutons but is, rather, a single, large mass of nearly uniform composition (King and Beikman, 1975). This statement should be viewed with caution, however, because large areas of the Idaho batholith are poorly known—or unknown. Imprecise dating places its intrusion at around 100–70 Myr. Both Peninsular Range and Idaho batholiths were probably formed by subduction-related magmatism, quite likely along the same convergent plate-boundary as Sierra Nevada rocks. Thus, the three largest Cretaceous batholiths of the Western Cordillera probably define a nearly continuous Andean-type magmatic arc.

Review of the dates of intrusive epochs in the Peninsular Range, Sierra Nevada, Klamath Mountains, and Idaho batholiths (see Table II) will reveal an interesting fact: while igneous activity in any given area is episodic, igneous activity along the whole Andean arc was nearly continuous from (at least) the onset of Inyo Mountains phase plutonism to the end of Idaho batholith emplacement. Thus, we may reaffirm our earlier point that tectonic activity along the Cordilleran margin was continuous rather than episodic.

iv. The Sevier Orogenic Belt.
To this point in the discussion, we have described the Cretaceous trench and subduction complex, fore-arc basin, and magmatic arc of western North America. The final elements of Cretaceous paleotectonics that remain to be discussed are the retroarc foreland basin and the fold-thrust belt of the back-arc area.

The Cretaceous retroarc foreland basin is a complex mass of cyclic, detrital molasse deposited on the western margin of the craton. Sediments were derived from the arc orogen and adjacent fold-thrust belt and carried in fluvial systems to the epeiric sea of the western interior. The stratigraphy and depositional history of this foreland basin were discussed earlier in this chapter in the section dealing with the craton.

The fold-thrust belt, between the arc and the foreland basin, is represented in the Southern Cordillera by the Sevier orogenic belt (see Fig. 8-44). Originally described by Armstrong (1968), the Sevier belt is a zone of major, east-directed thrust faults which shoved up strata of the old Paleozoic continental-margin prism to produce a major detrital sourceland from which molasse wedges grew eastward into the foreland basin. There is considerable stratigraphic evidence that thrusting and sedimentation were penecontemporaneous. For example, Indianola conglomerates shed from earlier thrust sheets are themselves overthrust by Paleozoic rocks. That these later thrusts are of Cretaceous age may be seen in the fact that allochthonous rocks are overlain with angular unconformity by Upper Cretaceous strata, e.g., the Price River Formation. Several authors (e.g., Burchfiel and Davis, 1975; Dickinson, 1976; as well as

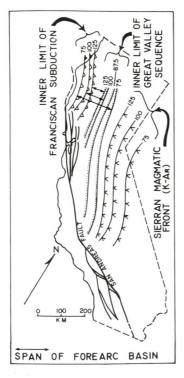

Fig. 8-43. Widening of the arc-trench gap during the Cretaceous and concomitant eastward overlap of the Great Valley Sequence. (From Ingersoll, 1978a, reproduced with permission of the Pacific Section, Society of Economic Paleontologists and Mineralogists.)

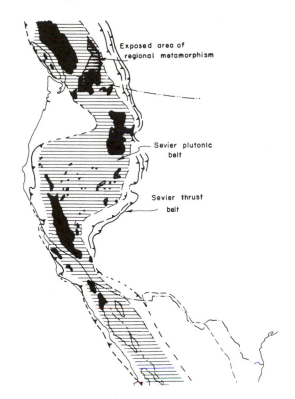

Fig. 8-44. Cordilleran orogenic belt during the Sevier orogeny. Sevier-age plutons shown in black; metamorphism related to the Sevier orogeny shown by horizontal ruling; barbed lines indicate Sevier thrusting. (From Burchfiel and Davis, 1975, *American Journal of Science*, Vol. 275-A, Fig. 5, p. 379.)

northward extension of the Sevier belt; see Fig. 8-44). It is instructive to compare Sevier deformation with deformation in the Allegheny belt (see Fig. 7-35) and the Ouachita belt (Fig. 7-43). The overall similarity is striking and probably hints that some generalities can be made about the response of a layered, anisotropic, sedimentary prism to unidirectional stress. In this regard, it is interesting to note that Burchfiel and Davis (1975) stated that Sevier thrusting in southeastern California occurred primarily within crystalline basement rocks (because the Paleozoic continental-margin prism is missing in that area due to the early Mesozoic truncation event). The crystalline rock behaved more isotropically than would a layered sedimentary prism and, consequently, thrusting is not primarily of the décollement type. Further, displacement along these faults is considerably less than displacement along Sevier décollement-faults to the north.

It remains for us to describe the mechanism of crustal shortening which caused Sevier thrusting and to relate it to Andean-arc activity. [This discussion is a blend of ideas discussed in Dickinson (1974b, 1976) and Burchfiel and Davis (1975).] In brief, Sevier deformation may be thought of as the result of incipient underthrusting of rigid continental crust beneath the plastic, magmatic arc. Recall that the North American plate, during the Cretaceous, was actively moving west- to northwestward, converging with an oppositely directed oceanic plate. The oceanic plate subducted beneath the continental plate and, partially melting, provided magmas which rose into the continental margin, causing it to grow thicker and rise isostatically. Intense heatflow accompanied magmatic activity, causing the arc orogen to heat up (which, of course, would also add an additional upward isostatic component). The fact that the arc and associated back-arc areas would be relatively hot is very important in our discussion because rocks there would behave in a more ductile manner than cooler, cratonic rocks. As the North American plate moved toward the arc, the effect was for rigid cratonic crust to be squeezed into and partially underthrust beneath the rising, thermally softened, arc orogen. One result was to squeeze mobilized, intensely metamorphosed rock of the arc orogen up and over the rigid crust that was being forced into the arc, resulting in an infrastructural belt, i.e., a belt of mobilized, deeply buried metamorphic rocks. Crustal shortening occurred deep beneath the rear of the arc orogen as the consequence of the squeezing of infrastructure rocks over rigid crust. Higher up, near the surface, this shortening was reflected by thrusting and folding. One may view this process *not* as thrusts being shoved eastward toward the craton but rather as continental crust being shoved westward *beneath* the sedimentary prism. East-directed Sevier thrusts are, therefore, somewhat the continental analogues of west-directed faults within the Franciscan Sequence.

Armstrong) have observed that the locus of Sevier deformation shifted eastward throughout the Cretaceous, paralleling the eastward migration of Sierra Nevada arc magmatism.

Sevier faults are major, low-angle thrusts (such as the Gass Peak thrust in southern Nevada to southwestern Utah or the Willard and Charleston thrusts in Northern Utah) which define a continuous zone of deformation from southeastern California to Idaho (> 800 km). Armstrong estimated total shortening along the Sevier belt to be at least 65 km and probably as much as 100 km. The thrusts are essentially lystric in nature and were considered by Armstrong to be rooted in a master décollement within Eocambrian detrital sediments near the bottom of the Paleozoic continental-margin prism (see Fig. 8-45). Sevier thrusts rise from the master décollement up through a series of tectonic ramps (in competent strata) and second-order décollements (in incompetent strata). Armstrong stated that major second-order décollement-zones include Middle Cambrian shales, Mississippian shales, and Middle Jurassic shales and evaporites. We should note here that most of the above discussion applies equally well to thrusting within the Columbian orogenic belt in western Canada (which is the

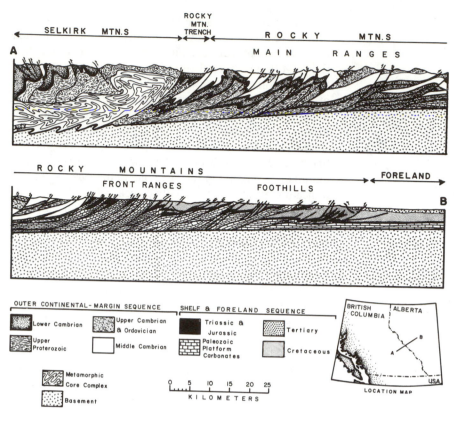

SELKIRK MTN.S   ROCKY
                                MTN.
                                TRENCH   R O C K Y   MTN.S

A   M A I N   R A N G E S

R O C K Y   M O U N T A I N S   FORELAND
     FRONT RANGES   FOOTHILLS

                                                                        B

OUTER CONTINENTAL-MARGIN SEQUENCE   SHELF & FORELAND SEQUENCE

Lower Cambrian   Upper Cambrian   Triassic &   Tertiary
                      & Ordovician   Jurassic

Upper   Middle Cambrian   Paleozoic   Cretaceous
Proterozoic   Platform
                                            Carbonates

Metamorphic
Core Complex   0   5   10   15   20   25

Basement   K I L O M E T E R S

BRITISH   ALBERTA
COLUMBIA

LOCATION MAP

Fig. 8-45. Deformation in the western Canadian Rocky Mountains associated with the Sevier orogeny. Note the master décollement above the Precambrian basement and the presence of second-order décollements higher in the sequence. Note also the lystric geometry of the faults. (Modified from cross section by Price and Mountjoy in Douglas, 1976.)

v. Later Cretaceous Tectonic Evolution. As time passed during the Cretaceous, the Franciscan trench was shifted westward as the subduction complex grew while the locus of arc magmatism moved eastward (see Fig. 8-43). Thus, the width of the arc-trench gap and, therefore, the Great Valley fore-arc basin, increased. The shift of the trench was simply due to progressive accretion of oceanic sediments to the subduction complex. Eastward migration of the arc is less easily explained; Dickinson (1976) suggested that it might have been due to flattening of the Benioff Zone, possibly caused by westward movement of the North American plate over the subducted oceanic lithosphere. This may also have been the cause of the eastward shift of Sevier deformation although Burchfiel and Davis (1975) argued that eastward Sevier migration could be explained by increase in the size of the infrastructural belt.

Summarizing our discussion of Southern Cordilleran orogenesis during the Cretaceous, we would stress that relatively simple, convergent-plate tectonics may explain the various structural and petrological events from California's Coast Ranges all the way to the front of the Sevier belt. Such extensive deformation seems, at first glance, to require a more complicated explanation. But, of course, the best example of an Andean-type orogen is the Andes, where tectonic activity today mirrors almost exactly Cretaceous activity in the western United States with the exception that no extensive fore-arc basin presently exists in the Andes.

D.2. The Northern Cordillera

It should come as no surprise that significantly different tectonic histories might complicate the development of a 5000-km-long continental margin or that the Northern Cordillera (consisting, for the purposes of this discussion, of Washington, western Canada, and Alaska) might have experienced sedimentary–tectonic events which bear little relationship to Southern Cordilleran events. The relatively straightforward story of Andean-arc tectonics developed in the previous section may be applied with confidence only to the California area. To the north, Zuni sedimentary–tectonic events were dominated by the accretion of displaced terranes and by deposition within a variety of successor basins.

The concept of successor basins in the Northern Cordillera has been elaborated by many geologists (some of whose ideas will be heard as we proceed); there is, however, an excellent overview of the subject by Eisbacher (1974) which we commend to the reader. In essence, a successor basin is a sedimentary basin developed *on top of* an older basin which has been strongly deformed and intruded; i.e., it is a new basin *succeeding* an older basin. As an illustration, consider a volcanic arc approaching and colliding with a continental margin. Following accretion of the arc to the continent, a new subduction zone develops; magmatism begins along a new arc. The old arc, consisting of volcanic and volcaniclastic rocks, perhaps with a few platform carbonates, all strongly deformed by collision, would subside rapidly because it would be cooling after cessation of magmatism. Into that subsiding area, the successor basin, would be carried detrital sediments from the buckled-up collisional-orogen on one side and/or from the new arc-orogen on the other. At a still later time, the successor basin may be strongly deformed itself by yet another collision.

From the preceding discussion, it is clear that one result of the accretion of displaced terranes (such as volcanic arcs) is the development of a kind of cyclic history in which arc orogens collide with the continent, are deformed, and subside, forming a successor basin which is, itself, later deformed by a new collision—and the cycle repeats. The perception of cycles in the tectonic development of a continental margin is not new, but had been dismissed by some as a misleading way to view structural developments, i.e., as the result of a rigidly deterministic set of processes and events. The model of successive displaced-terrane collisions provides a sort of redefined tectonic cycle, understandable in the context of plate tectonics and susceptible to considerable specific tectonic variation. It is, thus, a useful concept and much of the following discussion is molded around it.

D.2.a. Jurassic Events: Canada

During the latest Triassic, collisional accretion of the Stikine block (discussed in Chapter 7) to the outer margin of the North American block caused deformation of oceanic-crustal rocks (Cache Creek Group) and magmatic arc rocks (Takla Group) which had been caught between the colliding crustal elements (see Fig. 8-46). This deformational event is called the Inklinian orogeny and resulted in uplifts which shed detrital sediment both eastwards into a successor basin and westwards toward the ocean.

The main volcanic arc of Stikinia prior to collision had been located on the eastern side, associated with westward subduction of the Cache Creek oceanic lithosphere. Rocks of that arc comprise the Stuhini Group (Davis *et al.*, 1978). When collision ended Stuhini (and Takla) subduction, a new arc-trench complex was developed on the western side of Stikinia, associated with eastward subduction. Rocks of the new arc comprise the Hazelton Group. Tectonic activity associated with the Hazelton arc dominated the Canadian Cordillera throughout much of the Jurassic. The Hazelton Group consists of varicolored volcanic rocks, ranging in composition from basalt to rhyolite, as well as tuffs, volcaniclastics, and minor limestones (Tipper and Richards, 1976). They probably represent an archipelago of volcanic islands and low carbonate banks. The back-arc area, called the Hazelton trough by Tipper and Richards, received sediments not only from the arc but also from the uplifted Inklinian orogen to the east. South of the Stikine block,

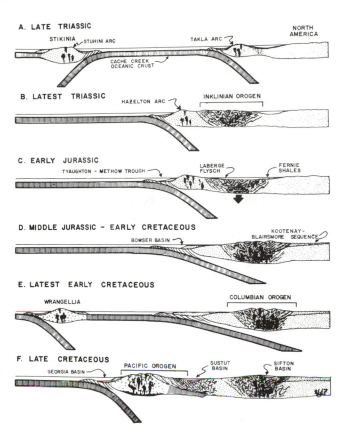

Fig. 8-46. Tectonic evolution of the Canadian Cordillera from the Late Triassic to the Late Cretaceous; see text for discussion.

volcanic activity, preserved today in the Nicola Group (see Chapter 7), had begun prior to Stikine accretion. Indeed, Davis *et al.* (1978) believed that the Nicola arc was continuous with the Takla arc farther north. Whereas Takla activity ceased following Stikine collision, the Nicola arc continued as before, a west-facing arc at the edge of the continent.

The trench and subduction complex which must have been associated with the Hazelton–Nicola arc has been difficult to recognize because of the great geological complexity of the area. One view (Dickinson, 1976) is that Hozameen Group ophiolites constitute oceanic crust which has been telescoped together with pelagic and hemipelagic sediments into a subduction complex.

More clearly understood is the fore-arc basin, whose deposits in southern and central British Columbia define the Tyaughton–Methow trough, which is composed of the Tyaughton Basin in the north and the Methow Basin in the south, separated today by faults (such as the Yalacom and Fraser faults; see Fig. 8-48). Earliest sediments of the trough (Jurassic to Early Cretaceous) are represented by the Newby and Ladner Groups which are primarily turbidite units, apparently deposited as submarine-fan complexes in a basin which deepened westward (Tennyson and Cole, 1978). Ladner rocks rest depositionally on ophiolitic rocks of the Coquihalla belt. To the east of the Tyaughton–Methow Trough, lying on Nicola basement, is the Ashcroft Formation whose conglomerates, sandstones, and shales appear to be the nearshore and subaerial facies of deeper-water Ladner rocks to the west. Taken together, Tyaughton–Methow deposits and the Ashcroft Formation probably represent a broad fore-arc basin. Volcaniclastic debris in both units indicates derivation from the Hazelton–Nicola arc. The south end of the Methow Basin is faulted off against the Ross Lake fault, but during the Jurassic, it must have continued farther south. Davis *et al.* (1978) correlated Tyaughton–Methow strata with the Nooksack Formation of the Northern Cascade Range in Washington; Nooksack and Ladner strata are petrographically similar and seem to record the same general historical development. Nevertheless, this correlation is highly debatable and dissent

abounds, revolving primarily around the nature of Cascades core complex, i.e., "Cascadia." We will return to this subject later in the section. Northern extensions of the fore-arc basin may be represented by deposits of the Bowser, Tantalus, and Dezadeash basins (see Fig. 8-47) which we will describe in more detail after discussion of the eastern successor basin.

Before Stikine-accretion, the Takla–Nicola arc had been located at the edge of the North American plate, separated from the shallow, continental shelf by a deep back-arc basin (Davis *et al.,* 1978; see Fig. 8-46). Following Stikine-accretion, large amounts of coarse detritus were carried out of the Inklinian orogen and deposited within the former back-arc basin. Resulting deposits of the Laberge Assemblage (Eisbacher, 1974) rest on deformed arc and back-arc rocks in western exposures. To the east, however, in

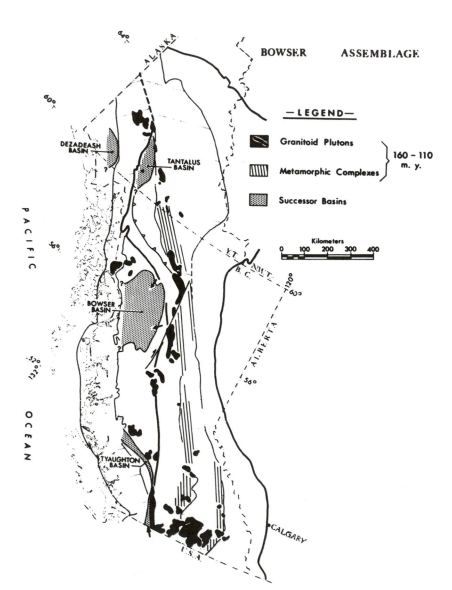

Fig. 8-47. Middle Jurassic successor basins in Canadian Cordillera and inferred contemporaneous uplifts of plutonic and metamorphic terranes; arrows show main paleocurrent directions. (From Eisbacher, 1974, reproduced with permission of the Society of Economic Paleontologists and Mineralogists.)

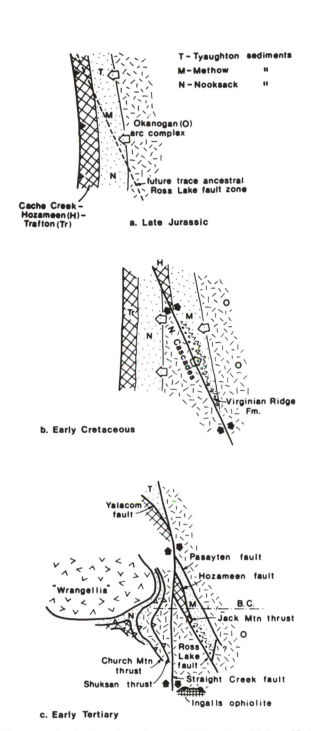

Fig. 8-48. Model of tectonic evolution of the Tyaughton–Methow–Nooksack fore-arc basin, showing transform-faulting origin of Cascadia displaced terrane and subsequent deformation due to collision with the Wrangellia terrane (see text for further discussion). (From Davis *et al.*, 1978, reproduced with permission of the Pacific Section, Society of Economic Paleontologists and Mineralogists.)

central and eastern exposures the contact with underlying strata is conformable, indicating that sedimentation continued there during deformation and uplift to the west. Laberge sediments were coarse gravels, graywacke sand, and mud; depositional environments range from high-energy, channeled, shallow-marine in the west (associated with the rugged coast of the Inklinian orogan) to deep-water slope and submarine-fan complexes. Deep-water deposits probably represent turbidity-current and debris-flow transportation. Laberge conglomerate clasts are composed of andesite, granodiorite, limestone, graywacke, chert, and quartzite, reflecting derivation from a complex volcanic–plutonic–sedimentary terrane.

Farther eastward, shallow continental-shelf limestones and quartz arenites of Triassic age are overlain by fine, dark shales, dark, argillaceous limestones, calcareous siltstones, and chert of the Fernie Formation. Thin layers of tuff are also present in the Fernie, probably blown eastward from the Hazelton–Nicola arc. Eisbacher (1974) said that Fernie sediments represent a starved basin along a continental slope and he suggested that Fernie deposits may be the distal facies of Laberge rocks. Thus, Laberge and Fernie deposits are considered to define a broad successor basin, called the Laberge–Fernie basin, which developed over former back-arc and continental-margin areas following the Inklinian orogeny.

During the later Jurassic, arc magmatism migrated eastward (in an analogous manner as Cretaceous, Sierra Nevada arc migration discussed earlier) and large plutons were emplaced into the Omineca crystalline belt (the infrastructural core of the Columbian orogen). Intrusion was accompanied by regional metamorphism and deformation during which rocks of the Laberge Assemblage were subjected to penetrative deformation associated with southwest-verging folds. This tectonic event is termed the Nassian orogeny (Douglas *et al.*, 1976) and is similar to the Nevadan orogeny discussed earlier. Uplift of the Omineca belt and Laberge rocks during the later Jurassic may be seen in the shallowing of the sea in the Fernie area as detritus from the orogen filled in the foredeep and by introduction of large volumes of detritus to the fore-arc basin system.

During the Late Jurassic, the fore-arc basins [Dezadeash, Tantalus, Bowser, Tyaughton–Methow (and Nooksack?); see Fig. 8-47] received abundant east-derived detrital sediment, called the *Bowser Assemblage* by Eisbacher (1974). The detrital composition of these sediments reflects derivation from supracrustal (i.e., volcanic and sedimentary) sources, indicating that the crystalline infrastructure of the orogen had not yet been unroofed. In eastern basin-areas, coarse fluvial and deltaic deposits dominated while western areas contain primarily graded graywackes and shales (flysch) deposited in submarine-fan environments.

D.2.b. Cretaceous Events: Canada

Cretaceous events in the Canadian Cordillera represent a combination of Andean-arc and microplate-accretion tectonics. Intrusion, deformation, and uplift within the Omineca belt continued and intensified into the Cretaceous, resulting in the Columbian orogeny which was analogous to the Sevier orogeny in the Southern Cordillera. In addition, during the early Late Cretaceous, a major displaced terrane called Wrangellia collided with the Northern Cordillera from southeastern Alaska to Vancouver. The collision ended Hazelton arc activity and led to the development of a new magmatic arc, recorded today by the Coast Plutonic Complex, along the seaward side of Wrangellia. It also resulted in new successor basins and decrease of orogenic activity in the Omineca belt. Several other tectonic events, especially possible rifting across the orogenic trend of the northwestern United States, further complicated the picture.

i. Early Cretaceous: The Columbian Orogeny.

The Omineca belt during the Early Cretaceous continued to be the site of intense magmatic activity. Major granitic and granodioritic plutons were emplaced during this time and volcanic rocks were erupted just northeast of the Omineca belt. Much of the Omineca belt was subjected to high-grade metamorphism due to intense compressive stresses associated with the convergent plate boundary and to great quantities of heat introduced by magmas rising off the Benioff zone. The result was a broad, mobile infrastructural zone (similar to that described for the Sevier orogen). As cooler, more rigid continental crust to the east moved toward the magmatic arc, it partially underthrust the hot, plastic infrastructure, causing crustal shortening at depth. [Eisbacher (1974) estimated up to 30% shortening perpendicular to the trend of the orogen.] The result was crustal thickening with concomitant isostatic uplift. "Underthrusting" also caused major décollement-type thrusting within the Paleozoic–Mesozoic continental-shelf prism (see Fig. 8-45). Columbian thrusting (the Alberta thrust belt) was penecontemporaneous with early Sevier thrusting; together, they produced a fold-thrust belt from Alaska to Mexico (see Fig. 8-44).

Naturally, uplift of the Omineca terrane and the fold-thrust belt provided a major detrital sourceland which supplied vast quantities of sediment both eastward and westward. Pouring eastward out of the rising orogen, sediments spread out into the northern Rocky Mountains foreland basin resulting in a great molasse wedge formed principally of Kootenay and Blairmore strata.

Orogenically derived sediments were also carried westward into the fore-arc system, where nonmarine environments dominated. In the Bowser Basin, for example, sediments of the Red Rose Formation record the progradation of nonmarine deposits westward over marine strata. Only in the Tyaughton–Methow Trough are marine sedimentary rocks of Lower Cretaceous age common. There, coarse, proximal deposits of eastern areas, e.g., the Jackass Mountain Group, grade westward into deeper-water, fine-grained rocks of the Harts Pass Group (Tennyson and Cole, 1978). These strata are considered by Tennyson and Cole to represent deposition in submarine fans. Note, however, that Douglas *et al.* (1976) stated that the Jackass Mountain Group was nonmarine. Regardless of the depositional environment, Lower Cretaceous rocks of the Tyaughton–Methow Trough are significant because of what they reveal about their provenance. Most Lower Cretaceous sandstones in the area are arkosic. Conglomerates contain coarse granitic clasts. Similar petrography characterizes sediments of the Bowser and Tantalus basins. Clearly, by the Early Cretaceous the Omineca infrastructure had be uncovered and was supplying detritus derived from an extensive, quartzose, plutonic–metamorphic terrane.

ii. Cascadia and Wrangellia.

It was during the later Early and early Late Cretaceous that the relatively simple, Andean-arc pattern described above began to be disrupted. Perhaps the first indication of a change appears in the Tyaughton–Methow Trough. As we have seen, Lower Cretaceous deposits of the trough were derived from a plutonic–metamorphic terrane to the east. But late Albian to Turonian strata of the Virginian Ridge Formation represent a *western* source area, probably a volcanic–sedimentary terrane. On the eastern side of the Tyaughton–Methow Trough, arkosic sandstones of the Winthrop Sandstone were apparently derived from the plutonic–metamorphic eastern terrane. Thus, during late Albian to Turonian time, the Tyaughton–Methow Trough received sediments from both eastern and western areas.

The obvious question is: From where were Virginian Ridge sediments derived? The Northern Cascades block (called "Cascadia") lies today just west of the Tyaughton–Methow trough and, since it was probably not there prior to Virginian Ridge time, one may posit that western-derived sediments came from Cascadia. But how did Cascadia arrive at its position off the Tyaughton–Methow trough? Naturally, there are several opinions. Hamilton (1978) argued that Cascadia was a "minicontinent," i.e., a displaced continental terrane, observing that Cascadia contains rocks as old as the Precambrian and had a complex geological history *before* collision with North America. He also observed that it bears some similarities to displaced-terrane rocks in California. Davis *et al.* (1978), on the other hand, believed the situation to have been more complicated; Figure 8-48 shows their model. Notice (as discussed previously) that Davis *et al.* regarded the Tyaughton, Methow, and Nooksack basins to have been a single, continuous, fore-arc basin during the Late Jurassic. By the late Early Cretaceous, in their view, a dextral wrench fault (probably a transform) had developed at a very low angle to the continental margin and had offset the southern part of the continental area, juxtaposing it to the west of the northern continental-margin area. Thus, in the scheme of Davis *et al.,* Cascadia was originally part of North America, located just southeastward of its present position.

A corollary of the Davis *et al.* model is proposed rifting in the Washington–Oregon–Idaho area (see Fig. 8-49). The basic idea is that the Cascadia block moved northwestward, opening a gap behind it and, thus, leading to the development of an active spreading center at a high angle to the trend of the Cordilleran orogen. This model is somewhat simplistic but seems to fit most of the data on the distribution of basement lithologies in the northwestern United States. The mechanics of such rifting are not immediately

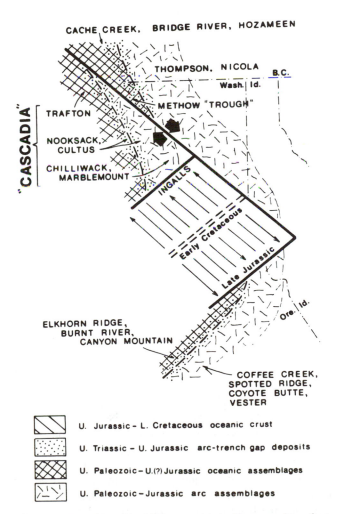

Fig. 8-49. Tectonic model of rifting associated with proposed transform-fault displacement of Cascadia. (From Davis *et al.,* 1978, reproduced with permission of the Pacific Section, Society of Economic Paleontologists and Mineralogists.)

obvious but may be related to oblique convergence of the western oceanic plate beneath the Cordilleran margin. If a sliver of North American crust became detached from the rest of the plate by transcurrent faulting in back-arc areas (related to oblique subduction), a spreading center would develop behind the sliver where it broke free from the continent.

There are several problems, especially of geometry, created by this rifting model. One such problem involves emplacement of the Seven Devils terrane in westcentral Idaho. The Seven Devils area is generally considered to have been a displaced terrane, probably a volcanic arc. Jones *et al.* (1977) regarded the Seven Devils terrane as a possible southern extension of Wrangellia (see below) but Davis *et al.* (1978) contended that the Seven Devils arc was distinct from Wrangellia and collided with the rifted continental margin during the later Early Cretaceous. They argued that magmatism of the Seven Devils arc was due to westward subduction beneath the approaching microplate. Following arc–continent collision, subduction resumed in an eastward direction beneath the western margin of the newly accreted terrane. Magmatism associated with this new subduction zone was responsible for the Idaho batholith. The problem with the rifting model described above is that one seemingly cannot reconcile Early Cretaceous spreading (as shown in Fig. 8-49) with arc convergence in the same area and at nearly the same time.

In 1977, Jones *et al.* described a major displaced terrane of the Northern Cordillera, which they named *Wrangellia* (from the Wrangell Mountains of southeastern Alaska; see Fig. 8-50). They presented stratigraphic comparisons of rocks in the Wrangell Mountains with rocks of Chichagof Island, the Queen Charlotte Islands, and Vancouver (see Fig. 8-51). With these comparisons, they were able to show gross stratigraphic correlations along the entire length of the terrane. Notice, for example, that in the Wrangell Mountains and Vancouver, Penno-Permian sedimentary rocks overlie andesitic rocks. In all areas, thick piles of pillowed and massive tholeiitic flows (Nikolai and Goon Dip Greenstones, Karmutsen Formation) comprise the bulk of the sections and are overlain by carbonate-platform strata. Thus, all of these areas appear to have undergone a similar history and therefore are considered to represent a continuous, narrow plate-sliver wedged into the Cordilleran collage.

The original location of Wrangellia is questionable. The Mesozoic sequence rests on the remains of an upper Paleozoic andesitic arc which probably formed at some distance from North America. Triassic flows from several different elements of Wrangellia have yielded paleomagnetic data indicating that they formed at 15°N or 15°S latitude (Jones *et al.,* 1977). Thus, Wrangellia must have moved northward *at least* 30° in latitude relative to the Canadian

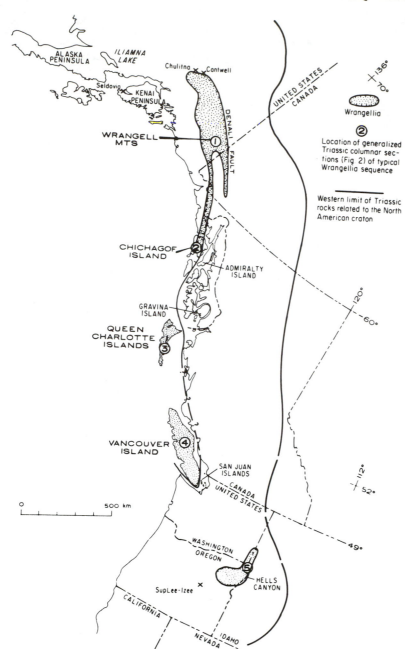

Fig. 8-50. Location map of major elements of the Wrangellia displaced terrane. (From Jones *et al.*, 1977; reproduced with permission of the *Canadian Journal of Earth Science.*)

Cordillera (based on Triassic paleopole locations). The time of Wrangellia's collision with North America is somewhat controversial although most authors (e.g., Jones *et al.*, 1977; Davis *et al.*, 1978) favor an early Late Cretaceous date. Wrangellia's collision had a number of important effects on Cordilleran tectonics during the Late Cretaceous and it is to these effects that we now turn.

iii. Late Cretaceous: A New Cycle.   The collision of Wrangellia ended the tectonic cycle, begun with Stikine collision, which was presided over by the Hazelton–

Omineca arc complex. With Wrangellia collision, the Bowser–Tyaughton–Methow fore-arc basin ceased to be active and underwent deformation. In the Bowser Basin, for example, a new sequence of upper Upper Cretaceous sediments was deposited over folded Bowser Group deposits. Sediments of this new sequence thus define a new successor basin, the Sustut Basin (see Fig. 8-52; Eisbacher, 1974). Disconnected from the Sustut Basin are similar, coarse deposits in the Sifton Basin to the east (see Fig. 8-52). The Sifton Basin probably represents an intermontane basin within the Omineca belt.

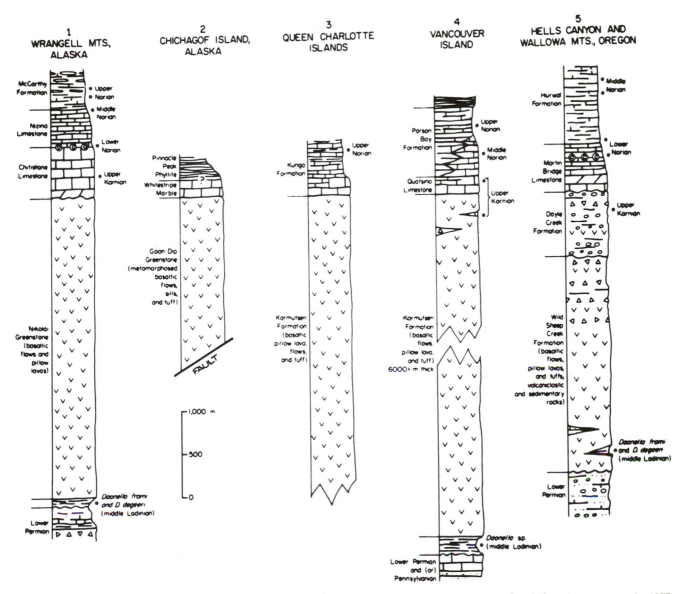

Fig. 8-51. Stratigraphic comparison of Wrangellia's disparate pieces; column numbers refer to location in Fig. 8-50. (From Jones *et al.*, 1977, reproduced with permission of the *Canadian Journal of Earth Science*.)

Wrangellia accretion also ended subduction along the precollision continental margin. Deprived of a subducting-plate source of magmas, tectonic activity in the Omineca area and thrusting along the Alberta thrust belt declined during the early Late Cretaceous. A reflection of this decline may be seen in the foreland basin where nonmarine, molasse deposition of the Kootenay–Blairmore assemblage gave way, during the Cenomanian, to marine deposition in which primarily fine-grained sediments (the Alberta assemblage of Eisbacher, 1974) accumulated.

The new cycle of arc tectonics and successor basins began with the development of a new subduction zone on the western margin of Wrangellia. Actually, the western Wrangellia subduction zone may have existed *prior* to collision; Davis *et al.* (1978) favored a "two arc" model, whereby Wrangellia, riding on a plate subducting eastward beneath North America, had an active volcanic arc on its western side due to eastward subduction beneath it. This would account for Jurassic plutonic activity in Wrangellia. At any rate, following collision *all* subduction was along the western zone and led to the development of a major belt of plutonic rocks, the Coast Plutonic Complex, located in westernmost British Columbia (see Fig. 8-52). This magmatic arc may have been continuous to the north, joining

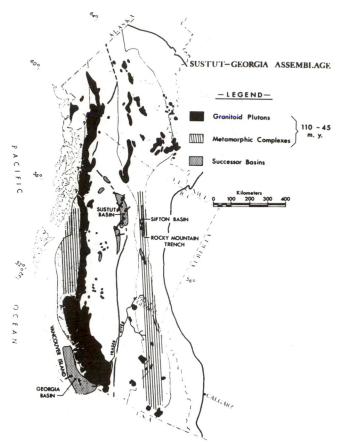

Fig. 8-52. Major Late Cretaceous sedimentary basins of the Canadian Cordillera. Granitoid plutonic bodies are part of Coast Plutonic complex; the Georgia Basin is believed to have been a fore-arc basin developed on the continental margin following accretion of Wrangellia; see text for further discussion. (From Eisbacher, 1974, reproduced with permission of the Society of Economic Paleontologists and Mineralogists.)

the Gravina–Nutzotin plutonic belt of southeastern Alaska (Dickinson, 1976). (Note, however, that such hypotheses are highly speculative given the large number of displaced terranes in Alaska. Mere similarity of ages of magmatism does not ensure that two arcs were continuous—or even related.)

In front of the Coast Plutonic Complex, a new fore-arc basin, called the Georgia Basin, developed on eastern Vancouver Island, beneath the Straits of Georgia, and on a few mainland areas (see Fig. 8-52; Dickinson, 1976; Eisbacher, 1974). Sedimentary rocks of the Georgia Basin in Vancouver, which comprise the Nanaimo and Comox Groups, lie on top of deformed Wrangellia rocks. Nanaimo and Comox rocks represent a complex of fluvial, deltaic, lagoonal, and barrier island environments.

Tectonic events associated with the new cycle of arc-magmatism continued into the Tertiary. Near the end of the Late Cretaceous, a major orogenic event began to affect the

eastern Cordillera in both the Southern and Northern Cordillera. This event, termed the Laramide orogeny, was principally a Tertiary affair and we will defer our discussion of it until the next chapter.

D.2.c. Alaska

The Cordilleran orogen in Alaska is a wide belt of displaced terranes of various shapes, sizes, crustal types, ages, original locations, and times of emplacement. We have already introduced the reader to the displaced terranes of Alaska but a short digression here might help to clarify the significance of the Alaskan "tectonic antique store." Figure 8-40 shows the aggregate of crustal slivers, lumps, wedges, and blocks that is the Alaskan Cordillera. Such a babble of terranes bespeaks a long and complex history involving numerous volcanic arcs, oceanic fracture zones, seamount chains, and so on riding the oceanic lithosphere toward the Northwest. During deposition of the Zuni Sequence, a fair number of displaced terranes were added to the continent, notably the Alexander terrane, Wrangellia, and the Peninsular terrane (see Fig. 8-40).

Other major elements of the Alaskan Cordillera are the arc-related volcanic and plutonic complexes extruded onto and intruded into the displaced terranes. Hudson (1979) described the major plutonic belts of Mesozoic age in southern Alaska and demonstrated their relative age relationships (see Fig. 8-53). In some cases, plutonic belts were emplaced within the displaced terrane before collisional accretion; in other instances, intrusion followed collision. The Kodiak–Kenai belt is the oldest, being of Upper Triassic–Lower Jurassic age. All of the rest are Middle Jurassic to Upper Cretaceous. Hudson considered these plutonic belts to represent four or five separate arc-complexes.

Along the Alaskan coast in the area south of Cook Inlet, rocks of the Upper Triassic to Lower Jurassic probably represent a volcanic island arc (Kirschner and Lyon, 1973). The Kamishak Formation, at the base of this sequence, is composed of fossiliferous limestone and chert, both primarily of a pelagic origin and both metamorphosed. Conformably overlying Kamishak strata is the Talkeetna Formation whose metamorphosed volcanic rocks, ultramafic masses, and flyschlike deposits of volcaniclastics probably record the main volcanic/sedimentary edifice of the arc-trench system. Associated with Talkeetna strata are dioritic to tonalitic plutons of Hudson's Kodiak–Kenai belt mentioned above (see Fig. 8-53). Again, we stress that this volcanic arc was probably *not* part of the Alaskan mainland during its formation; exactly where it was, however, is not known.

During the Middle Jurassic, a major orogenic event occurred in the area north of Cook Inlet. This event was possibly the result of a collision involving one of southern

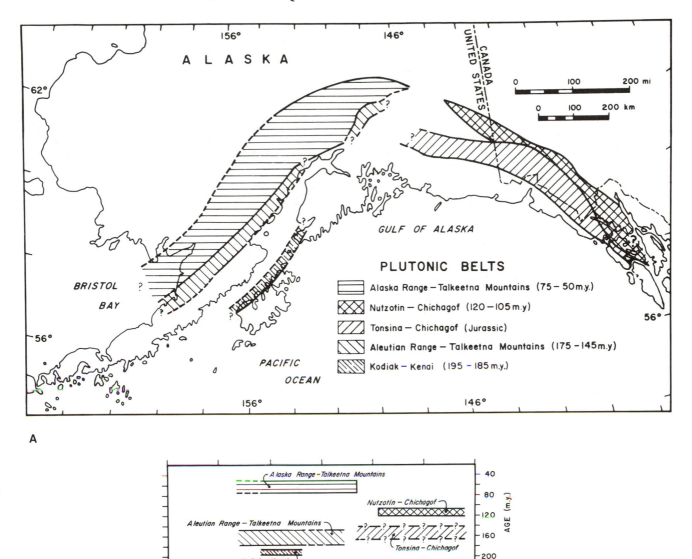

Fig. 8-53. (A) Location of Cook Inlet Basin and distribution of main Mesozoic plutonic belts of southern Alaska; note that several plutonic belts overlap each other. (B) Summary of the space–time relations of the plutonic belts shown in A; age ranges of individual belts may vary along its trend. (Modified from Hudson, 1979.)

Alaska's displaced terranes (Kirschner and Lyon, 1973). Following the collision, an Andean-type magmatic arc developed along the collisional orogen due to subduction of the oceanic plate beneath the new continental margin. Tectonic activity related to this collision and new Andean arc is termed the Alaska Range orogeny. The tectonic situation of the Alaska Range orogeny paralleled contemporary developments in the Sierra Nevada region of the Southern Cordillera, and in both areas a major episode of plutonic activity resulted. In southwest Alaska, large, elongate, batholithic complexes composed of tonalites, diorites, granodiorites, and so on were emplaced during the Alaska Range orogeny. These plutons comprise Hudson's Aleutian Range–Talkeetna Mountains plutonic belt (see Fig. 8-53). The Alaska Range orogeny was most intense in the southwestern part of Alaska, especially along the Aleutian peninsula and up into the southern Alaska Range itself. Intensity decreased eastward. Richter and Jones (1973), in characterizing the eastern Alaska Range, did not recognize any major Jurassic phase of plutonic activity. It may be noted, however, that the Tonsina–Chichagof belt (see Fig. 8-53) is of roughly the same age as the Aleutian Range–Talkeetna

Mountains belt and *may* represent an eastward continuation of the Alaska Range arc. (It may also represent a separate but contemporaneous volcanic arc.)

Associated with the Alaska Range orogeny, a large detrital wedge prograded southeastward, forming a sedimentary basin called the Cook Inlet Basin (see Fig. 8-53). The location of the Cook Inlet Basin (i.e., on the oceanward side of the arc) suggests that it occupied a fore-arc setting. Into this basin poured volcaniclastic sands and gravels derived from the arc to the northwest. These deposits comprise the Tuxedni Group. Kirschner and Lyon (1973) reported that 40% of clasts in Tuxedni conglomerates are diorite, indicating the unroofing of early Aleutian Range–Talkeetna Mountains plutons. Overlying Upper Jurassic sediments of the Chinitna and Naknek formations record the continuation of the unroofing process.

By the Early Cretaceous, the rate of tectonic activity had decreased in the Cook Inlet Basin and adjacent orogen. On the northwestern side of the basin, a shallow carbonate platform developed. Deposits of this platform comprise the Nelchina and Herendeen Limestones. To the southeast in the Cook Inlet Basin, deeper-water deposits of the Lower Cretaceous are represented by thick turbidite sequences of the Sunrise and Valdez groups.

During the early Late Cretaceous, Wrangellia collided with the Cordilleran margin of east Alaska, resulting in a major episode of deformation and magmatism. The Denali fault (see Fig. 8-50) is considered the line of suture between the two adjacent terranes and, therefore, the site of an ancient subduction zone. Slivers of ophiolitic rocks along the fault zone probably represent scraps of oceanic crust sheared up into the subduction complex prior to and during collision. Plutonic masses formed at this time comprise Hudson's Nutzotin–Chichagof belt (see Fig. 8-53). They range in size from small stocks to complex batholiths and are composed of diorites, granodiorites, and quartz monzonites. Richter and Jones (1973) said that intrusion of these plutons was "the most significant and widespread plutonic event" to affect the east Alaska area. (Note: east Alaska was *not* affected by plutonic activity during the Alaska Range orogeny discussed earlier.)

Following Wrangellia collision, a new magmatic-arc cycle began with the inception of subduction along the southern Alaska margin; again ensued the familiar pattern of arc magmatism and fore-arc sedimentation. Upper Cretaceous plutons associated with the new convergent boundary define the Alaska Range–Talkeetna Mountains belt (Hudson, 1979; see Fig. 8-53) which is the youngest Mesozoic plutonic belt in southern Alaska, with magmatism of this phase persisting into the Tertiary. Subduction was northwestward beneath the continent, as indicated by the northwestward increase in K_2O content of Alaska Range–Talkeetna Mountains plutons (Hudson, 1979).

Fore-arc basin sediments in the Cook Inlet area define the Matanuska Formation which consists of over 3000 m of marine sandstone and siltstone (Kirschner and Lyon, 1973). Coarse, quartzose Matanuska graywackes were deposited in the northwestern portion of the basin, suggesting that the detrital source consisted of a broad, relatively stable, plutonic–metamorphic terrane. Interestingly, a new and apparently unstable source appeared in the south. Coarse, polymictic conglomerates are interbedded with Matanuska siltstones in the southeastern area of Cook Inlet Basin. Composition of clasts in these conglomerates suggests a volcanic–metamorphic source, such as a volcanic-arc complex. Thus, another displaced terrane (possibly the Chugach terrane; see Fig. 8-40) was outboard from the continent during the Late Cretaceous, slowly approaching collision during the Tertiary.

E. The Innuitian Continental Margin

E.1. The Sverdrup Basin

The tectonic framework of the Sverdrup Basin during deposition of the Zuni Sequence was simple and straightforward. Maximum subsidence characterized the central region of the basin where marine shales were deposited. In basin-margin areas, subsidence was slower and, consequently, shallow-marine and/or nonmarine sands were deposited there. This pattern was established as early as the Permian and persisted, with several interruptions, until the Tertiary when the history of the Sverdrup Basin was terminated by the Eurekan orogeny.

E.1.a. Jurassic Stratigraphy

Lowermost Jurassic rocks (i.e., Hettangian Stage) are apparently absent from the Sverdrup Basin although they may comprise the top of the unfossiliferous Heiberg Formation (Upper Triassic). Thus, lowest Jurassic strata, the Sinemurian-age Borden Island Formation, rest on a paraconformity (Thorsteinsson and Tozer, 1976). The Borden Island is a thin marine unit composed of green and gray glauconitic sandstone with thin beds of red, iron-cemented sandstone. Its distribution is somewhat unusual, occurring around the margins of the basin but without correlative units in the basin center (see Fig. 8-54). This is different from most other Sverdrup Basin units which thicken toward the center.

Essentially normal depositional patterns were reestablished during Toarcian time. Basin-center deposits were primarily argillaceous, marine shales of the Savik Formation (Fig. 8-54) whereas correlative basin-margin rocks were marine sands. At the eastern end of the basin, these sands are termed the Jaeger Formation; at the western end,

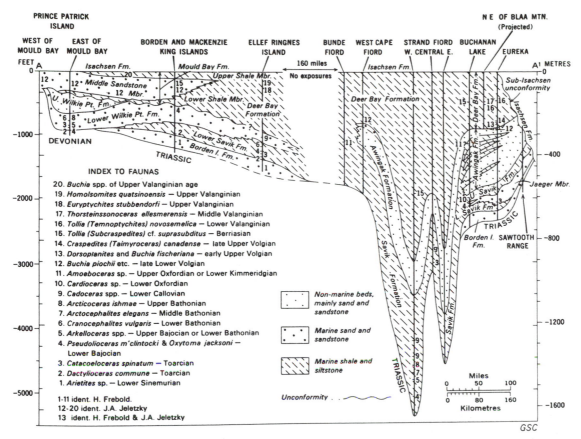

Fig. 8-54. East–west stratigraphic cross section of Jurassic strata in the Sverdrup Basin. Note that sandstone units are primarily confined to basin-margin areas where subsidence was limited and unconformities interrupt the sequence; shales characterize the basin-center sequence where subsidence was more rapid. (From Thorsteinsson and Tozer, 1976, reproduced with permission of the Minister of Supply and Services, Canada.)

they are called the lower Wilkie Point Formation. In both cases, the sands are quartzose with varying amounts of glauconite; they also contain iron-cemented beds and phosphatic nodules. Beginning probably during the Callovian, deltas began to grow into the basin from the south and east, probably reflecting an increase in the rate of sediment influx. The resulting mass of deltaic sediment, called the Awingak Formation in the east and the upper Wilkie Point Formation in the west, was confined mainly to the south side of the basin; little if any sediment entered the basin from the north at this time. During Oxfordian time, the Awingak delta system prograded rapidly northwestward, covering older Savik shales as much as 400 km into the basin center (see Fig. 8-54).

During the latest Jurassic, subsidence rates apparently increased (or sedimentation rate decreased) causing a return to the familiar basin-margin sand/basin-center shale pattern. The shale sequence, termed the Deer Bay Formation, straddles the Jurassic–Cretaceous boundary and contains a complete ammonite fauna from Portlandian to Valanginian Stages. Correlative sandstones on the southwestern margin

comprise the Mould Bay Formation. Local disconformities interrupt the Mould Bay so that its faunal record is much less complete than the Deer Bay record.

E.1.b. Cretaceous Stratigraphy

Plauchut (1973) stated that all Sverdrup Basin strata, from the lowest unit (Borup Fiord Formation; Mississippian) up through Deer Bay and Mould Bay formations, represent a single tectono-stratigraphic entity (which he called a sedimentary cycle) because they were all deposited within a localized, subsident basin. On the other hand, Plauchut argued, Cretaceous units were deposited much more widely, far outside the confines of the Sverdrup Basin; thus, they represent a different sedimentary cycle. It should be pointed out, however, that Cretaceous strata thicken appreciably within the basin indicating that it continued to be an active tectonic feature.

The first unit of Plauchut's second Sverdrup-Basin cycle is the Isachsen Formation, a widespread nonmarine unit composed of quartzose sandstone and conglomerate with subordinate shale, siltstone, and coal. Sandstones are commonly crossbedded and are arranged in fining-upward fluvial sequences containing leaf fossils and palynomorphs of Lower Cretaceous age. As shown in Fig. 8-55, Isachsen deposits can be traced as far west as the Alaskan border; indeed, lower strata of the nonmarine Ignek Formation in the Romanzof Mountains of northeastern Alaska may well represent a continuation of Isachsen-like deposition. To the east, the Isachsen has been correlated with coal-bearing strata on northern Baffin Island (Thorsteinsson and Tozer, 1976). Dawes and Soper (1973) reported a small amount of similar Cretaceous sedimentary rocks with leaf fossils from northern Greenland. If these disparate occurrences do, in fact, represent a nearly continuous mass of terrestrial deposits, they would suggest that, for the first time since long before the Late Devonian Ellesmere orogeny, much of the Innuitian continental margin behaved in a geologically similar manner. The nonmarine character of the sediments would seem to argue that the rate of subsidence approximately equaled the rate of sedimentation all along the continental margin.

Evidence that a plate-margin tectonic cause probably existed for this new depositional setting may be seen in the extensive basalt flows interbedded with Isachsen strata in northwestern Axel Heiberg Island. Similar mafic igneous activity persisted throughout Cretaceous time in the Axel Heiberg area and may have been related to extensive volcanic flows of the Kap Washington Group in North Greenland (of Cretaceous-Tertiary age) which is composed of up to 1500 m of rhyolitic lavas and tuffs. It may be that all of the various tectonic events described may be related to the opening of the Arctic Ocean (discussed later in this section).

Following Isachsen time, sea level rose, flooding much of the northern continental margin and resulting in deposition of Christopher Formation shales conformably over the Isachsen. Note in Fig. 8-55 that the Christopher is about as extensive as are strata of the Isachsen. Note also that Christopher lithologies and thickness do not seem to differ significantly within or without the Sverdrup Basin; this could indicate that the basin was not a significant geomorphological feature during this time. Christopher strata grade upward into the Hassel Formation, a somewhat variable quartzose sandstone and siltstone of probable Cenomanian age. Thorsteinsson and Tozer (1976) stated that the Hassel contains marine fossils in places but appears to be nonmarine on Ellesmere Island at the eastern end of the basin; they speculated that the Hassel might represent a deltaic sequence periodically flooded by the sea. On Axel Heiberg Island, a massive outpouring of basalt terminated Hassel deposition and resulted in the Strand Fiord Formation. The Strand Fiord is composed of up to 200 m of basalt and 70 m of pyroclastic breccia.

Upper Cretaceous strata comprise the Kanguk Formation, a dark-gray shale with minor sandstones and tuffs. The Kanguk is the youngest marine Mesozoic unit of the Arctic Archipelago and, unlike earlier Cretaceous units, is found only within the Sverdrup Basin. The Kanguk is unconformably overlain by nonmarine Lower Tertiary deposits of the Eureka Sound Formation.

E.2. Northern Alaska

The Jurassic and Cretaceous history of northern Alaska is considerably more complex than that of the Sverdrup

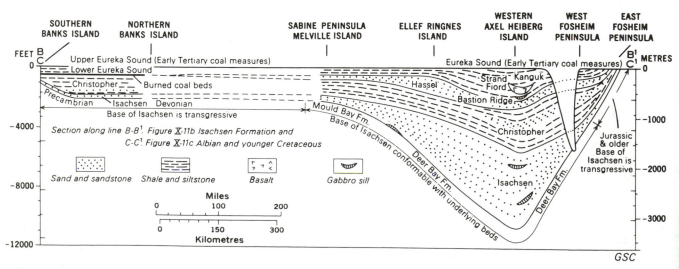

Fig. 8-55. East–west stratigraphic cross section of Cretaceous strata in the Sverdrup Basin and adjacent areas. Note that most Cretaceous units extend far to the west of the margin of the Sverdrup Basin. (From Thorsteinsson and Tozer, 1976, reproduced with permission of the Minister of Supply and Services, Canada.)

Basin because northern Alaska was intimately affected by the tectonic events of the Cordilleran continental margin. Especially in Cretaceous strata of the Brooks Range and North Slope may one see the interaction of different plate-margin regimes on either side of the narrow northern Alaska continental plate. Beginning during the Jurassic, rifting began to open the Arctic Ocean Basin to the north of the Brooks Range. At nearly the same time, or somewhat later, a collisional orogeny affected the Cordilleran margin south of the Brooks Range. Sediments derived from the orogen were deposited within the growing rifted-margin prism; obviously, this circumstance constituted a rather unusual tectonic situation.

E.2.a. Jurassic History

As discussed in the previous chapter, the Upper Triassic Shublik Formation records a low detrital source (the Beaufort uplift) to the north of the North Slope; water depth apparently increased toward the south. Following Shublik time, mild uplift and deformation affected much of the Brooks Range area, a harbinger of the greater structural disturbance to come. As a result, lowest strata of the Jurassic Kingak Shale were deposited over an angular unconformity (Detterman, 1973). The Kingak is a thick mass (up to 3000 m) of monotonous, dark shale with thin interbedded sandstone and siltstone beds. Kingak strata probably do not represent a continuous depositional-sequence but are interrupted at several horizons by disconformities or even, rarely, angular unconformities. Its deposition is considered to have spanned virtually the entire Jurassic.

The provenance of Kingak sediments has been a source of disagreement. Some authors, such as Brosgé and Tailleur (1970), have suggested that the northern source (i.e., the Beaufort uplift) continued to supply sediments for the Kingak. Others (e.g., Keller 1961) argued for a southern source, possibly the area of the present Brooks Range. Paradoxically, both answers seem to have been correct: lower Kingak strata coarsen toward the north, grading into glauconitic sand in the subsurface of the North Slope. On the other hand, upper Kingak strata contain beds of sandstone and siltstone which thicken and coarsen toward the south, probably reflecting earliest tectonic stirrings along the Cordilleran margin.

The lower Kingak grades southward into deep-water sediments and associated pillow lavas south of the Brooks Range (Brosgé and Tailleur, 1970). Richards (1974) said that these lavas represent the oldest, well-documented volcanic activity in the Alaskan Arctic sequence and, therefore, may have formed just following the onset of subduction in that area. Alternately, these volcanics might represent displaced terranes subsequently accreted to the Cordillera; nevertheless, their timing seems to coincide

nicely with initiation of deformation in Arctic Alaska, which began as early as Hettangian time (Detterman, 1973) as revealed by the angular unconformity between the Shublik and the Kingak.

By the Middle Jurassic, uplift was clearly affecting the Brooks Range area because coarse, conglomeratic detrital materials and tuffaceous sediments were deposited in that region. In the western Brooks Range area, near Cape Lisburne on the northwestern Alaskan coast, are upper Middle and Upper Jurassic graywackes, mudstones, and shales of the Ogotoruk and overlying Telavirak formations, considered by Detterman (1973) to represent flysch-type deposition. Farther north, coarse detrital sediments grade into fine, dark shales of the Kingak. The shift from a northern to a southern source for Kingak detrital sediments was considered by Richards (1974) to be the indication of a major tectonic change. Detterman reported that Bathonian and Callovian strata are missing from the Kingak and that the provenance shift may have occurred during that time.

E.2.b. Cretaceous History

By the beginning of the Cretaceous, an entirely new tectonic framework had developed in northern Alaska; Fig. 8-56 shows the main tectonic elements active during that time. Of primary interest is the Brooks Range orogen (or "geanticline"), developed as a result of Cordilleran-margin tectonics and composed of large, northward-directed thrust slices of ophiolites (Roeder and Mull, 1978) and deep-water sediments such as the Kagvik sequence. North of the Brooks Range orogen is the Colville foreland basin (or

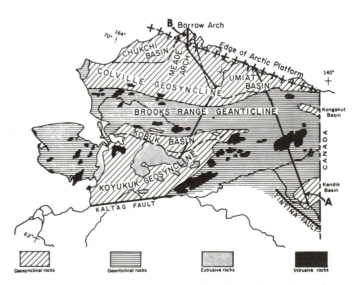

Fig. 8-56. The major tectonic elements of northern Alaska during the Cretaceous. (From Detterman, 1973, reproduced with permission of the American Association of Petroleum Geologists.)

"geosyncline") whose geometry and dynamics may have been due to some combination of tensional strain related to Arctic Ocean opening and compressive strain associated with the Cordilleran-margin orogeny. South of the Brooks Range orogen is the Koyukuk Basin whose lithologies are dominated by volcanics and volcaniclastics. The Koyukuk Basin was considered by Churkin *et al.* (1980) to be one of the displaced terranes of the Cordilleran collage.

Early Cretaceous (Neocomian) sedimentation reflected the instability of the region. Immature graywackes and mudstone turbidites of the Kisimilok Formation in the Cape Lisburne area conformably overlie the Jurassic Telavirak Formation, also a turbidite sequence. Similar turbidites of the Okpikruak Formation in central northern Alaska were deposited unconformably on Jurassic rocks (see Fig. 8-57). Equivalent strata in the Prudhoe Bay area also lie unconformably on eroded Jurassic strata and consist of thin, deep-water, argillaceous shales, probably a distal, offshore facies of Okpikruak turbidites. Morgridge and Smith (1972) referred to Okpikruak strata as clinothem deposits; thus, these rocks may define the newly formed rifted continental margin. Perhaps the unconformity at the base of the Cretaceous System may be related to uplift of the northernmost Alaska region just prior to the rapid-subsidence phase accompanying continental rifting. In the area of the Romanzof Mountains, nonmarine deposits comprise the Ignek Formation (Sable, 1977). Ignek strata, similar to underlying Kingak strata, probably represent episodic deposition interspersed with long periods of nondeposition. The resulting unit is shown by Detterman (1973) to span virtually the entire Cretaceous. Igneous activity, both intrusive and extrusive, and volcaniclastic sedimentation characterized events in the Early Cretaceous Koyukuk Basin.

The Aptian stage was considered by Eardley (1962) to have been the time of maximum orogenic activity in the Brooks Range orogen. Thus, it may have been the interval during which occurred the collision of the displaced terrane (the Yukon–Koyukuk terrane? Wrangellia?) with the Cordilleran continental margin.

Following the Aptian orogenic maximum, a great volume of coarse detrital sediments was shed both northward into the Colville Basin and southward into the Koyukuk Basin. Albian strata in the Colville Basin constitute a detrital wedge more than 3000 m thick. Directly adjacent to the Brooks Range orogen were deposited coarse detritus, predominantly polymictic conglomerates and immature sandstones; these rocks are termed the Fortress Mountain Formation (see Fig. 8-57). Toward the north, Fortress Mountain detrital sediments grade into sandstones and mudstones of the Torok Formation which themselves grade farther northward into fine-grained shales of the Oumalik and overlying Topagoruk formations. On the south flank of the Brooks Range orogen, Albian strata in the Koyukuk

comprise a thick volcaniclastic sequence with volcanic graywackes and igneous-pebble conglomerates.

By middle Albian time, both Colville and Koyukuk basins had been nearly filled by the massive detrital influx from the Brooks Range orogen. In the Colville Basin, this infilling is reflected by the Nanushuk Group (see Fig. 8-58), a major regressive sequence in which nonmarine rocks (the Chandler Formation) grade northward over littoral and shallow-marine strata (Tuktu, Grandstand, Ninuluk, and Umiat formations). Nanushuk deposition continued into the Cenomanian by which time nonmarine environments extended over virtually the entire Colville Basin.

Following Nanushuk Group deposition, much of the Colville Basin area was exposed to erosion. In some areas, especially in the northern part of the basin, virtually the entire Nanushuk Group was removed (see Fig. 8-58). When deposition resumed during Turonian time, it was greeted by a major modification of the Colville Basin. Figure 8-56 shows the basin divided into two smaller basins, the Umiat on the east and Chukchi on the west, by a prominent uplift termed the Meade Arch. Consideration of Fig. 8-58B will show that these features apparently had little effect on Nanushuk Group deposition but during ensuing Colville time, their influence was considerable. Upper Cretaceous strata of both nonmarine and marine affinities occur in the Umiat Basin whereas only minor amounts of Colville-equivalent, nonmarine strata are found in the Chukchi Basin. No Upper Cretaceous rocks exist on the Meade Arch.

The lowest unit of the Colville Group is the Seabee Formation, a fissile marine shale with bentonite beds and dark, micritic limestones. The Seabee sea was rapidly filled from the south by coarse, nonmarine detritus derived from the still-active Brooks Range orogen. These coarse detrital deposits are termed the Prince Creek Formation, and grade into marine rocks (sandstones, siltstones, shales, and tuff) of the Schrader Bluff Formation. In the Romanzof Mountains, nonmarine Ignek Formation strata are partly correlative with Prince Creek rocks. An unconformity is present at the top of the Cretaceous System throughout the area.

E.3. Early Opening of the Arctic Ocean

Intimately related to late Mesozoic and Cenozoic history of the northern continental margin is the history of the Arctic Ocean. Of particular importance are the tectonic events which resulted in its opening, because those events clearly must have left their marks in the surrounding continental margins. Thus, as we have shown earlier in this chapter for the Atlantic and Gulf continental margins, development of the northern continental margin must be viewed against its framework of global tectonics. In this section, after a brief primer on Arctic Ocean geography, we

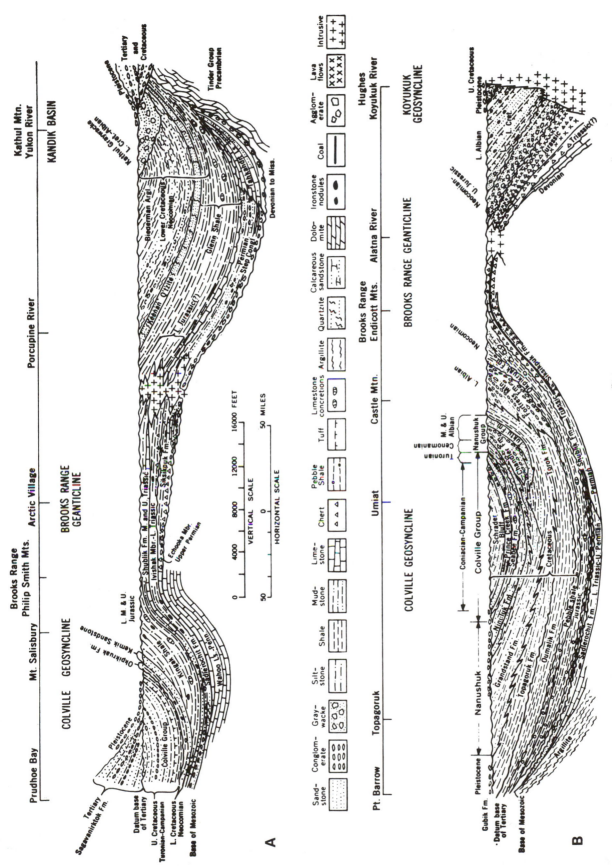

Fig. 8-57. Restored geological cross sections of Mesozoic units in northern Alaska; lines of section are shown in Fig. 8-56. (From: Detterman, 1973, reproduced with permission of the American Association of Petroleum Geologists.)

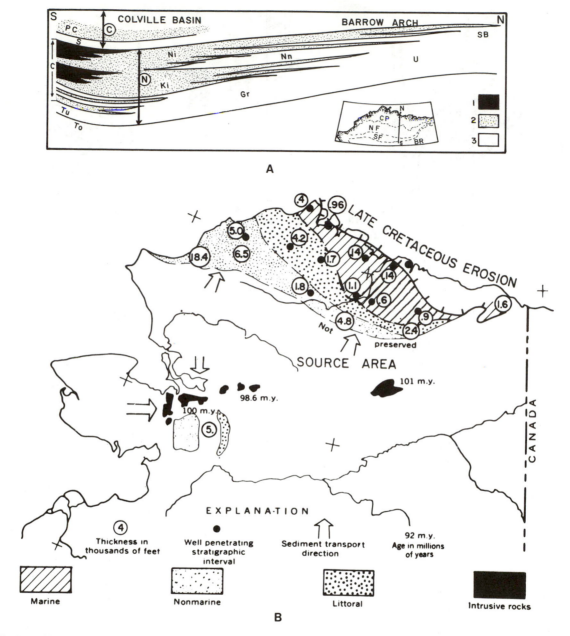

Fig. 8-58. Depositional environments and stratigraphic relations of Nanushuk Group strata in northern Alaska. (A) Stratigraphic cross section of Nanushuk and Colville Groups; symbols are as follows: circled C, Colville Group; circled N, Nanushuk Group; C, Chandler Formation; Gr, Grandstand Formation; Ki, Killik Tongue; Ni, Niakogon Tongue; Nn, Niniluk Formation; PC, Prince Creek Formation; SB, Schrader Bluff Formation; To, Torok Formation; Tu, Tuktu Formation; U, Umiat Formation; 1, terrestrial facies; 2, coastal facies; 3, offshore facies; on the index map, CP is Coastal Plain, NF is Northern Foothills of the Brooks Range, SF is Southern Foothills, and BR is the Brooks Range. (Redrawn and modified from Eardley, 1962.) (B) Paleoenvironmental map of Nanushuk strata. (From Detterman, 1973, reproduced with permission of the American Association of Petroleum Geologists.)

will describe the main hypotheses proposed to explain the early opening of the Arctic.

E.3.a. The Arctic Ocean Basin

Figure 8-59 illustrates major geographical and geological features of the Arctic Ocean and surrounding land-masses. The Arctic Basin is divided into two subbasins by the prominent Lomonosov ridge, which extends from northern Greenland to the Siberian shelf. The subbasin from the Lomonosov ridge to Alaska is termed the Amerasian Basin. The deep abyssal plain in the middle of the Amerasian Basin is named the Canada Basin. From the Lomonosov ridge to the Baltic shield is the Eurasian Basin. In the mid-

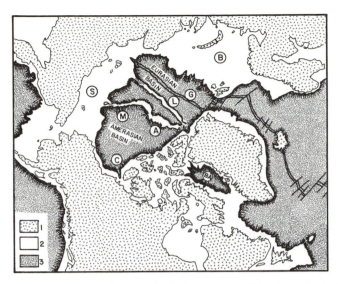

Fig. 8-59. Map of the major geomorphic features of the Arctic Ocean basin. Symbols are as follows: 1, continents; 2, continental shelf areas; 3, ocean basins; A, Alpha Cordillera; B, Barents Shelf; BA, Baffin Basin; C, Canada Basin; G, Gakkel Ridge (the active divergent plate boundary); L, Lomonosov Ridge; M, Mendeleyev Ridge; S, Siberian Shelf.

dle of the Eurasian Basin is an active spreading ridge, called the Gakkel Ridge, which is an extension of the North Atlantic Ridge into the Arctic Ocean Basin. The Gakkel ridge appears to extend into the Sadko Trough of the Siberian shelf; the Sadko Trough is seismically active and is considered a rift feature (Demenitskaya and Karasik, 1969). Two seismically active zones run southward and southeastward from the Sadko trough region, coinciding with the Verkhoyansk Mountains and Cherski Mountains, respectively.

The Lomonosov ridge is elongated roughly parallel to the edge of the Barents shelf (and both are parallel to the Gakkel ridge). It is considered by most workers (e.g., Wilson, 1963; Churkin, 1973; Herron *et al.,* 1974) to be a continental fragment rifted away from Eurasia during opening of the Eurasian Basin.

The Amerasian Basin is similar in many ways to the Eurasian Basin except it has no active spreading center. It does, however, have a ridge which runs through it parallel to the Lomonosov ridge. This ridge is termed the Alpha Cordillera near the Canadian side of the ocean and the Mendeleyev ridge on the Siberian side; we will refer to it as the Alpha–Mendeleyev ridge. The Alpha–Mendeleyev ridge has been the subject of conflicting interpretations, being considered either a fossil oceanic spreading center (analogous to the Gakkel ridge) or a fossil subduction zone (analogous to the Aleutian Archipelago). We will return to this question later in this section.

Several parts of northeastern Siberia are also important in the Arctic Ocean's history. We have already mentioned the Cherski Mountains, the most seismically active region of the USSR, which is considered to be the landward exten-

sion of the Gakkel spreading center (Churkin, 1973). Northeast of the Cherski Mountains is the Kolymski plate, a large and relatively stable tectonic element considered by some (e.g., Herron *et al.,* 1974) to be a continental block collisionally accreted against the Siberian platform. In addition, the Kolymski plate has been blamed for the Late Devonian collision which resulted in the Ellesmere orogeny. A recent reinterpretation (Andrews-Speed, 1981), however, has suggested that the Kolymski plate may actually be a collage of accreted volcanic-arc terranes (similar to the Sonomia plate). This reinterpretation is important in Arctic Ocean history and we will return to it later. To the east, the Kolymski plate is bordered by the Chukotski Peninsula. The deformed zone between Kolymski and Chukotski contains a number of ophiolitic bodies and is considered to represent a suture (called the Yuhzni Anyuy suture; Churkin *et al.,* 1980). Chukotski is believed to have been a continuous part of the sialic crust of North America since at least the Late Devonian.

E.3.b.   Opening the Arctic

Attempts to model Arctic Ocean history, using seafloor spreading, are problematical because of the lack of adequate data on the nature of the oceanic crust, on the character of the ridges which interrupt the bottom topography, and on the structure and stratigraphy of surrounding continental margins. Most syntheses of Arctic history agree that the ocean opened as the result of two separate events: the earlier opening of the Amerasian Basin (probably Late Jurassic to Cretaceous) and the later opening of the Eurasian Basin (during the Tertiary). Since the Eurasian Basin is, therefore, a Cenozoic feature, we will postpone discussion of its opening until the next chapter. Here, we will confine our discussion to Jurassic and Cretaceous events associated with the early history of the Amerasian Basin. Naturally, there are conflicting models of Amerasian Basin tectonics based on conflicting interpretations of the Alpha–Mendeleyev ridge.

Several authors (e.g., Vogt and Ostenso, 1970; Churkin, 1973) have interpreted the Alpha–Mendeleyev ridge as an extinct midocean spreading center (see Fig. 8-60). Advocates of this view point for evidence to low heat flow values along the ridge, suggestive of an inactive midoceanic ridge, transverse offsets interpreted as transform faults, and the existence of buried topography on the ridge which resembles the topography of present spreading ridges. If the Alpha–Mendeleyev ridge were, in fact, a divergent plate boundary, then it may have presided over the continental separation which opened the Amerasian Basin. In this scheme, a true rifted continental margin developed off northern Alaska whereas the continental margin north of Canada's Arctic Archipelago may have involved some component of transform movement (consider the geometry of spreading

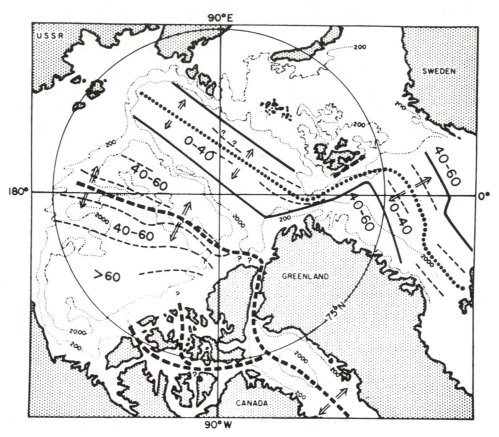

Fig. 8-60. Map of the Arctic Ocean Basin showing the Alpha–Mendeleyev ridge interpreted as an extinct spreading center; light dashed lines are inferred magnetic anomalies. Note that inferred spreading is shown to have ceased around 40 Myr. ago. (From Ostenso and Wold, 1973, reproduced with permission of the American Association of Petroleum Geologists.)

shown in Fig. 8-60). Spreading continued until about 40 Myr. ago when it apparently ceased. Tailleur (1973) proposed a modification of this model in which he suggested that earliest rifting may have begun along a spreading center striking due north from about the Alaska–Canada boundary (see Fig. 8-61). This hypothetical ridge is thought to have ceased activity around 60 Myr. ago, or at about the same time that spreading is considered to have commenced on the Alpha–Mendeleyev ridge.

There are several problems with these speculations, including the geometry of the inferred plate motions and lack of firm evidence for Tailleur's hypothetical ridge. But the major problem derives from the fact that the Alpha–Mendeleyev ridge may *not* have been a spreading center. Herron *et al.* (1974) observed that the elevation of the ridge is considerably higher than it should be were it composed solely of basaltic crust. It should, by now, have subsided nearly to the level of the surrounding abyssal plain. The fact that it has not may imply that it is not composed of basalt but rather some more felsic rock-type such as andesite. Second, Herron *et al.* pointed out that symmetrical magnet-

ic-anomaly striping of the oceanic crust, so striking in most ocean basins associated with a spreading center, is apparently lacking in the Amerasian Basin. A third argument is the presence on the ridge of Upper Cretaceous tuffs. These tuffs are not typical of midocean-ridge-type volcanism but are often associated with volcanic island-arc terranes.

Herron *et al.* interpreted the Alpha–Mendeleyev ridge as a volcanic-arc complex. They argued that the Amerasian Basin was initially opened during Jurassic time by the rifting and drift of the Kolymski plate away from its supposed location north of Canada and Alaska and toward a collision with Siberia (see Fig. 8-62). The nature of the rifting and the location of the spreading center which must have occasioned the rifting were not specified; nor was the relationship (if any) between the Kolymski plate and the Lomonosov ridge. This model also suffers from the reinterpretation of the Kolymski plate as an accreted mass of displaced arc-terranes (Andrews-Speed, 1981). If this were true, then the Kolymski ''plate'' may not have behaved as an independent tectonic element.

Following the opening of the Amerasian Basin,

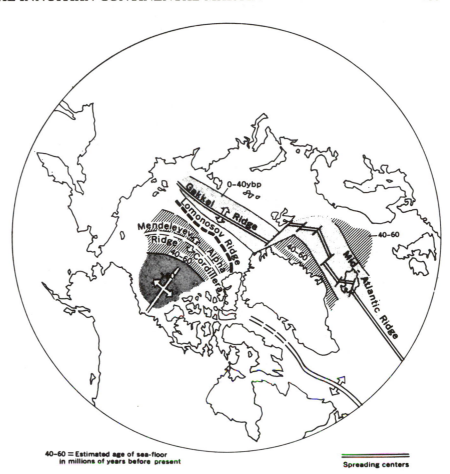

Fig. 8-61. Tailleur's seafloor spreading model of the Arctic Basin. Note that this scheme also interprets the Alpha–Mendeleyev ridge as a spreading center but also infers the presence of an earlier spreading ridge (at > 60 Myr. ago). 40–60, estimated age of sea-floor in Myr. before present; (=), spreading centers. (From Tailleur, 1973, reproduced with permission of the American Association of Petroleum Geologists.)

40–60 = Estimated age of sea-floor
in millions of years before present

Spreading centers

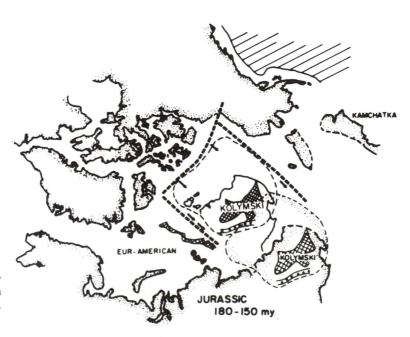

Fig. 8-62. Model of the opening of the Amerasian Basin by the rifting and drift of the Kolymski plate away from North America. (From Herron *et al.*, 1974, *Geology*, Fig. 3, p. 379.)

spreading of the Mid-Atlantic Ridge moved North America and Eurasia apart, rotating about a spreading pole located in northern Greenland. Of course, divergent motion of the two rigid plates in the Atlantic region would have resulted in convergent motion of the same two plates (see Fig. 8-63) in the Amerasian Basin on the other side of the spreading pole. Such convergence caused the initiation of subduction and the formation of a volcanic island-arc, represented today by the Alpha–Mendeleyev ridge. Subduction is believed to have continued into Early Tertiary time. Notice in Fig. 8-63 that a transform fault is inferred between the North American plate and the "Laramide plate." This fault cuts across eastern Alaska to the Cordilleran orogen, supposing right-lateral motion. It is interesting to compare this idea with the Churkin *et al.* (1980) conception of the Alaskan Cordilleran collage; consider that their "Porcupine megashear," although oriented somewhat differently from the transform fault shown by Herron *et al.*, could also be interpreted as a major right-lateral transform fault and, thus, may have been active in association with Alpha–Mendeleyev subduction.

The preceding discussion is intended in part to show that Arctic Ocean tectonics are still poorly understood. Major reinterpretations will, no doubt, be forthcoming. We will return to this discussion in the next chapter when we consider the opening of the Eurasian Basin and the Eurekan orogeny.

F. The Transition to Modern Invertebrates

F.1. Mesozoic Life, in General

The term "Mesozoic" means "middle life" and was originally based on assumptions that the overall biota represented a transition between archaic life of the Paleozoic and modern life of the Cenozoic. This concept is not invalid but it is grossly oversimplified. By the latest Mesozoic, the marine invertebrates were quite similar, at higher taxonomic levels, to modern invertebrates; however, at lower taxonomic levels, the differences are striking. On land, the Mesozoic tetrapods were vastly different from those of the earliest Cenozoic and there is little "transition" evident: indeed, the abruptness of the Cretaceous–Tertiary faunal change provides fertile ground for controversy (discussed in the next section). Further, the differences between Triassic organisms and Cretaceous organisms are no less profound than are those between the various eras.

In the following sections we make no formal attempt to describe all of Mesozoic life. Rather, focus will be on characteristic, distinctive, biostratigraphically important, and otherwise notable groups. Discussion is organized approximately by stratigraphic importance, abundance, and our personal interests.

F.2. Cephalopods
F.2.a. Ammonites

By far the most conspicuous marine invertebrates of Zuni time were those ammonoids which bore the complex type of sutures distinguished as "ammonite-type." These ammonites are distributed globally in Jurassic and Cretaceous strata and provide an unmatched biostratigraphic index suite.

To review briefly the history of the subclass Ammonoidea: the first ammonoids appeared in the Early Devonian, having most likely evolved from nautiloids similar to the common genus *Bactrites*. These ammonoids featured simply-flexed septa and wavy sutures of the goniatite type. In the Mississippian, ammonoids appeared with more complex septa yielding suture patterns in which the *lobes* (the posterior-directed folds) were subdivided into second-order lobes and *saddles* (the anterior-directed folds) while the saddles were simple; these ammonoids were of the ceratite type and dominate ammonoid faunas from Absarokan strata. In Early Permian time, a third major sutural variation evolved from the ceratite patterns and this is the ammonite-type pattern, featuring secondary lobes and saddles on both primary lobes and saddles, and in very advanced forms, featuring tertiary lobes and saddles which produce suture patterns of dazzling complexity. It is this complexity which provides much of the variation in Zuni-age ammonoids and which makes them such valuable index fossils as well as objects of beauty. Ammonite sutures are found on cephalopods ranging in age from Early Permian through Late Cretaceous but by far the majority of taxa lived from latest Triassic through early Late Cretaceous time.

Some shapes observed in Mesozoic ammonites were very aberrant, to the point of suggesting "racial senescence," but they may have actually been adaptations to specific habitats; the typical ammonoid shell shape is the tightly coiled, often convolute planispiral which served the animals well for swimming (apparent in modern *Nautilus*). The variously noncoiled and sometimes amorphous-appearing forms, called *heteromorphs*, may have been benthic and therefore did not need the balance of a symmetrical, coiled shell. Among the common heteromorphs are helical-spiral genera such as *Cochloceras* and *Turrilites*, U-shaped genera such as *Hamulina*, incomplete coiled genera such as *Scaphites*, and asymmetrical, almost knotted forms such as *Nipponites* (Fig. 8-64). Ward (1979) suggested that helically coiled ammonites were not benthic but rather were slow, balanced swimmers adapted to a variety of niches. The snail-like spiral shells may have facilitated distribution of cameral fluid (see below) among chambers and thus may have permitted more precise balancing.

The adaptive function of the complex sutures in ammonites is conjectural, but it probably relates to buoyancy and strengthening of the shell for deep diving. Clarkson

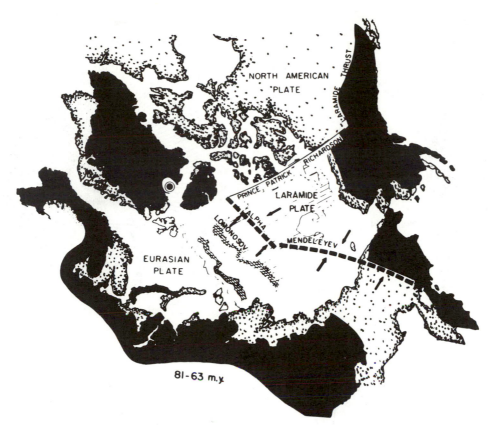

Fig. 8-63. Model of plate convergence in the Arctic which produced subduction associated with the inferred Alpha–Mendeleyev Volcanic Arc. Bull's eye in northern Greenland is pole of rotation of North America with respect to Eurasia. Black continental outline represents the assumed Cretaceous positions of the continents; stippled pattern represents present continental positions. (From Herron *et al.*, 1974, *Geology*, Fig. 4, p. 380.)

(1979) noted that ammonite shells are relatively thin compared to those of nautiloids and therefore the invagination of septa, as part of the ammonite pattern, would tend to give the shell greater rigidity and allow for deeper diving. In addition, it seems clear that the function of the shell itself is largely as a hydrostatic device (Arkell, 1957) since the animal's body in the adult occupies only a small part of the shell and the remainder is filled with gas and a quantity of *cameral fluid*. In the modern *Nautilus* (the only living shelled tetrabranch cephalopod), the gas and cameral fluid serve to balance and pressurize the shell for diving and emergence and it is safe to assume that ammonites functioned similarly. The complex sutures of ammonites may have partially served to allow precise changes in gas-to-fluid ratios in the main, unoccupied portion of the shell (*phragmocone*) and thus have allowed precise adjustments of buoyancy. Recent studies of tolerance to implosion under depth-pressure in *Nautilus pompilius* (Kanie *et al.*, 1980) suggest that roughly 785 m is the normal limit of diving for modern *Nautilus;* based on their shell morphology, it is probable that most ammonites could not dive as deep as *Nautilus.*

Two ammonoid groups (which had complex sutures but not of true ammonite type) were each apparently an-cestral to some ammonites and were deep-water forms like *Nautilus*. These ammonoids are called lytoceratids and phylloceratids and it seems likely that the diverse Zuni ammonoids evolved from these ancestral forms as the epeiric seas of the Jurassic and Cretaceous transgressed the world's cratons, creating major new habitats in greater abundances than had been present since the Permian. Arkell (1957) and Clarkson (1979) noted that ammonites apparently show *iterative evolution,* which is the process by which an ancestral stock periodically gives rise to sequences of offspring taxa while the ancestral form persists almost unchanged in a remote locality.

The stratigraphic record of ammonites seems strikingly ordered, with new forms appearing at almost regular intervals and disappearing quickly throughout most of the Zuni Sequence. But by the later Late Cretaceous, the rate of production of new taxa had dwindled greatly and by the end of the Maestrichtian, ammonoids were extinct. Indeed, virtually all shelled cephalopods were extinct by the latest Mesozoic since the nautiloids had been reduced to a single group of coiled forms and these had been apparently outcompeted through Zuni time by the ammonites. However, a relict nautilid population apparently regrouped in the Early

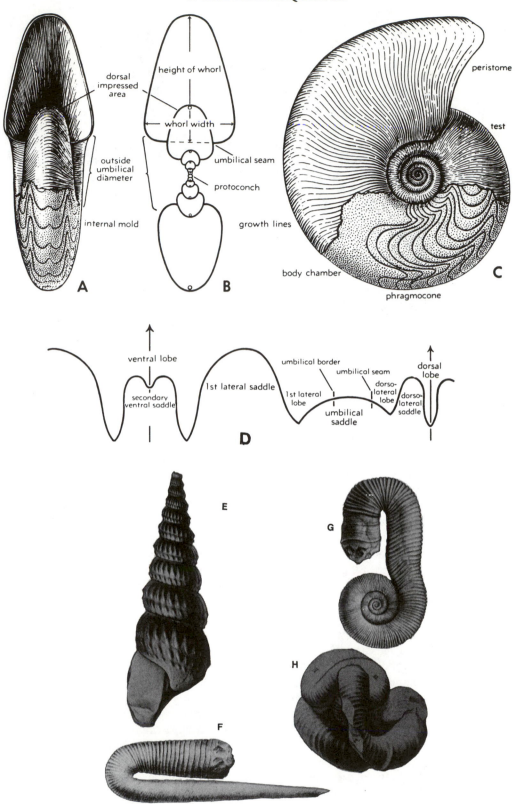

Fig. 8-64. Ammonoids, in general, and a set of Mesozoic heteromorphs. (A–D) Morphology of a representative goniatite ammonoid, *Manticoceras* (U. Devonian). (A) Ventral view; (B) cross-sectional view; (C) lateral view; (D) detail of the sutures, with saddles and lobes indicated (note arrows, which indicate the apertural direction). (E–H) Heteromorphic am- monites. (E) *Turrilites* (U. Cretaceous); (F) *Anahamulites* (L. Cretaceous); (G) *Macroscaphites* (L. Cretaceous); (H) *Nipponites* (U. Cretaceous). (All from the *Treatise on Invertebrate Paleontology*, courtesy of the Geological Society of America and the University of Kansas Press.)

Tertiary and survived to the present as *Nautilus*. Ward (1980) proposed that the end-Cretaceous extinction of ammonites opened new opportunities for nautiloids, since Tertiary nautiloids had (and have) shell forms like those of ammonites but unlike those of Mesozoic nautiloids. Presumably, post-Mesozoic nautiloids replaced ammonites.

F.2.b. Belemnites and Other Coleoids

The class Cephalopoda contains two major groups: the subclasses Tetrabranchiata (ammonoids, nautiloids, and relatives) and the Coleoidea (formerly called Dibranchiata, and including all other cephalopods). Most living coleoids are nonshelled, and therefore they logically have a poor fossil record. Coleoids do commonly have internal stiffeners such as the flexible ''pens'' of squids and the ''cuttlebone'' of *Sepia* but these are lightly mineralized structures which usually were not preserved. Coleoidea includes two orders: Octopoda and Decapoda. Octopods are the familiar, soft-bodied octopuses and *Argonauta*, which has eight arms and in the female secretes a paper-thin shell that functions as a brood pouch. The fossil record of octopods is almost nonexistent but rare specimens show that the group is at least as old as the Cretaceous.

The decapods have a generally poor record too but one group is actually superabundant. Decapods are ten-armed cephalopods which include the suborder Teuthoidea (squids); among the living forms is the common edible *Loligo* (''calamari,'' in Mediterranean restaurants), and the apparently rare, fantastic giant squids such as the 18-m *Architeuthis* (which is plausibly represented by individuals several times that size). Fortuitously preserved specimens of fossil squids are known from rocks as old as the Jurassic (Tasch, 1980). The second, living order of decapods is the Sepoidea, which includes forms with internal stiffeners such as the cuttlefish, *Sepia*, and a form with a coiled ''pen,'' *Spirula*. Despite the presence of internal calcitized structures, these cephalopods too have a poor fossil record, dating back to the Jurassic. The third decapod suborder, the Belemnoidea, includes the only common fossil coleoid group.

Belemnites possessed a stout, calcitic internal support which comprises commonly found fossils from later Mesozoic strata. The calcareous internal ''shell'' consisted of three parts: a heavy, posterior portion called the *rostrum*, which was in essence a bullet-shaped mass of solid calcite; a central-anterior cone, located in a cavity anterior to the bulk of the rostrum, called the *phragmocone;* and an extension far in front of the mass of the support called the *pro-ostracum.* Belemnites are extinct and thus the exact function of the skeleton is conjectural; but it probably served as a ballast and strengthening device. Why such a device was necessary is unknown, but the internal calcite must have served well for whatever purpose because belemnites are extremely abundant in many marine strata from the later

Triassic through the Cretaceous. Their stratigraphic range is unusual in that they persisted through the Mesozoic–Cenozoic boundary but disappeared in the Early Tertiary after a long decline throughout the later Cretaceous.

F.3. Bivalves

As we introduced in Chapter 7, bivalves dominated the Mesozoic benthic marine habitats just as they dominate modern marine habitats. We explained that mussels, pectens, numerous types of clams and oysterlike bivalves were present during the late Paleozoic and, although they were overshadowed by the diversity of Permian brachiopods, they made up much of the benthic fauna. During the Middle and Late Triassic Epochs, certain groups of bivalves initiated new radiations and comprise important index fossils for Zuni time, second only to ammonites. Indeed, they are of first-order importance in marginal, intertidal, and shallow-water environments beyond easy reach of the open sea, and in certain endemic, restricted faunas. The discussion below is organized nonsystematically because current bivalve taxonomy (indeed, the very name of the group!) is in flux and confusing to the nonspecialist, being based on a multiple set of characteristics (hinge, gill, dentition, and shell-microstructure).

F.3.a. Oysters

True oysters, i.e., pterioid bivalves of the suborder Ostreina, appeared in the Triassic and are distinguished by lacking hinge teeth, having inequal valves of which the left is commonly cemented to the substrate or to other shells, and by being generally asymmetrical but having shells which can close with virtually perfect sealing. It is the good seal around the gape of the shell that enables oysters such as modern species of *Crassostrea* to survive long periods of subaerial exposure within a sort of microenvironment enclosed within the shell. The attached habit enables oysters to build biohermal masses which, in turn, create additional benthic habitats. An additional set of oyster attributes are less obvious but extremely important in recognition of ''true'' oysters; among bivalves, only oysters have totally ''lost'' the muscular foot through specialization in the attached habit, and they have also ''lost'' the byssus, which is a feature of other pterioids (such as pectens). Morton (1967) noted that oysters evolved efficient mechanisms to expel sediment from within the valves, allowing them to inhabit silty and gravelly environments.

Early, Triassic oysters were not important benthic faunal elements and are represented best by the genus *Lopha* and several other forms with gryphaeate (i.e., like *Gryphaea;* see below) shapes (Cox, 1969). During Zuni time, however, oyster populations expanded and diversified, especially in the characteristic upper Mesozoic genera

Gryphaea and *Exogyra,* and in the long-ranging, surviving genera *Ostrea* and *Crassostrea* (Fig. 8-65). All of these common taxa featured massive shells with the lower, attached valve being inflated. In *Gryphaea* and *Exogyra* the lower valve features strong coiling reminiscent of gastropod coiling (such coiling is a persistent embryonic mollusk trait which is noticeable at the umbones of most modern bivalves); in *Gryphaea* the coiling is planispiral whereas in *Exogyra* it is helical-spiral. Both Mesozoic taxa featured unusually thick shells and Kauffman (1969) noted that the adaptive significance of such thick shells is obvious in high-energy, shallow environments (especially intertidal zones). In addition, he suggested that predation by boring organisms (especially gastropods) was rare in the thick-shelled Mesozoic oysters.

Some ostreids, notably species of *Exogyra,* are important index fossils for intervals of Zuni time. The species *E. costata,* for example, is the guide fossil for the early Maestrichtian on the Atlantic and Gulf Coastal Plains and its distribution is, remarkably, as wide as North Carolina to Mississippi. Other *Exogyra, Gryphaea,* and *Crassostrea* species are regional index taxa, especially on the coastal plains and especially in marginal-marine lithologies where cephalopods are uncommon.

Kauffman (1969) noted that *Exogyra* and *Gryphaea* lived detached from their substrate as adults and lay free on the bottom. They were apparently stabilized by the weight of their bottom valves and overall shape; it may have been that this mode of life led to evolution of the very thick shells of some taxa and that the protective aspects of the thick shells were secondary. The exogyrids disappeared at the Cretaceous–Tertiary boundary but *Gryphaea* persisted rela-

tively unchanged until the Eocene, when it too became extinct.

F.3.b.   Inoceramids, Trigoniids, and *Buchia*

These three, probably nonrelated taxa comprise large components of Zuni bivalve faunas and by their presence tend to distinguish these faunas from recent assemblages. *Inoceramus* has been mentioned previously in our discussion of the Niobrara Formation: it is the most characteristic bivalve in western interior Cretaceous strata and in some sequences reached enormous size and superabundance. The genus is quite distinctive, having a moderately thick shell with very strong concentric wrinkles on the surface. The inoceramids lacked hinge teeth and are classified as a distinct family within the Pterioida. Hattin (1965) observed that giant species of *Inoceramus* in the Niobrara Formation are commonly encrusted with larval oysters (called *spat*) and essentially they make solid substrates by their individual presence, much as the oysters do the same with their multitudes. *Inoceramus* is also found commonly in western Gulf Coast Cretaceous strata but is generally rare in the Atlantic Coastal Plain. The family Inoceramidae ranges from Early Permian through Late Cretaceous time, with questionable Early Tertiary representatives (McCormick and Moore, 1969); but the genus *Inoceramus* is of Jurassic and Cretaceous age. In western interior sequences, especially, *Inoceramus* is an important index fossil and many sequences have both *Inoceramus* and ammonite faunal zones defined.

Buchia is another common taxon within the Pectinacea, which in many ways resembles the mytilids (mus-

Fig. 8-65. Four characteristic Jurassic and Cretaceous bivalves. (A) *Exogyra* (Ripley Formation, Alabama); (B) *Gryphaea* (Sundance Formation, Wyoming); (C) *Trigonia* (Ripley Formation, Georgia); (D) *Inoceramus* (interior of incomplete specimen, Fort Hayes Member, Niobrara Chalk, Kansas). (A–C), approximately one-fifth life size; (D), approximately one-tenth life size. (DRS photographs.)

sels) by having asymmetrical, strongly *umbonate* (pointed) valves with strong concentric ribbing. The family Buchiidae is of Late Triassic through Cretaceous age but the genus *Buchia* is found only in Jurassic and Lower Cretaceous rocks. In western interior and Pacific sequences, *Buchia* is an important index genus and defines several zones in the Jurassic.

A third miscellaneous Mesozoic bivalve group, the Trigoniidae, comprise their own order which ranges from Devonian to Recent in age, with questionable Ordovician representatives [Moore (ed.), 1969]. Trigoniids feature trigonal outlines which are truncated posteriorly and with very strong umbonal ridges extending to the posterior angles of individual valves. Most trigoniids are heavily ornamented (Fig. 8-65) with varying types of ornamentation, and bear hinge teeth of the type called schizodont. Although trigoniids have a long stratigraphic range and are extant, in Mesozoic time they were very abundant whereas at present they are reduced to the single genus *Neotrigonia*. In Oxfordian through Meastrichtian strata, *Trigonia* is often the most abundant of bivalve genera present (although the other groups discussed are somewhat more spectacular and distinctive).

F.3.c.   Clams, in General

Briefly, and in order not to give the false impression that only the bivalves discussed above represent Zuni faunas, we must comment on the presence of a virtually modern "clam" component to Zuni faunas. As is often the case with character actors in theater, the ordinary and unspectacular are generally unobserved and unremembered. So too many common modern clams of various families trace their ancestry back to and preceding Zuni time. Among such taxa are *Crassatellites, Venericardia, Yoldia, Nucula, Pecten, Lucina, Arca,* and a host of others. Except locally, where *Trigonia, Inoceramus,* or one of the Mesozoic ostreids we have discussed are exceedingly well-represented, Zuni bivalve faunas overall still contain more common-looking "clams" and "scallops" than they do the "characteristic Mesozoic bivalves" discussed: in essence, it is a case of "*plus ça change, plus c'est la même chose.*"

F.3.d.   Rudists

A final bivalve group we will consider contains the most unusual of Mesozoic taxa. Rudists were a diverse group of unequal-valved bivalves which appeared in the Late Jurassic, flourished through the Cretaceous in the Tethyan region, and then became extinct at the end of the period. Their most striking feature was a tendency toward elongation of the lower valve with concomitant reduction of the upper; in this they converged on the morphology of horn

coral and certain Permian productid brachiopods. In addition, they were commonly reef-forming and effectively outcompeted coral in the Tethyan region during their time.

While most rudists were a few centimeters in overall size, some grew to nearly 2 m total length. In addition, the reef habit resulted in tall, erect growth in some species since, as Wilson (1975) stated, a premium exists in the competitive reef environment for rapid growth. The range of rudist morphologies is striking. Ancestral forms such as the Jurassic *Diceras* tended to resemble a ram's horns, with both valves being coiled. Typical genera showing the horn-coral morphology include the abundant genera *Hippurites* and *Radiolites* (Fig. 8-66). Further, some erect, elongate rudists were globose and barrel-shaped, whereas others were slender. Erect, slender forms were often intertwined in reef environments and comprised framework-builders.

The shells of attached rudists were often vesicular with open space possibly comprising a majority of the shell volume. Upper valves articulated with lowers by means of a very reduced set of hinge teeth: commonly, a single tooth in the lower valve and a pair of teeth in the upper (with corresponding sockets). These teeth were very large and interlocked in the bivalve such that the upper valve could only slide up and down, without rotating. This unique dentition is termed *pachyodont*. A further specialization is observed in several taxa, notably in *Radiolites*, where the upper valve is serrate (as in the Permian strophomenid brachiopod *Leptodus*). Presumably this allowed exchange of water in the bivalve with the upper valve down. Kauffman (1969) suggested that rudists may have had exposed mantle tissue containing *zooxanthellae* (symbiotic algae) in the manner of the modern "giant clam" *Tridacna:* if so, this points to further convergence on the habit of hermatypic coral.

The importance of rudists in the Cretaceous of the Tethyan region cannot be overemphasized. North American rudist populations and reefs are distributed in and around the Gulf of Mexico and on both Mexican coasts. There, they serve as important Cretaceous biostratigraphic fossils: the interested reader is directed to Wilson (1975), Skelton (1979), and Kauffman and Sohl (1974) for discussion of rudists and reefs.

F.4.   Microfossils

F.4.a.   Planktonic Algae: Coccolithophores, Dinoflagellates, and Silicoflagellates

In Zuni time, and most especially in the Cretaceous Period, several planktonic algal groups began their prominence in the oceans and left major fossil records. The Coccolithophyceae, a class within the Chrysophyta, appeared in the Late Triassic, diversified in great numbers of taxa in the Jurassic, and then underwent virtual explosive evolution in

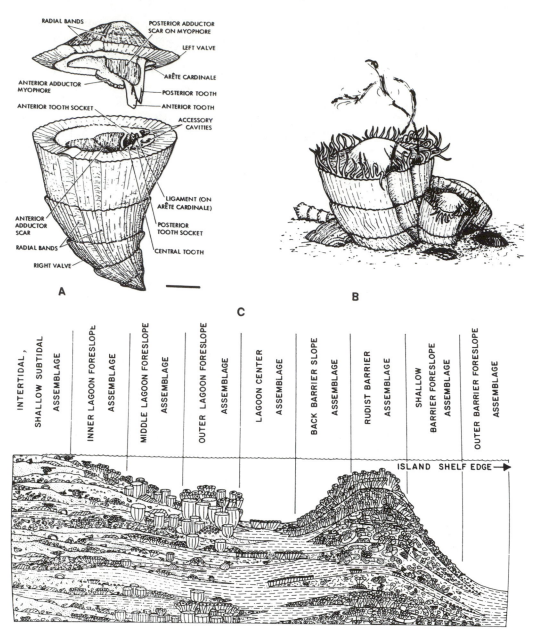

Fig. 8-66. Rudists and their reefs. (A) Labeled diagram of a representative rudist bivalve, *Radiolites* (bar = 1 cm); (B) a small group of *Radiolites* in inferred life positions; (C) a model of Caribbean rudist framework (i.e., biohermal-structural) distributions in the Cretaceous: note the variety of morphologies, including ram's-horn types (termed caprinids). (A and B, from Skelton, 1979; C, from Kauffman and Sohl, 1974.)

the Cretaceous (Brasier, 1980). Coccoliths are the individual plates of the coccolithophore, each 3–15 μm in diameter, which in bulk constitute very large proportions of the famous Cretaceous chalk formations such as the Niobrara and Austin. Coccolithophores each have numerous coccoliths encrusting their outer surface and the alga may be 20 μm in overall diameter. As coccolithophores die, their plates are shed, building chalky calcareous deposits but only in waters no deeper than the carbonate compensation depth; thus, chalk buildups are largely deposits of past epeiric seas. After Cretaceous time, coccoliths underwent intervals of depopulation, extinction, and rediversification, with a very sizable evolutionary burst in the Holocene, and a large number of taxa are present today. Among calcareous algal taxa classified with coccoliths are the discoasters, which lived in the upper Miocene and Pliocene Epochs and are distinguished by having stellate (star-shaped) plates with variable numbers of arms. An additional group of marine calcareous fossils are the pentagonal plates of *Braarud-osphaera*, which are known from Cretaceous through Recent sediments. Coccoliths, including discoasters, have served as important index fossils, especially in oceanic core samples and in deep-water carbonates. Additionally, ratios of cold-to-warm-water coccolith taxa have yielded paleoclimate measurements for the later Tertiary and Quaternary Periods.

A second important algal fossil group which reached prominence in the Mesozoic are the dinoflagellates, which constitute a separate division of Protista, the Pyrrhophyta. Living dinoflagellates range from 20 to 150 μm in length and take on a wide variety of shapes. Typically they have two flagella and irregularly suboval bodies with whiplike extensions. The cells are stiffened with cellulose and are not preserved in the adult stage; however, they have a resistant planktonic or benthic encysted stage which does fossilize. Dinoflagellate cysts are important microfossils and may become more so in the future. Oldest dinoflagellate cysts are of Rhaetian (uppermost Triassic) age but they are rare and of only one genus (Sarjeant, 1974). Beginning in the earliest Jurassic, they diversified greatly and, like the coccolithophores, reached an all-time peak of diversity in Upper Cretaceous seas. Unlike the coccoliths, dinoflagellates are not as diverse in Late Tertiary through Recent time as they were in the Cretaceous although there are over 1000 species known and they are very abundant (let the reader not forget that diversity and abundance are not necessarily related attributes). Dinoflagellates also are common freshwater algae and may be found in most drinking-water supplies. As with coccoliths, dinoflagellates can be used to determine paleotemperatures; additionally, they may also be used as water-depth and shoreline-distance indicators based on the known distributions of depth- and temperature-restricted taxa (Sarjeant, 1974).

A third planktonic algal group, the silicoflagellates, appeared first in the Early Cretaceous and persist to the present. Silicoflagellates are chrysophyte (green) algae comprising less than 20 known genera which secrete hollow, opaline siliceous skeletons. Like most planktonic algae, they are very small (20–100 μm in diameter). Silicoflagellates are not as common as are coccolithophores and dinoflagellates, and they are not generally used as index fossils. They reached their peak of diversification in the Miocene but are still represented by more than 50 species.

F.4.b. Mesozoic Foraminifera

Foraminiferan populations changed in makeup during Zuni time with the appearance of two suborders (Rotaliina and Miliolina) and with the emergence of planktonic taxa, largely from the rotaliids. Rotaliina are distinguished by having *hyaline* (i.e., glassy) tests which are perforate and commonly in the planispiral-coiled form (Fig. 8-67). A large number of morphological variants emerged from the basic rotaliid stock and the group today is the most numerous in taxa and absolute numbers.

Miliolids may have descended from the Late Paleozoic fusulinids and feature *porcelainous* (i.e., opaque, imperforate) tests. Many of the smaller taxa have straight or curved simple tubelike chambers arranged in parallel groups. The generic names of such forms are based on the number of tubes (e.g., *Triloculina*, *Quinqueloculina*). On the other hand, many large miliolids have fanlike, spiral, discoidal, and other very different shapes.

Rotaliids of the superfamily Globigerinacea constitute the most abundant planktonic foraminifera. Globigerinid rotaliids generally have a *trochospiral* pattern of chambers, which contributes to their buoyancy and enables planktonic existence. They thrived in the Late Cretaceous epicontinental seas and contributed very large proportions of the calcareous debris making up chalks. Planktonic foraminifera continue to thrive in the Holocene but their diversity is reduced somewhat from the Cretaceous (and subsequent Eocene and Miocene) diversity highs.

F.5. Echinoderms

F.5.a. Pelmatozoans

After virtual extinction of crinoids in the latest Permian, and after total extinction of blastoids at the same time, the Early Triassic seas were clearly short of pelmatozoans. Whether any Paleozoic crinoids survived the era boundary depends on taxonomic placement of a Rhaetian-age group (represented by the genus *Encrinus*); if these were inadunates, then they were relict Paleozoic crinoids. However, most authors believe they were articulate crinoids, the post-Paleozoic class, and their apparent resemblance to Paleozoic inadunates indicates ancestor–descendant relationship. By the latter identification, one might conclude that no known pelmatozoans survived the era boundary extinction (however, note the comment below).

The articulate crinoids appear in rocks from later Early Triassic time and this same crinoid class survives today and comprises the totality of post-Paleozoic Pelmatozoa. Articulate crinoids are quite different from their predecessors and in general show simplification of structures in the calyx and adaptations to deep-water and pelagic life rather than to epicontinental seas. It is probable that during the Permo–Triassic hiatus in crinoid history, remnant crinoid populations retreated to deeper waters and there adapted and diversified, to reappear in epicontinental strata with renewed later Early Triassic transgression.

Among the structural modifications of articulates are reduction of plates in the calyx to usually five each of

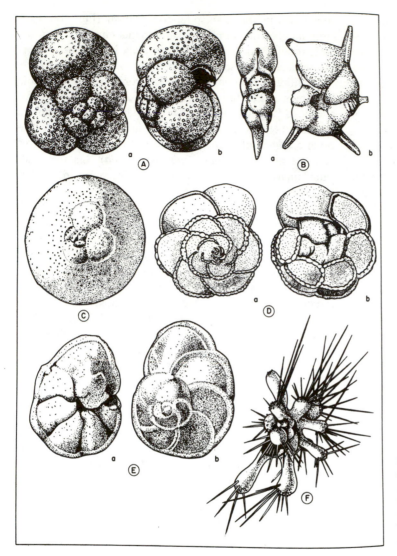

Fig. 8-67. Representative planktonic rotaliid foraminifera. (A) *Globigerina* (Paleocene to modern); (B) *Hantkenina* (Eocene); (C) *Orbulina* (Miocene to modern); (D) *Globotruncana* (U. Cretaceous); (E) *Globorotalia* (Paleocene to modern); (F) *Hastigerinella* (modern, shown with pseudopods extended). (From Tasch © 1980 by John Wiley & Sons, reprinted with permission of John Wiley & Sons.)

basals, infrabasals, and radials, while the arms simultaneously tend to be longer and more complex. The stems of articulates are typically short (under 50 cm) and most modern crinoids inhabiting waters less than 100 m deep are unstalked. Perhaps most unusual of articulate crinoid features is the existence of motile taxa; many Mesozoic and Cenozoic crinoids detached themselves from their stems during late larval stages and became free-living. Clarkson (1979) described behavior patterns for vagrant crinoids including the ability to hide in rocks by day and emerge by night to feed, ambulating on their *cirri* (small appendages near the base of the calyx). Among the better-known vagrant crinoids are the order Comatulida, which appear in Jurassic strata and may often be found in great abundance. Another stemless, motile group are the Uintacrinida, which lived only during the Late Cretaceous and are very abundant in limestone chalks from the western interior (such as the common *Uintacrinus* specimens in the Niobrara Formation from Kansas).

While stemmed articulates in the Recent are virtually restricted to deep ocean waters, in Zuni time they inhabited epeiric seas at shallow depths. The stems of crinoids are common fossils in many deposits from the western interior, many of which were likely formed in much less than 60 m of water. The stellate columnals of *Pentacrinites* (Fig. 8-68), for example, are common fossils in the Jurassic Sundance Formation in Wyoming. A minor, but locally common crinoid group of the Mesozoic, the order Roveacrinida, were ornate, lightly built forms that were not only unattached and free-living, but were also apparently planktonic-pelagic. The genus *Saccocoma* is a common representative of this group, found in apparent wind-collected deposits in many Jurassic lithographic limestones such as the Solenhofen in Bavaria.

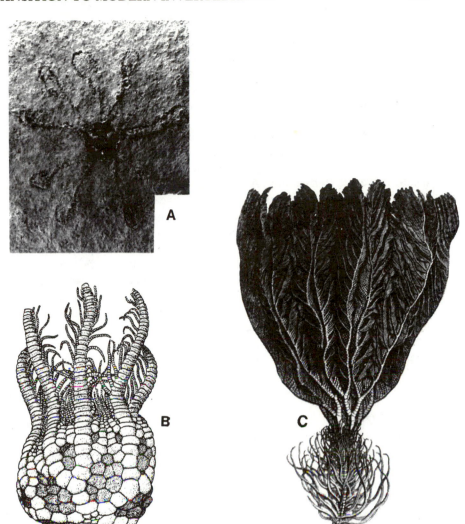

Fig. 8-68. Representative articulate crinoids. (A) *Saccocoma* (U. Jurassic), preserved in lithographic limestone (the Bavarian Solenhofen), a small, unattached form; (B) *Uintacrinus* (U. Cretaceous), another unattached crinoid very common in western interior North America, and reaching arm spans of 2 m; (C) *Pentacrinites* (Jurassic), a stemmed, attached articulate with star-shaped columnals, also very common in the western interior (notably the Sundance Formation). (A, DRS photograph; B and C, from the *Treatise on Invertebrate Paleontology*, courtesy of the Geological Society of America and the University of Kansas Press.)

F.5.b. Echinoids

Commonly known as "sea urchins" and "sand dollars," echinoids comprise a class of echinoderms featuring saclike bodies which are usually covered with spines, tube feet on the bottom surface which function in many ways and which make most echinoids motile, a water-vascular system which is largely involved with motion of the tube feet, and, in most groups, a five-part jaw apparatus, called "Aristotle's lantern," which enables the beast to break down food by the rasping and tearing action of five strong teeth in the lantern. Traditionally, the echinoids were classified in two main groups: irregular and regular forms, depending on whether or not they showed bilateral symmetry superimposed on the conventional echinoderm pentaradial symmetry. This classificatory scheme seems invalid since the irregular morphology has likely appeared several times independently in echinoid history (i.e., irregular echinoids

are *polyphyletic*) and a new taxonomy was used in the *Treatise on Invertebrate Paleontology* based on Durham and Melville (1957); unfortunately, this taxonomy is too detailed and complex to present here. Since we are interested primarily in post-Paleozoic echinoids, which were elements of a major radiation of new forms, we can simplify the taxonomy as follows: the subclass Perischoechinoidea includes exclusively Paleozoic taxa except for a single order, the Cidaroida, which survive to the present and were the ancestors of all post-Paleozoic echinoids. All Paleozoic echinoids were regular (i.e., pentaradially symmetrical) and possessed lanterns. All post-Paleozoic echinoids are placed in the subclass Euechinoidea, except for the surviving cidaroids, and euechinoids may be regular or irregular and include groups which do not have lanterns. Within the euechinoids are four superorders and at least 17 orders including the Clypeasteroida (sand dollars).

While the cidaroids survived the Permo-Triassic extinction, they did not evolve the diversity of Mesozoic taxa until the latest Triassic and Early Jurassic (Harland *et al.*, 1967). Some 13 orders of euechinoids appeared in the Jurassic or latest Triassic and three more appeared in the Cretaceous; all of the euechinoid superorders survive to the present as do most of the orders which appeared in the Mesozoic.

At least one taxon of irregular echinoids, the genus *Micraster,* received widespread attention because its Late Cretaceous evolutionary history is documented in minute detail. In the late 19th century, A. W. Rowe studied several thousand *Micraster* specimens through successive horizons in the English chalks. The clear evolutionary sequence in *Micraster* remains one of the best-known examples of gradual evolution.

Echinoids generally make excellent fossils since their skeletons are well-calcified. They are usually found as complete specimens but many strata contain echinoid spines as common fossils. As with other echinoderms, the plates making up the echinoid test as well as the spines are single calcite crystals. The reader interested in details of evolutionary trends in post-Paleozoic echinoids is referred to Kier (1974).

F.6. Bryozoa

Most Paleozoic bryozoans became extinct in latest Permian time. Harland *et al.* (1967) show one questionable, persistent line of fenestrate cryptostomes to survive into the Lower Triassic. In the later Triassic, only one family of ctenostomes and two or three families of cyclostomes are known to have been present. In the Jurassic, several new families of cyclostomes arose and by the earliest Cretaceous there were seven cyclostome families, which represents their all-time highest diversity. Cyclostomata survive to the present but only five families remain from the seven of the Cretaceous. Ctenostome bryozoans have membranous skeletons and so their fossil record is quite imperfect; the fossil record shows that a single family persisted through the early-Middle Cretaceous, and then disappeared. In the Miocene Epoch, a new group of ctenostome bryozoans appeared and persist to the present; presumably their ancestors were somewhere back in ctenostome history but for unknown reasons left no fossils or trace fossils.

The real story of post-Paleozoic bryozoan evolution deals with the order Cheilostomata, which appeared in the latest Albian and radiated almost explosively through the Late Mesozoic and Early Tertiary such that more than 81 families are defined in the *Treatise* (Bassler, 1953) within two suborders: the Anasca and the Ascophora. Harland *et al.* (1967) show more than 50 surviving cheilostome families. The excellent fossil record of cheilostomes is in part due to the relative youth of the order and to the generally excellent preservation of Upper Cretaceous and Tertiary marine strata; and it is in part due to the well-calcified skeletons of the organisms. Cheilostomes are thought commonly to have evolved from cryptostomes, and are distinguished from older bryozoan groups by having chitinous opercula covering individual zooecia (see Chapter 5 for morphological terms) and by having several types of specialized zooids (Fig. 8-69). In general, the Cheilostomata are highly varied and represent an eminently successful and dominant group in the modern seas.

The several types of specialized zooids present in cheilostomes are useful for classification of the order and they are also interesting examples of variability and specialization. Among these, *heterozooids* within a colony are projecting structures which resemble birds' heads and in fact are called *avicularia,* meaning, more or less, birds' heads. The "beak" of an avicularium is composed of the hinged operculum and sets of muscle serve to give the beak a definite "snap"; the avicularia apparently serve a defensive function for the bryozoan colony and provide a deterrent to colonizing and parasitizing organisms. Ryland (1970) stated that it is common to find nematodes grasped in avicularia. Other heterozooids include avicularia with highly elongate opercula which become slender and long enough to vibrate. These are called *vibracula* and they too may serve to prevent settling organisms from invading the bryozoan colony. Cheilostomes also develop egg broodchambers called *ooecia* which are readily distinguishable on examination of the colony. All of these specialized zooids, by their presence or absence among cheilostome groups, and by their frequency and arrangement within groups, serve as the basis of both classification and recognition of taxa.

F.7. Brachiopods

Most Paleozoic brachiopod groups became extinct at the Permo-Triassic boundary. Especially severe were the extinctions of flourishing productid and other strophomenid groups, which virtually characterized some Permian faunas. In the Early Triassic, only three strophomenid, one or two rhynchonellid, and four spiriferid families were present. By the later Early Jurassic, the spiriferids were extinct and only a single, persistent line of strophomenids were left among those orders, while the terebratulids expanded to two families and the rhynchonellids expanded to three. In general, brachiopods from the Jurassic to the present have not changed their population makeup dramatically except that terebratulids and rhynchonellids comprise virtually the entire modern articulate brachiopod population of the oceans, with two families of probable strophomenids surviving. Terebratulids did radiate slightly in later Tertiary time and are represented at present by seven families, whereas the

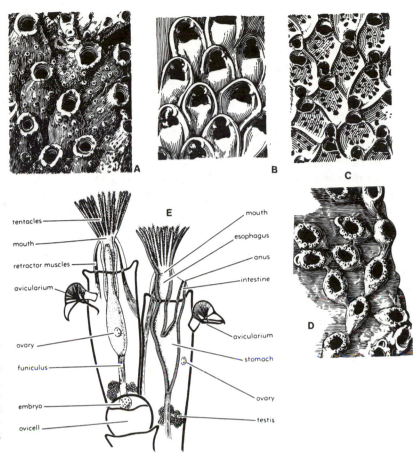

Fig. 8-69. Morphology of cheilostome bryozoa and representative taxa (A–D show the surface of colonies). (A) *Coleopora* (Miocene to modern); (B) *Floridinella* (Oligocene); (C) *Discoporella* (Miocene to modern); (D) *Allantopora* (Cretaceous to modern); (E) morphology of cheilostomes, showing several specialized zooids. (All from the *Treatise on Invertebrate Paleontology*, courtesy of the Geological Society of America and the University of Kansas Press.)

rhynchonellids have undergone no notable trends since the Late Cretaceous except for an interval of radiation during the Eocene. There are four Recent rhynchonellid families. Articulate brachiopods faded decidedly from importance in marine ecosystems after the Permian, even though they persist in fair numbers and are locally common in modern and Tertiary assemblages from tropical and subtropical waters.

Inarticulata contains the champions among persistent taxa. Harland *et al.* (1967) show the history of the three main groups of post-Mississippian inarticulates to be as follows: in Early Pennsylvanian time, the Lingulidae, Discinidae, and Craniidae were the surviving superfamilies. Since the Pennsylvanian and up to the present, all three superfamilies have survived. No new groups have emerged and few lower taxa within the three superfamilies have changed. The history of all three inarticulate groups is an amazing example of *bradytelic evolution* and all surviving inarticulates may be considered "living fossils." A fair number of species of *Lingula*, *Glottidia* (Lingulacea), *Philhedra* (Craniacea), and *Discina* (Discinacea) are present in modern oceans and the class Inarticulata seems to be in no danger of becoming extinct in the near future.

F.8. Arthropods: Decapod Crustaceans

Decapods are an order of the class Crustacea: in laymen's terms the group includes common shrimp, crayfish, lobsters, and crabs. Note the use of "common" with shrimp, which is intended to distinguish decapods of the shrimplike form from nonrelated taxa such as "pistol shrimp" (Stomatopoda) and "krill" (Euphausiacea). Characteristics which distinguish decapods include: presence of ten thoracic legs (hence the name), generally large size, carapace fused to all thoracic segments, and common development of claws. The group may have evolved from euphausiid "krill" (Glaessner, 1969) as suggested by the morphology of modern forms, but the fossil record of krill is virtually nonexistent and the idea is, therefore, unprovable. Shrimp morphology is the primitive format, with lobsters and crabs appearing after shrimp and with crabs representing the most specialized group. The size factor noted in the brief set of distinguishing characteristics of decapods is significant. Arthropods tend to be small (for example, there are great numbers of taxa among arthropods with individuals generally weighing less than 0.01 g) whereas decapods reach sizes to 4 m (spider crabs) and weights over 30 kg (lobsters).

The earliest-known fossil decapods appear in imprecisely dated strata from outside North America, of Permo-Triassic age. These oldest decapods include shrimp from the same family that includes the commercial modern genus, *Penaeus,* and representatives of a second family, the Erymidae, found in the Permo-Triassic of Siberia. During the Triassic, seven families appeared, including representatives of three suborders. The Jurassic, however, was the apparent time of major radiation of early decapods, with 22 families present by the end of the period including the Nephropidae (containing *Homarus,* the American lobster) and the first crabs in the infraorder Brachyura. Jurassic lobsters include excellent fossils of *Eryon* (Fig. 8-70) from the Solenhofen Limestone in Bavaria. Also found in Jurassic strata are the first representatives of the galatheid crabs, which in the Holocene occupy bathyal marine environments, and the first Callianassidae, a family of burrowing shrimp that live in marine intertidal zones in the Holocene. Callianasid shrimp (Fig. 8-70) create burrows which resem-

ble the ubiquitous *Ophiomorpha* burrows found in many Upper Cretaceous strata. During the Cretaceous, a diverse collection of crab families appeared while a number of the primitive shrimp groups became extinct. By the end of Cretaceous time, decapods had probably reached modern proportions of the marine fauna and probably occupied the same variety of habitats they now enjoy; however, the crab groups were still underrepresented relative to modern faunas. During the Tertiary, many new crab families appeared, especially within the Brachyrhyncha, Oxyrhyncha, and Cancridae; this Early Tertiary diversification appears to have been a virtual evolutionary "explosion." A final major event among decapods is shown by Glaessner (1969) to comprise a similar explosive diversification among the Anomura, a group which includes the deep-water crablike taxa. Five new families of anomurids appeared in the Holocene, along with certain representatives of doubtfully assigned fossil taxa; however, older representatives of these deep-water decapods may not have been preserved as fossils and the diversification may be more apparent than real.

G. Mesozoic Terrestrial Plants

G.1. Summary of Events among Terrestrial Floras at the Permo-Triassic Boundary

We last discussed terrestrial plants in Chapter 6, in context with the rise of arborescent forms in the Late Devonian and with the beginnings of the massive coal-swamp floras of the Mississippian and Pennsylvanian. In that same discussion we traced plant history in a general way through the late Paleozoic and noted that Permian floras, as known from a very imperfect fossil record, were quite similar to those of the Pennsylvanian. But, whereas the Pennsylvanian was a time of very favorable conditions for plant growth and preservation across much of the craton, Permian environments may have been far less favorable for life and certainly were less favorable for preservation of terrestrial plants.

At the era boundary, major changes in terrestrial plant communities occurred. By the Late Permian, all arborescent lycopods and sphenopsids were extinct; in fact, most important taxa from the Pennsylvanian coal swamps, such as *Lepidodendron, Sigillaria,* and *Calamites,* were gone by the Early Permian. These lycopods and sphenopsids were survived by herbaceous (i.e., small, nonwoody) members of the same groups, which persist to the present. Among other important Paleozoic plants, the so-called "preferns" of the order Coenopterales became extinct in Late Permian time whereas "true ferns" of the order Marattiales persisted into the Mesozoic and then evolved three new families from the single Paleozoic family. All three Mesozoic families are extant. The dominant order of "true ferns," the Filicales, comprised three families in the Paleozoic and each survived to the Recent, with more than a dozen additional Mesozoic

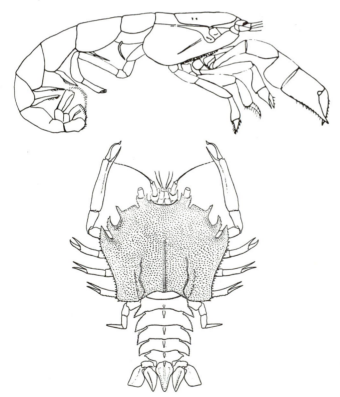

Fig. 8-70. Representative decapod crustaceans. (Above) *Callianassa,* a modern burrowing shrimp, which is probably very similar to the Late Cretaceous form which left the ubiquitous trace-fossil *Ophiomorpha.* (Below) *Eryon* (M. Jurassic to U. Cretaceous), an early form showing a rigid lobsterlike carapace. (Both from the *Treatise on Invertebrate Paleontology,* courtesy of the Geological Society of America and the University of Kansas Press.)

families added and another dozen Cenozoic families also added: the Filicales were in fact one of the dominant Mesozoic plant groups.

The "seed ferns" of the Paleozoic, the class Pteridospermopsida, declined drastically in diversity in the Late Pennsylvanian and almost disappeared at the end of the Permian; however, two orders of probable seed ferns are found in Triassic strata. Among these persistent groups are the Glossopteridales, which includes the genus *Glossopteris,* the characteristic element of the widespread Gondwanaland flora from Permo-Triassic time.

Finally, and to begin discussion of new Mesozoic floras, among Paleozoic arborescent gymnosperms, the Cordaitales became extinct at the end of the Permian, with the questionable exception of a few Triassic taxa assigned by some authors to the group, while other gymnospermous orders survived the era boundary and went on to dominate Mesozoic terrestrial floras.

G.2.   Mesozoic Gymnosperms

From the perspective of the botanist, the Mesozoic (especially the early Mesozoic, i.e., the Triassic and Jurassic) is the "Age of Gymnosperms" rather than the "Age of Reptiles" as the more common view would hold. Mesozoic gymnosperm populations were in part like those of modern floras, featuring conifers such as pines and *Sequoia;* but in part were quite different, being notably rich in cycads, ginkgophytes, araucarid conifers, and other groups now extinct or rare.

G.2.a.   Conifers

The Coniferales includes the most important of modern gymnosperms, including pine, fir, spruce, hemlock, cypress, yew, juniper, cedar, sequoia, redwood, larch, and other familiar trees. They are recognized readily by their so-called "softwood," cones, generally evergreen habit, and gummy resins. The largest known trees, past or present, are conifers. The order Coniferales was represented in the Paleozoic by a single family which appeared during the Late Pennsylvanian, and which can be distinguished from cordaitale gymnosperms by details of the female cones (conifers have cones with separate sexes, borne by the same tree). During the Triassic, four new conifer families arose and in the Jurassic five additional families evolved. During the Cretaceous, one new conifer family appeared and since that time little change in taxa has occurred. There are seven conifer families which survive almost unchanged from the Cretaceous, including the Pinaceae (e.g., pines, larch, spruce, hemlock) which appeared in the Early Cretaceous, and the Cupressaceae (cypress and junipers), which date to the Early Jurassic. Among the seven families are some less-

common, extant families including the Araucariaceae, represented today by two genera but in the Late Triassic comprising the imposing *Araucarioxylon* fauna of the Chinle Formation.

G.2.b.   Cycads and Bennettitales

These orders of Mesozoic gymnosperms featured palmlike foliage, with leaves reaching several meters' length in some groups, and with short, stout, leaf-scarred trunks (Fig. 8-71). The differences between cycads and

Fig. 8-71. Representative Mesozoic cycads. (A) A Lower Cretaceous cycad trunk, showing the leaf-scars in a honeycomblike pattern; (B) reconstruction of *Williamsonia* (U. Triassic through Jurassic), namesake of a cycad family; (C) reconstruction of *Leptocycas* (U. Triassic), a small, early cycad, known from relatively good material from North Carolina. (A, DRS photo; B, from Andrews, © 1961 by John Wiley & Sons, reprinted with permission from John Wiley & Sons; C, from Stewart, 1983, *Paleobotany and the Evolution of Plants,* courtesy of Cambridge University Press.)

Bennettitales center on the reproductive structures, which in cycads are *dioecious* cones (i.e., borne on separate male or female plants) whereas in Bennettitales were *monoecious* cones (i.e., borne on the same plant, as in conifers). Andrews (1961) suggested that the similarities of foliage between these groups obscure very real differences and that the groups are poorly related; however, Taylor (1981) classifies Bennettitales as the principal elements of the Cycadeoidophyta (a division), which suggests they are distinct from true cycads (Cycadophyta) but related to them.

Cycads and Bennettitales are the essence of popular conceptions of Mesozoic lush plant life. In fact, they were relatively small arborescent plants and may have been no more common than conifers and Filicale ferns; certainly the dominant trees of the upper stories in Mesozoic forests were conifers. Harland *et al.* (1967) showed the cycads to range back to the Late Triassic with certainty and to the Late Permian based on imperfect evidences. The Bennettitales also have Late Triassic occurrences but in addition their earliest reported occurrences are as early as the Pennsylvanian. The common fossils of these groups are leaves, followed by wood, cones, inflorescences, and other parts. In many respects, the flowers of the cycads are similar to those of angiospermous plants such as *Magnolia,* and cycads may have been the closest thing to pre-Cretaceous flowering plants, although the flowers of cycads lacked elements of angiosperm flowers (such as the carpels of *Magnolia;* Arnold, 1947).

Both groups probably evolved from seed ferns but the fossil record is inadequate to document the events; they are not common fossils except as foliage imprints in some Jurassic continental deposits. Some units, such as the Morrison Formation, have yielded numerous leaf, trunk, and fruit specimens, notable in the O. C. Marsh collections from Colorado. Andrews (1961) noted that most prolific bennettite collections have come from Lower Cretaceous nonmarine strata of the Black Hills where silicified trunks are present in the Dakota sandstones.

Bennettitales declined during the Early Cretaceous and became extinct in the late Campanian. Cycads similarly declined through the Cretaceous but persist to the present and are represented by nine genera and over 90 species including *Zamia floridana,* in central and southern Florida.

G.2.c.   Ginkgoes

The best-recognized "living fossil" plant is *Ginkgo biloba,* the sole surviving species (and genus) of the order Ginkgoales, which once were important trees of the Mesozoic forests. Ginkgoes have separate male and female trees (i.e., they are dioecious, like the cycads) and *cognoscenti* among ginkgophiles know not to plant females near their homes because the fruit are foul-smelling and unsightly when they fall from the trees. The separate sexes of ginkgo reflect a common Mesozoic mode of reproduction in gymnosperms.

Ginkgoes also feature notched, fan-shaped leaves borne in clusters of shoots from branches. The male inflorescences are catkins, whereas the female fruit are fleshy seeds. In growth pattern the ginkgo is very odd, growing straight up for perhaps 30 years before spreading; yet, as Andrews notes, terminal branches may grow half a meter in a single season. Its very distinctive characteristics easily separate it from all other living plants, including other gymnosperms.

The fossil record of ginkgophytes extends back to the Early Permian of France (Taylor, 1981). The taxonomy and relations among fossil ginkgoes are very poorly worked out and it is almost impossible to determine how many groups of ginkgoes were present at any time. The leaf in *Ginkgo* today is known to be quite variable and thus the variations in fossil ginkgo leaves may not truly indicate separate taxa; however, many genera and species of ginkgophytes are described, especially from uppermost Triassic to lowermost Jurassic rocks in East Greenland and numerous Jurassic and Cretaceous floras in the American western interior. Ginkgoes declined through the Cretaceous but persisted beyond the era boundary such that two genera were distributed widely in middle Tertiary time; and by the end of the Miocene, they were reduced to the single extant species which subsequently disappeared from all continents but Asia, before the intervention of modern man.

G.3.   Origin and Early Radiation of Angiosperms in the Cretaceous

Angiospermous plants exceed by far all other living plants in diversity and biomass. Angiosperms are defined as flowering plants which develop from seeds enclosed within an ovary (literally "vessel seeds" as opposed to the "naked seeds" that define gymnosperms). Because the significant distinction between higher plant groups is based on the seeds, one may begin to appreciate the long-standing paleobotanical mystery which is: When did the first angiosperms appear, from which group did they evolve, and most importantly, how did they diversify so rapidly in the late Mesozoic? Answers to these questions are unknown (Beck, 1976) because most fossils of putative early angiosperms are wood, leaf impressions, and pollen, none of which are securely diagnostic of the angiosperm condition. Also, it is unclear which, if any, of the known Mesozoic gymnospermous groups led to the angiosperms. The cycads, as stated previously, bear seeds that resemble slightly those of primitive angiosperms such as *Magnolia;* however, the differences are sufficiently great to eliminate cycads from consideration as direct angiosperm ancestors. No other likely ancestral forms are known.

What is known about the fossil record of angiosperms is that the earliest family appeared by the Neocomian (earliest Cretaceous) and was joined by three more during the Albian, by 28 families during the Cenomanian, and by many more throughout the later epochs of the Cretaceous such that fully 46 families were in existence at the close of the Mesozoic. Virtually all of these angiosperm families are still in existence and were joined by an even more diverse throng of Cenozoic families. Thus, there are more than 150 families extant which have known fossil records, and another 110 families extant with no fossil records (data from Harland *et al.*, 1967). The early history of angiosperms seems one of explosive evolution but it has also been suggested (Doyle, 1977) that the rate of evolution was comparable to that of other groups of organisms.

It is well beyond the scope of this book to describe the representatives of all those families of angiosperms present in the Cretaceous. Most Mesozoic fossil angiosperms are trees, since they produce great abundances of material and easily preserved structures, such as leaves and wood. Of *monocotyledonous* angiosperms (cattails, bamboo, grasses, and palms; defined as a subclass featuring flowering plants having an embryo contained in one seed leaf), the fossil record is much poorer than that of *dicotyledonous* angiosperms (all other flowering plants, including trees, having two seed leaves, also having net-veined leaves). It is apparent that the early fossil record of angiosperms is heavily biased toward groups which lived nearer to sea- or freshwater and against upland floras. The degree of this bias is undeterminable, but acknowledgment of the bias has led to a common belief among paleobotanists that the angiosperms must surely have evolved before their occurrence as fossils. It is possible that origin and considerable diversification of angiosperms occurred in, say, the Late Jurassic and earliest Cretaceous in regions remote from the very restricted sea of those times, and that fossils of those groups were not preserved until widespread deposition was initiated by Cretaceous transgression and subsequent regression.

To give the reader a glimpse into Cretaceous angiosperm life, the following modern plants have known Cretaceous relatives: sumacs and poison ivy (*Rhus*), birches (*Betula*), *Sassafras*, *Magnolia*, palms (*Palmoxyon*), oaks (*Quercus*), American plane or "sycamore" (*Platanus*), the rose family (*Pyrus*), poplars (*Populus*), breadfruit (*Artocarpus*), figs (*Ficus*), elms (*Ulmophyllum*), and, to add spice, cinnamon (*Cinnamomeides*).

G.4. New Mesozoic Insects

With the rise of angiosperms came new terrestrial insects. In the Jurassic, two very important orders appeared: Diptera (flies and mosquitoes) and Hymenoptera (wasps, bees, and ants). Both groups are strongly dependent on either flowering plants or warm-blooded animals during parts of their life cycle or for feeding; it is no coincidence that they appeared simultaneously with (or slightly before)

the rise of both angiosperms and modern mammalian groups. The presence of rare hymenopteran fossils in Jurassic strata, in fact, gives further evidence that angiosperms evolved in upland environments in the Jurassic and are not found due to some artifact of preservation. Mesozoic insects are found preserved in several fossil occurrences, including the Solenhofen and other Jurassic lithographic limestones, volcanic ash deposits, silicified trees, Cretaceous amber deposits in Canada, fine-grained lake deposits, and other means by which fragile, soft-bodied organisms could be preserved. Along with the dipterans and hymenopterans in the Jurassic appeared several less-important insects including the earwigs (Dermaptera) and the caddis flies (Trichoptera); walkingsticks and leaf insects (Phasmatodea) had appeared in the Triassic (data from Tasch, 1980). The last great radiation of insects occurred in the Early Tertiary with the appearance of mantids (Manteodea), butterflies and moths (Lepidoptera), and some less lofty creatures including fleas (Siphonaptera) and termites (Isoptera). Butterflies and fleas could only have appeared concomitant with flowers and abundant mammals (respectively) and their appearance may reflect renewed diversification of each in the Early Tertiary.

H. The Rise and Fall of Archosauria and Other Zuni Vertebrates

H.1. Origins and Varieties of Archosaurs

In our last discussion concerning archosaurian reptiles (Chapter 7), we showed that by later Early Triassic times advanced thecodont reptiles had achieved effective bipedal posture and had reached fairly large sizes. Thecodonts such as *Saltopsuchus* comprise the stem lineage which, while not being necessarily ancestral, represents forms similar to the probable ancestors of several orders of "ruling reptiles," including both saurischian and ornithischian dinosaurs and pterosaurs.

In the following discussion, the focus will be on the earliest representatives of the major dinosaur taxa, a few of the specializations which characterize dinosaurs and other archosaurs, and biostratigraphic details pertinent to the archosaurs. We will not attempt comprehensive description of the groups because these are so readily available at virtually any library and peripheral to the best use of space here.

H.1.a. Dinosaur Pelvises

Archosaurs of the two "dinosaur" orders possess one of two basic arrangements of bones in their pelvises, which evolved (probably independently) from the pelvis of thecodonts. In the saurischian pelvis ("lizard hip"), three radiat-

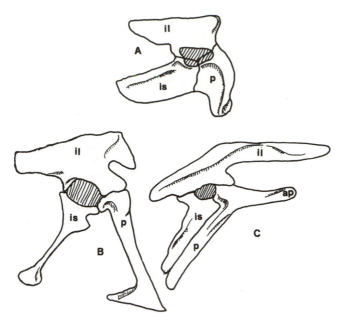

Fig. 8-72. Pelvises of archosaurs. (A) Thecodont type (also found in crocodilians); (B) saurischian type; (C) ornithischian type. Anterior is to the right. (Modified and redrawn from Colbert, 1980, and other sources.)

ing elements are present: ilium, ischium, and pubis, extending in diverging directions and leaving an open acetabulum into which the head of the femur articulates securely (Fig. 8-72). Although primitive, this type of pelvis was present in the largest dinosaurs and also in all carnivorous dinosaurs; taxa bearing this type of pelvis persisted until the end of the Maestrichtian along with bearers of the more highly evolved type. The alternate pelvis in dinosaurs was the ornithischian type (''bird hip'') which contained four radiating elements: the three mentioned above plus an additional anterior process (projection) on the pubis. This type of pelvis, as implied by the name, is similar to that of birds.

H.1.b. Earliest Dinosaurs

There are clear evolutionary lineages by which one may trace the progression from later Triassic taxa to the giants so well represented in the latest Jurassic Morrison Formation and other Jurassic strata. Since we will not discuss varieties of dinosaurs, beyond tabling the general taxonomic groups (Table III), the reader may wish to refer to Colbert (1980), Swinton (1970), or Glut (1982) for readable discussions of the dinosaurs mentioned in passing. The two major divisions among saurischians, sauropods and theropods (respectively herbivores and carnivores), apparently have separate origins from Triassic ancestors. The herbivores apparently stem from members of the infraorder Prosauropoda, represented in Europe by *Plateosaurus,* and in America by *Yaleosaurus (Anchisaurus).* These were rela-

tively large (to 6 m long), heavy-bodied creatures which feature good bipedal posture but with evidence of reversion to four-footed walking. In other words, in these prosauropods the drive toward bipedalism began to reverse, and we may be observing the progenitors of the gigantic ''brontosaurs.'' Important primitive characteristics additionally present in prosauropods were a five-digit count on fore and hind limbs (reduced in advanced dinosaurs), a thecodont pelvis, small skull fenestrae, and unspecialized teeth.

Early carnivorous dinosaurs trace their ancestry back to lightly built, fast-running saurischians called coelurosaurs (an infraorder). A common genus from the Triassic of North America is *Coelophysis,* known from the Chinle and Kayenta Formations, among others. Unlike prosauropods, coelurosaurs survived into the later Mesozoic and were in fact important members of the ruling reptiles. *Coelurus* (also called *Ornitholestes*), the infraorder's namesake, is well known from the Morrison Formation, and the ''ostrich mimic dinosaurs'' (e.g., *Ornithomimus*) were very common coelurosaurs of the Cretaceous and survived until the very end of the Maestrichtian. The giant carnivores (infraorder Carnosauria) are almost certain to have directly descended from Triassic coelurosaurs but intervening links are not known, largely because of the poor early Jurassic terrestrial record.

The earliest records of ornithischians, all of which were herbivorous, are poorer than those of saurischians. Colbert (1980) discussed three Upper Triassic ornithischian taxa, known from scanty remains, showing that the earliest

Table III. Annotated Summary of Higher Dinosaur Taxonomy

Order Saurischia (''lizard-hipped dinosaurs'')
  Suborder Theropoda (carnivorous dinosaurs)
    Infraorder Coelurosauria (slender ''running dinosaurs'')—Late Triassic to Late Cretaceous
    Infraorder Carnosauria (large to giant carnivores)—Early Jurassic to Late Cretaceous
    Infraorder (?) Dromaeosauria (small, very agile ''kicking dinosaurs'')—Early to Late Cretaceous
  Suborder Sauropodomorpha
    Infraorder Prosauropoda (semibipedal ancestral sauropods)—Middle to Late Triassic
    Infraorder Sauropoda (large to giant long-necked herbivores)—Early Jurassic to Late Cretaceous
Order Ornithischia (''bird-hipped dinosaurs'')
  Suborder Ornithopoda (stem ornithischians, including duckbilled and trachodonts)—Late Triassic to Late Cretaceous
  Suborder Stegosauria (stegosaurs, featuring erect dorsal armor)—Late Triassic (?) to Early Cretaceous
  Suborder Ankylosauria (armored dinosaurs)—Early to Late Cretaceous
  Suborder Ceratopsia (horned, frilled dinosaurs)—Late Cretaceous

ornithischians were small and lightly built, but already showed signs of specializing. *Heterodontosaurus,* from South Africa, appears to be an early camptosaur (an ornithopod, and a probable intermediate form leading to the very abundant duckbilled dinosaurs of the Cretaceous). *Fabrosaurus,* also from South Africa, is suggested by Colbert (1980) as an ornithopod. *Scutellosaurus* is a Colbert genus from the Kayenta Formation in northern Arizona. The fragmentary remains show armor in the skin, a very long tail, and evidence of reversion to quadrupedal posture. Colbert suggests it may be a very early armored dinosaur (i.e., of the ankylosaurs).

H.1.c. Important North American Dinosaur Assemblages

The known distribution of dinosaurs is undoubtedly very much skewed by the happenstance of preservation. The majority of specimens on display in museums are culled from a handful of formations; yet, the total number of taxa described is very large, with most specimens consisting of a scrap or two of bone. Glut (1982) presents a synoptic ''dictionary'' approach to cataloging dinosaur genera and demonstrates how varied and numerous they are.

In North America, three famous assemblages of dinosaurs are widely represented in museums and research institutions. The Morrison fauna is probably the most notorious of the world's dinosaur collections, and has filled many displays with the sauropods *Apatosaurus* (''*Brontosaurus*''), *Camarosaurus, Diplodocus,* and *Brachiosaurus;* the carnosaurs *Allosaurus* (*Antrodemus*) and *Ceratosaurus;* the ornithopod *Camptosaurus;* and the stegosaur *Stegosaurus* [see Dodson *et al.* (1980) for more taxa].

A second celebrated North American dinosaur assemblage contains no important skeletons at all; rather, it consists of numerous beautifully preserved trackways of a variety of dinosaurs. These are in the Trinity Group, at the Paluxy River, near Glen Rose, Texas. The beds are of Early Cretaceous age and show herding behavior in giant sauropods, predation on sauropods by carnosaurs (converging tracks), and the general impression that dinosaurs were at least locally and temporally common.

Several localities and units combine to make up the final ''notorious'' North American dinosaur assemblage. These are all centered around Montana, Wyoming, and southern Alberta, and include the Maestrichtian Laramie/Lance/Hell Creek/Old Man/Belly River/Judith River formations. Remains of the following dinosaurs are abundant: carnosaurs *Albertosaurus* and *Tyrannosaurus;* hadrosaurs *Edmontosaurus, Kritosaurus, Saurolophus, Parasaurolophus, Anatosaurus,* and *Lambeosaurus* (see Hopson, 1975); coelurosaurs *Ornithomimus* (*Struthiomimus*); ceratopsians *Triceratops, Anchiceratops,* and *Styracosaurus;*

and the ankylosaur *Euoplocephalus.* Many other less-common genera are also known from the Laramie equivalents.

Dinosaur fossils are widely distributed in North America and come from many additional settings. Notable among these second-order assemblages are Upper Triassic red beds in Nova Scotia, New Jersey, and Connecticut; Lower Jurassic coastal plain deposits in Massachusetts, New Jersey, and Connecticut; Lower Jurassic strata in Arizona; Lower Cretaceous strata in Wyoming, South Dakota, and British Columbia; and Upper Cretaceous strata from a very wide area and variety of lithologies. These latter include marine (coastal plain) strata in New Jersey, North Carolina, Delaware, Alabama, Mississippi [see Horner (1979) for a tabulation], and Georgia (Schwimmer, 1981); and nonmarine or interior-seaway strata in New Mexico, Colorado, Kansas, and other states adjacent to the Laramie and equivalent outcrop.

H.2. Pterosaurs

Along with the strikingly novel dinosaurs that evolved in the Mesozoic, flying reptiles of the order Pterosauria are probably the most unusual reptiles of all time. The earliest specimens of this order appear in Lower Jurassic strata from Europe, and these first taxa had already achieved apparently successful flight.

Two representative, primitive Jurassic taxa are *Dimorphodon* (Fig. 8-73) and *Rhamphorhynchus,* representing variations in the form of the skull and dentition. As a basic characteristic of pterosaurs, the all-important wing structure is derived from tremendous elongation and strengthening of the fourth digit of the manus, which became the wing support. The major wing-bones and most other sizable bones in the body had large hollows to save weight and the pelvis is quite different in structure compared to either order of dinosaur and to the thecodonts which probably gave rise to the pterosaurs. Other specializations evident in primitive pterosaurs include a relatively elongated neck, an even longer tail with a flattened distal portion in some species, and equally shortened main body, relatively short, weak hind limbs, and a relatively large head. The head is long and narrow in *Rhamphorhynchus,* but more typical of a primitive diapsid reptile in *Dimorphodon* (being of only moderate length). The teeth of *Rhamphorhynchus* are distinctive in that they project forward and cross each other when the mouth is closed; this has been presumed a fish-eating adaptation and indeed most pterosaur fossils are found in marine strata. *Dimorphodon* lacked forward-pointing teeth, and may have been insectivorous. Romer suggested that the common occurrence of pterosaurs in marine deposits suggests that they existed much like modern shore birds, flying far out to sea and nesting on land.

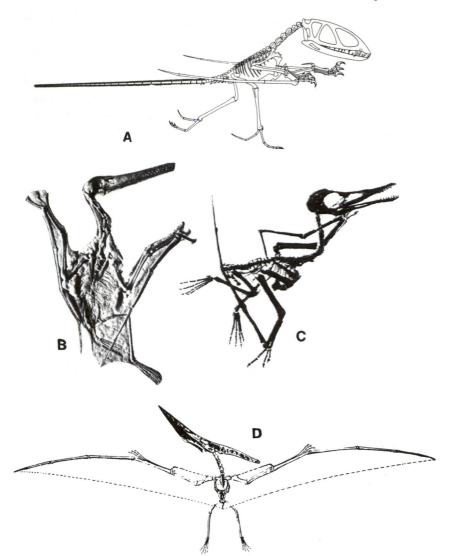

Fig. 8-73. Representative pterosaurs. (A) *Dimorphodon* (L. Jurassic), a rhamphorhynchoid with a relatively large skull (length about 20 cm), shown in running position. (B, C) *Pterodactylus*, from the Solenhofen Limestone (U. Jurassic); B is approximately 40 cm high, C approximately 10 cm high (the size of a sparrow). (D) Reconstruction of *Pteranodon* (U. Cretaceous), maximum wingspan to approximately 8 m. (A and D, from Padian, 1983; B and C, DRS photographs.)

Rhamphorhynchus and most early pterosaurs were small (chicken to sparrow sized) whereas the subsequent group of pterosaurs included some giants. Advanced pterosaurs appeared in the Middle Jurassic and became common by the Late Jurassic; this group is typified by *Pterodactylus,* one of the best-known pterosaurs due to excellent fossils from the Solenhofen Limestone in Bavaria. In pterodactyls, the tail became very short and the teeth in many genera were reduced drastically in size and number, disappearing from back to front. In the genus *Ctenochasma,* however, the teeth became very slender, long devices which were apparently used for straining food from water. In the Early Cretaceous, a specialized group of pterodactyls lost all their teeth and developed large bills, rather like those of pelicans. They also increased dramatically in size and some genera evolved elongate rudders from the bones of the skull. *Pteranodon* is the best known of large pterosaurs and

until recent years was considered the largest flying animal, reaching wingspans of up to 8 m (but with body weight approximately that of a large turkey). *Pteranodon* is found in many Upper Cretaceous chalk deposits, such as the Niobrara, from the western interior as well as three other continents.

A remarkable discovery in Texas in 1971 by D. A. Lawson in Big Bend Park was the humerus of a species of the pterosaur *Quetzalcoatlus* which proportions out to having a 12-m wingspan! This analysis is based on comparison with relatively complete specimens of *Quetzalcoatus* that reach to 6 m in wingspan (Langston, 1981). This also suggests that the body weight was in excess of 100 kg, making it several times as large as the largest *Pteranodon.*

Analysis of the flight of pterosaurs has occupied goodly parts of the paleontological literature over the past 100 years and continues. Stein (1975) did wind-tunnel and mechanical analyses of flight in *Pteranodon ingens* from Kansas and concluded that they could soar and glide, as has traditionally been believed, but he also claimed that the animal was "primarily adapted to slow, flapping flight and long flight endurance." This analysis contrasts with the common image of pterosaurs clambering up cliffs to begin their glides since Stein also showed that they could initiate flight from level ground. Padian (1983) went further and suggested that pterosaurs exclusively used active flight, rather than gliding, and were also agile on the ground. Padian's analysis of pterosaur morphology suggests that overall they were far more adept at various life functions than is generally perceived, and they should have been vigorous competitors with the first birds.

H.3. Crocodilians

Familiar as they are, crocodilians represent the sole surviving archosaur order. The group characteristics are largely modifications for semiaquatic life: elongate snout and body, propulsive tails, slight deemphasis of limbs for walking while retaining adequate strength to move on land, development of a secondary palate for air-breathing in water, and retention of a modified thecodont pelvis. Crocodilians are close in form to the phytosaurs (Chapter 7) except for the secondary palate; however, this suggests clearly separate evolutionary lineages of the groups.

Primitive crocodiles, such as *Protosuchus* from the Triassic, are little advanced beyond the phytosaur grade in development of the secondary palate. The Triassic crocodilians were heavily armored with scales (as are modern forms) and show crocodilian features in the postcranial skeletons. Jurassic time was the first flowering of several crocodile lines, most of which were dead ends. Romer (1966) noted that since most Jurassic strata are marine, one receives a biased suite of fossils and a poor record of the nonmarine taxa. Among marine crocodiles of the Jurassic were the Metriorhynchidae and Teleosauridae (Buffetaut, 1979). The former of the two groups were strongly adapted to open-sea life, with tails modified into virtual reverse-heterocercal types (as in some fish), formed by a downturned distal portion of the vertebrae and a secondary fin formed on the top. In *Metriorhynchus* the limbs were modified into flippers and the scaly armor was lost; it is likely that they could not walk on land. The teleosaurids were less specialized and retained functional walking legs and normal crocodile morphology. Both *Metriorhynchus* and *Teleosaurus* and their kin are representatives of the mesosuchian level of crocodile development (in contrast to the protosuchian level) featuring advanced development of the secondary palate and internal nostrils located far to the rear of the mouth; thus, they were truly crocodilian in head and body. Other mesosuchian groups were present in the

Jurassic and a Late Jurassic family featuring a relatively conservative genus, *Goniopholis,* may have been the progenitors of a large number of Cretaceous families. Crocodiles and the earliest alligators were abundant in Cretaceous waters and are among the commonest of fossil vertebrates found (for obvious reasons—their aquatic habit). Alligators differ from crocodiles chiefly in having blunt snouts and different patterns of interlocking in the teeth; gavials are a third group with very slender snouts and which are found in modern times only in India, Pakistan, and Burma. Gavials appeared in Tertiary time and were apparently distributed globally until the Miocene, at which time for unknown reasons they died out in Africa, North America, and Europe.

Most of the diverse late Mesozoic crocodilian families became extinct at the end of the Cretaceous but many new crocodiles appeared in the Early Tertiary and several groups persisted from the Cretaceous into the Tertiary (Buffetaut, 1979). The persistent groups include the Sebecidae, a mesosuchian family that survived until the Pleistocene (thus barely missing the chance to become living fossils). But, sebecids aside, some Late Cretaceous taxa and all the Tertiary crocodilians are placed in the suborder Eusuchia, which is distinctly progressive relative to the mesosuchians in having distinctive vertebral structure and in having secondary palates that completely separate the air passages from the mouth (a clear advantage to aquatic, air-breathing animals). The three modern crocodilian families are all eusuchians, possibly descended from a single stem stock.

Before leaving discussion of crocodilians, we should note that in the age of giants the crocodiles were not outdone. The Upper Cretaceous eusuchian *Deinosuchus,* common in the Atlantic and Gulf Coastal Plains, featured the largest-known skull length of 2 m (longer than *Tyrannosaurus*) and a probable body length in excess of 15 m. It lived in marginal-marine environments and may well have been a threat to dinosaurs venturing into the water.

H.4. Early Birds

The earliest fossil record of birds is, at the time of this writing, undergoing drastic re-examination in light of discoveries of Late Triassic bird-like fossils; these consist of two specimens from the Dockum Group in west Texas (refer to Chapter 7, Section B.7.e), discovered by S. Chatterjee of Texas Tech University. Since peer-reviewed publication about these fossils has not yet appeared, detailed discussion here is inappropriate; in general, the specimens represent about 70% of the animals, which were about the size of crows and appear to be somewhat more advanced in certain aspects than the next "oldest bird," *Archaeopteryx* (discussed below). The Triassic bird possessed a small *carina* (keel) on the sternum, relatively large brain and eyes, and a well-developed *furcula* (wishbone). Nevertheless, these Triassic fossils, to be assigned a new generic name *"Protoavis,"* show that the organisms retained several reptilian characteristics including teeth in the anterior jaws, fingers on the manus (rather than wings), and a long, multiboned tail (data from S. Chatterjee, paper read to Society of Vertebrate Paleontology, November 1986, and DRS, personal communication).

Assuming the formal report of *"Protoavis"* is accepted by the scientific community, the discovery suggests that birds *co-evolved with* dinosaurs during the mid- or Late Triassic from thecodont ancestry, rather than having *evolved from* dinosaurs during the Late Triassic or Early Jurassic. Further, the position of *Archaeopteryx* is probably off the mainstem of avian evolution, since *"Protoavis"* is more derived in the direction of modern birds than is *Archaeopteryx*.

Prior to 1986, the fossil record of birds began with the Jurassic (Oxford-Kimmeridge) species *Archaeopteryx lithographica* from the Bavarian Solenhofen Limestone. This species, known from six specimens (including two specimens with feather imprints and one comprising a single feather; Barthel, 1978) has been exhaustively studied. Its skeletal resemblance to a small coelurosaurian dinosaur is so striking that some specimens, lacking feathers, were originally classified as such. Among primitive features of *Archaeopteryx* are the following: the skull, though birdlike in lightness and closure of sutures, contains thecodont teeth; the tail is long and typically reptilian; the sternum is small with no carina present for attachment of flight muscles; overall, the bones are not fractionally as light as those of typical flying birds and they lack pneumatic structure; and the hands retain fingers rather than the strongly modified bones in later birds. In general, one could not expect a more primitive bird morphology in a flying creature.

The flying abilities of *Archaeopteryx* have aroused much controversy too. That it flew imperfectly, if it flew, is unequivocal, if for no other reasons than the lack of flight muscles (no carina), solid bones, and unmodified hands (Fig. 8-74). However, Feduccia and Tordoff (1979) make a strong case for active flight based on the feathers which are asymmetrically veined, like flying birds, rather than symmetrically veined such as is found in flightless modern birds. Ostrom (1974) and others contend that the feathers originated as an insect-gathering mechanism, and/or as a thermoregulatory device (i.e., insulation) with the flight function evolving secondarily. We observe that both viewpoints are compatible since one addresses the nature of flight in *Archaeopteryx,* and the other addresses the origin of flight and evolutionary processes.

The record of birds is somewhat better in the Cretaceous, particularly in the taxa from Upper Cretaceous North American chalks, such as the Niobrara. The best known of these late Mesozoic birds is *Hesperornis*, a

Fig. 8-74. Comparison of *Archaeopteryx*, with a modern pigeon. Distinctive features are in black and include the long tail, straight ribs, small breastbone (without carina), reduced occiput, and presence of unfused fingers on *Archaeopteryx*, among other features. (From Colbert © 1980 by John Wiley & Sons, reprinted with permission of John Wiley & Sons.)

flightless diving bird of fairly large size (roughly that of a crane). The flightless condition in *Hesperornis* is a secondary adaptation, since other, flighted birds are found in similar deposits and since the wings are almost entirely lost in *Hesperornis*. Primitive characteristics of *Hesperornis* include the presence of teeth in all but the anterior upper jaws and an unkeeled sternum (possibly a degenerate state). A fully flighted bird was also present in the Cretaceous of the western interior: *Ichthyornis*, a very small (robin-size) bird with a normal bird's tail, a long neck, and a full complement of teeth. *Ichthyornis* shows all skeletal features of modern birds except for teeth instead of a beak. Although bird fossils from the Lower Cretaceous are very rare, incomplete specimens of a flamingo-type bird, *Gallornis*, are found in Lower Cretaceous strata in Europe. *Gallornis* was toothless and of quite modern form, suggesting that the other known Cretaceous birds were specialized in having toothed jaws. Overall, it is apparent that by the Late Cretaceous, birds of nearly modern abilities, if not modern form, had appeared.

Since the fossil record of birds is so poor, the following paragraphs will summarize the remainder of the avian fossil record up to the present, including a brief summary of the taxonomy to make sense of events after the Cretaceous. Bird taxonomy is very simple at the higher levels: aside from *Archaeopteryx* (and, perhaps, "*Protoavis*"), which alone (or together) comprise the subclass Archaeornithes, all birds are placed in a single subclass Neornithes, which contains three superorders: Odontognathae (toothed Late Cretaceous flightless birds, like *Hesperornis*), Palaeognathae (flightless, usually large birds including ostriches, emus, and many giant extinct taxa of the Tertiary), and Neognathae (all remaining Late Cretaceous, Tertiary, and modern flying birds). The superordinal classification is based on morphology of the palatal bones in the skull, but ancillary differences are considerably more striking. Because palaeognathous birds are flightless, their wing, pelvic, leg, and other bones are greatly modified from the (presumed) ancestral flighted-bird condition. It is controversial whether the modern and later Cenozoic flightless taxa are monophyletic or not. Houde and Olson (1981) studied well-preserved Early Tertiary palaeognathous fossils and concluded tentatively that the group was once much more diverse, implying polyphyly.

By latest Cretaceous time, eight families in six orders of birds are known, nearly all flighted (neognathous) forms except the Hesperornithidae. The Paleocene record is very poor but apparently four orders of Neornithes survived, and by the latest Eocene, all but a very few modern orders had appeared. One must remember that the fossil record is very skewed toward shore birds and other taxa which live near water, since preservation is rare in woodland and upland environments. The small, fragile bones of most birds also disfavor fossilization and the record is similarly skewed toward large species. One may observe, for example, that roughly one-half of living bird species are in the order Passeriformes (perching birds); yet, they have a very poor fossil record because they are small, upland dwellers.

During the Tertiary, several large, predatory flightless birds left an abundant fossil record, most notably *Diatryma* of the Eocene and *Phororhacos* of the Miocene. These are placed in a distinct order because their relationships with other Paleornithes are undetermined. Several additional giant flightless taxa survived until later Holocene times and were victims of human overkills: these include the giant moas of New Zealand, *Dinornis,* and the elephant-bird of Madagascar, *Aepyornis*. A great number of flying birds have likewise suffered from human-caused extinction, including the passenger pigeons; but, generally, the flightless birds seem to succumb most effectively to humans (one may add the dodo and great auk to the large taxa mentioned above). A readable summary of the radiation of birds in the Tertiary is presented by Fisher (1967), and Romer (1966) discusses anatomical trends.

H.5. Marine Reptiles (Euryapsida) of the Zuni Sea

Euryapsida, introduced in Chapter 7, are unrelated to dinosaurs except by virtue of both groups belonging to the Reptilia; nevertheless, giant euryapsids were abundant in the Zuni seas during the "age of reptiles." In the following discussion, the reader will note that a third group of "giant marine reptiles," mosasaurs, are not mentioned. Mosasaurs were simply gigantic lizards and will be treated in the following subsection.

H.5.a. Plesiosaurs

Possibly evolving from Triassic nothosaurs (Chapter 7), in Jurassic and Cretaceous time plesiosaurs increased in size and diversity, culminating in forms such as *Elasmosaurus* of the Late Cretaceous, with more than 76 vertebrae in the neck and a total length over 16 m. The short-necked plesiosaurs (also called pliosaurs after the genus *Pliosaurus*) also became very large in the later Mesozoic; and Colbert (1980) stated that the huge genus *Kronosaurus* from Australia had a skull fully 4 m long! Short-necked plesiosaurs probably moved rapidly in the water whereas the long-necked forms are classically considered to have darted the head around snakelike, with the body serving as a sort of slow-moving base of operations. McFarland *et al.* (1979) cited studies that show the necks of long-necked forms to have been too stiff to behave in such fashion and suggest, rather, that the entire beast turned slowly and that the enormous length of the neck resulted in rapid movement of the head at the circumference of the arc. Morphological details in plesiosaurs worth noting are the apparent skeletal adaptations to provide equally strong forward *and* backward rowing motions in the long-necked taxa; from these features it is apparent that the body was literally rowed in the manner of a man-powered boat, with opposite strokes used to rotate the body and with the ability to move quickly in dead reverse. Since the paddles were relatively large, the plesiosaurs may have actually moved quickly, notwithstanding their size and bulk. The plesiosaurs, short- and long-necked, were quite successful and abundant in the Jurassic and were common even in latest Cretaceous seas; nevertheless, they succumbed along with the remnant dinosaurs to the end-Cretaceous mass extinction.

H.5.b. Ichthyosaurs

The first ichthyosaurs appear in the Late Triassic from many areas, and even the oldest forms are fully adapted to aquatic life. Ichthyosaurs featured the most extreme specializations for swimming (Fig. 8-75A): the pelvic girdle was vestigial, forelimbs were enlarged as paddles by the advent of both *hyperdactyly* (more than five digits) and *hyperphalangy* (many extra phalanges or joints in fingers); the tail terminated in a sharp downward bend in later forms, with a fleshy fin above forming a sharklike propulsive tail; the body was streamlined almost precisely as in porpoises; the snout was elongate and filled with simple pointed (i.e., fish-eating and catching) teeth; the eyes were very large; and the dorsal vertebrae lost interlocking projections (zygapophyses), resembling simple disks such as are found in many aquatic vertebrates. They clearly could not leave the seas and direct fossil evidence suggests they had live birth (in common with some living reptiles).

The assignment of ichthyosaurs to Euryapsida is quite tenuous. There is a temporal fenestra present in the proper euryapsid position, but in no other respects has an interrelationship between sauropterygians (i.e., plesiosaurs) and ichthyosaurs been demonstrated. The known anatomical details about ichthyosaurs come especially from the Solenhofen Limestone, where fin and flesh imprints prove the presumptive morphology derived from skeletal details.

Ichthyosaurs were especially common in the Jurassic. Two basic groups emerged: broad-finned (i.e., very hyperdactylous) taxa such as *Eurypterygius* and long-finned (i.e., very hyperphalangous) forms such as *Stenopterygius. Ichthyosaurus* (Fig. 8-75) was a very porpoise-like taxa and is the genus which includes many of the later long-finned species. The ichthyosaurs declined considerably after the latest Jurassic and are rare in the Cretaceous. Alone among the giant Mesozoic reptiles, ichthyosaurs disappeared long before the end of the Mesozoic. They were never morphologically varied to any great degree; however, the sizes ranged from a meter in length to over 10 m, and some taxa modified for mollusk-eating by development of buttonlike teeth.

H.6. Mosasaurs and Other Lepidosaurs

Among the most common and simultaneously impressive Mesozoic giants were mosasaurs. These lizards are in the same family as the modern Komodo monitors, which reach lengths of over 4 m and are the largest living reptiles. Mosasaurs were highly marine-adapted and, like plesiosaurs and ichthyosaurs, could not leave water: the pelvic girdle was virtually absent; the front limbs were modified into paddles by means of elongate digits; the tail was flattened and strengthened (as in crocodiles); the body was streamlined to a moderate degree; the head was somewhat elongate; teeth were slightly recurved, pointed, and the pterygoid generally featured rows of teeth (to prevent escape of fish). One of the most interesting features of mosasaurs is the joint in the lower jaw, which allowed parallel closure of the jaws rather than a more conventional scissorslike closure. This joint also allowed some fore-and-aft

motion which may have been sufficient to force fish into the gullet (see Russell, 1967). In addition to fish, there is direct evidence that mosasaurs ate ammonoids, since a specimen was found with multiple mosasaur tooth-imprints (Kauffman and Kesling, 1960). One rare mosasaur *Globidens* known from Georgia, Alabama, and Morocco featured buttonlike crushing teeth and may have been a mollusk- or turtle-eater.

Mosasaurs evolved only in the Cretaceous and were abundant during the Santonian through the latest Maestrichtian. All mosasaurs were large but the giants (e.g., *Tylosaurus proriger,* over 14 m long) generally are of Campanian and Maestrichtian ages. Fossils are especially abundant in marine chalks and shales in Kansas, South Dakota, Alabama, and adjacent areas, but traces of mosasaurs are distributed in North America virtually wherever marine Cretaceous rock is found. They were undoubtedly among the top predators in epicontinental marine environments and enormously successful right up to the very end of the Mesozoic.

Few lizards attract the attention in the fossil record as do mosasaurs. Following classification where the diapsid condition is present in two reptile subclasses, mosasaurs, lizards, and relatives are assigned to the subclass Lepidosauria (the other subclass is Archosauria). Lepidosaurs comprise three orders: Eosuchia, Squamata, and Rhynchocephalia.

Eosuchians date back to the Permian, notably in generalized forms such as the lizardlike *Youngina.* They differ from true lizards (squamates) in lacking the flexible jaw and skull structure which allows the characteristic accommodation to swallow large prey, as observed in snakes and lizards. Several eosuchian reptile groups persisted into the later Mesozoic and include an interesting aquatic genus, *Champsosaurus,* which converged on the morphology of crocodilians and phytosaurs. *Champsosaurus* is found in Cretaceous and Tertiary strata in North America and is one of the few Mesozoic reptiles to survive the era termination. It was also the last surviving eosuchian.

Squamates date to the Triassic and the first true lizards known are also among the most highly specialized! *Icarosaurus* and *Kuehneosaurus* (see Fig. 7-77) are Triassic flying lizards, which presumably glided on extended ribs, covered with skin, in the manner of the modern tropical "flying lizard," *Draco.* The record of squamates is generally poor through the Mesozoic, with the very notable exception of mosasaurs, but sufficient material shows they diversified through the Jurassic and that snakes derived from lizards during the Cretaceous. Romer (1966) believes snakes evolved from the varanoid lizards (the same family which includes mosasaurs), and the Booidea (boas) are the more primitive of two snake superfamilies (the others are termed Colubroidea).

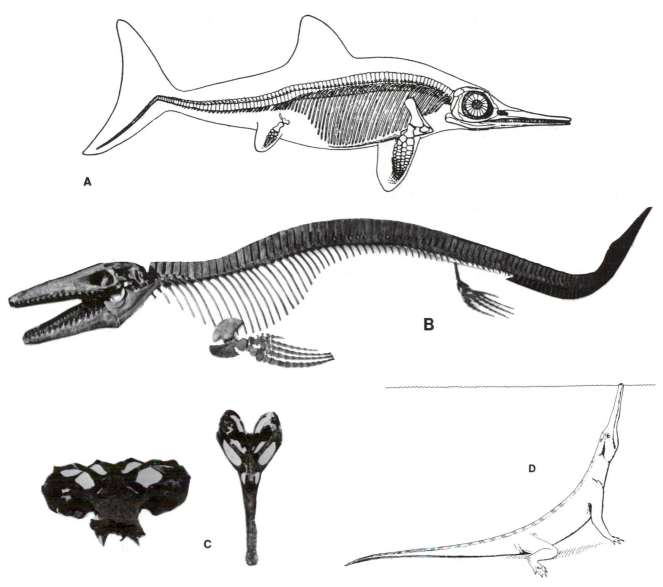

Fig. 8-75. Three distinctive Mesozoic aquatic/marine reptiles. (A) Reconstruction of *"Ichthyosaurus,"* the typical ichthyosaur genus into which many Jurassic forms are placed; lengths are quite variable, ranging from 2.5 to over 10 m for the adult. (B) A mounted skeleton of the very large mosasaur *Tylosaurus proriger*. (C, D) *Champsosaurus*, a eosuchian lizard, similar in size and shape to a small crocodile, which lived from the Late Cretaceous to the Early Tertiary. C shows anterior and dorsal views of the skull; D is a reconstruction of the animal in "snorkeling" position. (A, from Romer, A. S., *Vertebrate Paleontology*, 3rd ed., © 1966, University of Chicago Press; B, photograph reproduced with permission of the Texas Memorial Museum, Austin; C and D, from Erickson, 1985.)

The final lepidosaurian order, Rhynchocephalia, includes a living fossil, *Sphenodon*, found in coastal islands off New Zealand. Rhynchocephalia are differentiated from squamates by their retention of some eosuchian features including a functional pineal (third) eye, and in details of the skull bones. The group appeared in the Triassic, were never dominant, but did produce a group of sheep-sized, heavy-bodied herbivores (termed rhynchosaurs) during the Middle Triassic; and they declined to the single extant genus by the end of the Cretaceous.

H.7. Mesozoic Marine Turtles

In Chapter 7 we discussed the earliest turtles in the fossil record, from the Triassic, and noted that they were fully specialized with characteristic shells and beaks; however, they were apparently unable to withdraw the head in the manner of modern turtles. Jurassic turtles were somewhat advanced over this condition, featuring essentially modern skulls without even the vestiges of teeth (such as was present in the palates of the Triassic taxa), but still apparently unable to retract the head into the shell. This Jurassic turtle stock is assigned to the suborder Amphychelydia and some representatives of the group survived into the Tertiary and in one case the Quaternary, but none are extant. Turtles of essentially modern form appeared in the latest Jurassic and are assigned to the suborder Cryptodira; in these turtles the neck can be flexed in an S-shaped bend and the head can be withdrawn. In Late Cretaceous time a second modern (but now restricted) group appeared, the Pleurodira, which retract their heads by turning their necks sideways. In addition to this major morphological distinction between suborders, in pleurodires the pelvis is fused to the shell whereas in cryptodires it is free. Pleurodire fossils are very abundant in Upper Cretaceous shallow-marine strata of the Atlantic and Gulf Coastal Plains. They declined through the Tertiary and survive at present as remnant populations in South America and Madagascar, notably of the genus *Podocnemis.*

Among unusual Mesozoic turtles were some very big cryptodires including the largest, *Archelon,* from the Niobrara Formation in Kansas, which reached 4 m in length. Like some other very large marine turtles (such as the modern leatherback), *Archelon* featured a relatively light shell with mostly open space to its structure, covered by a leathery skin.

H.8.   The Earliest Mammals

H.8.a.   Reptile–Mammal Distinctions and Subdivisions of Mammalia

Mammals are usually considered animals of the Cenozoic Era, and rightly so because they dominate its fossil record overwhelmingly. However the first mammals date back to the very earliest age of dinosaurs, and they were apparently common throughout the long reign of the "ruling reptiles" prior to emerging as the dominant vertebrate class in the Paleocene Epoch. It is certain that mammals arose from advanced synapsid reptiles (Theriodontia) and certain genera within the theriodonts are so close to mammal-grade of development that their taxonomic placement is uncertain. The reptile–mammal evolutionary transition occurred during the late Middle or early Late Triassic, and at least five different mammal groups were present in the Late Triassic.

Distinguishing fossil reptiles of the most mammal-like groups from primitive mammals is not cut-and-dried because most mammalian characteristics are indeterminable from fossil evidence. Examination of Table IV shows that of the features listed as discriminating the classes, six—the presence/absence of hair, mammary glands, four-chambered heart, diaphragm, fat deposits, and endothermy (warm-

Table IV. Typical Reptilian versus Mammalian Features

Reptilian	Mammalian
Teeth single-cusped	Teeth (posterior) multicusped
Jaw articulation quadrate–articular	Jaw articulation squamosal–dentary
Multiboned jaw	Single-boned jaw
Internal nostrils open at anterior into mouth	Secondary palate present, internal nostrils at back of secondary palate
Middle ear contains stapes bone	Middle ear contains malleus, incus, stapes
Single occipital condyle	Double occipital condyle
Phalangeal formula 2 : 3 : 4 : 5 : 3	Phalangeal formula 2 : 3 : 3 : 3 : 3
Ectothermal metabolism	Endothermal metabolism
Three-chambered heart	Four-chambered heart
No mammary glands	Mammary glands present
Skin scaly	Skin hair- or fur-covered
Diaphragm absent	Diaphragm present
Little fat	Abundant body fat
Ovoviparous (some ovoviviparous or viviparous)	Viviparous (monotremes ovoviparous)

*a*Modified from McFarland *et al.* (1979).

blooded metabolism)—are determinable only from living specimens. The skeletal characteristics are generally ambiguous as an indication of the reptilian or mammalian affinities of a given taxon, because they may be found in one or another advanced reptiles. The most advanced of theriodonts, genera such as *Diarthrognathus* of the infraorder Ictidosauria, feature all mammalian characteristics except details of the jaw suspension. In mammals, the jaw hinge is formed by the squamosal (skull) and dentary (jawbone). In reptiles, there is a quadrate (skull) and articular (jaw) hinge. *Diarthrognathous* has the latter hinge with a partial squamosal–dentary contact.

Live (viviparous) birth is characteristic of most mammals and some reptiles. The basic subdivisions of Mammalia reflect embryonic development. Most living mammals bear live young internally throughout gestation; this is termed the *eutherian* (or placental) condition. Many living and fossil taxa show a more primitive system where the embryo develops partly internally and partly in an external pouch; this is termed the *metatherian* condition. The modern platypus and echidna from Australia feature egg-laying embryogeny, which may reflect a primitive mechanism of mammalian development termed the *prototherian* (monotreme) condition. Unfortunately, one cannot classify extinct mammals as proto-, meta-, or eutherians based on skeletal information except in certain circumstances, such as the presence of characteristic pelvic pouch bones in metatherians.

H.8.b. Subdivisions of Mesozoic Mammals

Table V shows the higher mammal taxa including living and extinct forms. Some of the assignments of Mesozoic taxa are very tenuous.

H.8.c. Triassic Mammals

Known Triassic mammals represent five taxa with unclear distribution among the higher taxa. The total collection of mammals known from the Late Triassic is: the families Morganucodontidae, Sinocodontidae, and Kuehneotheriidae, each represented by some quantity of material; the genus *Eozostrodon,* known from two teeth; and the family Haramyidae, known from a few teeth (data from Mills, 1971). Based on characteristics of the molars, which is the essential taxonomic criterion used with fossil mammals, morganucodonts, *Eozostrodon,* and sinocodonts were interrelated and might be placed (with considerable uncertainty) in the eotherian order Triconodonta. Triconodonts feature very primitive molars with three, aligned cusps that were obviously meat-processing structures. The known skeletons and skull material show that triconodonts were small (cat-size to mouse-size), very similar to advanced therapsids in overall form, and possessed large canine teeth.

The Haramyidae are very poorly known and seem to represent a line completely separate from the triconodonts. Mills (1971) doubted that they were mammalian and McFarland *et al.* (1979) placed them as a hypothetical offshoot of the cynodont reptiles. Until more haramyid material emerges, they are too poorly known to deal with. The final Triassic group, Kuehneotheriidae, were probably early symmetrodonts, the probable ancestors of pantotheres, and the most primitive therian mammals. In symmetrodonts, as in all higher mammals, the molars have cusps arranged in a nonlinear fashion: in the symmetrodonts, they were arranged in a symmetrical triangle.

Virtually all Triassic mammals have come from Euro-

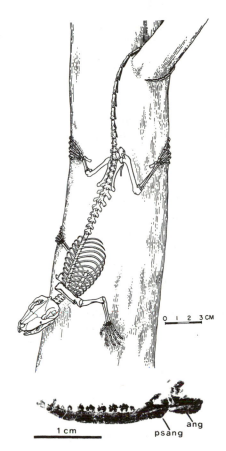

Fig. 8-76. Representative Mesozoic mammals. (Above) Reconstruction of a multituberculate, *Ptilodus,* in hypothetical position for descending a tree. (Below) The jaw of *Dinnetherium* (U. Triassic), a triconodont, from the Kayenta Formation in Arizona. (Above, from Jenkins and Krause, 1983; below, from Jenkins *et al.,* 1983; both reproduced from *Science,* © 1983, American Association for the Advancement of Science.)

pean and (more recently) Chinese deposits. Jenkins *et al.* (1983) reported the first Triassic mammals from North America, in the Kayenta Formation in Arizona. The few specimens represent a very tiny (shrew-sized) triconodont assigned to a new genus, *Dinnetherium* (Fig. 8-76). The overall impression of the Arizona fossils is that Triassic mammals were more diverse than thought, but the sparse fossil evidence very poorly reflects that diversity.

H.8.d. Jurassic Mammals

In Jurassic time, several new mammal groups joined the Triassic taxa mentioned, and one is faced with a complex set of relationships to comprehend. The new Jurassic groups are the pantotheres, docodonts, and multituberculates. Docodonts are the most poorly known, represented in the main by three genera based solely on tooth fossils from

Table V. Higher Taxonomy of Mammals[a]

Class Mammalia
  Subclass: Prototheria (external ontogenetic development)
    Infraclass: Eotheria (includes orders Docodonta and Triconodonta)—Triassic–Jurassic
    Infraclass: Allotheria (multituberculates)—Jurassic–Eocene
    Infraclass: Ornithodelphia (monotremes)—Holocene
  Subclass: Theria (internal ontogenetic development)
    Infraclass: Trituberculata (includes orders Symmetrodonta and Pantotheria)—Jurassic–Cretaceous
    Infraclass: Metatheria (marsupials)—Cretaceous–Holocene
    Infraclass: Eutheria (placentals)—Cretaceous–Holocene

[a]Modified from Colbert (1980).

North America and England. In *Docodon,* the molar cusps are arranged irregularly and with a distinct interlock between uppers and lowers. This has been considered generally a carnivorous feature probably not related to patterns known in other groups; however, Kermack and Kielan-Jaworowska (1971) consider them closely related to *Morganucodon* and *Sinocodon* and place all those genera in the Docodontia. These authors further observed that the time gap between docodonts and multituberculates (discussed below) allows docodonts to be ancestral to the latter. In summary, there is considerable uncertainty of the relations among these groups but there seems to be general agreement that morganucodonts, triconodonts, docodonts, and miscellaneous associated taxa form a related assemblage that probably was prototherian. It is in fact possible that from morganucodonts evolved both triconodonts and docodonts, which explains the confusing interrelationships in this assemblage.

The only additional prototherian group from the Mesozoic is also the most long-lived and best represented of Mesozoic mammals: the multituberculates. These were apparent herbivores (the first among mammals) with molars featuring double rows of cusps arranged in opposing rows in upper and lower jaws; in effect, these were griding teeth. The remaining skeleton in multituberculates is well known, featuring rodentlike incisors, no canines, strong, massive jaws, and a postcranial form much like that of the larger modern rodents (such as woodchucks). Jenkins and Krause (1983) showed that multituberculates were structurally adapted for tree-climbing. They reached large size (up to several kilograms) relative to the usually tiny Mesozoic mammals; and they survived for a very long time, from the Late Jurassic into the Eocene Epoch, which as Romer (1966) pointed out was an interval of some 100 Myr., far longer than the history of any other mammal. Despite their good fossil record, it is unclear whether multituberculates are in the ancestry of monotremes or whether they were an isolated, specialized, dead-end line.

Triconodonts are relatively well-represented in Jurassic strata and they were probably a common, small, predatory group (perhaps filling a niche similar to that of the modern weasels). Their stratigraphic range is very long, from the Upper Triassic to the Lower Cretaceous, and they apparently left no descendants.

Symmetrodonts are well-represented in the Upper Jurassic and in the Lower Cretaceous. It seems likely that *Kuehneotherium* of the Upper Triassic is ancestral to Jurassic symmetrodonts, and that symmetrodonts in turn are ancestral to the pantotheres and thence to the younger therian mammals (Simpson, 1971). Symmetrodont molars indicate carnivorous habit since the cusps of upper and lower teeth interlocked rigidly, which precludes grinding

capabilities. As they evolved through the middle Cretaceous, the cusps became longer and this apparent trend may have been an unfortunate specialization, since they did not survive into the Late Cretaceous. In form, the known postcranial skeletons, skulls, and jaws show them to be relatively primitive, very small carnivores of approximate large-rat size.

The final group of Jurassic mammals, pantotheres, were a possible offshoot of symmetrodonts, possessing molars with a three-cusped trigonal pattern but with the addition of a *talonid* or external element on the lowers and with a fourth large cusp inside the triangular-arranged cusps on the upper jaw. This condition is a fairly clear foreshadowing of molar structure in higher mammals; thus, pantotheres comprise one of the most evolutionarily important of Jurassic groups. Teeth aside, pantotheres were medium to small, carnivorous mammals, with poorly known postcranial morphology. There are sufficient jaw, skull, and tooth fossils to show that they were a diverse lot, with at least five families distributed widely in Europe and North America. Their known stratigraphic range is from Middle Jurassic through Lower Cretaceous but they are by far best known from the Upper Jurassic.

H.8.e. Cretaceous Mammals

By Early Cretaceous time mammalian faunas began to approach a "Tertiary" appearance. The Lower Cretaceous does not contain nearly as good a mammal (or for that matter, dinosaur) record as does the Upper Cretaceous, but evidence of triconodonts, multituberculates, and symmetrodonts has come from North America and Early Cretaceous pantotheres are known from Europe. In addition, the earliest known marsupials and placental mammals come from the Early Cretaceous.

Marsupials (or Metatheria) have small brains, embryogeny partially in external pouches (termed *marsupia*), primitive counts of teeth, digits, and other bones (i.e., few lost to specializations), and generalized molars of the *tribosphenic* type. In this molar, six shearing surfaces are present as a consequence of the arrangement of cusps (Crompton, 1971); the tribosphenic molar evolved from that of the pantothere in having an enlarged talonid process on the lower molar and a slight rearrangement of the cusps, producing the shearing surfaces.

In Lower Cretaceous Trinity Group strata from Texas are remains of an opossumlike marsupial called *Holoclemensia,* known from upper and lower molars. L. S. Russell (1975) observed that the upper molars are, in most respects, intermediate between those of pantotheres and opossums whereas the lowers are very similar to those of the modern opossum.

Also in the Paluxian (Lower Cretaceous) fauna from Texas are teeth from *Pappotherium,* the oldest eutherian mammal. The molars are sufficiently different from those of marsupials in the arrangement of cusps within the tribosphenic structure to show that they are eutherian but detailed description of those differences is beyond our scope of anatomical detail (see L. S. Russell, 1975). Eutherians comprise the overwhelming bulk of mammal taxa featuring full-term live birth, external mammae in females, and all the specializations of modern mammals. L. S. Russell (1975) discusses the Late Cretaceous mammals from North America; to summarize his observations, by Lance/Hell Creek/Laramie time, several placental mammal groups are known to have differentiated, including insectivores, ungulates (hoofed animals), and the first primate, *Purgatorius* (Van Valen and Sloan, 1965) from the Lance Formation. Marsupials are also common elements of the uppermost Cretaceous mammal faunas as are multituberculates; all other archaic Mesozoic types (triconodonts, docodonts, symmetrodonts, and pantotheres) had become extinct.

H.9. Hot-Blooded Dinosaurs and Related Notions

An impressive resurgence of interest in the biology of dinosaurs has occurred in recent years in part due to the set of notions which collectively have been popularized under the term "hot-blooded dinosaurs." The basic premises were presented in a few seminal papers by Bakker (e.g., 1972, 1975), and evidence based on inferred physiology and behavior of *Deinonychus* (Fig. 8-77) by its discoverer John Ostrom (e.g., 1969) was an important part of the overall idea. Additional input on the largely pro-hot-blood side was also provided by Ostrom (1974) on the relationship of birds to dinosaurs and on "herd" behavior in some dinosaurs (1972); and by de Ricqles (1968 and many other articles) who pointed up characteristics of dinosaur osteology that have been considered strong evidence of metabolic similarity to warm-blooded mammals.

H.9.a. Arguments

Robert T. Bakker and others suggest that dinosaurs maintained their body temperatures at a relatively constant, high level as do modern birds and mammals. That is the basic "hot-blooded dinosaur" premise, but with it go many behavioral and ecological implications. For example, Ostrom (1980) lists the following attributes of dinosaurs under the hot-blooded regime:

1. Posture and gait: upright, limbs move straight fore-and-aft, like mammals (unlike, say, crocodiles).
2. Blood pressure: high, heart must be four-chambered as in mammals and birds, not three-chambered as in reptiles.
3. Activity levels: high, as in mammals, not intermittently sluggish with cold temperature and active in warmth.
4. Feeding: requirement for high-caloric input to maintain warm-bloodedness.
5. Predator–prey ratios: large numbers of prey species, necessary to feed predators, relative to lesser needs for cold-blooded predators.
6. Bone histology: dinosaur compact bone features *Haversian systems,* which are also present in mammal bone and not in ectotherm (cold-blooded animal) bones. These provide means to transfer calcium and phosphorus through the tissues for rapid and heavy demand.
7. Biogeography: dinosaur fossils are found in latitudes that apparently were subpolar in later Mesozoic time, far beyond the tolerance range of extant reptiles.
8. Relationship with birds: evidence shows that *Archaeopteryx* was warm-blooded, as are modern birds, and because it was so very similar to coelurosaurian dinosaurs, the latter too were probably warm-blooded.

One conclusion reached by some aggressive proponents of the hot-blooded idea is that dinosaurs and birds

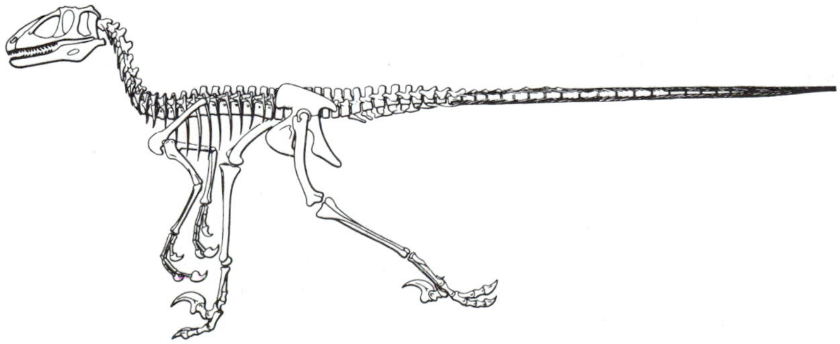

Fig. 8-77. Reconstruction of *Deinonychus,* an agile dinosaur with modifications of the tail suggesting it served as a support to allow employment of the huge claws on the pes. (From Colbert © 1980 by John Wiley & Sons, reprinted with permission of John Wiley & Sons.)

form a more closely related assemblage than do dinosaurs and other reptiles (including crocodilian archosaurs); they (e.g., Bakker and Galton, 1974) suggest that a class Dinosauria be erected to include the former groups, distinct from Reptilia, and that the class Aves be eliminated, and that living birds are, in fact, little dinosaurs! To evaluate the entire controversy is beyond our scope but we will share with the reader several of our observations based on the consensus view of recent authors.

One must begin this discussion with precise definitions of terms. ''Warm-blooded'' (or the unfortunate term, ''hot-blooded'') incorporates a number of types of metabolisms. For example, animals with regulated warm-blood are classified as *endotherms* if the core temperature is maintained by high metabolic rate, and further as *homeotherms* if the former condition is controlled within a few degrees (i.e., as in mammals and birds). In the common usage of ''warm-blooded'' then, one really means homeothermy. But there are other types of warm-bloodedness: for example, animals that use the sun's radiance selectively for internal temperature control are termed *heliotherms*. *Dimetrodon* (see Chapter 7) may have evolved its sail for that purpose. In addition, very large animals may maintain a constant internal temperature by means of great bulk and the inherent insulation of such a large body: such a mechanism is termed *inertial homeothermy* and the sheer bulk of larger dinosaurs suggests that as a likely condition for them. In opposition to the various ''warm-blooded'' systems, there are several ''cold-blooded'' (or ectothermal) metabolisms. Most typical is the state where an animal simply allows its core temperature to vary proportionally to the ambient temperature; such condition is termed *poikilothermy* and is characteristic of most vertebrates except birds and mammals. There is an additional condition called *heterothermy* in which core temperature varies more than in homeothermy but still the body is warm and regulated to some degree (terminology above modified from Ostrom, 1980).

It may be surprising to some readers to learn that certain fish, especially fast-swimming types such as tuna and mackerel sharks, have body temperatures that may be 10°C higher than their surrounding waters. Bees, too, generate considerable heat within their hives, and these fish and bees are ''warm-blooded'' in a real sense. But they are not homeothermal because their body temperature is dependent on muscular activity, not enzymatic control, and because their temperatures fluctuate widely. More importantly to this discussion, their metabolism is neither avian nor mammalian. We point up these examples to underscore that homeothermy is not the only extant warm-blooded state.

Let us now consider evidence why dinosaurs *should* have been warm-blooded. Endothermy (we will be less demanding and not require homeothermy for this discussion) allows an organism to be always ready to move quickly

without a ''warm-up'' period. It also allows properly insulated organisms to inhabit cold regions (high mountains and polar latitudes) and allows an overall higher proportional level of muscular performance. This last point applies at least to living animals; possibly other extinct systems may have also been highly efficient. Since dinosaurs were uncommonly large and thrived for much of the Mesozoic, it would seem that endothermy would explain their success. However, endothermy also is expensive metabolism requiring very large inputs of food for a given organism compared to the much smaller food requirements necessary for a same-sized poikilothermal organism. Examination of predator–prey ratios in both dinosaur and nondinosaur populations suggested to Bakker (1972) and others that proportions comparable to mammalian ratios were present; others, notably Farlow (1980), pointed out that predator–prey ratios become increasingly similar as the predator size increases and, since there are no living terrestrial predators of even close to dinosaur size, one cannot model existing ratios to those of dinosaurs. He also noted that cannibalism among carnivorous dinosaurs could skew paleobiological interpretations of their food sources as could insectivorous or even lizard-and-mammal-eating habits among smaller theropods.

Bone histology provides the second strongest argument in favor of dinosaur endothermy. The presence of Haversian systems is strongly in favor of similarity between birds, mammals, and dinosaurs. However, it is also true that some small mammals and birds lack Haversian systems and some cold-blooded modern reptiles possess them (these include some crocodilians and some turtles). In fact, impressive arguments have been made showing that Haversian systems in compact bone may be more closely related to size and growth rates rather than to temperature and activity. Tooth structure, in fact, has provided what is to some the strongest anti-hot-blood evidence. Johnston (1979) reported the clear evidence that dinosaur teeth contain rings like those found in crocodilians. Since such rings almost certainly reflect response to external temperature cycles (and are not found in mammals), this evidence alone might be considered ''cold-blooded murder'' of the ''hot-blooded dinosaurs.'' To our knowledge, no adequate refutation of this point has yet emerged.

Some aspects of the pro-warm-blooded arguments are compelling to us without requiring dinosaurs to have been endotherms. For example, that some dinosaurs may have been agile and quick is almost certain: skeletal evidence is clear that *Deinonychus* kicked with its ''terrible claw'' and balanced on its tail. Also, the tails of sauropods likely were held aboveground as they walked; such posture is indicated by their tracks and in fact the balance thus achieved would probably be mechanically less taxing than tail-dragging. That carnosaurs and coelurosaurs were rapid runners is very likely since there is good skeletal evidence that they had very muscular legs and must have been quicker than prey species. Known footprints also show clearly that sauropods moved in herds; carnivores may very likely have hunted in packs. We have already noted that sauropods may have evolved marvelously long necks in response to feeding from the high branches of gymnospermous plants; it has even been seriously proposed that they balanced on their tails while so feeding and we do not find this difficult to accept, even though it does require very strong myocardial tissue to generate the unusually high blood pressure necessary to raise a column of blood up that many meters of neck. Since we have no knowledge of dinosaur vascular systems, such inference is not beyond reason. (But we *do not* find in this proof of a four-chambered heart in dinosaurs.) A final point in this category of "accepted para-hot-blood arguments": it is clear that many dinosaurs had limbs rotated fully under their bodies, as do mammals and birds. The strain and strength limits of bone virtually require such posture.

This set of "mammalian" characteristics discussed above together has impressed many people that dinosaurs were therefore mammalian in metabolism. To us the evidence is strong that dinosaurs were convergent toward many mammalian characteristics because of their size and the times in which they lived, but that most of these attributes speak eloquently for the uniqueness of dinosaurs among reptiles. In essence, we follow the logically parsimonious conclusion that dinosaurs were like other reptiles, poikilothermal, but that their size and many known variations from modern reptiles allow for some different behaviors and features. It has been repeatedly pointed out (e.g., by Regal and Gans, 1980) that the sheer mass of dinosaurs was sufficient to maintain relatively constant internal temperatures; i.e., they would be inertial homeotherms. As such, they would reap some of the benefits of endothermy (e.g., quick start-up) without the cost in food. It would be virtually impossible for larger dinosaurs not to be warm-blooded and slow to respond internally to outside temperature change, considering their size.

To conclude this discussion, we will examine a few of the pithier peripheral controversies of the "hot-blood" thesis. Known dinosaur biogeography would be unusual for modern reptiles but their presence in subpolar latitudes is not really surprising considering that: (1) the Mesozoic was probably a much warmer and more equitable time than the present, and (2) global positions for landmasses, following the breakup of Pangaea II, are not known with precision sufficient to define flows of warm oceanic currents near landmasses nor the locations of specific sites within several degrees of rotation (equivalent to distances of hundreds of kilometers). In addition, nondinosaur reptiles are commonly found with northern dinosaur finds; it is unquestionable that those nondinosaur reptiles were poikilothermal.

The relationship of coelurosaurs to birds has aroused a number of subcontroversies, largely stemming from the very close morphological similarities in the remains of *Archaeopteryx* and some coelurosaurs. *Archaeopteryx* is very reptilian in form and it also may have been a fully functional bird (i.e., homeothermal, and with a four-chambered heart). Aggressive hot-bloods have maintained that *Archaeopteryx* was a dinosaur (as, they say, are all birds) and therefore its homeothermal nature implies homeothermy in all dinosaurs. This contention is logically circular in part and simply unfounded in others. Nevertheless, birds and dinosaurs are clearly related. We find no problem in accepting the idea that *Archaeopteryx* was a (literally) full-fledged bird with all the proper attributes. This suggests that evolution from reptilian to avian metabolism, with modification of the heart, accompanied evolution of feathers as part of the specializations accompanying flight.

H.10. The End–Mesozoic [Maestrichtian–Danian (?)] Mass Extinction

The most renowned of mass extinctions is certainly the terminal Cretaceous "end of the dinosaurs." In addition to terrestrial events which saw the demise of all dinosaurs, nearly all marine reptiles (save for crocodilians and turtles) were eliminated, and, in fact, all tetrapods weighing more than 10 kg (again, except turtles and crocodilians) became extinct (Bakker, 1977).

Major taxonomic upheavals also occurred among nonvertebrate animals and plants. Mass extinction of the abundant and characteristic Mesozoic bivalves is particularly evident in the disappearance of all rudists, *Inoceramus, Buchia, Exogyra,* and most *Gryphaea* and trigoniids. All ammonoids became extinct as did most nautiloids and all but a very few belemnoids. Among other megainvertebrates, five families of bryozoans, eight of echinoids, and the order of sphinctozoan sponges (important late Paleozoic through Mesozoic reef-builders) became extinct at the boundary. However, the above groups aside, benthic invertebrates were largely unaffected by the apparent mass extinction event.

Plankton suffered a diverse set of effects at the era boundary. The once-abundant Cretaceous calcareous nannoplankton were decimated. Coccolithophorids were reduced from approximately 500 to 100 species (Brasier, 1980) and only 13% of Cretaceous genera cross the boundary (Emiliani *et al.,* 1981). Planktonic foraminifera were similarly reduced, with loss of the characteristic Cretaceous families Rotaliporidae, Globotruncanidae, and Schakoinidae, along with genera from many other families (Loeblich and Tappan, 1964). However, noncalcareous plankton suffered less definitive losses: Emiliani *et al.* (1981) tabulated survival of 78% of dinoflagellate genera and 93% of radiolarian genera into the Tertiary; but Thierstein (1982) showed an 85% loss of radiolarian genera and a (comparable) 33% loss of dinoflagellate genera.

Terrestrial plants underwent no major taxonomic turnover, since only one characteristic Mesozoic group, Bennettitales, became extinct; but the abundant and characteristic Mesozoic cycads and ginkgoes declined to a vestigial population. Also, new conifers and ever-increasing numbers of angiosperms replaced many typical Mesozoic taxa. Such changes are not evident from study of taxonomic range charts (since most groups had survivors) but the proportional makeup of many regional forests in the Paleocene was significantly different from that of the Cretaceous.

Overall, our tabulation shows that taxa from all marine realms (plankton, nekton, and benthos) were affected by the end of Mesozoic mass extinctions, as were terrestrial elements. Forms which disappeared tended to be those apparently very well adapted and abundant during the later Mesozoic (although, admittedly, some groups declined prior to or during the Maestrichtian: e.g., sauropod and other dinosaurs, ichthyosaurs, and ammonites). Consequently, it may be stated that either: (1) some aspect of both marine and terrestrial environments changed drastically at the end of the Maestrichtian; (2) some extraordinarily lethal phenomenon was in operation for an interval at the end of the Maestrichtian; or (3) *stochastic* (i.e., random) events or changes coincidentally caused disappearance of several highly successful groups, giving the impression of a catastrophe among Mesozoic taxa. These conjectures are not mutually exclusive since both random and catastrophic elements may have been in operation for different groups; however, for there to have been a "mass extinction," at least several abundant groups must have disappeared at approximately the same time.

In the following discussions, we will offer alternative specific explanations for the observed events and analyze the most feasible. As with our previous discussion of "mass extinction" (that of the Permo-Triassic; Chapter 7), we need to first consider what is known of the time involved in the phrase "end of the Mesozoic." This ensnares one immediately into the "Danian" controversy.

H.10.a. Boundary Clays and the "Danian" Stage

European, and some American stratigraphers recognize a lowermost Tertiary "Danian" Stage, with a thin type section in Denmark. Most significantly, Danian strata overlie Maestrichtian strata with apparent conformity in several European regions, including Denmark. Therefore, it is reasonable to consider that the "end of the Mesozoic" may be traced to a specific horizon above the Maestrichtian and below the Danian rocks. However, even if this assumption were true, presence of precise bounding sequences may not

shed clear light on the nature of Cretaceous/Tertiary extinctions, as discussed below. One thing that may be said with some certainty: the stratigraphic interval, which may represent the era boundary, is quite precisely determinable in at least some regions. Therefore, if distinct changes at large scale can be tied to such a precisely determined bounding stratigraphic sequence, we may state that a real "mass extinction" occurred in a geologically brief span.

Confounding the attempt to tie faunas and strata at the precise boundary is the nature of environments represented by the known continuous Maestrichtian–Danian sections. Megafossils in the typical Danian are not abundant and those found are not diagnostic of either the Cretaceous or the Tertiary (i.e., not index taxa). Kauffman (1979) observed that few known boundary sequences formed in shallow-marine environments. Emiliani *et al.* (1981) stated that under the prevailing regressive conditions of the end-Cretaceous, no certain record remains from marine deposits formed in waters less than 10 m deep. Thus, much of the neritic biota of the latest Maestrichtian–earliest Danian may be unknown. On closer examination, one finds that nearly all known boundary sequences consist of either: (1) deep ocean sediments sampled in cores; (2) deep-water carbonates preserved largely in three regions (Gubbio, Italy/ Zumaya, Spain/Tampico, Mexico; data from Kauffman, 1979); or (3) part of the typical Danian section, widely exposed in Denmark. However, in the introduction to a symposium on the typical Danian, Birkelund and Bromley (1979) stated that each author has his or her own lower, middle, and upper Danian, and that there is no general acceptance of its boundaries. Further, the archetypical Danian section in the Stevns Klint region was shown by Surlyk (1979) to incorporate hardgrounds precisely at the boundary sequence. This suggests that an erosional hiatus is recorded there and may be present elsewhere in the Danian (despite apparent conformability). Overall, we have shown that some uncertainties exist about delimitation of the Maestrichtian–Danian boundary in the type area (where the only neritic–benthic habitats may be preserved).

The deeper-water boundary sequences feature a particularly interesting phenomenon at the precise bounding horizon. Smit (1979) and Alvarez *et al.* (1979) observed that a very thin, carbonate-free clay was present at the presumed boundary horizon in pelagic carbonate sequences from Spain and Italy. Further, Alvarez *et al.* (1980) showed that in Italy, this "boundary clay" was anomalously enriched in noble metals, principally iridium. Ganapathy (1980) reported similar enrichment in noble metals (iridium, osmium, gold, platinum, and others) in boundary sequences from Denmark, and these reports resurrected the school of "astronomical catastrophist" explanations for the end-Cretaceous extinctions, based on the hypothesis that the noble metal enrichment of the boundary clay represented meteoric debris from a very large object (e.g., an asteroid) that collided with Earth at the era boundary. At the time of this writing, there are virtually monthly reports of new noble-metal anomalies for Cretaceous–Tertiary boundary units, and older or younger strata.

H.10.b. Dinosaurs and the "End of the Cretaceous"

Since the uppermost Mesozoic Erathem is delimited in North America by the last appearance of dinosaur fossils, one must include dinosaur extinction with marine events in propounding an "end of the Cretaceous" scenario. Dinosaur events tend to be especially impressive because dinosaurs were impressive terrestrial tetrapods, just like humans. We empathize with their fate far more readily than with, say, that of echinoids. Thomas J. M. Schopf, in a review of Birkelund and Bromley, 1979 (*Science* **211**:571, and see Schopf, 1982), stated: "The problem of the extinction of the dinosaurs boils down to . . . what happened to a score of species inhabiting the river and floodplain habitats adjacent to the North American Western Interior seaway?" He attributes that putatively trivial event to a combination of sea-level and base-level lowering, seasonality change, and a reduction of habitat because of changed river regimes. Bakker (1977) stated that at the boundary, only a half-dozen dinosaur genera from the western interior remained and disappeared together: *Triceratops, Ankylosaurus, Edmontosaurus, Thescelosaurus, Leptoceratops,* and *Tyrannosaurus.* In addition, one may note that of the latest dinosaurs found, roughly 80% are *Triceratops;* that is to say, he suggests that diversity was very low among the last dinosaurs and populations of all but one genus were small. Russell

(1982) disagrees and shows that there is very imperfect knowledge of diversity and abundance of dinosaurs near the era boundary. He suggests that further explorations will likely reveal undescribed latest Maestrichtian taxa, and that they were globally distributed prior to the extinction event.

H.10.c. Hypotheses, Hypotheses

The terminal-Mesozoic extinctions have aroused more serious and spurious ideas than have any like event. Some can be dismissed quickly, such as ideas of interplanetary visitors carrying away all remaining dinosaurs, or an early, undiscovered human population killing too many. Among scholarly hypotheses are the following:

1. The climate randomly changed and became too cold, too hot, and/or too seasonal for dinosaurs.
2. The rise of angiosperms provided very favorable niches for insectivorous mammals which thrived, radiated quickly, and led to competition with small reptiles. These in turn no longer fed small dinosaurs and there was an effect up through the food webs to large dinosaurs.
3. The climate became too arid for large terrestrial animals.
4. Increasing mammal populations preyed heavily on dinosaur eggs, reducing populations below replacement levels (note that such a mechanism is in operation today on marine turtles).
5. Dinosaurs were at such low diversity and occupied such isolated, restricted basins by latest Maestrichtian time that their disappearance reflects a stochastic event.

Fig. 8-78. Hypothetical mode of combat among agile theropod dinosaurs. In this sketch, a *Ceratosaurus* (at right) balances on its tail to defend against two *Allosaurus. Ceratosaurus* approximately 5 m long; *Allosaurus* approximately 8 m long. (Illustration reproduced courtesy of Gregory S. Paul.)

6. Dinosaur eggshells became thin due to environmental conditions (reported in at least one genus; Erben *et al.*, 1979).

7. A terminal Mesozoic "greenhouse effect" enhanced a global warming trend. Atmospheric CO_2 release may have been triggered by Late Maestrichtian extinction of the huge coccolithophorid floras. The resulting warm conditions had catastrophic effects in oceans and on land (McLean, 1978).

8. The oceans became nutrient- and carbonate-poor as a consequence of peneplanated land surfaces, shrinking oceans, and with vast chalks deposited on continents. Consequently, a very shallow carbonate-compensation depth resulted which was coupled with a cloud-free atmosphere and extreme seasonality. The net effect was deleterious to marine carbonate-secreting organisms (e.g., ammonoids, coccoliths, foraminifera, echinoids, bivalves) and ultimately to land animals (Worsley, 1974).

9. The entire world ocean was covered by low-salinity water derived from spillover of the fresh-or-brackish Arctic Ocean which tectonically rejoined with the global ocean. This low-salinity water resulted in oxygen depletion below the surface layer and had terrible effects on the marine biota. This marine catastrophe would have caused climate change including drought and lowered temperature; these in turn would affect land animals and plants [in this form, see Gartner and McGuirk (1979); also see Thierstein and Berger (1978) and other "Arctic spillover" models cited in these papers].

10. A large, extraterrestrial object, probably an asteroid, collided with the Earth, initiating boundary-time. Two divergent scenarios derive from this hypothesis: Alvarez *et al.* (1980) postulated a continental asteroid collision, which threw up large quantities of dust and fragments of the asteroid, completely darkening the skies for an interval (they originally specified several years' duration for the darkness but more recent modifications call for much shorter intervals). The darkness interfered with photosynthesis and affected food webs among land and sea biotas, causing mass extinctions. Emiliani *et al.* (1981) accepted the concept of asteroid collision but proposed a model of collision with the ocean to best explain the data. Whereas the Alvarez *et al.* scenario involved darkness as the killing agent,

Emiliani *et al.* believe that higher surface temperature from water injected into the stratosphere was the agent of mass extinction.

As noted earlier, the "asteroid impact" concept is very much in vogue and reports of noble-metal enrichment in stratigraphic boundary sequences appear monthly. Iridium anomalies (implying extraterrestrial sources) have been found not only in marine strata (for a summary, see Alvarez *et al.*, 1982a), but also in Cretaceous–Tertiary swamp deposits in New Mexico (Orth *et al.*, 1981), and in nonmarine claystones in Montana (Bohor *et al.*, 1984). Iridium anomalies are also being reported from times other than the Cretaceous–Tertiary boundary: for example, in the Upper Devonian of Australia (Playford *et al.*, 1984), and in Eocene marine cores off Venezuela (Alvarez *et al.*, 1982b; Ganapathy, 1982). The situation has reached the point where paleontologists feel they must look for evidence of asteroid impacts where extinctions are noted [e.g., Orth *et al.* (1984) searched for Cambrian iridium anomalies and did not find them].

We believe the present high level of interest in the subject will make virtually anything written here obsolete by the time this book reaches press. At present, several main trends are evident in the "impact" school of extinction modeling which bear thought:

1. The sizable number of iridium anomalies reported shows the phenomenon is real but also commonplace through time.

2. The coincidence of the Cretaceous–Tertiary boundary with an iridium anomaly is also real; but the more such anomalies are found, the less will be the probability that there is cause-and-effect between astronomical impacts and the Cretaceous mass extinctions.

3. Geologically unusual events probably occurred at the end of the Maestrichtian. The "extinction" was real. However, the precise timing of all extinctions (terrestrial, benthic, epicontinental-marine, and so on) is not provably contemporaneous.

4. At present, the "asteroid impact" hypotheses are well supported by evidence. However, many questions are not answered by any reasonable hypothesis, such as why mammals were not affected (Tschudy and Tschudy, 1986), why turtles and crocodilians similarly survived, why plants were not affected (and they would be most strongly influenced, one would think, by atmospheric and solar effects), and why only certain among the many asteroid impacts on Earth had such major effects on life.

CHAPTER 9

The Tejas Sequence: Tertiary–Recent

A. Introduction

A.1. Overview of Tejas Events

At the end of the Maestrichtian Age, seas withdrew from North America and virtually every other continent. The Tejas episode began, as did others before, with high-standing continents; but unlike past times, the seas remained outside of the cratonic interior except for a brief transgression in the Paleocene which probably came from the Arctic Ocean and reached south as far as the Dakotas. The Tejas Sequence name derives from the site of the best-studied Cenozoic units, those of the Texas coastal plain, which feature a rich record of marine and marginal-marine strata (along with the Atlantic, Pacific, and adjacent Gulf coasts). In contrast, the cratonic interior also contains a very sizable Tejas record; but almost all units involved are nonmarine. In addition, substantial areas of the northern craton and Cordillera feature surficial Quaternary glacial deposits, which may obscure uppermost preglacial Tejas depositional events.

The Cenozoic history of the Atlantic and Gulf Coastal Provinces is enormously detailed but relatively simple, because no major tectonic events affected it (except in the Sierra Madre Oriental of Mexico). Continent-derived detrital sediments continued to accumulate on the eastern and southern margins in a variety of nonmarine, paralic, shelf, and slope environments. Atlantic Coastal Province areas received relatively little coarse detritus during the early Cenozoic, probably because the old Appalachian orogen had been eroded down to a low, nearly peneplaned surface. Consequently, lower Cenozoic deposits there are dominated by mudrocks and carbonates. Carbonates also continued to accumulate virtually free of terrigenous detritus in the Florida–Bahamas areas. In the Gulf Province. however, early Cenozoic sedimentation was overwhelmingly detrital in nature due to the flood of sediments derived from the Laramide orogen. Rapid sediment influx, especially in western and northwestern parts of the Gulf Province, resulted in the development of major, progradational sedimentary wedges, often dominated by deltaic systems. Later Cenozoic sedimentation in most Coastal Province areas was predominantly regressive, and detrital sediments began to encroach southward so that by the Pleistocene, carbonate deposition was limited principally to southern Florida. The Bahamas, cut off from the continent, continue even today as a carbonate province. Upper Cenozoic deposits of the Gulf Province are also detrital in character and reflect the onset and dominance of the Mississippi drainage system, whose sediments have completely filled the Mississippi Embayment and are now actively prograding into the Gulf of Mexico.

At the beginning of the Cenozoic, the Cordilleran continental margin was essentially a regular, Andean-type convergent plate-boundary although the very anomalous Laramide orogeny had just begun in the Rocky Mountains area. Laramide tectonics were dominantly vertical, characterized by up-and-down movements of basement blocks and concomitant drape-folding supracrustal strata; the cause of Laramide tectonics is not well understood but may be related to very-low-angle subduction of the Farallon plate beneath the western United States. Not long after Laramide events, the North American plate, in its westward drift, began to encounter and then override the divergent boundary between the Farallon and Pacific plates. The effects of this event were to initiate transform faulting, in those areas where North America was adjacent to the Pacific plate, while convergence continued in those other areas where the Farallon plate had not entirely been lost to subduction. The result was the development of the San Andreas and Queen Charlotte transform-fault systems and concomitant fragmentation of the continental margin in those areas to form a borderland terrane of fault-bounded basins and uplifts. Possibly related to the elaboration of transform tectonics in southern California was the beginning of block-faulting, volcanism, and crustal extension in the Basin and Range Province. Farther north, where the Farallon plate continued to subduct beneath North America, essentially normal arc-magmatism continued in the Cascades area, possibly accompanied by rotation of the Olympic Mountains block. These tectonic patterns persist today.

In the Innuitian area of North America, tectonic events were dominated by opening of the Eurasian Basin of the Arctic Ocean and the related rifting of Greenland away from North America. One result was the Eurekan orogeny, which ended the long history of the Sverdrup Basin. Following the Eurekan orogeny, sediments accumulated on the Arctic Coastal Plain and continental shelf and slope. The northern Alaskan continental margin was not affected by the Eurekan orogeny and continued, as during the Late Cretaceous, to develop a normal coastal plain and rifted continental margin.

The Quaternary Period, and especially its glacial stratigraphy, presents some of the most complex problems in detailed stratigraphic analysis known for the entire geological column; this is true despite the fact that the Quaternary is the most recent, shortest, and (clearly) uppermost stratigraphically of geological sequences. Much of the complexity derives from the nature of ice sheet deposition, which inherently disrupts and incorporates younger materials, thus obscuring older events with the younger. Then too, the glacial stratigraphic record involves many sedimentary forms which are not in accord with conventional stratigraphic principles and which include widely divergent and admixed stratigraphic styles; for example, a single ice advance and retreat may leave behind virtually all of the glacial erosional and depositional structures and deposits one learns about in school, or little but a wisp of loess and paleosol as evidence. Then too, a glacial interval may incorporate multiple advance and retreat cycles which leave variable evidences among diverse regions.

Terrestrial life of the Tertiary and Quaternary is radically different from that of the Cretaceous and before. The Cenozoic is certainly the "Age of Mammals" and many mammalian taxa originated in North America during the Tertiary. Generally the Tertiary marine fossil record shows forms often indistinguishable (except from the perspective of the specialist able to discriminate specific characters) from shells on modern beaches. Indeed, subdivisions of the Tertiary were based originally on percentages of extant mollusk species in fossil assemblages.

By late Quaternary time, North America hosted mammals from virtually all major orders. However, during the transition from Pleistocene to Holocene times at approximately 10,000 yr. B.P., a mass extinction event left the continent with a depauperate fauna containing few large carnivores and relatively few large ungulates. Humans first entered North America during the Quaternary and appear to have achieved some success.

A.2. On Cenozoic Time and Time–Stratigraphic Units, and "Lyellian Curves"

A.2.a. Nomenclature of the Periods

The Cenozoic Era has been subdivided using two contrasting geological time designations. The first was based on the 18th century time-stratigraphic units of Giovanni Arduino, recognizing an older "Tertiary System" and a younger "Quaternary System." The respective systems incorporated unconsolidated sediments at the top of most European sections, with the Quaternary portion representing glacial, volcanic, and recent alluvial materials. This "Tertiary/Quaternary" nomenclature, being based on an archaic misconception about the nature of strata, led many geologists to adopt an alternative time nomenclature with designation of "Paleogene/Neogene" Periods. Whereas the Tertiary encompasses all but the last 2.5 Myr. of Cenozoic time, the Paleogene comprises the first 40 Myr. and the Neogene the remaining 25 Myr. We prefer the Tertiary–Quaternary names for designation of Cenozoic time and strata, largely because the USGS uses such terminology and because the terms have precedence. In addition, as Krumbein and Sloss (1963) observed, the original meanings of "Tertiary" and "Quaternary" have long been ignored and they represent only abstract names in modern usage.

The universally adopted epochs of the Cenozoic Era (Paleocene, Eocene, and so on) also designate series in the American standard section, and will be the basis for subdivision of this chapter. Unfortunately, these are keyed to molluskan faunas of the coastal plains and are difficult to apply to continental strata of the interior. As a consequence, a second set of provincial ages has evolved, based on mammalian succession in North America (Fig. 9-1); fortunately, correlations among interior, coastal, and European time-stratigraphic units are generally clear.

A.2.b. Lyellian Curves

The historical basis for differentiating marine series of the Tertiary, originated with and is named for the champion of uniformitarianism; but, in truth, Charles Lyell, Paul Deshayes, and Heinrich Bronn independently came up with similar schemes approximately at the same time in history (in Stanley *et al.*, 1980). Lyell received common recognition for the concept because of his preeminence in early geology.

Lyellian curves are better termed "Lyellian percent-

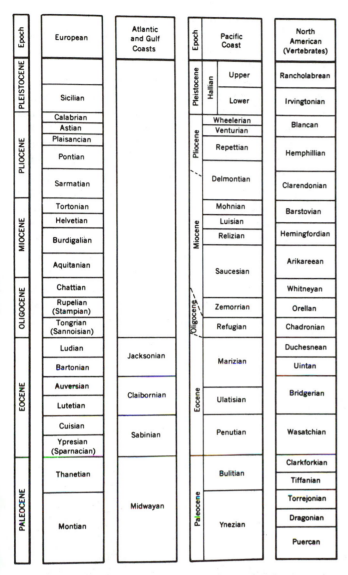

Epoch	European	Atlantic and Gulf Coasts	Epoch	Pacific Coast	North American (Vertebrates)
PLEISTOCENE			Pleistocene — Hallian	Upper	Rancholabrean
PLEISTOCENE	Sicilian		Pleistocene — Hallian	Lower	Irvingtonian
PLIOCENE	Calabrian		Pliocene	Wheelerian	Blancan
PLIOCENE	Astian		Pliocene	Venturian	Blancan
PLIOCENE	Plaisancian		Pliocene	Repettian	Hemphillian
PLIOCENE	Pontian		Pliocene	Repettian	Hemphillian
MIOCENE	Sarmatian		Miocene	Delmontian	Clarendonian
MIOCENE	Tortonian		Miocene	Mohnian	Barstovian
MIOCENE	Helvetian		Miocene	Luisian	Barstovian
MIOCENE	Burdigalian		Miocene	Relizian	Hemingfordian
MIOCENE	Aquitanian		Miocene	Saucesian	Arikareean
OLIGOCENE	Chattian		Oligocene	Saucesian	Whitneyan
OLIGOCENE	Rupelian (Stampian)		Oligocene	Zemorrian	Orellan
OLIGOCENE	Tongrian (Sannoisian)		Oligocene	Refugian	Chadronian
EOCENE	Ludian	Jacksonian	Eocene	Marizian	Duchesnean
EOCENE	Bartonian	Jacksonian	Eocene	Marizian	Uintan
EOCENE	Auversian	Claibornian	Eocene	Ulatisian	Bridgerian
EOCENE	Lutetian	Claibornian	Eocene	Ulatisian	Bridgerian
EOCENE	Cuisian	Sabinian	Eocene	Penutian	Wasatchian
EOCENE	Ypresian (Sparnacian)	Sabinian	Eocene	Penutian	Wasatchian
PALEOCENE	Thanetian	Midwayan	Paleocene	Bulitian	Clarkforkian
PALEOCENE	Thanetian	Midwayan	Paleocene	Bulitian	Tiffanian
PALEOCENE	Montian	Midwayan	Paleocene	Ynezian	Torrejonian
PALEOCENE	Montian	Midwayan	Paleocene	Ynezian	Dragonian
PALEOCENE	Montian	Midwayan	Paleocene	Ynezian	Puercan

Fig. 9-1. A comparison of Cenozoic time and time-stratigraphic units, including several North American regions, the European standard section, and North American land mammal ages. (From Kay and Colbert © 1965 by John Wiley & Sons, reprinted with permission of John Wiley & Sons.)

ages'' because they deal with percentages of fossil faunas having representatives of living species. In essence, one used Lyellian curves to determine the age of a particular stratum by calculating the percentage of living mollusk species therein. A percentage of, say, 50% falls within the Pliocene range of the curve. Older series contain smaller percentages of extant species. This approach is largely historical but is still occasionally in use to date strata.

Many authors (e.g., Kay and Colbert, 1965) have pointed out the circularity of logic inherent in such schemes and express distrust in their reliability. Particularly questionable are correlations by percentages of faunas from se-

quences along different oceans; e.g., the Atlantic and Pacific. Stanley *et al.* (1980) actually tested Lyellian percentages from suites in Japan and western North America, with such suites having been dated independently of the molluskan faunas (using nannofossil biostratigraphy and radiometric absolute dating). Surprisingly, they found good correspondence between the Lyellian percentages on the two continents and in both cases these percentages and the absolute dates matched the stratigraphic ranges for the faunas contained. They observed that the spread of values from data are fairly large and concluded that Lyellian curves should not be used for determinations of absolute age of rocks or faunas.

A.2.c. Quaternary Nomenclature

Historically, the Quaternary was the uppermost stratum in the European section, incorporating alluvium and Noachian flood ''diluvium'' above Tertiary fluvial terraces. Lyell, in 1839, coined the term ''Pleistocene'' for marine sequences in Italy and denoted the Plio-Pleistocene boundary at the first appearance of typical cold-water bivalves and foraminifera in fossil assemblages. Additionally, the youngest of discoasters are found in upper Pliocene marine nannoplankton and mark the boundary (this, however, was not known in Lyell's time). Using Lyellian curves, the Pleistocene is defined as the Cenozoic epoch with a fauna containing more than 70% of modern mollusks.

With understanding of the nature of continental glacial deposits, beginning in the 1830s [see Nilsson (1983) for a good discussion], it became evident that the ''Quaternary'' deposits and the ''Pleistocene'' marine sequences were largely synchronous and related by glacial events. Thus, the Quaternary emerged as the system above the Tertiary and the Pleistocene is a series within the Quaternary System characterized by glacial deposits.

Time and time-stratigraphic subdivision of the Quaternary Period is in a state of flux at the present time, largely because new understanding of pre-Wisconsinan history appears to contradict prior sequencing of glacial events. At the higher level, there is little controversy in terminology: the Quaternary Period/System includes two Epochs/Series, the Pleistocene and the Holocene. The Pleistocene encompasses the times of major ice advances, and the Holocene spans the past 10,000 yr. (see Nelson and Locke, 1981). Complexity and controversy enter with subdivision of the Pleistocene and delimitation of the Plio-Pleistocene boundary (to be discussed subsequently).

A.3. Global Paleogeography

Relative to older geological times, Cenozoic paleogeography is well known because the continental positions were similar to those of the present day, and because the history of drift is recorded in the magnetic anomaly patterns of the seafloor. Consequently, the major features of Cenozoic plate tectonics were reconstructed early in the history of plate tectonic theory.

Figure 9-2 shows modern global configurations, and Cenozoic seafloor spreading. Note that both the north and south Atlantic Oceans have opened considerably: as part of our discussion of Cordilleran tectonics, it will be shown that rapid plate convergence along North America's western margin, probably due to westerly drift, led to a very low-angle subduction geometry and to the somewhat unusual Laramide orogeny.

The old Gondwanaland assemblage had been sundered during the Mesozoic, with South America going one way and Austro-Antarctica going another. Africa was, however, still wedded to the Arabian landmass during the early Cenozoic and, possibly, to India. India became an independent continental block during the Paleocene, and began its northward movement; later, probably during the early Eocene, Australia and Antarctica parted company, shifting toward their present configurations. The timing of this event is constrained by the presence of upper Paleocene and lower Eocene flood basalts in Antarctica and the fact that the oldest magnetic anomaly between the two continents is dated at approximately 53 Myr.

By the middle Cenozoic, Africa was converging toward Eurasia, slowly closing the ancient Tethys Ocean. India, too, was approaching its collision with southern Asia which must have begun some time just prior to the Miocene, because Himalaya-derived molasse of Miocene age is present on both the Asian mainland and India. Among the most recent tectonic events is the development of a major rift system in East Africa. This rifting began the opening of the Red Sea, probably during the middle Miocene, and was caused by a hot spot under the Afar area of Ethiopia. In typical fashion of "hot spot"-type rifting, three "arms" developed, two of which opened to produce the Red Sea and the Gulf of Aden while the other failed to open, leaving the rift valley striking southward into Africa. This failed arm joined with the long and complex rift-system of East Africa, whose presence portends the development of a new continental margin.

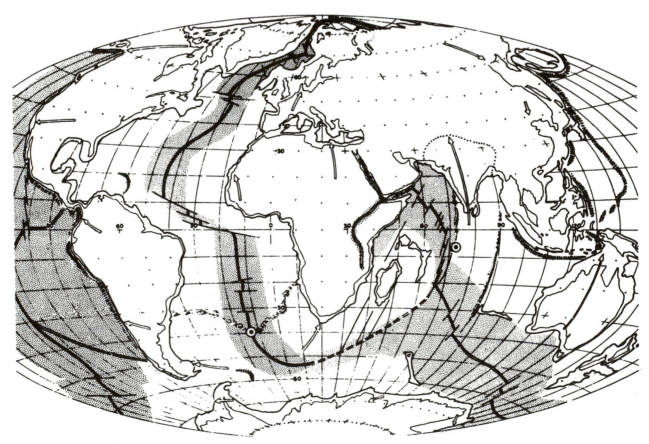

Fig. 9-2. Modern positions of the continents, showing the amount of Cenozoic seafloor spreading (in gray shading) which has occurred. Prevailing directions of plate motions are indicated by arrows. (From Dietz and Holden, 1970.)

B. Tertiary Cratonic Sedimentation

B.1. Overview

B.1.a. Summary

During Paleocene and Eocene times, major detrital basins developed on the flanks and between the Laramide mountains, in the present Rocky Mountain region. The earliest strata in these basins are typically coal-bearing cyclothems; however, by the Eocene a series of enormous lakes developed in some intermontane basins. Finely laminated, varved marls and organic-rich shales were formed in these lakes and include economically and paleontologically important resources. Adjacent to the Eocene lakes were alluvial plains where bright-colored. variegated detrital sediments were deposited. When the Eocene lakes dried up, the alluvial and fluvial variegated lithologies predominated across most Cordilleran basin and piedmont regions. Admixed with Paleocene through Oligocene cratonic strata are large quantities of volcanic ash, resulting from Laramide and subsequent tectonism. Among the notable products of this volcanism are the Eocene Yellowstone Park ''fossil forest'' beds, and the erodible, highly fossiliferous Oligocene strata of the Big Badlands of South Dakota.

During the Oligocene and in later times, cratonic sedimentation no longer occurred primarily in intermontane basins; rather, the western craton tilted upward and detritus was spread eastward from the eroding mountains. Deposits tend to form alluvial fans and aggrading fluvial channels reaching toward the center of the continent. Miocene and Pliocene strata comprise eastward-thinning wedges covering huge surface areas (partly because they are young and little-eroded). These later Tertiary deposits, especially those of Pliocene age, contain some of America's most important groundwater aquifers.

B.1.b. Tectonic Controls on Tertiary Cratonic Sedimentation

The Laramide orogeny yielded its greatest mountain ranges during latest Cretaceous through early Eocene times. Individual mountains culminated at differing times but, for example, during the Paleocene there were sizable ranges present in Wyoming, Montana, Colorado, northeastern Utah, Alberta, British Columbia, and New Mexico (see Fig. 9-7). Of particular importance to the craton's earliest Tertiary history was the Front Range of the Rockies, which typically shed sediments eastward into basins which formed at the feet of the mountains and which consequently contain most of the Paleocene continental stratigraphic record. During Eocene and later Tertiary times, intermontane *successor basins* commonly developed, comprising block-faulted grabens or blocks dropped by normal-faulted monoclines,

and becoming predominant receptacles for thick detritus shed from Laramide and older orogenic highlands (Fig. 9-3). The thickest terrestrial sequences were deposited in these basins along with very extensive and voluminous lake sediments.

The Laramide orogeny ended in the early Eocene and was followed by a prolonged period of planation with some gentle folding of strata adjacent to ranges. The folding seems to have mostly affected latest Paleocene strata. In later times, folding continued spasmodically, and in addition block-faulting occurred across a very wide area of the southwest. These blocks faults had vertical displacements of up to several thousand meters and produced one of the major North American physiographic features: the Basin-Range Province, containing a very sizable Late Tertiary stratigraphic record. Events there are discussed in conjunction with the western margin (Section D).

Basins east of the Front Ranges were especially important in the Paleocene and Eocene but were subsequently inactive. Regions farther east and generally south of those basins (i.e., in eastern Wyoming, South Dakota, Nebraska, Kansas, eastern Colorado, and Oklahoma) did not receive significant sedimentation from the Rockies until Oligocene time. In the Oligocene and Miocene, alluvial sedimentation on a broadly subsiding plain produced the deposits of the White River Group which now form the Badlands of South Dakota, and in Pliocene time a vast sheet of thin alluvial sediments, the Ogallala Formation, was deposited across a broad landscape from the panhandle of Texas to northernmost Nebraska. This veneer of Pliocene strata is the largest single-interval outcrop in North America and extends under the prairie states to comprise the most important aquifer in the United States. It is entirely possible that Oligocene and Miocene continental strata were once as extensive but have been eroded from the surface.

As a final point, it is clear that the Rocky Mountain region and points eastward show recent drainage which is superimposed on the Rocky Mountain structures. Apparently, rivers developed during late Miocene time, on Oligocene–Miocene basin fill, were unaffected initially by the underlying mountain structures. Once their courses were established, they continued competently along their way even as mountains were denuded and exposed. This extensive downcutting has yielded distinctive structures, including gaps through mountains in which no water flows while nearby a stream may cut through almost 1000 m of hard rock.

B.1.c. The Post–Zuni Unconformity

No conformable Maestrichtian–Paleocene (Danian) marine sequence are known in North America, and few are present anywhere else. In the previous chapter, we dis-

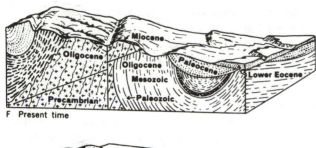

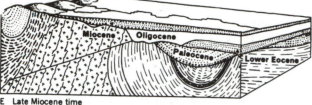

F Present time

E Late Miocene time

D Late Eocene time

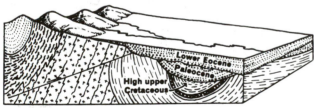

C Early Eocene time

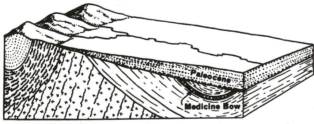

B Late Paleocene time

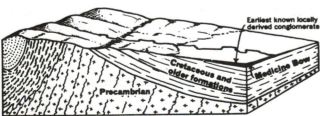

A Late Cretaceous time

cussed the boundary sequences and their relationship with the "Cretaceous extinctions" question. On the Atlantic and Gulf coastal plains, youngest Cretaceous sediments are regressive and capped by a distinct erosional surface, which in turn is overlain usually by Paleocene or Eocene sediment.

In the cratonic interior, there are, in contrast, numerous sections that shown no apparent sedimentary break at the era boundary and which feature relatively uniform lithologies extending above and below the boundary. Many such sections contain single formations which have beds in both eras: classical examples include the Hell Creek/Ft. Union transition in Montana, the Lance/Ft. Union transition in Wyoming, and the Denver Formation in eastern Colorado (see Fig. 8-23B). Typically in such units, uppermost Maestrichtian strata will contain dinosaur fossils and, within a short vertical distance, lowermost Paleocene beds will feature placental mammals (see Fig. 9-4). The intervening beds will typically contain Cretaceous palynomorphs and other plant materials and an absence of vertebrate signs. We have discussed this phenomenon in the previous chapter at some length.

Although one may not be certain in dealing with terrestrial strata that apparent conformity does in fact represent continuous sedimentation (because terrestrial deposition is inherently sporadic), paraconformable boundary sequences indicate that no significant tectonic changes of regional base-level occurred during the depositional interval (otherwise there would be some beveling and angular changes in deposition). Nichols *et al.* (1985) showed that the Beaverhead Conglomerate in Montana, a syntectonic deposit previously considered of Late Cretaceous through early Paleocene age, actually contains tectonically related material solely of Late Cretaceous age. Thus, the suggestion is that orogenic events locally were confined to Cretaceous time and were not occurring during the era-boundary interval.

B.2. Paleocene and Eocene Strata in the Intermontane Basins

All Early Tertiary (Paleocene–Eocene) strata in the western interior (Fig. 9-5) occur in a zone extending no farther east than the central Dakotas and terminating west-

Fig. 9-3. Reconstruction of the sequential development of the Front Range of the Rocky Mountains and the Laramie Basin in southeastern Wyoming from Late Cretaceous (A) to present (F). During Late Cretaceous through early Eocene time, primary relief produced by the Laramide orogeny results in cycles of sedimentation into the eastern basin. Late Eocene and subsequent times witnessed cycles of rejuvenation, thrust- and normal faulting, and differential erosion of both basinal and Laramide structures, culminating in the modern complex terrane. (From Kay and Colbert, 1965, after S. H. Knight, 1953, in Wyoming Geological Association 8th Guidebook, reprinted with permission of John Wiley & Sons.)

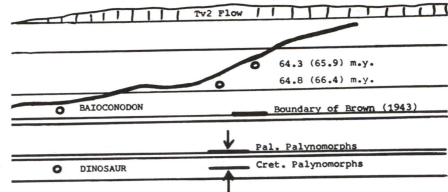

Fig. 9-4. Schematic diagram of the Cretaceous–Tertiary boundary section in the South Table Mountains, eastern Colorado. The heavy diagonal line indicates the ridge profile at the measured section. *Baioconodon*, found here above the era boundary, is a condylarth mammal. (Modified from Newman, 1979.)

ward at the Cordilleran thrust zone located west of the Rocky Mountains (Fig. 9-6). Few areas of Early Tertiary strata are present in Canada and these few lie in the Williston/Alberta basins.

As discussed in opening paragraphs of this chapter, virtually all interior Early Tertiary strata can be related directly to the Laramide orogeny. Deposits are concentrated in basins near or east of the Front Ranges of the Rocky Mountains and reflect mostly floodplain and swamp deposits for the Paleocene, and the same plus vast lake deposits for the Eocene. A single marine unit, the Cannonball Formation, represents the absolute last marine stratum in the North American interior. In our discussion below, we will draw heavily from King and Beikman (1978) for general concepts.

B.2.a. Paleocene Terrestrial Strata

The best-known and virtually generic Paleocene unit in the interior basins is the Fort Union Formation (locally elevated to group status), recognized in sections from the western Dakotas, eastern Montana, and northern Wyoming. Similar and probably related strata are described as: the Polecat Bench Formation in the Bighorn Basin, Wyoming; the Willow Creek/Porcupine Hills Formations in southwest Alberta; and the Ravenscrag Formation in southern Saskatchewan and parts of Alberta. In general, the Fort Union Formation and equivalent strata occupy the area of old Williston Basin and nearby regions. Paleocene strata outside the "Fort Union" region are found in the Denver, Raton, San Juan, Wasatch, and Uinta basins (see Fig. 9-7), all to be discussed below.

The Fort Union is a generally thick (to 1800 m), nonvolcanic, continental sequence which closely resembles the Lance and Hell Creek formations of the Maestrichtian. A typical section was described by Roehler (1979a) in southwestern Wyoming to contain a gray shale and gray siltstone,

with fine sandstones, gray and brown carbonaceous shales, and coal. Coal-bearing units are present largely in the lower parts of the formation which is typically 100 m thick. A regolith up to 3 m thick separates the lower Paleocene deposits in Roehler's section from the 330 m of the upper Fort Union section. The coals in the Fort Union are of great interest because they represent the single largest recoverable (by existing technology) fossil fuel deposit in the United States. Most of the coal is lignite or subbituminous and thus is of relatively-to very-low rank. However, it is present quite near the surface in many parts of Montana and the extreme western Dakotas, and it contains relatively little sulfur; therefore, what it lacks in heating value it makes up in abundance, "sweetness," and accessibility.

The Fort Union and related units represent terrestrial deposits formed after retreat of the Cretaceous seas, with abundant but not rapidly shed detritus supplied from the Rocky Mountains to the west. Coals in the Fort Union testify to development of swampy conditions in widespread areas of the detrital wedges shed from the Cordillera and suggest that the region centering around eastern Montana, down through western Wyoming, was of low relief during the Paleocene.

Other intermontane basins containing Fort Union strata are: the Piceance Basin in northwestern Colorado, The Washakie Basin in northwestern Colorado and adjacent Wyoming, the Green River Basin in southwestern Wyoming, the Wind River Basin in central Wyoming, the Powder River Basin in northeastern Wyoming (impinging on the southern Williston Basin), the Bighorn Basin in northwestern Wyoming, and several small basins in the region. All such basins contain strata reflecting alluvial and swampy environments which formed coals, variegated shales, and sandstones similar to those found in typical Fort Union strata described originally by Meek and Hayden for sections on the Yellowstone River near Buford, North Dakota.

It would not be worth the reader's time to elaborate on

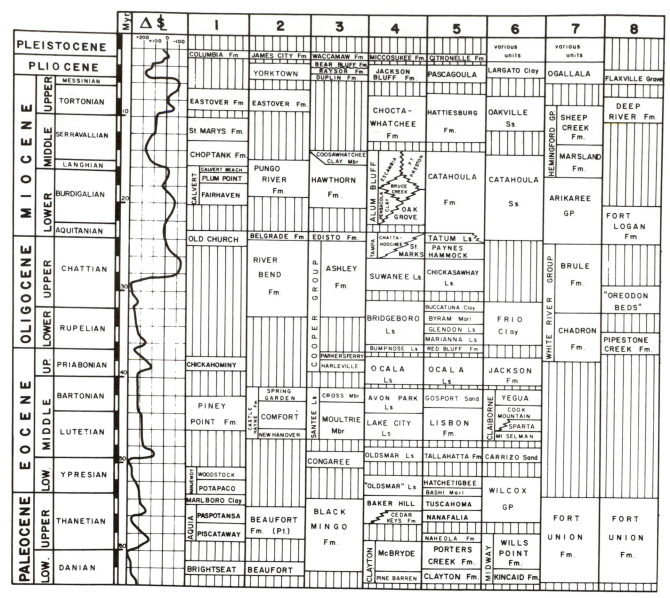

Fig. 9-5. Representative stratigraphic columns of Tejas deposits; most columns are synthetic, having been assembled from multiple sources and from different localities within the area represented. Sea-level curve from Vail and Hardenbol (1979). Areas are as follows: 1, Maryland; 2, North Carolina; 3, South Carolina; 4, Florida; 5, Alabama; 6, Texas; 7, Nebraska–South Dakota; 8, Montana; 9, Colorado Basin; 10, Utah; 11, southern Nevada and adjacent Arizona; 12, northeastern Nevada; 13, Santa Cruz Mountains, California; 14, central Oregon; 15, Cascade Range, western Oregon; 16, Alberta; 17, northern Alaska; 18, Sverdrup Basin, northern Canada.

strata from each intermontane basin of the Paleocene outside the Fort Union region: rather we will note their presence and highlight unusual or important sequences that differ from the typical Fort Union Formation. South of the Williston Basin region was the Denver Basin, which, as noted, contains a transitional Cretaceous–Paleocene unit, the Denver Formation, which is locally overlain or intergrades laterally with the Dawson Formation. The Denver is typically a fine-detrital unit whereas the Dawson is coarser, arkosic, and contains deposits which denote closer proximity to newly emerged mountains in the Front Range. Van Horn (1976) stated that Denver deposition was radiometrically dated (using interbedded volcanics) to have extended through most of Paleocene time. Upper beds of the Denver contain considerable quantities of volcanic tuff and tuffaceous sediments. In addition, latite lava flows are present in Tertiary parts of the Denver Basin, indicating sizable, local magmatism in the Front Range near Golden,

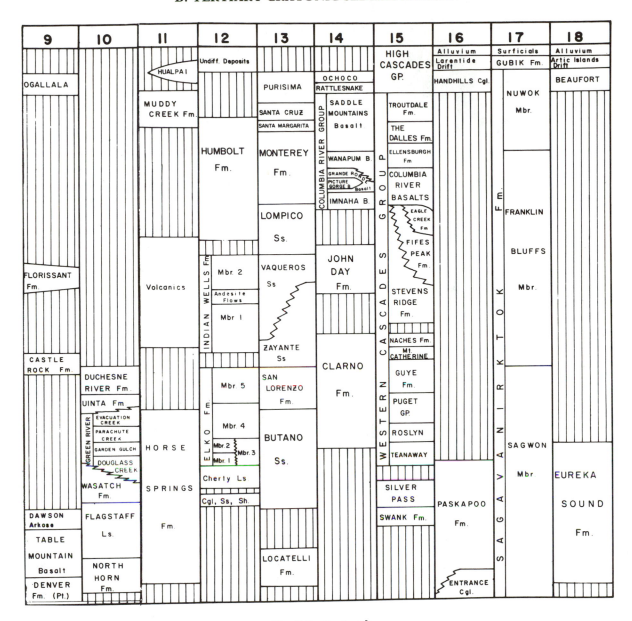

Fig. 9-5. (*Continued*)

Colorado. Numerous Paleocene mammals are preserved in tuff beds in the Denver Formation.

Farther south in Colorado and extending down to northcentral New Mexico, is the Raton Basin which, like the Denver Basin, contains a fine-grained Paleocene detrital unit, the Raton Formation, which intergrades laterally with or is overlain by a coarser detrital unit, the Poison Canyon Formation. The Raton is typically coal-bearing whereas the Poison Canyon, lying more to the north, contains volcanic debris. Since the strata coarsen northward, it is probable that detrital sediments came from the southernmost Front Ranges; however, the volcanics may have derived from the San Juan Mountains to the west and some detritus may have been shed from the Sangre de Cristo Mountains, to the east.

In northwestern New Mexico is the San Juan Basin, which continued to be a topographically low basin after the Cretaceous. In the San Juan Basin, the coal-bearing, Upper Cretaceous Kirtland/Fruitland formations are overlain by the Ojo Alamo Formation, an unfossiliferous unit of probable early Paleocene age. The Ojo Alamo is thin and conglomeratic, derived from streams eroding uptilted older strata in the basin. This is in turn overlain by the intertonguing Nacimiento and Animas formations, which are respectively a thick, variegated green-red shale and a thin, con-

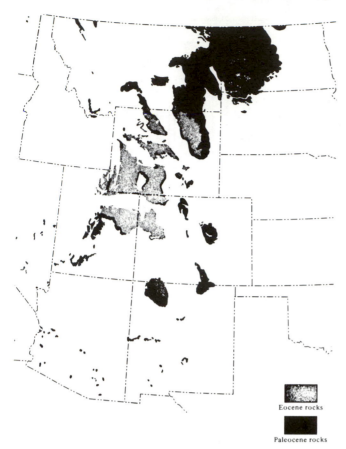

Fig. 9-6. Outcrop pattern of Paleocene and Eocene strata in the western interior of the United States. (Modified from King and Beikman, 1978.)

glomeratic sandstone. These latter two formations contain andesitic volcaniclastics derived from the nearby San Juan Mountains, which were active during the Paleocene, and they also contain sizable Paleocene mammal assemblages.

In the Uinta Basin, Utah, is another continuous Upper Cretaceous–Paleocene unit, the North Horn Formation, which was deposited throughout Paleocene time with sources from the Uinta and Wasatch Mountains in northernmost and central Utah. The North Horn is a variegated red-green shale unit with some coarse detrital sediments, especially toward the west, and was deposited mostly in fluvial environments. It grades northwestward into facies called the Current Creek Formation, deposited through Eocene time, composed of coarse fluvial gravels and shales. As noted in our previous discussion, the North Horn is of Cretaceous age in its lower beds and contains a *Triceratops* fauna. It also bears Paleocene mammals in upper beds. The North Horn is overlain by the youngest Paleocene to early Eocene age (Abbott, 1957) Flagstaff Limestone which is a white to gray, ledge-forming lacustrine unit. King and

Beikman (1978) noted that the Flagstaff contains a freshwater mollusk assemblage and, although it probably represents a very large lake, its outcrop is too incompletely known for the unit to achieve the recognition of the Eocene Green River Formation.

B.2.b. A Last Epicontinental–Marine Unit

In central North Dakota, northernmost central South Dakota, southwestern Manitoba, and a small part of adjacent Saskatchewan, a paralic-to-marine unit is present in the subsurface. Outcropping on the west flank of Turtle Mountain at the extreme north of North Dakota, it is called the Cannonball Formation, after the Cannonball River, North Dakota (and, coincidently, after the cannonball-shaped concretions which it contains). The Cannonball Formation intergrades with lower beds of the Fort Union Formation, but is of late Early Paleocene age (slightly younger than oldest Fort Union strata).

This marine unit, isolated among continental strata, is composed mostly of glauconitic sands and dark-green shales. It reaches 100 m in thickness and contains numerous mollusks and foraminifera (Stanton and Vaughan, 1920). These fossils yield information as to the source of the Cannonball but the information is ambiguous. King and Beikman (1978), and others, observed that the mollusks in the unit have affinities with those in the Midway Group in the Texas Coastal Plain; however, the foraminifera seem to have Arctic affinities (Fox, 1960, in McCrossan *et al.,* 1964). King and Beikman stated that Arctic waters transgressed to deposit the Cannonball; in contrast, Dott and Batten (1981) stated that an incursion of waters from the Gulf of Mexico was the source for the Cannonball. In truth, both extremes are unprovable since no other contemporary marine cratonic strata are known. We suggest the additional hypothesis that the Cannonball was deposited by a Paleocene remnant of the Late Cretaceous seaway which has gone undetected, probably due to erosion of intervening deposits. This remnant sea may have been located farther east than the older Cretaceous seaway, and a slight subsidence of the North Dakota Basin may have caused the shoreline to migrate westward, leaving the Cannonball. This hypothesis seems simpler and more plausible than the possibility of renewed transgression left in any adjacent region. It would also explain well the apparent mix of Arctic and Gulf faunas, since the Late Cretaceous sea clearly connected both oceans, and therefore its relict would share the same heritage.

B.2.c. Eocene Strata in the Intermontane Basins

During the Eocene, continental sedimentation into the eastern Front Range piedmont basins continued, generally

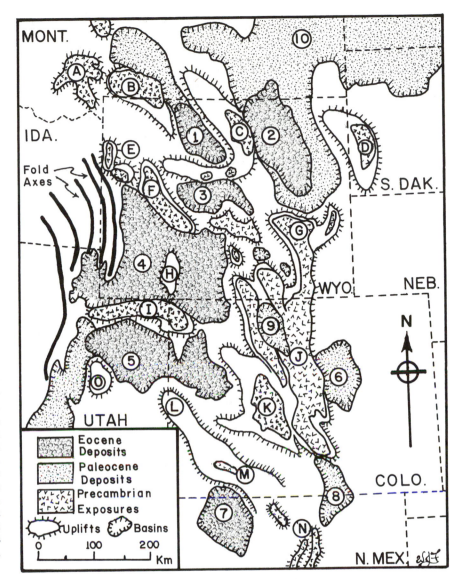

Fig. 9-7. The central Rocky Mountain region, showing principal uplifts and basins present during the Cenozoic Era. Basins indicated by deposits numbered as follows: 1, Bighorn Basin; 2, Powder River Basin; 3, Wind River Basin; 4, Green River and Washakie Basins; 5, Uinta and Piceance Basins; 6, Denver Basin; 7, San Juan Basin; 8, Raton Basin; 9, Park Basin; 10, Williston Basin; Uplifts are: A and B, Beartooth Mountains; C, Bighorn Mountains; D, Black Hills; E and F, Absaroka Range Volcanics; G, Laramie Range; H, Rock Springs Uplift; I, Uinta Mountains; J, Front Range; K, Sawatch Range; L, Uncompahgre Arch; M, San Juan Mountains Volcanics; N, Sangre de Cristo Mountains; O, San Raphael Swell.

at very high rates, but additional intermontane basins soon came to dominate the scene. The nature of continental sedimentation changed in the Eocene with coal swamps giving way to formation of extensive red beds on lower piedmont slopes and in floodplains. Some Paleocene forests gave way to prairies, and the first grasses appeared. With grasses came early grazing mammals. In general, more coarse detritus, red silt, sand, and conglomerate were deposited in intermontane basins. The environment had become generally more arid but locally, conditions evolved favoring the accumulation of very large lakes in basins. The largest of these lakes, Lake Gosiute, deposited the Laney Member of the Green River Formation.

The geographic extent of Eocene deposits is very different from that of the Paleocene. No Eocene strata are present on the craton in Canada, and virtually none are present in Montana and the Dakotas; the major sites are the Uinta Basin (northeastern Utah and Colorado), the Green River and Washakie basins (southwestern Wyoming), the Powder River Basin (northwestern Wyoming), the Wind River Basin (central Wyoming), and the San Juan Basin (northcentral New Mexico).

Three major lithological assemblages incorporate most of the described strata in these basins; they are, in ascending order, the Wasatch, Green River, and Bridger (Uinta) formations. Local names (sometimes given group status) are applied but the regional stratigraphic relations are quite clear. The Wasatch and Bridger/Uinta formations are predominantly of fluvial origin, whereas the Green River is lacustrine; however, lake sediments were deposited sporadically in both Wasatch and Bridger environments. Although the Wasatch/Green River/Bridger–Uinta succession is present in most regional basins, toward basin margins the Green River may be absent or pairs of the three units may intertongue.

The following discussion will treat the three Eocene generic units in sequence but the reader is cautioned to remember that although lake sediments are characteristic of the middle Eocene, and although fluvial sediments are characteristic of the early and late Eocene, still, lakes and rivers were present throughout the epoch (see Fig. 9-8).

i. The Wasatch Formation. This unit was described by Hayden (1869) in the Wasatch Mountains, Utah, on the west side of the Uinta Basin. It is thickest in that basin (reaching over 1800 m) but elsewhere is generally less than 800 m thick. The Wasatch Formation is a variably colored (but most often reddish) shale, sandstone, and conglomerate unit, arranged in fining-upward sequences typical of fluvial deposits. Coloration is one of the most distinctive Wasatch characteristics, especially in contrast to the rather drab Fort Union strata, and many olive and gray shales are bedded between red beds. In addition, lacustrine limestones are abundant and Picard (1957) suggested that these might be equivalent lithologically to the Flagstaff Limestone (discussed previously).

The Wasatch Formation is principally the product of aggrading streams on piedmont slopes, leaving sizable amounts of detritus in nearby basins (Murany, 1964). Some of the material may have been reworked by winds during particularly arid times in the early Eocene. In addition, bouldery strata in the Wasatch may be fanglomerates. Lake deposits in the Wasatch are generally dark gray and black, suggesting reducing environments.

The Wasatch terminology, as mentioned, is applied over a wide area to contemporary strata. King and Beikman (1978) observed that Wasatch units are recognized in Utah (outside the Wasatch Mountains), Wyoming, Colorado, and New Mexico. Other contemporary strata which are of similar origin include the Indian Meadows and Wind River formations in the Wind River Basin, Wyoming, the San Jose Formation in the San Juan Basin, New Mexico, and the Claron Formation in southwestern Utah (Rowley *et al.*, 1979; Cook *et al.*, 1975).

Numerous studies of the Wasatch and its many subunits have been done. A representative is an analysis of origin of variegated red beds in the Cathedral Bluffs Tongue of the Wasatch Formation in the northern Washakie Basin,

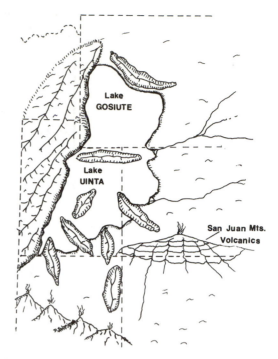

Fig. 9-8. Paleogeography of the southwestern craton during Green River time. (Redrawn and modified from Kay and Colbert, 1965.)

Wyoming (Braunagel and Stanley, 1977). Their study showed that variegated beds formed on very gently sloping alluvial plains adjacent to a very large lake (Lake Gosiute, discussed below). Green, fine-grained sediments derived from sediment-laden floodwaters in the lower floodplains while the red beds represent desiccation intervals, to be followed by subsequent floods. This model may apply widely to Wasatch sediments featuring variegated layers. The Wasatch Formation is excellently exposed in a number of localities, but nowhere better than Bryce Canyon National Park in southcentral Utah. There it is dissected into rugged and gorgeous badlands which show off the bright and varied colors to best advantage.

ii. The Green River Formation. Overlying and interfingering with Wasatch strata in the intermontane basins are thick and distinctive lake deposits collectively labeled the Green River Formation. These lake deposits cover almost 78,000 km^2 of area in southwestern Wyoming, northwestern Colorado, and northeastern Utah, and reach thicknesses to 1500 m. In addition, they contain the highly touted "western oil shale," the potentially largest hydrocarbon resource in the United States, although the material is marginally profitable to produce with current technology.

Classical study of the Green River Formation was done by Bradley (e.g., 1964) from whom we will extract much of the following discussion. The strata were deposited in two major lakes: Lake Gosiute in southwestern Wyoming (north of the Uinta Mountains), and Lake Uinta in northeastern Utah and adjacent Colorado. These lakes communicated at times to the east of the Uinta Mountains and at other times were isolated by a large structural feature called the Rock Springs Uplift in southcentral Colorado. The formation itself was named for exposures along the Green River in southernmost Wyoming, near Rock Springs, and it contains several members and tongues. Green River strata are easily recognized because they are very thin-bedded and usually varved, commonly light-colored (tan, green-brown, or gray), and of a very fine, creamy texture. Specific lithologies in the Green River feature some unique characteristics. For example, the finest of fish fossils come from the Green River shales and these fish are not rare finds but are, in fact, very common in some facies. Green River fossil fish (Fig. 9-9B) make their ways into rock shops and living rooms across the country. What is most impressive about these fossils is the preservation of delicate structures in exquisite detail, although the fish are flattened. In addition, plant material, insects, mollusks, reptiles, and mammals are found in some Green River shales, presumably representing nearshore habitats.

Another famous, and economically much more important feature of the Green River are the oil shale deposits, chiefly located in the Piceance Basin of northwestern Colorado. These "oil shales" are actually kerogen-rich, dolomitic or calcitic marls, which are especially concentrated in the Parachute Creek Member. Such shales will yield oil only on distillation, and the known quantity of kerogen-rich material in the Piceance basin more than equals the known remaining U.S. petroleum reserves.

Varved sediments allow estimations of a lake's life span. Each dark and light varve-pair represents a year, and each centimeter of strata contains a determinable number of varve-pairs. By simple mathematics, estimates of over 6.5 Myr. duration for Lake Gosiute have been made based on counts of varves in 100-m-thick units.

Bradley (1964) noted that Lake Gosiute changed size and characteristics greatly during its life span and members of the Green River Formation reflect those changes. In the early stage, the lake was large, fresh, and it had an outlet; during this time the Tipton Shale Member was deposited. Subsequently, the climate became drier and the lake shrank and became saltier. No outlet seems to have been present and commercially important quantities of trona (Na_2CO_3 $NaHCO_3 \cdot 2H_2O$), along with a host of other salts and minerals (Table I), were deposited in the middle Wilkins Peak Member. The Wilkins Peak occupies a restricted area surrounded by the Tipton Member, reflecting the outline of the shrinking lake. In latest Green River time, the climate again became wet and the lake expanded to its maximum size, depositing the marls and shales of the Laney Shale Member. Laney deposition continued through late Eocene time but the sedimentation rate exceeded the rate of subsidence and eventually the lake filled in. Lake Uinta had a similar history and in both Gosiute and Uinta lake beds, subsequent to Green River time, new and smaller lakes repeatedly appeared, as evident from lacustrine sediments in the overlying Bridger Formation.

Many studies have been done on specific components of the Green River sediments. Among these is the study by Johnson (1981) concerning the depth of Lake Uinta (which he reported to be over 300 m). Most geologists, however, have envisioned a shallow, playa-lake model for the Eocene lakes, with detritus derived from the Absaroka Mountains to the northwest. Stanley and Surdam (1978) described a transition from playa-type to freshwater lake at the Wilkins Peak-to-Laney Member transition, with deltaic rocks present in the Laney. In contrast to the proposal for Lake Uinta by Johnson, Stanley and Surdam suggest that Lake Gosiute in Laney time was predominantly shallow. Origin and concentration of the kerogen in Green River units are controversial (see Boyer, 1982), but most will agree that some combination of algae, other phytoplankton, fungal spores, and pollen, were the sources of organic material. Advocates of the playa model for much of Green River deposition have proposed several sets of conditions for oil-shale formation. Eugster and Hardie (1975), for example, proposed that accumulation of oil in the Wilkins Peak Member occurred when the central portion of Lake Gosiute was large (during relatively humid intervals) and masses of flocculent, benthic blue-green algae and fungi bloomed. During arid periods, the lake shrank, the plants declined, and trona and halite were deposited.

Studies of oil potential and methods of recovery from the Green River shales are legion. The interested reader is directed to Donnell (1961), Van West (1972), and Culbertson and Pitman (1973) for evaluations of the resource; and to Dinneen and Cook (1974) and Guthrie et al. (1979) for technology of future production.

iii. The Bridger and Uinta Formation.

With the demise of the great Eocene lakes, there was a return to widespread of deposition alluvial red beds. These units resemble the Wasatch Formation but, as Bradley stated, they contain much more volcanic ash and more gray and brown detritus. The latest Eocene unit in the Green River and Washakie basins is the Bridger Formation, whereas in the Uinta basin similar strata comprise the Uinta Formation. The Uinta is overlain by the Duchesne River Formation in

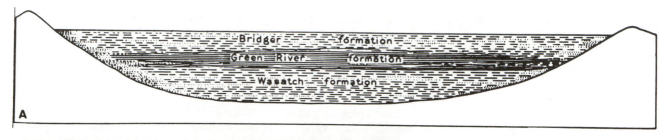

Fig. 9-9. (A) Bradley's classical model of the relationship among Bridger, Green River, and Wasatch strata. (B) A fossil fish, *Knightia,* which is extremely abundant in Green River shales. (C) Lateral view of a small section of typical Green River shale, showing the fine laminations and occasional pebbles or coprolites around which laminations are draped. (A and C, from Bradley, 1964; B, DRS photograph.)

the southern Uinta Mountains, and lower parts of the Duchesne River may be of late Eocene age.

The Bridger Formation has a wide areal extent and weathers to form badlands in central parts of the Green River and Washakie basins. It reaches over 700 m in thickness and consists largely of tuffaceous mudstones and sandstones, including volcanic ash beds with virtually no detritus. As mentioned, limestone and marl from ephemeral lakes also comprise important elements of the Bridger Formation; Bradley noted that these carbonates tend to be more resistant than most enclosing lithologies and form conspicuous benches in the badlands. Altered andesitic tuff minerals form striking colors, including deep blues, emerald greens, and clean whites in Bridger outcrops. Bradley stated that a likely source for most of this volcanic material, as with the Green River Formation, was the Absaroka Range in northwestern Montana.

The Uinta Formation is slightly younger than is the Bridger because Lake Uinta persisted longer than Lake Gosiute. Lithologically, it is similar but strata in the Uinta Basin contain less volcanic material. Bradley and others have tended to not differentiate Bridger and Uinta strata except in the central basin localities; and Bradley (1964) observed that along an 80-km stretch of the north flank of the Uinta Mountains, Wasatch, Green River, and Bridger–Uinta Formations are very conglomeratic and indistinguishable. The Duchesne River Formation is conformable with the Uinta but King and Beikman (1978) noted that it is coarser and marks a new cycle of regional sedimentation, more characteristic of the Oligocene. Uppermost volcanic materials in the Duchesne River yield Oligocene K/Ar dates.

Table I. Saline Minerals from the Green River Formation[a]

Shortite	$Na_2CO_3 \cdot 2CaCO_3$
Trona	$Na_2CO_3 \cdot NaHCO_3 \cdot 2H_2O$
Pirssonite	$Na_2CO_3 \cdot CaCO_3 \cdot 2H_2O$
Northupite	$Na_2CO_3 \cdot MgCO_3 \cdot NaCl$
Gaylussite	$Na_2CO_3 \cdot CaCO_3 \cdot 5H_2O$
Bromlite	$(Ca,Ba)\ CO_3$
Bradleyite	$Na_3PO_4 \cdot MgCO_3$

[a]From Bradley (1964).

iv. A Scattering of Additional Eocene Units and a Note on Eocene Fossils.

Although most Eocene strata lie within the basins we have discussed, scattered outliers of Eocene continental deposits occur on the western craton and may indicate formerly widespread strata. Most of these units are similar lithologically to the alluvial strata of the intermontane basins (i.e., the Wasatch/Bridger formations) and require only brief mention. King and Beikman noted that at the north end of the Sandia Mountains, near Albuquerque, is the Galisto Formation containing 900 m of variegated, partly tuffaceous sand and clay. The Galisto is of Wasatch age, based on vertebrate fossils. In parts of Texas near Big Bend Park, parts of southern New Mexico, and in local patches, are probable Eocene strata which are not clearly differentiable from Paleocene deposits and which are largely unnamed. In Arizona, King and Beikman mention there are more than a dozen small deposits considered of Eocene age. One unit, the Pantano Formation near Tucson, has been strongly deformed and thrust over gneiss complexes in the nearby Rincon and Catalina Mountains. Additional continental Eocene deposits are found in Montana (in the Lima area, above the Beaverhead Conglomerate), in Nevada, California, and Washington. Since we are discussing continental deposits in this section, virtually any area on the craton which experienced sedimentation at that time could contain preserved deposits.

The Absaroka Range, on the eastern border of Yellowstone National Park, northwestern Wyoming, not only provided volcanics for the Bridger and Green River formations, but also formed local volcaniclastic sedimentary rocks of Eocene age. The Lamar River Formation is one such unit and it includes a well-preserved "fossil forest." The classical work on the Lamar River forest was done by Dorf (e.g., 1964); among the very distinctive elements in the preserved Lamar River flora are many trees preserved upright, apparently in life position, with their roots in a putative paleosol. In addition, Dorf noted that at least 27 different layers of ash-buried forests are preserved and exposed (Fig. 9-10). Among the curious aspects of the "fossil forest" is the mix of species, including a few cold-loving (e.g., spruce and fir) and warmth-loving (e.g., breadfruit and magnolia). The presence of admixed cold-and-warm species has been troublesome. Fritz (1980) proposed that the Lamar River flora was not entirely autochthonous, and that one observes in it a mix of top-and-bottomland trees due to transport of parts of the flora down volcanic slopes. The high-elevation species would, of course, be cold-tolerant. Recent events and observations at Mt. St. Helens tend to confirm Fritz's model.

Vertebrate fossils abound in Eocene strata of the intermontane basins. The Uinta Basin deposits are especially rich with bones of a diverse set of mammals that roamed the region before, during, and after the great lakes time. Bones of early horses, rhinoceroses, and "uintatheres" have been recovered by the thousands, especially during the great bone hunts of the last century (see discussion in Chapter 8 of Cope and Marsh and the Morrison dinosaurs). We will discuss Tertiary mammals subsequently; but before continuing the reader should realize that environments such as those around Lake Gosiute and the adjoining alluvial plains provided varied and rich terrestrial habitats. The lake sediments and alluvial deposits also provided excellent tombs for the same animals and the badlands-type topography of Eocene strata in the intermontane basins provides excellent exposure of fossil bones.

B.3. Oligocene Strata in Wyoming and the Dakotas

The general nature of continental sedimentation changed after Eocene time in the Rocky Mountain region, notably in that basins between the Laramie Mountains (the focus of our previous discussions) were apparently filled, or nearly so, and after the late Eocene they received relatively little sediment. Strata from Oligocene and later Tertiary times are widespread in the intermontane basins, but are present as scattered outliers; either once-extensive continental detrital units have been eroded over most areas or deposition was localized. However, east of the Front Ranges of the Rocky Mountains and the basins at their feet, the Great Plains of the Dakotas and eastern Colorado/Wyoming/Montana apparently subsided in the Oligocene (and later times) and received considerable fluvial and alluvial detritus over a surface of eroded Cretaceous strata. Streams draining eastward across the plains eventually emptied into the Mississippi River drainage.

Climate changed regionally from the late Eocene into the Oligocene. Based on many studies of plant and vertebrate faunas in the respective strata, and based on additional studies of pollen and *sporomorphs* (i.e., spores from lower plants; see Frederiksen, 1980), it is apparent that the Oligocene was cooler and drier than the preceding epoch. However, Oligocene grasslands flourished, as did forests, and (since alligators are found in some Oligocene lake deposits) the climate probably could not have been cooler than that of, say, southern Georgia. Clark *et al.* (1967) suggested that prior to the Oligocene, a monsoonal air circulation prevailed over the Rockies and the Great Plains but, during the Oligocene, that pattern changed to a stronger hemispheric circulation system with prevailing westerlies.

As an additional change from Eocene to Oligocene, volcanic activity increased greatly in the Rockies and mountains farther to the west and south. This increase is reflected in the abundance of volcanic material present in Oligocene strata.

The classic and finest exposures of Oligocene strata are in the South Dakota Big Badlands (Fig. 9-11), where they

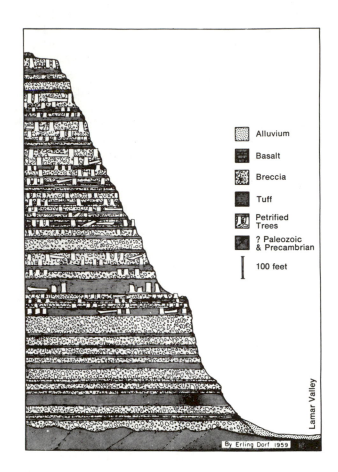

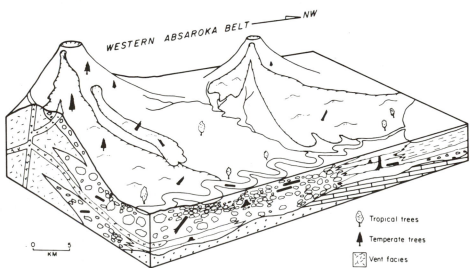

Fig. 9-10. (Above) Dorf's interpretation of the Yellowstone Eocene ''fossil forest'' beds above the Lamar River Valley. (Below) Fritz's model for origin of the Lamar River flora, assuming part of the fossil forest is allochthonous and derived from volcanic mudflows (lahars). (Above, from Banks, 1970, after Dorf; below, from Fritz, 1980.)

form the rugged topography that typifies a "badland." (We should point out that the severe erosion of the Dakota badlands is a post-Oligocene event, related to late Pliocene rejuvenation of the Rockies. Oligocene strata in the badlands did contribute to the badland topography by being poorly consolidated and easily eroded.)

The White River Group is the generic term for Oligocene continental deposits in the interior, named for exposures along the White River in South Dakota. This stratum dips below younger sediments to the south-southeast in Nebraska and eastern Colorado, but it is exposed in the Platte River valleys (North and South Platte Rivers). King and Beikman (1978) noted that the presence of outliers of the White River high up on mountains, as well as scattered in intermontane basins, suggests that these represent only remnants of a once-great sheet of strata. For much of the following discussion we will refer to Clark *et al.* (1967).

The White River Group in the Big Badlands consists of two formations, the Chadron and the Brule (Fig. 9-12), with a third, the Slim Buttes, present locally at the base of the section in the northwestern portion of the type region. The

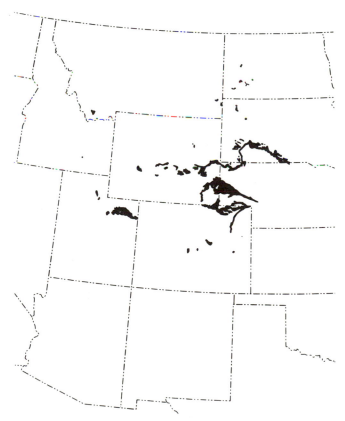

Fig. 9-11. Outcrop pattern of Oligocene deposits in the western interior of the United States. (Modified from King and Beikman, 1978.)

Slim Buttes Formation is either of very earliest Oligocene age, or perhaps of latest Eocene age, and is probably contemporaneous with the Duchesne River Formation of the Uinta Basin. It is a white to pale-green arkosic sandstone and clay unit, containing quartz and chert gravel beds, which forms discontinuous lenses of several kilometers areal extent and up to 35 m thick. Clark *et al.* (1967) have interpreted the Slim Buttes to be an initial Oligocene deposit formed by rapid erosion of the Black Hills, some 60–100 km to the northwest.

Following Slim Buttes deposition was a brief period of erosion, in which some erosion of the Slim Buttes ensued, followed by deposition of the Chadron and Brule formations throughout remaining Oligocene time. The Chadron Formation contains three members which are composed largely of pastel, multicolored sandstones, mudstones, and conglomeratic channel-fill sandstones. The Chadron reaches a thickness of 50 m and the upper contact with the Brule Formation is a distinctive limestone and intercalated-mudstone zone which may be as much as 1 m thick. The Brule is the thickest of the group, reaching a thickness of 140 m and containing two members. It is composed primarily of multicolored, pastel mudstones with siltstones, fine-grained sandstones, and channel-fillings which are sandy or conglomeratic. In general, the Chadron contains coarser-grained sediment than the Brule and the fossils are less well preserved. Volcanic ash beds are scattered throughout White River strata, generally present as thin layers altered to bentonite.

The Chadron/Brule formations are interpreted as deposits on a very gently sloping, broad plain dipping eastward from the Rockies. Sizable fluvial channels can be traced through the strata, extending many tens of kilometers from the mountains. Clark *et al.* (1967) observed that resumption of widespread fluvial deposition in the Great Plains during Oligocene time (after the apparent lack of the same in the Eocene) indicates factors of regional extent which overloaded streams east of the continental divide and caused them to aggrade and form sizable deposits. Among possible reasons for this renewed sedimentation, they listed: (1) the increase in volcanic ash volumes; (2) downwarp of the region; (3) eustatic sea-level change; (4) uplift of a barrier across the lower reaches of streams; and (5) climate change which would affect stream regimes. They opted for the last explanation, which fits speculations of a drier Oligocene climate, based on fossil evidence (see below).

White River strata, especially the Brule Formation, are rich in vertebrate fossils, notably those of horses and *Merycoidodon,* a primitive hoofed mammal of the group called oreodonts. Also present are huge, odd-toed hoofed animals called titanotheres, which resembled elephant-sized rhinoceroses. Along with terrestrial mammals are aquatic ver-

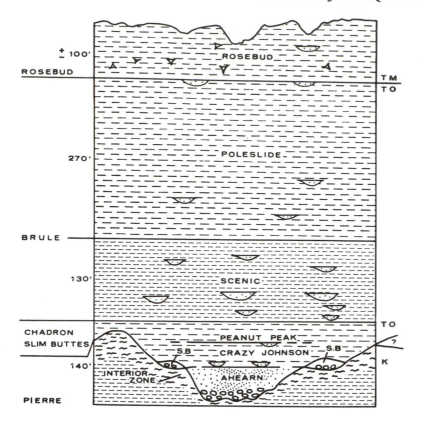

Fig. 9-12. (Above) Columnar section of the Big Badlands, showing Uppermost Cretaceous through Miocene strata. (Below) Molars of *Merycoidodon,* a very common fossil from the "oreodon" beds in the Brule Formation. (Above, from Clark *et al.,* 1967; below, DRS photograph.)

tebrates and invertebrates in lake and pond strata (such as the alligators mentioned above).

Retallack (1983) reported the presence of at least 87 fossil soils recognizable in a studied locality in Badlands National Park, South Dakota, of which the overwhelming majority were in Oligocene strata. The paleosols are evidence of depositional hiatuses with significant periods of subaerial erosion during White River time; and they substantiate vertebrate fossil evidence for sequential development of vast regional forest and grassland biotas. Prothero (1985) analyzed a mid-Oligocene extinction event among the vertebrates in nearby Wyoming, and concluded that the selective nature of extinctions suggests climatic/ecological causes for these events (rather than an extraterrestrial cause, such as was discussed in conjunction with Cretaceous/Tertiary extinctions in Chapter 8).

Outside the Great Plains region, outlying Oligocene strata reflect similar fluvial histories to White River strata. In Alberta and Montana, the Cypress Hills Conglomerate is a lower Oligocene unit derived from erosion of nearby ranges and containing a mix of sand and chert/quartz pebbles similar to that found in the Slim Buttes Formation. Southward in the Denver Basin, the Castle Rock Formation is very similar to the White River and may have been continuous with it. The Castle Rock overlies the Paleocene with an intervening ash-flow tuff bed. Other related Oligocene strata include the Wiggins Formation in the Wind River and Bighorn basins of Wyoming, the Farista Formation in the Raton Basin, and the Duchesne River Formation in the Uinta Basin.

A notably different unit of probable early Oligocene age is the Florissant Lake Beds, west of Colorado Springs in southern Colorado, occupying a small area but comprising a very important fossiliferous stratum. The Florissant Lake developed in a basin dammed by volcanic mudflow material from a volcanic field (now comprising the Cripple Creek mining area). Although vertebrate fossils are poorly represented in Florissant strata, insects and plants were beautifully preserved by heavy influx of volcanic ash into the lacustrine sediments (Fig. 9-13). King and Beikman (1978) noted that lakebed floral remains of existing species indicate that the lake formed at an altitude of approximately 1000 m, in a warm, subhumid climate, which contrasts with the 2750-m altitude of the present deposit: this demonstrates how strongly the region has been uplifted since the Oligocene. Classic study of the Florissant Lake Beds flora was done by MacGinitie (1953), who described over 100 plant species in the unit, including several surviving in the Orient (e.g., *Ailanthus*, which was successfully reintroduced to North America), and taxa known from upright tree stumps including *Metasequoia*.

Fig. 9-13. Insects from the Oligocene Florissant Lake Beds, southern Colorado. (Above) A hymenopteran, genus indeterminate; (below) an earwig, *Labiduromma*. (Both DRS photographs, specimens made available through the courtesy of Dr. Donald Baird, Princeton University.)

B.4. Miocene Strata of the Great Plains and Mountain Basins

Miocene time saw a continuation of the Oligocene sedimentary pattern (Fig. 9-14), i.e., sporadic sedimentation in intermontane basins in the Rocky Mountains, and widespread outwash-type deposition on the plains east of the mountains. A major difference, however, relates to climatic changes beginning in the Miocene and continuing (and culminating) through the Pleistocene. Cooling began at least as early as the early Miocene, and this mid-to-Late Tertiary climatic change encouraged the spread of grasslands in ever-widening areas of the Great Plains. Accompanying those grasslands were grazing ungulates, both odd-toed and even-toed, which left very abundant fossils including common species of *Merychippus*, the first of the large grazing horses, many camels, the common, odd *Moropus*, deer, cattle, and rhinoceroses. The Agate Springs Formation, a floodplain deposit from the upper Miocene of Nebraska, is famous for containing vast numbers of the bones of the rhinoceros *Diceratherium*.

Miocene strata in the Great Plains are assigned to the Arikaree and Hemingford groups. These cover larger areas than do the Oligocene units but they are still restricted to northwestern Nebraska, a small area of southwestern South Dakota, and southeastern Wyoming. Both the Arikaree and Hemingford are generally fine-grained detrital units, composed typically of light-colored sand and silt, which reflect continued erosion of the mountains to the west, but at relatively gradual rates. Lack of coarse material and arkosic beds in Miocene strata of the Great Plains suggests little topographic relief between mountains and high plains, pointing toward slow and limited uplift of the mountains during the Miocene.

Most notable about the Miocene deposits in the Great Plains is the penetrative nature of each succeeding deposit. For example, the Gering Formation, at the base of the Arikaree Group in Nebraska, commonly lies on and in a channeled surface of the Oligocene Brule Formation. In turn, succeeding Miocene formations fill channels in underlying Miocene units. These channel fills can be traced for long distances and speak clearly of the fluvial and alluvial nature of the deposits. A study by Stanley and Fagerstrom (1974) showed that low-gradient, braided streams were present during the early Miocene in Kansas, and that sizable accumulations of volcanic ash were major components of the sediment load. Bart (1975) discussed penecontemporaneous downward injection structures in such sediments and noted the presence of several types of structures characteristic of rapid loading of sediment with little bearing strength, which would be characteristic of ash beds. Strata

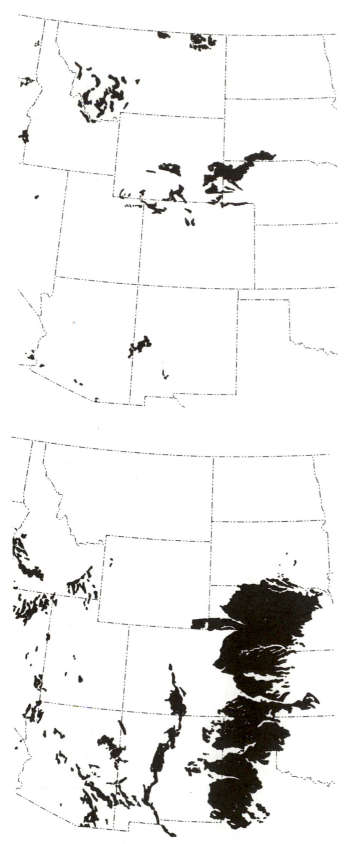

in the intermontane basins are strung along a series of basins extending down through Colorado and New Mexico and up through Wyoming. Additional deposits are present in Montana and very small areas of Saskatchewan. These basin deposits are generally isolated but are often thick; as King and Beikman (1978) noted, isolated Miocene basin units may attain thicknesses greater than those of typical Miocene sheet strata in Nebraska (reaching to 330 m). It is not clear whether intermontane basin deposits were formerly connected and subsequently isolated by erosion; however, based on the presence of scattered outliers of upper Miocene strata, King and Beikman believe this to be the case. Starting with the northernmost basin strata, in Saskatchewan, are small outcrops of the late Miocene-age Wood Mountain Gravels, an 18-m-thick stratum which McCrossan *et al.* (1964) consider to comprise reworked Cypress Hills Conglomerate, since they have in common a quartz/chert-pebble lithology. Farther south, in Montana, the only differentiated Miocene unit is the Flaxville Gravel, of very latest Miocene and early Pliocene age. The Flaxville forms a resistant caprock for plateaus in the region, overlying Paleocene and Cretaceous strata, and represents coarse detritus from mountains to the west mixed with locally derived finer detritus. The Flaxville is abundantly fossiliferous with numerous mammal remains. In the Uinta Mountains and Green River Basin are the Bishop and Browns Park formations, with the Bishop present in a small area of the Uinta Mountains where it forms the caprock for high terraces. The Bishop is a conglomerate of probable earliest Miocene age. The Brown Park is a widespread and thick unit (to 550 m) which contains quartzite conglomerate at the base, overlain by light-colored, tuffaceous sandstone which King and Beikman suggest may be of eolian or fluvial origin.

The thickest of Miocene deposits in the continental interior is found in central Wyoming, comprising the Split Rock Formation, which may have been deposited through all of Miocene time. It reaches to 830 m and consists of fine-grained silty sandstone with some coarser sands. Much of the sediment came from the northwest and some of the sands are eolian. The general lithological similarity and times of deposition of typical Arikaree/Hemingford strata on the Great Plains and the Split Rock farther to the west, suggest a common genesis of the units.

The series of Miocene strata which extend down along the trend of the Rocky Mountains into New Mexico, include a host of different formations in small areas. These southern Rocky Mountain Basin strata in the Miocene Series are

Fig. 9-14. Outcrop patterns of stratified deposits (i.e., excluding some alluvial channels and smaller unstratified deposits) in the western interior of the United States. (Above) Miocene outcrop; (below) Pliocene outcrop. (Both modified from King and Beikman, 1978.)

typically composed of bouldery conglomerates and large volumes of volcanic material, reflecting greater volcanic activity and perhaps greater uplift of the southern Rockies: for example, in the Raton Basin is the Devil's Hole Formation, composed of volcanic clasts and fragments of Precambrian rocks. Numerous other Rocky Mountains units are defined in New Mexico as are many that are unnamed. These intermontane basin strata are lithologically distinct from bolson–playa deposits within the Basin-Range Province farther to the west, and are generally older than the latest Miocene and Pliocene age of most Basin-Range strata. Such basin-full sequences in the desert environments of the Great Basin are relatively uniform and uncomplicated, and will be included with the tectonics discussion to follow.

B.5. Pliocene Sheet Deposits

The youngest strata of the Tertiary Period lie nearest to the surface (logically) and are also less eroded and reduced than are those of earlier times (also logically). Thus, there is a broad and extensive outcrop of Pliocene strata across the continental interior. This outcrop is the surface of a sheet deposit which is generally very thin, and which represents the last wave of erosion from the Rocky Mountains eastward onto the Great Plains.

The Pliocene of the Great Plains extends in a broad outcrop from southern South Dakota down to central Texas: in all, it covers 845,000 km². Virtually all of this sheet of sediment is assigned to the Ogallala Formation (or Group) which is typically 60 to 150 m thick but it thins eastward to a featheredge and thickens slightly westward. Remnants of probable Ogallala strata at the eastern base of the Rockies suggest that the formation reached farther west prior to erosion; similarly, it is likely that it extended farther east. The lithology is quite variable, including silt, sand, gravel, and minor quantities of volcanic ash. Since the unit is very young (at the top less than 5 Myr.), it is predictably poorly consolidated or unconsolidated. Parts are lime-cemented, the lime being caliche formed in the prevailing arid environments. The unconsolidated nature of the unit and its youth combined to make the Ogallala the largest and most important aquifer in the United States.

Typically, the Ogallala Formation lies on or just below the present land surface. It may be covered with a very thin veneer of Quaternary eolian sand, alluvium, basin-fill material, or, in some areas to the north, windblown loess. King and Beikman (1978) cited the work of Johnson (1901) who showed that Ogallala strata formed on a long eastward slope of the Rocky Mountains with stream gradients and climate virtually the same as in the Holocene (i.e., gradients of 1.2–1.4 m/km and semiarid climate). The Ogallala represents a mixture of alluvial, fluvial, bolson, and playa deposition across the western plains (see Diffendal, 1982) which ended with the change of climate with tectonic uplift of the plains. Vertebrate fossils are common in the formation, as are numerous fossils of plants and gastropods.

Notwithstanding the arid climate during Ogallala time, as the formation was deposited, water was incorporated with the buried sediment; and it is this connate water that has yielded irrigation water for much of the Great Plains states' agricultural "boom," and allows irrigation farming in normally arid regions. However, the aquifer is being rapidly depleted and once gone, the porewaters will take a very long time to replenish from groundwater recharge. It is also likely that replenished Ogallala groundwater will be saltier than the formation water due to migration of brine from underlying strata and due to the effects of leaching from saline surface conditions in the semiarid regions.

In some regions, local units within the Ogallala, or other formational names, are recognized. Among these units are the Batesland Formation, underlying the Ogallala in southwestern South Dakota, and the shaley, early Pliocene Moonstone Formation in the Wind River Basin in Wyoming. In Canada, there are numerous, Pliocene-age, coarse, detrital deposits found on mountain slopes and basins (McCrossan et al., 1964). However, one should consider that a gravel deposit of Pliocene age perched on a mountainside is only marginally a "formation," since it will likely be eroded off in short geological time; by analogy, one could scarcely consider a recent streambed-load to be a formation.

Finally, to culminate discussion of the Tertiary continental sequences, we must mention a series of widely distributed but areally restricted continental sheet deposits lying to the west of the Rocky Mountains, in northeastern Arizona and northwestern New Mexico. These are of approximate Ogallala age (early to middle Pliocene) and are largely assigned to the Bidahochi Formation, containing generally: (1) a lower lacustrine, mudstone unit, (2) a middle basaltic lava flow in many localities, and (3) an upper fluvial, crossbedded sandstone. King and Beikman (1978) stated that the fluvial parts of the Bidahochi were deposited by southwestward-flowing streams coming from the Rocky Mountains and from nearby volcanic mountains.

C. Tejas Sequence of the Atlantic and Gulf Coastal Province

During the Jurassic Period, the Atlantic and Gulf continental margins of North America were initiated by continental rifting. On the Atlantic continental margin, a rifted-margin prism developed as discussed in the previous chapter, but by the end of the Cretaceous the main subsidence phase of prism development was over. Consequently, Cenozoic deposits of the Atlantic Coastal Province, for all of their stratigraphic complexity, constitute only a veneer covering the main, Mesozoic mass of the sedimentary prism. For example, in the Baltimore Canyon Trough, of the total 12 km of sediment fill, only the upper 2 km is of Cenozoic age. At least one reason for thinner Cenozoic strata is the lack of a major source of detrital sediments.

In the northern Gulf of Mexico region, however, there was no such lack. The carbonate-dominated shelves that had surrounded the northern Gulf during Cretaceous time were overwhelmed early in the Cenozoic by a great mass of detrital sediments. These sediments were deposited in a series of major, progradational, sedimentary wedges, each one offlapping the last so that the continental margin grew hundreds of kilometers into the Gulf. The thickness of this great mass of Cenozoic deposits is over 12 km (i.e., equal to the *entire* sedimentary mass of the Baltimore Canyon trough)! The development of this major sedimentary prism was directly related to the Laramide orogeny in the Rocky Mountains region and to other orogenic events in the North American Cordillera, which supplied the detritus to feed the growing mass of sediments.

The Florida and Bahamas region, far removed from significant detrital influx, continued (as during the Cretaceous) to develop as a carbonate platform and bank complex.

C.1. The Paleocene Series

Returning seas of the early Paleocene flooded the Atlantic and Gulf Coastal Province, spreading marine deposits over much of that area. Indeed, in many regions of the Coastal Province, there is evidence that Paleocene strata may have extended even farther inland than they do today. In most eastern areas, detrital sedimentation was limited; thus, Paleocene units of the Atlantic Coastal Province are thin and discontinuous in outcrop. For the same reason, contemporaneous deposits of the eastern Gulf Coastal Province and the Florida–Bahamas area, are dominantly carbonates. In the western Gulf Coastal Province, however, Paleocene strata are thicker and principally detrital. In North America, Paleocene stage nomenclature is derived from Gulf Coastal Province strata. The lower Paleocene is termed the Midway Stage from the Midway Group. Overlying strata, including lowermost Eocene rocks, are represented by the Wilcox Group; unfortunately, the term Wilcox has been used in a variety of ways in the past and is not, therefore, a good choice for a time-stratigraphic term. Consequently, upper Paleocene and lowermost Eocene rocks are included in the Sabine Stage, named after the local, Louisiana designation for Wilcox rocks. European stage names (Danian and Thanetian) are also applied to

Coastal Province strata and we will use both systems, as appropriate, because they are not exactly correlative.

Atlantic Coastal Province. Paleocene strata of the Atlantic Coastal Province outcrop primarily in the Salisbury embayment and in scattered, smaller exposures farther north. In New Jersey and Delaware, for example, Paleocene deposits comprise the Hornerstown Sand and Vincentown Formation. The Hornerstown is a sandy, glauconitic unit, probably deposited on a shallow marine shelf, which disconformably overlies Cretaceous strata (Richards, 1967). The overlying Vincentown is also a marine unit but appears to have been deposited under deeper-water conditions. Olsson (1970) indicated that a disconformity separates the Hornerstown from the Vincentown; this hiatus may represent the mid-Paleocene sea-level fall shown in Fig. 8-3.

Similar relations may be seen in Maryland, where Paleocene sedimentary rocks are divided into the Brightseat Formation, Aquia Formation, and Marlboro Clay (Glaser, 1968). The Brightseat is correlated with the Hornerstown Formation but is less glauconitic and more fossiliferous. Nogan (1964) said that Brightseat deposits formed within the sublittoral zone, under approximately 80–90 m of water. More recently, however, Minard (1980) suggested an estuarine or lagoonal origin based on Brightseat depositional patterns and fossils. The Aquia rests disconformably on the Brightseat and is both faunally and lithologically similar to the Vincentown Formation. Minard posited an inner-shelf to nearshore depositional environment for the Aquia. Aquia deposition was basically regressive, occurring as the initially deep environment gradually shoaled throughout later Paleocene time (Glaser, 1968). The culmination of this regressive trend was deposition of the Marlboro Clay, whose thin but distinctive silver-gray deposits were formed under very shallow, brackish conditions.

South of the Chesapeake Bay, no Paleocene strata are exposed until they reappear in southwestern Georgia as the Midway Group of the Gulf Coastal Province. In the subsurface, however, Paleocene strata have been reported by Swain (1951) from deep wells in eastern North Carolina. These strata, termed the Beaufort Formation, are glauconitic, siliceous, sandy shales and marls overlain by fine-grained sandstone. The fossils and lithologies are similar to those of the Clayton Formation (of the Midway Group) in southwestern Georgia. In South Carolina, Beaufort strata appear to be absent but lower parts of the Black Mingo Formation may also be of Paleocene age (Richards, 1967).

Paleocene deposits are thicker and more continuous beneath the Atlantic continental shelf and slope. In the Georges Bank Basin, for example, Paleocene strata are almost 1 km thick and are composed of clayey silt with benthic foraminifers of outer-shelf origin (Poag, 1978). In the Baltimore Canyon trough, Paleocene deposits comprise the lower part of Schlee's (1981) seismic-stratigraphic Unit F; where sampled in deep wells, these deposits are dense, clayey limestone with middle- to outer-shelf microfossil assemblages. Similarly, in the Southeast Georgia embayment, Paleocene strata are primarily clayey, dolomitic, cherty limestones of relatively deep-water origin. The carbonate content of these shelf deposits increases progressively southward along the continental margin (just as it did in Cretaceous strata of the shelf); this trend continues farther southward into the Florida and Bahamas area.

Florida and the Bahamas. The broad, shallow platforms of Florida and the Bahamas during the Paleocene Epoch remained isolated from significant detrital-sediment influx by the Suwanee Channel and, thus, continued to be sites of widespread carbonate-sediment deposition. This may be seen in the fact that Paleocene strata of southern Georgia and northern Florida are mainly calcareous arenites and argillites whereas correlative rocks of the Florida Peninsula are shallow-water limestones of the Cedar Keys Limestone. These carbonates contain bioclastic, oolitic, and micritic facies associated in some areas with anhydrite and gypsum, suggesting supratidal environments scattered about on the shallow platform. The Cedar Keys Limestone is correlative with the Midway Group of the Gulf Coastal Province. As we will discuss in the next subsection, Paleocene strata of the Gulf area were considered until recently to be represented entirely by the Midway Group; recent revisions of the stratigraphic usage now include lower portions of the Wilcox Group in the Paleocene Series. Therefore, lower portions of the overlying Oldsmar Limestone are probably also of Paleocene age. Like the Cedar Keys, the Oldsmar is a shallow-marine carbonate unit, with lithologies ranging from calc-arenites to dolostones to evaporites (Applin and Applin, 1944).

It is interesting to note that coral bioherms are not found in Paleocene strata of the Florida–Bahamas area, even though they were plentiful during the preceding Cretaceous Period. Indeed, as Frost (1977) pointed out, coral bioherms are almost entirely absent in Paleocene rocks of the entire Caribbean region. This lack of coral reef-structures is no doubt a result of the fact that hermatypic scleractinian coral are likewise nearly absent in Caribbean Paleocene deposits. Perhaps this paucity is related to the global, end-of-Cretaceous "crisis" in the history of life during which many taxa were extinguished or greatly diminished in numbers. In any event, coral reefs did not reappear throughout the Caribbean until the early Eocene and were not abundant until the Oligocene, when they reached the zenith of their development (Frost, 1977).

Paleocene deposits of the southern Florida–Hatteras shelf, slope, and Blake Plateau Basin are also carbonates, mainly pelagic chalks of deep-water origin as indicated by their bathyal microfossil assemblages (Poag, 1978). Trun-

cating the top of this deep-water carbonate sequence is a major erosional unconformity (see Fig. 9-15). Paull and Dillon (1980) described this unconformity and argued that it formed during the latest Paleocene or early Eocene because the surface is covered by progradational clinoform strata of Eocene age. This unconformity may be traced over much of the continental margin east of Florida but is most strikingly preserved in two zones of deeply channeled topography now buried beneath Eocene strata: one of these deeply eroded zones trends northeasterly beneath the modern Florida–Hatteras slope and the other one trends more easterly across the central Blake Plateau, exiting the plateau just north of the Blake Spur (see Fig. 8-24). The character of this buried erosional surface is very much like that scoured into the present-day bottom of the Straits of Florida by the Gulf Stream current and is considered to have formed in the same way. Pinet *et al.* (1981) suggested that the Gulf Stream was initiated in this region during the later Paleocene, possibly as the result of the eustatic sea-level rise shown in Fig. 8-3. They said that the Paleocene Gulf Stream was able to run so far to the west of its present position because the Florida–Hatteras slope did not then exist; instead, the main slope off the Florida platform must have been farther westward. During this time, the current scoured out the northeast-trending zone mentioned above. Later, perhaps during the earliest Eocene, falling sea level caused the Gulf Stream's flow to be diverted eastward, carving the channeled surface beneath the central Blake Plateau. Pinet *et al.*, thus, considered the Gulf Stream to have originated (at least, as it exists today) during the later Paleocene; the reader may recall, however, that Sheridan *et al.* (1981) argued for a Cretaceous (Cenomanian) age for inception of the Gulf Stream. Perhaps the best comment on this subject is that the Gulf Stream, being the North Atlantic Ocean's major western boundary current, has existed since the North Atlantic existed and its actual location at any point in geological time was probably controlled by eustatic sea-level fluctuations.

Gulf Coastal Province. Paleocene deposits of the Gulf Coastal Province comprise the Midway Group and lower parts of the overlying Wilcox Group (see Fig. 9-16). At one time, only Midway rocks were considered to be of Paleocene age and Wilcox rocks were thought to be of lower Eocene age (e.g., Murray, 1961). More recently, based on further biostratigraphic evidence, lower Wilcox strata have been included in the Paleocene Series (e.g., Gibson, 1982).

In the eastern Gulf Province, Midway strata are divided into the Clayton Formation, Porters Creek Clay, and Naheola Formation. The Clayton is primarily a detrital unit in its type locality in up-dip parts of eastern Alabama, being composed of barrier-island and lagoonal deposits (Gibson, 1982). Down-dip, however, the Clayton is dominated by highly fossiliferous, shallow-water limestones; these car-

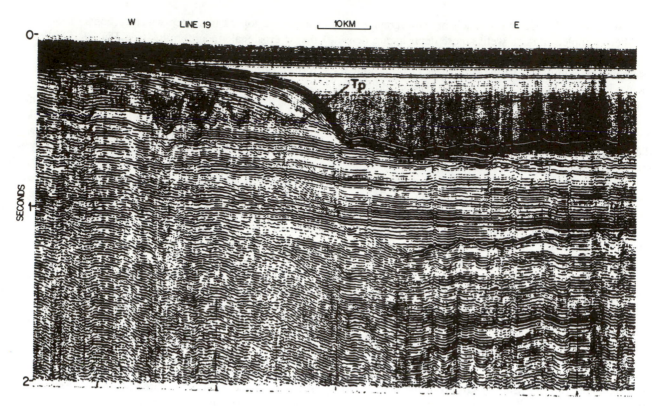

Fig. 9-15. Seismic-reflection profile of continental shelf strata offshore from South Carolina; notice the profound erosional surface marked "Tp." Strata below this surface are Paleocene and Mesozoic whereas clinoform deposits above it are of Eocene age (From Paull and Dillon, 1980, reproduced with permission of the American Association of Petroleum Geologists.)

bonates can be traced in the subsurface as far west as Louisiana. Even farther down-dip, in southern Georgia and northern Florida, the Clayton is composed of marly and chalky deposits which may represent an outer-shelf environment. The Porters Creek Clay is a distinctive dark, laminated claystone of shallow, open-shelf origin. The Porters Creek retains its lithological character all the way from the Georgia–Alabama area to southern Texas (where it is termed the Wills Point Formation; see below). The Naheola, which is not present in eastern Alabama and Georgia, differs from lower Midway units by being dominantly sandy instead of muddy or calcareous. Naheola sands are glauconitic and fossiliferous.

In Texas, Midway rocks comprise the Kinkaid, Wills Point, and Solomon Creek formations. The Kinkaid is similar to the Clayton Formation but is of greater detrital character with limestone constituting only a minor lithology. The overlying Wills Point, as mentioned above, is very similar to the Porters Creek Clay and, like that unit, represents an open marine shelf environment. The Solomon Creek is also a dark, argillaceous deposit, somewhat similar to the Wills Point but less limy.

Note, from the above discussion, that Midway strata in both eastern and western Gulf areas record a major marine transgression followed by regression. Subaerial exposure following regression caused considerable weathering and erosion of Midway (and older) rocks. In western Georgia, for example, the Porters Creek (and Naheola?) was stripped nearly away and a karst topographic surface developed on the Clayton. Indeed, in some up-dip areas the Clayton is represented only by a *terra rosso* residuum associated with scattered accumulation of ore-grade goethite, probably of epigenetic origin associated with karst formation

The Wilcox Group is of economic importance because it contains, especially in Texas and Louisiana, large accumulation of hydrocarbons (see, e.g., Chuber, 1972). In the eastern Gulf, Wilcox strata comprise the Nanafalia, Tuscahoma, and Hatchetigbee formations, of which only the latter is of Eocene age (Gibson, 1982). The Nanafalia was deposited on top of the post-Midway erosional surface during the later Paleocene transgression. Lower Nanafalia strata are mainly nonmarine but pass upwards into marginal-marine and fully marine strata. Marginal-marine deposits in the Nanafalia of western Georgia and eastern Alabama contain considerable amounts of bauxite which apparently developed during periods of exposure prior to Tuscahoma deposition (Cofer and Frederiksen, 1982). The lower Tuscahoma Formation is a strongly transgressive unit com-

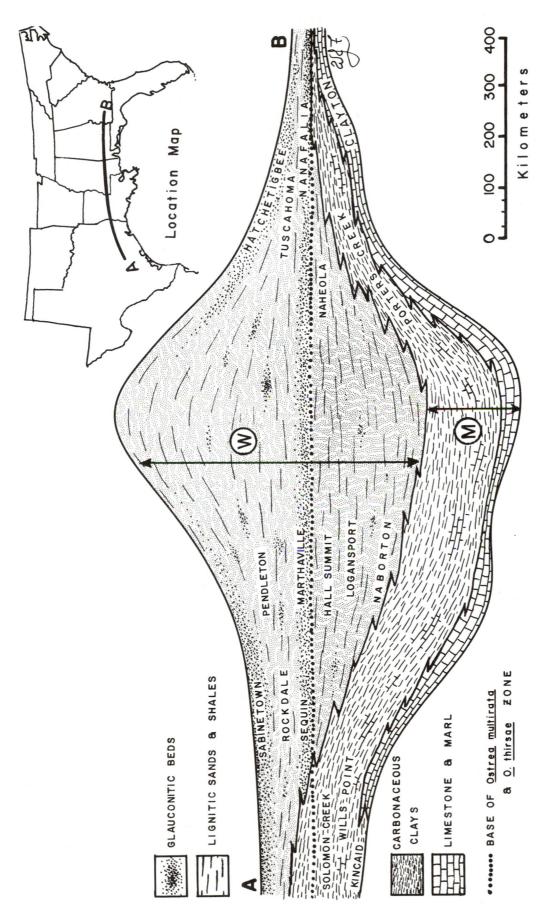

Fig. 9-16. Generalized stratigraphic cross section of Midway (M) and Wilcox (W) rocks of the Gulf Coastal Province. Note the thickening of both units in the Mississippi embayment. (Modified from Murray, 1961.)

posed of pebbly, glauconitic sand; upper Tuscahoma strata are composed of dark, laminated claystones and siltstones of lagoonal origin. Gibson (1982) said that a complex of lagoons probably extended across much of the eastern Gulf coast during this time. About 50 km north of the Fall Line in western Georgia is a small, fault-bounded pod of kaolinitic and bauxitic sediment of lower Wilcox age, based on its palynomorphs (Christopher *et al.*, 1980). These deposits not only demonstrate the greater original extent of Paleocene deposits but also provide one of the very few evidences of Cenozoic faulting in the eastern United States.

In the western Gulf region, lower Wilcox rocks comprise numerous formations, most of which are dominantly coarse, sandy detrital sediments. In large parts of its outcrop area in Texas and the Mississippi embayment, the Wilcox is nonmarine and not readily divisible into smaller stratigraphic units. In the subsurface, however, intertonguing marine and nonmarine strata allow an elaborate stratigraphy. Lowest Wilcox rocks in central Texas are called the Sequin Formation, which is a thin, fossiliferous, marine unit. Overlying the Sequin is the Rockdale Formation, which comprises the main mass of Texas Wilcox strata. The

Rockdale Formation is a generally sandy unit deposited in a complex of progradational deltas and associated fluvial systems (see Fig. 9-17). The result is a lithologically complicated unit over 300 m thick in outcrop and up to 1500 m thick in the subsurface. Clearly, lower Wilcox strata represent a major influx of detrital sediments. The origin of this great volume of detritus was probably the Laramide orogen in the Rocky Mountain region. Fisher (1969) showed that the major direction of sediment transport was from the northwest. Thus, the mid- to late-Paleocene timing of Wilcox deposition provides a reference for the onset of Laramide topographic uplift. During this time, the eastern Texas shelf edge prograded rapidly over 400 km eastward into the Gulf Basin (Woodbury *et al.*, 1973). Upper Wilcox units will be discussed in the next section.

C.2.   The Eocene Series

Eocene Coastal Province strata are parceled into the upper Sabine (Ypresian), Caliborne (Lutetian and Bartonian), and Jackson (Priabonian) Stages. This division

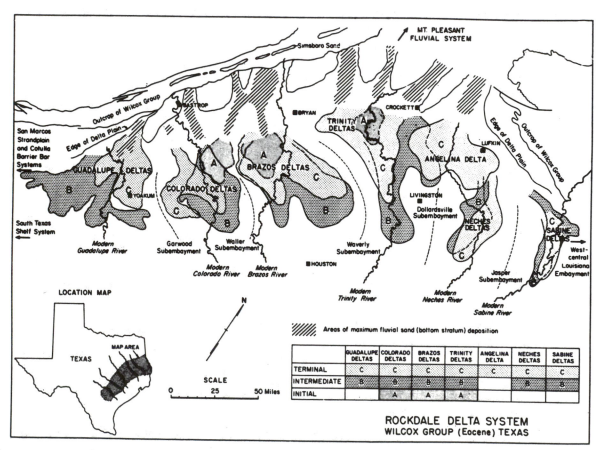

	GUADALUPE DELTAS	COLORADO DELTAS	BRAZOS DELTAS	TRINITY DELTAS	ANGELINA DELTA	NECHES DELTAS	SABINE DELTAS
TERMINAL	C	C	C	C	C	C	C
INTERMEDIATE	B	B	B	B		B	B
INITIAL		A	A	A			

ROCKDALE DELTA SYSTEM
WILCOX GROUP (Eocene) TEXAS

Fig. 9-17. Distribution of principal Rockdale system deltas showing their development through time. Note that the principal apparent source-direction was to the northwest. (From Fisher and McGowen, 1969, reproduced with permission of the American Association of Petroleum Geologists.)

roughly reflects three transgressive episodes (see Fig. 8-3). This, plus the inherent lithological complexity of western Gulf Province strata results in an elaborate Eocene stratigraphy. Generally, however, Eocene stratigraphic relations are regressive and uppermost Eocene facies-patterns in most Coastal Province areas are farther seaward than those of lower Eocene strata.

Atlantic Coastal Province. Eocene deposits of the Atlantic Coastal Province are less well-developed than in the Gulf region and they occupy a relatively limited outcrop area. Nevertheless, they form several academically and economically important units. Upper Sabinian strata comprise the Manasquan Formation of New Jersey, the Nanjemoy of Maryland and Virginia, and the Black Mingo of South Carolina. The Manasquan and Nanjemoy are both principally sandy units with varying amounts of silt, clay, and fossil debris; glauconite is also a common component of both units but is most abundant in the Nanjemoy of Virginia, where it constitutes the major component (Richards, 1967). Owens and Sohl (1973) showed that the character of Manasquan glauconites suggested an inner-shelf depositional environment. The greater glauconite content of the Nanjemoy probably reflects a deeper-water origin, possibly the middle to outer shelf. Sabinian strata are absent in the North Carolina and northern South Carolina, suggesting either that the Cape Fear Arch (see Fig. 8-24) was a positive feature at that time or that such strata were stripped off by pre-Claiborne erosion. In central and southern South Carolina, however, Sabinian sands and clays of marine to marginal-marine origin are present; termed the upper Black Mingo Formation, these deposits constitute the exposed, nearshore, detrital facies of offshore carbonates in the Southeast Georgia embayment (see below). Sabine time ended with a drop of sea level and an erosional hiatus separates Sabinian rocks from Claiborne strata in most areas.

Claibornian deposits are rare in Atlantic Province outcrop. The thin, glauconitic Shark River Formation of New Jersey and the upper Nanjemoy of Maryland and Virginia are the only exposed, post-Sabinian, Eocene strata north of North Carolina. During later Claiborne time, the sea flooded across the Cape Fear Arch, initiating deposition of calcareous and marly deposits of the Castle Hayne Limestone. The Castle Hayne, which extends up into the Jackson Stage, is a highly fossiliferous, shallow-water carbonate which has a very high porosity and, consequently, is one of the major aquifers of eastern North Carolina. In South Carolina and Georgia, Calibornian strata accumulated within the area influenced by the Southeast Georgia embayment. Marginal-marine and shallow-marine detrital deposits, especially across the Central Georgia Uplift (see Fig. 8-24), comprise the McBean Formation. To the east, i.e., toward the center of the embayment shelf, carbonates of the Santee Limestone were deposited. Santee carbonates are continuous with similar limey strata in the subsurface of eastern Georgia (Gohn *et al.,* 1982).

Jackson Stage strata are exposed only south of Virginia. Castle Hayne carbonates comprise the Jackson section in North Carolina but in South Carolina and eastern Georgia, the situation is much more complicated. In easternmost areas, carbonates of the Santee Limestone continued to be deposited (persisting into the early Oligocene). In western, marginal-marine areas, sands and muds of the Irwinton Sand and Barnwell Formation (South Carolina) and the Barnwell Group of Georgia (Clinchfield Sand, Dry Branch Formation, and Tobacco Road Sand; Huddlestun and Hetrick, 1982) all represent a diversified, high-energy shoreline and shelf complex. In central Georgia and extending southward into Florida is the Ocala Limestone (see below) and in western Georgia and Alabama are mudstones of the Yazoo Group. Taken together, all of these rocks may be considered as three major lithosomes: a sand lithosome (the Barnwell Group), a carbonate lithosome (the Ocala), and a mud lithosome (the Yazoo). These three lithosomes all interfinger in central and eastern Georgia (Huddlestun and Hetrick, 1982).

Eocene strata of the continental margin are mainly calcareous and argillaceous deposits of the outer shelf and upper slope. In the middle of the Baltimore Canyon trough, for example, is over 270 m of Eocene clay shale and chalky limestone of bathyal origin (Poag, 1978). Closer to shore, trough deposits of Eocene age are coarser and glauconitic; farther out, near the outer margin of the trough, Eocene strata contain small reeflike masses (Schlee, 1981). In the Southeast Georgia embayment are shallow-marine, bioclastic limestones correlative with and probably continuous with the outcropping Santee Limestone of South Carolina (Gohn *et al.,* 1982). Farther out, beneath the present Florida–Hatteras slope, Eocene beds are predominantly clayey, calcareous muds of deep-water origin. Paull and Dillon (1980) stated that a major progradation of the Florida–Hatteras shelf occurred during the Eocene so that, by the end of the epoch, the shelf edge was near its present position. Clinoform strata overlying the post-Paleocene unconformity (see Fig. 9-15) are the result of that progradation. The lack of Gulf Stream-related erosion during the progradation suggested to Paull and Dillon that the current must have been farther to the east at that time, possibly confined to the central Blake Plateau south of 31°N latitude, as posited by Pinet *et al.* (1981).

Florida and the Bahamas. Carbonate sediments continued to accumulate in the Florida region throughout the Eocene. The Peninsular Arch was still active, as shown both by thinning of Eocene strata across its top and by parallelism of isopachs and facies trends (Chen, 1965). Over the arch stretched the broad platform, on which shallow-water carbonates were deposited. Local evaporite deposits and areas of dolomitization imply the presence of scattered islands. The rock record of this platform comprises the upper Oldsmar Limestone (discussed earlier), the Lake City and Avon Park limestones (Claiborne Stage), and the Ocala Group (Jackson Stage). The Lake City is a porous, skeletal limestone, irregularly dolomitized, with scattered, minor gypsum and anhydrite. Partially correlative with the Lake City is the Tallahassee Limestone, which is found only in the panhandle region of Florida. The Tallahassee is cherty limestone with minor gypsum and a greater clay content than Lake City carbonates. The Avon Park is a chalky limestone and dolostone (Murray, 1961). The Ocala (divided in Florida into the Inglis, Williston, and Crystal River formations) is a highly fossiliferous, porous limestone recognized over the entire Florida platform as well as in southern Georgia and Alabama, It is correlative with the upper Santee Limestone of South Carolina and the Castle Hayne Limestone of North Carolina; with them, the Ocala constitutes a major carbonate lithosome from the Cape Fear Arch all the way to southern Alabama.

At the beginning of the Eocene, the Florida platform was separated from the southern Georgia mainland by the Suwanee Channel but by the end of the Eocene, the channel no longer existed or, at least, was not a significant barrier to detrital influx from the north (Chen, 1965). The infilling of the channel may have been related to the progradation of the Florida–Hatteras shelf, mentioned earlier.

It is not our intention to elaborate on Caribbean tectonics but mention should be made of the suggestion of Dickinson and Coney (1980) regarding Eocene events to the south of continental North America. The reader will recall our discussion of the opening of the Gulf of Mexico. By the end of the Cretaceous in Dickinson and Coney's model (see Figs. 8-32 and 9-18), both the Gulf and the proto-Caribbean Sea had been opened and a convergent plate-boundary existed all along the western margin of North America. This convergent boundary continued to the south, also bounding South America but between the two continents existed a gap, the details of which are entirely hypothetical. According to Dickinson and Coney, rocks of the subduction complex and magmatic arc associated with the convergent boundary that spanned that gap formed the tectonic embryo which would become Cuba, Hispaniola, and the Antilles. With no continental mass to act as "backstop," the Cuban–Antilles arc migrated relatively eastward as shown in Fig. 9-18. During the Eocene, movement of the northern portion of the arc was stopped by collision with the southwestern margin of the Florida–Bahama province. The southern end of the arch, however, continued to migrate, rotating clockwise as the proto-Caribbean widened and closed finally against the Venezuelan margin of South America to produce the Caribbean Coast Ranges. One result of the Cuba–Bahamas collision was the underthrusting of Bahamian

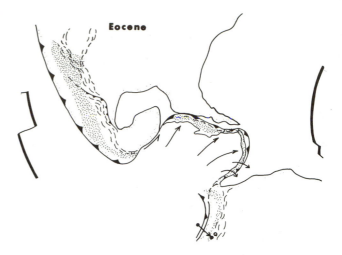

Fig. 9-18. Interpretative model of Eocene tectonics in the Gulf of Mexico region and the southern Cordillera; see text for discussion. (From Dickinson and Coney, 1980, reproduced with permission of the Louisiana State University School of Geosciences.)

limestones beneath ophiolitic rocks of Cuba, which Dickinson and Coney proposed to be the suture zone along which the Cuban–Antilles arc was welded to the Bahama platform.

Gulf Coastal Province. Eocene strata of the Gulf region are lithologically and stratigraphically complex (Figs. 9-19 and 9-20). Nevertheless, several generalizations can be made. For example, Eocene strata of the eastern Gulf Coastal Province are thin, predominantly marine, and composed mainly of carbonates. Correlative strata on the west, however, are much thicker, almost entirely of detrital character, and represent a variety of nonmarine and marginal-marine environments. Eocene strata of the Gulf Province also show much less influence of the Mississippi embayment than do older rocks; this may be seen by comparing the stratigraphic cross sections of Midway and Wilcox rocks (Fig. 9-16) with those of Claibornian and Jacksonian deposits (Figs. 9-19 and 9-20). Note that thickest Paleocene strata occur within the Mississippi embayment whereas thickest Eocene strata occur in the Texas and eastern Mexico area, i.e., adjacent to the Laramide orogen (to the northwest) and the Mexican Cordilleran orogen (to the west).

Upper Sabinian rocks of the eastern Gulf region comprise the Hatchetigbee Formation, of which the Bashi Marl is the lower member. The Bashi is the sole Hatchetigbee unit in the Chattahoochee Valley region of Georgia and Alabama (Gibson, 1982) and consists of glauconitic, fossiliferous, calcareous sand, often with large, spheroidal concretions. Gibson (1981) reported that the Bashi contains deposits of "sawdust sand," i.e., accumulations of coarse clay and silt floccules indicative of brackish to very shallow marine conditions; other Bashi rocks represent more open

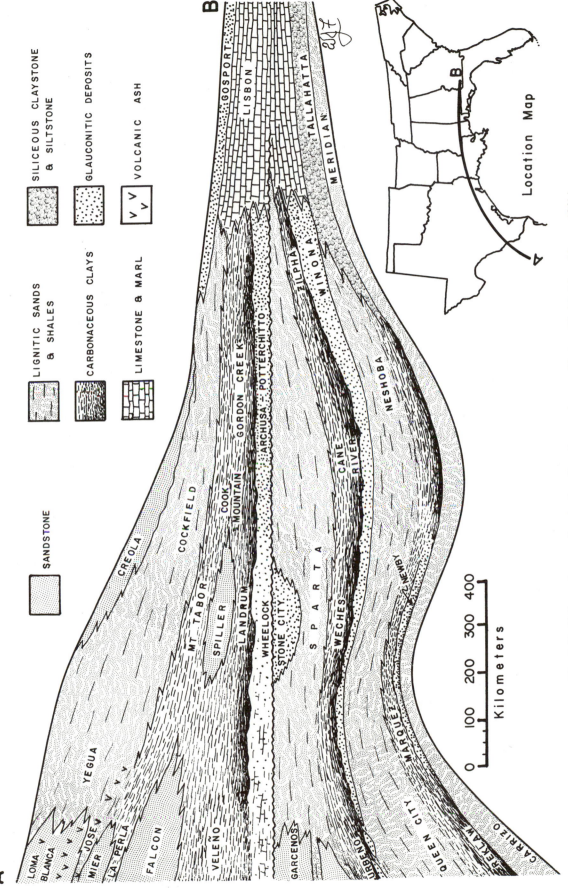

Fig. 9-19. Claiborne Stage stratigraphy from Georgia to northeastern Mexico. (Modified from Murray, 1961.)

SANDSTONE

LIGNITIC SANDS & SHALES

SILICEOUS CLAYSTONE & SILTSTONE

CARBONACEOUS CLAYS

GLAUCONITIC DEPOSITS

LIMESTONE & MARL

VOLCANIC ASH

Location Map

Kilometers

0   100   200   300   400

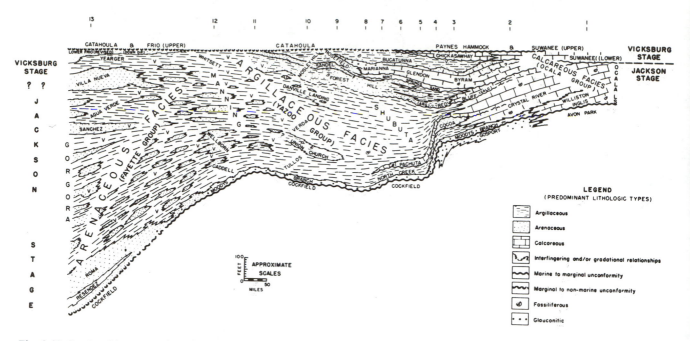

Fig. 9-20. Stratigraphic cross section of Jackson and Vicksburg Stages from northern Florida to the Rio Grande Embayment. Note the subdivision of Jackson strata into arenaceous (Fayette), argillaceous (Yazoo), and calcareous (Ocala) facies. (From Murray, 1961.)

marine conditions. The upper, unnamed member of the Hatchetigbee in Alabama and Mississippi and is composed of fine to medium sands and interbedded carbonaceous muds. Murray (1961) correlated the Meridian Sand of Mississippi with the Hatchetigbee but subsequent work has placed it as the basal member of Mississippi's Tallahatta Formation (basal Claiborne; see below).

Claiborne strata in the east are divided into Tallahatta, Lisbon, and Gosport formations. The Tallahatta Formation is a slightly calcareous, glauconitic, clayey sand and was deposited during a transgression following the retreat of the sea after Sabine time. Consequently, basal up-dip Tallahatta strata are gravelly, nonmarine deposits formed later in the transgression. Down-dip Tallahatta rocks are of wholly marine origin. The Lisbon is a calcareous, fossiliferous, glauconitic limestone, sandier in up-dip regions. It is correlative with the Lake City and Tallahassee Limestone of Florida and represents the initial Eocene extension of carbonate depostion into Georgia. The Gosport Sand is a richly fossiliferous unit composed of quartz and glauconite sand with associated thin, carbonaceous clays.

Most eastern Gulf strata of the Jackson Stage may be consigned to one of two major lithosomes: the Ocala carbonate lithosome or the argillaceous Yazoo Group. At the base of the Stage everywhere from Georgia to central Texas is the Moodys Branch Formation, a thin, transgressive unit composed generally of a lower, glauconitic, fossiliferous sand with associated clay beds and an upper calcareous and argillaceous sequence. The Moodys Branch rests disconfor-

mably on Claiborne rocks. Above the Moodys Branch in Georgia and eastern Alabama is the Ocala Limestone (correlative with the Ocala Group in Florida). In Georgia, the Ocala is a bioclastic, skeletal limestone composed principally of bryozoans, coralline algae, and large foraminifers; it has a porosity of up to 30% (Gelbaum and Howell, 1982). In western Alabama, the Ocala passes by facies change into the generally argillaceous Yazoo Group (see Fig. 9-20). The Yazoo is divided into the North Twistwood Creek Clay, Cocoa Sand, Pachuta Marl [considered by Murray (1961) to be a western tongue of Ocala rocks], and the Shubuta Clay. Yazoo strata are generally of marine or marginal-marine origin in their down-dip occurrences but become increasingly continental northward in the Mississippi embayment. Atop the Wiggins uplift in southwestern Alabama and southeastern Mississippi, the entire Jackson Stage is represented by carbonates, probably formed as an offshore carbonate bank (Murray, 1961). A regional unconformity caps the sequence.

In the western Gulf region, Eocene rocks are much thicker than in the east and are dominantly arenaceous and argillaceous. Upper Sabinian rocks comprise the Sabinetown Formation of the Texas Wilcox Group. The Sabinetown varies from a fossiliferous, glauconitic, calcareous sand of shallow marine origin in down-dip areas to a nonmarine deposit composed of fluvial fining-upward sequences in up-dip areas. Edwards (1981) described the deep subsurface Upper Wilcox in eastern Texas and concluded that it represents a major deltaic system (the Rosita system) which prograded rapidly to the Sabine shelf edge. During Rosita progradation, a number of down-to-the-basin-growth faults were active (see Fig. 9-21) with the result that Rosita system strata thicken from approximately 180 m in western areas to more than 900 m at their easternmost occurrence.

Murray (1961) included the Carrizo Formation of Texas with Wilcox strata but subsequent work has placed it within the Claiborne (King and Beikman, 1978). The Carrizo is a dominantly sandy, continental deposit correlated with the lower Tallahatta Formation in the eastern Gulf. The rest of the Claiborne (over 300 m thick in outcrop, increasing to almost 1800 m in the deep subsurface) is divided into three large-scale cyclic sequences (see Fig. 9-19). Each sequence is composed of a thin, lower, transgressive unit consisting of glauconitic, fossiliferous sandstone, a middle unit composed of dark argillaceous strata of deep-water origin, and a thick, upper, regressive unit composed of marginal-marine and continental deposits which often are carbonaceous and lignitic. The first cycle is composed generally of the Mount Selman Formation (consisting of the Reklaw, Queen City, and Weches members). The second cycle consists of the Sparta and Cane River formations and the third of the Cook Mountain and Yegua Formation. Stratigraphic cross sections across Claiborne strike show that each of these cycles represents a large regressive tongue of marginal-marine and nonmarine deposits separated from sub- and superjacent tongues by much thinner transgressive marine strata.

Throughout Claiborne time, massive amounts of sedi-ments were deposited on the western and northwestern margin of the Gulf of Mexico extending the shelf many tens of kilometers seaward. The reader may visualize the volume of strata involved by considering that the entire northern two-thirds of the Mississippi embayment is underlain by Claiborne strata and that the Texas Claiborne outcrops in a band almost 100 km wide across the state. Strata younger than Claiborne rocks outcrop in bands which approximately parallel the modern coastline whereas pre-Claiborne outcrops trend parallel to the inner margins of the Mississippi embayment. The great volume of Claiborne sediments testifies to the intensity of uplift and denudation in the Laramide orogenic area to the northwest. It should be noted, however, that other areas probably also yielded detritus during Claiborne time. McCarley (1981) conducted a mineralogical analysis of several Claiborne units (the Carrizo and Reklaw) and showed that an Ouachita source may also have provided some sediments at least episodically, to the prograding coastal complex.

Jackson strata of Texas comprise the Moody's Branch (discussed earlier), Caddell, Wellborn, McElroy, Manning, and Whitsett formations, all of which (except the Moodys Branch) comprise the Fayette Group. Taken together, Fayette rocks represent the proximal facies of the distal, clayey Yazoo Group mentioned earlier. Like Claiborne rocks, Jackson strata represent not a single depositional environment but rather a complex of fluvial, deltaic, barrier and back-barrier, and open marine deposits. Figure 9-22 illustrates the major sediment-dispersal systems of Jackson

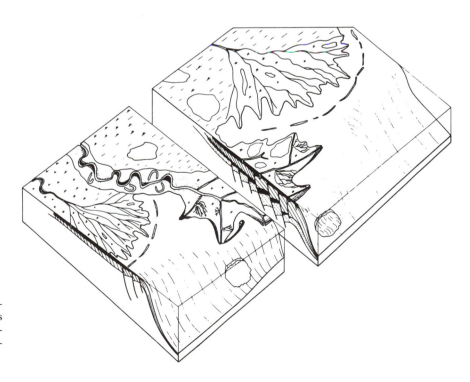

Fig. 9-21. Illustration of a typical Rosita-system delta, showing its relation to growth faults in the subsurface. (From Edwards, 1981, reproduced with permission of the American Association of Petroleum Geologists.)

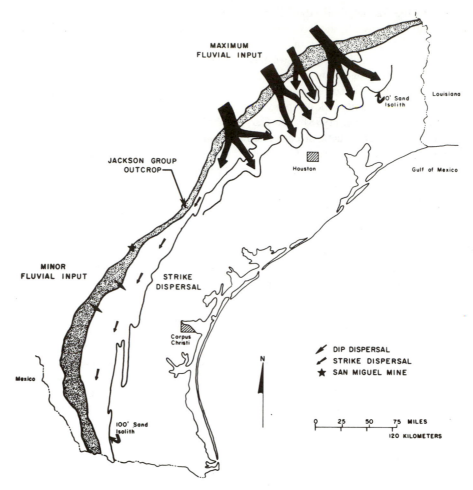

Fig. 9-22. Sand dispersal system of the Jackson Group in Texas. (From Snedden and Kersey, 1981, reproduced with permission of the American Association of Petroleum Geologists.)

environments. Notice that maximum fluvial input was from the northwest, corresponding with a Laramide source area.

C.3. The Oligocene Series

Oligocene strata are very poorly represented throughout the Atlantic and Gulf Coastal Province. The main reason for this underrepresentation is that a major eustatic fall of sea level occurred during the later Oligocene. The sea-level curve of Vail and Mitchum (1979; see our Fig. 8-3) indicates that this eustatic event was the most profound drop of sea level of the entire Cenozoic. Sea level was as much as 250 m below its present stand and virtually the entire coastal province out to the continental-shelf edge was exposed. Resultant erosion stripped away much of the lower Oligocene and even upper Eocene record, especially where these strata would have been thin anyway, i.e., on the Atlantic Province.

A second reason for the paucity of Oligocene strata

was the cessation of the Laramide orogeny. By the beginning of the Oligocene, the Laramide orogen had been eroded down to a low, hilly terrane and the supply of detritus to the coast, so great during earlier Cenozoic time, had slowed to a dribble.

Thickest Oligocene strata of the Coastal Province occur in the northern Gulf area around Louisiana and Mississippi where they comprise the Vicksburg Group and Chickasawhay Formation. These units have lent their names to Gulf Province provincial stages, the Vicksburgian being roughly equivalent to the European Rupelian and the Chickasawhayan correlative with the Chattian.

Atlantic Coastal Province. Oligocene strata do not outcrop on the Atlantic coastal plain north of South Carolina but they have been reported to occur in the subsurface of several areas. In eastern North Carolina, for example, between Cape Lookout and Cape Hatteras, well data reveal fine-grained sands of Oligocene age with a shallow-water benthic fauna (Poag, 1978). In eastern Maryland and New

Jersey, a glauconitic, clayey sand named the Piney Point Formation has recently been reclassified as Oligocene (Olsson *et al.*, 1980). Originally considered to be of late Eocene age, the Piney Point contains a late Oligocene planktonic fauna and is thought by Olsson *et al.* to be the basal member of a primarily Miocene transgressive sequence. This is a significant observation because it requires that the late Oligocene be a time of rising sea level rather than of falling, as shown by Vail and Mitchum's curve (Fig. 8-3). There is further stratigraphic support for this proposition in the fact that lower Oligocene strata are generally absent in Georgia, Alabama, and Mississippi but upper Oligocene beds, mainly of marine origin, are encountered in all three states. Olsson *et al.* further pointed out that an early Oligocene regression would correlate with severe climatic changes noted by other authors to have begun at the end of the Eocene.

Oligocene deposits of the continental margin are better developed than corresponding strata beneath the coastal plain. In the Baltimore Canyon Trough, for example, there is around 150 m of calcareous shale and claystone overlain by a dense, impermeable limestone, all of Oligocene age (Poag, 1978). Schlee (1981), however, showed that uppermost Oligocene strata of the Baltimore Canyon Trough are part of a series of progradational, deltaic wedges. If dating of these rocks is accurate, it poses an interesting problem: a regression must have been going on in the Baltimore Canyon area while, at the same time just to the west, the Piney Point transgression was transpiring. This problem is as yet unsolved. There is, however, general agreement that a major fall of sea level did occur at some time during the Oligocene and that it left erosional and/or nondepositional hiatuses in almost all sections containing deposits of the proper age. Grow *et al.* (1979) observed that fluvially transported detrital sediments must have bypassed shelf areas during the low stand of sea level, being funneled by submarine canyons down to abyssal depths.

In South Carolina, Oligocene strata comprise the Cooper Marl, a phosphatic marl which overlies the Santee Limestone. Cooper deposits are coarser and more detrital in the up-dip direction and grade-down-dip into shelf carbonates of the Southeast Georgia embayment. In Georgia, the Cooper Marl grades into the Suwanee Limestone. The Suwanee is a granular to nodular, highly recrystallized, partially dolomitized, cherty limestone with abundant foraminifers (Gelbaum and Howell, 1982). Between 30 and 60 m thick in Georgia, the Suwanee is of later Oligocene age. In up-dip areas of Georgia and adjacent South Carolina is exposed a residual, chert-block deposit, termed the Flint River Formation. The Flint River was apparently caused by dissolution of a cherty limestone and is probably the up-dip equivalent of the Suwanee (Richards, 1967).

Florida and the Bahamas. Oligocene strata of the Florida platform are mainly shallow-marine carbonates sim-

ilar to older rocks beneath them. They are thinner, however, and locally absent; for example, Oligocene strata are missing in northeastern Florida but do occur beneath thin inner continental shelf deposits where they are micritic and biomicritic rocks of shallow-water origin. Farther eastward, in the Blake Plateau Basin, Oligocene strata are composed of chalky oozes, some with volcanic ash (Schlee, 1977). On the West Florida shelf and slope, Oligocene rocks comprise parts of two depositional sequences, one of Eocene and lowest Oligocene age and the other of late Oligocene to Miocene age (Mitchum, 1978). Here again, one sees the effect of the Oligocene sea-level fall. West Florida slope deposits contain minor volcanic debris. The origin of this ash, as well as that of the Blake Plateau, may have been the Cuba–Antilles arc (see Fig. 9-18).

In southern-mainland Florida, deep subsurface Oligocene strata are generally classified as Suwanee Limestone but, in the panhandle region, the term Suwanee is used only for upper Jacksonian carbonates. Lower Jacksonian rocks of the Florida panhandle comprise the Marianna, Glendon, and Byram formations. These are primarily skeletal to chalky limestones with minor amounts of argillaceous insoluble residue. Marianna, Glendon, and Byram are all Gulf Province terms and we will return to them. Uppermost Oligocene strata of northern Florida comprise the Chickasawhay Formation.

Coral-reef deposits are common in Oligocene rocks of Florida and may even be found occasionally in the Georgia Suwanee. Indeed, as Frost (1977) has shown, the Oligocene was a time of major coral-reef growth all over the Caribbean region. At the end of the Oligocene, however, almost one-half of Caribbean hermatypic coral disappeared. Frost mentioned several possible causes for this extinction episode, such as changes in surface-current patterns due to tectonic alternations of Caribbean geometry, but concluded that a completely satisfactory explanation has yet to be proposed.

Gulf Coastal Province. In the Gulf region, too, Oligocene strata are rare. They are exposed in a narrow band from Georgia to Mississippi where they are lost beneath younger deposits of the Mississippi embayment. They reappear in Texas as the Catahoula Formation.

In the eastern Gulf, Jacksonian strata comprise two intertonguing lithosomes: an eastern, carbonate lithosome (shown in Fig. 9-20 to be a younger extension of the Eocene Ocala lithosome) and a western, argillaceous–arenaceous lithosome. Over the whole area, lowest Oligocene rocks are absent. In western Georgia, the whole Oligocene section is comprised of carbonates of the Suwanee Limestone, but in Alabama, Jackson strata are divided into the Marianna Limestone and Byram Formation. The Marianna is a glauconitic, fossiliferous marl in Alabama but becomes increasing detrital westward, finally passing into glauconitic sands of the Mint Springs Formation in western Mississippi. As discussed earlier, the Marianna becomes increasingly limy into Florida. The Alabama Byram Formation is divided into the lower, Glendon Limestone Member, a middle unnamed member, and the upper, Buccatuna Clay Member. The Glendon, which is considered a formation in Mississippi, is a well-indurated, fossiliferous limestone whose strata become increasing argillaceous upward into the middle, marly member. Byram clay content further increased upward into the Buccatuna, which is a dark, carbonaceous claystone. In Mississippi, lowest Jackson strata are sands and associated muds of the Forest Hill Formation, which is overlain by the Mint Springs, mentioned above. Overlying the Mint Springs and its eastern Mississippi equivalent, the Marianna, are Glendon, Byram, and Buccatuna formations.

Stratigraphic relations in the Texas Oligocene are inherently complicated but have been rendered virtually opaque by years of repeated reclassification. The only outcropping Oligocene strata in Texas are composed of the Frio Clay and the Catahoula Sandstone but Oligocene deposits thicken greatly in the subsurface and include a large number of stratigraphic units. Probably the most important of these units are the Vicksburg Formation, the Frio Formation (*not* the same as the Frio Clay), and part of the Anahuac Shale (the rest of which is Miocene). Figure 9-23 shows inferred stratigraphic relations of these units; notice the dramatic increase of stratigraphic thickness on the down-dip side of growth faults and the presence of salt domes in the Houston embayment. The Vicksburg Formation is a regressive sandy mudstone which grades eastward into shelf muds. Overlying the Vicksburg in up-dip areas is the Catahoula Sandstone, which everywhere onlaps the Vicksburg. The Catahoula Formation is a coarse nonmarine sequence dominated by fluvial deposits. The Frio Clay, of lacustrine origin, is now believed to correlate in part with both the Vicksburg and the lowest Catahoula. The Frio Formation is an entirely subsurface unit but is of major importance; it constitutes one of the great progradational, deltaic wedges of the western Gulf Tertiary System. In addition, it has proved to be a premier hydrocarbon reservoir, having produced over 6 billion barrels of oil and more than 60,000 cubic feet of natural gas. The Frio is a complex marginal-marine to marine unit and was considered by Galloway *et al.* (1982) to be the down-dip equivalent of Catahoula sands. Figure 9-24 shows the Galloway *et al.* reconstruction of major Frio–Catahoula depositional systems. Notice that the Houston and Rio Grande embayments were the sites of extensive fluvial and deltaic sedimentation whereas only minor fluvial deposits and barrier-island/strand-plain deposits occur across the top of the San Marcos Arch. The Anahuac Shale, also entirely subsurface, represents a major marine transgression during which dark shelf muds extended far up-dip (but not as far as had earlier Claiborne and Jackson regressive units). The age of the Anahuac has been the source of long debate (see

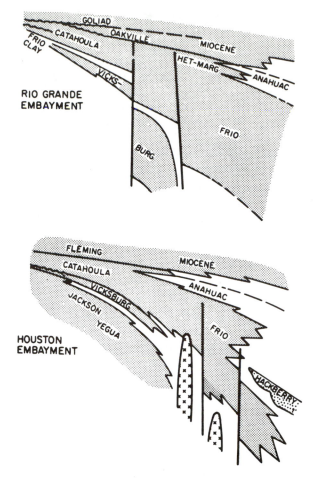

Fig. 9-23. Schematic stratigraphic cross sections (dip sections) of the Frio Formation and associated units of Texas. (From Galloway *et al.,* 1982, reproduced with permission of the American Association of Petroleum Geologists.)

summary of argument's history in Murray, 1961); it was considered entirely of Miocene age by Warren (1957) who defined the Anahuac Stage to be the lowest Miocene stage. Galloway *et al.* show the lower Anahuac to be of late Oligocene age (see Fig. 9-23).

C.4.   The Miocene Series

Miocene deposits are extensively exposed on the coastal plains from southern Texas to New Jersey, absent only where covered by Mississippi River alluvium and across the tops of the Peninsular Arch in Florida and the Cape Fear Arch in North Carolina. The Miocene was the longest epoch of Tertiary time (lasting almost 19 Myr.) and, consequently, the Miocene Series is extremely thick, especially beneath the Texas–Louisiana continental shelf and slope, where it exceeds 6.5 km. Miocene strata of the Atlantic Coastal Province are of primarily marine origin but, unlike

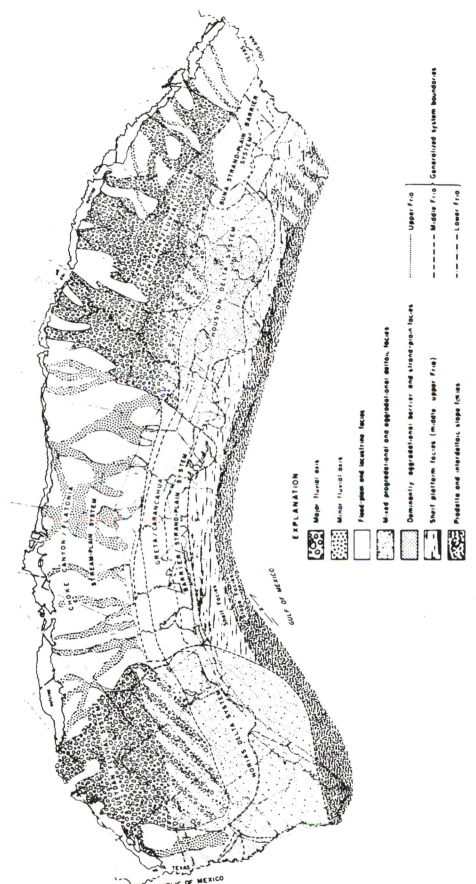

Fig. 9-24. Major depositional systems of the Frio Formation in Texas. Notice that fluvial and deltaic systems are well-developed in the Houston and Rio Grande Embayments whereas poorly developed stream and barrier deposits characterize the San Marcos Arch between the embayments. (From Galloway *et al.*, 1982, reproduced with permission of the American Association of Petroleum Geologists.)

Eocene and Oligocene marine strata of the same region, are dominantly detrital rather than calcareous. Exposed Gulf Province Miocene deposits are mainly of brackish to non-marine origin, passing down-dip in the subsurface to marine facies. Carbonate deposition continued in the Florida area.

Atlantic Coastal Province. The Miocene Series is the most extensive Tertiary unit of the northern Atlantic Coastal Province. Resulting from a major marine transgression (which began during the late Oligocene with deposition of the Piney Point Formation, discussed earlier), Miocene strata onlap the Piney Point and overstep older units all the way to the Piedmont of North Carolina and Virginia. In Virginia, Maryland, and Delaware, Miocene rocks comprise the Chesapeake Group, divided into Calvert, Choptank, St. Marys, and Yorktown formations. The Calvert is a shallow-marine deposit composed of a lower unit (the Fairhaven Member) of poorly sorted, argillaceous sands and diatomaceous earth and an upper unit (the Plum Point Marl) containing a rich marine fauna, especially in well-defined shell beds traceable over large areas (Glaser, 1968). Gibson (1962) described benthic foraminifers of the Calvert and concluded that they indicate a generally cool-water environment. Overlying Choptank strata represent a continuation of the shoaling trend begun late in Calvert time. The Choptank consists of very-shallow-marine clays and sands associated with shell beds. The St. Marys Formation overlies the Choptank conformably and represents a slightly deeper-water environment. In New Jersey, marine strata correlative with Calvert, Choptank, and St. Marys deposits are termed the Kirkwood Formation.

The Yorktown Formation, upper part of the Chesapeake Group, is a richly fossiliferous unit ranging in composition from glauconitic sand, to marly sand, to coquina. It is the most extensive single coastal-plain unit in southern Virginia and North Carolina. It is also the source of considerable controversy concerning its age. Originally thought to be entirely of Miocene age (e.g., LeGrande, 1961; Richards, 1967), the Yorktown has been shifted partially or completely up into the Pliocene. Hazel (1971), for example, concluded that of the three ostracod assemblage-zones in the Yorktown, two are upper Miocene and one is Pliocene. Akers (1972), using foraminiferal biostratigraphy, argued that the entire Yorktown is Pliocene, a conclusion with which King and Beikman (1978) concurred. In New Jersey, the Cohansey Sand is thought to be correlative with the Yorktown and probably represents a regressive barrier and back-barrier facies of offshore Yorktown sediments (Carter, 1978).

South of the Cape Fear Arch, Miocene strata are also well-developed but contain some striking differences from northern Miocene deposits. In South Carolina and Georgia, the Miocene is divided into the Tampa Limestone, Hawthorn Formation, Raysor and Duplin marls. The Tampa is a Florida unit traceable in the subsurface into southern Georgia and South Carolina. The Hawthorn is mainly a dolomitic limestone with up-dip sandy and gravelly facies. Hawthorn strata are remarkable, however, in that they contain considerable phosphate (also found in correlative strata beneath Georgia's inner continental shelf in the Southeast Georgia embayment; Poag, 1978) and fuller's earth (palygorskite–sepiolite). These materials, especially the clays, are rare and indicative of extraordinary environmental conditions during deposition and/or diagenesis. Weaver and Beck (1977) presented a thorough discussion of these sediments and concluded that palygorskite–sepiolites of the Southeast's Miocene were formed by the alteration of montmorillonite in a humid, subtropical to tropical environment. This conclusion is most interesting because, as mentioned earlier, Calvert Formation foraminifers seem to indicate cool-water conditions (Gibson, 1962). Weaver and Beck suggested that the major factor controlling palygorskite–sepiolite formation was oceanic surface-current patterns. Thus, the Gulf Stream, which channels water northward from Caribbean climes, may have been responsible for warming the Southeast while cooler waters influenced northern coastal province areas, probably because the Gulf Stream turned eastward in the area offshore from Cape Hatteras, as it does today. Weaver and Beck hypothesized that changes in surface-current patterns in the North Atlantic (as a result of continental drift, especially the closing off of the Mediterranean by collision of Europe and Africa at Gibraltar) ended the special, environmental conditions which allowed the formation of the clays. The abundantly fossiliferous Duplin Marl of North Carolina to Georgia and the Raysor Marl which underlies the Duplin in South Carolina are biostratigraphically equivalent to the Yorktown Formation and, like the Yorktown, have been ping-ponged back and forth across the Miocene–Pliocene boundary.

Florida and the Bahamas. Miocene strata are extensively exposed in both peninsular and panhandle Florida. They are also present in the subsurface over much of the Florida–Bahamas region except across the top of the Peninsular Arch. In most of these areas, Miocene deposits are lithologically complex but generally represent shallow-marine environments. Unlike strata of underlying Tertiary section, Miocene strata in Florida contain large amounts of detrital sediment as well as carbonates, especially in northern and eastern parts of the state, because of the absence of the Suwanee channel, which had acted during earlier epochs to present detrital influx. By the end of the Miocene, detrital sediments had been deposited in both eastern and western parts of the Florida Peninsula as far south as Miami.

At the base of Florida's Miocene section is the Tampa Limestone, whose carbonate facies (termed the St. Marks facies) is present over much of peninsular Florida and whose terrigenous detrital facies (the Chattahoochee facies) is confined mainly to northern parts of the state. Following Tampa time, detrital influx increased and silty, clayey, phosphatic deposits of the Hawthorn Formation (discussed above) spread down the eastern side of peninsular Florida. Similar deposits were reported by Paull and Dillon (1980) to occur beneath the Florida–Hatteras shelf south of the Southeast Georgia embayment. Detrital deposits also accumulated on the western side of the Peninsular Arch but were overshadowed by carbonate deposition, perhaps because the arch functioned as a geographic barrier to western dispersal of detrital sediments. In southern Florida, carbonates continued to form but are phosphatic. The resulting complex of detrital carbonate lithofacies is termed the Alum Bluff Group, generally of middle Miocene age. Overlying Alum Bluff strata in northern Florida are arenaceous and argillaceous carbonates of the Choctawhatchee Formation. Like the Alum Bluff, the Choctawhatchee is lithologically complicated with rocks distributed into a variety of facies (see Puri, 1953). The Choctawhatchee is disconformably overlain by regressive strata of the Jackson Bluff Formation (Dubar and Taylor, 1962). In South Florida, upper Miocene deposits comprise the Tamiami Formation, composed of sands, clays, and limestones overlying phosphatic carbonates of the Hawthorn Formation (Alum Bluff Group). Peck et al. (1979) discussed the Tamiami Formation in southwestern Florida and showed that it contains evidence of a major late Miocene (Messinian) regression. They correlated this regression with similar sequences at the base of the Duplin Marl in South Carolina and in the Jackson Bluff Formation of northern Florida. Farther north, the Yorktown Formation and Cohansey Sand are correlative with the Tamiami and also are regressive. Thus, there is considerable evidence of a significant, eustatic sea-level fall during the late Miocene (see Fig. 8-3). Peck et al. correlated this with the Messinian regression of Europe and with the hypothesized desiccation of the western Mediterranean Basin (and the ''Messinian salinity crisis''; see Adams et al., 1977). This eustatic event was considered by Peck et al. to have been caused by the severe late Miocene glaciation in Antarctica.

Gulf Coastal Province. Miocene strata of the Gulf Province have been the focus of considerable study, especially in the subsurface, and their stratigraphic relations, regional correlations, and formal classifications have all been controversial. Our review follows King and Beikman's (1978) discussion of the geological map of the United States except that we place the Catahoula Formation of the northwestern Gulf into the Oligocene Series instead of the Miocene, based on the more recent analysis by Galloway et al. (1982).

In the eastern Gulf Province, Miocene strata are primarily clayey and calcareous deposits of brackish-water or marine origin. In western Georgia, lower Miocene rocks are sandy carbonates of the Tampa Limestone, which is exposed in a narrow band in the southwestern part of the state. The Tampa is overlain by phosphatic sands and clays of the Hawthorn Formation which are, in turn, overlain by the Miccosukee Formation. From Alabama westward, exposed Miocene strata show progressively less marine influence. In Mississippi, for example, Miocene rocks comprise the fluvio-deltaic Hattiesburg and Pascagoula Clays. Of course, these strata pass down-dip into marine deposits of decreasing grain size farther away from the outcrop belt. In the subsurface of Louisiana are carbonate masses which formed as patch reefs on top of salt domes; continued upward migration of the salt diapirs punctured through the carbonate masses, forming small but highly porous hydrocarbon reservoirs (Forman and Schlanger, 1957).

The Pascagoula Clay in Mississippi and Louisiana contains a number of marine tongues pinching out northward, resulting in a generally cyclothemic deposit. Other cyclic strata have been reported in upper Miocene beds in the Gulf area (Murray, 1961). Peck et al. (1979) described cyclic deposition in the Tamiami Formation of southern Florida and suggested that the cause of such repetitive sedimentation was periodic, eustatic sea-level fluctuations. These fluctuations may have been related to advances and retreats of the Antarctic continental glacier.

In Texas, Miocene strata comprise the Oakville Sandstone and Fleming Formation. The Oakville Sandstone is a coarse to gravelly sandstone of fluvial origin. It contains the bones of numerous horses, camels, mastodons, and rhinoceroses associated with reworked Cretaceous marine shells, derived by erosion of Cretaceous strata in north-central Texas. The Fleming Formation is a fluvio-deltaic unit, finer-grained than the Oakville and containing both brackish-water invertebrates and mammal remains of latest Miocene–Pliocene age.

C.5. The Pliocene Series

Throughout the Atlantic and Gulf Coastal Province, exposed stratigraphic sequences (of virtually any age) are covered by a thin, discontinuous veneer of gravelly and sandy deposits referred to generally as surficials. Surficial deposits are commonly unfossiliferous but may rarely contain leaf imprints, lignitic materials, and, near coastal areas, marine or brackish-water mollusks. They range in age from Miocene to Recent but specific age determinations for given deposits are usually difficult due to the general lack of fossils. Consequently, correlations among surficial deposits are hypothetical at best and are often drawn on the basis of geomorphological features such as terraces, scarps, or other features. As a result, disagreements abound over correlations, age assignments, and even depositional environments. In the following paragraphs, we discuss the Pliocene component of coastal-plain surficial deposits but the reader should remember that virtually every unit mentioned is of questionable age.

Atlantic Coastal Province. Until recently, Pliocene surficial deposits on the northern coastal plain were not the subject of much study. They were given a variety of local formational names, e.g., Brandywine, Lafayette, Bryn Mawr, Sunderland, Beacon Hill, Pensauken, and others. In the last 10 to 15 years, an even more elaborate stratigraphy has developed as understanding of relative age relationships and facies patterns has increased. It is interesting to notice that the ranks of Pliocene stratigraphic units has decreased as they continue to be reassigned to either Miocene or Pleistocene Series. In the New Jersey–Delmarva Peninsula region, Miocene–Pliocene surficial strata comprise the Bridgeton, Pensauken, Beaverdam, and Walston formations. The Bridgeton and Pensauken are fluvial deposits, principally gravelly, crossbedded, arkosic sands of late Miocene age (and therefore correlative with deltaic and barrier Yorktown–Cohansey strata; Owens and Denny, 1979). The Beaverdam, too, is fluvial but is of later Pliocene age. Figure 9-25 shows these and other gravel deposits of the northern coastal plain and the inferred direction of sediment dispersal for each deposit. Note that most sediments were transported southwestward along the inner margin of the New Jersey coastal plain, then turned south- to southeastward in much the same fashion as do sediments transported by the modern Delaware and Susquehanna Rivers. Overlying the Beaverdam in the central Delmarva area is the Walston Silt, which is believed to be of marginal-marine origin, probably deposited in a back-barrier lagoon (Owens and Denny, 1979). Surficial deposits stratigraphically higher than the Walston are probably of Pleistocene age and will be considered later.

On the southern Atlantic coastal plain, Pliocene strata are similar to those described above but are stratigraphically less complicated. Pliocene surficial deposits of North and South Carolina comprise the fluvial Waccamaw Formation and, near the coast, the marginal-marine James City Formation. The Macks Formation south of Raleigh, North Carolina, contains late Miocene shells and may, therefore, be a Yorktown correlative. In central Georgia, surficial deposits in upland areas are unnamed; on the coast, however, there are thin calcareous clays and marly limestones called the Charlton Formation. Charlton fauna are of Pliocene age and indicative of brackish-water conditions (Richards, 1967).

On the Atlantic continental shelf, Pliocene deposits are primarily of marine origin. In the Baltimore Canyon Trough area, there is about 65 m of sand and silt with Pliocene-age inner-shelf benthic foraminifers (Poag, 1978). In the Southeast Georgia embayment, Pliocene strata are thinner (6–15 m) but are of similar character albeit more calcareous, especially on the outer shelf.

Florida and the Bahamas. In north Florida, Pliocene strata are also surficial deposits and are similar to Georgia and Alabama Pliocene deposits. The Miccosukee Formation (mentioned earlier in our discussion of the Georgia Miocene) is a fluvial surficial deposit of late Miocene and/ or Pliocene age. The Alachua Formation is composed of fluvial deposits as well as argillaceous residuum formed by carbonate dissolution. Sinkhole-fillings associated with the Alachua contain abundant terrestrial-vertebrate fossils. The Bone Valley Formation, a gravelly and marly surficial deposit in northcentral Florida, is correlative with the Alachua and also contains vertebrate fossils. It has commercially important phosphate deposits and may represent estuarine environments.

In south Florida, Pliocene deposits are not surficials but rather thin limestone strata of the upper Tamiami Formation and the Caloosahatchee Marl. The Caloosahatchee consists of clayey sand, marl, and molluskan shell beds; it is exposed in the Lake Okeechobee area and dips gently southward into the subsurface beneath Pleistocene carbonates.

Pliocene strata beneath Florida's eastern continental shelf are very thin and consist of quartzose sand in coastal areas and calcareous–argillaceous muds farther out. On the Blake Plateau, Pliocene strata are patchily distributed, being entirely absent in several large areas. Where they are present, Pliocene deposits of the Blake Plateau Basin are homogeneous, foraminifer–coccolith muds (Poag, 1978). On Florida's western shelf, Pliocene strata are thicker than on the eastern shelf but are also mainly carbonate muds, argillaceous in northern areas and virtually clay-free toward the south (Mitchum, 1978).

Gulf Coastal Province. Pliocene deposits exposed on the Gulf coastal plain are also surficial. On the other hand, relatively thick Pliocene strata are present beneath the outer continental margin south of eastern Texas, Louisiana, and Mississippi. No doubt, sediments to build this Pliocene sedimentary lens were carried to the site by the ancestral Mississippi River.

Pliocene surficial deposits of the northern Gulf region are termed the Citronelle Formation, which is principally a fluvial unit composed of disconformity-bounded, finingupward sequences. In its type area of southwestern Alabama, the Citronelle contains fossil plants and abundant vertebrate material. Otvos (1981) said that Citronelle sediments underlie upland areas between elevations of 9 to 16 m above mean sea level. It is interesting to note that the Citronelle lies atop a low-angle erosional surface but the dip of the Citronelle base is a lower angle than the dip of underlying strata. Thus, regional tilting must have occurred prior to Citronelle deposition. This is an important observation and we will return to it in our discussion of Pleistocene strata. While the Citronelle is mainly fluvial, there are coastal areas where Citronelle deposits are of lagoonal, estuarine, or paludal origin.

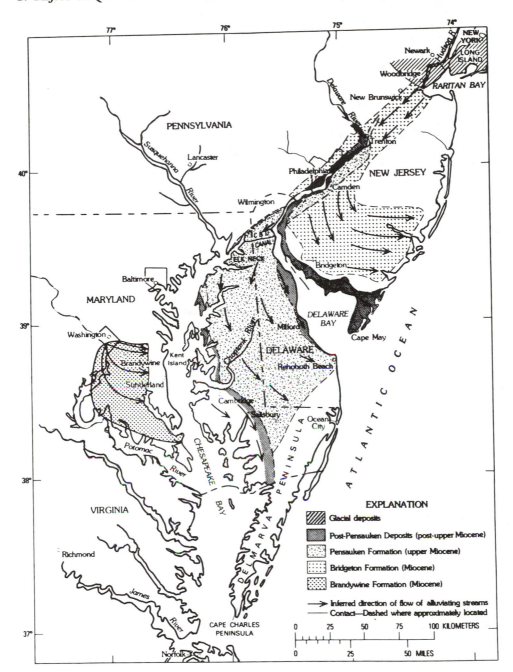

Fig. 9-25. Distribution and paleocurrent patterns of major surficial gravel sheets of the northern Atlantic Coastal Province. (From Owens and Denny, 1979.)

In eastern Texas, strata equivalent to the Citronelle are termed the Willis Sand. In southern Texas, however, Pliocene deposits comprise the Goliad Formation. The Goliad is similar to the Citronelle, also being composed mainly of fluvial fining-upward sequences. Basal Goliad gravels contain reworked Miocene vertebrates bones. Indigenous Pliocene vertebrate remains of the Goliad closely resemble those of the Ogallala Formation. The specific age relations

of Goliad and Willis strata are questionable but in the area just north of Texas's Colorado River, the Willis overlies the Goliad disconformably.

C.6.   Quaternary Deposits

Quaternary deposits of the Atlantic and Gulf Coastal Province are mostly unconsolidated or weakly consolidated detritus and may be divided roughly into three groups: (1) surficial deposits associated with fluvial and marine terraces; (2) recent stream alluvium; (3) marine deposits of coastal and continental shelf and slope areas. In addition, shallow-marine carbonates of Quaternary age are extensively exposed in southern Florida and throughout the Bahamas. Quaternary deposits are thin except in some continental shelf–slope areas and in the northern Gulf of Mexico, where large volumes of sediment were introduced by the Mississippi River. Although they are thin and extensively exposed, Quaternary strata have generated several major controversies, the most heated of which is the debate over the origin of low, sediment-veneered terraces which have been recognized on both Atlantic and Gulf coastal plains. These terraces have been ascribed to marine action during high (interglacial) stands of sea level, to marine action combined with tectonic uplift, and to fluvial action. In the following paragraphs, we will discuss this problem and describe modern stratigraphic reclassification of terrace-associated deposits. We will not discuss modern stream or shoreline deposits because they have received extensive study elsewhere.

Atlantic Coastal Province. Recognition of low terraces in coastal-plain areas goes back to the late 19th century (e.g., McGee, 1888; Shattuck, 1901) but elaboration of the concept awaited the work of C. W. Cooke in the early 1930s. Cooke (1930, 1931) recognized seven terraces in North Carolina and Virginia (called, from upper to lower, Brandywine, Coharie, Sunderland, Wicomico, Penholoway, Talbot, and Pamlico; see Fig. 9-26); other terraces recognized in southern coastal-plain areas include the Silver Bluff and Princess Anne, below the Pamlico terrace, and the Okefenokee, between the Sunderland and Wicomico. Cooke also recognized higher-angle "scarps" separating the terraces. These scarps vary in their degree of development, with several being distinct, showing up as pronounced lineaments on aerial and even space photos (e.g., the Orangeburg and Coats scarps of South and North Carolina and the Surry and Suffolk scarps of North Carolina and Virginia) whereas others are indistinct at best, and may even be figments (Flint, 1940). Cooke believed that each terrace and associated scarp formed by marine erosion and/or deposition during high sea levels. He assumed abso-

lute crustal stability during formation of the terraces and proceeded to correlate terrace features over great distances based solely on their elevation above sea level.

Subsequent work has cast considerable doubt on many of Cooke's original ideas. One problem, for example, concerns the ages of terrace deposits. Cooke considered them all to be of Pleistocene age and he held that sea-level fluctuations recorded by the terraces were due to glacio-eustasy. More recently, however, several terrace-associated deposits have been reassigned to the Pliocene or even Miocene (e.g., Beacon Hill, Pensauken, and Bridgeton formations, among others). Of greater significance is the fact that sediments associated with upper terraces in Virginia, Maryland, and New Jersey have been shown to be of fluvial, rather than littoral, origin (Hack, 1955; Schlee, 1957). Most damaging to the terrace concept, however, is the work of Winker and Howard (1977) who demonstrated that coastal areas have indeed been affected by tectonic deformation, albeit slight. Winker and Howard showed that scarps and relict beach-ridges from the Cape Fear River in North Carolina to south-central Florida are naturally grouped into three shoreline sequences (see Fig. 9-27). They termed these, from oldest to youngest, the Trail Ridge sequence (associated with Orangeburg and Coats scarps), the Effingham sequence (corresponding to the Surry scarp), and the Chatham sequence (corresponding to the Suffolk scarp). They then showed that elevations of these relict shorelines vary regularly along their trends, indicating slight differential uplift of the coast. Cronin (1981) also demonstrated coastal uplift and showed that the rate of uplift ranged from 1–3 cm/1000 yr. to a maximum of 5–10 cm/1000 yr. This uplift, Cronin suggested may be due to lithospheric flexure in response to sediment (and hydrological) loading of the continental shelf. Clearly, the fact of differential uplift renders inoperative any correlations based on elevation above sea level. For this reason, Winker and Howard (1977) recommended that Cooke's terrace names be abandoned except in those localities where they were originally defined. It may be noted, nevertheless, that eustatic sea-level fluctuations were probably the primary control on shoreline location and that tectonic factors were of secondary importance (Cronin, 1981).

We can illustrate the kind of modern sedimentolog-

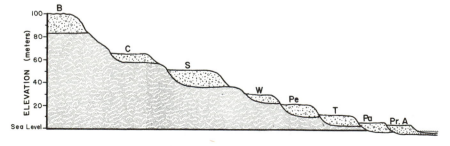

Fig. 9-26. The major terraces designated by Cooke (1931) in North Carolina. Symbols are as follows: B, Brandywine; C, Coharie; S, Sunderland; W, Wicomico; Pe, Penholoway; T, Talbot; Pa, Pamlico; Pr.A, Princess Anne.

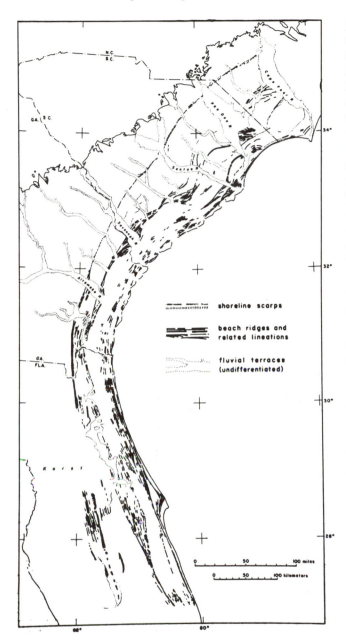

Fig. 9-27. Major surficial geomorphic features of the southern Atlantic Coastal Plain and their genetic interpretation. (From Winker and Howard, 1977, *Geology*, Fig. 1, p. 124.)

shoreline scarps

beach ridges and related lineations

fluvial terraces (undifferentiated)

ical/stratigraphic analysis currently being applied to Quaternary deposits of the Atlantic Coastal Province by considering briefly the work of J. P. Owens and colleagues (e.g., Owens and Denny, 1979; Owens and Minard, 1979). Pleistocene surficial deposits in the Delaware River valley comprise the so-called "Trenton gravels"; Owens and Minard subdivided these informally into the Spring Lake beds and overlying Van Sciver Lake beds. Both of these units were thought by Owen and Minard to have been de-

posited in environments generally similar to those of the modern Delaware River. For example, gravelly deposits of Sangamon-age Spring Lake beds north of Trenton, New Jersey, are considered to be of fluvial origin, sandy Spring Lake deposits in the Trenton area are probably of deltaic origin, and clayey Spring Lake silts in the Philadelphia area were probably deposited in a broad estuary like Delaware Bay. Correlative with these fluvial–estuarine deposits are marine strata of the Omar, Ironshire, and Kent Island formations, all of the central Delmarva Peninsula. The Omar, which is correlative with Spring Lake beds, is a sandy, barrier-related deposit which grades down-dip into open-shelf deposits of the Cape May Formation. Similar environments are recorded by the Ironshire Formation, which is correlative with Van Sciver Lake beds (late Sangamon age). The Ironshire, too, grades down-dip into the Cape May.

East of the Delmarva Peninsular, in the Baltimore Canyon Trough, Quaternary sediments are relatively thick and record similar environments as do subaerially exposed deposits. Poag (1978) said that inner-shelf deposits of Pleistocene age are up to 120 m of unconsolidated gravel, sand, and mud which represent marginal-marine and fluvial environments, obviously deposited during lower (glacial) stands of sea level. On the outer continental shelf, over 170 m of muddy sand and clay records deeper-water environments.

Although more will be said about them later in this chapter, we should mention in passing that Pleistocene deposits also comprise the main mass of Long Island, New York; and Cape Cod, Massachusetts. In both of these cases, sediments were deposited as terminal and ground moraines of the great Laurentide continental glacier.

Florida and the Bahamas. In northern Florida, Quaternary sediments are primarily detrital surficial deposits associated with relict shoreline features similar to those discussed above (e.g., Trail Ridge; see Fig. 9-27). In southern Florida, however, widespread Pleistocene limestones are exposed. As shown in Fig. 9-28, the surface of mainland south Florida is underlain by the Miami Limestone, composed of two facies. In their western areas, Miami deposits are entirely bioclastic sands associated with very abundant bryozoan colonies. Hoffmeister et al. (1968) interpreted these deposits, termed the bryozoan facies, to have accumulated in the warm, clear waters of a shallow, shelf lagoon similar to the shelf lagoon of the central Grand Bahamas Banks today. To the east, bryozoan-facies rocks pass upward into an oolitic calc-arenite termed the oolitic facies of the Miami Limestone. These deposits are strongly crossbedded and very well sorted; they probably represent current-washed, shelf-edge, oolitic shoals such as those which today border the Bahamian shelf lagoon on the west. A third carbonate lithology, termed the Key Largo Limestone, underlies the northern Florida Keys. The Key Largo

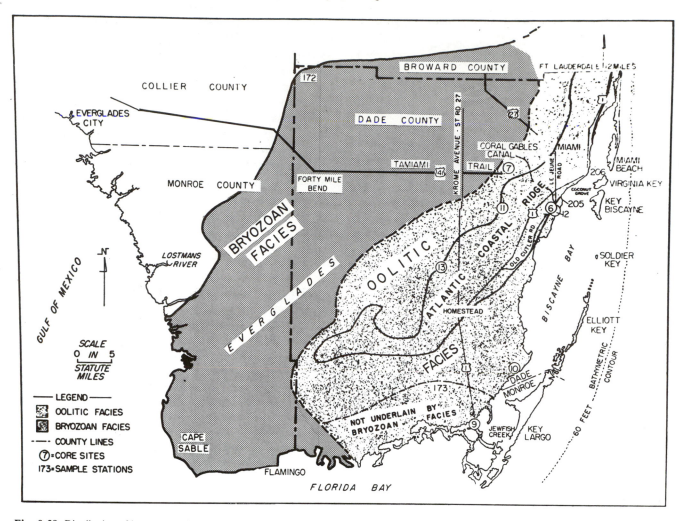

Fig. 9-28. Distribution of bryozoan and oolitic facies of the Miami Limestone of mainland Florida. (From Hoffmeister *et al.,* 1967.)

is composed almost exclusively of coral-reef deposits which must have developed along the outer edge of the Florida carbonate platform. The organisms whose fossilized remains compose the Key Largo are virtually identical to those which compose the living reefs east of the Florida Keys. In the southern Keys, Key Largo reef-rocks are overlapped by the oolitic facies of the Miami Limestone. Thus, it may be seen that oolitic sediments spread out with time, growing both westward into the bryozoan-facies lagoon and eastward over the Key Largo reef-complex. This expansion of the oolitic sheet was probably due to shoaling of the carbonate shelf during later Pleistocene time.

Gulf Coastal Province. Perhaps the major difference between Quaternary deposits of the Gulf versus the Atlantic Coastal Province is the profound influence of the Mississippi River system, which acquired an essentially modern aspect during the Miocene. During the Pleistocene, the vast Laurentide ice sheet provided enormous quantities of de-

trital sediment; this sediment was washed by meltwater streams from the glaciers, forming extensive, braided, outwash plains. In the absence of significant plant cover, glacially ground silt (so-called rock flour) was easily picked up by prevailing winds and carried eastward away from the cold, barren plains to be deposited on the eastern side of the Mississippi Valley as extensive sheets of loess. Coarser sediments were carried in streams to the Mississippi River and thence to the sea.

Alluvial deposits of the Mississippi River valley comprise a complex of stream terraces and related features stretching from southern Illinois to the Louisiana coast. During periods of falling sea level, the Mississippi and its tributaries were strongly rejuvenated and cut deeply into their floodplains, resulting in stream terraces along the margins of their valleys. Later, during periods of rising sea level, streams rapidly aggraded their floodplains. After four glacial epochs and (counting the Recent) four interglacials, the result of Mississippi River dynamics has been to create a series of stream terraces like giant stair-steps down to the current Mississippi floodplain (see Fig. 9-29). The upper terraces, called the Williana and Bentley by Fisk (1944), are relatively indistinct and their sediments poorly preserved. The Montgomery terrace (probably of Yarmouth age) is much better developed than earlier ones as is the Prairie terrace below it (which is probably of Sangamon age). Near the coast, terrace deposits pass gradationally into marginal-marine and then fully marine deposits, thickening progressively in the down dip direction. At the mouth of the Mississippi River, a great delta-complex has developed whose history has been characterized by shifts of the locus of major stream activity. Consequently, the greater Mississippi delta is composed of many individual subdeltas which were active for a time and then abandoned as fluvial activity shifted. Naturally, very large quantities of sediments were carried down the prodelta slope and out into the depths of the Gulf of Mexico. The result was the Mississippi fan (see Fig. 8-24), a huge mass of sediments up to 3 km thick and covering an area only slightly smaller than New England (Moore *et al.*, 1978).

Prominent scarps are also recognized on the northern Gulf coastal plain and their interpretation is as controversial as are those of the Atlantic region. Some of these features are clearly of marine origin but others are ambiguous and have been ascribed to faulting (cf. Willetts *et al.*, 1980, with Otvos, 1981). Given both the nature of Pleistocene glacio-eustasy and the fact that faulting is a well-known facet of Gulf-area stratigraphic development, our armchair deduction is that both processes were probably at work during deposition of Quaternary sedimentary sequences.

In coastal parts of Texas, Quaternary strata comprise a marginal-marine assemblage dominated by deltaic deposits but also containing barrier and back-barrier deposits as well. Because Texas rivers probably did not drain glaciated areas, they did not supply to the coast anywhere near the volume of detritus as did the Mississippi. Thus, cuspate

and/or lobate deltas, rather than bird's-foot deltas developed, and bed-load sediments were reworked from distributary-mouth bars to form continuous sheets of sand in front of the delta (instead of bar-finger sand bodies) and then redistributed into barrier islands. Correlation of Texas Quaternary strata with Mississippi valley terrace formations has proved difficult because depositional systems in the two areas were reacting to different rates of sediment supply. Nevertheless, a few correlations have been proposed. The Lissie Formation of the Houston area is considered to be correlative with Bentley Terrace and Montgomery Terrace formations and the Beaumont Formation equates with the Prairie Terrace Formation. It is interesting to notice that the slower rate of delivery of sediment to the Texas coast allowed wave and current activity to hold most sediments along the shore so that submarine fan complexes, such as the Mississippi fan, did not develop. Recent deposits of both Texas and Louisiana areas show that processes operating during the Pleistocene continue, somewhat modified, to operate today.

D. Tejas Sequence in the Cordillera

D.1. Overview

When last we considered the Cordilleran region of North America, we described the inception of an Andean-type-arc orogen along the western margin of the continent and showed how such a plate-tectonic setting could explain formation of Franciscan and Great Valley strata, intrusion of batholithic complexes such as the Sierra Nevada batholith, and events of the Sevier orogeny. We also described the effects of collisional accretion of displaced terranes, such as Wrangellia, into the Northern Cordillera. By the end of the Cretaceous, a trench and subduction zone lay just west of the North American continental margin from Mexico to Alaska.

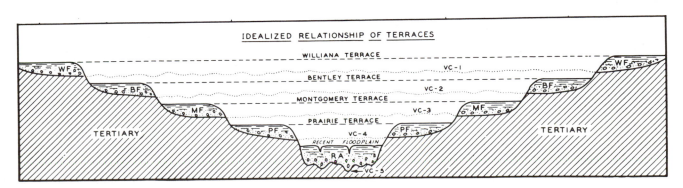

Fig. 9-29. Major Mississippi valley stream terraces and associated alluvial deposits as described by Fisk (1944). WF, Williana Terrace Formation; BF, Bentley Terrace Formation; MF, Montgomery Terrace Formation; PF, Prairie Terrace Formation; RA, Recent alluvium.

Beginning during the latest Cretaceous and running into the Eocene, a major orogenic event, termed the Laramide orogeny, affected the area of the present Rocky Mountains in both the United States and Canada. Whereas orogenic events in Canada were similar to earlier events (e.g., the Columbian orogeny), the Laramide orogeny in the United States was characterized by large-scale basement uplifts, major folding and faulting, intrusion of large granitic plutons, and minor volcanism. Laramide events are enigmatic both because of their location *within* the North American plate (rather than at the plate margin) and because of the somewhat unusual nature of the events themselves, e.g., fragmentation and uplift of the basement and associated folding of supracrustal strata.

During the Laramide orogeny in the Rocky Mountain region, continental-margin sedimentation continued in western and central California. Youngest Franciscan rocks are of Early Tertiary age. In the area of the Salinian block, a continental borderland terrane developed, with rapidly rising, positive crustal elements and adjacent subsiding basins. All along the Early Tertiary California coast, continental shelf sediments accumulated in the region just west of the Sierra Nevada terrane while continental slope and abyssal plain sediments were deposited in the region of the modern coast. In Washington and Oregon, also, subduction of a western oceanic plate beneath the continental margin dominated tectonic developments. The major effects of this subduction were arc magmatism and the accretion of oceanic blocks to the continental margin. In Alaska, too, convergent-plate tectonics continued to dominate geological activity on the continental margin. Of special interest is the subduction of an active spreading ridge (the Kula ridge) beneath the southern Alaskan margin at the Aleutian trench. Effects of this event were striking and dominated southern Alaska tectonics from Eocene to Miocene time.

During the Miocene, Cordilleran tectonics underwent a profound change. Just prior to that time, the modern San Andreas fault had begun its activity and crustal extension in the Basin and Range Province, which also persists today, occurred penecontemporaneously with inception of the San Andreas fault. Related events include extensive outpourings of basaltic lava, in the northwestern United States, forming the Columbia River basalt plateau. Similar basalts were also extruded in the southern Canadian Cordillera. Major basaltic volcanism during Pliocene time resulted in the Snake River Plain volcanics of southeastern Oregon and southwestern Idaho. Rejuvenation of the Sierra Nevada region led to the initiation of major stream-drainage patterns in the southwestern United States; these patterns, with some modification, persist today. Andesitic volcanism characterized western Oregon and Washington, suggesting continuation of subduction off the western continental margin; indeed, subduction and arc magmatism continue there today, as

evidenced by the spectacular eruptions of Mount St. Helens in 1980, 1981, and 1982.

It is clear from the foregoing discussion that Cenozoic tectonic events in the Cordillera were both numerous and complex. In this section, we will describe these events in detail and will attempt to integrate them into a coherent paleotectonic–paleogeographic picture using the logic of plate tectonics. Paleogeographic reconstructions of the Cordillera, especially for the Early Tertiary, are difficult because of major Cenozoic tectonic modifications of the original continental-margin area. These events "shuffled" disparate tectonic elements together into a confusing polyglot of terranes. Chief among these modifications are: (1) major lateral movements along wrench faults, such as the San Andreas fault, along which has occurred over 300 km of dextral slip (Nilsen and McKee, 1979); (2) large-scale rotations of tectonic blocks, e.g., the Coast Ranges of Washington and Oregon where paleomagnetic data indicate a clockwise rotation of 50–70° (Simpson and Cox, 1977); (3) accretion of displaced terranes, such as oceanic blocks in the Coast Range and Olympic Mountains of Oregon and Washington (Cady, 1975); and (4) east–west extension in the Basin and Range, which Hamilton and Myers (1966) estimated to have exceeded 100 km.

D.2. Early Cenozoic Paleogeography

D.2.a. Paleocene Events

The Paleocene record of the North American Cordillera is sparse and in many areas (especially the northwestern United States) appears to be absent (Nilsen and McKee, 1979). Thickest Paleocene strata occur in western California where they record a variety of continental-margin environments as well as deeper-water environments farther west. Thick sections of Paleocene terrestrial sediments occur in the Rocky Mountain area, originally deposited in rapidly subsiding, intermontane basins associated with Laramide uplifts. The central part of the Cordilleran orogen (referred to as the "Ancestral Cordillera" by Nilsen and McKee, 1979) was primarily emergent during the Paleocene and very little strata of that age occur in western Canada or in the area from northern Idaho and adjacent Washington down to the Peninsula Ranges in southern California and Mexico. Figure 9-30 shows the major features of western United States paleogeography during the Paleocene.

California.   In California, Paleocene sediments were deposited in several different paleogeographic situations. Youngest rocks of the Franciscan assemblage's coastal belt in northwestern California (see Fig. 9-31) are of Paleocene and Eocene age (Evitt and Pierce, 1975), suggesting that activity along the western subduction zone had continued in

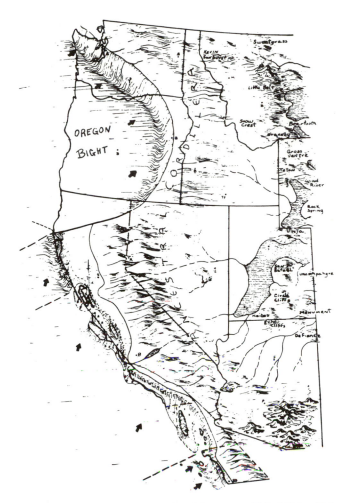

Fig. 9-30. Paleocene paleogeography of the western United States. Note the lack of volcanic activity west of the Laramide orogen. Compare the size of the Oregon Bight shown here with that shown in Fig. 9-33. (From Nilsen and McKee, 1979, reproduced with permission of the Pacific Section, Society of Economic Paleontologists and Mineralogists.)

Ancestral Cordillera during this time; see discussion of Laramide tectonics in following paragraphs). Nevertheless, the Sierra Nevada and areas farther east continued to be emergent, supplying feldspathic sediments toward the sea. In the Great Valley area, between the Franciscan trench and the emergent Sierra Nevada orogen, marine and nonmarine deposits accumulated on a relatively stable continental shelf. Terrestrial environments in the east passed westward into the marine environment, which gradually deepened farther to the west (see Fig. 9-30). The Great Valley area is divided into the Sacramento Basin (to the north) and the San Joachin Basin (to the south) by the Stockton Arch, an east–west uplift which was emergent during the Paleocene.

In westcentral and southwestern California, however, an entirely new tectonic situation had begun to develop during latest Cretaceous time which had profound effects on

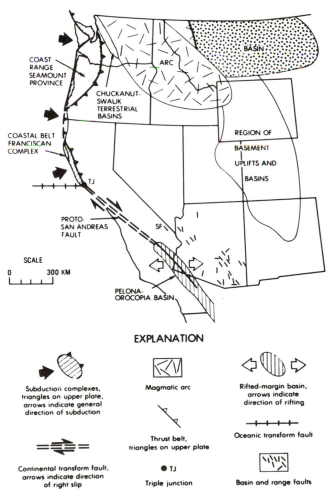

EXPLANATION

Subduction complexes, triangles on upper plate, arrows indicate general direction of subduction

Magmatic arc

Rifted-margin basin, arrows indicate direction of rifting

Continental transform fault, arrows indicate direction of right slip

Thrust belt, triangles on upper plate

Oceanic transform fault

● TJ
Triple junction

Basin and range faults

Fig. 9-31. Nilsen's (1978) speculative paleogeographic map of the Late Cretaceous Cordillera showing the proto-San Andreas fault; see text for further discussion. (Reproduced with permission of the Pacific Section, Society of Economic Paleontologists and Mineralogists.)

that area since the Late Cretaceous. Lower Tertiary Franciscan rocks, like Upper Cretaceous rocks associated with them, contain arc-derived andesitic detritus deposited in submarine fans and trench-slope basins. The reader may recall that Creataceous Franciscan rocks contain little arc-derived sediment because the outer arc (formed by uplift of the subduction complex) blocked dispersal of arc sediment from the fore-arc basin to the trench (see Fig. 8-42). The existence of arc sediments in Lower Tertiary Franciscan strata implies that the outer arc no longer functioned as a sediment barrier. Ingersoll (1978a) said that by the Paleocene, the fore-arc basin had been almost completely filled, and arc-derived sediments bypassed the fore-arc area, ending up in the trench (see Fig. 9-32). Magmatic activity in the Sierra Nevada region had ceased by the Paleocene (a reflection of the general absence of magmatic activity in the

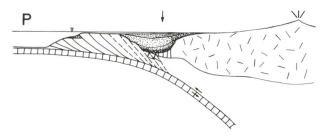

Fig. 9-32. Schematic cross section of the Paleocene fore-arc basin of western California; note that the Great Valley Sequence nearly fills the fore-arc basin and the Franciscan Subduction Complex forms a nearly flat-topped outer arc. (From Ingersoll, 1978a, reproduced with permission of the Pacific Section, Society of Economic Paleontologists and Mineralogists.)

subsequent continental-margin history. Large-scale dextral offset (as much as 200 km) was postulated by Nilsen (1978) to have occurred along a zone of wrench faults termed the *proto-San Andreas fault* (see Fig. 9-31). First proposed by Suppe (1970), the proto-San Andreas fault is poorly understood but may have been a continental transform fault (similar in general orientation, location, and sense of offset to the modern San Andreas fault). If it was in fact a transform fault, then it may have formed in response to complex plate interactions during the latest Cretaceous, in a similar fashion as was the modern San Andreas fault during the Miocene. An alternate possibility is that the proto-San Andreas fault may have been an intraplate fracture located *within* the fore-arc area and caused by the dextral-shear component of stresses associated with oblique subduction; this interpretation was favored by Dickinson (1979) and by Graham (1979).

But if the mechanism of proto-San Andreas faulting is unclear, its major effects are better understood. In addition to the large-scale dextral offset mentioned above, the other major effect was the development of a continental-borderland terrane within the Salinian block and in the area of the modern Transverse Ranges. A continental borderland is characterized by fragmentation of the continental margin into a series of relatively small, highly unstable crustal blocks whose relative vertical movements result in a series of rapidly subsiding basins and adjacent uplifts. The modern California continental margin is a well-known example (see Fig. 9-58). The cause of borderland fragmentation during the Paleocene is considered to have been horizontally directed shear stress within the crust due to lateral offset on the proto-San Andreas fault (Nilsen and McKee, 1979). Behaving in a brittle manner, the crust broke up into a series of large and small blocks which were jostled against each other as shearing continued, causing the vertical movements which typify a continental borderland. In some borderland-type basins of the Salinian block, deep-water environments

are inferred, whereas in adjacent basins only shallow-water deposits are found. The history of a specific basin may be very complicated, not only because of the diversity of basins but also because basinal activity was typically not constant with time; rather, subsidence rates varied considerably within a given basin. Indeed, in some instances, a formerly subsident basin may have subsequently become an uplift. In the Transverse Ranges, two major basins (the Santa Ynez on the east and the Sierra Madre on the west) were separated by a positive area called the San Raphael high, on which shallow-water algal limestone accumulated during the Paleocene. Figure 9-30 shows the San Raphael high as an isolated offshore block. This is because the two basins apparently merged and shoaled toward the north. Eastern parts of the Santa Ynez Basin during the Paleocene received primarily nonmarine and shallow-marine sediments probably deposited on a narrow, stable shelf (Nilsen and McKee, 1979). In most other parts of both basins, deep-water conditions prevailed and primarily turbidity-current sediments accumulated. Of arkosic composition, these sediments were probably derived from granitic, Sierra Nevada terranes to the north and east.

Washington and Oregon. In northernmost California, Oregon, and Washington, the general absence of Paleocene strata allows considerable latitude in paleogeographic reconstructions. Nilsen and McKee (1979) hypothesized that the whole area at that time was underlain by oceanic crust, possibly part of the Farallon plate (see p. 588 and Fig. 9-53), which subducted beneath North America along an arcuate trench convex toward the continental interior (see Fig. 9-30). Paleocene volcanic materials in western Montana lend support to this idea. Dickinson (1979) believed that the hypothetical Oregon–Washington trench was continuous with the Franciscan trench to the south (see Fig. 9-33) but Nilsen and McKee speculated the presence of a trench–trench transform fault between the two features (Fig. 9-30). In the northern Cascades, Paleocene fluvial deposits consisting of coal-bearing, conglomeratic sandstones outcrop in scattered areas around Bellingham, Washington. These deposits are termed the Chuckanut Formation (Misch, 1966) and were considered by Dickinson (1976) to represent a terrestrial fore-arc basin.

Canada. Paleocene strata are also sparse throughout much of the Canadian Cordillera because most of that area was emergent and undergoing erosion. A major exception was the nonmarine Sustut basin (see p. 476), which continued to receive detrital sediments from peripheral uplands. Paleocene strata are also rare in coastal areas of the Northern Cordillera. On the southeastern side of the Georgia basin (see p. 478), nonmarine deposits of the lower Burrard Formation may be Paleocene in age but there are no other apparently correlative strata on Vancouver Island. On the southern Queen Charlotte Islands, however, several fac-

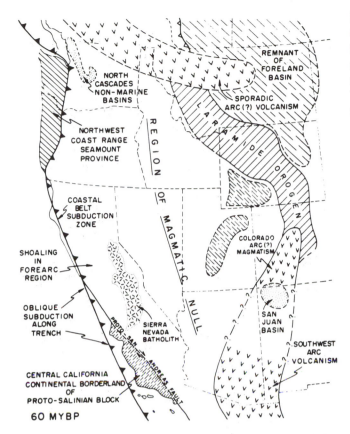

Fig. 9-33. Dickinson's (1979) paleotectonic map of the Cordillera during the Paleocene; contrast this interpretation of the proto-San Andreas fault with Nilsen's as seen in Fig. 9-31. (Reproduced with permission of the Pacific Section, Society of Economic Paleontologists and Mineralogists.)

late Mesozoic arc and arc-related rocks of the Yakutat Group which represents a displaced terrane (probably the Chugach terrane; see Fig. 8-40) accreted into the Cordilleran collage during the latest Cretaceous. It is possible that Cretaceous–Tertiary plutonic rocks intruding the Yakutat formed due to inception of subduction beneath the Chugach arc after accretion of the arc to the continent.

Farther west, in the area northeast of Cook Inlet, Kirschner and Lyon (1973) reported that terrestrial to estuarine sediments accumulated during the Paleocene in a successor basin which developed following accretion of the Chugach terrane. Called the Chickaloon Formation, this unit received lithic and volcanic detritus from the Chugach terrane to the south and arkosic sediments from plutonic terranes (probably the Alaska Range–Talkeetna Mountains plutonic belt; see Fig. 8-53) to the north. The reader may recall that the Upper Cretaceous Matanuska Formation also contains both granitic detritus from the north and arc-derived sediments from the south. Paleocene deposits of the Chickaloon probably represent a continuation of this same sediment-dispersal pattern following arc accretion. Chickaloon deposition continued into the Eocene.

To the north and west of the Cook Inlet Basin, large-scale plutonic activity commenced during latest Cretaceous time. This episode of magmatism, recorded by the Alaska Range–Talkeetna Mountains plutonic belt (Hudson, 1979; see Fig. 8-53), persisted until approximately 50 Myr. ago (middle Eocene) and was, therefore, concurrent with magmatism in both southeastern Alaska and the Aleutian Islands (see below). The composition of Alaska Range–Talkeetna Mountains plutons ranges from diorite to granite but granodiorite and quartz monzonite predominate.

The Bering Sea west of Alaska was the site of several interesting tectonic events. Before discussing these, however, a brief tour of Bering Sea geography is in order (Fig. 9-34). The Bering Sea lies between the Alaskan and Siberian mainlands to the northeast and northwest, respectively, and, on the south, the Aleutian Islands. The Aleutians are an active volcanic island-arc related to northwest-directed oblique subduction of the Pacific plate. The deep, Bering Sea Basin is divided into three smaller basins by two prominent bathymetric features. Bowers Ridge extends northward from the central Aleutian chain, curving strongly westward and isolating the Bowers Basin from the larger Aleutian Basin to the north and east. The Komandorsky Basin lies on the western side of the Bering Sea and is separated from the Aleutian Basin by the Shirshov Ridge. Cooper et al. (1976) studied these two ridges and concluded that they were formed by constructional volcanism (e.g., due to arc magmatism) rather than by simple uplift of the basaltic ocean-crust. The Bowers Ridge area was carefully described by Cooper et al. (1981) who demonstrated that it represents an extinct magmatic arc and subduction complex. The Bering

ies of the Massett Formation are of Paleocene age; Massett rocks are primarily volcanic; they consist of pyroclastic breccias with both rhyolitic and basaltic clasts, rhyolitic ash flows, tuffs, and dacitic to basaltic lava flows. These volcanic rocks are coeval with intrusive rocks of the Coast Plutonic Complex, whose activity migrated gradually eastward throughout the Paleocene (Dickinson, 1976). This magmatism seems to have been continuous with spasmodic igneous activity in Washington, Idaho, and Montana (see Fig. 9-33). This magmatic belt has been termed the Pacific Northwest arc (Dickinson, 1979) and probably represented the continuation, after the Cretaceous, of oceanic-plate consumption beneath the continental margin.

Alaska. In southern Alaska, rocks of Paleocene age are more common than in Canada. Alaskan Paleocene rocks record a magmatic arc which possibly was continuous with the Pacific Northwest arc. In the panhandle region of southeastern Alaska, Hudson et al. (1977) described a series of Late Cretaceous to Early Tertiary intrusions which they believed represent a major, regional plutonic event. These plutons, mostly of tonalitic composition, are intruded into

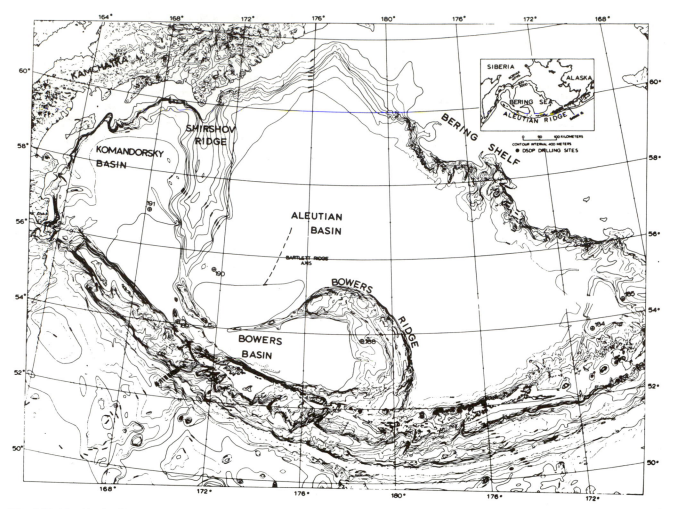

Fig. 9-34. The Aleutian Islands and the Bering Sea Basin. (From Cooper *et al.*, 1976, *Geological Society of America Bulletin*, Fig. 1, p. 1120.)

Shelf, which borders the Aleutian Basin to the north and east, is underlain by strongly deformed Mesozoic arc-related rocks and is believed to have been the site of plate consumption prior to latest Cretaceous time (see Fig. 9-35). The Aleutian Archipelago has probably been the site of plate consumption since the latest Cretaceous, which is the age of the older rocks found in the Aleutians (Marlow *et al.*, 1973). At the intersection of the Bering Shelf and the Aleutian arc is a broad submarine plateau called the Umnak Plateau.

During Late Cretaceous time, the Pacific Ocean Basin is believed to have been composed of at least four major lithospheric plates as shown in Fig. 9-35. Details of these plates' configurations and the exact orientations of their boundaries are speculative but the Farallon plate is believed to have subducted beneath the Cordilleran margin of North America south of Alaska while the Kula plate subducted beneath the Alaskan and Siberian continental margins. The

Pacific plate had no effect on North America until much later; the Phoenix plate had no direct effect on North America at all.

The following developmental model of the Bering Sea Basin and Aleutian island-arc is taken primarily from Cooper *et al.* (1976) and Ben-Avraham and Cooper (1980). The crust of the Bering Sea Basin is believed to have been part of the Kula plate as it subducted beneath the Bering Shelf during the Late Cretaceous (see Fig. 9-35). At some time near the end of the Cretaceous, subduction jumped southward to the approximate position of the modern Aleutian trench, thus preserving the unsubducted Bering Sea crust as a "fossil" oceanic plate. The cause of the jump-shift of subduction to the Aleutian area may have been collision of the Umnak plateau with the Bering Shelf. Such a collision could have "clogged" the old subduction zone, causing plate consumption to begin on the other south side of the colliding block. There are two ways in which this

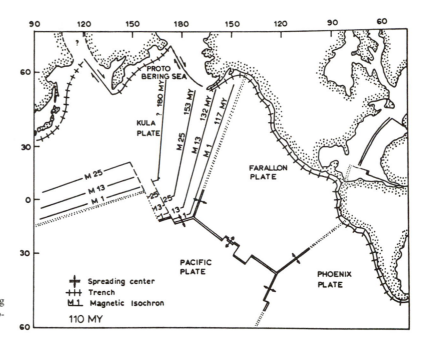

Fig. 9-35. Reconstruction of Pacific-basin plates during the Cretaceous. (Modified from Cooper *et al.*, 1976, *Geological Society of America Bulletin*, Fig. 3d, p. 1122.)

might have occurred. Figure 9-36 illustrates the first model, in which it is presumed that the Umnak plateau was an oceanic plateau (similar to western Pacific plateaus today) riding on the Kula plate as it moved generally northward. Bowers Ridge is considered in this model to have been a magmatic arc separated from the Kula plate by a subduction zone. With collision of the Umnak plateau and the Bering Shelf, subduction jumped south of Bowers Ridge and Umnak plateau as suggested above. Shirshov Ridge in this scheme is thought to have been formed by volcanic activity associated with a hot spot and localized by a major north-trending transform fault (similar to the Ninety-East Ridge in the Indian Ocean today).

In the alternate model (see Fig. 9-37) Shirshov Ridge, Bowers Ridge, and Umnak plateau are all considered to have been parts of a once-continuous magmatic arc developed just east of the Kamchatka Peninsula. Transform faulting (associated with a spreading ridge separating the Kula plate from a smaller plate to the west) fragmented the Shirshov–Bowers–Umnak arc into three blocks. These blocks moved differentially away from the Asian mainland, possibly due to back-arc spreading. When the Umnak plateau collided with the Bering Shelf, the new phase of plate consumption was localized along the transform fault just south of the Umnak block. At the present time, there is insufficient evidence to choose between these two models. Nevertheless, the effects of Late Cretaceous–Early Tertiary plate tectonics in the North Pacific are reasonably clear; these effects include inception of subduction in the Aleutian area and isolation of the Bering Sea Basin. The new Aleutian trench-arc system is believed to have been continuous

with the one just south of the Alaskan mainland and, therefore, continuous with the rest of the Pacific Northwest arc.

Ancestral Cordillera. The preceding paragraphs have dealt primarily with the paleogeography of continental-margin areas of the Cordillera. But mention must also be made of the Ancestral Cordillera and regions even farther east. The Ancestral Cordillera, as defined by Nilsen and McKee (1979), refers to the mountainous terrane, remnant of the Cretaceous Andean-type arc and Sevier fold-thrust belt, which stretched from the Peninsula Ranges in the south through the Sierra Nevada, up to eastern Oregon, western Idaho, and northeastern Washington, and on into the Columbian orogen in Canada (see Fig. 9-30). In this area, Paleocene strata are few and far between, being primarily poorly dated, fluvial and alluvial deposits at the bottoms of Eocene sections. For the most part, these deposits were formed in local, intermontane basins within the Peninsula Ranges and Sierra Nevada. This near-absence of sedimentary deposits of Paleocene age is, perhaps, understandable considering that the Ancestral Cordillera was, at that time, a mountainous area in which erosion (not deposition) was dominant. Sediments derived from within the Ancestral Cordillera were carried westward in drainage systems which fed the coastal areas of California. Of greater significance, however, is the apparent absence of Paleocene igneous rocks throughout virtually the entire western Cordillera of the United States (see Fig. 9-33). This has been termed the Paleocene magmatic null (Dickinson, 1976). The cause of this magmatic null is intimately related to tectonics of the Laramide orogeny; thus, we will defer further discussion until later in this section.

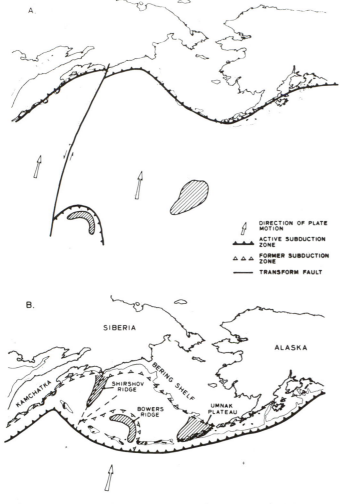

mention here that thick Paleocene (and Eocene) sections accumulated in rapidly subsiding basins associated with adjacent, rising, uplifts. Sediments to fill these basins were derived from both the adjacent, uplifted blocks and the Ancestral Cordillera to the west (see Fig. 9-30). Major Laramide basins include the Bighorn, Powder River, Wind River, Green River, Uinta, Denver, North Park, Raton, and San Juan (see Fig. 9-7; for further discussion of stratigraphic relations within these basins, see Section B).

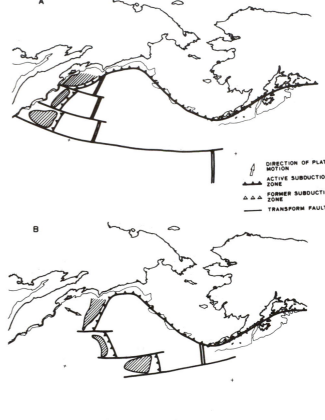

Fig. 9-36. Ben-Avraham and Cooper's first tectonic model of the origin of the Aleutian Ridge. (A) Late Mesozoic time; (B) early Tertiary time. (From Ben-Avraham and Cooper, 1980, *Geological Society of America Bulletin,* Fig. 9, p. 492.)

Of course, detrital sediments were also carried eastward out of the Ancestral Cordillera. In the foreland basin associated with the Canadian Cordillera, for example, Paleocene sediments formed the Belly River–Paskapoo Assemblage (Eisbacher *et al.,* 1974), a major molasse wedge which thins and fines eastwards away from the mountainous sourceland. In the foreland area of the United States Cordillera, however, a major (and somewhat unusual) tectonic development, which began during the latest Cretacous and persisted through most of the Eocene, disrupted "normal" foreland-basin sedimentation. This was the Laramide orogeny, which was responsible for major vertical movements of crustal blocks in the area of the present Rocky Mountains. We will discuss Laramide tectonism and its structural effects in the next few paragraphs but we should

Fig. 9-37. Ben-Avraham and Cooper's second tectonic model of the origin of the Aleutian Ridge. (A) Late Mesozoic time; (B) early Tertiary time. (From Ben-Avraham and Cooper, 1980, *Geological Society of America Bulletin,* Fig. 10, p. 493.)

D.2.b. The Laramide Orogeny

Throughout most of Paleozoic and Mesozoic time, the site of major, Cordilleran tectonic activity had shifted episodically westward. But even early tectonic activity, such as the Antler orogeny (Late Devonian–Mississippian), had occurred to the *west* of the modern Rocky Mountain region. That area had been the site of rather normal inner-continental-shelf sedimentation during much of the early and middle Paleozoic and was part of the Cordilleran foreland basin during most of the Mesozoic. During Mississippian and Pennsylvanian times, western areas of the Rocky Mountain region had been affected by intense subsidence resulting in enormously thick sedimentary basins, such as the Oquirrh Basin in Utah. This system of late Paleozoic basins and uplifts is termed the "Ancestral Rockies." But no major, compressive, tectonic activity had affected that area since some time during the Precambrian. Thus, the Rocky Mountain area could easily be considered to have occupied an intraplate location during much of Phanerozoic time. It is therefore surprising that this area was, during the latest Cretaceous and early Tertiary, the site of large-scale tectonic activity (see Fig. 9-31) complete with major structural deformation, uplift, and intrusive igneous activity. Exactly how these events, which define the Laramide orogeny, came to affect an area almost 1500 km away from the nearest plate boundary has been a major problem of North American Cordilleran geology. The problem, however, is even more complex because the nature of Laramide deformation itself is unusual, being considerably different from the styles of deformation commonly associated with either Andean-type or collisional-type orogens. Thus, the Laramide orogeny at first glance appears to resist interpretations based on plate-margin tectonics; indeed, it has been cited by some opponents of plate tectonics as a failure of plate theory. Within the past 10 years, however, a new approach to the "Laramide problem" has had some success in explaining not only the siting and nature of Laramide tectonics but also related events such as the Paleocene magmatic null mentioned earlier in this section. In the following paragraphs, we will describe major Laramide effects and present this recent speculative model for Laramide tectonics.

It should be noted at the outset that the term "Laramide orogeny" is sometimes misapplied to virtually any Cordilleran tectonic event which occurred during the interval from the latest Cretaceous to the Oligocene. For example, thrust faulting in the Columbian orogen of Canada during the early Tertiary represents only a continuation of the same style of tectogenesis which characterized the Cretaceous; nonetheless, it is referred to as the Laramide orogeny (Douglas *et al.*, 1976). The term has also been applied to coeval events in Alaska. In the following discussion, however, we will use the term "Laramide orogeny" in its "classical" sense, i.e., to refer to large-scale vertical movements of crustal blocks and related structural effects in the central Rocky Mountain region of the United States (see Fig. 9-33).

i. Laramide Effects. Probably the most characteristic feature of Laramide deformation was the fragmentation of the crust into fault-bounded, basement-cored blocks which underwent rapid and spectacular vertical movements. We have already mentioned the major Laramide basins; associated uplifted blocks include the Sweetgrass, Kevin Sunburst, Little Belt, and Beartooth uplifts of Montana; the Gros Ventre, Teton, Wind River, and Laramie uplifts of Wyoming; the Black Hills of South Dakota; the Uinta, San Rafael, Circle Cliffs, and Monument uplifts of Utah; the Front Range, Sierra Madre, Sawatch, White River, and Uncompahgre uplifts of Colorado; the Kaibab, Echo Cliffs, and Defiance uplifts of Arizona; and the Sangre de Cristo uplifts of New Mexico (see Figs. 9-7 and 9-30). Figure 9-38A is Matthews and Work's (1978) sketch of basement-block configuration in part of the Front Range uplift. Notice that bounding faults are very high angle to nearly vertical. A similar pattern was illustrated by Palmquist (1978) for uplifts in the Big Horn Mountains of Wyoming (Fig. 9-38B). Bounding faults are primarily of the normal, or gravity, type but some are high-angle reverse faults. The dominance of high-angle faulting indicates that the greatest principal stress in these areas during Laramide time was oriented approximately vertically. This is clearly different from the dominantly horizontal, compressive stresses encountered by plate-margin orogens.

An interesting corollary of Laramide uplifts is the style of deformation affecting supracrustal strata blanketing Laramide crustal blocks. As the blocks rose and subsided, supracrustal strata above the blocks were folded in much the same way that covers on a bed are folded when pushed up from underneath. The resultant style of deformation has been termed "drape folding" (Stearns, 1971) because it resembles folds of cloth "draped" over the basement blocks (see Fig. 9-39). The basement blocks, because they provided the vertical "push and pull" for folding, are termed forcing blocks. Notice in Fig. 9-40 that smaller, subsidiary faults splay off the bounding fault. Notice also that many of these subsidiary faults are of the reverse type even though the bounding fault of the forcing block is of the gravity type. Notice finally that several subsidiary faults die out before reaching the surface. This is because the mechanics of drape folding requires considerable internal slip along bedding planes; near the surface, the weight of overburden is relatively slight and resistance to bedding-plane slip is correspondingly low. But with increasing depth, stress normal to bedding increases so that bedding-slip becomes more difficult. Consequently, strata at depth cannot experience the

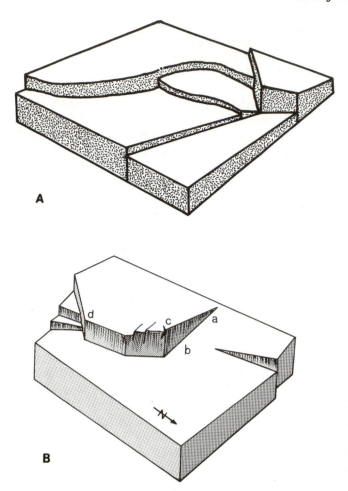

Fig. 9-38. Hypothetical reconstruction of basement blocks in the Laramide orogen. (A) The northern Front Range (redrawn and modified from Matthews and Work, 1978, Geological Society of America Memoir 151, Fig. 12g, p. 115); (B) the Big Horn Mountains of Wyoming (from Palmquist, 1978, Geological Society of America Memoir 151, Fig. 5, p. 132).

Igneous activity accompanied Laramide deformation. Although Laramide plutons are similar in composition to magmatic-arc-type plutons, they are generally smaller and do not coalesce to form great batholithic complexes such as the Jurassic–Cretaceous Sierra Nevada batholith of eastern California. A significant exception is the granitic Boulder batholith of western Montana which is over 100 km long by 30 km across at the surface. Timing of Boulder-batholith emplacement is constrained by the fact that it intrudes Upper Cretaceous lavas and fossiliferous tuffs and is nonconformably overlain by fossiliferous middle Eocene strata; radiometric dating puts its emplacement at about 71–82 Myr. ago, i.e., only slightly younger than the Upper Cretaceous volcanics it intrudes (King, 1977). Other Laramide plutons in Montana include stocks, laccoliths, and dikes of Eocene age in the Crazy Mountains, Little Belt Mountains, Sweetgrass Hills, and other areas (King and Beikman, 1978). In Colorado, Laramide magmatism is recorded in the Colorado Mineral Belt, which stretches from the Front Ranges to the San Juan Mountains. Andesitic volcanism characterized parts of this area during the Paleocene and Eocene, especially the San Juan volcanic field which lies over the Uncompahgre Uplift. The correspondence of Laramide magmatism with the area of basement uplifts, which had not experienced previous magmatic activity since the Precambrian, is particularly significant when one recalls that the Paleocene and Eocene were times of general magmatic quiescence throughout virtually the entire United States Cordillera west of the Laramide orogen.

same amount of bedding-slip as strata near the surface and must, therefore, fracture when pushed up from underneath (Stearns, 1978). Other factors which influence the relationship between folding and faulting include how strongly adjacent layers are welded to the forcing block and the physical properties of the supracrustal blanket (e.g., thin versus thick bedded, brittle versus ductile). Before leaving this subject, it is instructive to compare the drape style of folding (Fig. 9-40) with the rootless style of folding, in supracrustal strata affected by thin-skinned tectonics (e.g., Fig. 8-45). Here again, the difference between Laramide tectonics and the more common plate-margin tectonics is clearly apparent. [Readers who wish to pursue the subject of Laramide deformation should consult Matthews's (1978) symposium, from which much of the preceding information was taken.]

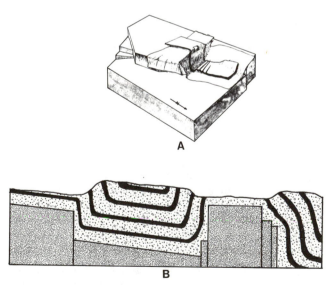

Fig. 9-39. Illustrations of drape folding. (A) Diagrammatic sketch showing "draping" of strata over basement blocks; (B) north–south structural cross section of the Bellview Dome in the northern Front Range. (A, from Palmquist, 1978, Geol. Soc. Amer. Mem. 151, Fig. 6, p. 132; B, redrawn from Matthews and Work, 1978, Geol. Soc. Amer. Mem. 151, Fig. 12d, p. 115.)

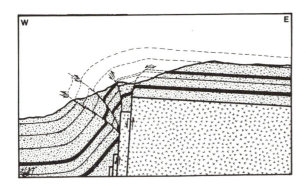

Fig. 9-40. Stratigraphic cross section across Rattlesnake Mountain in Wyoming. (Redrawn and modified from Stearns, 1971, Guidebook for 23rd Annual Field Conference, Wyoming Geological Association, Fig. 6, p. 131.)

A final observation is necessary before turning to interpretation of Laramide tectonics: whereas large-scale magmatism ceased throughout much of the United States during the Paleocene and Eocene, essentially normal magmatic activity continued virtually unabated north of the Laramide orogen in Canada and Alaska (the Pacific Northwest arc) and south of it in Mexico (the Southwest arc) as seen in Fig. 9-33. It is significant that thin-skinned thrusting also continued in both these areas, usually just east of the main arc region. From these observations, we may summarize that the common pattern of oceanic-plate subduction (and concomitant arc magmatism, infrastructure mobilization, and foreland-basin thrusting) continued in the Canadian and Mexican segments of the Cordillera from the Cretaceous on through the Eocene with no apparent interruption. Models of Laramide tectonics must, therefore, explain not only why events in the United States Cordillera were unusual but also why normal Andean-type orogenesis continued both northwest and south of the Laramide orogen.

ii. **Interpretation of Laramide Tectonics.** Most previous attempts to explain Laramide tectonics concentrated on the cause of crustal fragmentation and uplift. Proposed mechanisms have included uplift due to thrust-faulting (Fanshawe, 1939), simple vertical uplift (Eardley, 1951), wrench faulting (Stone, 1969), and fold-thrust uplift of the basement (Sales, 1968). In their discussion of Laramide tectonics, Burchfiel and Davis (1975) stated that crustal shortening in the Laramide area may have been as much as 100 km and that the only model of the four mentioned above which is compatible with this amount of shortening is that of Sales (1968). In Sales's model, strong compressive stress in the deep crust caused folding and thrusting and, consequently, led to the development of a thickening, tectonic "welt" in lower crustal levels. Thus, the vertical stresses which caused Laramide tectonics were due to the

upward "push" as the deep, tectonic "welt" grew and thickened. This model seems somewhat contrived but it does dovetail nicely with plate-tectonics speculations discussed in following paragraphs.

Sales's model only addresses the origin of the vertical stresses; it does not explain why those stresses were manifested in the Rocky Mountain region. Perhaps the most obvious explanation for siting of the Laramide orogeny is that basement structure was responsible. We have mentioned that much of the Laramide orogen occupies the same area as the "Ancestral Rockies" system of basins and uplifts which was subjected to major vertical movements during the late Paleozoic. Indeed, some Laramide features, e.g., the Uncompahgre and Front Range uplifts in Colorado, were actually reactivated structural elements originally formed during the late Paleozoic. Thus, there is reason to believe that some basement-control of Laramide siting was significant. But only some. Laramide areas to the north of Colorado were not part of the Ancestral Rockies system and in those areas, Laramide structures are strongly discordant with basement structures. This led Houston (1971), speaking of the Wyoming area, to state ". . . basement anistropy is not responsible for the major mountain structures developed in the Laramide" (p. 25). So, we are still left with the question of what did control localization of Laramide tectonics. Burchfiel and Davis (1975) argued that the major control was increased crustal ductility due to high rates of heat flow accompanying Laramide magmatism. The development of Sales's (1968) tectonic "welt" would clearly have been facilitated in that part of the crust already thermally weakened. On the other hand, as mentioned earlier, Laramide magmatism was seemingly rather minor; thus, one may argue that insufficient heat would have been available for the ductility model. Burchfiel and Davis reacted to this point by suggesting that the bulk of Laramide magmas never rose high enough in the crust to be exposed at the present-day surface. Although that is obviously a speculation, there is a clear congruence of Laramide uplifts with magmatic activity and that congruence suggests some genetic relationship, such as that postulated by Burchfiel and Davis.

Again, however, acceptance of Burchfiel and Davis's (1975) ideas leaves us with a question: namely, what was the cause of magmatism in the Rocky Mountain region? The answer to this question leads us to recent plate-tectonics models of the Laramide orogeny. The reader will recall that arc-related magmatism swept eastwards from the Sierra Nevada region during the Cretaceous (see Chapter 8). One may view the onset of magmatism in the Rocky Mountain region as merely a continuation of that eastward-migration trend. By the same token, the Paleocene magmatic null may be considered a result of continued widening of the arc-trench gap (see Fig. 8-43). Viewed in this way, Laramide

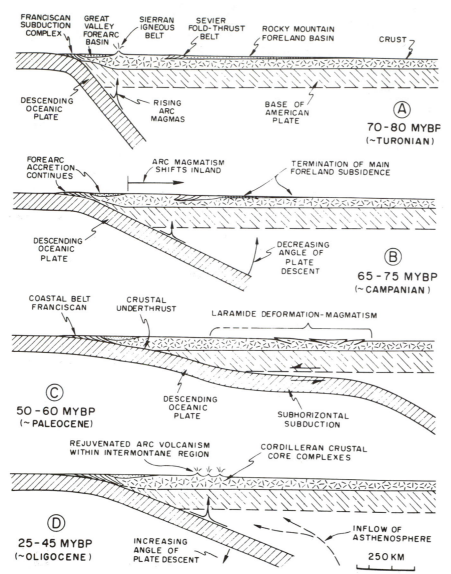

Fig. 9-41. Dickinson's plate-tectonic model of the Laramide orogeny; note the change of subduction dip-angle from relatively steep during the Cretaceous to subhorizontal during the Paleocene; note also that arc-type magmatism is suppressed by the Paleocene low subduction angle. (From Dickinson, 1979, reproduced with permission of the Pacific Section, Society of Economic Paleontologists and Mineralogists.)

magmatic activity may be seen as part of a continuum of arc magmatism reaching back to the Late Jurassic.

But all of this is no answer. *Why* did the arc-trench gap widen? That is, why did arc magmatism sweep eastwards? The emerging answer to those questions is that the dip angle of the subducting plate (the Farallon plate; see Fig. 9-35) may have become progressively more shallow with time until, by the Paleocene, it was less than 10°. Figure 9-41 shows Dickinson's (1979) model of this process. Note that subduction during the Turonian was hypothesized to have been relatively steep and arc magmatism was restricted essentially to the Sierra Nevada area; as subduction dip-angle decreased. the locus of arc magmatism moved eastwards (as did deformation in the Sevier fold-thrust belt). By Paleocene time, the subducting slab is believed to have dipped at such a low angle that it sheared against the base of the

continental lithosphere, thus generating the lower crustal compression called for by Sales's (1968) model of Laramide tectonics. Further, the actual site at which the subducting plate finally did plunge steeply down into the asthenosphere (in other words, the site of magma generation within the downgoing slab) was *beneath the Rocky Mountain area*. Notice that little or no magma would have been generated within the subhorizontal portion of the slab, thus accounting for the magmatic null west of the Laramide orogen. Magma rising into the crust beneath the Rocky Mountain area resulted in increased crustal ductility, which explains localization of the deep, tectonic "welt" and the congruence of basement uplifts and igneous activity.

Originally developed by Lipman *et al.* (1971), the idea of decreased subduction dip-angle during Laramide time has been advocated more recently by Burchfiel and Davis

(1975), Dickinson and Snyder (1978), and Dickinson (1979). Others have also used the concept to good advantage: Henyey and Lee (1976) used the low-angle subduction model to explain heat-flow history in the United States Cordillera; and Keith (1978) chronicled the change of dip angle from the Late Cretaceous to the later Tertiary by using the geochemical relationship between igneous-magma composition and depth of magma generation (see Fig. 9-42).

Difference in tectonics northwest and south of the Laramide orogen may have been due to transform faults which divided the Farallon plate into adjacent segments, which moved together even though they were decoupled from each other. Whereas the dip angle of the central, "Laramide segment" decreased with time, dips of segments both north and south of the "Laramide segment" remained steep so that normal Andean-type activity persisted. There are, however, no clear transform offsets at either end of the Laramide orogen although Sales (1968) did postulate a left-lateral couple north of the Lewis and Clark lineament in Montana, although this is a very controversial suggestion. The lack of apparent strike-slip faulting north and south of the Laramide area is disturbing but may be understood if one recalls that the North American lithosphere would have been a single, rigid plate; it was the Farallon plate that was segmented by faults. Thus, different tectonic conditions (e.g., in Canada versus Wyoming) should not, perhaps, be expected to be discordant across a knife-sharp line. This is because those conditions were being caused by differences in slab geometry 100 km or more beneath the thick North American plate. Rather, tectonic conditions in such a situation would probably change gradually from one style to the other across a broad and poorly defined zone. This is, in essence, what is seen at both ends of the Laramide orogen.

Of course, we are left with one fundamental question: why did the angle of subduction of the "Laramide segment" decrease so dramatically? One answer is that the rate of spreading of the North American plate increased during Late Cretaceous time and that North America consequently overrode the Farallon plate faster than it was being subducted into the asthenosphere. This is an attractive idea not only because it is intuitively pleasing but also because it attempts to relate evidence of spreading in the western Atlantic to Cordilleran tectonics. But, were this hypothesis strictly correct, then one must explain why Canadian and Mexican "segments" of the Farallon plate did not also experience decreased subduction dip angles. Another, and more promising idea lies in the work of Pilger (1981) on the Andean orogen of South America. In the Andes today are two areas of anomalously low-angle subduction (see Fig. 9-43). Notice in the figure that the observed distribution of earthquake foci is similar to the geometry of subduction envisioned by Dickinson (1979) for Laramide tectonics (cf. Fig. 9-41C with Fig. 9-43B). Pilger noted that both areas of low-angle subduction are adjacent to major aseismic ridges (the Nazca Ridge and the Juan Fernandez seamount chain) on the Nazca plate subducting beneath South America. In Pilger's view, the thickened crust of these ridges provides extra "buoyancy" to the subducting plate so that it does not sink as steeply into the asthenosphere as it would were the ridge not there. Application of this idea to Laramide tectonics seems a possibility: not only would subduction of an aseismic ridge on the Farallon plate explain the low subduction dip angle, but it would also explain differences in tectonic behavior along the various segments of the North American Cordillera *without* appeal to transform faults where there is no positive evidence of such.

Clearly, some of the ideas mentioned are virtual flights of wishful thinking. We caution the reader that much has yet to be learned about continental tectonics in general and the Laramide orogen in particular. Even the validity of application of plate-margin tectonics to continental interiors is

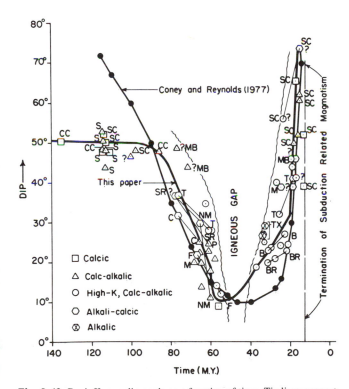

Fig. 9-42. Benioff zone dip-angle as a function of time. Tie lines connect different magmatic suites in the same geographic area. S = Salinian block, westcentral California; F = Front Range, Colorado; MB = Mojave block, southeastern California; SC = southern California batholith area; CC = central California batholith area; NM = southwestern New Mexico; BR = Black Range, southwestern New Mexico; B = Blue Range, eastcentral Arizona. Localities in southeastern Arizona: T = Tuscon area; C = Christmas area; SR = Santa Rita Mountains; P = Patagonia Mountains; M = Mammoth–Cooper Creek area; TX = western Texas. Question mark denotes questionable age or depth assignment. (From Keith, 1978, *Geology*, Fig. 3, p. 519.)

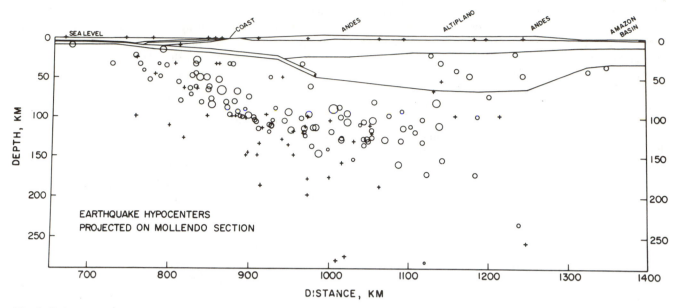

Fig. 9-43. Low-angle subduction in the modern Andean orogen. (A) Map of the Andean continental arc showing relation of low-angle subduction zones to the Nazca and Juan Fernandez Ridges (from Pilger, 1981, *Geological Society of America Bulletin*, Fig. 1, p. 448). (B) Earthquake hypo- centers defining the low-angle subduction zone beneath the modern Andean orogen in the general area of Mollendo, Peru; compare this illustration with Fig. 9-41C (from Couch *et al.,* 1981, *Geological Society of America Memoir 154,* p. 724, Fig. 15B).

open to debate. As Stearns said (1978, p. 32), "wholesale extrapolation of proved continental-edge systems into the interior of continents, with no geophysical justification, may be unwarranted."

D.2.c.   Eocene Paleogeography

In some respects, events of the Eocene Epoch in the Southern Cordillera were primarily a continuation of the Paleocene with the exception that a major marine transgression during the early Eocene flooded considerably more of the continental margin, leaving Eocene strata widely distributed (see Fig. 9-44). On the other hand, important tectonic differences characterized Eocene history of the Northern Cordillera, especially in Alaska, where the Kula spreading ridge began to interact with the Alaska–Aleutian plate-margin. In the Ancestral Cordillera, erosion had further lowered the old, Andean-type orogen, permitting deposition of nonmarine sediments in local, intermontane basins. At the same time, volcanism increased in intensity along the Washington–Idaho–Montana segment of the Pacific Northwest arc, whose locus migrated southward with time. And in the Central Rocky Mountains, Laramide tectonic activity continued in essentially the same manner as during the Paleocene.

California.   Eocene paleogeography of southern California in the area of the modern Peninsula Ranges dif- fered little from that of Paleocene times except that more of the narrow shelf was flooded by the Eocene transgression. Westward-flowing streams carried coarse, immature detritus from the southern Ancestral Cordillera to the shelf where it was redistributed by coastal and shelf currents. Based on their study of the La Jolla and Poway groups (lower and middle Eocene and upper Eocene, respectively) of the San Diego area, Lomar and Warme (1979) described the history of a portion of the southern California shelf (see Fig. 9-45). During earliest Eocene time, much of the area was subjected to erosion which beveled Cretaceous rocks; also during this time, a submarine canyon was carved into the edge of the shelf. As the sea flooded onto the shelf during the later early Eocene, a transgressive sheet-sand was deposited in a series of landward-migrating barrier-bars and back-barrier lagoons. At maximum transgression, marine currents reworked upper parts of the sand sheet into a sandy, shallow-marine shelf deposit. Sediments were funneled into the submarine canyon, finally filling it by the late middle Eocene. During the late Eocene, marine regression again exposed much of the shelf and rejuvenated the detrital source-terrane to the east. The result was a flood of detritus carried down the submarine canyon, which had been scoured out accompanying regression. On the shelf, conglomeratic debris accumulated in a major delta-system (the Stadium delta, Fig. 9-45D) while turbidity-current sediments, flushed down the canyon, built out the large Poway submarine fan (see Fig. 9-44).

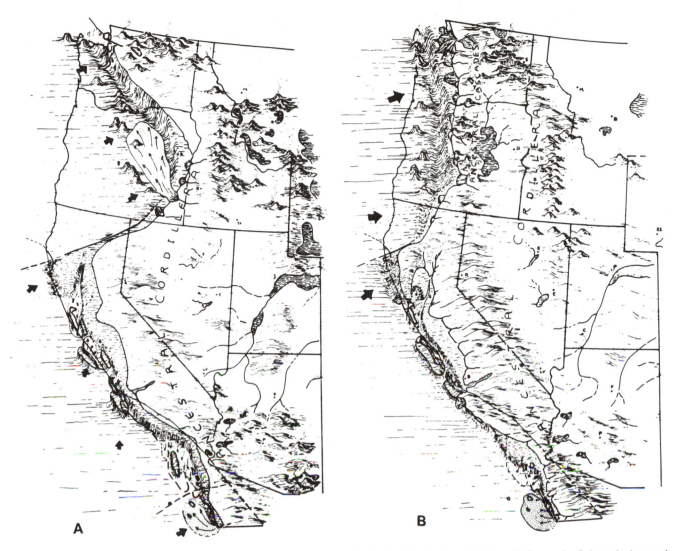

Fig. 9-44. (A) Early and middle Eocene paleogeography of the western United States. (B) Late Eocene paleogeography of the western United States. (Both from Nilsen and McKee, 1979, reproduced with permission of the Pacific Section, Society of Economic Paleontologists and Mineralogists.)

In the Sierra Madre and Santa Ynez basins, separated by the San Rafael high, deep-water deposition continued as sediment-laden turbidity currents rushed down the continental slopes. Farther north along the Salinian borderland terrane, other basins and uplifts were active, associated with continued movement along the proto-San Andreas fault. Figure 9-46 shows Graham and Berry's (1979) interpretation of central California paleogeography during the Eocene. Note the narrow shelf in the San Joachin Basin south of the Stockton Arch (which continued to be a positive feature). Nonmarine deposits accumulated in the Goler and Witnet basins adjacent to the Garlock fault. Rivers draining the Sierra Province carried immature detrital materials to the shelf where they were reworked and transported down several submarine canyons toward the deep ocean basin. At the bottoms of those submarine canyons, large turbidite-fan complexes developed in which flysch of the Cantua and Butano Sandstones, among others, was deposited. Shallow-water sediments are rare in the Salinian borderland of the early and middle Eocene but shallow-marine environments are recorded in the Gatchell Sandstone of the eastern San Joachin Basin. Graham and Berry hypothesized that Gatchell shelf-sands were the source of Cantua flysch, possibly carried by long shore drift to the mouth of the Cantua canyon and thence by turbidity currents to the depths. By late Eocene time, however, shallow environments were more common as regression and rejuvenation sent floods of detritus spilling westward onto the shelf and filling some of the smaller basins.

In the Sacramento Basin, north of the Stockton arch,

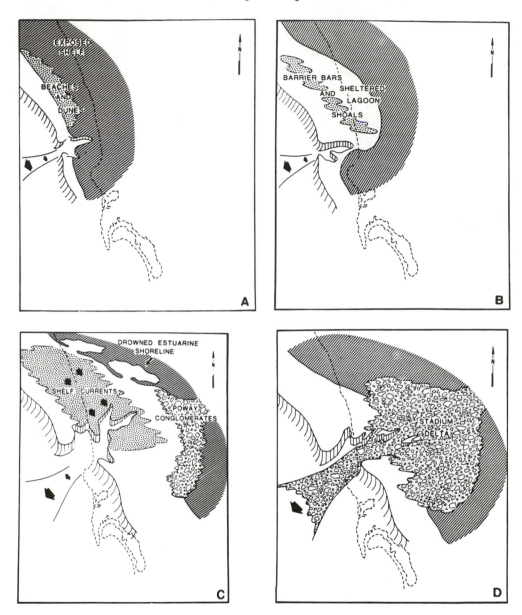

Fig. 9-45. Paleogeographic evolution of shelf sedimentary-environments near the San Diego area. (A) Early Eocene; (B) late early Eocene; (C) middle Eocene; (D) late middle to late Eocene; see text for further discus- sion. (From Lomar and Warme, 1979, reproduced with permission of the Pacific Section, Society of Economic Paleontologists and Mineralogists.)

early Eocene transgression resulted in a broad but very shal- low sea in which local, stagnant-bottom conditions devel- oped. Near the western side of the Sacramento Basin, a deep submarine canyon had developed during earliest Eo- cene time (possibly contemporaneous with erosion of the above-mentioned San Diego canyon). This so-called Mega- nos canyon had apparently filled with sediment by the late middle Eocene but renewed canyon-cutting concomitant with the late Eocene regression carved a new canyon, the Markley gorge, in the same general area (see Fig. 9-44). Turbidity currents washed down the Markley gorge to a

large submarine fan in the San Francisco Bay area. Upper Eocene strata are generally absent in northern California (Nilsen and McKee, 1979).

Washington and Oregon. It was during the late Eocene that tectonic events leading to the present-day geol- ogy of Washington and Oregon began to develop. Before discussing these events, however, it is necessary to tour the geography of modern Washington and Oregon and to de- scribe the nature of the Cordilleran orogen in that area prior to the middle Eocene. Figure 9-47 is a generalized geo- logical map of the U.S. Pacific Northwest. Of particular

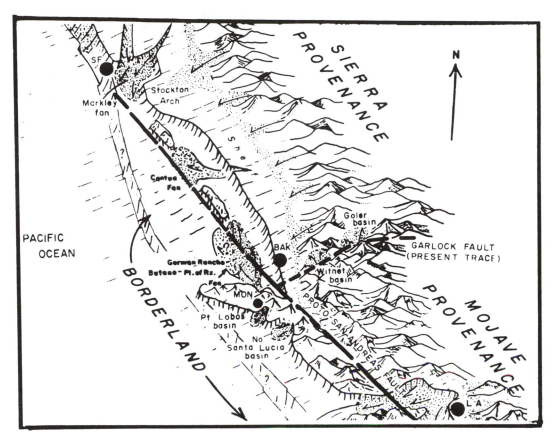

Fig. 9-46. Detailed paleogeographic map of the southern California borderland during the Eocene. (From Graham and Berry, 1979, reproduced with permission of the Pacific Section, Society of Economic Paleontologists and Mineralogists.)

importance in the following discussion is the Coast Range of extreme western Washington and Oregon (and including the Olympic Mountains of the Olympic Peninsula west of Puget Sound) and the Cascades Range, separated from the Coast Range by the long, discontinuous depression formed by the Puget Trough in Washington and the Willamette Valley in Oregon. The Cascades are a presently active Andean—type orogen starring some of North America's most famous volcanoes, e.g., Mt. St. Helens, Mt. Ranier, Mt. Baker, Lassen Peak, Mt. Adams, Mt. Hood, and Mt. Shasta. Miocene and younger lava-flows blanket over and obscure almost the entire pre-Miocene surface east of the Cascades except for areas such as the Blue and Ochoco Mountains which project up through the lavas.

Early Tertiary paleogeography of the Pacific Northwest is highly speculative, as we showed in the discussion of Paleocene events. Most authors today agree that an indentation of the continental margin (which we will term the Oregon Bight) characterized the area at that time, corresponding with the large concavity in the Ancestral Cordillera (see Fig. 9-44). However, disagreement exists over the size of the Oregon Bight (e.g., cf. Fig. 9-44 with 9-33).

A major point of contention is the Mesozoic–early Tertiary location of the Klamath Mountain terrane (of northern California) and the Coast Range block. Notice in Figure 9-44A that Nilsen and McKee show only ocean floor in the Klamath area of northern California; the Klamaths, in their opinion, constituted part of the Mesozoic subduction complex (correlative with the Sierran Foothills belt) lying as much as 500 km east of their present position (see Fig. 9-48A). The Coast Range block is believed to be composed of oceanic rocks (e.g., basaltic seamounts, pelagic sediments, and turbidites) which were accreted to the northeastern side of the Oregon Bight (as shown in Fig. 9-48B). Subsequently, the entire Coast Range block and adjacent Klamath terrane are believed to have been rotated into their present position. Evidence for this comes from paleomagnetism studies of volcanic rocks in the Coast Range which showed that their early Eocene paleopole positions are at angles as great as 50–70° to the expected Eocene field direction (obtained from Eocene rocks in other parts of the continent). This paleomagnetic discrepancy was first noted by Cox in 1957 and has since been confirmed by numerous other workers. In 1977, Simpson and Cox proposed that the

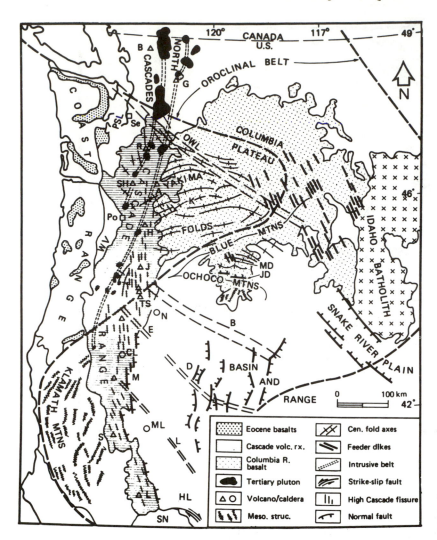

Fig. 9-47. Generalized geological map of the Pacific Northwest, showing major Cenozoic tectonic elements. *Cities:* Po = Portland; Se = Seattle. *Fault zones:* B = Brothers; D = Denio; E = Eugene; HL = Honey Lake; JD = John Day; K = Klickitat; L = Likely. *Landmarks:* MD = Monument dike swarm; PSL = Puget Sound lowland; SN = northern end of the Sierra Nevada; WV = Williamette Valley. *Volcanoes:* Z = Mount Adams; B = Mt. Baker; CL = Crater Lake; G = Glacier Peak; H = Mt. Hood; J = Mt. Jefferson; L = Lassen Peak; ML = Medicine Lake caldera; N = Newberry caldera; R = Mt. Rainier; S = Mt. Shasta; SH = Mt. St. Helens; TS = Three Sisters. (From Hammond, 1979, reproduced with permission of the Pacific Section, Society of Economic Paleontologists and Mineralogists.)

Coast Range block had rotated in a clockwise direction away from its original position, pivoting about a point within the Olympic Mountain area. The Klamaths, disposed at the end of the Coast Range block, were also transported westward, possibly shearing past the hypothetical transform fault shown by Nilsen and McKee to have formed the southeastern side of the Oregon Bight (Fig. 9-44). Having described pre-middle Eocene paleogeography of the Pacific Northwest, we can now discuss the Eocene geological history of that area. At the appropriate time, we will return to the subject of Coast Range block-rotation.

Subduction at the early Eocene trench along the northeast side of the Oregon Bight led to magmatism along the Pacific Northwest arc (called the Challis–Absaroka arc in Fig. 9-50). Fluvial strata, such as the Swauk Formation, continued to accumulate in the same fore-arc basin in the Northern Cascades in which the Paleocene Chuckanut For-

mation had been deposited. Along other segments of the fore-arc area, deltaic and other marginal-marine deposits accumulated. Buckovic (1979) described the development of deltaic–fluvial units within the Puget Group. Figure 9-49 illustrates his model; notice that extensive, subaerial, volcanic activity, represented by andesitic rocks of the Tukwila and Northcraft formations (later middle Eocene), occurred penecontemporaneously with sedimentation.

During the early Eocene, a large number of basaltic seamounts began to encounter the trench and to be thrust-intercalated together to form a massive pile of volcanic rocks ($> 250,000$ km^3; perhaps the largest volcanic mass of the Pacific Northwest; King and Beikman, 1978). As is typical of subduction-complex structure, the tops of these basaltic strata face eastwards but they become progressively younger toward the west (Cady, 1975). These rocks comprise the Roseburg, Umpqua, Siletz River, and Tillamook

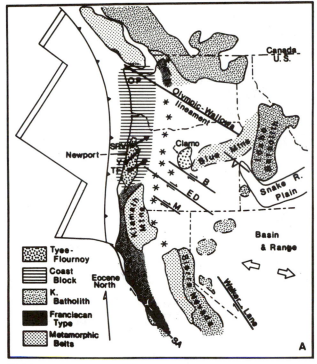

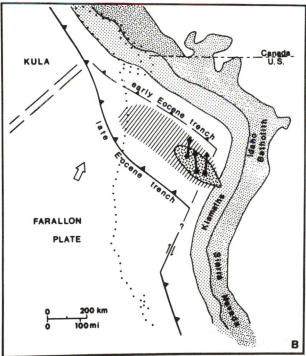

Fig. 9-48. (A) Generalized geological map of the northwestern United States showing orientation of the Siletz River volcanics (SRV), Yachats Basalt (Y), and Tyee and Flournoy Formations (TF). (B) Tectonic model to explain observed paleomagnetic orientations; note the jump of subduction from the early Eocene position adjacent to the Olympic–Wallowa lineament to the late Eocene position on the other side of the Coast Range block. (Both from Simpson and Cox, 1977, *Geology,* Figs. 1 and 5, pp. 585 and 588.)

Volcanics of Oregon, the Crescent Formation of Washington, and the Metchosin Volcanics of Vancouver. These rocks have been interpreted as oceanic-ridge tholeiites overlain by tholeiitic and alkalic basalts formed as seamounts and volcanic islands similar to the Hawaiian Islands (Cady, 1975). Simpson and Cox (1977) suggested that the seamount province may have formed over the hot spot now under the Yellowstone area of northwestern Wyoming. Associated with the volcanic strata are volcaniclastic rocks, pelagic micrites, and turbidity-current deposits (whose continentally derived sediment suggests that the seamounts did not form far from land). Early Eocene volcanics are succeeded in Oregon by thick and extensive turbidite deposits of the Tyee and Flournoy formations, which accumulated as a large submarine fan prograding northwestward from its Klamath Mountain source-area (see Fig. 9-44A). Farther north, Tyee-equivalent strata are thinner and composed primarily of pyroclastic debris.

At some time during the middle Eocene, subduction ceased along the early Eocene trench, possibly because the thick mass of Coast Range seamount rocks clogged up the subduction zone (Hammond, 1979). Subduction then jumped southwestwards to a new position on the other side of the Coast Range block (see Fig. 9-48B). One effect of this subduction jump was shift of arc magmatism from the Pacific Northeast (i.e., Challis–Absaroka) arc to the area just behind the Coast Range block. (This scenario may remind the reader of Triassic–Jurassic events in the Southern Cordillera.) The result was onset of western Cascades Range magmatism. In Oregon, much of the western Cascades is composed of the Little Butte Volcanic Series of upper Eocene to lower Miocene age. The Little Butte Series is a thick sequence of massive andesitic, dacitic, and rhyodacitic-flows, tuffs, and volcanic breccias (King and Beikman, 1978). In southern Washington, correlative rocks comprise the Ohanapecosh Formation which consists of > 4500 m of volcaniclastic strata and andesitic flows. Fluvial and deltaic sedimentation in the arc and fore-arc area continued apace with Cascades volcanism (see Fig. 9-44B). For example, the Tukwila and Northcraft formations of the Puget area (Fig. 9-49) were formed during this episode of volcanism. Plutonic activity accompanied Cascadian volcanism but was not, during Eocene time, very extensive. King and Beikman (1978) said that the earliest phase of intrusion of the granodioritic Chilliwack batholith in the Northern Cascades was probably Eocene in age.

Concurrently with the inception of volcanism in the Cascaes, clockwise rotation of the Coast Range-Klamath block began (see Fig. 9-50). Simpson and Cox (1977) hypothesized that back-arc spreading caused the Coast Range block to rift away from the continent along the *Olympic–Wallowa lineament* (OWL in Figs. 9-47 and 9-50), which is believed to mark a fundamental, crustal discontinuity sepa-

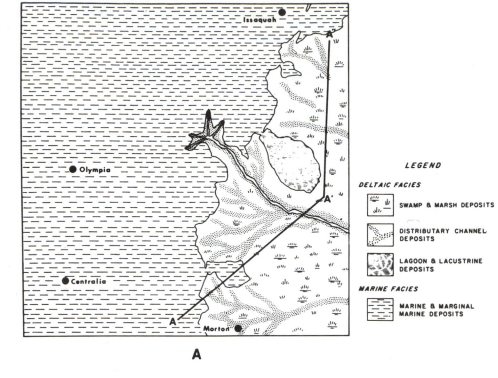

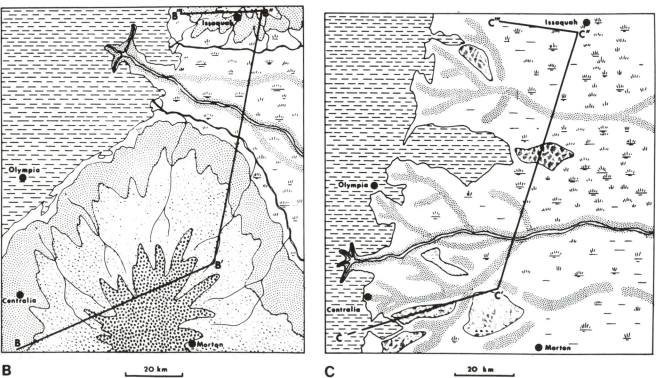

Fig. 9-49. Development of Eocene deltaic and volcanic deposits in west-central Washington (location of area shown in Fig. 9-44A). (A) Lower Puget sequence; (B) middle Puget sequence; (C) upper Puget sequence.

(From Buckovic, 1979, reproduced with permission of the Pacific Section, Society of Economic Paleontologists and Mineralogists.)

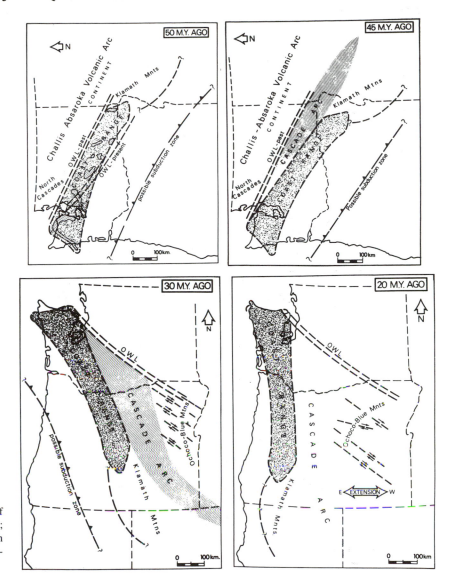

Fig. 9-50. Hammond's (1979) tectonic model of the Coast Range block and Cascades magmatic arc; see text for further discussion. (Reproduced with permission of the Pacific Section, Society of Economic Paleontologists and Mineralogists.)

rating continental crust to the northeast from more oceanic-like crust to the southwest (Skehan, 1965). Thus, the OWL may represent the line along which back-arc rifting began. Contributing to the rotation process may have been a change in rate and/or direction of subduction at the late Eocene trench. Simpson and Cox suggested several possible factors which may have influenced this presumed change; the most intriguing of these factors is the possible passage by the Oregon Bight area of the spreading ridge between the Kula and Farallon oceanic-plates (Atwater, 1970; see Fig. 9-53). As much as 50° of rotation must have occurred prior to the Oligocene, based on paleomagnetic study of Oligocene volcanics. It is possible, however, that rotation of the Coast Range block about the Olympic pivot was essentially ended by the close of the Eocene and that the remaining rotation

(about 20°) was caused by crustal extension in the Basin and Range during Miocene and later times (see Fig. 9-50).

Much of the area east of the Cascades is obscured by the Columbia River basalts so that tectonic details of the Eocene "back-arc" area are mostly speculative. Several recent wells penetrated the basalts, however, and showed the presence of nonmarine and deltaic sands (William Fritz, personal communication). The Blue and Ochoco Mountains are believed to have broken away from their original positions and dragged westward during Coast Range block-rotation. This motion was along major dextral transcurrent faults such as the Brothers, Eugene–Denio, and McLoughlin faults (Fig. 9-48A). The mechanics of this "breaking and dragging" process, however, remains elusive.

Ancestral Cordillera. In the mountainous terrane

east of the coastal areas discussed above, conditions during the Eocene had changed only a little since the Paleocene. Erosion dominated over much of this large area with the exception of a handful of small, scattered intermontane basins in which nonmarine sediments occurred. For example, in the region around Elko in northeast Nevada, fluvial and lacustrine sediments comprise the Elko Formation (Solomon *et al.*, 1979). The Elko is about 520 m thick and contains oil shale (of lacustrine origin) of possible commercial importance. The Elko Formation is considered by Solomon *et al.* to be correlative with the Uinta Formation of Utah. To the south, in eastcentral Nevada, the Sheep Pass Formation is similar to and possibly correlative with the Elko (Fouch, 1979). In western Montana, nonmarine strata such as the Sage Creek Formation (upper Eocene) contain a variety of mammal and plant remains which attest to a moderately warm, moist, and seasonal climate. Probably by the end of the Eocene, Ancestral Cordilleran mountains had been worn considerably down from their prior glory, by then possibly resembling the forested, rounded, wrinkled hills of the Blue Ridge in the Southern Appalachians or, perhaps, the Adirondack Mountains of upstate New York.

As discussed earlier, volcanic activity continued along the Pacific Northwest arc but had, by the middle Eocene, increased somewhat in intensity. The Challis Volcanics of Idaho, the Absaroka Volcanics of Wyoming and Montana, and the Clarno of central Oregon were erupted during the early and middle Eocene along the Pacific Northwest arc (hence the term Challis–Absaroka arc, mentioned previously). By the late Eocene, however, volcanic activity had migrated southwestward, perhaps in response to the jump-shift of subduction in the Oregon Bight or to a new regime of plate subduction ensuing after passage of the Kula–Farallon spreading ridge by the adjacent continental margin. Volcanism also continued in the Southwest arc area of Arizona and New Mexico. Throughout the rest of the Ancestral Cordillera, the magmatic null persisted.

The Laramide orogeny in the Central Rocky Mountain area also continued during the Eocene but by the end of the Eocene was beginning to wind down. We have elsewhere discussed Laramide-basin stratigraphy and Laramide tectonics. Thus, we move on.

Canada.   As during the Paleocene, the Eocene Canadian Cordillera was almost entirely emergent and undergoing erosion. Nonmarine deposition occurred in a few, small intermontane basins as well as within the long, broad valley between the Omineca Crystalline belt to the east and the Coast Plutonic belt to the west. This valley is called the Intermontane belt or the Columbian Zwischengebirge ("between the mountains"). During the Eocene, widespread, subaerial volcanic activity occurred in the Intermontane belt. In southern parts of the belt, near the Canada–United States border, Eocene volcanics and related sedimentary rocks comprise the Kamloops and Princeton Groups (among others), which have been dated both radiometrically (K–Ar: 49–45 Myr.; Douglas *et al.*, 1976) and paleontologically (plant fossils and mammal teeth) as middle Eocene. Volcanic rocks of these units are principally andesitic flows. In central parts of the Intermontane belt, rhyolitic flows and pyroclastics with subordinate andesites comprise the Ootsa Lake Group; in northern areas, coeval rocks of the Sloko Group are composed of dacitic and rhyolitic pyroclastics. Notice that this extensive area of volcanic activity, part of the Pacific Northwest arc, lies to the east of the Coast Plutonic belt, in which large-scale arc magmatism had occurred during the Late Cretaceous. Thus, arc magmatism had migrated eastwards with time. This is obviously similar to arc-migration trends previously noted in our discussions of Cordilleran tectonics.

Décollement-style thrusting and related deformation in the eastern Cordillera also continued into the Eocene, affecting areas farther to the east than during the Cretaceous. [This episode of deformation has been termed the Laramide orogeny by Canadian geologists (e.g., Douglas *et al.*, 1976); however, since we prefer to restrict that term to the Central U.S. Rocky Mountains, we will designate the Canadian event as the "Laramide" orogeny.] Regardless of its name, deformation during this event was considerable and many of the structural features of the eastern Cordillera in Canada were initiated (or attained their final form) at that time.

The "Laramide" orogeny in Canada is interpreted to have been an Andean-type orogenic event, a continuation of the Cretaceous Columbian orogeny. The oceanic (Kula?) plate, which had subducted during the middle Cretaceous beneath the newly accreted Wrangellia block, continued to plunge beneath the northwestern continental margin. The result was the now-familiar pattern of arc magmatism, infrastructure mobilization (represented in "Laramide" tectonics by regional metamorphism in the Omineca Crystalline belt), and thin-skinned deformation in the Rocky Mountain foreland thrust-belt. It is interesting to note that no Eocene strata have been reported from the foreland area, possibly because the whole area had been uplifted and subjected to erosion or because Eocene strata were removed by subsequent erosion.

Fore-arc sedimentation along the Canadian coast was very limited at this time, possibly because the rugged topography of the Insular belt dropped sharply westward into a deep and very narrow fore-arc basin. This basin, called the *Tofino Basin,* lay between the subduction complex just to the west (probably a continuation of the subduction complex comprising the Olympic Mountains) and the accreted Wrangellia block; there may have been virtually no continental shelf (Drummond, 1979). This interpretation is based upon analysis of the sedimentary petrology and stratigraphic relations of middle Eocene Lyre and upper Eocene Hesquiat and Escalente formations of the Tofino Basin. Deposits constituting these units principally record submarine-fan environments in which coarse, subaqueous-debris-flow sediments accumulated in proximal parts of the fan while turbidity-current deposition predominated in more distal areas (Cameron, 1979; Drummond, 1979; for an opposing view, see Jeletzky, 1975). Deposition of Hesquiat and Escalente sediments continued into the Oligocene. There are no other deposits of apparent Eocene age along the coast of western Canada.

Alaska. Subduction along the Aleutian trench had begun during latest Cretaceous to Paleocene time and, by the Eocene, arc volcanism was occurring along virtually the entire Aleutian archipelago. Formed at this time were volcanic and volcaniclastic units such as the Amchitka Formation of Amchitka Island, the Finger Bay Volcanics of Adak Island, and others, all of which have been grouped informally into the "early series" (DeLong et al., 1978). Possibly related to Aleutian volcanism was intrusion of numerous plutonic bodies in the Alaska Range–Talkeetna Mountains plutonic belt (Hudson, 1979; see Fig. 8-53). In the Cook Inlet area, deposition of the nonmarine Chickaloon Formation continued from Paleocene time.

During later Eocene time, a new tectonic development began to be manifested in both the Aleutian archipelago and along the southern Alaska coast. One indication of a change was decrease in the intensity of magmatic activity. DeLong et al. (1978) stated that the "early series" of the Aleutians is composed of two subdivisions: an earlier unit relatively rich in primary volcanic materials and a later, volcanics-poor unit. Volcanic activity in the Aleutians is believed to have ceased altogether about 45 Myr. ago. Similar events characterized the mainland where magmatic activity ceased in the Alaska Range–Talkeetna Mountains belt around 50 Myr. ago. Concurrent with shutdown of volcanic activity in the Aleutians, the archipelago was uplifted. This is indicated by a change in the nature of sedimentary rocks associated with "early series" volcanics: in lower volcanics-rich strata, sedimentary rocks are principally argillites, cherts, siltstones, and sandstones, all considered by DeLong et al. to be deep-water deposits; in overlying volcanics-poor rocks, the dominant sedimentary-rock type is conglomerate and sandstone of shallow-water origin. Uplift led, finally, to emergence of the entire archipelago. On all Aleutian Islands where it is present, the Eocene section is truncated by a major unconformity above which are Miocene or younger strata. In the Cook Inlet region of the mainland, deposition of the Chickaloon Formation ended with uplift of the whole area during the late Eocene (Kirschner and Lyon,

1973). Apparently accompanying uplift was widespread deformation and low-grade (greenschist) metamorphism. These affected both the Aleutian archipelago and the Alaskan mainland.

Thus, the magmatic and sedimentary history of the Aleutian–Alaskan magmatic arc was interrupted during the late Eocene; this interruption was accompanied by regional uplift, deformation, and low-grade metamorphism. All of these events are probably related to the slow but inevitable approach of the Kula Ridge (the spreading ridge between the Kula and Pacific plates) toward the Aleutian–Alaskan trench (DeLong et al., 1978). The actual arrival of the ridge at the trench was probably an Oligocene event.

D.2.d. Oligocene Paleogeography

The Oligocene Epoch lasted ~ 13.5 Myr (from ~ 38 to ~ 24.5 Myr.; Ness et al., 1980), i.e., less than either the Eocene before it (~ 17 Myr.) or the Miocene after it (~ 19.5 Myr). Further, the Oligocene was a time of major sea-level decline. (Indeed, as discussed in an earlier section, sea-level drop during the middle Oligocene is thought to have been the sharpest and most significant eustatic decline of the Cenozoic Era; sea level at that time was as low as 250 m below the present level; see Fig. 8-3). Thus, Oligocene deposits in the Cordillera are much less extensive than either Eocene or Miocene strata and, being primarily terrestrial, are commonly difficult to distinguish from younger terrestrial strata. Of course, good Oligocene sections do exist in the Cordillera, especially in Washington and Oregon where Cascades volcanism continued and shallow- to deep-marine environments existed in the fore-arc area. Nevertheless, there is a general paucity of Oligocene rocks in the Cordillera (see Fig. 9-51).

This paucity is unfortunate because the Oligocene was a time of fundamental change in the style of tectonics affecting western North America. Consider: (1) by the earliest Oligocene, Laramide deformation in the Central Rockies and "Laramide" events in Canada and Mexico had ceased; (2) a new phase of magmatic activity, recorded in the westward sweep of volcanism across the eastern Great Basin, began during the early Oligocene; (3) the Kula spreading-ridge was apparently subducted beneath the Aleutian–Alaskan trench; and (4) by the latest Oligocene, both the modern San Andreas fault in California and the Queen Charlotte fault north of Vancouver had begun their histories. The relationship of these events in time suggests a common (or related) cause but the details of this cause are still only poorly known.

California. In the Peninsula Ranges of southern California, only nonmarine deposits were formed during the Oligocene. Farther north, in the area of the Transverse

Fig. 9-51. Oligocene paleogeography of the western United States. (From Nilsen and McKee, 1979, reproduced with permission of the Pacific Section, Society of Economic Paleontologists and Mineralogists.)

and probably formed a cluster of low hills at the southern end of the Sacramento Basin. Almost entirely emergent itself, the Sacramento Basin was a broad, low valley acquiring stream-transported detritus from both the Sierran province to the east and the Franciscan terrane of the Coast Ranges to the west. Only in its southwestern area did the sea persist in the Sacramento Basin, deepening toward the Markley gorge, which continued to funnel sediments down to the submarine-fan complex in the San Francisco Bay area.

The San Andreas Fault. It was during the later Oligocene (~ 30 Myr. ago) that dextral offset began on the modern San Andreas fault (Atwater, 1970). The fault today is believed to be a complex boundary-transform fault connecting the divergent plate-boundary in the Gulf of California to the Gorda Ridge west of Oregon (see Fig. 9-52). Originally, these two ridge-segments were part of a single, continuous ridge separating the Pacific plate from the Farallon plate (see Fig. 9-53). But complex interactions as the North American plate encroached upon these oceanic plates led to the loss by subduction of most of the Farallon plate (except for several remaining fragments: the Juan de Fuca plate west of Washington and Oregon and the Cocos plate west of Central America; Fig. 9-52) and to the inception of the San Andreas fault. In the following paragraphs, we will describe the history of these events.

The reader will recall that, during the Late Cretaceous, three oceanic plates (the Kula, Farallon, and Pacific) comprised the Northern Pacific Basin. Whereas the configurations of the Kula Ridge (between the Kula and Pacific plates) and the Kula–Farallon Ridge are mainly speculative, that of the Farallon Ridge (between the Farallon and Pacific plates) is well constrained by magnetic-anomaly patterns and fracture zones of the Pacific plate (Fig. 9-52) and by the fact that segments of that ridge still exist today. From these data, we may reconstruct the shape of the Farallon Ridge with some accuracy to have had a number of major transform affects, especially major ones corresponding to the Mendocino and Murray Fracture Zones (see Fig. 9-52). During the early Tertiary, the North American plate gradually encroached on the Pacific-basin plates as shown in Fig. 9-53, with the Farallon plate subducting beneath the Southern Cordillera and the Kula plate beneath the Northern Cordillera. The relative motions of the North American, Farallon, and Kula plates were such that the position of the Kula–Farallon Ridge moved northward along the North American margin. The effects of this movement are speculative but, as we discussed earlier, some have argued that that change in rate of subduction and/or direction accompanying the sweep of the ridge past the area of the Oregon Bight may have caused (or allowed) the Eocene clockwise rotation of the Coast Range block (Simpson and Cox, 1977).

About 30 Myr. ago, the trench at the leading edge of

Ranges, the old Santa Ynez and Sierra Madre basins had become inactive and, over much of that area, a broad sheet of nonmarine sediment was deposited. This sheet, called the Sespe Formation, is over 1500 m thick and composed of red, arkosic debris with granite clasts, probably derived from the east, and clasts of Franciscan-terrane derivation which Nilsen and McKee (1979) said were possibly weathered from the local, reactivated San Raphael high. Toward the west, the Sespe grades into marine deposits of the Alegria and Gaviota formations (King and Beikman, 1978). The Salinian block apparently constituted a long island on which were deposited mainly fluvial sands and gravels. The Salinian block was separated from the mainland by a deep trough along the western edge of the San Basin in which were deposited muds of the Kreyenhagen Shale. In most other parts of the San Joachin Basin, nonmarine deposits prograded westward over lower Oligocene shallow-marine sediments. The Stockton Arch continued to be emergent

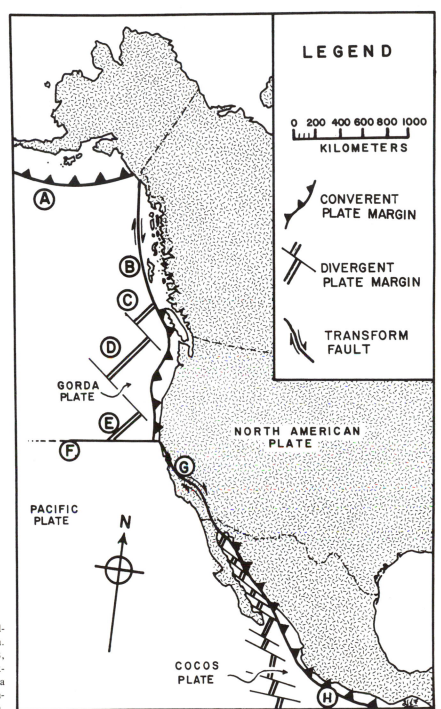

Fig. 9-52. Present-day configuration of tectonic elements along the western margin of North America. Symbols are as follows: A, Aleutian Trench; B, Queen Charlotte Islands transform Fault; C, Explorer Ridge; D, Juan de Fuca Ridge; E, Gorda Ridge; F, Mendocino Fracture Zone; G, San Andreas Transform Fault; H, Middle America Trench.

North America began to intersect the Farallon Ridge as shown in Fig. 9-53. The result was the formation of two triple-plate junctions: a fault–fault–trench type in the north, called the Mendocino triple junction (associated with the Mendocino Fracture Zone), and a ridge–trench–fault type in the south, called the Rivera triple junction, with a single transform fault (the San Andreas) connecting them. At first glance, the development of a transform fault from a ridge–trench intersection seems difficult to visualize. But one must remember that a ridge is considered in plate theory to be only a linear zone along which the lithosphere is thinnest and, consequently, weakest; it is *not* necessarily indicative

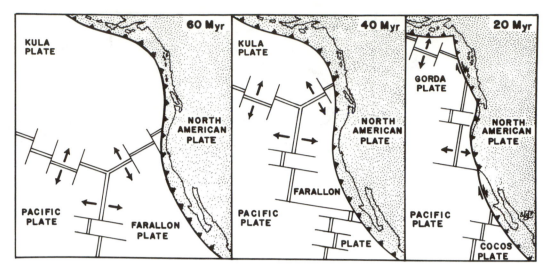

Fig. 9-53. Evolution of eastern Pacific-basin plates and the western portion of the North American plate during the Cenozoic; see text for discussion.

of any fundamental feature (e.g., a rising convective limb) of the mantle beneath it; thus, when a ridge encounters a subduction zone and dives into it, it dies (or becomes "stifled") in most situations. In the case of the Oligocene northeastern Pacific, the Pacific plate (against which the North American plate was juxtaposed by subduction of the intervening Farallon plate) was moving essentially parallel to the edge of North America. The absolute motion of any one plate is inconsequential in defining the character of a plate margin; it is the *relative* motion of adjacent plates that is crucial! Thus, the fact that the relative movement of the Pacific plate with respect to the North American plate was parallel to the western margin of the continent required the ensuing boundary to be a transform fault. And, as more of the Pacific plate was brought into juxtaposition with North America, the length of the fault must increase. [Further discussion may be found in McKenzie and Morgan (1969).] By the present time, only a small scrap of the Farallon plate (the Juan de Fuca plate) remains west of the United States. The San Andreas fault had (and continues to have) profound consequences for California tectonics; we will return to this topic in subsequent discussions.

Notice in Fig. 9-53 that a similar evolutionary pattern is envisioned for the Queen Charlotte transform fault from Vancouver to southeastern Alaska. The Queen Charlotte fault is believed to have developed during the earlier Oligocene or even late Eocene (Dickinson, 1976), i.e., earlier than the San Andreas, but the precise ages of inception are functions of the geometry of the plates.

Washington–Oregon.   Subduction of the Farallon plate beneath the North American continent continued in the Washington–Oregon region as did magmatic activity in the

Cascades Range (see Fig. 9-51). Cascades volcanism during the Oligocene was dominated by explosive eruptions which spread pyroclastic material over much of the area. Especially common in Oligocene units of the Little Butte Volcanic Series (which makes up the bulk of the western Cascades) are ash-flow tuffs, welded tuffs (i.e., ignimbrites), and volcanic breccias, bespeaking the violence which wracked the region. Fossil plants of middle Oligocene to early Miocene age associated with Little Butte volcanics record a gradually changing climate, with older floras indicating tropical–subtropical conditions and younger floras a cooler climate (King and Beikman, 1978). This may have been due to general climatic cooling or to an increase in the altitude at which the volcanics were erupted.

The massive quantities of pyroclastic debris generated by Cascades volcanism fairly overwhelmed the supply of other kinds of sediments in the Washington–Oregon area so that most Oligocene strata there are tuffaceous. In western Oregon and southwestern Washington, shallow marine sedimentation occurred in the broad fore-arc region associated with the Coast Range block. Resulting shallow-marine deposits of the Yaquina Formation (Oregon) and Lincoln Creek Formation (Washington) grade eastward into fossiliferous and coal-bearing, deltaic and marginal-marine strata (e.g., the Eugene Formation of Oregon). These, in turn, pass eastward into Little Butte volcanics. Deep-water conditions persisted in the Puget Bay area between the Cascades and the Olympic Peninsula (the northern end of the Coast Range block) as recorded by the Blakeley Formation. The Blakeley is a 2400-m-thick mass of conglomerate, sandstone, siltstone, and shale which McLean (1977) interpreted as a submarine-fan complex.

In eastern Oregon, a broad lake developed in the topographic basin between the Ancestral Cordillera to the east and the newly uplifted Cascades arc to the west (see Fig. 9-51). Deposits of this lake are termed the John Day Formation and consist primarily of about 600 m of tuffaceous siltstone and rhyolitic air-fall and ash-flow tuffs (King and Beikman, 1978). John Day strata contain a rich and well-preserved fauna of Oligocene mammals which has been famous since the turn of the century.

Ancestral Cordillera. By the beginning of the Oligocene, tectonic events associated with the Laramide orogeny had finally ceased and much of the Ancestral Cordillera was undergoing erosion. Laramide uplifts had been planed down, basins had been filled up, and over the entire area known today as the Rocky Mountains, a hilly, rolling landscape was all that remained of the Laramide orogen. Sediments derived from the Cordillera were carried by streams across the high plains toward the east, resulting in the White River Formation in South Dakota, Colorado, and Nebraska, the Castle Rock Conglomerate in the Denver Basin, the Florissant Lake Beds, and so on.

During Laramide times, volcanism had migrated eastward, affecting areas far inland from the continental margin. During the later Eocene and Oligocene, volcanic activity began to migrate back westward (see Fig. 9-54 and compare with Fig. 9-33). This prograde migration of arc volcanism is interpreted by Dickinson (1979) as a reflection of steepening of the subduction dip-angle (see Fig. 9-41), perhaps due to decrease in the rate of plate convergence (Oligocene spreading rates of the Mid-Atlantic ridge were slower than during earlier Tertiary time) or because the hypothetical aseismic ridge had been entirely consumed.

There are several important volcanic areas of the Ancestral Cordillera which formed during the Oligocene. Of these, perhaps the best known is the San Juan volcanic field of southwestern Colorado (see Fig. 9-55). Initial Oligocene activity (35–40 Myr.) consisted of voluminous flows and breccias of andesite, rhyodacite, and quartz latite. These rocks comprise the San Juan, Lake Fork, and Conejos formations. Succeeding these early Oligocene volcanics are more than 15 large sheets of rhyolitic, ash-flow tuff associated with large calderas. These, in turn, are overlain by lava flows such as the Fisher Quartz Latite of later Oligocene age (~ 25 Myr.). West of the San Juan field, arc volcanism also occurred in the eastern Great Basin (eastern Nevada and western Utah; see Fig. 9-54). Andesitic and dacitic lava-flows in western parts of this area pass eastward into a more felsic volcanic facies consisting primarily of rhyolitic ash-flow tuff. In the Virginia City, Nevada area, Oligocene andesitic lavas comprise the hostrock of the fabled Comstock Lode. The Needles Range Formation is an example of ash-flow strata in eastern Nevada and Utah (see Fig. 9-51); it consists of two to five flows traceable over 33,000 km!

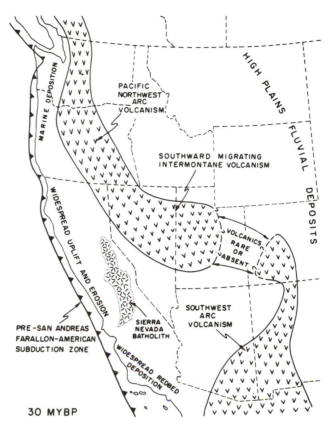

Fig. 9-54. Paleotectonic model of the United States Cordillera during the Oligocene. (From Dickinson, 1979, reproduced with permission of the Pacific Section, Society of Economic Paleontologists and Mineralogists.)

Violent volcanic ash-flows over such vast areas must have been truly awesome, like hundreds of Mount St. Helens' erupting at once. In southeastern Arizona and adjacent New Mexico, a large area, termed the *Mogollon Plateau* (see Fig. 9-51), was also volcanically active at this time. Here, too, violent ash-flow eruptions were associated with large calderas. Basaltic lava-flows were also extruded in this area, continuing on into Miocene and later epochs. From the above discussion and Fig. 9-54, it may be seen that Oligocene volcanism formed a nearly continuous zone down the entire length of the Cordillera except for a segment in western Canada (see below). This is taken as evidence that an almost-continuous subduction zone existed along the western margin of North America. As we will show in subsequent paragraphs, this pattern was interrupted during the later Oligocene and succeeding epochs by extensional tectonics in the Great Basin concurrent with activity of the San Andreas (and Queen Charlotte) fault.

Canada. The Oligocene history of western Canada is very poorly known because rocks of that age are almost entirely missing in the Canadian Cordillera. "Laramide" deformation had ceased by the Oligocene and the area was

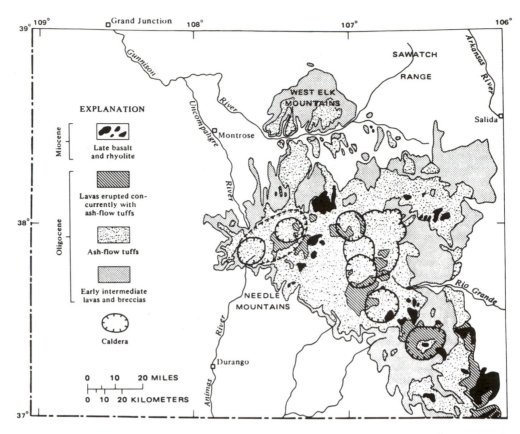

Fig. 9-55. Lower and middle Tertiary volcanic rocks in the San Juan Mountains of southwestern Colorado. (From King and Beikman, 1978.)

undergoing extensive erosion. Since "Laramide" tectonics in Canada were intimately related to arc magmatism, their cessation just prior to the beginning of the Oligocene is not surprising since magmatism had also ceased by that time (Dickinson, 1976). And, since both igneous and structural effects of "Laramide" Andean-type orogenic activity had ceased at approximately the same time, the most probable cause was cessation of subduction along the continental margin, probably due to inception of faulting along the Queen Charlotte fault. Unlike the San Andreas fault in California, movement of the Queen Charlotte fault probably began during later Eocene time (Chase and Tiffin, 1972), suggesting that interaction of the North American and Pacific plates had started earlier in the north than in the south. Some sedimentary rocks of the Tofino fore-arc basin are of Oligocene age, especially upper parts of the Hesquiat and Escalente formations and the Carmanah Formation (Douglas *et al.*, 1976).

Alaska. Oligocene rocks in Alaska, too, are relatively uncommon and, along the Aleutian archipelago, are virtually absent. This is because the entire Aleutian–Alaskan magmatic arc and adjacent areas were uplifted during the late Eocene and Oligocene and subjected to erosion,

resulting in unconformities which truncate all pre-Oligocene rocks in the region. Also formed at this time was the discontinuous, nearly planar, submarine surface (called the Ridge Shelf, Fig. 9-34) along the crest of the Aleutian arc which forms the platform off which the Aleutian Islands rise to the surface. The Ridge Shelf is believed to be a wave-cut platform planed across the top of the arc ridge during Oligocene uplift. No doubt arc uplift was aided in this by the very low stand of sea level during the middle Oligocene. There were several consequences of this arc-uplift event, perhaps the most interesting of which is the isolation of the Bering Sea from the World Ocean. Cut off from the Pacific by the uplifted Aleutian arc and from the Arctic Ocean by the Bering Shelf (which was left emergent by lowered Oligocene sea level), the Bering Sea must have been completely landlocked. DeLong *et al.* (1978) hypothesized that the Bering Sea may have become brackish at this time and that strongly euxinic bottom conditions developed. Further, lower sea level in the Bering Sea may have been responsible for initiation of the submarine canyons along the southwest margin of the Bering Shelf (e.g., Zemchug, Pribilof, and Bering canyons, Fig. 9-34, which DeLong *et al.* said may be the largest submarine canyons on Earth).

As discussed in the section on Eocene paleogeography, uplift was only one process at work in the Aleutian–Alaskan region during the Oligocene. Accompanying uplift were deformation, regional greenschist-grade metamorphism, and greatly reduced igneous activity. Only a few igneous rocks of this region are of Oligocene age, including gabbroic and diabasic intrusions on Agattu Island (DeLong et al., 1978) and several small plutons in the northeastern part of the Alaska Range–Talkeetna Mountains plutonic belt (Reed and Lamphere, 1973). By the middle and late Oligocene, however, uplift had ceased in the Cook Inlet Basin and a marine transgression ensued, during which the West Foreland Formation was deposited in the center of the basin. Primarily a polymictic conglomerate with coarse graywacke, tuffs, and local, subaerial, basaltic lava-flows (Kirschner and Lyon, 1973), the West Foreland Formation represents alluvial sedimentation in a broad valley flanked by mountains of moderate relief. Transgression persisted, reaching a maximum during the late Miocene. (As we will discuss in a subsequent section, Miocene volcanics and sediments are also common on the Aleutian Islands.) The reader will recall that the Oligocene was a time of general regression; thus, transgression in the Cook Inlet must be due to local (rather than eustatic) effects, representing crustal subsidence immediately following late Eocene–early Oligocene uplift.

A single, pleasantly parsimonious explanation for all of the above tectonic observations has been offered by De-Long et al. (1978), elaborating on an earlier suggestion by Grow and Atwater (1970). Simply stated, late Eocene and Oligocene tectonic activity in Alaska and the Aleutians is believed to have been caused by the approach to, arrival at, and subduction beneath the Aleutian trench of the still-active Kula ridge (see Fig. 9-53). The reasoning runs as follows: as a ridge approaches a trench, progressively younger crust flanking the ridge will be subducted. But the age of the crust is directly related to its elevation. Thus, as progressively younger, and therefore progressively higher, portions of the crust subduct, the result must be to cause uplift in the arc-trench complex. Eocene–Oligocene uplift in the Aleutian Islands is recorded by the change of "early series" rocks from deep-water environments to shallow-water environments to emergence. Also, paradoxical as it might seem, the subduction of a spreading ridge probably results in decrease of magmatic activity. DeLong et al. offered several possible explanations: (1) because the lithosphere directly associated with the ridge is still quite warm, it behaves much more plastically than cooler lithosphere so that frictional heating during subduction would be slight; (2) being warm, the down-going slab would have a lower-than-normal density so that it might sink into the athenosphere at a lower angle, thus causing magmatism to abandon the old arc. Higher levels of heat

flow from the warm, subducting ridge would probably result in an episode of thermally induced regional metamorphism, such as the greenschist-grade metamorphism in the Aleutians. A final line of evidence supporting ridge subduction may be seen in the magnetic-anomaly "stripes" of the Pacific plate south of the Aleutian trench. In contrast to the expected pattern of increasing crustal ages closer to the trench, just the opposite is observed, i.e., the age of the crust *decreases* toward the trench. Grow and Atwater (1970) argued that this was because the Pacific plate seen entering the trench today is actually the *trailing* flank of the Kula ridge, i.e., it was formed on the south side of the northward-moving ridge.

It is, as mentioned earlier, hypothetically uncommon for a spreading ridge to subduct beneath a trench. Presumably, when a ridge and trench intersect, the result is the evolution of some kind of triple-plate junction, such as those developed accompanying the history of the San Andreas (and Queen Charlotte) fault. DeLong et al. postulated that the reason the Kula ridge was physically subducted was that the ridge's approach was nearly parallel to the trench. In such a case, no triple-plate junction would form because the ridge and trench would meet not at a point but along a line. Of equal importance is the probability that the Pacific plate's relative motion was toward the Aleutian–Alaskan trench so that plate convergence continued. Crustal subsidence following ridge subduction, as seen in the Cook Inlet area, was probably due to the subduction of progressively older (and therefore cooler and bathymetrically lower) crust on the trailing flank of the ridge. Thus, as the Kula ridge was lost "down the tube," the arc-trench area subsided. By the Miocene, things had returned to normal, convergent-plate tectonics.

D.3. Late Cenozoic Paleogeography

As discussed in earlier parts of this section, early Cenozoic igneous and structural events were dominated by convergent-plate tectonics involving subduction, fore-arc basins, and arc magmatism. This general pattern of Andean-type orogenesis was complicated occasionally by variations in subduction dip-angle, accretion of oceanic seamounts, and so on, but was, nevertheless, the dominant tectonic *milieu* of the Cordillera from as far back as the Late Jurassic until the Oligocene, when North America began to overrun the spreading ridge between the Farallon and Pacific plates (see Fig. 9-53).

Late Cenozoic tectonics in the western United States were significantly different from those of earlier epochs and were characterized by an almost bewildering variety of disparate igneous and structural events. Chief among these events were: (1) large-scale extension in the Basin and Range Province; (2) major basaltic volcanism in the northwestern United States and adjacent areas of Canada; and (3) development of an elaborate system of basins, uplifts, and faults in southern California. In addition, "normal" convergent-plate tectonics persisted in the northern Sierra Nevada, the Cascades Range, and southwestern British Columbia as well as along the Aleutian–Alaskan trench-arc system. All of this tectonic complexity has provoked considerable speculation but little agreement among Cordilleran geologists. In the following discussions, we will describe the major tectonic events of Miocene, Pliocene, and later times and will review several hypotheses to account for them.

D.3.a. Miocene Events

It was during Miocene time that the familiar appearance of North America began to emerge from the scrabble of tectonic–geographic elements whose developments we have traced. We have shown that the modern Mississippi River drainage was probably initiated at that time, related to epeirogenic uplift in the Appalachians and Rocky Mountains. A coastal plain existed on the Atlantic and Gulf continental margins; the San Andreas fault presided over a continental-borderland complex of basins and uplifts; volcanic activity characterized the Cascades Range. In the Canadian Cordillera, Miocene deposits lie primarily in great valleys carved into the mountains, valleys which today continue to act as major controls of drainage. Of course, there were also many dissimilarities, especially in the Columbia Plateau and Basin and Range Provinces, but, were one to imagine the entire geological history of North America as a time-lapse film shot from high Earth-orbit, the Miocene would be the time when the appearance of the continent familiar to us becomes recognizable.

The Cordilleran continental margin of North America during the Miocene was segmented into five main areas (Fig. 9-53). An Andean-type orogen existed along the Southern Cordilleran margin of Central America and southern Baja California (which at that time was farther to the southeast and not separated from the Mexican mainland by the Gulf of California). This convergent plate boundary was underthrust by the Cocos plate, descendant of the Farallon plate. Along northern Baja and southern California, a borderland terrane developed in association with the San Andreas fault. The continental margin from central California to northern Vancouver was characterized by convergent-plate tectonics as the Juan de Fuca plate subducted eastwards. The western Canada continental margin was dominated by the Queen Charlotte fault and, along the Alaska–Aleutian margin, a third segment of plate convergence existed. In the discussions to follow, we will begin with a brief consideration of magmatic-arc areas, especially the northwestern United States, whose tectonic activity was

rather mundane compared with the excitement to the south and east. We will then consider the development of the southern California borderland, the Basin and Range Province, and the Columbia Plateau.

i. Pacific Northwest Arc. A magmatic arc associated with subduction of the Juan de Fuca plate stretched along the continent's edge from central California to the Queen Charlotte fault (see Fig. 9-56). At the beginning of the Miocene, magmatic activity in this Pacific Northwest arc ran in a narrow belt from south of the present Sierra Nevada northward through the Cascades Range and on into the Pemberton arc of southwestern British Columbia. As the Miocene progressed, however, the length of the arc decreased as the southern terminus moved northward [from south of the Garlock fault (see Fig. 9-60) in the early and middle Miocene to the latitude of San Francisco by the early Pliocene; Cole and Armentrout, 1979]. This trend has continued to the present: arc magmatism in California today is found only in the northern part of the state. The cause of this

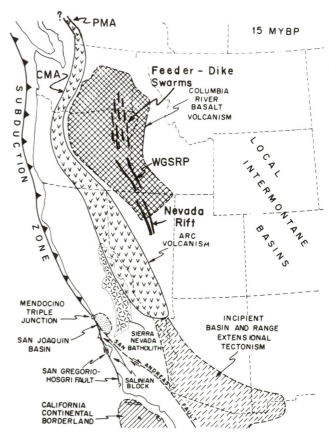

Fig. 9-56. Paleotectonic map of the western United States during the early Miocene. WGSNP is the western graben of the Snake River Plain; PMA is the Pemberton magmatic arc; CMA is the Cascades magmatic arc. (Modified from Dickinson, 1979, reproduced with permission of the Pacific Section, Society of Economic Paleontologists and Mineralogists.)

progressive northward cessation of arc magmatism was the concurrent increase in length of the San Andreas fault as progressively more of the Pacific plate was brought into juxtaposition with the North American plate.

Igneous activity along the Pacific Northwest arc was dominated by subequal volumes of silicic and andesitic volcanism in the form of lava flows, ash flows and falls, and coarser pyroclastic debris. In the area of Mt. Ranier, for example, upper Oligocene and early Miocene volcanic and volcaniclastic rocks comprise the Stevens Ridge Formation; the overlying Fife Peak Formation is of probable mid-Miocene age and consists of basaltic and andesitic volcanics (King and Beikman, 1978). These units are intruded by the Tatoosh Granodiorite (which forms the foundation of Mt. Ranier); the Tatoosh pluton was explosively unroofed during later Miocene time, spreading ash and other debris over a very large area. Other plutons, such as the large Snoqualmie Granodiorite northeast of Mt. Rainier, were also emplaced at this time. Their intrusion was accompanied by local folding of earlier Tertiary strata of the Cascades orogen. Also during later Miocene time, Cascades rocks and rocks just to the east were deformed into a set of relatively open folds whose axes run generally east–west and are broadly convex northward (see Fig. 9-47). This deformation is called the Yakima folding event. Apparently due to nearly north–south-directed, compressive stress, Yakima folding is probably related in some way to contemporaneous extension in the Basin and Range (more on this later).

The Coast Range block was being uplifted during the Miocene and marine sedimentation was restricted to a few, small, coastal embayments along the narrow continental shelf (such as the Newport embayment of western Oregon or the area around the mouth of the modern Columbia River; see Fig. 9-57). In the Newport area, over 1000 m of dark, organic-rich marine muds of the Nye Mudstone were deposited. At the same time, turbidite sands were being spread out on the ocean floor in front of the Columbia's mouth, resulting in the ancestral Astoria fan.

Far to the north, at the other end of the Queen Charlotte fault, similar volcanic and plutonic activity resumed along the Aleutian–Alaskan arc during the Miocene following subduction of the Kula Ridge. For example, on Amchitka Island (Fig. 9-34), lavas and granodioritic plutons yield K–Ar dates ranging from 15.8 to 8.9 Myr. (DeLong *et al.*, 1978). In the Cook Inlet area, the marine transgression which had begun during the late Oligocene continued into the Miocene, as recorded by fluvial and deltaic sediments of the Hemlock Conglomerate and Tyonek Formation (Kirschner and Lyon, 1973). In the Yakutat–St. Elias region northwest of Juneau, Miocene intrusive activity was widespread, resulting in numerous granitic plutons as well as both mafic and felsic dikes, sills, and plugs (Hudson *et al.*, 1977).

ii. California and the Continental Borderland.

For the purpose of the present discussion, we divide California into three areas: (1) the magmatic arc of the Sierra Nevada and regions to the north (discussed above); (2) the Great Valley and northern Coast Ranges, north and east of the San Andreas fault; and (3) the continental borderland of southern California.

The San Joachin and Sacramento basins of the Great Valley occupied a fore-arc setting during the early and middle Miocene and may be considered a conventional fore-arc-basin system, with a broadly uplifted outer ridge (the Franciscan terrane of the Coast Ranges) to the northwest and a gradually lengthening transform fault slicing it off at the south. At the beginning of the Miocene, sedimentation in the Great Valley was about equally divided between non-marine deposition, especially in the Sacramento Basin, and shallow-marine deposition in much of the San Joachin Basin. Rise of sea level, at least partly a result of the early middle Miocene eustatic sea-level rise (see Fig. 8-3), resulted in the flooding of much of the Great Valley (see Fig. 9-57). As a consequence, deep-water conditions (perhaps as deep as 2000 m; Cole and Armentrout, 1979) predominated in the western San Joachin Basin, where turbidite and pelagic sedimentation occurred. To the east, however, the sea gradually shoaled and shallow-marine conditions of the central San Joachin Basin graded eastward into fluvial deposits (e.g., the Mehrten and Zilch formations). Note in Fig. 9-57 (which represents late Miocene time) that the main detrital source for much of the Great Valley was the Sierran province to the east. Franciscan-derived detrital sediments present primarily in the northwestern Sacramento Basin, indicate that the northern Coast Ranges were also emergent at that time. Following the middle Miocene flooding, detrital sedimentation gradually filled in the basins leaving a generally regressive stratigraphic sequence throughout the Great Valley region.

The southern California borderland is an area of striking complexity, an abstract-expressionist's portrait of topography (Fig. 9-58). It is the type example of a continental borderland, that variety of continental margin characterized by numerous, small tectonic blocks adjoined by various kinds of faults and jostled together by fault-displacements. Some blocks are exclusively uplifted while others are exclusively depressed; some blocks are tilted, some rotated, and still others have complex histories in which different behaviors alternate. The reader will appreciate that the stratigraphy of such a terrane is extremely complex; thus, we will omit most stratigraphic detail and concentrate on regional tectonic development. [There are many papers on the California borderland; we direct the interested reader to the bibliographies in Crowell (1974a,b) and Luyendyk *et al.* (1980).]

The borderland terrane was originally formed (and

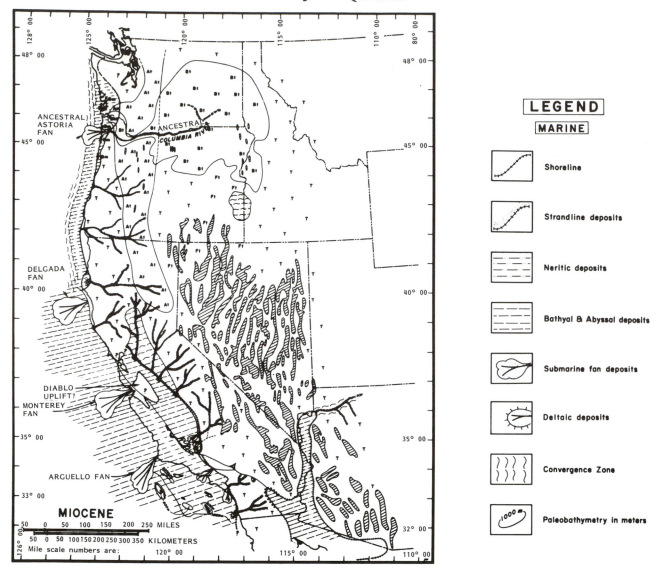

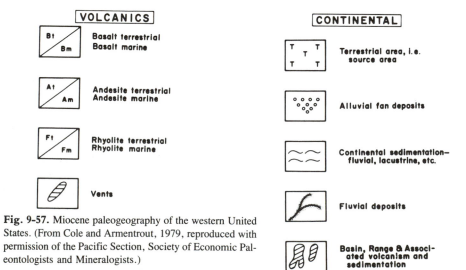

Fig. 9-57. Miocene paleogeography of the western United States. (From Cole and Armentrout, 1979, reproduced with permission of the Pacific Section, Society of Economic Paleontologists and Mineralogists.)

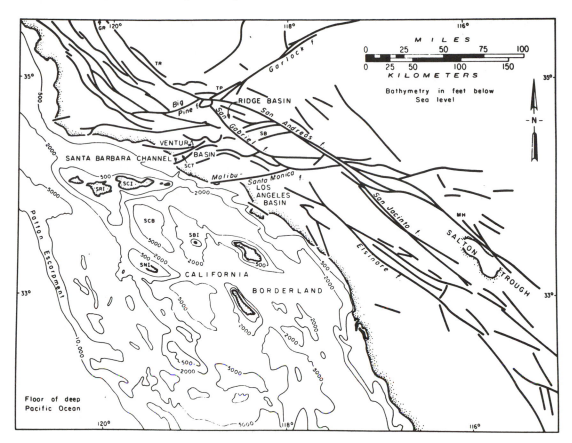

Fig. 9-58. Map of the modern southern California borderland and major onshore faults. Abbreviations: TP, Tejon Pass; SB, Soledad Basin; SCT, Santa Clara Trough; SRI, Santa Rosa Island; SCI, Santa Cruz Island; SCB, Santa Cruz Basin; SBI, Santa Barbara Island; SNI, San Nicolas Island; GR, Gabilan Range; TR, Tremblor Range; MH, Mecca Hills. (From Crowell, 1974a, reproduced with permission of the Society of Economic Paleontologists and Mineralogists.)

continues to be controlled by movements along the transform boundary between the North American and Pacific plates. In most discussions of this transform boundary, it is presumed that displacement occurred principally along the San Andreas fault. But Crowell (1974a) stated that of the 1500 km of relative movement of the Pacific plate past North America during the past 25 Myr., only about 300 km can be accounted for by dextral slip on the San Andreas fault. The remaining 1200 km of offset is to be found as much-smaller-scale displacements along the many borderland wrench faults (see Fig. 9-58). In Crowell's view, the basins and uplifts of the borderland owe their existence to this large-scale horizontal displacement and concomitant shearing.

Figure 9-59 shows the ways by which subsidence and uplift are ascribed to wrench faulting. Figure 9-59A illustrates the case in which a gentle curve exists along the trace of a wrench fault. The result of dextral shear along this fault would be twofold: shortening, or "crowding," along the convex-to-the-left part of the bend, with resultant gentle uplift, and stretching of the crust at the convex-to-the-right bend, causing subsidence. (Note that sinistral movement would reverse the location of uplift and subsidence). Ridge Basin, just north of the San Gabriel fault (Fig. 9-60), was considered by Crowell (1974a) to be due to such subsidence along a curved fault trace. In the case of a much sharper offset along the fault (Fig. 9-59B,C), a similar but more pronounced effect is produced. If the offset is to the left (B), dextral slip will cause uplift and strong deformation (characterized by compressive folding and thrust faulting); if the offset goes to the right (C), a rapidly subsiding, pull-apart basin will result. The Salton trough, along the San Andreas fault, may represent such a pull-apart basin. As the extent of pull-apart increases, lavas may work themselves up from below causing local plutonism, volcanism, and high levels of heat flow in basin centers. A final wrench-fault mechanism of uplift and subsidence is shown in Fig. 9-59D. In this case, a large number of splays have developed rather than a single fault; such a situation probably results when a relatively thin and weak plate-edge is subjected to transform shearing. These splays intersect to form a patchwork of fault-wedge blocks which, as they are sheared past one

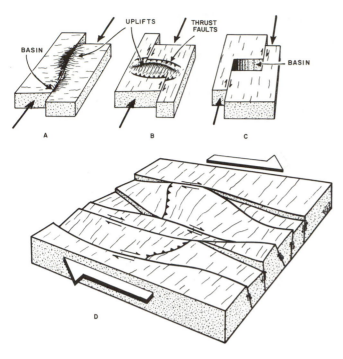

Fig. 9-59. Relation of diastrophism to faulting in continental-borderland uplifts and basins; see text for discussion. (Redrawn from Crowell, 1974a.)

another, may be forced up or down depending on their specific geometries. Alternately, they may be tilted so that one end rises while the other subsides. Deformation at block edge depends critically on the location of the edge relative to the other blocks; in some cases, a block margin will undergo compressive deformation including considerable thrust faulting; in other cases, extensional deformation will result while, in still others, simple shearing.

This final mechanism of basin/uplift activity has been elaborated and expanded by Luyendyk *et al.* (1980), who presented a geometric model to explain not only the origin of specific southern California basins and uplifts but also block rotations (revealed by magnetization directions of Miocene basaltic flows and shallow intrusions in the Transverse Ranges, the central Mojave area, and Catalina, among others; see Fig. 9-60). The geometric model of Luyendyk *et al.* is presented as Fig. 9-61. Notice that an initial pattern of fractures (perhaps caused during earliest shearing or perhaps controlled by basement inhomogeneities) resulted in a mosaic of independent blocks which were transported by shearing along the deformed zone as the Pacific and North American plates shifted past each other. The Transverse Range blocks are considered to have acted as a lever, forcing the adjacent blocks to shift past and jostle each other as

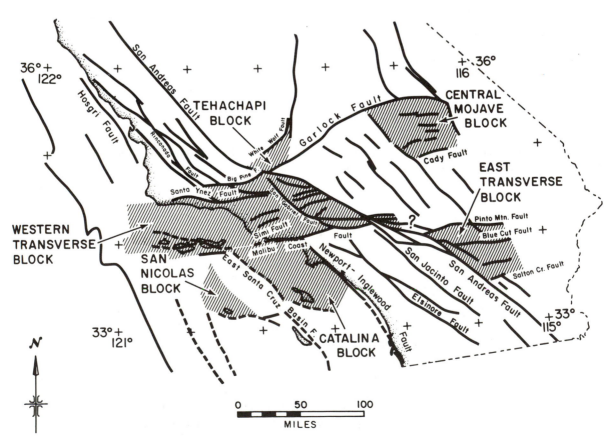

Fig. 9-60. Borderland tectonic-elements in the area of the Transverse Ranges; shaded areas have undergone clockwise rotation. (From Luyendyk *et al.*, 1980, *Geological Society of America Bulletin*, Fig. 2, p. 213.)

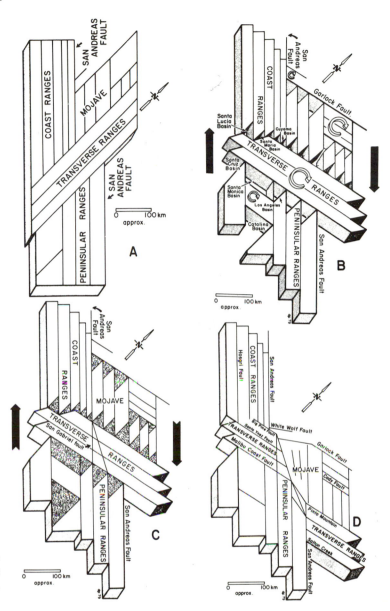

Fig. 9-61. Tectonic model of borderland-terrane evolution during the late Cenozoic; see text for discussion. (From Luyendyk *et al.*, 1980, *Geological Society of America Bulletin*, Fig. 3, pp. 214–215.)

the Transverse Range blocks themselves were rotated clockwise by the dextral shear. Notice that a consequence of these movements was the opening of a variety of triangular and rhomb-shaped pull-apart basins. Luyendyk *et al.* considered this to have been the origin of the offshore Santa Lucia, Santa Cruz, Catalina, and San Nicolas basins and the onshore Santa Maria, Cuyama, and Los Angeles basins.

Sedimentation within borderland basins varied considerably. Some basins were almost exclusively marine and may even have enjoyed deep-water conditions if nearby sources of detrital sediment were not present. Other basins were marginal to the sea and were, therefore, alternately emergent and submergent, depending on the whim of local tectonics and eustasy. A good example is the Monterey Bay area (south of San Francisco). This area was uplifted and deeply eroded during the Oligocene and then flooded at the

beginning of the early middle Miocene (Greene and Clark, 1979). The transgressive Lompico Sandstone was deposited in the Santa Cruz Mountains area as a result of this inundation. Overlying the Lompico are deep-water deposits of the Monterey Formation, whose mudstones, siliceous shales, diatomites, bedded cherts, and other siliceous sediments are widespread in the southern Coast Ranges. By the late Miocene, the Monterey Bay region was again emergent (briefly) and was then flooded again, resulting in a new transgressive sand, termed the Santa Margarita Sandstone. These sands were succeeded, apparently without interruption, by deeper-water muds, represented by the Santa Cruz Mudstone. The up-and-down dance of Monterey Bay continued into the Pliocene. Indeed, Monterey Bay (obviously) remains today partially flooded.

Still other borderland basins in southern California were

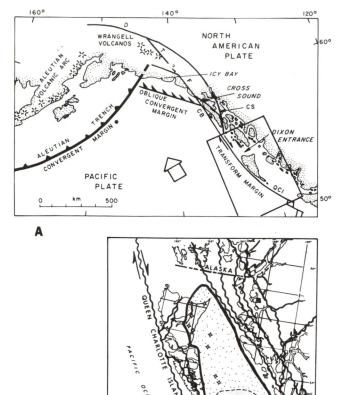

A

B

Fig. 9-62. (A) Late Cenozoic plate boundaries and related features in the Gulf of Alaska. D, Denali Fault; T, Totschund Fault; F, Fairweather Fault; CS, Chatham Strait Fault; CB, Chicagof–Baranof Fault; QCI, Queen Charlotte Islands Fault; box shows location of B. (Modified from Von Huene *et al.*, 1979, reproduced with permission of the Pacific Section, Society of Economic Paleontologists and Mineralogists.) (B) Distribution of facies in the Queen Charlotte Basin. (Modified from Drummond, 1979, reproduced with permission of the Pacific Section, Society of Economic Paleontologists and Mineralogists.)

(and are) entirely nonmarine. In the Mojave Desert, for instance, Miocene continental strata are named the Tropico Group and are composed of fluvial and alluvial deposits. King and Beikman (1978) stated that upper Miocene mammal remains are found in the nonmarine strata of the Barstow Formation, which is probably correlative with the Tropico. The nonmarine Punchbowl Formation (south of the San Andreas fault) and the similar Cajon Formation (90 km to the north of the fault) may represent the same original unit, subsequently displaced by faulting. Both of these are ex-

tremely thick: 1800 m for the Punchbowl and 2400 m for the Cajon. Clearly, subsidence of the Punchbowl-Cajon area must have been very rapid.

iii. Queen Charlotte Transform Zone.

Since the Oligocene, a tectonic environment similar to that of southern California has characterized the Northern Cordilleran continental margin from southeastern Alaska to an area just northwest of Vancouver Island. The Queen Charlotte transform zone, which is composed of the Queen Charlotte Islands, Chicagof–Baranof, and Fairweather faults (see Fig. 9-62), was probably formed by the same kind of plate interactions which caused the San Andreas fault. But much of the Queen Charlotte zone's length is currently underwater, under alpine glaciers, or exposed on the precipitous walls and cliffs of fjords; thus, much less is known of the Queen Charlotte zone. Von Huene *et al.* (1979) stated that the present episode of orogeny in the eastern Gulf of Alaska probably began sometime during the Miocene and has been dominated by transform motion.

As in southern California, subordinate splay-faulting and basin and uplift activity accompanied slippage of the Pacific plate past the North American plate. The Queen Charlotte Basin (Fig. 9-62B) contains about 3 km thickness of Miocene and Pliocene sediments, primarily nonmarine in the northwest (east of the modern Queen Charlotte Islands) and marine in the southeast. Galloway (1974) considered the Queen Charlotte Basin to be a fore-arc basin but the lack of an adjacent arc and its relation to the Queen Charlotte transform zone suggest the possibility that it is a borderland-type basin such as described above from southern California. Notice in Fig. 9-62 that several faults, such as the Chatham Strait fault, seem to splay off the Queen Charlotte transform zone in a manner analogous to the situation portrayed in Fig. 9-59D.

iv. The Basin and Range Province.

During the later Oligocene, extensional tectonics began to affect southern Nevada and adjacent areas. With time, the affected area increased so that crustal extension today characterizes the entire physiographic province known as the Basin and Range (see Fig. 9-63). The Basin and Range is sometimes confused with the Great Basin but they are *not* synonymous: the Basin and Range Province is that area of the American Southwest typified by numerous, subparallel mountain ranges and intervening basinal areas; the Great Basin is that area of California, Nevada, Utah, and Oregon, characterized by internal drainage (i.e., no streams flow out of the area to the sea). Possibly related to Basin and Range extensional tectonics was the development of the Rio Grande Rift, which runs almost due south from southcentral Colorado through New Mexico to westernmost Texas (the trans-Pecos region). In between the Rio Grande Rift and the

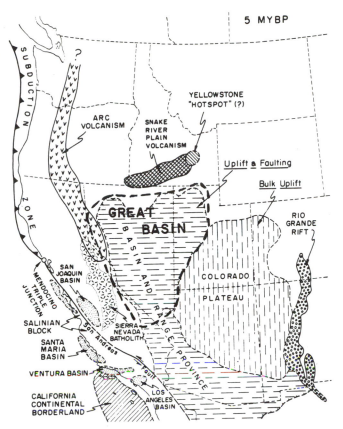

Fig. 9-63. Paleotectonic map of the western United States during the latest Miocene and early Pliocene. (Modified from Dickinson, 1979, reproduced with permission of the Pacific Section, Society of Economic Paleontologists and Mineralogists.)

lake deposits. These, along with erosional features such as steep, sediment-choked valleys, badlands topography, pediments, and inselbergs, define the characteristic arid-region physiography familiar to students of physical geology. Arid conditions, however, are relatively recent in the Basin and Range, probably having originated during the Pliocene. During the Miocene, in the absence of a high, mountain ridge in the Sierra Nevada (which acts today as a barrier to moisture-laden Pacific air masses, thus throwing a rain-shadow over the Great Basin), external drainage characterized most of the Basin and Range, with principal streams flowing westward to the Pacific.

It has long been known that the major cause of Basin and Range topography is normal faulting and, consequently, that the region's structure is a response to crustal extension. But the subsurface character of these faults and the nature of the lower crust in the region continue to be controversial. Three geometric models of Basin and Range faulting have been proposed (see Fig. 9-65): (1) horst-and-graben model, which explains uplifted ranges as horst blocks and basins as grabens; (2) tilted-block model, in which fault blocks are considered to have been tilted so that

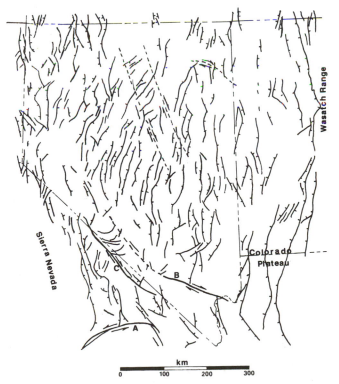

Fig. 9-64. Major faults in the Great Basin and adjacent areas of the Basin and Range Province. Normal faults are indicated by hachured lines, with hachures on the downthrown sides. Important strike-slip fault zones are labeled as follows: A, Garlock Fault; B, Las Vegas Fault Zone; C, Death Valley Fault Zone. (Modified and redrawn from King, 1977, and the Geologic Map of the United States, published by the USGS, 1974.)

Basin and Range is the broad, high Colorado Plateau which was essentially unaffected by late Cenozoic extension. The nature of extensional faulting in the Basin and Range and the causes of such faulting, both of which are controversial, will be considered in following discussions.

Figure 9-64 illustrates the major ranges and basins of the Great Basin. Notice that ranges are bounded on one side or the other by large, normal faults and that these faults show a regular change of orientation from NNW–SSE in northwestern Utah to NNE–SSW in northwestern Nevada. In southwestern Nevada and adjacent California, fault patterns are more complex but trend generally NW–SE (e.g., the Death Valley and Las Vegas fault zones). The result is a vast territory of grossly north–south-trending fault-block mountains and intervening lowlands. The lack of external drainage causes sediments derived from weathering of the mountains to accumulate in adjacent basins in a complex of depositional environments whose characteristics are strongly controlled by the aridity of the region. Thus, principal sediment-types include fanglomerates of alluvial fans and bajadas, ephemeral-stream sands and gravels, and playa-

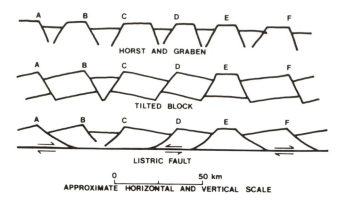

Fig. 9-65. The three main models of Basin and Range faulting. (From Stewart, 1980, *Geological Society of America Bulletin*, Fig. 2, p. 461.)

a given range and adjacent basin are underlain by a single block; and (3) lystric-fault model, which also calls upon tilted fault blocks but which further postulates that the controlling normal faults are lystric in nature and are rooted in a low-angle to horizontal décollement zone at some intermediate depth in the crust. The horst-and-graben model is generally discounted at the present time because of abundant evidence of tilting in the ranges and because ranges are not usually bounded on both sides by faults, as would be required by this scheme. Choosing between the other two models, however, is less easy and several authors have recently suggested that elements of both structural models may be found in the Basin and Range (Davis, 1980; Stewart, 1980).

Of course, the distinction between the two models depends rather critically on the specific response of the crust to extension; thus, it is useful to consider the deep crust of the Basin and Range. Eaton (1980) summarized such knowledge and presented the interpretive model shown in Fig. 9-66. Notice that he favors the lystric-fault model of upper-crust faulting and shows these faults to be rooted in a thin zone of ductile deformation. Crustal extension in levels below this ductile zone is considered to be due mainly to dike emplacement and stretching. Eaton also pointed out that the Basin and Range crust is thinner than that of surrounding areas (being no more than 25–30 km as opposed to 49–50 km for the Colorado Plateau and Great Plains provinces). Indeed, the lithosphere itself is thinner beneath the Basin and Range than "normal" continental crust ($\leq$ 65 km instead of $\sim$ 100 km). Further, heat flow in the Basin and Range is about twice as high as heat flow in the stable interior and eastern parts of North America. Combined with the relatively high permeability of Basin and Range rocks, this high heat-flow is responsible for a vigorous hydrothermal circulation system (see arrows in Fig. 9-66) which, in turn, promotes the epigenetic hydrothermal mineralization which is responsible for major commercial metals deposits

in Nevada and adjacent states. The cloud around this silver lining is the concomitant lack of hydrocarbon accumulations, which absence is also due to the vigorous hydrothermal circulation (Eaton, 1980).

High levels of heat flow are at least partly related to magmatic activity, which has characterized the Basin and Range throughout the late Cenozoic. During the early Miocene, violent felsic eruptions (which had persisted in the area since the Oligocene) spread ash-flow tuffs over much of Nevada. But volcanism of this phase ceased during the late early Miocene and subsequent volcanism was predominated by alkali-basaltic and andesitic lava flows. The oldest of these is dated at approximately 17 Myr. (i.e., contemporaneous with the onset of extensional tectonics); these oldest flows unconformably overlie the felsic tuffs of earlier volcanism. The basaltic and andesitic flows of the Basin and Range are closely related to volcanic rocks of the Snake River Plain to the north, which began to form concurrently.

The direction of crustal extension which caused Basin and Range faulting has also been debated. The present direction, according to Zoback and Thompson (1978), is approximately northwestward (N65°W–S65°E). But there is reason to believe that the Miocene extension direction was southwesterly (N68°E–S65°W, according to Zoback and Thompson). Evidence cited for this earlier stress regime includes the orientation of dike swarms in Nevada and the Columbia River Plateau and the trends of the western graben of the Snake River Plain and the Nevada rift, which all align if plotted on a Miocene palinspastic map (see Fig. 9-56).

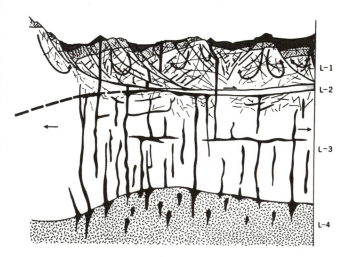

Fig. 9-66. Eaton's (1980) cross-sectional model of the continental crust beneath the Basin and Range. L-1 is faulted, surface layer, 8–15 km thick; L-2 is ductile intermediate layer, 0–3 km thick; L-3 is lower crustal layer, 10–20 km thick; L-4 is the mantle. Arrows in L-1 indicate thermally driven ground water circulation.

v. Columbia Plateau.   To the north of the Basin and Range is a most remarkable terrane characterized by voluminous and aerially extensive basaltic lava flows. This is the Columbia Plateau and the basalts are termed the Columbia River Basalt Group (see Fig. 9-56), divided (from oldest to youngest) into Picture Gorge, Imnaha, and Yakima Basalts. These rocks cover an area of approximately 300,000 km^2, their volume has been estimated at 200,000 km^3, and, near the center of the Plateau, they are over 3000 m thick. Davis (1980) said of the Columbia Plateau that it represents one of the most voluminous outpourings of basaltic lava in the entire geological record! It is comparable only to the Deccan Trap of India. Over much of the Plateau, basalts are essentially undeformed but in the west they were affected by the Yakima folding event (see Fig. 9-47).

The flows probably originated from long fissures (now represented by basaltic dike swarms) in southeastern Washington and northeastern Oregon. As mentioned previously, when these fissures are plotted on a Miocene palinspastic map, they align with other dike swarms and graben to define a zone of extension which runs approximately parallel to the Pacific Northwest arc (Fig. 9-56). This parallelism of extensional features in the rear-arc area to the trend of the arc itself is considered by Davis (1980) and others to provide a significant constraint on regional tectonic models. Spreading out from these fissures, the flows covered an area of low rolling-hills, blanketing the entire region except for a few upland areas such as the Blue Mountains of Oregon. Following each major flow, stream drainages were re-established, often resulting in spectacular gorges carved into the basalt. Swanson and Wright (1979) recognized three major, successive river systems on the Plateau during the Miocene. The Miocene climate is inferred (on the basis of plant fossils) to have been warmer and moister than today; consequently, weathering proceeded rapidly and well-developed soil profiles were developed on the tops of lava flows. Associated with these paleosols in some instances are the remains of forests which were covered by subsequent lavas. Occasionally, a lava flow would dam a river system empounding large lakes on the lava plain. Thus, lacustrine and fluvial deposits are interbedded with the basalts. Similarly, large lakes were empounded at the margins of the Plateau, resulting in lacustrine formations such as the Latah Formation of northeastern Washington and the Payette in eastern Oregon and southwestern Idaho.

The Columbia River basalts are uniformly tholeiitic in composition but there is a slight trend toward increasing silica content with time (SiO_2 content equals 50–47% in Picture Gorge rocks, 54–53% in middle Yakima rocks). Tholeiitic lavas typically have very low viscosities and the Columbia River basalts were no exception. Individual flows have been traced over 200 km without reaching their ends

(King and Beikman, 1978). Indeed, during Yakima extrusion, splendidly voluminous lavas poured into the ancestral Columbia River drainage and were channeled westward through the river's cleft in the Cascades Range and out to the Pacific where they met the sea.

There are other extensive areas of basaltic lava-flows in the Pacific Northwest which formed contemporaneously with the Columbia River flows. In southern Oregon, extensive outpourings of alkaline basalt comprise the Steens Basalt and similar, but less voluminous, flows in northern California are termed the Warner Basalt. In central British Columbia, an east–west-trending belt of alkaline basalts and associated volcanics of late Miocene to Quaternary age is termed the Anahim belt (Bevier *et al.*, 1979). One particularly interesting mass near the center of the Anahim belt defines the late Miocene Rainbow Range shield volcano which covers an area of 900 km^2. Clearly, a vast quantity of basalt was extruded in the Pacific Northwest during the Miocene.

The origin of all this basalt is debatable, with no single answer which fits all the data. [In fact, Dickinson (1979, p. 7) described Columbia Plateau lavas as "enigmatic."] The Anahim belt volcanics of British Columbia were considered by Bevier *et al.* (1979) to be due to the passage of North America over a "hot spot" within the sublithospheric mantle. Similarly, the Steens Basalt (as well as younger lavas of the Snake River Plain which lay to the east of Steens Basalt occurrences) is thought to have formed as the continent overrode a "hot spot" (Dickinson, 1979). In both of these cases, the principal reason for choosing the hot spot model is that oldest lavas occur in the west and basalt-ages become progressively younger along the lengths of the belts toward the east. In the case of the Steens and Snake River Plain basalts, the guilty "hot spot" is believed to have been the "Yellowstone hot spot," started at about 2.2 Myr. ago and currently beneath northwestern Wyoming (see Fig. 9-63). The Columbia River Group, however, shows no such age relationships. The above-mentioned orientation of feeder-dike swarms and the contemporaneity of volcanism and extensional tectonics in the Basin and Range suggest that events in both areas were related to a single, common factor. But what that factor might have been awaits understanding of Cordilleran regional tectonics during the Miocene.

vi. Speculations on Tectonic Models.   Few problems of Phanerozoic regional tectonics have engendered as many contrasting models in as short a time as has Miocene tectonics of the North American Cordillera. The essential facts are relatively clear and well known; data for interpretive models abound—indeed, the quantity of data is one of the biggest reasons that there are so many models. In the following discussion, we will describe the main models of Cordilleran tectonics during the late Cenozoic.

It has been suggested, for example, that the Basin and Range extension is caused by sublithospheric spreading associated with the Farallon ridge which was overrun by North America but which continues to spread down beneath the continent. But, as we have seen, divergent plate boundaries are not now believed to represent a fundamental zone of divergence in the sublithospheric mantle but rather a zone of weakness in the plate armor of the Earth.

It has been suggested that Basin and Range extension might represent inception of continental rifting. This may be true (although the geological situation in the Basin and Range is rather different from that of the East Africa–Red Sea rift which is the best extant example of a continental rift zone) but, even so, it does not explain *why* rifting began where and when it did. This suggestion also fails to explain contemporaneity of Basin and Range extension with San Andreas fault activity (not that contemporaneity *necessarily* equals causal equivalence) and the west-to-east movement of volcanic centers of the Snake River Plain and the Anahim belt.

A model with more promise was developed by Atwater (1970) who postulated that Basin and Range extension was caused by northwest–southeast-directed crustal "stretching" due to movement along the San Andreas fault. In this model, San Andreas faulting causes dextral shear across the western Cordillera, which Atwater called a "wide, soft boundary" between the two rigid plates. Opponents of this scheme argued that some extension features (e.g., the western graben of the Snake River Plain) lie far to the north of the Mendocino triple junction and, therefore, cannot be related to San Andreas faulting. This objection was countered by Livaccari (1979) who greatly expanded Atwater's model. Livaccari said that Miocene (and later) tectonic events in the western Cordillera were caused by distributive shearing related to the San Andreas fault. To understand this, imagine shearing a deck of cards; the "forced" cards (i.e., the ones on the outsides of the deck which are actually in contact with the shearing force) transfer shear to cards successively deeper in the deck, thereby *distributing* the shear throughout the deck. Figure 9-67 presents Livaccari's application of this idea to Miocene tectonics of the west. Notice that distributed shear, "forced" by movement along the San Andreas fault, is shown to have affected a large part of the western United States whereas similar shear associated with the Queen Charlotte transform zone had little effect on western Canada. This, Livaccari said, was because Alaska acted as a "buttress" to restrain deformation. As shearing and concomitant crustal extension proceeded, a conjugate pair of fracture-sets developed; Livaccari suggested that most Basin and Range normal faults, Columbia Plateau feeder dikes, and the western graben of the Snake

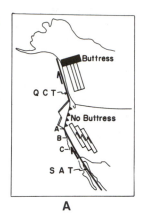
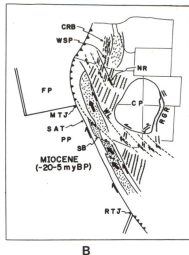

Fig. 9-67. (A) Schematic diagram illustrating Livaccari's card-deck model for the western United States and Canada. A, B, and C indicate, respectively, the positions of the Mendocino Triple Junction at 0, 5, and 20 Myr. ago; SAT, San Andreas Transform Fault; QCT, Queen Charlotte Transform Fault. (B) The major features of livaccari's tectonic model of evolution of the Basin and Range, the Rio Grande Rift, and the Columbia Plateau. Hachured lines indicate gravity faults; NR, Nevada Rift; WL, Walker Lane; RGR, Rio Grande Rift; CP, Colorado Plateau; RTJ, Rivera Triple Juncture; SAT, San Andreas transform; PP, Pacific plate; FP, Farallon plate; WSP, western graben of the Snake River Plain; CRB, feeder dikes of the Columbia River basalts, MTJ, Mendocino Triple Junction; SB, Salinian Block. (Both from Livaccari, 1979, *Geology,* Figs. 1 and 2, pp. 72 and 372.)

River Plain all represent one set of fractures and that features such as Walker lane, the Death Valley and Las Vegas fault zones, and the Nevada rift represent the other fracture set (Fig. 9-67B). Notice that the distributed shear concept allows San Andreas-forced deformation to occur in areas far north of the Mendocino triple junction for the same reason that a card pushed at only one point nevertheless will move as a coherent entity. An interesting corollary of this model is the possibility that shear across the southwest margin of the Colorado Plateau block caused it to rotate clockwise. The Colorado Plateau area may have behaved as a single block, rather than being fractured into Basin and Range topography, because it was cooler and thicker than crustal areas farther west. As the Colorado block rotated, the Rio Grande rift opened behind it and propagated northward as rotation continued. Thus, Livaccari's model explains a number of different features of Cordilleran tectonics.

There are, however, several problems with Livaccari's scheme. Perhaps the most serious of these is his interpretation of features such as the feeder-dike fissures, the western graben of the Snake River Plain, and the Nevada rift, as shear fractures. In rejecting Livaccari's model, Zoback and Thompson (1979) pointed out that fissures such as those of the eastern Columbia Plateau are usually oriented perpendicular to the direction of least principal stress (i.e., the extension direction). Livaccari (1979) countered that the fissures might represent extension gashes oblique to a local shear-direction and might then be rotated with continued deformation into their present position. While this is possible, it is hardly parsimonious. Further, extension gashes related to shear are usually short and in an *en échelon* pattern. The dike swarms of the Columbia Plateau are neither. Another problem with Livaccari's model is the total amount of shear necessary to produce the observed deformation. Crowell (1974a) estimated that about 300 km of slip has occurred along the San Andreas fault [King and Beikman (1978) say 480 km]. But if one sums the total amount of Basin and Range extension (estimates range from 75 to 200 km), plus net slip along fault zones such as Walker lane, the Death Valley and Las Vegas zones and others, plus extension in the Rio Grande rift, plus whatever shear stress was taken up in plastic and cataclastic deformation and released in earthquakes, then there seems to be more strain than can be accounted for by the available stress. Finally, there is the fact that initial extension-directions in the southern Basin and Range were southwesterly rather than northwesterly as shown by Livaccari. All of this would seem to weigh heavily against the distributive shear model. It should be noted, however, that considerable amounts of shear stress must have been exerted on the area east of the San Andreas fault, just as Livaccari has said; we only wonder if it was sufficient to be the *sole* cause of *all* Miocene tectonics in the United States Cordillera.

Another tectonic model (advocated by Davis, 1980, among others) postulates that opening of Columbia Plateau fissures, the western graben of the Snake River Plain, and the Nevada rift (all of which align on a Miocene palinspastic map and parallel the Pacific Northwest arc) was due to back-arc spreading. Davis suggested that acceleration of seafloor spreading rates during the middle Miocene may have caused intensified back-arc activity, reflected not only by crustal extension but also by the extrusion of copious tholeiitic lavas of the Columbia River Group. Notice in Fig. 9-56 that the orientation of the early Miocene subduction zone is shown to have been generally NNW–SSE; since back-arc spreading is usually perpendicular to the arc-trend, the main direction of extension due to back-arc spreading would have been WSW–ENE. Such an orientation is exactly that described by Zoback and Thompson (1978) from northern Nevada. As progressively more of the western plate margin was affected by transform movement along the San Andreas fault, however, dextral shear would cause clockwise rotation of the arc and, consequently, of the extension direction. This change may be seen by comparing the trench/arc orientation shown in Fig. 9-56 with that in Fig. 9-63. This model, therefore, explains extension in the

northern Basin and Range and Columbia Plateau volcanism but seems less able to explain events in the southern Basin and Range. It also avoids the penecontemporaneous onset of San Andreas faulting and Basin and Range extension.

A final scenario should be mentioned. Dickinson (1979) observed that one effect of the inception of the San Andreas fault was the termination of subduction along the length of the transform fault. As the Farallon plate continued to subduct both to the north and south of the San Andreas fault, a "slab window" opened and grew progressively wider as the San Andreas fault lengthened (see Fig. 9-68). Note that this slab-window effect is a direct consequence of plate-margin evolution which caused the San Andreas fault. Into the opening of the slab window, warm, plastic asthenosphere-material upwelled from below. This caused an upward-directed "push" from beneath the crust. Consider also that the crust above the slab window would be experiencing a "relaxation" from the compressive stress which had been occasioned by subduction *prior* to onset of the San Andreas fault. Thus, compressive stress oriented perpendicular to the San Andreas fault would have been decreased in the same area beneath which the slab window was developing. It is in the nature of faulting that the type and orientation of faults is controlled by the orientation of the stress field which causes the fault. When the greatest principal stress is vertical, normal faults ensue; the strikes of these normal faults are perpendicular to least principal stress. From the situation described above (i.e., asthenospheric upwelling causing a pronounced vertical stress and postsubduction "relaxation" causing least principal stress oriented perpendicular to the San Andreas fault), the predictable result would be normal faults striking NNW–SSE in the area over the slab window. No doubt, the consequent crustal thinning and extension in the Basin and Range was aided by the fact that the crust in the area was still warm and ductile due to arc-magmatic activity which had occurred there only a short time before.

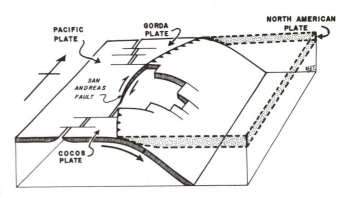

Fig. 9-68. Interpretative sketch showing development of "slab window" accompanying the growth of the San Andreas transform system.

As we said earlier, speculation abounds. It is our oft-stated attitude that such controversies are usually resolved by finding the common ground between opposing models. Thus, some synthesis of the above models may come closest to the truth. For example, an early phase of south-westerly extension in the Basin and Range may have begun due to the slab-window effect while concomitant San Andreas fault-related shear (i.e., distributive shear) led to the development of wrench-fault zone such as the Death Valley and Las Vegas zones. Further, distributive shear may have led to the development of an initial fracture-pattern which later controlled extensional features developed in both the slab-free area *and* the back-arc area farther north, where extension was due to back-arc spreading. Increased seafloor spreading rates during the middle Miocene could be responsible not only for increased back-arc activity but also for more rapid opening of the slab window and increased shear-rate on the San Andreas fault, thus accounting for the observed contemporaneity of these geographically disparate events. It is possible that extension in the slab-free area to the south proceeded at a slightly greater rate than back-arc spreading farther north; the result would have been a clockwise torque on the area west of the region under extension. Alternately, simple dextral shear related to San Andreas faulting may have caused the same clockwise torque. At any rate, the effect was to rotate the Coast Range block in Oregon and Washington, thus accounting for the remaining 10–20° of clockwise rotation described by Simpson and Cox (1977). This torque would probably also account for the north–south compression which caused the Yakima folding event. Admittedly, this synthetic scenario is based less on field data than on tectonic intuition and may well be as full of holes as the central Florida karst; nevertheless, it illustrates our belief that no one model discussed above can be dismissed *in toto* and that the best answer lies somewhere in between.

Associated with Basin and Range extension, the Colorado Plateau and the Rocky Mountains (site of the Laramide orogen) underwent epeirogenic uplift beginning during the Miocene and continuing today. One result was increase in the rate of erosion and concomitant rejuvenation of regional topography in both areas. On the Colorado Plateau, for instance, rejuvenation led to the development of such well-known scenic areas as Bryce Canyon, Glen Canyon, and (especially) the Grand Canyon of the Colorado River. In the Rocky Mountains, uplift rejuvenated a number of Laramide structures so that the resulting topography emphasizes the old basins and uplifts. Differential erosion of Laramide structures resulted in the rugged landscape which today typifies the Rocky Mountains. In addition to topographic rejuvenation, uplift in the Rocky Mountains caused the development of essentially modern drainage patterns in the western continental interior. With the establishment of the western continental divide, virtually the entire continental interior formed a single drainage basin: that of the Mississippi River. Eastward-flowing streams from the rising Rockies carried great quantities of detrital sediments. These sediments accumulated in front of the mountains, forming a great ramp from the flatlands of the Great Plains up to the Rockies. (The stratigraphy of this great mass of alluvial sediment was discussed earlier in this chapter.) On the western side of the Rocky Mountains, fewer streams developed because of the more arid conditions in the central Cordillera.

The cause of epeirogenic uplift in the Colorado Plateau and Rocky Mountains is not well understood. Contemporaneity of uplift with Basin and Range extension suggests that the two processes are related: possibly, extension in the Basin and Range might have caused compression of the crust on either side of the Basin and Range. In this regard, it is interesting to note that uplift of the Sierra Nevada area, just to the west of the Basin and Range, also began at about the same time as did uplift east of the Basin and Range. As we have observed before, contemporaneity does not automatically signify a causal relationship, although in this case the coincidence of uplift in the Sierras and the Colorado Plateau and Rocky Mountains with extension in the Great Basin seems too significant to ignore.

D.3.b. Plio-Pleistocene Events

The tectonic conditions which had been established in the Cordillera during the Miocene have continued, in the approximately 5 Myr. since the beginning of the Pliocene, to dominate geological developments in the region. Since we have already described those conditions in some detail, we limit the present discussion to new developments of the Pliocene and Pleistocene Epochs; in most cases, however, these new developments were related to the same tectonic situations as were those during the Miocene. Further, since this is a historical geology (rather than a neogeology) text, we will not describe Recent conditions except where they differ significantly from those of the past.

California.   Continental borderland tectonics, caused by shear stress related to the transform boundary between the Pacific and North American plates, dominated the Plio-Pleistocene history of southern California (see Fig. 9-69). Extreme rates of subsidence characterized some borderland basins at that time, such as the Ventura and Los Angeles basins (Fig. 9-58), both of which received predominantly marine sediments. In the Ventura Basin, for example, there is over 4500 m of Pliocene strata; since the Pliocene lasted approximately 3 Myr., this stratigraphic thickness indicates a subsidence rate in excess of 1.5 m/1000 yr. (without considering the effects of sediment compaction)! Pliocene sediments of the Ventura and Los Angeles basins are divided

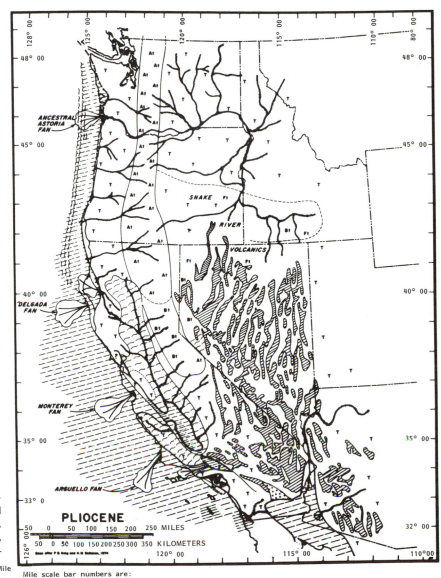

Fig. 9-69. Pliocene paleogeography of the western United States. Symbols are same as those used in Fig. 9-57. (From Cole and Armentrout, 1979, reproduced with permission of the Pacific Section, Society of Economic Paleontologists and Mineralogists.)

into the Repetto and Pica formations and are typified by turbidite sandstones and mudstones which are diatomaceous in some horizons. Natland and Kuenen (1951) said that the Ventura Basin was, at the beginning of the Pliocene, about 1200–1500 m deep but gradually filled. Both of these basins continued to receive marine sediments during the early Pleistocene. In the classic lower Pleistocene section of the Palos Verdes Hills (on the southwest side of the Los Angeles Basin), the main units are the Lomita Marl, Tims Point Siltstone, and San Pedro Sandstone. Similar deposits are found in the Ventura Basin. The lower Pleistocene strata of these basins (and other borderland basins as well) were subjected to moderate to strong deformation and uplift prior to the deposition of upper Pleistocene strata. This deformation event, termed the "Pasadenan orogeny" (King, 1977),

has been considered by some the most significant orogeny since the Cretaceous. But this event was only the most recent in a long series of locally strong deformations in the borderland terrane; thus, the Pleistocene episode probably does not warrant its own name.

Numerous other borderland basins of southern California received Pliocene and Pleistocene marine sediments. An important example was the Salton trough (Fig. 9-58), a northward extension of the newly formed Gulf of California. To the north, in the San Francisco area, both marine and nonmarine sediments accumulated. In the Santa Cruz Mountains west of the San Andreas fault, the marine Purisima Formation is composed of approximately 1800 m of Pliocene volcanic sandstone and mudstone (King and Beikman, 1978). On the east side of the fault is the Merced

Formation which resembles the Purisima but is somewhat younger, extending into the lower Pleistocene. The Merced was deformed and eroded prior to deposition of the upper Pleistocene sequence. In the Berkeley Hills east of San Francisco Bay, nonmarine sediments dominate the Plio-Pleistocene section (e.g., the volcanic Berkeley Group and overlying coarse, continental gravels).

Even farther north, in the northern Coast Ranges, borderland tectonics were not a significant factor and Pliocene-Pleistocene strata are principally composed of marine and marginal-marine sedimentary rocks deposited in local, coastal embayments. To the east, on the other side of the northern Coast Ranges, Plio-Pleistocene strata of the Sacramento and San Joachim valleys are exclusively nonmarine (Fig. 9-69). In addition to their detrital constituents, Pliocene units of these area (e.g., the Tuscan and Mehrten formations) also contain andesitic tuff and other pyroclastic debris. The presence of these volcanic materials argues that arc magmatism continued during the Pliocene in the area northwest of the Mendocino triple junction. Sediments of this age adjacent to the Sierra Nevada, which rose rapidly during the Pliocene to a height of nearly 4300 m, are primarily coarse deposits. Only in the southwestern part of the San Joachim Valley are there significant amounts of marine strata of Pliocene age. These include the Jacalitos and Etchegoin formations which are overlain by the freshwater Tulare Formation which contains a lower Pleistocene fauna. The Pliocene sea in the southwestern San Joachim Valley was probably open to the ocean only through a narrow strait as shown in Fig. 9-69; the position of this strait is now debatable since the whole western block has slipped farther northwestward.

Perhaps the most obvious physiographic change in the southern Cordillera during the Pliocene was the opening of the Gulf of California. Crouch (1979) discussed a model for the evolution of this region (see Fig. 9-70). According to Crouch, tectonic evolution of Baja California can be described as a series of shifts in the position of major plate-margin activity as the Rivera triple junction (Fig. 9-53) migrated southward. At about 18 Myr., the position of the spreading ridge "jumped" eastward to the middle of the Baja Peninsula. This jump caused faulting all along the length of the peninsula and may have been the cause of north–south rifting in the Sebastion Vizcaino Bay area of central Baja. By about 8 Myr. ago, the Rivera triple junction had reached the present tip of the Baja Peninsula and again jumped eastward, this time to its present position. Subsequent seafloor spreading has opened the Gulf of California. Marine sediments of Pliocene age occur in several regions around the margins of the rift, as well as in the Salton trough to the north (which was mentioned earlier). Spreading in this area continues today and the reader is invited to speculate on the future history of the "Baja block." Consider that just such an origin could have led to the formation of displaced terrances such as Wrangellia or the Alexander terrane.

Eastward shift of the main locus of transform faulting may also be seen in southern California. According to Crowell (1974b), the now-inactive San Gabriel fault (Fig. 9-58) was the main fault of the Pliocene transform margin. Ridge basin (Figs. 9-58 and 9-71) was a pull-apart basin formed at the convex-to-right bend in the San Gabriel fault (compare Fig. 9-71 with 9-59A). Later, the locus of faulting jumped to the modern San Andreas fault. As a result of this jump-shift of active faulting, the area of the Transverse Ranges was subjected to north–south compression (in the same manner as illustrated in Fig. 9-59B), causing complex faulting and uplift in that area. This tectonic situation continues today.

A final aspect of Californian geology during the Plio-Pleistocene should be mentioned. The old Sierra Nevada arc (which had been intermittently active since the Jurassic but which had been worn low in the time since the early to middle Miocene when it was abandoned by arc magmatism) was subjected to uplift beginning around 18 Myr. ago (Cole and Armentrout, 1979). The main episode of uplift, however, began during the Pliocene. Axelrod (1957) used fossil floras to show that the Sierra Nevada, during the latest Miocene and early Pliocene, had been a broad ridge with elevations of up to about 900 m. By the end of the Pliocene, the present elevation of over 4250 m had been attained. The concurrence of Sierra Nevada uplift with Basin and Range extension suggests that the two processes were related but the details of the relationship, if any, are uncertain.

Basin and Range. Crustal extension in the Basin and Range continued throughout the Plio-Pleistocene and persists today. But during the Pliocene, the climate of the Great Basin part of the Basin and Range began to change, becoming increasingly more arid as a result of the rising Sierra Nevada just to the west. The Sierra Nevada acted as a barrier to moist, westerly Pacific winds, thus casting a rain-shadow over the Great Basin. With increasing aridity, drainage in the region ceased to be external (except for major through-flowing streams such as the Colorado River, which arises beyond the limits of the Great Basin; the present Colorado River drainage began during the earliest Pliocene as the Gulf of California first began to open). Perhaps the most obvious effect of increased aridity was the onset of desert conditions with concomitant changes in flora, fauna, weathering style, and general topography. Of at least equal significance, however, is the fact that locally derived sediments in such a setting cannot be carried beyond the basin adjacent to their upland source; therefore, great thickness of detritus accumulated in the basins. These deposits are characterized by coarse alluvial sediments around basin margins and finer sediments, sometimes with

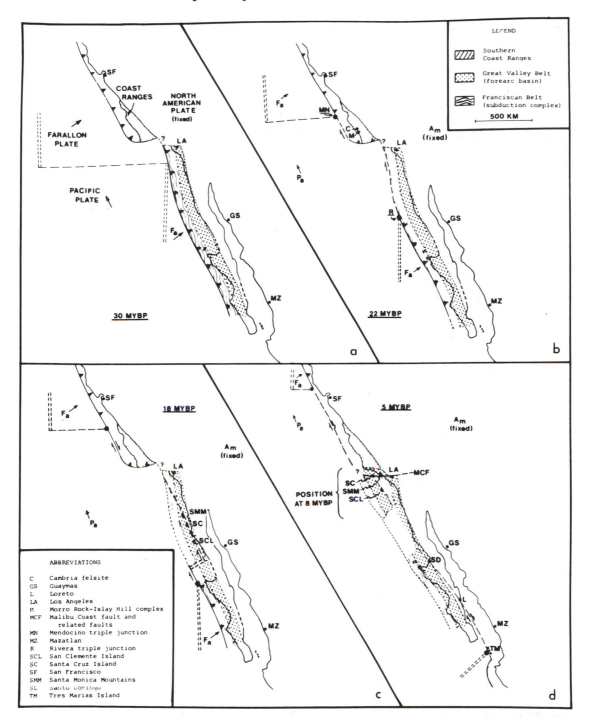

Fig. 9-70. Tectonic model of the evolution of the southern California borderland and Baja Peninsula; see text for discussion. (From Crouch, 1979, *Geological Society of America Bulletin*, Fig. 2, p. 342.)

playa evaporites, in the centers. Pliocene strata in the Basin and Range are given different formational names in different areas. For example, in New Mexico, such deposits are called the Santa Fe Formation; in southeastern Arizona, they are called the Gila Formation; in southern Nevada, the Muddy Creek Formation; in northeastern Nevada, the Humbolt Formation; and in Utah, the Salt Lake Formation. In many basins, these Pliocene units are entirely in the subsurface, being covered with nearly identical Quaternary deposits.

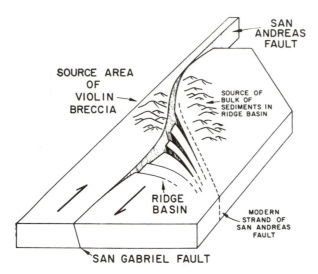

Fig. 9-71. Diagram showing the origin of Ridge Basin at a curve along the San Gabriel Fault prior to initiation of faulting on the present San Andreas Fault. (From Crowell, 1974b, reproduced with permission of the Society of Economic Paleontologists and Mineralogists.)

One important difference between Pliocene and Quaternary deposits of the Basin and Range is the presence in Quaternary strata of extensive lacustrine strata. The lakes in which these sediments were deposited were formed during periods of increased rainfall (so-called "pluvial" periods) associated with episodes of continental glaciation in the northern continental interior. Whereas pre-Wisconsin-age lacustrine deposits have been recognized in the Basin and Range, the most extensive lakes (called Lake Bonneville and Lake Lahontan) were formed during the Wisconsin Glacial. Lake Bonneville, at its maximum extent, covered almost 51,700 km² of western Utah and was probably about 335 m deep (King and Beikman, 1978). The Great Salt Lake today is but a shrunken remnant of the great Lake Bonneville.

An interesting feature of Quaternary tectonics of the Basin and Range is broad epeirogenic upwarping of the entire province. This upwarping has been greatest in central portions of the province such that basin floors in the central area stand 600 to 900 m higher than those on the eastern and western margins of the province. King and Beikman (1978) noted that this is the reason most Quaternary lakes occurred around the provinces' margins rather than in the center. Epeirogenic uplift is probably related to the high levels of heat flow and associated volcanic activity in the Basin and Range.

Pacific Northwest Arc. We turn finally to Plio-Pleistocene strata, mainly of volcanic and Lacustrine origin, in the northwestern United States from northern California to southwestern British Columbia and including rocks of the Snake River Plain. By the end of the Miocene, intrusions and extrusions of andesite and related felsic rocks had built the Cascades arc into a high, impressive, mountainous ridge. But at about 5 Myr. ago, a major change occurred in the nature of magmatic activity: andesitic and dacitic volcanism decreased considerably in intensity and long, north-south-oriented fissures opened along the crest and eastern slopes of the arc. From these fissures issued basaltic lavas which developed large shield and cinder coves, often aligned in linear patterns along the fissures. These rocks, and the relatively minor andesites and dacites, comprise the High Cascades Group (Hammond, 1979). The cause of crustal extension which led to fissuring and basalt eruptions is unclear. Hammond (1979) suggested that a decrease in the rate of subduction of the Juan de Fuca plate may have been responsible; alternately, fissuring and volcanism may represent an extensive of back-arc Basin and Range tectonics into the arc province itself.

Serrating the crest of the High Cascades are large, volcanic cones whose classic, snow-capped features adorn thousands of postcards. These include Mount Rainier, Mount St. Helens, Mount Hood, Mount Shasta, and others. One of these volcanos, "Mount Mazama" in Oregon, exploded in a splendidly violent eruption around 6000 yr. ago, the entire top being blown off and collapsing into a gaping crater. This great hole has partially filled with water and is now called Crater Lake. Lassen Peak in northern California erupted violently in 1914 and 1915 and, of course, Mount St. Helens has recently proven that dormant does not equal extinct. Fumarolic activity and active hot springs indicate that other areas, too, remain thermally active. Magmatic activity has also continued along the extension of volcanic arc in British Columbia, referred to as the Garibaldi arc.

In the Columbia Plateau region, behind the Cascades arc, eruption of the Yakima Basalt (upper unit of the Columbia River Group) persisted into the Pliocene but ceased soon thereafter. Upper Miocene and lower Pliocene sediments of the Ellensburg Formation were deposited in yet another large, lava-damned lake, this one on the west side of the Columbia Plateau, between it and the Cascades arc. In northern California, basaltic volcanism activity also continued and was similar to High Cascades Group activity to the north with the exception of greater amounts of felsic volcanism associated with it. Also, lacustrine strata occur, e.g., the Alturas Formation, containing tuffaceous and diatomaceous sediments.

Extensive volcanism also occurred on the Snake River Plain during the Plio-Pleistocene. The earliest activity is recorded by the Idavada Volcanics, which outcrop in a broad zone in southwestern Idaho. Idavada rocks consist principally of ash-flow tuffs and felsic lavas. Silicic volcanism changed gradually to mafic and the Idavada is overlain by the Banbury Basalt of the Idaho Group which contains middle Pliocene mammals in sedimentary interbeds.

Upper Idaho Group strata contain lower Pleistocene faunas. These, in turn, are overlain by Quaternary rocks of the Snake River Group. All of the above-mentioned activity along the Snake River Plain is believed to be related to the passage of North America over a deep hot-spot. The location of the hot spot today is thought to be under the Yellowstone National Park area of northwestern Wyoming where Quaternary volcanic rocks, and abundant hot springs, geysers, and other hydrothermal features indicate very high levels of heat flow.

This brings us to the end of our discussion of the Cenozoic development of the North American Cordillera. We have not examined events along the Queen Charlotte Islands transform system or the Alaska–Aleutian magmatic arc because the general tectonic setting has changed little since the late Miocene. Before turning to the north, however, we might note that the extremely complicated nature of Cordilleran tectonics portends some very interesting events in the future, especially when North America completely overruns the Juan de Fuca plate and, possibly, a transform boundary will run the entire length of the Cordilleran continental margin. Time will tell.

E.   The Innuitian Continental Margin

The Cenozoic history of North America's Innuitian margin, like its earlier history, is enshrouded in cold and ice so that only its broad outlines can be perceived. Those outlines, however, hint frustratingly of major tectonic developments. For example, the long history of the Sverdrup Basin was finally ended in the paroxysms of the mid-Tertiary Eurekan orogeny. Related to the Eurekan orogeny was a series of rifting events during which developed the gross configuration of the Queen Elizabeth Archipelago; these rifting events were probably associated with the separation of Greenland from North America. Erosion of Eurekan highlands yielded considerable amounts of detrital sediment which were deposited on the continental margin to form the Arctic Coastal Plain and adjacent sedimentary prism underlying the modern continental shelf, slope, and rise. The Cenozoic history of northern Alaska is less complicated than that of northern Canada because Alaska was largely unaffected by Eurekan deformation; consequently, geological events in northern Alaska were dominated by erosion of the ancestral Brooks Range and concomitant deposition of detritus on the Arctic Coastal Plain.

E.1.   Opening of the Eurasian Basin

The reader will recall from our discussion of the Zuni Sequence in the North that the Arctic Ocean is divided by the Lomonosov ridge into Amerasian and Eurasian basins

(see Fig. 8-59) and that the Amerasian Basin had been opened during the Mesozoic Era as a result of the Boreal Rifting event (Kerr, 1981). Although late stages of Boreal rifting persisted into early Tertiary time, most of the Tertiary was dominated by a different rifting event which led to the opening of the Eurasian Basin. Because this and related events greatly affected the development of the Innuitian margin, we will consider them before turning to the details of continental margin geology.

Just prior to the end of the Cretaceous, northern Europe, Greenland, and North America were closely adjacent to each other and a continuous continental shelf to their north was shared by all three. It is believed that the Lomonosov Ridge comprised the outer margin of that shelf. According to Pitman and Talwani's (1972) analysis, the North Atlantic ridge propagated northward during early Tertiary time with Europe and North America swinging away from each other like opening scissors-blades. The result was initiation of a divergent plate boundary beneath the northern continental shelf. Rifting along this new plate boundary split the Lomonosov Ridge off from the Barents Shelf and led to the formation of the Gakkel Ridge. From then to the present, the Gakkel Ridge has continued to be an active spreading ridge and has presided over the opening of the Eurasian Basin.

Intimately associated with extension of North Atlantic spreading into the Arctic was separation of Greenland from North America. Details of this event are poorly known and disagreement exists over the timing of rifting; nevertheless, there is general agreement that Greenland first moved with the European plate. Only later, possibly due to a jump-shift of seafloor spreading, did Greenland cease its differential drift with respect to North America (see Fig. 9-72). During Greenland's early Tertiary separation from North America, Baffin Bay and the Labrador Sea were opened as a continental rift propagated northward. By the middle Tertiary, this rift system had encountered the eastern Innuitian area and played a major role in the tectonic events which subsequently affected North America's northern margin.

E.2.   Cenozoic Events in Northern Canada

E.2.a.   Sverdrup Basin

The Sverdrup Basin, which had begun during the latest Mississippian, finally ceased to act as a subsident feature during the early Tertiary. Termination of subsidence was caused by the Eurekan orogeny, a relatively minor tectonic event whose greatest effects were felt in the eastern Sverdrup area, especially on Ellesmere and Axel Heiberg islands. Actually, the basin's orogenic climax was foreshadowed during latest Cretaceous and Paleocene time by mobilization of several arches within the basin. These arches (e.g., Corn-

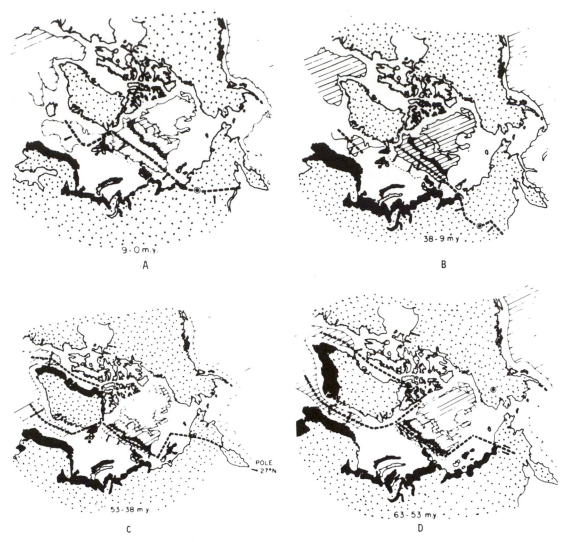

Fig. 9-72. Plate-tectonics model for opening of the Eurasian Basin, Arctic Ocean, during the Cenozoic. Hachures represent areas underlain by oceanic crust; crosshatch pattern represents the Lomonosov Ridge; dashed lines designate boundary between American and Eurasian plate; bull's-eye (in D) is the rotational pole of Eurasia's movement relative to North America. The amount of rotation of Eurasia and Greenland relative to North America is indicated by blackened areas, which show positions of continental blocks at the beginning of each interval represented. (From Herron *et al.*, 1974, *Geology*, Fig. 2, p. 378.)

wall and Princess Margaret arches; see Fig. 9-73) were positive features which subdivided the Sverdrup Basin into smaller subbasins and which shed detrital sediments into those subbasins to form the Eureka Sound Formation. The Eureka Sound was the last unit to be deposited in the Sverdrup Basin; it is composed of sandstone, shale, and coal seams and is considered to be of nonmarine origin. Adjacent to the intrabasin arches, Eureka Sound strata are very coarse-grained (even being termed a fanglomerate where associated with contemporaneous faults; Thorsteinsson and Tozer, 1976) and overlie older rocks with angular unconformity. In centers of subbasins, however, Eureka Sound rocks are finer-grained and conformably overlie the Upper Cretaceous Kanguk Formation.

Balkwill (1978) considered arch-subdivision of the Sverdrup Basin to be the first phase of the Eurekan orogeny. Kerr (1981) proposed that the cause of this event was related to the onset of rifting in the Baffin Bay area as the Greenland–North America divergent boundary propagated northward. Figure 9-73 illustrates Kerr's tectonic model; notice that Boreal rifting is shown still to have been active, causing oceanic crust of the Amerasian Basin to slip past the outer rim of the Sverdrup Basin along the Kaltag transform fault and leading to the rift-opening of the western Parry

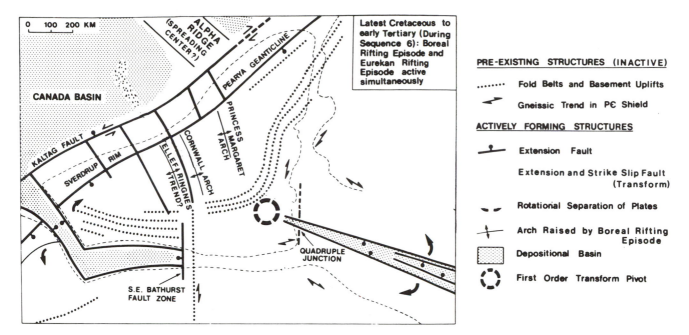

Fig. 9-73. Simultaneous development of Boreal Rifting event (associated with spreading of Canada-Basin oceanic crust past the Innuitian margin along the Kaltag Transform Fault) and the Eurekan Rifting event (associ-ated with opening of Baffin Bay) resulting in activation of arches within the Sverdrup Basin. (From Kerr, 1981.)

Channel. As rifting began in the ancestral Baffin Bay, stresses set up by that event interacted with shear stress associated with Kaltag-fault movement to mobilize the intrabasin arches (most of which were old basement features, not active since far back in the Paleozoic). Continuation and intensification of these stresses led to the main phase of the Eurekan orogeny.

E.2.b.   Eurekan Orogeny and Related Events

During the middle Eocene, the Eurekan orogeny entered its main deformational phase, characterized in the eastern Sverdrup Basin by compressional folding, thrust-faulting, and uplift (see Fig. 9-74). In the center of the Sverdrup Basin, the geometry of Eurekan folding was strongly influenced by diapiric intrusion of evaporite masses derived from Penno-Permian strata near the bottom of the basin. Evaporites may also have provided low-shear-strength horizons facilitating thrust-faulting (e.g., the Stolz fault on Axel Heiberg Island; Trettin, 1972). In areas to the south of the main compressional zone of the Eurekan orogen, deformation consisted of large-scale movements along high-angle normal and reverse faults as a result of which large basement blocks shifted vertically past each other. Folding of supracrustal strata above these vertically shifting basement blocks was of the drape-folding style, similar to folding associated with the Laramide orogeny. Igneous activity associated with the Eurekan orogeny was minor and limited primarily to felsic and subsidiary mafic volcanism along the northern margin of Greenland (the Kap Washington Group; Dawes and Peel, 1981). Regional metamorphism associated with the Eurekan orogeny has not been reported.

It is significant that tensional deformation was occurring to the southeast and west of the Sverdrup Basin at the same time that the compressional Eurekan orogeny was affecting central and eastern parts of the basin. As a result of further westward drift of Greenland, the Baffin Bay rift continued to enlarge. At the end of the modern Baffin Bay, resultant tensional-fracturing splayed out into three separate rifts, propagating westward to form the Lancaster Sound aulacogen, northward to form the Nares Strait rift basin, and northwestward to form the Jones Sound rift (between Ellesmere and Devon islands). This major episode of crustal rifting was termed the Eurekan Rifting event by Kerr (1981). In the western Queen Elizabeth Islands, tensional stress, possibly associated with the final throes of the Boreal Rifting event, led to the development of the Eglington graben and rejuvenation of the Prince Patrick Uplift (Trettin, 1972).

Kerr (1981) proposed a tectonic model of Eurekan deformation which genetically unites both compressional and tensional features of the orogeny (see Fig. 9-74). Kerr postulated that separation of Greenland from North America across the Labrador Sea–Baffin Bay rift was the prime cause of Eurekan events. As mentioned above, tensional

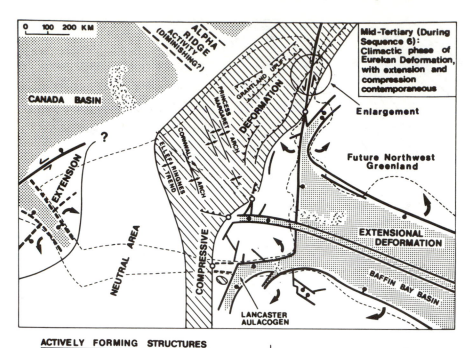

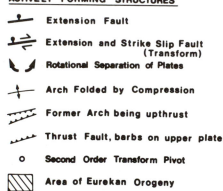

ACTIVELY FORMING STRUCTURES

- Extension Fault
- Extension and Strike Slip Fault (Transform)
- Rotational Separation of Plates
- Arch Folded by Compression
- Former Arch being upthrust
- Thrust Fault, barbs on upper plate
- o Second Order Transform Pivot
- Area of Eurekan Orogeny
- Depositional Basin

ENLARGEMENT, NARES STRAIT

Compression and incipient Strike Slip (Transpression)

Fig. 9-74. Kerr's (1981) tectonic model for deformational events associated with the climactic phase of the Eurekan orogeny; see text for discussion.

fracturing and consequent rifting splayed out from the head of Baffin Bay, possibly because that area served as the pivot about which the Baffin Bay rift opened (see Fig. 9-73). A number of lesser tensional fractures developed as complex conjugate faults associated with the major rifts and these served to fragment the area south of the main Eurekan orogen into decoupled basement blocks which were jostled against each other to produce the vertical crustal movements and supracrustal drape folding mentioned earlier.

As crustal extension pulled the Lancaster Sound and Nares Strait rifts apart, compression was set up in the northeastern Sverdrup Basin, which is located between these two rifts at the apex of the Baffin Bay rift. Uplift, such as along the Cornwall and Princess Margaret arches, probably was caused by compression of the Sverdrup block caught between the Baffin–Lancaster Sound–Nares Strait rifts and

the Boreal-rifting-related Parry Channel rift to the west. Major thrusting on Axel Heiberg and Ellesmere islands was due to transpression (i.e., compression and transform faulting) required by the geometry of interaction of adjacent crustal blocks (see Fig. 9-74).

The final phase of the Eurekan orogeny (during the Miocene) resulted when the Sverdrup block became completely separated by rift valleys from mainland continental crusts (see Fig. 9-75). The Lancaster Sound aulacogen broke through to the Parry Channel rift while the Nares Strait rift propagated entirely through the crust to the Arctic Ocean. With complete separation, the Sverdrup block became a separate continental-crustal element, capable of independent movement and no longer as susceptible to internal deformation. Further microplate interactions might have been expected but none occurred. This is probably because

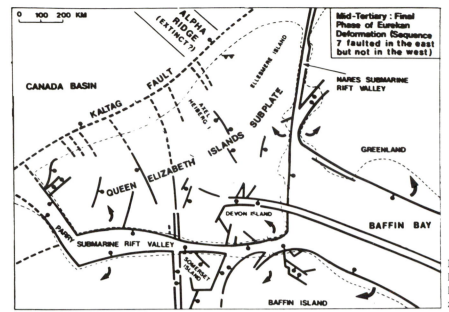

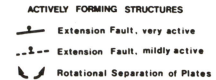

Fig. 9-75. Final phase of the Eurekan orogeny (early Miocene or later) when the Queen Elizabeth Islands subplate became completely surrounded by rift zones. (From Kerr, 1981.)

drift of Greenland away from North America had ceased, thus depriving the Innuitian area of the forcing jaw of the vise that had caused Eurekan orogenic deformations. During this time, erosion beveled the Eurekan orogen, production detrital sediments whose deposition on the northern continental margin formed the Arctic Coastal Plain.

E.2.c. The Arctic Coastal Plain

The sedimentary prism which underlies the modern continental margin of northern Canada (and Alaska; see below) probably began to develop as far back in time as the middle Mesozoic as a result of the opening of the Amerasian Basin. However, throughout much of its early history, apparently little sediment accumulated there because the Sverdrup Basin, which lay between the continental mainland and the continent's outer margin, acted as a trap for continent-derived detritus (Trettin, 1972). The Eurekan orogeny altered that pattern by causing the collapse of the Sverdrup Basin so that it no longer functioned as a sediment trap; obviously, uplifts associated with Eurekan deformation provided an additional source of detrital sediments. Thus, considerable volumes of sediment began to accumulate on the northern continental margin, forming the thick sedimentary prism which underlies the continental shelf, slope, and rise. Very little is known about this prism because most of it is beneath the frigid waters of the Arctic Ocean and has been explored only briefly with geophysical methods. The prism thickens from approximately 2.5 km under Ellef Ringnes Island to about 7.5 km under the continental rise (Trettin, 1972).

The only subaerially exposed unit of the continental-margin prism is the Beaufort Formation which underlies virtually the entire Arctic Coastal Plain of Canada. The Beaufort Formation is mainly of alluvial origin, except in scattered coastal areas where some Beaufort strata contain marine mollusks. Beaufort deposits consist of unconsolidated quartz sand, often highly crossbedded, with lesser amounts of gravel, silt, clay, and rare peat. A notable minor constituent of Beaufort sediments is wood (called "driftwood" by Trettin, 1972). Some of this wood is well-preserved and can be identified as spruce and pine. Palynomorphs include representatives of both these trees as well as birch, alder, and other floral elements of a cool, temperate forest (Thorsteinsson and Tozer, 1976). The Beaufort is considered to be of Miocene–Pliocene age (Hills, 1970) and is overlain by glacial deposits of Pleistocene age.

E.3. Northern Alaska

The area north of Alaska's Brooks Range was essentially unaffected by deformation associated with the Eurekan orogeny and, therefore, continued throughout the Tertiary to develop as it had during the later Cretaceous. Sediments derived from weathering and erosion in the Brooks Range were carried northward in numerous fluvial systems to be deposited on Alaska's North Slope and continental margin. These deposits comprise the Sagavaniktok Formation, which ranges in age from Paleocene to Pliocene. Sagavaniktok strata consist of poorly consolidated nonmarine and shallow-marine shale, sandstone, and conglomerate with minor carbonaceous shale, lignite, and bentonite (Grantz et al., 1981). Plant fossils preserved within the Sagavaniktok represent a cool-temperate climate (as do similar fossils in the Beaufort Formation of Canada). Nonmarine facies are developed mainly in southern North Slope areas adjacent to the Brooks Range and grade northward into marine facies near the modern shoreline and beneath the Western Beaufort Shelf.

Unconsolidated Pleistocene deposits of northern Alaska are termed the Gubik Formation (Dutro, 1981). The Gubik is composed of glacial deposits. Post-Wisconsinan deposits of the North Slope consist entirely of thin, unconsolidated surficial sediments, mainly of alluvial origin.

F. The Quaternary

F.1. Introduction

The most distinctive geological events occurring during the latter part of the Cenozoic Era were directly or indirectly related to the great continental glaciations. Beginning during the later Pliocene Epoch and extending through the entire Pleistocene Epoch, substantial portions of North America were intermittently covered with ice and then deglaciated, leaving complex sequences of proglacial, glacial, postglacial, and periglacial geological features over much of the continent's surface. The pre-late Wisconsinan glacial events (i.e., prior to ~ 27,000 yr. B.P.) are far less clear than are events of the last major glaciation which culminated at about 18,000 yr. B.P. Understanding of the timing and nature of pre-Illinoian glacial events (i.e., prior to ~ 400,000 yr. B.P.) is undergoing drastic reexamination at present, and any discussion presented in this book is likely to become archaic; therefore, we will be brief and noncommital in much of the subsequent discussion.

During periods of maximum ice buildup, the continent probably appeared as shown in Fig. 9-76. The areal extent of North American continental ice was at least 16 million km^2 (Nilsson, 1983). East of the Rocky Mountains, a single vast ice sheet covered Canada, parts of the Arctic Archipelago, and northern United States, reaching as far south as latitude 39° in the Mississippi River valley. This Laurentide ice sheet, as it is generally named, actually featured two centers of ice buildup: one to the east of Hudson Bay, in Nouveau Quebec–Labrador, and one to the west of Hudson Bay in the District of Keewatin, Northwest Territories (data from Shilts, 1980). To the northeast, the Laurentide ice sheet was continuous with the Ellesmere–Baffin glacier complex, which probably included large areas of thin sea-ice (Flint, 1971). The Ellesmere–Baffin glaciers connected to the north intermittently with the Greenland ice cap (England and Bradley, 1978), which is present today as one of two surviving continental glaciers. The present Greenland glacier is approximately 80% as extensive as it was during maximum glaciations.

To the west, the Laurentide ice sheet met the Cordilleran glacier complex. However, the nature of the juncture is controversial; either the ice masses intergraded at the eastern flank of the Rocky Mountains, or an ice-free corridor was present in between. The Cordilleran ice consisted of a variety of alpine and piedmont glaciers, which formed in and spread down from the Canadian Rockies, the Brooks and Aleutian ranges in Alaska, and the northern U.S. Rocky Mountains and Sierra Nevada.

The essential difference between conditions on the present Earth, with continental glaciers occupying only Greenland and Antarctica and with alpine glaciers at high elevations, and conditions during glacial maxima, is largely one of scale. During "ice ages," mean annual temperatures in northern continental interiors may have averaged 10°C cooler than at present. However, temperatures in equatorial regions may have been no more than 1°C cooler than at present, suggesting that enhanced seasonal and latitudinal

Fig. 9-76. North America during a maximum glacial interval (the Woodfordian advance of the late Wisconsinan). The hypothetical sea level is shown at the approximate present 100-m submarine contour; speculative ice margins are indicated by broken hachured lines. Lettered ice masses are: A, Cordilleran glacier complex; B, Baffin–Ellesmere Islands glacier complex; C, Greenland ice sheet. (Data from Flint, 1971; Shilts, 1980; Mickelson *et al.*, 1983.)

differences were characteristic of glacial episodes. Alpine glaciation occurred at lower elevations during glacial episodes, and Flint (1971) presented a simple curve relating latitude and elevation, and demonstrating the interactive nature of global temperature, altitude, and glacial ice (see Fig. 9-77).

The effects of late Quaternary glaciations extended well beyond the region directly covered by ice or meltwaters. The southerly extension of cold air masses affected global wind circulation and caused southerly migration of temperate cyclonic-storm belts associated with prevailing westerly winds. Consequently, greater amounts of rain fell in formerly (and presently) arid regions, including the Great Basin, where vast lakes formed within internally drained basins. These so-called ''pluvial lakes'' persisted well into Holocene time and left elevated desert beaches and the Great Salt Lake as relics.

The last Pleistocene glaciation, the Wisconsinan, was among the most pervasive in areal extent and, therefore, it left deposits covering most previously glaciated areas. Only in the upper Mississippi River valley and southern Great Lakes region can pre-Wisconsinan drift be identified on the surface; and, even there, the record is very spotty. Unlike most other erosional–depositional processes, glaciation typically obliterates much of the previous surficial sedimentary and structural evidence (see Gibbons *et al.,* 1984); and glaciers typically redeposit sediments in a largely unweathered state. Consequently, it requires the most painstaking study to determine whether a Wisconsinan moraine is composed of redeposited older till or represents a first glacial incursion into a given region.

F.2. Late Cenozoic Paleoclimatology

Glaciation obviously resulted from general climatic cooling during the late Pliocene and Pleistocene. The pattern of advances and retreats of continental glaciers probably involves additional factors, such as astronomical causes (discussed subsequently; see Section F.11), but these do not detract from the fundamental temperature drop which must have preceded or accompanied glaciation. The record of such global climatic cooling is best documented by two sources of data: oceanic strata in cores, and ice cores.

F.2.a. Ocean Cores

The upper portions of oceanic strata contain a sedimentary record which may reach back in relatively undisturbed state at least into the later Tertiary. Within these sediments are two excellent sources of paleotemperature data: (1) calcareous material (often foraminiferal tests) which yield $^{18}O/^{16}O$ ratios, and (2) species of planktonic foraminifera which show latitudinal and hence temperature preferences in modern oceans (and by implication, in the Pleistocene and Tertiary oceans).

The principle involved in oxygen isotope paleotemperature work is based on the relative rates of evaporation of the various oxygen isotopes. ^{16}O, being the lightest, evaporates fastest and, thus, enriches atmospheric water reservoirs, which, in turn, enrich glacial ice in the isotope. In consequence, seawater tends to become depleted in ^{16}O (or it may be viewed as being enriched in ^{18}O) during glacial intervals, and vice versa. As oxygen is taken up in $CaCO_3$ during formation of calcareous shells of organisms, two factors control the isotopic makeup: (1) isotopic composition of seawater, as explained above, and (2) ambient temperature of the seawater. The final result is a relatively sensitive indicator of past conditions of both the amount of glacial ice on Earth, and the ancient marine temperatures. Calculation of $^{18}O/^{16}O$ ratios is done against a standard (derived from a Cretaceous belemnite rostrum) and is done according to the formula:

$$\delta^{18}O = \text{ratio } (^{18}O/^{16}O) = \frac{^{18}O/^{16}O \text{ sample}}{^{18}O/^{16}O \text{ standard}} - 1$$

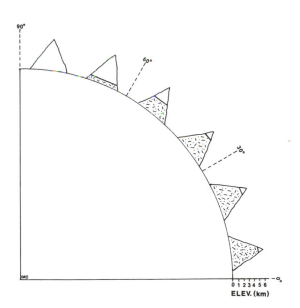

Fig. 9-77. Interrelationship of latitude, elevation, and iceline. Iceline, i.e., the condition under which annual snow retention exceeds ablation, is shown here to be related to latitude and elevation; thus, at high latitudes (e.g., 60°), under modern conditions, perennial ice may form at approximately 1500 m elevation. At lower latitudes, higher elevations are necessary to delimit iceline, reaching elevations over 6000 m near the equator. During a glacial event, the iceline migrates simultaneously lower in elevation and latitude, such that glacial ice may form at nearly 0 m elevation down to approximately latitude 45°. Conversely, during an especially warm interval, there may be no iceline at 0 m elevation even at the poles. (Based on concept in Flint, 1971.)

Paleotemperature data are obtained from fossil foraminifera rather handily, because many temperature-dependent taxa or morphological variants survive at present after long histories of Cenozoic existence. For example, *Globorotalia truncatulinoides* is a dominantly left-coiling form in the polar plankton, whereas in mid-to-moderately high latitudes it is found in a right-coiling variant. Therefore, given a Temperate Zone marine core, if the layers below the topmost contain *G. truncatulinoides* showing predominantly left-coiling, one may infer that the sediment was deposited under cooler regimes. The problem remains with both isotopic and foraminiferal data to determine the time represented by the data.

In recent years, paleomagnetic reversal chronology has been available to solve the ocean-core correlation problem. The practice is based on the known patterns of paleomagnetic reversals, which have been determined with reliability back to the late Middle Jurassic (see Palmer, 1983). Assuming the present magnetic polarity to be "positive," the Earth's magnetic poles have reversed irregularly but frequently (roughly one to five times per million years) in patterns which may be determined by remnant magnetism in a variety of geological media (oceanic iron-rich basalts are excellent, but many strata contain sufficient ferrous materials to yield remnant magnetism). If one can match the patterns of remnant magentism in a given core with published reversal chronologies, the dating of the core is straightforward. In practice, it is commonplace to see published marine core data simultaneously showing magnetic polarity dating, and foraminiferal and isotopic paleotemperature data.

All is not idyllic, however. Many marine sediments are compressed, which distorts paleomagnetic patterns and impairs dating. Then too, it is often mechanically difficult to obtain deep cores in many areas, and it has been frequently observed that there may be poor correspondence between ocean cores and continental glacial records, suggesting that low ocean temperatures may not correlate with glacial episodes (or, perhaps more logically, there is a significant time lag between global climatic cooling and the maximum advance of continental glaciers). Nevertheless, the oxygen isotopic record is sufficiently clear to define general cooling and warming trends through the Quaternary and well back into the Tertiary (see, e.g., Woodruff *et al.*, 1981).

F.2.b.   Ice Cores

A related method for determining paleoclimates uses ice cores, which are analyzed for their oxygen isotopic ratios in the H_2O molecule in the same manner as ocean core carbonates were analyzed. Ice cores have the advantage of yielding continental climate data, which are more directly applicable to glacial studies; but they have the dis-

advantage of incorporating generally shorter intervals of time than ocean cores. Among the first and best-known ice core projects was the Camp Century core, from northwestern Greenland (Dansgaard *et al.*, 1971). This core penetrated 1390 m and reached bedrock, incorporating minimally 120,000 yr. of ice accumulation. Unfortunately, it is not possible to absolutely date ice cores with confidence beyond approximately 100,000 yr. Ice core dating is done by extrapolation from calculated rates of accumulation compared with ice thickness; as the ice lies closer to the bottom, it undergoes plastic deformation and compaction, thereby distorting thickness-to-age ratios.

Analysis of the Camp Century core (Fig. 9-78) is interesting for several reasons. First, as mentioned, it provided among the first data on continental climates derived from fossil ice. Second, the paleotemperature record from the ice tended to corroborate the past 100,000-yr. history of

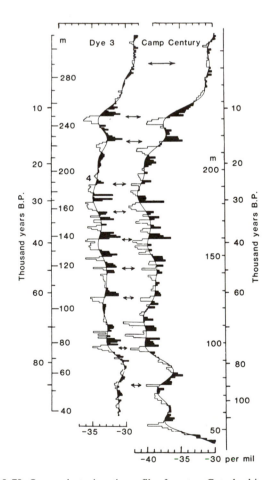

Fig. 9-78. Oxygen isotopic ratio profiles from two Greenland ice cores. The Camp Century core (right) penetrated older ice but the correspondence in isotopic ratios between the two sets of data reinforces the validity of isotopic paleotemperature analysis from ice. (From Dansgaard *et al.*, in *Science*, © 1982, American Association for the Advancement of Science.)

glaciations extrapolated from till-and-loess stratigraphy in Canada, Europe, and the northern United States. And finally, the ice core revealed at least one episode of severe cooling in the pre-Wisconsinan layers that was not evident from the glacio-stratigraphic record in lower latitudes. Ice core work continues at present, with additional published reports from southern Greenland (Dansgaard *et al.*, 1982) and the Ross ice shelf in Antarctica (Zotikov *et al.*, 1980).

F.2.c. Additional Paleoclimatic Data

Several additional sources of paleoclimatic data have been used in recent years to document the timing of glacial episodes of the late Cenozoic. Chief among these are pollen analyses, and studies of submerged and elevated beaches and terraces.

The basis of beach and terrace work is the assumption that glacial episodes produce eustatically lower global sea levels due to the uptake of water to form ice. To minimize the component of glacio-isostatic rebound, many such studies have dealt with low-latitude areas unencumbered by glacio-isostatic effects (see, e.g., Karrow and Bada, 1980). Cronin *et al.* (1981) went beyond study of the physical terrace/beach levels and used uranium and thorium dates from corals associated with Pleistocene marine deposits to derive a paleoclimate key for selected sites. Their work suggested that warm climates prevailed at 188,000, 120,000, 94,000, and 72,000 yr. B.P.

Pollen data from nonmarine sources have been especially valuable in analyzing late Quaternary paleoclimatology in nonglaciated areas. Since few, if any, major plant taxa at present have evolved subsequent to the later Pleistocene, correlation of the distribution of present versus past plant biomes yields good evidence of climatic change. Adam *et al.* (1981), for example, presented an estimated continuous 130,000-yr. pollen record from a California lake, dated by a combination of isotopic methods and correlation with marine core oxygen isotopes. They analyzed the interchange of oak and conifer species to determine temperature changes through time. Heusser and Shackleton (1979) correlated an estimated 150,000-yr. record of pollen from a nearshore ocean core with the marine oxygen isotope record, and showed a strong correlation between the ratios of hemlock and spruce pollen and the oxygen isotopic concentrations. Their work suggests that ocean temperatures and land-floral responses occur nearly simultaneously.

F.3. The Pliocene–Pleistocene Boundary Problem

As prior discussion noted (see Section A), the Pleistocene Epoch was originally designated as the Cenozoic inter-val featuring cold-water marine bivalves at the top of the Italian coastal stratigraphic section; however, through use the term became synonymous with the time of late Cenozoic glaciations. Detailed work during the past two decades has shown that substantial glaciations occurred in the late Pliocene, and probably range back into Miocene time. Consequently, one cannot automatically designate continental Pleistocene strata based on glacial deposits and, similarly, one cannot definitively date the Plio-Pleistocene boundary because its definition is equivocal.

As an example of the discord present in the boundary question, two fairly recent and authoritative publications specify very different Plio-Pleistocene dates. Palmer (1983), in a comprehensive North American time scale, stated that the boundary dated to 1.6 Myr. Nilsson (1983), in a comprehensive volume on the Quaternary, placed the boundary somewhere between 1 and 3 Myr., and listed eight lines of evidence showing the date to be no younger than 2 Myr. Boelstorff (1978), although now a dated source in such a volatile subject, specifically addressed the history of the problem and concluded that 2.5 ± 0.1 Myr. was the beginning of the Pleistocene. Two questions require solution before the boundary date can receive agreement: What comprises the nature of the epochal break? How can one recognize it in varied geological environments?

As noted above, Lyell's original definition can be used to recognize the Pleistocene in fossiliferous marine sequences. On the continents, a similar approach can be used with land mammal ages (see Section A), if one assumes absolute dating of the marine Pleistocene yields dates for the continental Pleistocene. The Irvingtonian mammal fauna, based on suites in California, is generally recognized as an early Pleistocene assemblage which includes the first North American appearance of jackrabbits (*Lepus*), mammoths (*Mammuthus*), and muskrats (*Ondatra;* data from Kurtén and Anderson, 1980). Savage and Russell (1983) presented a comprehensive global analysis of mammalian faunas through time, and assumed that the Blancan-to-Irvingtonian faunal change occurred at the Plio-Pleistocene boundary; their date for the event was 1.8 Myr.

F.4. Pre-Wisconsinan Laurentian Events

The classical upper Mississippi River valley Quaternary chronology, presented in Table II, shows three major glacial and interglacial intervals dating from the Plio-Pleistocene to the Wisconsinan. However, current studies, many awaiting publication, present an emerging impression that classical glacial stratigraphy is far too simple. It is now evident that the late Wisconsinan record shows a multitude of glacial advances and retreats, often spaced by less than 10,000 yr., and derived from local lobes of the Laurentide ice sheet. And, although the data preserved from pre-late

Table II. "Classical" Pre-Wisconsinan Glacial Chronology[a]

Glacial stage	Stadial	Stratotype (location)	Age (onset, yr. B.P.)
Wisconsinan Glacial		(See Table III)	100,000
Sangamonian Interglacial		Sangamon Soil (Illinois)	250,000 ± 50,000
Illinoian Glacial	Jubilean (or Buffalo Hart)	Radnor Till, Sterling Till (Illinois)	?
	Monican	Vandalia, Hulick, Ogle Tills (Illinois)	?
	Liman	Petersburg Silt, many tills (Illinois)	?400,000
Yarmouthian Interglacial		Yarmouth Soil (Iowa)	?600,000
Kansan Glacial		Nickerson, Cedar Bluffs, Clarkson Tills (Kansas)	1,000,000 ± 100,000
Aftonian Interglacial		Afton Soil (Iowa)	?
Nebraskan Glacial		Elk Creek Till, Iowa Point Till (Nebraska)	2,000,000 ± 500,000
		Pliocene Epoch	

[a]Data from Wright and Frey (1965), Flint (1971), Mahaney, ed., (1976).

Wisconsinan events are often poor, it is very reasonable to assume that multilobate ice margins and localized ice dynamics were features of the early Pleistocene and later Pliocene glaciations as well; therefore, one must assume that there is much greater complexity to early Quaternary glaciations than the available evidence implies.

Prior discussion highlighted some of the reasons glacio-sedimentary correlations are difficult. Quaternary specialists have long recognized that there were probably substantial errors in interregional glacio-stratigraphic correlations, and that tenuous arrangements are presented in even the midcontinent's "classical" Quaternary chronologies. Current evidence shows that glaciations extend well back before the earliest Pleistocene, and that the classical "Nebraskan" and "Kansan" deposits may be the scanty remains from a series of glaciations.

A given glacial advance and retreat ideally leaves many types of deposits in the diverse areas affected by the event. Immediately prior to glaciation, tundra conditions may prevail and fine-grained lacustrine and eolian sediments, especially varved lake beds, are characteristically deposited. Glacial ice leaves several types of tills in a variety of morainal and tabular morphologies (see Mickelson *et al.*, 1983, and Nilsson, 1983). In addition, various types of striations and lineations, exotic boulder deposits, eskers, and other characteristic structures and materials testify to the proximity of glacial ice. Several varieties of lacustrine deposits (e.g., in morainal lakes, kettle lakes, outwash lakes, periglacial lakes, and kame deltas), as well as outwash, may form during or after glaciation. Loess generally forms after a substantial area has been denuded by glaciation, but a given loess deposit (Fig. 9-79) may actually be a redeposition of older windblown peri- and post-glacial materials. Finally, with return of warmer climates, soil forms from all or part of the above-mentioned material, and tends to obscure part of the depositional record.

However, any or all glacial evidences may be lost to erosion or may never have formed in a given site from a

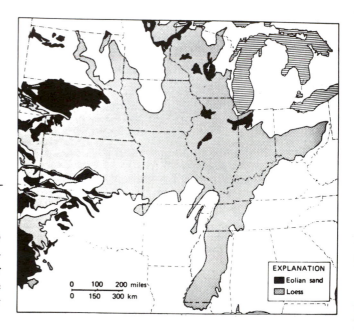

State / Stage	Nebraska	Iowa	Illinois	Indiana
Wisconsin	Bignell Loess / Peorian Loess / Gilman Canyon Loess	Wisconsin loess	Peoria Loess / Farmdale Silt / Roxana Silt	Peoria Loess Member silt bodies
Illinoian	Loveland Loess / Beaver Creek Loess / Grafton Loess	Loveland Loess	Loveland Silt / Petersburg Silt	unnamed loess
Kansan	Sappa Loess	unnamed loess	unnamed loess	Cagle Silt
Nebraskan				

Fig. 9-79. Areal extent of Quaternary loess and eolian sand deposits in the United States; and a table of eolian formations in the upper Mississippi River valley states which provided part of the basis for the "classical" Quaternary sequence. (From Flint, © 1971 by John Wiley & Sons, reprinted with permission of John Wiley & Sons.)

particular glacial event. In development of the Quaternary history for North America, virtually all types of glacial and periglacial deposits have been used as evidence of ice advances or retreats; a great amount of ice has often been inferred from slim traces of till and loess. Table II includes data on the key stratigraphic units which form the basis for the "classical" pre-Wisconsinan chronology; the remainder of this section will focus on Wisconsinan history, for which there exists better documentation. The reader should be warned, however, that we will generally avoid using names for specific glacial events, and that few specific formations will be described in the remaining discussion of the Pleistocene glaciations. This lack of specifics reflects the current flux in Quaternary nomenclature and stratigraphy, and the fact that glacial formations are inherently local in their extent.

F.5. Wisconsinan Events of the Laurentide Ice Sheet

F.5.a. Constructional Topography and ^{14}C Dating

Since the Wisconsinan glaciations were the last in North America, the preserved records are largely constructional (i.e., structure-forming) and surficial. Many relatively non-resistant structures (e.g., morainal ridges, kettles, eskers, drumlins, kames, deranged drainages) are preserved from the Wisconsinan, underscoring the recency of the last ice age events. It is a matter of conjecture whether the Wisconsinan glaciations exceeded earlier ice advances in areal extent; the preserved record is areally more extensive than that of past ice advances, but pre-Wisconsinan deposits reach south of the Wisconsinan edge in the midwestern United States, suggesting that some pre-Wisconsinan ice lobes were at least locally more pervasive.

Of prime importance to interpretation of the Wisconsinan is the impact of ^{14}C dating, which allows relatively accurate absolute ages to be determined as far back as 50,000 yr. B.P. and, with computer enhancement, to 75,000 yr. or more (Stuiver et al., 1978), using common materials. A substantial portion of the later Wisconsinan stratigraphy and paleoclimatology is based on radiocarbon dates from wood, peat, carbon-rich sediments (especially lake deposits), and preserved animal tissues. Even the maximum radiocarbon range may not encompass the entire span of Wisconsinan time, however, and just as Wisconsinan events are clearer than are those of prior glaciation, so too is the late Wisconsinan much better known than the earlier Wisconsinan.

F.5.b. Wisconsinan Chronology

As with the stage nomenclature of the entire Quaternary, one encounters classical Wisconsinan substage no-

menclature plus a series of revisions, many still in research. The "classical" sequences vary among the Great Lakes states and provinces, most of which have significant Wisconsinan deposits (notably Minnesota, Iowa, Illinois, Indiana, Wisconsin, Michigan, Ohio, Ontario, New York, and Quebec); however, the Illinois nomenclature might be considered the closest to a standard usage available and is as given in Table III.

Although many glacial and *interstadial* (i.e., a short nonglacial episode within a glacial stage) deposits may be correlated with the Illinois "classical" sequence and sequences in other states, about as many cannot be correlated. It is apparent that glacial advances were more frequent and localized than the scheme given in Table III suggests, with overlapping times of advances and retreats by various lobes of the ice front of the Laurentide ice sheet. A more flexible approach to Wisconsinan chronology divides the stage into three units, each potentially incorporating several ice advances and retreats, as shown in Table IV.

Indeed, even Table IV presents too gross a pattern of late Wisconsinan ice advances, as suggested by Mickelson et al. (1983); their tentative correlation chart of late Wisconsinan tills and end moraines shows eight advances and at least that number of interstadials and other categories of ice breakups during the post-Farmdalian (ca. 21,000 yr. B.P. to the Holocene). It is noted in their discussion that one-third of the ^{14}C dates used to delimit their sequence of events have substantial statistical error ranges.

It is beyond the scope of this discussion to analyze regional Wisconsinan stratigraphy because, as explained before, each Laurentide ice lobe probably featured a distinct erosional/depositional history, thereby making the concept of "representative units" such as have been exploited throughout this book, meaningless. Figure 9-76 represents North America at a maximum glacial substage, say that of the range 18,000–20,000 yr. (the "Woodfordian" in classical Illinois terminology). We have incorporated a recon-

Table III. "Classical" Wisconsinan Chronology, based on the Illinois Section[a]

Name and nature	Lithological basis	Approx. age (yr. B.P.)
Greatlakean Advance	Till and varved clay	10,000 to 11,500
Twocreekan Interstadial	Fossil wood and peat	11,800 to 12,500
Woodfordian Advance	Till and loess	12,500 to 22,000
Farmdalian Interstadial	Peat and reworked loess	22,000 to 28,000
Altonian Advance	Loess, till, soil	?30,000 to over 40,000

[a]Data from Flint (1971) and Nilsson (1983).

Table IV. Tripartite Wisconsinan Chronology[a]

Name and glacio/climatic trends	Time (yr. B.P.)
Late Wisconsinan	10,000 to 22,000
Major ice advance at 18,000 to 20,000 yr. B.P.	
Minor ice advances at 15,000, 12,500, and 11,500 yr. B.P.	
No significant temperate periods	
Middle Wisconsinan	22,000 to ?55,000
Moderate ice advance at ~33,000 yr. B.P.	
Predominantly cool-temperate at other times	
Early Wisconsinan	?55,000 to ?75,000
Major ice advance poorly dated between 55,000 and 65,000 yr. B.P.	
Minor ice advance poorly dated at ~75,000 yr. B.P.	
Probably cool-temperate interstadials 65,000 to 75,000 yr. B.P.	

[a]Data from Flint (1971), Nilsson (1983), Mickelson et al. (1983).

struction with maximum extent of Wisconsinan ice as shown by Flint (1971), while simultaneously showing the lobate nature to the ice margin, after Mickelson et al. (1983). Additionally, we have incorporated the concept of a twofold ice center in Canada as suggested by Shilts (1980).

F.6. The Cordilleran Glacier Complex

Alpine and piedmont glaciers developed on and around literally hundreds of mountains and ranges of the Cordillera, many reaching far south of the maximum southern extent of Laurentide ice. The Laurentide and Cordilleran perimeters were undoubtedly thin at the line of their juncture, and, indeed, there may have been an ice-free intervening region. Major subdivisions of the Cordilleran glacier complex include: the Brooks and Alaska ranges, the Canadian and northern U.S. Rocky Mountains, the Cascade Range, and the Sierra Nevada Range. In very general view, Flint (1971) described the configuration of the Cordilleran glaciers south of Alaska as two series of mountain glaciers (Rockies to the east and Cascades/Sierras to the west), feeding ice to piedmont and low valley glaciers in the intermontane basins. During maximum glacial intervals, ice may have covered a substantial portion of the northwestern continent, with the highest mountain peaks reaching above the ice mass by perhaps 100 m, comprising nunataks. However, ice-free lowlands undoubtedly existed, and may have been areally more extensive than glaciers across the Cordillera.

Having begged description of Laurentide events in all but cursory fashion, we assume the reader will easily under-

stand that the histories of hundreds of individual mountain glaciers are not attempted here. The Quaternary section in the Wind River Mountains, Wyoming, has been used as a reference section, with correlations made with the classical Laurentide sequence in Illinois/Wisconsin as given in Table V.

As with the situation in the Laurentide regions, current studies suggest that Table V, localized as it may be, is grossly oversimplified. Porter et al. (1983), dealing only with the later Wisconsinan, indicate that the Pinedale Glaciation left end moraines with dates of 70,000–60,000, 40,000–30,000, and 20,000–15,000 yr. B.P., thereby bringing the record in closer accord with modern interpretations from the Laurentide record, and also suggesting that the Pinedale Glaciation reached back to the early Wisconsinan. In addition, it should not surprise the reader that substantial numbers of Cordilleran glaciers survived well into the Holocene (indeed, alpine glaciers are common today in Alaska and in higher montane areas in Canada, Washington, Oregon, Montana, Wyoming, and Colorado). In the following subsections, we capsulize Quaternary glacial data from a few Cordilleran localities, largely to show the disjunct regional glacial histories. Having already explained why there need be no uniformity to alpine events in the Cordillera, we are able to use strictly local names for glaciations.

F.6.a. Canadian Rocky Mountains

Rutter (1976) discussed late Wisconsinan events in the Peace River Valley and vicinity, northeastern British Columbia. Preglacial materials (wood and fluvial deposits below glacial drift) yielded dates of 44,000 and 28,000 yr. B.P. Dates taken from plant material above the glacial drift center on 25,900 yr. B.P., suggesting a sizable regional glaciation between 25,900 and 28,000 yr. B.P., termed only "Early advance" by Rutter. This was followed by a glacial–interstadial–glacial sequence, spanning approximately 25,900 to 11,600 yr. B.P., termed the Portage Mountain advance. Deglaciation was underway by 9960 yr.

Table V. Generalized Chronology of the Cordilleran Glacier Complex[a]

Wind River Section	Wisconsin/Illinois	Time (yr. B.P.)
Pinedale Glac.	late Wisconsinan	ca. 15,000–20,000
Bull Lake Glac.	mid-to-early Wisconsinan	>30,000–>70,000
Sacagawea Glac.	Illinoian	>250,000, <400,000
Cedar Ridge Glac.	pre-Illinoian	? >600,000
Washakie Point Glac.	pre-Illinoian	? >1,000,000

[a]Data from Richmond (1965), Madole (1980), Nilsson (1983).

B.P. and evidence of substantial lake sedimentation is present. A minor glacial episode occurred well into the Holocene after 9280 yr. and sometime prior to 7470 yr. B.P., termed the Deserter's Canyon advance.

F.6.b. Cascade Mountains

For the Quaternary glaciations in the Pacific Northwest region of the Cordillera, Easterbrook (1976) and Waitt and Thorson (1983) reported the following: lobes of the lowland Cordilleran continental glaciers flowed down the Puget and Juan de Fuca Lowlands, with alpine glaciation forming in the Olympic and Cascade mountains. At least three early (i.e., pre-Fraser, in local usage) glaciations are documented in the Puget Lowlands, but firm dates are unavailable. Easterbrook terms these events the Orting and Stuck glaciations, the Stuck showing two advances and an interstadial, with the Alderton and Payallup interglacials intervening; however, Waitt and Thorson decline to assign names to the events, theorizing that multiple unrecognized glaciations are as likely to have occurred there as in other regions.

A sizable interglacial seems to be well documented for the interval between 20,000 and 32,000 yr. B.P. The final regional glaciation is termed the Fraser, spanning approximately 10,000 to 18,000 yr. B.P. This Fraser glaciation incorporated an initial alpine phase, followed by lowlands advance, retreat, and readvances. A minor readvance in the northern Cascade Range occurred during the early Holocene (Beget, 1981). Overall, fairly good correspondence may be noted between the Cascades Cordilleran and midwestern Laurentide chronology, with respect to the ca. 20,000-yr. major ice advance, and the 10,000-yr. deglaciation. However, relatively little relationship can be inferred between the Cascades and Peace River sequences.

F.6.c. Alaska

Not surprisingly, Alaska features both earlier glaciations than did the rest of North America, and a larger area of modern alpine glaciers than most of the continent. In general, alpine glaciation has always been more extensive in southern Alaska (the Aleutian, Alaska, and Boundary ranges) than in the northern strata (the Brooks Range); this is due to both the higher elevations in the southern area, and the greater moisture available in the south from the North Pacific and the Gulf of Alaska (Hamilton and Thorson, 1983).

Pewe (1976) described overall patterns in Alaska as follows: alpine glaciation began by the middle Miocene, with glaciers reaching down the southern ranges to the tidewater of the Gulf of Alaska, based on offshore deposits interpreted as drift. The Late Tertiary record is obscured, but marine deposits show the Bering land bridge was open intermittently during the Late Tertiary, suggesting intervals of glacial- and ice-free conditions. During the early Pleistocene, unnamed glaciations left deposits dated generally to 1.9 and 0.7 Myr. Locally, evidence exists for two or more pre-Illinoian stadials and a sizable interglacial period marked by high sea-level stands, dating between 250,000 and 100,000 yr. B.P.

Illinoian time (ca. 250,000 to 400,000 yr. B.P.) was apparently a major period of glaciation in Alaska, with snow lines approximately 500–600 m lower than at present, and virtually all highlands and coastal regions glaciated. Main ice-free regions included the Yukon and Kuskokwim River valleys, between major ranges and the North Slope. Pewe (1976) noted that the Yukon and Kuskokwin rivers wandered down and across the dry floor of the Bering Sea, which had been largely exposed by uptake of water in ice.

The Sangamonian (ca. 100,000 to 250,000 yr. B.P.) of the midwestern section, is evident in Alaska as a major interglacial interval, with summer temperatures probably higher than at present. But, by the early Wisconsinan, glacial conditions returned and spread to nearly all mountainous regions and many valleys. Pewe (1976) estimated that approximately 50% of Alaska, mostly in the central lowlands, remained ice-free and supported a tundra flora. The remaining Wisconsinan featured multiple cycles of advances and retreats in common with the other glaciated regions of the continent. And, as mentioned, the Holocene saw a higher snow line but continuation of many mountain glaciers.

Detailed discussion of the late Wisconsinan histories and deposits in the Brooks Range is presented by Hamilton (1982). A detailed southcentral Alaska glacial history and stratigraphy are given by Hamilton and Thorson (1983); and Porter et al. (1983) focus on late Wisconsinan stratigraphy in selected sites in the Brooks and Alaska Ranges.

F.7. A Case of Authentic "Catastrophism": The Channeled Scablands and the Lake Missoula Jökulhlaups

Beginning with the Bull Creek glaciations in the Rocky Mountains (i.e., pre-Wisconsinan time), the Okanogan, Purcell Trench, and possibly other lobes (see Fig. 9-80) of the Cordilleran glacier complex in southern British Columbia, dammed the Clark Fork River in Montana and northern Idaho to form Glacial Lake Missoula. This lake occupied 2500 km² and had a single outlet via the ice dam in northern Idaho, which broke periodically, resulting in episodic outburst floods (termed *jökulhlaups*) that crossed Washington state and eventually entered the Columbia River drainage to the Pacific Ocean. The discharge of these jökulhlaups was sufficient to form "gravel" bars with clasts several meters across, and to produce megaripples, deltas, and the spectacular erosion feature termed the "Channeled Scablands"

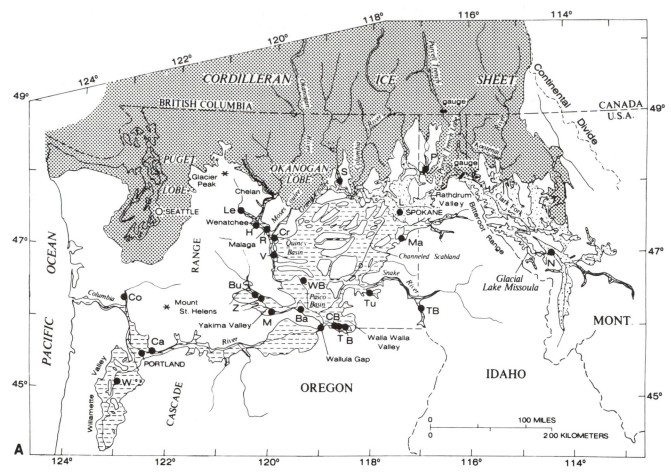

Fig. 9-80. The Pacific Northwest and ice-dam outbursts (jökulhlaups). (A) Regional map showing the channeled scablands and the ice lobes from the Cordilleran ice sheet which dammed Glacial Lake Missoula and other regional streams. (B) Local map of the Sanpoil River region during an ice-dam stage, which formed part of Glacial Lake Columbia west of Glacial Lake Missoula (see inset map). (A, from Waitt, 1985; B, from Atwater, 1984.)

to the west and south of Spokane. The Channeled Scablands feature numerous large, dry valleys (termed locally "coulees"), anastomosing small dry stream channels, dry waterfalls, and a number of oversize stream features now devoid of water. Most impressively, the Channeled Scablands are cut into solid basalt, and it is apparent that the various features formed in very short time: possibly in a handful of floods.

Bretz (1969, and numerous papers dating back to 1923) championed the flood-surge mechanism for the Channeled Scablands, but his idea was met with skepticism because it involved a distinctly nonuniformitarian event. However, in recent years, the catastrophic nature of the Scablands is generally accepted and the dominant question has focused on the number of floods involved as well as additional areas which may have experienced jökulhlaups. Richmond (1965) believed that only one pre-Wisconsinan and two Wisconsinan floods occurred, whereas Waitt

(1980) postulated tens of floods. Atwater (1984) supported Waitt's idea with evidence based on varves in the Sanpoil arm of Glacial Lake Columbia (see Fig. 9-80B), which formed to the west of Glacial Lake Missoula. Both Atwater (1984) and Waitt (1985) suggest that floods were regularly periodic through (at least) the late Wisconsinan; therefore, they are essentially "uniformitarian" events, even though of "catastrophic" proportions.

F.8. Retreat of Late Wisconsinan Glaciers

Since the maximum Wisconsinan glaciations occurred ca. 18,000 yr. B.P., one could reasonably consider all events postdating that time as part of the "retreat" of the glacial stage. Nevertheless, the last glacial advances in the United States and southern Canada ended by 10,000 yr. B.P. and, by ca. 9900 yr. B.P., the ice front had retreated from the conterminous United States. Most of southern

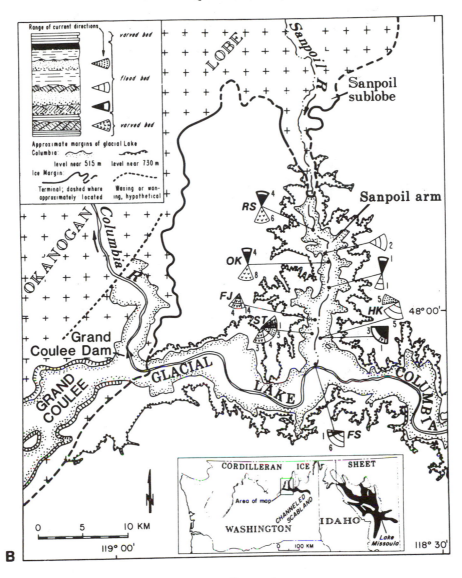

Fig. 9-80. (*Continued*)

Canada was ice-free ca. 8000 yr. B.P. (Fig. 9-81), and most remaining continental ice by that date centered around the Keewatin and Labrador regions from which it had originally spread (see Fig. 9-76).

Many localities featured substantial blocks of stagnant ice during and after retreat of glaciers. Innumerable and probably undocumentable lakes, swamps, deranged drainages, and other typical postglacial water features formed across the newly ice-free regions, largely as the result of poor drainage properties of typical glacial deposits. Tundra vegetation followed ice in deglaciated regions, to be replaced in turn by successions of forest biomes culminating in forest vegetation of modern latitude preferences. Establishment of forests and grasslands undoubtedly modified and obscured many glacial features.

Deglaciation of the western margin of the Laurentide ice sheet featured unusually large numbers of stagnant ice blocks, leaving as evidence the Prairie "Pothole" (actually kettlehole) region of southern Manitoba, Minnesota, and the eastern Dakotas. Deglaciation of the central Laurentide region (the Great Lakes and Mississippi River valley) seems to have featured a more general overland retreat, leaving sequences of recessional moraines and showing increasing isolation of ice lobes during retreat. Deglaciation of the eastern margins of the Laurentide ice sheet is difficult to characterize because several mountainous regions, especially the White, Adirondack, and Green ranges, featured alpine glaciers which continued after continental ice had ablated. Deglaciation of the Cordilleran glacier complex involved a multitude of local events, including many ranges which retained alpine glaciers long after the Holocene Epoch commenced.

F.9. Glacio-isostatic Rebound and Sea-Level Variations

The weight of glacial ice is sufficient to depress the Earth's crust by a ratio of

$$\text{ice thickness} : \text{crustal response} = 1 : 3.6$$

(Flint, 1971). Therefore, for a typical 2-km thickness of ice sheet, one would expect a crustal depression of 0.56 km. Upon removal of ice during and consequent to deglaciation, glaciated regions rebound at variable rates; indeed, many

regions of North America are still rebounding from the late Wisconsinan glaciations.

Glacio-isostatic rebound often resulted in crustal warpage, due to regionally variable rebound rates. Many regions experienced severe tilting during the Holocene and such tilting often affected regional streamflows by changing flow regimes and relationships to base level. For example, the southern Great Lakes region currently is undergoing measurable tilt to the north as southernmost portions rebound faster than the north (see Fig. 9-82).

Deglaciation also engenders rising sea levels and addi-

Fig. 9-81. Deglaciation of Canada. Approximate positions of ice fronts are shown for the indicated ages (in years B.P.; single digits, ×1000). Virtually complete deglaciation occurred by 6000 yr. B.P. Individual mountains in the Cordillera retained ice after dates shown. (Considerably simplified from Geological Survey of Canada Map 1257A, by V. K. Prest, 1969.)

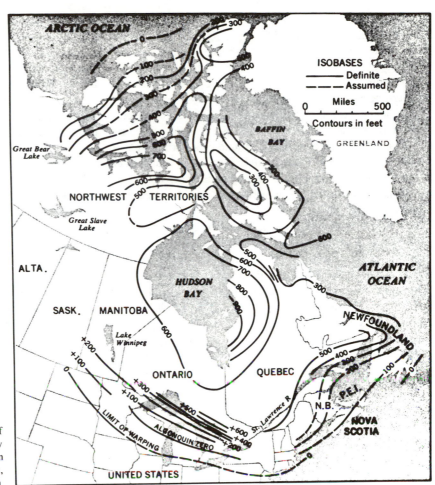

Fig. 9–82. Glacio-isostatic rebound since retreat of the Wisconsinan glaciers. Data are in feet and show rebound from depressed glaciated elevations. (From Kay and Colbert © 1965 by John Wiley & Sons, reprinted with permission of John Wiley & Sons.)

tional base-level effects on streams. Porter (1983) stated that estimates for total sea-level drop during the Wisconsinan range from 100 to 150 m, more than enough to expose large areas of the Atlantic and Gulf Coast continental shelves and to create a large subcontinent (Beringia) out of the now-submerged Bering Shelf. Flint (1971) noted that if the present Greenland and Antarctic ice caps melted, sea level would peak about 65 m above the present level.

The effects of relative sea level changes are especially evident in Pleistocene stream terraces, which derive from complex interactions of flow regimes, land depression, sediment loading, and sea level. Ages of terraces are often in seemingly anomalous sequence: for example, old terraces may be elevated high above younger valley fill, when a region is uplifted. But younger material may also be deposited above old terraces consequent to sea-level rise and superposition.

In coastal regions, isostatic and eustatic changes have produced numerous submerged beaches, deltas, barrier bars, and moraines. Submarine valleys extend far below sea level from river mouths on formerly exposed continental shelves; and this may be the origin of submarine canyons on

the continental slopes (e.g., Hudson Canyon, an extension of the Hudson River in New York). Fjord coasts, such as Maine's, are spectacular examples of alpine glacial scour followed by marine submergence after deglaciation.

F.10. The Wisconsinan Lakes

F.10.a. Lakes in the Glaciated Regions

The Great Lakes at the Canada–United States border, and the string of giant lakes extending across the Canadian Northwest (including Lake Athabaska, Great Slave Lake, and Great Bear Lake), are individually enormous bodies of fresh water. However, they are all remnants of late Wisconsinan lakes which were several times as large at their maximum times of existence.

The approximate configurations of the Great Lakes (and the Finger Lakes in New York, as well as many other northern lakes) were predetermined by bedrock structure prior to glaciation (Flint, 1971). Glacial scour widened and deepened valleys and other depressions in nonresistant bedrock, yielding the vast lakes. As an example, a glance at a geological map will show that Lakes Ontario, Superior, and

Michigan follow closely the strike of lower and middle Paleozoic strata around the Michigan Basin. Lake Ontario was carved from the soft Queenston Shale (Ordovician age; see Chapter 5) and closely follows the strike of the southward-dipping lower Paleozoic homocline which forms the northern portion of western New York.

Evidence for the presence of earlier, larger glacial lakes may be found in diverse areas. For example, elevated beaches and lake deposits may be found near Toronto, comprising the Don and Scarborough formations. Clark and Karrow (1984) discussed late Pleistocene water bodies in the St. Lawrence Lowlands in New York and identified four fossil lake levels.

The size of glacial lakes in Canada is truly impressive (Fig. 9-83). Lake Agassiz, which reached its maximum at approximately 12,500 yr. B.P., covered approximately 500,000 km² in the Red River drainage of Manitoba, Minnesota, Ontario, and the Dakotas. Lake Ojibway-Barlow, dating from approximately 10,500 to 8200 yr. B.P., occupied a large part of eastern Ontario and western Quebec. Lake Ojibway-Barlow ended when Hudson Bay deglaciated sufficiently for marine waters to enter the isostatically depressed region, forming a marine transgression termed the

Fig. 9-83. Principal meltwater lakes in Canada. Shown are the overall outlines of deposits in Canada, indicating maximum extent of Wisconsinan and Holocene meltwater lakes. Note that lakewater did not fill these entire areas contemporaneously. Areas in diagonal shading are ancient lake areas; black regions are present lakes and seaways; horizontal shading indicates marine transgression due to glacial depression. A, Glacial Lake Agassiz; OB, Glacial Lake Ojibway-Barlow; C, Champlainian Sea. (Data from Flint, 1947, 1971.)

Champlainian (or Tyrell) Sea; waters of the lake drained northward to the inland sea, and left only remnant, ephemeral lakes. The largest of all North American lakes was Lake McConnell, which Flint (1971) cited as extending across 9° of latitude in northwestern Canada. So large was this lake that its widely separated relict lakes (Lakes Athabaska, Great Slave Lake, and Great Bear Lake) are themselves enormous.

F.10.b. Pluvial Lakes in the Great Basin

While glaciation occurred in northern parts of the continent, the shifting of climate zones accompanying glaciation changed global wind patterns in latitudes far below the ice fronts. A distinctive consequence was voluminous rainfall in present (and pre-Pleistocene) desert basins, with formation of large, often salty lakes.

Lakes Bonneville and Lahontan (Fig. 9-84) were the largest pluvial lakes in North America, and their former extents are visible in high-standing beaches, far from the present lake shores in the case of the Great Salt Lake (a remnant of Lake Bonneville). At maximum, Lake Bonneville covered approximately 51,000 km² in northwestern Utah, and reached depths of 335 m. Lake Lahontan was actually a series of interconnected basins in northwestern Nevada, occupying approximately 22,000 km² and reaching depths of 213 m (data from Morrison, 1965).

The history of Lake Bonneville has been determined from lacustrine deposits with intervening soils; in brief, it originated from a series of smaller pre-Wisconsinan lakes, and it waxed and waned with glacial and interglacial events. Lake Bonneville remained considerably larger than the present Great Salt Lake until approximately 5000 yr. B.P., and the concentration of salts engendered by shrinkage from the maximum extent to the present basin easily explains the hypersalinity of the present lake. The history of Lake Lahontan was roughly similar, except that its configuration involving a number of disjunct basins makes interpretation of its history less clear.

F.11. Causes of Ice Ages

Among great geological "mysteries" is the cause of Pleistocene glaciations. Actually, three questions are involved: What caused the onset of glaciation? What caused repeated advances and retreats of glaciers? What lies in the future (i.e., is the present an interglacial or interstadial time)? The first question has a ready answer, while the others will receive the bulk of this discussion.

F.11.a. Why Glaciation in the First Place?

It is clear from multiple sources that global climate underwent erratic but progressive cooling from the Late

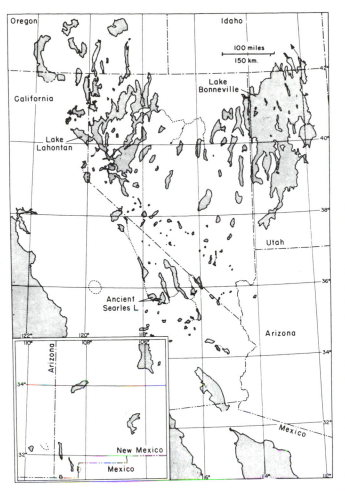

Fig. 9-84. Pluvial Lakes Bonneville, Lahontan, and others which formed in the Great Basin during the late Pleistocene. (From Flint © 1971 by John Wiley & Sons, reprinted with permission of John Wiley & Sons.)

tivity in the upper atmosphere with resulting decrease in net insolation)

2. Changes in continental position with concomitant changes in oceanic circulation and continental temperature–moisture regimes [e.g., the Ewing and Donn (1956) hypothesis]
3. Changes in the sun's thermal output
4. Changes in the Earth's astronomical position and the nature of its motions (e.g., the Milankovitch hypothesis, discussed below)

Categories 1 and 3 will not be discussed in detail because the hypothetical mechanisms do not lend themselves to testing with available evidence. For example, changes in atmospheric transparency can affect global temperatures following particular explosive and massive volcanism (as has been witnessed in historical times). However, volcanic events are individually short-lived and there is no evidence for unusual, cyclic volcanism during the Pleistocene such as would cause the known pattern of glacial advances and retreats. Solar fluctuations are a realistic possibility that may explain the long-term climate deterioration which brought on initial glaciation in the early Pleistocene, and which may be responsible for cyclicity of glaciations. Unfortunately, sunspot and solar flare cycles are poorly known and one can only speculate on their importance in Pleistocene glaciations. We tend to reject the suggestion that solar variation is the total cause of glaciation, since short-term sunspot cycles occur frequently and seem to have little relationship to modern climates. It is unknown whether long-term sunspot and flare cycles are sufficiently large to have larger effects.

Of the two remaining categories of glacial hypotheses, the geographical hypotheses (i.e., based on variations in continent–ocean positions) largely stem from the known, important effects of oceans on climate; for example, the Gulf Stream and Kuroshio (Japan) currents control coastal climates on eastern and western North America and distribute warm waters to northern latitudes into which they otherwise might not flow. Ewing and Donn (1956) proposed that ice ages occur when the Earth's rotational poles become thermally isolated (i.e., such as during the Cenozoic Era when Antarctica occupied the South polar region and the north polar region region became surrounded by land), thereby limiting regional circulation of seawater. When global oceanic heat is not circulated from warm to cold reservoirs, due to isolated polar regions, these regions become ice-bound, initiating conditions which may be termed preglacial. Pleistocene cyclicity within the Ewing and Donn mechanism resulted from alternation between an ice-covered and an ice-free Arctic Ocean; during ice-free stages, there would be moisture available at high latitudes for snowfall, causing continental glaciation. Glacial meltwater would lower Arctic Ocean temperatures, cause for-

Cretaceous to the Pliocene (see Section G.3). This climatic deterioration paved the way for Pleistocene glaciations by providing threshold conditions. It is apparent that by the late Pliocene, either a slight additional cooling, an increase in precipitation at higher latitudes, a change in ocean circulation, or some other relatively minor fluctuation could allow expansion of glaciers down from the high latitudes where they had already formed. The remaining discussion considers details of the minor fluctuations that brought glaciers to mid-latitudes.

F.11.b. Why Multiple Glaciations?

Herein lies the heart of the ice-ages controversy. Numerous hypotheses have been presented, which we will categorize as follows:

1. Changes in atmospheric transparency (e.g., due to volcanic or cosmic dust, causing increased reflec-

mation of sea ice, and eventually an Arctic ice cover. This would stop the precipitation and start the cycle over. At present, there is an Arctic ice cover and no major glaciation on North America.

The mechanism which would cause alternation between ice-free and ice-covered Arctic Ocean would be the opening and closing of the Arctic–Atlantic oceanic interchange, which would be dependent on sea level, which would in turn be dependent on the uptake of seawater in continental glaciers. In support of the hypothesis, recent studies suggest that the Arctic Ocean was ice-free episodically during the Pliocene and Pleistocene: Worsley and Herman (1980) showed that photosynthetic nannoplankton were living in the Arctic Ocean during various intervals of the Plio-Pleistocene, and Herman and Hopkins (1980) suggested that the Arctic perennial sea-ice cover did not form before approximately 7 Myr. These tend to reinforce the possibility that Arctic sea ice was cyclically variable during the ice ages, but they offer no firm proof that the Ewing–Donn mechanism caused overall glacial cyclicity.

In current, widespread favor are astronomical hypotheses for glacial cycling. Chief among these is the Milankovitch hypothesis, which was best described in Milankovitch (1941) but was preceded by earlier writings by Milankovitch and a considerable body of work by James Croll (numerous papers through the 1860s) and others. Basically, Milankovitch and others relate the known motions of the Earth in space with consequent effects on the total insolation received by the Earth. As it moves around the sun, the Earth undergoes three cyclic fluctuations of orbital and rotational movement with known periods: (1) obliquity (tilt of the axis), with a period of approximately 41,000 yr. and a range of values from 21.5° to 24.5°; (2) precession of the equinoxes (cyclicity of the relative position of the axis in space), with a 21,000-yr. period; and (3) ellipticity of the orbit, with a 91,800-yr. period and a range of values from near-perfect circle to 0.947 ratio of semiminor to semimajor axes of orbit. Milankovitch charted these motions for various latitudes and showed the predicted changes in insolation during the past 600,000 yr. that would result from calculated effects of the above cyclic fluctuations. He postulated basically that glacial events should correspond with low solar radiation peaks, and vice versa. Testing of this hypothesis has been hampered by imprecise dating of glacial events (especially those outside the effective range of radiocarbon). However, recent work (discussed below) with long ocean cores has renewed interest in the Milankovitch hypothesis and, in the view of many, tends to confirm elements of the hypothesis.

Milankovitch believed that the European glacial record correlated with his solar radiation curves, and most European geologists observed some correspondence. However, certain known glacial events on the North American conti-

nent were decidedly un-Milankovitch; for example, an interstadial occurred at 25,000 yr. B.P., which should have been a cool interval according to the Milankovitch curve. Similarly, the late Wisconsinan maximum at 18,000 yr. B.P. occurred 7000 yr. after the same Milankovitch low-radiation time. In general, for the past 40,000 yr., the radiocarbon record has shown considerable deviation from the temperature events predicted by Milankovitch. As of the mid-1960s, the Milankovitch hypothesis was largely out of favor with American geologists because of the apparent poor correspondence between his postulated late Pleistocene temperatures and the continental glacial record.

Renewed interest in the Milankovitch hypothesis arose with an oceanographic project called CLIMAP, whose primary focus was interpreting the ice age record from ocean cores. Preceding CLIMAP, studies of reef terraces in Caribbean and Pacific tropical islands revealed high terraces which yielded dates of 125,000, 105,000, and 82,000 yr. B.P. Since these were accretional terraces, they represented high sea-levels which must have been caused by virtually complete melting of polar ice. Modern followers of Milankovitch's hypothesis noted a correspondence between these dates and peaks in the 45° latitude curves by Milankovitch, which were due primarily to the 21,000-yr. precession effect.

The CLIMAP project documented another important cycle previously unnoticed, in paleotemperatures from ocean core data; this was a strong 100,000-yr. rhythm of maximum and minimum temperatures indicated by the oxygen isotope and paleontological data. The seminal paper by Hays *et al.* (1976) strongly emphasized the 100,000-yr. cycle and related it to the period of the orbital eccentricity. The apparent discrepancy between glacial timing from continental data and oceanic data was postulated to be due to the lag between continental glaciations and oceanic temperature change. Imbrie and Imbrie (1979) presented simplified curves showing the pattern of climate cycles in typical Indian Ocean cores, which feature a dominant 100,000-yr. cycle, and successively lesser 43,000-, 24,000-, and 19,000-yr. peaks, corresponding to axial tilt, and precession (the 19,000-yr. precession cycle is a second such cycle, subordinate to the major 21,000-yr. cycle).

Debate rages fiercely over these neo-Milankovitch observations. Critics point out that the correspondence between Milankovitch curves and continental glacial timing is, despite oceanographic data, poor. In addition, the 100,000-yr. cycle does not literally follow a period of orbital eccentricity (which, recall, was 91,800 yr.). Proponents of the orbital forcing concept generally attribute slop in fit of data to the poor continental record and lags in temperature correspondence between continental and ocean environments. Opponents are bothered by up to 7000-yr. lags. Most authors recognize strong periodicity in the

Pleistocene glacial record and are willing to attribute some part of the cyclicity of Pleistocene glaciations to astronomical forcing, as per Milankovitch; but many are uncertain why the relatively small variation in the Earth's eccentricity of orbit might cause the strongest climatic effects.

We will not take a stand on the subject, since it is still in active research; however, most current work seems to assume the orbital forcing concept is correct in explaining Pleistocene ice ages [see Covey (1984) for a summary]. If the orbital forcing concept is valid, it also allows predictions of climates. For example, the Earth should experience another major glacial maximum peaking approximately 23,000 yr. from now; a lesser "little ice age" is due approximately 2000 yr. hence (Imbrie and Imbrie, 1979). One complicating factor is anthropogenic (i.e., man-made) CO_2 which may (it too is controversial) engender sufficient "greenhouse-effect" heating to mask glacial trends due to orbital forcing. What is almost certain is that the Holocene is an interglacial episode, not a post-glacial time; therefore, one should expect future glaciations, regardless of cause.

G. Tertiary Invertebrates, Plants, and Paleoclimatology

G.1. Changes and Trends among Tertiary Invertebrates

All invertebrate taxa with substantial fossil records have been introduced by this point in the book. No major changes among higher taxa occurred within benthic marine invertebrates at the era boundary, except for those noted in the last chapter's discussion of end-Cretaceous extinctions. At lower taxonomic levels, the complexion of marine molluskan faunas is recognizably different for the different eras, featuring changes listed in Table VI. The absences of ammonites, *Inoceramus*, and exogyrines probably comprise the most substantial differences of Cenozoic assemblages.

Within the epochs of the Cenozoic, variations in generic and specific makeup of molluskan faunas generate the "Lyellian curves" introduced early in this chapter. The only higher megainvertebrate taxa which appeared for the first time in the Cenozoic were three new orders of insects: fleas (Siphonaptera), butterflies and moths (Lepidoptera), and termites (Isoptera). We observed previously that the appearance of fleas may coincide with the diversification of land mammals, and the butterflies may have coevolved with flowering plants (angiosperms). Termites may also have coevolved with angiosperms, but that is less evident.

The record of Cenozoic insects is very good, due to excellent preservation in European and Caribbean amber deposits, and in fine-grained lacustrine sediments especially

Table VI. Very Generalized Trends in Later Mesozoic versus Cenozoic Molluskan Communities[a]

Taxa	Mesozoic	Cenozoic
Bivalves		
Exogyra	Very abundant, moderately diverse	Extinct
Gryphaea	Very abundant, moderately diverse	Rare (extinct Eocene), few species
Inoceramus	Very abundant, diverse	Extinct
Trigoniids	Common, moderately diverse	Rare, few species
Rudists	Regionally abundant, diverse	Extinct
Ostreinids	Very abundant, moderately diverse	Very abundant, very diverse
Taxodonts	Abundant, moderately diverse	Common, reduced diversity
Heterodonts	Common, moderately diverse	Very abundant, very diverse
Pectens	Common, moderately diverse	Very abundant, very diverse
Cephalopods		
Ammonites	Very abundant, very diverse	Extinct
Nautiloids	Common, moderately diverse	Restricted numbers, few species
Belemnoids	Very abundant, moderately diverse	Rare (extinct Eocene), few species
Gastropods	No discernible trends at higher taxonomic levels, many changes within families and genera	

[a]Only notably changed taxa are listed.

notable from the American western deserts. Amber insect fossils from the Baltic seacoast of Germany, dating to the Eocene and Oligocene, are among the best known, and include many first occurrences in the fossil record. Brues (1951) experimented with sheets of sticky paper tacked to trees, in an attempt to simulate the environment under which amber-preserved insects were trapped in tree resins. He reported a strong bias toward small insects in the collections, probably because larger forms could escape entrapment by force. Additional important amber insect faunas are known from the Greater Antilles region, especially the island of Hispaniola: Wilson (1985) described a fauna from the Dominican Republic including 37 genera of Late Oligocene or early Miocene ants. Insects preserved in lake deposits include the very notable Florissant Lake Bed specimens in Colorado (discussed previously), of Eocene age.

G.2. Diversification of Marine Microfossil Taxa

Because marine microfossil taxa are so economically important in petroleum exploration in Cenozoic strata of the Gulf coast, and because many taxa diversified during the

Cenozoic Era, we will continue here the discussion of microfossil taxa initiated in Chapter 8. The following will be subdivided (somewhat artificially) into submicroscopic forms (nannoplankton) and those forms which are subvisible or barely visible (microfauna: largely foraminifera and ostracodes).

G.2.a. Nannoplankton

As noted in Chapter 8, coccolithophores appeared during the Jurassic, waxed into numerous taxa during the Cretaceous, and then declined dramatically during the Maestrichtian. However, during the Paleocene, many new species appeared, soon followed by more and more forms through the Early Tertiary, so that by the end of the Eocene the total number of known species (~ 375) rivals the Cretaceous peak of diversity (almost 500 species; data from Brasier, 1980, and Harland et al. 1967). The Eocene diversification of coccoliths included the discoasters; however, many discoaster species became extinct at the end of the Eocene as did a large number of other coccoliths, leading to a general Oligocene through Pleistocene decline in the total population. By the latest Pleistocene, less than 30 species remained; and yet there are over 350 extant species, indicating either: (1) poor preservation of Pleistocene taxa or (2) a recent explosive diversification. In terms of biostratigraphic value, one may quote Brasier (1980): "coccoliths define Cenozoic biostratigraphic zones that are the standard to which other groups are compared."

Silicoflagellates, which appeared during the Cretaceous, underwent an explosive radiation of taxa during the Miocene but shortly afterward declined to two species. They regrouped in the Quaternary and are common modern marine plankton. However, their appearance in fossil assemblages and cores is sporadic simply because most such lithologies are calcareous and sparsely populated by siliceous algae; as such, they have not yet been widely used as biostratigraphic fossils.

Diatoms, like silicoflagellates, have less wide-ranging distribution in sediments than do calcareous coccoliths; but they are, nevertheless, very abundant in both Tertiary and Quaternary deposits. Although they appeared during the Mesozoic, diatoms diversified into ever-increasing numbers of species through the Cenozoic and reached a maximum diversity in the Holocene (exceeding 3000 species). Earliest diatoms were *centric* (i.e., circular) but during the Paleocene *pinnate* (i.e., elongate, narrow) morphologies appeared. The presence of diatoms in nonmarine environments makes them useful in determining paleotemperatures of Tertiary lakes, since the oxygen in the siliceous tests show diagnostic $^{18}O/^{16}O$ ratios. Marine diatoms are infrequently used in biostratigraphy; however, thick accumulations of diatoms comprise the economically important substance diatomite, which is used as a filtering and polishing agent and abrasive. In a typical diatomite, over 6 million diatoms are present in each cubic centimeter (Brasier, 1980).

G.2.b. Foraminifera and Ostracodes

The most notable developments among foraminifera during the Cenozoic were the diversification of planktonic rotaliids, of the family Globigerinacea, and the evolution of enormous (for single-celled organisms) rotaliids, especially *Nummulites*. The globigerinids appeared in the Cretaceous but culminated in the Cenozoic (continuing to dominate the plankton at present) with the characteristic genera *Globigerina* and *Globorotalia* (see Fig. 8-67) among others. These planktonic forams feature globose chambers, often in loosely arranged spirals or in virtually free-form arrangements, with very perforate hyaline tests through which penetrate large numbers of pseudopods. The pseudopods help the organism buoy-up in the water column. *Globigerina* species, and other planktonic forams, are strongly zoned by latitudinal temperature differences and this phenomenon has allowed paleotemperature data to be extracted from fossiliferous marine cores.

Benthic foraminifera too persisted and diversified through the Cenozoic; it is beyond the scope of this discussion to deal with all benthic forams and the reader is directed to Loeblich and Tappan (1964). However, before dismissing benthic forams, we should mention the unusual large foraminifera of the Cenozoic, including four genera that survive to the present, which may be placed in a family named for the widely recognized genus *Nummulites*. Nummulitidae are rotaliid forams reaching sizes over 3 cm in diameter and usually taking the shape of a flattened, floppy disk with a radiate pattern superimposed on the upper surface (Fig. 9-85). They are extremely abundant in Early Tertiary deposits from the Tethyan region and survive today in restricted areas of the Indo-Pacific region. Among the better-known occurrences of Nummulitidae are the Eocene limestones which make up the Great Pyramid of Cheops at Giza, as well as other ancient Egyptian structures.

Ostracodes are biostratigraphic index fossils second only to foraminifera in importance in Cenozoic stratigraphy; ostracodes were introduced in Chapter 5. During Cenozoic time, they diversified significantly and this diversification apparently continued to the present. Most Cenozoic ostracodes belong to the order Podocopina, which dates back as far as the Ordovician and achieved dominance by the Triassic. Elements of two additional ostracode orders survive but they are represented by a handful of families. Detailed taxonomy of podocopinid taxonomy is beyond the scope here; the reader is directed to Benson et al. (1961) for detailed description and systematics.

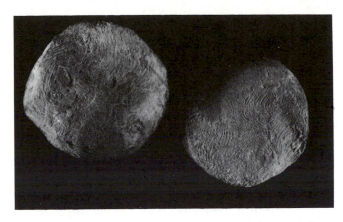

Fig. 9-85. *Nummulites,* the abundant Eocene large rotaliid foraminiferan (these from the Ocala Limestone in Florida; DRS photograph).

G.3.   Tertiary Climatology and the Diversification of Angiosperms

After the appearance of angiosperms in the Early Cretaceous (Chapter 8), there was a major Cretaceous radiation of families such that 66 were present by the end of the Maestrichtian. These include a variety of arborescent taxa and a smaller number of herbaceous taxa (possibly an artifact of preservation). Conspicuously absent in Cretaceous floras were grasses (however, a putative form of marsh reed has been reported from Upper Cretaceous beds in New Jersey; this and data above from Harland *et al.,* 1967).

Harland *et al.* reported that 62 angiosperm families appeared (or are reported to have appeared based on suggestive fossil evidence) during the Paleocene and Eocene, yet none of the Mesozoic families are known to have become extinct. Thus, one observes a twofold diversification of angiospermous plants (Cretaceous and Early Tertiary) with no time of mass mortality occurring between the radiations. It is therefore little wonder that modern floras are dominated by angiosperms. In the following discussion, we will briefly mention some well-studied floras from North America and describe Tertiary paleoclimates since distribution of plants is directly keyed to climate.

G.3.a.   Paleocene–Eocene Floras and Climates

Oldest Tertiary floras which have been studied occur in Midway Group strata from the Gulf Coastal Plain, notably in Louisiana. Wolfe (1978) stated that these indicate the presence of tropical, rain forest conditions at that latitude, which implies mean annual temperatures of around 27°C. Younger Paleocene floras from the same localities show a slight cooling trend to yield a ''paratropical'' rain forest. Wolfe explained, in the same article, that the physiognomy of leaves provides important clues to the temperature and

moisture conditions of growth for various plants. For example, thick (*coriaceous*) leaves correlate with high temperatures and moisture; smooth edges (in contrast to toothed or lobed margins) likewise are common tropical features. Tropical rain forest plants notably feature ''drip tips'' in lower-story (i.e., species below the canopy of tallest trees) plants as an additional characteristic; thus, one may diagnose even extinct taxa to their habitats by these and other features.

Paleocene floras from the interior, notably from the Fort Union Series, show parallel climatic conditions with the Coastal Plain material; but, since they are located farther to the north, they indicate that the North American climate may have been more equitable latitudinally relative to the present.

Eocene conditions are quite well-studied and show some changes from the Paleocene. Forests represented by fossils in the Green River sediments show moderate conditions but some contemporary interior assemblages show dry conditions favorable to the spreading of grasslands. Grasses (Graminae) appeared during the Eocene and with their diversification came a parallel, gradual rise of grazing mammals (such as the early horses, rhinoceroses, and artiodactyls). It is quite likely that both localized and widespread environments are represented within the interior floras. The Lamar River ''fossil forests'' in Yellowstone Park are somewhat anomalous, as explained previously, because both cool-temperate and tropical species are represented; however, Fritz's (1980) explanation for the seeming problem suggests that several thousand meters of elevation change is recorded in the ''fossil forest'' since part of the flora is allochthonous. Regardless of the correctness of Fritz's suggestion, subtropical conditions prevailed at low elevations, which is considerably different from the cool-temperate environment of the region today. Fossil floras on the Pacific coast, especially Washington and British Columbia, record cool-moderate conditions (similar to those of the Holocene) in middle Eocene time; and fossils from the Gulf coast (largely in the Mississippi Embayment) record drier-than-present, moderate to subtropical conditions. In summary, late Paleocene to middle Eocene environments, as recorded by plants, seem to be less variable latitudinally than at present (considering that the Gulf coast was drier, the interior warmer, and the Pacific coast almost identical to the present). Overall, Wolfe (1978) suggests that the mean annual range of North American temperatures in the middle Eocene was approximately half that of present and that absolute temperatures were several degrees warmer.

It is widely accepted that a gradual lowering of temperatures occurred from the Middle Eocene until the end of the epoch, at which time dramatic changes occurred. Rapid cooling occurred almost everywhere, indicated by a shift from broad-leaved evergreen foliage to deciduous foliage, and of greater effect, a striking increase in range of annual temperatures occurred (i.e., greater seasonality). A dramatic contrast in diversity can be observed between the Late Eocene Florissant flora of Colorado, including numerous species which probably could not survive in temperatures below −6°C, and the low-diversity floras in Oligocene strata, including predominantly cold-tolerant temperate species.

G.3.b. Oligocene–Pliocene Floras and Climates

With a major shift in climatic equitability at the end of the Eocene, a pattern of increased seasonality and latitudinal variation became established and prevailed throughout the remaining Tertiary (and into the Quaternary). Unfortunately, such a pattern makes description much more complicated, since each region must be dealt with separately; thus, we will not delve too deeply into the later Tertiary climates and floras. Upper Oligocene floras in the Bridger Creek Formation of the John Day Basin in Oregon, show forest species similar to those of the Holocene (e.g., redwood and oak) as well as some species found in other parts of the western continent but not in modern Oregon. In the western interior, the White River Formation of the Big Badlands reflects varied climates and environments but it notably includes abundant grassland mammal species, including horses, rhinoceroses, and oreodonts. One may assume that relatively drier conditions ensued there during the Oligocene relative to the Early Tertiary.

Post-Oligocene paleoclimates in North America were on the average probably no warmer than the Oligocene low-mean temperatures. Beginning in the early Miocene, a gradual and progressive cooling is evident and this cooling continued (with much variability) into the Pleistocene, when it culminated in the great ice ages (discussed earlier). The episodic Miocene-to-Holocene cooling events, with intervening warm periods, have been documented by various means. Spread of cold-tolerant forest species in Canada and the northern tier of American states through the later Tertiary is one means by which the cooling trend has been documented. Blackwelder (1981) documented episodic Late Tertiary cooling based on major depositional hiatuses in Atlantic Coastal Plain marine sequences. He suggested correlations between such hiatuses, reflecting eustatic sea-level drop, and the buildup of the Antarctic ice sheet (Fig. 9-86). Thus, such cool–warm intervals in the Tertiary long preceded the multicycles of the Pleistocene ice ages and may in fact be related (however, at higher mean temperatures and therefore not resulting in global ice ages).

Wolfe (1978) described a warming trend of the late Oligocene and a slight warming during the early to middle Miocene. Andrews (1961) distinguished two basic Miocene floras in America: (1) an Arcto-Tertiary flora in the northern Great Basin and Columbia Plateau, composed of temperate

hardwoods and conifers living under relatively humid conditions, and (2) the Madro-Tertiary flora of southeastern California and Mexico, consisting of semiarid taxa reflecting hot, dry conditions. The reader will note that neither is significantly different from floras and conditions of today.

There is good evidence of a severe late Miocene glaciation in Antarctica, probably accompanied by global cooling. Peck et al. (1979) discussed evidence for such based on regressive facies in southwestern Florida (the Tamiami Formation), and Herman and Hopkins (1980) related the same late Miocene major expansion of the Antarctic ice sheet to onset of alpine glaciation in high latitudes in North America. Additional effects of the late Miocene cooling and lowering of sea level were the isolation of the Mediterranean Sea and the prominence of the Bering land bridge between Asia and North America, isolating the Arctic Ocean once again. After the Miocene, global climates and floral patterns were virtually like those of the Holocene, on the average, but with continued warm and cool intervals. In the southwestern United States, Late Tertiary block faulting and increased uplift of the Sierras amplified the rain-shadow effect, enhancing desertification of the Great Basin and inland California.

H. Cenozoic Mammals

H.1. Introduction and Some Generalities about Mammalian Diversification

It was shown in the previous chapter that mammals evolved from therapsid reptiles very near the time of dinosaur evolution. Yet, for unknown reasons, they failed to dominate the vertebrate biota until the Paleocene. From a starting point of approximately 30 known Late Cretaceous genera, by the latest Paleocene there were 200 known mammalian genera and by the middle Eocene there were over 325 genera. This explosive radiation of mammal taxa was actually followed by a decline in diversity from the latest Eocene through the Oligocene, and subsequently by a renewed diversification through the Miocene–Pliocene culminating in a peak diversity of over 800 Pliocene genera [data from Gingerich (1977) based on Romer (1966)]. Section 8.c will discuss a Pleistocene/Holocene mass extinction event which terminated some 300 genera leading to the present depauperate global (especially North American) mammalian fauna.

One may speculate that combinations of climatic changes, competition among newly evolving forms, changing landmasses in global configurations, land bridges (e.g., the Central American and Bering Sea land bridges), and marine transgressions, cause fluctuations in mammalian diversity. It also seems likely that emergence of grasses in the Eocene strongly selected for certain taxa, especially among

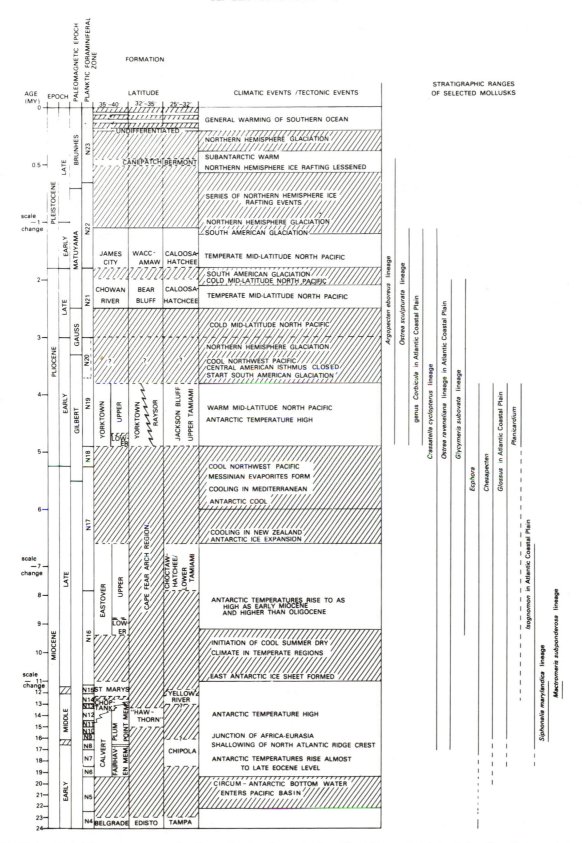

Fig. 9-86. Blackwelder's Miocene-to-present paleoclimate chart based on major depositional hiatuses and their relationship with Antarctic ice volumes. (Reproduced courtesy of the author.)

ungulates (hoofed animals), and that the deterioration of global temperatures toward the later Pliocene had an especially powerful selection effect on certain taxa. The fossil and paleoclimatological record yields many clues to these lines of speculation, but few firm conclusions are generally reached.

H.2.   Mammals of the Latest Cretaceous–Paleocene Transition

The youngest Cretaceous mammals in North America are designated the Lancian fauna (L. S. Russell, 1975) and comprise four marsupial and eight multituberculate genera (see Chapter 8), and six eutherian genera. Of the eutherian mammals, four are insectivores, one appears to be a distant ancestor of ungulates (the genus *Protoungulatum*), and the sixth is the oldest primate known, *Purgatorius*.

The dentition of many Mesozoic mammals show adaptations to insect-eating. The order Insectivora are the most primitive of extant (and fossil) eutherian mammals and probably descended directly from Cretaceous panthotheres; they also may be the common ancestors of all higher eutherian mammals. Although we will not attempt to describe the multitude of mammals of the Cenozoic, the key evolutionary stem groups (Fig. 9-87) will be worth special attention, beginning with the order Insectivora.

Insectivores are typically small, secretive, nocturnal or burrowing creatures, and are most notable for what they lack rather than possess. The teeth have a formula of 3 : 1 : 4 : 3 (meaning three incisors, one canine, four premolars, and three molars) in each quarter of the mouth, and this is close to the presumed ancestral state. Most higher taxa have a reduced formula (e.g., humans have a relatively primitive 2 : 1 : 2 : 3; rodents have a very specialized formula, typically 1 : 0 : 1 : 3). The molars are tribosphenic; in general, insectivores typically lack dental specializations. They lack auditory bullae (ear capsules), feature incomplete zygomatic arches (cheekbones), and possess a full primitive count of five digits in manus and pes. Among living insectivores, the shrews probably come closest in morphology to the Cretaceous–Paleocene ancestral forms. Representative North American Cretaceous insectivores include *Gypsonictops,* which Romer (1966) suggested may be distantly related to hedgehogs, and the shrewlike taxa *Procerberus,* which L.S. Russell (1975) suggests as a possible ancestor of most later insectivores and ungulates, based on its extremely primitive dentition.

Purgatorius, the earliest putative primate (Clemens, 1974), is represented by a single molar in Cretaceous strata and by tooth and jaw fossils from the early Paleocene. Reconstructing morphology from such scanty materials is tenuous but the animal was probably similar to the tupaias

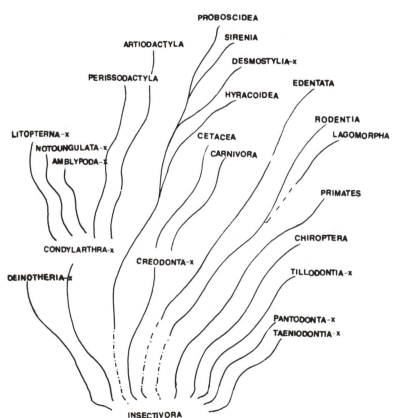

Fig. 9-87. A generalized evolutionary "bush" (trees imply to us long-term monophyly which is not the case with mammals) of placental mammals, omitting very minor orders. Extinct orders are marked with an "x."

(tree shrews) of Madagascar, which are the most primitive of extant primates. Several other tupaialike forms are known from the Paleocene as discussed subsequently, and this helps reinforce the identification of *Purgatorius* as a prosimian (the suborder of primates below the monkey-grade of specialization).

The final Late Cretaceous taxon of note is *Protoungulatum*, whose name suggests it to be the "first ungulate." At the earliest stages of specialization, ungulates (i.e., hoofed animals) did not show many of their typical characters, such as large size, distinctive limb structure, and very specialized teeth characterized by loss of canines, molarization of premolars, and development of infolded enamel to accommodate grazing and browsing. The ancestral ungulate group, condylarths, actually resembled carnivores more than hoofed animals; however, as we will detail subsequently, they did feature rudimentary hooves. Essentially, *Protoungulatum* of the Late Cretaceous and Paleocene featured dentition like that of condylarths, and condylarths were clearly ancestral to ungulates; hence, *Protoungulatum* was a remote ancestor of ungulates.

Early Paleocene faunas contain several "holdovers" from the Cretaceous (e.g., *Protoungulatum*, *Purgatorius*, many species of multituberculates, and several insectivore groups), and they include new (but still primitive) forms. Among these latter are (1) taeniodonts, which are a short-lived group of moderate-sized animals descended from insectivores and featuring rodentlike incisors, (2) several prosimian primates, (3) several types of condylarths, and (4) the earliest carnivores, including a more primitive group called creodonts and, in the middle Paleocene, earliest advanced carnivores or fissipeds. The occurrence of such mammals in North America is common; in fact, their presence in Lower Paleocene strata such as the Nacimiento, Fort Union, Polecat Bench, North Horn, Ravenscrag, and Denver formations, is sufficiently frequent to have biostratigraphic use.

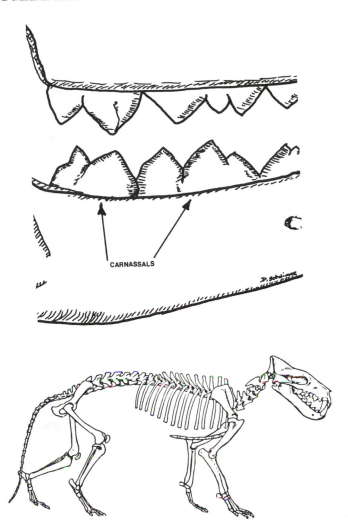

Fig. 9-88. Primitive carnivores. (Above) Diagram of carnassal molars (these drawn from a house cat) to show the shearing action. (Below) Reconstruction of the creodont *Hyaenodon* (u. Eocene to u. Oligocene). (Below, from Romer, A. S., *Vertebrate Paleontology*, 3rd ed., © 1966, University of Chicago Press.)

H.3. Archaic Eutherians of the Paleocene and Eocene

H.3.a. Creodontia and Carnivora

The appearance of carnivorous mammals marks a definitive shift in the general composition of mammalian biotas from insectivore/herbivore (Cretaceous) to a much more diverse set of ecological roles. Creodonts date from the Early Paleocene through the Pliocene (in very reduced numbers); Carnivora were a very minor presence until the Eocene and, of course, are diverse in the modern biota.

Dentition is the best means to recognize fossil carnivores. Even the primitive creodonts developed molars of the form termed *carnassals* (Fig. 9-88) which meet on the diagonal in a shearing action which effectively cuts tissue into manageable portions. This specialization is a derivative of the tribosphenic molar pattern and, indeed, some creodonts may have been insectivorous and detritivorous (carrion-eating) in part, as their teeth suggest. Other carnivore specializations include clawed toes, adaptations for fast running, and in Carnivora (at least), brain development suggesting the intelligence for pack-hunting behavior.

Creodonts may have been replaced in the later Tertiary (especially Oligocene and later times) because they were smaller-brained than Carnivora and possibly could not successfully kill the newly evolving ungulates of the later Tertiary. Two creodont families are recognized (Gingerich, 1980b): oxhyaenids (late Paleocene through Eocene) and hyaenodonts (early Eocene through Pliocene). *Oxhyaena* was approximately the size of a modern wolverine, whereas *Hyaenodon* was wolf-sized.

Carnivora are sufficiently familiar to need no description here. Romer (1966) describes the ancestral forms of the Eocene, and by the Oligocene carnivores of nearly modern abilities and morphologies were present. Two suborders are defined: Fissipeda (dog, cat, weasel, bear, racoon, and a few archaic families), and Pinnipeda (aquatic families: seals and walruses). Fissipeds date back to the Paleocene whereas pinnipeds have a surprisingly poor fossil record dating only to the Miocene.

H.3.b. Condylarths and Ungulate Characteristics

These are very essential creatures in higher mammalian evolution, because they were probably ancestral to all hooved mammals, including odd-toed ungulates (horses, rhinoceroses, tapirs, and several archaic groups), even-toed ungulates (pigs, camels, deer, oxen, sheep, and many others), subungulates (elephants and several additional forms), a number of extinct mammalian orders which evolved in apparent isolation through middle Tertiary time in South America (see Simpson, 1980), and even cetaceans (whales and smaller marine mammals). Further, condylarth morphology shows clear evidence of descent from a common ancestor with creodonts, i.e., Cretaceous or earlier insectivores.

Typical early condylarths, such as *Phenacodus* (Fig. 9-89) of the early Eocene, feature dentition with similarities to that of both insectivores and creodonts. However, the molars are neither carnassal nor tribosphenic, but rather a form termed *bunodont,* which features low, strong cusps adapted to crushing and prolonged chewing. Condylarths were generally sheep-sized but ranged from taxa as small as rabbits to taxa larger than the modern water buffalo. The most significant feature, aside from the dentition, is the small hooves present in condylarths, with an additional tendency toward enlargement of the central toes. Hooves are the feature which define the various ungulate groups and point most directly toward condylarth ancestry. In summary, condylarths provide a very excellent transitional group between the insectivores and ungulate orders. Without crea-

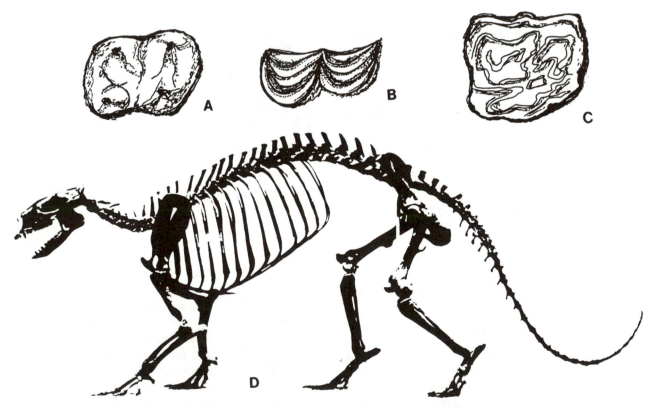

Fig. 9-89. Ungulate molars and a typical condylarth. (A–C) Occlusal views (i.e., grinding surface) of lower left molars. (A) Bunodont-type molar, found in many primitive ungulates (this drawn from *Phenacodus*), and, in modified form, in modern swine and primates (including humans); (B) selenodont-type molar, featuring crescentic enamel ridges, found in many ruminants (e.g., cervids, bovids, oreodonts), and see photograph in Fig. 9-12; (C) lophodont-type molar, featuring cross-crests of hard enamel, forming a complex set of grinding surfaces (this drawn from a fossil horse); (D) reconstruction of the skeleton of *Phenacodus,* a typical condylarth, showing admixture of primitive ungulate features (bunodont molars, small hooves, elongate tarsus, generally large size) and features derived from or with the creodonts (compare with skeleton of *Hyaenodon,* Fig. 9-88). (D courtesy of the Museum of Natural History, Princeton University.)

tures of condylarth morphology, the apparently enormous morphological differences between these ends of the mammalian spectrum would seem to preclude any relationship (consider the gross morphology of a shrew and cow to understand the obscurity of the evolutionary relationship).

Condylarths lived from Paleocene through early Oligocene time, and some taxa (e.g., *Mesonyx*) probably converged toward the carnivore condition and resembled bears in form and habit. The earliest true ungulates appear by the late Paleocene, and therefore probably evolved from condylarths very soon after the latter appeared. In the following paragraphs we will summarize the features and trends among major ungulate groups and certain taxa will be discussed in context later in the chapter.

Almost all ungulates are herbivores (with omnivory in some groups, e.g., swine, a secondary development) and the principal evolutionary modifications which characterized ungulates, after hooves, are molar and premolar teeth adapted to processing various types of vegetable food. We have already mentioned the basic bunodont molar, found in taxa as diverse as pigs and mastodonts. A refinement of the bunodont morphology is the *selenodont* molar, in which the cusps become crescentic and the enamel doubles back along the crescents, forming a very effective surface for prolonged chewing. Cattle, deer, and a common extinct (Oligocene) group called oreodonts feature this molar morphology. A different derivation of the basic ungulate molar is formed by ridges of enamel linking adjacent cusps, again producing a multiple set of hard surfaces. This is the *lophodont* molar, present in rhinoceroses and certain early horses. The most specialized molar features complex invaginations of the enamel, forming very effective grinding surfaces which are able to survive years of wear and the abrasion of grasses and other silica-bearing plants. This last type is the *hypsodont* molar found in modern horses and elephants.

Along with modification of the molars came a general loss of canine teeth (which are, however, retained in several groups, notably swine) and "molarization" of the premolars. In addition, in many ungulates the incisors became quite important as the major cropping elements; thus, a majority of ungulates feature a strong set of incisors, no canines, and a series of flat, reinforced grinding molars and premolars in their dental battery.

A second line of ungulate modifications concerns large size and running speed. With loss of canine teeth and adoption of herbivorous habits, ungulates coevolved as prey for the creodonts and carnivores. Herbivores tend to grow large, but size approaching even that of modern elephants is not certain protection for a defenseless animal; therefore, many ungulates specialized for swift escapes, protective herd behavior, extreme sensitivity to danger (i.e., via keen senses), and, in numerous groups, development of horns

and antlers. Of these specializations, running speed produced the most distinctive skeletal changes.

In all but the heaviest ungulates, the limbs are slim in proportion to other mammals. The main bones (humerus, radius/ulna; femur, tibia/fibula) are rarely elongated but the manus and pes feature extremely long metacarpals and metatarsals. The result is a three-part flexure with long bones offering leverage at each joint (e.g., hip, knee, ankle). One discovers in many fast-running ungulates that the longest part of the limbs are below the ankle or wrist joints. This extra lever adds the spring to the flight of a deer or antelope. Hooves, of course, facilitate running and long-distance walking.

H.3.c. Archaic North American Ungulates

The term "ungulates" encompasses a vast variety of descendants from condylarths. Certain groups of the Paleocene and Eocene may collectively be called "archaic" and probably were the main prey of archaic carnivores. Among the taxa included are two which occur commonly in North America: uintatheres and pantodonts.

Uintatheres are common in upper Eocene strata from the interior basins, including (not surprisingly) the Uinta Formation in the Uinta Basin and the Bridger Formation in the Washakie and Green River basins. They were probably the largest of early Tertiary mammals, reaching several metric tons (approximately the size of a very large rhinoceros), with disproportionately long skulls and perhaps the weirdest arrangement of horns and teeth known among mammals. The classical uintathere genus, *Uintatherium* (Fig. 9-90), featured: two very long upper canines projecting downward past its lower jaw; bony extensions of the lower jaw (probably to protect the canines); and fully six horns on the head (two over the nose, two over the eyes, two over the topmost skull). In addition, the shoulders were much higher than the rump, presenting an awkward backward slope to the body. Earlier uintatheres, dating as early as the late Paleocene, lacked the horns of *Uintatherium* but generally were large and featured the projecting canines. Although they were an evolutionary dead-end terminating in Oligocene time, fossils of uintatheres are disproportionately abundant because they lived in a time and place of extensive alluvial and lacustrine environments, producing numerous bone beds.

A related suborder, Pantodonta, includes primitive ungulates reaching the size of an Iowa hog, featuring heavy limbs, generalized browsing-type dentition, and distinctively, a pair of elongate canine teeth in Eocene genera such as *Coryphodon*. Pantodonts appeared during the middle Paleocene and survived into the Oligocene before becoming extinct. Unlike uintatheres which are largely restricted to North America (and found rarely in Asia), panto-

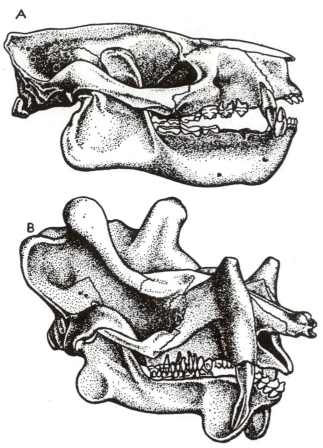

Fig. 9-90. Two archaic North American ungulates. (A) Skull of *Pantolambda,* a pantodont (which as a group were also present in most Holarctic continents); (B) skull of *Uintatherium,* the very bizarre, typical uintathere found abundantly in the Eocene deposits of the western interior. (Both from Romer, A. S., *Vertebrate Paleontology,* 3rd ed., © 1966, University of Chicago Press.)

donts have common occurrence in Europe, Asia, and North America.

H.4. Chiroptera (Bats)

The morphology of bats is too well known to require detailed description here; of major note are the wings developed from elongations of four fingers on each hand (in contrast to the elongate single fourth digit of pterosaurs, and the wing developed from the entire, nonelongate arm in birds). Additionally distinctive are the very small eyes, unique echolocating adaptations (i.e., a virtual "radar" system), leaf-shaped nose structure, and weak-hipped skeleton.

The dentition shows that bats evolved directly from insectivores; however, this close resemblance in the most commonly preserved structures also renders it difficult to distinguish bat fossils from those of insectivores. It is not known, for example, whether Paleocene bats have been found and misidentified as insectivores. The earliest fossils known to be bats are of Middle Eocene age and are found in the Green River shales and penecontemporaneous European strata. The American specimen, *Icaronycteris,* is marvelously well-preserved and shows that fully evolved bats were present by the Eocene; Romer suggested that they had evolved by the late Paleocene. Subsequent to the Eocene, bats are relatively rare as fossils until the Pleistocene, from which time we have abundant bat fossils from caves as well as masses of fossil bat guano.

H.5. Prosimians (Primates)

During the Paleocene, a number of prosimians joined *Purgatorius* in representing primates in North America. By Eocene time, advanced prosimians, possibly ancestral to modern lemurs and tarsiers, were present in the western interior.

Primates are largely unspecialized animals which, in the most primitive representatives, appear to be quite similar to insectivores; indeed, it is not clear whether some fossil taxa should be included in the order Insectivora or in the order Primates. The single, apparent evolutionary trend which diverted primates from insectivory is the arboreal way of life. Colbert (1980) emphasized that arboreal life requires good vision and especially good depth perception, good control of the hands, a tendency to develop mobile thumbs, and a keen sense of understanding surroundings (which may have led to selection for greater curiosity and intelligence). Olfactory sense (i.e., smell) was of little use in treetops and may have been suppressed early in primate evolution.

Most Paleocene primates (and the Upper Cretaceous specimens) are assigned to the suborder Plesiadapoidea, which contains three families which may be represented by *Plesiadapis* from the middle Paleocene in the western interior. Plesiadapids were a distinct, early primate group featuring rodentlike incisors, tribosphenic molars, and in contrast to later primates, bearing claws instead of nails on the digits.

In Eocene strata, advanced lemuroids are known from North America and these are assigned to the suborder Lemuroidea in the family Adapidae. Gingerich (1979) discussed Adapidae from the Middle Eocene and noted that six species among two genera were present in Bridger-age strata; and Gingerich (1980a) illustrated biostratigraphy in North America based on primate faunal zones. Typical of Eocene Adapidae are the genera *Notharctus* (Fig. 9-91) and *Smilodectes,* with species of *Notharctus* represented by the oldest complete skeletons of primates. The adapid-grade prosimians featured long limbs, reduced incisors and canine teeth, and development of an additional element, the *hypo-*

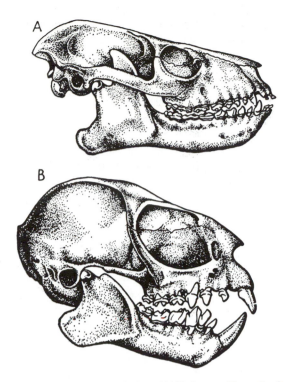

Fig. 9-91. North American prosimians. (A) *Notharctus* (Eocene), a lemur; (B) *Tetonius* (Eocene), a tarsier. Both creatures have skull lengths of only a few centimeters. (Both from Romer, A. S., *Vertebrate Paleontology*, 3rd ed., © 1966, University of Chicago Press.)

cone, on the molars creating an efficient omnivore tooth by modification of a bunodont-type molar.

A second Eocene primate group is probably ancestral to the living *Tarsius* of the East Indies and the Philippines. Early Cenozoic tarsiers were abundant in North America and Europe, including the widespread American genus *Tetonius* (Fig. 9-91). Like other tarsiers, the eyes were very large, the fingers and toes showed great prehensile abilities, and the brain had undergone some expansion. Szalay and Delson (1979) suggest that a group of tarsiers, the Omomyidae, are the probable ancestors of higher primates, and ultimately, apes and humans. However, after Eocene time, the diverse prosimian primates of North America declined rapidly, leaving only a handful of Oligocene omomyid tarsier genera, and a single known Miocene taxon with the Sioux-derived name *Ekgmowechashala* ("little cat man").

Primate history is naturally of great interest, since it leads to human history; but all higher primates are from the Old World, except for the Catarrhine, or New World, monkeys. These undoubtedly derived from North American prosimian ancestors who migrated during the Eocene and subsequently developed in isolation, but to date, prosimian fossils have not been found in South America. The reader is referred to Simons (1964) for a nontechnical discussion of prehuman primates, and to Szalay and Delson (1979) for a detailed treatment of all primate groups. One should note, however, that the field of paleoanthropology features more controversy on any given topic than does any other branch of science. To avoid entering the fray, we next discuss primates with the migration of Paleo-Indians back to North America during the late ice age (Section H.9).

H.6. Edentates (Sloths, Anteaters, Armadillos)

Although it is outside the scope of this book to examine the South American taxa of the Cenozoic, many of which owe their ancestry to North American taxa, several important South American forms migrated to the northern continent when land bridges were established by lowered sea levels. Chief among these are two taxa within the order Edentata (meaning "without teeth"): sloths and armadillos (and their relatives).

It is uncertain whether edentates comprise a monophyletic group. General characteristics include the reduction or loss of teeth, extra articulations between successive arches of the rear trunk vertebrae, often fused tibia and fibula, usually very massive pelvis, relatively small brains, a variable number of neck vertebrae (nearly all other mammals, except sirenians, have seven), and a tendency to develop skin armor, either solid, as in glyptodonts, jointed, as in armadillos, or disseminated, as in sloths.

The ancestors of South American edentates may be a late Paleocene North American taxon Paleodonta, which resemble unarmored armadillos. Paleodonta survived to the Oligocene, but all subsequent development of edentates was in South America (see Romer, 1966). During Pliocene time, a land connection with South America developed and representatives of two major edentate groups, Cingulata (armadillos) and Pilosa (sloths), wandered north and very successfully spread across the southern and western United States.

Glyptodonts were very large armadillos with unjointed shells and tails often featuring a club or spike-ball defense structure. They reached sizes as large as the largest bears, and survived until the latest Pleistocene. Armadillos, of the small modern format, also migrated from South America but they clearly survive to the present and are in fact currently expanding their range northward.

Sloths migrated to North America in the form of ground sloths, which were much larger than the tree sloths of the tropics and walked on their knuckles since they retained the very long claws found in arboreal sloths. They likely spent much of their time in a bipedal position, feeding on the low branches of trees. Sloths did not develop armor but the giant ground sloths possessed a type of bony pavement within the skin, perhaps serving as protection from smaller carnivores. Ground sloths, such as the common genera *Nothrotherium, Paramylodon,* and the largest taxon, *Megatherium,* spread widely in North America in the Plio-Pleistocene after reestablishment of the land bridge, and comprise some of the commonest fossils in such famous Quaternary sites as the La Brea tar pits in Los Angeles and Rampart Cave in the Grand Canyon. Rampart Cave, in fact, is floored with sloth dung as the dominant matrix.

H.7. Progressive North American Mammals

The term "progressive" is somewhat imprecise but here it is applied to mammalian orders which generally are larger-brained and more specialized than the typical Paleocene mammals and their direct descendants. Curiously, humans are primates, one of the "archaic" groups in this terminology. Nevertheless, all the progressive orders discussed below are extant, as are many of the archaic orders. Most of the progressive mammals appear first in the Eocene; however, the earliest rodents are of late Paleocene age and it is always possible that Paleocene forms of nearly any listed group may yet be found.

H.7.a. Rodents and Lagomorphs

Rodents are very rare in pre-Eocene strata, which is all the more surprising because of their abundance in the Holocene. At present over 1700 rodent species are known (Colbert, 1980) and they exceed all other mammals in terms of biomass. Diagnostic characteristics of the order Rodentia include: generally small size, herbivorous habit, possession of a single pair of strong incisors which grow continuously through the animal's life, and development of a jaw-hinging mechanism that allows front-to-back motion which enables use of the incisors as gnawing tools. Premolars are usually reduced in number and canines are always absent, while the molars may be high-crowned and flat; thus, a common rodent dental formula is 1:0:1:3.

There is no fossil record to show the ancestry of rodents. Earliest rodents appear in the Upper Paleocene of North American (Harland *et al.,* 1967) and are assigned to the genus *Paramys,* a squirrel-like, small animal which already possessed fully rodent-type incisors. *Paramys,* and its relatives in the Eocene, did possess a greater number of premolars than virtually all subsequent rodents and in that respect are primitive. The Paramyidae were abundant and diverse in the Eocene and are represented in the Holocene by the sewellel *Aplodontia.* From these ancestral rodents sprang directly the squirrels, woodchucks, and chipmunks (suborder Sciuromorpha), the "true" mice and rats (Myomorpha), and beavers (Castorimorpha), all appearing first in the Oligocene. The isolation of South America through most of the Tertiary allowed development of an endemic rodent population, much of which survives to the present; and with reestablishment of the Central America land bridge in the Pliocene, at least one South American rodent, the porcupine, successfully invaded North America.

Few rodents are very large but, during the Quaternary, beavers reached sizes as large as black bears. *Castoroides* (giant beavers) became extinct just after the "ice age" and may have been victim of climate change or of human predation. Not all beavers were aquatic dam-builders; some have left impressive fossils in the form of burrows in a distinctive corkscrew shape, found in Miocene strata of the western interior.

Rabbits, hares, and pikas are not rodents since they possess a double pair of incisors rather than the single incisors of rodents; in fact, they show a number of morphological details that suggest great distinction from rodents, including possession of two or three premolars, structurally different molars, and a simpler jaw musculature. Earliest rabbits are from the Upper Paleocene of Mongolia (Harland *et al.,* 1967) and other taxa appear in the Eocene and Oligocene of North America and Asia. Even early lagomorphs show the characteristic paired chisel-like incisors and the fenestrate structure of the snout portion of the skull; thus, it seems that lagomorphs were distinct from rodents in the Paleocene and remain a conservative group which have succeeded very well in the Pleistocene and Holocene.

H.7.b. Cetaceans

Perhaps the most difficult ancestor–descendant relationship for the casual observer to accept among mammals is that between mesonychid condylarths (e.g., *Mesonyx*) and cetaceans. Cetaceans are marvelously adapted to marine life and feature a morphology showing almost total transformation from that of land-living ancestors. In virtually all whales and porpoises, the axial skeleton is flexible, with numerous vertebrae; but the neck is short and the seven cervicals may be fused. The flukes (tail) are oriented horizontally and totally lack bone (which terminates with the vertebral column just anterior to the flukes). The dorsal fin is similarly formed by stiff tissue without bone. The pectoral limbs are modified to flippers and feature hyperphalangy (multiple joints) in common with those of plesiosaurs. The posterior limbs and pelvic girdle are almost entirely lost except for vestiges of the pelvis. In many cetaceans, the ribs are reduced in number, and in all taxa the skull features considerable modification of the ears (for the extraordinarily good underwater hearing of most forms) and nostrils.

The classification of whales is based on the dentition. Alone among mammals, modern toothed whales feature virtually undifferentiated (i.e., uniformly shaped and sized) teeth, which is almost certainly a secondary development. A report by Gingerich *et al.* (1983) has added considerable insight into the earliest history of whales and their tooth development. Fragments of an apparently amphibious creature, with dentition very similar to the oldest toothed whales, has come from the early Eocene of Pakistan (the genus *Pakicetus*). Although the entire postcranial skeleton is unknown, the presence of this skull material in terrestrial sediments, combined with the dentition clearly like that of later undoubted whales, suggests both the ancestral and amphibious nature of *Pakicetus*.

The oldest true (i.e., undoubtedly marine) whales are assigned to the suborder Archaeoceti and are slim, toothed whales of the later early Eocene through Miocene. The oldest archaeocete is *Anglocetus,* from England; however, whales only slightly younger have global distribution (as one might expect). In *Prozeuglodon,* a small archaeocete, additional evidence of condylarth ancestry is evident in the premolars and molars which bear vestiges of the condylarth shearing morphology (typical of the carnivorous forms such as *Mesonyx*). The most characteristic and earliest of the large whales is *Basilosaurus* (or *Zeuglodon*), widely distributed in upper Eocene strata and reaching 20 m in length.

Advanced whales first appear in the upper Eocene and increasingly in younger strata. The advanced toothed whales are called Odontoceti and include ancestors (and living taxa) of porpoises, orcas, narwhales, belugas, pilot whales, and the sperm whales. Oligocene and Miocene odontocetes were advanced over the Eocene archaeocetes in having the nasal passage (blowhole) migrated to the fully posterior position (Fig. 9-92). In this evolutionary change, the parietal bones were lost and the maxillae touch the supraoccipital bones. Additionally, odontocetes feature many more teeth than do archaeocetes, far exceeding the primitive eutherian count of 44 (as a secondary adaptation).

The oldest of known baleen whales (Mysticeti) are found in the lower Oligocene of New Zealand. Mysticetes differ from odontocetes in lacking teeth and in place possessing the unique sheets of ''whalebone'' which serve to sieve small crustaceans (typically of the Euphausiidae) as the primary source of food.

They also possess structural differences in the maxillae and superorbitals relative to toothed whales and it seems likely that baleen whales evolved from archaeocete ancestors rather than from odontocetes. Mysticetes are a low-diversity group, represented by only five extant genera divisible into two groups: rorquals (blue, finback, sei) and right whales (gray, Greenland right, humpbacks).

H.7.c. Perissodactyls

Living Perissodactyla include rhinoceroses, horses and their relatives, and tapirs. The order is characterized as

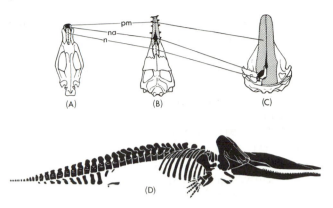

Fig. 9-92. Hypothetical evolution of the cetacean skull. (A) *Apterodon*, a mesonychid condylarth (Oligocene of Africa); (B) *Prozeuglodon*, an early Eocene cetacean; (C) *Aulophyseter*, a Miocene toothed whale. Bones and other features labeled are: n, nasals; na, external nares (nostrils); pm, premaxillary. (D) *Physeter*, the modern sperm whale, showing the extreme specialization in cetaceans. (Note that A is actually younger than B, suggesting the relationship is not directly ancestral.) (From Colbert © 1980 by John Wiley & Sons, reprinted with permission of John Wiley & Sons.)

including progressive ungulates in which there are an odd number of toes present and in which the balance of each foot normally passes through the middle toe. Most perissodactyls possess one or three toes on each foot, although some extinct taxa retained the primitive five toes. Otherwise, the typical ungulate characters described are found in most members of the group. Overall, perissodactyls seem to be an order that has fared rather badly in the Pleistocene and Holocene in the face of competition from the artiodactyls (see next subsection); of the three extant perissodactyl groups, only the horses, asses, and zebras are abundant, whereas rhinoceroses seem to be approaching extinction. In the Early–Middle Tertiary, however, odd-toed ungulates were diverse and widespread, and included the largest of land mammals among their ranks. The fossil record of horses and rhinoceroses is among the best of any terrestrial group.

All major perissodactyl lines appeared in the early Eocene. The horses are first found in Wasatch strata in the American western interior and in correlative units in Europe, whereas rhinoceroses appear in both of the above and also in China (Harland *et al.*, 1967). A third important perissodactyl group, Ancylopoda (or chalicotheres), became extinct in the Pleistocene but also appears first in lower Eocene units from North America and Europe.

Evolution of the horse has become one of the classics of paleontology. Since it is a subject of such renown, we will only describe the bare bones of the events. The first horse, *Hyracotherium* (or ''Eohippus'') of the lower Eocene, was roughly the size of a medium to small dog and had bunodont molars and small canine teeth with only a small *diastema* (i.e., space between teeth) behind the incisors. It possessed four toes on each foot but the fourth on the hind limbs, and to a lesser degree on the front limbs, was reduced and little used. The animal was a browsing forest inhabitant. Most mammal specialists consider *Hyracotherium* to be not only the ancestor of horses but also a reasonable ancestor of all odd-toed ungulates; however, for that latter distinction it would have to have appeared before the early Eocene (Paleocene relatives may have existed and for unknown reasons are undiscovered). The basic morphologies that characterize horses all stem from the ability to run swiftly. Although small, *Hyracotherium* may still have been the swiftest animal of its time.

At the end of Eocene time, *Hyracotherium* became extinct in Europe and all subsequent evolution of horses occurred in North America, leaving abundant fossils. In North America, other horse genera appeared during the Eocene (see Fig. 9-93), although all were four-toed and small. In the early Oligocene appeared large-dog-sized horses with three-toed manus and pes and with molars and premolars better adapted for grinding leaves (but still low-crowned and generally unsuited for grazing). *Mesohippus* best represents this group while in the middle and upper Miocene the slightly larger (but otherwise similar) *Miohippus* is the common horse genus.

After the Oligocene the apparent straight-line evolution (*monophyly*) ended and horses branched out into several groups. Some lineages increased in size while retaining three-toed feet and browsing habit (e.g., *Anchitherium*); others remained very much like *Miohippus*, but all such conservative lines became extinct by the later Pleistocene. The ancestry of modern horses lies through the Miocene taxa *Parahippus* (lower Miocene) and *Merychippus* (upper Miocene). In *Parahippus*, the side toes on feet were virtually unused and the molars and premolars began to show infolded enamel such as would be required by a grazing animal. *Merychippus* had high-crowned cheek teeth, a good cementum covering on the grinding teeth, and slightly larger size than previous horses; it was probably a true grazing animal living well on coarse siliceous prairie grasses in the spreading grasslands of the western interior.

In the Pliocene, descendants of *Merychippus* and related forms spread to Asia (via the Bering land bridge) and from there to most other continents (excluding Australia, Antarctica, and South America). Two lines of descent can be traced from *Merychippus*. The first includes gracile (i.e., slim), pony-sized, three-toed horses with excellently adapted grazing dentition, represented by the genus *Hipparion*. So many species and variants of European *Hipparion* have been described that Woodburne and Bernor (1980) suggested several genera of other groupings might be represented (i.e., a *Hipparion* "complex"). The other *Merychippus* descendants are represented by *Pliohippus*, a single-toed large horse. The hipparions died out in the late

Pliocene, after apparent success, while *Pliohippus* gave rise to *Equus*, the genus which includes zebras, asses, and the Pleistocene horses. *Equus* evolved in North America, spread across Asia, Africa, and Europe, and (curiously) became extinct in North America during the Pleistocene glaciations but survived well on the other continents. Horses in modern North America have all been introduced from Old World stock. A lineage of horses, descended from *Pliohippus*, reached South America during the Pleistocene and evolved as a distinctive group of short-legged, heavy taxa such as *Hippidium*. These became extinct in the Pleistocene and only the single genus *Equus* survives today among all the horses which once lived on six continents.

The Tertiary history of perissodactyls is by no means limited to horses; rhinoceroses were very abundant in the early and middle Cenozoic. The former abundance and diversity of rhinos are difficult to imagine, given their rarity and low diversity at the present time; additionally, modern rhinos are restricted to the tropics whereas most Tertiary rhino fossils are found in the Holarctic continents.

Earliest rhinos are of Eocene age and are found commonly in North American and Asian strata. They were of sheep size, slightly built, hornless, and apparently fast-running, browsing animals (rather more like modern horses in overall form than were the horses of the time!). These "running" rhinos survived into the Oligocene and are represented by the common *Hyracodon* from the Big Badlands strata. In the late Eocene appeared a second group of rhinos, amynodonts, which were heavier, hippolike, river-dwelling forms (living perhaps in the same manner as modern hippos), with short legs and still no horns. The Amynodonta were a dead-end line, surviving until the Miocene in Asia, while the hyracodonts or "running rhinos" themselves became extinct in the Miocene but gave rise to the modern rhinos.

The "true" rhinos are represented in the Oligocene by *Caenopus*, a large, hornless, short-legged form that, nonetheless, shows advanced dental characteristics (notably molarization of premolars), pointing toward ancestry of modern types. *Caenopus* also was among the earliest true rhino to show only three toes on manus and pes. Evolution of rhinos is complex following the *Caenopus* grade but, in general, several types of rhinos followed in the late Oligocene and succeeding epochs including: horned forms present by the Miocene, huge hornless taxa in the Oligocene–Miocene, short-legged, two-horned, pygmy, and a host of other morphologies. Romer (1966) suggested eloquently that in contrast to the stately "tree" of horse evolution with a main stem and a few branches, rhino evolution was "bush" with no main stem. We should note that again, like horses, rhinos were an abundant North American group through most of the Tertiary but disappeared from this continent during the Late Tertiary (the late Pliocene).

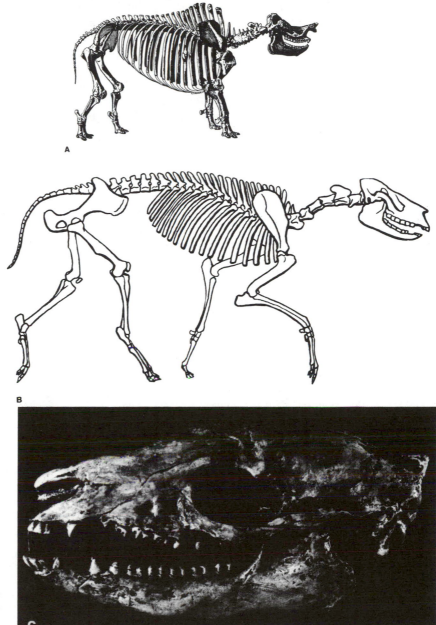

Fig. 9-93. Perissodactyl fossils. (A) *Brontops* (l. Oligocene), a large titanothere (not to scale; the specimen is approximately 3.5 m long); (B) *Hyracodon* (Oligocene), a lightly built "running rhinoceros" showing generalized ungulate morphology; (C) skull of *Epihippus* (u. Eocene), a very primitive North American horse. (A, from Romer, A. S., *Vertebrate Paleontology*, 3rd ed., © 1966, University of Chicago Press; B, from Colbert © 1980 by John Wiley & Sons, reprinted with permission of John Wiley & Sons; C, DRS photograph.)

They flourished in Eurasia as the large "woolly rhinoceroses" of the Pleistocene but are now reduced to five modern species.

Finally, there are two important, extinct perissodactyl groups that deserve discussion. The largest American odd-toed ungulates were the titanotheres, (also called brontotheres), which appeared in the Eocene as small forms similar to *Hyracotherium,* but which became larger and heavier through the Oligocene to culminate in impressive large genera such as *Brontotherium* from the Oligocene badlands faunas. These forms reached almost 3 m at the shoulders and had huge, low skulls featuring a curious "sag" in the middle and an even more curious forked, blunt horn over the nostrils. Their dentition never evolved beyond the bunodont stage and they became extinct in middle Oligocene time. However, during their brief span they were abundant and diverse, especially in the Great Plains.

Another unsuccessful but formerly abundant perissodactyl group were the chalicotheres, also appearing in the Eocene from "*Eohippus*"-like ancestors in North America, Europe, and Asia. Chalicotheres grew larger with time but retained a horselike morphology with the notable exception of possessing clawed feet. In addition, the teeth were like those of titanotheres (low crowned, bunodont) and their front limbs were longer than the rear, giving a backward slope to the carriage. They apparently used the claws for digging roots in a swampy environment. The best-known American genus is *Moropus,* a horse-sized and very horselike animal of the Miocene. Miocene chalicotheres were the last in North America but the group survived until the later Pleistocene in Europe and Asia.

H.7.d. Artiodactyls

Most modern ungulates are artiodactyls, i.e., even-toed ungulates. In this group the weight bearing on each limb passes between the two center toes (in a four-toed form) or between the two toes. In consequence of this balance, there are basic structural modifications of the upper and lower limb bones and in the manus and pes, which make it relatively easy to determine whether or not a fossil taxon was an artiodactyl. A very distinctive feature is the double-pulley shape of the *astragalus* (ankle bone) in artiodactyls, which allows flexible motion between the tibia and bones of the foot. Consequently, many odd-toed ungulates (especially goats, deer, pronghorn, and so on) can spring with great force because of this flexibility. As with perissodactyls, various types of dentition are present. In many artiodactyls, especially deer, the incisors are lost on the upper jaw and a fleshy hard surface comprises an anvil against which the lower incisors chop off forage.

Classification of artiodactyls is a classical taxonomist's dilemma, which we will avoid by generalizing: one can easily recognize an interrelated assemblage of cud-chewing taxa called ruminants (bovines, deer, tragulids), and a large group of swinelike taxa, including pigs, peccaries, and hippopotamuses. There are also numerous types of living and (especially) fossil camels, which are a cud-chewing group but of a more primitive lineage than the other ruminants. They seem to be much more closely related to, say, deer than they are to swine, but they are, nevertheless, distinct. Finally, there are several important extinct artiodactyl groups, including superficially piglike forms called entelodonts, and some common rather nondescript camel-like forms called oreodonts.

The earliest artiodactyls known come from the Eocene of North America, Europe, and Asia and show the even-

toed format fully developed (Rose, 1982). The early Eocene genus *Diacodexus* (Fig. 9-94) resembles the primitive extant chevrotain (a deerlike ruminant) far more than it does the suinid branch of artiodactyls, and Rose (1982) suggested that a bunodont-toothed ancestor of Suina may be found at some point in the Paleocene or earliest Eocene.

Among the earliest Suina are the entelodonts of late Eocene to Miocene age. These were first large artiodactyls, reaching the size of bison in *Dinohyus* of the Miocene, featuring two-toed feet, huge heads (over 1 m long), small eyes, small brains, canines modified as tusks, and as a distinctive feature, broad extensions of the zygomatic arches giving the head a wide-cheeked look. In fact, they resembled warthogs but were not true swine; rather, they were an early, large-bodied offshoot of the Suina which dead-ended in the Miocene.

While the entelodonts were evolving, more advanced

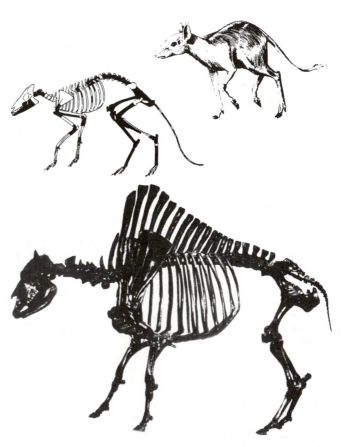

Fig. 9-94. (Above) The oldest-known artiodactyl, *Diacodexus* (l. Eocene), which shows a deerlike morphology. Actual preserved bone is shown in black on the skeletal reconstruction. (Below) A short-horned bison from the La Brea tar pits, showing typical morphology of the larger bovids: Note the very large neural spines on the dorsal vertebrae, for attachment of heavy muscles ("buffalo hump") to lift and brace the head. (Above, from Rose, in *Science,* © 1982, American Association for the Advancement of Science; below, DRS photograph.)

swine appeared in the late Eocene and are well-represented from the Oligocene onward. Pigs are Old World creatures whereas peccaries are of the New World. The difference between them is the greater adaptation for running in peccaries, with reduced side toes and longer limbs, and the shorter skulls with simple shearing molars and straight canines in peccaries. Pigs naturally tend to develop curving and sideways-growing tusks from the canines (a feature bred out of domestic hogs, which are probably descended from Asian pigs). Pigs and peccaries are intelligent animals with large brains. A variety of taxa are known from Oligocene to recent times and will not justify detailed description here.

Ruminant artiodactyls are cud-chewing forms that evolved the well-known complex stomach arrangement featuring the *rumen,* a large pouch in which bacteria break down cellulosic plant material such as tree bark and coarse grass; thus, ruminants are very well-fitted to extracting maximum energy from their forage. Most ruminants have selondont dentition and lack upper canines. In addition, as a group they tend to be long-legged and demonstrate herding behavior.

The more primitive of ruminants are tylopods, an infraorder including camels and llamas among extant groups, and including a host of extinct taxa. Features distinguishing tylopods from the advanced ruminants, called pecorans, include: separate tarsals and carpals, which are commonly fused in pecorans; simpler arrangement of the stomach in tylopods; and less modified dentition in tylopods. First tylopods appear in the upper Eocene of the Uinta Basin and this appearance includes a genus, *Poebrodon,* that is the first true camelid known (Harland *et al.,* 1967). Early tylopods were short-legged but did show selenodont molars and premolars, reduced canines, and other features which differentiate them from the Suina. Along with early camelids in the late Eocene and Oligocene arose many other tylopod stocks including cainotheres, anoplotheres, xiphodonts, oreodonts, and others.

Oreodonts, unlike others mentioned above, occurred and in fact were abundant in North America during the early Tertiary. They appeared in the late Eocene and flourished through Oligocene–Miocene times to dwindle drastically and finally disappear in the Pliocene. The oreodont body retained the stout, squat, piglike form of earliest artiodactyls throughout their tenure; however, they also developed selenodont dentition, showing that they were almost certainly ruminants. They are commonly referred to as "ruminating swine" and they comprise the majority of fossil mammals found in all Oligocene and Miocene strata from the western interior. The extremely fossiliferous beds from the Big Badlands are called the "Oreodon Beds," featuring the most common and characteristic genus, *Merycoidodon.* Their decline from virtual dominance of the ruminant fauna on this continent may be attributed to competition from pecoran artiodactyls.

While oreodonts were diversifying in the Oligocene, camelids were also evolving apace but as a group distinct from the oreodonts. Camels early in their evolution lost the two side toes of each foot, thus becoming the earliest two-toed artiodactyls, and featured elongate necks. *Poebrotherium,* a typical North American Oligocene taxon, was of the size of a very small sheep and featured virtually complete primitive ungulate dentition. Subsequent camelids diversified into various large and small forms, including a very long-necked, long-limbed, gracile genus of the Miocene, *Oxydactylus,* and a very large, giraffelike genus *Alticamelus,* of Pliocene age. These Middle and later Tertiary camels probably had hooves whereas modern camels, adapted to desert life, lost the hooves and rather adopted broad toes. Modern, Old World camels probably trace their history back to the Pliocene in North America, from where they spread during the Pliocene and Pleistocene, while the camelids became extinct in North America during the latest Pleistocene. Llamas also migrated from North America to South America during the Pleistocene. The distinguishing features of modern camels (e.g., the humps, ability to exist without frequent water supply, the long hair, and so on) are probably Pleistocene and Holocene specializations adopted by camels to cope with their specific Old World and South American environments.

The Pecora, comprising the suborder which includes all remaining even-toed ungulates, also appear in the upper Eocene, where they are very rare but become fairly common in Oligocene strata in North America. Among the most primitive pecorans are very small (hare-sized) deerlike animals called tragulids, which survive today as the oriental mouse deer or "chevrotain." The Eocene Old World genus *Archaeomeryx* was quite similar to the modern *Tragulus* of Asia and similar forms in Africa. In Oligocene and younger Tertiary strata of North America are found fossils from a line of tragulids which became of deer-size and feature extreme development of antlers. Antlers and horns (the former being shed annually, the latter being permanent) are very characteristic of artiodactyls; early North American tragulids featured four or six pairs of horns, such as in *Synthetoceros* of the Pliocene. These multihorned tragulids died out during the Pliocene but were followed by ever-diversifying horned ruminants including deer (Cevidae), goats, cattle (Bovidae), sheep, pronghorns, and giraffids. Again, the diversity of advanced artiodactyls is far beyond the scope of this book and the reader is referred to Romer (1966).

H.7.e. Subungulates

A final, major group of Tertiary mammals to be considered in this chapter may be a polyphyletic collection. The "subungulates" are several eutherian orders which share some of the following characteristics: the incisors or canines are typically enlarged, the remaining teeth are usually reduced greatly in number, the feet retain the primitive five digits, the clavicle is lost, and the molars and premolars tend to be bunodont or hypsodont in form. The most important of subungulates are proboscideans (i.e., elephants and their kin) but we will also discuss briefly the sirenids (manatees and dugongs).

The Proboscidea are a large and important group whose ancestry is known to trace back to the late Eocene of the Fayum (in Egypt) and other parts of Africa. Earliest proboscideans are called moeritheres, after the characteristic genus (Fig. 9-95). *Moeritherium* was a calf-sized animal that featured eyes set very far forward in the skull and enlarged upper and lower second incisors. The body was heavy-set with unusually stout legs. Moeritheres persisted through the early Oligocene, by which time earliest mastodonts had appeared. It is not completely certain that moeritheres are near the ancestry of higher proboscideans, but their morphology is close to that of a likely ancestral taxon.

Colbert (1980) noted that from the moerithere stage, proboscidean evolution probably went in two directions: to the elephantoids and to the dinotheres. Characteristics of both groups include the obvious large size, development of trunks, short necks, development of very large anterior teeth (usually tusks derived from the upper second incisors but also developed from other teeth), and great proportional increase in size of the head. The skull increased in length in the development of modern proboscideans but there was a secondary shortening of the jaws with great reduction of the lower jaw in modern elephants. Along the elephant lineage are three stages: the long-jawed mastodonts, short-jawed mastodonts, and elephants; however, as Colbert noted, these lineages commonly evolved side-by-side through the later Tertiary.

In the lower Oligocene Fayum deposits in Egypt are remains of primitive mastodonts such as *Phiomia*, the size and proportions of a small elephant, but with heavy, bunodont molars, and short tusks developed on upper and lower jaws. By Miocene time, mastodonts such as *Trilophodon*, a full-sized, heavily tusked proboscidean (with upper and lower tusks), had spread through the entire northern region, reaching North America by the late Miocene. There followed a very complex divergence of mastodonts, with many changes in the nature of the molars as the principal tool used to diagnose relationships.

One distinctive mastodont group of the Miocene–Pliocene in North America and Asia is the famous "shovel tusk" group in which the lower jaw flattened, elongated, and turned downward, presumably creating an effective

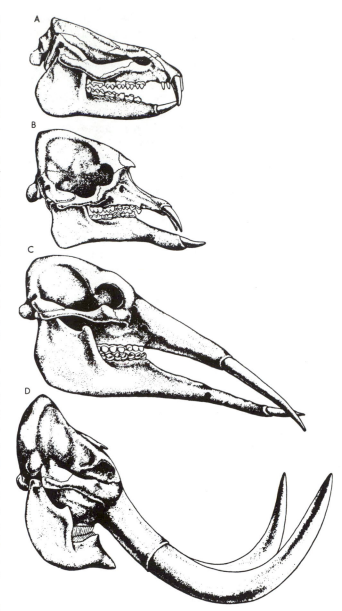

Fig. 9-95. Subungulates and a possible evolutionary lineage. (A) *Moeritherium* (u. Eocene to l. Oligocene, Egypt), a very primitive proboscidean, not very derived from possible condylarth ancestors; (B) *Phiomia* (l. Oligocene, Egypt), a bunomastodont; (C) *Gomphotherium* (Miocene, Europe and North America), a common mastodont; (D) *Mammuthus primigenius*, the woolly mammoth, a true elephant. (All from Romer, A. S., *Vertebrate Paleontology*, 3rd ed., © 1966, University of Chicago Press.)

shovel to dig roots. The American *Amebelodon* is one of the best-known genera. Another line of American mastodonts developed strongly cross-crested molars and lost the lower tusks, while simultaneously developing shortened lower jaws. This last characteristic parallels developments in the true elephants and gave this mastodont group an "advanced" appearance, culminating during the Pleistocene with the huge *Mammut americanum* which survived into the earliest Holocene and is a very common fossil in North America. In most respects, except the molars, it resembled elephants; however, it was a browser, living on tree leaves (determined directly from preserved stomach contents). The close resemblance to elephants was convergent because of the jaw developments. As with several other Pleistocene taxa we will highlight in the next section, the American mastodon may have succumbed to human predators.

Numerous additional mastodonts appeared on several continents, and the reader is referred to Colbert (1980) for more detailed discussion. The line that gave rise to elephants is uncertain but the chief characteristic, that of very-high-crowned, complexly infolded molars, seems to have evolved rapidly during the Pleistocene, probably stemming from bunodont mastodont ancestry. During the Pleistocene, the great mammoths (which are simply hairy elephants) were very abundant in the Northern Hemisphere and, like the American mastodon, survived until almost historical times.

A final group of subungulates to be considered are the sirenids and a possible fossil representative or collateral genus, *Demostylus*. Sirenids are dwellers of fresh water in the Americas, Africa, and Asia. They are strongly modified for aquatic life, featuring an almost entirely degenerate pelvis, hands modified as flippers, a fleshy swimming tail, and (uniquely among higher mammals) only six cervical vertebrae. The manatees of Africa and America reach 3.5 m in length and a weight of several hundred kilograms; however, a recently extinct northern species, the Steller's sea cow, weighed as much as two metric tons before mankind eliminated the group from the Bering Sea. Earliest sirenians are from middle Eocene strata in Jamaica, Europe, and Egypt. These were quite similar to later sirenians except for details of the dentition and presence of small but probably functional hind limbs and pelvis. A distinctive difference in manatees versus dugongs is that manatees develop their molars at the back of the jaws and these subsequently push forward during life as the forward molars are worn down; this feature is also characteristic of certain mastodonts and elephants.

An odd set of fossils come from large mammals that feature the same tooth-replacement mechanism and also show general morphology of primitive mastodonts and sirenids. *Demostylus* (and its relatives from Miocene units in Japan and the west coast of North America) was a large aquatic mammal with ancestry in some taxon of subungulate. However, the legs of *Demostylus* and others were well-developed and functional, showing that their occurrence in marine strata reflects a hippolike mode of life.

H.8. Quaternary Mammal Assemblages and Mass Extinction

H.8.a. The Irvingtonian Assemblage

Prior discussion (Section F.3) highlighted the Pliocene–Pleistocene boundary controversy and noted that the Blancan–Irvingtonian land mammal transition is generally considered to mark the epoch boundary in continental North American deposits (Table VII). The Irvingtonian assemblage dates between approximately 1.8 and 0.75 Myr., thereby spanning most of the pre-Illinoian Pleistocene, and was described originally from deposits in the vicinity of Irvington, Alameda County, California. Contemporary deposits are known from at least 35 additional sites in ten states and two provinces, including southwestern, Rocky Mountain, midwestern, and east-coast locations (Savage and Russell, 1983).

The Irvingtonian assemblage may be recognized by the first appearances of hares (*Lepus*), North American mammoths (*Mammuthus*), and muskrats (*Ondatra*). The fauna also contains many Asian immigrants, including new species of one-toed horses, saber-toothed cats, musk-oxen, brown bear, antelopes, and a variety of rodents. South American taxa also poured into the continent, including new glyptodonts, armadillos, and capybaras.

H.8.b. The Rancholabrean Assemblage

This fauna was named for the Rancho La Brea tar pits in Los Angeles, from which the bones of an impressive and beautifully preserved set of late Pleistocene mammals have been recovered. Animals became ensnared in the innocent-looking tar, and their struggles attracted predators and scavengers; there is, in fact, a bias in the numbers of individual animals toward forms which could not escape the tar. Kurtén and Anderson (1980) dated the assemblage from 0.75 Myr. to the present; however, others (e.g., Hibbard *et al.*, 1965) terminate the Rancholabrean several thousand

Table VII. Relationships among Pliocene–Pleistocene Epochs, Absolute Dates, and North American Land Mammal Ages[a]

Onset (Myr.)	Epoch	Land mammal age (distinctive genera)
0.01	Holocene	
0.07		Rancholabrean (*Bison*)
1.8	Pleistocene	Irvingtonian (*Mammuthus, Lepus*)
3.5		Blancan (*Geomys, Equus*)
5.3	Pliocene	Hemphilloblancan (*Megalonyx, Mammut*)

[a]Data from Kurtén and Anderson (1980), Savage and Russell (1983).

years before the present. Regardless, the assemblage encompasses the Illinoian and Wisconsinan Glacial Stages, and all intervening warm episodes.

The diagnostic taxon in the Rancholabrean is *Bison,* a Eurasian immigrant. In addition, the fauna includes other immigrants such as skunks, various rodents, caribou, bighorn sheep, bats, and, toward the end of Rancholabrean time, humans. Proboscideans were common and especially characteristic of the assemblage. The American mastodon, *Mammut americanum,* persisted through the entire course of the Pleistocene and survived to the end of the epoch. Elephants too were well-represented by species of *Mammuthus* (mammoths), mostly derived from Asian stock. *Mammuthus primigenius,* the woolly mammoth, inhabited taiga and tundra regions on both sides of the Bering Sea; the very large Columbian mammoth (*M. columbi*) was a survivor from the Irvingtonian; and the Jefferson's mammoth (*M. jeffersonii*) was probably a descendant of Irvingtonian *columbi* ancestors. All of the North American elephants became extinct approximately 11,000 yr. B.P. and were briefly survived by the American mastodons.

The Rancholabrean assemblage included several waves of immigration, in response to glacial-sea level fluctuations which caused the intermittent emergence of Beringia as a subcontinent between Asia and Alaska, and the opening and closing of the Panamanian land bridge to South America. Many of the later Wisconsinan taxa featured obvious adaptations to cold, such as wool or heavy fur, and large body-size (thermal efficiency increases in proportion to body volume, a cubed function; whereas surface area, a major factor in heat loss, increases in proportion to a squared function of size; therefore, increase in size tends to be thermally efficient). The Rancholabrean assemblage lost many taxa at the end of the Pleistocene, and additional taxa became extinct during the Holocene (several in historical times, at the hands of humans). Approximately one-half of the known Rancholabrean taxa survive at present.

H.8.c. Pleistocene Mammalian Extinctions

As noted in preceding sections, the present-day North American mammalian fauna is depauperate in comparison with the diversity of forms present in the late Wisconsinan. There are several distinctly "vacant" ecological niches, such as that of the large herbivore once filled by proboscideans, and there are other niches marginally filled by opportunistic forms: the coyotes as replacements for vanishing American wolves, saber cats, and other large predators (see Janzen and Martin, 1982).

The date for late Pleistocene "mass extinction" of North American mammals spans approximately 12,000 to 11,000 yr. B.P. Kurtén and Anderson (1980) suggested that the late Pleistocene event was of no greater magnitude than

that of the Pliocene-to-Pleistocene mammalian change (the Blancan-to-Irvingtonian mammal assemblage transition). They cited 138 species extinct at the end of Blancan time, versus 108 at the end of Rancholabrean time (and 89 extinct at the end of Irvingtonian time). Indeed, it is apparently the loss of large and spectacular creatures such as the woolly mammoths, glyptodonts, saber cats, native horses, camels, ground sloths, and others which makes the Pleistocene-to-Holocene transition seem so dramatic. One cannot fail to see a parallel with the end-Cretaceous mass extinction, where disappearance of the dinosaurs creates an impression of vast and dramatic change, which yet failed to hold up to a sober taxon-by-taxon tally of numbers gained and lost per time period. However, whereas we feel, at present, more comfortable with explanations for the Cretaceous extinctions based on stochastic processes (see Chapter 8), we believe evidence for a truly catastrophic loss of taxa exists in the case of Pleistocene mammals, since they were living not in a few restricted regions but, rather, were spread across the entire Holarctic and were abundant.

i.   The Nature of the Pleistocene/Holocene Faunal Changes.   A brief, vernacular tabulation of characteristic mammals which became extinct after the Pleistocene is as follows (at various taxonomic levels):

mammoths	ground sloths
mastodons	giant beavers
saber-toothed cats	dire wolves
glyptodonts	giant broad-horned bison
giant short-faced bears	American lion

This list does not point up abundances or diversities in each group; for example, there were several species of mammoths, saber-toothed cats, ground sloths, and glyptodonts living in the late Pleistocene, and most were widespread on the continent.

Mammals which disappeared from North America but survived elsewhere at the end of the Pleistocene include:

horses	yaks
camels	capybaras
peccaries	spectacled bears
tapirs	Asiatic antelopes

North America lost far more species and types of animals than did any other continent and this fact enhances the impression that specific extinction forces were at work in the latest Pleistocene. The magnitude of the loss of species is still being documented. For example, only recently have remains of a cheetahlike cat been reported (Martin *et al.,* 1977) and certainly other extinct taxa will be discovered.

ii.   Reasons for Pleistocene/Holocene Extinctions.   Two main lines of thought seek to explain the North American Pleistocene extinctions and changes:

1. They are the result of climatic change with the end of glaciation.
2. They are the result of human predation and overkill.

The end of the Pleistocene obviously featured climatic changes and unquestionably the ranges of many species changed. Forms such as the woolly mammoth were highly specialized for life at the margins of glaciers and on the tundra covering deglaciated areas; therefore, their loss may be explained as a case of overspecialization for glacial conditions. However, other mammoths roamed across the continent ranging (along with the American mastodon) to the extreme south. They too perished at the end of the Pleistocene and a simple change of climate cannot easily explain the disappearance of all proboscideans from North America (especially considering they survived in Asia and Africa). Similarly, saber-toothed cats and glyptodonts have been cited frequently as examples of overspecialized animals whose existence depended on fragile webs of circumstance, which might be broken by the end-Pleistocene changes. Yet, it may be shown that saber-toothed cats appeared several times during the Late Tertiary and Cenozoic, from independent lineages (Adams, 1981), and therefore their morphology and specialization should have been tested during varied environmental conditions. Glyptodonts likewise were abundant during the entire Plio-Pleistocene, and they apparently weathered series of glacial and interglacial events with vigor, until the end of the Pleistocene.

In consideration of the first suggestion for loss of Pleistocene species, deglaciation is a partial solution at best. The span during which most extinctions occurred, 12,000 to 11,000 yr. B.P., was a time featuring retreat of glaciers, yet it was no warmer than was the interval before the last major expansion of glaciers (before approximately 28,000 to 18,000 yr. B.P.), when most of the species mentioned were also present. In addition, the classical Rancholabrean fauna, from southern California, was sited far from glaciated lands yet it features many extinct species. True, changes with deglaciation in the North may have affected Californian climates; but there is also no compelling reason that the typical Rancholabrean fauna could not have migrated with changing climate.

The concept of human overkill is recurrent in the literature and has been championed especially by Martin (e.g., 1967) and Mosimann and Martin (1975). Arguments that man has outcompeted and eliminated many or all of the large and unique Pleistocene species are supported by three sets of observations: (1) the common, early occurrences of humans in North America are roughly coincident with the time of most species' disappearances, approximately between 15,000 and 11,000 yr. B.P.; (2) several mass-kill sites have been discovered and it is apparent that at least some Paleo-Indians drove herds of mammoths and bison over cliffs; and (3) in historical times, man has efficiently eliminated many species (e.g., passenger pigeons, great auks, dodos, Stellar's sea cow), and thus there is precedent for the hypothesis. The guilt-ridden reader is directed to Martin (1982) for further evidence of the human-overkill idea.

Arguing against human overkill as the exclusive agent of Pleistocene mass extinction are several lines of reasoning. First, mass kills are only proven for mammoths and bison. It is probable that other animals were hunted but no large kill-sites are known. Second, the human population does not seem to have been sufficiently large, based on the volume of archaeological finds, to have been able to kill off so many creatures [however, Mosimann and Martin (1975) presented a mathematical model purported to show the latter to have been feasible].

Third, if Paleo-Indians killed off so many species, why were other, equally esculent taxa not eliminated? And why were such dangerous forms as saber-toothed cats and dire wolves exterminated? Certainly there were more suitable prey. Finally, why were far fewer Pleistocene animals killed off in Eurasia? The human populations was, if anything, larger there.

In summary, there is no direct answer for the mystery of Pleistocene extinctions. The best approach at present is to consider each taxon individually and search for a cause, then to determine whether these assemble into a mosaic: a few examples will be given, and the reader is directed to Martin and Klein (1984) for many additional arguments. The mammoths, especially the woolly mammoths, may well have suffered both from climatic change (for the woolly species) and from being large, conspicuous prey for bands of hunters. Bison were also excellent prey and Kurtén and Anderson (1980) further suggested that the large, big-horned species may have interbred and assimilated into the smaller short-horned species. Bear-sized *Castoroides* beavers were such unusually large rodents that they strongly suggest overspecialization, and they may have been outcompeted in more temperate climates by *Castor* species. Ground sloths were very large, very sluggish creatures which were quite abundant during the Pleistocene. In Rampart Cave in the Grand Canyon, Arizona, Shasta ground sloth dung comprises a layer reaching over a meter in thickness. From this dung (which is stratiform and can be dated by layers) have come remains of a diverse flora representing plants eaten by the sloths (Hansen, 1978). If the ground sloths comprised the base of a predator–prey food web, then their elimination might have triggered decline of prey species in the typical Rancholabrean assemblage. Glyptodonts, too, were archaic, sluggish creatures that may have been naturally eliminated by the climatic change accompanying deglaciation. They may also have been easy prey for Paleo-Indian hunters armed with rocks or using cliff-falls.

H.9. Quaternary Man in North America

In Section H.5 it was shown that primates, after having first appeared, having initially diversified, and having radiated from and in North America, died out on this continent effectively by the late Eocene and totally in the Miocene. Primates did not reenter the continent until humans migrated in from Asia by way of Beringia.

As with so many prior discussions, this final topic of the book involves controversy. Specifically, there is debate on the timing of human arrival on the continent. Both skeletal and cultural evidence unambiguously confirm a human presence by 15,000 yr. B.P., as we noted previously with discussion of Pleistocene extinctions, and these remains are attributed to the so-called Paleo-Indian culture group. Kurtén and Anderson (1980) listed many of the best-known early Paleo-Indian finds, including Wilson Butte, Idaho (~ 14,500–15,000 yr. B.P.), Fort Rock Cave, Oregon (13,200 ± 720 yr.), Clovis, New Mexico (11,310 ± 240 yr.), and many others. In addition, finds have come from Tepexpan, New Mexico (11,000 yr.), Guitarrero Cave, Peru (12,000 yr.), and other Central and South American localities, showing that men contemporary with Paleo-Indians spread down through the southern Americas.

While no real controversy exists concerning the presence of humans in this hemisphere at about 15,000 yr. B.P. and thereafter, diametrical opinion and evidence have been brought to bear on reports of earlier humans here (some purported to date to over 100,000 yr. B.P.; e.g., Kraft and Thomas, 1976). Much of the problem stems from imprecision of dating methods for the interval concerned, and other uncertainties result from the nature of human burial. Where ritual burial is practiced by humans, such events often produce intrusive situations where materials are intruded into stratigraphically lower deposits. Thus, a recent burial might emplace bones and artifacts into Pleistocene (or, poten-

tially, even Precambrian!) strata. In other words, dating of enclosing sediments may not date the humans.

Among the best-known early human finds in North America are several sites in California: an occipital bone from the Los Angeles River yielded a radiocarbon date of 23,600 yr. B.P., whereas remains from Del Mar yielded 46,000 yr. B.P., measured by *amino acid racemization* (a technique by which the known rate of racemization of originally L-oriented amino acids in living tissue is used to date organic materials). Other California sites (Laguna and Yuba) have yielded ages of approximately 17,000 and 21,500 yr. B.P. by radiocarbon dating. Central and South American materials and fossils have been dated at 23,150 and 21,000 yr. B.P. in Mexico, and 14,000 to 19,000 yr. B.P. in Peru.

The oldest North American human artifacts include those discussed in Bada *et al.* (1974) from southern California, dated to 44,000 and 48,000 yr. B.P. by amino acid racemization, and those noted by Stanford *et al.* (1981), from the Yukon Territory and Colorado, dated over 50,000 yr. B.P. These last dates are based on human-worked bones from butchered animals.

Contrary opinion on the antiquity of man in North America was expressed by Bischoff and Rosenbaur (1981). They reported uranium series analyses of the Del Mar and other purportedly old human remains from California and showed dates no older than 11,000 yr. B.P. They further suggested that amino acid racemization dating is unreliable. Dumond (1980) and Borden (1979) provided good discussions of early man on the western border of North America, to which the reader is directed.

What seems apparent to us is that the 15,000 yr. B.P. and younger dates are indisputable and widespread, indicating Paleo-Indian and derivative cultures were firmly established by that date. There are too many early dates from the Pacific coast to ignore the additional likelihood that a pre-Paleo-Indian immigration before 25,000 yr. B.P., populated that part of the continent with a less-successful group of Americans who may have been replaced or interbred with latest Wisconsinan Paleo-Indians.

APPENDIX

Synoptic Taxonomy of Major Fossil Groups

Important fossil taxa are listed down to the lowest practical taxonomic level; in most cases, this will be the ordinal or subordinal level. Abbreviated stratigraphic units in parentheses (e.g., UCamb–Rec) indicate maximum range known for the group; units followed by question marks are isolated occurrences followed generally by an interval with no known representatives. Taxa with ranges to "Rec" are extant. Data are extracted principally from Harland *et al.* (1967), Moore *et al.* (1956 *et seq.*), Sepkoski (1982), Romer (1966), Colbert (1980), Moy-Thomas and Miles (1971), Taylor (1981), and Brasier (1980).

KINGDOM MONERA

DIVISION CYANOPHYTA

Class Cyanophyceae
    Order Chroococcales (Archean–Rec)
    Order Nostocales (Archean–Rec)
    Order Spongiostromales (Archean–Rec)
    Order Stigonematales (LDev–Rec)
    Three minor orders

KINGDOM PROTISTA

PHYLUM PROTOZOA

Class Rhizopodea
    Order Foraminiferida*
        Suborder Allogromiina (UCamb–Rec)
        Suborder Textulariina (LCamb–Rec)
        Suborder Fusulinina (Ord–Perm)
        Suborder Miliolina (Sil–Rec)
        Suborder Rotaliina (Miss? Perm–Rec)
Class Ciliata

*Foraminiferan taxonomy is of little meaning above the familial or generic level.

Class Ciliata (*cont.*)
    Order Spirotrichia (Tintinnida) (UOrd–Rec)
?Class *Insertae sedis*
    Order Chitinozoa (Proterozoic?, LOrd–UDev)
Class Actinopoda
  Subclass Radiolaria
    Order Polycystina
        Suborder Spumellaria (MCamb–Rec)
        Suborder Nasselaria (Dev–Rec)

KINGDOM ANIMALIA

PHYLUM PORIFERA

Class Hexactinellida
    Order Amphidiscophora (Miss–Rec)
    Order Hexactinosida (MTrias–Rec)
    Order Lyssacinosida (LCamb–Rec)
    Order Lychniscosida (UTrias–Rec)
Class Demospongia
    Order Monaxonida (MCamb–Rec)
    Order Lithistida (LOrd–Rec)
    Order Choristida (Miss–Rec)
Class Calcarea
    Order Sphinctozoa (UOrd–UCret)
    Order Pharetrones (LPerm–Rec)

Class Calcarea (*cont.*)
    Order Calcaronea (LJur–Rec)
    Order Heteractinida (MOrd–LPerm)
Class Sclerospongiae
    ?Order Stromatoporoidea (?LCamb, MOrd–UCret)
    Order Tabulospongidae (LOrd–Rec)

?PHYLUM ARCHAEOCYATHA

Class Regulares
    Order Monocyathida (LCamb–MCamb)
    Order Putapacyathida (LCamb)
    Order Ajacicyathida (LCamb–MCamb)
Class Irregulares
    Order Archaeocyathida (LCamb–MCamb)
    Order Rhizacyathida (LCamb–MCamb)
    Order Syringocnematida (LCamb)

?PHYLUM RECEPTACULITIDA (LOrd–MDev)*

PHYLUM CNIDARIA (= COELENTERATA)

Class Hydrozoa
    Order Trachylinida (MCamb?, Jur–Rec)
    Order Hydroida (MCamb–Rec)
    Order Siphonophorida (Ediacar–Rec)
    Order Spongiomorphida (LTrias–UJur)
    Order Milliporina (Danian–Rec)
    Order Stylasterina (Danian–Rec)
    Order Sphaeractinida (UPenn–UCret)
Class Scyphozoa
    Order Carybdeida (UJur–Rec)
    Order Coronatida (LCamb–Rec)
    Order Semaeostomatida (UJur–Rec)
    Order Lithorhizostomatida (UJur)
    Order Rhizostomatida (UJur–Rec)
    ?Order Conulariida (LCamb–UTrias)
Class Anthozoa
    Order Stolonifera (LCret–Rec)
    Order Alcyonacea (LCamb?, MSil–Rec)
    Order Gorgonacea (UOrd–Rec)
    Order Pennatulacea (Ediacar?, Trias–Rec)
    Order Tabulata (LOrd–UPerm)
    Order Rugosa (MCamb?, MOrd–UPerm)
    Order Heterocorallia (MDev–LPenn)
    Order Scleractinia (MTrias–Rec)

PHYLUM BRYOZOA (= ECTOPROCTA, POLYZOA)

Class Stenolaemata
    Order Cyclostomata (LOrd–Rec)
    Order Trepostomata (LOrd–UPerm)

*Position uncertain, may alternatively be a class of Porifera, or a subdivision of Chlorophyta (Plantae).

Class Stenolaemata (*cont.*)
    Order Cystoporata (LOrd–UPerm)
    Order Cryptostomata (LOrd–LTrias)
    Order Fenestrata (MOrd–LTrias)
Class Gymnolaemata
    Order Ctenostomata (LOrd–Rec)
    Order Cheilostomata (UJur–Rec)

PHYLUM BRACHIOPODA

Class Inarticulata
    Order Lingulida (LCamb–Rec)
    Order Acrotretida (LCamb–Rec)
    Order Obolellida (LCamb–MCamb)
    Order Paterinida (LCamb–MOrd)
    Order Kutorginida (LCamb–MCamb)
Class Articulata
    Order Othida
        Suborder Orthidina (LCamb–UDev)
        Suborder Clitambonitidina (LOrd–UOrd)
        Suborder Triplesiidina (MOrd–MSil)
    Order Pentamerida (MCamb–UDev)
    Order Strophomenida
        Suborder Strophomenidina (LOrd–UTrias)
        Suborder Chonetidina (UOrd–LJur)
        Suborder Productidina (LDev–LTrias)
        Suborder Oldahaminidina (Penn–UTrias)
    Order Rhynchonellida (MOrd–Rec)
    Order Atrypida (MOrd–UDev)
    Order Spiriferida (MOrd–MJur)
    Order Terebratulida (USil–Rec)
    Order Thecideida (LJur–Rec)

PHYLUM MOLLUSCA

Class Amphineura
    Order Palaeoloricata (UCamb–UCret)
    Order Neoloricata (MOrd–Rec)
Class Monoplacophora
    Order Tryblidoidea (MCamb–Rec)
    Order Cyrtonellida (LCamb–MDev)
Class Gastropoda
    Order Archaeogastropoda (UCamb–Rec)
    Order Bellerophontida (UCamb–Rec)
    Order Mesogastropoda (LOrd–Rec)
    Order Neogastropoda (LCret–Rec)
    Order Entomotaeniata (MDev–Rec)
    Order Cephalaspidea (Miss–Rec)
    Seven minor fossil orders
Class Scaphopoda
    Order Dentalioida (MOrd–Rec)
    Order Siphonodentalioida (LPerm–Rec)
Class Rostroconchia
    Order Ribeirioida (LCamb–UOrd)

Class Rostroconchia (*cont.*)
   Order Ischyrinioida (LOrd–UOrd)
   Order Conocardioida (LOrd–UPerm)
  Class Bivalvia (= Pelecypoda, Lamellibranchiata)
   Order Fordilloida (LCamb)
   Order Solemyoida (LDev–Rec)
   Order Nuculoida (LOrd–Rec)
   Order Mytiloida (LOrd–Rec)
   Order Praecardioida (LOrd–UTrias)
   Order Arcoida (LOrd–Rec)
   Order Pterioida (LOrd–Rec)
   Order Modiomorphoida (LOrd–UPerm)
   Order Trigonioida (LDev–Rec)
   Order Unionoida (LTrias–LJur)
   Order Veneroida (LOrd–Rec)
   Order Myoida (LPerm–Rec)
   Order Hippuritoida (LSil–UCret)
   Order Pholadomyoida (MOrd–Rec)
Class Cephalopoda
 Subclass Nautiloidea
   Order Ellesmerocerida (UCamb–UOrd)
   Order Orthocerida (LOrd–UTrias)
   Order Actinocerida (LOrd–UPenn)
   Order Endocerida (LOrd–LSil)
   Order Nautilida (LDev–Rec)
   Six minor orders
 Subclass Bactritoidea (LOrd–UPerm)
 Subclass Ammonoidea
   Order Goniatitida (MDev–UPerm)
   Order Ceratitida (UPerm–UTrias)
   Order Phylloceratida (LTrias–UCret)
   Order Lytoceratida (LJur–UCret)
   Order Ammonitida (LJur–UCret)
 Subclass Coleoidea
   Order Belemnoida (LJur–Eocene)
   Order Octopoda (UCret–Rec)
   Order Sepiida (UJur–Rec)
   Order Teuthida (UTrias–Rec)
   Three minor orders
?Class Tentaculitida (LOrd–UDev)
?Class Hyolitha (LCamb–UPerm)

PHYLUM ANNELIDA

Class Polychaeta
   Order Errantia (Ediacar–Rec)
   Order Sedentaria (Ediacar–Rec)

PHYLUM ARTHROPODA

Subphylum Trilobitomorpha

Class Trilobitoidea (LCamb–MCamb, LDev?)
Class Trilobita

Class Trilobita (*cont.*)
  Order Agnostida
   Suborder Agnostina (LCamb–UOrd)
   Suborder Eodiscina (LCamb–MCamb)
  Order Olenellida (Redlichiida)
   Suborder Olenellina (LCamb–MCamb?)
   Suborder Redlichiina (LCamb–MCamb)
  Order Corynexochida (LCamb–UCamb)
  Order Ptychopariida
   Suborder Ptychopariina (LCamb–UOrd)
   Suborder Asaphina (UCamb–UOrd)
   Suborder Illaenina (LOrd–UDev)
   Suborder Harpina (UCamb–UDev)
   Suborder Trinucleina (LOrd–USil)
  Order Proetida (UCamb–UPerm)
  Order Phacopida
   Suborder Cheirurina (UCamb–MDev)
   Suborder Calymenina (LOrd–MDev)
   Suborder Phacopina (LOrd–UDev)
  Order Lichida (LOrd–UDev)
  Order Odontopleurida (LOrd–UDev)

Subphylum Chelicerata

Class Merostomata
 Subclass Xiphosuria
   Order Aglaspida (LCamb–MOrd)
   Order Limulida (LDev–Rec)
   Order Eurypterida (LOrd–UPerm)
   Two minor orders
Class Arachnida (MSil–Rec)

Subphylum Pycnogonida (LDev–Rec)

Subphylum Mandibulata

Class Myriopoda
 Subclass Diplopoda (MSil–Rec)
 Subclass Arthropleurida (Penn)
 Subclass Chilopoda (MSil–Rec)
Class Crustacea*
 Subclass Branchiopoda (LDev–Rec)
 Subclass Ostracoda
   Order Archaeocopida (LCamb–LOrd)
   Order Leperditicopida (LOrd–MDev)
   Order Paleocopida (LOrd–UPerm)
   Order Myodocopida (MOrd–Rec)
   Order Podocopida (MOrd–Rec)
 Subclass Cirripeda (Penn–Rec)
 Subclass Malacostraca
   Order Decapoda (UPerm–Rec)
   Order Isopoda (UPerm–Rec)
   Order Amphipoda (Eocene–Rec)
Class Insecta (= Hexapoda)[†]

*Crustacea incorporates many additional arthropods of MCamb and younger ages whose affinities are indeterminate.
[†]Insecta includes more than 43 orders, most with a sparse fossil record.

Class Insecta (*cont.*)
    Subclass Apterygota (LDev–Rec)
    Subclass Pterygota (Penn–Rec)

PHYLUM ECHINODERMATA

insertae sedis
Class Homalozoa (= Carpoidea, Calcichordata) (LCamb–MDev)
Class Helicoplacoidea (LCamb)

Subphylum Pelmatozoa

Superclass Cystoidea
    Class Eocrinoidea (LCamb–USil)
    Class Paracrinoidea (MOrd–UOrd)
    Class Diploporita (LOrd–MDev)
    Class Rhombifera (LOrd–UDev)
    Class Blastoidea (LOrd–UPerm)
Superclass Crinozoa
    Class Crinoidea
        Subclass Camerata (LOrd–UPerm)
        Subclass Inadunata (LOrd–MTrias)
        Subclass Flexibilia (MOrd–UPerm)
        Subclass Articulata (LTrias–Rec)

Subphylum Eleutherozoa

Class Edrioasteroidea (LCamb–Penn)
Class Asteroidea (LOrd–Rec)
Class Ophiuroidea (LOrd–Rec)
Class Holothuroidea (MCamb?, LOrd–Rec)
Class Echinoidea
    Subclass Perischoechinoidea (MOrd–Rec)
    Subclass Euechinoidea (UTrias–Rec?)

PHYLUM CONODONTA

Order Paraconodontida (Ediacar?, LCamb–MOrd)
Order Conodontophorida (UCamb–UTrias

PHYLUM HEMICHORDATA

?Class Graptolithina
    Order Dendroidea (MCamb–LPenn)
    Order Tuboidea (UCamb–MSil)
    Order Cameroidea (LOrd–UOrd)
    Order Stolonoidea (LOrd)
    Order Graptoloidea (LOrd–LDev)
    Two minor orders
Class Pterobranchia (MCamb–Rec)
Class Enteropneusta (MCamb?, Rec)

PHYLUM CHORDATA

Subphylum Cephalochordata
Class Acrania (MCamb?, Rec)

Subphylum Vertebrata (= Craniata)

Infraphylum Agnatha
Superclass Cephalaspidiomorphi (MOrd–Rec)
Superclass Pteraspidiomorphi (UCamb–UDev)

Infraphylum Gnathostomata
Superclass Elasmobranchiomorphi
    Class Placodermi
        Order Rhenanida (LDev–UDev)
        Order Ptyctodontida (MDev–UDev)
        Order Petalichthyida (LDev–MDev)
        Order Arthrodira (LDev–UDev)
        Order Antiarcha (MDev–Miss)
    Class Chondrichthyes
        Subclass Elasmobranchii
            Order Cladoselachii (MDev–UPerm)
            Order Pleuracanthodii (UDev–UTrias)
            Order Selachii (UDev–Rec)
            Order Batoidea (LJur–Rec)
        Subclass Holocephali (Miss–Rec)
Superclass Teleostomi
    Class Acanthodii (MSil–Penn)
    Class Osteichthyes
        Subclass Sarcopterygii
            Infraclass Crossopterygii
                Superorder Rhipidistia (MDev–Perm)
                Superorder Actinistia (= Coelacanthi) (UDev–Rec)
        Subclass Actinopterygii
            Infraclass Chondrostei (MDev–Rec)
            Infraclass Holostei (UPerm–Rec)
            Infraclass Teleostei (UTrias–Rec)
Superclass Tetrapoda
    Class Amphibia
        Subclass Labyrinthodontia
            Order Ichthyostegalia (UDev–Miss)
            Order Temnospondyli (Miss–UTrias)
            Order Anthracosauria (Miss–UPerm)
        Subclass Lepospondyli
            Order Nectridea (LPenn–LPerm)
            Order Aistopoda (Miss–LPerm)
            Order Microsauria (Miss–LPerm)
        Subclass Lissamphibia (LTrias–Rec)
    Class Reptilia
        Subclass Anapsida
            Order Captorhinomorpha (LPenn–MPerm)
            Order Chelonia (MTrias?, UTrias–Rec)
            Order Mesosauria (LPerm)

Class Reptilia (*cont.*)
  Subclass Lepidosauria
      Order Eosuchia (UPerm–Eocene)
      Order Squamata (UTrias–Rec)
      Order Rhynchocephalia (LTrias–Rec)
  Subclass Euryapsida
      Order Araeoscelidia (LPerm–UCret)
      Order Sauropterygia (LTrias–UCret)
      Order Placodontia (LTrias–UTrias)
  Subclass Ichthyopterygia (MTrias–UCret)
  Subclass Archosauria
      Order Thecodontia (UPerm–UTrias)
      Order Crocodilia (UTrias–Rec)
      Order Pterosauria (LJur–UCret)
      Order Saurischia (UTrias–UCret)
      Order Ornithischia (UTrias–UCret)
  Subclass Synapsida
      Order Pelycosauria (UPenn–MPerm)
      Order Therapsida (MPerm–LJur)
Class Aves
  Subclass Archaeornithes (UTrias?, UJur)
  Subclass Neornithes
      Superorder Odontognathae (UCret)
      Superorder Paleognathae (Eocene–Rec)
      Superorder Neognathae (LCret–Rec)
Class Mammalia
  Subclass Prototheria
    Infraclass Eotheria (UTrias–LCret)
    Infraclass Allotheria (UJur–Eocene)
    Infraclass Ornithodelphia (= Monotremata) (Pleist–Rec)
      Subclass Theria
    Infraclass Trituberculata (MJur–LCret)
    Infraclass Metatheria (UCret–Rec)*
    Infraclass Eutheria (LCret?, MCret?, UCret–Rec)

KINGDOM PLANTAE
Subkingdom Thallophyta

DIVISION CHRYSOPHYTA

Class Chrysophyceae
    Order Coccolithineae (LJur–Rec)
    Order Chrysomonadales (Silicoflagellata) (LCret–Rec)
    Order Bacillariophyceae (Diatomacea) (LCret–Rec)

DIVISION PYRRHOPHYTA

Class Dinophyceae (LSil?, LPerm–Rec)

*Metatheria includes the single order of marsupials; Eutheria includes 22 or more orders of placental mammals.

?DIVISION ACRITARCHA*

DIVISION RHODOPHYTA

    Family Solenoporaceae (LCamb–Rec)
    Family Corallinaceae (Penn–Rec)

DIVISION CHLOROPHYTA

Class Chlorophyceae
    Order Siphonales (Proter?, UCamb–Rec)
    Order Dasycladales (LCamb–Rec)†

Subkingdom Tracheophyta

?DIVISION RHYNIOPHYTA (USil–UDev)

DIVISION ZOSTEROPHYLLOPHYTA (LDev)

DIVISION TRIMEROPHYTOPHYTA (LDev–UDev)

DIVISION LYCOPHYTA

    Order Protolepidodendrales (LDev–UDev)
    Order Lepidodendrales (Miss–UPenn)
    Order Lycopodiales (UDev–Rec)
    Order Sigillariales (Miss–LPerm)
    + ten minor orders

DIVISION SPHENOPSIDA

    Order Hyeniales (LDev–MDev)
    Order Pseudoborniales (UDev)
    Order Sphenophyllales (UDev–UPerm)
    Order Equisetales (Miss–Rec)

DIVISION PTERIDIOPHYTA

Class Cladoxylopsida (MDev–Miss)
Class Coenopteridopsida (LDev–UPerm)
Class Filicopsida
    Order Marattiales (LPenn–Rec)
    Order Filicales (Miss–Rec)
    + four minor orders

DIVISION GYMNOSPERMATOPHYTA

Class Progymnospermopsida (MDev–Miss)
Class Pteridospermopsida
    Order Medullosales (Miss–UPerm)

*There are 13 informal subdivisions of acritarchs, ranging from Proterozoic to Recent, but there is no recognized taxonomy.
†Dasycladales may incorporate the questionable animal phylum Receptaculitida.

Class Pteridospermopsida (*cont.*)
    Order Callistophytales (MPenn–UPenn)
    Order Calymopityales (UDev–Miss)
    Order Glossopteridales (UPenn–LTrias)
    Order Archaeopteridales (UDev–LPerm)
    Order Diplopteridales (UDev–Penn)

DIVISION CYCADOPHYTA

Class Cycadopsida
    Order Cycadales (LPerm?, UTrias–Rec)
    Order Bennettitales (UTrias–UCret)

DIVISION GINKGOPHYTA (LPerm–Rec)

DIVISION CONIFEROPHYTA

Class Cordaitopsida (LPenn–UPerm)
Class Coniferopsida (LPenn–Rec)
Class Taxales (LJur–Rec)

DIVISION ANTHOPHYTA (= ANGIOSPERMAE)*

*Includes all flowering plants, with over 300 extant families. Earliest definite records from the LCret, but questionable fossils date to the UTrias.

References

Aalto, K. R., 1971, Glacial marine sedimentation and stratigraphy of the Toby Conglomerate (Upper Proterozoic), southeastern British Columbia, northwestern Idaho and northeastern Washington. *Can. J. Earth Sci.* **8**, 753–787.

Aalto, K. R., 1981, Multistage mélange formation in the Franciscan Complex, northernmost California. *Geology* **9**, 602–607.

Abbott, W., 1957, Tertiary of the Uinta basin. Intermountain Assoc. Pet. Geol., 8th Annu. Field Conf., Guidebook, pp. 102–109.

Acomb, L. J., Mickelson, D. M., and Evenson, E. B., 1982, Till stratigraphy and late glacial events in the Lake Michigan Lobe of eastern Wisconsin. *Geol. Soc. Am. Bull.* **93**, 289–296.

Adam, D. P., Sims, J. D., and Throckmorton, C. K., 1981, 130,000-yr. pollen record from Clear Lake, Lake County, California. *Geology* **9**, 373–377.

Adams, C. G., *et al*, 1977, The Messinian salinity crisis and evidence of late Miocene eustatic changes in the world ocean. *Nature* **269**, 383–386.

Adams, D. B., 1981, Nine lives of the sabercat. *Science* **81**, 42–47.

Adams, J. F., 1962, Foreland Pennsylvanian rocks of Texas and eastern New Mexico. *In* Branson, C. C. (ed.), Pennsylvanian System in the United States—A symposium. AAPG, Tulsa, pp. 372–384.

Ahr, W. M., 1973, The carbonate ramp: An alternative to the shelf model. *Trans. Gulf Coast Geol. Soc.* **23**, 221–225.

Akers, W. H., 1972, Planktonic foraminifera and biostratigraphy of some Neogene formations, northern Florida and Atlantic Coastal Plain. *Tulane Stud. Geol. Paleontol.* **9**(1–4).

Alberstadt, L. P., Walker, K. R., and Zurawski, R. P., 1974, Patch reefs in the Carters Limestone (Middle Ordovician) in Tennessee, and vertical zonation in Ordovician reefs. *Geol. Soc. Am. Bull.* **85**, 1171–1182.

Allison, C. W., 1975, Primitive fossil flatworm from Alaska: New evidence bearing on ancestry of the Metazoa. *Geology* **3**, 649–652.

Allmendinger, R. W., Brewer, J. A., Brown, L. D., Kaufman, S., Oliver, J. E., and Houston, R. S., 1982, COCORP profiling across the Rocky Mountain Front in southern Wyoming, part 2: Precambrian basement structure and its influence on Laramide deformation. *Geol. Soc. Am. Bull.* **93**, 1253–1263.

Alvarez, L. W., Alvarez, W., Asaro, F., and Michel, H. V., 1980, Extraterrestrial cause for the Cretaceous–Tertiary extinction. *Science* **208**, 1095–1108.

Alvarez, W., Alvarez, L. W., Asaro, F., and Michel, H. V., 1979, Anomalous iridium levels at the Cretaceous/Tertiary boundary at Gubbio, Italy: Negative results of tests for a supernova origin. *In* Christensen, W. K., and Birkelund, T. (eds.), *Cretaceous–Tertiary Boundary Events*, Vol. II. University of Copenhagen, Denmark, p. 69.

Alvarez, W., Alvarez, L. W., Asaro, F., and Michel, H. V., 1982a,

Current status of the impact theory for the terminal Cretaceous extinction. *Geol. Soc. Am. Spec. Pap.* **190**, 305–316.

Alvarez, W., Asaro, F., Michel, H. V., and Alvarez, L. W., 1982b, Iridium anomaly approximately synchronous with terminal Eocene extinctions. *Science* **216**, 886–888.

Amsden, T. W., Caplan, W. M., Hilpman, P. L., McGlasson, E. H., Rowland, T. L., and Wise, O. A., Jr., 1967, Devonian of the southern midcontinent area, United States. *In* Devonian System, Int. Symp. Calgary, Alberta, *Soc. Pet. Geol.* **1**, 913–932.

Anderson, E. J., Goodwin, P. W., and Sobieski, T. H., 1984, Episodic accumulation and the origin of formation boundaries in the Helderberg Group of New York State. *Geology* **12**, 120–123.

Andrews, H. N., 1961, *Studies in Paleobotany*. Wiley, New York.

Andrews–Speed, C. P., 1981, The case against a Phanerozoic Kolyma plate in the northeastern USSR. *Geology* **9**, 174–177.

Anhaeusser, C. R., 1973, The evolution of the early Precambrian crust of southern Africa. *Philos. Trans. Roy. Soc. London* **273**, 359–388.

Anhaeusser, C. R., 1975, Precambrian tectonic environments. *Annu. Rev. Earth Planet. Sci.* **3**, 31–53.

Anhaeusser, C. R., 1978, The geological evolution of the primitive Earth—Evidence from the Barberton Mountain Land. *In* Tarling, D. H. (ed.), *Evolution of the Earth's Crust*. Academic Press, New York, pp. 71–106.

Anhaeusser, C. R., Roering, C., Viljoen, M. J., and Viljoen, R. P., 1968, The Barberton Mountain Land—A model of the elements and evolution of an Archean fold belt. *Trans. Geol. Soc. Afr.* **71**(annexure), 225–254.

Antoine, J. W., Martin, R. G., Jr., Pyle, T. G., and Bryant, W. R., 1974, Continental margins of the Gulf of Mexico. *In* Burk, C. A., and Drake, C. L. (eds.), *The Geology of Continental Margins*. Springer-Verlag, Berlin, pp. 683–693.

Applin, P. L., and Applin, E. R., 1944, Regional subsurface stratigraphy and structure of Florida and southern Georgia. *Am. Assoc. Pet. Geol. Bull.* **28**, 1673–1753.

Arden, D. D., Beck, B. F., and Morrow, E. (eds.), 1982, 2nd symposium on the Geology of the southeastern Coastal Plain. *Ga. Geol. Surv. Inf. Circ.* **53**.

Arkell, W. J., 1957, Mesozoic Ammonoidea. *In* Moore, R. C. (ed.), *Treatise on Invertebrate Paleontology,* Part L, Mollusca 4. Geological Society of America and University of Kansas Press, Lawrence, pp. 81–129.

Arkle, T., Jr., 1959, Monongahela Series, Pennsylvanian System and Washington and Greene Series, Permian System of the Appalachian basin. Geological Society of America, Guidebook for field trips, Pittsburgh meeting, Field trip 3, pp. 115–141.

Arkle, T., Jr., 1969, The configuration of the Pennsylvanian and Dunkard (Permian?) strata in West Virginia; a challenge to classical concepts.

In Donaldson, A. C. (ed.), *Some Appalachian Coals and Carbonates: Models of Ancient Shallow-Marine Deposition.* W. Va. Geol. Surv. pp. 55–87.

Armentrout, J. M., Cole, M. R., and TerBest, H., Jr. (eds.), 1977, Cenozoic paleogeography of the western United States. SEPM, Pacific Section, Pacific Coast Paleogeography Symp. No. 3.

Armstrong, R. L., 1968, Sevier orogenic belt in Nevada and Utah. *Geol. Soc. Am. Bull.* **79**, 429–458.

Armstrong, R. L., and Dick, H. J. B., 1974, A model for the development of thin overthrust sheets of crystalline rock. *Geology* **2**(1), 35–40.

Armstrong, R. L., and Suppe, J., 1973, Potassium–argon geochronometry of Mesozoic igneous rocks in Nevada, Utah, and southern California. *Geol. Soc. Am. Bull.* **84**, 1375–1392.

Arnold, C. A., 1947, *An Introduction to Paleobotany.* McGraw–Hill, New York.

Atherton, E., and Palmer, J. E., 1979, The Mississippian and Pennsylvanian (Carboniferous) System in the United States—Illinois. *U.S. Geol. Surv. Prof. Pap.* **1110-L**.

Atwater, B. F., 1984, Periodic floods from glacial Lake Missoula into the Sanpoil arm of glacial Lake Columbia, northeastern Washington. *Geology* **12**, 464–467.

Atwater, T. M., 1970, Implications of plate tectonics for the Cenozoic tectonic evolution of North America. *Geol. Soc. Am. Bull.* **81**, 3513–3536.

Awramik, S. M., Schopf, J. W., and Walter, M. R., 1983, Filamentous fossil bacteria 3.5×10^9 years old from the Archaean of Western Australia. *Precambrian Res.* **20**, 357–374.

Axelrod, D. I., 1957, Late tertiary floras and the Sierra Nevada uplift (California–Nevada). *Geol. Soc. Am. Bull.* **68**, 19–45.

Bada, J. L., Schroeder, R. A., and Carter, G. F., 1974, New evidence for antiquity of Man in North America deduced from amino acid racemization. *Science* **184**, 791–793.

Baer, A. J., 1981, A Grenvillian model of Proterozoic plate tectonics. *In* Kröner, A. (ed.), *Precambrian Plate Tectonics.* Elsevier, Amsterdam, pp. 353–385.

Bailey, E. H., Irwin, W. P., and Jones, D. L., 1964, Franciscan and related rocks and their significanc in the geology of western California. *Calif. Div. Mines Geol. Bull.* **183**.

Bailey, T. L., Evans, F. G., and Adkine, W. S., 1945, Revision of stratigraphy of part of Cretaceous in Tyler basin, northeast Texas. *Am. Assoc. Pet. Geol. Bull.* **29**, 170–186.

Baird, D., 1978, *Pneumatoarthrus* Cope, 1870, not a dinosaur but a sea-turtle. *Proc. Acad. Nat. Sci. Philadelphia* **129**(4), 71–81.

Baird, D., and Case, G. R., 1966, Rare marine reptiles from the Cretaceous of New Jersey. *J. Paleontol.* **40**(5), 1211–1215.

Baird, G. C., 1979, Lithology and fossil distribution, Francis Creek Shale in northeastern Illinois. *In* Nitecki, M. H. (ed.), *Mazon Creek Fossils.* Academic Press, New York, pp. 41–68.

Bakker, R. T., 1972, Anatomical and ecological evidence for endothermy in dinosaurs. *Nature* **238**, 81–85.

Bakker, R. T., 1975, Dinosaur renaissance. *Sci. Am.* **232**(4), 57–78.

Bakker, R. T., 1977, Tetrapod mass extinctions—A model of the regulation of speciation rates and immigration by cycles of topographic diversity. *In* Hallam, A. (ed.), *Patterns of Evolution as Illustrated by the Fossil Record.* Elsevier, Amsterdam, pp. 439–468.

Bakker, R. T., 1980, Dinosaur heresy—Dinosaur renaissance. *In* Thomas, R. D. K., and Gibson, E. C. (eds.), A cold look at the warm–blooded dinosaurs. *AAAS Symp.* **28**, 351–462.

Bakker, R. T., and Galton, P. M., 1974, Dinosaur monophyly and a new class of vertebrates. *Nature* **248**, 168–172.

Baldwin, J. E., 1973, The Moenkopi Formation of north–central Arizona: An interpretation of ancient environments based upon sedimentary structures and stratification types. *J. Sediment. Petrol.* **43**(1), 92–106.

Balkwill, H. R., 1978, Evolution of Sverdrup basin, Arctic Canada. *Am. Assoc. Pet. Geol. Bull.* **62**, 1004–1028.

Ball, M. M., 1967, Carbonate sand bodies of Florida and the Bahamas. *J. Sediment. Petrol.* **37**, 556–591.

Bamber, E. W., and Copeland, M. J., 1976, Carboniferous and Permian faunas. *Geol. Surv. Can. Econ. Geol. Rep.* **1**, 623–632.

Banks, H. P., 1970, *Evolution and Plants of the Past.* Wadsworth, Belmont, Calif.

Banks, H. P., Leclercq, S., and Hueber, F. M., 1975, Anatomy and morphology of *Psilophyton dawsonii*, sp. n. from the late Lower Devonian of Québec (Gaspé) and Ontario, Canada. *Paleontogr. Am.* **8**, 75–127.

Baragar, W. R. A., 1970, The igneous succession and chemical characteristics of the igneous rocks. *Can. Geol. Surv. Pap.* **70–40**, pp. 81–84 and 116–122.

Baragar, W. R. A., and Scoates, R. F. J., 1981, The Circum–Superior belt: A Proterozoic plate margin? *In* Kröner, A. (ed.), *Precambrian Plate Tectonics.* Elsevier, Amsterdam, pp. 296–330.

Barghoorn, E. S., 1971, The oldest fossils. *In* Laporte, L. F. (ed.), *Evolution and the Fossil Record.* Freeman, San Francisco, pp. 44–56.

Barghoorn, E. S., and Schopf, J. W., 1966, Microorganisms 3–billion years old from the Precambrian of South Africa. *Science* **152**, 758–763.

Barghoorn, E. S., and Tyler, S. A., 1965, Microorganisms from the Gunflint Chert. *Science* **147**, 563–577.

Barrell, J., 1908, Relations between climate and terrestrial deposits. *J. Geol.* **16**, 159–190, 255–295, 363–384.

Barrell, J., 1915, Central Connecticut in the geologic past. *Conn. State Geol. Nat. Hist. Surv. Bull.* **23**.

Bart, H. A., 1975, Downward injection structures in Miocene sediments, Arikaree Group, Nebraska. *J. Sediment. Petrol.* **145**, 944–950.

Barthel, K. W., 1978, Solenhofen. Ott Verlag, Thun, Switzerland.

Bassett, H. G., and Stout, J. G., 1967, The Devonian of western Canada. *In* Devonian System, Int. Symp. Calgary, Alberta, *Soc. Petr. Geol.*, **1**, 717–752.

Bassler, R. S., 1941, A supposed jellyfish from the Pre–Cambrian of the Grand Canyon. *Proc. U.S. Nat. Hist. Mus.* **48**, 519.

Bassler, R. S., 1953, Bryozoa. *In* Moore, R. C. (ed.), *Treatise on Invertebrate Paleontology*, Part 6. Geological Society of America and University of Kansas Press, Lawrence.

Bates, R. L., 1955, Permo–Pennsylvanian formations between Laramie Mountains, Wyom. and Black Hills, S. D. *Am. Assoc. Pet. Geol. Bull.* **39**(10), 1979–2002.

Bathurst, R. G. C., 1971, *Carbonate Sediments and Their Diagenesis.* Elsevier, J. Amsterdam.

Beach, A., 1976, The interrelations of fluid transport, deformation, geochemistry and heat flow in early Proterozoic shear zones in the Lewisian complex. *Philos. Trans. R. Soc. London Ser. A* **280**, 579–604.

Bearce, D. N., Black, W. W., Kish, S. A., and Tull, J. F., 1982, Tectonic studies in the Talledaga and Carolina Slate Belts. *Geol. Soc. Am. Sp. Paper* **191**, p. 164.

Beck, C. B. (ed.), 1976, *Origin and Early Evolution of Angiosperms: A Perspective.* Columbia University Press, New York, pp. 1–11.

Beerbower, J. R., 1963, Morphology, paleoecology, and phylogeny of the Permo-Pennsylvanian amphibian *Diploceraspis*. *Bull. Mus. Comp. Zool. Harvard Univ.* **130**(2), 31–108.

Beget, J. E., 1981, Early Holocene glacier advance in the North Cascade Range, Washington. *Geology* **9**, 409–413.

Bell, R. T., 1976, Preliminary notes on the Hurwitz Group, Padlei map area, Northwest Territories. *Geol. Surv. Canad. Pap.* **69–52**.

Belt, E. S., 1968, Post-Acadian rifts and related facies, eastern Canada. *In* Zen, E., White, W. S., and Hadley, J. B. (eds.), *Studies of Appalachian Geology: Northern and Maritime.* Wiley–Interscience, New York, pp. 95–113.

Ben–Avraham, Z., and Cooper, A. K., 1980, Early evolution of the Bering Sea by collision of oceanic rises and North Pacific subduction zones. *Geol. Soc. Am. Bull.* **92**, 485–595.

Ben–Avraham, Z., Nur, A., and Cox, A., 1981, Continental accretion: From oceanic plateaus to allochthonous terranes. *Science* **213**, 47–54.

Benson, R. H., *et al.*, 1961, *Treatise on Invertebrate Paleontology*, Part Q (Arthropoda 3) (Ostracoda). Geological Society of America and University of Kansas Press, Lawrence.

Bentley, R. D., and Neathery, T. L., 1970, Geology of the Brevard Fault zone and related rocks of the Inner Piedmont of Alabama. Ala. Geol. Soc. Field Trip Guidebook, p. 119.

Bergeron, R., 1957, Late Precambrian rocks of the northern shore of the St. Lawrence River and of the Mistassini and Otish Mountains area, Quebec. *In* Gill, J. E. (ed.), *The Proterozoic in Canada.* R. Soc. Can. Pap. **2**, 124–131.

Berggren, W. A., and Hollister, C. D., 1974, Paleogeography, paleobiogeography, and the history of circulation in the Atlantic Ocean. *Soc. Econ. Paleontol. Mineral. Spec. Publ.* **20**, 126–186.

Berkner, L. V., and Marshall, L. C., 1965, On the origin and rise of oxygen concentration in the Earth's atmosphere. *J. Atmos. Sci.* **22**, 225.

Berman, D. S., 1977, A new species of *Dimetrodon* (Reptilia, Pelycosauria) from a non-deltaic facies in the Lower Permian of north-central New Mexico. *J. Paleontol.* **51**(1), 108–115.

Berry, B. N., 1977, Some Siluro-Devonian patterns in the western United States. *In* Stewart, J. H., Stevens, C. H., and Fritsche, A. E. (eds.), Paleozoic paleogeography of the western United States. Pacific Section, SEPM, Los Angeles, pp. 241–249.

Berry, W. B. N., 1960, Graptolite faunas of the Marathon region, West Texas. Tex. Bur. Econ. Geol. Publ. 6005, p. 103

Besairie, H., 1967, The Precambrian of Madagascar. *In* Rankama, K. (ed.), *The Precambrian,* Vol. 3. Wiley–Interscience, New York, pp. 133–142.

Bevier, M. L., Armstrong, R. L., and Souther, J. G., 1979, Miocene peralkaline volcanism in west-central British Columbia—Its temporal and plate-tectonics setting. *Geology* **7**, 389–392.

Bird, J. M., 1969, Middle Ordovician gravity sliding—Taconic region. *Am. Assoc. Pet. Geol. Mem.* **12**, 670–686.

Bird, J. M., and Dewey, J. F., 1970, Lithosphere plate: Continental margin tectonics and the evolution of the Appalachian orogen. *Geol. Soc. Am. Bull.* **81**, 1031–1059.

Birkelund, T., and Bromley, R. G. (eds.), 1979, *Cretaceous–Tertiary Boundary Events,* Vol. I. University of Copenhagen, Denmark.

Bischof, F., 1875, Die steinsalzwerke bie Strassfurt. Halle, Pfeffer, p. 95.

Bischoff, J. L., and Rosenbaur, R. J., 1981, Uranium series dating of human skeletal remains from the Del Mar and Sunnyvale sites, Calif. *Science* **213**, 1003–1005.

Bissell, H. J., 1974, Tectonic control of late Paleozoic and early Mesozoic sedimentation near the hinge line of the Cordilleran miogeosynclinal belt. *Soc. Econ. Paleontol. Mineral. Spec. Publ.* **22**, 83–97.

Bissell, H. J., and Barker, H. Y., 1977, Deep–water limestones of the Great Blue Formation (Mississippian) in the eastern part of the Cordilleran miogeosyncline in Utah. *Soc. Econ. Paleontol. Mineral. Spec. Publ.* **23**, 171–186.

Black, L. P., Gale, N. H., Moorbath, S., Pankhurst, R. J., and McGregor, V. R., 1971, Isotopic dating of very early Precambrian amphibolite facies gneisses from the Godthaab district, West Greenland. *Earth Planet. Sci. Lett.* **12**, 245–259.

Black, R. F., 1976, Quaternary geology of Wisconsin and contiguous Upper Michigan. *In* Mahaney, W. C. (ed.), *Quaternary Stratigraphy of North America.* Dowden, Hutchinson & Ross, Stroudsburg, Pa., pp. 93–118.

Black, W. W., and Fullagan, P. D., 1976, Avalonian ages of metavolcanics and plutons of the Carolina Slate Belt near Chapel Hill, N.C. *Geol. Soc. Am. Abstr. Prom.* **8**, 136.

Blackwelder, B. W., 1981, Late Cenozoic marine deposition in the United States Coastal Plain related to tectonism and global climate. *Paleogeogr. Paleoclimatol. Paleoecol.* **34**, 87–114.

Blatt, H., Middleton, G., and Murray, R., 1980, *Origin of Sedimentary Rocks,* 2nd ed. Prentice–Hall, Englewood Cliffs, N.J.

Boardman, R. S., *et al.,* 1983, *Treatise on Invertebrate Paleontology,* Part G, Bryozoa (rev.), Vol. 1. Geological Society of America and University of Kansas Press, Lawrence.

Bobyarchick, A. R., 1980, The Eastern Piedmont fault system and its relationship to Alleghenian tectonics in the Southern Appalachian. *J. Geol.* **89**, 335–347.

Boellstorff, J., 1978, North American Pleistocene stages reconsidered in light of probable Pliocene–Pleistocene continental glaciation. *Science* **202**, 305–307.

Bohor, B. F., Foord, E. E., Modreski, P. J., and Triplehorn, D. M., 1984, Mineralogic evidence for an impact event at the Cretaceous–Tertiary boundary. *Science* **224**, 867–869.

Borden, C. E., 1979, Peopling and early cultures of the Pacific Northwest. *Science* **203**, 963–972.

Bostock, H. H., 1971, Geological notes on Aquatuk River map–area,

Ontario, with emphasis on the Precambrian rocks. *Geol. Surv. Can. Pap.* **70–42**.

Boucot, A. J., 1962, Appalachian Siluro-Devonian. *In Some Aspects of the Variscan Fold Belt,* 9th Inter–University Geol. Congr. Manchester University Press, pp. 155–163.

Boucot, A. J., 1968, Silurian and Devonian of the Northern Appalachians. *In* Zen, E., White, W. S., and Hadley, J. B. (eds.), *Studies of Appalachian Geology: Northern and Maritime.* Wiley–Interscience, New York, pp. 83–94.

Boucot, A. J., 1983, Does evolution take place in an ecological vacuum? *J. Paleontol.* **57**(1), 1–31.

Boucot, A. J., and Johnson, J. G., 1967, Paleogeography and correlation of Appalachian Province Lower Devonian sedimentary rocks. *Tulsa Geol. Soc. Dig.* **35**, 35–87.

Bouma, A. H., Moore, G. T., and Coleman, J. M. (eds.), 1978, Framework, facies, and oil–trapping characteristics of the upper continental margin. AAPG, Studies in Geology No. 7.

Bowen, D. Q., 1978, *Quaternary Stratigraphy.* Pergamon Press, Elmsford, N.Y.

Boyer, B. W., 1982, Green River laminates: Does the Playa–lake model really invalidate the stratified–lake model? *Geology* **10**, 321–324.

Bradley, D. C., 1983, Tectonics of the Acadian orogeny in New England and adjacent Canada. *J. Geol.* **91**, 381–400.

Bradley, W. H., 1964, Geology of the Green River Formation and associated Eocene rocks in southwestern Wyoming and adjacent parts of Colorado and Utah. *U.S. Geol. Surv. Prof. Pap.* **496–A**.

Branson, C. C., 1962a, Pennsylvanian System of the mid-continent. *In* Branson, C. C. (ed.), Pennsylvanian System in the United States—A symposium. AAPG, Tulsa, pp. 431–460.

Branson, C. C. (ed.), 1962b, Pennsylvanian System in the United States—A symposium. AAPG, Tulsa.

Brasier, M. D., 1980, *Microfossils.* Allen & Unwin, London.

Braunagel, L. H., and Stanley, K. O., 1977, Origin of variegated redbeds in the Cathedral Bluffs tongue of the Wasatch Formation (Eocene), Wyoming. *J. Sediment Petrol.* **47**, 1201–1219.

Brenner, R. L., and Davies, D. K., 1974, Oxfordian sedimentation in the western interior United States. *Am. Assoc. Pet. Geol. Bull.* **58**(3), 407–428.

Bretsky, P. W., Jr., 1969, Ordovician benthic marine communities in the Central Appalachians. *Geol. Soc. Am. Bull.* **80**, 193–212.

Bretsky, P. W., Jr., and Lorenz, D. M., 1970, An essay on genetic adaptive strategies and mass extinctions. *Geol. Soc. Am. Bull.* **81**, 2449–2456.

Bretz, J. H., 1969, The Lake Missoula floods and the channeled scablands. *J. Geol.* **77**, 505–543.

Briden, J. C., Turnell, H. B., and Watts, D. R., 1984, British paleomagnetism, Iapetus Ocean, and the Great Glen Fault. *Geology* **12**, 428–431.

Bridgwater, D., and Collerson, K. D., 1976, The major petrological and geochemical characters of the 3,600 m.y. Uivak gneisses from Labrador. *Contrib. Mineral. Petrol.* **54**, 43–59.

Bridgwater, D., Watson, J., and Windley, B. F., 1973, The Archean craton of the North Atlantic region. *Philos. Trans. R. Soc. London Ser. A* **273**, 493–512.

Bridgwater, D., McGregor, V. R., and Myers, J. S., 1974, A horizontal tectonic regime in the Archean of Greenland and its implications for early crustal thickening. *Precambrian Res.* **1**, 179–198.

Bridgwater, D., Keto, L., McGregor, V. R., and Myers, J. S., 1976, Archean gneiss complex of Greenland. *In* Escher, A., and Watt, W. S. (eds.), *Geology of Greenland,* Geol. Surv. Greenland, Copenhagen, pp. 19–75.

Bridgwater, D., Collerson, K. D., and Myers, J. S., 1978, The development of the Archaean gneiss complex of the North Atlantic region. *In* Tarling, D. H. (ed.), *Evolution of the Earth's Crust.* Academic Press, New York, pp. 19–69.

Briggs, D. F., Gilbert, M. C., and Glober, L., III, 1978, Petrology and regional significance of the Roxboro Metagranite, North Carolina. *Geol. Soc. Am. Bull.* **89**, 511–521.

Briggs, G. (ed.), 1974a, Carboniferous of the Southeastern United States. *Geol. Soc. Am. Spec. Pap.* **148**.

Briggs, G., 1974b, Carboniferous depositional environments in the

Ouachita Mountains–Arkoma basin area of southeastern Oklahoma. *Geol. Soc. Am. Spec. Pap.* **148**, 225–239.

Bristol, H. M., and Howard, R. H., 1974, Sub–Pennsylvanian valleys in the Chesterian surface of the Illinois basin and related Chesterian slump blocks. *Geol. Soc. Am. Spec. Pap.* **148**, 315–336.

Brocoum, S. J., and Dalziel, I. W. D., 1974, The Sudbury Basin, the Southern Province, the Grenville Front, and the Penokean Orogeny. *Geol. Soc. Am. Bull.* **85**, 1571–1580.

Brooks, J., and Muir, M. D., 1971, Morphology and chemistry of the organic insoluble matter from the Onverwacht Series Precambrian chert and the Orgueil and Murray carbonaceous meteorites. *Grana* **11**, 9–14.

Brosgé, W. P., and Dutro, J. T., Jr., 1973, Paleozoic rocks of northern and central Alaska. *Am. Assoc. Pet. Geol. Mem.* **19**, 361–375.

Brosgé, W. P., and Tailleur, I. L., 1970, Depositional history of northern Alaska. *In* Adkison, W. L., and Brosgé, W. P. (eds.), Proceedings of the geological seminar on the North Slope of Alaska. Los Angeles, Pacific Section, AAPG, pp. D–1–D–17.

Brown, L. F., Jr., 1969, North Texas (eastern shelf) Pennsylvanian delta systems. *In* Delta systems in the exploration for oil and gas—a research colloquium. Univ. Texas, Austin, Bur. Econ. Geol., pp. 40–53.

Brown, L. F., Jr., 1973, Depositional systems in the Cisco Group of north–central Texas. *In* Brown, L. F., Jr., Cleaves, A. W., II, and Erxleben, A. W., Pennsylvanian depositional systems in north–central Texas; a guide for interpreting terrigenous clastic facies in a cratonic basin. Univ. Texas, Austin, Bur. Econ. Geol. Guidebook **14**, 57–73.

Brues, C. T., 1951, Insects in amber. *Sci. Am.* **185**(11), 56–61.

Buckovic, W. A., 1979, The Eocene deltaic system of west–central Washington. *In* Armentrout, J. M., Cole, M. R., and TerBest, H., Jr. (eds.), Cenozoic paleogeography of the western United States. SEPM, Pacific Section, Pacific Coast Paleogeography Symp. **3**, pp. 147–163.

Buehler, E. J., and Tesmer, I. H., 1963, Geology of Erie County, N.Y. *Buffalo Soc. Nat. Sci. Bull.* **21**(3).

Buffetaut, E., 1979, The evolution of the crocodilians. *Sci. Am.* **241**(4), 130–144.

Buffler, R. T., Shaub, F. J., Watkins, J. S., and Worzel, J. L., 1979a, Anatomy of the Mexican Ridges, southwestern Gulf of Mexico. *Am. Assoc. Pet. Geol. Mem.* **29**, 319–327.

Buffler, R. T., Watkins, J. S., and Dillon, W. P., 1979b, Geology of the offshore Southeast Georgia embayment, U.S. Atlantic continental margin, based on multichannel seismic reflection profiles. *Am. Assoc. Pet. Geol. Mem.* **29**, 11–25.

Buffler, R. T., Watkins, J. S., Shaub, F. J., and Worzel, J. L., 1980, Structure and early history of the deep central Gulf of Mexico basin. *In* Pilger, R. H., Jr. (ed.), The Origin of the Gulf of Mexico and the Early Opening of the Central North Atlantic Ocean. Proc. Symp., LSU, Baton Rouge, pp. 3–16.

Bulman, O. M. B., 1970, *Treatise on Invertebrate Paleontology*, Part V, Graptolithina (rev.). Geological Society of America and University of Kansas Press, Lawrence.

Burchfiel, B. C., and Davis, G. A., 1972, Structural framework and evolution of the southern part of the Cordilleran orogen, western United States. *Am. J. Sci.* **272**, 97–118.

Burchfiel, B. C., and Davis, G. A., 1975, Nature and controls of Cordilleran orogenesis, western United States: Extensions of an earlier synthesis. *Am. J. Sci.* **275–A**, 363–396.

Burk, C. A., and Drake, C. L. (eds.), *The Geology of Continental Margins*. Springer–Verlag, Berlin.

Burke, K., and Kidd, W. S. F., 1978, Were Archean continental geothermal gradients much steeper than those of today? *Nature* **272**, 240–241.

Burke, K., Dewey, J. F., and Kidd, W. S. F., 1976, Precambrian paleomagnetic results compatible with contemporary operation of the Wilson cycle. *Tectonophysics* **33**, 287–299.

Butler, H., 1961, Devonian deposits of Central East Greenland. *In* Raasch, G. O. (ed.), *Geology of the Arctic,* 1, University of Toronto Press, pp. 188–196.

Butler, J. R., 1972, Age of Paleozoic regional metamorphism in the Car-

olinas, Georgia, and Tennessee, Southern Appalachians. *Am. J. Sci.* **272**, 319–333.

Butler, J. R., and Ragland, P., 1979, A petrographic survey of plutonic intrusions of the Piedmont, southeastern Appalachians, U.S.A. *Contrib. Mineral. Petrol.* **24**, 164–190.

Cady, W. M., 1975, Tectonic setting of the Tertiary volcanic rocks of the Olympic Peninsula, Washington. *U.S. Geol. Surv. Res.* **3**, 575–582.

Caldwell, W. G. E. (ed.), 1975, The Cretaceous System in the western interior of North America. *Geol. Assoc. Can. Spec. Pap.* **13**.

Calvin, M., 1969, *Chemical Evolution.* Oxford University Press, New York.

Cambray, F. W., 1978, Plate tectonics as a model for the environment of deposition and deformation of the early Proterozoic (Precambrian X) of northern Michigan. *Geol. Soc. Am. Abstr. Progr.* **10**, 376.

Cameron, B. E. B., 1979, Early Cenozoic paleogeography of Vancouver Island, British Columbia [abstract]. *In* Armentrout, J. M., Cole, M. R., and TerBest, H., Jr. (eds.), Cenozoic Paleogeography of the western United States. SEPM, Pacific Section, Pacific Coast Paleogeography Symp. No. 3, p. 326.

Camfield, P. A., and Gough, D. I., 1977, A possible Proterozoic plate boundary in North America. *Can. J. Earth Sci.* **14**, 1229–1238.

Campbell, G. G., 1978, Some problems in hominid classification and nomenclature. *In* Jolly, C. J. (ed.), *Early Hominids of Africa,* St. Martin's Press, N.Y., pp. 567–582.

Card, K. D., Church, W. R., Franklin, J. M., Frarey, M. J., Robertson, J. A., West, G. F., and Young, G. M., 1972, The Southern Province. *Geol. Assoc. Can. Spec. Pap.* **11**, 335–380.

Carey, S. W., 1958, A tectonic approach to continental drift. *In* Carey, S. W. (ed.), Continental Drift—A symposium. Tasmania University, Hobart, pp. 177–355.

Carozzi, A. V., 1964, Complex ooids from Triassic lake deposits, Virginia. *Am. J. Sci.* **262**, 231–241.

Carpenter, F. M., and Burnham, L., 1985, The geological record of insects. *Annu. Rev. Earth Planet. Sci.* **13**, 297–314.

Carrasco-V., Baldomero, 1977, Albian sedimentation of submarine autochthonous and allochthonous carbonates, east edge of the Valles, San Luis Potosi Platform, Mexico. *Soc. Econ. Paleontol. Mineral. Spec. Publ.* **25**, 263–272.

Carrington, T. J., 1967, Talladega Group of Alabama. *In* Ferm, J. C., Ehrlich, R., and Neathery, T. L., *A Field Guide to Carboniferous Detrital Rocks in Northern Alabama.* Annual field trip of the Coal Div., Geol. Soc. Am., and Ala. Geol. Soc., pp. 24–27.

Carroll, R. L., 1980, The hyomandibular as a supporting element in the skull of primitive tetrapods. *In* Panchen, A. L. (ed.), *The Terrestrial Environment and the Origin of Land Vertebrates.* Academic Press (Systematics Assoc.), New York/London, pp. 293–317.

Carroll, R. L., and Gaskill, P., 1978, The Order Microsauria. *Mem. Am. Philos. Soc.* **126**.

Carter, C. H., 1978, A regressive barrier and barrier–protected deposit; depositional environments and geographic setting of the Later Tertiary Cohansey Sand. *J. Sediment. Petrol.* **48**, 933–950.

Cavanaugh, M. D., and Seyfert, C. K., 1977, Apparent polar wander paths and the joining of the Superior and Slave provinces during early Proterozoic time. *Geology* **5**, 207–211.

Cebull, S. E., and Shurbet, D. H., 1980, The Ouachita belt in the evolution of the Gulf of Mexico. *In* Pilger, R. H., Jr. (ed.), The Origin of the Gulf of Mexico and the Early Opening of the Central North Atlantic Ocean. Proc. Symp., LSU, Baton Rouge, pp. 17–26.

Chamberlain, C. K., 1971, Bathymetry and paleoecology of Ouachita geosyncline of southeastern Oklahoma as determined from trace fossils. *Am. Assoc. Pet. Geol. Bull.* **55**, 34–50.

Chamberlain, C. K., 1978, A guidebook to the trace fossils and paleoecology of the Ouachita geosyncline, Guidebook, S.E.P.M. field trip, Tulsa, Oklahoma, 68 p.

Chamberlain, C. K., and Clark, D. L., 1973, Trace fossils and conodonts as evidence for deep–water deposits in the Oquirrh basin of central Utah. *J. Paleontol.* **47**, 663–682.

Chamberlain, J. A., Jr., 1978, Mechanical properties of coral skeleton: Compressive strength and its adaptive significance. *Paleobiology* **4**(4), 419–435.

Chapman, C. A., 1968, A comparison of the Maine coastal plutons and the magmatic central complexes of New Hampshire. *In* Zen, E., White, W. S., and Hadley, J. B. (eds.), *Studies of Appalachian Geology: Northern and Maritime*. Wiley–Interscience, New York, pp. 385–396.

Chase, C. G., and Gilmer, T. H., 1973, Precambrian plate tectonics: The midcontinent gravity high. *Earth Planet. Sci. Newslett.* **21**, 70–78.

Chase, R. L., and Tiffin, D. L., 1972, Queen Charlotte fault zone, British Columbia. 24th Int. Geol. Congr., Sect. 8, pp. 17–27.

Chen, C. S., 1965, The regional lithostratigraphic analysis of Paleocene and Eocene rocks of Florida. *Fla. Geol. Surv. Geol. Bull.* **45**.

Chown, E. H., and Caty, J. L., 1973, Stratigraphy and paleocurrent analysis of the Aphebian clastic formations of the Mistassini–Otish Basin. *Geol. Assoc. Can. Spec. Pop.* **12**, 49–71.

Chowns, T. M. and William, C., 1983, Pre–Cretaceous rocks beneath the Georgia Coastal Plain–regional implications. *In* Gohn, G. (ed.) Studies related to the Charleston, South Carolina earthquake of 1886–tectonics and seismicity. *U.S. Geol. Surv. Prof. Pap.* **1313**, L1–L42.

Chowns, T. M., and Elkins, J. E., 1974, The origin of quartz geodes and cauliflower cherts through the silicification of anhydrite nodules. *J. Sediment. Petrol.* **44**, 885–903.

Christensen, W. K., and Birkelund, T. (eds.), 1979, *Cretaceous–Tertiary Boundary Events*, Vol. II. University of Copenhagen, Denmark.

Christie-Blick, N., 1982, Upper Proterozoic and Lower Cambrian rocks of the Sheeprock Mountains, Utah: Regional correlation and significance. *Geol. Soc. Am. Bull.* **93**, 735–750.

Christopher, R. A., 1982, Palynostratigraphy of the basal Cretaceous units of the eastern Gulf and southern Atlantic Coastal Plains. *In* Arden, D. D., Beck, B. F., and Morrow, E. (eds.), 2nd symposium on the geology of the southeastern Coastal Plains. *Ga. Geol. Surv. Info. Circ.* **53**, 10–23.

Christopher, R. A., Owens, J. P., and Sohl, N. F., 1979, Late Cretaceous polynomorphs from the Cape Fear Formation of North Carolina. *Southeast. Geol.* **20**, 145–159.

Christopher, R. A., Prowell, D. C., Reinhardt, J., and Markewich, H. W., 1980, The stratigraphic and structural significance of Paleocene pollen from Warm Springs, Georgia. *Palynology* **4**, 105–123.

Christopher, Y. E., 1975, The depositional setting of the Manville group (Lower Cretaceous) in southwestern Saskatchewan. *Geol. Assoc. Can. Spec. Pap.* **13**, 523–552.

Chronic, J., 1979, Colorado (*In* the Mississippian and Pennsylvanian (Carboniferous) Systems in the United States). *U.S. Geol. Surv. Prof. Pap.* **1110-V**.

Chronic, J., McCallum, M. E., Ferris, C. S., Jr., and Eggler, D. H., 1969, Lower Paleozoic rocks in diatremes, southern Wyoming and northern Colorado. *Geol. Soc. Am. Bull.* **80**, 149–156.

Chuber, S., 1972, Milbur (Wilcox) field, Milam and Burleson Counties, Texas. *Am. Assoc. Pet. Geol. Mem.* **16**, 399–405.

Church, W. R., 1967, The occurrence of kyanite, andalusite and kaolinite in lower Proterozoic (Huronian) rocks of Ontario. Abstr. Int. Meet. Geol. Can., Kingston, Ontario, pp. 14–15.

Churkin, M., Jr., 1973, Geologic concepts of Arctic Ocean Basin. *Am. Assoc. Pet. Geol. Mem.* **19**, 485–499.

Churkin, M., Jr., 1974, Paleozoic marginal ocean basin–volcanic arc systems in the Cordilleran foldbelt. *Soc. Econ. Paleontol. Mineral. Spec. Publ.* **19**, 174–192.

Churkin, M., Jr., and Eberlein, G. D., 1977, Ancient borderland terranes of the North American Cordillera: Correlation and microplate tectonics. *Geol. Soc. Am. Bull.* **88**, 769–786.

Churkin, M., Jr., Nobleberg, W. J., and Huie, C., 1979, Collision-deformed Paleozoic continental margin, western Brooks Range, Alaska. *Geol.* **7**, 379–383.

Churkin, M., Jr., Carter, C., and Trexler, J. H., Jr., 1980, Collision-deformed Paleozoic continental margin of Alaska—Foundation for microplate accretion. *Geol. Soc. Am. Bull.* **91**, 648–654.

Cisne, J. L., 1974, Trilobites and the origin of arthropods. *Science* **186**, 13–18.

Ciurca, S. J., 1973, Eurypterid horizons and the stratigraphy of the Upper Silurian and Lower Devonian of western New York State. *In* Guidebook to 45th Annual Meeting, Rochester. N.Y. State Geol. Soc., pp. D–1–D–12.

Clark, J., Beerbower, J. R., and Kietzke, K. K., 1967, Oligocene and sedimentation, stratigraphy, paleoecology and paleoclimatology in the Big Badlands of South Dakota. *Fieldiana Geol. Mem.* **5**.

Clark, P., and Karrow, P. F., 1984, Late Pleistocene water bodies in the St. Lawrence Lowland, N.Y. and regional correlations. *Geol. Soc. Am. Bull.* **95**, 805–813.

Clarke, J. W., 1952, Geology and mineral resources of the Thomaston Quadrangle, Georgia. *Ga. Geol. Surv. Bull.* **59**.

Clarkson, E. N. K., 1979, *Invertebrate Paleontology and Evolution*. Allen & Unwin, London.

Clarkson, E. N. K., and Levi-Setti, R., 1975, Trilobite eyes and the optics of Des Cartes and Huygens. *Nature* **254**, 663–667.

Cleaves, A. W., II, 1975, Upper Desmoinesian–lower Missourian depositional systems (Pennsylvanian), north–central Texas. Ph.D. dissertation, University of Texas, Austin.

Clemens, W. A., 1974, *Purgatorius*, an early paromomyid primate (Mammalia). *Science* **184**, 903–905.

Clendening, J. A., 1969, The base of the Permian in the Appalachians: Ninety years of controversy. *In* Donaldson, A. C. (ed.), *Some Appalachian Coals and Carbonates: Models of Ancient Shallow–Water Deposition*. W. Va. Geol. Surv., pp. 229–233.

Cloos, E., 1972, Experimental imitation of the upturned Precambrian surface along the Blue Ridge, Maryland and Virginia. *In* Leasing, Peter and others (eds.), *Appalachian Structures: Origin, Evolution, and Possible Potential for New Exploration Frontiers*. Morgantown, W. Va., West Virginia University and West Virginia Geol. Econ. Surv., pp. 17–37.

Cloos, E., and Pettijohn, F. J., 1973, Southern border of the Triassic basin west of York, Pennsylvania: Fault or overlap? *Geol. Soc. Am. Bull.* **84**, 523–536.

Cloos, M., 1980, Numerical modeling of mixing and uplift in flow mélanges of the Franciscan subduction complex, California. *Geol. Soc. Am. Abstr. Progr.* **12**(7), 404.

Cloud, P. E., 1968, Premetazoan evolution and the origins of the Metazoa. *In* Drake, E. T. (ed.), *Evolution and Environment*. Yale University Press, New Haven, Conn., pp. 1–72.

Cloud, P. E., 1976, Beginnings of biospheric evolution and their biogeochemical consequences. *Paleobiology* **2**, 351–387.

Cloud, P. E., and Glaessner, M. F., 1982, The Ediacarian Period and System: Metazoa inherit the Earth. *Science* **218**, 783–792.

Cloud, P. E., Wright, J., and Glover, L., III, 1976, Traces of animal life from 620–million–year old rocks in North Carolina. *Am. Sci.* **64**, 396–406.

Coates, D. R., 1976, Quaternary stratigraphy of New York and Pennsylvania. *In* Mahaney, W. C. (ed.), *Quaternary Stratigraphy of North America*. Dowden, Hutchinson & Ross, Stroudsburg, Pa., pp. 65–90.

Cobban, W. A., 1951, Colorado Shale of central and northwestern Montana and equivalent of Black Hills. *Am. Assoc. Pet. Geol. Bull.* **35**(10), 2170–2198.

Cobban, W. A., 1977, Characteristic marine molluscan fossils from the Dakota Sandstone and intertongued Mancos Shale, west–central New Mexico. *U.S. Geol. Surv. Prof. Pap.* **1009**.

Cobban, W. A., and Reeside, J. B., Jr., 1952a, Correlation of the Cretaceous formations of the western interior of the U.S. *Geol. Soc. Am. Bull.* **63**, 1011–1044.

Cobban, W. A., and Reeside, J. B., Jr., 1952b, Frontier Formation, Wyoming and adjacent areas. *Am. Assoc. Pet. Geol. Bull.* **36**(10), 1913–1961.

Cobban, W. A., Erdmann, C. E., Lemke, R. W., and Maughan, E., 1976, Type sections and stratigraphy of the members of the Black Leaf and Marias River Formations (Cretaceous) of the Sweetgrass Arch, Montana. *U.S. Geol. Surv. Prof. Pap.* **974**.

Cofer, H. E., Jr., and Frederiksen, N., 1982, Paleoenvironment and age of kaolin deposits in the Andersonville district, Georgia. *Ga. Geol. Surv. Inf. Circ.* **53**, 24–37.

Colbert, E. C., 1980, *Evolution of the Vertebrates*, 3rd ed. Wiley, New York.

Cole, M. R., and Armentrout, J. M., 1979, Neogene paleogeography of the western United States. *In* Armentrout, J. M., Cole, M. R., and TerBest, H., Jr. (eds.), Cenozoic paleogeography of the western

United States, SEPM, Pacific Section, Pacific Coast Paleogeography Symp. No. 3, pp. 297–323.

Collins, H. R., 1979, Ohio. *In* The Mississippian and Pennsylvanian (Carboniferous) Systems in the U.S. *U.S. Geol. Surv. Prof. Pap.* **1110**, E1–E26.

Collinson, C., 1967, Devonian of the north–central region, U.S. *In* Devonian System, Int. Symp. Calgary, Alberta, *Soc. Pet. Geol.*, 1, 933–997.

Collinson, J. W., and Hasenmueller, W. A., 1978, Early Triassic paleogeography and biostratigraphy of the Cordilleran miogeosyncline. *In* Howell, D. G., and McDougall, K. A. (eds.), Mesozoic paleogeography of the western United States. SEPM, Pacific Section, Pacific Coast Paleogeography Symp. No. 2, pp. 1750–1807.

Colton, G. W., 1970, The Appalachian Basin–its depositional sequences and their geologic relationships. *In* Fisher, F. W., Pettijohn, F. J., and Reed, J. C., Jr. (eds.), *Studies of Appalachian Geology.* Wiley–Interscience, New York, 5–47.

Compagno, I. J. V., 1973, Interrelationships in living elasmobranchs. *In* Greenwood, P. H., Miles, R. S., and Patterson, C. (eds.), *Interrelationships of Fishes.* Academic Press, New York, Vol. 53 (Suppl. 1), pp. 15–61.

Cooper, A. K., Marlow, M. S., and Ben–Avraham, A., 1981, Multichannel seismic evidence bearing on the origin of Bowers Ridge, Bering Sea. *Geol. Soc. Am. Bull.* **92**, 474–484.

Cooper, A. K., Scholl, D. W., and Marlow, M. S., 1976, Plate tectonic model for the evolution of the eastern Bering Sea basin. *Geol. Soc. Am. Bull.* **87**, 1119–1126.

Cooper, B. N., 1966, Geology of the salt and gypsum deposits in the Saltville area, Smyth and Washington Counties, Virginia. Northern Ohio Geol. Soc., 2nd Salt Symp. **1**, 11–34.

Cooper, B. N., 1968, Profile of the folded Appalachians of western Virginia. Missouri Univ. Tolla J. No. 1, pp. 27–64.

Cooper, G. A., and Grant, R. E., 1972, 1974, 1975, 1976, Permian brachiopods of West Texas, Parts I–IV. *Smithson. Contrib. Paleobiol.* No. 14, 15, 19, 21.

Copeland, C. W., 1972, Upper Cretaceous Series in central Alabama. *In* Tolson, J. S. (ed.), Guide to Alabama geology: Guidebook for field trips. 21st Annu. Meet. Southeast. Geol. Soc. Am., pp. 2(1)–2(24).

Cornell, W. C., and LeMone, D. V., 1979, (response to) A terminal Cretaceous greenhouse. *Science* **206**, 1428–1429.

Cornet, B., 1975, Palynological evidence for the Late Triassic position of the Newark Supergroup. Abstr., Triassic Conf., Wesleyan University.

Cornet, B., and Traverse, A., 1975, Palynological contributions to the chronology and stratigraphy of the Hartford basin in Connecticut and Massachusetts. *Geoscience and Man*, Vol. XI, pp. 1–33.

Couch, R., Whitsett, R., Huehn, B., and Briceno–Guarupe, L., 1981, Structures of the continental margin of Peru and Chile. *In* Kulm, L. D., Dymond, J., Dasch, E. J., Hussong, D. M., and Roderick, R. (eds.), Nazca Plate: Crustal formation and Andean convergence. *Geol. Soc. Am. Mem.* **154**, 703–726.

Condie, K. C., 1981, *Archean Greenstone Belts*. Elsevier, Amsterdam.

Condie, K. C., 1982, Plate-tectonics model for Proterozoic continental accretion in the southwestern U.S. *Geology* **10**, 37–42.

Condie, K. C., 1984, Archean geotherms and supracrustal assemblages. *Tectonophysics* **105**, 29–41.

Condie, K. C. and Budding, A. J., 1979, Geology and geochemistry of Precambrian rocks, central and south–central Mexico. *N. M. Bur. of Mines and Mineral Res. Mem.* **35**.

Condie, K. C., Macke, J. E., and Reimer, T. O., 1970, Petrology and geochemistry of Early Precambrian graywackes from the Fig Tree Group, South Africa. *Geol. Soc. Am. Bull.* **81**, 2759–2775.

Conway Morris, S., and Fritz, W. H., 1980, Shelly microfossils near the Precambrian–Cambrian boundary, Mackenzie Mountains, northwest Canada. *Nature* **286**, 381–384.

Conway Morris, S., and Whittington, H. B., 1979, The animals of the Burgess Shale. *Sci. Am.* **241**(1), 122–133.

Cook, H. E., and Enos, P. (eds.), 1977, Deep-water carbonate environments. *Soc. Econ. Paleontol. Mineral. Spec. Publ.* **23**.

Cook, T. D., and Bally, A. W., 1975, *Stratigraphic Atlas of North and Central America*. Princeton University Press, Princeton, N.J.

Cooke, C. W., 1930, Correlation of coastal terraces. *J. Geol.* **38**, 577–589.

Cooke, C. W., 1931, Seven coastal terraces in the Southeastern United States. *Wash. Acad. Sci.* **21**, 500–513.

Cooke, F. A., Albaught, D. S., Brown, L. D., Kaufman, S., Oliver, J. E., and Hatcher, R. D., Jr., 1979, Thin–skinned tectonics in the crystalline southern Appalachians; COCORP seismic–reflection profiling of the Blue Ridge and Piedmont. *Geology* **7**, 563–567.

Covey, C., 1984, The Earth's orbit and ice ages. *Sci. Am.* **250**(2), 58–66.

Cowan, D. S., 1974, Deformation and metamorphism of the Franciscan subduction zone complex northwest of Pacheco Pass, California. *Geol. Soc. Am. Bull.* **85**, 1623–1634.

Coward, M. P., and James, P. R., 1974, The deformation patterns of two Archean greenstone belts in Rhodesia and Botswana. *Precambrian Res.* **1**, 235–258.

Coward, M. P., Francis, P. W., Graham, R. H., Myers, J. S., and Watson, J., 1969, Remnants of an early metasedimentary assemblage in the Lewisian Complex of the Outer Hebrides. *Proc. Geol. Assoc.* **80**, 387–408.

Coward, M. P., Lintern, B. C., and Wright, L. I., 1976, The pre-cleavage deformation of the sediments and gneisses of the northern part of the Limpopo belt. *In* Windley, B. F. (ed.), *The Early History of the Earth*. Wiley, New York, pp. 323–330.

Cowie, J. W., 1967, Life in Precambrian and early Cambrian times. *In* Harland, W. B., *et al.* (eds.), *The Fossil Record*. Geological Society of London, pp. 17–35.

Cowie, J. W., 1971, The Cambrian of the North American Arctic regions. *In* Holland, C. H. (ed.) *Cambrian of the New World*. Wiley–Interscience, London, pp. 325–384.

Cox, A., 1957, Remanent magnetism of lower to middle Eocene basalt flows from Oregon. *Nature* **179**, 685–686.

Cox, L. R., 1969, General features of Bivalvia. *In* Moore, R. C. (ed.), 1969, *Treatise on Invertebrate Paleontology*, Pt. N, Vol. 1, Bivalvia. Geological Society of America and University of Kansas Press, Lawrence, pp. 2–129.

Craig, L. C., and Conners, C. W. (eds.), 1979, Paleotectonic investigations of the Mississippian System in the U.S. *U.S. Geol. Surv. Prof. Pap.* **1010**.

Craig, L. C., and Varnes, K. L., 1979, History of the Mississippian System—An interpretive summary. *U.S. Geol. Surv. Prof. Pap.* **1010**, 371–406.

Crimes, T. P., and Anderson, M. M., 1985, Trace fossils from late Precambrian–Early Cambrian strata of southeastern Newfoundland (Canada): Temporal and environmental implications. *J. Paleontol.* **59**(2), 310–343.

Crompton, A. W., 1971, The origin of the tribosphenic molar. *In* Kermack, D. A., and Kermack, K. A. (eds.), *Early Mammals*. Academic Press (Linnean Society), London, pp. 65–88.

Cronin, T. M., 1981, Rates and possible causes of neotectonic vertical crustal movements of the emerged southeastern U.S. Atlantic Coastal Plain. *Geol. Soc. Am. Bull.* **92**, 812–833.

Cronin, T. M., Szabo, B. J., Ager, T. A., Hazel, J. A., and Owens, J. P., 1981, Quaternary climates and sea levels of the U.S. Atlantic Coastal Plain. *Science* **211**, 233–240.

Crosby, E. J., and Mapel, W. J., 1975, Central and West Texas. *In* Paleotectonic investigations of the Pennsylvanian System in the U.S. *U.S. Geol. Surv. Prof. Pap.* **853**(I), 197–232.

Cross, A. T., and Schemel, M. P., 1956, Geology of the Ohio River Valley in West Virginia. *W. Va. Geol. Surv.* **22**(I).

Crouch, J. K., 1979, Neogene tectonic evolution of the California continental borderland and western Transverse Range. *Geol. Soc. Am. Bull.* **90**, 338–345.

Crowell, J. C., 1974a, Origin of Late Cenozoic basins in southern California. *Soc. Econ. Paleontol. Mineral. Spec. Publ.* **22**, 190–204.

Crowell, J. C., 1974b, Sedimentation along the San Andreas fault, California. *Soc. Econ. Paleontol. Mineral. Spec. Publ.* **19**, 292–303.

Culbertson, W. C., and Pitman, J. K., 1973, Oil shale. *In* U.S. minerals. *U.S. Geol. Surv. Prof. Pap.* **820**, 497–505.

Curray, J. R., 1965, Late Quaternary history, continental shelves of the U.S. *In* Wright, H. E., Jr., and Frey, D. G. (eds.), *The Quaternary of the United States*. Princeton University Press, Princeton, N.J., pp. 723–735.

Cushing, E. J., 1967, Late Wisconsin pollen stratigraphy and the glacial sequence in Minnesota. *In* Cushing, E. J., and Wright, H. E., Jr. (eds.), *Quaternary Paleoecology*, Vol. 7. Yale University Press, New Haven, Conn., pp. 59–88.

Cushing, E. J., and Wright, H. E., Jr. (eds.), 1967, *Quaternary Paleoecology*, Vol. 7, Proceedings of Congress INQUA. Yale University Press, New Haven, Conn.

Cushman, J. A., 1950, *Foraminifera*. Harvard University Press, Cambridge, Mass.

D'Allura, J. A., Moores, E. M., and Robinson, L., 1977, Paleozoic rocks of the northern Sierra Nevada: Their structural and paleogeographic implications. *In* Stewart, J. H., Stevens, C. H., and Fritsche, A. E., (eds.), *Paleozoic paleogeography of the western United States*. SEPM, Pacific Section, Los Angeles, pp. 395–408.

Dallmeyer, R. D., Wright, J. E., Secor, D. T., Jr., and Snoke, A. W., 1986, Character of the Alleghanian orogeny in the Southern Appalachians: Part II. Geochronological constraints on the tectonothermal evolution of the eastern Piedmont in South Carolina. *Geol. Soc. Am. Bull.* **97**, 1329–1344.

Dana, J. D., 1895, *Manual of Geology*, 4th ed. New York.

Danner, W. R., 1977, Paleozoic rocks of northwest Washington and adjacent parts in British Columbia. *In* Stewart, J. H., Stevens, C. H., and Fritsche, A. E., (eds.), Paleozoic paleogeography of the western United States. SEPM, Pacific Section, Los Angeles, pp. 481–501.

Dansgaard, W., Johnsen, S., Clausen, H. B., and Langway, C. C., Jr., 1971, Climatic record revealed by the Camp Century ice core. *In* Turekian, K. K. (ed.), *Late Cenozoic Glacial Ages*. Yale University Press, New Haven, Conn., pp. 37–56.

Dansgaard, W., Clausen, H. B., Gundestrup, N., Hammer, C. U., Johnsen, S., Kristinsdottir, P. M., and Reeh, N., 1982, A new Greenland deep ice core. *Science* **218**, 1273–1277.

Dapples, E. C., 1955, General lithofacies relationships of St. Peter Sandstone and Simpson Group. *Am. Assoc. Pet. Geol. Bull.* **39**, 444–467.

Darby, D. G., 1982, The early vertebrate *Astraspis*, habitat based on a lithologic association. *J. Paleontol.* **56**(5), 1187–1196.

Davidson, A., 1972. The Churchill Province. *Geol. Assoc. Can. Spec. Pap.* **11**, 382–433.

Davies, F. B., 1975, Origin and ancient history of gneisses older than 2,800 Myr in Lewisian Complex. *Nature* **258**, 589–591.

Davies, G. R., 1977a, Carbonate–anhydrite facies relationships, Otto Fiord Formation (Mississippian–Pennsylvanian), Canadian Arctic archipelago. *In* Fisher, J. H. (ed.), Reefs and evaporites—Concepts and depositional models. AAPG, Studies in Geology No. 5, pp. 145–167.

Davies, G. R., 1977b, Turbidites, debris sheets and truncation structures in upper Paleozoic deep-water carbonates of the Sverdrup basin, Arctic Archipelago. *Soc. Econ. Paleontol. Mineral. Spec. Publ.* **23**, 221–247.

Davies, G. R., and Nassichuk, W. W., 1975, Subaqueous evaporites of the Carboniferous Otto Fiord Formation, Canadian Arctic Archipelago: A summary. *Geology* **3**(5), 272–278.

Davis, G. A., 1980, Problems of intraplate extensional tectonics, western United States. *In* Continental tectonics. Geophysics Study Committee, Studies in Geophysics, pp. 84–95.

Davis, G. A., Monger, J. W. H., and Burchfiel, B. C., 1978, Mesozoic construction of the Cordilleran "collage," central British Columbia to central California. *In* Howell, D. G., and McDougall, K. A. (eds.), Mesozoic paleogeography of the western United States. SEPM, Pacific Section, Pacific Coast Paleogeography Symp. No. 2, pp. 1–32.

Davis, M. B., 1967, Late-glacial climate in northern United States: Comparison of New England and the Great Lakes region. *In* Cushing, E. J., and Wright, H. E., Jr. (eds.), Quaternary Paleoecology, Vol. 7. Yale University Press, New Haven, Conn., pp. 11–44.

Davis, W. M., 1886, Triassic Formation of the Connecticut Valley. *U.S. Geol. Surv.* 7th Annu. Rep., pp. 455–490.

Dawes, P. R., and Peel, J. S., 1981, The northern margin of Greenland from Baffin Bay to the Greenland Sea. *In* Nairn, A. E. M., Churkin, M., Jr., and Stehli, F. G. (eds.), *The Ocean Basins and Margins*, Vol. 5. Plenum Press, New York, pp. 201–264.

Dawes, P. R., and Soper, N. J., 1973, Pre-Quaternary history of North Greenland. *Am. Assoc. Pet. Geol. Mem.* **19**, 117–134.

Dawley, R. M., Zawiskie, J. M., and Cosgriff, J. W., 1979, A ravisuchid thecodont from the Upper Triassic Popo Agie Formation in Wyoming. *J. Paleontol.* **53**(6), 1428–1430.

DeBoer, J., 1967, Paleomagnetic–tectonic study of Mesozoic dike swarms in the Appalachians. *J. Geophys. Res.* **72**, 2237–2250.

DeBoer, J., and Snider, F. G., 1979, Magnetic and chemical variations of Mesozoic diabase dikes from eastern North America: Evidence for a hotspot in the Carolinas. *Geol. Soc. Am. Bull.* **90**, 185–198.

Deevy, E. S. J., 1965, Pleistocene non-marine environments. *In* Wright, H. E., Jr., and Frey, D. G. (eds.), *The Quaternary of the United States*. Princeton University Press, Princeton, N.J., pp. 643–652.

DeGraw, H. M., 1975, The Pierre–Niobrara unconformity in western Kansas. *Geol. Assoc. Can. Spec. Pap.* **13**, 589–606.

DeLong, S. E., Fox, P. J., and McDowell, F. W., 1978, Subduction of the Kula Ridge at the Aleutian Trench. *Geol. Soc. Am. Bull.* **89**, 83–95.

Demenitskaya, R. M., and Karasik, A. M., 1969, The active rift system of the Arctic Ocean. *Tectonophysics* **8**, 345–351.

Dennison, J. M., 1971, Petroleum related to Middle and Upper Devonian deltaic facies in the Central Appalachians. *Am. Assoc. Pet. Geol. Bull.* **55**, 1179–1193.

Dennison, J. M., 1970, Silurian stratigraphy and sedimentary tectonics of southern West Virginia and adjacent Virginia. *In* Silurian stratigraphy, Central Appalachian Basin. Field Conf., Appalachian Geol. Soc., April 17–18, pp. 2–33.

Dennison, J. M., 1976, Gravity tectonic removal of cover of Blue Ridge anticlinorium to form Valley and Ridge province. *Geol. Soc. Am. Bull.* **87**, 1470–1476.

Dennison, J. M., and Head, J. W., 1975, Sea-level variations interpreted from the Appalachian basin Silurian and Devonian. *Am. J. Sci.* **275**, 1089–1120.

Dennison, J. M., and Textoris, D. A., 1966, Stratigraphy and petrology of Devonian Tioga ash fall in northeastern United States [abstr.]. Geol. Soc. Am. Progr. 1966 Annu. Meet., p. 52.

Dennison, J. M., and Wheeler, W. H., 1975, Stratigraphy of Precambrian through Cretaceous strata of probable fluvial origin in southeastern United States and their potential as uranium host rocks. *Southeast. Geol. Spec. Publ.* **5**.

de Ricqles, A. J., 1968, Récherches paléohistologiques sur les os longs de tetrapodes. I. Origine du tissue osseux plexiforme des dinosaurien sauropodes. *Annu. Paleontol. Vertebr.* **54**, 133–145.

Desmond, A. T., 1977, *The Hot–Blooded Dinosaurs*. Dial, New York.

Detterman, R. L., 1973, Mesozoic sequence in Arctic Alaska. *Am. Assoc. Pet. Geol. Mem.* **19**, 376–387.

Dewey, J. F., 1969, Evolution of the Appalachian/Caledonian orogen. *Nature* **22**, 124–129.

Dewey, J. F., and Bird, J. M., 1970, Mountain belts and the new global tectonics. *J. Geophys. Res.* **75**, 2625–2647.

Dewey, J. F., and Burke, K., 1973, Tibetan, Variscan, and Precambrian basement reactivation: products of continental collision. *J. Geol.* **81**, 683–692.

Dewey, J. F., and Kidd, W. S. F., 1974, Continental collisions in the Appalachian–Caledonian orogenic belt: Variations related to complete and incomplete suturing. *Geology* **2**, 543–546.

Dickinson, W. R. (ed.), 1974a, Tectonics and sedimentation. *Soc. Econ. Paleontol. Mineral. Spec. Publ.* **22**.

Dickinson, W. R., 1974b, Plate tectonics and sedimentation. *Soc. Econ. Paleontol. Mineral. Spec. Publ.* **22**, 1–27.

Dickinson, W. R., 1976, Sedimentary basins developed during evolution of Mesozoic–Cenozoic arc–trench system in western North America. *Can. J. Earth Sci.* **13**, 1268–1287.

Dickinson, W. R., 1977, Paleozoic plate tectonics and the evolution of the Cordilleran continental margin. *In* Stewart, J. H., Stevens, C. H., and Fritsche, A. E. (eds.), Paleozoic paleogeography of the western United States. SEPM, Pacific Section, Los Angeles, pp. 137–152.

Dickinson, W. R., 1979, Cenozoic plate tectonic setting of the Cordilleran region in the United States. *In* Armentrout, J. M., Cole, M. R., and TerBest, H., Jr. (eds.), Cenozoic paleogeography of the western United States. SEPM, Pacific Section, Pacific Coast Paleogeography Symp. No. 3, pp. 1–13.

Dickinson, W. R., and Coney, P. J., 1980, Plate tectonic constraints on the origin of the Gulf of Mexico. *In* Pilger, R. H., Jr. (ed.), The origin of the Gulf of Mexico and the early opening of the central North Atlantic Ocean. Proc. Symp., LSU, Baton Rouge, pp. 27–36.

Dickinson, W. R., and Snyder, W. S., 1978, Plate tectonics of the Laramide orogeny. *Geol. Soc. Am. Mem.* **151**, 355–366.

Dietz, R. S., 1972, Geosynclines, mountains, and continent building. *Sci. Am.* **226**, 30–38.

Dietz, R. S., 1964, Sudbury structure as an astrobleme. *J. Geol.* **72**, 412–434.

Dietz, R. S., and Holden, J. C., 1970, Reconstruction of Pangaea: Breakup and dispersion of continents, Permian to present. *J. Geophys. Res.* **75**(26), 4939–4955.

Dietz, R. S., Holden, J. C., and Sproll, W. P., 1970, Geotectonic evolution and subsidence of Bahama Platform. *Geol. Soc. Am. Bull.* **81**, 1915–1928.

Diffendal, R. F., Jr., 1982, Regional implications of the geology of the Ogallala Group, Upper Tertiary of southwestern Morrill County, Nebraska, and adjacent areas. *Geol. Soc. Am. Bull.* **93**, 964–976.

Dillon, W. P., Paull, C. K., Buffler, R. T., and Fail, J. P., 1979, Structure and development of the northern Blake Plateau: Preliminary analysis. *Am. Assoc. Pet. Geol. Mem.* **29**, 27–41.

Dimroth, E., 1972, The Labrador geosyncline revisited. *Am. J. Sci.* **272**, 487–506.

Dimroth, E., 1981, Labrador geosyncline: Type example of early Proterozoic cratonic reactivation. *In* Kröner, A. (ed.), *Precambrian Plate Tectonics.* Elsevier, Amsterdam, pp. 331–352.

Dimroth, E., and Kimberley, M. M., 1976, Precambrian atmospheric oxygen: Evidence in the sedimentary distributions of carbon, sulfur, uranium, and iron. *Can. J. Earth Sci.* **13**, 1161–1185.

Dinneeen, G. V., and Cook, G. L., 1974, Oil shale and the energy crisis. *Technol. Rev.* **76**(3), 26–22.

Dixon, G. H., 1967, Permian System in Northeastern New Mexico and Texas–Oklahoma panhandles. *U.S. Geol. Surv. Prof. Pap.* **515**, 65–84.

Dodd, J. R., 1964, Environmentally controlled variations in the shell structure of a pelecypod species. *J. Paleontol.* **38**(6), 1065–1071.

Dodson, P., Behrensmeyer, A. K., Bakker, R. T., and McIntosh, J. S., 1980, Taphonomy and paleoecology of the dinosaur beds of the Jurassic Morrison Formation. *Paleobiology* **6**(2), 208–232.

Doe, T. W., and Dott, R. H., Jr., 1980, Genetic significance of crossbedding with examples from the Navajo and Weber sandstones of Utah. *J. Sediment. Petrol.* **50**(3), 793–812.

Donaldson, A. C., 1974, Pennsylvania sedimentation of central Appalachians. *Geol. Soc. Am. Spec. Pap.* **148**, 47–78.

Donaldson, J. A., 1970, Labrador subprovince. *Geol. Surv. Can. Econ. Geol. Rep.* No. 1, pp. 101–104.

Donnell, J. R., 1961, Tertiary geology and oil shale resources of Piceance Creek Basin, between the Colorado and White Rivers, northwestern Colorado. *U.S. Geol. Surv. Bull.* **1082-L**, 842–877.

Dorf, E., 1964, The petrified forests of Yellowstone Park. *Sci. Am.* **210**(4), 106–114.

Dott, R. H., Jr., and Batten, R. L., 1981, *Evolution of the Earth*, 3rd ed. McGraw–Hill, New York.

Dott, R. H., Jr., and Roshardt, M. A., 1972, Analysis of cross–stratification orientation in the St. Peter Sandstone in southwestern Wisconson. *Geol. Soc. Am. Bull.* **83**, 2589–2596.

Dott, R. H., Jr., and Shaver, R. H. (eds.), 1974, Modern and ancient geosynclinal sedimentation. *Soc. Econ. Paleontol. Mineral. Spec. Publ.* **19**.

Douglas, R. J. W. (ed.), 1976, Geology and economic minerals of Canada. *Geol. Surv. Can. Econ. Geol. Rep.* No. 1 (3 volumes).

Douglas, R. J. W., *et al.*, 1976, Geology of western Canada. *Geol. Surv. Can. Econ. Geol. Rep.* No. 1, pp. 366–488.

Doyle, J. A., 1977, Patterns of evolution in early angiosperms. *In* Hallam, A. (ed.), *Patterns of Evolution as Illustrated by the Fossil Record.* Elsevier, Amsterdam, pp. 501–566.

Droste, J. B., and Shaver, R. H., 1977, Synchronization of deposition: Silurian reef–bearing rocks on Wabash Platform, with cyclic evaporites of Michigan basin. *In* Fisher, J. H. (ed.), *Reefs and evaporites—Concepts and depositional models.* AAPG, Studies in Geology No. 5, pp. 93–109.

Droste, J. B., Shaver, R. H., and Lazor, J. D., 1975, Middle Devonian paleogeography of the Wabash platform, Indiana, Illinois, and Ohio. *Geology* **3**, 269–272.

Drummond, J. M., 1979, Factors influencing the distribution and facies of Cenozoic marine sediments along the west coast of British Columbia, Canada. *In* Armentrout, J. M., Cole, M. R., and TerBest, H., Jr. (eds.), Cenozoic paleogeography of the western United States. SEPM, Pacific Section, Pacific Coast Paleogeography Symp. No. 3, pp. 53–61.

Drummond, K. J., 1974, Paleozoic Arctic margin of North America. *In* Burk, C. A., and Drake, C. L. (eds.), *The Geology of Continental Margins.* Springer-Verlag, Berlin, pp. 797–810.

Drummond, K. J., 1974, Paleozoic Arctic margin of North America. *In* Burk, C. A., and Drake, C. L. (eds.), *The Geology of Continental Margins.* Springer-Verlag, Berlin, pp. 797–810.

Drury, S. A., 1978, Basic factors in Archean geotectonics. *In* Windley, B. F., and Naqri, S. M. (eds.), *Archean Geochemistry.* Elsevier, Amsterdam, pp. 3–23.

Dubar, J. R., and Taylor, D. S., 1962, Paleoecology of the Choctawhatchee deposits, Jackson Bluff, Florida. *Gulf Coast Assoc. Geol. Soc. Trans.* **12**, 349–376.

Duff, P. M., and Walton, E. K., 1962, Statistical basis for cyclothems: A quantitative study of the sedimentary succession in the East Pennine coalfield. *Sedimentology* **1**, 235–255.

Duff, P. M., Hallam, A., and Walton, E. K., 1967, *Cyclic Sedimentation.* Elsevier, Amsterdam.

Dumond, D. E., 1980, The archaeology of Alaska and peopling of America. *Science* **209**, 984–991.

Dumple, E. T., 1920, The geology of east Texas. *Univ. Texas, Austin, Bull.* **1869** (Dec. 10, 1918).

Dunbar, C. O. and Rodgers, J., 1957, *Principles of Stratigraphy.* Wiley, New York.

Dunham, J. B., and Olson, E. R., 1978, Diagenetic dolomite formation related to Paleozoic paleogeography of the Cordilleran miogeocline in Nevada. *Geology* **6**, 556–559.

Durham, C. O. Jr., 1957, The Austin Group in Central Texas. Ph.D. dissertation, Columbia University, Department of Geology.

Durham, J. W., and Caster, K. E., 1963, Helicoplacoidea: A new class of echinoderms. *Science* **140**, 820–822.

Durham, J. W., and Melville, R. C., 1957, A classification of echinoids. *J. Paleontol.* **31**, 242–272.

Dutch, S. I., 1983, Proterozoic structural provinces in the north–central United States. *Geology* **11**, 478–481.

Dutro, J. T., Jr., 1981, Geology of Alaska bordering the Arctic Ocean. *In* Nairn, A. E. M., Churkin, M., Jr., and Stehli, F. G. (eds.), *The Ocean Basins and Margins*, Vol. 5. Plenum, New York, pp. 21–36.

Eardley, A. J., 1951, *Structural Geology of North America.* Harper & Row, New York.

Eardley, A. J., 1962, *Structure of North America*, 2nd ed. Harper, New York.

Easterbrook, D. J., 1976, Quaternary geology of the Pacific northwest. *In* Mahaney, W. C. (ed.), *Quaternary Stratigraphy of North America.* Dowden, Hutchinson & Ross, Stroudsburg, Pa., pp. 441–463.

Eaton, G. P., 1980, Geophysical and geological characteristics of the crust of the Basin and Range Province. *In Continental Tectonics, Studies in Geophysics*, National Academy of Sciences, pp. 96–113.

Edmunds, W. E., *et al.*, 1979, Pennsylvania and New York. *U.S. Geol. Surv. Prof. Pap.* **1110–B**.

Edwards, M. B., 1981, Upper Wilcox Rosita delta system of South Texas: Growth–faulted shelf-edge deltas. *Am. Assoc. Pet. Geol. Bull.* **65**, 54–73.

Eisbacher, G. H., 1974, Evolution of successor basins in the Canadian Cordillera. *Soc. Econ. Paleontol. Mineral. Spec. Publ.* **19**, 274–291.

Eisbacher, G. H., Carrigy, M. A., and Campbell, R. B., 1974, Paleodrainage pattern and late–orogenic basins of the Canadian Cordillera. *Soc. Econ. Paleontol. Mineral. Spec. Publ.* **22**, 143–166.

Eldredge, N., 1974, Stability, diversity, and speciation in Paleozoic epeiric seas. *J. Paleontol.* **48**(3), 541–548.

Eldredge, N., 1977, Trilobites and evolutionary patterns. *In* Hallam, A. (ed.), *Patterns of Evolution as Illustrated by the Fossil Record.* Elsevier, Amsterdam, pp. 305–332.

Eldredge, N. and Gould, S. J., 1972, Punctuated equilibria: An alternative to phyletic gradualism. *In* Schopf, T. J. M. (ed.) *Models in Paleobiology.* Freeman, Cooper & Co., San Francisco, pp. 82–115.

Ells, G. D., 1979, The Mississippian and Pennsylvanian (Carboniferous) System in the United States—Michigan. *U.S. Geol. Surv. Prof. Pap.* **1110–J**.

Elston, D. P., and McKee, E. H., 1982, Age and correlation of the late Proterozoic Grand Canyon disturbance, northern Arizona. *Geol. Soc. Am. Bull.* **93**, 681–699.

Emiliani, C., Kraus, E. B., and Shoemaker, E. M., 1981, Sudden death at the end of the Mesozoic. *Earth Planet. Sci. Lett.* **55**, 317–334.

Engel, A. E. J. and Kelm, D. L., 1972, Pre–Permian global tectonics: A tectonic test. *Geol. Soc. Am. Bull.* **83**, 2325–2340.

Engel, A. E. J., *et al.*, 1968, Alga–like forms in Onverwacht Series, South Africa: Oldest recognized lifelike forms on Earth. *Science* **161**, 1005–1008.

England, J., and Bradley, R. S., 1978, Past glacial activity in the Canadian High Arctic. *Science* **200**, 265–269.

Englund, K. J., 1974, Sandstone distribution patterns in the Pocahontas Formation of southwest Virginia and southern West Virginia. *Geol. Soc. Am. Spec. Pap.* **148**, 31–45.

Enos, P., 1977, Tamabra limestone of the Poza Rica Trend, Cretaceous, Mexico. *Soc. Econ. Paleontol. Mineral. Spec. Publ.* **23**, 273–314.

Erben, H. K., Hoefe, J., and Wedepohl, K. H., 1979, Paleobiological and isotopic studies of eggshells from a declining dinosaur. *Paleobiology,* **5**(4), 380–414.

Erickson, B. R., 1985, Aspects of some anatomical structures of *Champsosaurus* (Reptilia, Eosuchia). *J. Vertebr. Paleontol.* **5**(2), 111–127.

Eriksson, K. A., and Truswell, J. F., 1978, Geological processes and atmospheric evolution in the Precambrian. *In* Tarling, D. H. (ed.), *Evolution of the Earth's Crust.* Academic Press, New York, pp. 219–238.

Ermanovics, I. F., and Davison, W. L., 1976, The Pikwitonei granulites in relation to the northwestern Superior Province of the Canadian Shield. *In* Windley, B. F. (ed.), *The Early History of the Earth.* Wiley, New York, pp. 331–347.

Ervin, C. P., and McGinnis, L. D., 1975, Reelfoot Rift: Reactivated precursor to the Mississippi Embayment. *Geol. Soc. Am. Bull.* **86**, 1287–1295.

Erxleben, A. W., 1973, Depositional systems in the Canyon Group of north–central Texas. *In* Brown, L. F., Jr., Cleaves, A. W., II, and Erxleben, A. W., Pennsylvanian depositional systems in north–central Texas; A guide for interpreting terrigenous clastic facies in a cratonic basin. Univ. Tex., Austin, Bur. Econ. Geol. Guidebook **14**, 43–56.

Erxleben, A. W., 1975, Depositional systems in the Canyon Group (Pennsylvanian System), north–central Texas. Univ. Tex., Austin, Bur. Econ. Geol. Rep. Invest. **82**.

Escher, A., Escher, J. C., and Watterson, J., 1975, The reorientation of the Kangamiut dike swarm, West Greenland. *Can. J. Earth Sci.* **12**, 158–173.

Escher, A., Jack, S., and Watterson, J., 1976, Tectonics of the North Atlantic dike swarm. *Philos. Trans. R. Soc. London Ser. A* **280**, 529–539.

Estes, R., 1964, Fossil vertebrates from the Late Cretaceous Lance Formation, eastern Wyoming. *University of California Publ. Geol. Sci.* **49**.

Eugster, H. P., and Hardie, L. A., 1975, Sedimentation in an ancient playa–lake complex: The Wilkins Peak Member of the Green River Formation of Wyoming. *Geol. Soc. Am. Bull.* **86**, 319–334.

Evernden, J. F., and Kistler, R. W., 1970, Chronology of emplacement of Mesozoic batholith complexes in California and western Nevada. *U.S. Geol. Surv. Prof. Pap.* **216**.

Evitt, W. R., and Pierce, S. T., 1975, Early Tertiary ages from the coastal belt of the Franciscan Complex, northern California. *Geology* **3**, 433–436.

Ewing, M., 1971, Late Cenozoic history of the Atlantic basin. *In* Turekian, K. K. (ed.), *Late Cenozoic Glacial Ages.* Yale University Press, New Haven, Conn., pp. 565–573.

Ewing, M., and Donn, W. L., 1956, A theory of ice ages. *Science* **123**, 1061–1066.

Faill, R. J., 1973, Tectonic development of the Triassic Newark–Gettysburg Basin in Pennsylvania. *Geol. Soc. Am. Bull.* **84**, 725–740.

Fanshawe, J. R., 1939, Structural geology of Wind River Canyon area, Wyoming. *Am. Assoc. Pet. Geol. Bull.* **23**, 1439–1492.

Farlow, J. O., 1980, Predator/prey biomass ratios, community food webs and dinosaur physiology. *In* Thomas, R. D. K., and Olson, E. C. (eds.), A cold look at the warm–blooded dinosaurs. AAAS Symp. 28, pp. 55–84.

Fassett, Y. E., and Hinds, J. S., 1971, Geology and fuel resources of the Fruitland Formation and the Kirtland Shale of the San Juan Basin, New Mexico and Colorado. *U.S. Geol. Surv. Prof. Pap.* **676**.

Fay, R. O., *et al.*, 1979, The Mississippian and Pennsylvanian Systems in the United States—Oklahoma. *U.S. Geol. Surv. Prof. Pap.* **1110–R**.

Fedonkin, M. A., 1981, White Sea biota of the Vendian (Precambrian, non–skeletal fauna of the Russian Platform North). *USSR Acad. Sci. Geol. Inst. Proc.* **342**, 1–100.

Feduccia, A., and Tordoff, H. B., 1979, Feathers of *Archaeopteryx:* Asymmetric veins indicate aerodynamic function. *Science* **203**, 1021–1022.

Feldman, R. M., and Palubniak, D. S., 1975, Paleoecology of Maestrichtian oyster assemblage in the Fox Hills Formation. *Geol. Assoc. Can. Spec. Pap.* **13**, 221–234.

Ferguson, H. G., Miller, S. W., and Roberts, R. J., 1951, Geology of the Winnemucca quadrangle, Nevada. U.S. Geol. Surv. Quad. Map (GQ–12), scale 1:125,000.

Ferm, J. C., 1974, Carboniferous environmental models in eastern United States and their significance. *Geol. Soc. Am. Spec. Pap.* **148**, 79–95.

Ferm, J. C., and Cavaroc, V. V., Jr., 1968, A nonmarine sedimentary model for the Allegheny rocks of West Virginia. *Geol. Soc. Am. Spec. Pap.* **106**, 1–19.

Ferm, J. C., Milici, R. C., and Eason, J. E., 1972, Carboniferous depositional environments in the Cumberland Plateau of southern Tennessee and northern Alabama. *Tenn. Div. Geol. Rep. Invest.* **33**.

Ferrigno, K. F., and Walker, K. R., 1973, Stop 1: Holston Formation and Chapman Ridge Formation; field trips 1 and 2: Stratigraphy and depositional environments in the Valley and Ridge at Knoxville. *Tenn. Div. Geol. Bull.* **70**, 117–122.

Finks, R. M., and Toomey, D. F., 1969, The paleoecology of Chazyan (Lower Middle Ordovician) "reefs" or "mounds." *In* Guidebook, 41st Annu. Meet. N.Y. State Geol. Soc., pp. 93–120.

Fisher, D. C., and Nitecki, M. H., 1982, Standardization of the Anatomical orientation of receptaculitids. *Paleontol. Soc. Mem.* **13**.

Fisher, G. W., Pettijohn, F. J., Reed, J. C., Jr., and Weaver, K. N., (eds.), 1970, *Studies in Appalachian Geology: Central and Southern.* Wiley–Interscience, New York.

Fisher, J., 1967, Fossil birds and their adaptive radiation. *In* Harland, W. B., *et al.* (eds.), *The Fossil Record.* Geological Society of London, pp. 133–154.

Fisher, J. H. (ed.), 1977, Reefs and evaporites—Concepts and depositional models. AAPG, Studies in Geology No. 5.

Fisher, W. L., 1969, Facies characterization of Gulf Coast basin delta systems, with some Holocene analogies. *Gulf Coast Assoc. Geol. Soc. Trans.* **19**, 239–261.

Fisher, W. L., and McGowen, J. H., 1969, Depositional systems in Wilcox Group (Eocene) of Texas and their relation to occurrence of oil and gas. *Am. Assoc. Pet. Geol. Bull.* **53**, 30–54.

Fisher, W. L., and Rodda, P. U., 1969, Edwards Formation (Lower Cretaceous), Texas: Dolomitization in a carbonate platform system. *Am. Assoc. Pet. Geol. Bull.* **53**, 55–72.

Fisk, H. N., 1944, Geological investigations of the alluvial valley of the lower Mississippi River. Corps of Engineers, Mississippi River Commission.

Flawn, P. T., and Muehlberger, W. R., 1970, The Precambrian of the United States of America: South–central United States. *In* Rankama, K. (ed.), *The Precambrian*, Vol. 4. Wiley–Interscience, New York, pp. 73–143.

Flessa, K., and Imbrie, J., 1973, Evolutionary pulsations: Evidence from Phanerozoic diversity patterns. *In* Tarling, D. H., and Runcorn, K. (eds.), *Implications of Continental Drift to the Earth Sciences.* Academic Press, New York, pp. 247–285.

Flint, R. F., 1940, Pleistocene features of the Atlantic Coastal Plain. *Am. J. Sci.* **238**, 757–787.

Flint, R. F., 1947, *Glacial Geology and the Pleistocene Epoch.* Wiley, New York.

Flint, R. F., 1971, *Glacial and Quaternary Geology.* Wiley, New York.

Folk, R. L., and McBride, E. F., 1976, The Caballos Novaculite revisited, (I): Origin of novaculite members. *J. Sediment. Petrol.* **46**, 659–669.

Fontaine, W. M., and White, I. C., 1880, The Permian or Upper Carboniferous flora of West Virginia and southwestern Pennsylvania. 2nd Geol. Surv. Pa., Report of Progress.

Ford, T. D., 1979, The history of the study of the Precambrian rocks of Charnwood Forest, England. *In* Kupsch, W. O., and Sarjeant, W. A. S. (eds.), *History of Concepts in Precambrian Geology*, Paper 19, pp. 65–80.

Forman, M. J., and Schlanger, S. O., 1957, Tertiary reef and associated limestone facies from Louisiana and Guam. *J. Geol.* **65**, 611–627.

Fortier, Y. O., 1957, The Arctic Archipelago. *In* Geology and economy of Canada. *Geol. Surv. Can.*, Econ. Ser. No. 1, 4th ed., pp. 393–442.

Fouch, T. D., 1979, Character and paleographic distribution of Upper Cretaceous (?) and Paleogene nonmarine sedimentary rocks in east-central Nevada. *In* Armentrout, J. M., Cole, M. R., and TerBest, H., Jr. (eds.), Cenozoic paleogeography of the western United States. SEPM, Pacific Section Pacific Coast Paleogeography Symp. No. 3, pp. 97–111.

Fox, S. W., and Dose, K., 1977, *Molecular Evolution and the Origin of Life.* Dekker, New York.

Franks, P. C., 1975, The transgressive–regressive sequence of the Cretaceous Cheyenne, Kiowa and Dakota Formations of Kansas. *Geol. Assoc. Can. Spec. Pap.* **13**, 469–522.

Franks, P. C., 1980, Models of marine transgressions—Example from Lower Cretaceous fluvial and paralic deposits, north–central Kansas. *Geology* **8**, 56–61.

Fraser, G. S., 1976, Sedimentology of a Middle Ordovician quartz arenite–carbonate transition in the upper Mississippi Valley. *Geol. Soc. Am. Bull.* **86**, 833–845.

Fraser, J. A., Hoffman, P. F., Irvine, T. N., and Mursky, G., 1972, The Bear Province. *In* Price, R. A., and Douglas, R. J. W. (eds.), Variations in Tectonic Style in Canada. *Geol. Assoc. Can. Spec. Pap.* **11**, 454–503.

Frazier, W. J., 1975, Celestite in the Mississippian Pennington Formation, central Tennessee. *Southeast. Geol.* **16**, 241–248.

Frazier, W. J., 1982, Sedimentology and paleoenvironmental analysis of Upper Cretaceous Tuscaloosa and Eutaw Formations in western Georgia. *In* Arden, D. D., Beck, B. B., and Morrow, E. (eds.), 2nd Symposium on the Geology of the Southeastern Coastal Plain, Ga. Geol. Surv. Info. Circ. 53, pp. 39–52.

Frazier, W. J., and Taylor, R. S., 1980, Facies changes and paleogeographic interpretations of the Eutaw Formation (Upper Cretaceous) from western Georgia to central Alabama. *In* Tull, J. F. (ed.), Field trips for the southeastern section. *Geol. Soc. Am.*, Birmingham, Ala., pp. 1–27.

Frederiksen, N. O., 1980, Mid–Tertiary climate of southeastern United States: The sporomorph evidence. *J. Paleontol.* **54**, 728–239.

Freeman, W. E., and Visher, G. S., 1975, Stratigraphic analysis of the Navajo Sandstone. *J. Sediment. Petrol.* **45**(3), 651–668.

French, B. M., 1970, Possible relations between meteorite impact and igneous petrogenesis as indicated by the Sudbury structure, Ontario, Canada. *Bull. Volcanol.* **34**, 466–517.

Frezon, S. E., and Dixon, G. H., 1975, Pennsylvanian System in Texas panhandle and Oklahoma. *U.S. Geol. Surv. Prof. Pap.* **853**, 177–195.

Friedman, G. M., and Sanders, J. E., 1978, *Principles of Sedimentology.* Wiley, New York.

Fritz, W. H., 1971, Geological setting of the Burgess Shale. Proceedings of the North American Paleontological Convention, (I), pp. 1155–1170.

Fritz, W. J., 1980, Reinterpretation of the depositional environment of the Yellowstone "fossil forests." *Geology* **8**, 309–313.

Frost, S. H., 1977, Cenozoic reef systems of Caribbean—Prospects for paleoecologic synthesis. AAPG, Studies in Geology No. 4, pp. 93–110.

Frye, J. C., Willman, H. R., and Black, R. F., 1965, Outline of glacial geology of Illinois and Wisconsin. *In* Wright, H. E., Jr., and Frey, D. G. (eds.), *The Quaternary of the United States.* Princeton University Press, Princeton, N.J., pp. 43–61.

Gabrielse, H., 1966, Tectonic evolution of the northern Canadian Cordillera. *Can. J. Earth Sci.* **4**, 271–298.

Gabrielse, H., 1972, Younger Precambrian of the Canadian Cordillera. *Am. J. Sci.* **272**, 521–536.

Gadd, N. R., 1976, Quaternary stratigraphy in southern Quebec. *In* Mahaney, W. C. (ed.), *Quaternary Stratigraphy of North America.* Dowden, Hutchinson & Ross, Stroudsburg, Pa., pp. 133–158.

Gaffney, E. S., 1975, A revision of the side–necked turtle *Taphrosphys sulcatus* (Leidy) from the Cretaceous of New Jersey. *Am. Mus. Novit.* No. 2571.

Galloway, W. E., 1974, Deposition and diagenetic alteration of sandstone in northeast Pacific arc–related basins: Implications for graywacke genesis. *Geol. Soc. Am. Bull.* **85**, 319–390.

Galloway, W. E., Hobday, D. K., and Magara, K., 1982, Frio Formation of Texas Gulf Coastal Plain: Depositional systems, structural framework, and hydrocarbon distribution. *Am. Assoc. Pet. Geol. Bull.* **66**, 649–688.

Ganapathy, R., 1980, A major meteorite impact on the Earth 65 million-years–ago: Evidence from the Cretaceous–Tertiary boundary clay. *Science* **209**, 921–923.

Ganapathy, R., 1982, Evidence for a major meteorite impact on the Earth 34 million years ago: Implication for Eocene extinctions. *Science* **216**, 885–886.

Gartner, S., and McGuirk, J. P., 1979, Terminal Cretaceous extinction: Scenario for a catastrophe. *Science* **206**, 1272–1276.

Gelbaum, C., and Howell, J., 1982, The geohydrology of the Gulf trough. *Ga. Geol. Surv. Inf. Circ.* **53**, 140–153.

Gibb, R. A., 1983, Model for suturing of Superior and Churchill plates: An example of double indentation tectonics. *Geology* **11**, 413–417.

Gibb, R. A., and Walcott, R. J., 1971, A Precambrian suture in the Canadian Shield. *Earth Planet. Sci. Lett.* **10**, 417–422.

Gibbons, A. B., Megeath, J. D., and Pierce, K. L., 1984, Probability of moraine survival in a succession of glacial advances. *Geology* **12**, 327–330.

Gibson, G. G., Tetter, S. A., and Fedonkin, M. A., 1984, Ediacarian fossils from the Carolina Slate Belt, Stanly County, N.C. *Geology* **12**, 387–390.

Gibson, T. G., 1962, Benthonic foraminifera and paleoecology of the Miocene deposits of the Middle Atlantic Coastal Plain. Unpublished Ph.D. dissertation, Princeton University.

Gibson, T. G., 1981, Facies changes of lower Paleogene strata. *In* Reinhardt, J., and Gibson, T. G. (eds.), Upper Cretaceous and Lower Tertiary of the Chattahoochee River Valley, western Georgia and eastern Alabama. Ga. Geol. Soc., 16th Annu. Field Trip, Guidebook. pp. 19–28.

Gibson, T. G., 1982, Paleocene to middle Eocene depositional cycles in eastern Alabama and western Georgia. *Ga. Geol. Surv. Inf. Circ.* **53**, 53–63.

Gill, J. R., and Cobban, W. A., 1966, The Red Bird Section of the Upper Cretaceous Pierre Shale in Wyoming. *U.S. Geol. Surv. Prof. Pap.* **393A**.

Gill, J. R., and Cobban, W. A., 1973, Stratigraphy and geologic history of the Montana Group and equivalent rocks, Montana, Wyoming, and North and South Dakota. *U.S. Geol. Surv. Prof. Pap.* **776**.

Gill, J. R., Merewether, E. A., and Cobban, W. A., 1970, Stratigraphy and nomenclature of some Upper Cretaceous and Lower Tertiary rocks in southcentral Wyoming. *U.S. Geol. Surv. Prof. Pap.* **667**.

Gingerich, P. D., 1977, Patterns of evolution in the mammalian fossil record. *In* Hallam, A. (ed.) *Patterns of Evolution as Illustrated by the Fossil Record.* Elsevier, Amsterdam, pp. 469–500.

Gingerich, P. D., 1979, Phylogeny of middle Eocene Adapidae (Mammalia, Primates) in North America: *Smilodectes* and *Notharctus. J. Paleontol.* **53**, 153–163.

Gingerich, P. D., 1980a, Evolutionary patterns in early Cenozoic mammals. *Annu. Rev. Earth Planet. Sci.* **8**, 407–424.

Gingerich, P. D., 1980b, *Tytthaena parrisi*, oldest–known oxhyaenid (Mammalia, Creodonta) from the late Paleocene of western North America. *J. Paleontol.* **54**(3), 570–576.

Gingerich, P. D., Wells, N. A., Russell, D. A., and Shah, S. M. I., 1983, Origin of whales in epicontinental remnant seas: New evidence from the early Eocene of Pakistan. *Science* **220**, 403–406.

Glaessner, M. F., 1961, Precambrian animals. *Sci. Am.* **204**(3), 72.

Glaessner, M. F., 1969, Arthropoda. *In* Moore, R. C. (ed.), *Treatise on Invertebrate Paleontology,* Part R 4. Geological Society of America and University of Kansas Press, Lawrence.

Glaessner, M. F., 1971, Geographic distribution and time range of the Ediacara fauna. *Geol. Soc. Am. Bull.* **82**, 509.

Glaser, J. D., 1968, Coastal plain geology of southern Maryland. Md. Geol. Surv. Guideb. No. 1.

Glaser, J. D., 1968a, Provenance, dispersal, and depositional environment of Triassic sediments in Newark–Gettysburg basin. *Pa. Geol. Surv.,* 4th Ser., Bull. 43.

Glibson, A. Y., 1976, Stratigraphy and evolution of primary and secondary greenstones: Significance of data from Shields of the southern hemisphere. In Windley, B. F., (ed.), 1984, *The Evolving Continents,* 2nd edition. Wiley, New York, pp. 257–277.

Glick, E. E., 1975, Arkansas and northern Louisiana. *U.S. Geol. Surv. Prof. Pap.* **853**, 157–175.

Glick, E. E., 1979, Arkansas. *U.S. Geol. Surv. Prof. Pap.* **1010**, 125–145.

Glikson, A. Y., 1972, Early Precambrian evidence of a primitive ocean crust and island nuclei of sodic granite. *Geol. Soc. Am. Bull.* **83**, 3323–3344.

Glover, L., III, and Sinha, A. K., 1973, The Virgilian deformation, a late Precambrian to Early Cambrian(?) orogenic event in the central Piedmont of Virginia and North Carolina. *Am. J. Sci.* **273–A**, 234–251.

Glut, D. F., 1982, *The New Dinosaur Dictionary.* Citadel Press, Secaucus, N.J.

Gohn, G. S., *et al.,* 1982, A stratigraphic framework for Cretaceous and Paleogene margins along the South Carolina and Georgia coastal sediments. *Ga. Geol. Surv. Inf. Circ.* **53**, 64–74.

Goodwin, A. M., 1956, Facies relations in the Gunflint Iron Formation. *Econ. Geol.* **51**, 565–595.

Goodwin, A. M., 1968, Archean protocontinental growth and early crustal history of the Canadian Shield. Proc. 23rd Int. Geol. Congr., Prague, **1**, 69–89.

Goodwin, A. M., 1973, Archean iron–formations and tectonic basins of the Canadian Shield. *Econ. Geol.* **68**, 915–933.

Goodwin, A. M., 1978, The nature of Archean crust in the Canadian Shield. *In* Tarling, D. H. (ed.), *Evolution of the Earth's Crust.* Academic Press, New York, pp. 175–218.

Goodwin, A. M., and Ridler, R. H., 1970, The Abitibi orogenic belt. *Geol. Surv. Can. Pap.* 70-40, pp. 1–24.

Gordon, M., Jr., and Stone, C. G., 1973, Correlation of Carboniferous rocks of the Ouachita geosyncline with those of the adjacent shelf. *Geol. Soc. Am. Abstr. Progr.* **5**, 259.

Gordon, M., Jr., and Stone, C. G., 1977, Correlation of the Carboniferous rocks of the Ouachita trough with those of the adjacent foreland. *In* Stone, C. G. (ed.), Symposium on the geology of the Ouachita Mountains. Ark. Geol. Comm., Vol. 1, pp. 70–91.

Gould, S. J., and Calloway, C. B., 1980, Clams and brachiopods—Ships that pass in the night. *Paleobiology* **6**(4), 383–396.

Gould, S. J., and Katz, M., 1975, Disruption of ideal geometry in the growth of receptaculitids: A natural experiment in theoretical morphology. *Paleobiology* **1**(1), 1–20.

Graham, S. A., 1979, Tertiary paleotectonics and paleogeography of the Salinian block. *In* Armentrout, J. M., Cole, M. R., and TerBest, H., Jr. (eds.), Cenozoic paleogeography of the western United States. SEPM, Pacific Section, Pacific Coast Paleogeography Symp. No. 3, pp. 45–51.

Graham, S. A., and Berry, K. D., 1979, Early Eocene paleogeography of the central San Joaquin Valley. *In* Armentrout, J. M., Cole, M. R., and TerBest, H., Jr. (eds.), Cenozoic paleogeography of the western United States. SEPM, Pacific Section, Pacific Coast Paleogeography Symp. No. 3, pp. 119–127.

Graham, S. A., Dickinson, W. A., and Ingersoll, R. V., 1975, Himalayan–Bengal model for flysch dispersal in the Appalachian–Ouachita System. *Geol. Soc. Am. Bull.* **86**, 273–286.

Grantz, A., Eittreim, S., and Whitney, O. T., 1981, Geology and physiography of the continental margin north of Alaska and implications for the origin of the Canadian basin. *In* Nairn, A. E. M., Churkin,

Mr., Jr., and Stehli, F. G. (eds.), *The Ocean Basins and Margins,* Vol. 5. Plenum, New York, 439–492.

Grasso, T. X., 1973, Stratigraphy of the Genesee Gorge at Rochester. *In* Guidebook, 45th Annu. Meet., Rochester. N.Y. State Geol. Soc., pp. I–1–I–19.

Gray, H. H., 1979, The Mississippian and Pennsylvanian (Carboniferous) System in the United States—Indiana. *U.S. Geol. Surv. Prof. Pap.* **1110–K**.

Gray, J., and Boucot, A. J., 1978, The advent of land plant life. *Geology* **6**, 489–492.

Gray, J., Massa, D., and Boucot, A. J., 1982, Caradocian land plant micro–fossils from Libya. *Geology* **10**, 197–201.

Green, D. H., 1972, Magmatic activity as the major process in the chemical evolution of the Earth's crust and mantle. *Tectonophysics* **13**, 47–71.

Green, D. H., 1975, Genesis of Archean peridotitic magmas and constraints on Archean geothermal gradients and tectonics. *Geology* **3**, 15–18.

Green, D. H., 1981, Petrogenesis of Archean ultramafic magmas and implications for Archean tectonics. *In* Kröner, A. (ed.), *Precambrian Plate Tectonics.* Elsevier, Amsterdam, pp. 469–489.

Green, D. H., and Ringwood, A. E., 1963, Mineral assemblages in a model mantle composition. *J. Geophys. Res.* **68**, 937–945.

Green, D. H., and Ringwood, A. E., 1968, Genesis of the calc–alkaline igneous rock suite. *Contrib. Mineral. Petrol.* **18**, 105–162.

Greene, H. G., and Clark, J. C., 1979, Neogene paleogeography of the Monterey Bay area, California. *In* Armentrout, J. M., Cole, M. R., and TerBest, H., Jr. (eds.), Cenozoic paleogeography of the western United States. SEPM, Pacific Section, Pacific Coast Paleogeography Symp. No. 3, pp. 277–296.

Griffith, L. S., Pitcher, M. G., and Rice, G. W., 1969, Quantitative analysis of a Lower Cretaceous reef complex. *Soc. Econ. Paleontol. Mineral. Spec. Publ.* **14**, 120–138.

Grow, J. A. and Atwater, T., 1970, Mid–Tertiary tectonic transition in the Aleutian arc: *Geol. Soc. Am. Bull.,* **81**, 3715–3722.

Grow, J. A., Mattick, R. E., and Schlee, J. S., 1979, Multichannel seismic depth sections and interval velocities over outer continental shelf and upper continental slope between Cape Hatteras and Cape Cod. *Am. Assoc. Pet. Geol. Mem.* **29**, 27–41, 65–83.

Grow, J. A. *et al.,* 1983, Representative multichannel seismic reflection profiles over the U.S. Atlantic continental margin. *In* Bally, A. W. (ed.), Seismic expression of structural style. AAPG, Studies in Geology No. 15, **2**: 2.2.3–1–2.2.3–19.

Guthrie, H. D., Blum, E. H., and Braitsch, R. J., 1979, Commercialization strategy report for oil shale. U.S. Department of Energy, Report TID–28845.

Gwinn, V. E., 1964, Thin–skinned tectonics in the Plateau and northwestern Valley and Ridge provinces of the Central Appalachians. *Geol. Soc. Am. Bull.* **75**, 863–900.

Habicht, J. K. A., 1979, Paleoclimate, paleomagnetism, and continental drift. AAPG, *Studies in Geology* No. 9.

Hack, J. T., 1955, Geology of the Brandywine area and origin of the upland of southern Maryland. *U.S. Geol. Surv. Prof. Pap.* **267–A**.

Hadley, J. H., 1970, The Ocoee Series and its possible correlatives. *In* Fisher, G. W., Pettijohn, F. P., Reed, J. C., Jr., and Weaver, K. N. (eds.), *Studies of Appalachian Geology: Central and Southern.* Wiley–Interscience, New York, pp. 247–259.

Haidutov, I. S., 1976, A greenstone belt–basement relationship in the Tanganyika Shield. *Geol. Mag.* **113**, 53–60.

Hale, L. A., and Van de Graaf, F. R., 1964, Cretaceous stratigraphy and facies patterns, northeastern Utah and adjacent areas. Intermountain Assoc. Pet. Geol. 13th Annu. Field Conf. Guidebook, pp. 115–138.

Hall, W. E., Batchelder, J. N., and Douglass, R. C., 1974, Stratigraphic section of the Wood River Formation, Blaine County, Idaho. *U.S. Geol. Surv. J. Res.* **2**(1), 89–95.

Hallam, A., 1975, *Jurassic Environments.* Cambridge University Press, London.

Hallam, A. (ed.), 1977, *Patterns of Evolution as Illustrated by the Fossil Record.* Elsevier, Amsterdam.

Hallgarth, W. E., 1967, Permian System of Western Colorado, southern

Utah, and northwestern New Mexico. *U.S. Geol. Surv. Prof. Pap.* **515**, 175–202.

Hamblin, W. K., 1961, Paleogeographic evolution of the Lake Superior region from Late Keweenawan to Late Cambrian time. *Geol. Soc. Am. Bull.* **72**, 1–18.

Hamilton, T. D., 1982, A late Pleistocene chronology for the southern Brooks Range: Stratigraphic record and regional significance. *Geol. Soc. Am. Bull.* **93**, 700–716.

Hamilton, T. D., and Thorson, R. M., 1983, The Cordilleran Ice Sheet in Alaska. *In* Wright, H. E., Jr., and Porter, S. C. (eds.), *Late Quaternary Environments of the United States,* Vol. 1. University of Minnesota Press, Minneapolis, pp. 38–52.

Hamilton, W., 1978, Mesozoic tectonics of the western United States. *In* Howell, D. G., and McDougall, K. A. (eds.), Mesozoic paleogeography of the western United States. SEPM, Pacific Section, Pacific Coast Paleogeography Symp. No. 2, pp. 33–70.

Hamilton, W., and Myers, W. B., 1966, Cenozoic tectonics of the western United States. *Rev. Geophys.* **4**, 509–547.

Hammond, P. E., 1979, A tectonic model for evolution of the Cascade Range. *In* Armentrout, J. M., Cole, M. R., and TerBest, H., Jr. (eds.), Cenozoic paleogeography of the western United States. SEPM, Pacific Section, Pacific Coast Paleogeography Symp. No. 3, pp. 219–237.

Hampton, M. A., 1972, The role of subaqueous debris flow in generating turbidity currents. *J. Sediment. Petrol.* **42**, 775–793.

Hancock, J. M., 1967, Some Cretaceous–Tertiary marine faunal changes. *In* Harland, W. B., *et al.* (eds.), *The Fossil Record.* Geological Society of London, pp. 91–104.

Hancock, J. M., 1975, The sequence of facies in the Upper Cretaceous of northern Europe compared with that in the Western Interior. *In* Caldwell, W. G. E. (ed.), 1975, The Cretaceous System in the western interior of North America. *Geol. Assoc. Can. Spec. Pap.* **13**, 83–118.

Hansen, R. M., 1978, Shasta ground sloth food habits, Rampart Cave, Arizona. *Paleobiology* **4**, 302–319.

Hansen, W. R., 1965, Geology of the Flaming Gorge area, Utah–Colorado–Wyoming. *U.S. Geol. Surv. Prof. Pap.* **490**.

Harland, W. B., 1974, The Precambrian boundary. *In Cambrian of the British Isles, Norden, and Spitsbergen.* Wiley, New York, p. 15.

Harland, W. B., *et al.* (eds.), 1967, *The Fossil Record.* Geological Society of London.

Harland, W. B., Cox, A. V., *et al.,* 1982, *A Geologic Time Scale.* Cambridge University Press, London.

Harrington, H. J., 1959, Classification (of trilobites). *In* Moore, R. C. (ed.), *Treatise on Invertebrate Paleontology,* Part O(1) Arthropoda. Geological Society of America and University of Kansas Press, Lawrence, pp. 145–170.

Harris, L. D., 1976, Thin–skinned tectonics and potential hydrocarbon traps—illustrated by a seismic profile in the Valley and Ridge Province of Tennessee. *U.S. Geol. Surv. J. Res.* **4**(4), 379–386.

Harris, L. D., 1979, Similarities between the thick–skinned Blue Ridge anticlinorium and the thin–skinned Powell Valley anticline. *Geol. Soc. Am. Bull.* **90**, 525–539.

Harris, L. D., and Bayer, K. C., 1979, Sequential development of the Appalachian orogen above a master decollement—A hypothesis. *Geology* **7**, 568–572.

Harris, L. D., and Milici, R. C., 1977, Characteristics of thin–skinned style of deformation in the southern Appalachians, and potential hydrocarbon traps. *U.S. Geol. Surv. Prof. Pap.* **1018**.

Harrison, J. E., 1972, Precambrian Belt basin of northwestern United States: Its geometry, sedimentation, and copper occurrences. *Geol. Soc. Am. Bull.* **83**, 1215–1240.

Harrison, J. E., Griggs, A. B., and Wells, J. D., 1974, Tectonic features of the Precambrian Belt basin and their influences on post–Belt structures. *U.S. Geol. Surv. Prof. Pap.* **866**.

Hartman, C. M., and Banks, H. P., 1980, Pitting in *Psilophyton dawsonii,* an Early Devonian trimerophyte. *Am. J. Bot.* **67**(3), 400–412.

Hartman, W. D., and Goreau, T. F., 1970, Jamaican coralline sponges: Their morphology, ecology and fossil relatives. *Sym. Zool. Soc. London* **25**, 205–243.

Hartshorn, J. H., 1976, Quaternary problems in southern New England. *In*

Mahaney, W. C. (ed.), *Quaternary Stratigraphy of North America.* Dowden, Hutchinson & Ross, Stroudsburg, Pa., pp. 65–90.

Haskett, G. T., 1959, Niobrara Formation of northwest Colorado. *In* Guidebook, 11th Annu. Field Conf. Rocky Mountain Assoc. Geol., 1959, pp. 46–50.

Hass, W. H., 1962, Conodonts. *In* Moore, R. C. (ed.), *Treatise on Invertebrate Paleontology,* Part W, Miscellanea. Geological Society of America and University of Kansas Press, Lawrence, pp. 1–83.

Hasson, K. O., 1971, Lithostratigraphy of the Grainger Formation (Mississippian) in northeast Tennessee. Ph.D. dissertation, University of Tennessee.

Hasson, K. O., and Dennison, J. M., 1974, The Pokejoy Member, a new subdivision of the Mahantango Formation (Middle Devonian) in West Virginia, Maryland, and Pennsylvania. *W. Va. Acad. Sci. Proc.* **46**, 78–86.

Hatcher, R. D., Jr., 1971, Structural, petrologic, and stratigraphic evidence favoring a thrust solution to the Brevard problem. *Am. J. Sci.* **270**, 177–202.

Hatcher, R. D., Jr., 1972, Developmental model for the southern Appalachians. *Geol. Soc. Am. Bull.* **83**, 2735–2760.

Hatcher, R. D., Jr., 1978, Tectonics of the western Piedmont and Blue Ridge, Southern Appalachians: Review and speculations. *Am. J. Sci.* **278**, 276–304.

Hatcher, R. D., Jr., and Zeitz, I., 1980, Tectonic implications of regional aeromagnetic and gravity data from the Southern Appalachians. *In* Wones, D. (ed.), Proceedings of the IGCP Caledonides Orogen Project—The Caledonides in the U.S.A. Va. Polytech. Inst. Mem. **2**, 235–244.

Hatcher, R. D., Jr., Howell, D. E., and Talwani, P., 1977, Eastern Piedmont fault system; speculation on its extent. *Geology* **5**, 636–640.

Hattin, D. E., 1965, Upper Cretaceous stratigraphy, paleontology and paleoecology of western Kansas. Field Conf. Guidebook, Annu. Meet., Kansas City, Mo., 1965, pp. 1–28.

Hattin, D. E., 1967, Stratigraphic and paleoecologic significance of macroinvertebrate fossils in the Dakota Formation (Upper Cretaceous) of Kansas. *In* Essays in Paleontology and Stratigraphy. Univ. Kans. Spec. Publ. **2**, 570–589.

Hattin, D. E., 1975, Stratigraphic study of the Carlile–Niobrara (Upper Cretaceous) unconformity in Kansas and northeastern Nebraska. *Geol. Assoc. Can. Spec. Pap.* **13**, 195–210.

Haworth, R. T., Lefort, J. P., and Miller, H. G., 1978, Geophysical evidence for an east-dipping Appalachian subduction zone beneath Newfoundland. *Geology* **6**, 522–526.

Hayden, F. V., 1869, 3rd Annual Report. *U.S. Geol. and Geograph. Surv. of Terr.*

Hayes, P. T., 1970a, Mesozoic stratigraphy of the Mule and Huachuca Mountains, Arizona. *U.S. Geol. Surv. Prof. Pap.* **658-A**.

Hayes, P. T., 1970b, Cretaceous paleogeography of southeastern Arizona and adjacent areas. *U.S. Geol. Surv. Prof. Pap.* **658-B**.

Hays, J. D., and Pitman, W. C., 1973, Lithospheric plate motions, sea level changes and climatic and ecological consequences. *Nature* **246**, 18–22.

Hays, J. D., Imbrie, J., and Shackleton, N. J., 1976, Variations in Earth's orbit: pacemaker of the ice ages. *Science* **194**, 1121–1132.

Hazel, J. E., 1971, Ostracode biostratigraphy of the Yorktown Formation (upper Miocene and lower Pliocene) of Virginia and North Carolina. *U.S. Geol. Surv. Prof. Pap.* **704**.

Heaton, M. J., 1980, The Cotylosauria: A reconsideration. *In* Panchen, A. L. (ed.), *The Terrestrial Environment and the Origin of Land Vertebrates.* Academic Press (Systematics Assoc.), London/New York, pp. 497–551.

Heaton, M. J., and Reisz, R. R., 1980. A skeletal reconstruction of the Early Permian captorhinid reptile *Eocaptorhinus laticeps* (Williston). *J. Paleontol.* **54**(1), 136–143.

Helwig, J., 1974, Eugeosynclinal basement and a collage concept of orogenic belts. *Soc. Econ. Paleontol. Mineral. Spec. Publ.* **19**, 359–376.

Henderson, J. B., 1977, Archean geology and evidence of ancient life in the Slave structural province, Canada. *In* Ponnamperuma, C. (ed.),

Chemical Evolution of the Early Precambrian. Academic Press, New York, pp. 41–54.

Henyey, T. L., and Lee, T. C., 1976, Heat flow in Lake Tahoe, California–Nevada, and the Sierra Nevada–Basin and Range transition. *Geol. Soc. Am. Bull.* **87**, 1179–1187.

Herman, G., and Barkell, C. A., 1957, Pennsylvanian stratigraphy and productive zones, Paradox salt basin. *Am. Assoc. Pet. Geol. Bull.* **41**(5), 861–881.

Herman, Y., and Hopkins, D. M., 1980, Arctic oceanic climate in late Cenozoic time. *Science* **209**, 557–562.

Herron, S. D., Jr. and Wheeler, W. H., 1964, The Cretaceous formations along the Cape Fear River, North Carolina. Atlantic Coastal Plain Geological Association field guide, 5th annu. field excursion, Oct. 1964.

Herron, E. M., Dewey, J. F., and Pitman, W. C., III, 1974, Plate tectonics model for the evolution of the Arctic. *Geology* **2**, 377–380.

Heusser, L. E., and Shackleton, N. J., 1979, Direct marine–continental correlation: 150,000–year oxygen isotope–pollen record from the North Pacific. *Science* **204**, 837–839.

Heywood, W. W., 1961, Geological notes, northern district of Keewatin. *Geol. Surv. Can. Pap.* **61–18**.

Heywood, W. W., 1967, Geologic notes, northern District of Keewatin and southern Melville Peninsula, District of Franklin, Northwest Territories. *Geol. Surv. Can. Pap.* **66–40**.

Hibbard, C. W., *et al.*, 1965, Quaternary mammals of North America. *In* Wright, H. E., Jr., and Frey, D. G. (eds.), *The Quaternary of the United States.* Princeton University Press, Princeton, N.J. pp. 509–525.

Hickman, C. P., Jr., Roberts, L. S., and Hickman, F. M., 1984, *Integrated Principles of Zoology.* Times Mirror/Mosby College Publ., St. Louis.

Higgins, M. W., 1972, Age, origin, regional relations, and nomenclature of the Glenarm Series, Central Appalachian Piedmont: A reinterpretation. *Geol. Soc. Am. Bull.* **83**, 989–1026.

Higgins, M. W., *et al.*, 1984, A brief excursion through two thrust stacks that comprise most of the crystalline terrane of Georgia and Alabama. 19th Annu. Field Trip Ga. Geol. Soc.

Hill, D., 1972, *Treatise on Invertebrate Paleontology,* Part E, Archaeocyatha (rev.). University of Kansas Press, Lawrence.

Hills, F. A., and Houston, R. S., 1979, Early Proterozoic tectonics of the central Rocky Mountains, North America. *Contrib. Geol.* **17**, 89–109.

Hills, L. V., 1970, Stratigraphy of Bearfoot Formation along western margin of Canadian Arctic islands. *Am. Assoc. Pet. Geol. Bull.* **54**, 2486.

Hintze, L. F., and Robison, R. A., 1975, Middle Cambrian stratigraphy of the House, Wah Wah, and adjacent ranges in western Utah. *Geol. Soc. Am. Bull.* **86**, 881–891.

Hobday, D. K., 1974, Beach– and barrier–island facies in the Upper Carboniferous of northern Alabama. *Geol. Soc. Am. Spec. Pap.* **148**, 209–223.

Hoffman, P. F., 1973, Evolution of an early Proterozoic continental margin: The Coronation geosyncline and associated aulacogens of the northwest Canadian Shield. *Philos. Trans. R. Soc. London* **273**, 547–581.

Hoffman, P. F., and Bowring, S. A., 1984, Short–lived 1.9 Ga continental margin and its destruction, Wopmay orogen, northwest Canada. *Geology* **12**, 68–72.

Hoffman, P. F., Dewey, J. F., and Burke, K., 1974, Aulacogens and their genetic relation to geosynclines, with a Proterozoic example from Great Slave Lake, Canada. *Soc. Econ. Paleontol. Mineral. Spec. Publ.* **19**, 38–55.

Hoffman, P. F., Card, K. D., and Davidson, A., 1982, The Precambrian: Canada and Greenland. *In* Palmer, A. R. (ed.), Perspectives in regional geological synthesis: Planning for geology of North America. *Geol. Soc. Am., DNAG Spec. Publ. No. 1,* pp. 3–6.

Hoffmann, H. J., and Jackson, G. D., 1969, Precambrian (Aphebian) microfossils from Belcher Islands, Hudson Bay. *Can. J. Earth Sci.* **6**, 1137–1144.

Hoffmeister, J. E., and Multer, H. G., 1968, Geology and origin of the Florida Keys. *Geol. Soc. Am. Bull.* **79**, 1487–1502.

Holst, T. B., 1984, Evidence for nappe development during the early Proterozoic Penokean Orogeny, Minnesota. *Geology* **12**, 135–138.

Hon, R., Acheson, D., III, and Shulman, J., 1981, Geochemical and petrological correlation of Acadian magmatic rocks in northwest and northcentral Maine. *Geol. Soc. Am. Abstr. Progr.* **13**, 138.

Hopkins, D. M., Matthews, J. V. Jr., Schweger, C. E., and Young, S. B., (eds.), 1982, *Paleoecology of Beringia.* Academic Press, New York.

Hopson, J. A., 1975, The evolution of cranial display structures in hadrosaurian dinosaurs. *Paleobiology* **1**, 21–43.

Horne, J. C., Ferm, J. C., and Swinchatt, J. P., 1974, Depositional model for the Mississippian–Pennsylvanian boundary in northeastern Kentucky. *Geol. Soc. Am. Spec. Pap.* **148**, 97–114.

Horner, J. R., 1979, Upper Cretaceous dinosaurs from the Bearpaw Shale (marine) of south–central Montana with a checklist of Upper Cretaceous dinosaur remains from marine sediments in North America. *J. Paleontol.* **53**(3), 566–577.

Horodyski, R. J., and Bloeser, B., 1978, 1400 million–year–old shale facies microbiota from the Lower Belt Supergroup, Montana. *Science* **198**, 396–398.

Hotz, P. E., and Willden, R., 1964, Geology and mineral deposits of the Osgood Mountains quadrangle, Humbolt County, Nevada. *U.S. Geol. Surv. Prof. Pap.* **431**.

Houde, P., and Olson, S. L., 1981, Paleognathous carinate birds from the Early Tertiary of North America. *Science* **214**, 1236–1237.

House, M. R., 1967, Fluctuations in the evolution of Paleozoic invertebrates. *In* Harland, W. B., *et al.* (eds.), *The Fossil Record.* Geological Society of London, pp. 41–54.

Houston, R. S., 1971, Regional tectonics of the Precambrian rocks of the Wyoming Province and its relationship to Laramide structure. *In* Wyoming Geol. Assoc., 23rd Annu. Field Conf., Wyoming Tectonics Symp., Guidebook, pp. 19–27.

Howell, D. G., and McDougall, K. A. (eds.), 1978, Mesozoic paleogeography of the western United States. SEPM, Pacific Section, Pacific Coast Paleogeography Symp. No. 2.

Howell, D. G., Suchecki, R. K., and Callahan, R. K. M., 1977, The Cowhead Breccia: Sedimentology of the Cambro–Ordovician continental margin, Newfoundland. *Soc. Econ. Paleontol. Mineral. Spec. Publ.* **23**, 125–154.

Hubert, J. F., Reed, A. A., and Carey, P. J., 1976, Paleogeography of the East Berlin Formation, Newark Groups, Connecticut Valley. *Am. J. Sci.* **276**, 1183–1207.

Hubert, J. F., Suchecki, R. K., and Callahan, R. K. M., 1977, The Cow Head Breccia: Sedimentology of the Cambro–Ordovician continental margin. *Soc. Econ. Paleontol. Mineral. Spec. Publ.* **25**, 125–154.

Huddlestun, P. F., and Hetrick, J. H., 1982, Upper Eocene stratigraphy of eastern Georgia [abstract]. *Ga. Geol. Surv. Inf. Circ.* **53**, 75.

Hudson, T., 1979, Mesozoic plutonic belts of southern Alaska. *Geology* **7**, 230–234.

Hudson, T., Plafker, G., and Turner, D. L., 1977, Intrusive rocks of the Yakutat–St. Elias area, south–central Alaska. *U.S. Geol. Surv. J. Res.* **5**, 155–172.

Hughes, C. J., 1970, The late Precambrian Avalonian orogeny in Avalon, southeast Newfoundland. *Am. J. Sci.* **269**, 183–190.

Hughes, N. F., and Smart, J., 1967, Plant–insect relationships in Paleozoic and later time. *In* Harland, W. B., *et al.* (eds.), *The Fossil Record.* Geological Society of London, pp. 107–118.

Huh, J. M., Briggs, L. I., and Gill, D., 1977, Depositional environments of pinnacle reefs, Niagaran and Salinan, northern shelf, Michigan basin. *In* Fisher, J. H. (ed.), Reefs and evaporites—Concepts and depositional models. AAPG, *Studies in Geology* No. 5, pp. 1–21.

Humphris, C. C., Jr., 1978, Salt movement on continental slope, northern Gulf of Mexico. *In* Bouma, A. H., Moore, G. T., and Coleman, J. M. (eds.), Framework, facies, and oil–trapping characteristics of the upper continental margin, AAPG, *Studies in Geology* No. 7, pp. 69–85.

Hunter, R. E., 1970, Facies of iron sedimentation in the Clinton Group. *In* Fisher, G. W., Pettijohn, F. J., Reed, J. C., Jr., and Weaver, K. N.

(eds.), *Studies in Appalachian Geology: Central and Southern.* Wiley–Interscience, New York, pp. 101–124.

Imbrie, J., 1955, Quantitative lithofacies and biofacies study of the Florena Shale (Permian) of Kansas. *Am. Assoc. Pet. Geol. Bull.* **39**(5), 649–670.

Imbrie, J., and Imbrie, K. P., 1979, Ice ages. Enslow Publ., Short Hills, N.J.

Imlay, R. W., 1980, Jurassic paleobiogeography of the conterminous United States in its continental setting. *U.S. Geol. Surv. Prof. Pap.* **1062**.

Ingels, J. J. C., 1963, Geometry, paleontology, and petrography of Thornton Reef complex, Silurian of northeastern Illinois. *Am. Assoc. Pet. Geol. Bull.* **47**, 405–440.

Ingersoll, R. V., 1978a, Paleogeography and paleotectonics of the late Mesozoic forearc basin of northern and central California. *In* Howell, D. G., and McDougall, K. A. (eds.), Mesozoic paleogeography of the western United States. SEPM, Pacific Section, Pacific Coast Paleogeography Symp. No. 2, pp. 471–482.

Ingersoll, R. V., 1978b, Evolution of the Late Cretaceous forearc basin, northern and central California. *Geol. Soc. Am. Bull.* **90**, 813–826.

Ireland, H. A., 1965, Regional depositional basin and correlations of the Simpson Group. *Tulsa Geol. Soc. Dig.* **33**, 74–89.

Irvine, T. N. and Findlay, T. C., 1972, Alpine—type peridotite with particular reference to the Bay of Islands igneous complex. Can. Dept. Energy, Mines, and Res. Earth Phys. Br. Pub. **42**, No. 3, pp. 97–128.

Irving, E., and McGlynn, J. C., 1976, Proterozoic magneto-stratigraphy and the tectonic evolution of Laurentia. *Philos. Trans. R. Soc. London Ser. A* **280**, 433–468.

Irving, E., and McGlynn, J. C., 1981, On the coherence, rotation and paleolatitude of Laurentia in the Proterozoic. *In* Kröner, A. (ed.), *Precambrian Plate Tectonics.* Elsevier, Amsterdam, pp. 561–598.

Irwin, W. P., 1972, Terranes of the western Paleozoic and Triassic belt in the southern Klamath Mountains, California. *U.S. Geol. Surv. Prof. Pap.* **800-C**, pp. C103–C111.

Irwin, W. P., 1977, Review of Paleozoic rocks of the Klamath Mountains. *In* Stewart, J. H., Stevens, C. H., and Fritsche, A. E. (eds.), Paleozoic paleogeography of the western United States. SEPM, Pacific Section, Los Angeles, pp. 441–454.

Irwin, W. P., Jones, D. L., and Kaplan, T. A., 1978, Radiolarians from pre-Nevadan rocks of the Klamath Mountains, California and Oregon. *In* Howell, D. G., and McDougall, K. A. (eds.), Mesozoic paleogeography of the western United States. SEPM, Pacific Section, Pacific Coast Paleogeography Symp. No. 2, pp. 303–310.

Jackson, G. D., 1971, Operation Penny Highlands, south–central Baffin Island. *In* Report of activities, Part A: April to October, 1970. *Geol. Surv. Can. Pap.* 70–1, pp. 138–140.

Jackson, G. D., and Taylor, F. C., 1972, Correlation of major Aphebian rock units in the northeastern Canadian Shield. *Can. J. Earth Sci.* **9**, 1650–1669.

James, H. L., 1954, Sedimentary facies of iron formation. *Econ. Geol.* **49**, 235–293.

James, H. L., 1972, Stratigraphic Commission note 40; Subdivisions of Precambrian, an interim scheme to be used by the U.S. Geological Survey. *Am. Assoc. Pet. Geol. Bull.* **56**, 1128–1133.

James, N. P., 1981, Megablocks of calcified algae in the Cow Head Breccia, western Newfoundland: Vestiges of a Cambro–Ordovician platform margin. *Geol. Soc. Am. Bull.* **92**, 799–811.

James, W. C., 1980, Limestone channel storm complex (Lower Cretaceous) Elkhorn Mountains, Montana. *J. Sediment. Petrol.* **50**(2), 447–456.

Jansa, L. F., and Wade, J. A., 1975, Geology of the continental margin off Nova Scotia and Newfoundland. *In* Van der Linden, W. J. M., and Wade, J. A. (eds.), Offshore geology of eastern Canada. Geol. Surv. Can. Pap. 74–30, pp. 55–105.

Janzen, D. H. and Martin, P. S., 1982, Neotropical anachronisms: The fruits the gomphotheres ate. *Science,* **215**, 19–27.

Jefferies, R. P. S., 1984, Locomotion, shape, ornament, and external ontogeny in some mitrate calcichordates. *J. Vertebr. Paleontol.* **4**(3), 292–319.

Jeletzky, J. A., 1975, Age and depositional environment of the lower part of Escalante Formation, western Vancouver Island, British Columbia (92E). *Geol. Surv. Can. Pap.* 75–1C, pp. 9–16.

Jell, P. A., 1974, Faunal provinces and possible planetary reconstruction of the Middle Cambrian. *J. Geol.* **82**, 319–350.

Jenkins, F. A., Jr., and Krause, D. W., 1983, Adaptations for climbing in North American multituberculates (Mammalia). *Science* **220**, 712–715.

Jenkins, F. A., Jr., Crompton, A. W., and Downs, W. R., 1983, Mesozoic mammals from Arizona: New evidence on mammalian evolution. *Science* **220**, 1233–1235.

Jenkins, R. J. F., 1985, The enigmatic Ediacaran (late Precambrian) genus *Rangea* and related forms. *Paleobiology* **11**(3), 336–355.

Johnson, G. A. L., 1978, European plate movements during the Carboniferous. *In* Tarling, D. H. (ed.), *Evolution of the Earth's Crust.* Academic Press, New York, pp. 343–360.

Johnson, J. G., 1970, Taghanic onlap and the end of North American Devonian provinciality. *Geol. Soc. Am. Bull.* **81**, 2077–2105.

Johnson, J. G., and Potter, E. C., 1975, Silurian (Llandovery) downdropping of the western margin of North America. *Geology* **3**, 331–334.

Johnson, K. E., 1968, Sedimentary environment of Stanley Group of the Ouachita Mountains of Oklahoma. *J. Sediment. Petrol.* **38**, 723–733.

Johnson, K. G., 1976, Alluvial and tidal facies of the Catskill deltaic system. *In* Johnson, J. H. (ed.), Guidebook to Field Excursions at the 48th Annu. Meet. of the New York State Geological Association, pp. B–8–1–B–8–26.

Johnson, K. G., and Friedman, G. M., 1969, The Tully clastic correlatives (Upper Devonian) of New York State: A model for recognition of alluvial, dune (?), tidal, nearshore (bar and lagoon), and offshore sedimentary environments in a tectonic delta complex. *J. Sediment. Petrol.* **39**, 451–485.

Johnson, R. C., 1981, Stratigraphic evidence for a deep Eocene Lake Uinta, Piceance Creek basin, Colorado. *Geology* **9**, 55–62.

Johnson, W. H., 1976, Quaternary stratigraphy in Illinois: Status and current problems. *In* Mahaney, W. C. (ed.), *Quaternary Stratigraphy of North America.* Dowden, Hutchinson & Ross, Stroudsburg, Pa., pp. 161–196.

Johnston, P. A., 1979, Growth rings in dinosaur teeth. *Nature* **278**, 635–636.

Jones, D. L., Silberling, N. J., and Hillhouse, J., 1977, Wrangellia—A displaced terrane in northwestern North America. *Can. J. Earth Sci.* **14**, 2565–2577.

Judson, S., 1965, Quaternary processes in the Atlantic Coastal Plain and Appalachian highlands. *In* Wright, H. E., Jr., and Frey, D. G. (eds.), *The Quaternary of the United States.* Princeton University Press, Princeton, N.J., pp. 133–136.

Kanie, Y. Fukuda, Y., Nakayama, H., Seki, K., and Hattori, M., 1980, Implosion of living *Nautilus* under increased pressure. *Paleobiology* **6**(1), 44–47.

Kapp, U. S., 1974, Mode of growth of middle Chazyan (Ordovician) stromatoporoids, Vermont. *J. Paleontol.* **48**(6), 1235–1240.

Karig, D. E., 1974, Evolution of arc systems in the western Pacific. *Annu. Rev. Earth Planet Sci.* **2**, 51–75.

Karl, H. A., 1976, Depositional history of Dakota Formation (Cretaceous) sandstones, southeastern Nebraska. *J. Sediment. Petrol.* **46**(1), 124–131.

Karlstrom, K. E., Flurkey, A. J., and Houston, R. S., 1983, Stratigraphy and depositional setting of the Proterozoic Snowey Pass Supergroup, southeastern Wyoming: Record of an early Proterozoic Atlantic–type cratonic margin. *Geol. Soc. Am. Bull.* **94**, 1257–1274.

Karrow, P. F., and Bada, J. L., 1980, Amino acid racemization dating of Quaternary raised marine terraces in San Diego County, California. *Geology* **8**, 200–204.

Karson, J., and Dewey, J. F., 1978, Coastal complex, western Newfoundland: An Early Ordovician oceanic fracture zone. *Geol. Soc. Am. Bull.* **89**, 1037–1049.

Kauffman, E. G., 1969, Form, function and evolution. *In* Moore, R. C. (ed.), *Treatise on Invertebrate Paleontology,* Part N(1), Mollusca 6. Geological Society of America and University of Kansas Press, Lawrence, pp. 129–203.

Kauffman, E. G., 1979, The ecology and biogeography of the Cretaceous–Tertiary extinction event. *In* Christensen, W. K., and Birkelund, T.

(eds.), *Cretaceous–Tertiary Boundary Events,* Vol. II. University of Copenhagen, Denmark, pp. 29–37.

Kauffman, E. G., and Kesling, R. V., 1960, An Upper–Cretaceous ammonite bitten by a mosasaur. *Mus. Paleontol. Univ. Contrib.* **15**(9), 1005–1018.

Kauffman, E. G., and Sohl, N. F., 1974, Structure and function of Antillean rudist frameworks. *Ver. Naturforsch. Ges.* Basel 84(1), 399–467.

Kaula, W. M., 1975, The seven ages of a planet. *Icarus,* **26**, 1–15.

Kay, M., and Colbert, E. H., 1965, *Stratigraphy and Life History.* Wiley, New York.

Kazmierczak, J., 1984, Favositid tabulates: Evidence for poriferan affinity. *Science* **225**, 835–837.

Kearey, P., 1976, A regional model of the Labrador trough, northern Quebec, from gravity studies, and its relevance to continental collision in the Precambrian. *Earth Planet. Sci. Lett.* **28**, 371–378.

Keen, C. E., and Keen, M. J., 1974, Continental margins of eastern Canada and Baffin Bay. *In* Burk, C. A., and Drake, C. L. (eds.), *The Geology of Continental Margins.* Springer-Verlag, Berlin, pp. 381–390.

Keith, S. B., 1978, Paleosubduction geometries inferred from Cretaceous and Tertiary magmatic patterns in southwestern North America. *Geology* **6**, 516–521.

Kellberg, J. M., and Grant, L. F., 1956, Coarse conglomerates of the Middle Ordovician in the southern Appalachian Valley. *Geol. Soc. Am. Bull.* **67**, 697–716.

Keller, A. S., Morris, R. H., and Detterman, R. L., 1961, Geology of the Shaviovik and Sagavanirktok Rivers region, Alaska. *U.S. Geol. Surv. Prof. Pap.* **303-D**, D169–D222.

Keller, G. R., and Cebull, S. E., 1973, Plate tectonics and the Ouachita System in Texas, Oklahoma, and Arkansas. *Geol. Soc. Am. Bull.* **83**, 1659–1666.

Keller, G. R., and Shurbet, D. H., 1975, Crustal structure of the Texas Gulf Coastal Plain. *Geol. Soc. Am. Bull.* **86**, 897–910.

Kent, H. C., 1968, Biostratigraphy of Niobrara–equivalent part of Mancos Shale (Cretaceous) in northeastern Colorado. *Am. Assoc. Pet. Geol. Bull.* **52**(11), 2098–2115.

Kepferle, R. C., 1977, Stratigraphy, petrology, and depositional environment of the Kenwood Siltstone Member, Borden Formation (Mississippian), Kentucky and Indiana. *U.S. Geol. Surv. Prof. Pap.* **1007**.

Kermack, D. A., and Kermack, K. A. (eds.), 1971, *Early Mammals.* Academic Press (Linnean Society), London.

Kermack, K. A., and Kielan-Jaworowska, F., 1971, Therian and nontherian mammals. *In* Kermack, D. A., and Kermack, K. A. (eds.), *Early Mammals.* Academic Press (Linnean Society), London, pp. 103–116.

Kerr, J. W., 1967, Devonian of the Franklin miogeosyncline and adjacent central stable region, arctic Canada. *In* Devonian System, Int. Symp., Calgary. *Alberta Soc. Petrol. Geol.* **1**, 677–692.

Kerr, J. W., 1981, Evolution of the Canadian Arctic islands: A transition between the Atlantic and Arctic Oceans. *In* Nairn, A. E. M., Churkin, M., Jr., and Stehli, F. G. (eds.), *The Ocean Basins and Margins,* pp. 105–199.

Kershaw, S., 1981, Stromatoporoid growth form and taxonomy in a Silurian biostrom, Gotland. *J. Paleontol.* **55**(6), 1284–1295.

Kesling, R. V., and Graham, A., 1962, *Ischadites* is a dasycladacean alga. *J. Paleontol.* **36**(6), 943–952.

Ketner, K. B., 1977, Late Paleozoic orogeny and sedimentation, southern California, Nevada, Idaho, and Montana. *In* Stewart, J. H., Stevens, C. H., and Fritsche, A. E. (eds.), Paleozoic paleogeography of the western United States. SEPM, Pacific Section, Los Angeles, pp. 363–369.

Kidd, J. T., and Neathery, T. L., 1976, Correlation between Cambrian rocks of the southern Appalachian geosyncline and the interior low plateaus. *Geology,* **4**(12), 767–769.

Kier, P. M., 1974, Evolutionary trends and their functional significance in the post–Paleozoic echinoids. *Paleontol. Soc. Mem.* **5**, (II).

Kier, R. S., Brown, L. F., Jr., and McBride, E. F., 1979, The Mississippian and Pennsylvanian (Carboniferous) Systems in the United States—Texas. *U.S. Geol. Surv. Prof. Pap.* **1110–S**.

King, L. H., and MacLean, B., 1970, Seismic–reflection study, Orpheus gravity anomaly. *Am. Assoc. Pet. Geol. Bull.* **54**, 2007–2031.

King, P. B., 1948, Geology of the Guadalupe Mountains, West Texas. *U.S. Geol. Surv. Prof. Pap.* **215**.

King, P. B., 1955, Orogeny and epirogeny through time. *Geol. Soc. Am. Spec. Pap.* **62**, 723–739.

King, P. B., 1970, The Precambrian of the United States of America: Southeastern United States. *In* Rankama, K. (ed.), *The Precambrian,* **4**. Wiley–Interscience, New York, pp. 1–71.

King, P. B., 1976, Precambrian geology of the United States: An explanatory text to accompany the geologic map of the United States. *U.S. Geol. Surv. Prof. Pap.* **902**.

King, P. B., 1977, *Geological Evolution of North America.* Princeton University Press, Princeton, N.J.

King, P. B., and Beikman, H. M., 1976, The Paleozoic and Mesozoic rocks: A discussion to accompany the geologic map of the United States. *U.S. Geol. Surv. Prof. Pap.* **903**.

King, P. B., and Beikman, H. M., 1978, The Cenozoic rocks: A discussion to accompany the geologic map of the United States. *U.S. Geol. Surv. Prof. Pap.* **904**.

King, P. B., and Zeitz, I., 1978, The New York–Alabama lineament: Geophysical evidence for a major crustal break in the basement beneath the Appalachian Basin. *Geology* **6**, 312–319.

Kirschner, C. E., and Lyon, C. A., 1973, Stratigraphic and tectonic development of Cook Inlet petroleum province. *Am. Assoc. Pet. Geol. Mem.* **19**, 396–407.

Kish, S. A., Fullagar, P. D., Snoke, A. W., and Secor, D. T., Jr., 1978, The Kiokee Belt of South Carolina (part I): Evidence for late Paleozoic deformation and metamorphism in the Southern Appalachian Piedmont. *Geol. Soc. Am. Abstr. Progr.* **10**, 172–173.

Kistler, R. W., 1978, Mesozoic paleogeography of California: A viewpoint for isotope geology. *In* Howell, D. G., and McDougall, K. A. (eds.), Mesozoic paleogeography of the western United States. SEPM, Pacific Section, Pacific Coast Paleogeography Symp. No. 2, pp. 75–84.

Kjellesvig–Waering, E. N., 1972, *Brontoscorpio anglicus:* A gigantic lower Paleozoic scorpion from central England. *J. Paleontol.* **46**(1), 39–42.

Klapper, G., and Bergstrom, S. M., 1984, The enigmatic Middle Ordovician fossil *Archaeognathus* and its relations to conodonts and vertebrates. *J. Paleontol.* **58**(4), 949–976.

Klein, G. deV., 1962, Triassic sedimentation, Maritime Provinces, Canada. *Geol. Soc. Am. Bull.* **73**, 1127–1146.

Klein, G. deV., 1969, Deposition of Triassic sedimentary rocks in separate basins, eastern North America. *Geol. Soc. Am. Bull.* **80**, 1825–1832.

Klepper, M. R., Weeks, R. A., and Ruppel, E. T., 1957, Geology of the southern Elkhorn Mountains, Jefferson and Broadwater counties, Montana. *U.S. Geol. Surv. Prof. Pap.* **292**.

Klitgord, K. D., and Behrendt, J. C., 1979, Basin structure of the U.S. Atlantic margin. *Am. Assoc. Pet. Geol. Mem.* **29**, 84–112.

Klitgord, K. D., and Grow, J. A., 1980, Jurassic seismic stratigraphy and basement structure of western Atlantic magnetic quiet zone. *Am. Assoc. Pet. Geol.* **64**, 1659–1680.

Kluth, C. F., and Coney, P. J., 1981, Plate tectonics of the Ancestral Rocky Mountains. *Geology* **9**, 10–15.

Knauth, L. P., 1979, A model for the origin of chert in limestone. *Geology* **7**, 274–277.

Knoll, A. H., and Barghoorn, E. S., 1977, Archean microfossils showing cell division from the Swaziland System of South Africa. *Science* **198**, 396–398.

Knoll, A. H. and Rothwell, G. W., 1981, Paleobotany: Perspectives in 1980. *Paleobiology,* **7**(1), 7–35.

Konish, K., 1959, Stratigraphy of Dakota Sandstone, northwestern Colorado. *In* Guidebook, 11th Annu. Field Conf. Rocky Mountain Assoc. Geol., 1959, pp. 30–32.

Kottlowski, F. E., 1962, Pennsylvanian rocks of southwestern New Mexico and southeastern Arizona. *In* Pennsylvanian System in the U.S. Tulsa, pp. 331–371.

Kozlowski, R., 1949, Les graptolites et quelques nouveaux groups d'animaux du Tremadoc de la Pologne. *Paleontol. Pol.* **3**, 1–235.

Kraft, J. C., and Thomas, R. A., 1976, Early man at Holly Oak, Delaware. *Science* **192**, 756–761.

Krauskopf, K. B., 1968, A tale of ten plutons. *Geol. Soc. Am. Bull.* **79**, 1–17.

Kreisa, R. D., and Bambach, R. K., 1973, Environments of deposition of the Price Formation (Lower Mississippian) in its type area, southwestern Virginia. *Am. J. Sci.* **273-A**, 326–342.

Kröner, A., 1985, Evolution of the Archean continental crust. *Annu. Rev. Earth Planet. Sci.* **13**, 49–74.

Krumbein, W. C., and Sloss, L. L., 1963, *Stratigraphy and Sedimentation.* Freeman, San Francisco.

Krynine, P. D., 1950, Petrology, stratigraphy, and origin of the Triassic sedimentary rocks of Connecticut. *Conn. State Geol. Nat. Hist. Surv. Bull.* **73**.

Kues, B. S., Lehman, T., and Rigby, J. K., Jr., 1980, The teeth of *Alamosaurus sanjuanensis*, a Late Cretaceous sauropod. *J. Paleontol.* **54**(4), 864–886.

Kummel, B., 1954, Triassic stratigraphy of southeastern Idaho and adjacent areas. *U.S. Geol. Surv. Prof. Pap.* **254–H**, 165–189.

Kummel, B., and Teichert, C., 1970, Stratigraphy and paleontology of the Permian–Triassic boundary beds, Salt Range and Trans-Indus Ranges, West Pakistan. *In* Kummel, B., and Teichert, C. (eds.), Stratigraphic boundary problems. University of Kansas, Department of Geology, Spec. Publ. 4, pp. 1–110.

Kurtén, B., and Anderson, E., 1980, *Pleistocene Mammals of North America.* Columbia University Press, New York.

Lambert, R. St. J., 1976, Archean thermal regimes, crustal and upper mantle temperatures, and a progressive evolutionary model for the Earth. *In* Windley, B. F. (ed.), *The Early History of the Earth.* Wiley, New York, pp. 377–403.

Langston, W., Jr., 1981, Pterosaurs. *Sci. Am.* **244**, n. 2, 122–136.

Lanham, U. N., 1973, *The Bone Hunters.* Columbia University Press, New York.

Laporte, L. F., 1969, Recognition of a transgressive carbonate sequence within an epeiric sea; Helderberg Group (Lower Devonian) of New York State. *Soc. Econ. Paleontol. Mineral. Spec. Publ.* **14**, 54–75.

Larson, E. R., and Langenheim, R. L., Jr., 1979, The Mississippian and Pennsylvanian (Carboniferous) System in the U.S.—Nevada. *U.S. Geol. Surv. Prof. Pap.* **1110–BB**.

Larue, D. K., 1981, The Chocolay Group, Lake Superior region, U.S.A.: Sedimentologic evidence for deposition in basinal and platformal settings on an early Proterozoic craton. *Geol. Soc. Am. Bull.* **92**, 417–435.

Larue, D. K., and Sloss, L. L., 1980, Early Proterozoic sedimentary basins of the Lake Superior region: Summary. *Geol. Soc. Am. Bull.* **91**(1), 450–452.

Lawson, A. C., 1913, The petrographic designation of alluvial–fan formations. *Calif. Univ. Dep. Geol. Bull.* **7**(15), 325–334.

Leeper, W. S., 1963, Interpretation of primary bedding structures in Mississippian and Upper Devonian rocks of southeastern Somerset County, Pennsylvania. *Pa. Topogr. Geol. Surv. Bull. G* **39**, 165–181.

LeGrand, H. E., 1961, Summary of geology of Atlantic Coastal Plain. *Am. Assoc. Pet. Geol. Bull.* **45**, 1557–1571.

LePichon, X., and Fox, P. J., 1971, Marginal offsets, fracture zones and the early opening of the North Atlantic. *J. Geophys. Res.* **76**, 6294–6308.

Lewis, R. Q., Sr., and Taylor, A. R., 1979, The Science Hill Sandstone Member of the Warsaw Formation and its relation to other clastic units in south-central Kentucky. *U.S. Geol. Surv. Bull.* **1435–D**.

Licari, G. R., 1978, Biogeology of the Late Pre-Phanerozoic Beck Springs Dolomite of eastern California. *J. Paleontol.* **52**(4), 767–792.

Lilly, H. D., 1966, Late Precambrian and Appalachian tectonics in the light of submarine exploration of the Great Bank of Newfoundland and in the Gulf of St. Lawrence: Preliminary view. *Am. J. Sci.* **264**, 569–574.

Lindholm, R. C., 1969, Carbonate petrology of the Onondaga Limestone (Middle Devonian), New York: A case for calcisiltite. *J. Sediment. Petrol.* **39**, 268–275.

Lindsley–Griffin, N., 1977, Paleogeographic implications of ophiolites: The Ordovician Trinity Complex, Klamath Mountains, California. *In* Stewart, J. H., Stevens, C. H., and Fritsche, A. E. (eds.), Paleozoic paleogeography of the western United States. SEPM, Pacific Section, Los Angeles, pp. 409–420.

Lipman, P. W., Prostka, H. J., and Christiansen, R. L., 1971, Evolving subduction zones in the western United States as interpreted from igneous rocks. *Science* **174**, 821–825.

Little, H. W., 1960, Nelson map area, west half, British Columbia. *Can. Geol. Surv. Mem.* **308**.

Livaccari, R. F., 1979, Late Cenozoic tectonic evolution of the western United States (and comment). *Geology* **7**, 72–75, 371–373.

Lochman–Balk, C., 1971, The Cambrian of the craton of the U.S. *In* Holland, C. H. (ed.), *Cambrian of the New World.* Wiley, New York, p. 79.

Lochman–Balk, C., and Wilson, J. L., 1958, Cambrian biostratigraphy in North America. *J. Paleontol.* **32**(2), 312–350.

Loeblich, A. R., Jr., and Tappan, H., 1964, *Treatise on Invertebrate Paleontology,* Part C, Protista 2. Geological Society of America and University of Kansas Press, Lawrence.

Lomar, J. M., and Warme, J. E., 1979, An Eocene shelf margin: San Diego County, California. *In* Armentrout, J. M., Cole, M. R., and TerBest, H., Jr., (eds.), Cenozoic paleogeography of the western United States. Pacific Section, Pacific Coast Paleogeography Symp. No. 3, pp. 165–175.

Longwell, C. R., 1933, Eastern New York and western New England. Int. Geol. Congr. Guidebook 1.

Lowe, D. R., 1975, Regional controls on silica sedimentation in the Ouchita system. *Geol. Soc. Am. Bull.* **86**, 1123–1127.

Lowe, D. R., 1980, Stromtolites 3400 myr-old from the Anchean of Western Australia. *Nature* **284**, 441–443.

Lowe, D. R., and Knauth, L. P., 1977, Sedimentology of the Onverwacht Group (3.4 billion years), Transvaal, South Africa, and its bearing on the characteristics and evolution of the early Earth. *J. Geol.* **85**, 699–723.

Lowenstam, H. A., 1961, Mineralogy, $^{18}O/^{16}O$ ratios, and strontium and magnesium contents of recent and fossil brachiopods and their bearing on the history of oceans. *J. Geol.* **69**, 241.

Ludlum, J. C., 1959, Rock salt, rhythmic bedding, and salt–crystal impressions in the Upper Silurian limestones of West Virginia. *Southeast. Geol.* **1**(1), 22–32.

Luyendyk, B. P., Kamerling, M. J., and Terres, R., 1980, Geometric model for Neogene crustal rotations in southern California. *Geol. Soc. Am. Bull.* **91**, 211–217.

McBride, E. F., 1962, Flysch and associated beds of the Martinsburg Formation (Ordovician), Central Appalachians. *J. Sediment. Petrol.* **32**, 39–91.

McBride, E. F., and Folk, R. L., 1977, The Caballos Novaculite, (II): Chert and shale members and synthesis. *J. Sediment. Petrol.* **47**, 1261–1286.

McCall, G. J. H. (ed.), 1977, *The Archean: Search for the Beginning.* Dowden, Hutchinson & Ross, Stroudsburg, Pa.

McCarley, A. B., 1981, Metamorphic terrane favored over Rocky Mountains as source of Claiborne Group, Eocene, Texas Coastal Plain. *J. Sediment. Petrol.* **51**, 1267–1276.

McCave, I. N., 1973, The sedimentology of a transgression: Portland Point and Cooksburg Members (Middle Devonian), New York State. *J. Sediment. Petrol.* **43**, 484–504.

McCrossan, R. G., Glaister, R. P., Austin, G. H., and Nelson, S. J., 1964, Stratigraphic Atlas of Western Canada. Alberta Soc. Pet. Geol.

McElhinny, M. W. and McWilliams, M. O., 1977, Precambrian geodynamics—a paleomagnetic view. *Tectonophysics,* **40**, 137–159.

McFarland, W. N., Pough, F. H., Cade, T. J., and Heiser, J. B., 1979, *Vertebrate Life.* Macmillan Co., New York.

McGee, W. J., 1888, Three formations of the Middle Atlantic slope. *Am. J. Sci. 3rd Ser.* **35**, 120–143, 328–330, 367–388, 448–466.

McGlynn, J. C., 1976, The Southern Province. *In* Douglas, R. J. W. (ed.), 1976, Geology and economic minerals of Canada. *Geol. Surv. Can. Econ. Geol. Rep.* No. 1, pp. 108–119.

McGlynn, J. C., 1976a, Slave Province, in Chapter IV, Geology of the Canadian Sheild. *In* Douglas, R. J. W. (ed.), 1976, Geology and economic minerals of Canada. *Geol. Surv. Can. Econ. Geol. Rep.* No. 1, 71–76.

McGlynn, J. C., 1976c, Superior Province. *In* Douglas, R. J. W. (ed.),

1976, Geology and economic minerals of Canada. *Geol. Surv. Can. Econ. Geol. Rep.* No. 1, 54–71.

McGlynn, J. C., and Henderson, J. B., 1970, Archean volcanism and sedimentation in the Slave structural province. *Geol. Surv. Can. Pap.* **70–40**, pp. 31–44.

McGlynn, J. C., and Irving, E., 1975, Paleomagnetism of early Aphebian diabase dikes from the Slave structural province, Canada. *Tectonophysics* **26**, 23–38.

McGregor, V. R., 1973, The early Precambrian gneisses of the Godthab district, West Greenland. *Philos. Trans. Roy. Soc. London Ser. A* **273**, 343–358.

McGregor, V. R. and Mason, B., 1977, Petrogenesis and geochemistry of metabasaltic and metasedimentary enclaves in the Amitsog gneisses, W. Greenland. *Am. Mineral.* **62**, 887–904.

McIver, N. L., 1970, Appalachian turbidites. *In* Fisher, G. W., Pettijohn, F. J., Reed, J. C., Jr., and Weaver, K. N. (eds.), *Studies in Appalachian Geology: Central and Southern.* Wiley–Interscience, New York, pp. 69–81.

McIver, N. L., 1972, Cenozoic and Mesozoic stratigraphy of the Nova Scotia shelf. *Can. J. Earth Sci.* **9**, 54–70.

McKee, E. D., 1954, Stratigraphy and history of the Moenkopi Formation of Triassic age. *Geol. Surv. Am. Mem.* **61**.

McKee, E. D., 1967, Arizona and western New Mexico. *U.S. Geol. Surv. Prof. Pap.* **515**, 203–223.

McKee, E. D., and Crosby, E. J. (coords.), 1975, Paleotectonic investigations of the Pennsylvanian System in the United States, part I. *U.S. Geol. Surv. Prof. Pap.* **853**.

McKee, E. D., *et al.* (eds.), 1967, Paleotectonic investigations of the Permian System in the United States. *U.S. Geol. Surv. Prof. Pap.* **515**.

McKenzie, D. P. and Morgan, W. J., 1969, Evolution of Triple junctions. *Nature,* **224**, 125–133.

McKenzie, D. P., and Weiss, N. O., 1975, Speculations on the thermal and tectonic history of the Earth. *Geophys. J. R. Astron. Soc.* **42**, 131–174.

McKerrow, W. S., and Ziegler, A. M., 1971, The Lower Silurian paleogeography of New Brunswick and adjacent areas. *J. Geol.* **79**, 635–646.

McLaughlin, J. C., 1979, *Archosauria.* Viking, New York.

McLean, D. M., 1978, A terminal Mesozoic "greenhouse": Lessons from the past. *Science* **201**, 401–406.

McLean, H., 1977, Lithofacies of the Blakeley Formation, Kitsap County, Washington: A submarine fan complex. *J. Sediment. Petrol.* **47**, 78–88.

McLelland, J., and Isachsen, Y., 1980, Structural synthesis of the southern and central Adirondacks: A model for the Adirondacks as a whole and plate-tectonics interpretations: Summary. *Geol. Soc. Am. Bull.* **91**(I), 68–72.

McMenamin, M. A. S., 1982, A case for two late Proterozoic–earliest Cambrian faunal province loci. *Geology* **10**, 290–292.

McMenamin, M. A. S., Awramik, S. M., and Stewart, J. H., 1983, Precambrian–Cambrian transition problem in western North America: Part II, Early Cambrian skeletonized fauna and associated fossils from Sonora, Mexico. *Geology* **11**(4), 227–230.

McNair, A. H., 1961, Relations of the Parry Islands fold belt and Cornwallis Islands fold, eastern Bathurst Island, Canadian Archipelago. *In* Raasch, G. O. (ed.), Geology of the Arctic. Proc. 1st Int. Symp. Arctic Geol. **1**, 421–426.

McPhee, J., 1980, Annals of the former world, Basin and Range I. *The New Yorker* **LVI**(35), 58–136.

MacGinitie, H. D., 1953, Fossil plants of the Florissant beds, Colorado. *Carnegie Inst. Wash. Contrib. Paleontol. Publ.* **599**.

MacGregor, A. M., 1951, Some milestones in the Precambrian of southern Africa. *Proc. Geol. Soc. S. Afr.* **54**, 27–71.

MacLachlan, M. E., 1967, Oklahoma. *U.S. Geol. Surv. Prof. Pap.* **515**, 85–96.

Mack, G. H., Thomas, W. A., and Horsey, C. A., 1983, Composition of Carboniferous sandstones and tectonic framework of Southern Appalachian–Ouachita orogen. *J. Sediment. Petrol.* **53**, 931–946.

Madole, R. F., 1980, Time of Pinedale deglaciation in north–central Colorado: Further considerations. *Geology* **8**, 118–122.

Mahaney, W. C. (ed.), 1976, *Quaternary Stratigraphy of North America.* Dowden, Hutchinson & Ross, Stroudsburg, Pa.

Maher, H. D., Palmer, A. R., Secor, D. T., and Snoke, A. W., 1981, New trilobite locality in the piedmont of South Carolina, and its regional implications. *Geology* **9**, 34–36.

Mallory, W. W., 1958, Pennsylvanian coarse arkosic redbeds and associated mountains in Colorado. *In* Pennsylvanian rocks of Colorado and adjacent areas. Rocky Mountain Assoc. Geol., Annu. Field Trip Guidebook.

Mancini, E. A., and Benson, D. J., 1980, Regional stratigraphy of Upper Jurassic Smackover carbonates of southwest Alabama. *Trans. Gulf Coast Assoc. Geol. Soc.* **30**, 151–165.

Manger, W. L., and Saunders, W. B., 1980, Lower Pennsylvanian (Morrowan) ammonoids from the North American mid–continent. *J. Paleontol. Mem.* **10**.

Manspeizer, W., 1981, Early Mesozoic basins of the Central Atlantic passive margins. In Bally, A. W. *et al.* (eds.), Geology of passive continental margins: History, structure and sedimentologic record (with special emphasis on The Atlantic margin). AAPG, Education Course Note Series No. 19, pp. 4–1–4–60.

Manspeizer, W., Puffer, J. H., and Cousminer, H. L., 1978, Separation of Morocco and eastern North America: A Triassic–Liassic stratigraphic record. *Geol. Soc. Am. Bull.* **89**, 901–920.

Manton, S. M., 1977, *The Arthropoda.* Oxford, University Press (Clarendon), London.

Maples, C. G., and Waters, J. A., 1984, Algal–archaeocyathan patch reefs from the Cartersville mining district, Georgia. *Southeast. Geol.* **24**(4), 159–167.

Marcher, M. V., and Sterns, R. G., 1962, Tuscaloosa Formation in Tennessee. *Geol. Soc. Am. Bull.* **73**, 1365–1386.

Marlow, M. S., Scholl, D. W., Buffington, E. C., and Alpha, T. R., 1973, Tectonic history of the central Aleutian arc. *Geol. Soc. Am. Bull.* **84**, 1555–1574.

Marshall, L. G., Butler, R. F., Drake, R. E., and Curtis, G. H., 1981, Calibration of the beginning of the age of mammals in Patagonia. *Science* **212**, 43–45.

Marshall, L. G., Webb, S. D., Sepkoski, J. J., Jr., and Raup, D. M., 1982, Mammalian evolution and the great American interchange. *Science* **215**, 1351–1357.

Marshall, L. G., *et al.,* 1979, Calibration of the great American interchange. *Science* **204**, 272–279.

Martin, L. D., Gilbert, B. M., and Adams, D. B., 1977, A cheetah–like cat in the North American Pleistocene. *Science* **195**, 981–982.

Martin, P. E. and Klein, R. G. (eds.), 1984, *Quaternary Extinctions.* The University of Arizona Press, Tucson.

Martin, P. S., 1967, Prehistoric overkill. *In* Martin, P. S., and Wright, H. E., Jr. (eds.), *Pleistocene Extinctions: The Search for a Cause.* Yale University Press, New Haven, Conn., pp. 75–120.

Martin, P. S., 1982, The pattern and meaning of Holarctic mammoth extinction. *In* Hopkins, D. M., Matthews, J. V., Jr., Schweger, C. E., and Young, S. B. (eds.), *Paleoecology of Beringia.* Academic Press, New York, pp. 399–408.

Martin, P. S., and Klein, R. G. (eds.), 1984, *Quaternary Extinctions.* University of Arizona Press, Tucson.

Martin, R. G., 1978, Northern and eastern Gulf of Mexico continent margin: Stratigraphy and structural framework. *In* Bouma, A. H., Moore, G. T., and Coleman, J. M. (eds.), Framework, facies, and oil–trapping characteristics of the upper continental margin. AAPG, Studies in Geology No. 7, pp. 21–42.

Martin, R. G., and Bouma, A. H., 1978, Physiography of Gulf of Mexico. *In* Bouma, A. H., Moore, G. T., and Coleman, J. M. (eds.), Framework, facies, and oil–trapping characteristics of the upper continental margin. AAPG, Studies in Geology No. 7, pp. 3–19.

Mather, W. W., 1838, Natural history of New York, Div. 4, Geology of New York, Part I, comprising the geology of the first geological district, Albany.

Matthews, R. K., 1984, *Dynamic Stratigraphy,* 2nd ed. Prentice–Hall, Englewood Cliffs, N.J.

Matthews, S. C., and Missarzhevsky, V. V., 1975, Small shelly fossils of late Precambrian and Early Cambrian age: A review of recent work. *J. Geol. Soc. London* **131**, 289–304.

Matthews, V., III (ed.), 1978, Laramide folding associated with basement block faulting in the western United States. *Geol. Soc. Am. Mem.* **151**.

Matthews, V., III, and Work, D. F., 1978, Laramide folding associated with basement block faulting along the northeastern flank of the Front Range, Colorado. *Geol. Soc. Am. Mem.* **151**, 101–124.

Matti, J. C., Murphy, M. A., and Finney, S. C., 1974, Summary of Silurian and Lower Devonian basin and basin–slope limestones, Copenhagen Canyon, Nevada. *Geology* **2**, 575–577.

Maughan, E. K., 1967, Eastern Wyoming, eastern Montana and the Dakotas. *U.S. Geol. Surv. Prof. Pap.* **515**, 129–156.

Maxwell, J. C., 1975, Anatomy of an orogen. *Geol. Soc. Am. Bull.* **85**, 1195–1204.

May, P. R., 1971, Pattern of Triassic–Jurassic diabase dikes around the North Atlantic in the context of predrift position of the continents. *Geol. Soc. Am. Bull.* **82**, 1285–1291.

Meckel, L. D., 1967, Origin of Pottsville conglomerates (Pennsylvanian) in The Central Appalachians. *Geol. Soc. Am. Bull.* **78**, 223–258.

Meckel, L. D., 1970, Paleozoic alluvial deposition in the central Appalachians: A summary. *In* Fisher, G. W., Pettijohn, F. J., Reed, J. C., Jr., and Weaver, K. N. (eds.), *Studies in Appalachian Geology: Central and Southern*. Wiley–Interscience, New York, pp. 49–67.

Melton, W., and Scott, H. W., 1973, Conodont–bearing animals from the Bear Gulch Limestone, Montana. *Geol. Soc. Am. Spec. Pap.* **141**, 31–65.

Miall, A. D., 1973, Regional geology of northern Yukon. *Bull. Can. Pet. Geol.* **21**, 81–116.

Miall, A. D., 1976, Devonian geology of Banks Island, Arctic Canada, and its bearing on the tectonic development of the circum–Arctic region. *Geol. Soc. Am. Bull.* **87**, 1599–1608.

Mickelson, D. M., Clayton, L., Fullerton, D. S., and Borns, H. W., Jr., 1983, The Late Wisconsin Glacial Record of the Laurentide Ice Sheet in the United States. *In* Wright, H. E., Jr., and Porter, S. C. (eds.), *Late Quaternary Environments of the United States*, Vol. 1. University of Minnesota Press, Minneapolis, pp. 3–37.

Milankovitch, M., 1941, Kanon der Erdbestrahlung und Seine anwerdung auf des eiszeitproblem [Laws of Earth's solar radiation and their applications to the problem of ice ages]. Acad. R. Serbe, ed. Spec. **133**, 633.

Milici, R. C., 1974, Stratigraphy and depositional environments of Upper Mississippian and Lower Pennsylvanian rocks in the southern Cumberland Plateau of Tennessee. *Geol. Soc. Am. Spec. Pap.* **148**, 115–133.

Milici, R. C., 1975, Structural patterns in the Southern Appalachians: Evidence for a gravity slide mechanism for Alleghanian deformation. *Geol. Soc. Am. Bull.* **86**, 1316–1320.

Milici, R. C., *et al.*, 1979, The Mississippian and Pennsylvanian (Carboniferous) Systems in the United States—Tennessee. *U.S. Geol. Surv. Prof. Pap.* **1110-G**.

Milici, R. C., and Wedow, H., Jr., 1977, Upper Ordovician and Silurian stratigraphy in Sequatchie Valley and parts of the adjacent Valley and Ridge, Tennessee. *U.S. Geol. Surv. Prof. Pap.* **996**.

Miller, R. H., 1978, Early Silurian to Early Devonian conodont biostratigraphy and depositional environments of the Hidden Valley Dolomite, southeastern California. *J. Paleontol.* **52**(2), 323–344.

Miller, R. H., and Walch, C. A., 1977, Depositional environments of Upper Ordovician through Lower Devonian rocks in the southern Great Basin. *In* Stewart, J. H., Stevens, C. H., and Fritsche, A. E. (eds.), Paleozoic paleogeography of the western United States. SEPM, Pacific Section, Los Angeles, pp. 165–180.

Mills, J. R. E., 1971, The dentition of *Morganucodon*. *In* Kermack, D. A., and Kermack, K. A. (eds.), *Early Mammals*. Academic Press (Linnean Society), London, pp. 29–64.

Minard, J. P., 1980, Geology of the Round Bay quadrangle, Anne Arundel Co., Md. *U.S. Geol. Surv. Prof. Pap.* **1109**.

Misch, P., 1966, Tectonic evolution of the northern Cascade of Washington State—A west–Cordilleran case history. *In* Symposium on the tectonic history, mineral deposits of the western Cordillera in British Columbia and in neighboring parts of the U.S.A. *Can. Inst. Min. Metal. Spec.* **8**, 101–148.

Mitchum, R. M., Jr., 1978, Seismic stratigraphic investigation of West Florida Slope, Gulf of Mexico. *In* Bouma, A. H., Moore, G. T., and Coleman, J. M. (eds.), Framework, facies, and oil–trapping characteristics of the upper continental margin. AAPG, Studies in Geology No. 7, pp. 193–223.

Molnar, P., and Tapponnier, P., 1977, The collision between India and Eurasia. *Sci. Am.* **236**(4), 30–41.

Money, P. L., 1968, The Wollaston Lake fold–belt system, Saskatchewan–Manitoba. *Can. J. Earth Sci.* **5**, 1489–1504.

Monger, J. W. H., 1977, Upper Paleozoic rocks of the western Canadian Cordillera and their bearing on Cordilleran evolution. *Can. J. Earth Sci.* **14**, 1832–1859.

Monger, J. W. H., Souther, J. G., and Gabrielse, H., 1972, Evolution of the Canadian Cordillera: A plate–tectonic model. *Am. J. Sci.* **272**, 577–602.

Monroe, W. H., 1947, Stratigraphy of outcropping Cretaceous beds of southern states. *Am. Assoc. Pet. Geol. Bull.* **31**, 1817–1824.

Moorbath, S., 1975, The geological significance of early Precambrian rocks. *Proc. Geol. Assoc.* **86**, 259–279.

Moorbath, S., 1977, The oldest rocks and the growth of continents. *Sci. Am.* **236**(3), 92–104.

Moorbath, S., and Pankhurst, R. J., 1976, Further Rb–Sr age and isotope evidence for the nature of the late Archean plutonic event in West Greenland. *Nature* **262**, 124–126.

Moorbath, S., O'Nions, R. K., Pankhurst, R. J., Gale, N. H., and McGregor, V. R., 1972, Further rubidium–strontium age determinations on the very early Precambrian rocks of the Godthab district, West Greenland. *Nature Phys. Sci.* **240**, 78–82.

Moore, C. H., and Druckman, Y., 1981, Burial diagenesis and porosity evolution, Upper Jurassic Smackover, Arkansas and Louisiana. *Am. Assoc. Pet. Geol.* **65**, 597–628.

Moore, G. T., *et al.*, 1978, Mississippi fan, Gulf of Mexico—Physiography, stratigraphy and sedimentational patterns. *In* Bouma, A. H., Moore, G. T., and Coleman, J. M. (eds.), Framework, facies, and oil–trapping characteristics of the upper continental margin. AAPG, Studies in Geology No. 7, pp. 155–191.

Moore, G. T., *et al.*, 1979, Investigation of Mississippi fan, Gulf of Mexico. *Am. Assoc. Pet. Geol. Mem.* **29**, 383–402.

Moore, R. C., 1936, Stratigraphic classification of the Pennsylvanian rocks of Kansas. *Kans. Geol. Surv. Bull.* **22**, 1–256.

Moore, R. C., 1950, Late Paleozoic cyclic sedimentation in central United States. *In* Rhythm in sedimentation, Reports of the 18th Int. Geol. Congr., (IV) Proc. Sect. C, pp. 5–16.

Moore, R. C. (ed.), 1956 (and continuing), *Treatise on Invertebrate Paleontology*. Geological Society of America and University of Kansas Press, Lawrence, multi–volumes.

Moore, R. C., 1958, *Introduction to Historical Geology*. McGraw–Hill, New York.

Moore, R. C. (ed.), 1969, *Treatise on Invertebrate Paleontology*, Pt. N, Vol. 1, Mollusca 6, Bivalvia. Geological Society of America and University of Kansas Press, Lawrence.

Moore, R. C., Lalicker, C. G., and Fischer, A. G., 1952, *Invertebrate Fossils*. McGraw–Hill, New York.

Moorhouse, W. W., and Beales, F. W., 1962, Fossils from the Animike, Port Arthur, Ontario. *Trans. R. Soc. Can.* **56**, 97–100.

Moran, S. R., *et al.*, 1976, Quaternary stratigraphy and history of North Dakota, southern Manitoba, and northwestern Minnesota. *In* Mahaney, W. C. (ed.), *Quaternary Stratigraphy of North America*. Dowden, Hutchinson & Ross, Stroudsburg, Pa., pp. 37–50.

Morey, G. B., 1972, The middle Precambrian-Minnesota River Valley. *In* Sims, P. K., and Morey, G. B. (eds.), Geology of Minnesota: A centennial volume. Minn. Geol. Surv., St. Paul, pp. 204–262.

Morey, G. B., 1973, Stratigraphic framework of middle Precambrian rocks in Minnesota. *Geol. Assoc. Can. Spec. Pap.* **12**, 211–249.

Morey, G. B., 1978, Lower and middle Precambrian stratigraphic nomenclature for east–central Minnesota. *Minn. Geol. Surv. Rep. Invest.* **21**.

Morey, G. B., and Sims, P. K., 1976, Boundary between two Precambrian W terranes and its geologic significance. *Geol. Soc. Am. Bull.* **87**, 141–152.

Morgridge, D. L., and Smith, W. B., Jr., 1972, Geology and discovery of Prudhoe Bay field, eastern Arctic slope, Alaska. *Am. Assoc. Pet. Geol. Mem.* **16**, 489–501.

Morris, H. T., Douglass, R. C., and Kopf, R. W., 1977, Stratigraphy and microfaunas of the Oquirrh Group in the southern east Tintic Mountains, Utah. *U.S. Geol. Surv. Prof. Pap.* **1025.**

Morris, R. C., 1971, Classification and interpretation of disturbed bedding types in Jackfork flysch rocks (Upper Mississippian), Ouachita Mountains, Arkansas. *J. Sediment. Petrol.* **41,** 410–424.

Morris, R. C., 1974a, Carboniferous rocks of the Ouachita Mountains, Arkansas: A study of facies patterns along the unstable slope and axis of a flysch trough. *Geol. Soc. Am. Spec. Pap.* **148,** 241–279.

Morris, R. C., 1974b, Sedimentary and tectonic history of the Ouachita Mountains. *Soc. Econ. Paleontol. Mineral. Spec. Publ.* **22,** 120–142.

Morris, S. C., and Whittington, H. B., 1979, The animals of the Burgess Shale. *Sci. Am.* **241**(1), 122–133.

Morrison, R. B., 1965, Quaternary geology of the Great Basin. *In* Wright, H. E., Jr., and Frey, D. G. (eds.), *The Quaternary of the United States.* Princeton University Press, Princeton, N.J., pp. 265–286.

Morton, J. E., 1967, *Molluscs.* Hutchinson University Library, London.

Mosimann, J. E., and Martin, P. S., 1975, Simulating overkill by Paleoindians. *Am. Sci.* **63,** 781–820.

Mossop, G. D., 1980, Geology of the Athabasca oilsands. *Science* **207,** 145–152.

Mount, J. F., Gevirtzman, D. A., and Signor, P. W., III, 1983, Precambrian–Cambrian transition problem in western North America. Part I, Tommotian fauna in the southwestern Great Basin and its implications for the base of the Cambrian System. *Geology* **11**(4), 224–226.

Moy-Thomas, J. A., and Miles, R. S., 1971, *Paleozoic Fishes.* Chapman & Hall, London.

Mudge, M. R., 1967, Midcontinent (Permian of). *U.S. Geol. Surv. Prof. Pap.* **515,** 93–126.

Muehlberger, W. R., 1980, The shape of North America during the Precambrian. *In* The National Research Council, Continental tectonics: Studies in geophysics. National Academy of Sciences, pp. 175–183.

Muir, M. D., and Grant, P. R., 1976, Micropaleontological evidence from the Onverwacht Group, South Africa. *In* Windley, B. F. (ed.), *The Early History of the Earth.* Wiley, New York, pp. 595–604.

Muller, E. H., 1965, Quaternary geology of New York. *In* Wright, H. E., Jr., and Frey, D. G. (eds.), *The Quaternary of the United States.* Princeton University Press, Princeton, N.J., pp. 99–112.

Multer, H. G., and Hoffmeister, J. E., 1968, Subaerial laminated crusts of the Florida Keys. *Geol. Soc. Am. Bull.* **79,** 183–192.

Munthe, J., and Coombs, M. C., 1979, Miocene dome–skulled chalicotheras (Mammalia, Perissodactyla) from the western United States: A preliminary discussion of a bizarre structure. *J. Paleontol.* **53,** 77–91.

Murany, E. E., 1964, Wasatch Formation of the Uinta Basin. *In* 13th Annu. Field Conf. Guidebook: Rocky Mountain Assoc. Pet. Geol., pp. 145–155.

Murray, G. E., 1961, *Geology of the Atlantic and Gulf Coastal Province of North America.* Harper & Row, New York.

Nairn, A. E. M., Churkin, M., Jr., and Stehli, F. G. (eds.), 1964, *The Ocean Basins and Margins,* 5. Plenum, New York.

Natland, M. L., and Kuenen, P. H., 1951, Sedimentary history of the Ventura basin, California and the action of turbidity currents. *Soc. Econ. Paleontol. Mineral. Publ.* **2,** 76–107.

Naylor, R. S., 1968, Origin and regional relationships of the core–rocks of the Oliverian domes. *In* Zen, E., White, W. S., and Hadley, J. B. (eds.), *Studies in Appalachian Geology: Northern and Maritime.* Wiley–Interscience, New York, pp. 231–240.

Nelson, A. R., and Locke, W. W., III, 1981, Quaternary stratigraphic usage in North America: A brief survey. *Geology* **9,** 134–137.

Nelson, B. K., and DePaolo, D. J., 1982, Crust formation age of the North American midcontinent. *Geol. Soc. Am. Abstr. Progr.* **14**(7), 575.

Nelson, H. F., Brown, C. W., and Brineman, J. H., 1962, Skeletal limestone classification. *Am. Assoc. Pet. Geol. Mem.* **1,** 224–252.

Ness, G., Levi, S., and Couch, R., 1980, Marine magnetic anomaly timescales for the Cenozoic and Late Cretaceous: A precis, critique, and synthesis. *Rev. of Geophys. and Space Phys.* **18,** 753–770.

Newell, N. D., 1956, Catastrophism and the fossil record. *Evolution* **X**(1), 97–101.

Newell, N. D., 1967, Revolutions in the history of life. *Geol. Soc. Am. Spec. Pap.* **89,** 63–91.

Newell, N. D., 1971, An outline history of tropical organic reefs. *Am. Mus. Novit.* **2465.**

Newell, N. D., and Boyd, D. W., 1970, Oyster–like Permian bivalves. *Bull. Am. Mus. Nat. Hist.* **143**(4), 221–281.

Newell, N. D., Rigby, J. K., Purdy, E. G., and Thurber, D. L., 1959, Organism communities and bottom facies, Great Bahama Bank. *Bull. Am. Mus. Nat. Hist.* **117,** 180–228.

Newman, K. R., 1979, Cretaceous/Paleocene boundary in the Denver Formation at Golden, Colorado, U.S.A. *In* Christensen, W. K., and Birkelund, T. (eds.), *Cretaceous–Tertiary Boundary Events,* Vol. II. University of Copenhagen, Denmark, pp. 246–248.

Nichols, D. J., Perry, W. J., Jr., and Haley, J. C., 1985, Reinterpretation of the palynology and age of Laramide syntectonic deposits, southwestern Montana, and revision of the Beaverhead Group. *Geology* **13,** 149–153.

Nichols, K. M., and Silberling, N. J., 1977, Depositional and tectonic significance of Silurian and Lower Devonian dolomites, Roberts Mountains and vicinity, east–central Nevada. *In* Stewart, J. H., Stevens, C. H., and Fritsche, A. E. (eds.), Paleozoic paleogeography of the western United States. SEPM, Pacific Section, Los Angeles, pp. 217–240.

Niem, A. R., 1976, Patterns of flysch deposition and deep–sea fans in the lower Stanley Group (Mississippian), Ouachita Mountains, Oklahoma and Arkansas. *J. Sediment. Petrol.* **46,** 633–646.

Nilsen, T. H., 1977, Paleogeography of Mississippian turbidites in south–central Idaho. *In* Stewart, J. H., Stevens, C. H., and Fritsche, A. E. (eds.), Paleozoic Paleogeography of the western United States. SEPM, Pacific Section, Los Angeles, pp. 275–299.

Nilsen, T. H., 1978, Late Cretaceous geology of California and the problem of the Proto-San Andreas fault. *In* Howell, D. G., and McDougall, K. A. (eds.), Mesozoic paleogeography of the western United States. SEPM, Pacific Section, Pacific Coast Paleogeography Symp. No. 2, pp. 559–573.

Nilsen, T. H., and McKee, E. H., 1979, Paleogene paleogeography of the western United States. *In* Armentrout, J. M., Cole, M. R., and Ter-Best, H., Jr. (eds.), *Cenozoic paleogeography of the western United States.* SEPM, Pacific Section, Pacific Coast Paleogeography Symp. No. 3, pp. 257–276.

Nilsson, T., 1983, *The Pleistocene.* Reidel, Dordrecht.

Nitecki, M. H. (ed.), 1979a, *Mazon Creek Fossils.* Academic Press, New York.

Nitecki, M. H., 1979b, Mazon Creek fauna and flora: A hundred years of investigation. *In* Nitecki, M. H. (ed.), *Mazon Creek Fossils.* Academic Press, New York, pp. 1–12.

Nogan, D. S., 1964, Foraminifera, stratigraphy, and paleoecology of the Aquia Formation of Maryland and Virginia. Cushman Foundation for Foraminiferal Res. Spec. Publ. No. 7, p. 50.

North, B. R., and Caldwell, W. G. E., 1975, Foraminiferal faunas in the Cretaceous System of Saskatchewan. *Geol. Assoc. Can. Spec. Pap.* **13,** 303–332.

North, F. K., 1971, The Cambrian of Canada and Alaska. *In* Holland, C. H. (ed.), *Cambrian of the New World.* Wiley, New York, p. 219.

Nurmi, R. D., 1978, Use of well logs in evaporite sequences. *In* Dean, W. E., and Schreiber, B. C. (eds.), Marine evaporites. SEPM, Short Course No. 4, pp. 144–176.

Nurmi, R. D., and Friedman, G. M., 1977, Sedimentology and depositional environments of basin–arc evaporites, Lower Salinan (Upper Silurian), Michigan basin. *In* Fisher, J. H. (ed.), Reefs and evaporites—Concepts and depositional models. AAPG, Studies in Geology No. 5, pp. 23–52.

Obradovich, J. D., and Cobban, W. A., 1975, A time–scale for the Late Cretaceous of the western interior of North America. *Geol. Assoc. Can. Spec. Pap.* **13,** 31–54.

Ocola, L. C., and Meyer, R. P., 1973, Central North American rift system, 1: Structure of the axial zone from seismic and gravimetric data. *J. Geophys. Res.* **23,** 5173–5194.

Odom, I. E., 1975, Feldspar–grain size relations in Cambrian arenites, upper Mississippi Valley. *J. Sediment. Petrol.* **45**(3), 636–650.

Odom, L., 1976, Was Florida a part of North America in the Lower Paleozoic? *Geol. Soc. Am. Abstr. Progr.* **8**(2), 237–238.

Ogden, J. G., III, 1967, Radiocarbon and pollen evidence for a sudden change in climate in the Great Lakes region approximately 10,000–years–ago. *In* Cushing, E. J., and Wright, H. E., Jr. (eds.), *Quaternary Paleoecology*, Vol. 7. Yale University Press, New Haven, Conn., pp. 117–127.

Ojakangas, R. W., and Matsch, C. L., 1980, Upper Precambrian (Eocambrian) Mineral Fork Tillite of Utah: A continental glacial and glaciomarine sequence. *Geol. Soc. Am. Bull.* **91**, 495–501.

Okulitch, V. J., 1955, Archaeocyatha. *In* Moore, R. C. (ed.), *Treatise on Invertebrate Paleontology*, Part E, Archaeocyatha. Geological Society of America and University of Kansas Press, Lawrence.

Oliver, W. A., Jr., DeWitt, W., Jr., Dennison, J. M., Hoskins, D. M., 1967, Devonian of the Appalachian basin, United States. *In* Oswald, D. H. (ed.), Devonian System, Int. Symp. Alberta Soc. Pet. Geol., **1**, 1001–1040.

Olson, E. C., 1971, *Vertebrate Paleozoology*. Wiley–Interscience, New York.

Olson, P. E., 1975, The Newark Supergroup. Triassic Conf., Wesleyan Univ., Abstr.

Olson, P. E., and Galton, P. M., 1977, Triassic–Jurassic tetrapod extinctions: Are they real? *Science* **197**, 983–986.

Olson, P. E., Cornet, B., and Thomson, K. S., 1978, Cyclic change in Late Triassic lacustrine communities. *Science* **201**, 729–733.

Olsson, R. K., 1970, Paleocene planktonic foraminiferal biostratigraphy and paleozoogeography of New Jersey. *J. Paleontol.* **44**, 589–597.

Olsson, R. K., Miller, K. G., and Ungrady, T. E., 1980, Late Oligocene transgression of Middle Atlantic Coastal Plain. *Geology* **8**, 549–554.

Oriel, S. S., Myers, D. H., and Crosby, E. J., 1967, West Texas Permian basin region. *U.S. Geol. Surv. Prof. Pap.* **515**, 21–64.

Orpen, J. L., and Wilson, J. F., 1981, Stromatolites at 3,500 Myr and a greenstone–granite unconformity in the Zimbabwean Archaean. *Nature* **291**, 219–221.

Orth, C. J., *et al.*, 1981, An iridium anomaly at the palynological Cretaceous–Tertiary boundary in northern New Mexico. *Science* **214**, 1341–1343.

Orth, C. J., *et al.*, 1984, A search for iridium abundance anomalies at two Late Cambrian biomere boundaries in western Utah. *Science* **223**, 163–165.

Osman, R. W., and Whitlatch, R. B., 1978, Patterns of species diversity: Fact or artifact? *Paleobiology* **4**(1), 41–54.

Ostenso, N. A. and Wold, R. J., 1973, Aeromagnetic evidence for origin of Arctic Ocean basin. *In* Pitcher, M. G. (ed.), Arctic geology. Am. Assoc. Pet. Geol. Mem. **19**, pp. 506–516.

Ostrom, J. H., 1969, Osteology of *Deinonychus antirrhopus*, an unusual theropod from the Lower Cretaceous of Montana. *Peabody Mus. Nat. His. Yale Univ. Bull.* **30**, 1–165.

Ostrom, J. H., 1972, Were some dinosaurs gregarious? *Palaeogeogr. Palaeoclimatol. Palaeoecol.* **11**, 287–301.

Ostrom, J. H., 1974, *Archaeopteryx* and the origin of flight. *Q. Rev. Biol.* **49**, 27–47.

Ostrom, J. H., 1980, The evidence in endothermy in dinosaurs. *In* Thomas, R. D. K., and Olson, E. C. (eds.), A cold look at the warm-blooded dinosaurs. AAAS Symp. **28**, 15–54.

Otvos, R. H., 1981, Tectonic lineaments of Pliocene and Quaternary shorelines, Northeast Gulf Coast. *Geology* **9**, 398–404.

Overstreet, W., and Bell, H., III, 1965, The crystalline rocks of South Carolina. *U.S. Geol. Surv. Bull.* **1183**.

Owens, J. P., and Denny, C. S., 1979, Upper Cenozoic deposits of the central Delmarva Peninsula, Maryland and Delaware. *U.S. Geol. Surv. Prof. Pap.* **1067**.

Owens, J. P., and Minard, J. P., 1979, Upper Cenozoic sediments of the lower Delaware Valley and the northern Delmarva Peninsula, New Jersey, Pennsylvania, Delaware and Maryland. *U.S. Geol. Surv. Prof. Pap.* **1067-D**.

Owens, J. P., and Sohl, N. F., 1973, Glauconites from New Jersey–Maryland Coastal Plain: Their K–Ar ages and application in stratigraphic studies. *Geol. Soc. Am. Bull.* **84**, 2811–2838.

Padian, K., 1983, A functional analysis of flying and walking in pterosaurs. *Paleobiology* **9**(3), 218–239.

Page, L. R., 1968, Devonian plutonic rocks in New England. *In* Zen, E., White, W. S., and Hadley, J. B. (eds.), *Studies in Appalachian Geology: Northern and Maritime.* Wiley–Interscience, New York, pp. 371–383.

Palmer, A. R., 1960, Early Late Cambrian stratigraphy of the United States. *J. Wash. Acad. Sci.* **50**, 8.

Palmer, A. R., 1962, Comparative ontogeny of some opisthoparian, gonatoparian, and proparian Upper Cambrian trilobites. *J. Paleontol.* **36**(1), 87–97.

Palmer, A. R., 1968, Cambrian trilobites of east–central Alaska. *U.S. Geol. Surv. Prof. Pap.* **559–B**.

Palmer, A. R., 1969, Cambrian trilobite distributions in North America and their bearing on the Cambrian paleogeography of Newfoundland. *Am. Assoc. Pet. Geol. Mem.* **12**, 139–144.

Palmer, A. R., 1971a, The Cambrian of the Great Basin and adjacent areas, western United States. *In* Holland, C. H. (ed.), *Cambrian of the New World.* Wiley, New York, p. 1.

Palmer, A. R., 1971b, The Cambrian of the Appalachians and eastern New England regions, eastern United States. *In* Holland, C. H. (ed.), *Cambrian of the New World.* Wiley, New York, p. 169.

Palmer, A. R., 1983, The decade of North American geology, 1983 geologic time scale. *Geology* **11**, 503–504.

Palmer, A. R., and Rozanov, A. Y., 1976, Archaeocyatha from New Jersey: Evidence for an intra–Cambrian unconformity in the north–central Appalachians. *Geology* **4**, 773–774.

Palmquist, J. C., 1978, Laramide structures and basement block faulting: Two examples from the Big Horn Mountains, Wyoming. *Geol. Soc. Am. Mem.* **151**, 125–138.

Palvides, L., 1981, The central Virginia volcanic–plutonic belt: An island arc of Cambrian(?) age. *U.S. Geol. Surv. Prof. Pap.* **1231–A**.

Panchen, A. L., 1977, The origin and early evolution of tetrapod vertebrae. *In* Andrews, S. M., Miles, R. S., and Walker, A. D. (eds.), *Problems in Vertebrate Evolution.* Academic Press, New York, pp. 289–318.

Panchen, A. L. (ed.), 1980, *The Terrestrial Environment and the Origin of Land Vertebrates.* Academic Press (Systematics Assoc.), London.

Paul, C. R. C., 1977, Evolution of the primitive echinoderms. *In* Hallam, A. (ed.), *Patterns of Evolution as Illustrated by the Fossil Record.* Elsevier, Amsterdam, pp. 123–158.

Paull, C. K., and Dillon, W. P., 1980, Structure, stratigraphy, and geologic history of Florida–Hatteras shelf and inner Blake Plateau. *Am. Assoc. Pet. Geol.* **64**, 339–358.

Peck, D. M., *et al.*, 1979, Late Miocene glacial–eustatic lowering of sea level: Evidence from the Tamiami Formation of south Florida. *Geology* **7**, 285–288.

Pedersen, K., Sichko, M., Jr., and Wolff, M. P., 1976, Stratigraphy and structure of Silurian and Devonian rocks in the vicinity of Kingston, N.Y. *In* Johnson, J. H. (ed.), Guidebook to field excursions, 48th Annu. Meet. New York State Geol. Assoc., pp. B–4–1–B–4–27.

Pepper, J. F., DeWitt, W., Jr., and Demarest, D. F., 1954, Geology of the Bedford Shale and Berea Sandstone in the Appalachian basin. *U.S. Geol. Surv. Prof. Pap.* **259**.

Pessagno, E. A., Jr., 1969, Upper Cretaceous stratigraphy of the western Gulf Coast area of Mexico, Texas, and Arkansas. *Geol. Soc. Am. Mem.* **111**.

Peterson, F., and Pipiringos, G. N., 1979, Stratigraphic relations of the Navajo Sandstone to middle Jurassic Formations, southern Utah and northern Arizona. *U.S. Geol. Surv. Prof. Pap.* **1035–B**.

Pettijohn, F. J., 1975. *Sedimentary Rocks*, 3rd ed. Harper & Row, New York.

Pewe, T. L., 1976, Late Cenozoic history of Alaska. *In* Mahaney, W. C. (ed.), *Quaternary Stratigraphy of North America.* Dowden, Hutchinson & Ross, Stroudsburg, Pa., pp. 493–506.

Pfeil, R. W., and Read, J. F., 1980, Cambrian carbonate platform marginal facies, Shady Dolomite, southwestern Virginia, U.S.A. *J. Sediment. Petrol.* **50**(1), 91–116.

Phillips, W. E. A., Stillman, C. J., and Murphy, T. A., 1976, A Caledonian plate tectonic model. *J. Geol. Soc. London* **132**, 579–609.

Picard, M. D., 1957, Green River and Lower Uinta Formations—Subsurface stratigraphic changes in central and eastern Uinta basin, Utah. *In* Guidebook, 8th Annu. Field Congr., Intermontane Assoc. Pet. Geol., pp. 116–130.

Pilger, R. H., Jr. (ed.), 1980, The origin of the Gulf of Mexico and the early opening of the central North Atlantic Ocean. Proc. Symp., LSU, Baton Rouge.

Pilger, R. H., Jr., 1981, Plate reconstructions, aseismic ridges, and low-angle subduction beneath the Andes. *Geol. Soc. Am. Bull.* **92**, 448–456.

Pinet, P. R., Popenoe, P., and Nelligan, D. F., 1981, Gulf Stream: Reconstruction of Cenozoic flow patterns over the Blake Plateau. *Geology* **9**, 266–270.

Piper, J. D. A., 1976, Definition of pre–2000 m.y. apparent polar movements. *Earth Planet. Sci. Lett.* **28**, 470–478.

Pipiringos, G. N., and Imlay, R. W., 1979, Lithology and subdivisions of the Jurassic Stump Formation in southeastern Idaho and adjoining areas. *U.S. Geol. Surv. Prof. Pap.* **1035-6**.

Pipiringos, G. N., and O'Sullivan, R. B., 1978, Principal unconformities in Triassic and Jurassic rocks, western interior United States—A preliminary survey. *U.S. Geol. Surv. Prof. Pap.* **1035-A**.

Pitcher, M. G., 1964, Evolution of Chazyan (Ordovician) reefs of eastern United States and Canada. *Bull. Can. Pet. Geol.* **12**, 632–691.

Pitcher, M. G. (ed.), 1973, Arctic geology. *Am. Assoc. Pet. Geol. Mem.* **19**.

Pitman, W. C., III, and Talwani, M., 1972, Sea–floor spreading in the North Atlantic. *Geol. Soc. Am. Bull.* **83**, 619–646.

Plauchut, B. P., 1973, Geology of Sverdrup basin—Summary. *Am. Assoc. Pet. Geol. Mem.* **19**, 76–82.

Playford, P. E., *et al.,* 1984, Iridium anomaly in the Upper Devonian of the Canning Basin, western Australia. *Science* **226**, 437–439.

Poag, C. W., 1978, Stratigraphy of the Atlantic continental shelf and slope of the United States. *Annu. Rev. Earth Planet. Sci.* **6**, 251–280.

Poag, C. W., 1980, Foraminiferal stratigraphy, paleoenvironments, and depositional cycles in the outer Baltimore Canyon Trough. *In* Scholle, P. A. (ed.), Geologic studies of the COST No. B–3 well, United States Mid–Atlantic continental slope area. *U.S. Geol. Surv. Circ.* **833**, 44–65.

Ponnamperuma, C. (ed.), 1977, *Chemical Evolution of the Early Precambrian.* Academic Press, New York.

Poole, F. G., 1974, Flysch deposits of Antler foreland basin, western United States. *Soc. Econ. Paleontol. Mineral. Spec. Publ.* **22**, 58–82.

Poole, F. G., and Stewart, J. H., 1964, Chinle Formation and Glen Canyon Sandstone in northeastern Utah and western Colorado. *U.S. Geol. Surv. Prof. Pap.* **501–D**, 30–39.

Poole, F. G., and Sandberg, C. A., 1977, Mississippian paleogeography and tectonics of the western United States. *In* Stewart, J. H., Stevens, C. H., and Fritsche, A. E. (eds.), Paleozoic paleogeography of the western United States, Pacific Section, SEPM, Pacific Coast Paleogeography Symp. No. 1, pp. 67–85.

Poole, F. G., Sandberg, C. A. and Boucot, A. J., 1977, Silurian and Devonian paleogeography of the western United States. *In* Stewart, J. H., Stevens, C. H., and Fritsche, A. E. (eds.), Paleozoic Paleogeography of the western United States, Pacific Section, SEPM, Pacific Coast paleogeography.

Poole, F. G., *et al.,* 1967, Devonian of the southwestern United States. *In* Devonian System, Int. Symp. Calgary, *Alberta Soc. Pet. Geol.* **1**, 879–912.

Porter, S. C., 1983, Introduction. *In* Wright, H. E., Jr., and Porter, S. C. (eds.), *Late Quaternary Environments of the United States,* Vol. 1. University of Minnesota Press, Minneapolis, pp. xi–2.

Porter, S. C., Pierce, K. L., and Hamilton, T. D., 1983, Late Wisconsin mountain glaciation in the western United States. *In* Wright, H. E., Jr., and Porter, S. C. (eds.), *Late Quaternary Environments of the United States,* Vol. 1. University of Minnesota Press, Minneapolis, pp. 71–114.

Potter, P. E., and Pryor, W. A., 1961, Dispersal centers of Paleozoic and later clastics of the upper Mississippi Valley and adjacent areas. *Geol. Soc. Am. Bull.* **72**, 1195–1250.

Potter, P. E., and Siever, R., 1956, Sources of basal Pennsylvanian sediments in the Eastern Interior basin; pt. 1, crossbedding. *J. Geol.* **64**, 225–244.

Potter, A. W., Hotz, P. E., and Rohr, D. M., 1977, Stratigraphy and inferred tectonic framework in the eastern Klamath Mountains, northern California. *In* Stewart, J. H., Stevens, C. H., and Fritsche, A. E.,

(eds.), Paleozoic paleogeography of the western United States, Pacific Section, SEPM, Pacific Coast Paleogeography Symp. No. 1, pp. 421–440.

Pratt, L. M., Phillips, T. L., and Dennison, J. M., 1978, Evidence of nonvascular land plants from the Early Silurian (Llandoverian) of Virginia, U.S.A. *Rev. Paleobot. Palynol.* **25**, 121–149.

Prothero, D. R., 1985, Mid–Oligocene extinction event in North American land mammals. *Science* **229**, 550–551.

Prouty, W. F., 1923, Geology and mineral resources of Clay County. *Ala. Geol. Surv. Cty. Rep.* **1** (Spec. Rep. 12).

Pryor, W. A., 1960, Cretaceous sedimentation in upper Mississippi embayment. *Am. Assoc. Pet. Geol.* **44**, 1473–1504.

Pryor, W. A., and Sable, E. G., 1974, Carboniferous of the Eastern Interior basin. *Geol. Soc. Am. Spec. Pap.* **148**, 281–313.

Purdy, E. G., Pusey, W. C., III, and Wantland, K. F., 1975, Continental shelf of Belize—Regional shelf attributes. *In* Purdy, E. G., Pusey, W. C., III, and Wantland, K. F. (eds.), Belize shelf—carbonate sediments, clastic sediments, and ecology. AAPG, Studies in Geology No. 2, pp. 1–40.

Puri, H. S., 1953, Contribution to the study of the Miocene in the Florida panhandle. *Fla. Dept. Conserv. Geol. Bull.* **36**, Part 1.

Rackoff, J. S., 1980, The origin of the tetrapod limb and the ancestry of tetrapods. *In* Panchen, A. L. (ed.), *The Terrestrial Environment and the Origin of Land Vertebrates.* Academic Press (Systematics Assoc.), London, pp. 255–289.

Rainwater, E. H., 1967, Resumé of Jurassic to Recent sedimentation history of the Gulf of Mexico basin. *Trans. Gulf Coast Assoc. Geol. Soc.* **17**, 179–210.

Ramsey, J. G., 1963, Structural investigations in the Barberton Mountain Land, eastern Transvaal. *Trans. Geol. Soc. S. Afr.* **66**, 353–398.

Ramsey, J. G., 1967, *Folding and Fracturing of Rocks.* McGraw–Hill, New York.

Randazzo, A. F., Swe, W., and Wheeler, W. H., 1970, A study of tectonic influence on Triassic sedimentation—The Wadesboro basin, central Piedmont. *J. Sediment. Petrol.* **40**, 998–1006.

Rankin, D. W., 1968, Volcanism related to tectonism in the Piscataquis volcanic belt, an island arc of Early Devonian age in north–central Maine. *In* Zen, E., White, W. S., and Hadley, J. B. (eds.), *Studies in Appalachian Geology: Northern and Maritime.* Wiley–Interscience, New York, pp. 355–369.

Rankin, D. W., 1970, Stratigraphy and structure of Precambrian rocks in northwestern North Carolina. *In* Fisher, G. W., Pettijohn, G. W., Reed, J. C., Jr., and Weaver, K. N. (eds.), *Studies in Appalachian Geology: Central and Southern.* Wiley–Interscience, New York, pp. 227–245.

Rankin, D. W., 1975, The continental margin of eastern North America in the Southern Appalachians: The opening and closing of the Proto–Atlantic ocean. *Am. J. Sci.* **275-A**, 298–336.

Rankin, D. W., Espenshade, G. H., and Shaw, K. W., 1973, Stratigraphy and structure of the metamorphic belt in northwestern North Carolina and southwestern Virginia: A study from the Blue Ridge across the Brevard fault zone to the Sauratown Mountains anticlinorium. *Am. J. Sci.* **273-A**, 1–40.

Rasetti, F., 1963, Middle Cambrian ptychoparioid trilobites from the conglomerates of Québec. *J. Paleontol.* **37**, 575.

Rast, N., and Stringer, P., 1980, A geotraverse across a deformed Ordovician ophiolite and its Silurian cover, northern New Brunswick. *Tectonophysics* **69**, 221–245.

Raup, D. M., 1976, Species diversity in the Phanerozoic: An interpretation. *Paleobiology* 2(4), 289–297.

Raup, D. M., 1977, Species diversity in the Phanerozoic: Systematists follow the fossils. *Paleobiology* 3(3), 328–329.

Raup, D. M., 1979, Size of the Permo–Triassic bottleneck and its evolutionary implications. *Science* **206**, 217–218.

Raup, D. M., and Stanley, S. M., 1978, *Principles of Paleontology,* 2nd ed. Freeman, San Francisco.

Raven, P. A., and Axelrod, D. I., 1975, History of the flora and fauna of Latin America. *Am. Sci.* **63**, 420–429.

Read, C. B., and Mamay, S. H., 1964, Upper Paleozoic floral zones and floral provinces of the United States. *U.S. Geol. Surv. Prof. Pap.* **454–K**, K1–K34.

Read, H. H., and Watson, J., 1975, *Introduction to Geology,* Vol. 2, Part I. Halsted Press, New York.

Reed, B. L., and Lamphere, M. A., 1973, Plutonic rocks of Alaska–Aleutian Range batholith. *Am. Assoc. Pet. Geol. Mem.* **19**, 421–428.

Reed, J. C., Jr., and Morgan, B. A., 1971, Chemical alteration and spilitization of the Catoctin greenstones, Shenandoah National Park. *Va. J. Geol.* **79**, 529–548.

Reeside, J. B., Jr., and Cobban, W. A., 1960, Studies of the Mowry Shale (Cretaceous) and contemporary formations in the United States and Canada. *U.S. Geol. Surv. Prof. Pap.* **355**.

Reeves, C. C., Jr., 1976, Quaternary stratigraphy of the High Plains. *In* Mahaney, W. C. (ed.), *Quaternary Stratigraphy of North America.* Dowden, Hutchinson & Ross, Stroudsburg, Pa., pp. 213–234.

Regal, P. J., and Gans, C., 1980, The revolution in thermal physiology: Implications for dinosaurs. *In* Thomas, R. D. K., and Olson, E. C. (eds.), A cold look at the warm–blooded dinosaurs. AAAS Symp. **28**, 167–188.

Regnell, G., 1966, Edrioasteroids. *In* Moore, R. C. (ed.), *Treatise on Invertebrate Paleontology,* Part 3(1), Echinodermata. Geological Society of America and University of Kansas Press, Lawrence, pp. 136–173.

Reif, D. M., and Slatt, R. M., 1979, Red bed members of the Lower Triassic Moenkopi Formation, southern Nevada: Sedimentology and paleogeography of a muddy tidal flat deposit. *J. Sediment. Petrol.* **49**(3), 869–890.

Reinemund, J. A., 1955, Geology of the Deep River coal field of North Carolina. *U.S. Geol. Surv. Prof. Pap.* **246**.

Reinhardt, J., 1977, Cambrian off–shelf sedimentation, Central Appalachians. *Soc. Econ. Paleontol. Mineral. Spec. Publ.* **23**, 83–112.

Reinhardt, J., 1981, Upper Cretaceous stratigraphy and depositional environments. *In* Reinhardt, J., and Gibson, T. G. (eds.), Upper Cretaceous and Lower Tertiary of the Chattahoochee River Valley, western Georgia and eastern Alabama. Ga. Geol. Soc., 16th Annu. Field Trip, Guidebook, pp. 2–8.

Reinhardt, J., and Gibson, T. G. (eds.), 1981, Upper Cretaceous and Lower Tertiary of the Chattahoochee River Valley, western Georgia and eastern Alabama. Ga. Geol. Soc., 16th Annu. Field Trip, Guidebook.

Repetski, J. E., 1978, A fish from the Upper Cambrian of North America. *Science* **200**, 529.

Retallack, G. J., 1983, A paleopedological approach to the interpretation of terrestrial sedimentary rocks: The mid–Tertiary fossil soils of Badlands National Park, South Dakota. *Geol. Soc. Am. Bull.* **94**, 823–840.

Rhodes, F. H. T., 1967, Permo–Triassic extinctions. *In* Harland, W. B., *et al.* (eds.), *The Fossil Record.* Geological Society of London, pp. 57–76.

Ricard, L. V., 1962, Late Cayugan (Upper Silurian) and Helderbergian (Lower Devonian) stratigraphy in New York. *N.Y. State Mus. Sci. Serv. Bull.* **386**.

Rice, C. L., Sable, E. G., Dever, G. R., Jr., and Kehn, T. M., 1979, The Mississippian and Pennsylvanian (Carboniferous) Systems of the United States—Kentucky. *U.S. Geol. Surv. Prof. Pap.* **1110–F**.

Rich, J. L., 1934, Mechanics of low–angle overthrust faulting as illustrated by Cumberland thrust block, Virginia, Kentucky, and Tennessee. *Am. Assoc. Pet. Geol. Bull.* **18**(12), 1584–1596.

Rich, M., 1977, Pennsylvanian paleogeographic patterns in the western United States. *In* Stewart, J. H., Stevens, C. H., and Fritsche, A. E. (eds.), Paleozoic paleogeography of the western United States. SEPM, Pacific Section, Los Angeles, pp. 87–111.

Richards, H. G., 1967, Stratigraphy of Atlantic Coastal Plain between Long Island and Georgia. Review. *Am. Assoc. Pet. Geol. Bull.* **51**, 2400–2429.

Richards, H. G., 1974, Tectonic evolution of Alaska. *A.A.P.G. Bull.,* **58**, 79–105.

Richardson, E. S., Jr., 1956, Pennsylvanian invertebrates of the Mazon Creek area, Illinois—Introduction; Insects; Marine fauna. *Fieldiana Geol.* **12**(1–3), 3–67.

Richmond, G. M., Fryxell, R., Neff, G. E., and Weiss, P. L. 1965, Glaciation of the Rocky Mountains. *In* Wright, H. E., Jr., and Frey, D. G. (eds.), *The Quaternary of the United States.* Princeton University Press, Princeton, N.J., pp. 217–230.

Richter, D. H., and Jones, D. L., 1973, Structure and stratigraphy of eastern Alaska Range, Alaska. *Am. Assoc. Pet. Geol. Mem.* **19**, 408–420.

Rickards, R. B., 1977, Patterns of evolution in the graptolites. *In* Hallam, A. (ed.), *Patterns of Evolution as Illustrated by the Fossil Record.* Elsevier, Amsterdam, pp. 333–358.

Roberts, A. E., 1979, Northern Rocky Mountains and adjacent plains region. *U.S. Geol. Surv. Prof. Pap.* **1010**, 221–247.

Roberts, D., and Gale, G. H., 1978, The Caledonian–Appalachian Iapetus Ocean. *In* Tarling, D. H. (ed.), *Evolution of the Earth's Crust.* Academic Press, New York, pp. 255–342.

Roberts, R. J., Holtz, P. E., Gilluly, J. and Ferguson, H. G., 1958, Paleozoic rocks of north–central Nevada. *Am. Assoc. Pet. Geol. Bull.* **42**, 2813–2857.

Robertson, J. A., 1961, Geology of Townships 143 and 144. *Ont. Dep. Mines Geol. Rep.* **4**.

Robertson, J. A., 1963, Geology of townships 155, 156, 161, and 162. *Ont. Dep. Mines Geol. Rep.* **13**.

Robertson, J. A., 1971, A review of recently acquired geological data, Blind River–Elliot Lake area. *Ont. Dep. Mines North. Aff. Misc. Pap.* **45**.

Robertson, F. and Marshall, F. C., 1975, *Historical Geology–Manual of Laboratory Exercises,* 3rd ed. Burgess Publ. Co., Minneapolis.

Robinson, P., and Hall, L., 1979, Tectonic synthesis of southern New England. *In* Wones, D. (ed.), The Caledonides in the U.S.A. Blacksburg, VPI, pp. 73–82.

Robison, R. A., 1960, Lower and Middle Cambrian stratigraphy of the eastern Great Basin. *In* Geology of east–central Nevada. Intermontane Assoc. Pet. Geol., Annu. Field Conf., **11**, pp. 43–52.

Robison, R. A., 1964, Upper Middle Cambrian stratigraphy in western Utah. *Geol. Soc. Am. Bull.* **75**, 995–1010.

Robison, R. A., 1972, Mode of life of agnostid trilobites. 24th Int. Geol. Congr., Sect. 7, p. 33.

Rodgers, J., 1956, The known Cambrian deposits of the southern and central Appalachian mountains. El Sistema Cambrico, su paleogeografia y el problema de su base 2, part 2, 20th Int. Geol. Congr., pp. 353–385.

Rodgers, J., 1970, *The Tectonics of the Appalachians.* Wiley–Interscience, New York.

Rodgers, J., 1971, The Taconic orogeny. *Geol. Soc. Am. Bull.* **82**, 1141–1178.

Rodgers, J., 1972, Latest Precambrian (Post–Grenville) rocks of the Appalachian region. *Am. J. Sci.* **272**, 507–520.

Rodgers, J., 1981, The Merrimack synclinorium in northeastern Connecticut. *Am. J. Sci.* **281**, 176–186.

Rodgers, J., 1982, The life history of a mountain range—The Appalachians. *In* Hsü, K. J. (ed.), *Mountain Building Processes.* Academic Press, New York, pp. 229–241.

Roeder, D., and Mull, C. G., 1978, Tectonics of Brooks Range Ophiolites, Alaska. *Am. Assoc. Pet. Geol. Bull.* **62**, 1696–1702.

Roehler, H. W., 1979a, Geology of the Cooper Ridge N.E. Quadrangle, Sweetwater Co., Wyoming. *U.S. Geol. Surv. Prof. Pap.* **1065–B**.

Roehler, H. W., 1979b, Geology and energy resources of the Sand Butte Rim N.W. Quadrangle, Sweetwater Co., Wyoming. *U.S. Geol. Surv. Prof. Pap.* **1065–A**.

Rogers, H. D., 1838, Second annual report on the geological exploration of the state of Pennsylvania. Harrisburg, Pa.

Rolfe, W. E. I., 1980, Early invertebrate terrestrial faunas. *In* Panchen, A. L. (ed.), *The Terrestrial Environment and the Origin of Land Vertebrates.* Academic Press (Systematics Assoc.), London, pp. 117–155.

Romer, A. S., 1952, Late Pennsylvanian and Early Permian vertebrates of the Pittsburg–West Virginia region. *Ann. Carnegie Mus.* **33**, 47–112.

Romer, A. S., 1966, *Vertebrate Paleontology,* 3rd ed. University of Chicago Press, Chicago.

Root, S. I., 1970, Structure of the northern terminus of the Blue Ridge in Pennsylvania. *Geol. Soc. Am. Bull.* **81**, 815–830.

Roscoe, S. M., 1973, The Huronian Supergroup, a Paleoaphebian succession showing evidence of atmospheric evolution. *Geol. Assoc. Can. Spec. Pap.* **12**, 31–47.

Rose, K. D., 1982, Skeleton of *Diacodexis,* oldest known artiodactyl. *Science* **216**, 621–623.

Ross, C. A., 1978, Late Pennsylvanian and Early Permian sedimentary rocks and tectonics setting of the Marathon geosyncline. SEPM Publ. 78–17, pp. 89–93.

Ross, C. A., 1979, Late Paleozoic collision of North and South America. *Geology* **7**(1), 41–44.

Ross, C. P., 1970, The Precambrian of the United States: Northwestern United States, the Belt Series. *In* Rankama, K. (ed.), *The Precambrian*, Vol. 4. Wiley–Interscience, New York, pp. 145–252.

Ross, R. J., 1977, Ordovician paleogeography of the western United States. *In* Stewart, J. H., Stevens, C. H., and Fritsche, A. E. (eds.), Paleozoic paleogeography of the western United States. SEPM, Pacific Section, Los Angeles, pp. 19–38.

Rowland, S. M., 1984, Were there framework reefs in the Cambrian? *Geology* **12**, 181–183.

Rowley, D. B., and Kidd, W. S. F., 1981, Stratigraphic relationships and detrital composition of the medial Ordovician flysch of western New England: Implications for the tectonic evolution of the Taconic orogeny. *J. Geol.* **89**, 199–218.

Rowley, P. D., Steven, T. A., Anderson, J. J., and Cunningham, C. G., 1979, Cenozoic stratigraphy and structural framework of southwestern Utah. *U.S. Geol. Surv. Prof. Pap.* **1149**.

Rudwick, M. J. S., 1970, *Living and Fossil Brachiopods*. Hutchinson University Library, London.

Ruppel, S. C., and Walker, K. R., 1984, Petrology and depositional history of a Middle Ordovician carbonate platform: Chickamauga Group, northeastern Tennessee. *Geol. Soc. Am. Bull.* **95**(5), 568–583.

Russell, D. A., 1967, Systematics and morphology of American mosasaurs (Reptilia, Sauria). *Peabody Mus. Nat. Hist. Yale Univ. Bull.* **23**.

Russell, D. A., 1982, Paleontological consensus on the extinction of the dinosaurs? *Geol. Soc. Am. Spec. Pap.* **190**, 401–406.

Russell, E. E., 1975, Upper Cretaceous Selma equivalents in Tennessee. *In* Sterns, R. G. (ed.), Field trips in west Tennessee. Southeastern Geol. Soc. Am., Tenn. Div. Geol., Rep. Invest., No. 36, pp. 16–24.

Russell, I. C., 1878, On the physical history of the Triassic Formation in New Jersey and Connecticut. *N.Y. Acad. Sci. Ann.* **1**, 220–254.

Russell, L. S., 1975, Mammalian faunal succession in the Cretaceous System of western North America. *Geol. Assoc. Can. Spec. Pap.* **13**, 137–162.

Rutten, M. G., 1971, *The Origin of Life by Natural Causes*. Elsevier, Amsterdam.

Rutter, N. W., 1976, Multiple glaciation in the Canadian Rocky Mountains. *In* Wright, H. E., Jr., and Frey, D. G. (eds.), *The Quaternary of the United States*. Princeton University Press, Princeton, N.J., pp. 409–440.

Ryer, T. A., 1983, Transgressive–regressive cycles and the occurrence of coal in some Upper Cretaceous strata of Utah. *Geology* **11**(4), 207–210.

Ryland, J. S., 1970, *Bryozoans*. Hutchinson University Library, London.

Sable, E. G., 1977, Geology of the western Romanzof Mountains, Brooks Range, northeastern Alaska. *U.S. Geol. Surv. Prof. Pap.* **897**.

Saleeby, J. B., Goodin, S. E., Sharp, W. D., and Busby, C. J., 1978, Early Mesozoic paleotectonic–paleogeographic reconstruction of the southern Sierra Nevada region. *In* Howell, D. G., and McDougall, K. A. (eds.), Mesozoic paleogeography of the western United States. SEPM, Pacific Section, Pacific Coast Paleogeography Symp. No. 2, pp. 311–336.

Sales, J. K., 1968, Crustal mechanics of Cordilleran foreland deformation: A regional and scale–model approach. *Am. Assoc. Pet. Geol. Bull.* **52**, 2016–2044.

Salop, L. I., 1977, *Precambrian of the Northern Hemisphere*. Elsevier, Amsterdam.

Salop, L. I., and Scheinmann, Y. M., 1969, Tectonic history and structure of platforms and shields. *Tectonophysics* **7**, 565–597.

Sandberg, C. A., and Mapel, W. J., 1967, Devonian of the northern Rocky Mountains and plains. *In* Devonian System, Int. Symp. Calgary, Alberta, *Soc. Pet. Geol.* **1**, 843–877.

Sanders, J. E., 1963, Late Triassic tectonic history of northeastern United States. *Am. J. Sci.* **261**, 501–523.

Sanders, J. E., 1968, Stratigraphy and primary sedimentary structures of fine–grained, well–bedded strata, inferred lake deposits, Upper Tri-

assic, central and southern Connecticut. *Geol. Soc. Am. Spec. Pap.* **82**, 265–305.

Sandford, B. V., 1967, Devonian of Ontario and Michigan. *In* Devonian System, Int. Symp. Calgary, Alberta. *Soc. Pet. Geol.* **1**, 973–999.

Sando, W. J., 1976, Mississippian history of the northern Rocky Mountains region. *U.S. J. Geol. Surv. Res.* **4**, 317–338.

Sando, W. J., Mackenzie, G., and Dutro, J. T., 1975, Stratigraphy and geologic history of the Amsden Formation (Mississippian and Pennsylvanian) of Wyoming. *U.S. Geol. Surv. Prof. Pap.* **848–A**.

Sarjeant, W. A. S., 1974, *Fossil and Living Dinoflagellates*. Academic Press, New York.

Savage, D. E., and Russell, D. E., 1983, *Mammalian Paleofaunas of the World*. Addison–Wesley, Reading, Mass.

Schafer, J. P., and Hartshorn, J. H., 1965, The Quaternary of New England. *In* Wright, H. E., Jr., and Frey, D. G. (eds.), *The Quaternary of the United States*. Princeton University Press, Princeton, N.J., pp. 113–127.

Schamel, S., Hanley, T. B., and Sears, J. W., 1980, Geology of the Pine Mountain Window and adjacent terranes in the Piedmont province of Alabama and Georgia. Guidebook Southeast. Sect. Geol. Soc. Am. 29th Annu. Meet.

Schlee, J. S., 1957, Upland gravels of Southern Maryland. *Geol. Soc. Am. Bull.* **68**, 1371–1409.

Schlee, J. S., 1977, Stratigraphy and Tertiary development of the continental margin east of Florida. *U.S. Geol. Surv. Prof. Pap.* **581–F**.

Schlee, J. S., 1981, Seismic stratigraphy of Baltimore Canyon Trough. *Am. Assoc. Pet. Geol. Bull.* **65**, 26–53.

Schlee, J. S., Dillon, W. P., and Grow, J. A., 1979, Structure of the continental slope off the eastern United States. *Soc. Econ. Paleontol. Mineral. Spec. Publ.* **27**, 95–117.

Schluger, P. R., 1973, Stratigraphy and sedimentary environments of the Devonian Perry Formation, New Brunswick, Canada, and Maine, U.S.A. *Geol. Soc. Am. Bull.* **84**, 2533–2548.

Schopf, J. M., 1975, Pennsylvanian climate in the United States. *U.S. Geol. Surv. Prof. Pap.* **853**, 23–31.

Schopf, J. W., 1968, Microflora of the Bitter Springs Formation, Late Precambrian, central Australia. *J. Paleontol.* **42**, 651–688.

Schopf, J. W., 1975, Precambrian paleobiology: Problems and perspectives. *Annu. Rev. Earth Planet. Sci.* **3**, 213–249.

Schopf, J. W., and Oehler, D. Z., 1976, How old are the eucaryotes? *Science* **193**, 47–49.

Schopf, J. W., and Walter, M. R., 1983, Archaean microfossils: New evidence of ancient microbes. *In* Schopf, J. W. (ed.), *Earth's Earliest Biosphere*. Princeton University Press, Princeton, N.J., pp. 214–239.

Schopf, T. J. M., 1982, Extinction of the dinosaurs, a 1982 understanding. *Geol. Soc. Am. Spec. Pap.* **190**, 415–422.

Schuchert, C., 1923, Sites and nature of the North American geosynclines. *Geol. Soc. Am. Bull.* **34**, 151–229.

Schuepbach, M. A., and Vail, P. R., 1980, Evolution of outer highs on divergent continental margins. *In* Continental Tectonics, Studies in Geophysics. National Academy of Sciences, pp. 50–61.

Schultz, L. G., Tourtelot, H. A., Gill, J. R., and Boerngen, J. G., 1980, Composition and properties of the Pierre Shale and equivalent rocks, northern Great Plains region. *U.S. Geol. Surv. Prof. Pap.* **1064–B**.

Schwartz, R. M., and Dayhoff, M. O., 1978, Origins of procaryotes, eucaryotes, mitochondria, and chloroplasts. *Science* **199**, 395–403.

Schweichert, R. A., 1976, Early Mesozoic rifting and fragmentation of the Cordilleran orogen in the western U.S.A. *Nature* **260**, 586–591.

Schweichert, R. A., 1978, Triassic and Jurassic paleogeography of the Sierra Nevada and adjacent regions, California and western Nevada. *In* Howell, D. G., and McDougall, K. A. (eds.), Mesozoic paleogeography of the western United States. SEPM, Pacific Section, Pacific Coast Paleogeography Symp. No. 2, pp. 361–384.

Schweichert, R. A., and Cowan, D. S., 1975, Early Mesozoic tectonic evolution of the western Sierra Nevada, California. *Geol. Soc. Am. Bull.* **86**, 1329–1336.

Schwimmer, D. R., 1975, Quantitative taxonomy and biostratigraphy of Middle Cambrian trilobites from Montana and Wyoming. *Math. Geol.* **7**(2), 149–165.

Schwimmer, D. R., 1981, A distinctive Upper Cretaceous fauna, 3–4 meters below the Blufftown–Cusseta contact in the Chattahoochee River Valley. *In* Reinhardt, J., and Gibson, T. G. (eds.), Upper

Cretaceous and Lower Tertiary of the Chattahoochee River Valley, western Georgia and eastern Alabama. *Ga. Geol. Soc., 16th Annu. Field Trip, Guidebook*, pp. 81–88.

Scotese, C. R., *et al.*, 1979, Paleozoic base maps. *J. Geol.* **87**, 217–277.

Scott, G. R., and Cobban, W. A., 1959, So–called Hygiene Group of northeastern Colorado. *In* Guidebook, 11th Annu. Field Conf., Rocky Mt. Assoc. Geol., pp. 124–131.

Scott, K. R., Hayes, W. E., and Fietz, R. P., 1961, Geology of the Eagle Mills Formations. *Trans. Gulf Coast Assoc. Geol. Soc.* **18**, 124–165.

Scott, R. W., 1970, Stratigraphy and sedimentary environments of Lower Cretaceous rocks, southern western interior. *Am. Assoc. Pet. Geol. Bull.* **54**(7), 1225–1244.

Sears, J. W., and Price, R. A., 1978, The Siberian connection: A case for Precambrian separation of the North American and Siberian cratons. *Geology* **6**, 267–270.

Sears, J. W., Graff, P. J., and Holden, G. S., 1982, Tectonic evolution of lower Proterozoic rocks, Uinta Mountains, Utah and Colorado. *Geol. Soc. Am. Bull.* **93**, 990–997.

Sears, S. O., and Lucia, F. J., 1979, Reef–growth model for Silurian pinnacle reefs, northern Michigan reef trend. *Geology* **7**, 299–302.

Secor, D. T., Jr., Samson, S. L., Snoke, A. W., and Palmer, A. R., 1983, Confirmation of the Carolina Slate Belt as an exotic terrane. *Science* **221**, 649–651.

Secor, D. T., Jr., *et al.*, 1986, Character of the Alleghanian orogeny in the Southern Appalachians: Part I. Alleghanian deformation in the eastern Piedmont of South Carolina. *Gol. Soc. Am. Bull.* **97**, 1319–1328.

Secor, D. T., Jr., Snoke, A. W., and Dallmeyer, R. D., 1986, Character of the Alleghanian orogeny in the Southern Appalachians: Part III. Regional tectonic relations. *Geol. Soc. Am. Bull.* **97**, 1345–1353.

Seiders, V. M. and Wright, J. E., 1977, Geology of the Carolina Slate Belt in the Asheboro, North Carolina area. Geol. Soc. Am. Southeast. Sect., Guidebook, North Carolina Dept. of Nat. and Econ. Res., Geol. and Min. Res. Sect.

Seilacher, A., 1953, Studien zur Palichnologie, I. Uber die Methoden der Palichnologie. *Neuer Jahib. Geol. und Paläontol.* **96**, 421–452.

Sellars, R. T., 1967, The Siluro–Devonian rocks of the Ouachita Mountains. *In* Toomey, D. F. (ed.), Silurian–Devonian rocks of Oklahoma and environs. *Tulsa Geol. Soc. Digest* **35**, 231–241.

Sepkoski, J. J., Jr., 1979, A kinetic model of Phanerozoic taxonomic diversity. II. Early Phanerozoic families and multiple equilibria. *Paleobiology* **5**, 222–251.

Sepkoski, J. J., Jr., 1982, A compendium of fossil marine families. *Milwaukee Public Museum Contrib. in Biol. and Geol.* No. 51.

Sevon, W. D., 1969, The Pocono Formation in northeastern Pennsylvania. Guidebook for the 34th Annual Field Conference of Pennsylvania Geologists, Harrisburg, Bur. Topogr. Geol. Surv.

Seyfert, C. K., 1980, Paleomagnetic evidence in support of a middle Proterozoic (Helikian) collision between North America and Gondwanaland as a cause of the metamorphism and deformation in the Adirondacks: Summary. *Geol. Soc. Am. Bull.* **91**(I), 118–120.

Seyfert, C. K., and Sirkin, L. A., 1979, *Earth History and Plate Tectonics: An Introduction to Historical Geology*, 2nd ed. Harper & Row, New York.

Shackleton, R. M., 1976, Pan–African structures. *Phil. Trans. R. Soc. London*, **A280**, 491–497.

Shanmugam, G., and Walker, K. R., 1978, Tectonic significance of distal turbidites in the Middle Ordovician Blockhouse and lower Sevier Formations in east Tennessee. *Am. J. Sci.* **278**, 551–578.

Shatsky, N. S., 1935, On the tectonics of the Arctic. *In* Trans., 1st Geol. Inv. Conf., Moscow, **1**, 149–168.

Shattuck, G. B., 1901, The Pleistocene problem of the North Atlantic Coastal Plain. *Johns Hopkins Univ. Circ.* **20**, 69–75; *Am. Geol.* **28**, 87–107.

Shaver, R. H., 1974, Silurian reefs of northern Indiana: Reef and interreef macrofaunas. *Am. Assoc. Pet. Geol. Bull.* **58**(6), 934–956.

Shaw, E. W., and Harding, S. R. L., 1949, Lea Park and Belly River Formations of east-central Alberta. *Am. Assoc. Pet. Geol. Bull.* **33**(4), 487–499.

Shaw, F. C., 1969, Stratigraphy of the Chazy Group (Middle Ordovician) in the northern Champlain Valley. N.Y. State Geol. Assoc. Guideb., 41st Annu. Meet., pp. 81–92.

Shawe, D. R., 1976, Geologic history of the Slick Rock district and vicinity, San Miguel and Dolores Counties, Colorado. *U.S. Geol. Surv. Prof. Pap.* **576-E**.

Shear, W. A., *et al.*, 1984, Early land animals in North America: Evidence from Devonian age arthropods from Gilboa, N.Y. *Science* **224**, 492–494.

Sheehan, P. M., 1977, Species diversity in the Phanerozoic: A reflection of labor by systematists? *Paleobiology* **3**(3), 325–327.

Sheldon, R. P., and Carter, M. D., 1979, Williston basin region. *U.S. Geol. Surv. Prof. Pap.* **1010**, 249–271.

Sheldon, R. P., Cressman, E. R., Cheney, T. M., and McKelvey, V. E., 1967, Middle Rocky Mountains and northeastern Great Basin. *U.S. Geol. Surv. Prof. Pap.* **515**, 157–174.

Sheridan, R. E., 1974, Atlantic continental margin of North America. *In* Burk, C. A., and Drake, C. L. (eds.), *The Geology of Continental Margins*. Springer-Verlag, Berlin, pp. 391–407.

Sheridan, R. E., Crosby, J. T., Bryan, G. M., and Stoffa, P. L., *et al.*, 1981, Stratigraphy and structure of southern Blake Plateau, northern Florida Straits, and northern Bahama Platform from multichannel seismic reflection data. *Am. Assoc. Pet. Geol. Bull.* **65**, 2571–2593.

Shideler, G. L., 1969, Dispersal patterns of Pennsylvanian sandstones in the Michigan basin. *J. Sediment. Petrol.* **39**(3), 1229–1237.

Shideler, G. L., 1970, Provenance of Johns Valley boulders in late Paleozoic facies, southeastern Oklahoma and southwestern Arkansas. *Am. Assoc. Pet. Geol. Bull.* **54**, 789–806.

Shilts, W. W., 1980, Flow patterns in the central North American ice sheet. *Nature* **286**, 213–218.

Shilts, W. W., Cunningham, C. M., and Kaszycki, C. A., 1979, Keewatin Ice Sheet—Re–evaluation of the traditional concept of the Laurentide Ice Sheet. *Geology* **7**, 537–541.

Shurbet, D. H., 1964, The high–frequency S phase and structure of the upper mantle. *J. Geophys. Res.* **69**, 2065–2070.

Siegel, B. Z., 1977, *Kakabekia*, a review of its physiological and environmental features and their relationship to its possible ancient affinities. *In* Ponnamperuma, C. (ed.), *Chemical Evolution of the Early Precambrian*. Academic Press, New York, pp. 143–154.

Siever, R., 1977, Early Precambrian weathering and sedimentation: An impressionistic view. *In* Ponnamperuma, C. (ed.), *Chemical Evolution of the Early Precambrian*. Academic Press, New York, pp. 13–23.

Signor, P. W., III, 1978, Species richness in the Phanerozoic: An investigation of sampling effects. *Paleobiology* **4**(4), 394–406.

Signor, P. W., III, and Brett, C. E., 1984, The mid–Paleozoic precursor of the Mesozoic marine revolution. *Paleobiology* **10**(2), 229–246.

Sigsby, R. J., 1976, Paleoenvironmental analysis of the Big Escambia Creek–Jay-Blackjack Creek Field area. *Trans. Gulf Coast Assoc. Geol. Soc.* **26**, 258–278.

Silberling, N. J., 1975, Age relationships of the Galconda thrust fault, Sonoma Range, north-central Nevada. *Geol. Soc. Am. Spec. Pap.* **163**.

Silberling, N. J., and Roberts, R. J., 1962, Pre-Tertiary stratigraphy and structure of northwestern Nevada. *Geol. Soc. Am. Spec. Pap.* **72**.

Silver, L. T., 1978, Precambrian formations and Precambrian history in Cochise County, southeastern Arizona. *In* Callendar, J. F., Wilt, J. C., and Clemmons, R. E. (eds.), Land of Cochise. N.M. Geol. Soc., 29th Field Conf. Guideb., pp. 157–163.

Silver, L. T., and Schultz, P. H. (eds.), 1982, Geological implications of impacts of large asteroids and comets on Earth. *Geol. Soc. Am. Spec. Pap.* **190**.

Simons, E. L., 1964, The early relatives of Man. *Sci. Am.*, **211**(1), 50–62.

Simpson, F., 1975, Marine lithofacies and biofacies of the Colorado Group (Middle Albian to Santonian) in Saskatchewan. *Geol. Assoc. Can. Spec. Pap.* **13**, 553–587.

Simpson, G. G., 1971, Concluding remarks: Mesozoic mammals revisited. *In* Kermack, D. A., and Kermack, K. A. (eds.), *Early Mammals*. Academic Press (Linnean Society), London, pp. 181–198.

Simpson, G. G., 1980, *Splendid Isolation: The Curious History of South American Mammals*. Yale University Press, New Haven, Conn.

Simpson, R. W., and Cox, A., 1977, Paleomagnetic evidence for tectonic rotation of the Oregon Coast Range. *Geology* **5**, 585–589.

Sims, P. K., 1980, Boundary between Archean greenstone and gneiss terranes in northern Wisconsin and Michigan. *Geol. Soc. Am. Spec. Pap.* **182**, 113–124.

Sims, P. K., Card, K. D., Morey, G. B., and Peterman, Z. E., 1980, The Great Lakes tectonic zone—A major crustal structure in central North America. *Geol. Soc. Am. Bull.* **91**, 690–698.

Skehan, J. W., 1965, The Olympic–Wallowa lineament: A major deep-seated tectonic feature of the Pacific Northwest (abs.). *Am. Geophsys. Union Trans.*, **46**, 71.

Skehan, J. W., *et al.*, 1978, Significance of fossiliferous Middle Cambrian rocks of Rhode Island to the history of the Avalonian microcontinent. *Geology*, **6**, 694–698.

Skelton, P. W., 1979, Preserved ligament in a radiolitid rudist bivalve and its implication of mantle marginal feeding in the group. *Paleobiology* **5**(2), 90–106.

Skipp, B., Sando, W. J., and Hall, W. E., 1979, The Mississippian and Pennsylvanian (Carboniferous) Systems in the United States—Idaho. *U.S. Geol. Surv. Prof. Pap.* **1110-AA**.

Sloss, L. L., 1963, Sequences in the cratonic interior of North America. *Geol. Soc. Am. Bull.* **74**, 93–114.

Sloss, L. L., 1984, The greening of stratigraphy 1933–1983. *Annu. Rev. Earth Planet. Sci.* **12**, 1–10.

Sloss, L. L., and Speed, R. C., 1974, Relationships of cratonic and continental-margin tectonic episodes. *Soc. Econ. Paleontol. Mineral. Spec. Publ.* **22**, 98–119.

Sloss, L. L., Dapples, E. C., and Krumbein, W. C., 1966, *Lithofacies Maps*. Wiley, New York.

Sloss, L. L., Krumbein, W. C., and Dapples, E. C., 1949, Integrated facies analysis. *Geol. Soc. Am. Mem.* **39**, 91–123.

Smith, J., 1979, The Cretaceous/Tertiary transition in the Barranco Del Gradero, Spain. *In* Christensen, W. K., and Birkelund, T. (eds.), *Cretaceous–Tertiary Boundary Events*, Vol. II. University of Copenhagen, Denmark, pp. 156–163.

Smith, D. L., 1977, Transition from deep- to shallow-water carbonates, Paine Member, Lodgepole Formation, central Montana. *Soc. Econ. Paleontol. Mineral. Spec. Publ.* **25**, 187–201.

Smithson, S. B., Murphy, D. J., and Houston, R. S., 1971, Development of an augen gneiss terrain. *Contrib. Mineral. Petrol.* **33**, 184–190.

Snedden, J. W., and Kersey, D. G., 1981, Origin of San Miguel Lignite deposit and associated lithofacies, Jackson Group, South Texas. *Am. Assoc. Pet. Geol. Bull.* **65**, 1099–1109.

Snoke, A. W., and Secor, D. T., Jr., 1982, The Eastern Piedmont fault system and its relationship to Alleghenian tectonics in the Southern Appalachians: A discussion. *J. Geol.* **90**, 209–213.

Sohl, N. G., and Kauffman, E. G., 1964, Giant Upper Cretaceous oysters from the Gulf Coast and Caribbean. *U.S. Geol. Surv. Prof. Pap.* **483-H**.

Sohn, I. E., 1979, Non-marine ostracodes in the Lakota Foundation (Lower Cretaceous) from South Dakota and Wyoming. *U.S. Geol. Surv. Prof. Pap.* **1069**.

Solomon, B. J., McKee, E. H., and Andersen, D. W., 1979, Stratigraphy and depositional environments of Paleogene rocks near Elko, Nevada. *In* Armentrout, J. M., Cole, M. R., and TerBest, H., Jr. (eds.), Cenozoic paleogeography of the western United States. SEPM, Pacific Section, Pacific Coast Paleogeography Symp. No. 3, pp. 75–88.

Sonnenfeld, P., Hudec, P. P., Turek, A., and Boon, J. A., 1977, Base metal concentration in a density–stratified evaporite pan. *In* Fisher, J. H. (ed.), Reefs and evaporites—Concepts and depositional models. AAPG, *Studies in Geology* No. 5, pp. 181–187.

Speed, R. C., 1971, Permo–Triassic continental margin tectonics in western Nevada. *Geol. Soc. Am. Abstr. Progr.* **3**, 197.

Speed, R. C., 1977a, An appraisal of the Pablo Formation of presumed late Paleozoic age, central Nevada. *In* Stewart, J. H., Stevens, C. H., and Fritsche, A. E. (eds.), Paleozoic paleogeography of the western United States. SEPM, Pacific Section, Los Angeles, pp. 315–324.

Speed, R. C., 1977b, Excelsior Formation, west–central Nevada: Stratigraphic appraisal, new dimensions, and paleogeographic interpretations. *In* Stewart, J. H., Stevens, C. H., and Fritsche, A. E. (eds.), Paleozoic paleogeography of the western United States. SEPM, Pacific Section, Los Angeles, pp. 325–336.

Speed, R. C., 1977c, Island–arc and other paleogeographic terranes of late Paleozoic age in the western Great Basin. *In* Stewart, J. H., Stevens, C. H., and Fritsche, A. E. (eds.), Paleozoic paleogeography of the

western United States. SEPM, Pacific Section, Los Angeles, pp. 349–362.

Speed, R. C., 1978, Paleogeographic and plate tectonic evolution of the early Mesozoic marine province of the western Great Basin. *In* Howell, D. G., and McDougall, K. A. (eds.), Mesozoic paleogeography of the western United States. SEPM, Pacific Section, Pacific Coast Paleogeography Symp. No. 2, pp. 253–270.

Speed, R. C., and Sleep, N. H., 1980, Antler orogeny: A model. *Geol. Soc. Am. Abstr. Progr.* **12**(7), 527.

Speed, R. C., 1979, Collided Paleozoic microplate in the western United States. *J. Geol.* **87**, 279–292.

Speed, R. C., McMillan, J. R., Poole, F. G., and Kleinhampl, F. J., 1977, Diablo Formation, central–western Nevada; composite of deep and shallow water upper Paleozoic rocks. *In* Stewart, J. H., Stevens, C. H., and Fritsche, A. E. (eds.), Paleozoic paleogeography of the western United States. SEPM, Pacific Section, Los Angeles, pp. 301–314.

Spieker, E. W., 1949, Sedimentary facies and associated diastrophism. *Geol. Soc. Am. Mem.* **39**, 55–81.

Sprinkle, H. J., 1976, Classification and phylogeny of Pelmatozoan echinoderms. *Syst. Zool.* **25**, 83–91.

Stalker, A. M., 1976, Quaternary stratigraphy of the southwestern Canadian prairies. *In* Mahaney, W. C. (ed.), *Quaternary Stratigraphy of North America*. Dowden, Hutchinson & Ross, Stroudsburg, Pa., pp. 381–408.

Stanford, D., Bonnichsen, R., and Morlan, R. E., 1981, The Ginsberg experiment: Modern and prehistoric evidence of a bone–flaking technology. *Science* **212**, 438–440.

Stanley, G. D., 1981, Early history of scleractinian corals and its geological significance. *Geology* **9**, 507–511.

Stanley, K. O., and Fagerstrom, J. A., 1974, Miocene invertebrate trace fossils from a braided river environment, western Nebraska, U.S.A. *Palaeogeogr. Palaeoclimatol. Palaeoecol.* **15**(1), 63–82.

Stanley, K. O., and Surdam, R. C., 1978, Sedimentation on the front of Eocene Gilbert–type deltas, Washakie Basin, Wyoming. *J. Sediment. Petrol.* **48**, 557–573.

Stanley, K. O., Jordan, W. M., and Dott, R. H., Jr., 1971, New hypothesis of Early Jurassic paleogeography and sediment dispersal for western United States. *Am. Assoc. Pet. Geol. Bull.* **55**(1), 10–19.

Stanley, S. M., 1973, An ecological theory for the sudden origin of multicellular life in the late Precambrian. *Proc. Natl. Acad. Sci. USA* **70**(5), 1486–1489.

Stanley, S. M., 1976, Fossil data and the Precambrian–Cambrian evolutionary transition. *Am. J. Sci.* **276**, 56–76.

Stanley, S. M., Addicott, W. O., and Chinzer, K., 1980, Lyellian curves in paleontology: Possibilities and limitations. *Geology* **8**, 422–426.

Stanton, T. W., and Vaughan, T. W., 1920. The fauna of the Cannonball marine member of the Lance Formation. *U. S. Geol. Surv. Prof. Paper* **128-A**.

Stearn, C. W., 1980, Classification of Paleozoic stromatoporoids. *J. Paleont.* **54**(5), 881–902.

Stearn, C. W., Carroll, R. L., and Clark, T. H., 1979, *Geological Evolution of North America*, 3rd ed. Wiley, New York.

Stearns, D. W., 1971, Mechanisms of drape folding in the Wyoming Province. *In* Wyo. Geol. Assoc., 23rd Annu. Field Conf., Wyo. Tectonics Symp., Guidebook, pp. 125–143.

Stearns, D. W., 1978, Faulting and forced folding in the Rocky Mountains foreland. *Geol. Soc. Am. Mem.* **151**, 1–37.

Stein, R. S., 1975, Dynamic analysis of *Pteranodon ingens*: A reptilian adaptation to flight. *J. Paleontol.* **49**(5), 534–548.

Stekk, C. R., 1975, Basement control of Cretaceous sand sequences in western Canada. *Geol. Assoc. Can. Spec. Pap.* **13**, 427–440.

Stevens, C. H., 1977, Permian depositional provinces and tectonics, western United States. *In* Stewart, J. H., Stevens, C. H., and Fritsche, A. E. (eds.), Paleozoic paleogeography of the western United States. SEPM, Pacific Section, Los Angeles, pp. 113–135.

Stevenson, J. S., 1971, Origin of the Onaping Tuff, Sudbury, Ontario. Geol. Assoc. Can., Mineral. Assoc. Can., Abstr. Progr., Sudbury, p. 68.

Stewart, J. H., 1972, Initial deposits in the Cordilleran geosyncline: Evidence of a late Precambrian continental separation. *Geol. Soc. Am. Bull.* **83**, 1345–1360.

Stewart, J. H., 1976, Late Precambrian evolution of North America: Plate tectonics implication. *Geology* **4**, 11–15.

Stewart, J. H., 1980, Regional tilt patterns of late Cenozoic Basin–Range fault blocks, western United States. *Geol. Soc. Am. Bull.*, Pt. 1, v. 91, pp. 460–464.

Stewart, J. H., and Poole, F. G., 1974, Lower Paleozoic and uppermost Precambrian Cordilleran miogeosyncline, Great Basin, western United States. *Soc. Econ. Paleontol. Mineral. Spec. Publ.* **22**, 28–57.

Stewart, J. H., and Suczek, C. A., 1977, Cambrian and latest Precambrian paleogeography and tectonics in the western United States. *In* Stewart, J. H., Stevens, C. H., and Fritsche, A. E. (eds.), Paleozoic paleogeography of the western United States. SEPM, Pacific Section, Los Angeles, pp. 1–17.

Stewart, J. H., Poole, F. G., and Wilson, R. F., 1972, Stratigraphy and origin of the Chinle Formation and related Upper Triassic strata in the Colorado Plateau region. *U.S. Geol. Surv. Prof. Pap.* **690**.

Stewart, J. H., Stevens, C. H., and Fritsche, A. E. (eds.), 1977b, Paleozoic paleogeography of the western United States. SEPM, Pacific Section, Los Angeles.

Stewart, J. H., McMillan, J. R., Nichols, K. M., and Stevens, C. H., 1977, Deep–water upper Paleozoic rocks in north–central Nevada— A study of the type area of the Havallah Formation. *In* Stewart, J. H., Stevens, C. H., and Fritsche, A. E. (eds.), Paleozoic paleogeography of the western United States. SEPM, Pacific Section, Los Angeles, pp. 337–347.

Stewart, W. N., 1983, *Paleobotany and the Evolution of Plants.* Cambridge University Press, London.

St. Jean, J., 1973, A new Cambrian trilobite fauna from the Piedmont of North Carolina. *Am. J. Sci.* **273–A**, 196–216.

St. Julien, P., and Hubert, C., 1975, Evolution of the Taconic orogen in the Quebec Appalachians. *Am. J. Sci.* **275-A**, 337–362.

Stockwell, C. H., 1964, Fourth report on structural provinces, orogenies, and time classification of rocks of the Canadian Precambrian Shield. *Can. Geol. Surv. Pap.* **64–17**, Pt. II, pp. 1–21.

Stockwell, C. H., 1972, Revised Precambrian time scale for the Canadian Shield. *Can. Geol. Surv. Pap.* **72–52**.

Stockwell, C. H., 1976, Introduction to Chapter IV, Geology of the Canadian Shield. *Geol. Surv. Can. Econ. Geol. Rep.* No. 1, pp. 44–54.

Stokes, W. L., 1952, Lower Cretaceous in Colorado Plateau. *Am. Assoc. Pet. Geol. Bull.* **36**(9), 1766–1776.

Stokes, W. L., Peterson, J. A., and Picard, M. D., 1955, Correlation of Mesozoic formations of Utah. *Am. Assoc. Pet. Geol. Bull.* **39**(10), 2003–2019.

Stone, D. S., 1969, Wrench faulting and Rocky Mountain tectonics. *Mt. Geol.* **6**, 67–79.

Stose, A. J., and Stose, G. W., 1944, Geology of Carroll and Frederick Counties. *In* The physical features of Carroll County and Frederick County. Md. Dept. of Geology, Mines and Water Res., 11–131.

Stott, D. F., 1975, The Cretaceous System in northeastern British Columbia. *Geol. Assoc. Can. Spec. Pap.* **13**, 427–441.

Stowe, C. W., 1974, The older tonalite gneiss complex in the Selukwe area, Rhodesia. *Geol. Soc. S. Afr. Spec. Publ.* **3**, 85–96.

Strong, D. F., Dickson, W. L., O'Driscoll, C. F., Kean, B. F., and Stevens, R. K., 1974, Geochemical evidence for an east–dipping Appalachian subduction zone in Newfoundland. *Nature* **248**, 37–39.

Stuiver, M., Huesser, C. J., and Yang, I. C., 1978, North American glacial history extended to 75,000 years ago. *Science* **200**, 16–21.

Sues, H. D., 1980, A pachycephalosaurid dinosaur from the Upper Cretaceous of Madagascar and its paleobiological implications. *J. Paleontol.* **54**, 954–962.

Suppe, J., 1970, Offset of late Mesozoic basement terranes by the San Andreas fault system. *Geol. Soc. Am. Bull.* **81**, 3253–3258.

Surlyk, F., 1979, Guide to Stevns Klint. *In* Birkelund, T., and Bromley, R. G. (eds.), *Cretaceous–Tertiary Boundary Events,* Vol. I. University of Copenhagen, Denmark, pp. 164–170.

Sutherland, P. K., and Henry, T. W., 1977, Carbonate platform facies and new stratigraphic nomenclature of the Morrow Series (Lower and Middle Pennsylvanian), northeastern Oklahoma. *Geol. Soc. Am. Bull.* **88**, 425–440.

Suttner, L. J., 1969, Stratigraphic and petrographic analysis of Upper Jurassic–Lower Cretaceous Morrison and Kootenai Formations, southwest Montana. *Am. Assoc. Pet. Geol. Bull.* **53**(7), 1391–1410.

Sutton, J., 1978, Proterozoic of the North Atlantic. *In* Tarling, D. H. (ed.), *Evolution of the Earth's Crust.* Academic Press, New York, pp. 239–254.

Swain, F. M., 1951, Cenozoic Ostracoda, Pt. 1 of Ostracoda from wells in North Carolina. *U.S. Geol. Surv. Prof. Pap.* **234-A**.

Swann, D. H., 1963, Classification of Genevievian and Chesterian (Late Mississippian) rocks of Illinois. *Ill. State Geol. Surv. Rep. Invest.* **216**.

Swann, D. H., 1964, Mississippian rhythmic sediments of the Mississippi Valley. *Am. Assoc. Pet. Geol. Bull.* **48**, 637–658.

Swanson, D. A. and Wright, T. L., 1979, Source area and distribution of major units in the Columbia River Basalt Group. *Geol. Soc. Am. Abstr.* **11**, No. 3, 131.

Sweeney, J. F., 1977, Subsidence of the Sverdrup basin, Canadian Arctic Islands. *Geol. Soc. Am. Bull.* **88**, 41–48.

Sweet, W. C., 1985, Conodonts, those fascinating little whatzits. *J. Paleontol.* **59**(3), 485–494.

Swift, D. J. P., 1968, Coastal erosion and transgressive stratigraphy. *J. Geol.* **76**, 444–456.

Swift, D. J. P., Heron, S. D., Jr., and Dill, C. E., Jr., 1969, The Carolina Cretaceous: Petrographic reconnaissance of a graded shelf. *J. Sediment. Petrol.* **39**, 18–33.

Swinton, W. E., 1970, *The Dinosaurs.* Allen & Unwin, London.

Szalay, F. S. and Delson, E., 1979, *Evolutionary History of the Primates.* Academic Press, New York.

Szaniawski, H., 1982, Chaetognath grasping spines recognized among Cambrian protoconodonts. *J. Paleontol.* **56**(3), 806–810.

Tailleur, I., 1973, Probable rift origin of the Canada Basin. *Am. Assoc. Pet. Geol. Mem.* **19**, 526–535.

Tappan, H., 1976, Possible eucaryotic algae (Bangiophycidae) among early Proterozoic microfossils. *Geol. Soc. Am. Bull.* **87**, 633–639.

Tapponier, P., and Molnar, P., 1976, Slip–line field theory and large scale continental tectonics. *Nature* **264**, 319–324.

Tarling, D. H. (ed.), 1978a, *Evolution of the Earth's Crust.* Academic Press, New York.

Tarling, D. H., 1978b, The first 600 million years. *In* Tarling, D. H. (ed.), *Evolution of the Earth's Crust.* Academic Press, New York, pp. 1–17.

Tarney, J., Dalziel, I. W. D., and de Wit, M. J., 1976, Marginal basin 'Rocas Verdes' complex from S Chile: A model for Archean greenstone belt formation. *In* Windley, B. F. (ed.), *The Early History of the Earth.* Wiley, New York, pp. 131–146.

Tasch, P., 1980, *Paleobiology of the Invertebrates,* 2nd ed. Wiley, New York.

Taylor, M. E., 1966, Precambrian mollusk–like fossils from Inyo County, California. *Science* **153**, 3732.

Taylor, M. E., 1977, Late Cambrian of western North America: Trilobite biofacies, environmental significance, and biostratigraphic implications. *In* Kauffman, E. G., and Hazel, J. E. (eds.), *Concepts and Methods of Biostratigraphy.* Dowden, Hutchinson & Ross, Stroudsburg, Pa., pp. 397–425.

Taylor, T. N., 1981, *Paleobotany.* McGraw-Hill, New York.

Teichert, C., and Moore, R. C., 1964, Classification and stratigraphic distribution. *In* Moore, R. C. (ed.), *Treatise on Invertebrate Paleontology,* Part K, Mollusca 3. Geological Society of America and University of Kansas Press, Lawrence, pp. 94–106.

Tennyson, M. E., and Cole, M. R., 1978, Tectonic significance of upper Mesozoic Methow–Pasayten sequence, northeastern Cascade Range, Washington and British Columbia. *In* Howell, D. G., and McDougall, K. A. (eds.), Mesozoic paleogeography of the western United States. SEPM, Pacific Section, Pacific Coast Paleogeography Symp. No. 2, pp. 33–70.

Terasmae, J., and Dreimanis, A., 1976, Quaternary stratigraphy of southern Ontario. *In* Mahaney, W. C. (ed.), *Quaternary Stratigraphy of North America.* Dowden, Hutchinson & Ross, Stroudsburg, Pa., pp. 51–63.

Thayer, P. A., 1970, Stratigraphy and geology of the Dan River Triassic basin, North Carolina. *Southeast. Geol.* **12**, 1–32.

Thierstein, H. R., 1982, Terminal Cretaceous plankton extinctions: A critical reassessment. *Geol. Soc. Am. Spec. Pap.* **190**, 385–400.

Thierstein, H. R., and Berger, W. H., 1978, Injection events in ocean history. *Nature* **276**, 461–466.

Thomas, J. J., Shuster, R. D., and Bickford, M. E., 1984, A terrane of 1,350- to 1,400-m.y.-old silicic volcanic and plutonic rocks in the buried Proterozoic of the mid-continent and in the Wet Mountains, Colorado. *Geol. Soc. Am. Bull.* **95**, 1150–1157.

Thomas, M. D., and Gibb, R. A., 1977, Gravity anomalies and deep structure of the Cape Smith fold belt, northern Ungava, Québec. *Geology* **5**, 169–172.

Thomas, R. D. K., and Olson, E. C. (eds.), 1980, A cold look at the warm–blooded dinosaurs. AAAS Symp. **28**.

Thomas, W. A., 1966, Late Mississippian folding of a syncline in the western Appalachians, West Virginia and Virginia. *Geol. Soc. Am. Bull.* **77**, 473–494.

Thomas, W. A., 1973, Southwestern Appalachian structural system beneath the Gulf Coastal Plain. *Am. J. Sci.* **273-A**, 372–390.

Thomas, W. A., 1974, Converging clastic wedges in the Mississippian of Alabama. *Geol. Soc. Am. Spec. Pap.* **148**, 187–207.

Thomas, W. A., 1977, Evolution of Appalachian–Ouachita salients and recesses from reentrants and promontories in the continental margin. *Am. J. Sci.* **277**, 1233–1278.

Thomas, W. A., 1979, Mississippian stratigraphy of Alabama. *U.S. Geol. Surv. Prof. Pap.* **1110**, I1–I22.

Thomas, W. A., and Cramer, H. R., 1979, Georgia. *U.S. Geol. Surv. Prof. Pap.* **1110**, H1–H37.

Thomas, W. A., Tull, J. F., Bearce, D. N., Russell, G., and Odom, A. L., 1980, Geologic synthesis of the southernmost Appalachians, Alabama and Georgia. *In* Wones, D. R. (ed.), Proceedings of IGCP Project 27, Caledonide orogen, 1979 Meeting. *Blacksburg, Va. Inst. Dep. Geol. Sci. Mem.* No. 2, pp. 83–90.

Thomasson, M. R., 1959, Late Paleozoic stratigraphy and paleotectonics of central and eastern Idaho. Ph.D. thesis, University of Wisconsin, Madison.

Thompson, J. B., and Norton, S. A., 1968, Paleozoic regional metamorphism in New England and adjacent areas. *In* Zen, E., White, W. S., and Hadley, J. B. (eds.), *Studies in Appalachian Geology: Northern and Maritime.* Wiley–Interscience, pp. 319–327.

Thompson, M. L., 1964, Fusulinacea. *In* Moore, R. C. (ed.), *Treatise on Invertebrate Paleontology,* Part C, Protista 2. Geological Society of America and University of Kansas Press, Lawrence, pp. 358–436.

Thomson, A. F., and McBride, E. F., 1964, Summary of the geologic history of the Marathon geosyncline. *SEPM Publ.* **64–9**, pp. 52–60.

Thomson, A. F., and Thomasson, M. R., 1969, Shallow to deep water facies development in the Dimple Limestone (Lower Pennsylvanian), Marathon region, Texas. *Soc. Econ. Paleontol. Mineral. Spec. Publ.* **14**, 57–78.

Thomson, J. E., 1957, Geology of the Sudbury Basin. *Ont. Dep. Mines* **65**, Pt. 3.

Thomson, K. S., 1977, The pattern of diversification among fishes. *In* Hallam, A. (ed.), *Patterns of Evolution as Illustrated by the Fossil Record.* Elsevier, Amsterdam, pp. 377–404.

Thomson, K. S., 1980, Ecology of Devonian lobe–finned fishes. *In* Panchen, A. L. (ed.), *The Terrestrial Environment and the Origin of Land Vertebrates.* Academic Press (Systematics Assoc), London, pp. 187–224.

Thorsteinsson, R., and Tozer, E. T., 1976, Geology of the Arctic Archipelago. *Geol. Surv. Can. Econ. Geol. Rep.* No. 1, pp. 547–590.

Tipper, H. W., and Richards, T. A., 1976, Jurassic stratigraphy and history of north–central British Columbia. *Can. Geol. Surv. Bull.* **270**.

Tipper, J. C., 1979, Rarefaction and rarefiction: The use and abuse of a method in paleoecology. *Paleobiology* **5**(4), 423–434.

Tomlinson, C. W., and McBee, W. D., Jr., 1959, Pennsylvanian sediments and orogenies of Ardmore district, Oklahoma. *In* Petroleum geology of southern Oklahoma—A symposium. *Ardmore Geol. Soc.* **2**, 3–52.

Torrey, T. W., 1967, *Morphogenesis of the Vertebrates,* 2nd ed. Wiley, New York.

Towe, K. M., 1978, *Tentaculites,* evidence for a brachiopod affinity? *Science* **201**, 626–628.

Tozer, D. C., 1972, The concept of a lithosphere. *Geophys. Int.* **13**, 363–388.

Tremblay, L. P., 1968, Preliminary account of the Goulburn Group, Northwest Territories, Canada. *Geol. Surv. Can. Pap.* **67–8**.

Trettin, H. P., 1967, Devonian of the Franklin eugeosyncline. *In* Devonian System, Int. Symp. Calgary, Alberta, *Soc. Pet. Geol.* **1**, 693–701.

Trettin, H. P., 1972, The Innuitian Province. *Geol. Assoc. Can. Spec. Pap.* **11**, 83–179.

Trettin, H. P., 1973, Early Paleozoic evolution of northern parts of Canadian Arctic Archipelago. *Am. Assoc. Pet. Geol. Mem.* **19**, 57–75.

Tschudy, R. H. and Tschudy, B. D., 1986, Extinction and survival of plant life following the Cretaceous/Tertiary boundary event, Western Interior, North America. *Geology,* **14 (8)**, 667–670.

Tull, J. F., 1978, Structural development of the Alabama Piedmont northwest of the Brevard Zone. *Am. J. Sci.* **278**, 442–460.

Tull, J. F., 1982, Stratigraphic framework of the Talledega slate belt, Alabama Appalachians. *Geol. Soc. Am. Spec. Publ.* **191**, 3–18.

Tull, J. F., and Stowe, S. H., 1980, The Hillabee Greenstone: A mafic volcanic complex in the Appalachian Piedmont of Alabama. *Geol. Soc. Am. Bull.* **91**(I), 27–36.

Turekian, K. K. (ed.), 1971, *Late Cenozoic Glacial Ages.* Yale University Press, New Haven, Conn.

Tweto, O., and Lovering, T. S., 1977, Geology of the Minturn 15–minute quadrangle, Eagle and Summit Counties, Colorado. *U.S. Geol. Surv. Prof. Pap.* **956**.

Vail, P. R., and Hardenbol, J., 1979, Sea–level changes during the Tertiary. *Oceanus* **22**, 71–79.

Vail, P. R., and Mitchum, R. M., Jr., 1979, Global cycles of relative changes of sea level from seismic stratigraphy. *Am. Assoc. Pet. Geol. Mem.* **29**, 469–472.

Valentine, J. W., 1977, General patterns of metazoan evolution. *In* Hallam, A., (ed.), *Patterns of Evolution as Illustrated by the Fossil Record.* Elsevier, Amsterdam, pp. 27–58.

Valentine, J. W., and Moores, E. M., 1972, Global tectonics and the fossil record. *J. Geol.* **80**, 167–184.

Valentine, J. W., Hedgecock, D., Zumwalt, G. S., and Ayala, F. J., (eds.), 1973, Mass extinctions and genetic polymorphism in the "killer clam," *Tridacna. Geol. Soc. Am. Bull.* **84**, 3411–3414.

Van Andel, T. H., 1975, Mesozoic/Cenozoic calcite compensation depth and the global distribution of calcareous sediments. *Earth Planet. Sci. Lett.* **26**, 187–194.

Van de Graaf, F. R., 1972, Fluvial deltaic facies of the Castlegate Sandstone (Cretaceous), east-central Utah. *J. Sediment. Petrol.* **42**(3), 558–571.

Van der Voo, R., Mauk, F. J., and French, R. B., 1977, Permian–Triassic continental configuration and the origin of the Gulf of Mexico. *Geology* **5**, 177–180.

Van Horn, R., 1976, Geology of the Golden Quadrangle. *U.S. Geol. Surv. Prof. Pap.* **872**.

Van Houten, F. B., 1964, Origin of redbeds: Some unresolved problems. *In* Nairn, A. E. M., (ed.), *The Problems in Paleoclimatology.* Interscience Publishers, New York, pp. 647–659.

Van Houten, F. B., and Bhattacharyya, D. P., 1982, Phanerozoic oolitic ironstones. *Annu. Rev. Earth Planet. Sci.* **10**, 441–458.

Van Schmus, W. R., 1972, Geochronology of Precambrian rocks in the Penokean foldbelt subprovince of the Canadian Shield (abstr.) *18th Annu. Inst. on Lake Superior Geology,* (I), Pap. **32**, Houghton, Mich.

Van Schmus, W. R., 1980, Chronology of igenous rocks associated with the Penokean Orogeny in Wisconsin. *Geol. Soc. Am. Spec. Pap.* **182**, 159–168.

Van Schmus, W. R., and Bickford, M. E., 1981, Proterozoic chronology and evolution of the midcontinent region, North America. *In* Kröner, A. (ed.), *Precambrian Plate Tectonics.* Elsevier, Amsterdam, pp. 260–296.

Van Valen, L., and Sloan, R. E., 1965, The earliest primates. *Science* **150**, 743–745.

Van West, F. P., 1972, Green River oil shale. *In* Geologic atlas of the Rocky Mountain region. Rocky Mountain Assoc. Pet. Geol., Denver.

Veizer, J., 1983, Geologic evolution of the Anchean–early Proterozoic Earth. In Schopf, J. W. (ed.), *Earth's Earliest Biosphere: Its Origin and Evolution*. Princeton University Press, pp. 240–259.

Vigrass, L. W., 1968, Geology of Canadian heavy oil sands. *Am. Assoc. Pet. Geol. Bull.* **52**(10), 1984–1999.

Viljoen, M. L., and Viljoen, R. P., 1969, A collection of 9 papers on many aspects of the Barberton granite-greenstone belt, South Africa. *Geol. Soc. S. Afr. Spec. Publ.* **2**.

Vogt, P. R., 1973, Early events in the opening of the North Atlantic. *In* Tarling, D. H., and Runcorn, S. K. (eds.), *Implications of Continental Drift to the Earth Sciences*. Academic Press, New York, pp. 693–712.

Vogt, P. R., and Johnson, G. L., 1973, Magnetic telechemistry of oceanic crust? *Nature* **245**, 373–375.

Vogt, P. R., and Ostenso, N. A., 1970, Magnetic and gravity profiles across the Alpha Cordillera and their relation to Arctic sea–floor spreading. *J. Geophys. Res.* **75**, 4925–4937.

Von Brunn, V., and Hobday, D. K., 1976, Early Precambrian tidal sedimentation in the Pongola Supergroup of South Africa. *J. Sediment. Petrol.* **46**, 670–679.

Von Huene, R., Moore, G. W., Moore, J. C., and Stephens, C. D., 1979, Cross–section, Alaska Peninsula–Kodiak Island–Aleutian Trench: Summary: *Geol. Soc. Am. Bull., (I)*, **90**, 427–430.

Voorhies, M. R., and Thomasson, J. R., 1979, Fossil grass authoecia within Miocene rhinoceros skeletons: Diet in an extinct species. *Science* **206**, 331–333.

Waage, K. M., 1975, Deciphering the basic sedimentary structure of the Cretaceous System in the western interior. *Geol. Assoc. Can. Spec. Pap.* **13**, 55–82.

Waitt, R. B., Jr., 1980, About forty last–glacial Lake Missoula jökulhlaups through southern Washington. *J. Geol.* **88**, 653–679.

Waitt, R. B., Jr., 1985, Case for periodic, colossal jokulhlaups from Pleistocene glacial Lake Missoula. *Geol. Soc. Am. Bull.* **96**, 1271–1286.

Waitt, R. B., Jr., and Thorson, R. M., 1983, The Cordilleran Ice Sheet in Washington, Idaho, and Montana. *In* Wright, H. E., Jr., and Porter, S. C. (eds.), *Late Quaternary Environments of the United States*, Vol. 1. University of Minnesota Press, Minneapolis, pp. 53–70.

Walcott, C. D., 1910, Abrupt appearance of the Cambrian fauna in the North American continent: Cambrian geology and paleontology II. *Smithson. Misc. Collect.* **57**, 1–16.

Walker, J. C. G., *et al.*, 1983, Environmental evolution of the Archean–early Proterozoic Earth. *In* Schopf, J. W. (ed.), *Earth's Earliest Biosphere: Its Origin and Evolution*, Princeton University Press, pp. 260–290.

Walker, K. R., 1964, The stratigraphy and petrography of the Price–Pocono Formation in a portion of southwestern Virginia. M.S. thesis, University of North Carolina, Chapel Hill.

Walker, K. R., and Laporte, L., 1970, Congruent fossil communities from Ordovician and Devonian carbonates of New York. *J. Paleontol.* **44**, 938–939.

Walker, R. G., 1971, Nondeltaic depositional environments in the Catskill clastic wedge (Upper Devonian) of central Pennsylvania. *Geol. Soc. Am. Bull.* **82**, 1305–1326.

Walker, R. G., and Sutton, R. G., 1967, Quantitative analysis of turbidites in the Upper Devonian Sonyea Group, N.Y. *J. Sediment. Petrol.* **37**, 1012–1022.

Wallace, C. A., and Crittenden, M. D., Jr., 1969, The stratigraphy, depositional environments, and correlation of the Precambrian Uinta Mountains Group, western Uinta Mountains, Utah. *In* Lindsay, J. B. (ed.), Geologic Guidebook of the Uinta Mountains, Utah's maverick range. Intermountain Assoc. Geol. 16th Annu. Field Conf., pp. 126–141.

Walper, J. L., 1980, Tectonic evolution of the Gulf of Mexico. *In* Pilger, R. H., Jr. (ed.), The origin of the Gulf of Mexico and the early opening of the central North Atlantic Ocean. Proc. Symp., LSU, Baton Rouge, pp. 87–98.

Walper, J. L., and Rowett, C. L., 1972, Plate tectonics and the origin of

the Caribbean Sea and the Gulf of Mexico. *Trans. Gulf Coast Assoc. Geol. Soc.* **22**, 105–116.

Walter, M. R., 1983, Archaean stromatolites: Evidence of the Earth's earliest benthos. *In* Schopf, J. W. (ed.), *Earth's Earliest Biosphere*. Princeton University Press, Princeton, N.J., pp. 187–213.

Walter, M. R., Goode, A. D. T., and Hall, W. D. M., 1976, Microfossils from a newly discovered Precambrian stromatolitic iron formation in Western Australia. *Nature* **261**, 221–223.

Wanless, H. R., 1955, Pennsylvanian rocks of the Eastern Interior basin. *Am. Assoc. Pet. Geol. Bull.* **39**(9), 1753–1820.

Wanless, H. R., 1975, Appalachian region. *U.S. Geol. Surv. Prof. Pap.* **853**, 17–62.

Wanless, H. R., and Bell, K. G., 1975, New England. *In* McKee, E. D. and Crosby, E. J. (coords.), Paleotectonic investigations of the Pennsylvanian System in the United States (I), *U.S. Geol. Surv. Prof. Pap.* **853**, 9–15.

Wanless, H. R., and Shepard, R. P., 1936, Sea level and climatic changes related to late Paleozoic cycles. *Geol. Soc. Am. Bull.* **47**, 1177–1206.

Wanless, H. R., and Weller, J. M., 1932, Correlation and extent of Pennsylvanian cyclothems. *Geol. Soc. Am. Bull.* **43**, 1003–1016.

Wanless, H. R., Tubbs, J. B. Jr., Gednetz, D. E., and Weiner, J. L., 1963, Mapping sedimentary environments of Pennsylvanian cycles. *Geol. Soc. Am. Bull.* **74**, 437–486.

Ward, P., 1979, Functional morphology of Cretaceous helically–coiled ammonite shells. *Paleobiology* **5**(4), 415–423.

Ward, P., 1980, Comparative shell shape distributions in Jurassic–Cretaceous ammonites and Jurassic–Tertiary nautilids. *Paleobiology* **6**(1), 32–44.

Wardlaw, B. R., Collinson, J. W., and Maughan, E. K., 1979, Stratigraphy of Park City Group equivalents (Permian) in southern Idaho, northeastern Nevada, and northwestern Utah. *U.S. Geol. Surv. Prof. Pap.* **1163–C**, 9–16.

Warner, J., 1979, Continental nuclei and a terrestrial magma ocean. *In* Mueller, P. A., and Wooden, J. L. (organizers), 1979 Field Conf. U.S. Natl. Comm. Archean Geochem., University of Florida, p. A–16.

Warren, A. D., 1957, The Anahuac and Frio sediments in Louisiana. *Trans. Gulf Coast Assoc. Geol. Soc.* **7**, 221–238.

Watkins, J. S., *et al.*, 1978, Occurrence and evolution of salt in deep Gulf of Mexico. *In* Bouma, A. H., Moore, G. T., and Coleman, J. M. (eds.), Framework, facies and oil-trapping characteristics of the upper continental margin. AAPG, Studies in Geology No. 7, pp. 43–65.

Watson, J. V., 1978, Precambrian thermal regimes. *Philos. Trans. R. Soc. London Ser. A* **288**, 431–440.

Wayne, W. J., and Zumberge, J. H., 1965, Pleistocene geology of Indiana and Michigan. *In* Wright, H. E., Jr., and Frey, D. G. (eds.), *The Quaternary of the United States*. Princeton University Press, Princeton, N.J., pp. 63–84.

Weaver, C. E., and Beck, K. C., 1977, Miocene of the S.E. United States: A model for chemical sedimentation in a peri–marine environment. *Sediment. Geol.* **17**(1 & 2).

Weigand, P. W., and Ragland, P. C., 1970, Geochemistry of Mesozoic dolerite dikes from eastern North America. *Contrib. Mineral. Petrol.* **29**, 195–214.

Weimer, R. J., 1959, Upper Cretaceous stratigraphy, Colorado. *In* Guidebook, 11th Annu. Field Conf. Rocky Mountain Assoc. Geol., 1959, pp. 9–16.

Weimer, R. J., and Hoyt, J. H., 1964, Burrows of *Callianassa major* Say, geologic indicators of littoral and shallow neritic environments. *J. Paleontol.* **38**, 761–767.

Weimer, R. J., and Land, C. B., 1975, Maestrichtian deltaic and interdeltaic sedimentation in the Rocky Mountain region of the U.S. *Geol. Assoc. Can. Spec. Pap.* **13**, 633–666.

Weir, J. D., 1949, Marine Jurassic formations of southern Alberta plains. *Am. Assoc. Pet. Geol. Bull.* **33**(4), 547–563.

Weller, J. M., 1930, Cyclical sedimentation of the Pennsylvanian Period and its significance. *J. Geol.* **38**, 97–135.

Wells, J. W., 1963, Coral growth and geochronometry. *Nature* **197**, 948–950.

Wheeler, H. E., 1963, Post–Sauk and pre–Absaroka stratigraphic patterns in North America. *Am. Assoc. Pet. Geol. Bull.* **47**, 1497–1526.

Wheeler, J. O., and Gabrielse, H., 1972, The Cordilleran structural province, *Geol. Assoc. Can. Spec. Pap.* **11**, 1–82.

Wheeler, W. H., and Textoris, D. A., 1978, Triassic limestone and chert of playa origin in North Carolina. *J. Sediment. Petrol.* **48**, 765–776.

Whisonant, R. C., 1974, Petrology of the Chilhowee Group (Cambrian and Cambrian(?)) in central–eastern and southern Tennessee. *J. Sediment. Petrol.* **44**(1), 228–241.

Whisonant, R. C., 1977, Lower Silurian Tuscarora (Clinch) dispersal patterns in western Virginia. *Geol. Soc. Am. Bull.* **88**, 215–220.

White, W. S., 1966, Geologic evidence for crustal structure in the western Lake Superior basin. *In* Steinhart, J. S., and Smith, T. J. (eds.), The earth beneath the continents. Am. Geophys. Union, Geophys. Monogr. **10**, 28–41.

Whittington, H. B., and Hughes, C. P., 1972, Ordovician geography and faunal provinces deduced from trilobite distribution. *Proc. R. Acad. London Ser. B* **263**, 235–278.

Wickenden, R. T. D., 1953, Mesozoic stratigraphy of the eastern plains, Manitoba and Saskatchewan. *Can. Geol. Surv. Mem.* **239**.

Wickham, J., Roeder, D., and Briggs, G., 1976, Plate tectonics models for the Ouachita foldbelt. *Geology* **4**(3), 173–176.

Willetts, C. F., Staheli, A. C., and Manley, F. H., 1980, Mississippi coastal ridges: Origin and analysis. *Geol. Soc. Am. Abstr. Progr.* **12**, 549.

Williams, J. S., 1962, Pennsylvanian System in central and northern Rocky Mountains. *In* Branson, C. C. (ed.), Pennsylvanian System in the United States–A symposium. AAPG, Tulsa, pp. 159–187.

Williams, A., and Hurst, J. M., 1977, Brachiopod evolution. *In* Hallam, A. (ed.), *Patterns of Evolution as Illustrated by the Fossil Record.* Elsevier, Amsterdam, pp. 79–122.

Williams, E. G., and Bragonier, W. A., 1974, Controls of Early Pennsylvanian sedimentation in western Pennsylvania. *Geol. Soc. Am. Spec. Pap.* **148**, 135–152.

Williams, G. D., and Stelck, C. R., 1975, Speculations on the Cretaceous paleogeography of North America. *Geol. Assoc. Can. Spec. Pap.* **13**, 1–20.

Williams, H., and Hatcher, R. D., Jr., 1982, Suspect terranes and accretionary history of the Appalachian orogen. *Geology* **10**, 530–536.

Williams, H., and Stevens, R. K., 1974, The ancient continental margin of eastern North America. *In* Burk, C. A., and Drake, C. L. (eds.), *The Geology of Continental Margins.* Springer–Verlag, Berlin, pp. 781–796.

Wilson, E. O., 1985, Invasion and extinction in the West Indian ant fauna: Evidence from the Dominican amber. *Science* **229**, 265–267.

Wilson, H. D. B., 1956, Structure of lopoliths. *Geol. Soc. Am. Bull.* **67**, 289–300.

Wilson, J. L., 1975, *Carbonate Facies in Geologic History.* Springer–Verlag, Berlin.

Wilson, J. T., 1963, Hypothesis of Earth's behavior. *Nature* **198**, 925–929.

Wilson, P. C., 1962, Pennsylvanian stratigraphy of Powder River basin and adjoining area. *In* Branson, C. C. (ed.), Pennsylvanian System in the United States—A symposium. AAPG, Tulsa, pp. 117–158.

Windley, B. F. (ed.), 1976, *The Early History of the Earth.* Wiley, New York.

Windley, B. F., 1977, *The Evolving Continents.* Wiley, New York.

Windley, B. F., 1984, *The Evolving Continents,* 2nd. edition. Wiley, New York.

Windley, B. F., and Bridgwater, D., 1971, The evolution of Archean low– and high–grade terrains. *Geol. Soc. Aust. Spec. Publ.* **3**, 33–46.

Winker, C. D., and Howard, J. D., 1977, Correlation of tectonically deformed shorelines on the southern Atlantic Coastal Plain. *Geology* **5**, 123–127.

Woese, C. R., and Fox, G. E., 1977, Phylogenetic structure of the prokaryotic domain: The primary kingdoms. *Proc. Natl. Acad. Sci. USA* **74**, 5088–5090.

Wolfe, J. A., 1978, A paleobotanical interpretation of Tertiary climates in the Northern Hemisphere. *Am. Sci.* **66**, 694–716.

Wood, D. S., 1966, The Rhodesian basement. 10th Annual Report of the Institute of African Geology, University of Leeds, pp. 18–19.

Wood, D. S., 1973, Patterns and magnitudes of natural strain in rocks. *Philos. Trans. R. Soc. London Ser. A* **274**, 373–382.

Woodburne, M. O., and Bernor, R. L., 1980, On superspecific groups of some Old World hipparonine horses. *J. Paleontol.* **54**(6), 1319–1348.

Woodbury, H. O., *et al.,* 1973, Pliocene and Pleistocene depocenters, outer continental shelf Louisiana and Texas. *Am. Assoc. Pet. Geol.* **54**, 2428–2439.

Woodland, B. G., and Stenstrom, R. C., 1979, The occurrence and origin of siderite concretions in the Francis Creek Shale (Pennsylvanian) of northeastern Illinois. *In* Nitecki, M. H. (ed.), *Mazon Creek Fossils.* Academic Press, New York, pp. 69–103.

Woodrow, D. L., Fletcher, F. W., and Ahrnsbrak, W. F., 1973, Paleogeography and paleoclimate at the deposition sites of the Devonian Catskill and Old Red facies. *Geol. Soc. Am. Bull.* **84**, 3051–3064.

Woodruff, F., Savin, S. M., and Douglas, R. G., 1981, Miocene stable isotope record: A detailed deep Pacific Ocean study and its paleoclimatic implications. *Science* **212**, 665–668.

Woodward, H. P., 1957, Chronology of Appalachian folding. *Am. Assoc. Pet. Geol. Bull.* **41**, 2312–2327.

Worsley, T. R., 1974, The Cretaceous–Tertiary boundary event in the ocean. *In* Hay, W. W. (ed.), *Studies in Paleo–oceanography,* SEPM Special Publ. 20, Tulsa, Oklahoma pp. 94–125.

Worsley, T. R., and Herman, Y., 1980, Episodic ice–free Arctic ocean in Pliocene and Pleistocene time: Calcareous nannoplankton evidence. *Science* **210**, 323–325.

Worzel, J. L., and Burk, C. A., 1979, The margins of the Gulf of Mexico. *Am. Assoc. Pet. Geol. Mem.* **29**, 403–419.

Wright, H. E., Jr., 1971, Late Quaternary vegetational history of North America. *In* Turekian, K. K. (ed.), *Late Cenozoic Glacial Ages.* Yale University Press, New Haven, Conn., pp. 425–464.

Wright, H. E., Jr., 1976, Ice retreat and revegetation in the western Great Lakes area. *In* Mahaney, W. C. (ed.), *Quaternary Stratigraphy of North America.* Dowden, Hutchinson & Ross, Stroudsburg, Pa., pp. 119–132.

Wright, H. E., Jr., and Frey, D. G. (eds.), 1965, *The Quaternary of the United States.* Princeton University Press, Princeton, N.J.

Wright, H. E., Jr., and Porter, S. C. (eds.), 1983, *Late Quaternary Environments of the United States,* Vol. 1. University of Minnesota Press, Minneapolis.

Wright, H. E., Jr., and Ruhe, R. V., 1965, Glaciation of Minnesota and Iowa. *In* Wright, H. E., Jr., and Frey, D. G. (eds.), *The Quaternary of the United States.* Princeton University Press, Princeton, N.J., pp. 29–42.

Wright, L. A., *et al.,* 1974, Precambrian sedimentary environments of the Death Valley region, eastern California. *In* Guidebook—Death Valley region, California and Nevada, Field Trip No. 1. Geol. Soc. Am., Cordilleran Sect., 70th Annu. Meet., Las Vegas, pp. 27–35.

Wynne–Edwards, H. R., 1972, The Grenville Province. *Geol. Assoc. Can. Spec. Pap.* **11**, 264–334.

Wynne–Edwards, H. R., 1976, Proterozoic ensialic orogenesis: The millipede model of ductile plate tectonics. *Am. J. Sci.* **276**, 927–953.

Yochelson, E. L., 1963, Problems of the early history of the Mollusca. 16th Int. Congr. Zool. Proc., Washington, D. C., p. 187.

Young, G. M. (ed.), 1973a, Huronian stratigraphy and sedimentation. *Geol. Assoc. Can. Spec. Pap.* **12**.

Young, G. M., 1973b, Tillites and aluminous quartzites as possible time markers for Middle Precambrian (Aphebian) rocks of North America. *Geol. Assoc. Can. Spec. Pap.* **12**, 96–126.

Young, G. M., and Church, W. R., 1966, The Huronian System in the Sudbury district and adjoining areas of Ontario; a review. *Geol. Assoc. Can. Proc.* **17**, 65–82.

Young, K., 1963, Upper Cretaceous ammonites from the Gulf Coast of the United States. *Tex. Univ. Publ.* **6304**.

Youngquist, W., 1967, Fossil systematics. *In* Teichert, C., and Yochelson, E. L. (eds.), Essays in paleontology and stratigraphy. University of Kansas, Press, Dept. Geol. Spec. Publ. **2**, 57–62.

Zangerl, R., and Richardson, E. S., Jr., 1963, The paleoecological history of two Pennsylvanian black shales. *Fieldiana Geol. Mem.* **4**.

Zen, E., 1968, Nature of the Ordovician orogeny in the Taconic area. *In* Zen, E., White, W. S., and Hadley, J. B. (eds.), *Studies in Appalachian Geology: Northern and Maritime*. Wiley–Interscience, New York, pp. 129–139.

Zen, E., 1972, The Taconide Zone and the Taconic orogeny in the western part of the northern Appalachian orogen. *Geol. Soc. Am. Spec. Pap.* **135**.

Zen, E., White, W. S., and Hadley, J. B. (eds.), 1968, *Studies in Appalachian Geology: Northern and Maritime*. Wiley-Interscience, New York.

Zhuravleva, I. T., 1970, Marine faunas and Lower Cambrian stratigraphy. *Am. J. Sci.* **269**, 417–445.

Zoback, M. L., and Thompson, G. A., 1978, Basin and Range rifting in northern Nevada: Clues from a mid–Miocene rift and its subsequent offsets. *Geology* **6**, 111–116.

Zoback, M. L., and Thompson, G. A., 1979, Comment on Late Cenozoic tectonic evolution of the western United States. *Geology* **7**, 370–371.

Zotikov, I. A., Zagorodnov, V. S., and Raikovsky, J. V., 1980, Core drilling through the Ross ice shelf (Antarctica) confirmed basal freezing. *Science* **207**, 1462–1463.

Stratigraphic Index

This index lists lithostratigraphic units discussed in text and/or included in the stratigraphic-correlation diagrams in each of the Phanerozoic chapters. Stratigraphic nomenclature follows standard usage; however, where several hierarchical levels or descriptive terms are applied to a unit, such as Formation vs. Group or Formation vs. Sandstone, the appellation most often appearing in text is listed followed by additional common usages. Consecutive page numbering in this and subsequent indexes does not necessarily imply continuous discussion of a topic; rather, it may signify the item's appearance in parts of the bracketed pages.

Taxonomic Index

All fossil and living taxa discussed or tabulated in the text are listed here following the most commonly recognized systematic term. Because suffixes of taxonomic terms (e.g., -ida, -acea, -ina, -oida, -oidea) generally correspond to their assigned hierarchical level, a particular taxonomic name may vary in appearance in this book relative to other publications: and, possibly, between our usage in the Synoptic Taxonomy (Appendix A) and references in text. To accommodate these inevitable discrepancies, this index lists all internal variations in usage for organisms and trace fossils, even where our coverage consists of only a listing in the synoptic taxonomy. Taxa discussed in the text using vernacular terms (e.g., horses, saber-tooth cats, rudists) are listed here following the most appropriate formal taxonomic usage. The reader may wish to use a standard dictionary to look up the appropriate Linnean entry; however, most taxa of general interest have sufficiently well-known names and are here sufficiently cross referenced to allow easy access (e.g., horses may be found under Perissodactyla, *Equus,* or many older genera—in any case, the pertinent discussion will be found nearby). For a few groups whose vernacular names are very different from the formal systematic usage, we supply entries in the Subject Index with cross references to this index.

Subject Index

The nature of this index is largely self-evident; however, several usages or absences bear noting. Topics are omitted if they feature both extensive text coverage and clearly demarcated entries in the Table of Contents; such topics include most continental regions, physiographic provinces, time and time-stratigrapic intervals, and other major entities. Index entries are given where references to the above are scattered through the book. All references to stratigraph-

ic units *per se* appear in the Stratigraphic Index but references to sedimentary basins appear herein. References to fossil organisms are similarly omitted here and placed in the Taxonomic Index; exceptions are those important taxa with common names not readily related to their formal taxonomic name (e.g., mosasaurs are listed under Squamata; dinosaurs include several taxa, none conventionally termed ''Dinosauria'').